ESTUARINE AND COASTAL MODELING

PROCEEDINGS OF THE SIXTH INTERNATIONAL CONFERENCE

November 3–5, 1999
New Orleans, Louisiana

EDITED BY
Malcolm L. Spaulding
H. Lee Butler

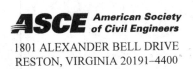

ASCE *American Society of Civil Engineers*

1801 ALEXANDER BELL DRIVE
RESTON, VIRGINIA 20191–4400

Abstract: This volume contains the Proceedings of the 6th International Conference on Estuarine and Coastal Modeling held in New Orleans, Louisiana, November 3-5, 1999. The conference included oral and poster sessions describing model development, testing, application, skill assessment, interpretation, and visualization of results. Four special sessions were held in conjunction with the American Meteorology Society, Committee on Meteorology and Oceanography of the Coastal Zone on the development and operation of coastal forecast systems. There were two poster sessions featuring model application and demonstration of model visualization and systems integration. Each paper in this publication was presented orally or at a poster session at the conference and was peer reviewed prior to publication. The paper authors represent a well balanced mix of professionals in the field from private industry, government, and academic institutions both from the US and abroad.

Library of Congress Cataloging-in-Publication Data

Estuarine and coastal modeling : proceedings of the 6th international conference, November 3-5, 1999, New Orleans, Louisiana / edited by Malcolm L. Spaulding, H. Lee Butler.
    p. cm.
  Includes bibliographical references and index.
  ISBN 0-7844-0504-2
    1. Estuaries—Mathematical models—Congresses. 2. Coast changes—Mathematical models—Congresses. 3. Hydrodynamics—Mathematical models—Congresses. 4. Water quality—Mathematical models—Congresses. 5. Sediment transport—Mathematical models—Congresses. I. Spaulding, Malcolm L. II. Butler, H. Lee. III. International Conference on Estuarine and Coastal Modeling (6th : 1999 : New Orleans, La.)

GC96.5 E74 2000
551.46'09--dc21   00-033176

Foreword

This conference represents the sixth in a biennial series to explore the development, testing, application, calibration, validation, and visualization of predictions from estuarine and coastal models. Application of models to problems in hydrodynamics, water quality and sediment transport were given. Attendance at the meeting was 175 and included representatives from both the US and many foreign countries. In addition Estuarine and Coastal Modeling conference participants could also attend sessions of a meeting sponsored by the American Meteorology Society, Committee on Meteorology and Oceanography of the Coastal Zone (AMS/MOCZ) being held at the same time. Attendance was predominantly from government and academic engineers and scientist, but also included active participation by industry professionals.

The goal of the conference, as for the past conferences in the series, was to bring together a diverse group of model researchers, users, and evaluators to exchange information on the current state-of-the-art and practice in marine environmental modeling. The primary focus was on application of models to bays, sounds, lagoons, estuaries, embayments, bights, and coastal seas. The models were addressed at solving engineering and environmental impact assessment problems and also at better understanding circulation and pollutant transport in these near shore waters. Model applications to address regulatory requirements for facility sighting and operation were also presented.

The conference included 12 oral sessions and 2 poster sessions held over the two and one half day meeting period. Four special sessions were also held in conjunction with the American Meteorology Society, Committee on Meteorology and Oceanography of the Coastal Zone (AMS/MOCZ) on the development and operation of coastal forecast systems. Papers from both poster and oral sessions are included in the conference proceedings. Each paper in the proceedings was presented at the meeting, subjected to two external peer reviews, and accepted, if appropriate and after revision, by the proceedings editors.

The enthusiastic support of the organizing and advisory committees, whose names are listed below, and Joseph Pittle and his staff at the University of Rhode Island Conference and Special Programs Development Office contributed greatly to the success of the meeting. Special thanks are also due to Frank Aikman who organized and chaired the special, joint sessions on coastal forecast systems. Thanks are also extended to the many other individuals who generously served as session chairs and moderators.

ORGANIZING COMMITTEE:

Professor Keith W. Bedford, Ohio State University, Columbus, OH.
Dr. Alan F. Blumberg, HydroQual Inc., Mahwah, NJ.
Mr.. H. Lee Butler, Veri-Tech, Vicksburg, MS.
Dr. Ralph T. Cheng, U.S. Geological Survey, Menlo Park, CA.
Professor Malcolm L. Spaulding, University of Rhode Island, Narragansett, RI.

ADVISORY COMMITTEE:

Dr. Frank Aikman, NOAA, National Ocean Service, Silver Spring, MD.
Dr. Mark Dortch, U.S. Army Corp of Engineers, Vicksburg, MS.
Dr. Billy Johnson, U.S. Army Corp of Engineers, Vicksburg, MS.
Professor Nikolaos Katopodes, University of Michigan, Ann Arbor, MI.
Professor Ian King, University of California, Davis, CA.
Professor Wilbert Lick, University of California, Santa Barbara, CA.
Professor Y. Peter Sheng, University of Florida, Gainsville, FL.
Dr. Richard P. Signell, U.S. Geological Survey, Woods Hole, MA.
Dr. Peter E Smith, U.S. Geological Survey, Sacramento, CA.
Professor Guus Stelling, Delft University of Technology, The Netherlands
Dr. Craig Swanson, Applied Science Associates, Inc, Narragansett, RI.
Dr. Roy A. Walters, National Institute for Water and Atmospheric Research, New Zealand

Special thanks are extended to Dolores Provost of the University of Rhode Island, Ocean Engineering Department who generously provided her time to perform the many administrative tasks necessary to plan the meeting, solicit and distribute the abstracts, and produce the conference proceedings. Thanks are also due to Larry Simoneau who developed and maintained the conference web site.

Given the very enthusiastic response of the conference participants and the record setting attendance, planning is currently in progress for the 7 th International Estuarine and Coastal Modeling Conference to be held in the fall 2001.

Malcolm L. Spaulding
Narragansett, RI

H. Lee Butler
Vicksburg, MS

Conference Co-Chairmen

Contents

Thursday, November 4, 1999

Session 1: General Joint Session
No papers were submitted for this session.

Session 2A: Finite Volume and Adjoint Methods

Control of Estuarine Salinity Using the Adjoint Method 1
Michael Piasecki and Brett F. Sanders

An Unsteady Finite Volume Circulation Model 17
Matthew C. Ward

Session 2B: Predicting Impacts at Sensitive Sites

Modeling the Impacts of Dredged Material Placement in Upper Chesapeake Bay 34
Billy H. Johnson, Hans R. Moritz, Allen M. Teeter, Carl F. Cerco, and
Harry V. Wang

**A Multi-Size, Multi-Source Formulation for Determining Impacts of Sediments
on Near-Shore Sensitive Sites** 59
Panagiotis Velissariou, Vasilia Velissariou, Yong Guo, and Keith W. Bedford

Session 2C: Simulating Thermal Processes

**Modeling Three-Dimensional, Thermohaline-Driven Circulation
in the Arabian Gulf** 74
Cheryl Ann Blain

A Numerical Simulation of the Japan/East Sea (JES) Seasonal Circulation 94
Peter C. Chu, Jian Lan, and Hilbert Strauhs

Effect of Heat Flux on Thermocline Formation 114
Joseph S. Helfand, David P. Podber, and Michael J. McCormick

Session 3A: Hydrodynamic Model Formulation

**A Comparison of Two Three-Dimensional Shallow-Water Models Using
Sigma-Coordinates and z-Coordinates in the Vertical Direction** 130
M.D.J.P. Bijvelds, J.A.Th.M. Van Kester, and G.S. Stelling

**Three-Dimensional Hydrostatic and Non-Hydrostatic Modeling of Seiching
in a Rectangular Basin** 148
XinJian Chen

**A Three-Dimensional Non-Hydrostatic Model for Free Surface Flows:
Development, Verification, and Limitations** 162
H. Weilbeer and J.A. Jankowski

Session 3B: Sediment Transport Processes I

**Developing a Capability to Forecast Coastal Ocean Optics: Minerogenic
Scattering** 178
Timothy R. Keen and Robert H. Stavn

A Large Domain Convection Diffusion Based Finite Element Transport Model 194
Norman W. Scheffner

Modeling Sand Bank Formation Around Tidal Headlands..209
Richard P. Signell and Courtney K. Harris

Session 3C: Models for Environmental Assessment

Proposed Third Crossing of Hampton Roads, James River, Virginia:
Feature-Based Criteria for Evaluation of Model Study Results...................................223
J.D. Boon, A.Y. Kuo, H.V. Wang, and J.M. Brubaker

Application of VIMS HEM-3D to a Macro-Tidal Environment......................................238
Sung-Chan Kim, Jian Shen, Chang Shik Kim, and Albert Y. Kuo

Simulation of Tunnel Island and Bridge Piling Effects in a Tidal Estuary...................250
Harry V. Wang and Sung-Chan Kim

Session 4A: Hydrodynamic Model

Monotonicity Preserving Model for Free-Surface Flow with Sharp Density Fronts.....270
Scott F. Bradford and Nikolaos D. Katopodes

Generation of a Two-Dimensional Unstructured Mesh for the East Coast
of South Carolina...290
Scott C. Hagen and Robert J. Bennett

Internal Characteristics of Two-Layer Stratified Flow Systems Under Wind Stress....304
Koji Kawasaki, Jong Sung Yoon, and Keiji Nakatsuji

The Mary River Estuary: A New Approach to Modeling of Marshes and Over-Bank
Areas 316
Ian P. King and David K. Williams

Session 4B: Three-Dimensional Water Quality Simulations

Modeling Water Quality for Dredged Material Disposal...335
Mark Dortch, Beth Fleming, and Barry Bunch

A Framework for Integrated Modeling of Coupled Hydrodynamic–Sedimentary–
Ecological Processes..350
Y. Peter Sheng

Session 4C: Waves and Wave-Induced Currents

Application of a Barotropic Hydrodynamic Model to Nearshore Wave-Induced
Circulation...363
Mark Cobb and Cheryl Ann Blain

Circulation and Sediment Transport in the Vicinity of the Hudson Shelf Valley..........380
Courtney K. Harris and Richard P. Signell

Inverse Flow Modelling of Waves and Currents in the Surf Zone................................395
David H. Henderson, David Johnson, and Graham J.M. Copeland

Session 5A: Three-Dimensional Hydrodynamic Model Simulations I

Three-Dimensional Modeling of Temperature Stratification and Density-Driven
Circulation in Lake Billy Chinook, Oregon 411
Zhaoqing Yang, Tarang Khangaonkar, Curtis DeGasperi, and Kevin Marshall

Session 5B: Simulating Shelf-Scale Processes I

Sources of Uncertainty for Oil Spill Simulations...426
Eric Anderson and Henry Rines

Session 5C: Plume Simulations

**Topographic and Wind Influences on the Chesapeake Bay Outflow Plume
and the Associated Fronts** 443
    Patrick C. Gallacher, Michael Schaferkotter, and Paul Martin

Modeling Surface Trapped River Plumes: A Sensitivity Study 452
    Jason Hyatt and Richard P. Signell

**Typical and Extreme Responses of Chesapeake Bay and Its Coastal Plume
to Riverine Forcing** 466
    Jerry L. Miller, Patrick Gallacher, Michael Schaferkotter, and Paul Martin

Session 6: Hydrodynamics and GIS

**High Performance Estuarine and Coastal Environmental Modeling: The CH3D
Example** 470
    Justin R. Davis and Y. Peter Sheng

**Use of Membrane Boundaries to Simulate Fixed and Floating Structures
in GLLVHT** 485
    Venkat S. Kolluru, Edward M. Buchak, and Jian Wu

Using GIS as an Interface for Three-Dimensional Hydrodynamic Modeling...............501
    I. Morin, K. Hickey, and M. Greenblatt

Numerical Simulation of Wave Propagation in the Entrance of the Tagus Estuary.....510
    Filipa S.B.F. Oliveira

Intrusions of the Loop Current over the Mississippi Bight 526
    Germana Peggion

**The Use of GIS in Three-Dimensional Hydrodynamic Model Pre- and
Post-Processing for Feature-Specific Applications** 538
    G. McAllister Sisson, John D. Boon, and Katherine L. Farnsworth

**Three-Dimensional Curvilinear Modelling of Wind-Induced Flows Through
the North Channel of the Irish Sea** 549
    Emma F. Young, John N. Aldridge, and Juan Brown

Thursday, November 4, 1999

Session 7: Overview of Papers in Session 12
No papers were submitted for this session.

Session 8A: Transport of Organisms

Development of a Long-Term Predictive Model of Water Quality in Tokyo Bay..........564
    Jun Sasaki and Masahiko Isobe

A Three-Dimensional Shear Dispersion Model Applied to Georges Bank....................581
    Zhigang Xu, Charles G. Hannah, and John W. Loder

Session 8B: Modeling Enclosed Bays

**Hydrodynamic Modeling for the 1998 Lake Michigan Coastal Turbidity
Plume Event** 597
    Dmitry Beletsky, David J. Schwab, Michael J. McCormick, Gerald S. Miller,
    James H. Saylor, and Paul J. Roebber

Numerical Modeling and Field Experiments of Wind-Induced Currents in Enclosed Bay..614
Tadashi Fukumoto and Takehiro Nakamura

An Isopycnic Model Study of the Black Sea......................................630
Gokay M. Karakas, Alec E. James, and Alaa M.A. Al-Barakati

Session J1-A: Development and Operation of Coastal Forecast Systems

The Coastal Marine Demonstration Project......................................646
L.J. Walstad, G. J. Szilagyi, F. Aikman, L.C. Breaker, F. Klein, J.S. D'Aleo, and J.T. McQueen

Session 9A: Three-Dimensional Hydrodynamic Modeling II

A Hydrodynamic Model Calibration Study of the Savannah River Estuary with an Examination of Factors Affecting Salinity Intrusion..........663
Daniel L. Mendelsohn, Steven Peene, Eduardo Yassuda, and Steven Davie

The Gaspé Current and Cyclonic Motion over the Northwestern Gulf of St. Lawrence...686
Jinyu Sheng

Session 9B: Innovative Computational Techniques

Quadtree Grids for Dispersion and Inverse Flow Models....................705
G.J.M. Copeland, R.D. Marchant, and A.G.L. Borthwick

A Parallel Finite Volume Scheme for Two Dimensional Hydrodynamic Flows..........721
Prasada Rao and Scott A. Yost

Integrating High Performance Computing Strategies with Existing Finite Element Hydrodynamic Codes...738
Scott A. Yost and Prasada Rao

Session J1-B: Development and Operation of Coastal Forecast Systems

Lake Michigan Forecast System and the Prospects for a Sediment Transport Model...755
Yi-Fei Philip Chu, Keith W. Bedford, and David J.S. Welsh

The Tampa Bay Nowcast−Forecast System......................................765
Mark Vincent, David Burwell, and Mark Luther

Session 10A: Two-Dimensional Hydrodynamic Model Applications

Navigation Study for Matagorda Bay, Texas....................................781
David J. Mark, Dennis W. Webb, and Nicholas C. Kraus

Hydrodynamic Modeling of a Sea-Breeze Dominated Shallow Embayment, Baffin Bay, Texas...795
Adele Militello

Session 10B: Hydrodynamic Model Skill Assessments

Tidal Inlet Circulation: Observations, Model Skill, and Momentum Balances..........811
James L. Hench and Richard A. Luettich, Jr.

Modeling of Tide-Induced Circulation near a Strait............................827
Masahide Ishizuka, Keiji Nakatsuji, and Shuzo Nishida

Application of Quantitative Model: Data Calibration Measures to Assess Model
Performance 843
    Malcolm Spaulding, Craig Swanson, and Daniel Mendelsohn

Session J1-C: Development and Operation of Coastal Forecast Systems

Demonstration of a Nowcast–Forecast System for Galveston Bay 868
    Richard A. Schmalz, Jr.

Skill Assessment Methods for NOS Forecast Systems.......................................884
    Kurt W. Hess and Thomas F. Gross

Coastal Wave Measurement and Forecast System: Preliminary Results
and Model Selection...899
    Michael S. Bruno, Kelly L. Rankin, and Thomas O. Herrington

Session 11A: Sediment Transport Processes II

Effects of Bed Coarsening on Sediment Transport...915
    Craig Jones and Wilbert Lick

Session 11B: Data Assimilation and Model Skill

Mixing in a Small Tidal Estuarine Plume..931
    M. Pritchard, D.A. Huntley, and T.J. O'Hare

Improving Coastal Model Predictions through Data Assimilation 947
    J. Craig Swanson and Matthew C. Ward

Parameter Estimation for Subtidal Water Levels Using an Adjoint Variational
Optimal Method..964
    Aijun Zhang, Eugene Wei, and Bruce Parker

Session J1-D: Development and Operation of Coastal Forecast Systems

Development of the High-Resolution Data-Assimilating Numerical Model
of the Monterey Bay 980
    I. Shulman, C.-R. Wu, J.K. Lewis, J.D. Paduan, L.K. Rosenfeld, S.R. Ramp,
    M.S. Cook, J.C. Kindle, and D.-S. Ko

Session 12 (Posters): Environmental Aspects

Modeling Residence Times: Eulerian versus Lagrangian...................................995
    David Burwell, Mark Vincent, Mark Luther, and Boris Galperin

Pearl River Estuary Pollution Project (PREPP): An Integrated Approach.................1010
    Jay-Chung Chen, Ophelia Lee, and Chan Wai Man

Implementation of Vertical Acceleration and Dispersion Terms in an Otherwise
Hydrostatically Approximated Three-Dimensional Model......................................1019
    John Eric Edinger and Venkat S. Kolluru

Modeling Hydrodynamic and Sediment Processes in Morro Bay............................1035
    Zhen-Gang Ji, Michael R. Morton, and John M. Hamrick

An Ocean Model Applied to the Chesapeake Bay Plume......................................1055
    Paul J. Martin

Simulation of the Oil Trajectory and Fate of the Diamond Grace Spill
in Tokyo Bay..1070
    K. Okamoto

A Multi-Estuarine Model in Long Island Sound 1084
    Frederick E. Schuepfer, Guy A. Apicella, and Robert O'Neill

Friday, November 5, 1999

Session 13A: Three-Dimensional Hydrodynamic Model Simulations III

Three-Dimensional Hydrodynamic Model of an Estuary in Nova Scotia 1100
    Ken Hickey, Isabelle Morin, Marcia Greenblatt, and Gavin Gong

Modeling the Circulation in Penobscot Bay, Maine 1112
    Huijie Xue, Yu Xu, David Brooks, Neal Pettigrew, and John Wallinga

A Three-Dimensional Modeling Study of the Estuarine System: An Application
to the Satilla River 1128
    Lianyuan Zheng and Changsheng Chen

Session 13B: Simulating Shelf-Scale Processes II

Spatial Decorrelation Scales in Coastal Regions: A Precursor to More Accurate
Data Assimilation... 1150
    James K. Lewis, Alan F. Blumberg, and B. Nicholas Kim

Numerical Modeling Experiments of the East (Japan) Sea Circulation
and Mesoscale Eddy Generation Based on POM-ES... 1167
    Young Jae Ro

Session 13C: Morphodynamic Modeling

Simulating Morphodynamical Processes on a Parallel System....................................... 1182
    Hartmut Kapitza and Dieter P. Eppel

MORWIN: Collaborative Modeling of Coastal Morphodynamics.............................. 1192
    Rainer Lehfeldt and Volker Barthel

An Improved Formulation for the Bed Shear Stress in Morphodynamic
Simulations... 1206
    Andreas Malcherek

Session 14A: Three-Dimensional Hydrodynamic Model Simulations IV

Modeling Tidal Circulation in a Barrier-Island Estuary: Apalachicola Bay.............. 1216
    W. Huang, K. Jones, T.S. Wu, and H. Sun

A PC-Based Visualization System for Coastal Ocean and Atmospheric Modeling..... 1233
    Hongqing Wang, Kai-Hon Lau, Wai-Man Chan, and Lai-Ah Wong

Session 14B: Two-Dimensional Water Quality Model Applications

Calibration Performance of a Two-Dimensional, Laterally-Averaged
Eutrophication Model of a Partially Mixed Estuary... 1244
    James D. Bowen

Modelling the Humber Estuary Catchment and Coastal Zone....................................... 1259
    Roger Proctor, Jason Holt, John Harris, Alan Tappin, and David Boorman

Development of a Waste Load Allocation Model for the Charleston Harbor Estuary.
Part III: Project Application.. 1275
    Eduardo A. Yassuda, Steven J. Peene, Steven R. Davie,
    Daniel L. Mendelsohn, and Tatsu Isaji

Indexes

Subject Index.. 1294

Author Index 1297

CONTROL OF ESTUARINE SALINITY USING THE ADJOINT METHOD

Michael Piasecki [1] M. ASCE and Brett F. Sanders [2] M. ASCE

ABSTRACT

This paper presents a new and efficient method to control the upstream migration of saltwater in well mixed estuaries. The underlying goal is to prevent adverse effects on salinity-sensitive biota and to simultaneously allow for maximum freshwater diversions. The conflicting objectives are formulated as an optimization problem where control is achieved by optimizing the fresh water diversion rate as a function of time subject to environmental constraints, in this case maximum allowable salinity concentrations. Salinity migration is formulated as a one-dimensional problem including continuity, momentum, and a transport equation for salinity. A monitoring station is placed in the brackish water zone to measure current salinity and to compare it with desired levels. The differences serve as a quality measure for the current diversion strategy and to solve the adjoint set of equations. The solution vector of the adjoint set of equations is used to determine gradients which report changes in quality measure with respect to control action. These gradients are utilized by a quasi-Newton gradient search method. Several diversion scenarios are tested by varying the number of diversion points. It is shown that for one and two diversion points rapid convergence to the optimal solution is obtained, rendering the selected scenarios suitable for real-time adaptive control. For more than 2 diversion nodes the optimization problem becomes too complex for the gradient search and a global search algorithm has to be employed in order to identify optimal withdrawal rates.

[1] Assistant Professor, Department of Civil and Architectural Engineering, Drexel University, Philadelphia, PA 19104

[2] Assistant Professor, Department of Civil and Environmental Engineering, University of California, Irvine, CA 92697

I - INTRODUCTION

Throughout the late 1980s and early 1990s the western part of the United States suffered from a prolonged drought. Among the many problems that the drought caused was the inland migration of saline waters in estuaries. This migration was worsened by the increased withdrawal of fresh water from declining stream flows in order to meet the public and private demand for drinking and irrigation water. While the upstream increase in salinity affected water quality at municipal drinking water intakes (Denton and Briggs 1997), similar shifts in the position of the brackish water zone have been known to adversely impact inland biota that are intolerant of salinity increases (Goldman and Horne 1983).

The problem of managing salinity intrusion into coastal streams is typical of systems of open channel waterways that must be operated to meet various, often competing, objectives. A typical set of conflicting objectives is to maintain low salinity levels at freshwater withdrawal points (or sensitive habitats) on the one side, and to supply a sufficient amount fresh water to municipalities and agriculture on the other. This conflict constitutes a major challenge to a river management system and requires a tool to systematically make real-time control decisions. Specifically, this tool needs to be able to maximize the amount of water withdrawn, subject to the constraint that salinity levels have to be maintained at a regulated level.

Tools recently developed for water resources management are able to assist in balancing competing demands on the nation's surface waters. The adjoint sensitivity method allows use of a numerical flow model to optimize control actions subject to stipulated constraints. The technique has been used to optimize the diversion of water from rivers to mitigate the impact of floods (Sanders and Katopodes 1999). Additionally, this approach has been used to optimize the location and strength of pollutant sources in estuaries (Katopodes and Piasecki 1996) and to optimize the mass loading rate in rivers and estuaries (Piasecki and Katapodes 1997a,b). Adjoint methods also have been used for parameter identification problems (Panchang and O'Brien 1989, Das and Lardner 1992, Chertok and Lardner 1996).

This paper presents a novel approach to optimize the diversion of fresh water from a well-mixed estuary with a strong tidal influence. The adjoint problem is introduced to compute the sensitivity of an objective function, which measures the upstream penetration of saline water, to changes in the diversion rates. The sensitivity information provided by the adjoint problem solution is used as gradient information in a quasi-Newton optimization algorithm. Subsequently, the time variation in the diversion rate that minimizes the objective function is identified.

II - MATHEMATICAL MODEL

The Direct Problem

A system of cross-sectionally averaged equations can be applied to describe flow in a class of estuaries where mixing occurs rapidly, and pressure in the vertical direction remains hydrostatic. Such an estuary has been defined as *well-mixed* by Fischer et al. (1979). The system of one-dimensional equations comprised of continuity, momentum, and salt transport equations appear in its per-unit-width formulation as follows (Odd 1981),

$$\frac{\partial \mathbf{U}}{\partial x} + \frac{\partial \mathbf{F}}{\partial x} + \mathbf{S} = 0 \tag{1}$$

where

$$\mathbf{U} = \begin{pmatrix} h \\ uh \\ Sh \end{pmatrix} \qquad \mathbf{F} = \begin{pmatrix} uh \\ u^2h + \frac{1}{2}gh^2\rho/\rho_o \\ uhS - hD_L\frac{\partial S}{\partial x} \end{pmatrix} \qquad \mathbf{S} = \begin{pmatrix} q_l \\ -gh(S_o - S_f) + u_l q_l \\ Sq_l \end{pmatrix}$$

and h is the depth of flow, U is the depth-averaged velocity in the channel, ρ is the fluid density and is assumed to be solely a function of salt concentration S, ρ_o is a reference fluid density, S_o is the bed slope, S_f is the friction slope given by the Manning equation, q_l is the lateral outflow per stream-wise length and unit width of channel, u_l is the velocity in the stream-wise direction with which the lateral outflow leaves the channel, and D_L is the longitudinal dispersion coefficient. Here, the fluid density is assumed to be a function of salt concentration, independent of temperature, as follows

$$\rho = k_1 + k_2 S \tag{2}$$

where $k_1 = 999\,\mathrm{kg/m^3}$, and $k_2 = 0.755\,\mathrm{kg/(m^3 ppt)}$. This relation is valid at salinities up to 35 ppt.

The computational system that is used for the evaluation of freshwater diversion assumes a known discharge q_u with salinity $S = 0$ at the upstream boundary of the channel. The channel has constant slope and bed roughness. The freshwater is diverted from the channel a distance x_d from the upstream boundary. At the downstream boundary ($x = L$), it is assumed that the depth, $h(L,t)$, varies periodically as a result of tidal oscillations. In addition, it is assumed that an infinite reservoir of saline water with $S = S^{ocean}$ is present at the terminus of the channel, resulting in a finite dispersive flux into the channel at all times and an advective flux during flood tides. A sketch of the problem formulation, including boundary conditions, is presented in Figure 1.

The location of the step in the salinity gradient shown in Figure 1 is controlled by the diversion rate q_l. It is assumed that other controlling factors as upstream

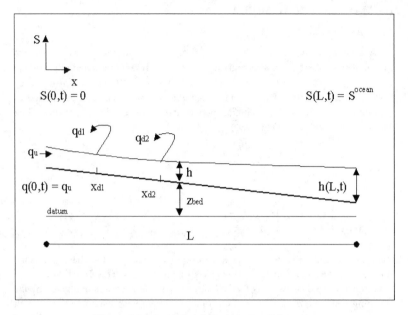

Figure 1: Schematic of Computational Domain

inflow, dispersion parameter, water levels at downstream end, etc, are known a priori and are not part of the control vector. An objective function involving the salinity concentrations at a target location, x_T, is to be minimized. The objective function can be constrained to prevent the salinity step from moving beyond a desired location in the upstream direction.

The objective function is written in terms of a measuring function, r, as follows,

$$J = \int_0^T \int_0^L r(h, q, S; x, t) \, dx \, dt \tag{3}$$

where $q = uh$ is the discharge per unit width in the channel. The measuring (or response) function is specified to minimize the amount of time salinity at the target location exceeds a design, or critical, value. Two commonly used alternatives exist that can be used. They read as follows

$$r(h, q, S; x, t) = \frac{1}{2} w(x, t) \left[S - S^c(t) \right]^2 \delta(x - x_T) \tag{4}$$

at all times during the simulation, or

$$r(h, q, S; x, t) = \begin{cases} \frac{1}{2} w(x, t) \left[S - S^c(t) \right]^2 \delta(x - x_T) & \text{if} \quad S(x_T, t) > S^c(t) \\ 0 & \text{otherwise} \end{cases} \tag{5}$$

where $S^c(t)$ is a critical salinity value, $w(x,t)$ is a weighing function that can be used to decrease or increase the influence of pre-specified target values in the response function, and $\delta(x - x_T)$ is the Dirac-Delta function that is used to identify the target locations. The weighting function can be used to scale monitored values of $S - S^c(t)$ that are larger than 0, therefore indicating an exceedance of the allowable tolerance, and as a result receive a higher level of importance than values that are less than 0. The latter would indicate that the salinity has been pushed to far downstream, and, as result, decreases potentially higher withdrawal rates. Both formulations will be used in this work and referred to in a later section.

The Adjoint Problem

Optimization can be carried out using a variety of techniques. Among them gradient search methods are known to be extremely efficient, provided the solution domain of the functional, Eq. 3, remains free of local minima and saddlepoints. When this is not true global search algorithms are needed. They perform searches on a pre-defined grid and then narrow their focus on promising areas. This search methodolgy normally results in greatly improved solution vectors (in our case diversion rates) but comes at a considerable computational cost. We will therefore prefer to use a gradient search algorithm like the Conjugate Gradient or quasi-Newton search algorithm and only switch to a global search method, like the Simulated Annealing algorithm (da Conceicao Cunha and Sousa 1999, Press et al. 1990), in case convergence cannot be achieved by the gradient methods.

A quasi-Newton search algorithm known as CONMIN (Shanno and Phua 1980) is employed to optimize the solution vector, q_l, which contains discrete values for the lateral outflow. This algorithm has successfully been applied in a variety of parameter optimization problems, (Das and Lardner 1992, Piasecki and Katopodes 1997b, and Sanders and Katopodes 1999). The vector q_l is selected to be the best solution that minimizes the objective function J. The objective function is comprised of an m-dimensional parameter space containing the value of lateral outflow at every 10 time levels in the direct problem solution, i.e., $m = N_T/10$, where N_T is the number of timesteps used in the simulation. Using a subset of the lateral outflow at all time levels was found to speed up the optimization procedure by reducing the dimension of the parameter space while retaining the ability to resolve the time variation of the lateral outflow. The required input to CONMIN includes an initial guess for the solution vector of the objective function in addition to the gradient, g, of the objective function with respect to the solution vector.

The computation of the gradient, g or $\frac{\partial J}{\partial q_l}$, of the objective function with respect to the control variables (in our case the diversion rates q_l at locations x_d) can be done in a variety of ways. A trial-and-error approach is to change the diversion rates one at a time and measure the corresponding change in the objective function.

However, this is not very computationally efficient. A more suitable and much more efficient way to compute the gradient is to use an adjoint formulation of the direct problem, Eq. 1. The adjoint methodology is based on the duality principle, (Ciarlet 1989, Marchuk 1995), which establishs that any linear functional J can be computed by using its argument, here it is S, or its adjoint S^*, as long as they are twice differentiable in both space and time and are members of a Gilbert space. For further clarification the interested reader is refered to (Ciarlet 1989 and Marchuk 1995).

The adjoint problem is derived by multiplying each of the governing equations with a Lagrangian multiplier, and then integrating the resulting expressions over space and time. Since the resulting equations are of the same type as the right hand side of Eq. 3 and are equal to zero, they can be added to the objective function. The resulting sum of all equations is called a Lagrangian, $\mathcal{L}$, and must consequently be equal to zero. It appears as,

$$
\begin{aligned}
\mathcal{L} = & \ J + \int_0^T \int_0^L \left\{ \phi \left[\frac{\partial h}{\partial t} + \frac{\partial q}{\partial x} + q_l \right] \right. \\
& + \psi \left[\frac{\partial q}{\partial t} + \frac{\partial}{\partial x} \left(\frac{q^2}{h} + \frac{\rho g h^2}{2\rho_o} \right) - gh(S_o - S_f) + u_l q_l \right] \\
& \left. + \zeta \left[\frac{\partial(hS)}{\partial t} + \frac{\partial(qS)}{\partial x} - \frac{\partial}{\partial x} \left(hD_L \frac{\partial S}{\partial x} \right) + q_l S \right] \right\} dx \, dt
\end{aligned}
\tag{6}
$$

Following integration-by-parts the spatial and temporal derivatives acting on the flow variables, h, q, S, are passed onto the adjoint variables, ϕ, ψ, ζ, respectively. Since we are interested in knowing how a pertubation in q_l changes values of $\mathcal{L}$, the variation of the Lagrangian with respect to its state variables h, q, and S, and the time dependent parameter q_l needs to be taken. The state varibales need to be included in the variation, because they themselves are a function of q_l. Finally grouping terms with common variationals, Eq. 6 becomes,

$$
\begin{aligned}
\delta \mathcal{L} = & \ \frac{\partial J}{\partial q_l} \delta q_l + \int_0^T \int_0^L \left\{ \left[-\frac{\partial \phi}{\partial t} + \left(\frac{q^2}{h^2} - \frac{\rho g h}{\rho_o} \right) \frac{\partial \psi}{\partial x} - g(S_o - S_f)\psi \right. \right. \\
& \left. + gh \frac{\partial S_f}{\partial h} \psi - S \frac{\partial \zeta}{\partial t} - S \frac{\partial}{\partial x} \left(D_L \frac{\partial \zeta}{\partial x} \right) + \frac{\partial r}{\partial h} \right] \delta h \\
& + \left[-\frac{\partial \psi}{\partial t} - \frac{\partial \phi}{\partial x} - \frac{2q}{h} \frac{\partial \psi}{\partial x} + gh \frac{\partial S_f}{\partial q} \psi - S \frac{\partial \zeta}{\partial x} + \frac{\partial r}{\partial q} \right] \delta q \\
& + \left[-h \frac{\partial \zeta}{\partial t} - q \frac{\partial \zeta}{\partial x} - \frac{\partial}{\partial x} \left(hD_L \frac{\partial \zeta}{\partial x} \right) + q_l \zeta - \frac{gh^2}{2\rho_o} \frac{\partial \psi}{\partial x} \frac{d\rho}{dS} + \frac{\partial r}{\partial S} \right] \delta S
\end{aligned}
$$

$$+ \left[\phi + u_l \psi + S\zeta \right] \delta q_l \Bigg\} \, dx \, dt$$

$$+ \int_0^L \left[(\phi + \zeta S)\delta h + \psi \delta q + \zeta h \delta S \right]_0^T \, dx$$

$$+ \int_0^T \left[\left(\frac{\rho g h}{\rho_o} \psi - \frac{q^2}{h^2} \psi - \zeta D_L \frac{\partial S}{\partial x} + S D_L \frac{\partial \zeta}{\partial x} \right) \delta h \right.$$

$$\left. + \left(\phi + \frac{2q}{h} \psi + \zeta S \right) \delta q + \left(\zeta q + h D_L \frac{\partial \zeta}{\partial x} + \psi \frac{g h^2}{2\rho_o} \frac{d\rho}{dS} \right) \delta S \right]_0^L \, dt \qquad (7)$$

where the adjoint equations are given as follows,

$$-\frac{\partial \phi}{\partial t} + \left(\frac{q^2}{h^2} - \frac{\rho g h}{\rho_o} \right) \frac{\partial \psi}{\partial x} - g(S_o - S_f)\psi + gh \frac{\partial S_f}{\partial h} \psi + \frac{Sq}{h} \frac{\partial \zeta}{\partial x}$$

$$+ \frac{SD_L}{h} \frac{\partial h}{\partial x} \frac{\partial \zeta}{\partial x} - \frac{Sq_l}{h} \zeta + \frac{Sgh}{2\rho_o} \frac{d\rho}{dS} \frac{\partial \psi}{\partial x} - \frac{S}{h} \frac{\partial r}{\partial S} + \frac{\partial r}{\partial h} = 0$$

$$-\frac{\partial \psi}{\partial t} - \frac{\partial \phi}{\partial x} - \frac{2q}{h} \frac{\partial \psi}{\partial x} + gh \frac{\partial S_f}{\partial q} \psi - S \frac{\partial \zeta}{\partial x} + \frac{\partial r}{\partial q} = 0$$

$$-\frac{\partial \zeta}{\partial t} - \frac{q}{h} \frac{\partial \zeta}{\partial x} - \frac{1}{h} \frac{\partial}{\partial x} \left(h D_L \frac{\partial \zeta}{\partial x} \right) + \frac{q_l}{h} \zeta - \frac{gh}{2\rho_o} \frac{d\rho}{dS} \frac{\partial \psi}{\partial x} + \frac{1}{h} \frac{\partial r}{\partial S} = 0 \qquad (8)$$

The gradient of J with respect to lateral diversions q_l at time t_d and location x_d therefore becomes,

$$\left. \frac{\partial J}{\partial q_l} \right|_{x_d, t_d} = - \left[\phi + u_l \psi + S\zeta \right]_{x_d, t_d} \qquad (9)$$

which, in fact, can be computed by solving the direct and adjoint problem only once each per iteration step. This constitutes a remarkable improvement compared to the time consuming repetitive approach using the direct problem only.

In order to compute the sensitivities given by Eq. 9, appropriate initial and boundary conditions must be specified for the adjoint problem. Initial conditions are specified at the final time, T, to yield a well-posed problem, as follows,

$$\phi(x, T) = 0 \qquad \psi(x, T) = 0 \qquad \zeta(x, T) = 0 \qquad (10)$$

Following the approach introduced by (Sanders 1997), the number of boundary conditions imposed in the adjoint problem at $x = 0$ and $x = L$ corresponds to the number of characteristics leaving each boundary. Additionally, the adjoint Riemann variables along the characteristics entering the solution domain are set to zero. The characteristic wave speeds in the adjoint problem are given by

$\lambda_1 = -u - a$, $\lambda_2 = -u$, and $\lambda_3 = -u + a$ where $a = (gh)^{1/2}$. Hence, at the upstream boundary, where $0 < u < a$, the boundary condition is given by,

$$\phi + (u - a)\psi + S\zeta = 0 \tag{11}$$

At the downstream boundary, where tidal oscillations occur, the appropriate boundary condition is given by,

$$\phi + (u + a)\psi + S\zeta = 0 \quad \text{and} \quad \zeta = 0 \qquad \text{for} \quad u > 0 \tag{12}$$

$$\phi + (u + a)\psi + S\zeta = 0 \qquad \text{for} \quad u \leq 0 \tag{13}$$

Because the initial conditions for the adjoint problem are specified at the final time, T, integration is performed in the reverse time direction, $\tau = T - t$. In this direction the adjoint equations appear as,

$$\frac{\partial \phi}{\partial \tau} + \left(\frac{q^2}{h^2} - \frac{\rho g h}{\rho_o}\right)\frac{\partial \psi}{\partial x} - g(S_o - S_f)\psi + gh\frac{\partial S_f}{\partial h}\psi + \frac{Sq}{h}\frac{\partial \zeta}{\partial x}$$
$$+ \frac{SD_L}{h}\frac{\partial h}{\partial x}\frac{\partial \zeta}{\partial x} - \frac{Sq_l}{h}\zeta + \frac{Sgh}{2\rho_o}\frac{d\rho}{dS}\frac{\partial \psi}{\partial x} - \frac{S}{h}\frac{\partial r}{\partial S} + \frac{\partial r}{\partial h} = 0$$
$$\frac{\partial \psi}{\partial \tau} - \frac{\partial \phi}{\partial x} - \frac{2q}{h}\frac{\partial \psi}{\partial x} + gh\frac{\partial S_f}{\partial q}\psi - S\frac{\partial \zeta}{\partial x} + \frac{\partial r}{\partial q} = 0$$
$$\frac{\partial \zeta}{\partial \tau} - \frac{q}{h}\frac{\partial \zeta}{\partial x} - \frac{1}{h}\frac{\partial}{\partial x}\left(hD_L\frac{\partial \zeta}{\partial x}\right) + \frac{q_l}{h}\zeta - \frac{gh}{2\rho_o}\frac{d\rho}{dS}\frac{\partial \psi}{\partial x} + \frac{1}{h}\frac{\partial r}{\partial S} = 0 \tag{14}$$

or in matrix form

$$\frac{\partial \mathbf{\Phi}}{\partial \tau} + \mathbf{A_\Phi}\frac{\partial \mathbf{\Phi}}{\partial x} + \mathbf{B_\Phi}\frac{\partial}{\partial x}\left(hD_L\frac{\partial \mathbf{\Phi}}{\partial x}\right) + \mathbf{C_\Phi}\mathbf{\Phi} + \mathbf{D_\Phi} = 0 \tag{15}$$

where

$$\mathbf{\Phi} = \begin{pmatrix} \phi \\ \psi \\ \zeta \end{pmatrix} \qquad \mathbf{A_\Phi} = \begin{pmatrix} 0 & \frac{q^2}{h^2} - \frac{\rho g h}{\rho_o} + \frac{Sgh}{2\rho_o}\frac{d\rho}{dS} & \frac{Sq}{h} + \frac{SD_L}{h}\frac{\partial h}{\partial x} \\ -1 & -\frac{2q}{h} & -S \\ 0 & -\frac{gh}{2\rho_o}\frac{d\rho}{dS} & -\frac{q}{h} \end{pmatrix} \tag{16}$$

$$\mathbf{B_\Phi} = \begin{pmatrix} 0 & 0 & 0 \\ 0 & 0 & 0 \\ 0 & 0 & -1/h \end{pmatrix} \tag{17}$$

$$\mathbf{C_\Phi} = \begin{pmatrix} 0 & -g(S_o - S_f) + gh\frac{\partial S_f}{\partial h} & \frac{-Sq_l}{h} \\ 0 & gh\frac{\partial S_f}{\partial q} & 0 \\ 0 & 0 & \frac{q_l}{h} \end{pmatrix} \qquad \mathbf{D_\Phi} = \begin{pmatrix} -\frac{S}{h}\frac{\partial r}{\partial S} + \frac{\partial r}{\partial h} \\ \frac{\partial r}{\partial q} \\ \frac{1}{h}\frac{\partial r}{\partial S} \end{pmatrix} \tag{18}$$

Both sets of equations, i.e. Eqs.(1) and (15), are solved using a Godunov-type finite volume method that is second-order accurate in both space and time. For more detail on the development of the numerical integration schemes the interested reader is referred to (Sanders and Piasecki 1999).

III - CONTROL OF FRESHWATER DIVERSIONS

The flow domain used for all test series has a length of $L = 20\,km$ and a bottom slope of $S_o = 0.0002$. The total simulation time is $T = 43,200\,s$ (12 hrs). The downstream boundary conditions are comprised of a tidal wave with period $T_w = 21,600\,s$ (6 hrs), wave amplitude $A_w = 1\,m$, and salinity concentration $S^{ocean} = 35\,ppt$. The longitudinal dispersion coefficient is selected to be $D_L = 100\,m^2/s$, and the upstream flow boundary condition is $q(0,t) = 0.35\,m^2/s$. The flow domain is subdivided into $N = 200$ elements of length $\Delta x = 100\,m$. The simulation is carried out for $N_T = 4000$ steps ($\Delta t = 10.8\,s$). The diversion location is $x_d = 4\,km$ (cell 40). The diversion location is close to the upstream boundary in order to ensure that no saline water is being withdrawn and to seperate the target and diversion locations. The initial conditions used for this series are shown in Fig.2 while a plot of the salinity is shown in Fig.3.

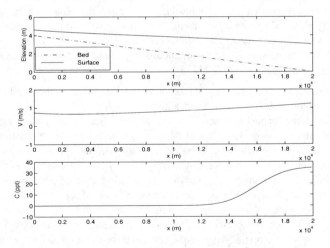

Figure 2: Initial Conditions for Test Channel

Single Diversion Node

The performance of the quasi-Newton optimization procedure using adjoint

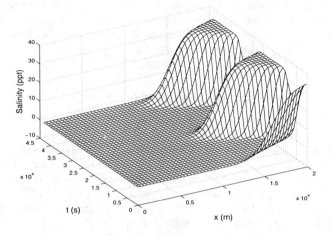

Figure 3: Initial Salinity as a Function of Space and Time

sensitivities is first examined by considering a single diversion node. Additionally, the target location, i.e. the node at which the objective function is measured, is placed at varying locations along the brackish zone. The latter variation is intended to identify preferred monitoring locations. In all runs performed the measurement function from Eq. 5 is used. This one is less demanding than the one presented in Eq. 4 as it merely requires the concentration history to drop below a constant maximum value, as illustrated in Fig. 4.

Prior to implementing the control procedure, an upper bound on the withdrawal rate was determined. It was found that a diversion of $0.25 \, m^2/s$ is the maximum allowable for the channel to remain wet over the entire tidal cycle. This diversion represents about 70% of the inflow at the upstream boundary. (Note that in this finite-volume code, which is based on conservation over the cell not at a point, the lateral outflow acts over one cell length $\Delta x = 100 \, m$. Hence, an outflow of $0.25 \, m^2/s$ corresponds to $q_l = 0.0025 \, m^3/s$ per m of channel width per m of outflow section.) A constant critical salinity concentration S^c was also determined for each target location prior to optimization. The concentration was computed as the peak salinity concentration at each location prior to any fresh water diversions upstream. Hence, the optimized diversion rates, in theory, will not allow the salinity concentrations at the target location to exceed the salinity levels experienced prior to any diversions along the channel reach.

The target location is first placed near the upstream end of the brackish water zone using $x_T = 12\,km$ (cell 120). The concentration at this cell in the absence of diversions ranges from $\sim 0.0\,ppt$ to $0.8\,ppt$. In order to allow for a non-zero diversion vector the critical concentration is set to a value of $S^c = 1.0\,ppt$. Using an initial estimate for the solution vector given by a constant rate of $0.200\,m^2/s$, the optimization procedure is applied. A solution is achieved after just 8 iterations, requiring just 5 minutes of computational time on a desktop P-II workstation with 200MHz processor. The final, initial, and maximum allowable concentration histories are shown in Fig. 4. The corresponding initial guess and optimal outflow vector are presented in Fig. 5.

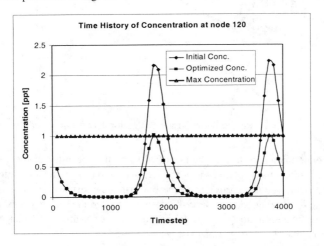

Figure 4: Resulting Time History of Concentration for 1 Diversion Node

The observable return of the solution vector to its initial guess at the end of the simulation time is a typical characteristic of the adjoint method and can not be avoided. However, it can easily be taken into account be extending the total simulation time in order to move this section beyond the time-period of interest. It is clear from the variation in lateral outflow present in the solution vector that substantial changes in the lateral outflow rate must be implemented to achieve the desired objective. These changes are not trivial, in fact they would be extremely difficult to predict given any other less sophisticated method. Additionally, various initial guesses were tested to investigate whether the selection of the start vector influences the outcome of the search. While the quality of the initial guess was found to influence the rate of convergence, all trials resulted in the same output vector. This suggests that the solution domain (comprised of 400 parameters) displays a monotone convex surface and therefore allows for the existence of

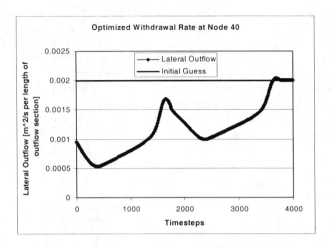

Figure 5: Resulting Withdrawal Rate for 1 Diversion Node

an absolute minimum that can be identified using the suggested gradient search method. Additionally, placement of the target location in the middle (kilometer 14) and at the end (kilometer 16) of the brackish zone revealed that it is best to place the monitoring point towards the upstream end so the search algorithm can make use of the salinity gradient. It appears that the sensitivity calculated from the tip or midsection of the brackish zone is less sensitive to small changes in salinity concentrations.

Multiple Diversion Nodes

In practice, water is diverted at more than just one node. As a result, a scenario is selected where two or more nodes are involved in the overall withdrawal situation. The modest increase in diversion node numbers is based on the assumption that the complexity increases substantially for each node added. This is due to the nature of withdrawal action that continuously sends waves through the flow domain as a reaction to each change in diversion rate at each node. Furthermore, for multiple diversion nodes it is necessary to use the more restricting measurement function from Eq. 4. The desired concentration history is being computed by assuming no diversion and then using the resulting history as a base-function. This base function is then multiplied by a constant factor larger than one in order to open a window within which optimization can take place. For the following test it was assumed that peak values for salinity were allowed to exceed the ones of the base function by ~40%.

As in the previous section, the monitoring location is at channel kilometer 12, while diversion points are placed at kilometer 4 and 8. The initial conditions and critical salinity history are identical to the previous section. At each diversion point, the diversion is described using 100 parameters, which corresponds to the diversion rate change every 40 time steps. This is done to reduce the parameter space without compromising the flexibility in the temporal space. Initially, all parameter values are set to a uniform diversion rate of $0.0004 m^2/s/m$, a value on the low side in comparison to the rates used in the previous test.

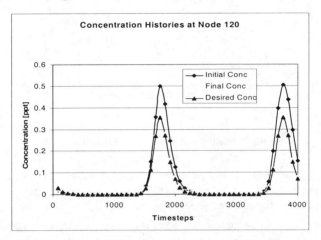

Figure 6: Resulting Time History of Concentration

Fig. 6 shows the critical concentration history in addition to the concentration histories resulting from the initial and optimal parameter estimates. The match between the optimized system response and the critical system response is nearly perfect, which is indicated by the very small residual functional of 0.034 as compared to the value resulting from the initial guess of $668.66 \frac{gr^2 s}{m^5}$.

Fig. 7 shows the optimized diversion rates for each diversion point. At both diversion points, the diversions are reduced from the initial guess. The reduction occurs earlier at kilometer 4 than kilometer 8. This is due to the prolonged travel time associated with information moving between kilometer 4 and 12, versus 8 and 12. If a diversion point closer to the monitoring location is chosen instead, then the lag time would become even smaller. Additionally, when the impact of the control action at kilometer 4 passes kilometer 8, the diversion rate at 8 tries to counterbalance the action at kilometer 4 with a slight increase in the diversion rate. A further decrease in the diversion rate at point 4 together with more arriving

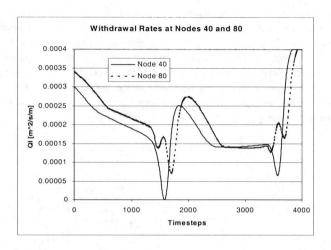

Figure 7: Resulting Withdrawal at Kilometer 4 and 8

sensitivity information from the target point at kilometer 12 forces the diversion at kilometer 8 to drastically reduce outflow. The same reactive pattern can be observed when the second peak travels upstream.

The uniqueness of the solution was examined by repeating the optimization process using different initial estimates for the diversion rates. Specifically, initial diversions of 0.0002 and $0.00005 m^2/s/m$ were tested. In all cases the resulting withdrawal pattern remained the same, with satisfactory agreement among the descrete withdrawal values. The pattern is characterized by a brief shut down of the diversion near high tide, followed by a substantial increase and then a gradual decrease. As in the previous section the diversion rates in the final portion of the time series are unaffected by optimization. In all cases, the search algorithm found the general shape of the solution vectors quickly (in less than 10 function calls). The remainder of the execution time, about 80%, was spent by the search algorithm fine-tuning the result.

To further examine the applicability of the proposed algorithm a number of test runs were performed with the number of diversion nodes increased to 3, 4 and 5. For either attempt the quasi-Newton-method failed. For the 3-node scenario, either using 10 or 100 parameters per node, the simulated annealing method was able to identify an acceptable solution vector. However, the required CPU time was excessive (about 16hrs) and because of that reason no runs were performed for the 4- or 5-node scenario using the simulated annealing method.

IV - CONCLUSIONS

A mathematical model and numerical solution method capable of identifying the optimal diversion rate are presented. The best location at which the salinity concentration can be constrained is near the upstream boundary of the brackish water zone. Optimization is achieved using a quasi-Newton gradient search algorithm. Gradient information is supplied by the solution of the adjoint problem that is derived and presented here. The numerical solution for solving the basic and adjoint set of equations is facilitated using a second order accurate finite-volume method. The performance of the optimization method in a series of control problems reveals that sensitivities are well resolved by the adjoint problem solution. In addition, the efficiency of the overall method, made possible by the rapid evaluation of the sensitivities by the adjoint problem, yields a technique that is capable of determining a control action in real-time. Hence, this control algorithm can be coupled with a water quality forecasting model to provide engineers with on-line recommendations for control actions that achieve a desired operational goal.

The performance of the control algorithm depends on the number of diversion nodes involved. For one- and two-node setups the algorithm proves to be ideally suited to mitigate adverse water quality conditions such as the upstream migration of saline water in real-time. The increase to three or more nodes scenarios renders the shape of the functional domain extremely complex. The complexity arrives from the inherent presence of many local minima and saddle points that apparently characterize the solution domain. This leads to a deterioration of the performance of the gradient search methods and the advantage of using the adjoint method becomes less pronounced. In this case a global search algorithm is necessary to find acceptable solution vectors, however, at great computational costs.

APPENDIX I - REFERENCES

Ciarlet, P.G. (1989) "Introduction to Numerical Linear Algebra and Optimisation", Cambridge Texts in Applied Mathematics, Cambridge University Press.

Chertok, D.L. and Lardner, R.W., (1996) "Variational Data Assimilation for a Nonlinear Hydraulic Model," Applied Mathematical Modeling, Vol. 20, pp. 675-682.

da Conceicao Cunha, M., and Sousa, J. (1999) "Water Distribution Network Design Optimization: Simulated Annealing Approach" J. of Water Resources Planning and Management, ASCE, Vol. 125, No. 4, July/August 1999, pp. 215-221.

Das, S.K. and Lardner, R.W., (1992) "Variational Parameter Estimation for a Two-Dimensional Numerical Tidal Model," International Journal for Numerical Methods in Fluids, Vol. 15, pp. 313-327.

Denton, R.A. and Briggs, D.A. (1997) "Incorporating a New Salinity-Outflow Algorithm Into an Operations Model for the Central Valley, California," Proceedings of the XXVII IAHR Congress, Theme B. p. 611-616.

Fischer, H.B., List, E.J., Koh, R.C.Y., Imberger, J., and Brooks, N.H., (1979) "Mixing in Inland and Coastal Waters," Academic Press.

Goldman, C.R. and Horne, A.J. (1983) "Limnology," McGraw Hill.

Katopodes, N.D. and Piasecki, M., (1996) "Site and Size Optimization of Contaminant Sources in Surface Water Systems," ASCE J. of Environmental Engineering, Vol. 122, 1996, pp. 917-923.

Marchuk, G.I., (1995) "Adjoint Equations and Analysis of Complex Systems," Kluwer, 1995.

Odd, N., (1981) "The predictive ability of one-dimensional estuary models," Transport Models for Inland and Coastal Waters, H.B. Fischer, ed., pp. 39-62.

Panchang, V.G. and O'Brien, J.J., (1989) "On the Determination of Hydraulic Model Parameters Using the Adjoint State Formulation," Modeling Marine Systems, Vol. I, pp. 5-18, A.M. Davies, ed., CRC Press.

Piasecki, M. and Katopodes, N.D., (1997a) "Control of Contaminant Releases in Rivers, I: Adjoint Sensitivity Analysis," ASCE J. of Hydraulic Engineering, Vol. 123, pp. 486-492.

Piasecki, M. and Katopodes, N.D., (1997b) "Control of Contaminant Releases in Rivers, II: Optimal Design," ASCE J. of Hydraulic Engineering, Vol. 123, pp. 493-503.

Press, W.H., Flannery, B.P., Teukolsky, S.A., Vetterling, W.T., (1990) Numerical Recipes, Cambridge University Press.

Sanders, B.F., (1997) "Control of Shallow-Water Flow Using the Adjoint Sensitivity Method," Ph.D Thesis, Department of Civil and Environmental Engineering, University of Michigan, 1997.

Sanders, B.F. and Katopodes, N.D., (1999) "Active Flood Hazard Mitigation Part I: Bidirectional Wave Control", ASCE J. of Hydraulic Engineering, Vol. 125, No. 10, October 1999, pp. 1057-1070.

Sanders B.F., Piasecki M., "Adaptive Control of Salinity in Well-Mixed Estuaries", Journal of Hydraulic Engineering ASCE, in review , January 2000.

Shanno, D.F. and Phua, K.H. (1980) "Remark on algorithm 500, a variable metric method for unconstrained minimization," ACM Trans. Math. Softw., Vol. 6, pp. 618-622.

An Unsteady Finite Volume Circulation Model

Matthew C. Ward[1]

Abstract

A finite-volume formulation of the two-dimensional, unsteady shallow water equations was developed to predict sea surface elevation and currents in estuarine and coastal waters on a system of unstructured grids. The finite volume methodology integrates the governing equations over discrete spatial volumes and conservation of mass and momentum. The approach allows for accurate discretization of complex regions without resorting to coordinate transformations. The governing equations are expressed as a Helmholtz equation in terms of sea-surface elevation and the vertically averaged conservation of momentum equations and solved using a semi-implicit solution methodology that overcomes the time step restriction of standard explicit schemes.

Model predictions were compared to the analytic solutions of a standing wave in a rectangular closed basin of constant depth, periodic forcing in rectangular basins of constant and linearly varying depth, and wind setup in closed basins of varying geometry. Each test included a sensitivity study of the model solution accuracy to grid Courant number, which ranged from 1 to as high as 13, resulting in maximum errors on the order of 5% for the surface elevation and velocity. The model was applied to Greenwich Bay, RI to evaluate its ability to predict tidally driven flow in an estuary of complex geometry and bathymetry. The simulations conducted were in good agreement with available surface elevation data and consistent with shallow water theory.

---

[1] Hydrodynamic Modeler, Applied Science Associates, Narragansett, RI

Introduction

To understand tidal and storm surge dynamics in coastal and estuarine waters, investigators typically employ hydrodynamic models derived from the two-dimensional shallow water equations (SWE). The presence of irregular coastlines, islands, man-made structures and the dynamics of coastal boundary layers, encountered in these regions, often require large variations in the spatial discretization of the computational domain. Two approaches, in common use, address this issue: 1) the finite element method (FEM) on unstructured grids and 2) the finite difference method (FDM) using curvilinear or boundary fitted grids. The finite element method, in conjunction with unstructured grids, provides a convenient method to describe complex domains, by allowing the governing equations to be solved in a Cartesian coordinate system without having to resort to complex coordinate transforms. Models employing the finite difference method, on curvilinear grids, provide great flexibility in representing complex regions and allow for the application of well-known and tested differencing schemes. The drawback, of this approach, is that the equations of motion are considerably more complex than their orthogonal counterparts due to the transformation from rectangular to general curvilinear space.

An alternative to the FEM and the FDM is the finite volume method (FVM). In this approach the computational domain is divided into discrete control volumes, similar to the FEM. However, the FVM employs integration of the governing equations over each control volume, in contrast to the weighted residual or variational approaches of the FEM and to the Taylor series expansions of the FDM. The volumetric integral equations express the conservation of the dependent variables for each control volume, just as the differential equation expresses conservation for an infinitesimal control volume.

Recently, several investigators have developed finite volume models of the shallow water equations (Zhao et al., 1994,1996; Zhou, 1995; Chippada et al., 1996; Mingham and Causon, 1997; Hu et al., 1997). Zhao et al. (1994) developed an explicit model of the SWE, on a grid composed of triangular and quadrilateral elements, to investigate river basin systems. In order to capture hydraulic shock waves, the numerical fluxes were posed as a Riemann problem and solved using the Osher scheme. Continuing this work, Zhao et al. (1996) presented three approximate Riemann solvers, flux vector splitting, flux difference splitting and the Osher scheme, to model hydraulic shock waves. The Osher scheme was found to be the most accurate and the flux vector splitting the least accurate for cases involving one and two dimensional dam break flows. Again, as in the previous work, the model employed a fully explicit time scheme. Chippada et al. (1996) presented a model, employing an explicit time scheme, based upon a vorticity formulation of the SWE on unstructured grids composed of triangular elements.

Mingham and Causon (1997) developed an explicit flow solver for the SWE based on the FVM on boundary fitted grids. The model employed an upwind scheme of the Godunov type using cell-centered collocated data which was extrapolated to the cell interfaces using the MUSCL (Monotonic Upwind Scheme for Conservative Laws) reconstruction approach and slope limiters to limit the occurrence of spurious oscillations. The model performed well for test cases involving one-dimensional dam break flow, steady state hydraulic jump, oblique bore wave reflection, and tidally induced flow in a narrow channel. Hu et al., (1997) developed a model similar to that of Mingham and Causon (1997) except using a two-stage Hancock time integration scheme. This model was applied to determine the effect of the construction of a stormwater tank on a sea wall.

The use of explicit time schemes, in the above models, requires relatively small timesteps and are limited to Courant numbers generally no greater than 1, in accordance with the Courant-Friedrichs-Lewy (CFL) stability condition. Thus, the use of explicit time schemes leads to computationally expensive simulations when grid resolutions are refined and water depths are relatively deep.

The purpose of this paper is to present the development and testing of a semi-implicit finite volume formulation of the shallow water equations on unstructured triangular grids. The semi-implicit method (Madala and Piasek, 1977) involves the solution of a Helmholtz equation in terms of the sea surface elevation. This formulation permits an iterative technique to be employed for the solution of the sea surface elevation and an explicit calculation of the velocities, thereby removing the CFL stability condition constraint. Also, the use of unstructured grids allows the system of governing equations to be solved on a Cartesian coordinate system without resorting to coordinate transforms.

The paper first presents the governing equations in rectangular coordinates and the associated boundary conditions. This is followed by the presentation of the discretization of the governing equations, the solution methodology, and model testing for selected cases where analytic solutions are available. The model is then applied to simulate tidal circulation in Greenwich Bay and the results compared with available observations.

Governing Equations

The depth averaged shallow water equations on a rectangular coordinate system (Kowalik and Murty, 1993), where y is positive north and x is positive east can be expressed, respectively, as

$$\rho_0 \left(\frac{\partial v}{\partial t} + u \frac{\partial v}{\partial x} + v \frac{\partial v}{\partial y} + fu \right) = -\frac{\partial P_a}{\partial y} - g\rho_0 \frac{\partial \eta}{\partial y} + \frac{\tau_{sy}}{D} - \frac{\tau_{by}}{D} + \rho_0 N_h \left(\frac{\partial^2 v}{\partial x^2} + \frac{\partial^2 v}{\partial y^2} \right), \qquad (1a)$$

$$\rho_0\left(\frac{\partial u}{\partial t}+u\frac{\partial u}{\partial x}+v\frac{\partial u}{\partial y}-fv\right)=-\frac{\partial P_a}{\partial x}-g\rho_0\frac{\partial \eta}{\partial x}+\frac{\tau_{sx}}{D}-\frac{\tau_{bx}}{D}+\rho_0 N_h\left(\frac{\partial^2 u}{\partial x^2}+\frac{\partial^2 u}{\partial y^2}\right),\qquad (1b)$$

where u and v are the vertically averaged velocities, f is the Coriolis parameter, D is the total depth, P_a is the atmospheric pressure, η is the sea surface elevation, ρ_0 is the sea water density, N_h is the horizontal eddy viscosity, and τ_s and τ_b are the surface and bottom stresses, respectively. The surface and bottom stresses are related to the wind velocity and depth averaged velocities using quadratic drag laws and are expressed, respectively, as

$$\tau_{sx}=C_s\rho_a W_x\left(W_x^2+W_y^2\right)^{1/2}\ ,\ \ \tau_{sy}=C_s\rho_a W_y\left(W_x^2+W_y^2\right)^{1/2}\ ,\qquad (3a,b)$$

$$\tau_{bx}=C_b\rho_0 u\left(u^2+v^2\right)^{1/2}\ ,\ \ \tau_{by}=C_b\rho_0 v\left(u^2+v^2\right)^{1/2}\ ,\qquad (3c,d)$$

where C_s and C_b are the surface and bottom drag coefficients, W is the wind speed with the subscripts indicating direction, and ρ_a is the density of the air. To complete the set of equations, the vertically integrated equation of water mass conservation is given by

$$\frac{\partial \eta}{\partial t}+\frac{\partial(uD)}{\partial x}+\frac{\partial(vD)}{\partial y}=0\cdot\qquad (4)$$

Two types of boundary conditions are typically used for the above system of equations. On a closed boundary (land) there is no momentum flux and the normal velocities are set to zero thus, $\vec{V}\cdot\vec{n}=0$. At an open boundary either the vertically averaged velocity or the sea surface elevation is specified as a function of time.

This set of second order, non-linear equations is of mixed hyperbolic-parabolic type. However, in most situations the second order viscous terms are negligible, reducing the equations to first order, and will be omitted in the proceeding sections.

Finite Volume Discretization

This section presents the finite volume discretization on unstructured grids of the system of equations developed in the preceding section. The choice of unstructured grids stems from the complicated coastlines often encountered while investigating coastal and estuary dynamics. Traditional solution methodologies based on Cartesian grids give a staircase approximation to the boundaries when the boundaries are not parallel to the coordinate axes. This can lead to erroneous flows in these regions. In addition, Cartesian grids do not give the user flexibility

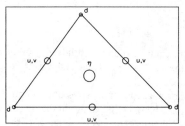

Figure 1: Typical triangular element.

in concentrating mesh points without enriching the entire mesh and suffering a subsequent decrease in computational speed. Unstructured gridding methodologies overcome these two problems while allowing computations to be performed in a Cartesian coordinate system.

The computational grid is composed of triangular elements of the form shown in Figure 1. The sea surface elevation is located at the centroid of the element, the bottom depth (d) at the vertices, and the vertically averaged velocities at the midpoints of the element's sides. This grid arrangement has three primary advantages, 1) there is no ambiguity associated with the direction of the normal vectors, 2) it allows a clear definition of the boundary conditions, and 3) it provides high resolution of the bathymetry relative to the number of elements.

The vertically averaged continuity equation (4) can be expressed in the following form,

$$div\left(D\vec{V}\right) + \frac{\partial \eta}{\partial t} = 0 \, . \tag{5}$$

Integrating Equation 5 over the control volume, shown in Figure 1, produces the following volumetric integral equation,

$$\iiint_{V} div\left(D\vec{V}\right) dV + \frac{\partial}{\partial t} \iiint_{V} \eta \, dV = 0 \, . \tag{6}$$

By applying Gauss's divergence theorem the volumetric integral, containing the divergence of the velocity field, is transformed into a surface integral,

$$\iint_{S} \left(D\vec{V} \bullet \hat{n}\right) dS + \frac{\partial}{\partial t} \iiint_{V} \eta \, dV = 0 \, . \tag{7}$$

This in turn allows for the spatial discretization of the vertically averaged continuity equation,

$$A_c \frac{\partial \eta}{\partial t} + \sum_{i=1}^{3} \left(D\vec{V} \bullet \hat{n} \right)_i l_i = 0, \tag{8}$$

where A_c is the area of the control volume, i is the index of the cell's sides, $\vec{V}$ is the velocity vector, l_i is the length of a cell side and $\hat{n}$ is the outward directed normal vector to a cell side.

In order to evaluate the total depth at the sides of the control volume, a simple arithmetic average is applied between the values of the surface elevation located at the centroids of the neighboring cells and the centroid of the control volume. The values of the bottom depth are averaged over the cell's sides. The full spatial discretization is then expressed as,

$$A_c \frac{\partial \eta}{\partial t} + \sum_{i=1}^{3} \overline{D}_i J_i = 0, \text{ where } J_i = \left(\vec{V} \bullet \hat{n} \right)_i l_i = \left(un_x + vn_y \right)_i l_i, \tag{9}$$

$\overline{D}$ is the average total depth, and n_x and n_y are the outward directed normal vectors in the x and y directions, respectively. In order to complete the formulation, Equation 9 is discretized with respect to time, resulting in the solution for sea surface elevation,

$$\eta_c^{n+1} = \eta_c^{n-1} - \frac{\Delta t}{A_c} \sum_{i=1}^{3} \overline{D}_i^n \left(J_i^{n+1} + J_i^{n-1} \right), \tag{10}$$

where n+1 is the time level to be predicted, n is the current time level, and n-1 is the previous time level.

The control volume for the solution of the vertically averaged velocities is somewhat more complicated than that used for the sea surface elevation. Since, the velocities are defined at the sides of the cells, it is convenient to define the

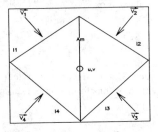

Figure 2: Control volume for velocities.

control volume for the velocities as shown in Figure 2 (Chippada et al., 1996). Note that the elements adjacent to the side form the control volume, resulting in overlapping control volumes.

The integration of Equation 1a over the control volume shown in Figure 2 produces the following volumetric integral equation,

$$\frac{\partial}{\partial t} \iiint_V v \, dV = -\iiint_V u(\hat{i} \bullet \vec{\nabla} v) \, dV - \iiint_V v(\hat{j} \bullet \vec{\nabla} v) \, dV - f \iiint_V u \, dV$$

$$-g \iiint_V (\hat{j} \bullet \vec{\nabla} \eta) \, dV - \frac{1}{\rho_o} \iiint_V (\hat{j} \bullet \vec{\nabla} P_a) \, dV + \frac{1}{\rho_o} \iiint_V \frac{\tau_{sy} - \tau_{by}}{D} \, dV \quad (11)$$

Due to its similarity to Equation 1a, the discretized form of Equation 1b is presented at the end of this section. The resulting spatial discretization of the equation of motion in the y-direction is

$$A_m \frac{\partial v_m}{\partial t} = -u_m \sum_{j=1}^{4} (n_x v)_j l_j - v_m \sum_{j=1}^{4} (n_y v)_j l_j - A_m f u_m$$

$$-g \sum_{j=1}^{4} (n_y \eta)_j l_j - \frac{1}{\rho_o} \sum_{j=1}^{4} (n_y P_a)_j l_j + A_m \frac{1}{\rho_o} \frac{(\tau_{sy} - \tau_{by})_m}{D_m} \quad , \quad (12)$$

where A_m is the area of the velocity control volume, j is the index of the sides forming the control volume, and u_m, v_m, $\overline{D}_m$, are the velocities in the x and y directions and the average total depth, respectively, at the mid-point of the side surrounded by the control volume. Next the equation is discretized with respect to time, resulting in the solution of the velocity in the y-direction,

$$v_m^{n+1} = VOLD + \frac{\Delta t}{A_m \theta} [CONVY + CORRY + PRESSY + SEAY + STRESSY], \quad (13)$$

where

$$VOLD = v_m^n - \frac{(1-\theta)}{\theta} [v_m^n - v_m^{n-1}]$$

$$CONVY = -u_m^n \sum_{j=1}^{4} (n_x l)_j v_j^n - v_m^n \sum_{j=1}^{4} (n_y l)_j v_j^n ,$$

$$CORRY = -A_m f u_m^n ,$$

$$PRESSY = -\frac{1}{\rho_o} \sum_{j=1}^{4} (n_y l)_j P_{aj}^n ,$$

$$SEAY = -\phi \sum_{j=1}^{4} g(n_y l)_j \eta_j^{n+1} - \omega \sum_{j=1}^{4} g(n_y l)_j \eta_j^n - (1 - \phi - \omega) \sum_{j=1}^{4} g(n_y l)_j \eta_j^{n-1} ,$$

$$STRESSY = A_m \frac{1}{\rho_o} \frac{\left(\tau_{sy} - \tau_{by}\right)^n_m}{\overline{D}^n_m} \, ,$$

and θ, ϕ, and ω are weighting factors used to control the order of the time scheme approximation (Kinnmark, 1986). The approximation of the unsteady and sea surface elevation terms are second order accurate and centered around $t+(\theta-1/2)\Delta t$ if the following condition on ϕ and ω is maintained: $\omega + 2\phi = 1/2 + \theta$.

The equation describing motion in the x-direction is developed through the same procedure as above, and is given by:

$$u_m^{n+1} = UOLD + \frac{\Delta t}{A_m \theta}\left[CONVX + CORRX + PRESSX + SEAX + STRESSX\right], \tag{14}$$

where

$$UOLD = u_m^n - \frac{(1-\theta)}{\theta}\left[u_m^n - u_m^{n-1}\right]$$

$$CONVX = -u_m^n \sum_{j=1}^{4}\left(n_x l\right)_j u_j^n - v_m^n \sum_{j=1}^{4}\left(n_y l\right)_j u_j^n \, ,$$

$$CORRX = A_m f v_m^n \, ,$$

$$PRESSX = -\frac{1}{\rho_o} \sum_{j=1}^{4}\left(n_x l\right)_j P_{aj}^n \, ,$$

$$SEAX = -\phi \sum_{j=1}^{4} g\left(n_x l\right)_j \eta_j^{n+1} - \omega \sum_{j=1}^{4} g\left(n_x l\right)_j \eta_j^n - \left(1 - \phi - \omega\right)\sum_{j=1}^{4} g\left(n_x l\right)_j \eta_j^{n-1} \, ,$$

$$STRESSX = A_m \frac{1}{\rho_o} \frac{\left(\tau_{sx} - \tau_{bx}\right)^n_m}{\overline{D}^n_m} \, .$$

The set of discretized equations developed in this section are second order accurate in time and first order in space. The lower order spatial accuracy is due to the simple arithmetic averaging of variables across the cell interfaces, which is equivalent to a first order central difference scheme.

Solution Methodology

The first step in the computation of the sea surface elevation and the vertically averaged velocities is to derive a Helmholtz equation in terms of the sea surface elevation. Upon substitution of Equations 13 and 14 into Equation 10 the Helmholtz equation becomes,

$$\eta_c^{n+1} = \frac{\eta_c^{n-1} - \frac{\Delta t}{A_c} \sum_{i=1}^{3} \overline{D}_i^n \left(u_i^* + v_i^* + J_i^{n-1} + NSEAXR_i + NSEAYR_i \right)}{1 + \frac{\Delta t}{A_c} \sum_{i=1}^{3} \overline{D}_i^n \left(NSEAXL_i + NSEAYL_i \right)},$$ (15)

where

$$u_i^* = \left(n_x l \right)_i \left\{ UOLD + \frac{\Delta t}{A_m} [CONVX + CORRX + PRESSX + OSEAX + STRESSX] \right\}_i,$$

$$v_i^* = \left(n_y l \right)_i \left\{ VOLD + \frac{\Delta t}{A_m} [CONVY + CORRY + PRESSY + OSEAY + STRESSY] \right\}_i,$$

$$NSEAXL = \left(n_x l \right)_i \left\{ \frac{\Delta t}{A_m \theta} \left[\frac{-\phi g}{2} \left\langle \left(n_x l \right)_{i1} + \left(n_x l \right)_{i2} \right\rangle \right] \right\},$$

$$NSEAYL = \left(n_y l \right)_i \left\{ \frac{\Delta t}{A_m \theta} \left[\frac{-\phi g}{2} \left\langle \left(n_y l \right)_{i1} + \left(n_y l \right)_{i2} \right\rangle \right] \right\},$$

$$NSEAXR = \left(n_x l \right)_i \left\{ \frac{\Delta t}{A_m \theta} \left[-\phi g \left\langle \frac{1}{2} \left(n_x l \right)_{i1} \eta_{1N}^{n+1} + \frac{1}{2} \left(n_x l \right)_{i2} \eta_{2N}^{n+1} + \left(n_x l \right)_{i3} \eta_{i3}^{n+1} + \left(n_x l \right)_{i4} \eta_{i4}^{n+1} \right\rangle \right] \right\},$$

$$NSEAYR = \left(n_y l \right)_i \left\{ \frac{\Delta t}{A_m \theta} \left[-\phi g \left\langle \frac{1}{2} \left(n_y l \right)_{i1} \eta_{1N}^{n+1} + \frac{1}{2} \left(n_y l \right)_{i2} \eta_{2N}^{n+1} + \left(n_y l \right)_{i3} \eta_{i3}^{n+1} + \left(n_y l \right)_{i4} \eta_{i4}^{n+1} \right\rangle \right] \right\}.$$

The formulation of the Helmholtz equation allows for the use of a semi-implicit solution methodology (Madala and Piasek, 1977) in which the sea surface elevation is solved implicitly using a Gauss-Siedel, point-by-point iterative scheme with successive over-relaxation. Once the solution for the sea surface elevation is available the velocities are solved explicitly. The semi-implicit approach is not hindered by the CFL stability condition thus allowing the use of large time steps which in turn increases computational efficiency. It should be noted that although the model's stability is maintained as the timestep increases the accuracy of the solution should be expected to decrease.

Model Testing

In order to evaluate the model's formulation and performance, several test cases were chosen for which analytical solutions were available for comparison. All the cases presented below neglect the convective acceleration and Coriolis terms.

The first case evaluated was a standing wave in a frictionless rectangular basin 1000m long with a constant depth of 40m. The analytical solution is presented in

Figure 3: Characteristic rectangular basin grid.

Sorenson (1993). The simulation was conducted on a mesh of right triangles (Figure 3) with normal velocities at the basin walls set to zero and at a Courant number of 1.7. The spatial scale used to determine the Courant number was one-third the height of the triangle. The simulation was initialized with a half cosine wave for the sea surface elevation and zero velocities. Figure 4a-b shows the solution for the surface elevation and velocity as a function of distance along the basin at 1/10 increments of the standing wave period. The agreement with the analytical solution is excellent throughout the entire simulation with maximum errors on the order of 4% for both the sea surface elevation and velocity. Further model tests show that finer spatial and temporal resolution reduce this error.

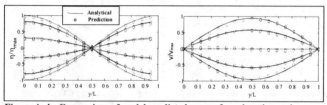

Figure 4a-b: Comparison of model predicted sea surface elevation and velocity for a standing wave in a closed basin.

In order to assess the performance of the semi-implicit technique in allowing time steps in excess of the CFL criteria, a sensitivity study of the error in the model predicted sea surface elevation as a function of the Courant number was conducted. The Courant number was varied from 1.7 to 13.6 by maintaining a constant grid size and varying the timestep. The percent error after 5 basin oscillation cycles for Courant numbers of 1.7, 3.4, 4.5, 6.8, 9.1, and 13.6 was approximately 2.5%, 3.2%, 3.5%, 4.2%, 4.5%, and 5.5%, respectively. The percent error was calculated by averaging the error, from an element far enough removed from the boundary, four cells, to minimize the possibility of boundary effects, over an entire oscillation period. For comparison a purely explicit technique becomes unstable for a Courant number greater than 1. The present model maintains stability for high Courant numbers. The accuracy decreases however as the Courant number increases.

Next the model's ability to simulate the response to periodic forcing was tested. The basin and grid configuration were identical to those for the standing wave case, except for the basin being open at one end. The surface elevation at the open end was forced using a cosine wave 1m in amplitude with a period of 50.5s. The analytical solution is given in Ippen (1966). Figure 5a-b shows the solution for the surface elevation and velocity as a function of distance along the basin at 1/10 increments of the forcing period. Again, the agreement with the analytical solution is excellent throughout the entire simulation with maximum errors of 3% and 7% for the sea surface elevation and velocity, respectively. As an additional test, the basin was rotated ±45° and ±135° and forced with the same conditions as above. The results showed no appreciable deviation from the original basin orientation. Thus, confirming the model's insensitivity to grid orientation.

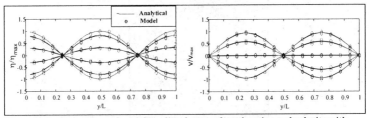

Figure 5a-b: Comparison of model predicted sea surface elevation and velocity with the analytic solution for periodic forcing in a basin open at one end.

In order to study the effects of varying bottom depth, a simulation was conducted involving periodic forcing in a frictionless rectangular basin with bottom depths of 60 m and 10 m at the open and closed ends, respectively, resulting in a linear bottom slope of 0.05. The analytical solution is given in Lynch and Gray (1978). Figure 6a-b shows the solution for the surface elevation and velocity as a function of distance along the basin at 1/10 increments of the forcing period. The agreement with the analytical solution is similar to that of the periodic forcing of a flat bottom basin with maximum errors on the order of 4% and 7% for the sea surface elevation and velocity, respectively.

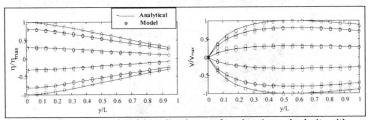

Figure 6a-b: Comparison of model predicted sea surface elevation and velocity with the analytic solution for periodic forcing in a rectangular basin open at one end with a bottom slope of 0.05.

The final test conducted was wind setup in a closed rectangular and irregular basin (Fig. 7) with a constant depth of ten meters, length of 50km, and a bottom friction factor (C_b) of 2×10^{-3}. The simulation was performed using a constant, uniform wind stress of $0.4N/m^2$, in a northerly direction. The model was started from rest and driven by constant wind stress until steady state was achieved. The predicted sea surface slope for both cases (4×10^{-6}) was in agreement with the analytical solution presented in Horikawa (1978) which is independent of geometry.

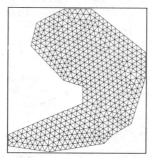

Figure 7: Grid configuration for wind setup in an irregular basin.

Model Application

The model was applied to predict tidal circulation in Greenwich Bay (Figure 8). The bay is located in north-western Narragansett Bay and contains a central basin and four small coves (Warwick, Brush Neck, Apponaug, and Greenwich). The central basin has a mean depth of 2.7m while the coves have mean depths on the order of 1 to 2m. A distinct bathymetric feature of the bay is an entry channel, near its mouth, with a mean depth of 10m which decreases in depth towards the west. Due to its small width (5 km) the Coriolis terms were not included in the computations.

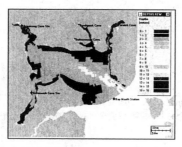

Figure 8: Greenwich Bay.

Simulations were performed using an unstructured grid (Figure 9) composed of 110 cells. The grid is coarse, designed only to represent the dominant characteristics of the bay, with cells ranging in area from $2 \times 10^4 m^2$ to $4 \times 10^5 m^2$. Depth data was derived from NOAA bathymetric data (NOAA/NOS, Chart #13224) by averaging data within a 20m radius around each grid node.

The model was forced, at the open boundary, by a time series consisting of the 17 major tidal constituents. The time series algorithm was developed by the Department of Ocean Engineering at the University of Rhode Island in order to simulate tidal forcing within Greenwich Bay for any calendar month or year. The Department of Ocean Engineering maintains a YSI/ENDECO 6000 sensor system moored near the open boundary from which surface elevation data was collected in order to develop the tidal component time series.

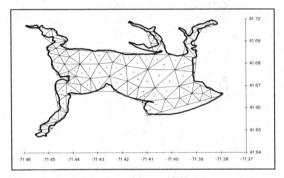

Figure 9: Computational Grid for Greenwich Bay application.

The simulation was conducted using a bottom drag coefficient of 0.003. Initialized with a surface elevation of 0.5m and zero velocities and spun-up over a period of 4 tidal cycles. Figure 10 shows the maximum predicted flood tide on 20

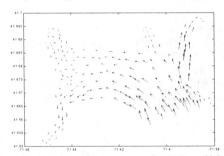

Figure 10: Model predicted maximum flood tide on 20 Spet. 1996.

Sept. 1996. The predicted currents are relatively uniform across the central basin
and decrease markedly upon entry into Brush Neck, Apponaug, and Greenwich
coves. The current speeds are greatest in areas of steep bathymetric gradients,
evident in the area surrounding the entry channel, near the open boundary. The
maximum ebb currents for 20 Sept. 1996 are similar to those of the flood except
in the opposite direction.

Figure 11 shows a comparison between the model predicted and observed sea
surface elevations at Greenwich Cove from 19 Sept. 1996 to 22 Sept. 1996. The
data was provided by the Department of Ocean Engineering, which maintains a
data collection buoy, similar to the one near the open boundary, in this cove. The
predictions are generally in good agreement with the observations. The tidal
phasing and tidal peak amplitudes are accurately represented. However,
predictions of individual tidal troughs are in error by as much as 20cm. This can
be attributed to the appropriate non-linear dynamics not being captured in the
coves due to the under resolution of the bathymetry in these areas.

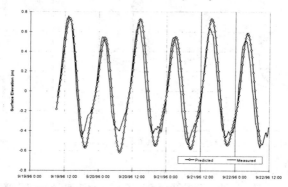

**Figure 11: Comparison between model predicted and observed sea surface
elevation at Greenwich Cove from 19 Sept. to 22 Sept. 1996.**

To investigate the effect of bottom friction, simulations were conducted with
varying quadratic bottom drag coefficients ranging from 0.001 to 0.005 and at a
maximum Courant number of 2. The results for both sea surface elevation and
velocity proved to be insensitive to these variations in drag coefficient. This is to
be expected due to the length scale of Greenwich Bay (5km) being much less than
the scale of the tidal wavelength, which is on the order of 400km.

Discussion

An unsteady vertically averaged finite control volume circulation model was developed for application in estuarine and coastal waters. The model's formulation and performance were tested against a series of problems for which analytical solutions were available. The model was then applied to Greenwich Bay.

Predicted sea surface elevation and currents were in excellent agreement with the analytical solutions for a standing wave in a closed constant depth rectangular basin; tidal forcing in a constant depth rectangular basin; tidal forcing in a linearly sloped rectangular basin; and wind setup in closed rectangular and irregular basins of constant depth. The results of the simulation conducted for Greenwich Bay showed the predicted sea surface elevation to have the proper phase and tidal peak amplitudes, when compared to available data from Greenwich Cove. Errors in the tidal troughs were as much as 20cm and were attributed to inadequate bathymetric resolution in the coves. Although the results were in good agreement with available data and theory, future investigations of this region should attempt to collect surface elevation and velocity data from all the coves and central basin in order to provide a greater level of comparison and validation.

A major premise for the development of this model was to remove the restriction of the CFL stability condition on the timestep through the implementation of a semi-implicit solution methodology. In order to address this issue a Courant number study was conducted for the standing wave test case. The resulting predictions showed excellent stability and reasonable accuracy for Courant numbers as high as 13.

The finite control volume, semi-implicit methodology developed in this paper shows promise and deserves further development. The method is stable and reasonably accurate for high Courant numbers in regular geometry. Focus needs to be placed on acquiring similar results in problems involving complex geometry and bathymetry. Future investigations will involve gridding schemes that produce high quality smooth grids e.g., boundary fitted grids or mixed quadrilateral and triangular grids.

Acknowledgments

The author wishes to thank Professors Frank M. White and Malcolm L. Spaulding of the University of Rhode Island for their support while completing this work.

References

Chippada, S., C.N. Dawson, M.L. Martinez, and M.F. Wheeler, 1996. Numerical Simulation of the Shallow Water Equations, 1996 Fluids Engineering Division Conference Volume 3.

Horikawa, K., 1978. Coastal Engineering - An Introduction to Ocean Engineering, University of Tokyo Press.

Hu, K., C.G. Mingham, and D.M. Causon, 1997. Numerical Simulation of Tidal Induced Circulation and Waves in Harbours in Computer Modelling of Seas and Coastal Regions III edited by J.R. Acinas and C.A. Brebia, pp. 325-334.

Ippen, A., 1966. Esturay and Coastline Hydrodynamics, McGraw Hill.

Kinnmack, I., 1986. The Shallow Water Equations: Formulation, Analysis, and Application, Springer and Verlag, New York.

Kowalik, Z., and Murty, T.S., 1993. Numerical Modelling of Ocean Dynamics, World Scientific Publishing Co. Pte. Ltd., Singapore.

Lynch, D.R., and Gray, W.G., 1978. Analytical Solutions for Computer Flow Model Testing, Journal of the Hydrodynamics Division, Proceedings of the American Society of Civil Engineers, Vol.104, No.HY10, pp.1409-1428.

Madala, R.V., and S.A. Piacsek, 1977. A Semi-Implicit Numerical Model for Baroclinic Oceans, Journal of Computational Physics, 23, pp. 167-178.

Mingham, C.G., and D.M. Causon, 1997. On High Resolution Finite Volume Modelling of Discontinuos Solutions of the Shallow Water Equations in Computer Modelling of Seas and Coastal Regions III edited by J.R. Acinas and C.A. Brebia, pp. 43-52.

Sorensen, R.M., 1993. Basic Wave Mechanics for Coastal and Ocean Engineers, John Wiley and Sons, Inc., New York.

Zhao, D.H., H.W. Shen, G.Q. Tabios III, J.S. Lai, and W.Y. Tan, 1994. Finite-Volume Two-Dimensional Unsteady-Flow Model for River Basins, Journal of Hydraulic Engineering, Vol. 120, No. 7, pp. 863-883.

Zhao, D.H., H.W. Shen, J.S. Lai, and G.Q. Tabios III, 1996. Approximate Riemann Solvers in FVM for 2D Hydraulic Shock Wave Modelling, Journal of Hydraulic Engineering, Vol. 122, No. 12, pp. 692-702.

Zhou, J.G., 1995. Velocity-Depth Coupling in Shallow Water Flows, Journal of Hydraulic Engineering, Vol. 121, N0. 10, pp. 717-724.

MODELING THE IMPACTS OF DREDGED MATERIAL
PLACEMENT IN UPPER CHESAPEAKE BAY

by

Billy H. Johnson[1], Member ASCE: Hans R. Moritz[2]; Allen M. Teeter[3]; Carl F. Cerco[4]; and Harry V. Wang[5]

ABSTRACT

The potential for placed material to leave the disposal site and the impact on water quality of placing 18 million cubic yards (cu yds) of dredge material at Site 104 in Upper Chesapeake Bay over a five year period have been determined through numerical modeling. The placement method modeled is bottom disposal from a spilt-hull barge. The material is fine-grained maintenance material to be dredged from several upper bay channels. The models used were STFATE (Short-Term FATE), MDFATE (Multiple Disposal FATE), CH3D (Curvilinear Hydrodynamics in three dimensional (3D)), and CEQUAL-ICM (Corps Engineers QUALity -Integrated Compartment Model). STFATE computes the short-term fate of material as it falls through the water column and impacts the bottom. MDFATE builds mounds and subsequently computes the erosion of material from the mounds, consolidation of the mounds, and slumping of the mounds. CH3D is a 3D numerical hydrodynamic model that was applied on a relatively fine grid of the entire upper Chesapeake Bay to provide bottom shear stresses and currents to MDFATE. CEQUAL-ICM is a 3D eutrophication water quality model of the

1, 3, 4 - U.S. Army Engineer Research and Development Center, Vicksb;urg, MS
2 - U.S. Army Engineer District, Portland, Portland, OR
5 - Virginia Institute of Marine Science, Gloucester Point, VA

entire Chesapeake Bay. Erosion tests were performed on samples of the material to be dredged to provide insight on the erosional properties of the material. Results from the modeling were that the total "potential" sediment losses would be about 17%. About 3-4% of the 17% of material potentially lost from the site is material that will never be deposited. The remainder is material that will be eroded over time after placement on the bottom. No significant changes were computed in the dissolved oxygen at Site 104 after placement of the dredged material.

INTRODUCTION

Site 104 is a previously used open-water, dredged material placement site that is located in the upper Chesapeake Bay (Figure 1). At the request of the Maryland Port Administration (MPA), the U.S. Army Engineer District, Baltimore, is evaluating Site 104 for placement of approximately 18 million cu yds of clean dredged material over a five year period. The dredged material that is proposed for open-water placement at Site 104 will be dredged from federal navigation channels including the Craighill Entrance, Craighill Channel, Craighill Angle, Craighill Upper Range, Cutoff Angle, Brewerton Channel Eastern Extension, Swan Point Channel, Tolchester Channel, and the Southern Approach Channel to the C&D Canal (Figure 1).

Three questions that have been posed in the evaluation of the site are: (1) How much of the material placed at the site has the potential to leave the site? (2) If material leaves the site, where will it ultimately be deposited? (3) What will be the water quality impact of placing material at the site? To aid in addressing questions 1 and 3, the Maryland Environmental Service (MES) under the sponsorship of the MPA requested that the Coastal and Hydraulics Laboratory of the U.S. Army Engineer Research and Development Center at Vicksburg, MS conduct a numerical modeling study. To address question 2 would require application of a 3D sediment transport model of much of the Chesapeake Bay. The development of such a model was beyond the scope of this study.

The study utilized existing full bay 3D hydrodynamics and water quality models and dredged material placement models developed under the Corps Dredging Research Program (DRP). In addition, an upper bay 3D hydrodynamics model was developed. The manner in which these models were applied and results from the study are presented.

NUMERICAL MODELS

The following models were applied to answer the questions of how much of the dredged material placed at the site has the potential to leave the site and what will be the impact on water quality at the site as a result of material placement.

CH3D (Curvilinear Hydrodynamics in 3D) - CH3D is a numerical model for computing 3D hydrodynamics in coastal, estuarine, and riverine environments. As its name implies, CH3D makes computations on a curvilinear or boundary-fitted grid. Physical processes impacting circulation that are modeled include tides, wind, density effects (salinity and temperature), freshwater inflow, turbulence, and the effect of the earth's rotation. Details concerning the equations solved and numerical techniques employed can be found in Johnson et al. (1991).

Under the Chesapeake Bay Program (CBP), a CH3D model of the entire Chesapeake Bay was developed and applied to provide flow fields for the years of 1985-94 to the 3D water quality model of the full bay discussed below. The full bay plan form numerical grid is shown in Figure 2. Grid cell dimensions are 2-4 km. A maximum of 20 layers represents the vertical, with each layer being 1.52 m thick except for the top layer which varies with the tide. For the Site 104 study, a CH3D model was developed on a refined gird of the upper Chesapeake Bay (Figure 3). Typical grid cell dimensions are 300-1000 m.

CEQUAL-ICM (Corps of Engineers Quality-Integrated Compartment Model) - Cereo and Cole (1994) developed a 3 D water quality model of the full Chesapeake Bay that was driven by flow fields provided by CH3D. The model calculates 22 state variables. These include components of the nitrogen, carbon, phosphorous, and silica cycles. In addition, there is a fully predictive benthic sediment diagensis sub-model. This sub-model was used to provide information on characteristics of the material proposed for dredging from the upper bay channels and to predict how newly deposited dredged material would interact with the water column. The end product from the model is a computation for the dissolved oxygen over the water column. The model makes computations in such a manner that flow fields computed on both structured and unstructured grids can be imported to drive model transport.

STFATE (Short-Term FATE) - STFATE is a model for computing the fate of material placed from either a split-hull barge or a hopper dredge (Johnson and Fong 1995). The model computes the movement of the material from the moment it is injected into the water column until the material is either deposited on the sea floor or transported out of the grid covering the site. The computations consist of three phases. The first is the convective descent of the dredged material cloud through the water column, during which the cloud grows as a result of the entrainment of ambient water. Normally, the descending cloud of material strikes the bottom with a resulting spreading of the material on the sea floor (Phase 2). Finally, material in suspension is transported and diffused by the ambient flow field (Phase 3).

MDFATE (Multiple Disposals FATE) - MDFATE is a numerical model that simulates open-water placement activities impacting both the short and long-term morphology of dredged material placed on the sea floor. MDFATE is used for planning and managing the use of open- water placement sites by bridging the gap between the modeling of individual placement events and tracking a myriad of placements that occur within a site over the duration of the site's operative cycle. The MDFATE model has evolved from several earlier concepts (Moritz and Randall 1992,1995).

MDFATE defines an open-water disposal site (ODMDS) in terms of a numerical grid and incorporates modified versions of stand-alone computer models to simulate bathymetry change resulting from a series of placement cycles. A version of the STFATE model discussed above and a model developed under the DRP called LTFATE (Long-Term FATE) (Scheffner et al. 1995) are coupled within the MDFATE simulation.

Once a particular ODMDS grid has been created, MDFATE can be used to simulate a given placement operation that may extend over months and consist of hundreds of placements. The entire placement operation is divided into separate week-long episodes over which long-term fate processes governing dredged material behavior on the sea floor are simulated using a modified version of LTFATE. Results are modeled in a cumulative manner. Long-term processes include self-weight consolidation, sediment erosion by waves and currents, and mound slumping.

Within each week-long episode, a modified version of STFATE simulates short-term fate processes that govern each placement operation, with the changing bathymetry accounted for in the computation of the cumulative mound distribution arising from each placement. Short-term processes are those that influence placed material up to the point at which all energy possessed by the material is expended. An assumption in the MDFATE model is that other than the placed dredged material, no sedimentary material enters the ODMDS.

STUDY APPROACH

Results from the MDFATE, STFATE, and CEQUAL-ICM models provided the information required to assess the potential for sediment loss and the water quality impact of dredged material placed at Site 104. However, a critical component of the study was the development of the upper bay 3D hydrodynamic model to provide bottom shear stresses and vertically-averaged currents to MDFATE. Verification of the hydrodynamic model used tide, currents, salinity, and temperature data collected during October 1992 - September 1993 (Fagerburg 1995) along with data collected in July 1997 by Sanford (1997) and data collected by Hamilton and Boicourt (1984). Verification of the 3D hydrodynamic model is presented by Johnson, et al. (in preparation).

Based on the availability of data and the wide range of environmental forcings that occurred during 1993 in the upper Chesapeake Bay, the decision was made to use hydrodynamics from that year for input to MDFATE. The initial year of 1993 hydrodynamics was run with the existing bathymetry at Site 104 (Figure 4). After running MDFATE for the first year of the dredged material placement schedule shown in Table 1, CH3D was rerun with the next bathymetry predicted by MDFATE at the end of the first year of placement. The new hydrodynamics were then used in MDFATE to simulate the second year of placement. These steps continued until the entire 5 years of the placement schedule was completed. Thus, the effect of the reduction of the seabed elevation was included in the hydrodynamic modeling.

Table 1. Five year placement schedule		
Year	Volume of Dredged Material (cu yds)	Placement Duration (Days starting from 1 Nov)
1	2,502,000	139
2	4,500,000	125
3	3,024,000	168
4	6,408,000	168
5	2,016,000	112

As can be seen from Table 1, the total volume of material placed varied from one year to the next. The placement schedule was based on the State of Maryland's anticipated dredging needs. To maximize the capacity of Site 104 without creating excessively high mounds of placed material, locations of the individual placements were determined using a placement grid. The construction of the placement plan used in MDFATE is discussed later.

In addition to currents and bottom shear stresses, MDFATE requires values for the erosion parameters (critical shear stress and erosion rate constant) in the erosion model presented below.

$$E = M(\frac{\tau}{\tau_c} - 1) \qquad \text{If } \tau \text{ is larger than } \tau_c$$

$$E = 0.0 \qquad \text{If } \tau \text{ is less than } \tau_c$$

where

E = rate of erosion in kg/sq m/sec
M = erosion rate constant in kg/sq m/sec
τ = bottom shear stress in Pascal (Pa)
τ_c = critical shear stress for erosion in Pa

To provide reliable values for these parameters, sediment cores were taken from the various channels to be dredged and combined to form a composite material. As discussed later, this material was then subjected to erosion tests in the laboratory to provide values for the erosion parameters.

MDFATE provides an estimate of the amount of material placed on the sea floor that will be eroded from the placement mounds and the amount of material near the bottom that doesn't settle out before leaving the site. To provide an estimate of losses to the water column during descent of the dredged material through the water column, STFATE runs were made. These tests were made for several water depths representative of the depths within Site 104. Existing depths at the site range from about 45 ft to 75 ft.

As previously noted, in addition to the issue concerning the potential for placed material to leave the site, the impact on water quality due to placing material within Site 104 was also addressed. Existing CEQUAL-ICM and CH3D models for the full Chesapeake Bay were employed to provide insight on water quality impacts. Full bay hydrodynamics from a previous study were available for the years of 1985-87 along with nutrient loads. To simulate 5 years of placement operations, a 5 year sequence was constructed using hydrodynamics and nutrient loads for 1985, 86, 87, and then 1985 and 86 again. Nutrient characteristics of the material expected to be dredged from the upper bay channels were determined and specified for the material placed at Site 104 during the times of placement shown in Table 1. Three runs, each of which was a continuous 5 year simulation, were made; namely, a base run in which no material was placed, a run in which no direct release of nutrients to the water column was allowed, and a run in which all of the nutrients were assumed to be directly released to the water column. Because of the coarseness of the full bay grid (Figure 2), Site 104 is represented by only one cell. A 3D water quality model on the upper bay grid shown in Figure 3 does not exist.

PLACEMENT PLANS

Placement plans for locating the individual placements were determined by assuming Years 1, 3 and 5 would have 5 placements each day (each placement is 3600 cu yds) and Years 2 and 4 would have 10 placements each day. Buffers of 500 ft along the east and west boundaries of the site and 1200 ft at the southern boundary of the site (placement was not allowed in the buffers) were instituted to prevent the transport of dredged material by currents or bottom surge motion beyond the formal site boundary during placement operations. Placement plans were developed for each year with the goal of minimizing both vertical accumulation on the seabed and the surface area of accumulation. Vertical accumulation was important since the minimum water depth to be allowed over placements mounds is 45 ft. Minimizing the surface area of accumulation was important since the amount of material eroded increases with the surface area of exposure. The placement plan for the first year is shown in Figure 5. It can be seen that the first year placement plan creates a berm at the southern boundary to constrain the bottom transport of material during placement operations in the succeeding years.

EROSION TESTS

Sediment samples used in the erosion tests were a composite of six core samples collected from the Craighill Angle, Cutoff Angle, Brewerton Eastern Extension, Tolchester, and Swan Point Channels (Figure 1). Core lengths roughly matched the thickness of channel deposits that would normally be dredged during maintenance dredging. The volume of each channel sample represented in the overall composite was proportional to the annual maintenance dredging that takes place in each channel (Table 2).

Table 2. Percent of Annual Maintenance Material from each Channel	
Craighill Entrance	11.25%
Craighill Channel	5%
Craighill Angle	25%
Craighill Upper Range	2.5%
Cutoff Angle	13.75%
Brewerton Eastern Extension	25%
Swan Point	2.5%
Tolchester	15%

The composite represented 81.25% of material proposed for placement at Site 104. The remaining 18.75% is from the Craighill Entrance, Craighill Channel, and Craighill Upper Range (Figure 1). Sediment characteristics for these channel reaches reported by EA Engineering, Science, and Technology (1996) indicated larger sand fractions and lower moisture contents.

Percent moisture by weight, bulk wet density (BWD), particle size distribution, and organic content were determined for the composite sediment. Results for the composite were 1.251 g/cu cm BWD, 67.44 percent moisture, and 10.1 percent organics. The particle size distributions for the individual sediment cores, as well as for the composite, are given in Table 3.

Table 3. Sediment-Core Particle Size μm			
Channel	Mean	Median	Std Dev
Craighill Angle	10.05	5.75	13.1
Cutoff Angle	10.36	5.61	14.2
Brewerton Extension	12.45	7.12	15.5
Tolchester	11.25	6.56	13.9
Swan Point	11.1	6.12	14.4
Composite	10.88	6.02	14.2

The composite sediment was slurried one part sediment to three parts of site water to represent expected dilution shortly after placement. The slurry was then added to erosion cylinders, and erosion tests were performed after 1, 2, 4, 7, and 8 days of settling. It was expected that in 7 days the settled material would regain most of its shear strength in the field. Composite channel material was also tested in its original undiluted condition to represent the lowest expected erodibility at the placement site.

The erosion experiments were carried out in a particle-entraiment simulator (PES). The PES is a portable device proposed by Tsai and Lick (1986) and used extensively for testing undisturbed core samples. An oscillating grid in the PES generates turbulence above the sample to simulate a shear stress. Equivalent shear stresses were increased stepwise during each erosion test and the concentration of suspensions were monitored over time. From these tests an erosion threshold of τ_c = 0.219 Pa and an erosion rate constant of 11.9 g/sq m/min were determined for input to MDFATE for material that had been in place at the site for less than one week.

The original composite material, without addition of water, was also tested to provide information on material that had been in place at the site for longer than one week. Shear stresses up to 0.6 Pa were applied without the erosion threshold being reached. Thus, based upon tests of similar material from other studies, the critical shear stress for material older than one week used in MDFATE was assumed to be 0.7 Pa and the erosion rate constant was taken to be 50 g/sq m/min.

PHYSICAL FATE MODELING RESULTS

Results from the STFATE simulations showed that even for placement in water depths of 70 ft, less than 1 percent of the disposed material will be stripped during descent through the water column. This percentage agrees qualitatively with observations of Tavolaro (1984) and Truitt (1988). Model results indicate that because of the action of ambient turbulence very little of this fine suspended material will be deposited within the site, regardless of where the placement takes place.

MDFATE results provide information on the amount of placed material that has the potential to leave the site and the bathymetry of the site after placement of the dredged material. Basic hydrodynamic and sediment erosion data required as input by MDFATE have previously been discussed. These data determine the amount of placed material that is eroded. The areal and vertical configuration of aggregate dredged material mounds at the site are controlled by a specified shearing angle and postsheared angle of the dredged material. Geometrically, the extent to which the material accumulates is limited by the steepest angle at which the material can sustain itself before gravity forces the material to slump and redistribute down slope. The slumped sediment comes to rest when some equilibrium angle is reached. The limiting angle of respose (shearing angle) is the steepest angle the material can attain before slumping. The postsheared angle defines the slope of the slumped material after it has come to rest. At site 104, the shearing angle was assumed to vary between 1.7 and 2.0 degrees. These values are based on assessments of similar types of dredged material placed at the New York Dump Site (Clausner et al. 1998). The angle at which slumping stops (postsheared angle) once it has begun was estimated to be 1.6 - 1.9 degrees.

MDFATE also accounts for consolidation of the placed material. Consolidation data required in the finite strain concept (Cargill 1985 and Poindexter-Rollings 1990) used in MDFATE were also based on data developed by Clausner et al. (1998) for the New York Dump Site. Time constraints placed on the Site 104 study did not allow for consolidation tests to be conducted with the proposed dredged material. With all of the processes noted above being simulated, the cumulative distribution of placed material at Site 104 computed by MDFATE after 5 years of placement is shown in Figure 6. From Figure 7, it can be seen that mounds have been created that have less than 45 ft of water depth over them.

With the buffers previously discussed and the 45 ft clearance criterion, the site capacity is only 14 million cu yds. Thus, the placement of 18 million cu yds results in excessive mound heights. If the site is selected for open-water placement of dredged material, management decisions will be required. Either some material must be allowed to be transported from the site during placement operations (remove the buffers), or the minimum water depth allowed after placement must be decreased, or less than 18 million cu yds should be placed at the site.

During the 5 years of computations, MDFATE also computed the cumulative mass of placed material that was eroded from the bottom each year along with the amount of material in each bottom surge that was swept out of the site before being depositing on the seabed. Table 4 lists the cumulative amount of material not deposited during and after the completion of the bottom surge associated with each individual placement operation. It can be seen that for the 5 years of placement 3.3% of the disposed material is not initially deposited in the site.

Table 4. Material not Deposited for the Split-Hull Barge Placement	
Year	Percent
1	2.2
2	2.4
3	3.1
4	4.5
5	3.1
5-year avg	3.3

The percent of material not deposited varies for each placement operation since the loss is highly dependent on the ambient currents that exist during the operation, as well as the water depth at the location of the placement operation.

Table 5 lists the computed percent of material disposed at Site 104 that will be eroded. Virtually all of the erosion takes place immediately after deposition of the material. Recall that the erosion model is different for material more than one week old, e.g., $\tau_c = 0.7$ Pa, whereas, material in place less than one week has a critical shear stress of 0.22 Pa. It can be seen that the total amount of eroded material ranges from a low of about 9.5% for Years 2 and 4 to a high of 18.9% for

Year 3. The primary reason for the difference is due to the placement plan implemented for each year. Year 3 had a placement plan that resulted in more spreading of the material, resulting in an increased surface area of accumulation which then resulted in the total amount of material eroded being greater. When viewing these results, it should be remembered that the modeling results do not provide information on how much of this material will be carried out of the site by the ambient currents. However, both field data and hydrodynamic modeling results clearly show that the residual currents near the bottom of Site 104 are directed to the north. Thus, the long-term transport of resuspended material will be to the north.

Table 5. Percent of Material placed at Site 104 that will be eroded	
1	16.0
2	9.5
3	18.9
4	9.5
5	15.6
5-year avg	12.6

The total potential sediment loss is the sum of sediment stripped during descent through the water column (less than 1%), material that does not deposit within the site during the bottom surge nor afterwards (3.3%), and material that is eroded after initial deposition on the bottom (12.6%). These results indicate a total of about 17% of the total solids (mass of dredged material) placed at Site 104 over the 5 years could potentially leave the site.

WATER QUALITY MODEL RESULTS

The water quality modeling aspect of the study employed the 3D Chesapeake Bay model previously discussed. The model code and parameter values were identical to the model described by Cerco and Cole (1994). Three years of hydrodynamics were available from the full Chesapeake Bay hydrodynamic model. Since dredging operations were proposed for 5 years, a synthetic sequence of years 1985, 86, 87, 85, and 86 was created. It should be noted that unlike the MDFATE simulations, the hydrodynamics were only computed with the existing bathymetry, i.e., the effect of the site being filled was not considered.

As previously noted, two limiting approaches were simulated. In the first approach, complete elutriation was assumed. All ammonium, nitrate, phosphate, and chemical oxygen demand (COD) associated with the dredged material was released to the water column while particulates were deposited on the bottom. Release was to two model layers corresponding to the lower 10 ft of the water column. In the second approach, no elutriation was assumed. All particulate and dissolved substances contained in the dredged material were placed directly on the bottom at Site 104. Properties of benthic sediments computed by the model at the dredged sites were examined and used to create an average composite sediment (Table 6). This composite sediment was placed directly on the bottom for the no elutriation run. For the complete elutriation run, dissolved and sorbed substances were released to the water column at the rates shown in Table 7. 1

Table 6. Bulk Characteristics of Composite Sediment	
Substance	Density, g/cu m
Labile particle organic carbon	46.7
Refractory particulate organic carbon	759
Inert particulate organic carbon	99,000
Labile particulate organic nitrogen	7.58
Refractory particulate organic nitrogen	168
Inert particulate organic nitrogen	1,410
Labile particulate organic phosphorous	0.41
Refractory particulate organic phosphorous	10.7
Inert particulate organic phosphorous	146
Ammonium	9.79
Nitrate	0.077
Total inorganic phosphorous	174
Chemical oxygen demand	3.08

Table 7. Release (kg/day) by Complete Elutriation					
Substance	Year 1	Year 2	Year 3	Year 4	Year 5
Ammonium	134	242	162	324	108
Nitrate	2	2	2	2	2
Phosphate	2,396	4,312	2,876	5,750	1,916
COD	42	76	50	102	34

For the case of no elutriation, the model computed a slightly enhanced sediment-ammonium release during the dredging season (Nov-Mar) when subsurface sediments from the dredging sites are deposited at the sediment-water interface at Site 104. A second feature computed was an enhanced sediment-phosphorous release during periods of bottom-water anoxia in the summer. The enhanced release occurs because the dredged material placed at the site contains more inorganic phosphorous than material already at Site 104. However, the impact was restricted to bottom waters immediately over the sediments. Since vertical stratification tends to trap the additional phosphorous in bottom waters below the photic zone, little or no stimulation of algae in surface waters occurs. Therefore, no noticeable impact on dissolved oxygen in the bottom waters overlying Site 104 as a result of the placement of dredged material was computed.

Results from the complete elutriation run showed that since the phosphate is released to the water column, elevated concentrations were computed near the bottom during the placement season, but not at other times. The phosphorous has a slight stimulatory effect on the spring algal bloom but no apparent effect on summer chlorophyll levels due to the timing of the placements. The phosphorous released by elutriation is dispersed before the summer algal population appears. The releases by elutriation of ammonium, nitrate, and COD are minor, so no influence on water column nitrogen or dissolved oxygen were apparent from the computations.

SUMMARY AND CONCLUSIONS

A previously used open-water disposal site in upper Chesapeake Bay called Site 104 is being considered for the placement of 18 million cu yds of maintenance material from the upper bay navigation channels. Several numerical models were applied to provide information on how much of the dredged material proposed for placement at Site 104 has the potential to leave the site and to provide information on the impact on water quality at the site due to the placed material. STFATE provided information on how much of the disposed material would be stripped during descent through the water column. CH3D provided full bay hydrodynamics to drive 3D water quality computations in the CEQUAL-ICM model and upper bay hydrodynamics to provide bottom shear stresses and vertically-averaged flow fields in the MDFATE model. Final Site 104 bathymetry and the amount of material that has the potential to leave the site during the five years of placement were computed by MDFATE.

Laboratory erosion tests of a composite sample of the material to be dredged resulted in the specification of a critical shear stress of about 0.22 Pa and an erosion rate constant of about 12 g/sq m/min for the placed material within the first week after placement. After being on the bottom for more that one week, a critical shear stress of 0.7 Pa and an erosion constant of 50 g/sq m/min were assumed for the placed material.

Modeling results from STFATE and MDFATE indicate that up to 17% of the placed material has the potential to leave the site. Of this total, about 1% will be stripped from the dredged material as it descends through the water column, about 3.3% will remain in the water column after impact with the bottom, and about 12.6% will be resuspended after initial deposition on the seabed. However, these results only represent the potential for material to leave the site. It is likely that much of the resuspended material will be redeposited within the site before actually leaving the site. In addition, since the long-term net flow near the bottom of the site is to the north, the net transport of resuspended material will be to the north.

Results from the water quality model for both the no elutriation and the complete elutriation runs show that there will be virtually no impact on dissolved oxygen in the water column over Site 104. Thus, although there will be some impact on near bottom concentrations of variables such as phosphorous, this impact has negligible impact on the overall water quality at the site with regard to water column dissolved oxygen concentrations.

ACKNOWLEDGMENTS

The study described herein was funded by the Maryland Environmental Service under contract to the Maryland Port Administration. Permission to publish this paper was granted by the Chief of Engineers.

REFERENCES

Cargill, K. W. 1985. "Consolidation of soft layers by finite strain analysis," Miscellaneous paper GL-82-3, U.S. Army Engineer Waterways Experiment Station, Vicksburg, MS.

Cerco, C. F. and Cole, T. M. 1994. "Three-dimensional eutrophication model of Chesapeake Bay," Technical Report EL-94-4, U.S. Army Engineer Waterways Experiment Station, Vicksburg, MS.

Clausner, J. E., Lillycrop, L. S., McDowell, S. E., and May, B. 1998. "Overview of the 1977 Mud Dump Capping Project," Proceedings Xvth World Dredging Congress 1998. CEDA. Amsterdam, The Netherlands.

EA Engineering, Sciences, and Technology. 1996. "FY 1995 sediment sampling and chemical analysis for Baltimore Harbor and Chesapeake Bay, MD," Draft Report prepared for the Baltimore District, Hunt Valley, MD.

Fagerburg, T. 1995. "Delaware Bay field data report," Technical Report HL-95-1, U.S. Army Engineer Waterways Experiment Station, Vicksburg, MS.

Hamilton, P. and Boicour, W. 1984. "Long-term salinity, temperature, and current measurements in Upper Chesapeake Bay," Contract NO. P81-81-04, Submitted to Department of Natural Resources, Tawes State Office Building, Annapolis, MD, by Science Applications, Raleigh, NC.

Johnson, B. H., Kim, K. W., Heath, R. E., Hsieh, B. B., and Butler, H. L. 1991. "Development and verification of a three-dimensional numerical hydrodynamic, salinity, and temperature model of Chesapeake Bay," Technical Report HL-91-7, U.S. Army Engineer Waterways Experiment Station, Vicksburg, MS.

Johnson, B. H. and Fong, M. T. 1995. "Development and verification of numerical models for predicting the initial fate of dredged material disposed in open water: Report 2, Theoretical developments and verification results," Technical Report DRP-93-1, U.S. Army Engineer Waterways Experiment Station, Vicksburg, MS.

Johnson, B. H., Moritz, H. R., Teeter, A. M. Cerco, C. F., and Wang, H. V. In preparation. "Modeling the fate and water quality impact of the proposed dredge material placement at Site 104," Technical Report CHL-99-, U. S. Army Engineer Research and Development Center, Vicksburg, MS.

Moritz, H. R. and Randall, R. E. 1992. "Users guide for open water disposal management simulation," Contract Number DACW39-90-K-0015, final report to U.S. Army Engineer Waterways Experiment Station, submitted through the Texas A&M Research Foundation, College Station, TX.

_____. 1995. "Simulating dredged material placement at open water disposal sites," Journal of Waterway, Port, Coastal, and Ocean Engineering, 121(1).

Poindexter-Rollings, M. E. 1990. "Methodology for analysis of subaqueous mounds," Technical Report D-90-2, Environmental Laboatory, U.S. Army Engineer Waterways Experiment Station, Vicksburg, MS.

Sanford, L. 1997. Personal communication. Horn Point Laboratory, Cambridge, MA.

Scheffner, N. W., Thevenot, M., Tallent, J. and Mason, J. 1995. "LTFATE: A model to investigate the long-term fate and stability of dredged material disposal sites; User's guide," Instruction Report DRP 95-1, U.S. Army Engineer Waterways Experiment Station, Vicksburg, MS.

Tavolaro, J. F. 1984. "A sediment budget study of clamshell dredging and ocean disposal activities in the New York Bight," Environmental Geology and Water Science 6(3), 133-140.

Truitt, L. C. 1988. "Dredged material behavior during open-water disposal," Journal of Coastal Research 4(3).

Tasi, C. H. and Lick, W. 1986. "A portable device for measuring sediment resuspension," Jr. Great Lakes Research 12(4), 314-321.

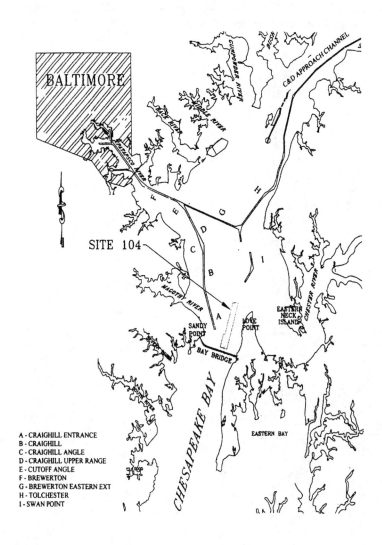

A - CRAIGHILL ENTRANCE
B - CRAIGHILL
C - CRAIGHILL ANGLE
D - CRAIGHILL UPPER RANGE
E - CUTOFF ANGLE
F - BREWERTON
G - BREWERTON EASTERN EXT
H - TOLCHESTER
I - SWAN POINT

Figure 1. Upper Chesapeake Bay navigation channels

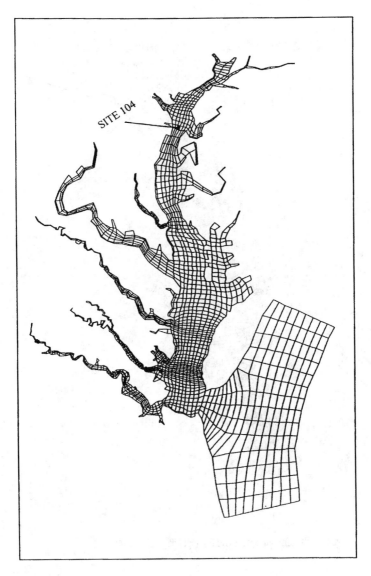

Figure 2. Planform numerical grid for full bay CH3D model

Figure 3. Planform numerical grid for upper bay CH3D model

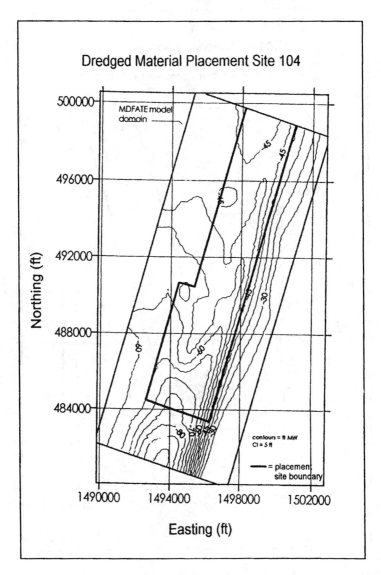

Figure 4. Existing bathymetry at Site 104

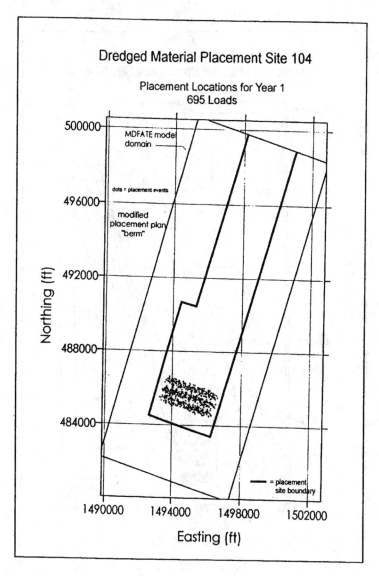

Figure 5. Placement plan for the first year of the placement schedule

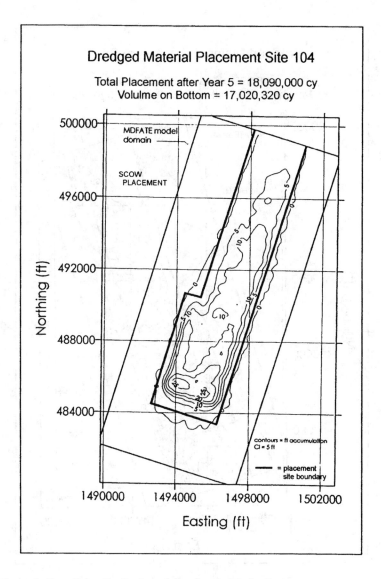

Figure 6. Cumulative distribution of placed material after 5 years

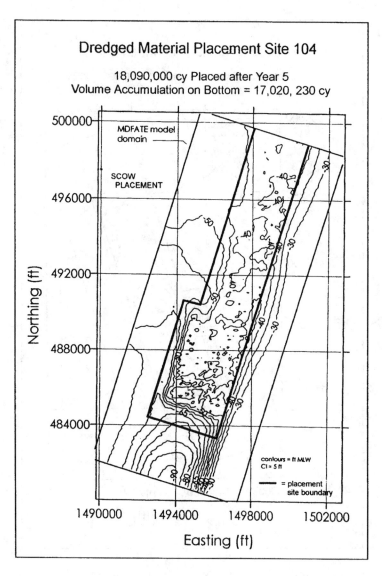

Figure 7. Final bathymetry at Site 104 after 5 years of placement

A Multi-Size, Multi-Source Formulation for Determining Impacts of Sediments on Near-Shore Sensitive Sites

Panagiotis Velissariou[1], Vasilia Velissariou, Yong Guo and Keith W. Bedford

Abstract

The unconfined placement of dredged material raises the issue of potential adverse impacts on sensitive near-shore areas (e.g. water intakes) located some distance away from the placement site. Two management questions arise: first, whether the material from the placement site is actually transported to the sensitive area; and second what is the relative intensity of this load compared to other sites.

In an effort to evaluate the relative impact of different sediment sources on sensitive near-shore areas we considered a multi-grain size, multi-source sediment transport model formulation. Calculations were performed using the CH3D circulation model coupled with a sediment model known as the CH3D-SED, which includes a suspended sediment and a mobile-bed sediment module.

The location of interest has been the Toledo, Ohio, water intake area in the western basin of Lake Erie. The sediment disposal site is located a few miles northeast of the sensitive site. The four sediment sources selected are: the Maumee River, the Detroit River, the entrained bottom sediments and the sediments originating from the disposal site. Three representative grain sizes, one from each sediment size class (sand, silt, clay), were assigned for each source.

The model was run on a 2 km spatial grid for the test period from April 1, 1997 to December 31, 1997. Model outputs include: 3-D velocity, temperature and concentration (all grain sizes) fields, vertically integrated velocities and grain size distributions at the lake bottom, from which hourly trajectory maps and time traces are generated.

---

[1]Department of Civil & Environmental Engineering and Geodetic Science, The Ohio State University, Columbus, Ohio, 43210. E-mail: Velissariou.1@osu.edu.

A first evaluation of the model results implies that the major contributor to suspended sediment load is the bottom sediment, which overwhelms the sensitive site. The impact from the disposal site seems to be minimal while more significant contributions are observed from the river inputs. In fairly calm weather conditions the finer sediment classes are the major contributors to the suspended sediment loading, while during storm events the resuspension and transport of sands is also significant.

Introduction

This paper summarizes some of the first findings of a research project funded under the US Army Corps of Engineers cooperative agreement whose objectives were to determine: a) if dredged sediments disposed at the new unconfined sediment disposal site at the western basin of Lake Erie are actually being transported to the Toledo water intake area under various meteorological conditions, and b) to compare the relative intensities of the water intake area particle load from the disposal site to those water intake area loads originating from other sediment sources such as local resuspension and tributary inputs.

The Toledo, Ohio, water intake area, is located 12600 m south to southeast the disposal site. The disposal site is located at 41°48.6' N, and 83°17.0' W, in water of 6.1 m to 7.0 m depth relative to IGLD Low Water Datum (Figure 1).

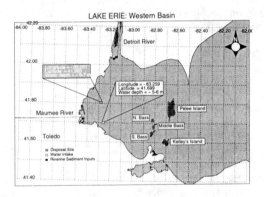

Figure 1: Location of the disposal and water intake sites at the western basin of Lake Erie.

To answer the above questions a multi-grain, multi-source sediment transport formulation was used. Three sediment sources were selected to represent the sediments originating from the disposal site, from the Maumee River, and from the Detroit River. A fourth sediment source selected was associated with the entrainment and transport of the lake bottom sediments. The sediments of each source are represented using three size classes (sand, silt, and clay). The sediment particle diameters for these classes are set to vary slightly among the different sources. This approach allows both the identification of the sources of the sediment material transported at the water intake area and, the evaluation of the relative intensity of the sediment load transported from each source, to the total load transported.

Model Selection

The calculations were performed using the CH3D circulation model coupled with a sediment model, known as CH3D-SED model (Spasojevic and Holly, 1994. The decision to engage CH3D-SED in this project came out of its ability to model sediments of various grain sizes and its ability to model the sediments originating from different sources.

CH3D is a three dimensional, non-linear primitive equation circulation model. The equations governing the circulation of the lake include the continuity equation, the momentum equations and the conservation equation for the thermal energy. CH3D-SED describes the advection and turbulent diffusion of the suspended sediments, as well as the entrainment of the bottom sediments and the evolution of the lake bottom (as bottom sediments are being transported). The sediment equations of the model describe the behavior of a non-uniform sediment mixture, which is represented by an appropriate number of sediment size classes.

Using the concept of an elemental volume ΔV that includes the upper layer of the bed and the bed surface, and assuming a uniform sediment size distribution within this volume, the conservation of mass equations for each size class are given by (Spasojevic and Holly, 1994):
For each size class separately:

$$\rho_s(1-p)\frac{\partial(\beta E_m)}{\partial t} + \nabla\vec{q}_b + S_e - S_d - S_F = 0 \tag{1}$$

For the sum of all size classes:

$$\rho_s(1-p)\frac{\partial E_m}{\partial t} + \sum(\nabla\vec{q}_b + S_e - S_d - S_F) = 0 \tag{2}$$

with the constraint: $\sum\beta = 1$.

In the above equations, p is the porosity of the bed material and ρ_s is the density of the sediment (both assumed to be constant); β represents the fraction of the mass of one particular size class over the mass of all sediment particles in the elemental volume; and $\vec{q}_b$ is the bedload mass flux expressed as a two-dimensional vector parallel to the bed surface. The bedload is calculated in CH3D-SED by using an empirical relation proposed by Van Rijn (1984a). For one particular size class the bedload flux is given as:

$$(q_b)_s = 0.053 \cdot \rho_s \cdot \sqrt{(s-1)gD_s} \frac{D_s}{D_{*s}^{0.3}} \left[\frac{u_*^2 - u_{*c}^2}{u_{*c}^2} \right]^{2.1} \tag{3}$$

where, D_s is the particle diameter; D_{*s} is the dimensionless particle diameter defined as:

$$D_{*s} = D_s \left[\frac{\sqrt{(s-1)g}}{v^2} \right]^{1/3} \tag{4}$$

u_* is the bed shear velocity; and u_{*c} is the critical shear velocity.

The source term, S_e represents the entrainment of the bed sediments into the water column, S_d represents the settling of suspended sediments and S_F describes the exchange of sediment particles between the active layer elemental volume and an elemental volume immediately underneath, called the active stratum elemental volume (Spasojevic and Holly, 1994).

Similarly, the mass conservation equations for an elemental active stratum volume can be written as:
For each size class separately:

$$\rho_s(1-p)\frac{\partial}{\partial t}\left[\beta_s(z_b - E_m)\right] + S_F = 0 \tag{5}$$

For the sum of all size classes:

$$\rho_s(1-p)\frac{\partial}{\partial t}(z_b - E_m) + S_F = 0 \tag{6}$$

with the constraint: $\sum \beta_s = 1$, where, z_b is the bed elevation and β_s is the active stratum fraction of the mass of one particle size class over the mass of all sediment particles.

The advection and turbulent diffusion of each particular size class of the suspended sediment can be expressed in mathematical form as:

$$\frac{D(\rho C)}{Dt} = \frac{\partial}{\partial x}\left[D_h \frac{\partial(\rho C)}{\partial x} \right] + \frac{\partial}{\partial y}\left[D_h \frac{\partial(\rho C)}{\partial y} \right] + \frac{\partial}{\partial z}\left[D_v \frac{\partial(\rho C)}{\partial z} \right] + \frac{\partial}{\partial z}(\rho C w_f) \tag{7}$$

In the above equation, C is the ratio of the mass of one particular sediment size class to the mass of all size classes within an elemental volume ΔV, ρ is the density of the sediment-water mixture, represented by all size classes, w_f is the settling velocity of the sediment particles and D_h, D_v are the horizontal and vertical mass diffusion coefficients.

The hydrodynamic part of the CH3D-SED model provides information about the fluid velocities, water depths and temperature changes that are required input for the sediment part. The sediment model in return provides information about the changes of the bed elevation, the bed surface roughness due to the changes of the bed-surface size distributions and changes of the density of the water-sediment mixture (Spasojevic and Holly, 1994).

Data Analysis

The availability of the extensive databases of the Great Lakes Forecasting System (GLFS) made it a natural choice to use these databases to obtain the meteorological data required by CH3D-SED. The database chosen was the year 1997, and for the ice-free period between April 1 to December 31, 1997. The meteorological data obtained from GLFS were the wind speed and temperature fields, at the free water surface of Lake Erie. The hourly wind speed data are adjusted to reflect: a) a common anemometer height and b) the over-water conditions. Further manipulation of the wind and the temperature data is not required for use in CH3D-SED. From these data the model internally calculates the wind stresses.

The daily flow rates of the three tributaries considered in this study were obtained from the U.S.G.S databases (1998). The daily flow rates for Maumee and Niagara rivers were available for the simulation year 1997 and were directly used as model input. Measured data for the Detroit River are sparse and not available for the simulation year. Considering the fact that the flow rates of Detroit river over the years have a very close resemblance and small value variations (G. F. Koltun, 1990), the latest available data from the USGS database (USGS Water Resources of the United States) were used, which are the data from the year 1977. The flow rate data are shown in Figure 2.

The sediment modeling requires the inclusion of all the major sediment sources. The sediment sources considered in Lake Erie are the bottom sediments (source 1), the disposal site unconfined sediments (source 2) and the riverine sediment inputs (sources 3 and 4). Despite the fact that shore erosion is extremely important in sediment forecasting, and a very significant sediment source, it was not considered in this study simply because the purpose is to determine the relative intensity of the impacts of the other four sources on the water intake site.

Qualitative information about the bottom sediments and their grain size distribution for Lake Erie was obtained by Thomas et al. (1976) who used both sediment sampling and acoustic profiling to examine 275 sampling locations all over the lake. Their results show a distribution of the bottom sediments based upon four basic types which are identified as follows: a) sand and/or gravel (S),

b) post-glacial mud (M), c) soft gray mud with some sand (SM) and d) glacial
sediments (GL).

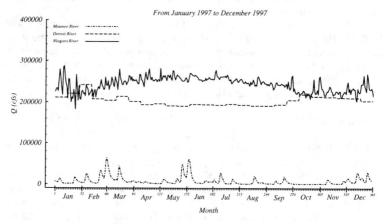

Figure 2: Daily flow rates for the Maumee, Detroit and Niagara Rivers for the
simulation year 1997.

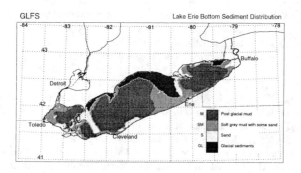

Figure 3:   Distribution of the bottom sediment types for Lake Erie.

Following the qualitative distribution of the bottom sediments described above,
each grid point at the bottom of the lake, depending upon its location, is assigned
one of these types (Figure 3). With this information, the analysis continues with
the identification of the fractions representing the sediment grain sizes.

Quantitative information about the bottom sediments of Lake Erie has been
obtained from the technical report prepared by Herdendorf et al., 1978. The

results were reported as the percentages of sand, silt and clay contents of the samples (according to the Wentworth sediment grade scale).

To estimate the percent of sand, silt and clay contained in each one of the four sediment types the samples taken within the region of each sediment type were first identified and then the average percent of each size class for each sediment type was calculated using the following formula:

$$F_{ji} = \frac{1}{N_i} \sum_{k=1}^{N_i} f_{jk} \qquad (8)$$

where F_{ji} is the average percent of each size class for each sediment type; i, is each sediment type (M, S, GL, SM); N_i is the number of samples corresponding to each sediment type; f_{jk} is the percent of each size class for each sample N_i; and j, is each size class (sand, silt, clay). The results of this analysis are shown in Table 1.

Table 1: Definition of the sediment fractions for the sediment types in Lake Erie.

Sediment Type	No. of Samples N	Sand	Silt	Clay
M	759	0.0 %	70.0 %	30.0 %
SM	137	70.0 %	20.0 %	10.0 %
S	361	97.0 %	2.0 %	1.0 %
GL	410	55.0 %	35.0 %	10.0 %

The average sediment concentration for the disposal site used in this study is 35 mg/L (Bedford et al., 1999). This value was calculated by adjusting the data collected during a field study project contacted the summer of 1996 (Fan and Bedford, 1998). Grain size distributions for the bottom sediments at the disposal site have been reported by the Toledo Harbor Planning Group in 1998. The average of four samples showed a 96.2 % of silts and clays and a 3.8 % of sands and gravel. Further analysis on the silts and clays at the disposal site performed by the Automatic Particle Size Analyzer (HIAC Model-320), at the Coastal Engineering Laboratory at O.S.U, revealed that a 72.2% are silts and 24% are clays (Bedford et al., 1999).

Detailed suspended sediment concentration data for both Maumee and Detroit rivers are sparse to non-existent and the main source of sediment data for the Maumee river used in this study is the annual USGS Water-Data Report of Ohio for the water year October 1991 to September 1992 (US Geological Survey, 1993). Using the suspended sediment concentration and flow rate data obtained for Maumee River for the water year 1992 the following linear relationship was obtained:

$$C = 0.015 \cdot Q + 12.116 \qquad (9)$$

where C (mg/L) is the daily average suspended sediment concentration and Q (ft^3/s) is the daily average flow rate for the Maumee River. Using Equation (9) and the flow rates for the water year 1997, the daily average suspended sediment concentrations for the Maumee River were obtained. Detailed suspended sediment concentration data for the Detroit River are not available. Kemp et al., 1976, have reported the sediment loadings from various sources in Lake Erie. Using their findings an average coefficient $\bar{f} = 0.77$ which reflects the relative loading between the Detroit and the Maumee Rivers was calculated. The following equation was used to determine the daily suspended sediment concentrations for the Detroit River:

$$C_D = \bar{f} \cdot C_M \cdot \frac{Q_M}{Q_D} \qquad (10)$$

where, C (mg/L) is the average daily concentration, Q (ft^3/s) is the average daily flow rate, and the subscripts "D" and "M" denote the Detroit and Maumee Rivers respectively.

In addition to suspended-sediment loads, the particle-size distribution of the suspended sediment for the two rivers needed to be estimated. Available data for the Detroit River describe a suspended sediment mixture with 87 to 100 percent particles classified as silt and clay (US Geological Survey, 1975b). An average of 6 percent by weight for the sand size class was used in the present study for the Detroit River. An average of 7.5 percent by weight for the sand size class has been estimated for the Maumee River (Toledo Harbor Planning Group in 1993). Finally, a 30 to 70 percent contribution from the clay and silt size classes respectively has been assumed for both the Detroit and the Maumee Rivers.

Numerical Domain, Boundary and Initial Conditions

Lake Erie is 388 km long and 92 km wide with a southwest to northeast alignment. In order to establish the "x" coordinate axis along the longitudinal axis of the lake the flow domain is rotated by 27.33° clockwise. The resolution of the numerical grid used in CH3D-SED is 2x2 km, which yields 209 grid points in the "x" direction and 57 grid points in the "y" direction (normal to the "x" direction). All the land grid points are assigned with a water depth equal to zero, so they can be identified during the model calculations.

In the vertical direction CH3D-SED uses the σ-coordinate system in order to accommodate for the depth variation throughout the lake. In this study fourteen grid points in the vertical direction were used, resulting in thirteen irregularly spaced vertical slices in the (x, y, z)-coordinate system. The free surface is identified at $\sigma = 0$, while the lake bottom is identified at $\sigma = -1$.

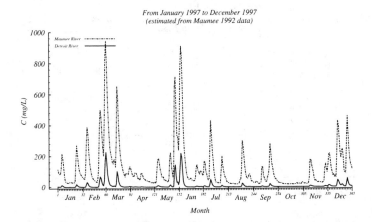

Figure 4: Estimated concentrations of the suspended sediments for Maumee and Detroit Rivers (simulation year 1997).

The boundary conditions can be divided into three categories: a) meteorological boundary conditions, b) hydrodynamic boundary conditions, and c) sediment boundary imposed at the river boundaries and at the bottom of the lake.

The meteorological boundary conditions imposed at the free water surface of the lake are the 2-D wind and temperature fields. The wind velocities, required for the above calculations, were obtained from the Great Lakes Forecasting System (GLFS). CH3D-SED has been slightly modified to accept space and time varying temperature at the free water-surface for each time step. These data are readily available from GLFS for the whole simulation period. The riverine boundary conditions require the definition of the flow-rates, imposed at the river boundaries. The boundary conditions for the inflow and the outflow boundaries are simply the daily averaged flow rates, Q.

The simulation of the sediment distribution in Lake Erie required the definition of twelve sediment size classes (Table 2). These twelve sediment classes were equally divided among the four sediment sources and the grain sizes were selected to represent the typical sediment particle sizes found in Lake Erie. While the flow-rates for the hydrodynamic part of the simulation were specified at all three tributaries, in the case of the sediment boundary conditions, the Niagara River was treated as an open boundary.

Table 2   Definition of the twelve sediment size classes and their fractions for the four sediment sources in Lake Erie.

Location	Size Class	Grain Diameter (μ)	Grain Size Fractions (%)
Lake Bottom (Source #1)	1	150	
	2	50	Refer to Table 1
	3	4	
Disposal Site (Source #2)	4	149	3.8
	5	49	72.2
	6	3.7	24.0
Maumee River (Source #3)	7	148	7.5
	8	48	65.0
	9	3.3	27.5
Detroit River (Source #4)	10	147	6.0
	11	47	65.0
	12	3	29.0

The daily concentration C (mg/L) of each sediment class was found by multiplying the fraction for that class with the average daily concentrations at each tributary. The concentration profiles at the tributaries (fourteen vertical grid points) were assumed constant for each sediment class.

Both the flow velocities and the concentrations were initialized to be zero everywhere in the lake. The concentration profile of the suspended sediment at the disposal site was initialized with a constant value of 35 mg/L (all sediment classes). For each sediment class, the constant concentration profile was determined by multiplying the total concentration of 35 mg/L with the fraction of that class. The bottom sediments in Lake Erie were assigned three sediment classes throughout the lake, but for each grid point, a different class fraction distribution was assigned. For the grid point corresponding to the disposal site the class fractions used are the same as the ones used for the suspended sediment.

Results and Discussion

This study covers, a one year ice-free test period, from April 1, 1997 to December 31, 1997. The model results include: a) the full 3-D velocity field, b) the 2-D vertically averaged velocity field, c) the 3-D temperature field, d) the 3-D sediment concentration field, e) the bottom sediment fraction distributions and, f) the changes of the bottom elevation. From these results, contour maps and time traces of the variables have been generated. Figures 5 and 6 present contour maps of the water surface wind speed, and the total suspended mass of the "global sediments" originating from the lake bottom and the disposal site. The term "global sediments" is used inclusively here and describes all the sediment classes assigned to the sediment source (sand, silt and clay).

The units used in the total suspended mass contour maps are metric tons per unit depth (tons/m). The total suspended mass is calculated using the equation:

$$\text{TSSM} = 10^{-6} \cdot \frac{\overline{C} \cdot V_{eff}}{d} = 10^{-6} \cdot \frac{\overline{C} \cdot A_{eff} \cdot d}{d} = 10^{-6} \cdot \overline{C} \cdot A_{eff} \tag{11}$$

where TSSM, is the total suspended mass of the sediments (tons/m), $\overline{C}$, is the vertically averaged concentration (mg/L), V_{eff}, is the effective volume (m^3) where the total suspended mass is calculated, A_{eff}, is the corresponding effective area (m^2), and d, is the water depth (m). The effective area, A_{eff}, is defined as the area of the horizontal square extending half a grid point from the grid point where $\overline{C}$ is calculated. For the grid resolution used for this simulation it is: $A_{eff} = 4 \text{ km}^2$.

The plots in Figure 7 present the time traces of the "relative intensity" of the "global sediments" originating from the different sources. The term "relative intensity" is defined as the ratio of the "global sediments" from one source to the "global sediments" from all the sources and is dimensionless (expressed as %).

The average relative intensity for the simulation period of the TSSM originating from the disposal site is approximately 0.5 %, while the corresponding seasonal peaks do not exceed the 3.5 % mark. The average relative intensity for the simulation period of the TSSM originating from the lake bottom is approximately 90 %, and that of the TSSM originating from the Detroit and the Maumee Rivers is approximately 9.5 %. From early June to late September the riverine contributions increase significantly. The contribution of the sediments originating from the disposal site is fairly constant throughout the simulation period, but occasional peaks are observed during storm events. In any case, these peaks do not exceed the 3.5 % mark.

Overall the impact at the water intake site of the sediments originating from the disposal site is fairly small (during storm events) to insignificant.

Acknowledgments

This paper is the result of the ongoing US Army Corps of Engineers contract agreement no. DACW39-95-K-0018. This support is sincerely appreciated. The thoughts presented in this paper represent the authors' own and in no way are to be interpreted as the official position or policy of the supporting agency. The authors would also like to thank Dr. Yi-Fei Philip Chu, Dr David Welsh and Sean O'Neil for their help and valuable suggestions throughout this research project.

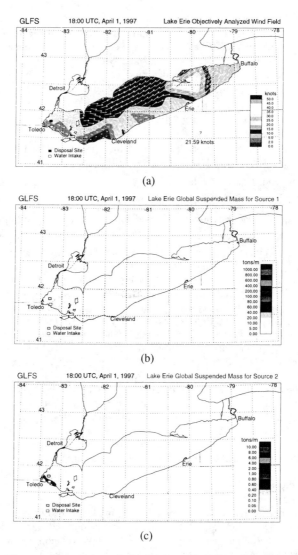

(a)

(b)

(c)

Figure 5: Water-surface wind speed distribution (a), and horizontal distributions of
the global suspended mass of the sediments originating from (b) source 1
(lake bottom), and (c) source 2 (disposal site) for April 1, 1997.

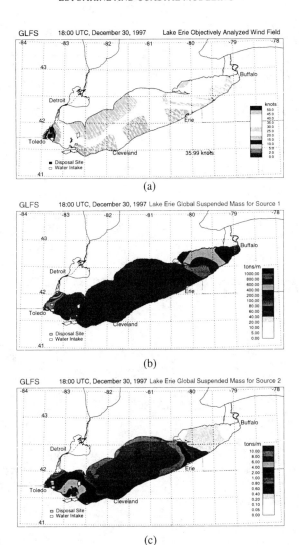

Figure 6: Water-surface wind speed distribution (a), and horizontal distributions of the global suspended mass of the sediments originating from (b) source 1 (lake bottom), and (c) source 2 (disposal site) for December 30, 1997.

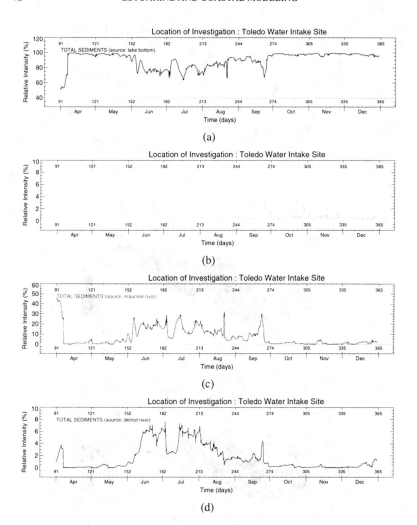

Figure 7:   Relative intensity, percent of the total suspended sediment mass from all sources, at the water intake site of the global sediments originating from: a) the lake bottom, b) the disposal site, c) the Maumee River and d) the Detroit River.

References

1. K. W. Bedford, P. Velissariou, V. Velissariou and Yong Guo, 1999. *A Multi-Size, Multi-Source Formulation for Determining the Impacts of Sediments on Near-Shore Sensitive Sites.* Report no. 3, US Army Corps of Engineers, Contract DACW39-95-K-0018, September 1999.

2. S. C. Fan and K. W. Bedford, 1998. *Data Analysis of the 1996 ARMS Field Experiments in Western Basin of Lake Erie.* Report no. 1, US Army Corps of Engineers, Contract DACW39-95-K-0018, January 1998.

3. The Great Lakes Forecasting System (GLFS) Database. Internal Database Constructed Yearly from the Great Lakes Nowcasts. Web Address: http://superior.eng.ohio-state.edu/.

4. C. E. Herdendorf, D. B. Gruet, M. A. Slagle and P. B. Herdendorf, 1978. *Descriptions of Sediment Samples and Cores from the Michigan and Ohio Waters of Lake Erie.* Clear Technical Report, no. 85, January 1978.

5. A. L. W. Kemp, R. L. Thomas, C. I. Dell and J. M. Jaquet, 1976. *Cultural Impact on the Geochemistry of Sediments in Lake Erie.* J. Fish. Res. Board Can., 33:440-462, 1976.

6. G. F. Koltun, 1990. *A Statistical Approach to Characterizing Bottom Shear Stress Entrainment and Related Hydrologic and Climatic Phenomena in the Detroit River, Trenton Channel.* MS Thesis, The Ohio State University, Department of Civil Engineering, 1990.

7. M. Spasojevic, F. M. Holly Jr., 1994. *Three-Dimensional Numerical Simulation of Mobile-Bed Hydrodynamics.* Contract Report HL-94-2, US Army Corps of Engineers, Waterways Experiment Station, August 1994.

8. R. L. Thomas, J. M. Jaquet, A. L. W. Kemp and C. F. M. Lewis, 1976. *Surficial Sediments of Lake Erie.* J. Fish. Res. Board Can., 33:385-403, 1976.

9. Toledo Harbor Planning Group, 1993. *Long-Term Dredged Material Management Plan within the Context of Maumee River Watershed. Sediment Management Strategy.* Executive Committee Phase 3 Report with Environmental Assessment, December 1993.

10. US Geological Survey, 1975b. *Water Resources Data for Michigan-1974, Part 2.* Water Quality Records: US Dept. Int., Geol. Survey-Water Resources Div., 165 p., 1975.

11. US Geological Survey, 1993. *U.S. Geological Survey Water-Data Report, OH-92-2.* Water Resources Data, Ohio Water Year 1992, Volume 2, St. Lawrence River Basin and Statewide Project Data, Ohio 1993.

12. USGS Water Resources of the United States, 1998. *Database accessible via the World Wide Web.* Web Address: http://www.usgs.gov.

13. L. C. van Rijn, 1984a. *Sediment Transport, Part I: Bed Load Transport.* Journal of Hydraulic Engineering, ASCE, Vol. 110, No 10, pp. 1431-1456, 1984.

Modeling Three-Dimensional, Thermohaline-Driven Circulation in the Arabian Gulf

Cheryl Ann Blain[1]

ABSTRACT

The 3-D finite element circulation model, QUODDY4, is applied to investigate the dynamical composition of circulation in the Arabian Gulf during the winter and summer seasonal extremes. The QUODDY4 model is a free surface, time domain model that includes tidal dynamics, wind- and buoyancy-driven flows and level 2.5 advanced turbulence closure. Forcing for the region takes the form of multiple tidal constituents, initial climatological temperature and salinity fields, and seasonal mean wind stress. A series of simulations that incorporate increasing dynamic complexity and combinations of forcing illuminate the contribution of individual processes (e.g. tidal rectification, baroclinic pressure gradients, and wind driven circulation) to the overall circulation pattern observed in the Gulf. Brief qualitative comparisons to published observations of currents, temperature and salinity indicate simulated solutions reasonably agree with observations but highlight the need for realistic evaporation and heat flux forcing.

INTRODUCTION

To date there have been only a few attempts at modeling the thermohaline-driven circulation in the Arabian Gulf (e.g. Horton et al., 1992; Chao et al., 1992;

---

[1] Oceanographer, Oceanography Division, Naval Research Laboratory, Stennis Space Center, MS 39529-5004

Lardner et al., 1987); none of these efforts have involved the same level of resolution and geometric detail as the study herein nor have they focused on understanding the dynamical partitioning of the circulation. Also lacking have been observations of Gulf circulation that span the seasonal cycle and/or include significant spatial coverage. Measurements taken during the Mt. Mitchell expedition (Reynolds, 1993) sample much of the Gulf but extend only from February - June, 1992. Recent measurements taken as part of the Strait of Hormuz Experiment (Johns and Zantopp, 1999) span a longer period, Dec. 1996 – May 1998, but are limited to a single mooring. Still questions remain as to how the dominant forces in the region drive the underlying circulation throughout the Gulf at seasonal time scales. In particular understanding is needed concerning the role of tides, wind, and baroclinic pressure gradients on the total circulation.

Application of a fine resolution (3-5 km) numerical model in the Gulf can begin to answer some of these questions. A series of simulations that incorporates increasing dynamic complexity and combinations of forcing form the basis of this study. The set of experiments includes barotropic, baroclinic diagnostic, baroclinic prognostic, and composite seasonal computations for summer and winter mean conditions. The modeled circulation is partitioned by forcing to illuminate the contribution of tides, baroclinic pressure gradients and wind to the overall circulation in the Gulf. Limited comparison to published observations of temperature and salinity in the Strait and mean currents in the Gulf reinforce the validity of model predictions.

MODEL DESCRIPTION

The circulation model applied to the Arabian Gulf is the Dartmouth College Model, QUODDY4 (Lynch and Werner, 1987; Lynch and Werner, 1991). This model is a time marching simulator based on the 3-D hydrodynamic equations subject to the conventional Boussinesq and hydrostatic assumptions. A wave-continuity form of the mass conservation equation (Lynch and Gray, 1979; Kinnmark, 1985), designed to eliminate numerical noise at or below two times the grid spacing, is solved in conjunction with momentum conservation and transport equations for temperature and salinity. The sub-grid scale dissipation is represented by an eddy viscosity (diffusivity) formulation, parameterized in terms of stratification plus turbulent kinetic energy and a turbulent mixing length, both of which evolve at the macroscale. This formulation is described by the well-known Mellor-Yamada 2.5 level turbulence closure (Mellor and Yamada, 1982) with improvements by Galperin et al. (1988) and Blumberg et al. (1992). This turbulence closure scheme accounts for processes occurring over the vertical extent of the water column such as diffusion, shear production, buoyancy production and dissipation.

At the water surface kinematic boundary conditions are imposed. Wind stress is a source of horizontal momentum, and heat flux is treated as a restoring condition relative to equilibrium temperature. At the bottom boundary fluid is constrained to follow topography while the bottom stress serves as a sink of momentum. No buoyancy flux through the bottom is permitted.

The method of solution is based on a discretization that uses Galerkin finite elements with nodal quadrature. Solution of the continuity and momentum equations involves three sequential steps at each time level. First the wave equation is solved semi-implicitly for elevation at three time levels. The vertical structure of horizontal velocity is then determined from the horizontal momentum equations using a semi-implicit formulation at two time levels. Lastly the vertical component of the continuity equation determines the vertical velocity. Conservation equations for heat and turbulent quantities, i.e. twice the kinetic energy and twice the kinetic energy times a turbulent length scale, are solved using Galerkin finite elements. Dissipation terms are handled implicitly. Density is vertically interpolated onto pre-defined level surfaces on which the baroclinic pressure gradient is computed. This is followed by re-interpolation of the baroclinic pressure gradient to the vertical mesh and vertical integration. The vertical mesh is a time-dependent mesh that tracks the evolution of the free surface. A fixed number of vertical nodes beneath each horizontal point are redistributed in time over the water column above the pre-specified resolution of boundary layer.

The QUODDY4 model is dynamically equivalent to the often used Princeton Ocean model (Blumberg and Mellor, 1987). The advantage of the current model lies in its finite element formulation that allows for greater flexibility in representing geometric complexity and strong horizontal gradients in either bathymetry and/or velocity.

ARABIAN GULF MODEL

Mesh and Bathymetry

The model domain encompasses the Arabian Gulf, extends through the Strait of Hormuz and truncates on the Gulf of Oman shelf at depths of 200m (Fig. 1). Bathymetric depths are interpolated from a fine resolution data set, DBDB-V, compiled by the Naval Oceanographic Office (1996). For the Arabian Gulf, this database provides two-minute resolution. The Arabian Gulf is a very shallow basin with an average depth of approximately 50m. Bathymetry is asymmetrical about the NW-SE axis of the basin with a deep channel (80m) parallel to this axis and positioned offshore from the Iranian coast. A broad shallow shelf off the coast of Oman and the United Arab Emirates contrasts the steep slope and short shelf on the Iranian side of the Gulf.

The domain is discretized with three-dimensional finite elements. The horizontal mesh contains 33322 elements (17440 nodes) that are projected vertically up from bottom. Spatial resolution of the mesh is designed to adequately capture the speed of gravity waves. A resolution of at least 3 km is maintained in waters less than 40m depth; for the remainder of the Gulf spatial resolution is approximately 6 km with 8 km nodal spacing near the open boundary. In the vertical dimension, twenty non-uniform linear elements (21 nodes) vary sinusoidally over the water depth with increasing resolution (1 m) at the surface and bottom Ekman layers. An identical number of linear vertical elements exist everywhere in space. Admittedly, 21 vertical nodes is the minimum threshold of acceptable resolution but was selected to facilitate formulation of a reasonable computational problem.

Forcing

External forcing is comprised of tides, surface wind stress, and heat flux for which sea surface temperature is a surrogate. A climatological density field serves as an initial condition. The source of bi-monthly climatological field serves

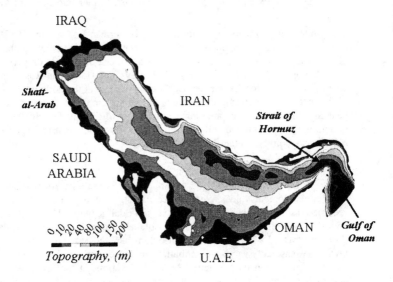

Figure 1. The Arabian Gulf model domain. Shown are bathymetric contours at 10, 20, 40, 60, 80, and 100m.

as an initial condition. The source of bi-monthly climatological hydrography is the MODAS 2.0 Synthetic Ocean Environment (Harding et al., 1998). This climatology was established utilizing data from 1920 to the present that resides in the U.S. Navy's Master Oceanographic Observation Data Set (MOODS) (Bauer, 1982). The MODAS hydrography is output at 37 standard depths and has a horizontal resolution of 0.125 degree. Bi-monthly estimates represent three-month averages centered on the 15th of each respective month.

MODAS-derived surface and near-bottom temperature, salinity and density for January (winter) and July (summer) appear in Figs. 2a and 2b. For the month of July surface heating leads to highly stratified temperature profiles throughout the Gulf; temperatures can range over 20 degrees throughout the water column. At depth, more saline water is present in the northern Gulf due to the sinking of surface water subject to significant evaporation. The signature of freshwater inflow from the Shatt-al-Arab river is evident in the surface salinity. From the hydrography, less saline water enters from the Gulf of Oman through the Strait of Hormuz. The winter hydrography contrasts sharply with that of summer. Generally waters are well-mixed over the entire Gulf. River inflow in the northwest is evident in both temperature and salinity fields as a plume of cool, fresh water that extends along the Arabian coast, further into the Gulf than was seen in summer.

The wind stress climatology is taken from Hellerman-Rosenstein (1983). Their calculations were based upon surface observations obtained during the period 1870 through 1976 and an application of the speed-dependent drag coefficient formulation of Bunker (1976). Fig. 3 displays the Hellerman-Rosentein wind stress forcing data for the Arabian Gulf model domain at the horizontal resolution of the data set (2 degrees). Mean winds in January are the strongest winds of the year and blow from the predominant northwest direction. During the summer period, winds weaken and shift to a more westerly component in response to the SW monsoon in the southern Gulf.

Boundary Conditions

At the open ocean boundary, temperature and salinity are held constant at their climatological values. Sea surface elevation at the offshore boundary is specified as a composite of tidal and sub-tidal signals. The tidal elevation boundary condition components are obtained from the Grenoble database (LeProvost et al., 1994). Tidal contributions are considered to be seasonally invariant. Seasonally varying contributions to the sea surface elevation include the effects of wind and baroclinicity. Residual elevations at the offshore boundary due to baroclinic effects are derived from the density field to produce zero-geostrophic bottom velocity normal to the boundary. This specification allows cross-boundary flows while assuming a near-bottom "level of no motion". The

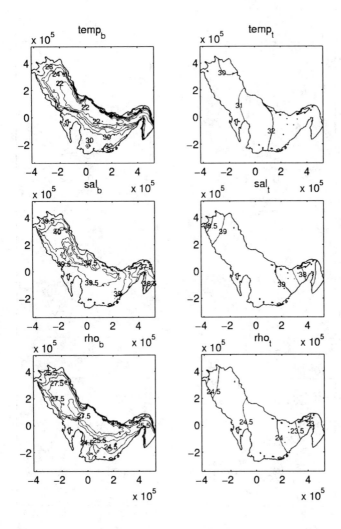

Figure 2a. MODAS 2.0 Hydrography. Shown are the near-bottom (left) and surface (right) temperature (top), salinity (middle), and density (bottom) for the month of July.

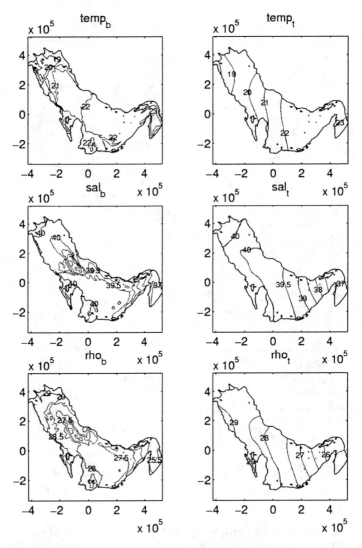

Figure 2b. MODAS 2.0 Hydrography. Shown are the near-bottom (left) and surface (right) temperature (top), salinity (middle), and density (bottom) for the month of January.

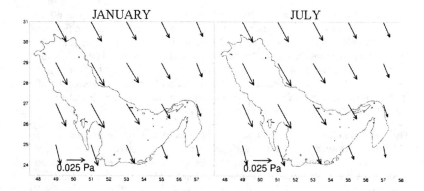

Figure 3. Summer and winter seasonal mean wind stress. Hellerman-Rosenstein wind stress over the Arabian Gulf for the months of January and July.

seasonal wind contribution to elevation is obtained from simulations over a larger domain, forced by wind stress alone. The linearized model of Lynch and Werner (1991) is used for these simulations and results are sampled at the seaward boundary of the model mesh. To this is added the baroclinic signal, calculated as described above, and the tidal elevation.

At the sea surface, the climatological wind stress is applied as the natural boundary condition for the momentum equation. Surface temperature is nudged to climatology on a one day time scale as a surrogate for the seasonal heating; the effects of evaporation are not considered explicitly. A quadratic stress condition ($C_d = 0.005$) is implemented at a height of 1m above the seabed. Within the context of this bottom stress parameterization, the selected value for C_d dictates a roughness height of 0.0035m (Lynch et al., 1996). No heat or salt exchange with the seabed is assumed. At landward boundaries a conventional free-slip condition is employed with no normal transport of heat, mass, or momentum.

SIMULATION STRATEGY

A series of simulation experiments are designed to understand the computed circulation in terms of forcing mechanisms active in the Arabian Gulf environment. Initially a barotropic tidal run, assuming homogeneous conditions, is used to assess the tidal response of the basin, examine sensitivity to the bottom stress coefficient, and validate predicted tidal elevations against available station data. Barotropic simulations are 10 days in length and use a time step of 60 sec. These simulations are not connected in any way to the diagnostic computations described below. Conceptually these simulations are considered to be seasonally invariant, an assumption not challenged in this work. Barotropic tidal rectification describes the residual circulation or that portion of the current field remaining after the extraction of the primary tidal components from the solution. The low frequency residual circulation is important only in the sense that it can interact with the baroclinic-forced component of the flow.

The set of simulations described below is applied to each of the months January and July, the winter and summer seasonal extremes, respectively. A time step of 120 sec. is implemented and deemed appropriate for the low frequency motions modeled. A diagnostic simulation subject to tide and density forcing is initialized with climatological mass fields and a fluid at rest. The simulation proceeds for 10 days, i.e. January 5-15, with density held constant at the initial state. In these simulations the density structure is presumed known and constitutes a fixed baroclinic pressure gradient. Such diagnostic computations represent a balance between reduced physics and reduced computational burden. First order effects of 3-D flow over topography are captured in these experiments. The resulting residual circulation is described as baroclinic tidal rectification and indicates the importance of stratification on the computed eddy viscosity coefficients.

The third type of simulation, termed prognostic, includes the transport of heat and salt in tidal time. Initialization is with the climatological mass field in addition to seasonal pressure and currents derived diagnostically. The model is driven by tides and residual elevation at the open boundary and extends in time for an additional 10 days beyond the diagnostic calculation, i.e. January 16-25. Prognostic calculations allow the evolution of tidal time variations of the velocity and density fields. In these simulations the density adjustment to a changing baroclinic pressure gradient is evident. The final simulation, named seasonal, is identical to the prognostic calculations with the exception that wind forcing is now activated, i.e. the simulation runs 10 days beyond diagnostic computations, January 16-25.

This simulation strategy, used in previous studies by Naimie et al. (1994) and Naimie (1996), allows for short-term prognostic adjustment of the climatological initial mass fields to tidal-time advection and mixing. Such an adjustment facilitates the development of proper mixing fronts, the removal of database anomalies associated with the forcing, and the sharpening of stratification features that are necessarily blurred in the climatology. At the same time, the short-term nature of the simulation preserves the large-scale climatological signal embedded in the mass field initalization. These calculations should be considered diagnostic at inter-seasonal time-scales, but prognostic at intra-seasonal scales.

The 10 day length of the simulations allow spin-up of nonlinear coupling between the 3-D tidal-time variations in density, turbulent mixing processes, and current structure. All solutions are sampled hourly in the final nine days of the run period. The residual circulation is then determined by fitting these hourly samples in a least-squares sense to a spectrum that includes modeled tidal constituents and the mean circulation, using the tidal package of Foreman, (1977, 1978). The 10 day simulation period is sufficient to accurately separate four primary tidal constituents, M_2, S_2, K_1, O_1.

MODEL RESULTS

Barotropic Tidal Rectification

Computed co-tidal charts in the Arabian Gulf for four primary tidal constituents are presented in Fig. 4. The tidal signal is largely a mixed diurnal and semi-diurnal response. Tidal elevations range upwards to 1 meter. In comparison to measured elevations at 94 International Hydrographic Office stations, root mean square errors on a constituent-wise basis average less than 9 cm for elevation and less than 18 degrees for phase. A more detailed discussion of the tidal response in the Arabian Gulf is given by Blain (1998).

Generally the tidal residual circulation is quite small with values less than 2 cm/s.

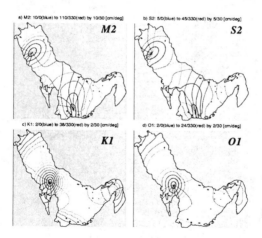

Figure 4. Computed co-tidal charts for four primary tidal constituents, M_2, S_2, K_1, O_1.

The primary feature of the stream function for the depth-averaged velocity, shown in Fig. 5, is an anti-clockwise gyre in the northeast section of the basin. Intensification of the current occurs along the eastern boundary of the gyre in response to shallower water and the peninsula protrusion off the Iranian coast. It is possible that the no flow condition at the coastal boundary may prevent more realistic flows such as flooding/drying leading to an artificially high residual circulation at this location. Other notable residual currents occur around the Musandem peninsula (the bend in the Strait of Hormuz leading into the Gulf of Oman) and in the Strait itself. Amplification of the residual circulation is highly localized and can be traced largely to complexities in the geometry (e.g. islands) or very shallow bathymetry.

Summer Mean Circulation

Baroclinic Tidal Rectification – Over a majority of the Gulf during summer months, evaporation far exceeds the input of freshwater from precipitation and river inflow (e.g. Sultan and Elghribi, 1996; Abdelrahman and Ahmad, 1995).

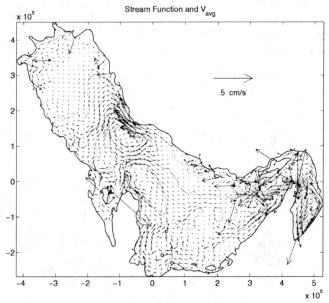

Figure 5. Barotropic tidal rectification. Depth-averaged velocity vectors on contours of the stream function computed from a barotropic, homogenous simulation.

This imbalance creates a density gradient along the axis of the Gulf with denser water residing at the northern end of the Gulf. The resulting pressure gradient drives freshwater inflow from the Gulf of Oman into the Arabian Gulf through the Strait of Hormuz. Water flowing northward is deflected by Coriolis forcing toward the Iranian coast. As depths become shallower and temperatures cool (relatively), the water becomes more saline and dense due to these factors and continued evaporation as reflected in the hydrography. Denser water then sinks in the northern Gulf and flows down into the deep trough located in the central Gulf. Eventually this dense, saline water exits the Strait of Hormuz as a deep outflow (e.g. Lardner et al., 1987; Abdelrahman and Ahmad, 1995).

The stream function for depth-averaged velocity (Fig. 6) computed from the diagnostic simulation clearly exhibits the described reverse estuarine flow pattern and coincides with published currents measurements (e.g. Hunter, 1986; Reynolds, 1993; Abdelrahman and Ahmad, 1995). A cyclonic gyre of circulation extends the length of the Gulf with inflow through the northern half of the Strait of Hormuz. In comparing surface and near-bottom velocities, this inflow is confined to the upper layers of the water column and is approximately 10-20 cm/s, a value in good agreement with Abdelrahman and Ahmad (1995). The excursion of the freshwater inflow is 200-300 km along the Iranian coast and also matches published observations (Abdelrahman and Ahmad, 1995). At the northern reach of this excursion, the current re-circulates creating a strong gyre embedded within the larger Gulf-length gyre. This feature has not been reported previously though there is nothing to suggest that it does not exist. The remaining coastal current moves northward and is enhanced by the tidal residual as it passes off the Iranian coast. Flow returns down the central Gulf. This deep outflow is evident in plots of the near-bottom velocity (not shown). As deduced by comparing the MODAS July surface and bottom salinities, the high salinity shallow waters off the Oman coast, likely caused by increased evaporation, sink and flow out of the Gulf at depth. A compensating freshwater inflow from the Gulf of Oman enters through the Strait hugging the southeast coast. In the southeast region of the Gulf a strong clockwise circulation gyre forms (Fig. 6) as the saline water from the central Gulf flows toward the region of freshwater influx along the Oman coast.

Prognostic Evolution of the Mass Variables – The stream function and depth-averaged velocities (Fig.7) form a circulation pattern that is significantly more developed than that of the diagnostic calculations. The main cyclonic gyre widens and intensifies in both its southern and northern regions of circulation, extending even farther to the north. Outflow expands across the entire Strait. The increased intensity of the circulation demonstrates the importance of time-dependant

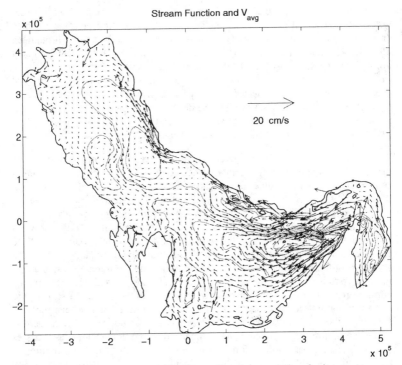

Figure 6. Baroclinic tidal rectification. Depth-averaged velocity vectors on contours of the stream function computed from a summer diagnostic (constant stratification) simulation.

baroclinic effects. The clockwise circulation in the southern Gulf is far weaker than in the diagnostic case. An incoming surface current remains along the southeast coast but deeper water flows in a southerly direction from the main channel to satisfy continuity. This direction of the deep flow is a clear departure from the purely diagnostic calculations, as is the small counter-clockwise circulation within the Strait itself.

Summer Seasonal Circulation – The addition of wind forcing is the only dynamical difference between the seasonal and prognostic simulations. The northwest wind sets up a coastal current flowing southward along the Arabian coast as shown in the computed stream function plotted in Fig. 8. Generally the northwest wind spreads the cyclonic gyre westward over the length of the Gulf.

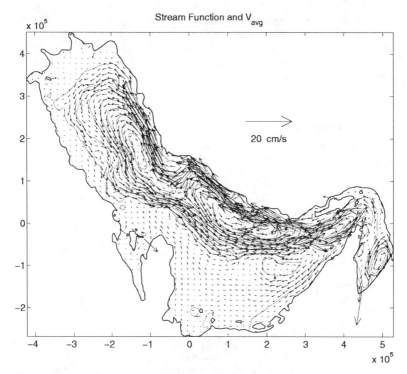

Figure 7. Baroclinic tidal rectification. Depth-averaged velocity vectors on contours of the stream function computed from a summer prognostic (time-dependent stratification) simulation.

The westerly component of the wind breaks up the anti-cyclonic gyre in the southeast portion of the basin. Water is pushed southward onto the shallow U.A.E. and Oman shelf, then west along the coast and out through the Strait. The presence of the wind appears to intensify the circulation gyre present in the Strait.

Winter Mean Circulation

The winter seasonal mean depth-averaged currents and associated stream function are shown in Fig. 9. Conceptually the dynamics of the circulation follow that of the summer months but the magnitude and structure is significantly varied. The two-gyre (cyclonic/anti-cyclonic) structure of the currents is maintained but the currents are far weaker. The northern inflow through the Strait along the Iranian coast does not penetrate as far into the Gulf due to the reduced stratification of the water column and the counteracting strong northwesterly winds in winter. These winds spread the cyclonic gyre over the central Gulf and lead to the formation of several small internal re-circulation gyres. The northwest winds also forms a strong coastal current along the Arabian coast. The clockwise circulation in the southern Gulf becomes elongated as the wind pushes the flow to the south toward the shoreline. The circulation gyre within the Strait is much intensified.

Qualitative Model-Data Comparisons

The very recent Strait of Hormuz Experiment (Johns and Zantopp, 1999) offers the first direct time series observations of the current and water mass

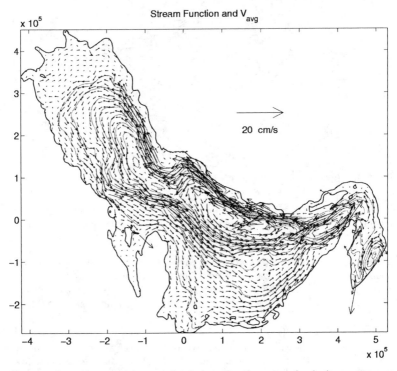

Figure 8. Summer seasonal mean circulation. Depth-averaged velocity vectors on contours of the stream function computed from a summer prognostic (time-dependent stratification) simulation that includes wind and tidal forcing.

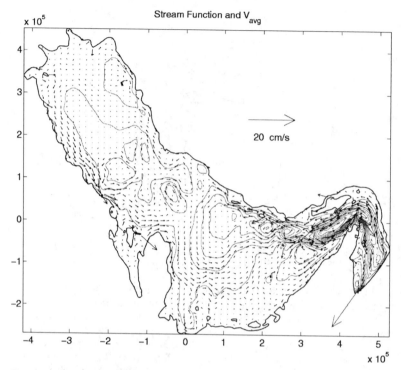

Figure 9. Winter seasonal mean circulation. Depth-averaged velocity vectors on contours of the stream function computed from a winter prognostic (time-dependent stratification) simulation that includes wind and tidal forcing.

structure in the Strait. Hydrographic stations were occupied along a line crossing the southern Strait of Hormuz shown in Fig. 10. Observations taken from 6 shipboard CTD profilers on 29 July 1997 are qualitatively compared to the computed temperature and salinity at a nearby transect also shown in Fig. 10. Warm surface water present along the southern side of the Strait in the observations is also seen in the model-generated temperature field. Both observed and simulated temperatures are 30-31 degrees C. At the same location is a pocket of very salty water observed to be nearly 40 PSU. The model also contains saltier water at this location but the salinity values are fresher than those measured, approximately 37.9 psu. This warm, salty water is the result of highly evaporative conditions off the Oman coast. Proceeding down through the water column, the observations indicate a high salinity wedge banked against the southern side of

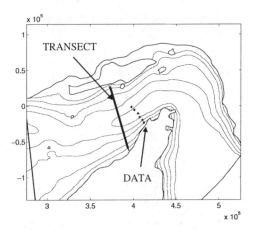

Figure 10. Locations of 6 CTD locations for 29 July, 1997 and a nearby model transect in the Strait of Hormuz.

the Strait. Such a wedge is perceptible in the model but it does not continue to the bottom of the water column as seen in the observations. Generally the horizontal and vertical circulation patterns are well-represented by the model. However, model profiles of temperature and most markedly salinity lose the strong stratification seen in the climatological values. Sources for this error are thought to be largely due to the lack of evaporative flux in the model as configured. Secondly there may be contributions to over-mixing by the Mellor-Yamada 2.5 turbulence closure scheme and/or the advective scheme. Comparisons between the climatological temperature and salinity and the observations do however suggest that prognostic evolution of the mass fields is important.

CONCLUSION

It is well-known that the tidal influence on the overall circulation in the Arabian Gulf is minimal; a fact confirmed by the present model results. In only a few localized areas (e.g. Iranian coast and the Strait of Hormuz) are tidal residuals large enough to enhance the dominant density-driven flow. Currents derived from baroclinic diagnostic computations in which the density structure remains as imposed by climatology reinforce the existence of the often-hypothesized reverse estuarine flow in the Gulf. A prognostically evolving baroclinic pressure gradient noticeably intensifies the main cyclonic circulation. In summer the baroclinic pressure gradient is the primary mechanism for flow throughout the Gulf. Wind effects during summer are limited to the Arabian coast and the very shallow waters in the southern Gulf. Wind over the central Gulf merely enhances the cyclonic gyre of circulation.

Winter circulation is far more controlled by wind forcing. The northern Gulf is well-mixed with wind-generated, southward flowing coastal currents particularly off the Arabian coast. Freshwater inflow from the Gulf of Oman is limited to a region near the Strait where influence of the baroclinic pressure gradient is a maximum. Over the eastern portion of the Gulf wind and density-driven influences are comparable.

Limited initial comparisons to published observations in the Gulf and the Strait itself indicate that model predictions are in reasonable agreement with observed conditions. Obviously, more extensive model-data comparisons are necessary. The lack of vertical structure in the temperature and salinity profiles regardless of the background prescribed indicates an imbalance of salt and heat. The inclusion of time and spatially variable evaporation and heat flux at the surface will likely improve the computed profiles of temperature and salinity but data available to compute such fluxes is sparse. The influx of freshwater from the northern rivers must also be taken into account. Overall, the finite element model QUODDY4 produces realistic circulation patterns over the Arabian Gulf at the seasonal extremes of January and July. This model has also proved to be a useful tool for understanding the dynamics of the region by examination of a series of process-oriented simulations.

Acknowledgements: The author is grateful to W. E. Johns for sharing his observational data; C. E. Naimie for the use of the Dartmouth College circulation models and ongoing discussions regarding the modeling. This work is supported by the Office of Naval Research through the NOMP Program, Award No. N0001499WX30113.

REFERENCES

Abdelrahman, S. M. and F. Ahmad, 1995. A note on the residual currents in the Arabian Gulf, *Continental Shelf Research,* 15(8), 1015-1022.

Bauer, R., 1982. Functional description: Master Oceanographic Observation Data Set (MOODS), Compass Systems, Inc.

Blain, C. A., 1998. Barotropic and tidal residual circulation in the Arabian Gulf, *Estuarine and Coastal Modeling, Proceedings of the 5th International Conference,* M. L. Spaulding and A. F. Blumberg, eds., 166-180, American Society of Civil Engineers.

Blumberg, A. and G. L. Mellor, 1987. A description of a three-dimensional coastal ocean circulation model, in *Three-Dimensional Coastal Models*, N. S. Heaps, ed., American Geophysical Union, Washington, D. C., Coastal and Estuarine Series, 4, 1-16.

Blumberg, A. F., B. Galperin, and D. J. O'Conner, 1992. Modeling vertical structure of open channel flows, *ASCE Journal of Hydraulic Engineering*, 118, 1119-1134.

Bunker, A. F., 1976. Computations of surface energy flux and annual air-sea interaction cycles of the North Atlantic Ocean, *Monthly Weather Review*, 104, 1122-1140.

Chao, S-Y., T. W. Kao, and K. R. Al-Hajri, 1992. A numerical investigation of circulation in the Arabian Gulf, *J. Geophys. Res.*, 97(C7), 11219-11236.

Foreman, M. G. G., 1977. Manual for tidal heights analysis and prediction, Pacific Marine Science Report 77-10, Institute for Ocean Sciences, Patricia Bay, Sidney, B.C.

Foreman, M. G. G., 1978. Manual for tidal currents analysis and prediction, Pacific Marine Science Report 78-6, Institute for Ocean Sciences, Patricia Bay, Sidney, B.C.

Galperin, B., L. H. Kantha, S. Hassid, and A. Rosati, 1988. A quasi-equilibrium turbulent energy model for geophysical flows, *Journal of Atmospheric Science*, 45, 55-62.

Harding, J., R. Preller, R. Rhodes, 1998. Coastal ocean prediction at the Naval Research Laboratory, *Proceedings of the 2nd Conference on Coastal and Oceanic Prediction and Processes*, 11-16 January, 1998, Phoenix, AZ, 108-114, American Meteorological Society.

Hellerman, S. and M. Rosenstein, 1983. Normal monthly wind stress over the world ocean with error estimates, *J. Phys. Oceanogr.*, 13(7), 1093-1104.

Hunter, J., 1986. A physical oceanography of the Arabian Gulf: A review and theoretical interpretation of previous observations, *Marine Environment and Pollution, Proceedings of the First Arabian Gulf Conference on Environment and Pollution*, Environmental Protection Council, Kuwait.

Johns, W. E. and R. J. Zantopp, 1999. Data report for the Strait of Hormuz experiment December 1996 – March 1998, Technical Report 99-001, University of Miami, Miami, FL.

Kinnmark, I. P. E., 1985. *The Shallow Water Wave Equations: Formulation, Analysis and Application*, Ph.D. Dissertation, Department of Civil Engineering, Princeton University.

Lardner, R. W., W. J. Lehr, R. J. Fraga, and M. A. Sarhan, 1987. Residual circulation in the Arabian Gulf, I. Density-driven flow, *Arabian J. Science and Engr.*, 12(3), 341-354.

LeProvost, C., M. L. Genco, F. Lyard, P. Vincent, and P. Canceil, 1994. Spectroscopy of the world ocean tides from a finite element hydrodynamic model, *J. Geophys. Res.,* 99(C12), 24777-24798.

Lynch, D.R. and W.G. Gray, 1979: A Wave equation model for finite element tidal computations, *Comp. Fluids,* 7, 207-228.

Lynch, D. R. and F. E. Werner, 1987. Three-dimensional hydrodynamics on finite elements, Part I: linearized harmonic model, International Journal for Numerical Methods in Fluids, 7, 871-909.

Lynch, D. R. and F. E. Werner, 1991. Three-dimensional hydrodynamics on finite elements, Part II: non-linear time-stepping model, International Journal for Numerical Methods in Fluids, 12, 507-533.

Lynch, D. R., F. E. Werner, D. A. Greenberg, and J. W. Loder, 1992. Diagnostic model for baroclinic and wind-driven circulation in shallow seas, Continental Shelf Research, 12, 37-64.

Lynch, D. R., J. T. C. Ip, C. E. Naimie, and F. E. Werner, 1996. Comprehensive coastal circulation model with application to the Gulf of Maine, Continental helf Research, 16(7), 875-906.

Mellor, G. L. and T. Yamada, 1982. Development of a turbulence closure model for geophysical fluid problems, Reviews of Geophys. Space Phys., 20, 851-875.

Naimie, C. E., J. W. Loder, and D. R. Lynch, 1994. Seasonal variation of the three-dimensional residual circulation on Georges Bank, J. Geophys. Research, 99(C8), 15967-15989.

Naimie, C. E., 1996. Georges Bank residual circulation during weak and strong stratification periods: Prognostic numerical results, , J. Geophys. Research, 101(C3), 6469-6486.

Naval Oceanographic Office, 1996: Data base description for digital bathymetric data base - Variable resolution (DBDB-V), Draft Report, Naval Oceanographic Office, Stennis Space Center, MS.

Reynolds, R. M., 1993: Physical oceanography of the Gulf, Strait of Hormuz, Gulf of Oman-Results from the Mt. Mitchell Expedition, Marine Pollution Bulletin, 100, 35-59.

Sultan, S. A. R. and N. M. Elghribi, 1996. Temperature inversion in the Arabian Gulf and Gulf of Oman, Continental Shelf Research, 16(12), 1521-1544.

A numerical simulation of the Japan/East Sea (JES) seasonal circulation

Peter C. Chu, Jian Lan, and Hilbert Strauhs
Naval Postgraduate School, Monterey, CA 93943

Abstract

The seasonal variability of the Japan/East Sea (JES) circulation was studied numerically using the Princeton Ocean Model (POM) with horizontal resolution varying from 11.54 to 18.53 km and 15 sigma levels conforming to a relatively realistic bottom topography. The model was integrated using climatological monthly mean wind stresses, heat and salt fluxes as surface forcing and observational oceanic inflow/outflow at open boundaries. The seasonally averaged effects of isolated forcing terms are presented and analyzed from the following experiments: 1) non-linear effects removed, 2) no lateral transport at open boundary, and 3) wind effects removed. Major currents are simulated reasonably well compared to observations. The nonlinear advection does not affect the general circulation pattern evidently, but does affect the formation of the mesoscale eddies, especially the Ulleung/Tsushima Basin (UTB) eddy (all seasons) and the Japan Basin (JB) cyclonic gyre (spring). The lateral boundary forcing enhances (weakens) the JES volume transport in the summer (winter). The wind forcing is the most important factor (80%) for generating the JB cyclonic gyre. Besides, it drives the Liman Current and damps the East Korean Warm Current in the winter, and generates the UTB eddy, and eddies along the Japan Nearshore Branch (JNB) in all seasons. However, it has almost no effect on the JNB currents for all seasons.

1 Introduction

The Japan Sea, known as the East Sea in Korea, has a steep bottom topography (Fig. 1) that makes it a unique semi-enclosed ocean basin overlaid by a pronounced monsoon surface wind. The Japan/East Sea, hereafter referred to as JES, covers an area of 10^6 km^2. It has a maximum depth in excess of 3,700 m, and is isolated from open oceans except for small (narrow and shallow) straits, which connects the JES with the North Pacific through the Tsushima/Korean and Tsugaru Straits and with the Okhotsk Sea through the Soya and Tatar Straits. The JES contains three major basins called the Japan Basin (JB), Ulleng/Tsushima Basin (UTB), and Yamato Basins (YB), and a high central seamount called the Yamato Rise (YR). The JES has great scientific interest as a miniature prototype ocean. Its basin-wide circulation pattern, boundary currents, Subpolar Front (SPF), mesoscale eddy activities and deep water formation are similar to those in a large ocean.

Numerical studies on the JES circulations started in early 1980. Various types of models were used such as the multi-layer model (Sekine, 1986, 1991; Kawabe, 1982b; Yoon, 1982a,b; Seung and Nam, 1992a,b; Seung and Kim, 1995), the Modular Ocean Model (MOM) (Kim and Yoon, 1994; Holloway et al., 1995; Kim and Yoon, 1996; Yoshikawa et al., 1999), rigid-lid z-level model (Yoshikawa et al., 1999), the Miami Isopycnal Coordinate Model (MICOM) (Seung and Kim, 1993) and the Princeton Ocean Model (POM) (Mooer and Kang, 1995; Chu et al., 1999c,d). Most of the numerical efforts are concentrated on simulating basin-wide circulation, the Tsushima Warm Current (TWC) bifurcation, and formation of the intermediate waters. However, there is lack of studies on dynamical mechanisms for the formation of seasonal variability of the JES circulation.

In this study, we use the POM to investigate driving mechanisms for the formation of the seasonal variabilities of the JES circulation including the subpolar front (SPF) meandering and eddies, the TWC bifurcation and its effect on the formation of mesoscale eddies in the UTB and the YB, and the Liman Cold Current (LCC) and its penetration into the southwestern waters along the Korean coast. The control run, forced by the climatological monthly wind stress, heat and fresh water fluxes, is designed to best simulate reality against which each experiment is compared. In the experiments, various external and internal factors are modified and the resulting circulation patterns and magnitudes compared to the control run results. Specifically, we estimate the contribution, in terms of volume transport and circulation patterns, of non-linear advection, wind forcing and lateral boundary transport to the ocean features identified in the control results. From this, we can estimate the relative importance of these factors to the seasonal variability of the JES circulation.

2   JES Current Systems

Most of the nearly homogeneous water in the deep part of the basin is called the Japan Sea Proper Water (Moriyasu, 1972) and is of low temperature and low salinity. Above the Proper Water, the TWC, dominating the surface layer, flows from the East China Sea through the Tsushima/Korean Strait and carries warm water from the south. The LCC flows in the JES from the Okhotsk Sea through the Tatar Strait, carries cold fresh water from the north, and becomes the North Korean Cold Current (NKCC) after reaching the North Korean coast (Yoon, 1982). Both currents turn eastward to flow roughly along 40°N latitude, forming the SPF between the Tsushma Warm Water and the cold and fresh water from the north (Fig. 2).

Uda (1934) was the first one to sketch the JES general circulation from limited observational data. The TWC separates north of 35°N into two branches into a western and an eastern channel (Kawabe, 1982a,b; Hase et al., 1999) and flows through the western channel, called the East Korea Warm Current (EKWC), and closely follows the Korean coast until it separates near 37°N into two branches. The eastern branch follows the SPF to the western coast of Sapporo Island, and the western branch moves northward and forms a cyclonic eddy at the Eastern Korean Bay (EKB). It flows through the eastern channel which closely follows the Japanese Coast, called the Nearshore Branch (NB) by Yoon (1982a). More accurately, we call it the Japan Nearshore Branch (JNB). The JNB is usually weaker than the EKWC. The strength of the Tsushima Current at both channels reduces with depth.

The NKCC meets the EKWC at about 38°N with some seasonal meridional migration. After separation from the coast, the NKCC and the EKWC converge and form a strong front that stretches in a west-east direction across the basin. The NKCC makes a cyclonic recirculation gyre in the north but most of the EKWC flows out through the outlets. The formation of NKCC and separation of EKWC are due to local forcing by wind and buoyancy flux (Seung, 1992). Large meanders develop along the front and are associated with warm and cold eddies. Readers may find qualitative depiction from a textbook written by Tomczak and Godfrey (1994).

Chu et al. (1999b) identified major features of the three-dimensional circulation and the volume transport from the Navy's 0.5°×0.5° global monthly climatological temperature and salinity data set using the P-vector method (Chu, 1995; Chu et al., 1998b). The transport pattern is largely determined by the upper layer circulation and characterized by a large-scale cyclonic recirculation gyre, in which the EKWC and the JNB take part, as the inflow-outflow system, and the NKCC in the North. At a few hundred kilometers off the separation area, the EKWC makes an anticyclonic gyre. The gyre becomes stronger as the EKWC develops. On the other hand, the northern cyclonic gyre is very deep and is most significantly in the winter strengthened by the wind and buoyancy flux. The gyre, or the southward coastal current related to it, is deep enough to intrude southward beneath the EKWC most of the time. Seung also confirmed the summertime presence of counter-current beneath the JNB. North of the SPF there exists a cyclonic gyre in the JB usually called the JB gyre.

3   Seasonal Variation of Atmospheric Forcing

The Asian monsoon strongly affects the thermal structure of the JES (Chu et al., 1998a). During the winter monsoon season, a very cold northwest wind blows over the JES as a result of the Siberian High Pressure System. By late April, numerous frontally-generated events occur making late April and May highly variable in terms of wind speeds and number of clouds. During this period storms originating in Mongolia may cause strong, warm westerlies. By late May and early June, the summer surface atmospheric low pressure system begins to form over Asia. Initially this low pressure system is centered north of the Yellow Sea producing westerly winds. In late June, this low begins to migrate to the west setting up the southwest monsoon that dominates the summer months. The winds remain variable through June until Manchurian low pressure system strengthens. Despite the very active weather systems, the mean surface wind speed over the JES in summer is between 3 and 4 m/s, which is weaker than in winter. By July, however, high pressure (the Bonin High) to the south and the low pressure over Manchuria produce southerly winds carrying warm, moist air over the East China Sea/Yellow Sea. In the summer, warm air and strong downward net radiation stabilize the upper layer of the water and causes the surface mixed layer to shoal. October is the beginning of the transition back to winter conditions. The southerly winds weaken and let the sea surface slope reestablish the winter pattern.

The datasets used were the objectively analyzed fields of surface marine climatology and anomalies of fluxes of heat, momentum, and fresh water. The fields are derived from individual observations in the Comprehensive Ocean-Atmosphere Data Set (COADS) from 1945 to December 1989 and are analyzed on a 1° by 1° grid (da Silva et al., 1994).

4   The Numerical Ocean Model

4.1   Model Description

Coastal oceans and semi-enclosed seas are marked by extremely high spatial and temporal variability that challenge the existing predictive capabilities of numerical simulations. The POM is a time dependent, primitive equation circulation model on a three dimensional grid that includes realistic topography and a free surface (Blumberg and Mellor, 1987). River outflow is also not included. However, the seasonal variation in sea surface height, temperature, salinity, circulation and transport are well represented by the model. From a series of numerical experiments, the qualitative and quantitative effects of non-linearity, wind forcing and lateral boundary transport on the JES are analyzed, yielding considerable insight into the external factors affecting the region oceanography. The model results were sampled every thirty days.

The model contains $94 \times 100 \times 15$ horizontally fixed grid points. The horizontal spacing is 10' latitude and longitude (approximately 11.54 to 15.18 km in the zonal direction and 18.53 km in the latitudinal direction) and 15 vertical sigma coordinate levels are used. The model domain is from $35.0°$ N to $51.5°$ N, and from $127.0°$ E to $142.5°$ E. The bottom topography is the smoothed data from the Naval Oceanographic Office Digital Bathymetry Data Base 5 minute by 5 minute resolution (DBDB5). The horizontal diffusivities are modeled using the Smagorinsky (1963) form with the coefficient chosen to be 0.2 for this application. The bottom stress τ_b is assumed to follow a quadratic law

$$\tau_b = \rho_0 C_D |\mathbf{V}_b| \mathbf{V}_b \tag{1}$$

where ρ_0 ($= 1025$ kg/m^3) is the characteristic density of the sea water, $\mathbf{V}_b$ is the horizontal component of the bottom velocity, and C_D is the drag coefficient which was determined by the similarity theory.

4.2 Surface Forcing Functions

The atmospheric forcing for the JES application of the POM includes mechanical and thermohaline forcing. The wind forcing is depicted by

$$\rho_0 K_M \left(\frac{\partial u}{\partial z}, \frac{\partial v}{\partial z} \right)_{z=0} = (\tau_{0x}, \tau_{0y}) \tag{2}$$

where (u, v) and (τ_{0x}, τ_{0y}) are the two components of the water velocity and wind stress vectors, respectively. The wind stress at each time step is interpolated from monthly mean climate wind stress from COADS (1945-1989). We interpolated the COADS wind stress of the resolution of $1° \times 1°$ onto the model grid of the resolution of $10'$.

Surface thermal forcing is depicted by

$$K_H \frac{\partial \theta}{\partial z} = \alpha_1 \left(\frac{Q_H}{\rho C_p} \right) + \alpha_2 C(\theta_{OBS} - \theta) \tag{3}$$

$$K_S \frac{\partial S}{\partial z} = -\alpha_1 S(P - E) + \alpha_2 C(S_{OBS} - S) \tag{4}$$

where θ_{OBS} and S_{OBS} are the observed potential temperature and salinity, c_p is the specific heat, Q_H is the net surface heat flux (downward positive), P is the precipitation rate, and E is evaporation rate. The relaxation coefficient C is the reciprocal of the restoring time period for a unit volume of water. The parameters (α_1, α_2) are $(0,1)$-type switches: $\alpha_1 = 1$, $\alpha_2 = 0$, would specify only flux forcing is applied; $\alpha_1 = 0$, $\alpha_2 = 1$, would specify that only restoring type forcing is applied.

Chu (1989) pointed out the importance of using compatible surface wind and thermal boundary conditions. Chu et al. (1998c) further found that the restoring surface boundary condition does not exist anywhere in the world ocean. Therefore, in this study, the surface thermohaline forcing is determined solely by the flux forcing, that is, $\alpha_1 = 1$, $\alpha_2 = 0$ in (3)-(4). The mixing coefficients K_M, K_H, and K_S were computed using a level two-turbulence closure scheme (Mellor and Yamada, 1982).

4.3 Lateral Boundary Forcing

Closed lateral boundaries, i.e., the modeled ocean bordered by land, were defined using a free slip condition for velocity and a zero gradient condition for temperature and salinity. No advective or diffusive heat, salt or velocity fluxes occur through these boundaries.

At open boundaries, the numerical grid ends but the fluid motion is unrestricted. Uncertainty at open boundaries makes marginal sea modeling difficult. Three approaches,

Month	Jan	Feb	Mar	Apr	May	Jun	Jul	Aug	Sep	Oct	Nov	Dec
Soya	-0.32	-0.12	-0.20	-0.15	-0.20	-0.35	-0.55	-0.57	-0.57	-0.6	-0.6	-0.4
Tatar	0.2	0.2	0.2	0.2	0.2	0.2	0.2	0.2	0.2	0.2	0.2	0.2
Tsugaru	-0.68	-0.38	-0.60	-0.45	-0.60	-1.05	-1.65	-1.73	-1.73	-1.8	-1.8	-1.2
Tsushima	0.7	0.3	0.6	0.4	0.6	1.2	2.0	2.1	2.1	2.2	2.2	1.4

Table 1: Monthly variation of volume transport (Sv) at the lateral open boundaries. The positive/negative values mean inflow/outflow.

local-type, inverse-type, and nested basin/coastal modeling, are available for determining the open boundary condition (Chu et al., 1997). Here, we take the local-type approach, i.e., to use the radiative boundary condition with specified volume transport. When the water flows into the model domain, temperature and salinity at the open boundary are likewise prescribed from the Navy's climatological data. When water flows out of the domain, the radiation condition was applied,

$$\frac{\partial}{\partial t}(\theta, S) + U_n \frac{\partial}{\partial n}(\theta, S) = 0 \tag{5}$$

where the subscript is the direction normal to the boundary.

Warm water enters the JES through the Tsushima/Korean Strait with the TWC from the East China Sea, and cold water enters the JES through the Tatar Strait with the Liman Current from the Sea of Okhotsk (Uda, 1934). The water exits the JES through the Tsugaru and Soya straits. We use a more recent estimation of the monthly mean volume transport, reported by Chu et al. (1999b), through the Tsushima/Korean Strait with the annual average of 1.3 Sv, a maximum of 2.2 Sv in October, and a minimum of 0.3 Sv in February. The total inflow transport through the Tatar and Tsushima/Korean straits should be the same as the total outflow transport through the Tsugaru and Soya straits. We assume that 75% of the total inflow transport should flow out of the JES through the Tsugaru Strait, and 25% through the Soya Strait. The monthly volume transports at open boundaries are listed in Table 1.

4.4   Initial Conditions and Initialization

The model was integrated with all three components of velocity (u, v, w) initially set to zero, and with temperature and salinity specified by interpolating climatology data to each model grid point. The model year consists of 360 days (30 days per month), day 361 corresponds to 1 January. It was found that 90 days were sufficient for the model kinetic energy to reach quasi-steady state under the imposed conditions. In order to first capture the winter monsoon to summer monsoon transition, the model was started from 30 October, and run to 30 January next year for spin-up. After that day (30 January), the model was run another three years for each experiment. The monthly mean data are used for comparison.

4.5   Mode Splitting

For computational efficiency, the mode splitting technique (Blumberg and Mellor, 1987) is applied with a barotropic time step of 25 seconds, based on the Courant-Friederichs-Levy (1928) computational stability (CFL) condition and the external wave speed; and a baroclinic time step of 900 seconds, based on the CFL condition and the internal wave speed.

4.6 Volume Transport Streamfunction

To investigate the JES circulation as a whole, we integrate the velocity vertically from the surface $(z = 0)$ to the bottom $(z = -h)$

$$(U, V) = \int_{-h}^{0} (u, v) dz \tag{6}$$

Due to the continuity the volume transport stream function (Ψ) is defined by

$$U = -\frac{\partial \Psi}{\partial y}, \quad V = \frac{\partial \Psi}{\partial x} \tag{7}$$

and satisfies the Poisson equation

$$(\frac{\partial^2}{\partial x^2} + \frac{\partial^2}{\partial y^2})\Psi = \frac{\partial V}{\partial x} - \frac{\partial U}{\partial y} \tag{8}$$

For each time instance, we solve the two-dimensional Poisson equation (10) with the given boundary conditions, that is, zero transport at the rigid boundaries and known transport at the open boundaries (see Table 1).

5 The Simulated Circulation (Control Run)

5.1 General Description

The simulated surface velocity field (Fig. 3) coincides with the earlier description of the JES circulation presented in Section 2. The TWC separates at the Tsushima/Korea Strait into western and eastern branches. Flow through the western channel (i.e., EKWC) closely follows the Korean coast until it separates near 38°N into two branches. The eastern branch follows the SPF to the western coast of Sapporo Island, and the western branch, moves northward and forms a cyclonic eddy at the Eastern Korean Bay (EKB). Flow through the eastern channel (i.e., JNB) is a little weaker than through the western channel. The simulated LCC carries fresh and cold water along the Russian coast and becomes the NKCC at the North Korean coast. The simulated NKCC meets the EKWC at about 38°N. After separation from the coast, the NKCC and the EKWC converge to a strong front that runs in the west-east direction across the basin.

5.2 JB Cyclonic Gyre

A large-scale cyclonic recirculation gyre over the JB is simulated with a strong seasonal variation. This gyre is easily identified by the volume transport streamfunction (Fig. 4). The JB gyre is the strongest and recirculates 8 Sv in the winter. It weakens and retreats northward in the spring and summer. In the fall, the cyclonic JB gyre disappears and weak anticyclonic eddies appears.

5.3 LCC

The LCC is a southwestward current following along the Russian coast. It bifurcates into two branches near the east Russian bight (134°E, 42°N): the western branch flows along the Russian-Korean coast and becomes the NKCC. The eastern branch flows southeastward, then turns eastward at 41.5°N, and becomes the south flank of the JB gyre. The LCC has a strong seasonal variation with the maximum speed in the winter (Fig. 3a) and the minimum speed in the summer (Fig. 3c).

5.4   EKWC

EKWC, western branch of TC north of the Tsushima/Korea Strait, acts as a western boundary current and has a strong seasonal variability. At 37°N, the EKWC v-velocity component, which is practically the direction of the current on that position, varies from 0.30 m/s in winter to 0.42 m/s in summer (Fig. 5). The width of EKWC is around 60 km all year round. However, the depth is around 1,400 m in the summer and 800 m in the winter.

5.5   EKWC - NKCC Confluence

The (northward) overshooting of the EKWC near the Korean Bight at 37.5°N is stronger in the winter than in the summer (Fig. 3). The overshot EKWC leaves the Korean coast and moves northward. It converges with the southward flowing NKCC, at 39°N, 130°E, and forms a current meandering toward the east along the SPF. We may call it the SPF Current (PFC).

5.6   JNB

JNB, the eastern branch of TC north of the Tsushima/Korea Strait at 135°E, has a maximum eastward component of 0.18 m/s in the winter (Fig. 6). A counter current beneath it is simulated with a westward component of 0.09 m/s. In the summer, the counter current remains almost unchanged, but the JNB increases to 0.21 m/s (Fig. 6).

5.7   UTB Eddy

The western branch of the warm TC (i.e., EKWC) moves northward along the southern part of Korean coast and separates from the coast after approaching the East Korean Bay (EKB). It keeps its northward motion until meeting the NKCC near 40°N, meanders southeastward, and forms a warm-core anticyclonic eddy (Fig. 3). The simulated anticyclonic UTB eddy is strongest in the summer (Fig. 3c). The center is located at 38.5°N 130°E. The size of the eddy is around 150 km. tangential velocity is around 0.4 m/s.

6   Driving Mechanisms

We analyzed the results of the four experiments to identify the driving mechanisms for the JES circulation.

6.1   Effects of Nonlinearity (Run 1 - Run 2)

In the first sensitivity study, the nonlinear advection terms were removed from the dynamic equations. Otherwise, the same parameters as the control run were used. The figures and analysis were performed after computing the difference between control and non non-linear terms runs. The differential surface vector velocities (Fig. 7) do not show evidence for general circulation patterns except a couple of mesoscale eddies, and especially the UTB eddy, which indicates that the non-linearity does not change the general circulation pattern except for the UTB anticyclonic eddy. However, our computation shows that the non-linearity causes a noticeable change in the volume transport.

6.2   Wind-Induced Circulation (Run 1 -Run 3)

Compared to the surface velocity vector map for the control run (Fig. 3), the surface differential current vectors ΔV (Fig. 8) clearly show the wind effects on the JES circulation: (a) driving the LCC in the winter, (b) damping the EKWC especially in the winter,

(c) generating the UTB eddy in all the seasons, and (d) generating eddies along the JNB. The wind has almost no effect on the occurrence of the JNB for all seasons.

The winter monsoon winds blow from the northwest to the southeast over the JES surface and drives the Ekman flow in the upper ocean to the right of the wind direction. Such a surface current moves southward at the western coast of the JES and strengthens the LCC and weakens the EKWC. The summer monsoon winds blow from the south and southeast to the northwest with much smaller speeds. Thus, the summer wind effects on reducing LCC and strengthening EKWC are quite weak.

The winds drive mesoscale eddies especially near UTB and along the west coast of Honshu. The surface UTB eddy, occurring on the differential velocity vector map (Fig. 8), has the similar swirl velocity (anticyclonic) with same order of magnitudes (maximum value of 0.7 m/s) as the control run (Fig. 3). This means that the wind effect is in fact a key factor for generating the UTB anticyclonic eddy. Furthermore, there is no evident current along the west coast of Honshu on the ΔV map (Fig. 8), indicating that wind forcing is not a major factor for maintaining JNB.

6.3 Inflow/Outflow Induced Circulation (Run 1 - Run 4)

The third sensitivity study used the control run equations and forcing but closed all open lateral boundaries, preventing the transport of mass, heat or salinity through the Tsushima, Tsugaru, Soya, and Tatar Straits. With inflow or outflow, the JES vertically integrated circulation pattern is more pronounced in the summer than in the winter. This is because Ψ (control - no transport) pattern (Fig. 9) is quite similar to Ψ_C (control run) (Fig. 4) in the summer than in the winter. Increased circulation in the summer by the lateral transport generally leads to greater horizontal and vertical variabilities of the current structure. In the winter, Ψ has positive values (2 Sv), and Ψ_C has negative values in JB. This suggests that the lateral transport reduces the JB cyclonic gyre by 2 Sv in the winter.

7 Conclusions

(1) The JES circulation was simulated in this study by the POM model under the seasonal surface flux forcing. The northern JES (i.e., the Japan Basin) is occupied by a cyclonic gyre (called the JB gyre) and the southern JES is characterized as a multi-eddy structure. The volume transport streamfunction has a double-gyre structure with negative values (cyclonic) in the northern JES and positive values (anticyclonic) in the southern JES. The simulated JB gyre is the strongest and recirculates 8 Sv in the winter. It weakens and retreats northward in the spring and summer. In the fall, the cyclonic JB gyre disappears and weak multi-anticyclonic eddies appear.

(2) The current systems such as the LCC, EKWC, NKCC, and JNB are simulated reasonably well. The simulated LCC has a maximum southward component (0.21 m/s), occurring near the surface in the winter with a width of 100 km and extending to a depth of 800 m. It weakens to a minimum of 0.15 m/s in the summer and fall, and shrinks in size to a width of 60 km and depth of 400 m.

The simulated EKWC varies from 0.42 m/s (summer) to 0.30 m/s (winter). The width of the EKWC is around 60 km all year round. However, the depth is around 1,400 m in the summer and 800 m in the winter. The (northward) overshot EKWC leaves the Korean coast moving northward and converges with the southward flowing NKCC, at 40°N, 130°E, and forms a current meandering toward east along the SPF.

The simulated JNB has a maximum eastward component of 0.24 m/s in the winter and weakens to 0.09 m/s in the summer. This is consistent with the earlier observational study. The effect of coastal geometry, such as the Noto Peninsula on the JNB, is also simulated.

The simulated UTB warm-core anticyclonic eddy has a western branch of the TWC (i.e., EKWC) which moves northward along the southern part of the Korean coast and separates from the coast after approaching the East Korean Bay. It keeps its northward motion until meeting the NKCC near $40°$ N, meanders southeastward, and forms a warm-core anticyclonic eddy. This eddy is strongest in the summer and its center is located at $38.5°$ N $130°$ E. The size of the eddy is around 150 km. The tangential velocity is around 0.4 m/s.

(3) The nonlinear advection does not affect the general circulation pattern evidently, but does affect the formation of the mesoscale eddies, and especially the UTB eddy (all seasons) and the JB cyclonic gyre (spring).

(4) The model wind effects on the JES circulation are more pervasive than those of non-linear dynamic effects. The winter winds cause a strong basin-wide JES cyclonic gyre with 8 Sv recirculation in the northern JES and the summer winds drive a weak nearly basin-wide JES anticyclonic gyre. Thus, the wind forcing is the most important fact (80%) for the generation of the JB cyclonic gyre. Besides, the winds also influence the surface circulation such as driving the LCC (winter), damping the EKWC (winter), generating the UTB eddy (all seasons), and generating eddies along the JNB (all seasons). The wind has almost no effect on the occurrence of the JNB for all seasons.

(5) The model boundary-forcing enhances (weakens) the JES volume transport in the summer (winter). It has very weak effects on the occurrence of the LCC except in the winter, when the boundary-forcing accounts for 30% of the LCC at $46°$ N. It weakens the JB cyclonic gyre by 2 Sv (25%) in the winter. Besides, the boundary forcing also influences the surface circulation such as driving the UTB eddy in all the seasons (50% in the winter), generating the EKWC (50-100%) in the winter and 21% in the summer), and generating the JNB and eddies along the JNB.

(6) Future studies should concentrate on less simplistic scenarios. Realistic lateral transport should be included and the use of extrapolated climatological winds needs to be upgraded to incorporate synoptic winds to improve realism. Finally, the assumption of quasi-linearity that allowed us to use simple differences to quantify the effect of external forcing needs to be rigorously tested. It is very important to develop a thorough methodology to perform sensitivity studies under the highly non-linear conditions that may exist in the littoral environment.

8   Acknowledgments

The authors wish to thank George Mellor and Tal Ezer of the Princeton University for most kindly proving us with a copy of the POM code and to thank Chenwu Fan and Yuchun Chen for programming assistance. This work was funded by the Office of Naval Research NOMP Program, the Naval Oceanographic Office, and the Naval Postgraduate School.

References

[1] Blumberg, A., and G. Mellor, 1987: A description of a three dimensional coastal ocean circulation model. Three-Dimensional Coastal Ocean Models, edited by N.S. Heaps, *Amer. Geophys. Union*, Washington D.C., 1-16.

[2] Cai, W.J., and P.C. Chu, 1998: A thermal oscillation under a restorative forcing. *Q. J. R. Meteorol. Soc.*, **124**, 793-809.

[3] Chu, P.C., 1989: Relationship between thermally forced surface wind and sea surface temperature gradient. *Pure Appl. Geophys.*, **130**, 31-45.

[4] Chu, P.C., 1995: P-vector method for determining absolute velocity from hydrographic data. *Marine Tech. Soc. J.*, **29**(3), 3-14.

[5] Chu, P.C., C.W. Fan, and L.L. Ehret, 1997: Determination of open boundary conditions from interior observational data. *J. Atmos. Oceanic Tech.*, **14**, 723-734.

[6] Chu, P.C., Y.C. Chen, S.H. Lu, 1998a: Temporal and spatial variabilities of Japan Sea surface temperature and atmospheric forcings. *J. Oceanogr.*, **54**, 273-284.

[7] Chu, P.C., C.W. Fan, and W.J. Cai, 1998b: P-vector inverse method evaluated using the Modular Ocean Model (MOM). *J. Oceanogr.*, **54**, 185-198.

[8] Chu, P.C., Y.C. Chen, and S.H. Lu, 1998c: On Haney-type surface thermal boundary conditions for ocean circulation models. *J. Phys. Oceanogr.*, **28**, 890-901.

[9] Chu, P.C., S.H. Lu, and Y.C. Chen, 1999a: A coastal atmosphere-ocean coupled system (CAOCS) evaluated by an airborne expandable bathythermograph survey in the South China Sea, May 1995. *J. Oceanogr.*, **55**, 543-558.

[10] Chu, P.C., J. Lan, and C.W. Fan, 1999b: Japan/East Sea (JES) seasonal circulation and thermohaline variabilities, Part 1, Climatology. *J. Phys. Oceanogr.*, submitted.

[11] Chu, P.C., S.H. Lu, and Y.C. Chen, 1999c: Circulation and thermohaline structures of the Japan/east Sea (JES) and adjacent seas simulated by a nested basin/coastal model, p.108-111, *Proc. CREAMS'99*, Fukuoka, Japan, Jan. 26-28, 1999.

[12] Chu P. C., Y. Chen, S. Lu, 1999d: Japan/East Sea (JES) SPF Meandering and Eddy Shedding in May 1995, p 11-13, *Proc. CREAMS'99*, Fukuoka, Japan, Jan. 26-28, 1999.

[13] Courant, R., K.O. Friedrichs, and H. Levy, 1928: Uber die partiellen differenzengleichungen der mathematischen physik. *Math. Annalen,* **100**, 32-74.

[14] Hase, H., J.-H. Yoon, W. Koterayama, 1999: The Branching of the Tsushima Warm Current along the Japanese coast., p 19-22, *Proc. CREAMS'99*, Fukuoka, Japan, Jan. 26-28, 1999.

[15] da Silva, A.M., C.C. Young, and S. Levitus, 1994: Atlas of Surface Marine Data 1994. *Tech. Rep. Geosci.*, 94, University of Wisconsin-Milwakee, 83pp.

[16] Hirose, N., C.H. Kim, and J.H. Yoon, 1996: Heat budget in the Japan Sea. *J. Oceanogr., Soc. Jpn.*, **52**, 553-574.

[17] Holloway, G.T. Sou and M. Eby, 1995. Dynamics of circulation of the Japan Sea. *J. Mar, Res.*, **53**, 539-569.

[18]Kawabe, M., 1982a: Branching of the Tsushima Current in the Japan Sea, Part I: Data analysis. *J. Oceanogr. Soc.*, **38**, 95 - 107.

[19]Kawabe, M., 1982b: Branching of the Tsushima Current in the Japan Sea, Part II: Numerical experiment. *J. Oceanogr. Soc.*, **38**, 183 - 192.

[20]Kim, C.-H. and J.-H. Yoon, 1994: A numerical study on the seasonal variation of the Tsushima Warm Current along the coast of Japan. Proceedings of the CREAMS Third Workshop. Seoul, Korea, pp. 73-79.

[21]Kim. C.-H. and J.-H. Yoon, 1996: Modeling of the wind-driven circulation in the Japan Sea using a reduced gravity model. *J. Oceanogr.*, **52**, 359-373.

[22]Mellor, G., and T. Yamada, 1982: Development of a turbulence closure model for geophysical fluid problems. *Rev. Geophys. Space Phys.*, **20**, 851-875.

[23]Mooers, C. N.K. and H.-S. Kang (1995). Initial spin-up of a Sea of Japan numerical circulation model. In: (Eds. A.S. Alekseev and N.S. Bakhvalov) Advanced Mathematics: Computations and Applications, NCC Publisher, Novosibirsk; pp. 350-357.

[24]Ro. Y. J.. 1999. Numerical Experiments of the Meso - Scale Eddy Activities in the East (Japan) Sea, p 116-119, *Proc. CREAMS'99*, Fukuoka, Japan, Jan. 26-28, 1999.

[25]Smagorinsky, J.. General circulation experiments with the primitive equations, I. The basic experiment, *Mon. Wea. Rev.*, **91**, 99-164, 1963.

[26]Sekine, Y. , 1986: A numerical experiment on the seasonal variation of the oceanic circulation in the Japan Sea. In: Oceanography of Asian Marginal Seas (Ed., K. Takano) Elsevier Oceanography Series, 54:113-128.

[27]Sekine. Y., 1991: Wind-driven circulation in the Japan Sea and its influence on the branching of the Tsushima Current. *Prog. Oceanogr.*, **17**, 297-312.

[28]Seo. J. W.. 1998: Research on the sea surface winds and heat flux in the East Asian Marginal Seas. Ph.D Thesis, Hanyang University.

[29]Seung, Y.H., 1992: A simple model for separation of East Korean Warm Current and formation of the North Korean Cold Current. *J. Oceanol. Soc. Korea*, **27**, 189-196.

[30]Seung. Y.H. and K. Kim, 1989: On the possible role of local thermal forcing on the Japan Sea circulation. *J. Oceanol. Soc. Korea*, **24**: 1-14.

[31]Seung, Y.H., S.Y. Nam, and S.R. Lee, 1990: A combined effect of differential cooling and topography on the formation of Ulleung Warm Eddy. *Bull. Korean Fish. Soc.*, **22**, 375-384.

[32]Seung, Y.-H. and S.-Y. Nam, 1991: Effect of winter cooling of subsurface hydrographic conditions off Korean Coast in the East (Japan) Sea. In: Oceanography of Asian Marginal Seas (Ed., K. Takano) Elsevier Oceanography Series, pp. 163-178.

[33]Seung. Y.H. and S.Y. Nam, 1992a: A numerical study on the barotropic transport of the Tsushima Warm Current. *La mer*, **30**, 139-147.

[34]Seung, Y.H. and S.Y. Nam, 1992b: A two-layer model for the effect of cold water formation on the East Korean Warm Current. *Bull. Korean Fish. Soc.*, **25**, 65-72.

[35]Seung, Y.-H. and K. Kim, 1993: A numerical modeling of the East Sea circulation. *J. Oceanol. Soc. Korea*, **28**, 292-304.

[36]Seung, Y.H. and K.J. Kim, 1995: A multilayer model for dynamics of upper and intermediate layer circulation of the East Sea. *J. Oceanol. Soc. Korea*, **30**, 227-236.

[37]Tomczak, M., and J.S. Godfrey, 1994: Regional Oceanography: An Introduction. Pergamon, Tarrytown, New York, pp.180-181.

[38]Uda, M., 1934: The results of simultaneous oceanographic investigartions in the Japan Sea and its adjacent waters in May and June. *J. Imp. Fish. Exp. Sta.*, **5,** 57-190 (in Japanese).

[39]Yoon, J.-H., 1982a: Numerical experiment on the circulation in the Japan Sea, Part I. Formation of the East Korean Warm Current. *J. Oceanogr. Soc. Japan*, **38**, 43-51.

[40]Yoon, J.-H. 1982b: Numerical experiment on the circulation in the Japan Sea, Part II. Influence of seasonal variations in atmospheric conditions on the Tsushima Current. *J. Oceanogr. Soc. Japan*, **38**, 81-94.

[41]Yoon, J.-H. 1982c: Numerical experiment on the circulation in the Japan Sea Part, III. Mechanism of the Nearshore Branch of the Tsushima Current. *J. Oceanogr. Soc. Japan*, **38**, 125-130.

[42]Yoshikawa, Y., T. Awaji, and K. Akitomo, 1999: Formation and circulation processes of intermediate water in the Japan Sea. *J. Phys. Oceanogr.*, **29,** 1701-1722.

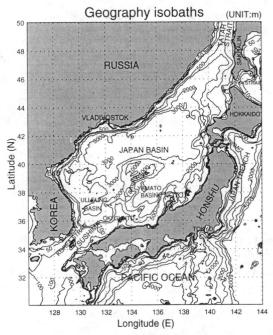

Fig. 1. Geography and isobaths showing the bathymetry (m) of the Japan/East Sea (JES).

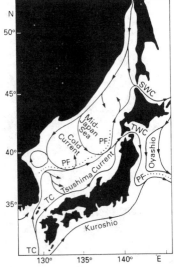

Fig. 2. Schematic map of surface current systems (after Tomczak and Godfrey, 1994).

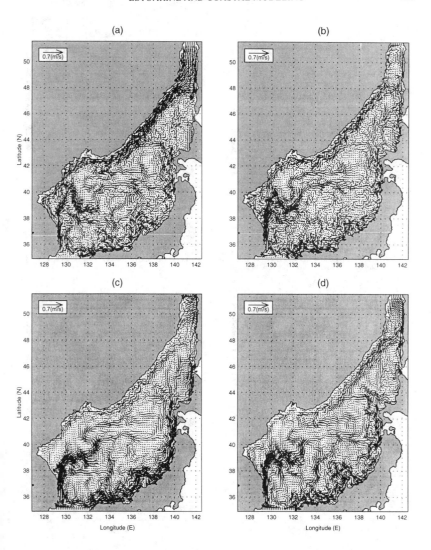

Fig. 3. Simulated surface circulation from the control run for (a) January, (b) April, (c) July and (d) October.

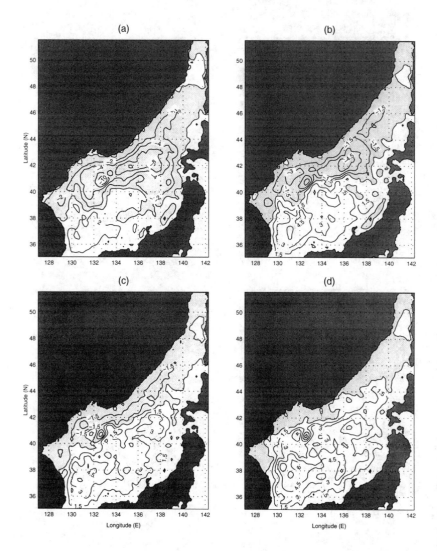

Fig. 4. Simulated volume transport (Sv) from the control run for (a) January, (b) April, (c) July and (d) October.

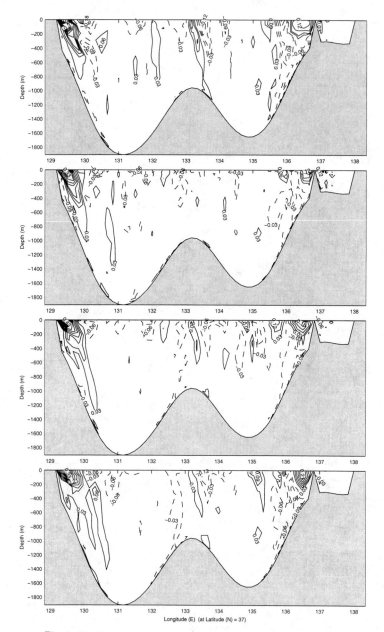

Fig. 5. Simulated zonal cross-section of v-velocity component (m/s) along 37°N from the control run for (a) January, (b) April, (c) July and (d) October.

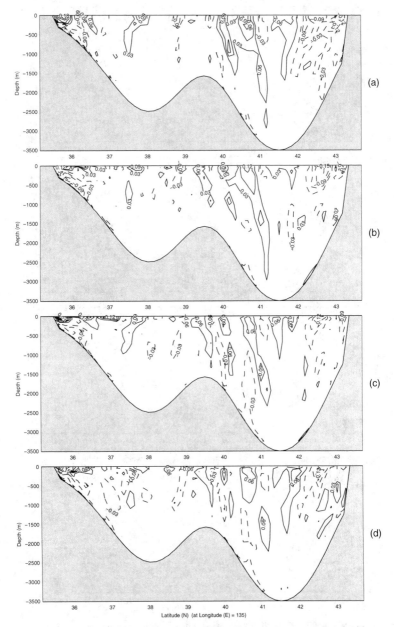

Fig. 6. Simulated latitudinal cross-section of u-velocity component (m/s) along 135°E from the control run for (a) January, (b) April, (c) July and (d) October.

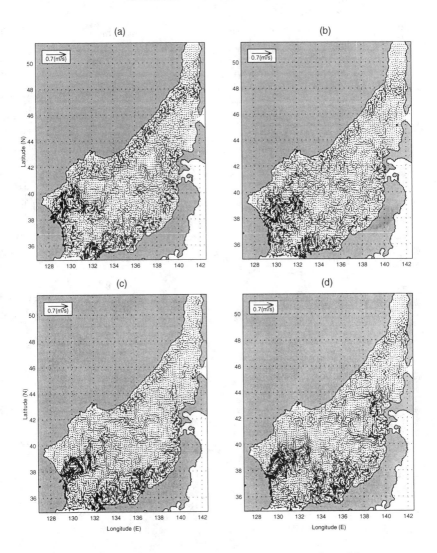

Fig. 7. Surface circulation caused by the non-linear effect: (a) January,
(b) April, (c) July and (d) October.

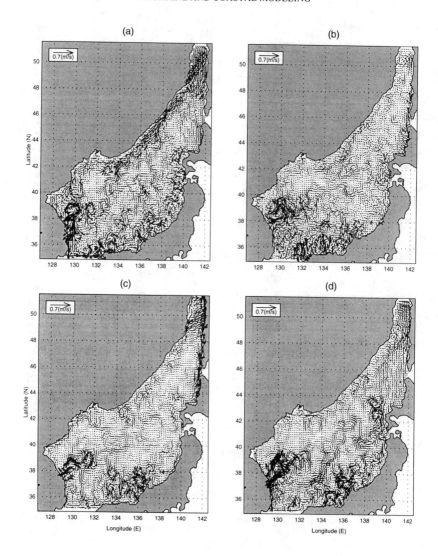

Fig. 8. Surface circulation driven by the winds: (a) January, (b) April, (c) July and (d) October.

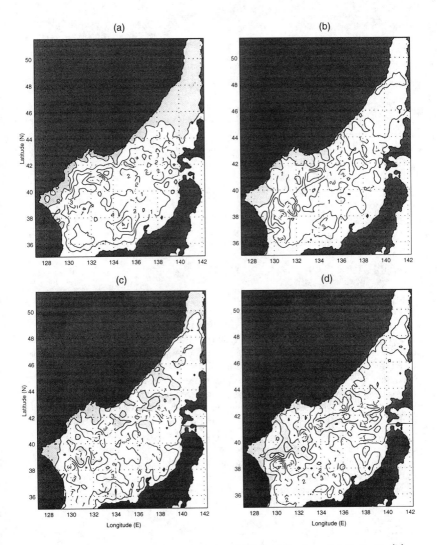

Fig. 9. Volume transport driven by the lateral forcing: (a) January, (b) April, (c) July and (d) October.

EFFECT OF HEAT FLUX ON THERMOCLINE FORMATION

Joseph S. Helfand[1], David P. Podber[1], and Michael J. McCormick[2]

ABSTRACT

A numerical model based upon Deardorff (1980) was used to simulate heat transfer between a lake and the atmosphere. The model includes a lake-atmosphere heat exchange submodel, with penetration of solar radiation. Heat and momentum are transferred in the vertical with a one-dimensional model that disregards horizontal processes. Atmospheric forcing to drive the models was taken from NOAA Marobs buoy 45007 in the south central region of Lake Michigan, and from the nearby land based stations at Milwaukee, WI, and Muskegon, MI. Three simulations were run and compared to field data collected by NOAA GLERL in 1996. The simulations focused on the formation of the thermocline from June 9 to August 1. The first simulation used land-based atmospheric data as inputs, the second used buoy data supplemented with land data where needed. The results were compared to the field data and showed that using terrestrial data can lead to results that are plausible, but inaccurate. The third simulation used buoy and land data as before, but adjusted the incoming solar radiation to produce a better fit to the observed data. The adjustment apparently compensated for deficiencies in the land-based data.

INTRODUCTION

Numerical modeling of large lakes, estuaries, coastal waters, and gulfs has advanced to the point that it is now standard to run seasonal simulations with highly resolved spatial (1~10 km) and temporal (30~90 s) scales. The accurate

---

[1] Limno-Tech, Inc. 501 Avis Drive, Ann Arbor, MI 48108
[2] Great Lakes Environmental Research Laboratory, NOAA, 2205 Commonwealth Blvd., Ann Arbor, MI 48105

computation of the ocean (lake)-atmosphere heat exchange is necessary to compute the development of the thermal structure (i.e., thermocline location) accurately over a seasonal scale. The density stratification associated with thermal stratification affects the circulation patterns produced by the atmospheric input of momentum, which is responsible for the circulation in non-tidal systems and the residual circulation in tidal systems.

A vertical direction, one-dimensional numerical model for velocity, temperature, and turbulence was encoded. The vertical mixing model was the one-equation turbulence model of Deardorff (Deardorff, 1980). The relatively simple 1-equation model of Deardorff was chosen instead of the more sophisticated 2 and 2 ½ equation models (k-ε, or Mellor and Yamada (1974)), to test whether or not a simple mixing scheme is sufficient. Future work is planned to compare the quality of the model results using a suite of models and vertical spacings that are more applicable to a three dimensional model application. A four component heat flux model was used, including incoming solar radiation (short wave), net long wave radiation, evaporative or latent heat, and sensible heat.

Various simulations were run to analyze the sensitivity of the model results to heat flux parameters and data. In particular the effect of land-based versus water-based atmospheric forcings was examined. The model results are compared to the field data, in the form of time series temperatures at various depths, gathered in Lake Michigan by the National Oceanic and Atmospheric Administration (NOAA) Great Lakes Environmental Research Lab (GLERL). The field data used for comparison consisted of velocity and temperature (thermistor) data from a location in the central part of the southern basin of Lake Michigan. This location (Figure 1) was chosen to minimize the advective contribution of heat from the nearshore region, although some advective effects can still be seen in the data.

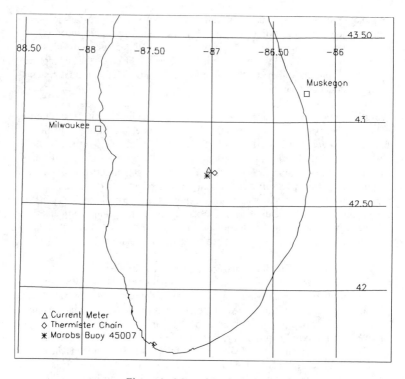

Figure 1: Map of Study Location

DATA

In all cases except for rainfall, linear interpolation of the actual data was used to create hourly data series. For rainfall, hourly precipitation amounts were used as described below. Figure 2 examines air temperatures averaged by the hour over the model period. It is apparent that the land-based data show a clear diurnal cycle, whereas the water-based data show no cycle, in addition to being significantly lower. As will be shown below, data differences between land and water may result in large differences in model results.

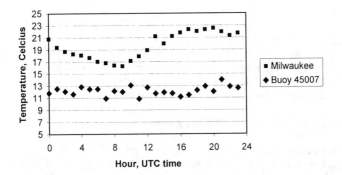

Figure 2. Averaged Air Temperatures by Hour at the Milwaukee Airport and Buoy 45007 for the simulation period of June 9 – August 1, 1996

Land-based Data

Land-based data came from Muskegon Airport, 43.167 N latitude and 86.25 W longitude. Data included time, wind speed and direction, air temperature, dewpoint temperature, cloud cover (as a fraction), precipitation, and pressure. Precipitation amounts were converted into hourly presence/absence rainfall data for subsequent use in a regression equation for cloud cover attenuation of incoming solar radiation.

Water-based Data

Water-based data came from instruments on a buoy at the study location (42.70 N, 87.00 W), or from surface-moored thermistors at fixed depths nearby (42.697 N, 86.970 W). Atmospheric data included time, wind speed and direction, air temperature, and pressure adjusted to sea level. Pressures were readjusted to an elevation of 198.12 meters (650 feet) using a standard equation (Iribarne and Godson, 1981). Dewpoint temperatures were calculated from the air temperature and an assumed relative humidity of 80 percent, using a curve-fit equation for water vapor pressure as a function of temperature. Thermistor temperatures were collected at depths of 1.3, 3.5, 5.1, 6.7, 8.3, 9.1, 10.0, 11.4, and 13.7 meters at intervals of 15 minutes. Total water depth was about 155 meters. The thermistor data record began June 8, 1996 at hour 2300 EST.

METHODS

Main components of the model are 1) a model for conservation of momentum and heat, 2) a turbulence model which controls mixing, thereby controlling the exchange of heat between water layers at different depths, and 3) a heat flux model, which controls the exchange of heat at the upper boundary (the water surface). The lower boundary is controlled by a zero flux condition.

Governing Equations for Momentum and Heat

The governing equations for momentum and heat are simplified by the assumption that all exchanges occur in the vertical, since all horizontal gradients are assumed to be zero. Effects due to the earth's rotation are neglected since only one direction in the horizontal is addressed. Also vertical advective fluxes are assumed to be zero, so that the vertical fluxes of momentum and heat are due solely to turbulent motions in the fluid. The equations are as follows:

Momentum:

$$\frac{\partial u}{\partial t} = \frac{\partial}{\partial z}(K_m \frac{\partial u}{\partial z})$$

Heat (Temperature):

$$\frac{\partial T}{\partial t} = \frac{\partial}{\partial z}(K_h \frac{\partial T}{\partial z}) + \frac{q_s}{\rho c_p}$$

where u(z,t) is the horizontal velocity [m/s], T(z,t) is the temperature [°C], K_m is the vertical eddy viscosity [m²/s], and K_h is the vertical eddy diffusivity [m²/s]. K_m and K_h are given by the following relationships, that are explained in the Turbulent Mixing section: $K_m = C_1 \lambda q + K_0$, and $K_h = C_2 K_m$. q_s [W/m²/m] is the rate of absorption of the penetrating solar radiation which varies with depth, ρ is the density of water [kg/m³], and c_p is the heat capacity of water [J/kg/K].

The boundary conditions for the two governing equations are given as:

Surface: $K_m \frac{\partial u}{\partial z}\Big|_s = \frac{\tau_s}{\rho}$, where τ_s is the surface wind stress [Pa],

$K_h \frac{\partial T}{\partial z}\Big|_s = \frac{q_{hf}}{\rho c_p}$, where q_{hf} is the surface heat flux [W/m²].

Bottom: $u_b = 0.$ $\left.\dfrac{\partial T}{\partial z}\right|_b = 0.$

Turbulent Mixing in the Vertical

The vertical mixing of momentum and heat was based on the Deardorff (Deardorff, 1980) model, which was also applied more recently by Denbo and Skyllingstad (Denbo and Skyllingstad, 1996). This method can be characterized as a one-equation turbulent kinetic energy (tke) model with an empirically determined mixing length and dissipation rate. The mixing length is based on density stratification, and the dissipation of tke is computed by using the mixing length, empirical constant, and tke$^{-3/2}$.

The governing equation from Deardorff has been modified by dividing through by one-half of the square root of the tke to avoid the computational burden of taking the square root (although this burden is minimal in the one-dimensional application used in this study). The governing equations were also simplified because this study is one-dimensional. The equations are given below:

$$\frac{\partial q}{\partial t} = \frac{K_m}{2q}[P] \;+\; \frac{\beta g}{2q}K_h\frac{\partial T}{\partial z} \;+\; \frac{\partial}{\partial z}(C_{dg}K_m\frac{\partial q}{\partial z}) \;-\; \frac{\varepsilon}{2q}$$

$$\qquad\quad\; \vee \qquad\qquad \vee \qquad\qquad\quad \vee \qquad\qquad\quad \vee$$

shear production   buoyant production/   vertical diffusion   dissipation rate
                              dissipation   of turbulence

where $q^2 = k = (\overline{u'}^2 + \overline{v'}^2 + \overline{w'}^2)/2.0$, and u',v',w' are the turbulent fluctuations in the u,v,w velocities respectively; $K_m = C_{km}\lambda q + K_0$ is the vertical eddy viscosity [m^2/s], C_{km} is an empirical coefficient, λ is the mixing length, $K_0 = 1.0 \times 10^{-6}$ is the background mixing, $K_h = (C_{hl} + C_{hs}\dfrac{\lambda}{\Delta s})K_m$ is the vertical eddy diffusivity [m^2/s], C_{hl} and C_{hs} are empirical coefficients, Δs is the vertical grid spacing, P is the production of turbulence due to shear in the mean flow; in tensor notation it is written as $P = (\dfrac{\partial u_i}{\partial x_j} + \dfrac{\partial u_j}{\partial x_i})\dfrac{\partial u_i}{\partial x_j}$, which simplifies to $P = (\dfrac{\partial u}{\partial z})^2$ for one dimension using standard notation, $\beta = 8.75 \times 10^{-6}$ (T + 9.0) for T in °C., and β has units [1/°C].

The mixing length, λ, is computed differently for stable and unstable conditions, as given below:

$$
\begin{array}{cc}
\text{Stable Fluid} & \text{Unstable Fluid} \\
\end{array}
$$

$$
\lambda = \left|
\begin{array}{l}
l = C_l q \left(\dfrac{g}{\rho_0} \dfrac{\partial \rho}{\partial z} \right)^{-1/2} \\
\lambda = \min(l, \Delta s)
\end{array}
\right. \qquad \qquad \lambda = \Delta s
$$

The dissipation rate ε, is given by

$\varepsilon = \dfrac{C}{\lambda} q^3$, where the constant $C = \left(C_{\varepsilon l} + C_{\varepsilon s} \left(\lambda \big/ \Delta s \right) \right) C_{wall}$.

The empirical constants were assigned the values described in Deardorff, as follows:

$C_{dg} = 2.0$, $C_{\varepsilon l} = 0.19$, $C_{\varepsilon s} = 0.51$, $C_l = 0.76$, $C_{km} = 0.1$, $C_{hl} = 1.0$, $C_{hs} = 2.0$,
$C_{wall} = \{1.0$ away from solid boundaries, 3.9 next to solid boundaries and at the surface.$\}$.

The boundary conditions imposed on the governing equation are as follows:

Surface conditions for q and P : $\dfrac{\partial^2 q}{\partial z^2} = 0$. $P\big|_s = (\dfrac{\partial u}{\partial z}\big|_s)^2 = \left[\dfrac{\tau_s}{\rho K_m} \right]^2$, where τ_s is

the wind stress. Bottom conditions are $\dfrac{\partial q}{\partial z} = 0$ and $P = \varepsilon$ (following ASCE, 1988).

Heat Flux

The heat flux model was derived from a 1988 heat flux model developed by McCormick of the Great Lakes Environmental Research Laboratory as applied to Lake Erie (Kelley, 1995). The model uses standard meteorological observations and surface water temperatures as model input. Model output consists of two types of components generated by four submodels. The surface flux (equivalent to q_{hf} in the above governing equation boundary conditions) is the sum of the fluxes from the net long wave radiation, evaporative or latent heat, and sensible heat submodels. The penetrating flux (the source of q_s in the above governing equations) comes from the incoming solar radiation (short wave).

The heat flux model uses a heat balance approach in which the total heat flux is equal to the sum of the shortwave radiation, long wave radiation, sensible heat flux and latent heat flux. The terms, in watts/meter^2, are positive for energy flowing into the lake. In equation form, the model is:

$$H_{total} = H_{SR} + (H_{LR} + H_S + H_L) = \int_{z=0}^{z=b} q_s \, dz + q_{hf}$$

where H_{total} is the net heat flux with the atmosphere, H_{SR} is the net incoming short wave radiation, H_{LR} is the net long wave radiation, H_S is the sensible heat transfer, and H_L is the latent heat transfer.

H_{SR}, Short Wave Radiation

The short wave radiation is a function of clear sky radiation. Clear sky radiation is predicted by a regression equation from the solar zenith angle, and is modified by cloud cover. The resulting equation follows Cotton (1979):

$$H_{SR} = (SR_c)(CLD)$$

where SR_c is the net incoming short wave radiation for clear sky conditions, and CLD is the "cloud effect" term. SR_c is computed from the equation:

$$SR_c = a_0 + a_1\cos(Z) + a_2(\cos(Z))^2 + a_3(\cos(Z))^3$$

where a_0, a_1, a_2, and a_3 are empirical coefficients, and Z is the solar zenith angle (defined as the angular distance of the sun from local vertical). Z is computed following Guttman and Matthews, 1979. CLD is computed from the equation:

$$CLD = c_0 + c_1 S_c + c_2(S_c)^2 + c_3(S_c)^3 + c_4(RN)$$

where S_c is the opaque cloud cover amount, c_0, c_1, c_2, c_3, and c_4 are empirical coefficients, and RN = 0 if it is not raining, RN = 1 if it is.

The attenuation of the short wave radiation in the water column is described by the Beer's law equation:

$$I_{z,\lambda} = I_{0,\lambda} e^{-K_{e,\lambda} z}$$

where $I_{z,\lambda}$ is the light intensity [W/m^2] at depth z, z is the depth [m], $K_{e,\lambda}$ is the extinction coefficient [m^{-1}], $I_{o,\lambda}$ is the incoming solar radiation at the surface, and λ is the wavelength [nm]. The effect on the computed light attenuation of using low resolution or high resolution extinction coefficients was investigated. It was found that the low resolution of just two wavelength bands, infrared and visible/ultraviolet, gave very similar results to using the high resolution of many wavelength bands. Therefore the model was implemented with two wavelength bands, infrared and visible/ultraviolet, with extinction coefficients of 2.85 and 0.296 respectively. The light attenuation is converted to heat as q_s.

H_{LR}, Long Wave Radiation

The net long wave radiation is a function of the water temperature, air temperature and cloud cover. The equation used follows Wyrtki (1965):

$$H_{LR} = -S_{SB}\,\varepsilon\,(T_W^4(0.39-0.05e_a^{1/2})[1.0-kS_c^2] + 4T_W^3(T_W-T_A))$$

where S_{SB} is the Stefan-Boltzmann constant, 5.67051×10^{-8} watts/(m^2 K^4), ε is the emissivity assumed equal to 0.97, T_W is the water temperature [K], e_a is the vapor pressure of air [millibars], k is a parameter that increases linearly with latitude from 0.5 at the equator to 0.8 at 70° of latitude, and T_A is the air temperature [K].

H_S, Sensible Heat

The sensible heat flux is a function of the air-water interface temperature differential, the air density, heat capacity, and the wind speed. The calculation uses a bulk transfer approach according to McBean et al. (1979):

$$H_S = \rho_a\,CpC_HU(T_W-T_A)$$

where ρ_a is the density of air [kg/m^3], Cp is the specific heat of air at constant pressure [J/(kg K)], C_H is the bulk heat transfer coefficient for momentum, U is the wind speed [m/s], T_W is the water temperature, and T_A is the air temperature.

H_L, Latent Heat

The latent heat flux (due to evaporation and condensation) is a function of air density, water latent heat of vaporization, wind speed, water temperature, and air humidity. The equation used (Kelley, 1995) is:

$H_L = \rho_a L_V C_D U(q_A - q_W)$

where L_V is the latent heat of vaporization at the temperature of the water surface [J/kg], C_D is the bulk transfer coefficient for momentum, q_A is the specific humidity of the air (ratio of water vapor mass to total humid air mass), and q_W is the specific humidity of the air at the water surface.
The q_A and q_W functions are calculated by:

$q_A = 0.62185e_A/(P - 0.37815e_A)$
$q_W = 0.62185e_W/(P - 0.37815e_W)$

where e_A is the saturation vapor pressure [millibars], P is the air pressure, [millibars], and e_W is the saturation vapor pressure of water at T_W, [millibars].

C_D and C_H, Bulk Transfer Coefficients

The dimensionless bulk transfer coefficients for momentum (C_D) and heat (C_H) are used in the calculations of the latent and sensible heat transfer terms, described above. Both C_D and C_H are dependent on the stability of the lowest few meters of the atmospheric boundary layer, the height of wind speed measurement and the wind speed. These coefficients also include the effects of surface roughness. C_D is determined from the equation:

$$C_D = \frac{u_*^2}{u_z^2}$$

where u_* is the friction velocity [m/s], and u_z is the wind speed at height z. C_H is calculated as a function of the dimensionless stability height using parameters developed in the calculation of C_D. A profile method is used to estimate u_* (Schwab, 1978; Lui and Schwab, 1987). The method is based on Monin-Obukhov similarity theory (Monin and Obukhov, 1954) and on the works of Businger et al. (1971), Long and Shaffer (1975), Panofsky (1963), Paulson (1970), and Smith and Banke (1975).

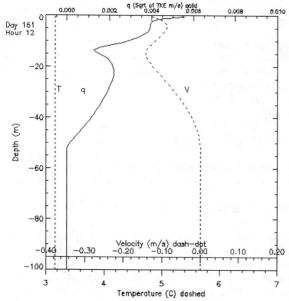

Figure 3: Turbulence Variables vs. Depth

RESULTS

Overview of Runs

Model runs were done for the period Julian day 161 (June 9, 1996), hour 0400 UTC, to day 214 (August 1, 1996), hour 0000 UTC. Figure 3 shows an example of turbulence-related variables (temperature, velocity, and q) versus depth. Figure 4 shows an example of average penetrating solar radiation power input estimated by the heat flux model versus depth.

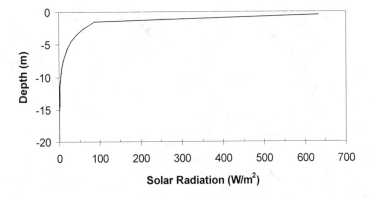

Figure 4: Power Input vs. Depth

Data Choice, Calibration, and Comparisons to Data

A series of three sets of model runs were done, resulting in a steadily improving fit of model outputs to observed data. The first set used only land-based atmospheric forcings (from Muskegon Airport), and did not employ any calibration of model to data (other than original literature empirical constants). The second set used the available atmospheric forcings recorded at the buoy, along with the land-based cloud cover and rainfall (which were not recorded at the buoy). The third set used the same forcings as the second, except that solar input was reduced by 50 percent for the first 17 days in accordance with the observation that the observed water temperature data indicated a markedly lower heat input during this period. During the first 17 days there are small diurnal variations in the surface water temperature that can be explained by a radiative heat flux of 30-40 W/m^2. It is surmised that the small amount of radiative heat flux was due to a thicker cloud cover over the lake than was reported at the Muskegon airport. Fog over the lake is a common spring time feature, and was accounted for by increasing the cloud cover where needed.

The first two sets are compared to data in Figure 5, which shows temperature vs. time at thermistor depths. The land-based forcing produces results which are plausible but inaccurate. The buoy+land forcing produces more accurate results. It is apparent, especially at greater depths, that some features of the data (e.g. high-frequency variability due to internal waves) are due to horizontal processes which the present vertical application cannot resolve.

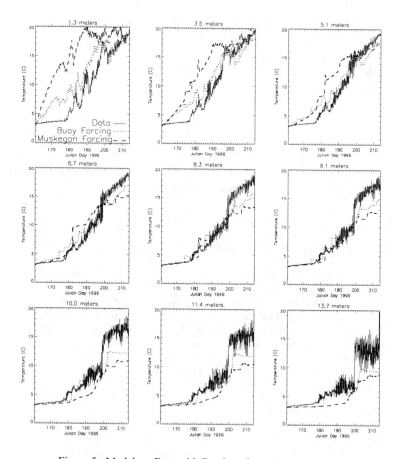

Figure 5: Model vs. Data with Forcings from Two Locations

The third set is compared to data in Figure 6. The adjustment made in this third set apparently compensates for deficiencies in the land-based cloud cover data during this period, resulting in a much-improved fit of model to data.

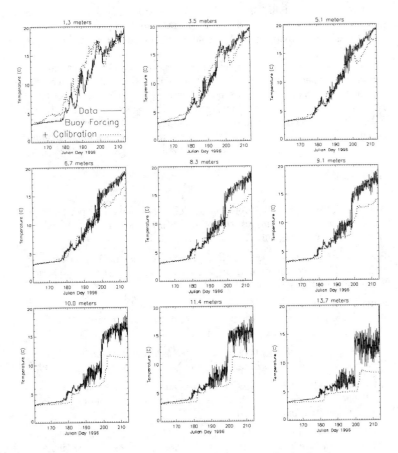

Figure 6: Model vs. Data with Calibration of Solar Input

CONCLUSIONS

Currently available numerical models are able to simulate thermocline formation accurately, given accurate forcing functions. The predictive capabilities of the models are limited by the accuracy of the available data. The use of terrestrial data will lead to thermal structures that may be plausible but inaccurate.

The top few meters of the lake tend to equilibrate with the atmosphere and are insensitive to horizontal processes. This is not true for lower layers, which are strongly influenced by internal waves and the collapse of the thermal bar.

The level one turbulence closure scheme of Deardorff tended to inhibit mixing across the thermocline to a greater extent than data indicates it ought to be. Some sensitivity analysis was performed on the empirical coefficients reported by Deardorff, though not reported in this article. The results from those analyses indicate that it is possible to adjust those coefficients to better simulate site-specific physics. It should be noted that, even with no modification to Deardorff's empirical coefficients, the model performed well.

REFERENCES

ASCE Task Committee on Turbulence Models in Hydraulic Computations: 1988. *Turbulence modeling of surface water flow and transport: parts i-v.* Journal of Hydraulic Engineering, vol. 114, no. 9, pages 970-1073.

Businger, J.A.; Wyngaard, J.C.; Izumi, Y.; Bradley, E.F.: 1971. *Flux-Profile Relationships in the Atmospheric Surface Layer.* Journal of the Atmospheric Sciences, vol. 28, pages 181-189.

Cotton, Gerald F.: 1979. *ARL Models of Global Solar Radiation, Appendix VI.* Air Resources Laboratories, Environmental Research Laboratories, U.S. Department of Commerce, pages 165-184.

Deardorff, J.W.: 1980. *Stratocumulus-Capped Mixed Layers Derived From a Three-Dimensional Model.* Boundary-Layer Meteorology, vol. 18, pages 495-527.

Denbo, D.; Skyllingstad, E.: 1996. *An Ocean Large-Eddy Simulation Model with Applications to Deep Convection in the Greenland Sea.* Journal of Geophysical Research, vol. 101, pages 1095-1110.

Guttman, Nathaniel B.; Matthews, J. David.: 1979. *Computation of Extraterrestrial Solar Radiation, Solar Elevation Angle, and True Solar Time of Sunrise and Sunset, Appendix II.* National Climatic Center, Environmental Data and Information Service, U.S. Department of Commerce, pages 47-54.

Iribarne, J.V., Godson, W.L.: 1981. *Atmospheric Thermodynamics.* 2nd Edition. 259 pages.

Kelley, John Gormley Walsh: 1995. *One-Way Coupled Atmospheric-Lake Model Forecasts For Lake Erie.* Dissertation. Ohio State U., 450 pages.

Liu, Paul C.; Schwab, David J.: 1987. *A Comparison of Methods for Estimating u* *From Given u_z and Air-Sea Temperature Differences.* Journal of Geophysical Research, vol. 92, no. C6, pages 6488-6494.

Long, Paul E. Jr.; Shaffer, Wilson A.: 1975. *Some Physical and Numerical Aspects of Boundary Layer Modeling.* NOAA Tech. Memo. NWS TDL-56, 37 pages.

McBean, G.A.; Bernhardt, K.; Bodin, S.; Litynska, Z.; Van Ulden, A.P.; Wyngaard, J.C.: 1979. *The Planetary Boundary Layer (Technical Note No. 165).* World Meteorological Organization, no. 530, 200 pages.

Mellor, G.; Yamada, T.: 1974. *A Hierarchy of Turbulence Closure Models for Planetary Boundary Layers.* Journal of the Atmospheric Sciences, vol. 31, pages 1791-1806.

Monin, A.S.; Obukhov, A.M.: 1954. *Basic Laws of Turbulent Mixing in the Atmosphere Near the Ground.* Tr. Akad. SSSR Geofiz. Inst. vol. 24, pages 163-187.

Paulson, C.A.: 1970. *The Mathematical Representation of Wind Speed and Temperature Profiles in the Unstable Atmospheric Surface Layer.* J. Appl. Meteor., vol. 9, pages 857-861.

Panofsky, H.A.: 1963. *Determination of Stress From Wind and Temperature Measurements.* Quat. J. Meteorol. Soc., vol. 89, pages 85-94.

Schwab, David J.: 1978. *Simulation and Forecasting of Lake Erie Storm Surges.* Monthly Weather Review, vol. 106, no. 10, pages 1476-1487.

Smith, S.D., Banke, E.G.: 1975. *Variation of the Sea Surface Drag Coefficient With Wind Speed.* J. R. Met. Soc., vol. 101, pages 665-673.

Wyrtki, Klaus: 1965. *The Average Annual Heat Balance of the North Pacific Ocean and Its Relation to Ocean Circulation.* Journal of Geophysical Research, vol. 70, no. 18, pages 4547-4559.

ACKNOWLEDGMENT

The authors would like to acknowledge the Donald C. Cook Nuclear Plant, Bridgman, Michigan for its support for the development of the heat flux algorithms.

[1]A comparison of two 3D shallow-water models using sigma-coordinates and z-coordinates in the vertical direction

M.D.J.P. Bijvelds[1], J.A.Th.M. Van Kester[2] and G.S. Stelling[1]

Abstract

In this paper two 3D shallow-water models are presented, using different vertical coordinate systems. The first one uses σ–coordinates (Phillips, 1957), the second one z-coordinates. Numerical problems related to both vertical coordinate systems are discussed. Improvements to minimize the numerical errors related to these problems are presented. The improvements are shown on the basis of two simple test cases and for the Haringvliet estuary, which is situated in the Netherlands. At this moment the estuary is subject to a changing control of the sluices to introduce tidal influence, which may have a significant effect on the intrusion length of the salt wedge and on the ecology of the estuary.

Introduction.

In coastal seas, estuaries and lakes stratified flow occurs in combination with steep topography. 3D numerical modelling of the hydrodynamics and water quality in these areas requires accurate treatment of the vertical exchange processes. The existence of vertical stratification influences the turbulent exchange of heat, salinity and passive contaminants. The accuracy of the discretization of the vertical exchange processes is determined by the vertical grid system. The vertical grid should:

- resolve the boundary layer near the bottom to allow an accurate evaluation of the bed stress
- be fine around pycnoclines
- avoid large truncation errors in the approximation of strict horizontal gradients.

The two commonly used vertical grid systems in 3D shallow-water models, see Figure 1: the "z"-coordinate system (ZCM) and the "sigma" coordinate system (SCM) both do not meet all the requirements. The ZCM model has horizontal coordinate lines which are (nearly) parallel with isopycnals, but the bottom is usually not a coordinate line and is represented as a staircase (zig-zag boundary). The staircase leads to inaccuracies in the approximation of the bed stress and the horizontal advection near the bed (Fortunato & Baptista, 1994). A flux along the bed is split into a horizontal and vertical part, which leads to numerical cross-wind diffusion in the transport equation for matter. On the other hand, the SCM is boundary fitted but will not always have enough resolution around the pycnocline. The SCM may give significant errors in the approximation of strictly horizontal gradients (Leendertse, 1990), (Stelling & Van Kester, 1994). This was the reason that more "generalised" vertical grid systems (Deleersnijder & Ruddick, 1992) or adaptive σ grids (Zegeling et al., 1998) were introduced in 3D circulation models, but these generalized vertical grids may become locally unsmooth introducing new truncation errors. First, we define the vertical grid for both the SCM and the ZCM.

In many 3D circulation (ocean) models the σ-transformation is employed (Cheng & Smith, 1990). This terrain-following vertical coordinate system has several advantages compared to

[1] Delft Univ. of Techn., Faculty of Civil Eng. and Geosciences, p.o. box 5048, 2600 GA Delft, The Netherlands and WL | Delft Hydraulics, p.o. box 177, 2600 MH Delft, The Netherlands
[2] WL | Delft Hydraulics, p.o. box 177, 2600 MH Delft, The Netherlands

the Cartesian coordinate system. In this paper the σ-transformation for the SCM is defined as follows:

$$\sigma = \frac{z-\zeta}{H} \tag{1}$$

where z is the Cartesian vertical coordinate, ζ is the free surface elevation and H is the total water depth given by $H=d+\zeta$ with d the bottom depth (measured downward). Grid spacing in the SCM model are constructed by lines of constant σ and are determined by Eq. (1). In a finite-difference model, due to the σ-transformation, the number of computational nodes in the vertical direction is constant over the entire computational domain. The relative layer thickness $\Delta\sigma$ does not depend on the horizontal coordinates x and y. This makes it impossible to locally refine the grid around pycnoclines in regions with steep bed topography, unless some adaptive sigma grid method is applied (Zegeling *et al.*, 1998). Moreover, the σ-transformation gives rise to unnecessarily high grid resolution in shallow areas (tidal flats) and possibly insufficient grid resolution in deep parts (holes) of the computational domain, see Figure 1. At tidal flats at low tide the mapping may even become singular. The numerical scheme may become non-convergent in these areas due to hydrostatic inconsistency (Haney, 1991).

Several 3D z-coordinate models were presented in the past. The vertical grid system is based on horizontal surfaces with constant z intersecting the water column, see Figure 1. The layers of a z-coordinate model should not be confused with layers of constant density in stratified flows. The layer thickness in a z-coordinate model is defined as the distance between two consecutive surfaces and is independent of space and time for an intermediate layer. The layer thickness of the top layer is defined as the distance between the free surface and the first horizontal surface, and may vary in space and time. In most of the z-coordinate models the vertical position of the free surface $z = \zeta(x,y,t)$ is restricted to only one layer, e.g. (Leendertse & Liu, 1975), imposing a lowerbound to the thickness of the top layer. Only recently, z-coordinate models have been developed where the free surface moves through the vertical grid (Casulli & Cheng, 1992). The vertical index k of the top layer of neighbouring horizontal grid cells may vary. In that case, in contrast with the SCM model, fluxes may be defined at cell faces that do not necessarily have a "wet" neighbouring grid cell. The thickness of the bottom layer is the distance between the bottom $z = -d(x,y)$ and the first horizontal surface above the bed. Many of the z-coordinate models do also not allow for variable grid spacing near the bed. This restriction may lead to a coarse "zig-zag" (staircase) representation of the bottom topography, which introduces errors in the propagation speed of the free surface waves and the flow field in the bottom boundary layer. Let $z = z_k$ be strict horizontal surfaces, where k is an integer indicating the layer index. In the present model the vertical grid spacing Δz_k is defined by:

$$\Delta z_k \big(x, y, t\big) = \min\big[\zeta(x, y, t), z_k\big] - \max\big[-d(x, y), z_{k-1}\big] \tag{3}$$

Taking into account variable grid sizes near the bed and allowing the free surface to move through the vertical grid introduces a lot of book keeping and makes the free surface boundary elaborate to treat in the numerical method. The gridpoints that are "wet" are determined every half time step. A computational cell is set "wet" whenever $\Delta z_{i,j,k} > 0$. Since the grid spacing near the bed and free surface may vary as a function of space and time, velocity points on the staggered grid of two adjacent grid cells may be situated at different vertical positions. Formally, this leads to additional terms in the discretized equations but this is not accounted for in the ZCM model. The variation of the free surface and bed topography is smooth in most areas, which justifies the neglect of the cross terms involved.

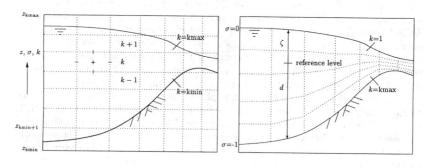

Figure 1: vertical computational grid z-coordinate model (left) and the σ–model (right).

To compare the effect of the different truncation errors on the flow field and the turbulent mixing coefficients for a ZCM and SCM, it is necessary to have two 3D shallow water models which only differ in the vertical coordinate system. In this paper two 3D shallow-water models are compared using σ-coordinates and z-coordinates, respectively which are based on the same numerical method. The 3D shallow-water equations are discretized on a staggered grid (Arakawa C-grid). The shallow-water equations (SWE) are solved by an ADI-type of factorization for the barotropic pressure (Stelling, 1983). Both the horizontal components of the velocity vector, u and v, are computed once in a full time step Δt. The vertical advection and viscosity term are integrated fully implicitly in order to avoid an excessive small time step imposed by the relatively small vertical grid spaces. The vertical grid space may vanish due to drying and flooding of shallow areas. The horizontal advection and viscosityterms are integrated explicitly. The timestep is restricted by the Courant-Friedrichs-Lewy condition for horizontal advection.

Central differences are employed for the discretization of advection in vertical direction. The horizontal advective terms are approximated by upwind discretization along streamlines (Van Eijkeren *et al.*, 1993). Although this method is formally of first order accuracy, it was found to be much less diffusive than the standard first order upwind method (Bijvelds, 1997).

The vertical eddy viscosity and eddy diffusivity are computed by the standard k-ε turbulence closure model (Rodi, 1984). It uses two partial differential equations to compute the transport of turbulent kinetic energy and energy dissipation rate. The production term only depends on the vertical gradients of the horizontal velocity. The presence of stratification is taken into account by the buoyancy flux.

A finite volume approach is used for the discretization of the scalar transport equation, which ensures mass conservation. To circumvent time step restrictions imposed by the small vertical grid size in the SCM model in drying areas, implicit time integration is used for the vertical derivatives in the transport equation. For horizontal derivatives explicit approximations are used. In horizontal and vertical direction, diffusion is discretized using central differences. Horizontal advective terms are computed by the Van Leer II TVD scheme and vertical advection is computed using a first order upwind method in order to obtain positive and monotonic solutions. It is noted that in the ZCM model the time step restriction due to an explicit treatment of the vertical advection is not as severe as in the SCM model. Apart from the free surface, the vertical grid spacing is not a function of time. In the ZCM near the free surface a threshold for the vertical grid spacing is introduced that allows the use of explicit schemes for vertical advection. This is favorable since it allows for the use of accurate positive monotonic schemes in vertical direction, that are easy to implement.

First, two major numerical problems of both coordinate systems were improved in the present models. Firstly, problems concerning the pressure gradient in the σ-coordinate model are coped with by application of a non-linear filter as presented by (Stelling & Van Kester, 1994). The tendency of this filter to systematically underestimate gradients in case of horizontally moving gravity currents, as reported by (Slørdal, 1997), is removed by extension of the non-linear filter in such a way that it still fulfils the min-max principle.

Secondly, the inaccuracies related to the stair case boundary representation of the bed in the ZCM model are removed by simple adjustments of the determination of the bed shear stress and the advection near solid vertical walls.

For the test cases presented, a significant improvement is obtained. The comparison of the SCM with the ZCM is based on the improved implementation of both models.

Aspects of the σ-coordinate model.

Further numerical problems may arise in stratified areas. In general, the numerical approximation of horizontal gradients in areas with steep bed topography in SCM models introduces large truncation errors. For stratified flows over continental slopes, the currents induced by the truncation error in the baroclinic pressure term may be as large as the typical physical velocities (Walters & Foreman, 1992).

The σ-transformation is not angle preserving. Transformation of the horizontal diffusion term introduces cross terms that are proportional with the slopes $\partial H / \partial x$ and $\partial H / \partial y$. In the implementation of the horizontal diffusion term these cross-derivative terms are often neglected, assuming mild slopes, in order to simplify the implementation and to anticipate problems concerning accuracy, stability and monotonicity (Mellor and Blumberg, 1985). On sloping bottoms this simplification may lead to significant errors as shown in (Paul, 1994) and (Podber & Bedford, 1994).

An approach to minimize the truncation errors in computing the horizontal diffusive fluxes and the horizontal pressure gradient force was presented in (Stelling & Van Kester, 1994). In this method (SVK) the horizontal diffusive fluxes and baroclinic pressure gradients are approximated in Cartesian coordinates by defining rectangular finite volumes around the sigma coordinate grid points, see Figure 1. This finite volume approach does not have to satisfy the hydrostatic consistency condition (Haney, 1991),

$$\left| \frac{\sigma}{H} \frac{\partial H}{\partial x} \right| \Delta x \leq \Delta\sigma \qquad (2)$$

which is severely restrictive for the grid, especially in areas with a rapidly varying bathymetry.

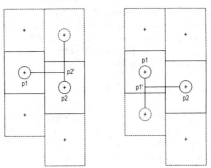

Figure 2: Cartesian finite volumes constructed on the basis of the σ-grid, used to approximate strict horizontal gradients.

Since the centre of the finite volumes is not at the same vertical level, a z-interpolation of the scalar concentration p is needed to compute horizontal derivatives. The values obtained from this interpolation are indicated by p_1' and p_2' respectively, in Figure 2. A non-linear filter is applied to the two consistent gradients, $s_1 = (p_2' - p_1)/\Delta x$ and $s_2 = (p_2 - p_1')/\Delta x$,

> if $s_1 * s_2 < 0$ then
> $$\frac{\Delta p}{\Delta x} = 0$$
> else
> $$\frac{\Delta p}{\Delta x} = \text{sign}(s_1) * \min(\text{abs}(s_1, s_2))$$
> endif

which results in a horizontal gradient which is used in the baroclinic pressure gradient in the momentum equation. In (Slørdal, 1997) it was stated that the above approximation may produce errors of the same sign which leads to a systematic underestimation of the baroclinic pressure term. This underestimation can be ascribed to the non-linear filter which selects the minimum of the two gradients under consideration. This limiter is fully analogue to the minmod limiter used for the construction of monotone advection schemes (Hirsch, 1990). Since the same approximation is used for the horizontal diffusion flux, it is important to ensure that the difference operator is positive definite in order to get physically realistic solutions. The maximum and minimum of a variable being transported by diffusion do not increase or decrease (min-max principle). By taking the minimum of the gradients, (Stelling & Van Kester, 1994) show that the min-max principle is fulfilled. (Beckers et al., 1997) show that any nine-point consistent linear discretization of the horizontal diffusion on the σ-grid does not fulfill the min-max principle.

By introducing an additional approximation of the horizontal gradient in the filter algorithm defined by, $s_3=(p_2-p_1)/\Delta x$, the stringent conditions of the minimum operator can be relaxed somewhat. The drawback of underestimation reported by (Slørdal, 1997) can be minimized without loosing that the method fulfills the min-max principle. This third gradient s_3, which is consistent for $\min(s_1,s_2)<s_3<\max(s_1,s_2)$, has point-to-point transfer properties and therefore leads to a positive scheme for sufficiently small time steps. Let s_4 be a consistent approximation of the horizontal gradient s_4 which is more accurate than taking the minimum of the absolute value of the two slopes s_1 and s_2. (Slørdal, 1997) used:

$$s_4 =(s_1+s_2)/2.$$

The following non-linear approach is both consistent and assures the min-max principle:

> if $s_1 * s_2 < 0$ then
> $$\frac{\Delta p}{\Delta x}=0$$
> elseif $\mathrm{abs}(s_4)<\mathrm{abs}(s_3)$ then
> $$\frac{\Delta p}{\Delta x}=\mathrm{sign}(s_4)*s_4$$
> elseif $\min(s_1,s_2)<s_3<\max(s_1,s_2)$ then
> $$\frac{\Delta p}{\Delta x}=\mathrm{sign}(s_3)*s_3$$
> else
> $$\frac{\Delta p}{\Delta x}=\mathrm{sign}(s_1)*\min(\mathrm{abs}(s_1,s_2))$$
> endif

Application of a linear interpolation formula instead of this non-linear filter may introduce wiggles and a small persistent artificial vertical diffusion (except for linear vertical density distributions). Due to the related artificial mixing, stratification may disappear entirely for long term simulations, unless the flow is dominated by the open boundary conditions. This artificial mixing is prevented using the SVK method (Stelling & Van Kester, 1994), but the interpolation algorithm requires a binary search to find the indices of neighbouring grid boxes, which is time consuming.

Aspects of the z-coordinate model

the transport model

A key component of any circulation model is the scalar transport equation. It is not trivial to develop a numerical transport algorithm which ensures mass conservation and monotony for a z-coordinate model where the vertical index of the free surface layer varies in space and time. We introduce an approach based on a fractional step method. The water column is divided into a "3D" and a "2D" part, see Figure 3. In this Figure the free surface at the old time level t and the new time level $t+1/2\Delta t$ are indicated by a solid line and a dotted line, respectively. The 3D part of the water column consists of the cells (finite volumes) which are situated entirely under the free surface at both time levels. The 2D cells, indicated by the hatched area, are formed by the remaining "wet" cells at position (i,j), and reduce to one cell in the case the free surface remains within the same cell during the time step $\Delta t / 2$. In the latter case, the

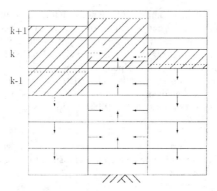

entire column consists effectively of 3D cells. This will be the case most of the time since water level variations are commonly slow in estuarine areas. The concentrations at the new time level, $t + 1/2\Delta t$, are computed on the basis of "3D" and "2D" fluxes, indicated by solid and dotted arrows in Figure 3.

Figure 3: "3D" and "2D" fluxes, indicated by solid and dotted arrows

The stepwise representation of the bed topography can distort the bed shear stress and thereby affect the flow (Sheng, 1983). The bed shear stress τ_{bed} is computed both in the ZCM model as in the SCM model by a quadratic friction law:

$$\frac{\tau_{bed}}{\rho} = u_*^2 = \frac{g}{C_{3D}^2} u_b^n u_b^{n+1} \tag{4}$$

where ρ is the density of the fluid, u_* is the friction velocity, g is the gravitational acceleration and u_b^n is the horizontal velocity in the first grid point above the bed at time step n. The 3D Chezy value C_{3D} is obtained from the logarithmic law of wall

$$C_{3D} = \frac{\sqrt{g}}{\kappa} \ln\left[\frac{1/2\Delta z_{k\min}}{z_0}\right] \tag{5}$$

where κ is the Von Kármán constant, $1/2\Delta z_{k\min}$ is half the grid spacing of the grid cell near the bed (the horizontal velocity is located at $1/2\Delta z_{k\min}$) and z_0 is the roughness height. This dependence on the grid size near the bed, which can vary strongly in space, degrades the computation of the bed shear stress. This can easily be circumvented by using the full grid spacing $\Delta z_{k\min} = z_k - z_{k-1}$ (see Figure 1) or by computing the bed shear stress from the velocity at layer with index $kmin+1$ instead of $kmin$. In contrast with the first adjustment, the latter method is fully consistent with the above equations for the bed shear stress, since a logarithmic velocity profile is assumed in the derivation of these expressions and therefore the bed shear stress can be computed at any height in principle. Both adjustments yield significant improvement of the determination of the bed shear stress, as will be shown later.

The staircase representation of the bed and the free surface also leads to a local degradation of the truncation error of the spatial discretization, since only first-order approximations can be used at these artificial boundaries. Moreover, this stair-case topography is known to introduce artificial friction near the boundaries. By disregarding advection near closed boundaries if the flow is directed from the wall, the development of an artificial boundary layer can be prevented (Stelling, 1983).

Lastly, the ZCM model provides a good basis for the implementation of a non-hydrostatic module since the Cartesian grid guarantees favorable matrix properties (Casulli & Stelling, 1998), which makes solution of the system of equations relatively easy and cheap.

Numerical models: σ vs. Cartesian

Although quite some attention has been paid to the problems related to the σ-transformation, very little consistent comparison of both approaches is available from the literature. Quite often, it is unclear whether differences between results from numerical simulations and observations can be attributed to the discretization, the schematization of the geometry, the turbulence model that is used or the coordinate system. The ZCM and SCM are developed for intercomparison of results of numerical experiments. Similar discretizations for time and space are used in both models, in order to prevent differences stemming from the method of integration of the partial differential equations. Such a comparison allows us to draw more definite conclusions about the performance of the two coordinate systems, without interference of other uncertainties.

Testcases

During the development of the models, several testcases were used to test the consistency between the two models. Both for flows with nearly uniform depth and oscillating tidal flow over a horizontal bed, results of both models were in good agreement with each other (Bijvelds, 1998). However, differences between the models in situations where the bed is no longer horizontal are observed. These differences are mostly related to the inherent disadvantages of the models. This implies that the stair case representation of the bathymetry is the source of most errors in the ZCM model. In the SCM model the baroclinic pressure gradient and horizontal diffusion term mostly give rise to errors. In the following section, the deficiencies of the models will be addressed. The effectiveness of the solutions to these problems, as presented in previous sections, is shown on the basis of the following testcases.

Flow over a linearly sloping bed

In this section two different flows over a linearly sloping bed are considered. The slope of the bed is equal to 0.1, see Figure 4b. First a steady flow is considered to compare different methods for determining the bed shear stress. In the second case the flow is driven by buoyancy effects only, i.e. the inflow at the open boundaries is set equal to zero. Inviscid and viscous flow are studied subsequently. For the latter situation computational results will be compared to measurements available from literature.

Bed shear stress

To test the different methods for the determination of the bed shear stress in the ZCM model, a steady flow is imposed at one open boundary. The mean velocity is equal to 0.5 m/s and the Chezy value equals 55 $m^{1/2}/s^2$. In the vertical direction 25 and 50 uniformly distributed layers are used in the ZCM model and SCM model respectively. The dependence of the bed shear stress on the grid spacing of the lowest grid cell is clearly shown in Figure 4a, where a bed shear stress coefficient is plotted as a function of horizontal position x. The grid spacing of the lowest cell $\Delta z_{k\,min}$ and the depth are plotted in the lower Figure. It is observed that the bed shear stress varies strongly if computed on the basis of $\Delta z_{k\,min}$. By using the full grid spacing $\Delta z_{k\,min} = z_k - z_{k-1}$ (indicated by z-method 2) or involving an extra grid cell at kmin+1 in the determination for the bed shear stress coefficient (indicated by z-method 1), smooth solutions are obtained. Both approximations yield a bed shear coefficient that is in good agreement with the coefficient obtained with the SCM model. In this steady flow, the method based on a vertical grid spacing near the bed of $\Delta z_{k\,min} = z_k - z_{k-1}$ appears to yield a smooth distribution of the bed shear stress coefficient. It is noted that large variations of the vertical grid spacing in vertical direction may result in inaccurate discretization of derivatives in this direction.

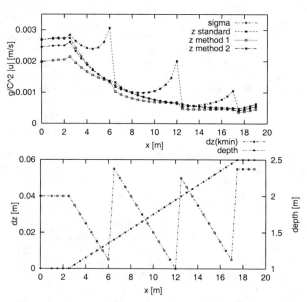

Figure 4: a) Computed friction coefficient for the SCM model, the standard ZCM model and two adjustments to the standard ZCM model (see text for explanation) b) Vertical grid spacing in the lowest cell $\Delta z_{k\,min}$ of the ZCM model and water depth as a function of x.

Pressure gradient

We consider the flow as sketched in Figure 5. Since for this situation both isopycnals and stream-lines are aligned with the computational mesh in the SCM model, the SCM model is beforehand suited best for this flow. The number of grid cells in x-direction is equal to 80. In vertical direction the number of grid cells in the ZCM is equal to 100 whereas only 50 layers are defined in the SCM model. The ratio $\Delta x / \Delta z$ is typically larger than 10, which is characteristic for shallow-water models The increase of the number of active grid cells in the ZCM case is only of the order of 20 % since not all grid cells are included in the computation. Time step restrictions are imposed by advective velocities and the courant number based on the maximal occurring advective velocities is approximately equal to 0.85.

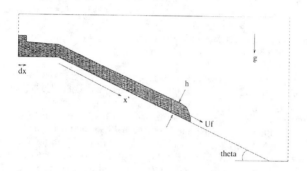

Figure 5: Gravity current on a slope. The excess density is kept constant in time over a distance dx.

First we consider the case of a gravity current that runs down a slope in absence of bed friction and viscosity, see Figure 5. For this simplified flow, the momentum equation is given by:

$$\frac{dU_f^2 h}{dx} = g' h \sin\theta \tag{6a}$$

where g' is the reduced gravity

$$g' = \sqrt{\frac{\Delta\rho}{\rho}} g \tag{6b}.$$

Equation (6a) describes an accelerating flow in the presence of density differences and a sloping bed. Although an analytical solution is available for this equation, it is hard to compare numerical results to this analytical solution since several assumptions that are made to obtain the analytical solution, such as a constant velocity over the layer depth h of the excess density fluid, are not fulfilled in the numerical model.

In agreement with Eq. (6), the gravity current in the SCM model accelerates as it runs down the slope. However, at some point the velocity becomes constant which might be due to the lack of resolution in the vertical direction since the density current becomes thinner as it accelerates. The acceleration is not reproduced in the ZCM model with 80x100 grid points which is due to excessive numerical cross-wind diffusion. Although it is not shown in the Figure 6, implementation of the Van Leer II scheme in both horizontal and vertical direction instead of in horizontal direction only, showed an increase of the front speed in the ZCM model. This indicates the need for accurate transport schemes when the flow is not aligned with the grid. In absence of bed friction and viscosity, hardly any (artificial) vertical diffusion is present in the SCM model, retaining the density difference and hence the driving force. This is caused by the relatively low vertical transformed velocities in the SCM model that are used for solving the transport equation which in its turn is caused by the fact that the grid layers and the flow coincide. An increase of the horizontal resolution in the ZCM model by a factor 4 is needed to obtain a converged solution.

The errors in computing the horizontal pressure gradient in the SCM model are revealed by comparing results of the SCM model with the reference solution of the ZCM model (320x100 grid points). Computing the pressure gradient along σ-planes is shown to overestimate the horizontal pressure gradient reflected by a front speed which is too high. The front speed as computed with both the standard and the improved SVK method is in good agreement with the reference solution. Although it is clear from a mathematical point of view that the new non-linear filter will generally increase the computed horizontal pressure gradients, no difference can be observed between the results of the old and new non-linear filter. Using 80x50 grid points is sufficient in the SCM model; increasing the number of grid points to 320x50 hardly influences the front speed.

It is noted that both in the SCM and ZCM models most of the diffusion takes place near the front (salt wedge) of the density current. Here unphysical large vertical velocities occur due to the use of the hydrostatic pressure assumption in both models.

Including bed friction and viscosity in the model diminishes the differences between the SCM and ZCM models. From measurements (Britter & Linden, 1980) it was concluded that for gravity currents on slopes larger than 5° with sufficiently large Reynolds numbers, the front velocity is constant. Acceleration due to gravity is balanced by bed friction and entrainment and the front speed U_f can be described by:

$$\frac{U_f}{(g'q)^{1/3}} = 1.5 \pm 0.2 \tag{7}$$

where $g'q$ is the buoyancy flux. This relation is plotted in Figure 6 together with the numerical results. The upper and lower limit of Eq. (7) are depicted by the dashed lines. Effects of turbulence is accounted for by using a k-ε turbulence model in which buoyancy effects are included. In agreement with the previous case, the results in the ZCM model (80x100 grid) are dictated by artificial diffusion. Hardly any difference can be observed between the inviscid and the viscous case for this grid resolution. Again, increasing the number of grid points to 320x80 showed to give grid-converged results which are similar to the SCM model with the (improved) SVK method. Although some acceleration of the flow can be observed in the numerical models, results are in reasonably good agreement with the experiments.

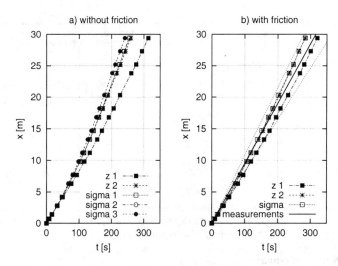

Figure 6: Front position as a function of time. Grids in the SCM model consist of 80x50 grid cells. a) no bed friction and viscosity: z 1: 180x100, z 2: 320x100, sigma 1: original Stelling & Van Kester method, sigma 2: improved Stelling & Van Kester method, sigma 3: discretization along sigma-planes. b) z 1: 80x100, z 2: 320x100, sigma improved Stelling & Van Kester method.

Advection near boundaries

In the ZCM model, artificial vertical closed boundaries are introduced by the stair case representation of the bottom topography. This means that the discretization of the advection terms necessarily locally reduces to first order approximations. The correct treatment of advection at the boundaries is essential to avoid the creation of an artificial boundary layers (Stelling, 1983). The creation of an artificial boundary layer can easily be observed in Figure 7. For the frictionless gravity current described in the previous section, vertical velocity profiles are drawn at a certain position at the slope. Since friction is neglected, we expect the largest velocities to occur near the bed. This type of behaviour is only observed if advection is neglected near the artificial vertical boundaries, when the flow is directed from the closed boundary. If in this case advection is not neglected, artificially generated friction will harm the solution and reduce the velocity of the front.

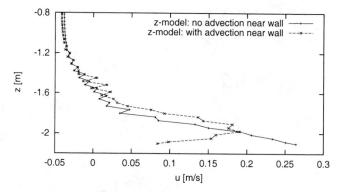

Figure 7: Vertical profiles of horizontal velocity at steady state at 2/3 of slope: ZCM model with advection near solid boundary and ZCM model without advection near solid boundary. The wiggles are due to the absence of viscosity in the model.

The Haringvliet estuary

The Haringvliet estuary is situated in the south-west of the Netherlands. A large part of the Haringvliet estuary was cut off from the tidal influence of the sea several decades ago, by the construction of a dam. Because of this dam, the area considered can be characterized by two regions: a rather shallow part at the sea side of the sluices, of which a part dries during low tide and a region behind the sluices, where deeper channels and some very deep holes are present, see Figure 8. Induced by a changing opion with respect to the ecology of the system, there are plans to introduce tidal influence on the estuary by changing the control of the sluices. The effects of this measure on the hydrodynamics and particularly on the position of the intruding salt wedge, is studied with a 3D numerical model. In this paper the influence of the control of the sluices is disregarded and attention is focussed only on differences between the results of the two models for the situation with open sluices.

The forcing of the model consists of a water level boundary at the sea side and a combined discharge and water level boundary at the river side. At both boundaries the salt concentration is prescribed and the Thatcher-Harleman return time is set equal to zero. From convergence checks (Zijlema, 1999) it was concluded that the increase of the number of layers to more than 8 in the SCM model hardly influenced the computational results. Therefore, only results of a 8-layer SCM model are presented here. Because of the deep holes, the number of fixed layers in the ZCM model is larger than 8. Two calculations were carried out using 11 and 16 layers, respectively. The increase in computational effort compared to the 8-layer SCM model is only minor since only a few extra "wet" cells are added in the vertical in the deep areas in the model. The horizontal computational grid consists of 100x150 grid cells.

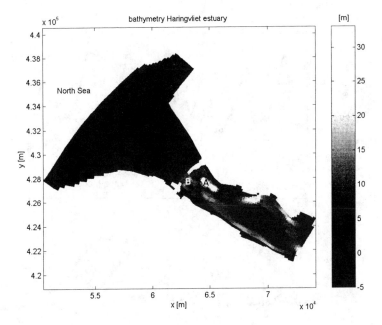

Figure 8: Bathymetry of the Haringvliet estuary.

We first consider the free surface elevation as a function of time at position A, see Figure 9. Hardly any difference between the water level in the SCM model (with the standard SVK approach) and the ZCM model (16 layers) can be observed. Since this monitoring station is situated far away from the water level boundary, the ZCM model seems capable of modelling tidal oscillations equally well as the SCM model.

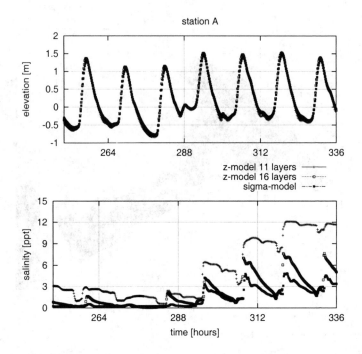

Figure 9: Time series of a) Free surface elevation and b) salinity near the bed in station A.

The time series of salinity near the bed show a strong dependence on the number of layers in the ZCM model. Clearly, taking 11 layers, which causes the shallow parts of the estuary to be underresolved, is insufficient. Increasing the number of layers to 16 leads to similar results as for the SCM model, although some higher values for salinity can be observed in the ZCM model. The latter is also the case in station b. In Figure 10, vertical salinity profiles are plotted at different times for the ZCM and the SCM model. From this figure it can be concluded that the salt wedge tends to intrude somewhat further in the ZCM model although the differences are only of minor importance.

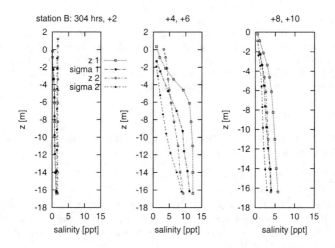

Figure 10: Vertical salinity profiles at station B at different times. The index 1 and 2 in the legend refers to the first and second time mentioned above the figure, respectively.

Conclusions

Two 3D shallow-water models are presented that are designed for estuarine and coastal flows. In regions with strongly varying bathymetry, the flow direction and the orientation of isopycnals generally do not coincide. Under these circumstances, cross-wind diffusion in the transport equation may lead to inaccurate results. As a general rule it may be stated that accurate transport schemes are a prerequisite for flows that are not aligned with the computational mesh, which will be the case in most simulations.

The above is confirmed by the test case of a gravity current on a sloping bed, without any external forcing. The front speed of the gravity current in the z-model was shown to be strongly dependent upon the grid resolution and the advection scheme that was used. This is caused by the fact that in this model the grid lines are not aligned with the flow, which results in cross-wind diffusion. In the σ–model, numerical diffusion was of minor importance due to the small vertical transformed velocities. Comparing the grid-converged solutions of the z-model with results of the σ-model showed that the SVK method for the computation of horizontal gradients, is superior to the discretization of the baroclinic pressure gradient along σ–planes. Although the proposed adjustment of the original SVK method did not yield different results, it is clear from a mathematical point of view that the extension of the non-linear filter should improve results under certain conditions. The better performance of the SCM model (less grid cells needed for the same accuracy as the ZCM model) is due to the fact that this test case is perfectly suited for the SCM model. Although not reported in this paper, it is important to note that the above is not true in general; there are situations in which the ZCM model will perform equally well or better than the SCM model, such as the simaulation of wind-driven flow in stratified lakes with a steep topography .

Application of explicit transport schemes in vertical direction is possible in the ZCM model without severe restrictions on the time step. This allows the implementation of accurate TVD schemes. This is an advantage of the ZCM model since the accuracy of the numerical model in stratified regions often depends on the vertical transport scheme adopted in the model. Moreover, it was shown that problems related to the stair case representation of the bed in the ZCM model are easy to solve. It can therefore be concluded that coping with problems related to the z-coordinate approach is easier than constructing methods to minimize the inherent drawbacks of the SCM model for steep topography. Results of the application of both models to the Haringvliet estuary in the Netherlands, showed only minor differences.

Acknowledgements

The ministry of Public Works and Transport (RWS/RIKZ) in the Netherlands is greatly acknowledged for making available the MOHA model of the Haringvliet estuary.

References

Beckers, J.M., Burchard H., Campin, J.M., Deleersnijder E., and Mathieu, P.P., 1998: Another Reason Why Simple Discretizations of Rotated Diffusion Operators Cause Problems in Ocean Models: Comments on "Isoneutral Diffusion in a z-Coordinate Ocean Model". J. Phys. Oc. 28, pp. 1552-1559.

Britter, and Linder, , 1980: The motion of the front of a gravity current traveling down an incline. J. Fl. Mech. 80(3), pp. 531-543

Bijvelds, M.D.J.P., 1997: Recirculating steady flows in harbours: comparison of numerical computations to laboratory experiments. Tech. Rep. 1-97, Delft Univ. of Techn.

Bijvelds, M.D.J.P., 1998: A three-dimensional fixed grid model for shallow-water flow. Tech. Rep. 6-98, Delft Univ. of Techn.

Casulli, V. and Cheng, R.T., 1992: Semi-implicit finite difference methods for three-dimensional shallow water flow. Comp. Math. Appl., 27 pp. 99-112

Casulli, V. and Stelling, G.S., 1998: Numerical Simulation of 3D Quasi-Hydrostatic Free-Surface Flows. J. of Hydr. En. July, pp. 678-688

Cheng, R.T. and Smith, P.E., 1990: A survey of three-dimensional numerical estuarine models. Estuarine and coastal modeling (proc. conf.), Ed. Spaulding, M.L., pp. 1-15.

Deleersnijder, E. and Ruddick, K.G., 1992: A generalized vertical coordinate for 3D marine models. Bulletin de la Societe Royale des Sciences de Liege, 61 pp. 489-502

Eijkeren, J.C.H. Van, Haan, B.J. De, Stelling, G.S. and Stijn, Th.L. Van, 1993: Linear upwind biased methods. Numerical methods for advection-diffusion problems, Eds. Vreugdenhil, C.B. and Koren, B.,55-91.

Fortunato, A.B. and Baptista A.M., 1994: Modeling near-bottom advective acceleration in surface water models. Computational methods in water resources X, Ed. Peters, A. *et al.* pp. 1045-1052.

Haney, R.L.: 1991: On the pressure gradient force over steep topography in sigma coordinate ocean models. J. Phys. Oc. 21, pp. 610-619.

Hirsch, C.: 1990: Numerical Computation of internal and external flows. Wiley, New York, 1990.

Leendertse, J.J., 1990: Turbulence modeling of surface water flow and transport: part IV[a]. J. Hydr. Eng. 116(4), pp. 603-606

Leendertse, J.J. and Liu, S.K., 1975: A three-dimensional model for estuaries and tidal seas. II Aspects of computation. Tech. Rep. R-1764-OWRT Rand Corporation.
Mellor, G. and Blumberg, A., 1985: Modelling vertical and horizontal diffusivities with the sigma coordinate system. Monthly Weather Rev. 11, pp. 1379-1383

Paul, J.F., 1994: Observations related to the use of the sigma coordinate transformation for estuarine and coastal modeling studies. Estuarine and coastal modeling III (proc. conf.), Eds. Spaulding, M.L., Bedford, K., Blumberg, A., Cheng, R. and Swanson, C., pp. 336-350.

Phillips, N.A., 1957: A coordinate system having some special advantages for numerical forecasting. J. of Meteorology, 14, pp. 184-185.

Podber, D.P. and Bedford, K.W., 1994: Tributary loading with a terrain following coordinate system. Estuarine and coastal modeling III (proc. conf.), Eds. Spaulding, M.L., Bedford, K., Blumberg, A., Cheng, R. and Swanson, C., pp. 475-488.

Rodi, W., 1984: Turbulence models and their application in hydraulics - a state of the art review. International Association for Hydraulics Research.

Sheng, Y.P., 1983: Mathematical modeling of three-dimensional coastal currents and sediment dispersion. Tech. Rep. CERC-33-2. U.S. Army Corps Engineers.

Slørdal, H., 1997: The pressure gradient force in sigma co-ordinate ocean models. Int. J. Num. Meth. Fl. 24, pp. 987-1017.

Stelling, G.S., 1983: On the construction of computational methods for shallow-water problems. Ph. D. thesis, Delft Univ. of Techn.

Stelling, G.S. and Van Kester, J.A.Th.M., 1994: On the approximation of horizontal gradients in sigma co-ordinates for bathymetry with steep bottom slopes. Int. J. Num. Meth. Fl. 18, pp. 915-935.

Walters, R.A. and Foreman, M.G.G., 1992: A 3D finite element model for baroclinic circulation on the Vancouver Island continental shelf. J. Mar. Sys. 3.

Zegeling, P., Borsboom, M. and Van Kester, J.A.Th.M., 1998: Adaptive moving grid solutions of a shallow-water transport model with steep vertical gradients. Proc.12th Int. Conf. on Computational Methods in Water Resources, Vol. 2, pp. 427--434.

Zijlema, M., 1999: personal communications.

Three-Dimensional, Hydrostatic and Non-Hydrostatic Modeling of Seiching in a Rectangular Basin

XinJian Chen[1]

Abstract

When vertical acceleration terms have a comparable effect on flow movement as the gravity acceleration term, the hydrostatic pressure assumption used by most hydrodynamic models is questionable. In these cases, a fully hydrodynamic (non-hydrostatic) model, which solves all three momentum equations, needs to be used.

This paper presents a three-dimensional, non-hydrostatic model, which treats the hydrostatic assumption for pressure as a special case. The model has two solution approaches: an Euler-Lagrangian method with a pressure-correction procedure and a semi-implicit method with a fractional step scheme. Both solution approaches were used to study two seiche oscillation cases, in which effects of non-hydrostatic pressure on flow are not negligible. The first oscillation is caused by a free surface setup, while the second one is driven by a horizontal density gradient. The model was run with and without the hydrostatic assumption for pressure. Model results of the hydrostatic and non-hydrostatic runs show that the velocity field and surface elevation with the hydrostatic assumption are different from those without the hydrostatic assumption. Model results also indicate that the hydrostatic simulation tends to cause model results nonlinear in a faster manner than the non-hydrostatic simulation does, with larger fluctuations of velocity and surface elevation.

Introduction

The hydrostatic pressure assumption has been used in most hydrodynamic models. In most cases, this assumption is valid because the horizontal gradients of hydrostatic pressure is much larger than those of non-hydrostatic pressure. However, there are also many cases where this assumption may be questionable. One example is the flow induced by a very large horizontal density gradient. Other examples include short wave flow, flow controlled by a structure, the near-field of brine disposal from a

---

[1]SWFWMD, 7601 Highway 301 North, Tampa, Fla. 33637

desalination plant, flow field near the uptake point of a water withdrawal system, etc.. In all these cases, effects of non-hydrostatic pressure on flow may be comparable to hydrostatic pressure, and thus can not be neglected in model simulations.

There exist only a limited number of 3-D models which are capable of including non-hydrostatic pressure in the simulation. Casulli's semi-implicit model is one of them (Casulli and Stelling, 1995 and 1998). Mahadevan et al. (1996) developed a non-hydrostatic mesoscale ocean model which used the similar solution procedure as that of Casulli's model, although the former used a control volume method, while the latter used a finite difference method. Both models employed a fractional step scheme to first solve an intermediate velocity field and the final free surface elevation using hydrostatic pressure and then solve an elliptic (Poisson) equation for the non-hydrostatic portion of the pressure field. The velocity field at the n+1-th time step was finally solved after non-hydrostatic pressure was obtained.

This study developed a three-dimensional, fully hydrodynamic model which treats the hydrostatic assumption for pressure as a special case. Two different solution approaches are used in the model. The first is an Euler-Lagrangian model which solves the whole set of Reynolds Averaged Navier-Stokes (RANS) equations by first guessing a pressure field and then correcting the pressure field by forcing the velocity field to satisfy the continuity equation. The second solution procedure splits pressure into barotropic, baroclinic, and non-hydrostatic components, and also uses a fractional step scheme. It first solves RANS equations without the vertical eddy viscosity term and then updates the final velocity and surface elevation at the vertical turbulent mixing step. The model was used to study two seiche oscillation cases, in which the non-hydrostatic pressure effects on flow is comparable to that of hydrostatic pressure and thus can not be ignored. The model was run with and without the hydrostatic pressure assumption using the two solution approaches. Model results show that the velocity field and surface elevation with the hydrostatic assumption are different from those without the hydrostatic assumption. Model results also indicate that the hydrostatic simulation tends to cause model results nonlinear in a faster manner than the non-hydrostatic simulation does, with larger fluctuations of velocity and surface elevation.

In the following, a brief description of model formulation is presented, followed by simulations of two seiching cases with and with out the hydrostatic pressure assumption. Model results are then discussed and conclusions are drawn from the model results. The terms non-hydrostatic simulation and (fully) hydrodynamic simulation used in the following text are simulations where the pressure term includes all the pressure components, including hydrostatic pressure and non-hydrostatic pressure.

Model Formulation

Governing Equations:

The model solves the following continuity and momentum equations:

$$\frac{\partial u_i}{\partial x_i} = 0 \tag{1}$$

$$\frac{du_l}{dt} = \frac{\partial u_l}{\partial t} + u_m \frac{\partial u_l}{\partial x_m} = -\frac{1}{\rho}\frac{\partial p}{\partial x_l} + \phi_l + \frac{\partial}{\partial x_m}(A\frac{\partial u_l}{\partial x_m}) \tag{2}$$

where, t is time, $x_l (l=1,2,3)$ represents the Cartesian coordinates in three directions (x_1 and x_2 are in the west-east and south-north directions, respectively, while x_3 is in the vertical direction) u_l represents velocity in x_l-direction, ρ is density, p is pressure, $\phi = (f u_2, -f u_1, -g)$, A represents eddy viscosity, and f and g are Coriolis parameter and the gravitational acceleration, respectively.

The model also solves the following transport equation of concentration:

$$\frac{\partial c}{\partial t} + u_m \frac{\partial c}{\partial x_m} = \frac{\partial}{\partial x_m}(B\frac{\partial c}{\partial x_m}) + S + R \tag{3}$$

where c is concentration, S denotes source/sink terms, R represents reaction terms and B is the eddy diffusivity.

With consideration of the exchange with the atmosphere, integration of Equation (1) yields

$$\frac{\partial h}{\partial t} = -\frac{\partial}{\partial x_1}(\int_{h_0}^h u_1 dx_3) - \frac{\partial}{\partial x_2}(\int_{h_0}^h u_2 dx_3) + r \tag{4}$$

where, h is the surface elevation, h_o is the bottom elevation, and r is the net rain intensity (rainfall minus evaporation).

In Equations (2) and (3), eddy viscosity A and eddy diffusivity B in horizontal directions are calculated using the Sub-Grid Scale (SGS) model (Smagorinsky, 1963), while those in the vertical direction are calculated by solving the turbulent kinetic energy equation from the velocity gradient (Chen, 1994).

Boundary conditions specified in the vertical direction are shear stresses. At the free surface, wind shear stresses are used. At the bottom, the log-layer distribution of velocity is used to calculate the bottom shear stress:

$$\tau_l = \rho\left[\frac{\kappa}{\ln(z_b/z_o)}\right]u_l\sqrt{u_1^2 + u_2^2} \qquad l = 1, 2 \tag{5}$$

where κ is the von Karman constant (0.41), u_1 and u_2 are horizontal velocity calculated at a level z_b near the bottom, $z_o = k_s/30$ and k_s is the bottom roughness.

Density in Equation (2) and (5) is a function of temperature (T) and salinity (s)

$$\rho = \rho(T,s) \tag{6}$$

The pressure gradient term in Equation (2) includes both the hydrostatic pressure gradient and non-hydrostatic pressure gradient. The pressure can be calculated as follows:

$$p = p_h + p_n = g\int_{x_3}^h \rho dx_3 + \int_{x_3}^h \rho a_3 dx_3 = \int_{x_3}^h \rho G dx_3 \tag{7}$$

where p_h and p_n are hydrostatic pressure and non-hydrostatic pressure, respectively, a_3 represents all the vertical acceleration terms contributing to non-hydrostatic pressure, and $G = g + a_3$.

Numerical Schemes for the Continuity and Momentum Equations :

Two solution approaches using the finite difference method were developed to solve Equations (1) and (2). The first one uses an Euler-Lagrangian method with a pressure-correction procedure, and the second one utilizes a semi-implicit method with a fractional step scheme. Both finite difference solution approaches were developed in a Cartesian grid system with N_1, N_2, and N_3 grids in the west-east, south-north, and vertical directions, respectively. A brief description of the two solution procedure is presented below, and more details can be found in Chen (2000).

1. Euler-Lagrangian Method with a Pressure-Correction Procedure

For the Euler-Lagrangian method, a colocated arrangement of model variables was used, where all variables are at the center of the grid cell. Now, let us consider a water particle whose location at the n+1-th time step is at the center of the cell (i, j, k) with the coordinates of (x_1, x_2, x_3). At the n-th time step, the location of this water particle was at

$$x_l^* = x_l - \hat{u}_l \Delta t \qquad (8)$$

where $\hat{u}_l$ is the average velocity along the streamline segment $x_l^* \to x_l$.

However, without knowing the velocity distribution at the n+1-th time step, $\hat{u}_l$ in Equation (8) is an unknown. Therefore, x_l^* is an unknown unless velocities at the n+1-th time step are solved. Thus, an iteration process has to be employed to solve this problem. Let us use u_l^n as an initial guess of $\hat{u}_l$ for the iteration, where the superscript n denotes the n-th time step. An initial estimate of x_l^* can be calculated from Equation (8) and the water particle velocity at the n-th time step (at x_l^*), u_l^* can be estimated based on x_l^* and the velocity distribution at the n-th time step.

Once u_l^* is estimated, the following Euler-Lagrangian scheme can be used to solve the momentum equations:

$$\frac{u_{l,i,j,k}^{n+1} - u_{l,i,j,k}^*}{\Delta t} = -\frac{1}{\rho^n}\frac{\Delta p^*}{\Delta x_l} + \mathbf{H}(u_l^n) + \phi_l^n + \frac{1}{\Delta x_{3_k}}(A_{v_k}^n \frac{u_{lk+1}^{n+1} - u_{lk}^{n+1}}{\Delta x_{3k+1/2}} - A_{v_{k-1}}^n \frac{u_{lk}^{n+1} - u_{lk-1}^{n+1}}{\Delta x_{3k-1/2}}) \qquad (9)$$

where Δx_l is the spacing in the x_l direction; the subscripts i, j, and k are grid indexes (for clarity, subscripts i and j are omitted on the right hand side of the equation) in x_1-, x_2-, and x_3-directions, respectively; p^* is the average pressure along the path of $x_l^* \to x_l$, and $\mathbf{H}$ is an finite difference operator for the horizontal viscous terms. This average pressure is the total pressure including hydrostatic pressure and non-hydrostatic pressure. It is also an unknown. One can use the pressure at the n-th time step as an initial guess of p^* and then correct the pressure by forcing the final velocity field satisfying the continuity equation.

From Equation (4), the free surface at the new time step can be calculated:

$$\frac{h_{i,j}^{n+1} - h_{i,j}^n}{\Delta t} = -\frac{\Gamma_{1i+1/2,j,k}^{n+1} - \Gamma_{1i-1/2,j,k}^{n+1}}{\Delta x_1} - \frac{\Gamma_{2i+1/2,j,k}^{n+1} - \Gamma_{2i-1/2,j,k}^{n+1}}{\Delta x_2} + r \qquad (10)$$

where,

$$\Gamma_1 = \int_{h_0}^{h} u_1 dx_3, \qquad \Gamma_2 = \int_{h_0}^{h} u_2 dx_3$$

The velocity field solved from Equation (9) does not necessarily satisfy the continuity equation. Let u_l^{n-1} calculated from Equation (9) be u_l^{n*}, let the final u_l, which satisfies the continuity equation at the n+1-th time step, be u_l^{n-1}, and let the difference between the two be $\delta u_l = u_l^{n+1} - u_l^{n+*}$. To get u_l^{n+1}, the final pressure, p^{**}, has to be sought. Let $p'=p^{**}-p^*$, then from Equation (9) we have

$$\frac{\delta u_l}{\Delta t} = -\frac{1}{\rho^n}\frac{\partial p'}{\partial x_l} + \frac{\partial}{\partial x_3}(A\frac{\partial \delta u_l}{\partial x_3}) \tag{11}$$

Inserting this expression to Equation (2) leads to a Poisson equation for p':

$$\nabla^2 p' = \frac{\rho}{\Delta t}U' - \rho\frac{\partial}{\partial x_3}(A\frac{\partial U'}{\partial x_3}) \tag{12}$$

where p' is the correction to the pressure field p^*, and U' is the divergence of u_l^{n+*}.

Equation (12) is a seven-diagonal matrix system, which is symmetric and nonnegative everywhere. It can be solved using the conjugate gradient method with incomplete Cholesky preconditioning (Golub and van Loan, 1990). Once p' is solved, the final velocity field can be computed from Equation (11). However, u_l^{n+1} so calculated may still not be the final solution of Equations (1) and (2) because u_l^n was used as the initial guess for $\hat{u}_l$ in Equation (8). A better guess of $\hat{u}_l$ would be $(u_l^n+u_l^{n+1})/2$. Using this new estimate of $\hat{u}_l$ in Equation (8) and repeating the above steps again mean an iteration process involving solving the Poisson equation for p' in each iteration. This is very time consuming. Another relatively simple alternative for correcting pressure was also built in the model and explained below.

From the continuity equation, the vertical component, u_3, can be calculated from the two horizontal components, u_1 and u_2:

$$\frac{u_{3i,j,k}^{n++} - u_{3i,j,k-1}^{n++}}{\Delta x_{3k}} = -\frac{u_{1i+1/2,j,k}^{n+*} - u_{1i-1/2,j,k}^{n+*}}{\Delta x_{1i}} - \frac{u_{2i,j+1/2,k}^{n+*} - u_{1i,j-1/2,k}^{n+*}}{\Delta x_{2j}} \tag{13}$$

where u_3^{n++} denotes the new approximation of the vertical velocity which, together with u_1 and u_2, satisfies the continuity equation.

Let u_3 calculated from Equation (9) be u_3^{n+*} and let $\delta u_3^+ = u_3^{n++} - u_3^{n+*}$. If $\delta u_3 = 0$, then velocity fields calculated from Equation (9) automatically satisfy the continuity equation and would be the final solution. If $\delta u_3 \neq 0$, then the following finite difference equation relating δu_3 and p' can be obtained from Equation (9):

$$\frac{\delta u_3^+}{\Delta t} = -\frac{1}{\rho^n}\frac{\Delta p'}{\Delta x_3} + \mathbf{V}(\delta u_3^+) \tag{14}$$

where $\mathbf{V}$ is an finite difference operator for the vertical viscosity term.

Once p' is solved, a new estimate of the pressure field is sought and takes the following form:

$$p_{i,j,k}^{**} = \mathbf{I}_z(\rho^n G) = \rho_{i,j,k}^n G_{i,j,k}\frac{\Delta x_{3k}}{2} + \sum_{kk=k+1}^{k_m^*-1}\rho_{i,j,kk}^n G_{i,j,kk}\Delta x_{3kk} + \rho_{j,k_m^*}^n G_{i,j,k_m^*}\eta_{i,j}^{n+*} \tag{15}$$

where $\mathbf{I}_z$ is an operator integrating its argument from a level z to the free surface, k^*_m is the intermediate vertical grid index number of the free surface layer, $\eta_{i,j}^{n+*}$ is the intermediate top layer thickness calculated from Equation (10), and

$$G_{i,j,k} = g + \frac{u_{3i,j,k}^{n++} - u_{3i,j,k}^*}{\Delta t} - \mathbf{H}(u_3^n) - \mathbf{V}(u_3^{n++}) \tag{16}$$

G in the above equation (16) includes gravitational acceleration, vertical acceleration of the water particle and the viscosity terms. Therefore, the new pressure field (p^{**}) calculated using Equation (15) is fully hydrodynamic and includes not only hydrostatic pressure, but also non-hydrostatic pressure. Using $(u_l^n + u_l^{n*})/2$ as a new guess for $\hat{u}_l$ and p^{**} as a new guess for the average pressure along the streamline segment x_l^* $\rightarrow x_l$, the above steps can be iterated to find the true velocity and pressure field at the new time step.

2. Semi-Implicit Method with a Fraction Step Scheme

For the semi-implicit method, a staggered grid system was used, which stores u_1, u_2, and u_3 at the centers of the east, north, and top faces of the grid cell, respectively, and the remaining variables at the middle of the grid cell. Similar to Casulli and Stelling (1998), pressure is split into three components:

$$\nabla p = \rho_h g \nabla h + g \int_z^h \nabla \rho d\xi + \nabla q \tag{17}$$

where ρ_h is the density at the water surface. The first two terms on the right hand side of the above equation are the barotropic and baroclinic components and the third term is the non-hydrostatic term. To obtain a numerical scheme with better stability, a fractional step scheme was derived and is explained below.

Two steps are used in the model. In the first step, the vertical eddy viscosity term is not included in the difference equations and an intermediate velocity field, $\hat{u}_l$, is solved. Then, an implicit method is used for the vertical eddy viscosity term in the second step to obtain the final velocity field, u_l^{n+1}. All three pressure gradient components in Equation (17) are included in the first step, in which the semi-implicit method is used for the barotropic term and the remaining terms are treated explicitly, including the non-hydrodynamic term. The finite difference equations are as follows:

$$\frac{\ddot{u}_l^{n+1} - u_l^n}{\Delta t} = -\frac{\rho_h g}{\rho \Delta x_l}[\theta(h_{i+1,j}^{n+1} - h_{i,j}^{n+1}) + (1-\theta)(h_{i+1,j}^n - h_{i,j}^n)] - \frac{g}{\rho}\mathbf{I}_z(\frac{\partial \rho}{\partial x_l})$$
$$- \frac{1}{\rho}\frac{(q_{i+1,j}^n - q_{i,j}^n)}{\Delta x_l} + \mathcal{H}u_l^n + \phi_l^n \tag{18}$$

or

$$\ddot{u}_l^{n+1} = -\frac{\theta g \Delta t}{\Delta x_l}(h_{i+1,j}^{n+1} - h_{i,j}^{n+1}) + \mathbf{L}(u_l) \tag{19}$$

where θ varies between 0 (fully explicit) and 1 (fully implicit), $\mathcal{H}$ is a finite difference operator including the convective terms and horizontal eddy viscosity terms, and $\mathbf{L}$ includes all the explicit terms in Equation (18).

Combining Equation (19) with Equation (10) results in a finite difference equation for h^{n-1}:

$$\frac{h_{i,j}^{n+1} - h_{i,j}^n}{\Delta t} = -\frac{\theta g \Delta t}{\Delta x_1}(H_{i-1/2,j}\frac{h_{i,j}^{n+1} - h_{i-1,j}^{n+1}}{\Delta x_{1_{i-1/2}}} - H_{i+1/2,j}\frac{h_{i+1,j}^{n+1} - h_{i,j}^{n+1}}{\Delta x_{1_{i+1/2}}}) + \mathbf{I}_{h_{0_{i-1/2,j}}}[\mathbf{L}_{i-1,j}(u_1)] - \mathbf{I}_{h_{0_{i+1/2,j}}}[\mathbf{L}_{i,j}(u_1)]$$
$$- \frac{\theta g \Delta t}{\Delta x_2}(H_{i,j-1/2}\frac{h_{i,j}^{n+1} - h_{i,j-1}^{n+1}}{\Delta x_{2_{j-1/2}}} - H_{i,j+1/2}\frac{h_{i,j+1}^{n+1} - h_{i,j}^{n+1}}{\Delta x_{2_{j+1/2}}}) + \mathbf{I}_{h_{0_{i,j-1/2}}}[\mathbf{L}_{i,j-1}(u_2)] - \mathbf{I}_{h_{0_{i,j+1/2}}}[\mathbf{L}_{i,j}(u_2)] \tag{20}$$

where H is the water depth.

Equation (20) represents $N_1 \times N_2$ linear equations with the same number of unknowns. It forms a five-diagonal matrix system which is symmetric and nonnegative everywhere. The conjugate gradient method with incomplete Cholesky preconditioning can be used to solve the five-diagonal system.

After h^{n+1} is solved, $\ddot{u}_l$ ($l=1$, 2) are calculated from Equation (18) and the final solution for u_l^{n+1} ($l=1$, 2) can be solved in the second step as follows:

$$\frac{u_l^{n+1} - \ddot{u}_l^{n+1}}{\Delta t} = \frac{1}{\Delta x_{3_k}}(A_{v_k}^n \frac{u_{l_{k+1}}^{n+1} - u_{l_k}^{n+1}}{\Delta x_{3_{k+1/2}}} - A_{v_{k-1}}^n \frac{u_{l_k}^{n+1} - u_{l_{k-1}}^{n+1}}{\Delta x_{3_{k-1/2}}}) \qquad l=1, 2 \qquad (21)$$

and u_3^{n+1} is solved from the continuity equation.

It should be noted that the final u_1^{n+1} and u_2^{n+1} lead to a new surface elevation which may not be the same as h^{n+1} calculated from Equation (20). Therefore, h^{n+1} needs to be updated using Equation (10) once the final u_1^{n+1} and u_2^{n+1} are available. However, it can be proven that it is not necessary to do such update if the following relationship between two divergences holds:

$$\nabla \cdot \vec{\tau}_w = \nabla \cdot \vec{\tau}_b \qquad (22)$$

where $\vec{\tau}_w = \tau_{wx}\mathbf{i} + \tau_{wx}\mathbf{j} + 0\mathbf{k}$ is wind shear stress and $\vec{\tau}_b = \tau_{bx}\mathbf{i} + \tau_{bx}\mathbf{j} + 0\mathbf{k}$ is the bottom shear stress.

Finally, the non-hydrostatic portion of the total pressure for the n+1-th time step can be calculated from the momentum equation for the vertical velocity component. It includes the vertical acceleration term and the horizontal and vertical viscosity terms. This non-hydrostatic pressure field will be save for solving the velocity field at the next time step (n+2-th time step).

Numerical Scheme for the Transport Equation:

Numerical scheme for the transport equation of concentration (e.g., salinity, suspended sediment, etc.) Takes the following form:

$$\frac{c^{n+1} - c^n}{\Delta t} = \mathbf{H}_s(c^n) - \mathbf{W}(c^{n+1}) + \frac{1}{\Delta z_k}(B_{v_{i,j,k}}^n \frac{c_{i,j,k+1}^{n+1} - c_{i,j,k}^{n+1}}{\Delta x_{3_{k+1/2}}} - B_{v_{i,j,k-1}}^n \frac{c_{i,j,k}^{n+1} - c_{i,j,k-1}^{n+1}}{\Delta x_{3_{k-1/2}}}) \qquad (23)$$

where $\mathbf{H}_s$ is a finite difference operator for the horizontal advective terms and diffusive terms and $\mathbf{W}$ is a finite difference operator for the vertical advective term using the upwind scheme. Equation (23) constitutes $N_1 \times N_2$ linear equations with the same number of unknowns. It forms a triangle matrix system, which can be easily solved with the Thomas Algorithm.

Two Seiche Oscillation Cases

Using the numerical schemes discussed above, a three-dimensional lake and estuarine simulation system (LESS3D) has been developed by the author. The model was used to simulate two seiche oscillation cases in a rectangular basin to investigate the differences between the hydrostatic and non-hydrodynamic simulations. In the first seiche oscillation case, a surface setup was initially assumed and thus the oscillation is caused by the barotropic pressure gradient. In the second case, the oscillation is caused by the horizontal density gradient, or the baroclinic pressure gradient. In both cases, no wind is applied at the free surface and the bottom roughness is very small (z_0=0.00001).

In the first seiche oscillation case, the rectangular basin has a length of 63 meters and a water depth of 4 meters. At time = 0, water is still, but the free surface has the following sinusoidal form:

$$h = H + 10\sin(\frac{x - L}{2L}\pi)$$

where H is the average water depth (400 centimeters), L is the length of the basin, h is the surface elevation in centimeters, and the x-axis is pointing to the right from its origin at the bottom left corner of the basin.

A Cartesian grid system was used to simulate this oscillation case. The grid spacings in the horizontal directions are either 2 or 3 meters, while the vertical grid spacing varies between 10cm and 30cm. The model allows the free surface to travel between layers, so that there is no need to make the top layer thick enough to cover the whole range of free surface variation. For example, the free surface which was in the k-th grid at time t can fall down to the k-1-th grid at time $t+\Delta t$. In other words, numbers of vertical grids for each water columns varies with time in the computation and the vertical grid index number of the free surface layer (k_m in Equation (15)) is updated each time when the surface elevation is updated. For the free surface layer and the bottom layer, the exact layer thicknesses were used in the computation, instead of the grid spacings of these two layers.

In the second seiching case, the rectangular basin has a length of 40 meters and a depth of 5 meters. At time = 0, water in the basin is still and the free surface elevation is the same everywhere. Assuming that there exist a gate at the half length of the basin which separates salty water from freshwater, salinity in the basin has the following distribution at time ≤ 0 (see Figure 1):

$$s = \begin{cases} 0 \text{ ppt} & x \leq 0 \text{ m} \\ 20 \text{ ppt} & x > 20 \text{ m} \end{cases}$$

where s is salinity in ppt.

At time = 0, the gate was lifted and water in the basin would be first overturned clockwise by the initial density gradient and then sloshes back and forth in the basin. As time goes on, salinity in the basin will be mixed by the circulation and the baroclinic effect will be diminished and water in the basin will slowly calm down. Eventually, salinity in the basin will be fully mixed with a value of 10 parts per thousand everywhere.

A Cartesian grid system similar to that used for the first seiche oscillation was used for this oscillation case with the horizontal grid spacing of 2 meters and vertical grid spacing of 10 to 30 centimeters.

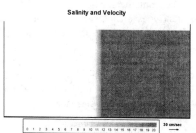

Figure 1 Salinity and velocity distributions at t=0 for the second oscillation.

Model Results of Hydrostatic and Non-Hydrostatic Simulations

Although the two oscillation cases are physically two-dimensional (the Coriolis effect is neglected), the 3-D model described above was used to simulate the oscillations. True three-dimensional oscillations were also simulated and model results for these 3-D problems can be found elsewhere (Chen, 2000) and are not presented here. For the two seiching cases, the model was run using both the Euler-Lagrangian method and the semi-implicit method with and without non-hydrostatic pressure. It was found that the two solution approaches (Euler-Lagrangian method with a pressure-correction procedure and a semi-implicit method with a fractional scheme) yield similar model results. Comparisons of model results using the two solution procedures are documented in Chen (2000), and omitted here. The main purpose of this study is to compare models results of hydrostatic and non-hydrostatic simulations.

For the first oscillation case, which is caused by the initial free surface setup, the celerity of the gravity wave is about 6.3 m/s. Therefore, the period for the oscillation is about 20 seconds (the basin length is 63 meters). The model was run for 400 seconds (about 20 cycles) and time series of surface elevations and velocities at two locations were printed out as outputs. Also printed out were the velocity distributions every 2 seconds during the last cycle.

Figure 2 shows simulated time series of surface elevations and velocities during the last 100 seconds (about five cycles) using the hydrostatic pressure assumption, while Figure 3 shows the corresponding model results using fully hydrodynamic pressure, including the non-hydrostatic component. The two locations shown in Figures 2 and 3 are at $x = 3$ m (solid lines) and $x = 41$m (dashed lines), respectively, with a depth of 2 m below the mean surface level for the former and 0.5 m for the letter. Because the oscillation is two-dimensional and the Coriolis effect was set to zero in the simulations, simulated velocities in y-direction are zero and are not shown in the two figures.

Figures 4 through 7 are simulated velocity fields at time = 378, 384, 390, and 392 seconds. The top graphs in each figures are the model results of the hydrostatic simulation, while the bottom graphs are those of the non-hydrostatic simulation. The free surfaces at the four time levels are also shown in these figures.

Comparison of Figures 2 and 3 suggests that model results using hydrostatic pressure contain more higher mode oscillations than those of non-hydrostatic simulation as time goes on. Overall, simulated free surface elevation and velocity with the hydrostatic pressure assumption are more nonlinear, with larger fluctuations. It can also be seen from Figures 2 and 3 that simulated oscillation using the hydrostatic pressure assumption has a phase lead in relative to the model results of non-hydrostatic simulation. For example, the last surface elevation peak at $x = 3$ m (solid lines) occurred at time = 392 seconds for the hydrostatic simulation, while the last surface elevation peak at the same location occurred at time = 396 seconds for the non-hydrostatic simulation. As can be seen from the velocity plots, the vertical velocity is comparable with the horizontal velocity during the course of seiching. For areas where the surface elevation gradient in x-direction is very small (for example, near the walls at the two ends), effects of non-hydrostatic pressure gradient will be significant and thus can not be neglected.

Figures 4 through 7 further confirm that model results of the hydrostatic pressure simulation are more nonlinear than those of the non-hydrodynamic simulation. It can also be seen from Figures 4 through 7 that the simulated oscillation using fully hydrodynamic pressure swings back and forth with a center near the free surface at the half length of the basin, whereas the simulated oscillation using hydrostatic pressure could have two swing centers. One of the swing centers in the hydrostatic simulation results is a primary one, while the other is secondary. The primary swing center migrates back and forth in the basin with a velocity equal to the celerity of the gravity wave. This again indicates that higher mode oscillations are significant in the model results of the hydrostatic simulation.

Figures 8 through 11 are simulated salinity and velocity distributions at time= 10,30, 90, and 1000 seconds for the second oscillation case. Again, the top graphs in each figures are the model results of hydrostatic simulation, while the bottom graphs show the model results of hydrodynamic simulation. As expected, both simulations give similar velocity and salinity distribution patterns where the fresh water moves to the right in the top layer and salty water moves to the left in the bottom layer under a negative pressure gradient in x-direction. A gyre centered at the middle of the basin is formed by an overturn of the water in the basin. At the beginning, the intensity of the circulation is rather strong and the direction of the circulation is clockwise. Later, the circulation alters direction and becomes weaker because the mixing of salinity gradually diminishes the horizontal density gradient.

Differences between the top graph and bottom graphs can be seen in Figures 8 through 11, especially in the top layer where simulated velocity without non-hydrostatic pressure has a higher vertical component than that of the non-hydrostatic simulation. As a result, simulation with non-hydrostatic pressure produces a much smoother free surface than that produced by the hydrostatic simulation. This, in turn, causes a slight difference between the two velocity fields below the top layer.

Conclusions

A three-dimensional, fully hydrodynamic model has been developed. The model solves the whole set of RANS equations with two different numerical solution approaches. The first uses an Euler-Lagrangian method with a pressure-correction procedure, and the second one is a semi-implicit method with a fractional step scheme. Both methods have been used to simulate two seiche oscillation cases in rectangular basins. The first seiching case is caused by a free surface setup, while the second one is caused by the horizontal density gradient. In spite of the obvious differences between the two solution approaches, no significant differences are seen in model results when using the two solution approaches to solve the same problem. However, differences exist between the model results of the hydrostatic and non-hydrostatic (fully hydrodynamic) simulations. Generally, the hydrostatic simulation tends to cause model results nonlinear in a faster manner than the non-hydrostatic simulation does, with larger fluctuations of velocity and surface elevation, while the non-hydrostatic simulation can preserve the linearity better. For example, in the first seiching case where the oscillation was caused by a free surface setup, higher mode oscillations can be clearly seen in the model results of the hydrostatic simulation, while these higher mode oscillations are not significant in the non-hydrostatic simulation. Simulated oscillation using the hydrostatic pressure assumption seems to have a phase lead in relative to that without such assumption.

Although the hydrostatic and non-hydrostatic simulations do not give exactly the same velocity fields and surface elevations, model results of salinity distributions are similar for the two simulations in the second oscillation case. This can be explained by the fact that the main difference of the velocity fields for the two simulations only occurs at the top layer. Casulli and Stelling (1998) found that the hydrostatic simulation would produce a sharp density front, which is unreal. However, this sharp front did not appear in the hydrostatic simulation in this study. Two factors might be responsible for the discrepancy between the hydrostatic simulation of density front in this study and that in Casulli and Stelling (1998). First, turbulent mixing was included in this study, while it was not included in their simulation. Second, this study used a much coarser grid size than that used in Casulli and Stelling (1998). According Stelling (1999), a very fine grid size has to be used to produce the sharp density front in the hydrostatic simulation. Future study with a very fine grid size and a very low viscosity needs to be done to confirm their finding.

References

Casulli, V. and G.S. Stelling, 1995, Simulation of Three-dimensional, Non-hydrostatic Free-surface Flows for estuaries and Coastal Seas, Proc. Of the 4th Int. Conf. On Estuarine and Coastal Modeling, San Diego, California (eds. M.L. Spaulding and R.T. Cheng), ASCE pp1-12.

Casulli, V. and G.S. Stelling, 1998, Numerical Simulation of 3D Quasi-Hydrostatic Free-Surface Flows, *Journal of Hydraulic Engineering*, 124(7), 678-686.

Chen, X.-J., 1994: Effects of Hydrodynamics and Sediment Transport Processes on Nutrient Dynamics in Shallow Lakes and Estuaries, Ph.D. Dissertation, University of Florida, Gainesville.

Chen, X.-J., 2000: A Three-Dimensional Lake and Estuarine Simulation System (LESS3D): Model Formulation, Technical Report (in preparation), Southwest Florida Water Management District.

Golub, G.H. and C. van Loan, 1990, Matric computations, John Hopkins University Press, Baltimore.

Mahadevan, A., J. Oliger, and R. Street, 1996, A nonhydrostatic mesoscale ocean model. Part II: Numerical Implementation, J. Physical Oceanography, 26(9), 1881-1900.

Smagorinsky, J., 1963, General circulation experiments with primitive equations, I. The basic experiment, Monthly Weather Review, 91(3) 99-164.

Stelling, G.S., 1999, Personal Communication.

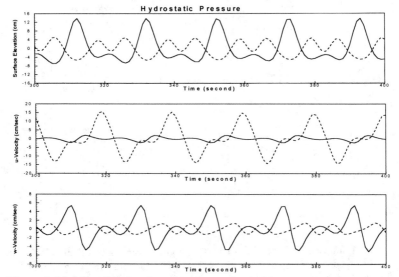

Figure 2 Model results of surface elevations and velocities at two locations. Non-hydrostatic pressure was not included in the simulation.

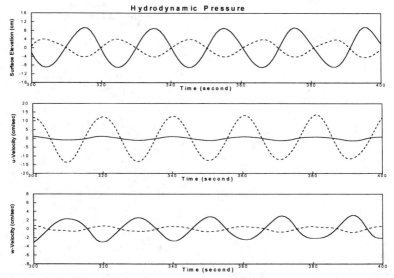

Figure 3 Model results of surface elevations and velocities at two locations. Non-hydrostatic pressure was included in the simulation.

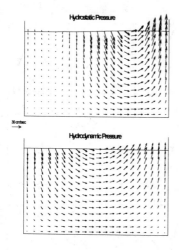

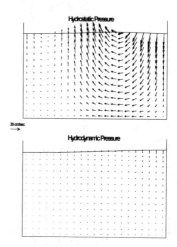

Figure 4 Simulated velocity fields with hydrostatic pressure (top graph) and fully hydrodynamic pressure (bottom graph) at time = 378 seconds.

Figure 5 Simulated velocity fields with hydrostatic pressure (top graph) and fully hydrodynamic pressure (bottom graph) at time = 384 seconds.

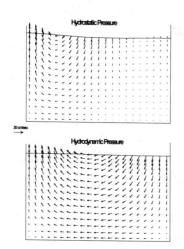

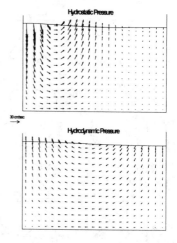

Figure 6 Simulated velocity fields with hydrostatic pressure (top graph) and fully hydrodynamic pressure (bottom graph) at time = 390 seconds.

Figure 7 Simulated velocity fields with hydrostatic pressure (top graph) and fully hydrodynamic pressure (bottom graph) at time = 392 seconds.

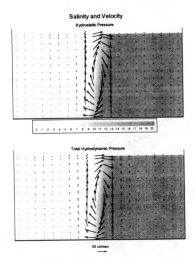

Figure 8 Simulated salinity and velocity for Case 2 with hydrostatic pressure (top) and hydrodynamic pressure (bottom) at t=10 seconds.

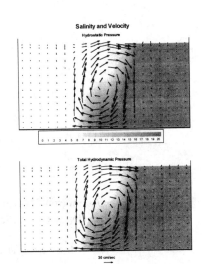

Figure 9 Simulated salinity and velocity for Case 2 with hydrostatic pressure (top) and hydrodynamic pressure (bottom) at t=30 seconds.

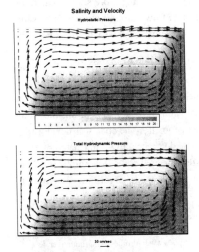

Figure 10 Simulated salinity and velocity for Case 2 with hydrostatic pressure (top) and hydrodynamic pressure (bottom) at t=90 seconds.

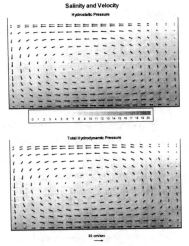

Figure 11 Simulated salinity and velocity for Case 2 with hydrostatic pressure (top) and hydrodynamic pressure (bottom) at t=1000 sec.

A Three–Dimensional Non–Hydrostatic Model for Free Surface Flows – Development, Verification and Limitations

H. Weilbeer[1] and J. A. Jankowski[2]

Abstract

The theoretical background of a new finite–element non–hydrostatic model for simulation of free surface flows based on the fractional step method and pressure decomposition is presented. One of the verification cases concerning the solitary wave propagation is provided. Further developments concerning more sophisticated turbulence modelling for practical applications as flows around structures or scour formation are discussed and illustrated with preliminary, but very promising results.

Introduction

The standard version of the code used for simulations presented in this paper, **Telemac3D**, solves numerically the three–dimensional shallow water equations for incompressible free surface flows. The typical application domains of this model are geophysical free surface flows with complex geometry. The code was developed in Laboratoire National d'Hydraulique (Electricité de France, EDF) in Chatou by Paris, based on the experience gained with a two–dimensional code **Telemac2D** (Galland, 1991). A description of the three–dimensional, hydrostatic algorithm (**Telemac3D**) is given by Janin et al. (1997).

[1] Institute of Fluid Mechanics and Computer Applications in Civil Engineering, University of Hannover, Appelstr. 9A, D–30167 Hannover, Germany

email: weilbeer@hydromech.uni–hannover.de   http://www.hydromech.uni–hannover.de

[2] Present adress:Federal Waterway Engineering and Research Institute, Kußmaulstraße 17, D–76187 Karlsruhe, Germany

email: jacek.jankowski@baw.de   http://www.baw.de

Telemac3D uses the finite element method for the numerical solution. The mesh consists of a constant number of tiers of prismatic elements over the whole computational domain, so that advantage of the σ–transformation can be taken. The interpolation functions are linear, which is a good compromise between the approximation accuracy and the required computational effort. The algorithm is based on the *operator splitting* technique which yields a significantly modular algorithm structure. The code is fully vectorised using the element–by–element method and can be parallelised by domain decomposition. For solution of the linear equations appearing in various algorithm steps, iterative solvers are used.

The non–hydrostatic version of the code removes the previously existing limitations due to the hydrostatic approximation which is intrinsic in the three–dimensional shallow water equations. Two important aspects are addressed. First, the vertical acceleration in free surface incompressible flows is taken into consideration by solving the vertical momentum conservation equation. Second, the free surface computation method allows simulation of its movements without limitations typical for shallow water equations, as the restriction to long waves or to gentle slopes of the free surface and the bottom. In following the theoretical background of the non–hydrostatic model is outlined.

Non–hydrostatic model formulation

The time (t) dependend hydrodynamic equation set to be solved consists of the three–dimensional equation of motion (Navier–Stokes), the continuity equation, the transport equation for a tracer (temperature, salinity, passive effluent concentration), equation of state and an equation for the free surface position. It is formulated for geophysical free surface flows in the non–inertial, orthogonal Cartesian co–ordinate system (x, y, z) connected with the surface of the earth. z points vertically upward in the direction of $-\mathbf{g}$, the acceleration of gravity. The dependend variables are velocity $\mathbf{u} = (u, v, w)$ and the free surface position S, as well as the pressure p. In order to simplify the basic form of the equations, a number of assumptions, approximations and simplifications is met. The flow is assumed to be incompressible, so that the density ϱ can be obtained from a separate equation of state. The variations of density $\Delta\varrho$ around an average flow density ϱ_0 are assumed to be small, so that the Boussinesq approximation is valid. The eddy viscosity (or diffusivity) concept is introduced to deal with the fluid turbulence. Leaving the description of the free surface to the next sections, the set of the governing equations can be written as follows:

$$\frac{\partial \mathbf{u}}{\partial t} + \mathbf{u}\nabla\mathbf{u} + 2\Omega \times \mathbf{u} = -\frac{1}{\rho_0}\nabla p + \frac{\rho}{\rho_0}g + \nabla \cdot (\mathbf{v}\nabla\mathbf{u}) \qquad (1)$$

$$\nabla \cdot \mathbf{u} = 0 \qquad (2)$$

$$\frac{\partial T}{\partial t} + \mathbf{u}\nabla T = \nabla \cdot (\mathbf{v}_T\nabla T) + q_T \qquad (3)$$

$$\rho = \rho(T, s, c) \qquad (4)$$

In the equation of state (4) the density is a function of transported *active tracers*, as temperature T, salinity s and/or suspended matter concentration c, but not of the pressure. The tracer transport is described using the transport equations (the equation for temperature (3) is cited above) including a source term q_T. Various turbulence models can be applied in order to obtain the values for eddy viscosity $\mathbf{v}$ and eddy diffusivity $\mathbf{v}_T$. The non–inertiality of the co–ordinate system is taken into consideration by introducing the Coriolis force term, where Ω is the angular velocity of the earth's rotation relative to the inertial system fixed to the distant stars.

For free surface tracking the *height function* method is applied. This technique is based on treating the free surface directly as a moving boundary. The distance between the interface and a given reference level is calculated from a separate equation. Therefore, it requires that the free surface can be represented by a single–value function (*height function*) S (x, y, t) with respect to one of the co–ordinate directions. This approach offers a simple and robust method of simulating environmental free surface flows. However, the restriction to single valued functions exclude some classes of flows, as e.g. breaking surfaces, bubbles, drops.

Two most widely used equations for tracking the free surface, the *kinematic boundary condition* and the *conservative free surface equation* are implemented. The first equation is obtained from kinematic conditions concerning the free surface particles:

$$\frac{\partial S}{\partial t} + u_s\frac{\partial S}{\partial x} + v_s\frac{\partial S}{\partial y} - w_s = 0 \qquad (5)$$

where the suffix *s* indicates the velocity components at the free surface. The latter equation is obtained from integration of the continuity equation over the depth from the bottom z = –B(x,y) to the free surface z = S(x,y,t), using the kinematic boundary condition (5) and the impermeability condition at the bottom (similar to (5)). The obtained *conservative form of the free surface equation* is:

$$\frac{\partial S}{\partial t} + \frac{\partial}{\partial x} \int_{-B}^{S} u dz + \frac{\partial}{\partial y} \int_{-B}^{S} v dz = 0 \tag{6}$$

Both equations mentioned above are hyperbolic. The main advantage of the conservative equation is that it includes the proper boundary conditions at the bottom and at the free surface. This approach brings a method of finding the free surface location while automatically satisfying the mass conservation criterion. However, the kinematic boundary condition allows easier implementation of some specific boundary conditions, as e.g. non–reflecting ones.

The treatment of pressure is one of the most characteristic and important features of the realised algorithm and is discussed in detail. The main idea of the new non–hydrostatic model is to decompose the pressure into two physically interpretable parts, the *hydrostatic* and *hydrodynamic* pressures, the latter treated as a form of a correction to the former. In contrast to the internal flows, in the free surface flows the pressure terms in the momentum conservation equation (1) can be separated into two terms, consisting of the *hydrostatic pressure* p_H, which can be explicitly computed, and the *hydrodynamic (motion) pressure* π, which can be found e.g. by solving a Poisson pressure equation. Therefore, in following, the global pressure p is decomposed into a sum:

$$p = p_H + \pi \tag{7}$$

The hydrostatic pressure p_H can be computed from the free surface elevation and the local fluid density $\varrho(x,y,z)$ field by integrating the equation of hydrostatics

$$\frac{\partial p_H}{\partial z} = -\rho g \tag{8}$$

in the water column, which yields:

$$p_H = \int_z^S \rho g dz = \int_z^S (\rho_0 + \Delta\rho) g dz = \rho_0 g(S - z) + \rho_0 g \int_z^S \frac{\Delta\rho}{\rho_0} dz \tag{9}$$

As a result of this decomposition, free surface gradients independent of the density (*barotropic part*), horizontal gradients of the pressure resulting from density differences (*baroclinic part*) and gradients of the hydrodynamic pressure appear in the horizontal equations of (1) for u and v. In the vertical equation (for w) only the vertical gradient of the hydrodynamic pressure remains. Equation (1) transforms to:

$$\frac{\partial \mathbf{u}}{\partial t} + \mathbf{u}\nabla\mathbf{u} = -g\nabla_H S - g\nabla_H\left[\int_z^S \frac{\Delta\rho}{\rho_0} dz\right] - \frac{1}{\rho_0}\nabla\pi - 2\Omega \times \mathbf{u} + \nabla \cdot (\mathbf{v}\nabla\mathbf{u})$$

(10)

The hydrodynamic pressure π can be found from a pressure Poisson equation. Due to the algorithm structure of the model described here, a special form of this equation is applied, namely the *pressure Poisson equation from fractional step formulation*. This equation is obtained from the *time–discretised* form of the Navier–Stokes equations. In the solution algorithm the velocity time derivative is treated explicitly and can be split into:

$$\frac{\partial \mathbf{u}}{\partial t} = \frac{\mathbf{u}^{n+1} - \tilde{\mathbf{u}}}{\Delta t} + \frac{\tilde{\mathbf{u}} - \mathbf{u}^n}{\Delta t}$$

(11)

where $\tilde{\mathbf{u}}$ is an intermediate solution for the velocity field, which does not need to satisfy the incompressibility condition. In this way equation (10) can be transformed into an equation containing all terms but hydrodynamic pressure gradients and a second one containing them exclusively:

$$\frac{\tilde{\mathbf{u}} - \mathbf{u}^n}{\Delta t} + \mathbf{u}\nabla\mathbf{u} = -g\nabla_H S - g\nabla_H\left[\int_z^S \frac{\Delta\rho}{\rho_0} dz\right] - \frac{1}{\rho_0}\nabla\pi - 2\Omega \times \mathbf{u} + \nabla \cdot (\mathbf{v}\nabla\mathbf{u})$$

(12)

$$\frac{\mathbf{u}^{n+1} - \tilde{\mathbf{u}}}{\Delta t} = -\frac{1}{\rho_0}\nabla\pi$$

(13)

Taking into consideration that the resulting field $\mathbf{u}^{n+1}$ must fulfil the incompressible continuity equation (2), the following form of the Poisson equation for the hydrodynamic pressure can be derived from (13):

$$\nabla^2\pi = \frac{\rho_0}{\Delta t}\nabla \cdot \tilde{\mathbf{u}}$$

(14)

Consequently, the equation (13) can be used to find the final (and divergence–free) velocity field at the time level n+1.

The boundary conditions for the Poisson equation for the hydrodynamic pressure (14) require attention. For example, it should be realised that setting the imposed (Dirichlet) value of zero for hydrodynamic pressure at a boundary has the physical meaning of applying purely hydrostatic pressure there. Hydrostatic pressure means no fluid motion, or that hydrostatic approximation is assumed to be valid. The hydrostatic approximation $\pi = 0$ is an acceptable boundary condition for open *inflow boundary sections*, where the velocity is thoroughly defined and its divergence is zero. For viscous flows, at the open boundaries and at free surface, the dynamic boundary conditions can be implemented:

$$\pi = \rho \mathbf{vu} \cdot \nabla \mathbf{u}_n \qquad (15)$$

where outside pressures and stresses (e.g. wind) are neglected for simplicity and **n** represents the normal vector to the boundary. Physical reasoning yields that for those *outflow sections* where $\mathbf{n} \cdot \nabla \mathbf{u}_n = 0$, the expression $\mathbf{n} \cdot \nabla \pi = 0$ is also valid. A Neumann boundary condition for the pressure Poisson equation (14) at the *solid walls* or at the *bottom* can be obtained from the equations (13) and the impermeability condition for the final velocity $\mathbf{n} \cdot \mathbf{u}^{n+1} = 0$, which yields:

$$\mathbf{n} \cdot \nabla \pi = \frac{\rho_0}{\Delta t} \bar{\mathbf{u}}_n \qquad (16)$$

Condition (16) imposes another constraint upon the hydrodynamic pressure field to be obtained from (14). Namely, it should provide not only the final solenoidal field but also the fulfillment of the impermeability condition at the solid walls. The disadvantage of this method lies in fact that it requires a very good approximation of pressure derivatives at boundaries. However, the values of $\bar{\mathbf{u}}_n$ are set to 0 by already applying the incompressibility boundary condition in the hydrostatic part of equation (12). In this case, the boundary condition (16) transforms to $\mathbf{n} \cdot \pi = 0$.

The solution algorithm

The algorithm applied in this model is based on the family of methods known under the common name of *decoupled methods*, where the solution of equation (1) is obtained in consecutive stages. The main idea of these methods is to solve sequentially a number of smaller, linear equation systems instead of an iterative solution of a larger, usually non-linear and slowly converging one. The free surface equation is also solved in a separate step. A number of these methods appear under various names, as *splitting schemes* (Galland (1991)), *fractional step schemes* (Quartapelle (1993)), *projection methods* (Chorin (1968), Gresho (1990), Shen (1993)) or *pressure methods* (Bulgarelli (1984), Casulli (1995)).

The solution is obtained in subsequent stages (*fractional step*) treating equations split into parts which have well–defined mathematical properties, so that the most adequate methods for a given differential operator type can be used (*operator–splitting*). In the particular case of the finite–element method, the decoupled algorithm structure allows application of equal–order linear interpolation functions for all variables. It is presumed, that the *Ladyzhenskaya–Babuska–Brezzi condition* (LBB–condition, Brezzi (1991)) is circumvented by the fact that the decoupled methods do not require an incompressibility condition in the form of an explicit equation $\nabla \cdot \mathbf{u} = 0$ occurring in the global equation set. The incompressibility is asymptotically achieved by the convergence of a pressure equation solution.

The time derivatives of the variables are split into fractional steps with respect to the mathematical operator properties and treated with appropriate numerical methods:

$$\frac{\partial \mathbf{u}}{\partial t} = \frac{\mathbf{u}^{n+1} - \mathbf{u}^d}{\Delta t} + \frac{\mathbf{u}^d - \mathbf{u}^a}{\Delta t} + \frac{\mathbf{u}^a - \mathbf{u}^n}{\Delta t} \tag{17}$$

In the first step of the non–hydrostatic algorithm the hydrodynamic pressure is excluded and only the hydrostatic pressure part is taken into consideration. In this stage (advection–diffusion), the intermediate solution for the velocity field $\mathbf{u}^d = \bar{\mathbf{u}}$ is obtained:

$$\frac{\mathbf{u}^d - \mathbf{u}^n}{\Delta t} + \mathbf{u}\nabla\mathbf{u} = \nabla \cdot (\nu\nabla\mathbf{u}) + \mathbf{F_u} \tag{18}$$

where the source terms $\mathbf{F_u}$ are:

$$\mathbf{F_u} = -g\nabla_H S^n - g\nabla_H \left[\int_z^{S^n} \frac{\Delta\rho^n}{\rho_0} dz \right] - 2\Omega \times \mathbf{u}^n + \mathbf{q_u} \tag{19}$$

According to the operator–splitting scheme, this stage is realised in two substeps, advection (hyperbolic) step and diffusion (parabolic) step, when the method of characteristics for advection, and semi–implicit standard Galerkin FEM for diffusion is required:

$$\frac{\mathbf{u}^a - \mathbf{u}^n}{\Delta t} + \mathbf{u} \cdot \nabla\mathbf{u} = 0 \tag{20}$$

$$\frac{\mathbf{u}^d - \mathbf{u}^a}{\Delta t} = \nabla \cdot (\nu\nabla\mathbf{u}) + \mathbf{F_u} \tag{21}$$

When the semi–implicit streamline upwind Petrov–Galerkin (SUPG) FEM for advection is chosen, the advection and diffusion steps are realised simultaneously. In general, the intermediate velocity field $\mathbf{u}^d$ is not solenoidal and its divergence is non–zero. Its value yields the source term for the (elliptic) Poisson equation for the hydrodynamic pressure (14), which is solved in the next algorithm stage (*continuity stage*) using standard Galerkin FEM with appropriate boundary conditions.

Consequently, the intermediate solution yielded by the advection–diffusion steps is corrected by the non–hydrostatic component. This component is computed from the hydrodynamic pressure gradients (equation (13)) under the assumption that the resulting final velocity must be divergence–free in the entire domain (incompressible flow) and satisfies appropriate boundary conditions:

$$\mathbf{u}^{n+1} = \mathbf{u}_d - \frac{\Delta t}{\rho_0} \nabla \pi \tag{22}$$

In this step the formal *velocity projection* (to the space of divergence–free vectors) is performed.

Finally, the free surface is found solving alternatively (5) or (6), which are hyperbolic equations solved with the method of characteristics or the semi–implicit SUPG FEM. The position of the advancing surface is tracked using a Lagrangian approach, where the mesh is adapted to the position of the free surface at each time step (so–called σ–mesh).

Model verification

Verification of the algorithm has been performed using a number of benchmark test cases covering the targeted model application domain. They include free surface and internal waves, sub– and supercritical channel flow over a steep ramp, wind– and buoyancy–driven currents (Jankowski (1998)). One of the most simple, but impressive test cases is solitary wave propagation in a long channel. The solitary wave, being a non–linear wave of finite amplitude, cannot be described properly in the framework of the shallow water equations. A solitary wave is a single elevation of water surface above an undisturbed surrounding, which is neither preceded nor followed by any free surface disturbances. Neglecting dissipation, as well as bottom and lateral boundary shear, a solitary wave travels over a horizontal bottom without changing its shape and velocity. The accuracy of the model can be evaluated by comparing the amplitude and celerity of the wave with its theoretical values, as well as by observing the conservation of the wave profile as it travels.

There are numerous analytical studies of this form of non–linear finite–amplitude wave. The first approximation provided by Laitone (1960) is the most frequently used for comparative studies. For a vertical section of an infinitely long channel of an undisturbed depth h (z=0 at the undisturbed surface), the following approximate formulae for velocity components u, w, free surface elevation η, pressure p and wave celerity c of a solitary wave with a height of H are valid:

$$u = \sqrt{gh}\frac{H}{h}\operatorname{sech}^2\left[\sqrt{\frac{3}{4}\frac{H}{h^3}}(x - ct)\right] \tag{23}$$

$$w = \sqrt{3gh}\left(\frac{H}{h}\right)^{\frac{3}{2}}\left(\frac{z}{h}\right)\operatorname{sech}^2\left[\sqrt{\frac{3}{4}\frac{H}{h^3}}(x - ct)\right]\tanh\left[\sqrt{\frac{3}{4}\frac{H}{h^3}}(x - ct)\right] \tag{24}$$

$$\eta = h + H\operatorname{sech}^2\left[\sqrt{\frac{3}{4}\frac{H}{h^3}}(x - ct)\right] \tag{25}$$

$$p = \rho g(\eta - z) \tag{26}$$

$$c = \sqrt{g(H + h)} \tag{27}$$

Following the test cases provided by Ramaswamy (1990), a solitary wave described by the formulae (23)–(27) is applied in a long channel as an initial condition, and the behaviour of the solution is observed thereafter. The simulation is performed in a finite domain, so that care must be taken choosing the initial position of the wave crest in the channel. The effective wave length λ concept is applied. λ is equal to twice the length between the wave crest and a point where the free surface elevation is $\eta(x)=0.01H$. According to Laitone:

$$\lambda = 6.9\sqrt{\frac{h^3}{H}} \tag{28}$$

A long channel 600m long and 6m wide, with a constant depth of h=10 m is taken. The mesh is 6 elements wide and 600 long, with a resolution in the direction parallel to the channel axis of 1 m. It consists of 4210 nodes and 7206 elements. The three–dimensional mesh has 11 equidistantly distributed levels. Inviscid flow without shear on the walls and bottom is assumed. All boundaries are impermeable. As the initial condition the hydrostatic approximation given by formulae (23)–(27) is applied, with a wave height of H=2m (H=0.2h) and the initial crest position is $\lambda/2 = 80$m away from the channel end, according to (28). The time step is taken as constant, $\Delta t=0.1$s, and the simulation time 40s (Courant number in the direction of wave propagation from 0.2 to about 1.0 at the wave crest).

For computation of the free surface elevation in the non–hydrostatic case the semi–implicit (Crank–Nicholson coefficient $\theta=0.55$) or implicit SUPG methods based on the kinematic boundary condition or the conservative free surface equation are applied. The hydrodynamic pressure is set to zero at the free surface and free Neumann BCs at all other boundaries are imposed.

A comparison between the hydrostatic and the non–hydrostatic solution is provided in figure 1. In contrast to the non–hydrostatic model, the hydrostatic one does not conserve the shape and amplitude of the solitary wave as it travels.

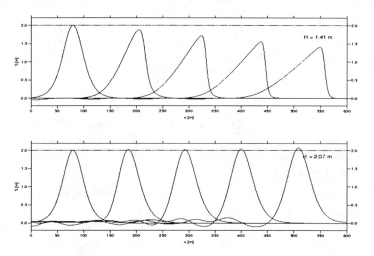

Figure 1: Solitary wave propagation in a long channel with a flat bottom. Above: hydrostatic solution. Below: non–hydrostatic solution found using free surface computation method based on the kinematic BC solved with implicit SUPG. The free surface profiles shown for t=0s, 10s, 20s, 30s, and 40s.

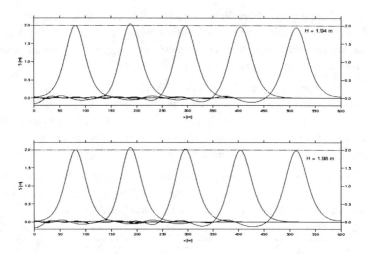

Figure 2: Solitary wave propagation in a long channel with a flat bottom for free surface computation method based on conservative free surface equation solved with SUPG. Above: implicit, below: semi–implicit θ=0.55. The free surface profiles shown for t=0s, 10s, 20s, 30s, and 40s.

The free surface schemes based on the kinematic BC show dispersive properties with ever–growing oscillations of the free surface behind the wave, while the conservative free surface algorithm is not sensitive to such effects. They are also much more sensitive to the influence of implicitness factor θ than the free surface conservative equation (Figure 2).

There are only a few restrictions limiting the application domain of the non–hydrostatic free surface model presented in this paper. Due to the σ–mesh structure, where the mesh nodes must be situated exactly along a vertical line, not all arbitrarily choosen three–dimensional geometries can be reproduced. Because of the limitations of the height function method, which requires that the free surface must be described by a single–valued function, breaking waves cannot be simulated.

Nevertheless, the ability to model three–dimensional hydrodynamical processes in the vicinity of structures is achieved for several geometries. But, in contrast to the simulations used for verification of the non–hydrostatic code which were carried out with simplest turbulence models, this aspect attracts more attention here. The second part of this paper sketches the preliminary considerations and first experience concerning the more sophisticated turbulence modeling.

Hydrodynamics in the vicinity of structures

Hydrodynamical processes in the vicinity of structures must be investigated with larger care for details. In technically relevant problems, such flows are always three–dimensional and indicate spatially strongly varying coherent turbulent structures. The horizontal and vertical scales of these flows may be of the same order of magnitude. Due to the large vertical velocities and accelerations hydrodynamical models based on the hydrostatic assumption cannot be applied.

The horseshoe vortex at the toe of a body which is induced by the stagnation pressure at the front, and the coherent turbulent structures (flow separation, periodic vortex shedding) at its rear belong to the most important and most interesting physical phenomena of flows around structures (i.e. vertical piles and abutments). They are also of practical interest. Typical technical problems directly connected with these flows are e.g. the flow and/or wave induced dynamic load of a structure or vibrations induced by oscillating loads or alternate separation processes (fluid–structure interaction). Furthermore the scour processes induced by these flows in the vicinity of structures are of vital importance in civil engineering.

Turbulence Modeling

The modeling of these phenomena is demanding for the hydrodynamical–numerical model, in particular turbulence modeling. Two totally different approaches are possible. On one hand statistical turbulence models based on the Reynolds averaged Navier–Stokes equations (RANS) are popular; on the other, large eddy simulations (LES) based on spatial filtered Navier–Stokes equations have aroused growing interest lately.

Statistical turbulence models are often applied for this task, but they are not suitable because they yield both the periodic and the turbulent fluctuations and therefore overpredict the turbulent stresses. In the case of LES the turbulent effects, which are reproducible by a given mesh resolution, are directly simulated. Only the subgrid scale turbulent stresses are taken into consideration by a turbulence model. LES concept is very promising, but seems to be too expensive for technically relevant Reynold numbers. Recently Very Large Eddy Simulations (VLES) are under development (Speziale (1998)) which can possibly close the gap between RANS and LES.

In order to gather more experience in modeling turbulent phenomena in the vicinity of structures flow around a circular cylinder is simulated, because of the broad availability of the results from experimental (e.g. Sumer (1997)) and numerical investigations (e.g. Fröhlich (1998), Breuer (1998)) for comparisons.

Examples

As mentioned above in the standard version of **Telemac3D** the turbulent viscosities can be set to constant values, or obtained from a mixing length model or a k–ε turbulence model. In the first test case an experiment was simulated which was carried out in a large wave flume. A solitary wave with a wave height of H=0,80m passes a vertical pile (diameter d=0.70m) (Figure 3). This is one of the most straightforward cases from numerous experimental series, whereby the forces induced by breaking and non–breaking waves on the vertical pile were measured. In the model, in a first attempt constant eddy viscosities for horizontal momentum exchange were chosen, and a mixing length model for vertical viscosities. As a result two symmetrical recirculation zones occur in the rear of the cylinder, but neither flow separation nor horseshoe vortex (Sumer (1997)) were observed. The forces from the solitary wave on the pile are not yet calculated.

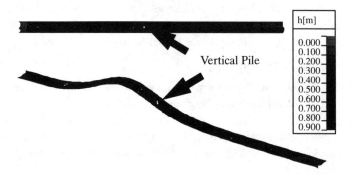

Figure 3: Solitary wave in the Large Wave Flume in Hannover.

In an next step two–dimensional simulations of a circular cylinder (d=0.10m) in a stationary flow field were carried out with **Telemac2D** in order to test different components of LES. The mesh resolution is refined nearby the walls, and no–slip boundary conditions for the velocities at the cylinder are prescribed. A two–dimensional version of the Smagorinski turbulence model is applied. It seems that the two–dimensional simulations gives a good representation of the quasi–two–dimensional mechanisms of the flow separation and the periodic vortex shedding motion (Bouris (1999), Sun (1996)). The Strouhal number of this period is in a good agreement with experimental data (St=0.21), as well as the Strouhal number of a square cylinder (St=0.14), for exact the same modelling conditions. The occurrence of a vortex street depends strongly on the grid resolution (Figure 4).

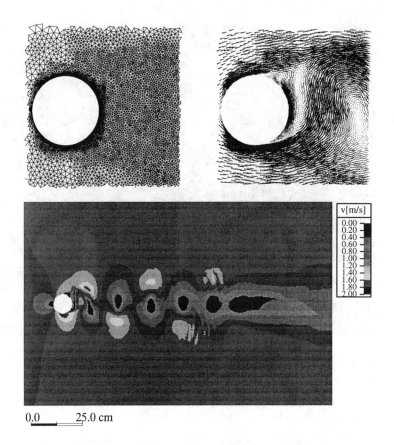

0.0   25.0 cm

Figure 4: Zoom of the computational mesh and resulting flow velocities of a two–dimensional flow around a circular cylinder (Re=100.000, St=0.21).

The last example is included in order to illustrate the direction of future developments (Figure 5). In the same model setting as described above the flow field is calculated with the three–dimensional hydrostatic code. Sediment transport is calculated as well. A horseshoe vortex occurs which is responsible for the scour pattern in the front of the cylinder. However, by first application tests using the non–hydrostatic version, the horseshoe vortex disappears, probably due to incompatible boundary conditions. The future developments regarding three–dimensional turbulence modeling will hopefully improve the results, so it will be possible to simulate flow and sediment transport in the vicinity of structures under nonstationary conditions.

Figure 5: Horseshoe vortex in front of a circular cylinder and calculated scour pattern. The flow field and the related sediment transport is calculated using the standard code of **Telemac3D**.

Conclusions

The model presented in the paper has been developed for dealing with free surface flows, where the hydrostatic approximation is not appropriate. Only a few less important restrictions concerning the free surface shape and the boundary geometry remain. The model has been thoroughly verified using typical examples from its aimed application domain. The further developments are concentrated on more sophisticated turbulence modelling, with the accent on large eddy simulation (LES). Introductory tests using the two–dimensional model version show that this approach yields promising results. However, turbulence is a three–dimensional phenomenon, so that the fully three–dimensional approach is required.

References

[1]   Bouris, D., Bergeles, G., 1999. 2D LES of vortex shedding from a square cylinder. *J. Wind Eng. Ind. Aerodyn.*, Vol. 80, 31–46.

[2]   Breuer, M., 1998. Large eddy simulation of the subcritical flow past a circular cylinder: Numerical and modeling aspects. *Int. J. for Numerical Methods in Fluids*, Vol. 28, 1281–1302.

[3]   Brezzi, F., Fortin, M., 1991. Mixed and hybrid finite element methods. Springer–Verlag, Berlin.

[4]   Bulgarelli, U., Casulli, V., Greenspan, D., 1984. Pressure methods for the numerical solution of free surface fluid flows. Pineridge Press, Swansea, U.K.

[5]   Casulli, V., Stelling, G.S., 1995. Simulation of Three–Dimensional, Non–Hydrostatic Free–Surface Flows for Estuaries and Coastal Seas. Estuarine and Coastal Modeling, Proceedings of the 4th International Conference, 1–12.

[6]   Chorin, A., 1968. Numerical solution of the Navier–Stokes equations. *Math. Comp.*, 22, 745–762.

[7]   Fröhlich, J., Rodi, W., Kessler, Ph., Parpais, S., Bertoglio, J.P. und Laurence, D., 1998. Large Eddy Simulation of Flow around Circular Cylinders on Structured and Unstructured Grids, in E.H. Hirschel (ed.): Notes on Numerical Fluid Mechanics, Vieweg–Verlag.

[8]   Galland, J.–C., Goutal, N., Hervouet, J.–M., 1991. TELEMAC: A new numerical model for solving shallow water equations. *Adv. Water Resources*, 14 (3), 138–148.

[9]   Gresho, P., 1990. On the theory of semi–implicit projection methods for viscous incompressible flows and its implementation via finite element method that also introduces a nearly consistent mass matrix. Part 1: Theory. *International Journal for Numerical Methods in Fluids*, 11, 587–620.

[10]   Janin, J.–M., Marcos, F., Denot, T., 1997. Code TELEMAC–3D–Version 2.2. Note théorique. Tech. Rep. HE–42/97/049/B, Electricité de France (EDF–DER), Laboratoire National d'Hydraulique.

[11]   Jankowski, J.A., 1998. A non–hydrostatic model for free surface flows. Dissertation, Institut für Strömungsmechanik und Elektronisches Rechnen im Bauwesen, Universität Hannover.

[12]   Laitone, E., 1960. The second approximation to cnoidal and solitary waves. *Journal of Fluid Mechanics*, 9, 430–444.

[13]   Quartapelle, L., 1993. Numerical solution of the incompressible Navier–Stokes equations. Birkhäuser, Berlin.

[14]   Ramaswamy, B., 1990. Numerical simulation of unsteady viscous free surface flow. *Journal of Computational Physics*, 90, 396–430.

[15]   Shen, J., 1993. A remark on the projection–3 method. *International Fournal for Numerical Methods in Fluids*, 16, 249–253.

[16]   Speziale, C. G., 1998. Turbulence Modeling for Time–Depending RANS and VLES: A Review. *AIAA Journal*, Vol. 36, No.2.

[17]   Sumer, B.M., Christiansen, N., Fredsøe, J., 1997. The horseshoe vortex and vortex shedding around a vertical wall–mounted cylinder exposed to waves. *J. Fluid Mechanics*, Vol. 332, 41–70.

[18]   Sun, X., Dalton, C., 1996. Application of the LES method to the oscillating flow past a circular cylinder. *Journal of Fluids and Structures*, Vol. 10, 851–872.

Developing a Capability to Forecast Coastal Ocean Optics: Minerogenic Scattering

Timothy R. Keen[1] and Robert H. Stavn[2]

Abstract

We are developing a detailed three-dimensional optical hindcast for a study site at Oceanside, California during October 1995. This study uses three steps to evaluate the prediction of coastal scattering: (1) a bio-optical model computes organic and inorganic (minerogenic) scattering coefficients using measured scattering data and in-situ chlorophyll measurements; (2) a numerical model calculates suspended sediment profiles. (3) Model-predicted distributions of sand-size sediment are used to calculate the scattering coefficients, which are then compared to values calculated by the bio-optical model. This study demonstrates that the near-bed optical field is strongly influenced by minerogenic particles even when meteorological forcing is weak. Furthermore, the time-dependent optical field is highly dependent on the variation of the suspended sediment concentrations above the bed.

1. Introduction

We are investigating the problem of the parameterization of optical Type II water, which Gordon and Morel (1983) define to be significantly affected by minerogenic matter. The optical properties of Type II, coastal ocean water change rapidly because of surface waves, advection, and internal waves or solitons (Weidemann et al., 1997a, b, 1998). This physical forcing resuspends bottom sediments. Furthermore, the optical effects of minerogenic scatterers and organic scatterers differ because of differences in the volume scattering function (Stavn and Weidemann, 1994; 1997; Weidemann et al., 1995). All of this has profound effects on

---

[1] Naval Research Laboratory, Oceanography Division, Stennis Space Center, MS 39529

[2] Department of Biology, University of North Carolina at Greensboro, Greensboro, NC 27402-6174

attempts to invert radiances and irradiances to obtain optical properties and concentrations of materials such as suspended sediment and chlorophyll. Unfortunately, data on suspended minerogenic matter are not routinely collected and this presents special difficulties in studying coastal waters.

One approach to this problem is to simulate the hydrodynamic and sedimentological nearshore environment using numerical models, which utilize bottom sediment properties and wind and wave forcing, to predict the three-dimensional (3D) concentration of resuspended sediment. This approach applies fundamental principles of physics to the problem of modeling and simulation. Thus, we are developing a "bottom up" model of optical variability, which complements "top down" approaches that focus on the activities of living cells as they are fueled by solar radiation.

2. Approach

This study utilizes two models. The first is a coupled bio-optical model, which uses observed chlorophyll to calculate the organic contribution to total scattering. The second model is a coupled, sedimentation-optical numerical model (TRANS98; Keen and Glenn, 1998), which predicts minerogenic scattering coefficients. There are two components to this model: (1) a sedimentation model for predicting the profile of suspended sediments; and (2) a light-scattering model to calculate the scattering coefficient for a given distribution of inorganic sediment grains.

A fundamental assumption of this approach is that biological and minerogenic scattering add linearly to the total scattering B_T field measured by in-situ instruments (Jerlov, 1976).

$$b_T = b_o + b_m \qquad (1)$$

where b_o is organic scattering and b_m is minerogenic scattering. The bio-optical model is used to calculate the biological scattering coefficients. These are then subtracted from the measured total scattering coefficients. The resulting values of minerogenic scattering coefficients are treated as "ground truth". The sedimentation-optical model predictions are then compared to the "ground truth" estimates.

2.1. Bio-optical model

The model utilized here is a simple parameterization based on chlorophyll concentration first described by Weidemann et al. (1995). This parameterization was derived for Type I, clear ocean water, and is applied here with the assumption that Type II water can be described optically by summation of the organic and minerogenic components. It has already been demonstrated that the absorption coefficient for Type II waters can be predicted from this chlorophyll parameterization (Stavn and Weidemann, 1997). What we are investigating here is the fact that Type I model parameterizations cannot predict the total scattering component of the optical field, primarily because minerogenic matter is not accounted for in Type I optical parameterizations. The scattering coefficient determined from the bio-optical model is totally organic and divided into an algal component b_a, a bacterial component b_{ba}, and an organic detritus component b_{od}:

$$b_o = b_a + b_{ba} + b_{od} \qquad (2)$$

The algal scattering is simply the specific scattering coefficient b^* (m^2/mg) at 532 nm (10^{-9} m) multiplied by the chlorophyll concentration [Chl] expressed as (mg Chl)/m^3. Specific scattering coefficients were obtained from Ahn et al. (1992) and Bricaud et al. (1988). The bacterial scattering is obtained from an estimate of bacterial cell numbers per m^3 parameterized to chlorophyll concentration (Cole et al., 1988). The cell numbers are then used to calculate the bacterial total scattering coefficient from Mie theory, assuming a mean cell diameter of 0.59 μm (Bohren and Huffman, 1983, McCartney, 1976). The cell number estimates used here were 0.63 x 10^{12} cells/m^3 for [Chl] less than 1.0 mg/ m^3 while the estimate was 0.91 x 10^{12} $[Chl]^{0.52}$ cells/m^3 for [Chl] greater than or equal to 1.0 mg/ m^3. The calculation of the scattering coefficient for organic detritus was carried out with the following formula adapted from the results of Prieur and Sathyendranath (1981):

$$b_{od} = \frac{[(a_{od}/a_{od}{}^{*\prime}) + 0.00029]}{0.42} \qquad (3)$$

where a_{od} is the absorption coefficient for suspended organic detritus at 532 nm, obtained from Roesler (1995), while $a_{od}{}^{*\prime}$ is the specific absorption coefficient for suspended organic detritus at 532 nm, normalized to the scattering coefficient for organic detritus (Prieur and Sathyendranath, 1981). We were fortunate to have Roesler's direct measurements for organic detritus absorption, which allowed the use of the simple and direct equation (3). The formula was a normalization of the scattering coefficient of organic detritus to a large suite of chlorophyll and hydrosol scattering data by Prieur and Sathyendranath (1981). The calculations from this formula were near the middle of a range established for standard chlorophyll normalization for organic detritus.

2.2. Sedimentation model

The wave-current interaction bottom boundary layer model (BBLM) described by Keen and Glenn (1998) is used to calculate suspended sediment profiles. This model is an extended version of the suspended-sediment-stratified BBLM of Glenn and Grant (1987). The model computes sediment concentrations for a number of size bins. The model uses a logarithmic axis stretching for the vertical dimension. The resulting resolution near the bed is less than 0.01 m.

A critical value in calculating the suspended sediment profile is the reference concentration, given by

$$c_n = c_{bn} \frac{\gamma_0 S_n}{1 + \gamma_0 S_n} \tag{4}$$

where c_{bn} is the bed concentration, γ_0 is a resuspension coefficient, and S_n is the excess skin friction (Glenn and Grant, 1987). Recent work in sandy substrates suggests that γ_0 has a value of 0.003 (Styles, 1998). The bed concentration for a specific size class is given initially by the assumed grain size distribution. If the resuspension depth exceeds the active layer thickness, c_{bn} is reduced to simulate armoring. New profiles of currents, sediment concentrations, and the stability parameter are then calculated at the grid point. This reduction procedure is repeated until the resuspension depth of the size class does not exceed the active layer.

The concentration of suspended sediment above the bed is dependent on the eddy diffusivity K_s. An important parameter therefore is the ratio of the eddy diffusivity to the eddy viscosity, $\beta = K_s/K_{mo}$. Hill et al. (1988) examined this parameter and suggested that it might be size dependent. They derived the following relationship:

$$\beta = 1.45 \left(\frac{w_n}{ku_*} \right)^{0.33} \tag{5}$$

where w_n is fall velocity of size n, k is von Karman's constant, 0.4, and u_* is the shear velocity. This formulation was examined for the California study site. The BBLM uses an eddy diffusivity parameter $\gamma = 1/\beta$. A fixed value of 0.74 is typically used but recent work also suggests a size dependence (Styles, 1998).

2.3. Optical scattering model

The optical scattering model is applied only to the suspended sediment distributions predicted by TRANS98. Since no silt or clay is present in the Oceanside study area, we assume the sediment is quartz. Further, we assume that these siliceous particles are approximately spherical. Thus, Mie theory can be applied (Shifrin, 1988). We use a slightly modified version of Wiscombe's Fortran code (Wiscombe, 1979). Since we are dealing with a distribution of sizes, we apply a polydisperse Mie calculation (McCartney, 1976). The wavelength used in this investigation is 532 nm. We determine the total scattering cross section (m^2) for every sediment size class that is in suspension:

$$\sigma_P = 2\pi \int_0^\pi \sigma_P(\theta) \sin\theta \, d\theta \tag{6}$$

where σ_p is the total scattering cross section for a single particle. We then determine the efficiency factor, the ratio of the scattering cross section to the projected cross sectional area of a particle.

$$Q_{sc} = \frac{\sigma_p}{\pi r^2} \tag{7}$$

Q_{sc} is the efficiency factor and r is the radius of the particle. We then determine the total scattering coefficient for particles of a given radius:

$$b = N\sigma_p = N\pi r^2 Q_{sc} \tag{8}$$

where b is the total scattering coefficient and N is the concentration of particles per m^3. When we have several size-classes (subscripted with i):

$$b = \sum_i N_i \pi r_i^2 Q_{sci} \tag{9}$$

In addition to the polydisperse Mie calculation outlined above, we have a simple regression relation for minerogenic scattering, based on the Mie size parameter and an average scattering efficiency (Haltrin et al, 1999). The regression is based on Mie variables calculated from the Wiscombe code. The scattering coefficients for this relation are compared with a polydisperse Mie calculation by Stavn et al. (1998).

The regression model is coupled to the sedimentation model through the 3D size-fractionated suspended sediment distributions predicted by TRANS98.

3. The Oceanside Study

The study area (figure 1) is located along the southwestern California coast of the U. S. Both optical and hydrographic surveys were conducted between 17 and 27 October 1995. Ocean conditions were mild during the field study (figure 2), with wave heights less than 1 m and surface currents less than 0.5 m/s. There were thus no significant meteorological or oceanographic events to dominate sediment resuspension.

3.1. Field measurements

Oceanographic data collection included ADCP profiles made while the ship was underway. Continuous timeseries of surface currents (solid line in figure 2), temperature, and bottom pressure near the coast were also measured. The pressure data were analyzed for the surface wave field (dotted line in figure 2) and astronomical tides. Time series of wind speed and direction were measured at the Oceanside Harbor (see figure 1 for location).

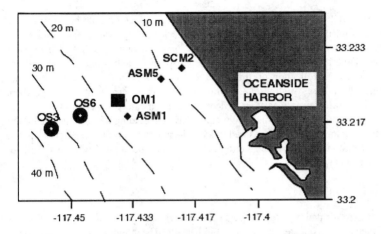

Figure 1. Map of study area and sedimentation-optical model grid.

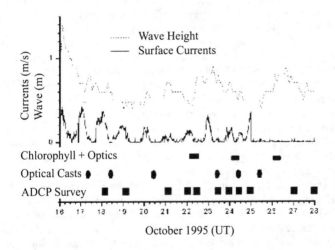

Figure 2. Overview of the Oceanside field study. The wave height was measured at the coast. The surface currents are indicative of coastal flow during the field effort. The ADCP surveys lasted approximately 4 hours. Optical casts were measured at some location within he field area during the time interval indicated. The same applies to the Chlorophyll + Optics measurements. The sizes of the symbols are indicative only.

The ADCP records for two intervals were selected to generate bottom currents for the sedimentation model. The profiles were binned at 1-m intervals. Figure 3 shows the currents from the lowest bin for the surveys beginning at 0000 UT October 18 and 0830 UT October 24. Each survey lasted approximately 4 hours.

Optical data collection occurred at the stations shown in figure 1. Times are indicated by ellipses in figure 2. Station OM1 was a permanent mooring where timeseries were also measured. Measurements included total scattering and absorption at nine wavelengths (Stavn et al., 1998), including 532 nm. In addition to scattering, chlorophyll was measured at some locations and times (see figure 2). The bio-optical model described in section 2.1 could only be applied when a full set of measurements was made.

3.2. Modeling approach

The sedimentation grid (figure 1) has a horizontal resolution of 200 m. The bathymetry was collected during the field study. The bottom currents and waves required to drive the TRANS98 model can be supplied by either timeseries or fields. This study uses two-dimensional (2D) fields for currents and timeseries for waves. Bottom currents come from ADCP surveys (figure 3) or 2D fields (figure 4) predicted by the Princeton Ocean Model (POM). The ADCP data are treated as instantaneous. Bottom current fields were interpolated to the sedimentation model grid. For this study, the ADCP fields are assumed to be more accurate, and were used to evaluate the sedimentation model sensitivity. The wave field was uniform over the model grid but time varying, using the timeseries data from the coastal pressure gauge (figure 2).

The southern California (SOCAL) POM model (Keen and Murphy, 1999) uses a Cartesian grid with cells of 903 m and 759 m along the x and y axes, respectively. The circulation model was spun up for 48 hrs and the hindcast began at 0000 UT 17 October. It was then run for the duration of the field experiment, ending on 0000 UT 28 October. The initial temperature and salinity fields were derived from a mean profile of all CTD data from 17 October. The winds measured at Oceanside Harbor were merged with the Navy Operational Global Atmospheric Prediction System (NOGAPS) wind analysis fields to produce wind fields for POM.

The sedimentation model contains 20 different size classes (bins) of sediment, ranging from 2 μm (10^{-6} m) to 3306 μm in diameter. Three sediment distributions were used (Inman, 1953). Distribution 1 has a mean of 49 μm, a standard deviation (SD) of 380 μm, and is located to a depth of 4 m. Distribution 2 has a mean of 110 μm, SD of 750 μm, and it is found to a depth of 30 m. The deepest distribution, with a mean of 82 μm and SD equal to 640 μm, is located in depths up to 100 m. No computations were performed for depths greater than 100 m.

The TRANS98 model can be run in either resuspension or advection mode. When resuspension mode is selected, no advection terms are computed and only equilibrium profiles are calculated. This study uses resuspension mode.

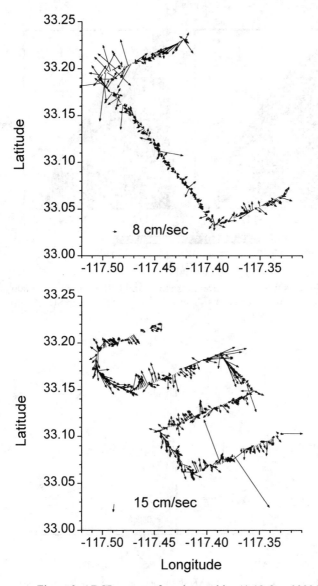

Figure 3. ADCP currents from lowest bin. A) 18 Oct. 0000 UT.
B) 24 Oct. 0832 UT.

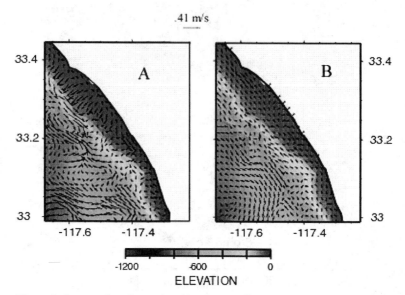

Figure 4. Currents from lowest level in the SOCAL POM. A) 18 October 0000 UT.
B) 24 October 0800 UT. The maps show a subregion of the entire SOCAL grid.

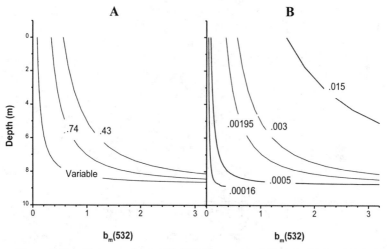

Figure 5. Sensitivity results for TRANS98 at station SCM2 for 24 October 0800 UT.
A) Eddy diffusivity parameter γ; resuspension coefficient coefficient = 0.003.
B) Resuspension coefficient γ_o; $\gamma = 0.43$.

4. Coupled sedimentation-optical model results

The sediment-optical model can only be compared to observations when both scattering and chlorophyll measurements were made, on 21, 24 and 26 October (see figure 2). Furthermore, model parameters were only adjusted on days when ADCP fields were available, limiting the best-run times to 21 and 24 October.

4.1. Model Sensitivity to γ and γ_o

There are no adjustable parameters in the optical scattering model discussed in section 2.3. Bottom type, wave fields, and steady currents are not adjustable model parameters. However, some physical parameters within the BBLM and sedimentation models are not well constrained although the model values are based on relevant field or laboratory studies. We will limit our tests to resuspension parameters that can be modified with some justification, and which are constrained by recent work. Thus, model sensitivity to the resuspension coefficient γ_o and the diffusivity parameter γ is examined.

The eddy diffusivity parameter γ has been proposed to be either constant, above unity, or below unity by various workers (see Hill et al., 1988 for a discussion). Note that Hill et al. use the inverse, β. The BBLM initially used a value of 0.74, which was determined from a stratified atmospheric boundary layer (Glenn and Grant, 1987). This value produces a moderate scattering profile (figure 5a), with b_m exceeding 1 m^{-1} within 1 m of the bottom. The smaller value of 0.43 suggested by Styles (1998) produces a much larger scattering profile. In light of the wide variation in γ reported by Styles (1998), we also examine the variable formulation in equation 5. This approach computes a different value for each size class in suspension at every time and location. The sediment at Oceanside comprised very fine sand to silt, with γ between 10 and 0.7 at station SCM2 at approximately 0800 UT October 24. The resulting scattering profile in figure 5a is lower than for either constant γ.

Field estimates of γ_o vary from 1.6×10^{-5} to 1.5×10^{-2}. The large range can be partly attributed to measurement uncertainty (Hill et al., 1988). Hill et al. proposed a value of 1.3×10^{-4}, which is smaller than the value of 1.8×10^{-3} suggested by Wikramanayake and Madsen (1992). Using the smallest γ_o reported in the literature, the model-predicted mineral scattering b_m at station SCM2 for 0800 UT October 24 (figure 5b) is 0.1 m^{-1}, at a height of 1 m above the bed. Using some of the larger reported values increases scattering, with a maximum profile when γ_o equals 0.015. Simulations for October 23 (not shown), when waves were significantly higher (see figure 2), reveal that for γ_o larger than 1.95×10^{-3}, suspended sediment stratification dampens turbulence and decreases scattering coefficients above the wave boundary layer. Consequently, predicted scattering above 6 m is lower for the largest γ_o than for a value of 5×10^{-4}.

4.2. Comparison with the bio-optical model

Before examining the sedimentation-optical model predictions of minerogenic scattering, it is useful to compare this model with the bio-optical model described in section 2.1. The optical measurements made at Oceanside were highly variable, in part because of environmental factors such as river inflow, local wind and turbulent mixing, internal waves, advection, and biological processes. In addition, there were some problems with the optical instruments. Consequently, the number of stations with quality optical, biological, and physical measurements is limited. The comparison is limited to model simulations using the ADCP currents only.

Separate profiles of b_o and b_m can be calculated using the bio-optical model. The scattering profiles at station OM1 (figure 6a) at approximately 0800 UT, October 24, are nearly equal below 8 m. The minerogenic scattering ranges from 0.2 m^{-1} at 9 m, to 0.35 m^{-1} at the surface. Note that there is no significant increase near the bed as would be expected with sediment resuspension. The sedimentation-optical model (dotted line) predicts a smooth distribution that matches the bio-optical model at 3 m. The quartz curve (solid line in figure 6a) from station ASM1 is more typical of a resuspension scattering profile. Note the similarity to the sedimentation model curve. The wave and currents from October 21 (figure 2) were weak and minerogenic scattering above the bottom was reduced (figure 6c). The TRANS98-predicted coefficients are within 30% of the bio-optical model values and follow a similar vertical distribution.

Based on this comparison with the bio-optical model, it seems that the suggested values for the resuspension coefficient and eddy diffusivity parameter from Styles (1988) are reasonable for the Oceanside coast. Thus, we use γ equal to 0.43 and γ_o equal to 3×10^{-3} in the following simulations.

4.3. Predicted minerogenic scattering

Forecasting the mineral backscattering requires the input of forecast current and wave fields. Herein we use a current field only, since measured waves from the field study supply the oscillatory currents to the BBLM. The "forecast" currents are supplied by the SOCAL POM bottom currents in figure 4. The currents from the ADCP (figure 3) are used for comparison. This discussion focuses on stations SCM2, ASM5, OM1, and OS6 (see figure 1 for locations).

October 18: Moderate currents and waves

The bottom currents from the northernmost end of the ship track (figure 3a) were about 0.15 m s^{-1} and the significant wave height was about 0.6 m. The resulting b_m at SCM2 (solid line in figure 7a) exceeds 1 m^{-1} to within 2 m of the surface. The scattering that is predicted with the POM currents (dashed line) is half this. Because of the effect of water depth on wave orbital parameters, scattering decreases by 40% at station ASM5 (figure 7b). At station OM1 (figure 7c), the currents from POM are

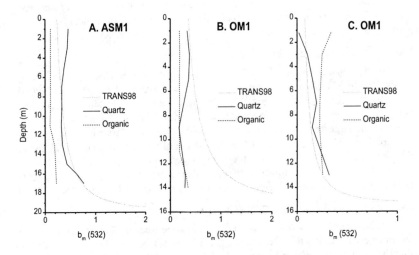

Figure 6. Comparison of scatter coefficients. A) Station ASM1 at 24 Oct. 0800 UT.
B) Station OM1 at 24 Oct. 0800 UT. C) Station OM1 at 21 Oct. 2030 UT.

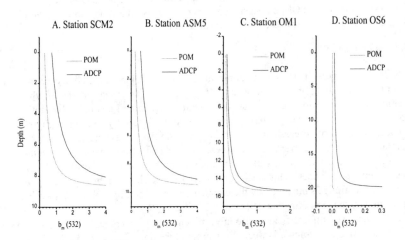

Figure 7. Scattering Coefficients at 18 Oct. 0000 UT.

larger relative to the ADCP currents and scattering is thus closer for the two simulations. Near-bottom scattering coefficients are nonetheless less than 0.1 m^{-1}. At station OS6 (figure 7d), the scattering calculated using the POM currents is nil.

October 23: Moderate currents and large waves

The significant wave height on 23 October was almost 1 m. Consequently, the b_m profiles at SCM2 (figure 8a) are very similar using both the ADCP and POM currents. The ADCP-generated profile is very near that from 18 October as well. This comparison reflects the wave dominance of the coastal bottom boundary layer. The effect of the larger waves is even more noticeable seaward. At station OM1 (figure 8c), predicted scattering exceeds 0.4 m^{-1} at a depth of 10 m, compared to 0.25 m^{-1} at this depth on 18 October. Despite the larger waves, the underprediction of bottom currents by POM results in lower scattering at this station. This error in POM currents is even greater at station OS6 (figure 8d), where no sediment is in suspension.

October 24: Weak currents and moderate waves

The surface currents near the coast (figure 2) were weak on 24 October and wave heights were about 0.6 m. The bottom currents from the ADCP's (figure 3b) were slightly lower on 18 October whereas the POM currents (figure 4b) were much lower. Because of the weaker flow, ADCP-driven scattering (figure 9a) was reduced even at station SCM2. The currents from POM did not decrease as much and scattering changed little. The decline in b_m using the ADCP currents was slightly less at ASM5 (compare figures 8b and 9b). This trend reversed at OM1 (figure 9c), where larger currents were encountered. This effect is further evidenced at station OS6 (figure 9d). The currents from the POM hindcast do not increase seaward and no similar pattern is predicted. Instead, scatter decreases to zero at station OS6.

5. Conclusions

This paper presents the results of recent efforts to develop a predictive capability for the optical field in coastal water. This effort has focused on the southern California coast. Part of the work has been the development of a coupled bio-optical model to calculate organic scattering from a measurable quantity like chlorophyll. A second direction of research has been the prediction of resuspended sand due to combined wave and current action. These two related efforts have been brought together to evaluate the validity of computing minerogenic scattering using a physics-based numerical model. This evaluation has been carried out by comparing the scattering coefficients predicted by the coupled sedimentation-optical model to values calculated by the bio-optical model.

The near-bed optical field is strongly influenced by minerogenic particles even when meteorological forcing is weak. Furthermore, the time-dependent optical field is highly dependent on the variation of the suspended sediment concentrations above the bed. This study shows that the coupled sedimentation-optical model can reasonably

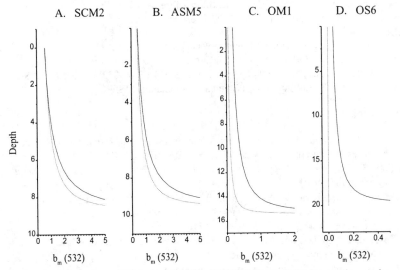

Figure 8. Scattering coefficients at 23 Oct. 0800 UT. The solid lines are the results using the ADCP currents. The dotted lines are the results using the SOCAL POM currents.

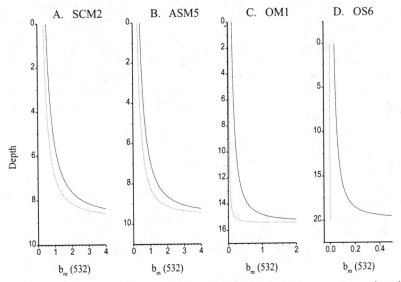

Figure 9. Scattering coefficients at 24 Oct. 0800 UT. The solid lines are the results using the ADCP currents. The dotted lines are the results using the SOCAL POM currents.

predict minerogenic scattering but, to develop a forecast capability, it is necessary to improve the hydrodynamic and wave models used to drive it.

Acknowledgments. The first author was funded by Program Element 62435N sponsored by the Office of Naval Research. The second author wishes to acknowledge the support of ONR Grant No. N00014-97-1-0872 and a Navy-ASEE Summer Faculty Fellowship held at Stennis Space Center, MS.

References

Ahn, Y. H., Bricaud, A., and A. Morel, Light backscattering efficiency and related properties of some phytoplankters, *Deep Sea Res.*, *39*, 1835-1855, 1992.

Bohren, C. F., and D. R. Huffman, *Absorption and Scattering of Light by Small Particles*, John Wiley, New York, 1983.

Bricaud, A., Bédhomme, A. L., and A. Morel, Optical properties of diverse phytoplanktonic species: experimental results and theoretical interpretation, *J. Plankt. Res.*, *10*, 851-873, 1988.

Cole, J. J., S. Findlay, and M. L. Pace, Bacterial production in fresh and saltwater ecosystems: a cross-system overview, *Mar. Ecol. Prog. Ser.*, *43*, 1-10, 1988.

Glenn, S. M. and W. D. Grant, A suspended sediment correction for combined wave and current flows, *J. Geophys. Res.*, *92*, 8244-8246, 1987.

Gordon, H. R., and A. Morel, Remote assessment of ocean color for interpretation of satellite visible imagery, in *Lecture Notes on Coastal and Estuarine Studies*, R.T. Barber, C.K. Mooers, M.J. Bowman, and B.Zeitschel, eds., 1-113, Springer Verlag, New York, 1983.

Haltrin, V. I., Shybanov, E. B., Stavn, R. H., and A. D. Weidemann, Light scattering and backscattering coefficients by quartz particles suspended in seawater, Int. Geosc. and Remote Sensing Symposium, Hamburg, Germany, 1420-1422, 1999.

Hill, P. S., R. M. Nowell, and P. A. Jumars, Flume evaluation of the relationship between suspended sediment concentration and excess boundary shear stress, *J. Geophys. Res.*, *93*, 12499-12509, 1988.

Inman, D. L., Areal and seasonal variations in beach and nearshore sediments at La Jolla, Calif., U.S. Beach Erosion Board Tech. Mem. No. 39, 1-82, 1953.

Jerlov, N. G., *Marine Optics*, Elsevier/North-Holland, New York, 231 pp., 1976.

Keen, T. R., and S. M. Glenn, Resuspension and advection of sediment during Hurricane Andrew on the Louisiana continental shelf, *5th Int. Conf. Estuarine and Coastal Model., Proc.*, 481-494, 1998.

Keen, T. R., and S. J. Murphy, Developing a relocatable coastal ocean forecast model. *Oceanology '99*, Singapore, April 27-29, 1999. Conf. Proc., 47-57, 1999.

McCartney, E. J., Optics of the Atmosphere, Scattering by Molecules and Particles, John Wiley & Sons, New York, 1976.

Prieur, L., and S. Sathyendranath, An optical classification of coastal and oceanic waters based on the specific spectral absorption curves of phytoplankton pigments, dissolved organic matter, and other particulate materials, *Limnol. Oceanogr., 26,* 671-689, 1981.

Roesler, C, Data report for Grant # N00014-95-1-G909. The influence of particulate and dissolved material on water clarity in the littoral zone, submitted to A. D. Weidemann, Naval Res. Lab., Code 7331, Stennis Space Center, MS 39529, 24 December, 1995.

Shifrin, K. S., *Physical Optics of Ocean Water,* 1-285, Am. Inst. Phys., New York, 1988.

Stavn, R. H., Keen, T. R., Haltrin, V. I., and A. D. Weidemann, Coastal ocean optics: hindcasting optical properties and variability from predicted minerogenic concentrations, Ocean Optics IV, Kailua-Kona, Hawaii, 10-13 Nov. 1998, Sponsors: ONR & NASA, Proc., 1998.

Stavn, R. H., and A. D. Weidemann, Coastal optical water type 2: modeling and minerogenic scattering, in *Ocean Optics XIII,* Steven G. Ackleson, Robert Frouin, Editors, Proc. SPIE 2963, 38-48, 1997.

Stavn, R. H., and A. D. Weidemann, Geometrical light field parameters for improving remote sensing estimates of the backscattering in the marine hydrosol, in *Ocean Optics XII,* Jules S. Jaffe, ed., Proc. SPIE 2258, 202-209, 1994.

Styles, R. B., A continental shelf bottom boundary layer model: development, calibration, and applications to sediment transport in the Middle Atlantic Bight, Ph.D. Thesis, Rutgers University, New Brunswick, New Jersey, 261 pp., 1998.

Weidemann, A. D., W. S. Pegau, and L. A. Jugan, The influence of internal waves on coastal optical properties, Aquatic Sciences Meeting, Am. Soc. of Limnol. and Oceanogr., Santa Fe, New Mexico, February 10-14, 1997a.

W eidemann, A. D., W. S. Pegau, L. A. Jugan, and T. E. Bowers, Tidal influences on optical variability in shallow water, in *Ocean Optics XIII,* Steven G. Ackleson, Robert Frouin, eds., Proc. SPIE 2963, 320-325, 1997b.

Weidemann, A. D., R. Holyer, J. Sandidge, W. S. Pegau, and L. A. Jugan, Detection of subsurface internal waves via remote sensing. Fifth Intl. Conf. for Remote Sensing of Marine and Coastal Environments, ERIM International, San Diego, Calif., October 1998.

Weidemann, A. D., R. H. Stavn, J. R. V. Zaneveld, and M. R. Wilcox, Error in predicting hydrosol backscattering from remotely sensed reflectance, *J. Geophys. Res., 100,* 13,163-13,177, 1995.

Wikramanayake, P. N., and O. S. Madsen, Calculation of suspended sediment transport by combined wave-current flows. Contract report DRP-92. Final report for the Dredging Research Program. U. S. Army Corps of Eng., Coastal Waterways Research Center, Vicksburg, Miss., 148 pp., 1992.

Wiscombe, W. J., Mie scattering calculations: advances in technique and fast, vector-speed computer codes, pp. 1-62, National Center Atmos. Res. Tech. Note TN-140+STR, 1979.

A Large Domain Convection Diffusion Based
Finite Element Transport Model

Norman W. Scheffner[1], Member, ASCE

Abstract

This paper describes the development and implementation of a convection-diffusion equation based transport model. The model has been applied to simulate both salinity distribution and intrusion as well as wave and current induced sediment transport and local bathymetry change. The current field is provided through coupling with the large domain long wave hydrodynamic model ADCIRC (ADvanced CIRCulation). The ADCIRC model is an unstructured grid finite element model which solves the Generalized Wave-Continuity Equation (GWCE) formulation of the governing equations in conjunction with the primitive momentum equations. The model was specifically developed to: 1) represent very large computational domains with the capability of providing high resolution details for both complex coastlines and areas of high bathymetric gradients, 2) properly represent all pertinent physics of the fully nonlinear governing equations, and 3) provide accurate and efficient simulations for intra-tidal periods extending from months to years.

The transport component of ADCIRC is formulated in the context of the full convection/diffusion equation written in a non-conservative form. The code is fully vectorized, uses a Crank-Nicolson time splitting scheme, and is solved with a preconditioned biconjugate gradient iterative solver. Transport is applied in two modes, fine grained sediment transport with associated erosion and deposition and salinity intrusion. Source and sink terms for sediment transport are formulated as a function of bed shear stress imbalance with defined critical shear stresses for erosion

[1]Senior Research Hydraulic Engineer, US Army Engineer Research and
Development Center's Waterways Experiment Station
3909 Halls Ferry Road, Vicksburg, MS 39180, USA

and deposition. The bed shear stresses are computed according to the formulation of Christoffersen and Jonsson for a combined wave-current environment. Salinity distribution is computed as a function of prescribed initial conditions, fixed offshore values, and defined salinity input at river sources.

An application of hydrodynamic and transport results are demonstrated for Willapa Bay, a coastal estuary located in the state of Washington on the west coast of the United States. The model is solved over a 17,661 node computational grid on a CRAY computer. Results show the overall hydrodynamic, salinity, and sediment modeling approach to be computationally efficient and accurate.

Introduction

The Waterways Experiment Station of the U.S. Army Engineer Research and Development Center (ERDC) is one of the primary governmental agencies responsible for providing guidance for dredging or structural enhancements to the coastal zone of the United States. Because of this mission, the ERDC is supported by a High Performance Computation (HPC) facility housing CRAY computer support. As a result of this available computational power, large domain circulation and transport model applications are routinely used to provide design criteria for construction and maintenance of coastal protection structures and navigation channels. The application capabilities presented in this paper employ large domain hydrodynamic and transport models to develop engineering solutions to practical problems. The presented application addresses proposed structural and dredging schemes to stabilize the inlet to Willapa Bay, a coastal estuary located on the Pacific coast of the State of Washington.

Two components of ADCIRC are presented in this paper, the hydrodynamic formulation and the transport capability incorporated in the model. Both flow field and transport capabilities are used to determine impacts resulting from each of the proposed stabilization schemes. In the following sections, the hydrodynamic formulation is discussed first, followed by a description of the transport subroutine. Finally, an application of the coupled model to the Willapa Bay project is presented.

Hydrodynamics

Tide and wind field forced circulation are computed by the large domain long-wave hydrodynamic model ADCIRC (ADvanced CIRCulation, Luettich, et al 1992). The ADCIRC model is an unstructured grid finite element long-wave hydrodynamic model which was developed under the 6-year U.S. Army Corps of Engineer funded Dredging Research Program (DRP). The model was specifically developed to: 1) represent very large computational domains with the capability of providing high resolution details for both complex coastlines and for areas of high bathymetric gradients, 2) properly represent all pertinent physics of the fully nonlinear governing equations, and 3) provide accurate and efficient simulations for intra-tidal periods extending from months to years.

In two dimensions, model formulation begins with the depth-averaged shallow-water equations for conservation of mass and momentum subject to incompressibility and hydrostatic pressure approximations. The Boussinesq approximation, where density is considered constant in all terms but the gravity term of the momentum equation, is also incorporated in the model. Using the standard quadratic parameterization for bottom stress and omitting baroclinic terms and lateral diffusion and dispersion, the following set of conservation statements in primitive, non-conservative form and expressed in a spherical coordinate system are incorporated in the model (Flather 1988; Kolar et al. 1994):

$$\frac{\partial \zeta}{\partial t} + \frac{1}{R\cos\phi}\left[\frac{\partial UH}{\partial \lambda} + \frac{\partial (UV\cos\phi)}{\partial \phi}\right] = 0 \tag{1}$$

$$\frac{\partial U}{\partial t} + \frac{1}{r\cos\phi} U\frac{\partial U}{\partial \lambda} + \frac{1}{R}V\frac{\partial U}{\partial \phi} - \left[\frac{\tan\phi}{R}U + f\right]V =$$

$$-\frac{1}{R\cos\phi}\frac{\partial}{\partial \lambda}\left[\frac{P_s}{\rho_0} + g(\zeta - \eta)\right] + \frac{\tau_{s\lambda}}{\rho_0 H} - \tau_* U \tag{2}$$

$$\frac{\partial V}{\partial t} + \frac{1}{r\cos\phi} U\frac{\partial V}{\partial \lambda} + \frac{1}{R}V\frac{\partial V}{\partial \phi} - \left[\frac{\tan\phi}{R}U + f\right]U =$$

$$-\frac{1}{R}\frac{\partial}{\partial \phi}\left[\frac{P_s}{\rho_0} + g(\zeta - \eta)\right] + \frac{\tau_{s\phi}}{\rho_0 H} - \tau_* V \tag{3}$$

where t represents time, λ and ϕ are degrees longitude (east of Greenwich is taken positive) and degrees latitude (north of the equator is taken positive), η is the free-surface elevation relative to the geoid, U and V are the depth-averaged horizontal velocities, R is the radius of the Earth, $H = \zeta + h$ is the total water column depth, h is the bathymetric depth relative to the geoid, $f = 2\Omega \sin\varphi$ is the Coriolis parameter, Ω is the angular speed of the Earth, P_s is the atmospheric pressure at the free surface, g is the acceleration due to gravity, η is the effective Newtonian equilibrium tide potential, ρ_0 is the reference density of water, $\tau_{s\lambda}$ and $\tau_{s\phi}$ are the applied free-surface stress, and τ_* is given by the expression $C_f(U^2 + V^2)^{1/2}/H$ where C_f equals the bottom friction coefficient which can be specified as either linear or nonlinear (Luettich et al 1992).

The momentum equations (Equations 2 and 3) are differentiated with respect to λ and t and substituted into the time differentiated continuity equation (Equation 1) to develop the following Generalized Wave Continuity Equation (GWCE):

$$\frac{\partial^2 \zeta}{\partial t^2} + \tau_0 \frac{\partial \zeta}{\partial t} - \frac{1}{R\cos\phi} \frac{\partial}{\partial \lambda} \left[\frac{1}{R\cos\phi} \left(\frac{\partial(HUU)}{\partial \lambda} + \frac{\partial(HUV\cos\phi)}{\partial \phi} \right) - UVH \frac{\tan\phi}{R} \right]$$

$$\left[-2\omega \sin\phi HV + \frac{H}{R\cos\phi} \frac{\partial}{\partial \lambda} \left(g(\zeta - \alpha\eta) + \frac{P_s}{P_0} \right) + \tau_* HU - \tau_0 HU - \frac{\tau_{s\lambda}}{\rho_0} \right]$$

$$-\frac{1}{R} \frac{\partial}{\partial \phi} \left[\frac{1}{R\cos\phi} \left(\frac{\partial(HVV)}{\partial \lambda} + \frac{\partial(HVV\cos\phi)}{\partial \phi} \right) + UUH \frac{\tan\phi}{R} + 2\omega\sin\phi HU \right]$$

$$-\frac{1}{R} \frac{\partial}{\partial \phi} \left[\frac{H}{R} \frac{\partial}{\partial \phi} \left(g(\zeta - \alpha\eta) + \frac{P_s}{\rho_0} \right) + (\tau_* - \tau_0)HV - \frac{\tau_{s\phi}}{\rho_0} \right]$$

$$\text{(4)}$$

$$-\frac{\partial}{\partial t} \left[\frac{VH}{R}\tan\phi \right] - \tau_0 \left[\frac{VH}{R}\tan\phi \right] = 0$$

The ADCIRC-2DDI model solves the GWCE (Equation 4) in conjunction with the primitive momentum equations given in Equations 2 and 3.

The ADCIRC model uses a finite-element algorithm in solving the defined governing equations over complicated bathymetry encompassed by irregular sea and shore boundaries. This algorithm allows for extremely flexible spatial discretizations over the entire computational domain and has demonstrated excellent stability characteristics. The advantage of this flexibility in developing a computational grid is that larger elements can be used in the open ocean regions where less resolution is needed whereas smaller elements can be applied in the nearshore and estuary areas where finer resolution is required to resolve hydrodynamic details.

Transport - General

The transport subroutine incorporated in the ADCIRC model provides a capability for modeling any substance or property which can be represented in the flow field by the convection/diffusion equation shown below:

$$H \frac{\partial C}{\partial t} + H \left[U \frac{\partial C}{\partial x} + \frac{\partial C}{\partial y} \right] = \frac{\partial}{\partial x} \left(HD_x \frac{\partial C}{\partial x} \right) + \frac{\partial}{\partial y} \left(HD_y \frac{\partial C}{\partial y} \right) + E - D$$

where C is depth averaged concentration, t represents time, x,y are cartesian directions, U,V are depth averaged horizontal velocities, D_x, D_y are horizontal diffusion coefficients, and E,D are terms representing sources and sinks.

The transport capability of ADCIRC uses a Crank-Nicolson time splitting scheme such that concentration is weighted equally at time steps t and t+1. Therefore, concentration can appear on both sides of the equation. Because some source and sink terms are written as a function of concentration, the model can construct the following three forms of computational matrices: 1) source/sink does not contain the dependent variable, e.g., salinity, 2) source/sink contains two terms, one with the dependent variable and one without, e.g., sediment, and 3) source/sink has one term which contains the dependent variable, e.g., temperature.

Transport – Salinity/Sediment

Computation of the salinity distribution requires no source/sink term, only the specification of some initial salinity field and any specified boundary salinity. Computation of sediment transport however, requires separate source and sink functions. Because of the various transport formulations, the sediment application of the transport capability is described below.

The sediment transport module incorporated in ADCIRC provides a capability of simulating transport of fine-grained cohesive sediments as a function of both waves and currents. Many sediment transport equations require near bottom velocities, but the methods incorporated in ADCIRC were developed and work well using a depth integrated velocity reflective of conditions outside the wave and current boundary layers. Unlike near-bottom velocities, these velocities are not significantly affected by bottom roughness. This is an advantage in regions where bottom roughness is unknown or continually changing.

The method used to compute bottom stresses due to combined current and wave action can be described as follows. Large water bodies such as bays, oceans, and estuaries are subject to current velocities generated by tides and wind as well as by wave generated orbital velocities. As a result, the bottom shear stress depends on the relative magnitude of current- and wave-induced velocities and the angle between them.

Bottom shear stress due to currents only is usually assumed to be quadratic in the velocity term. For example,

$$\tau_{cb} = \frac{\rho f_c U^2}{2} \qquad\qquad (5)$$

where τ_{cb} is the near bottom shear stress due to currents only, ρ is the density of water, f_c is the current related friction factor, and U is the current velocity outside the boundary layer. For pure currents, f_c ranges from 0.002-0.005, depending upon the bed roughness. For waves only, a similar equation for wave-induced shear stress is written as:

$$\tau_{wb} = \frac{\rho f_w u^2}{2} \qquad\qquad (6)$$

where u is the near bottom orbital velocity for the wave and f_w is the wave-related friction factor. The coefficient f_w ranges from 0.002-0.05, depending on u and the wave period T_s.

The method for calculating shear stress under a combined wave/current regime incorporated in ADCIRC is based on the following formulation reported by Christoffersen and Jonsson (1985):

$$\tau_{bm} = \frac{1}{2} \rho f_w u_{wbm}^2 m \qquad\qquad (7)$$

where τ_{bm} is the maximum bed shear stress due to waves and currents, u_{wbm} is the amplitude of the bottom orbital velocity at the top of the wave boundary layer and m is a nonlinear function of the ratio τ_{cb} and the maximum value of τ_{wb}.

The amplitude of the bottom orbital velocity at the top of the wave boundary layer (or maximum bottom orbital velocity) is calculated from linear wave theory by:

$$u_{wbm} = \frac{H_s\, gk\, T_s}{4\pi \cosh(hH_s)} \qquad (8)$$

where H_s is the wave height, k is the local wave number, and h is the total water depth. The nonlinear function m is defined as:

$$m = (1 + \sigma^2 + 2\sigma \,|\cos(\delta - \alpha)|\,)^{0.5}$$

where δ and α represent the angles of current velocity and wave orbital velocity. The value of $\cos(\delta-\alpha)$ is generally taken to be unity. The function σ is defined as:

$$\sigma = \frac{\tau_{cb}}{\tau_{wbm}} = \left[\frac{f_c}{f_w}\right]\left[\frac{U}{u_{wbm}}\right]^2$$

with τ_{cb} defined according to Equation 5 and τ_{wbm} idefined as the maximum value of τ_{wb} of Equation 6 using maximum orbital velocity values defined from Equation 8.

A commonly used method of relating erosion to shear stress (Parthenaides 1962) has been incorporated into ADCIRC. This version relates erosion as a function of shear stress to some exponential power. The equation for the erosion rate E is:

$$E = A_0 \left(\frac{\tau_{bm} - \tau_{cr}}{\tau_{cr}}\right)^M$$

where A_0 and M are site specific parameters, τ_{bm} is the shear stress due to currents and waves of Equation 7, and τ_{cr} is the site specific critical shear stress below which no erosion occurs.

Another commonly applied approach to deposition has also been incorporated into the transport formulation (Krone 1962). In the adopted formulation, deposition of suspended material is defined as follows (Van Rijn 1993):

$$D = \left(\frac{\tau_{bm} - \tau_{cd}}{\tau_{cd}} \right) W_s \frac{C}{H}$$

where τ_{cd} is the site specific critical shear stress below which deposition occurs and W_s is the depositional fall velocity.

The wave field contributions for the shear stress computations are provided via the STeady-state spectral WAVE model STWAVE (Resio 1987) which calculates a steady state wave height, period, and direction distribution over the hydrodynamic model computational domain. Input to the STWAVE model are global wind fields provided by the U.S. Navy Fleet Numerical Meteorology and Oceanography Center's database of wind fields. Wave field contributions are made not only to the erosion terms of the transport equation but can be included as radiation stress contributions to the momentum equations of the ADCIRC model.

Erosion and deposition terms are computed from the hydrodynamic flow field and used in the transport equation to compute erosion and deposition within the computational grid. This new total depth is then used to define depth for the next calculation of the hydrodynamic flow field. Currently, a transport and bathymetry change update is performed every 10 hydrodynamic time steps. Optimization of the 10:1 update was not investigated. In the following section, an example simulation of model capabilities is presented.

Model Application – Willapa Bay, Washington, U.S.

Willapa Bay, shown in Figure 1, is located on the Pacific coast of the United States in the state of Washington. The entrance channel into the bay has historically been unstable with primary channels alternating between a north and middle location. The goal of the present investigation is to determine the feasibility of various dredging and/or structural alternatives proposed to stabilize a single entrance channel, thus minimizing maintenance dredging. Sediments vary from coarse sands which deposit in the channel to fine grained silts and clays which are both deposited in the channel and transported and deposited within the back bay areas.

The coupled hydrodynamic, sediment transport, and bathymetry change model is being used to analyze and evaluate the effectiveness of various proposed alternatives. The domain of the 17,661-node computational grid is shown in the upper portion of Figure 2. The highly resolved study area is provided in the lower half of the figure. Flexibility of the unstructured grid can be quantified by a minimum node-to-node spacing on the order of 100 m, a maximum of approximately 80 km, and a ratio of largest to smallest element area of 1.9×10^6.

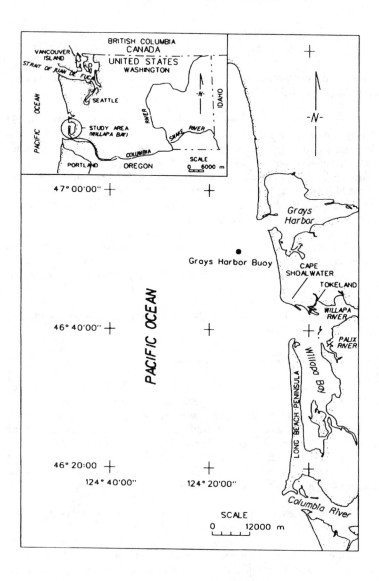

Figure 1 Location map for Willapa Bay, Washington

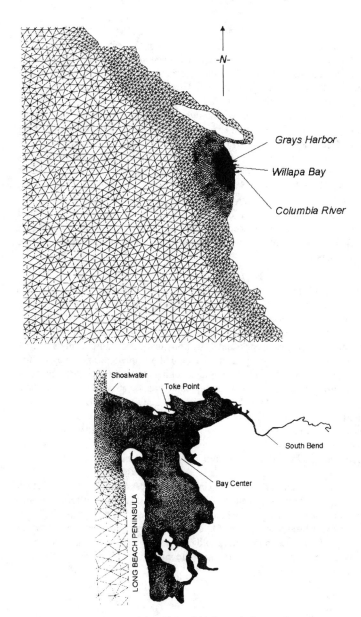

Figure 2 Full computational domain and high resolution study area

The model was run with an 8-constituent (K_1, O_1, P_1, Q_1, M_2, S_2, N_2, and K_2) tidal surface displacement boundary condition along the open ocean boundary. Tidal constituent data were extracted from the tidal constituent database of Le Provost et al (1994). Shoreline and bathymetry information were obtained from the NOAA medium-resolution World Vector Shoreline database and the NOS 1- by 1-deg bathymetry data for the Pacific Ocean, supplemented by high-density shoreline and bathymetric information from NOS and Seattle District surveys through 1996 in the vicinity of Willapa Bay. Verification was made to both surface elevation time series reconstructed from existing harmonic analyses and to prototype surface elevation and current data collected for the project. All comparisons of model results to collected or harmonically reconstructed data were considered very good as shown in some typical comparisons of Figure 3 in which Station 1 is for Toke Point shown in Figure 2 and Station 14 is located in the Willapa River near Toke Point. A quantitative verification to salinity and transport was beyond the scope of this project. However, computed salinity values approximated published observed values and computed erosion/deposition patterns in the area of the raised dike (option 3D shown in Figure 4) were very similar to erosion and deposition patterns measured at the project.

The goal of this study is to determine feasibility of various dredging and/or structural alternatives proposed to stabilize an entrance channel. To accomplish this, the sponsor initially defined 17 design alternatives for evaluation. Of these, 12 alternatives were further evaluated for: 1) impact to wave fields affecting navigability, 2) historical morphological trends, and 3) hydrodynamic and transport impacts such as maximum/minimum velocity, peak discharge, and erosion/deposition magnitudes. These 12 options, listed in Table 1, include dredging alternatives for both the north and middle channels as well as the construction of training structures to direct flow into either the north or middle channel. Figure 4 shows locations of the various alternatives. A rigorous analysis of each alternative was beyond the time and cost constraints of the project, therefore a screening level approach was adopted to reduce the number of alternatives to a manageable number of scenarios which were considered functionally viable and economically feasible to implement.

The hydrodynamic and transport elimination process was initially based solely on the impacts of maximum ebb and flood currents and discharge. Alternatives that had no significant impact or that generated large changes in bay circulation were eliminated from further consideration. Other study groups concurrently evaluated the 12 alternatives from a wave and morphological perspective. The conclusion of all three analyses was that further evaluation of the impacts of Alternatives 3A and 4A was warranted.

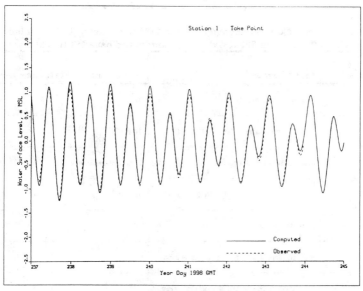

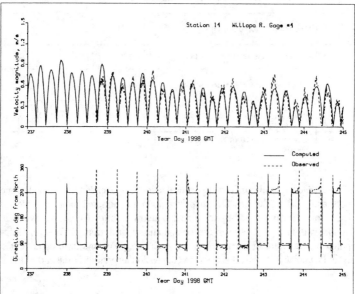

Figure 3 Typical comparisons of surface elevation and velocity

Table 1
Initial Dredging Volumes Assuming a Static 1998 Bathymetry (Unless noted otherwise, channels are 500 ft wide at the bottom and have 1:3 side slopes)

Alternative	Short Description	Volume (in units of 1,000,000 cubic yards)
3A	Straight North Channel	0.8
3B	Curved North Channel, 500 to 2500 ft width	0.04
3C	Extreme curved North Channel	Somewhat greater than 0.2
3D	Raise SR105 dike with straight North Channel	0.8
3E	North Channel with jetty	0.8
3F	3A orientation, 38 x 1000 ft channel dimensions	5.7
3G	3B orientation, 38 x 1000 ft channel dimensions	4.2
4A	Middle Channel (bar dredging)	2.2
4B	Bay Center training structure with 4A	2.2
4C	Dredging across whole entrance with 4B	9.4
4D	Willapa River training structure with 4A	2.6
4E	4A orientation with 38 x 1000 ft channel dimensions	12.3

The detailed hydrodynamic/transport evaluation was extended to include sediment transport and salinity change. The modeling approach adopted simulates the tidal circulation for a 10-day period beginning on 23 August. A normal condition wave field distribution was interpolated over the computational grid from STWAVE output. Computed metrics were differences between each alternative and base conditions for salinity and cumulative sediment erosion and deposition magnitude at the end of the 10-day period.

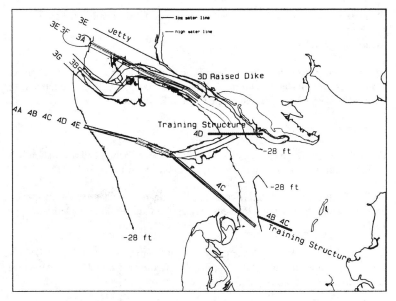

Figure 4 Channel and structure locations for alternatives listed in Table 1

Hydrodynamic and transport comparison of both dredging alternatives 3A and 4A showed no significant change in circulation, sediment transport (erosion and deposition), or salinity. The conclusion of all project analyses for this first phase of the project was that computed impacts are local and well within the variability of the system. Therefore, changes such as options 3A and 4A would be acceptable from a wave, geomorphic, hydrodynamic, and transport perspective. These results provided the necessary input to allow the sponsor to provide funding to investigate additional dredging alternatives. This additional scope of work involves detailed evaluation of the impacts of 9 alternative options.

Conclusions

The models and modeling procedures described in this paper demonstrate a capability for simulating hydrodynamics, sediment transport and associated bathymetry change, and salinity intrusion over very large computational domains. Because the modeled area is very large with respect to the study area of interest, boundary conditions are easily prescribed without adverse impact to the study area. The application to Willapa Bay described herein demonstrates use of existing numerical models to evaluate the physical impact of a large number of construction alternatives. By eliminating options that are clearly unacceptable according to either environmental or economic criteria, a small number of functionally viable and economically feasible alternatives can be considered for further detailed analysis.

Acknowledgements

The research described in this paper was performed at the US Army Engineer Waterways Experiment Station in Vicksburg, Mississippi, U.S.A. Permission to publish the material was granted by the Chief of Engineers.

References

Christoffersen, J. B., and Jonsson, I. G. (1985). "Bed friction and dissipation in a combined current and wave motion," Ocean Engineering 12(5), 387-423.

Flather, R. A. (1988). "A numerical model investigation of tides and diurnal-period continental shelf waves along Vancouver Island," Journal of Physical Oceanography 18,115-139.

Kolar, R. L., Gray, W. G., Westerink, J. J., and Luettich, R. A. (1994). "Shallow water modeling in spherical coordinates: Equation formulation, numerical implementation, and application," Journal of Hydraulic Research 32(1), 3-24.

Krone, R. B. (1962). "Flume Studies of the Transport of Sediment in Estuary Shoaling Processes," Technical Report, Hydraulic Engineering Laboratory, University of California, Berkeley, Calif.

Le Provost, C., Genco, M.L., Lyard,F., Vincent, P., and Canceill, P. (1994). "Spectroscopy of the World Ocean Tides from a Hydrographic Finite Element Model," Journal of Geophysical Research 99(C12):24,777-24,797.

Luettich, R. A., Westerink, J. J., and Scheffner, N. W. (1992). "ADCIRC: an advanced three-dimensional circulation model for shelves, coasts, and estuaries; Report 1, Theory and methodology of ADCIRC-2DDI and ADCIRC-3DL," Technical Report DRP-92-6, USAE Waterways Experiment Station, Vicksburg, MS.

Parthenaides, E. (1962). "A Study of Erosion and Deposition of Cohesive Soils in Salt Water," Thesis presented the University of California, Berkeley, Calif in 1962 in partial fulfillment of the requirements for the degree of Doctor of Philosophy.

Resio, D. T. (1987). Shallow-Water Waves. I: Theory. J. Wtrway, Port, Coast, and Oc Engrg. ASCE 1987; 113(3): 264-281.

Van Rijn, L. C. (1993). Principles of sediment transport in rivers, estuaries, and coastal seas. Aqua Publications, Amsterdam, Holland.

Modeling Sand Bank Formation Around Tidal Headlands

Richard P. Signell[1]
Courtney K. Harris[1]

Abstract

Sandbanks are often found in the vicinity of coastal headlands around which tidal flows are strong enough to generate significant tidally-forced residual eddies, typically with scales of 2-10 km. One popular hypothesis is that these sandbanks are generated by a "tidal stirring" mechanism in which the inward-directed pressure gradient associated with these residual eddies produces an inward-directed movement of sand near the seabed. This hypothesis predicts asymmetric sandbank formation when planetary vorticity is significant compared to the relative vorticity of the residual eddies. This mechanism is tested with a numerical sediment transport model, using idealized symmetrical coastline geometry and tidal forcing that represents conditions similar to regions where these tidal headland sandbanks are known to occur. For both suspended and bedload simulations, we find that nearly symmetric sandbanks form, and that the sediment transport patterns that are responsible for building and maintaining the banks are due more to the patterns of shear stress and sediment flux that occur over the course of the tidal cycle rather than to the characteristics of the tidally-averaged residual fields. We also find that sediment supply can be an important factor in controlling the nature of the resulting sandbanks.

Introduction

In an intriguing paper, Pingree (1979) suggested that the formation of sand banks commonly found in the vicinity of coastal headlands could be due to "tidal stirring"or "teacup effect" associated with tidal residual eddies. Just as tea leaves accumulate in the center of a stirred cup of tea due to the inward directed pressure gradient, so sand might accumulate in the center of a residual eddy "stirred" by the tidal flow. Using a depth-averaged tidal model, Pingree showed that for the case of "the Shambles," a 5

[1] U.S. Geological Survey, Woods Hole Field Center, Woods Hole, MA, 02543-1598. Email: rsignell@usgs.gov, ckharris@usgs.gov. This paper is also available with color figures and supplemental movies on the Web (http://smig.usgs.gov/SMIG/features_0300/headland.html).

km sand bank found just offshore of Portland Bill in the United Kingdom, the center of the bank coincided with the center of the cyclonic residual eddy found on the eastern side of the headland.

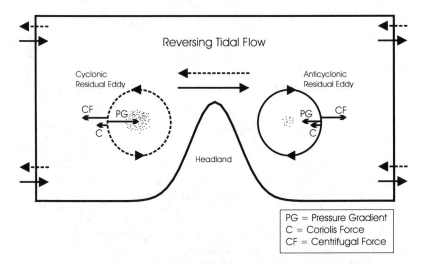

Figure 1. Cartoon of the "tidal stirring" hypothesis for sandbank generation around tidal headlands, modified from Pingree (1979). Reversing tidal flows generate symmetric residual eddies on either side of the headland. The inward-directed pressure gradient PG, however, is larger in the cyclonic eddy because it must balance the Coriolis force C in addition to the centrifugal force CF. In the anticyclonic eddy, the Coriolis force helps balance the centrifugal force and thus the pressure gradient is smaller. The hypothesis states that this asymmetry in pressure gradient would result in asymmetry in sandbank generation, since sand moving down-pressure gradient near the bottom would feel a stronger force in the cyclonic eddy. The larger stippled region inside the cyclonic eddy indicates this pictorially.

An equally strong anticyclonic residual eddy was found on the western side of the headland, but the sand bank associated with this eddy was much smaller. Pingree argued that the asymmetry in sand bank formation was due to the Earth's rotation. In the cyclonic eddy, the Coriolis force acting outward requires a larger inward directed pressure gradient to drive the centripetal acceleration (Figure 1). Conversely, in the anticyclonic eddy, the Coriolis force is directed toward the center of the eddy, thus requiring a smaller pressure gradient. Thus, near the bottom, where the pressure gradient balances stress divergence, more sand would be directed toward the center of the cyclonic eddy than toward the center of the anticyclonic eddy. In fact, if the residual vorticity of the anticylonic eddy were less than the planetary vorticity f, the pressure gradient at the bottom would be directed outward and no sand bank would form. This mechanism for generation of asymmetric banks around tidal headlands has remained popular (e.g. Dyer and Huntley, 1999).

Here we use idealized sediment transport simulations to explore sand bank formation around headlands by suspended and bed load and to test the "tidal stirring" hypothesis. We initialize the bed with a finite depth layer of uniform sediment, run a periodic tidal model with limiting cases of pure suspended load and pure bed load, then examine the resulting erosional and depositional patterns in relation to the tidal residual eddies.

Modeling Approach

For our modeling study we use the 3D sediment transport model ECOMSED, developed by HydroQual, Inc.[2], which is an extension of the Blumberg and Mellor (1987) circulation model.

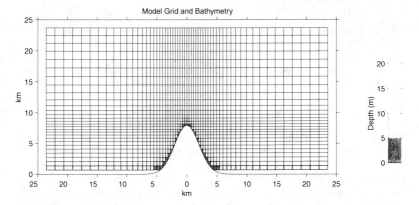

Figure 2. The headland model domain, grid and bathymetry used for this study. The tide is forced at the western boundary and a gravity wave radiation condition is applied at the eastern boundary.

For our idealized headland model domain, we utilize the base case of Signell and Geyer (1991) because the hydrodynamics have been thoroughly studied and it is a close match to the situation near the Shambles in terms of tidal flows, water depths and the scale of the headland. The domain is 50 x 25 km, with an 8 km width, 8 km amplitude Gaussian shaped headland located in the middle of the southern wall (Figure 2). The water depth is 20 m, except near the southern boundary, where it decreases to a minimum of 2 m over a distance of 3 km. The model grid spacing is

[2] Trade names are for identification purposes only, and do not constitute endorsement by the U.S. Geological Survey

variable, with 1-2 km spacing near the open boundaries, and 0.3-0.4 km spacing near the headland.

There are open boundaries at the west and east and a free-slip wall at the north. The tide is specified as a periodic normal velocity at the western open boundary, and a gravity radiation condition is applied at the eastern open boundary to allow the tidal wave to leave the domain. In the base case, an M_2 (12.42 hour period) tidal current with amplitude of 0.5 m s^{-1} is specified. Although Signell and Geyer (1991) used a curvilinear grid, we found that our Cartesian grid results closely match theirs. Table 1 summarizes other model characteristics and selected parameters.

Internal time step (DTI)	51.750 seconds
External time step (DTE)	5.175 seconds
Background vertical mixing (UMOL)	5×10^{-6} m^2 s^{-1}
Bottom roughness length z_0 (Z0)	0.003 m
Smagorinsky coefficient (HORCON)	0.05
Coriolis frequency f (COR)	1.0×10^{-4} s^{-1}

Table 1. Adjustable model parameters. The FORTRAN variable names are shown in parentheses for the benefit of POM/ECOM users.

To model erosion and deposition due to bed load transport of sediment, we use the method of Van Rijn (1984a), where the rate is proportional to the excess shear stress to the 2.1 power. The convergence and divergence of bedload transport determines the erosion and deposition rate of the sediment from the bed.

To model erosion and deposition due to suspended load transport of sediment, we use a concentration-based approach, where the sediment is treated as a 3D tracer field with a vertical fall velocity w_s. We consider only a single size class of noncohesive sediment, with constant settling w_s. The advection of suspended sediment is computed using a high-order advection scheme (Margolin and Smolarkiewicz, 1989), and vertical diffusion of sediment is determined from the Mellor and Yamada (1982) level 2.5 turbulence closure model with the Galperin et al. (1988) extensions.

The bottom boundary condition for suspended sediment concentration is specified as a sediment flux between the bed and the water column. This sediment flux depends on excess shear stress, fall velocity, and sediment concentration in the lowest layer.

If the bottom shear velocity u_* is less than the threshold for movement u_{*crit} or less than the setting velocity w_x then only settling occurs and the flux to the bed is $w_x * C_1$ where C_1 is the concentration in the lowest layer. If u_* exceeds w_x and u_{*crit}, the sediment concentration in the lowest layer is assumed to be in equilibrium (upward diffusion balancing downward settling). The equilibrium horizontal sediment flux in the lower layer is then computed in a manner similar to Van Rijn (1984b), essentially integrating the product of the Rouse profile and the law-of-the-wall velocity profile over the thickness of the lowest sigma layer. The vertical flux at the bed is then specified as the difference between this equilibrium flux and the flux currently computed by the model ($C_1 * U_1$), where U_1 is the velocity in the lowest layer and C_1 is the concentration in the lowest layer obtained from the previous time step. If the model flux is lower than equilibrium, erosion occurs so that the equilibrium flux is matched. If the model flux is higher than equilibrium, deposition occurs. Thus it is possible to have deposition even when u_* exceeds the critical movement and suspension thresholds. The cumulative impact of the vertical fluxes to and from the bed is tracked through erosional and depositional changes to the bed thickness. For simplicity, no waves, active layers, or multiple size classes were included.

The model is spun up from a state of rest for four tidal cycles before the sediment transport is initiated. The initial sediment bed is given a constant thickness of 3 m. As material is eroded and deposited, the bed thickness changes, and these modifications to the bed are allowed to change the overall depth in the model so that the developing bed features may affect the hydrodynamics. The present model configuration does not allow for formation of spits, islands, or other changes to the coastline because it cannot handle drying of previously wet grid cells.

Suspended Load Simulation

To determine what type of sand banks might develop from material that can be suspended and advected by the tidal flow, we initialized the bed with a 3-m thick layer of coarse silt, with a grain size of 30 microns, a settling velocity w_s =0.06 cm/s and a threshold of movement of $u_{*_{crit}}$ =0.86 cm/s. Because the settling velocity is less than the threshold of movement, any flow energetic enough to move coarse silt can suspend it. Coarse silt will therefore always move in suspension, and the bedload contribution can be neglected.

Insight into suspended sediment transport for this case can be gained by looking at snapshots of the tidal flow; suspended sediment; shear velocity magnitude; erosion rate; and net change in the bed thickness at key points during the 1st tidal cycle after sediment transport is initiated (Figures 3 and 4). Just after maximum ebb (Figure 3), the westward flow is accelerating as it approaches the tip of the headland and then separates at the headland tip, forming a cyclonic eddy on the westward side of the headland. The shear stress is sufficient to resuspend material over a large region surrounding the headland, and the rate of erosion is particularly strong at the headland tip where maximum shear stress occurs. Deposition is starting to occur downstream of the headland tip, where the flow is decelerating and the sediment capacity of the lower layer is therefore decreasing. The bed thickness is just starting to decrease at the headland tip and increase where the tidal flows decelerate to the west of the headland tip.

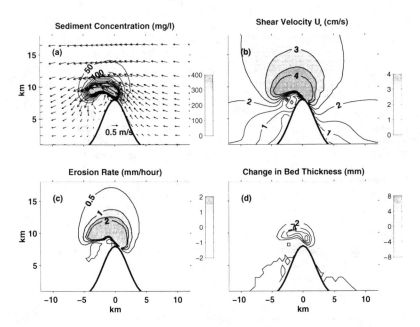

Figure 3. Suspended load sediment transport characteristics near maximum ebb (1.3 hours after sediment transport was initiated): (a) Depth-averaged sediment concentration and near-bed (1 m) velocity vectors; (b) Shear velocity magnitude; (c) Erosion rate; (d) Net change in bed thickness. At maximum flood and ebb, high sediment concentrations are found near the tip of the headland, where strong shear velocities cause rapid erosion of the bed. Eroded material is advected away from the headland tip.

At slack before flood (Figure 4), the cyclonic eddy is fully developed, shear stresses have decreased, so the sediment that was eroded from the headland tip during ebb is continuing to fall out of suspension, building the bed thickness on the western side. At the headland tip, however, the flood tide is already beginning and this eastward flow is starting to pick up sediment. This sediment is transported and deposited in the anticyclonic eddy that forms on the eastward side of the headland, mirroring the process occurring during the ebb tide. An animation of these variables over the tidal cycle adds considerable insight into the nature of the resuspension and transport process (http://smig.usgs.gov/SMIG/features_0300/headland_movie1.html).

The mean erosion/deposition rate over the first two tidal cycles indicates that the coarse silt is being eroded from the headland and deposited in banks forming near the centers of the residual eddies (Figure 5a). The deposition rates, however, are

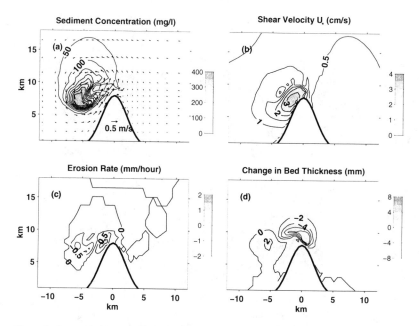

Figure 4. Suspended load sediment transport characteristics near slack water before flood (3.8 hours after sediment transport was initiated): (a) Depth-averaged sediment concentration and near-bed (1 m) velocity vectors; (b) Shear velocity magnitude; (c) Erosion rate; (d) Net change in bed thickness. Just 2.5 hours after the snapshot shown in Figure 3, the sediment resuspended during maximum ebb is being wrapped around a transient eddy. The shear velocities are weak, and the material is rapidly depositing due to the decreased carrying capacity of the flow. This erosion at maximum flood and ebb and deposition at slack before ebb and flood results in sand bank generation on both sides of the headland.

nearly identical on either side of the headland, indicating nearly symmetric sand bank formation. While the tidal stirring hypothesis predicts that the bank on the left hand side of the headland should grow much faster (in the Northern Hemisphere), the simulation actually shows slightly greater growth of the sandbank to the right of the headland. Investigation revealed that this is the slight frictional damping of the tide that occurs as it propagates eastward from the western open boundary.

As a further check as to the negligible influence of the Earth's rotation, the test case was run a second time using a Coriolis frequency $f = -1 \times 10^{-4}$ s^{-1}, valid for the Southern Hemisphere. Figure 5b confirms that the development of the banks is not being significantly impacted by the Earth's rotation.

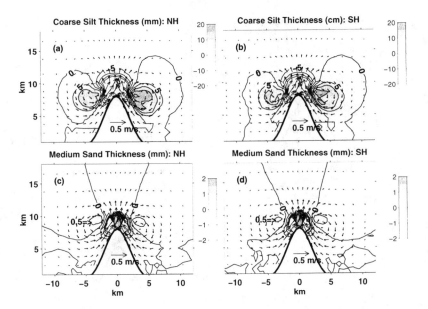

Figure 5. Net change in bed thickness over the 1st two tidal cycles following sediment transport initiation and near-bottom (1 m) residual currents: (a) Coarse silt (suspended load) case in the northern hemisphere; (b) Coarse silt (suspended load) case in the southern hemisphere; (c) Medium sand (bedload) case in the northern hemisphere; (d) Medium sand (bedload) case in the southern hemisphere. The nearly symmetric sand bank formation in both the suspended load and bedload cases, and in both the northern and southern hemisphere indicates that the Earth's rotation is not a significant factor in the generation of these banks. The slightly increased rate of growth of the bank to the right of the headland is due to slight decay of the forced tidal wave as it propagates eastward from the western boundary, which generates slightly more resuspension on the eastward flowing phase of the tide.

If we now focus on the long-term development of the sand banks instead of the first few tidal cycle patterns of erosion and deposition, we find that the source supply of sediment plays an important role in sand bank evolution. Within 1.5 years, the initial 3 m of sediment is completely removed from a roughly 25-km² region off the tip of the headland. As the sediment source near the headland tip is depleted, each bank rapidly shifts shoreward from the center of the residual eddy to a more stable position along the shoreward side of the residual eddy (Figure 6a). These banks continue to build in nearly the same locations, reaching heights of 4-5 m in 15 years when the simulation died (Figure 6b). Unfortunately, the banks were still slowly growing when the simulation died, and therefore the equilibrium condition (if one exists for this case) could not be established. At the time when the simulation died, the 4-5 m banks had only a modest impact on the structure and magnitude of the residual flow. An animation of the long-term development of the sandbanks clearly shows the rapid depletion of bed material near the tip and subsequent slow evolution of the banks (http://smig.usgs.gov/SMIG/features_0300/headland_movie2.html).

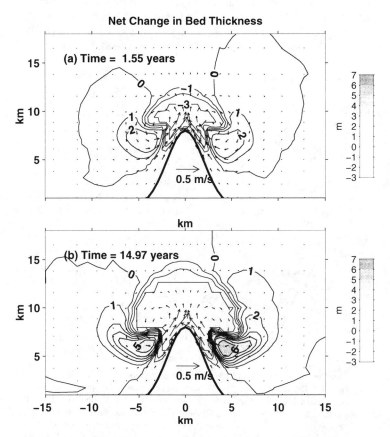

Figure 6. Long-term changes in bed thickness and near-bottom (1 m) residual currents for the coarse silt (suspended load) case: (a) after 1.5 years; (b) after 15 years. Although the banks initially form near the centers of the residual eddies, they migrate away from the headland tip as the sediment supply is exhausted. After 1.5 years, all of the original 3 m of sediment has been eroded from a roughly 25-km^2 area in the vicinity of the headland tip, and the banks are found immediately adjacent to the erosional region. After 15 years the banks are nearly in the same location, but the region from which the material has been completely removed has grown to more than 50 km^2. The banks are nearly 5 m high, but their growth rate has slowed dramatically due to the lack of available sediment from the erosional area in the vicinity of the headland.

Bedload Simulation

We next investigate sand bank formation by bedload transport. The model was
initiated with a 3-m thick layer of medium sand with a grain size of 250 microns, a
setting velocity of $w_s = 3.0$ cm/s and a threshold of movement $u_{*crit} = 1.38$ cm/s.
Because the tidal flows are sufficient to move medium sand but not to suspend it, it
will always move as bedload and suspended load can be neglected.

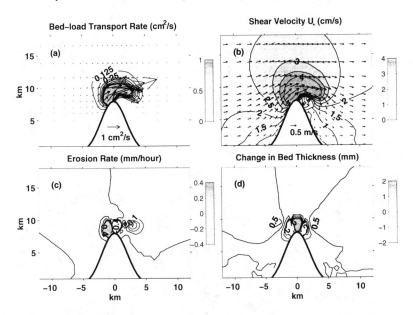

Figure 7. Bedload sediment transport characteristics at maximum flood: (a) Bedload
transport rate (cm²/s); (b) Shear velocity magnitude and near-bed (1 m) velocity vectors; (c)
Erosion rate (mm/hour); (d) Net change in bed thickness (mm).   Strong bedload transport
and shear velocities are found upstream and offshore as the flow separates from the headland
tip. This causes rapid erosion on either side of the tip, but strong convergence of sand
directly off the headland tip. As the shear velocities decrease downstream from the headland
tip, bedload convergence of sand results in rapid deposition and sandbank generation.

Snapshots of the bedload transport rate, the shear velocity, the erosion rate and the net
change to the bed again illustrate the mechanism of sand bank formation by bedload
transport (Figure 7). At maximum flood, as in the suspended load case, erosion rates
are highest near the tip of the headland where the flow speeds are greatest. Unlike
the suspended load case, however, the erosion rate depends only on the divergence of
the transport rate. While there is a region of deposition to the east of the headland
near where the sandbank formed in the suspended load case, there is also a small

region of strong deposition just offshore of the headland tip. There are regions of strong erosion along the upstream side of the headland and in the vicinity of the headland tip immediately adjacent to the small depositional zone. A movie of these fields (http://smig.usgs.gov/SMIG/features_0300/headland_movie3.html) clearly shows migration of the zone of deposition from the tip of the headland downstream as the flood tide progresses. Most of the deposition and erosion occurs during maximum flood and ebb, when the bottom stress gradients are strongest.

The mean erosion/deposition rate for the bedload case over the first two tidal cycles indicates that medium sand will be both eroded and deposited from the headland tip and will also be deposited in banks forming near the centers of the residual eddies, just as in the suspended load case (Figure 5c). The bank positions are slightly closer to the headland tip than they were for the suspended load case since there is no water column advection of sediment in this case. The banks are again nearly symmetric and the effects of the Earth's rotation are negligible (Figure 5d).

The long-term evolution of the bedload generated sand banks could not be determined since the strong deposition near the headland tip caused the run to terminate within a few weeks. Future work is needed to determine if the inclusion of wave effects might redistribute the sand resulting from this strong deposition or whether this process might actually build spits, a phenomenon that the current model is not capable of representing because water cells cannot dry.

Discussion

In our simulations, we get nearly symmetric sandbanks forming around the headland for both the suspended load and bedload-dominated cases. In both cases, the sandbanks form within the region encompassed by the residual eddies. Is the teacup effect responsible? If we calculate the vorticity of the residual eddies, it is only about $3f$. Thus in the conceptual "tidal stirring" or "teacup" model, the mean pressure gradient would balance $4f$ in the cyclonic eddy and $2f$ in the anticyclonic eddy. If this mean pressure gradient was responsible for driving flow toward the centers of the eddies to produce sandbanks, we should see sandbanks in the cyclonic eddy develop about twice as fast as in the anticyclonic eddy. We see no such effect. For such a nonlinear process as sediment transport, examining the nearly symmetric erosion and deposition patterns that occur over a tidal cycle forms a better conceptual model. These reveal that the gradients in stress and the downstream transport of material resuspended near the headland tip are the dominant factors in sandbank formation.

If the teacup effect does not explain why the asymmetric sandbanks observed off of Portland Bill, what does? It is possible that meteorological conditions provide the explanation. Pingree (1979) shows that the response to a strong wind event from the west should result in preferential deposition on the Shambles. It is also possible that the sediment supply is such that it favors asymmetric development. Ultimately, the supply of sand needs to be linked to the wave-dominated nearshore zone, which is not considered here. Still more mechanisms for asymmetric sand bank formation could be the presence of persistent non-tidal residual flows caused by winds or density gradients or strong ebb or flood dominance in the large-scale tidal flow.

These simulations have explored sandbank formation only in a small part of parameter space. Previous work has shown that the nature of the flow around tidal headlands depends on the ratios of three important length scales: the tidal excursion, the length scale of the headland, and the frictional spin-down scale (Signell, 1989; Signell and Geyer, 1991). Flow separation occurs when the tidal excursion is at least as large as the headland length scale and at least as small at the frictional length scale. When the flow separates, then the ratio of the tidal excursion to the frictional length scale determines whether eddies formed on one phase of the tide can maintain their vorticity long enough to interact with eddies formed on the following phase of the tide. In the case explored here, which represents conditions commonly found in nature, all three length scales are about the same magnitude, so flow separation occurs, but eddies that form on one phase of the tide are spun down by friction before they can interact with eddies formed on the following phase of the tide. Since these parameters control the nature of the circulation, it is expected that they would also affect the structure of the resulting sandbanks. The length scale over which sediment falls from suspension and the length scale over which the critical suspension threshold is exceeded are important additional parameters that will affect the characteristics of sand bank generation. A more thorough study is necessary to map out the types of sandbanks that form in this parameter space.

The fact that all the relevant length scales for our test case have about the same magnitude provides an explanation for why the sandbanks are observed to initially form near the centers of the tidal residual eddies, an observation that would appear consistent with the teacup stirring mechanism. A more appropriate explanation, however, is that both vorticity and suspended load are generated near the headland tip, and are advected together downstream. The length scale that they are advected is proportional to the tidal excursion. If the sediment falls out of suspension over the same length scale that the vorticity decays, then the sandbank will coincide with the residual eddy. For the bedload case, the length scale of the headland dictates the length scale of shear stress variability, and if this scale is comparable to the tidal excursion and the frictional length scale, the bedload-generated bank will also coincide with the residual eddy.

Conclusions

A three-dimensional model of sediment transport around tidal headlands is capable of generating sandbanks due to both suspended load and bedload. The sandbanks that form in a case similar to conditions at Portland Bill are symmetric about the headland and are not strongly influenced by the Earth's rotation, indicating that the "tidal stirring" or "teacup effect" of Pingree (1979) is not controlling the generation of these banks. While the suspended load case and the bedload case both initially generate sandbanks near the centers of the tidal residual eddies, this is due to the fact that the controlling length scales all have comparable magnitudes, and other regions of parameter space may be expected to behave differently. The results are also sensitive to sediment supply. In our suspended load case, the 3-m sediment initial layer of silt was rapidly depleted near the headland tip, and the sandbanks migrated shoreward, reaching a relatively stable position on the shoreward side of the residual eddy within about 1.5 years. The banks continued to grow throughout the duration of the 15-year run.

While the suspended load and bedload cases both predict sandbank formation on either side of the headland, the suspended load case produced pure erosion in the vicinity of the headland tip, while the bedload case indicates a small region of strong deposition just off the headland tip. Future work is required to determine if this spit building mechanism would be significant in the presence of waves.

These simulations have demonstrated the great utility of numerical models in exploring the complex, sometimes nonintuitive aspects of sediment transport, even in fairly simple hydrodynamic situations. A thorough parameter study varying the tidal excursion, frictional length scale and sediment characteristics would likely yield additional insightful results, as would a study examining the interaction and sorting of sediment for mixed grain size beds.

Acknowledgements

Thanks to Alan Blumberg and Pravi Shrestha for allowing us to use the ECOMSED code and for many helpful discussions regarding its use. Thanks also to Tim Keen and John Hamrick for providing insight into the merits and difficulties of different sediment transport modeling approaches.

References

Blumberg, A. F. and G. L. Mellor, 1987. A description of a three-dimensional coastal model. In *Three-Dimensional Coastal Ocean Models*, AGU Coastal and Estuarine Sciences Series, Washington, DC. **4**, 1-16.

Dyer, K. R. and D. A. Huntley, 1999. The origin, classification and modeling of sand banks and ridges, *Cont. Shelf Res.*, **19**, 1285-1330.

Galperin, B., Kantha, L. H., Hassid, S. and Rosati A., 1988. A quasi-equilibrium turbulent energy model for geophysical flows, *J. Atmosph. Sci.*, **45**, 55-62.

Margolin, L. G. and Smolarkiewicz, P. K., 1989. Antidiffusive velocities for multipass donor cell advection, *Technical Report UCID 21866*, Lawrence Livermore National Laboratory, Livermore, CA, 44 pp.

Mellor, G. and Yamada, T., 1982.   Development of a turbulence closure model for geophysical fluid problems, *Rev. Geophys. Space Phys.*, **20**, 851-875.

Pingree, R. D., 1979. The formation of the Shambles and other banks by tidal stirring of the seas, *J. Mar. Biol. Ass. U.K.*, **58**, 211-226.

Signell, R. P., 1989. Tidal dynamics and dispersion around coastal headlands, Ph.D thesis, MIT/WHOI Joint Program in Physical Oceanography, Woods Hole, MA, 159 pp.

Signell, R. P. & Geyer, W. R., 1991. Transient eddy formation around headlands, *J. Geophys. Res.*, **96** (C2), 2561-2575.

Van Rijn, L. C., 1984a. Sediment transport, part I: Bedload transport, *J. Hydraulic Eng.*, **110** (10), 1431-1456.

Van Rijn, L. C., 1984b. Sediment transport, part II: Suspended load transport, *J. Hydraulic Eng.*, **110** (11), 1613-1641.

Proposed Third Crossing of Hampton Roads, James River, Virginia: Feature-based Criteria for Evaluation of Model Study Results*

J. D. Boon[1], A.Y. Kuo[1], H.V. Wang[1], and J.M. Brubaker[1]

ABSTRACT

Advancements in GIS-controlled grid development and the design of feature-specific information systems (data files linked to a geographically referenced object or "feature") have greatly increased the value of numerical models used in so-called "what-if" development scenarios. This is particularly true for port and harbor development where resource managers as well as planners require detailed information on the environmental results expected for a given design, or alternatives to that design, in advance of the project being built. Morphological and structural features include shoreline extensions, bridges and tunnel islands, and modifications to navigation channels. Unlike structural design criteria, criteria for assessing structure-induced changes in large-scale circulation, sedimentation, and related physical characteristics of coastal estuaries are still evolving. In our search for useful criteria, physical attributes (e.g., water level, current, and salinity) were viewed as components of specific hydrodynamic features (e.g., tidal fronts, stratified layers, eddies). In Hampton Roads, one of the best known hydrodynamic features is a tidal intrusion front which is believed to be an essential pathway for recruitment of shellfish larvae in the James River estuary. Other potentially important features include: 1) a residual eddy at the mouth of the Elizabeth River, an important tributary of the James, 2) turbulence-induced changes in vertical salinity structure and sedimentation potential in the vicinity of

---

[1] Virginia Institute of Marine Science
School of Marine Science
College of William and Mary
Gloucester Point, Virginia 23062 USA

bridge pilings. These features are sensitive to spring-neap, perigean-apogean variations in tidal range requiring intermediate-term model simulations (simulation periods on the order of a month).

INTRODUCTION

The Port of Hampton Roads in Virginia, not unlike many others competing in the global economy, is approaching the next millennium with expansion in mind (Figures 1 and 2). The proposed expansion is expected to take place in three steps:

1. Expansion of port facilities, especially container cargo handling and storage areas
2. Construction of bridge-tunnel systems to expand rail and highway access
3. Deepening of navigational channels to expand vessel access

Regardless of the order in which they are accomplished, all three steps will have an influence in one degree or another on the circulation and other physical characteristics of the host body of water, the lower James River and its tributaries. This paper examines the results of a hydrodynamic modeling study recently conducted to address environmental questions associated with step 2; i.e., a proposed new bridge-tunnel crossing of Hampton Roads (Boon et al., 1999). For the Hampton Roads study, we used a version of the VIMS three-dimensional Hydrodynamic Eutrophication Model, HEM-3D. The hydrodynamic component of this model (Hamrick, 1996) was coupled with a revised sedimentation module (Boon et al., 1999). Model pre- and post-processing tasks were completed with the aid of a Geographic Information System, ArcView® GIS (Sisson et al., this volume). In addition to these tools, the need for guiding criteria to evaluate results from this and similar model studies in estuarine environments has proven to be exceptionally important. Model validation is the first step in the development of these criteria.

MODEL VALIDATION

An appropriate and properly validated computational model offers one of the few means of obtaining any predictive evidence, in advance of a planned modification at a site, for the changes that may occur as a result. More importantly, the model may provide further evidence indicating whether the changes can be considered significant or not.   Environmental Impact Studies (EIS) or Environmental Assessments (EA) have stressed the importance of the word 'significant' with the result that its use demands both objectivity and specificity about what is being claimed. Taking a pragmatic view toward model validation, Dee (1995) proposed that

"Validation of a computational model is the process of formulating and substantiating explicit claims about the applicability and accuracy of computational results, with reference to the intended purposes of the model as well as to the natural system it represents."

Our first explicit claim in the present study is that we substantiate the river-induced gravitational circulation and the major astronomical components of tide and tidal current in the model. No claims are made regarding variations in current or water level caused by other forcings; e.g., wind-driven circulation. For validation purposes, we compared predicted and observed tides as represented by nine tidal harmonic constituents at selected points in the model domain. The output of the model forced by these nine constituents was compared to the amplitudes and phases of the same nine constituents extracted from water level recordings using a least squares method. The average agreement was approximately a 3-4 cm difference in amplitude and a time equivalent phase difference of less than 20 minutes.

The above values are considered spatially representative because tidal amplitude and phase do not vary abruptly over short distances (1-5 km) in the lower James River. This is not the case with tidal currents whose speed and direction vary in three, rather than two, spatial dimensions. Current directions are spatially least variable at depth in straight channel sections where predicted and observed comparisons are least likely to be affected by any offset between the point of model prediction and the point of current observation. Consequently, tidal current comparisons were made using channel-parallel (flood and ebb) components of the current at selected locations such as the James River Bridge (Figure 2). An example of the results for surface and bottom currents there is shown in Figure 3. For the surface current in this example, an RMS error of about 6.7 cm/s is expected due to phase difference alone and about 9.9 cm/s due to the combination of amplitude and phase. For the bottom current in the channel, the corresponding RMS errors are 4.2 cm/s and 5.1 cm/s, respectively. The overall error in predicting the surface and bottom current is about 10 percent of the maximum current. We assume that a similar accuracy prevails in other regions of the model domain.

PREPARATION OF MODEL FOR TESTING

Development plans for a new bridge-tunnel crossing of Hampton Roads required model testing and evaluation of three alternative designs, each compared to the Base Case (existing conditions). These were provided to us as CAD drawings which we imported into ArcView® GIS. Using the GIS, bridge sections on pilings and tunnel islands on shoal bottom were precisely located within the geo-referenced cells of the model grid (horizontal cell dimensions: 370 m x 370 m). Flow impedance or blocking parameters were then assigned to the data base linked to the edited cells (Boon et al., 1999; Wang and Kim, this volume). For the purposes of this paper, only one of the bridge-tunnel designs, Alternative 9 (Figure 4a), will be discussed in detail.

Alternative 9 calls for the construction of two new tunnels, one running south from Newport News Point under the Newport News Channel and the other running west from Norfolk under the Norfolk Harbor Reach at the Elizabeth River entrance. Bridge roadways on pilings connect the tunnels as shown in Figures 4a and 4b. To determine the EIS/EA relevant changes expected for Alternative 9, model simulations were designed that would incorporate the expected range of estuarine conditions most likely to produce such changes. Previous studies (Kuo et al., 1988; 1990; Shen et al., 1999) have pointed to salinity stratification as one of the most important factors governing the behavior of three-dimensional circulation and material transport in the lower James River. The state of stratification in turn is principally governed by variation in tidal range at the mouth and fresh water discharge at the head of the river estuary.

To incorporate the relationship between tidal range and stratification, we chose a three–constituent tide for model input. Using available harmonic amplitudes for the M_2, S_2, and N_2 semidiurnal tidal constituents at Hampton Roads, we conducted model runs simulating both the spring-neap ($M_2 \pm S_2$) and apogean-perigean ($M_2 \pm N_2$) range cycles over a 30-day period. Combining all three constituent amplitudes with appropriate phasing enabled larger than average (perigean-spring) and smaller than average (apogean-neap) extremes in range to be simulated in the correct sequence during a single 30-day run. However, three separate runs were required to simulate historically representative rates of high, average, and low river discharge.

FEATURE-BASED COMPARISONS

Tidal Intrusion Front - One of the most distinctive features of the circulation in the lower James River is a tidal intrusion front that regularly forms in a small area off Newport News Point adjacent to the Newport News Channel (Figure 4a). This feature first appears as a line of convergence between opposing currents during the early part of the flood cycle and moves westward with time. Incoming saline water from the Chesapeake Bay meets and subducts beneath fresher James River water at the front and then moves upstream in the deep channel, transporting shellfish larvae upstream in the process (Shen et al., 1999). This feature is considered to be responsible for exceptionally high levels of oyster recruitment in the James (Ruzecki and Hargis, 1988) and was a focal point of concern in our evaluations. The relevant model results are shown in a display of current and salinity along a transect running normal to the front (Figures 5a and 5b). As noted in Figure 5b, the predicted effects of Alternative 9 are minor and limited primarily to the surface layer. Following the line of current convergence in model transect plots for Alternative 9 at successive times revealed no breakdown in the front itself, only a change in its rate of movement from east to west during the advance of the flood stage.

Tidal Mixing Induced By Bridge Pilings - One of the effects anticipated from the insertion of an array of regularly-spaced bridge pilings into an estuarine water column is turbulence-induced vertical mixing (Wang and Kim, this volume). The vertical salinity gradient near these structures will tend to weaken, or may be eliminated, because of this effect. Comparison of model results for the Base Case with Alternative 9 confirms that surface salinity increases while bottom salinity decreases in the areas surrounding the added bridge sections. This is particularly true during maximum ebb within an east-west strip immediately north of Craney Island (Figures 6a, 6b) where a predicted salinity change of about 4 ppt occurs in both surface and bottom layers. Most important in terms of biological functioning of species intolerant to salinity change, horizontal salinity gradients and the limits of salt intrusion were unaffected by the added structures of Alternative 9 in any of the model runs.

Tidal Prism of the Elizabeth River Basin - The Elizabeth River, a commercially exploited tributary of the James River, is divided into three relatively short branches with limited freshwater stream inflow. Circulation and flushing ability of this semi-enclosed basin depends on its tidal prism (volume of water entering or leaving the system in a half tidal period). We tested the obviously important question: will the addition of structures near the mouth of the Elizabeth River cause a significant reduction in tidal prism? Model tests integrating the flow past a transverse transect at the river entrance indicated no discernible reduction.

Non-tidal Transport including Eddies - Integrating the velocity field over a tidal period yields the non-tidal or Eulerian residual transport. One of the distinctive features of estuarine circulation is the presence of layered residual flows along the longitudinal axis, ideally involving landward bottom flow of saltier water and seaward surface flow of fresher water separated by a layer of zero net motion along which mixing occurs. This is the baroclinic or gravitational circulation. In addition, horizontal circulation cells, or eddies, can occur in association with current advection near certain types of shoreline and bathymetric features. Testing with the James River model, in fact, suggested that a large, clockwise eddy should exist at the Elizabeth River mouth with maximum development during apogean-neap tides and with little or no development during perigean-spring tides (Figure 7, Base Case). The presence of an eddy at the entrance to the Elizabeth River had not been previously observed but a subsequent 25-hour survey using a broad-band Acoustic Doppler Current Profiler (ADCP) during an apogean-neap tide confirmed its existence (Figure 8). Model tests for Alternative 9, however, showed a non-tidal eddy at this location that was considerably weakened and poorly organized during apogean-neap tide (Figure 7). The potential impact of the latter result is unknown; The importance of eddies and eddy-like features to the delineation of material dispersion or convergence zones in estuaries has only recently been investigated using numerical models (Hood et al., 1999; Shen et al., 1999; Shen and Kuo, 1999).

Sediment Deposition Potential - Sedimentation, although not unique to estuaries, is nevertheless an important and distinguishing feature of these systems. On the whole, estuaries tend to trap sediments originating from both landward and seaward sources. The James River is typical in having a well-developed turbidity maximum and fine-grained suspended sediments (silty sands and sandy silts) are uniformly abundant throughout the lower sections of the river. Locally, sediment deposition is a function not only of source material availability but hydrodynamic conditions favoring accumulation at the bed. We developed a measure for the latter using model-computed values of bed shear stress. The percentage of total time over a thirty-day model run during which the predicted shear stress fell below a critical value permitting deposition was defined as the sediment deposition potential. Sedimentation is often episodic and the deposition potential thus defined permits a degree of smoothing that enables relative comparisons to be made over different regions of the bottom area within the model domain. It is consistent with our objective, to assess whether a significant change in fundamental processes governing sediment deposition will occur as a result of the proposed construction. In that event, detailed studies utilizing a higher-level sedimentation-erosion model may be required.

The critical shear stress value allowing deposition of suspended sediments of varying grain size depends on a number of factors. For the expected population of fine sands and silts, a typical value of 0.1 pascal was considered adequate to evaluate change, and not absolute value, in bottom shear stress over a reasonably large area. After computing sediment deposition potential for the Base Case, we observed that areas of high potential coincided well with known distributions of fine-grained bottom sediments for the Lower James River (Nichols et al., 1991). The changes predicted for Alternative 9, as compared to the Base Case, were small and insignificant when compared to predicted variation in sediment deposition potential due to the tested variation in river discharge.

CONCLUSIONS AND RECOMMENDATIONS

The availability of GIS software and high-resolution satellite imagery has enabled precise geo-referencing of estuarine morphological and man-made structural features in parallel with the development of fine-scale, three-dimensional hydrodynamic model grids and associated model validation data. It is now possible to compare both near-field and far-field predictions of tide, current, and salinity within the model domain with spatially and temporally fixed measurement data collected with equal precision. The value of the added capability to Environmental Impact Studies and Environmental Assessments, as required by the U.S. National Environmental Policy Act (NEPA), should be clearly apparent. However, it should be equally apparent that increasing amounts of information require an improved set of criteria for their evaluation. We submit that feature-based criteria will prove exceptionally useful in this regard.

Feature-based criteria for evaluating model study results in coastal estuaries are suggested as a means to improve standard but often vague references to environmental quality. As a result of recent advances in estuarine research, a number of biological, physical and chemical processes have been linked to specific hydrodynamic features, either influencing the features or being influenced by them. Thus, it is important to know when, and to what degree, a feature is likely to be changed in terms of one or more of its definitive attributes. In this paper we have suggested tidal intrusion fronts, vertical and horizontal salinity gradients, tidal prisms, and residual current features including eddies. Predicted changes in these and other fundamental features of coastal estuaries are a focal point for concern and a starting point for further investigation as required. Improved field observations are needed to validate model predictions regarding the temporal and spatial behavior of specific estuarine features.

REFERENCES CITED

Boon, J.D., H.V. Wang, S.C. Kim, A.Y. Kuo and S.M. Sisson, 1999. Three dimensional hydrodynamic-sedimentation modeling study, Hampton Roads crossing, lower James River, Virginia. *Special Report No. 354 in Applied Marine Science and Ocean Engineering, Virginia Institute of Marine Science, Gloucester Point, 36p, 3 appendices.*

Dee, D.P., 1995. A pragmatic approach to model validation. In: D.R. Lynch and A.M. Davies (eds.), *Quantitative Skill Assessment for Coastal Ocean Models*, Coastal and Estuarine Studies, American Geophysical Union, Washington, DC, 47:1-13.

Hamrick, J. M., 1996. User's manual for the environmental fluid dynamics computer code. *Special Report in Applied Marine Science and Ocean Engineering No. 331*, Virginia Institute of Marine Science, Gloucester Point, VA, 223p.

Hood, R.R., H.V. Wang, J.E. Purcell, E.D. Houde, and L.W. Harding, 1999. Modeling particles and pelagic organisms in Chesapeake Bay: convergent features control plankton distributions. *Journal of Geophysical Research*, 104:1223-1243.

Kuo, A. Y., R. J. Byrne, J. M. Brubaker, and J. H. Posenau, 1988. Vertical transport across an estuary front, pp.93-109. In J. Dronkers and W. van Leussen (eds.), *Physical Processes in Estuaries*. Springer-Verlag, Berlin.

Kuo, A. Y., R. J. Byrne, P. V. Hyer, E. P. Ruzecki, and J. M. Brubaker, 1990. Practical application of theory for tidal-intrusion fronts. *Journal of Waterway, Port, Coastal, and Ocean Engineering* 116(3): 341-361.

Nichols, M.N., S.C. Kim, and C.M. Brouwer, 1991. Sediment Characterization of the Chesapeake Bay and its Tributaries, Virginian Province. *NOAA National Estuarine Inventory: Supplement*. VIMS contract report, 83p.

Ruzecki, E. P., and W. J. Hargis, Jr., 1988. Interaction between circulation of the estuary of the James River and transport of oyster larvae , p. 255-278. In B. J. Neilson, A. Y. Kuo, and J. M. Brubaker (eds.), *Estuarine Circulation*. The Humana Press, Clifton, N. J.

Shen, J., J.D. Boon and A.Y. Kuo, 1999. A modeling study of a tidal intrusion front and its impact on larval dispersion in the James River estuary, Virginia. *Estuaries,* 22(3a): 681-692.

Shen, J., and A.Y. Kuo, 1999. Numerical investigation of an estuarine front and its associated eddy. *Journal of Waterway, Port, Coastal, and Ocean Engineering* 125(3): 127-135.

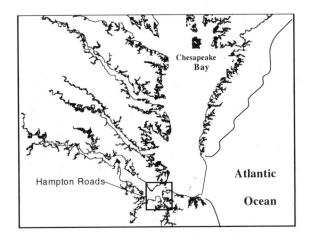

Figure 1. Location of Hampton Roads, Virginia

Figure 2. Existing Features, James and Elizabeth Rivers

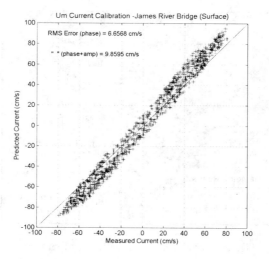

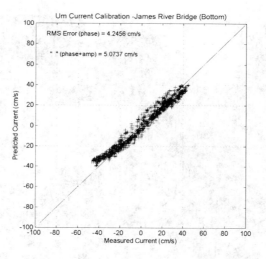

Figure 3. Predicted versus measured current, James River Bridge.

Figure 4a. Bridge-tunnel structure, proposed Alternative 9 crossing, Hampton Roads.

Figure 4b. Bridge-tunnel structure, proposed Alternative 2 crossing, Hampton Roads

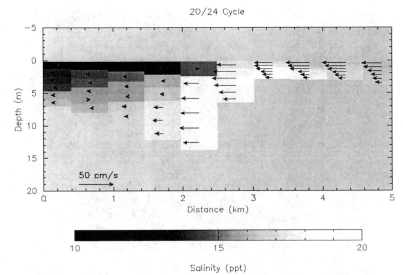

Figure 5a. Current and salinity transect crossing tidal front, Base Case.

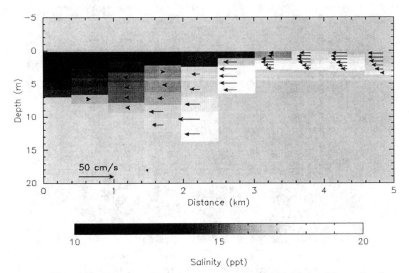

Figure 5b. Current and salinity transect crossing tidal front, Alternative 9.

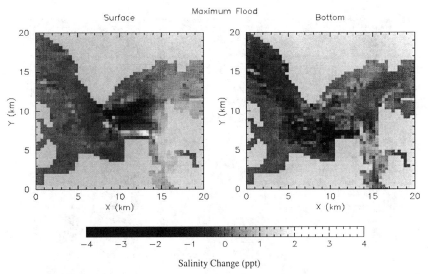

Figure 6a. Salinity change, Alternative 9, during maximum flood, apogean-neap tide, mean river inflow.

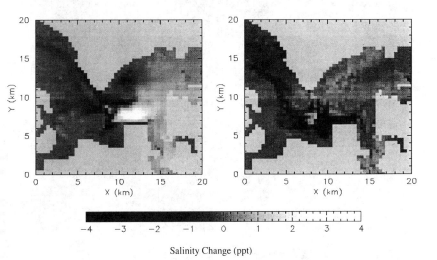

Figure 6b. Salinity change, Alternative 9, during maximum ebb, apogean-neap tide, mean river inflow.

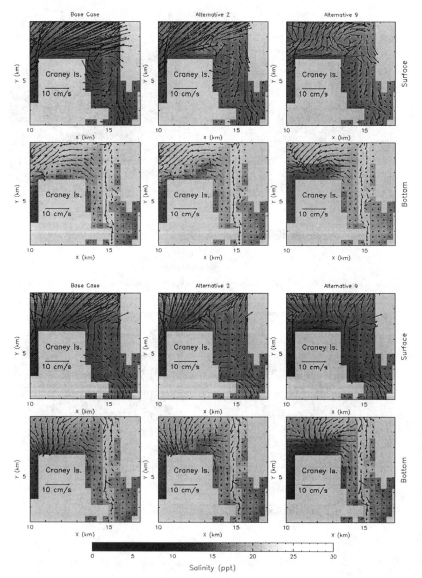

Figure 7. Tidally-averaged horizontal current and salinity; Apogean-neap tide (upper 6 panels) and perigean-spring tide (lower 6 panels), simulation comparison for Base Case, Alternative 2 and Alternative 9, Elizabeth River Entrance.

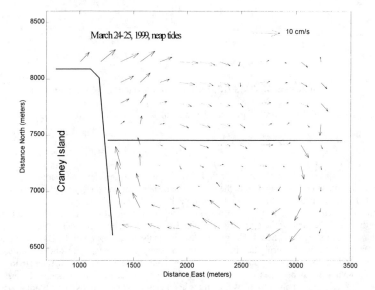

Figure 8. Measured surface current, 25-hour tidal average using a vessel-mounted ADCP at the entrance to the Elizabeth River.

Application of VIMS HEM-3D to a Macro-Tidal Environment

Sung-Chan Kim[1], Jian Shen[2], Chang Shik Kim[3], and Albert Y. Kuo[4]

Abstract

VIMS (Virginia Institute of Marine Science) HEM-3D (3-Dimensional Hydrodynamic-Eutrophication Model) was applied to the coastal water encompassing the Kyunggi Bay of Korea. This first phase of study was to demonstrate the suitability of the model to a macro-tidal environment. The Kyunggi Bay receives freshwater input, together with its associated pollutants from the Han River, one of the four major rivers in Korea with its drainage basin including the metropolitan capital city of Seoul. The mean tide range in the inner portion of the Bay is on the order of 6 meters and the spring tide range may reach as high as 11 meters. In addition there are extensive tidal flats in the inner portion of the Bay, which become exposed at low tide. Up to 5 % of the Kyunggi Bay becomes 'dry' at low tide. Therefore the accurate simulation of the wet-and-dry process and the mass conservation of the computation scheme are two important issues in this model application.

The hydrodynamic portion of HEM-3D uses the EFDC (Environmental Fluid Dynamics Code). It employs an internal-external mode splitting procedure to separate the internal shear or baroclinic mode from the external free surface gravity wave. The procedure determining wetting and drying of a cell is conducted during the process of solving the external mode, i.e., solving surface elevation and the vertically-averaged horizontal velocity components. At the beginning of solving the external mode at each time step, each dry cell is checked against a preset criterion to determine if it has reached the potential to become a wet cell. If the criterion is met, the cell is put back into computation and solving of the external mode is conducted. After solving the external mode, each wet cell is checked against preset criteria to determine if it becomes dry. The drying cells are taken out of computation and the same time step computation is repeated. This procedure is iterated until no new cell becomes dry. This improves the computational stability and ensures mass conservation over each step of computation. When a cell becomes dry, the non-zero value of water depth is maintained in the cell and the transport across all cell faces is blocked while the kinetic processes of water quality computation continue to be executed at each time step.

---

[1] Research Assistant Professor, Virginia Institute of Marine Science, School of Marine Science, College of William and Mary, Gloucester Point, VA 23062, e-mail: sunkim@vims.edu

[2] Environmental Engineer, Tetra Tech, Inc., 10306 Eaton Place, Suite 340, Fairfax, VA 22030, e-mail: shenji@tetratech-ffx.com

[3] Senior Scientist, Korea Ocean Research and Development Institute, 1270 Sadong, Ansan 425-170, South Korea, e-mail: surfkim@kordi.re.kr

[4] Professor, Virginia Institute of Marine Science, School of Marine Science, College of William and Mary, Gloucester Point, VA 23062, e-mail: kuo@vims.edu

The hydrodynamic portion of the model was preliminarily calibrated with mean tide ranges at six tidal stations along the coast. The model results compare very well with prototype observations. The model also reproduced the characteristic propagation of a tidal wave along the coast in the south-north direction. The area of tidal flats exposed at low tide was accurately portrayed. The utility of the water quality model was demonstrated with a hypothetical source of pollutants discharging from the Han River.

Introduction

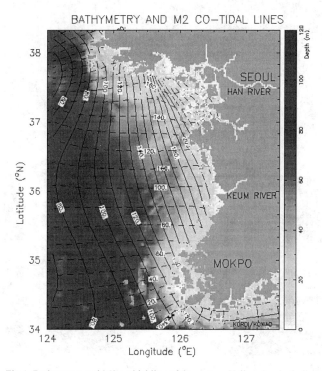

Fig 1. Bathymetry and M2 cotidal line of the eastern Yellow Sea including the Kyunggi Bay.

The Kyunggi Bay is located in the eastern part of the Yellow Sea. A major tributary—the Han River— flows into the Bay from the west coast of the Korean Peninsula. The Bay is a typical macro-tidal region with an average tidal range of 6 m and the maximum tidal range of 9.54 m (according to Korea Hydrography Office tide record, 1989). M2 is the most dominant constituent. At Inchon, located at the center of the coastline in the bay, the M2 amplitude is 2.92 m whereas the S2, K1 and O1 amplitudes are 1.13 m, 0.26 m, and 0.39 m, respectively (Choi, 1980). Fig 1 shows the location map with bathymetry and M2 cotidal lines. The Bay is comprised of complicated coastlines and many islands. High tidal range results in expansive tidal flats along the coastline in the Bay. Fig 2 is a satellite image showing these complicated coastlines with extensive tidal flats in the Bay. About 5 % of Kyunggi Bay is exposed during low tide. A modeling of the Bay should include the proper representation of the tidal flats.

Fig 2. A satellite image of the Kyunggi Bay.

VIMS has developed the HEM-3D model—a general purpose three-dimensional Hydrodynamic-Eutrophication Model. It is uniquely suited to estuaries and coastal seas. The model has been applied to a wide range of environmental studies in the Chesapeake Bay and its tributaries (e.g., Shen et al., 1997; Hamrick et. al., 1995; Hamrick, 1992a, 1992b) and other systems (e.g., Kim et al., 1997; Moustafa and Hamrick, 1994). One of the unique features of the HEM-3D model is the handling of the wet-and-dry scheme. In this study, we applied the HEM-3D model to the Kyunggi Bay to investigate the capability of the model's tidal flat representation. A brief description of the model and the wet-and-dry scheme is given. The application follows and then discussion is given.

HEM-3D model

The VIMS HEM-3D (Three-Dimensional Hydrodynamic-Eutrophication Model) consists of two portions: the hydrodynamic and water quality portions. The hydrodynamic portion of the model is essentially the EFDC (Environmental Fluid Dynamics Computer Code, Hamrick, 1992), which can be run alone. The transport calculations of the water quality portion of the model are integrated with the hydrodynamic portion, and must be run together with the latter. However the kinetic processes of the water quality portion of the model are decoupled from the transport processes (Park and Kuo, 1996) such that they can be computed with a time step and spatial domain that are different from those of the hydrodynamic portion of the model. The water quality model (Park et al., 1995) simulates nutrient cycling, phytoplankton dynamics and the dissolved oxygen budget. It has 21 water quality state variables, with the kinetic processes mostly the same as those of the Chesapeake Bay three-dimensional water quality model, CE-QUAL-ICM (Cerco and Cole, 1994). The model has been described in the reports by Hamrick (1992) and Park et al. (1995). Fig 3 shows the overall schematics of the HEM-3D model.

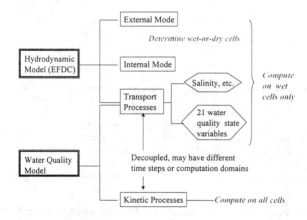

Fig 3. Schematics of the HEM-3D model.

The EFDC uses an internal-external mode splitting procedure to separate the internal shear or baroclinic mode from the external free surface gravity wave. The model uses a staggered grid structure. The horizontal grid system and the location of variables within the grid are shown in Fig. 4. The surface elevation (η) and total water depth (H) are defined at the center of each cell together with density (ρ), salinity (s), temperature (T), and other dissolved substance concentration (C). The velocity components, in both north (V) and east (U) directions, as well as transport fluxes are defined at the faces of each cell.

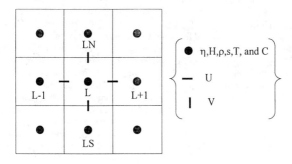

Fig 4. The horizontal grid pattern, location and indexing of variables.

Wet-and-dry scheme

The processes of determining if a cell is 'wet' or 'dry' are performed immediately prior to and after each solving of the external mode where the surface elevation, and thus the total water depth, is computed. Three pre-determined parameters are used to implement the criteria for assigning the wet-or-dry status of a cell: for the wet depth, H_{wet}, the dry depth, H_{dry}, and the minimum number of time steps, $N_{drystep}$, a cell has to remain dry once it is determined dry. An intermediate depth, H_{wd}, equal to the mean of wet and dry depths, is also used to determine if a cell is dry. The hydrodynamic and physical transport of water quality computations are performed over a domain including only the wet cells, while the kinetic process computations are performed for all cells.

The schematic diagram in Fig 5 demonstrates the procedure of determining wetting and drying of a cell. Before solving the external mode, each dry cell is checked as to whether it has remained dry for $N_{drystep}$ time steps. If it has, the cell is assigned as a wet cell and put into the computation domain. If it has not, the time step counter for the cell remaining dry is increased by one, and the cell remains excluded from the computation domain.

After solving the external mode and the new depth of each wet cell is calculated. Each cell is then tested against the drying criteria to determine its wet or dry status. This is done in two stages. First, a cell is considered a dry cell if the depth at the cell center is less than H_{dry}. A dry cell will be excluded from the computation domain in the next iteration of the external mode solution.

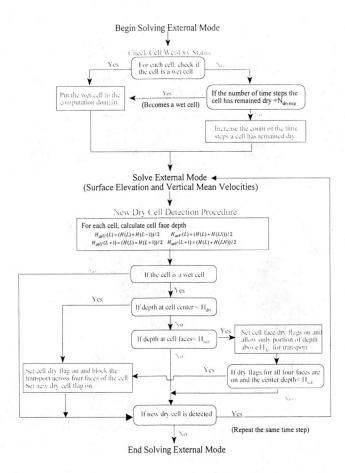

Fig 5. Procedure for wet-or-dry cells determination.

To enhance the stability of the computation scheme, a second set of drying critera is implemented for each cell with a computed center depth greater than H_{dry}. If the computed depths at all four faces of the cell are less than H_{wet} and the center depth is less than H_{wd}, the cell is also considered a dry cell and excluded from computation domain. Furthermore, a cell is considered only partially wet if any of the cell face depths are less than H_{wet}. Under this condition, water is allowed to flow through only the portion of that cell face or faces above the dry depth in the next iteration of computation.

If any new dry cell is detected, the solution of the external mode is repeated, excluding all new and old dry cells from the computation domain. The new dry cells are assigned the original depths at the beginning of the time step and no transport is allowed through any cell face to assure that mass conservation is maintained. The solving of the external mode for a given time step is repeated until no more new dry cell results from the last iteration. Then the model continues with the solution of the internal mode and the transport portion of the water quality model, with the computation domain including only the wet cells. However, the computation of kinetic processes in the water quality portion of the model is performed over the dry as well as wet cells. For the model application to the Kyunggi Bay, the following values are used in the wet-or-dry criteria: $H_{wet} = 15$ cm, $H_{dry} = 10$ cm, and $N_{drystep} = 6$.

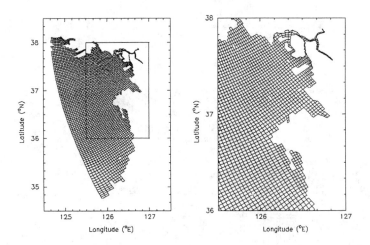

Fig 6. HEM-3D model grid for Kyunggi Bay, Korea.

A computational grid was generated utilizing a 78 by 96 matrix of orthogonal horizontal grid cells to cover the model domain. The total number of active surface water cells is 3122. Vertically, 6 layers (KC) are set in σ-coordinates (evenly spaced as 1/6 of the total depth). The generated curvilinear grid has a higher resolution near Inchon (~ 1 km) and lower resolution at the open boundary and towards the south (~ 3 km) (Fig 6). The Han River was included as a major connecting estuary at the eastern boundary. Both the north and south boundaries were land cells. A western open boundary with 90 cells was set along the co-tidal line of the M2 tide with a 1.2 m amplitude. The phase was assumed to increase by 2 degrees between adjacent western open boundary cells from south to north. Along the western open boundary, salinity and temperature were set as 30 psu and 25 °C. The hydrodynamic simulation also included the simulation of a conservative substance. Its concentration was assumed to be zero at the open boundary. A constant flow rate at the Han River head was assumed with a conservative concentration of 20 mg/l. The total simulation span was 10 days with a time step of 72 seconds.

Model Results

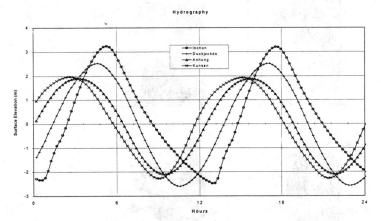

Fig 7. Surface elevation time series at selected stations.

Fig 7 shows a one-day portion of surface elevation time series at the selected cells representing important coastal locations. Propagation of tides from south (Kunsan) to north (Inchon) is apparent. The following table summarizes the tidal range calculated at selected stations compared with existing observations. Reasonable agreement is seen as higher M2 tide range at Inchon (~ 6 m) than at Kunsan (~ 4 m). In general peak flood current leads high tide and the lead increased from south to north, suggesting that the tide became characteristically standing waves toward the coastal zone near the entrance of the Han River.

Station	Observation (m)	Model (m)	HT lag (hour)	LT lag (hour)
Daechungdo	1.97	2.02	12 .84	11.21
Inchon	5.80	5.73	12.42	12.42
Duckjuckdo	4.94	5.09	11.79	9.94
Sungapdo	4.69	4.54	11.42	9.30
Auhung	4.20	3.99	10.57	8.88
Kunsan	4.20	4.21	9.94	8.46
Wonsando	4.50	4.36	9.73	8.88

Table 1. Predicted versus observed tide range and phase. Lag times are calculated from the model simulation.

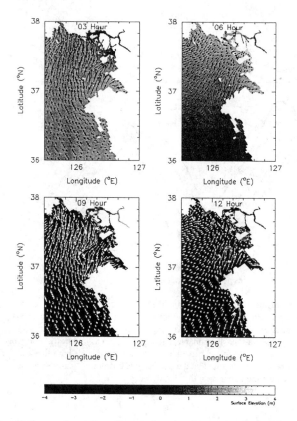

Figure 8. Temporal variation of sea surface elevations and surface currents

Time variation of the spatial distribution of surface elevation (Fig 8) shows not only the phase relationships but also the successful implementation of the wet-and-dry scheme. At hour 12, the dry cells were represented by a gray color. The dry cells represent qualitatively well the tidal flats in the Kyunggi Bay. Quantitative comparison of the exposed area was not attempted because the wet-and-dry scheme is sensitive to bathymetry and the model bathymetry was only visually checked with nautical charts.

Water quality was also simulated to test the HEM-3D model for demonstration purposes. All the conditions were hypothetical. After 7 days, chlorophyll was concentrated along the shallow coastal zone near the entrance of the Han River (Fig 9). Ammonium nitrogen in the same region was consumed by uptake and thus resulted in low concentration.

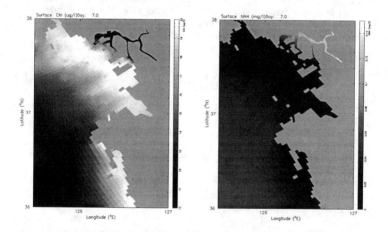

Fig 9. Surface chlorophyll and ammonium nitrogen distribution after 7 days.

Summary and Conclusions

The HEM-3D model was setup for the Kyunggi Bay, Korea. The model was forced at the western open boundary by the M2 tide with an amplitude half that of the mean tidal range. The calculated tidal ranges agree well with available field observations at 7 selected locations. The model results indicate that the tidal wave propagates in a general south to north direction. The tidal current has a phase lead with respect to the tidal height. The phase lead increases as the tide approaches the shoreline toward the northeast. Since the Kyunggi Bay is a macro-tidal sea with extensive subtidal flats, the proper simulation of wet-and-dry cells is of the uttermost importance. The successful calibration with tidal range suggests that the wet-and-dry scheme used in the HEM-3D model is suitable. One may also achieve the same tidal calibration without wet-and-dry scheme. However, the most important feature in the Kyunggi Bay is the tidal flats which can be represented only through wet-and-dry scheme. It is not uncommon that the computations become unstable because of wet-and-dry. The scheme adopted by the HEM-3D appears not subject to the instability. Under the normal flow condition, the influence of freshwater discharge from the Han River on the salinity distribution is limited to a small area around the river mouth. The grid layout of the current model is not fine enough to properly simulate the salinity distribution. A higher resolution grid is required.

Since there is insufficient water quality data for the calibration of the water quality portion of the model, the model was run with some hypothetical input conditions. The values of calibration coefficients were based on those of other systems such as the Chesapeake Bay. The results indicate that the model has been properly set up for the Kyunggi Bay. However, the results should be used only for demonstration purposes. It is hoped that more water quality and hydrographic data would enhance the operation of the HEM-3D model for the Kyunggi Bay and other coastal waters of Korea.

References

Cerco, C. F., and T. Cole, 1993. Three-dimensional eutrophication model of Chesapeake Bay. *J. Environ. Engnr.*, 119, 1006-1025.

Choi, B. H., 1980. A tidal model of the Yellow Sea and the Eastern China Sea. *KORDI Report* 80-02. Korea Ocean Research and Development Institute. 72 pp.

Hamrick, J. M., 1992a. A Three-Dimensional Environmental Fluid Dynamics Computer Code: Theoretical and Computational Aspects. The College of William and Mary, Virginia Institute of Marine Science. Special Report 317, 63 pp.

Hamrick, J. M., 1992b. Estuarine environmental impact assessment using a three-dimensional circulation and transport model. *Estuarine and Coastal Modeling, Proceedings of the 2nd International Conference*, M. L. Spaulding *et al*, Eds., American Society of Civil Engineers, New Yrok, 292-303.

Hamrick, J. M., A. Y. Kuo, and J. Shen, 1995. Mixing and dilution of the Surry Nuclear Power Plant cooling water discharge into the James River. A report to Virginia Power Company, Richmond. The College of William and Mary, Virginia Institute of Marine Science, 76 pp.

Kim, S.-C., L. D. Wright, J. P.-Y. Maa, and J. Shen, 1998. Morphodynamic responses to extratropical meteorological forcing on the inner shelf of the Middle Atlantic Bight: wind waves, currents, and suspended sediment transport. (*ed* by M.L. Spaulding and A.F.

Blumberg) *Estuarine and coastal modeling, Proceedings of 5th International Conference*, ASCE, 456-466

Moustafa, M. Z., and J. M. Hamrick, 1994. Modeling circulation and salinity transport in the Indian River Lagoon. *Estuarine and Coastal Modeling, Proceedings of the 3rd International Conference*, M. L. Spaulding *et al*, Eds., American Society of Civil Engineers, New York, 381-395.

Park, K. and A. Y. Kuo, 1996. A multi-step computation scheme decoupling kinetic processes from physical transport in water quality models. *Water Research*, 30, 2255-2264

Park, K., A. Y. Kuo, J. Shen, and J. M. Hamrick, 1995. A three-dimensional hydrodynamic-eutrophication model (HEM-3D): Description of water quality and sediment process submodels. VIMS Special Report in Applied Marine Science and Ocean Engineering # 327. 102 pp.

Shen, J., G. M. Sisson, A. Y. Kuo, J. D. Boon, and S.-C. Kim, 1998. Three-dimensional numerical modeling of the tidal York River system, Virginia. (*ed* by M.L. Spaulding and A.F. Blumberg) *Estuarine and coastal modeling, Proceedings of 5th International Conference*, ASCE, 495-510

Simulation of Tunnel Island and Bridge Piling Effects in a Tidal Estuary

Harry V. Wang[1] and Sung-Chan Kim[2]

Abstract

The passageways across the tidal estuary frequently entail structures such as bridge piers, pilings and tunnel islands in the midst of a water body. A structure generates interference to the flow and the disturbances can potentially be carried away by the tidal flow, creating a far field effect. In this paper, we use a three-dimensional numerical model in an attempt to simulate the effects of the tunnel island and the pilings on primary physical characteristics in the tidal estuary. The tunnel islands considered are comparable to the grid size and are treated either as islands or the thin wall boundaries. The bridge pilings, on the other hand, are spatially distributed and individually smaller in size compared to the grid dimension, and thus are parameterized by the porosity parameters. The porosity is defined as the water fraction of the piling area projected onto a constant x, y, z plane; and is used to quantify the collective effects of individual pilings on a control volume of numerical grid scale. Dynamically, the main effects that structures exert onto the flow are constriction, impedance (friction effect), and the production of turbulence energy; that tied to main flow properties such as volume flux, momentum balance and the turbulent mixing. The numerical model was modified, tested, and used to simulate the potential effect of highway structures in the lower James River. Overall, the results show that the effects of planned highway structures on the flow are local in nature, which does not significantly affect the major features in the lower James River; however, noticeable effects are identified.

---

1   Assistant professor, Virginia Institute of Marine Science. Gloucester Point, VA 23062;
    Tel:804-684-7215; email: wang@vims.edu
2. Research associate professor, Virginia Institute of Marine Science. Tel:804-684-7359; email:
    sunkim@vims.edu

1. Introduction

Many large ports and cities are located near coastal bays and estuaries. In order to support the highway infrastructure, man-made structures such as bridges and tunnels are frequently built across the tidal estuary. Most of the bridges are braced by piers or piling directly into the water body. The tunnel, on the other hand, is underground, but in many cases is built in conjunction with the bridge. It thus requires man-made islands be constructed on the shallow side of the water body to provide its entrance and exit. One such example is the Chesapeake Bay Bridge Tunnel located at the mouth of the Chesapeake Bay in the East Coast of the US.

From a hydrodynamic point of view, the bridge piling and tunnel island built in the middle of the water are destined to generate interference to an otherwise natural environment. Frictional effect is one, constriction and turbulent mixing are examples of others. The important question is whether the interference, in a broad context, creates an adverse effect to the environment. Stigebrandt (1992) uses an analytical model to assess the possible salinity change due to bridge piling in the Baltic Sea. Miller and Valle-Levinson (1996) measured the pile-induced stratification change in an existing pier and compared with the bottom and wind induced change. Their conclusion is that the effects induced by bridge piling are substantially less than the naturally occurring one. In an attempt to conduct environmental assessment, Boon et. al. (1999) suggest that impacts should not be evaluated solely on time series in fixed and selected stations, as was conducted traditionally. Spatial features such as the front and eddy should also be identified and considered because, in many instances, they are associated with important biological and chemical processes.

The difficulties in dealing with assessment of man-made-structure effects are two-fold. First, the effects cannot be measured until a real structure is actually built; second, the signals induced by man-made structure are likely to be intertwined with the natural variability and hard to separate. Three-dimensional numerical models have been successfully applied in the estuaries (Oey et. al., 1985; Johnson et. al., 1993; Casulli and Cheng, 1993). The simulated flow, salinity, temperature and water elevation under natural conditions are routinely output from a model. If the man-made structure such as bridge piling and the tunnel island can be properly formulated into the numerical model, then the model will possess a capability of simulating the man-made effect in a virtual reality (without actually building one) and can be analyzed separately from the natural forcing. With recent advances on GIS (Sisson et. al., 1999), the spatial features from model can also be identified robustly.

The purpose of the paper is to formulate the tunnel island and the piling effect into a three-dimensional numerical model. The goal is not driven by the interest of the near-field phenomena such as scouring effects on the structure; rather, it is intended to be used for far-field environmental impact evaluation.

2. Model description

The model used for the study is the HYSED3D-VIMS three-dimensional hydrodynamic model, a variation of the EFDC (Environmental Fluid Dynamics Code), originally developed by Hamrick

(1996). The model solves the three-dimensional, vertically hydrostatic, free surface equations of motions for a variable density fluid. The dynamically coupled transport equations for turbulent kinetic energy, turbulent length scale, salinity and temperature equations are also solved. The two-parameter turbulent transport equation implemented is the Mellor-Yamada lever 2.5 closure scheme. The model uses an orthogonal curvilinear coordinate in the horizontal and sigma vertical coordinate. The time integration employed is the second-order accurate three-time level finite difference scheme with an internal-external mode splitting procedure.

Two features are of particular interest in this study. 1) tunnel islands, whose dimension is about 1 km long and slightly less than 370 m wide and 2) bridge pilings, whose individual pier diameter is 1.37 m and which are spaced 24 m apart, spanning through a 5 to 10 km stretch of highway pavement. Since the dimension of the tunnel island is comparable to the grid resolution (on the order of 370 meter), it can either be treated as a thin-wall boundary (if the shape is slender) or as land cells if it occupies a whole grid cell or more. This thin-wall boundary can be masked in either east, west, south, north or both faces of a cell so that no flow is allowed to pass through the boundary and the resultant flow is forced to follow around the boundary. The dimension of the bridge piling, on the other hand, is much smaller than the grid resolution, and thus needs to be parameterized.

We use porosity to parameterize the piling effect. The porosity is defined as the water fraction of the piling area in the constant x, y, z cross-section plane. When incorporated into the mass balance of a control volume, it gives the continuity equation:

$$\frac{\partial}{\partial t}(\phi_z \zeta) + \frac{\partial}{\partial x}(\phi_x Hu) + \frac{\partial}{\partial y}(\phi_y Hv) + \frac{\partial}{\partial z}(\phi_z w) = 0$$

where ζ is the water surface elevation; H is the local depth of the water column; u, v, w are the horizontal and vertical velocities; ϕ_x, ϕ_y, ϕ_z are porosity in x, y and z planes and its values can be estimated from the actual design. The typical values of ϕ_x, ϕ_y, ϕ_z are 0.94, 0.94 and 0.99, but degenerate to 1.0 when the pilings are non-existent. The mass conservation equation for a dissolved or suspended material with concentration C is thus:

$$\frac{\partial}{\partial t}(\phi_z C) + \frac{\partial}{\partial x}(\phi_x HuC) + \frac{\partial}{\partial y}(\phi_y HvC) + \frac{\partial}{\partial z}(\phi_z wC) = \frac{\partial}{\partial z}\left(\phi_z H^{-1} k_z \frac{\partial}{\partial z} C\right) + (\phi_z S_c)$$

where k_z is the eddy diffusivity; S_c is the source or sink term.

In the momentum equation, a resistance term is incorporated to account for additional boundary shear stress introduced around the piling. The x- momentum equation reads:

$$\frac{\partial}{\partial t}(\phi_z\,Hu)+\frac{\partial}{\partial x}(\phi_x\,Huu)+\frac{\partial}{\partial y}(\phi_y\,Hvu)+\frac{\partial}{\partial z}(\phi_z\,wu)-f\,\phi_z\,H\,v$$

$$=-\phi_z H\frac{\partial}{\partial x}(g\zeta+p)+\frac{\partial}{\partial z}\left(\phi_z H^{-1}N_z\frac{\partial u}{\partial z}\right)-c_p\frac{B_p}{L_p^2}H\left(u^2+v^2\right)^{0.5}u$$

where the terms on the left represent the inertial terms and the Coriolis force terms, and those on the right, the pressure gradient term, the vertical shear stress term and lastly the resistance term due to the piling. N_z represents eddy viscosity, c_p the drag coefficient, B_p the projected piling width and L_p the separation length scale for the piling density. The resistance term is applied throughout the water column as opposed to the bottom shear stress, which is applied only at the bottom as a bottom boundary condition. The y–momentum equation (not shown) can be derived similarly.

The turbulence formulation used is the standard Mellor-Yamada turbulence closure scheme, as extended by Galperin et. al. (1988) and Blumberg et. al. (1992). The modifications were made to include an additional turbulence energy production and dissipation induced by the bridge piling. The turbulent energy equation reads:

$$\frac{\partial}{\partial t}(\phi_z\,Hq^2)+\frac{\partial}{\partial x}(\phi_x\,Huq^2)+\frac{\partial}{\partial y}(\phi_y\,Hvq^2)+\frac{\partial}{\partial z}(\phi_z\,wq^2)=\frac{\partial}{\partial z}(\phi_z H^{-1}A_q\frac{\partial}{\partial z}q^2)$$

$$+2\phi_z H^{-1}A_v((\frac{\partial u}{\partial z})^2+(\frac{\partial v}{\partial z})^2)+2c_p\frac{B_p}{L_p^2}H\left(u^2+v^2\right)^{1.5}+2\phi_z\,gA_b\frac{\partial b}{\partial z}$$

$$-2\phi_z H(B_1 l)^{-1}q^3$$

Where q is turbulent intensity, b is buoyancy, ϕ_x, ϕ_y, ϕ_z are porosity; C_p, B_p, and L_p are drag coefficient, projected width and separation length for the piling, as defined before. B_1 is constant derived from experiment (Blumberg et. al. 1992).

Essentially, what the bridge piling effect brought forth and exerted on the water body are the following dynamical effects: (1) constriction (2) impedance (frictional effect) and (3) turbulent mixing. Of which (1) is parameterized by porosity, (2) by water column quadratic resistance and (3) by additional turbulent energy production and dissipation due to flow-structure interaction.

3. Model verification

The three-dimensional numerical model described above was applied in James River. The model domain starts from the river mouth adjacent to the Chesapeake Bay, and extends as far as · Richmond, Virginia, covering a 150-km stretch of the River. It uses a Cartesian grid with 370 m by 370 m resolution, totaling about 27,500 cells. Figure 1 shows the grid in the lower James

River near Hampton Road.

The model was calibrated using a comprehensive set of field data collected during 1985-1990 (Kuo et. al, 1988, 1990). The calibration involved the adjustment of bottom friction to achieve an acceptable agreement between the measurement and the model simulated surface elevation and the horizontal velocity at selected locations. Examples of astronomic tidal height comparisons are shown in Figure 2; which shows the surface water elevation comparison between modeled and measured at (a) the James River Bridge and (b) Newport News Point. The comparison was very good; the expected phase error was about 5-8 minutes and the amplitude error of 3-4 cm. The current data from the surface and bottom layers of the main river channel at the James River Bridge are also compared, as shown in Figure 3 (a) and (b). The flow component compared are those in the direction parallel to the channel axis. The RMS error for predicted surface current is about 7-8 cm/sec on the surface and 4-5 cm/s at the bottom. When compared over a full tidal cycle, the amplitude estimation errors are less than 10% and the phase error are less than 20 minutes.

A special laboratory experiment was conducted in the Waterways Experiment Station to test the bridge piling and tunnel island effects on the shoaling effect in the Newport News-Portsmouth region (Fang, et. al. 1972). The experiment used gilsonite that was injected through a pipe following the main channel from just upstream of the James River Bridge for 2- 4 tidal cycle continuously. A grid was designed in the physical model domain to provide an area from which the gilsonite could be vacuumed and the major feature of the shoaling could be seen from the data collected. The numerical data for the deposition of gilsonite in each grid area are tabulated. Table 1 shows the distribution of deposition or scouring under the I-664 bridge-tunnel configuration (resembling configuration 2A in the table). Figure 4 provides the graphical representation of the table; the larger-than-one ratios given in the figures show the increased deposition while the less-than-one ratio show the scouring, both attributed to the bridge-tunnel configuration. The measured data suggest that erosion occurs immediately around the tunnel island and bridge piling, while deposition occurs in the main channel as well as in the north side of Craney Island, downstream of the structure.

We set up a numerical model experiment whose purpose was to mimic the physical model experiment. Sediment potential was determined by the percentage of time during which bottom stress falls below the critical bottom stress. The critical stress (0.1 Pascals) was chosen based on Maa et. al. (1993) who recorded measurements in the Lower Chesapeake Bay. The calculated distribution of sedimentation potential for scenarios with and without bridge piling-tunnel conditions is shown in Figure 5. When comparing the two scenarios, one immediately detects that the sediment potential is lower around the area where the bridge pier and tunnel were located. That means more erosion in the region. On the other hand, the sediment potential increases in the main channel downstream of the structure and also in the area north of Craney island; that means, for both of the regions, there is a tendency for deposition.

The results are highly consistent between numerical model simulation and the measurement from gilsonite experiment, at least qualitatively. Both experiments showed erosion occurred in the immediate vicinity of the bridge tunnel and bridge piling, while deposition occurred on the downstream side of the structure, both in the main channel and north side of Craney Island. The kinship of results from both numerical and physical models thus lend the support to the formulation instituted for simulating tunnel island and bridge piling effects in the numerical model.

4. Model application

The model was finally applied to evaluate the third crossing highway infrastructure in Hampton Roads, Virginia. The purpose is to determine whether the planned highway infrastructure will have an impact on the primary physical characteristics in the lower James River.

Design plans

There are three engineering plans for highway infrastructures, namely, Alternative 1, Alternative 2 and Alternative 9. In this paper, we focus primarily on impacts from Alternative 2 and Alternative 9. Figure 6 (a) show the design for Alternative 2, which connects Norfolk Harbor to the Craney Island across the Elizabeth River mouth. Figure 6(b) shows Alternative 9, which starts from Newport News and branches into two highways at the third crossing: one running from north to south continuing to Portsmouth and the other running from west to east continuing toward Craney Island. Alternative 2 and Alternative 9 are both causeways with which the bridges and the tunnel islands are constructed over in the shallow water and tunnels are built underneath the deep navigation channel. Both Alternative 2 and Alternative 9 have tunnel islands at both ends; whereas the former crosses the Elizabeth River and the latter crosses the James River. Most of the bridge piling are fairly straight and regular in spacing (24 m), with the exception at the intersection. The intersection is where highways of different branches merge and the pilings are denser there than in the straight section. The ϕ_x, ϕ_y, ϕ_z values are around 0.94, 0.94 and 0.99 in straight section, but reduced to 0.90, 0.90 and 0.97 at the intersection where dense pilings exist.

Scenario runs

In order to simulate the design plans realistically, three river flows condition, namely, high, medium and low river discharge were implemented. Also for simulating spring-neap tide, three tidal components (i.e., M_2, S_2, N_2) were synthesized. This is comparable to the traditional approach using 9 components when runs are conducted within 67 tidal cycles. In all, four plans (including base case), three river flow and three tidal condition totaling 36 scenarios, were conducted and the results were analyzed.

Results

Physical characteristics of primary concerns are: tidal elevation, currents, salinity, sedimentation and a special feature – tidal front. Following are the main conclusion from the modeling results for each of the components:

A. Tidal elevation: No discernable change in simulated tidal height was noted at any of the nine tide stations selected for comparison. None of the comparison evidenced any structure induced change in tidal height related to variation in tidal range or river inflow.

B. Currents: Spatial change in instantaneous surface and bottom currents is quite small and limited to a few highly local changes (e.g., in the vicinity of the bridge tunnels for Alternative 9). For individual stations, change of current in time history is apparent. Figure 7 shows the map for stations used for comparison. Station C1 at the entrance to the James River evidenced a slight difference in simulated currents for Alternatives 1 and 2. Station C2 at the entrance to the Elizabeth River demonstrated a more noticeable change in the surface current time histories for Alternatives 2 and 9 during all nine combinations of tidal range and river inflow. In addition to changes in vector curve, there are small differences in the strength of the current maximum (stronger flood and weaker ebb). This is suggestive of a residual current change when the duration of flood and ebb are approximately equal. At stations C3 (Newport News Channel) and C5 (Newport News Point) there is no change in bottom current. However, it is apparent that the surface current at station C3 possesses an ebb residual which is strengthened by Alternative 9 while the current range remain constant. At station C5, current range at mean tide (120 cm/sec) decreases by approximately 12 cm/sec (10%) while the ebb current residual again increases by a slight amount. These observations suggest a change in the direction of the surface current or a current divergence at station C5, which would produce a reduction on current range in the direction aligned with the channel axis.

C. Salinity: Changes in salinity were primarily limited to the vicinity of bridge pilings and the regions with measurable salinity stratification and enhanced vertical mixing. This is particularly noticeable for bridge pilings north of Craney Island. As expected, surface salinity increases and bottom salinity decreases in these regions as a result of turbulence-induced vertical mixing. The mixing is most intense during perigean-spring tides and least intense during apogean-neap tides.

D. Sedimentation: Figure 8 shows the sediment potential during the high flow. The results for Alternative 1, show little change except in limited areas. Runs for Alternatives 2 and 9, illustrate a more definite decrease in sedimentation potential along the immediate route of the bridge pilings north of Craney Island and near the Elizabeth River mouth. Similar results were produced when the freshwater inflow conditions were changed. Simulated release of tagged or traceable sediment particles with characteristic grain sizes and settling velocities indicate that only medium silt (15.6 – 31.3 μm) and finer grained sediment will reach the lower James River from known source regions upstream under the mean freshwater inflow condition. However, considerably more of this sediment and some coarser-grained material can reach the lower James during high river inflow. The change that occurs under natural occurring extreme condition thus outweighs the change due to structures added to by any of the Alternatives.

E. Special feature—tidal front: Further evaluation of potential impact in the hydrodynamics of lower James River is guided by the aforementioned study of the specific tidal front system located just below Newport News Point. Described as a tidally induced salinity front, this s system develops during the flood in each tide cycle as higher salinity water from the bay flows westward across Hampton Flats and converges with lower salinity (lower density) river water still ebbing to the southeast in the main channel. Because of the difference in density, the bay water subducts beneath river water at the front and travels upstream at the bottom of the channel (Kuo et al. 1988; 1990). This unique feature was recognized as an important mechanism for enhancing larval recruitment (Ruzecki and Hargis, 1988). Along the transect perpendicular to the front, the front movement during high river flow, as shown in Figure 9, is examined. The frontal interfaces are drawn through current convergence points at successive half hour intervals. The results reveal no substantive change in the position of the tidal front. However, in Alternative 9, it was noted that the shape of the frontal surface was changed in the surface layer and the phase was slightly delayed.

5. Conclusion

A three-dimension hydrodynamic and sedimentation model, HYSED, was modified to include the capability of simulating bridge tunnel and piling effect, in addition to the standard functions which provides information of water elevation, current, bottom shear stress and salinity field. The model was used to assess the effects of design infrastructure of highway crossing in the lower James River. The model uses a full range of hydrodynamic and hydrologic conditions expected for the prototype system, including tide of maximum and minimum range and extremes in freshwater inflow. In each test, model simulated properties were compared between the existing condition and that of the designed plans to determine the hydrodynamic response to the design structure.

Overall, the model results show that effect of the bridge tunnel and piling are local in nature; they are confined to the vicinity of the structure built. For example, at the location in the west and north of Craney where, under Alternative 9, a highway branches out from a junction directed west to east toward Norfolk. The impacts along the highway are noticeable and felt by hydrodynamic characteristics in currents, salinity and sediment potential. Another area of concern is the mouth of the Elizabeth River. There, a small change in the lower James River can have potential effect to the Elizabeth River proper, for the latter is a tributary of the former. The change of eddy structure at the mouth of Elizabeth River discussed by Boon et. al. (1999) is a good example. Further investigations are deemed to be necessary in order to thoroughly understand the full-fledged impact.

Acknowledgement

The authors appreciated the supports from Michael Baker Inc. and Virginia Department of Transportation.

References:

Blumberg, A. F., B. Galperin, and D.J. O'Connor (1992): Modeling vertical structure of open channel flows. Journal of Hydraulic Engineering, 118: 1119-1134.

Boon, J. D., A. Y. Kuo, H. V. Wang and J. M. Brubaker (1999): Proposed third crossing of Hampton Roads, James River, Virginia: feature-based criteria for evaluation of model study results. Submitted to ASCE 6th international conference on estuarine and coastal modeling, New Orleans, LA.

Casulli, V. and R. T. Cheng (1993): Semi-implicit finite difference methods for three-dimensional shallow water flow. Inter. J. for numerical Methods in Fluids, Vol. 15, p 629-648.

Fang, C. S., B. J. Neilson, A. Y. Kuo, R. J. Byrne, and C. S. Welch, (1972): Physical and geological studies of the proposed bridge-tunnel crossing of Hampton Roads near Craney Island. Special Report in Applied Marine Science and Ocean Engineering, No. 24, Virginia Institute of Marine Science, Gloucester Point, VA pp. 199-241.

Galperine, B., L. H. Kantha, D. Hassid, and A. Rosati (1998): A quasi-equilibrium turbulent energy model for geophysical flow. Journal of Atmospheric Science, 45: 55-62.

Hamrick, J. M. (1996): User's manual for the environmental fluid dynamics computer code. Special report in Applied Marine Science and Ocean Engineering No. 331, Virginia Institute of Marine Science, Gloucester Point, VA, 223 pp.

Johnson, B. H., K. W. Kim, R. E. Heath, B. B. Hsieh, and H. L. Butler (1993): Validation of a three-dimensional hydrodynamic model of Chesapeake Bay. Journal of Hydraulic Engineering, 119: 2-20.

Kuo, A. Y., R. J. Byrne, J. M. Brubaker, and J. H. Posenau (1988): Vertical transport across an estuary front, pp. 93-109. In J. Dronkers and W. van Leussen (eds.), Physical Processes in Estuaries, Springer-Verlang, Berlin.

Kuo, A. Y., R. J. Byrne, P. V. Hyer, E. P. Rusecki, and J. M. Brubaker (1990): Practical application of theory for tidal-intrusion fronts. Journal of Waterway, Port, Coastal, and Ocean Engineering, 116(3): 341-361.

Maa, J. P-Y, C. H. Lee, F. J. Chen (1995): VIMS sea carousel: bed shear stress measurements.

Marine Geology, 129: 129-136.

Miller, J. L. and A. Valle-Levinson (1996): The effect of bridge piles on stratification in the lower Chesapeake Bay. Estuaries, vol. 19, no.3, pp 526-539.

Oey, L. –Y., G. Mellor, R. Hires (1985): Three-dimensional simulation of the Hudson-Raritan estuary. Part III: salt flux analyses. Journal of Physical Oceanography 14: 629-645.

Ruzecki, E. P., and W. J. Hargis, Jr.(1988): Interaction between circulation of the estuary of the James River and transport of oyster larvae, pp. 255-278. In B. J. Neilson, A. Y. Kuo and J. M. Brubaker (eds.), Estuarine Circulation. The Humana Press, Clifton, N. J.

Sisson, G. M., J. D. Boon, K. L. Farnsworth (1999): The use of GIS in 3D Hydrodynamic model pre- and post-processing for feature-specific applications. Submit to ASCE 6th international conference on estuarine and coastal modeling, New Orleans, LA.

Stigebrandt, A (1992): Bridge-induced flow reduction in sea straits with reference to effects of a planned bridge across Öresund, Ambio 21:130-134.

Table 1. Gilsonite Study data

Area	Study 2X Raw	Pct.	Ratio	Study 2A Raw	Pct.	Ratio	Study 2B Raw	Pct.	Ratio
TLIN	44087.	100.00	1.00	44224.	100.00	1.00	45073.	100.00	1.00
1	125.	0.28	1.00	45.	0.10	0.35	70.	0.15	0.54
2	2100.	4.76	1.00	2060.	4.65	0.97	1990.	4.41	0.92
3	350.	0.79	1.00	160.	0.36	0.45	147.	0.32	0.41
4	505.	1.14	1.00	860.	1.94	1.69	705.	1.56	1.36
5	500.	1.13	1.00	1130.	2.55	2.25	595.	1.32	1.16
6	0.	0.00	0.00	0.	0.00	0.00	0.	0.00	0.00
7	715.	1.62	1.00	560.	1.26	0.78	520.	1.15	0.71
8	505.	1.14	1.00	270.	0.61	0.53	190.	0.42	0.36
9	430.	0.97	1.00	153.	0.34	0.35	198.	0.43	0.45
10	195.	0.44	1.00	93,	0.21	0.47	165.	0.36	0.82
11	520.	1.17	1.00	525.	1.18	1.00	500.	1.10	0.94
12	1085.	2.46	1.00	1115.	2.52	1.02	1186.	2.63	1.06
13	243.	0.55	1.00	240.	0.54	0.98	82.	0.18	0.33
14	280.	0.63	1.00	240.	0.54	0.85	40.	0.08	0.13
15	285.	0.64	1.00	150.	0.33	0.52	157.	0.34	0.53
16	285.	0.64	1.00	58.	0.13	0.20	126.	0.27	0.43
17	285.	0.64	1.00	310.	0.70	1.08	286.	0.63	0.98
18	175.	0.39	1.00	142.	0.32	0.80	131.	0.29	0.73
19	0.	0.00	0.00	0.	0.00	0.00	0.	0.00	0.00
20	242.	0.54	1.00	245.	0.55	1.00	136.	0.30	0.54
21	220.	0.49	1.00	253.	0.57	1.14	140.	0.31	0.62
22	155.	0.35	1.00	150.	0.33	0.96	192.	0.42	1.21
23	190.	0.43	1.00	165.	0.37	0.86	186.	0.41	0.95
24	1030	2.33	1.00	1222.	2.76	1.18	1115.	2.47	1.05
25	285.	0.64	1.00	420.	0.94	1.46	260.	0.57	0.89
26	265.	0.60	1.00	140.	0.31	0.52	255.	0.56	0.94
27	420.	0.95	1.00	175.	0.39	0.41	170.	0.37	0.39
28	210.	0.47	1.00	303.	0.68	1.43	210.	0.46	0.97
29	749.	1.69	1.00	650.	146	0.86	625.	1.38	0.81
30	233.	0.52	1.00	295.	0.66	1.26	200.	0.44	0.83
31	353	0.80	1.00	165.	0.37	0.46	282.	0.62	078
32	175.	0.39	1.00	62.	0.14	0.35	168.	0.37	0.93
33	315.	0.71	1.00	152.	0.34	0.48	450.	0.99	1.39
44AC	305.	0.69	1.00	430.	0.97	1.40	358.	0.79	1.14
J52C	500.	1.13	1.00	590.	1.33	1.17	715.	1.58	1.39
1X2C	0.	0.00	0.00	0.	0.00	0.00	0.	0.00	0.00
9EC	445.	1.00	1.00	510.	1.15	1.14	290.	0.64	0.63
7C	685.	1.55	1.00	298.	0.67	0.43	170.	0.37	0.24
1X7C	0.	0.00	0.00	0.	0.00	0.00	0.	0.00	0.00
9C	655.	1.48	1.00	640.	1.44	0.97	940.	2.08	1.40
9AC	605.	1.37	1.00	510.	1.15	0.84	720.	1.59	1.16
2AC	973.	2.20	1.00	850.	1.92	0.87	820.	1.81	0.82
FSC	660.	1.49	1.00	600.	1.35	0.90	490.	1.08	0.72
6AC	1855.	4.20	1.00	1870.	4.22	1.00			
COLL	20108.	45.60	1.00	18804.	42.51	0.93	17631.	39.11	0.85

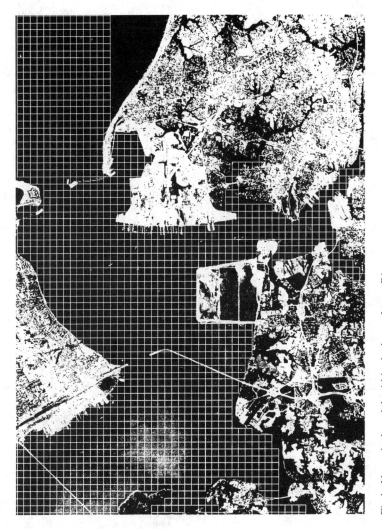

Fig 1. Numerical model grid in the lower James River.

(a)

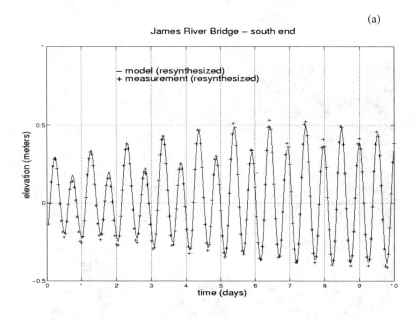

(b)

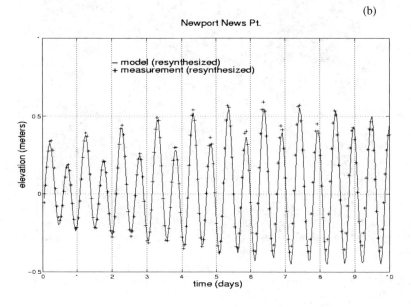

Fig 2. Surface water elevation comparison between modeled and measured
(a) at James River Bridge and (b) at Newport News Point.

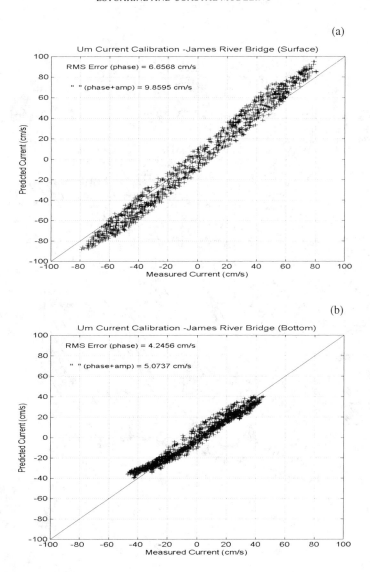

Fig 3. The current velocity comparison between modeled and measured
(a) at surface layer and (b) at bottom layer.

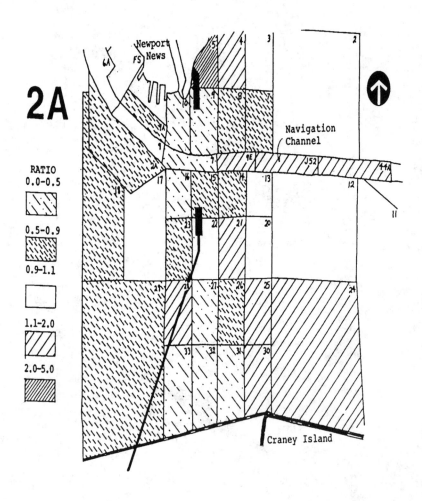

Fig 4. Gilsonite distribution measured in the physical model experiment.

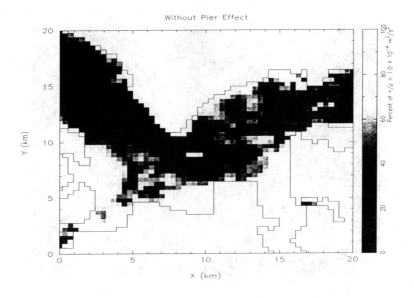

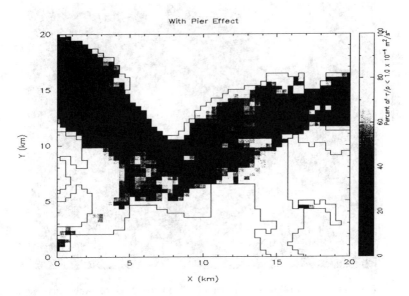

Fig 5. Distribution of sediment potential, with and without bridge piling effect

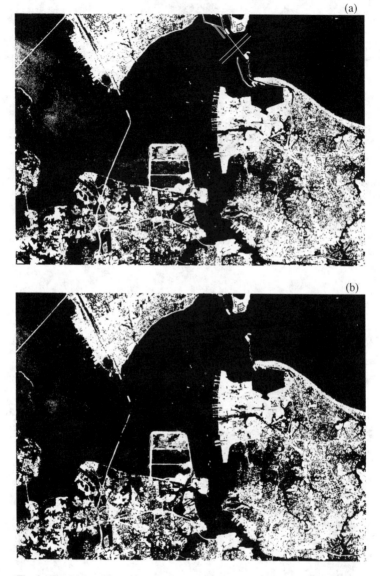

Fig 6. The highway crossing design (a) Alternative 2 and (b) Alternative 9.

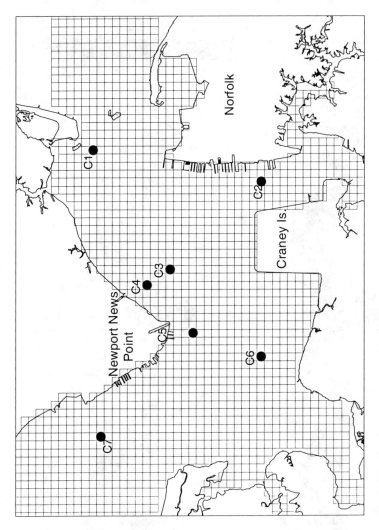

Fig 7. Map for the current stations used for comparison.

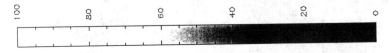

Percent of Times Below Critical Shear Stress

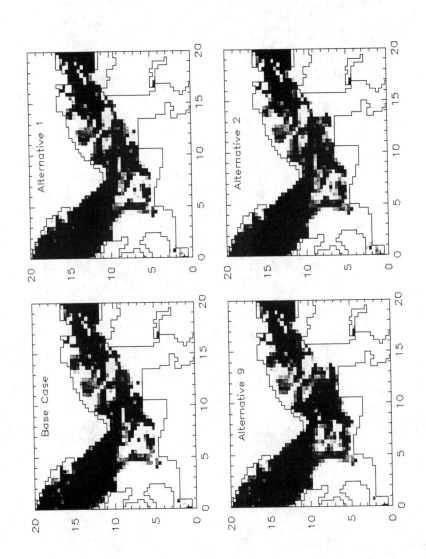

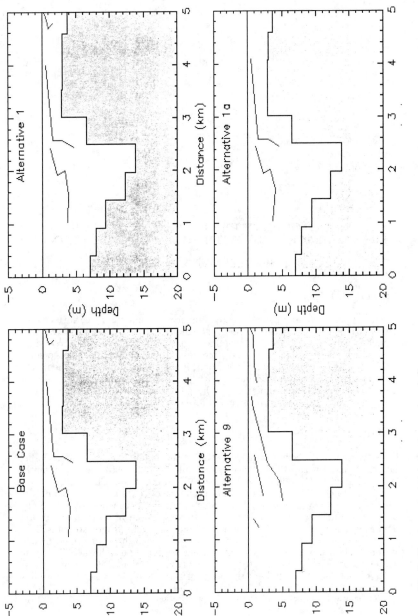

Fig 9. Front movement during mean river inflow.

MONOTONICITY PRESERVING MODEL FOR FREE-SURFACE FLOW WITH SHARP DENSITY FRONTS

Scott F. Bradford[1] Nikolaos D. Katopodes[2]

ABSTRACT

The development of an unsteady, three-dimensional model for free surface flow with sharp density fronts is presented. The model solves the Reynolds averaged Navier-Stokes equations on a curvilinear grid. A monotonicity preserving, second-order accurate in space and time finite volume scheme is used to solve the governing equations in a semi-implicit, fractional step manner. The model is applied to a variety of problems with large surface deformations, abrupt transitions in shoreline geometry, and sharp density fronts in order to evaluate its performance and illustrate its capabilities and robustness.

INTRODUCTION

Density fronts have been extensively studied through field observations (Valle-Levinson and Lwiza, 1996, Marmorino and Trump, 1996) and with mathematical models (Kapolnai, Werner, and Blanton, 1996, Chao, 1998, Shen and Kuo, 1999.) The widespread interest in such fronts is primarily because they act as a dynamic barrier that hinders cross-front transport of pollutants, sediments, and biota. In addition, such fronts play an important role in underwater acoustics in the nearshore environment. The fronts are caused by more buoyant water intruding into denser water, which occurs when heated water is released from a power plant or when fresh or brackish water flows into saline water. They are characterized by a strong horizontal flow convergence, with a length scale typically of the order of 1 km, and associated downwelling

---

[1]Mechanical Engineer, Image Science and Applications Branch, Code 7261, Naval Research Laboratory, Washington D.C. 20375

[2]Professor, Department of Civil and Environmental Engineering, University of Michigan, Ann Arbor, MI 48109

and intense mixing. The vertical front scale is typically a few meters and the downward velocity has been estimated to be as high as 0.15 m/s in the James river estuary (Marmorino and Trump, 1996). In addition, the density of water can change drastically within a few meters both laterally and vertically. For example, Garvine and Monk (1974) observed the salinity to vary from 12 ppt to 24 ppt in less than 1 m of depth at the Connecticut river front. The location of the front also changes with time and is governed primarily by tides, fresh/cool water input, wind, and domain geometry. Despite the frontal movement and intense mixing, the fronts have been observed to exist for several days with no sign of diminishing in strength.

Such sharp, relatively small scale flow features are difficult to simulate numerically. The treatment of advection and time integration greatly affects the stability and accuracy of the model. First order accurate methods are overly diffusive and severely smear discontinuities, while second order methods may be unstable for problems with rapid shoreline or bathymetric variations, strong shear or upwelling/downwelling, large free surface movements, or sharp fronts in density or temperature. In the present study, a semi-implicit, fractional-step scheme is used to solve the unsteady, incompressible, Navier-Stokes equations for stratified, nonhydrostatic, free-surface flows. The model closely follows the work of Casulli and Stelling (1998), but is fully second-order accurate in space and time except in the vicinity of discontinuities. The pressure is split into hydrostatic and hydrodynamic components and integrated implicitly with the Crank-Nicolson method. The semi-implicit treatment greatly improves the model's robustness and eliminates time-step restrictions associated with the movement of gravity waves. Advection and diffusion are treated with an explicit predictor-corrector scheme, which simplifies the model but introduces stability restrictions on the time step. The monotone, upstream scheme for conservation laws (MUSCL) is used to discretize the advection terms of the momentum and scalar transport equations to achieve second-order accuracy in space (Van Leer, 1979). Flux limiting is used to prevent the development and growth of numerical oscillations. This treatment is shown to yield superior results over traditional methods such as the upwind, Lax-Wendroff, and centered difference discretizations, with minimal additional computational overhead. The model is applied to various idealized problems with large free surface deformations, sharp density fronts, and abrupt shoreline transitions in order to illustrate the improvement.

THE GOVERNING EQUATIONS

The foundation of the model consists of the Reynolds averaged Navier-Stokes equations,

$$\frac{\partial \mathbf{u}}{\partial t} + \frac{\partial \mathbf{f}}{\partial x} + \frac{\partial \mathbf{g}}{\partial y} + \frac{\partial \mathbf{h}}{\partial z} = \mathbf{s} \tag{1}$$

where $\mathbf{u} = (u \ \ v \ \ w)^T$ is the vector of fluid velocities in the x, y, and z directions, respectively, and

$$\mathbf{f} = \begin{pmatrix} u^2 - \tau_{xx} + p + gh \\ uv - \tau_{xy} \\ uw - \tau_{xz} \end{pmatrix} \qquad \mathbf{g} = \begin{pmatrix} uv - \tau_{xy} \\ v^2 - \tau_{yy} + p + gh \\ vw - \tau_{yz} \end{pmatrix}$$

$$\mathbf{h} = \begin{pmatrix} uw - \tau_{xz} \\ vw - \tau_{yz} \\ w^2 - \tau_{zz} + p \end{pmatrix} \qquad \mathbf{s} = \begin{pmatrix} -\frac{g}{\rho_0} \int_z^h \frac{\partial \rho}{\partial x} dz^* + fv \\ -\frac{g}{\rho_0} \int_z^h \frac{\partial \rho}{\partial y} dz^* - fu \\ 0 \end{pmatrix}$$

The shear stresses, normalized by a reference density ρ_0, are defined as (Pedlosky, 1987)

$$\tau_{xx} = 2D_h \frac{\partial u}{\partial x} \qquad\qquad \tau_{yy} = 2D_h \frac{\partial v}{\partial y} \qquad\qquad \tau_{zz} = 2D_v \frac{\partial w}{\partial z}$$

$$\tau_{xy} = D_h \left(\frac{\partial u}{\partial y} + \frac{\partial v}{\partial x} \right) \qquad \tau_{xz} = D_v \frac{\partial u}{\partial z} + D_h \frac{\partial w}{\partial x} \qquad \tau_{yz} = D_v \frac{\partial v}{\partial z} + D_h \frac{\partial w}{\partial y}$$

The term h represents the deviation of the free surface from the mean water level, $p = p_d/\rho_0$ where p_d is the dynamic pressure component. D_h and D_v denote the horizontal and vertical turbulence eddy diffusivities, respectively, and f is the Coriolis parameter. Several assumptions are implicit in Eqs.(1), which most notably include the Boussinesq approximation, Boussinesq eddy viscosity assumption, and constant atmospheric pressure. In addition, it has been assumed that turbulence anisotropy exists only between the vertical and horizontal directions.

In general, the free surface moves and therefore the domain of interest deforms with time. From a computational perspective, this can be handled by using a fixed numerical grid or a deforming grid with an upper boundary that moves with the free surface. The latter procedure is adopted in this study, which unfortunately prevents a direct calculation of the local fluid acceleration, $\partial \mathbf{u}/\partial t$. However, the Lagrangian fluid acceleration, $D\mathbf{u}/Dt$, can be computed and therefore the local acceleration may be determined as,

$$\frac{\partial \mathbf{u}}{\partial t} = \frac{D\mathbf{u}}{Dt} - w_d \frac{\partial \mathbf{u}}{\partial z} \tag{2}$$

Note that it is assumed that the grid moves in the z direction only and w_d denotes the z component of the local velocity of the domain, i.e. $w_d = dz/dt$. This result can be substituted into Eq. (1) to yield

$$\frac{D\mathbf{u}}{Dt} - w_d \frac{\partial \mathbf{u}}{\partial z} + \frac{\partial \mathbf{f}}{\partial x} + \frac{\partial \mathbf{g}}{\partial y} + \frac{\partial \mathbf{h}}{\partial z} = \mathbf{s} \tag{3}$$

In addition, the fluid is assumed to be incompressible, which is expressed as

$$\frac{\partial u}{\partial x} + \frac{\partial v}{\partial y} + \frac{\partial w}{\partial z} = 0 \tag{4}$$

An expression for h is derived by integrating Eq. (4) in the z direction. Application of Leibnitz Rule, along with impermeable bottom and kinematic free surface boundary conditions, yields

$$\frac{\partial h}{\partial t} + \frac{\partial}{\partial x}\left(\int_{-d}^{h} u\,dz\right) + \frac{\partial}{\partial y}\left(\int_{-d}^{h} v\,dz\right) = 0 \tag{5}$$

An alternative approach is to directly solve the kinematic free surface boundary condition for the free surface location. Nichols and Hirt (1973) used this technique with some success, but they needed to append artificial diffusion terms to the kinematic free surface equation to prevent numerical instability. The approach outlined here is computationally stable without introducing additional numerical parameters.

Transport equations are solved for the fate and movement of scalars including salinity, thermal energy, and turbulence quantities and an equation of state is used to relate the density of water to salinity and temperature.

THE NUMERICAL SCHEME

The finite volume approach is used to discretize the governing equations. This method has the attractive feature of conserving mass and momentum locally, even in the presence of discontinuities. This is not guaranteed by finite difference or finite element methods. The domain is divided into hexagonal computational cells indexed with j, k, l. A staggered grid is utilized, with all scalars defined at cell centers and velocities located at the cell faces, in order to avoid checkerboard solutions. A computational cell is illustrated in Fig. 1. Eqs. (3) are written in generalized coordinates as

$$\frac{Du_\xi}{Dt} + F_\xi + H_\xi + P_\xi = s_\xi \qquad \frac{Du_\eta}{Dt} + F_\eta + H_\eta + P_\eta = s_\eta$$

$$\frac{Du_\zeta}{Dt} + F_\zeta + P_\zeta = s_\zeta \tag{6}$$

where

$$F_\xi = \frac{\partial}{\partial \xi}(u_\xi^2 - \tau_{\xi\xi}) + \frac{\partial}{\partial \eta}(u_\xi u_\eta - \tau_{\xi\eta}) + \frac{\partial}{\partial \zeta}(u_\xi u_\zeta - \tau_{\xi\zeta}) - w_d\zeta_z\frac{\partial u_\xi}{\partial \zeta}$$

$$F_\eta = \frac{\partial}{\partial \xi}(u_\xi u_\eta - \tau_{\xi\eta}) + \frac{\partial}{\partial \eta}(u_\eta^2 - \tau_{\eta\eta}) + \frac{\partial}{\partial \zeta}(u_\eta u_\zeta - \tau_{\eta\zeta}) - w_d\zeta_z\frac{\partial u_\eta}{\partial \zeta}$$

$$F_\zeta = \frac{\partial}{\partial \xi}(u_\xi u_\zeta - \tau_{\xi\zeta}) + \frac{\partial}{\partial \eta}(u_\eta u_\zeta - \tau_{\eta\zeta}) + \frac{\partial}{\partial \zeta}(u_\zeta^2 - \tau_{\zeta\zeta}) - w_d\zeta_z\frac{\partial u_\zeta}{\partial \zeta}$$

$$H_\xi = g\left(l_{\xi\xi}\frac{\partial h}{\partial \xi} + l_{\xi\eta}\frac{\partial h}{\partial \eta}\right) \qquad H_\eta = g\left(l_{\xi\eta}\frac{\partial h}{\partial \xi} + l_{\eta\eta}\frac{\partial h}{\partial \eta}\right)$$

$$P_\xi = l_{\xi\xi}\frac{\partial p}{\partial \xi} + l_{\xi\eta}\frac{\partial p}{\partial \eta} + l_{\xi\zeta}\frac{\partial p}{\partial \zeta} \qquad P_\eta = l_{\xi\eta}\frac{\partial p}{\partial \xi} + l_{\eta\eta}\frac{\partial p}{\partial \eta} + l_{\eta\zeta}\frac{\partial p}{\partial \zeta}$$

$$P_\zeta = l_{\xi\zeta}\frac{\partial p}{\partial \xi} + l_{\eta\zeta}\frac{\partial p}{\partial \eta} + (l_{\zeta\zeta} + \zeta_z^2)\frac{\partial p}{\partial \zeta}$$

The generalized velocities and source terms are defined as

$$u_\xi = u\xi_x + v\xi_y \qquad u_\eta = u\eta_x + v\eta_y \qquad u_\zeta = u\zeta_x + v\zeta_y + w\zeta_z$$

$$s_\xi = s_x\xi_x + s_y\xi_y \qquad s_\eta = s_x\eta_x + s_y\eta_y \qquad s_\zeta = s_x\zeta_x + s_y\zeta_y + s_z\zeta_z$$

The terms multiplying the velocities and source terms are the grid transformation metrics, which are computed by assuming a linear mapping from (x, y, z) space to (ξ, η, ζ) space. ξ is in the direction of contiguous j indices, while η and ζ are in the directions of k and l indices, respectively. Also note that s_x, s_y, and s_z represent the source terms of the x, y, and z momentum equations, respectively. The shear stresses are now defined in terms of u_ξ, u_η, and u_ζ and are,

$$\tau_{\xi\xi} = 2D_h\left(\frac{\partial u_\xi}{\partial \xi}l_{\xi\xi} + \frac{\partial u_\xi}{\partial \eta}l_{\xi\eta} + \frac{\partial u_\xi}{\partial \zeta}l_{\xi\zeta}\right) \qquad \tau_{\eta\eta} = 2D_h\left(\frac{\partial u_\eta}{\partial \xi}l_{\xi\eta} + \frac{\partial u_\eta}{\partial \eta}l_{\eta\eta} + \frac{\partial u_\eta}{\partial \zeta}l_{\eta\zeta}\right)$$

$$\tau_{\zeta\zeta} = 2D_h\left(\frac{\partial u_\zeta}{\partial \xi}l_{\xi\zeta} + \frac{\partial u_\zeta}{\partial \eta}l_{\eta\zeta} + \frac{\partial u_\zeta}{\partial \zeta}l_{\zeta\zeta}\right) + 2D_v\frac{\partial u_\zeta}{\partial \zeta}\zeta_z^2$$

$$\tau_{\xi\eta} = D_h\left\{\frac{\partial}{\partial \xi}\left(u_\xi l_{\xi\eta} + u_\eta l_{\xi\xi}\right) + \frac{\partial}{\partial \eta}\left(u_\xi l_{\eta\eta} + u_\eta l_{\xi\eta}\right) + \frac{\partial}{\partial \zeta}\left(u_\xi l_{\eta\zeta} + u_\eta l_{\xi\zeta}\right)\right\}$$

$$\tau_{\xi\zeta} = D_h\left\{\frac{\partial}{\partial \xi}\left(u_\xi l_{\xi\zeta} + u_\zeta l_{\xi\xi}\right) + \frac{\partial}{\partial \eta}\left(u_\xi l_{\eta\zeta} + u_\zeta l_{\xi\zeta}\right) + \frac{\partial}{\partial \zeta}\left(u_\xi l_{\zeta\zeta} + u_\zeta l_{\xi\zeta}\right)\right\} + D_v\frac{\partial u_\xi}{\partial \zeta}\zeta_z^2$$

$$\tau_{\eta\zeta} = D_h\left\{\frac{\partial}{\partial \xi}\left(u_\eta l_{\xi\zeta} + u_\zeta l_{\xi\eta}\right) + \frac{\partial}{\partial \eta}\left(u_\eta l_{\eta\zeta} + u_\zeta l_{\eta\eta}\right) + \frac{\partial}{\partial \zeta}\left(u_\eta l_{\zeta\zeta} + u_\zeta l_{\eta\zeta}\right)\right\} + D_v\frac{\partial u_\eta}{\partial \zeta}\zeta_z^2$$

where

$$l_{\xi\xi} = \xi_x^2 + \xi_y^2 \qquad l_{\xi\eta} = \xi_x\eta_x + \xi_y\eta_y \qquad l_{\xi\zeta} = \xi_x\zeta_x + \xi_y\zeta_y$$

$$l_{\eta\eta} = \eta_x^2 + \eta_y^2 \qquad l_{\eta\zeta} = \eta_x\zeta_x + \eta_y\zeta_y \qquad l_{\zeta\zeta} = \zeta_x^2 + \zeta_y^2$$

Note that it is assumed that the computational direction ζ is parallel to the z direction, which results in $\xi_z = \eta_z = 0$ for all time. In addition, it has been assumed that the grid metric terms are locally constant and can be taken inside derivative operators in the advection and diffusion terms.

Eq. (4) in generalized coordinates is

$$\frac{\partial u_\xi}{\partial \xi} + \frac{\partial u_\eta}{\partial \eta} + \frac{\partial u_\zeta}{\partial \zeta} = 0 \tag{7}$$

while Eq. (5) becomes

$$\frac{\partial h}{\partial t} + \frac{\partial}{\partial \xi}\left(\int_{-d}^{h} u_\xi dz\right) + \frac{\partial}{\partial \eta}\left(\int_{-d}^{h} u_\eta dz\right) = 0 \tag{8}$$

Time Integration

The advection and viscous terms are integrated in time using an explicit predictor-corrector method similar to the approach used by Bell et al. (1989) for confined incompressible flow problems. This approach is stable without resorting to smoothing filters required by other second order schemes such as the leapfrog method (Blumburg and Mellor, 1987). The pressure and free surface terms are discretized with the standard θ method, which yields,

$$\frac{u_\xi^{n+1} - u_\xi^n}{\Delta t} + F_\xi^{n+1/2} + \theta(H_\xi + P_\xi)^{n+1} + (1 - \theta)(H_\xi + P_\xi)^n = s_\xi^{n+1/2}$$

$$\frac{u_\eta^{n+1} - u_\eta^n}{\Delta t} + F_\eta^{n+1/2} + \theta(H_\eta + P_\eta)^{n+1} + (1 - \theta)(H_\eta + P_\eta)^n = s_\eta^{n+1/2}$$

$$\frac{u_\zeta^{n+1} - u_\zeta^n}{\Delta t} + F_\zeta^{n+1/2} + \theta P_\zeta^{n+1} + (1 - \theta)P_\zeta^n = s_\zeta^{n+1/2} \tag{9}$$

where the superscripts $n + 1$ and n denote the new and old time levels, respectively. The implicit treatment of the pressure terms is analogous to the SIMPLER scheme proposed by Patankar (1980) for confined incompressible flow and greatly improves the robustness of the model.

Computing the Predictor

First, predictor velocities at $n+1/2$ are computed explicitly with known values at the previous time level,

$$u_\xi^{n+1/2} = u_\xi^n - \frac{\Delta t}{2} (F_\xi + H_\xi + P_\xi - s_\xi)^n$$

$$u_\eta^{n+1/2} = u_\eta^n - \frac{\Delta t}{2} (F_\eta + H_\eta + P_\eta - s_\eta)^n$$

$$u_\zeta^{n+1/2} = u_\zeta^n - \frac{\Delta t}{2} (F_\zeta + P_\zeta - s_\zeta)^n \tag{10}$$

These velocities are then used to estimate the advection, diffusion, and source terms at the $n + 1/2$ time level.

Computing the Free Surface Elevation

Next, the new free surface position is calculated by integrating Eq. (8) in time with the θ method for a cell j, k, which yields,

$$h_{j,k}^{n+1} + \theta \Delta t \sum_{l=1}^{n_l} \Big[(\tilde{u}_\xi^{n+1} \Delta z)_{j+1/2,k,l} - (\tilde{u}_\xi^{n+1} \Delta z)_{j-1/2,k,l} +$$
$$(\tilde{u}_\eta^{n+1} \Delta z)_{j,k+1/2,l} - (\tilde{u}_\eta^{n+1} \Delta z)_{j,k-1/2,l} \Big] = R \tag{11}$$

where

$$R = h_{j,k}^n - \Delta t(1-\theta) \sum_{l=1}^{n_l} \Big[(u_\xi^n \Delta z)_{j+1/2,k,l} - (u_\xi^n \Delta z)_{j-1/2,k,l} +$$
$$(u_\eta^n \Delta z)_{j,k+1/2,l} - (u_\eta^n \Delta z)_{j,k-1/2,l} \Big]$$

and the integrals have been evaluated with the trapezoidal rule. Note that nl denotes the number of cells in the vertical direction and $j - 1/2$ and $j + 1/2$ denote the left and right cell faces, respectively. The front and back faces are denoted by $k-1/2$ and $k+1/2$. The velocities $\tilde{u}_\xi^{n+1}$ and $\tilde{u}_\eta^{n+1}$ are approximate velocities computed by assuming $p = 0$, i.e.,

$$\tilde{u}_\xi^{n+1} = \tilde{u}_\xi - g\theta\Delta t H_\xi^{n+1} \qquad \tilde{u}_\eta^{n+1} = \tilde{u}_\eta - g\theta\Delta t H_\eta^{n+1} \qquad (12)$$

where

$$\tilde{u}_\xi = u_\xi^n - \Delta t\, (F_\xi - s_\xi)^{n+1/2} - g(1-\theta)\Delta t H_\xi^n$$
$$\tilde{u}_\eta = u_\eta^n - \Delta t\, (F_\eta - s_\eta)^{n+1/2} - g(1-\theta)\Delta t H_\eta^n$$

For completeness, $\tilde{u}_\zeta^{n+1}$ is also given as

$$\tilde{u}_\zeta^{n+1} = \tilde{u}_\zeta = u_\zeta^n - \Delta t\, (F_\zeta - s_\zeta)^{n+1/2} \qquad (13)$$

Eqs. (12) can be substituted in Eq. (11) to yield,

$$h_{j,k}^{n+1} - g(\theta\Delta t)^2 \Big\{ (H_\xi^{n+1}D)_{j+1/2,k} - (H_\xi^{n+1}D)_{j-1/2,k} +$$
$$(H_\eta^{n+1}D)_{j,k+1/2} - (H_\eta^{n+1}D)_{j,k-1/2} \Big\} = \tilde{R} \qquad (14)$$

where

$$\tilde{R} = R - \theta\Delta t \sum_{l=1}^{n_l} \Big[(\tilde{u}_\xi \Delta z)_{j+1/2,k,l} - (\tilde{u}_\xi \Delta z)_{j-1/2,k,l} +$$
$$(\tilde{u}_\eta \Delta z)_{j,k+1/2,l} - (\tilde{u}_\eta \Delta z)_{j,k-1/2,l} \Big]$$

and D denotes the total water depth at a cell face. The free surface gradients are discretized with centered differences and therefore Eq. (14) can be rearranged to the form,

$$a_1 h_{j-1,k-1}^{n+1} + a_2 h_{j-1,k}^{n+1} + a_3 h_{j-1,k+1}^{n+1} + a_4 h_{j,k-1}^{n+1} + a_5 h_{j,k}^{n+1} +$$
$$a_6 h_{j,k+1}^{n+1} + a_7 h_{j+1,k-1}^{n+1} + a_8 h_{j+1,k}^{n+1} + a_9 h_{j+1,k+1}^{n+1} = \tilde{R} \qquad (15)$$

where

$$a_1 = -g(\theta\Delta t)^2 \frac{(l_{\xi\eta}D)_{j-1/2,k} + (l_{\xi\eta}D)_{j,k-1/2}}{4} \qquad a_3 = g(\theta\Delta t)^2 \frac{(l_{\xi\eta}D)_{j-1/2,k} + (l_{\xi\eta}D)_{j,k+1/2,l}}{4}$$

$$a_7 = g(\theta\Delta t)^2 \frac{(l_{\xi\eta}D)_{j+1/2,k} + (l_{\xi\eta}D)_{j,k-1/2}}{4} \qquad a_9 = -g(\theta\Delta t)^2 \frac{(l_{\xi\eta}D)_{j+1/2,k} + (l_{\xi\eta}D)_{j,k+1/2}}{4}$$

$$a_2 = -g(\theta\Delta t)^2 \left\{ (l_{\xi\xi}D)_{j-1/2,k} + \frac{(l_{\xi\eta}D)_{j,k-1/2} - (l_{\xi\eta}D)_{j,k+1/2})}{4} \right\}$$

$$a_4 = -g(\theta\Delta t)^2 \left\{ (l_{\eta\eta}D)_{j,k-1/2} + \frac{(l_{\xi\eta}D)_{j-1/2,k} - (l_{\xi\eta}D)_{j+1/2,k})}{4} \right\}$$

$$a_6 = -g(\theta\Delta t)^2 \left\{ (l_{\eta\eta}D)_{j,k+1/2} + \frac{(l_{\xi\eta}D)_{j+1/2,k} - (l_{\xi\eta}D)_{j-1/2,k})}{4} \right\}$$

$$a_8 = -g(\theta\Delta t)^2 \left\{ (l_{\xi\xi}D)_{j+1/2,k} + \frac{(l_{\xi\eta}D)_{j,k+1/2} - (l_{\xi\eta}D)_{j,k-1/2})}{4} \right\}$$

$$a_5 = 1 + g(\theta\Delta t)^2 \{(l_{\xi\xi}D)_{j-1/2,k} + (l_{\xi\xi}D)_{j+1/2,k} + (l_{\eta\eta}D)_{j,k-1/2} + (l_{\eta\eta}D)_{j,k+1/2}\}$$

Assembly of Eq. (15) over the entire surface domain creates a nine-diagonal matrix that is solved by a preconditioned biconjugate gradient method for the values of h^{n+1} (Press et al. 1992). Peric (1990) found this method to be more robust, and more computationally expensive, than several other common methods such as Stone's strongly implicit procedure. In this study, the matrix is preconditioned simply by multiplying it by its trace. The grid metric terms are then updated using the new free surface elevations and $\tilde{u}_\xi^{n+1}$, $\tilde{u}_\eta^{n+1}$, and $\tilde{u}_\zeta^{n+1}$ are computed.

Note that if the horizontal grid is orthogonal, the matrix is reduced to a more computationally tractable penta-diagonal form. Peric (1990) investigated the loss of accuracy when neglecting such terms in confined incompressible flow simulations and found an insignificant loss in accuracy if the angle between the ξ and η directions is between 45 and 135 degrees.

Computing the Dynamic Pressure

The simulation can continue to the next time step if only a hydrostatic solution is desired. However, for nonhydrostatic flows, the velocities can be corrected to satisfy momentum conservation including p. This is analogous to the fractional step methodology purposed by Chorin (1968) for confined, incompressible flow. In this case, the final velocities are defined as

$$u_\xi^{n+1} = u_\xi^* - \theta\Delta t P_\xi^{n+1} \quad u_\eta^{n+1} = u_\eta^* - \theta\Delta t P_\eta^{n+1} \quad u_\zeta^{n+1} = u_\zeta^* - \theta\Delta t P_\zeta^{n+1} \tag{16}$$

where

$$u_\xi^* = \tilde{u}_\xi^{n+1} - (1-\theta)\Delta t P_\xi^n \quad u_\eta^* = \tilde{u}_\eta^{n+1} - (1-\theta)\Delta t P_\eta^n$$

$$u_\zeta^* = \tilde{u}_\zeta^{n+1} - (1-\theta)\Delta t P_\zeta^n$$

The final velocities are required to satisfy the discrete incompressibility constraint,

$$u_{\xi j+1/2,k,l}^{n+1} - u_{\xi j-1/2,k,l}^{n+1} + u_{\eta j,k+1/2,l}^{n+1} - u_{\eta j,k-1/2,l}^{n+1} + u_{\zeta j,k,l+1/2}^{n+1} + u_{\zeta j,k,l-1/2}^{n+1} = 0 \tag{17}$$

Substituting Eqs. (16) into Eq. (17) and centrally discretizing the pressure gradients yields,

$$b_1 p_{j-1,k,l-1}^{n+1} + b_2 p_{j,k-1,l-1}^{n+1} + b_3 p_{j,k,l-1}^{n+1} + b_4 p_{j,k+1,l-1}^{n+1} + b_5 p_{j+1,k,l-1}^{n+1} \ +$$
$$b_6 p_{j-1,k-1,l}^{n+1} + b_7 p_{j-1,k,l}^{n+1} + b_8 p_{j-1,k+1,l}^{n+1} + b_9 p_{j,k-1,l}^{n+1} + b_{10} p_{j,k,l}^{n+1} \ +$$
$$b_{11} p_{j,k+1,l}^{n+1} + b_{12} p_{j+1,k-1,l}^{n+1} + b_{13} p_{j+1,k,l}^{n+1} + b_{14} p_{j+1,k+1,l}^{n+1} \ +$$
$$b_{15} p_{j-1,k,l+1}^{n+1} + b_{16} p_{j,k-1,l+1}^{n+1} + b_{17} p_{j,k,l+1}^{n+1} + b_{18} p_{j,k+1,l+1}^{n+1} + b_{19} p_{j+1,k,l+1}^{n+1} = R_p \quad (18)$$

where

$$b_1 = -\theta \Delta t \frac{l_{\xi\zeta(j-1/2,k,l)} + l_{\xi\zeta(j,k,l-1/2)}}{4} \qquad b_2 = -\theta \Delta t \frac{l_{\eta\zeta(j,k-1/2,l)} + l_{\eta\zeta(j,k,l-1/2)}}{4}$$

$$b_4 = \theta \Delta t \frac{l_{\eta\zeta(j,k+1/2,l)} + l_{\eta\zeta(j,k,l-1/2)}}{4} \qquad b_5 = \theta \Delta t \frac{l_{\xi\zeta(j+1/2,k,l)} + l_{\xi\zeta(j,k,l-1/2)}}{4}$$

$$b_6 = -\theta \Delta t \frac{l_{\xi\eta(j-1/2,k,l)} + l_{\xi\eta(j,k-1/2,l)}}{4} \qquad b_8 = \theta \Delta t \frac{l_{\xi\eta(j-1/2,k,l)} + l_{\xi\eta(j,k+1/2,l)}}{4}$$

$$b_{12} = \theta \Delta t \frac{l_{\xi\eta(j+1/2,k,l)} + l_{\xi\eta(j,k-1/2,l)}}{4} \qquad b_{14} = -\theta \Delta t \frac{l_{\xi\eta(j+1/2,k,l)} + l_{\xi\eta(j,k+1/2,l)}}{4}$$

$$b_{15} = \theta \Delta t \frac{l_{\xi\zeta(j-1/2,k,l)} + l_{\xi\zeta(j,k,l+1/2)}}{4} \qquad b_{16} = \theta \Delta t \frac{l_{\eta\zeta(j,k-1/2,l)} + l_{\eta\zeta(j,k,l+1/2)}}{4}$$

$$b_{18} = -\theta \Delta t \frac{l_{\eta\zeta(j,k+1/2,l)} + l_{\eta\zeta(j,k,l+1/2)}}{4} \qquad b_{19} = -\theta \Delta t \frac{l_{\xi\zeta(j+1/2,k,l)} + l_{\xi\zeta(j,k,l+1/2)}}{4}$$

$$b_3 = -\theta \Delta t \left(l_{\zeta\zeta(j,k,l-1/2)} + \frac{l_{\xi\zeta(j-1/2,k,l)} - l_{\xi\zeta(j+1/2,k,l)} + l_{\eta\zeta(j,k-1/2,l)} - l_{\eta\zeta(j,k+1/2,l)}}{4} \right)$$

$$b_7 = -\theta \Delta t \left(l_{\xi\xi(j-1/2,k,l)} + \frac{l_{\xi\eta(j,k-1/2,l)} - l_{\xi\eta(j,k+1/2,l)} + l_{\xi\zeta(j,k,l-1/2)} - l_{\xi\zeta(j,k,l+1/2)}}{4} \right)$$

$$b_9 = -\theta \Delta t \left(l_{\eta\eta(j,k-1/2,l)} + \frac{l_{\xi\eta(j-1/2,k,l)} - l_{\xi\eta(j+1/2,k,l)} + l_{\eta\zeta(j,k,l-1/2)} - l_{\eta\zeta(j,k,l+1/2)}}{4} \right)$$

$$b_{11} = -\theta \Delta t \left(l_{\eta\eta(j,k+1/2,l)} + \frac{l_{\xi\eta(j+1/2,k,l)} - l_{\xi\eta(j-1/2,k,l)} + l_{\eta\zeta(j,k,l+1/2)} - l_{\eta\zeta(j,k,l-1/2)}}{4} \right)$$

$$b_{13} = -\theta \Delta t \left(l_{\xi\xi(j+1/2,k,l)} + \frac{l_{\xi\eta(j,k+1/2,l)} - l_{\xi\eta(j,k-1/2,l)} + l_{\xi\zeta(j,k,l+1/2)} - l_{\xi\zeta(j,k,l-1/2)}}{4} \right)$$

$$b_{17} = -\theta \Delta t \left(l_{\zeta\zeta(j,k,l+1/2)} + \frac{l_{\xi\zeta(j+1/2,k,l)} - l_{\xi\zeta(j-1/2,k,l)} + l_{\eta\zeta(j,k+1/2,l)} - l_{\eta\zeta(j,k-1/2,l)}}{4} \right)$$

$$b_{10} = \theta \Delta t (l_{\xi\xi(j-1/2,k,l)} + l_{\xi\xi(j+1/2,k,l)} + l_{\eta\eta(j,k-1/2,l)} +$$
$$l_{\eta\eta(j,k+1/2,l)} + l_{\zeta\zeta(j,k,l-1/2)} + l_{\zeta\zeta(j,k,l+1/2)})$$

and

$$R_p = u^*_{\xi j+1/2,k,l} - u^*_{\xi j-1/2,k,l} + u^*_{\eta j,k+1/2,l} - u^*_{\eta j,k-1/2,l} + u^*_{\zeta j,k,l+1/2} - u^*_{\zeta j,k,l-1/2}$$

Assembly of Eq. (18) over the entire domain yields a nineteen-diagonal system of equations that are again solved by the preconditioned biconjugate gradient method for the values of p^{n+1}. With the dynamic pressures known, Eqns. (16) can then be solved for the final velocities at the $n+1$ time level. If the vertical and horizontal discretizations are orthogonal to each other, the matrix reduces to a seven-diagonal form. However, contrary to the horizontal discretization, the vertical discretization may not be designed to maintain orthogonality with the horizontal grid when the domain has drastic bathymetric variations or there are large free surface deformations. The effect of neglecting these terms will be illustrated below.

Scalar Transport

After the hydrodynamics are computed, the equations governing the fate and transport of scalar quantities are solved. The transport equation in curvilinear coordinates for an arbitrary scalar, ϕ, is

$$\frac{D\phi}{Dt} + \frac{\partial f_\xi}{\partial \xi} + \frac{\partial f_\eta}{\partial \eta} + \frac{\partial f_\zeta}{\partial \zeta} = s_\phi \tag{19}$$

where

$$f_\xi = u_\xi \phi - D_h^\phi \left(l_{\xi\xi} \frac{\partial \phi}{\partial \xi} + l_{\xi\eta} \frac{\partial \phi}{\partial \eta} + l_{\xi\zeta} \frac{\partial \phi}{\partial \zeta} \right)$$

$$f_\eta = u_\eta \phi - D_h^\phi \left(l_{\xi\eta} \frac{\partial \phi}{\partial \xi} + l_{\eta\eta} \frac{\partial \phi}{\partial \eta} + l_{\eta\zeta} \frac{\partial \phi}{\partial \zeta} \right)$$

$$f_\zeta = u_\zeta \phi - D_h^\phi \left(l_{\xi\zeta} \frac{\partial \phi}{\partial \xi} + l_{\eta\zeta} \frac{\partial \phi}{\partial \eta} + l_{\zeta\zeta} \frac{\partial \phi}{\partial \zeta} \right) - (D_v^\phi \zeta_z + w_d)\zeta_z \frac{\partial \phi}{\partial \zeta}$$

D_h^ϕ and D_v^ϕ are the horizontal and vertical eddy diffusivities, respectively, and s_ϕ represents a source/sink term.

First, a predictor is computed as,

$$\phi_{j,k,l}^{n+1/2} = \phi_{j,k,l}^n - \frac{\Delta t}{2} \left(\frac{\partial f_\xi}{\partial \xi} + \frac{\partial f_\eta}{\partial \eta} + \frac{\partial f_\zeta}{\partial \zeta} - s_\phi \right)^n \tag{20}$$

Then the corrector is computed as

$$\phi_{j,k,l}^{n+1} = \phi_{j,k,l}^n - \Delta t \left(\frac{\partial f_\xi}{\partial \xi} + \frac{\partial f_\eta}{\partial \eta} + \frac{\partial f_\zeta}{\partial \zeta} - s_\phi \right)^{n+1/2} \tag{21}$$

Spatial Discretization

An upwind biased expression for the advection flux of some quantity ψ, which could be a velocity component for example, is written as

$$f = u_\perp \psi = \frac{1}{2}\left(\overline{u}_\perp(\psi_L + \psi_R) - |\overline{u}_\perp|(\psi_R - \psi_L)\right) \tag{22}$$

where $u_\perp$ denotes the velocity perpendicular to the cell face and the subscripts L and R refer to reconstructed quantities to the left and right of the cell face, respectively. For example, for the $j + 1/2, k, l$ face of cell j, k, l,

$$\psi_R = \psi_{j+1,k,l} - \tfrac{1}{2}\overline{\Delta\psi}_{\xi j+1,k,l} \qquad\qquad \psi_L = \psi_{j,k,l} + \tfrac{1}{2}\overline{\Delta\psi}_{\xi j,k,l}$$

This technique is termed the Monotone Upstream Scheme for Conservation Laws (MUSCL) and was first proposed by Van Leer (1979) for solving the compressible Euler equations. The term $\overline{\Delta\psi}_\xi$ denotes some average gradient of ψ in the ξ direction. For example, the beta family of averages is

$$\overline{\Delta\psi} = \begin{cases} sign(a)min(max(|a|,|b|),\beta min(|a|,|b|)) & ab > 0 \\ 0 & ab \le 0 \end{cases} \tag{23}$$

where $1 \le \beta \le 2$. When computing the gradients in the ξ direction, $a = \psi_{j,k} - \psi_{j-1,k}$ and $b = \psi_{j+1,k} - \psi_{j,k}$. Analogous expressions can be deduced for the gradients in the other directions. In this study, $\beta = 2$ is used in all simulations, which yields the relatively less dissipative Superbee average. Setting $\overline{\Delta\psi} = 0$ at extrema tends to flatten smooth waves, particularly as $\beta \to 2$. This effect can be reduced by selecting smaller values of β, at the expense of introducing additional numerical dissipation into the solution. The diffusion terms, along with the surface and pressure gradients, are discretized with space-centered differences. The source terms are evaluated at the cell centers. Since the proposed model is nonlinear, even for linear problems, it is difficult to perform a precise stability analysis. However, because the model uses explicit time-stepping, then the sum of the Courant numbers in the three computational directions must be less than or equal to one in each cell. Also, the sum of the diffusion numbers in the three directions must be less than or equal to 0.5 in each cell. These conditions are necessary, but not sufficient to preserve the stability of the solution.

MODEL APPLICATIONS

Solitary Wave Run-up

This first application was performed to test the non-hydrostatic capabilities of the model and to illustrate the effect of neglecting the previously mentioned terms in the dynamic pressure equation. The surface elevation, velocities, and dynamic pressure for a solitary wave were specified as initial conditions in a 40 m long channel with a vertical downstream wall and still water depth, $d = 2$ m. The expressions developed by Chappelear (1962) were used for this purpose. The crest of the wave was initially at x = 20 m and wave heights $H = 0.2 - 0.8$ m were specified. The domain was discretized with 50 cells lengthwise and 20 cells in the vertical direction. The time step varied from 0.1 s for the smaller wave height cases to 0.05 s for the highest wave height case. The bottom was treated as a free slip boundary and turbulence was neglected, i.e. $D_h = D_v = 0$.

The predicted run-up, R, of the wave on the vertical wall was compared with experimental data presented by Chan and Street (1970). Fig. 2 compares the measured and computed run-up for varying initial wave height. The run-up was computed by solving the nineteen-diagonal and the simpler seven-diagonal forms of the dynamic pressure equation. From this figure it is seen that both predictions closely matched the observed R for low values of H/d, which indicates that grid distortion was not very severe. For increasing values of H/d, both predictions were slightly less than the measurements, which could be improved by increasing the grid resolution. As expected, solving the nineteen-diagonal system yielded slightly better predictions of R as the grid distortion increased. Surprisingly, both predictions converged to the observed value as $H/d \rightarrow 0.4$. This is in fact a transition region at which point a higher order solitary wave theory must be employed to obtain accurate initial conditions for the model (Chan and Street, 1970). The velocity divergence errors introduced by the current initial conditions may have masked any improvement gained by solving the nineteen-diagonal system. Also, the angle between the ξ and ζ directions ranged from 58 to 122 degrees for the $H/d = 0.4$ case, which is within the acceptable range given by Peric (1990). Therefore, the differences between the solutions should be small.

Fig. 3 shows contours of p and flow vectors case computed with the seven-diagonal form of the pressure equation for the $H/d = 0.4$ at $t = 1.5$ s and $t = 4.5$ s, while Fig. 4 shows the same quantities computed by solving the full system of equations. A comparison of these figures shows that there is a small difference in the predictions of p and the flow fields. As a result, the predicted run-up was similar for both approaches in this case.

Sudden Density Current Release

This test illustrates the accuracy and robustness of the monotonicity preserving scheme for simulating flows with sharp density fronts. The predicted speed of an intruding density current, initiated by the release of saline water in a channel filled with fresh water, is compared with experimental measurements made by Huppert and Simpson (1980). The channel was 9.6 m long with a gate located 0.39 m from the upstream end and was filled with water to a depth of 0.15 m. Salt was added to the water upstream of the gate to achieve two initial salt concentrations of 0.01156 and 0.03642. The domain was discretized with 50 cells along the channel length and 10 cells in the vertical direction, while the time step was 0.2 s for the low concentration case and 0.1 s for the high concentration case. A no-slip bottom condition was specified, although similar results were obtained with a quadratic bottom friction assumption with drag coefficients in the range of 0.01-0.02, which is a reasonable range for small scale laboratory experiments.

Constant values of $D_h = D_v = 1 \times 10^{-5}$ m^2/s were used in the simulation, which was chosen because it yielded reasonable predictions of the advance rate of the density current front using the second order accurate model. Fig. 5 shows the solutions for the high salt case at $t = 12$ s and $t = 120$ s using first order upwind differencing and fully implicit time stepping and Fig. 6 shows the analogous solutions computed with the MUSCL/Crank-Nicolson methodology. The additional numerical diffusion introduced by the upwind differencing has smeared out the head structure of the density front, which reduced the salt concentration in the head and thus led to an underprediction of the front intrusion speed as shown in Fig. 7. The head structure in Fig. 6 also qualitatively agrees with the experimental observations made by Huppert and Simpson (1980) and Rottman and Simpson (1983). Fig. 8 shows a sketch by Rottman and Simpson (1983) that summarizes the general shape of the dense underflow and the head structure is clearly evident. In addition, the current speeds predicted by the upwind method are somewhat lower than the MUSCL solution, which acts to reduce the shear stress at the upper boundary of the plume.

Geophysical Application

The final example is a geophysical application of an idealized release of fresh water into a coastal body of water with a salinity equal to 0.035. This problem is the extension of the previous case to three dimensions. Plan views of the domain are shown in Fig. 9 at the $z = -14.5$ m and $z = -0.5$ m levels and the depth was assumed to be a constant 15 m everywhere. The domain was assumed to be located at a mid northern latitude with $f = 7.3 \times 10^{-5}$ s^{-1}. The bottom was again treated as no-slip while $D_V = 1 \times 10^{-5}$ m^2/s and $D_H = 1 \times 10^{-5}$ m^2/s were arbitrarily specified. The lateral grid spacing was 1000 m and 8 cell levels were used in the vertical direction. The fresh water body was to the west and all boundaries were treated as free-slip walls. The salt water body was to the east and its northern, southern, and eastern boundaries were treated as open and the western boundary was specified as a free-slip wall. The channel boundaries were free-slip walls as well. At $t = 0$ s the two bodies were allowed to interact and the resulting contours of salinity and velocity vectors are shown in Fig. 9 at $t = 30000$ s. The solution was computed with the MUSCL/Crank-Nicolson approach with $\Delta t = 20$ s. From this figure it is seen that the salt water flow intruded into the fresh water body in the lower layers, while the light water spilled out over the salt water in the upper layers. The low salinity plume was beginning to bend to the right and attach to the coast due to the effect of the Coriolis acceleration. The model maintained a sharp density front and the solution was fairly smooth and stable despite the neglect of horizontal mixing.

CONCLUSIONS

A model that solves the three-dimensional, Reynolds averaged Navier-Stokes equations has been developed and presented. The model was shown to be robust and yielded accurate results for the run-up of a solitary wave on a vertical wall and the sudden release of saline water into fresh water. The previous problems were solved with Law-Wendroff and centered difference approximations to the advection terms but were found to be unstable. The neglect of the terms in the dynamic pressure equation was shown to introduce little error in the predicted wave run-up. In this case, it was particularly important to use $\theta = 0.5$ for the integration of the free surface equation. Using the more diffusive value $\theta = 1$ lead to a greater underprediction of the run-up. For the density current problem, the use of the MUSCL scheme for the advection terms was more important than the particular value of θ. For this problem, the upwind method yielded overly diffusive solutions and underpredicted the advance of the density current.

Future developments include the addition of a turbulence model, such as the one developed by Mellor and Yamada (1982) for geophysical flows. Also, the time integration of the vertical mixing terms needs further examination in order to relax the severe stability constraint this term can place on the time step. Finally, piecewise quadratic reconstruction is also being investigated. Such an approach should perform better than the piecewise linear MUSCL scheme used in the present study, particularly for problems involving waves and local maxima/minima in the state variables.

REFERENCES

Bell, J.B., Colella, P., and Glaz, H.M. (1989), "A Second-Order Projection Method for the Incompressible Navier-Stokes Equations", Journal of Computational Physics, Vol. 85, pp. 257–283.

Blumburg, A.F. and Mellor, G.L. (1987), "Description of a Three-Dimensional Coastal Ocean Circulation Model", In *Three-Dimensional Coastal Ocean Models*, American Geophysical Union, Washington D.C., pp. 1–16.

Casulli, V. and Stelling, G.S. (1998), "Numerical Simulation of 3D Quasi-Hydrostatic, Free-Surface Flows", Journal of Hydraulic Engineering, Vol. 124, pp. 678–686.

Chappelear, J.E. (1962), "Shallow-Water Waves,", Journal of Geophysical Research, Vol. 67, pp.4693–4704.

Chorin, A.J. (1968), "Numerical Solution of the Navier-Stokes Equations,", Mathematics of Computation, Vol. 23, pp. 341–354.

Chao, S. (1998), "Hyperpycnal and Buoyant Plumes from a Sediment-Laden River,", Journal of Geophysical Research, Vol. 103, pp. 3067–3081.

Galperin, B., Kantha, L.H., Hassid, S., and A. Rosati (1988), "A Quasi-Equilibrium Turbulent Energy Model for Geophysical Flows," Journal of the Atmospheric Sciences, Vol. 45, pp. 55–62.

Garvine, R.W. and Monk, J.D.(1974), "Frontal Structure of a River Plume," Journal of Geophysical Research, Vol. 79, pp. 2251–2259.

Huppert, H.E. and Simpson, J.E. (1980), "The Slumping of Gravity Currents," Journal of Fluid Mechanics, Vol. 99, pp. 785–799.

Kapolnai, A., Werner, F.E., and Blanton, J.O. (1996), "Circulation, Mixing, and Exchange Processes in the Vicinity of Tidal Inlets: A Numerical Study," Journal of Geophysical Research, Vol. 101, pp. 14253–14268.

Marmorino, G.O. and Trump, C.L. (1996), "High-Resolution Measurements Made Across a Tidal Intrusion Front," Journal of Geophysical Research, Vol. 101, pp. 25661–25674.

Mellor, G.L. and Yamada, T. (1982), "Development of a Turbulence Closure Model for Geophysical Fluid Problems," Reviews of Geophysics and Space Physics, Vol. 20, pp. 851–875.

Nichols, B.D. and Hirt, C.W., (1973), "Calculating Three-Dimensional Free Surface Flows in the Vicinity of Submerged and Exposed Structures," Journal of Computational Physics, Vol. 12, pp. 234–246.

Patankar, S.V. (1980), Numerical Heat Transfer and Fluid Flow, McGraw-Hill Book Company, New York.

Pedlosky, J. (1987), Geophysical Fluid Dynamics, Springer-Verlag, New York.

Peric, M. (1990),"Analysis of Pressure-Velocity Coupling on Nonorthogonal Grids," Numerical Heat Transfer, Part B, Vol. 17, pp. 63–82.

Press, W.H., Teukolsky, S.A., Vetterling, W.T., and Flannery, B.P. (1992), Numerical Recipes in Fortran, Cambridge University Press, New York.

Rottman, J.W. and Simpson, J.E. (1983),"Gravity Currents Produced by Instantaneous Releases of a Heavy Fluid in a Rectangular Channel," Journal of Fluid Mechanics, Vol. 135, pp. 95–110.

Shen, J. and Kuo, A.Y. (1999),"Numerical Investigation of an Estuarine Front and Its Associated Eddy," Journal of Waterway, Port, Coastal, and Ocean Engineering, Vol. 125, pp. 127–135.

Valle-Levinson, A. and Lwiza, K.M.M. (1995), "The Effects of Channels and Shoals on Exchange between the Chesapeake Bay and the Adjacent Ocean," Journal of Geophysical Research, Vol. 100, pp. 18551–18563.

Van Leer, B. (1979), "Towards the Ultimate Conservative Difference Scheme. V. A Second Order Sequel to Godunov's Method," Journal of Computational Physics, Vol. 32, pp. 101–136.

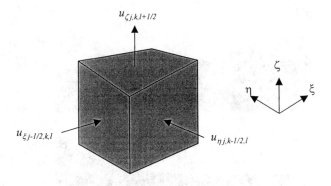

Figure 1: Sketch of a computational cell

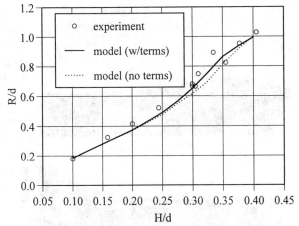

Figure 2: Comparison of predicted and measured solitary wave run-up on a vertical wall

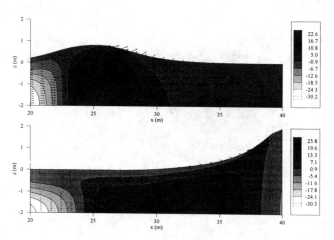

Figure 3: Contours of p and flow vectors at $t=1.5$ s and $t=4.5$ s computed with the seven-diagonal pressure equation.

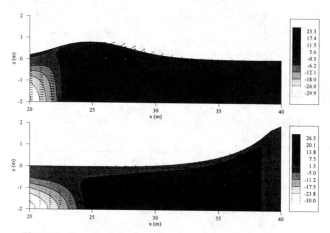

Figure 4: Contours of p and flow vectors at $t=1.5$ s and $t=4.5$ s computed with the nineteen-diagonal pressure equation.

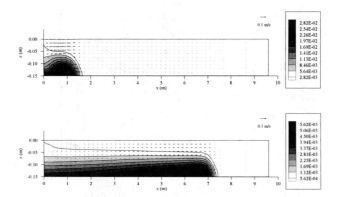

Figure 5: First order accurate contours of salinity and flow
vectors at t=12 s and t=120 s for an initial salt concentration
equal to 0.03642.

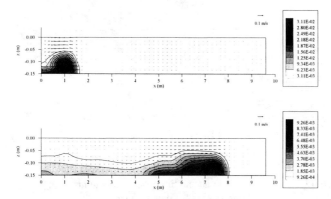

Figure 6: Second order accurate contours of salinity and flow
vectors at t=12 s and t=120 s for an initial salt concentration
equal to 0.03642.

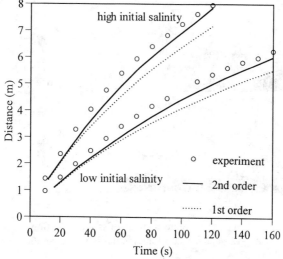

Figure 7: Comparison of predicted and measured front position.

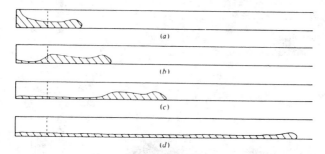

Figure 8: Sketch by Rottman and Simpson (1983) summarizing their experimental observations of dense underflows.

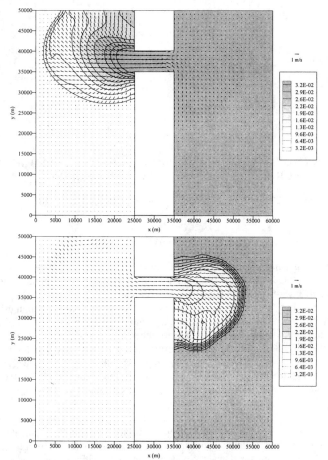

Figure 9: Contours of salinity and flow vectors at t=30000 s at z=-14.5 m (top) and z=-0.5 m (bottom)

GENERATION OF A TWO-DIMENSIONAL UNSTRUCTURED MESH FOR THE EAST COAST OF SOUTH CAROLINA

Scott C. Hagen[1] (ASCE Assoc. Member) and Robert J. Bennett[1]

ABSTRACT

The following paper describes the application of ADCIRC, a state-of-the-art, finite element model for coastal circulation, to perform real-time predictions of downstream tide-stage hydrographs for the Waccamaw River. Two major tasks are accomplished by this study. 1) An unstructured two-dimensional mesh is developed for a coastal region comprising the entire East Coast of South Carolina. 2) Real-time simulations are performed to generate tide-stage hydrographs that will be used by the National Weather Service River Forecast System for future forecasts. One of the resulting stage hydrographs will serve as a downstream boundary condition for the dynamic wave flood routing model (DWOPER) and will be applied by the NWS Southeast River Forecast Center.

A localized truncation error analysis of a linear form of the two-dimensional momentum equation is utilized to develop the unstructured mesh. Element sizes are varied such that truncation errors are driven to a constant level throughout the domain. An algorithm for boundary node placement is defined. Thus, when the boundary node procedure is combined with the generation of interior elements a foundation is provided for full reproduction of unstructured grids.

The approach described herein for generating downstream boundary conditions for DWOPER serves as a pilot study. This pilot study provides a basis that can be extended to the entire East Coast of the United States. Under full implementation, real-time predictions of tide-stage hydrographs can be incorporated with hurricane and tropical storm surge predictions, which will enhance National Weather Service River Forecast System efforts.

---

[1] University of Central Florida, Orlando, FL, U.S.A., shagen@mail.ucf.edu

Recent advances in surface water modeling have permitted the development and successful implementation of coastal ocean circulation models for increasingly larger domains (Blain et al., 1994; Foreman, 1986; Kinnmark, 1984; Kolar et al., 1996; Luettich et al. 1992; Lynch and Gray, 1979; Lynch, 1983; Westerink and Gray, 1991; Westerink et al., 1994). While a large domain increases the predictive capabilities of coastal ocean models (Blain et al., 1994; Westerink et al., 1994), it complicates the process of computational node placement. Large domains require a strategic placement of nodes in order to maintain acceptable levels of local and global accuracy for a given computational cost. However, the actual gridding of larger, more complex domains has not received the attention it deserves. In the following, the process of computational node placement will be discussed and a method of grid generation will be presented that more successfully couples the physics, as represented by discrete equations, underlying tidal flow and circulation to the mesh generation process.

Larger domains warrant a method of gridding that utilizes unstructured meshes, e.g., the finite element method, which allows for spatially-varying levels of discretization. Since, in general, shallower water has a higher localized wave number content than deeper water, higher resolution will be required in shallow water regions. Furthermore it has been shown that the computed response is highly sensitive to grid resolution in regions with steep bathymetric gradients (Hagen, 1998; Hagen et al., 2000; Luettich and Westerink, 1995; Westerink et al., 1992). Two-dimensional (2-D) response structures associated with intricate shorelines, 2-D topography, amphidromes (the intersection of all phase lines and a point at which all cotidal lines meet) and resonant bays also require local refinement of grids. Conversely, deep ocean waters usually result in large expanses with more slowly varying response structures in space, which can utilize a coarser level of resolution. These considerations indicate that variably-graded meshes are needed, which are easily implemented with the finite element method.

The method of production of variably-graded meshes is currently poorly defined, imprecise and *ad hoc*. It is a tedious process at best. Since no robust criterion or node spacing routine exists that incorporates the aforementioned physical characteristics and subsequent responses into the mesh generation process, modelers are left to rely on their knowledge of particular domains and their intuition. The mesh generation process results in an unstructured grid that is inherently irreproducible.

We utilize a localized truncation error analysis (LTEA), an *a posteriori* error estimation procedure, to define local limits on element sizes and then interpret these requirements with a hierarchical technique. The LTEA is of the actual discrete equations and includes approximations to the variables being simulated and their derivatives (Hagen, 1998; Hagen et al., 2000). Thus our LTEA-based approach directly couples the estimated truncation errors to the actual mesh generation process. We present our basis for a reproducible approach by describing an example mesh generation for the entire coast of South Carolina.

2. SOUTH CAROLINA MODEL

2.1 South Carolina Coastal Domain

The objective of this paper is to demonstrate a new approach to two-dimensional, finite element mesh generation. We choose the coastal region of South Carolina (Figure 1) because it provides an illustrative example for our mesh generation procedure. In addition, a mesh is needed to generate a tide stage hydrograph at the downstream end of the Waccamaw River for the National Weather Service's Southeast River Forecast Center.

The project area is located in the northern region of the South Atlantic Bight along the southeast coast of the United States. The Waccamaw River drains the coastal areas of southern North Carolina and northern South Carolina. The river leaves Lake Waccamaw in North Carolina and flows southward through Conway, South Carolina. From there, the river flows southward to the confluence with the Great Pee Dee and Black Rivers, through Winyah Bay, and into the Atlantic Ocean as shown in Figure 1.

For the purposes of the present paper and the tidal study, a grid domain for the area surrounding the Waccamaw River coastal region is defined. The domain is chosen to be large enough to include the area surrounding Charleston, South Carolina where historical tidal stage data is available. Further details on boundary definition, bathymetry and the entire finite element mesh generation approach will be provided in section 3.

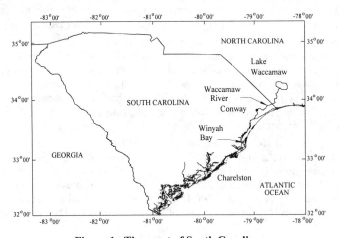

Figure 1. The coast of South Carolina.

2.2 Finite Element Model

The computations that are performed to generate a 2-D LTEA-based grid for the South Carolina domain are realized with a finite element model of the linearized shallow water equations. There are two main reasons that justify using a linear form of the shallow water equations. First, the concept of mesh generation that is based on multiple orders of the truncation error series is in the early stages of research. Simplicity, here in the form of linearized shallow water equations, facilitates a clear understanding of the details and implications of this theory. Second, shallow water modeling of a long-wave process in a large basin is weakly nonlinear. Because the nonlinear-term contribution is minimal, examination of the truncation error associated with the linear form of the shallow water equations should produce a finite element grid that will be suited for nonlinear simulations.

Two-dimensional shallow water equations are comprised of depth-integrated formulations of primitive continuity and momentum equations. For this paper, the continuity equation is formulated in the generalized wave continuity equation form (GWCE) (Kinnmark, 1984; Luettich et al., 1992). The linearized 2-D GWCE is given by:

$$\frac{\partial^2 \eta}{\partial t^2} + \tau_o \frac{\partial \eta}{\partial t} - g\left[\frac{\partial}{\partial x}\left(h\frac{\partial \eta}{\partial x}\right) - \frac{\partial}{\partial y}\left(h\frac{\partial \eta}{\partial y}\right)\right] - (\tau - \tau_o)\left[\frac{\partial}{\partial x}(uh) - \frac{\partial}{\partial y}(vh)\right] = 0 \quad (1)$$

and the 2-D, linearized, non-conservative momentum equations are expressed as:

$$\frac{\partial u}{\partial t} + g\frac{\partial \eta}{\partial x} + \tau u = 0 \quad (2)$$

$$\frac{\partial v}{\partial t} + g\frac{\partial \eta}{\partial y} + \tau v = 0 \quad (3)$$

where t = time, x and y = spacial coordinates, η = the deviation of the free surface from the geoid, u = velocity in the x-direction, v = velocity in the y-direction, τ_o = a weighting parameter in the GWCE, which controls the contribution from primitive continuity, g = gravitational acceleration, h = depth below the geoid and τ = bottom friction coefficient.

3. MESH GENERATION

The following sub-sections describe four major tasks that are performed to produce an unstructured grid for the South Carolina model domain.

3.1 Boundary Definition

World Vector Shoreline data is downloaded from the US Geologic Survey Internet site at http://crusty.er.usgs.gov/coast/getcoast.html.

Table 1. Segment end points for the Waccamaw River boundary (Figure 2).

Longitude (degrees west)	Latitude (degrees north)
79.247929	33.353808
79.100298	33.522266
79.142248	33.500265
79.247635	33.395684
79.242502	33.432060
79.279098	33.352561

Shoreline data with a coarse resolution of 1:250,000 (1:250) is extracted for the area bounded by 32 and 34 degrees north latitude and 78 and 81.5 degrees west longitude (see Figure 1). This corresponds to the entire shoreline of the South Carolina Coast plus a small section of the North Carolina coast up to Cape Fear. The data consists of nodes (longitude and latitude locations) that demarcate land/ocean boundaries.

Since the boundaries for the 1:250 resolution do not extend far enough into the estuary where the final tide stage hydrograph is desired, three segments of 1:70,000 (1:70) resolution data for the Waccamaw River are extracted to create a shoreline for this region. Table 1 provides the end points of each these segments. The 1:70 data is used to illustrate an algorithm for boundary node definition.

Figures 2 a and b display the raw and the refined shoreline data from the 1:70 resolution for the Waccamaw River boundary. Since the total number of nodes/elements in the finite element grid is a function of the boundary node resolution, the number of nodes on the boundary is minimized while maintaining physical geometry characteristics. In addition, the spacings between the nodes are required to be relatively uniform to assure that the triangular elements formed adjacent to the boundary nodes are approximately equilateral.

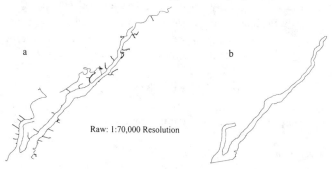

a

b

Raw: 1:70,000 Resolution

Figure 2. Raw and refined Waccamaw River boundary.

To accomplish both objectives, a minimum/maximum spacing routine is implemented. The routine deletes nodes that are closer together than a specified minimum distance and adds a node(s) when the spacing between two nodes is greater than 1.75 times the minimum spacing distance. A minimum distance of 0.002° is chosen for this region of the grid domain. In addition, all tributaries with a width of less than 0.002° are deleted. The final shoreline for the Waccamaw River boundary is displayed in Figure 2 b.

The minimum/maximum spacing routine is also applied to the 1:250 resolution data. Here a minimum spacing of 0.005° is utilized. The refined Waccamaw River boundary is then incorporated. Finally, an ocean boundary is generated by spinning an arc of node points. The center between the two end land boundary points is established at 79.38335° west, 33.02820° north and an ocean boundary node is placed at intervals equal to an arc length of 0.00834°. The final boundary is presented in Figure 3. Depth contours, from the National Geophysical Data Center (1999), are also included in Figure 3.

Figure 3. South Carolina boundary with depth contours (in meters).

3.2 Structured Mesh

The next step is to generate a structured grid. This is done in our application using polar coordinates. First, a center is established at 79.38335° west, 33.02820° north. (Note, this is the same location as is used for defining the ocean boundary.) Second, 198 arcs are swung beginning at 135° and ending before a complete revolution is made. Each arc radius, r, is incremented by $dr = 1.59031°/199$, where $1.59031°$ = distance from the center to the sea boundary. A node is placed every $d\theta = dr/r$ as each arc is swung.

In addition, nodes are placed within the refined Waccamaw River boundary (Figure 2 b). After the initial nodes from the previous routine are removed from within the Waccamaw River boundary, a center is established at 79.253024° west, 33.352297° north. Here 73 arcs are swung beginning at 45° and ending at 135°. Each arc radius, r, is incremented by $dr = 0.1931736°/74$. A node is placed every $d\theta = dr/r$ as each arc is swung. All boundary and interior nodes are then triangulated and bathymetry from the National Geophysical Data Center (1999) is interpolated at each node point to complete a <u>structured</u> mesh with a total of 69,816 nodes.

3.3 Localized Truncation Error Analysis

3.3.1 LTEA Formulation

The development of the spatial truncation error associated with solutions from any given structured or unstructured grid would result in a set of equations that would be tedious to work with because of the potential irregularity of the grid spacing. More importantly, the resulting equations would not lend themselves to an algorithm that promotes a domain-wide, constant truncation error, which is the basis for adjusting the local spacing. Herein the truncation error associated with a linear, harmonic form of the primitive momentum equations, (2) and (3), is developed for the nodes of an assumed regular triangular mesh (Figure 4).

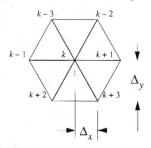

Figure 4. A typical interior node, k, and the surrounding nodes.

It is assumed that the finite element grid that will be generated will have equilateral, triangular elements on a local scale (as represented by Figure 4). The assumption permits: 1) an *a priori* estimation of truncation error for the finite element grid that is being developed; 2) that Δ_y may be expressed as a function of Δ_x, i.e., $\Delta_y = \sqrt{3}\Delta_x$.

Note that the assumption of equilateral triangular elements would be valid locally throughout a given finite element grid. However, since the grids that are generated will have irregular spacings, the assumption can not be valid globally. This does lead to some dependence on *pseudo* 2-D grid generation because the grid design is constrained to use equilateral triangles.

Because the truncation error is developed for a specific configuration, a valence of six (Figure 4), the truncation error series may be summed, truncated and solved together for Δ, noting that $\Delta = \Delta_x$. This provides an estimation of the second and fourth-orders of truncation error associated with the discrete form of the linear, harmonic, non-conservative momentum equations, (2) and (3), on the interior nodes of an equilateral triangular grid (Hagen, 1998):

$$\tau_{ME} = \Delta^2 \left[\left(\frac{\hat{i}\varpi + \tau}{2} \right) \left(\frac{\partial^2 \hat{u}_k}{\partial x^2} + \frac{\partial^2 \hat{v}_k}{\partial x^2} + \frac{\partial^2 \hat{u}_k}{\partial y^2} + \frac{\partial^2 \hat{v}_k}{\partial y^2} \right) + \right.$$

$$\frac{g}{2} \left(\frac{\partial^3 \hat{\eta}_k}{\partial x^3} + \frac{\partial^3 \hat{\eta}_k}{\partial x^2 \partial y} + \frac{\partial^3 \hat{\eta}_k}{\partial x \partial y^2} + \frac{\partial^3 \hat{\eta}_k}{\partial y^3} \right) \right] +$$

$$+ \Delta^4 \left[\left(\frac{\hat{i}\varpi + \tau}{8} \right) \left(\frac{\partial^4 \hat{u}_k}{\partial x^4} + \frac{\partial^4 \hat{v}_k}{\partial x^4} + 2\frac{\partial^4 \hat{u}_k}{\partial x^2 \partial y^2} + 2\frac{\partial^4 \hat{v}_k}{\partial x^2 \partial y^2} + \frac{\partial^4 \hat{u}_k}{\partial y^4} + \frac{\partial^4 \hat{v}_k}{\partial y^4} \right) + \right.$$

$$+ \frac{g}{24} \left(\frac{22}{10} \frac{\partial^5 \hat{\eta}_k}{\partial x^5} + \frac{\partial^5 \hat{\eta}_k}{\partial x^4 \partial y} + 2\frac{\partial^5 \hat{\eta}_k}{\partial x^3 \partial y^2} + 6\frac{\partial^5 \hat{\eta}_k}{\partial x^2 \partial y^3} + 3\frac{\partial^5 \hat{\eta}_k}{\partial x \partial y^4} + \frac{9}{5} \frac{\partial^5 \hat{\eta}_k}{\partial y^5} \right) \right] \quad (4)$$

where $\hat{u}, \hat{v}$ and $\hat{\eta}$ are the complex amplitudes of u, v and η evaluated at node k (the center node in Figure 4), g = gravitational acceleration, τ = bottom friction coefficient, $\hat{i} = \sqrt{-1}$, and ϖ = the response frequency.

Central difference approximations for a regular finite difference grid $(dx = dy)$ are applied to estimate the partial derivatives of equation (4). These approximations are carefully developed such that the estimate of the second and third-order partial derivatives have a leading-order accuracy of order four and the estimate of the fourth and fifth-order partial derivatives have a leading-order accuracy of order two (Hagen, 1998).

3.3.2 LTEA Computations

Hydrodynamic calculations are performed with ADCIRC-2DDI, a two-dimensional depth integrated circulation code (Luettich et al., 1992; Westerink et al., 1994). The simulation utilizes linear, Galerkin finite elements in two dimensions, with triangular elements, and employs a constant bottom friction coefficient and GWCE weighting parameter of 0.0004. A no-flow boundary condition is enforced at all land boundaries and open ocean boundaries are forced with the M_2 tidal constituent. 30 days of real time are simulated with the structured grid, which is described in section 3.2, to ensure that a dynamic steady-state is achieved. A time step of 30 s is used. In addition, a hyperbolic ramping function (Luettich et al., 1992) is imposed during the first two days.

Harmonic solutions from the structured mesh simulation are employed to compute equation (4), with $\Delta = 900$ meters, and produce local truncation error estimates. Note that use of the central difference approximations does not permit an estimate of the local truncation error up to the boundaries.

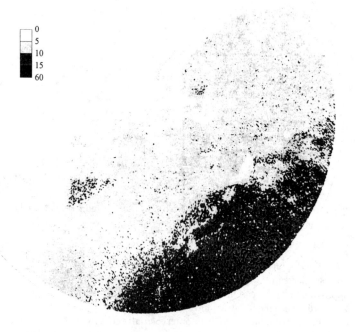

Figure 5. Maximum allowable radii (in kilometers).

3.4 Maximum Allowable Radii and Unstructured Mesh

A scalar value, which represents a radius of maximum allowable node spacing, is computed at interior nodes of the structured grid by setting equation (4) equal to the peak local truncation error value, 7.90×10^{-6} m/s^2. This complex quadratic is solved for Δ with the minimum real root selected as the scalar value (Hagen, 1998). The scalar value represents a radius of maximum allowable node spacing. This procedure is carried out for interior nodes of the structured grid.

Figure 5 presents a contour plot of the maximum allowable radii for the South Carolina domain. Note that use of the central difference approximations does not permit an estimate of the local truncation error up to the boundaries. The local node spacing requirements are changed as a result of forcing the truncation error to be constant, with the one exception being the node where the peak local truncation error is attained. Maximum allowable radii range from 0.9 to 55.89 kilometers.

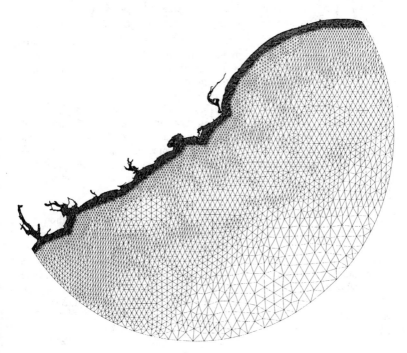

Figure 6. Unstructured mesh for the South Carolina domain.

The maximum allowable radii are employed to produce an unstructured mesh (Horstmann, 1998; Hagen et al., 1999) which is displayed in Figure 6. The final unstructured grid is reduced to 10,013 nodes from the 69,816 nodes of the structured mesh. The nearshore region has not been modified and maintains the element configuration of the structured mesh. Comparison of Figures 3 and 6 shows that the LTEA-based procedure captures important characteristics of the bathymetry, by successfully coupling the physics, as represented by discrete equations, underlying tidal flow and circulation to the mesh generation process.

4. SIMULATIONS

Fully nonlinear, hydrodynamic calculations are performed with ADCIRC-2DDI, a two-dimensional depth integrated circulation code (Luettich et al., 1992; Westerink et al., 1994). The simulation employs a constant bottom friction coefficient of 0.003, a GWCE weighting parameter of 0.009 and an eddy viscosity of 0.0. A no-flow boundary condition is enforced at all land boundaries and open ocean boundaries are forced with the M_2, M_4, M_6, O_1, N_2, S_2, K_1, and $STEADY$ tidal constituents. 90 days of real time are simulated with the unstructured grid (Figure 6), beginning at 12:00 AM (GMT) on January 1, 1998. A time step of 10 s is used. In addition, a hyperbolic ramping function (Luettich et al., 1992) is imposed during the first five days.

Figure 7 displays the historical and modeled tide elevations for a 15-day period at the end of March, 1998. When comparing the modeled results to the historical data, it is important to realize that the historical data results from all actual astronomical *and* meteorological tidal forcings present at that time while the modeled results include only the astronomical tide forcings. Figure 7 indicates that the modeled results estimate the tidal elevations reasonably well at Charleston, SC.

One the basis of the successful historical calculations, further simulations are performed to produce a series of tide-stage hydrographs for a region including Winyah Bay (Figure 1) and the entrance of the Waccamaw River. (Results are not included in this paper in the interests of brevity.) One of the resulting stage hydrographs will serve as a downstream boundary condition for the dynamic wave flood routing model (DWOPER), which is a finite difference approximation to the St. Venant equations, and will be applied by the NWS Southeast River Forecast Center.

5. DISCUSSION AND CONCLUSIONS

We have demonstrated how to generate a unique unstructured mesh for coastal and ocean circulation modeling. The truncation error-based approach to generating local node spacing requirements incorporates an *a posteriori* estimation of flow variables and their derivatives, which enable the resulting mesh to model shallow water flow accurately and efficiently. While the process lays the groundwork for achieving reproducible finite element grids, refinements must occur before a generalized application is produced.

At present, central difference approximations are used to estimate partial derivatives in the truncation error series. As a result, local truncation error can only be estimated on the main interior for an existing grid. A different technique must be developed that will permit truncation error estimation, and thereby maximum allowable radii computations, to be carried out up to and including the boundary.

In addition, it is possible to refine the mesh grading by the application of multiple transition zones. Transition zones (as applied in the transition between the nearshore structured grid and the LTEA-based portion of the interior) will enable the required spacing to be followed more accurately in the final mesh design and will promote more desirable element shapes.

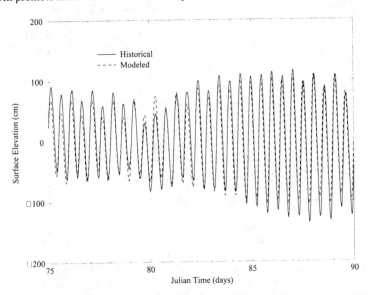

Figure 7. Charleston historical data and modeled results, 3/16/98 - 3/31/98

ACKNOWLEDGEMENTS

This paper is funded in part from University Cooperation for Atmospheric Research Subaward No. UCAR S98-98908 pursuant to National Oceanic and Atmospheric Administration Award No. NA67WD0097. The views expressed herein are those of the author(s) and do not necessarily reflect the views of NOAA, its sub-agencies, or UCAR.

REFERENCES

Blain, C.A., J.J. Westerink and R.A. Luettich Jr., "The influence of domain size on the response characteristics of a hurricane storm surge model," *J. geophys. research*, **99**, **C9**, 18,467-18,479 (1994).

Foreman, M.G.G., "An accuracy analysis of boundary conditions for the forced shallow water equations," *J. Comput. Phys.*, **64**, 334-367 (1986).

Hagen, S.C. "Finite element grids based on a localized truncation error analysis," *Ph.D. Dissertation*, Department of Civil Engineering and Geological Sciences, University of Notre Dame, IN, 1998.

Hagen, S.C., O. Horstmann, and R.J. Bennett, "A reproducible approach to unstructured mesh generation for shallow water models, "*Proceedings of the 8th International Meshing Roundtable*," Sandia National Laboratories, Albuquerque, NM, 1999.

Hagen, S.C., J.J. Westerink and R.L. Kolar, "One-dimensional finite element grids based on a localized truncation error analysis," *Int. J. Numer. Meth. Fluids*, **32**, 241-261 (2000).

Hagen, S.C., J.J. Westerink, R.L. Kolar and O. Horstman, "Two-dimensional, Unstructured Mesh Generation for Shallow Water Models: Comparing a Wavelength-based Approach to a Localized Truncation Error Analysis," *Int. J. Numer. Meth. Fluids*, In Review, January 2000.

Horstmann, O. "Adaptive grids for hydroengineering based upon predefined construction segments," Advances in Hydro-Science and -Engineering, Vol. III, K.P. Holz, et al. (eds.), *Proceedings of the 3rd International Conference on Hydroscience and Engineering*, 1998.

Kinnmark, P.E., "The shallow water wave equations: formulation, analysis and application," *Ph.D. Dissertation*, Department of Civil Engineering, Princeton University, NJ 1984.

Kolar, R.L., W.G. Gray, and J.J. Westerink, "Boundary conditions in shallow water models - an alternative implementation for finite element codes," *Int. J. Numer. Meth. Fluids*, 22, 603-618 (1996).

Luettich Jr., R.A., J.J. Westerink and N.W. Scheffner, "ADCIRC: an advanced three-dimensional circulation model for shelves, coasts and estuaries, Report 1: theory and methodology of ADCIRC-2DDI and ADCIRC-3DL," *Technical Report* DRP-92-6, Department of the Army, 1992.

Luettich Jr., R.A. and J.J. Westerink, "Continental shelf scale convergence studies with a barotropic tidal model," in D.R. Lynch and A.M. Davies (eds.), *Quantitative Skill Assessment for Coastal Ocean Models*, A.G.U., **47**, 349-371, 1995.

Lynch, D.R. and W.G. Gray, "A wave equation model for finite element tidal computations," *Comput. Fluids*, **7**, 207-228 (1979).

Lynch, D.R. "Progress in hydrodynamic modeling, review of U.S. contributions, 1979-1982," *Rev. Geophys. Space Phys.*, **21**, 741-754 (1983).

National Geophysical Data Center Coastal Relief Model, Vol. 02, version 1.0, US South East Atlantic Coast, Boulder, CO, 1999.

Westerink, J.J. and W.G. Gray, "Progress in surface water modeling," *Rev. Geophys.*, **29**, 210-217 (1991).

Westerink, J.J., J.C. Muccino and R.A. Luettich Jr., "Resolution requirements for a tidal model of the Western North Atlantic and Gulf of Mexico," T. F. Russell et al. (eds.), *Proceedings of the IX International Conference on Computational Methods in Water Resources*, 1992.

Westerink, J.J., R.A. Luettich Jr., J.K. Wu and R.L.Kolar, "The influence of normal flow boundary conditions on spurious modes in finite element solutions to the shallow water equations," *Int. J. Numer. Meth. Fluids*, **18**, 1021-1060 (1994).

Westerink, J.J., R.A. Luettich Jr. and J.C. Muccino, "Modeling tides in the western north atlantic using unstructured graded grids," *Tellus*, **46A**, 178-199 (1994).

Westerink, J.J., C.A. Blain, R.A. Luettich Jr., and N. W. Scheffner, "ADCIRC: an advanced three-dimensional circulation model for shelves, coasts and estuaries, Report 2: user's manual for ADCIRC-2DDI," *Technical Report DRP-92-6*, Department of the Army (1994).

Internal Characteristics of Two-Layer Stratified Flow Systems under Wind Stress

Koji Kawasaki[1], Jong Sung Yoon[2], and Keiji Nakatsuji[3]

Abstract

Two-dimensional numerical and hydraulic experiments were performed to study hydrodynamics of density interface and mixing processes of stratified water not only during the wind event but also after the release of the wind stress. According to the results of both numerical and hydraulic experiments, the internal characteristics of two-layer stratified flow system under the influence of wind stress can be classified mainly into three regimes in terms of the Wedderburn number. The internal flow after the release of wind forcing has a strong influence on the mixing process at a stratified interface. It is found that the density interface after the release of wind forcing changes with almost the same period as the internal seiche in a closed strongly stratified flume. In case of weak stratification in the flume with a slope, return flow and strong vortices occur over the slope after the wind stops. The results of numerical computation are confirmed to be in agreement with those of hydraulic experiment.

INTRODUCTION

When wind blows offshore in stratified estuaries and lakes, it is well known that the stratified interface is inclined in the reverse direction to the gradient of the water surface and turbulent entrainment takes place across the density interface. Upwelling phenomenon of the bottom water often occurs to compensate the surface water transported by the wind stress. In the head of Tokyo Bay, the color of surface water sometimes changes to milky-blue or milky-green when the offshore northeast wind blows during a few days in late summer or autumn. Such a phenomenon is referred to as 'aoshio' in Japanese. The 'aoshio' takes place as a result of the

---

[1] Research Associate, Dept. of Civil Eng., Osaka Univ., 2-1 Yamadaoka, Suita, Osaka 565-0871, Japan

[2] Associate Professor, Dept. of Civil Eng., Inje Univ., 607 O Bang-Dong, Kimhae, KyongNam, Korea

[3] Professor, Dept. of Civil Eng., Osaka Univ., 2-1 Yamadaoka, Suita, Osaka 565-0871, Japan

upwelling of anoxic bottom water. Since the 'Aoshio' phenomenon causes lots of fishes and shellfishes to perish, the fisheries have incurred damage. On the other hand, when the wind suddenly stops, the inclined interface causes an internal seiche or density current that propagates dynamically at the interface. In the head of Tokyo Bay, the internal wave is observed to propagate counterclockwise along the coastline after the release of wind forcing. The propagation speed corresponds to that of an internal Kelvin wave. Also, according to observations in Biwa Lake, the occurrences of strong flow near the bottom and the abrupt change of density interface were confirmed after the wind stops (Hayami et al., 1996).

Hence, it is very important from the viewpoint of environmental management in estuaries or lakes to understand the effects of wind stress on stratified water. Spigel and Imberger (1980) argued that the mixed-layer dynamics in lakes or reservoirs during the wind event could be classified into four regimes in terms of the Wedderburn number *We*. Furthermore, Thompson and Imberger (1980) explained based on the results of numerical experiments that the stratified interface could approach the water surface in the case of *We* < 3 or 4. However, the internal characteristics of stratified flow systems after the release of wind stress have not been examined sufficiently.

In the present study, hydraulic and numerical experiments for a closed two-layer stratified flume are conducted to understand the internal characteristics of stratified flow systems under the influence of wind stress. Based on the results of both numerical and hydraulic experiments, we discuss the characteristics of density interface variation, internal flow, and mixing process before and after the release of wind forcing in a two-layer stratified flume.

NUMERICAL SIMULATION MODEL

(a) Governing equations and boundary conditions

The governing equations consist of the continuity equation, the horizontal and vertical momentum equations under Boussinesq's approximation, and the diffusion equation of fluid density. The k-ε turbulence model is used to evaluate the eddy viscosity coefficient $v_t*(= c_u k*^2/\varepsilon*)$, as given by Eqs.(5) and (6). Note that the dependent variables in the following equations are dimensionless ones based on dimension analysis (Yoon , 1994).

$$\frac{\partial u*}{\partial x*} + \frac{\partial w*}{\partial z*} = 0 \tag{1}$$

$$\frac{\partial u*}{\partial t*} + u*\frac{\partial u*}{\partial x*} + w*\frac{\partial u*}{\partial z*} = -\frac{\partial p*}{\partial x*} + \frac{\partial}{\partial x*}\left\{\left(\frac{1}{R_e}+v_t*\right)\frac{\partial u*}{\partial x*}\right\} + \frac{\partial}{\partial z*}\left\{\left(\frac{1}{R_e}+v_t*\right)\frac{\partial u*}{\partial z*}\right\} \tag{2}$$

$$\frac{\partial w*}{\partial t*} + u*\frac{\partial w*}{\partial x*} + w*\frac{\partial w*}{\partial z*} = -\frac{\partial p*}{\partial z*} + \frac{\partial}{\partial x*}\left\{\left(\frac{1}{R_e}+v_t*\right)\frac{\partial w*}{\partial x*}\right\} + \frac{\partial}{\partial z*}\left\{\left(\frac{1}{R_e}+v_t*\right)\frac{\partial w*}{\partial z*}\right\} + R_i\rho* \tag{3}$$

$$\frac{\partial \rho^*}{\partial t^*} + u^* \frac{\partial \rho^*}{\partial x^*} + w^* \frac{\partial \rho^*}{\partial z^*} = \frac{\partial}{\partial x^*}\left\{\left(\frac{1}{R_e \cdot S_c} + \frac{v_t^*}{\sigma_t}\right)\frac{\partial \rho^*}{\partial x^*}\right\} + \frac{\partial}{\partial z^*}\left\{\left(\frac{1}{R_e \cdot S_c} + \frac{v_t^*}{\sigma_t}\right)\frac{\partial \rho^*}{\partial z^*}\right\} \tag{4}$$

$$\frac{\partial k^*}{\partial t^*} + u^* \frac{\partial k^*}{\partial x^*} + w^* \frac{\partial k^*}{\partial z^*} = \frac{\partial}{\partial x^*}\left\{\left(\frac{1}{R_e} + \frac{v_t^*}{\sigma_k}\right)\frac{\partial k^*}{\partial x^*}\right\} + \frac{\partial}{\partial z^*}\left\{\left(\frac{1}{R_e} + \frac{v_t^*}{\sigma_k}\right)\frac{\partial k^*}{\partial z^*}\right\}$$
$$+ 2 v_t^*\left\{\left(\frac{\partial u^*}{\partial x^*}\right)^2 + \left(\frac{\partial w^*}{\partial z^*}\right)^2\right\} + v_t^*\left(\frac{\partial u^*}{\partial z^*} + \frac{\partial w^*}{\partial x^*}\right)^2$$
$$- R_i \frac{v_t^*}{\sigma_t}\frac{\partial \rho^*}{\partial z^*} - \varepsilon^* \tag{5}$$

$$\frac{\partial \varepsilon^*}{\partial t^*} + u^* \frac{\partial \varepsilon^*}{\partial x^*} + w^* \frac{\partial \varepsilon^*}{\partial z^*} = \frac{\partial}{\partial x^*}\left\{\left(\frac{1}{R_e} + \frac{v_t^*}{\sigma_\varepsilon}\right)\frac{\partial \varepsilon^*}{\partial x^*}\right\} + \frac{\partial}{\partial z^*}\left\{\left(\frac{1}{R_e} + \frac{v_t^*}{\sigma_\varepsilon}\right)\frac{\partial \varepsilon^*}{\partial z^*}\right\}$$
$$+ c_{\varepsilon 1}\frac{\varepsilon^*}{k^*}(P_r^* + G^*)(1 + c_{\varepsilon 3}R_f) - c_{\varepsilon 2}\frac{\varepsilon^{*2}}{k^*} \tag{6}$$

where, $t^* = t/t_0 = t/(H/u_*)$, $(x^*, z^*) = (x/H, z/H)$, $(u^*, w^*) = (u/u_*, w/u_*)$,

$k^* = k/u_*^2$, $\varepsilon^* = \varepsilon H/u_*^3$, $\rho^* = (\rho - \rho_1)/\rho_0$, $p^* = p/(\rho_0 u_*^2)$

$$P_r^* = 2v_t^*\left\{\left(\frac{\partial u^*}{\partial x^*}\right)^2 + \left(\frac{\partial w^*}{\partial z^*}\right)^2\right\} + v_t^*\left\{\left(\frac{\partial u^*}{\partial z^*} + \frac{\partial w^*}{\partial x^*}\right)^2\right\}$$

$$G^* = R_i \frac{v_t^*}{\sigma_t}\frac{\partial \rho^*}{\partial z^*}$$

x^* and z^* are the horizontal and vertical axes in Cartesian coordinate. u^* and w^* are the velocities in the respective directions of x^* and z^*. t^* is the time, H is the overall water depth, p^* is the pressure, ρ^* is the fluid density, ρ_0 is the referential fluid density, ρ_1 is the fluid density in the upper layer, k^* is the turbulence energy, and ε^* is the turbulence energy dissipation ratio. Re is the Reynolds number ($= u_* h_1/v_0$) ; u_* : the wind-driven friction velocity at the water surface, h_1 : the initial thickness of upper layer, v_0 : the kinematic viscosity coefficient, and R_i is the Richardson number ($= g\Delta\rho h_1/\rho_0 u_*^2$) ; g : the gravity acceleration, $\Delta\rho$: the initial density difference between upper and lower layers. Sc is the Schmidt number, and has a value of 8.3 in this work. σ_t ($=1.0$) is the turbulent Schmidt number, and R_f is the flux Richardson number. $(c_u, \sigma_k, \sigma_\varepsilon, c_{\varepsilon 1}, c_{\varepsilon 2}, c_{\varepsilon 3}) = (0.09, 1.0, 1.3, 1.44, 1.92, 1.0)$.

Next, the boundary conditions are explained briefly as the following : The water surface is assumed to be a rigid lid in the present study. This assumption is valid under the condition that the water surface does not billow largely. Also, it is confirmed from hydraulic experiments that the water surface elevations were small for all experimental conditions. The wind-driven friction velocity is set as the surface boundary condition so that the wind-driven momentum can be provided at the water surface. At other boundaries, the log law was adopted.

(b) Computational algorithm

Figure 1 shows the flow chart of the present numerical simulation model. After setting of the initial and boundary conditions, the momentum equations are used to calculate the first approximation of velocities at the next time step. Then, the velocity components u^* and w^*, the pressure correction p', the turbulence energy k^*, and the turbulence energy dissipation ratio ε^* are calculated by means of the SIMPLE (Semi-Implicit Method for Pressure-Linked Equation) method (Patanker, 1980), which is one of non-hydrostatic models. After that, the boundary conditions are set. However, the calculated velocity components do not satisfy the continuity equation in general. An iteration procedure is, thus, performed to satisfy the continuity equation at reasonable criteria. Stable numerical results can be obtained by repeating the above-mentioned procedures under suitable boundary conditions at each time step.

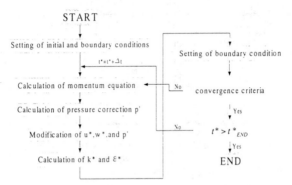

Fig. 1 Flow chart of the numerical simulation model

(c) Calculation conditions

We considered two kinds of computational domains with and without a slope at the windward side, as shown in Fig. 2. The positive direction x-axis is taken windward, and the vertical z-axis is taken positive upward with its origin being on the bottom at the leeward end. The computational domain is taken as $L = 5.4$m and $H = 0.3$m in the respective directions of x and z. Mesh sizes in the directions of x and z are $\Delta x = 0.075$m and $\Delta z = 0.01$m, respectively. The time increment Δt is set constant to be 0.005s at every time step.

Calculation conditions are shown in Table 1, where u_* is the wind-driven friction velocity at the water surface and its value was determined from the experimental results. h_1 is the initial thickness of upper layer, H is the overall water depth, $\Delta\rho$ is the initial density difference between upper and lower layers, and We is the Wedderburn number defined as the product of the Richardson number R_i and the aspect ratio of the stratified water flume $2h_1 / L$. The parameter

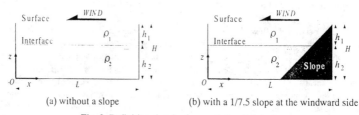

(a) without a slope   (b) with a 1/7.5 slope at the windward side

Fig. 2 Definition sketch of computational domain

Table 1 Calculation conditions

Run	u_* (m/s)	h_1 (m)	H (m)	$\Delta\rho$ (Kg/m³)	We
1	6.5×10^{-3}	0.1	0.3	38	39.6
2	6.5×10^{-3}	0.1	0.3	12	13.1
3	6.5×10^{-3}	0.1	0.3	7	5.8
4	6.5×10^{-3}	0.1	0.3	4	3.4
5	6.5×10^{-3}	0.1	0.3	1	0.8

settings are the same as those of the hydraulic experiments. In the present study, five experiments were conducted by changing $\Delta\rho$. The water surface boundary condition after the release of the wind forcing is set as $u_* = 0$.

HYDRAULIC EXPERIMENTS

Experimental equipment is composed of a main flume (600cm in length, 115cm in depth, and 15cm in width) and a wind tunnel (the maximum rotation : 1500rpm) with six shuttlecocks, as shown in Fig. 3. Hydraulic experiments were conducted in stratified flow systems with two layers,

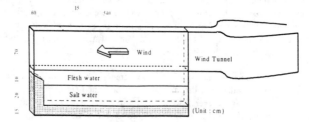

Fig. 3 Schematic of experimental equipment

fresh water over salt water. The stratification depends on the density of the salt water. The initial thickness of the upper layer h_1 was kept constant at 10cm for all experiments. Mean velocities in the upper layer were measured using a Laser Doppler Velocimeter (LDV). Two 35mm cameras and a video camera were employed to observe the movement of the density interface and mixing processes. The density profiles were measured using a salinometer.

RESULTS AND DISCUSSIONS

Figures 4 and 5 show the time variation of the density interface during the wind event and after the release of wind stress, respectively. The horizontal and vertical axes (t^*, z^*) indicate the dimensionless lapsed time after the start of wind forcing or wind stopping, and the dimensionless vertical coordinate of the density interface. The observation points are the windward end ($x^*=18.0$), the center ($x^*=9.0$), and the leeward end ($x^*=0.0$) of flume. The computational density interface is defined as the average value of initial fluid densities in two layers.

According to the results of both numerical and hydraulic experiments, the pattern of a two-layer stratified flow system under the influence of the wind stress can be classified mainly into three regimes in terms of the Wedderburn number We. The characteristics of three regimes, (a) $We=39.6$ and 13.1, (b) $We=5.8$ and 3.4, and (c) $We=0.8$, are as follows :

(a) Case of $We = 39.6$ and 13.1

As shown in Fig. 4(a), the density interface variation at each point is small because the stratification due to the density difference is extremely strong, and the wind stress hardly influences on the vertical mixing or entrainment at the density interface. On the other hand, it is found from Fig. 5 (a) that the density interface oscillates periodically after the wind stops. The oscillation periods T_c in case of $We=39.6$ and 13.1 are $T_c/t_0=1.5$ and 2.8, respectively. The internal seiche period of fundamental mode T_i is given in Eq. (7) derived from the theory of two-layer system. Applying Eq. (7) to the case of $We=39.6$ and 13.1, $T_i/t_0=1.5$ and 2.7 are obtained. These values are in good agreement with the numerical results. In other words, the density interface after the release of wind forcing changes with almost the same period as the internal seiche in a closed strongly stratified flume.

$$T_i = \frac{2L}{\sqrt{\Delta\rho \, h_1 \, h_2 / (\rho_0 H)}} \qquad (7)$$

In order to understand the internal flow after the release of wind forcing in the case of strong stratification, the computed vertical profiles of horizontal velocity at the center of the flume ($x^*=9.0$) are shown in Fig. 6. The different lines represent the horizontal velocities at various time steps since the start of wind forcing. At the time just before the wind stops ($t^*=2.17$), the flow in the same direction as the wind stress occurs in the range of $z^*>0.88$. The horizontal velocity at the water surface is 5.1 times as the wind-driven friction velocity. In $0.6<z^*<0.88$, the windward flow

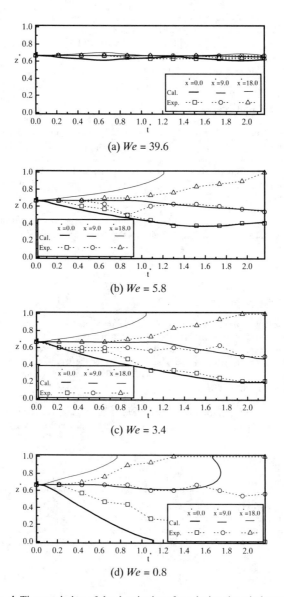

Fig. 4 Time variation of the density interface during the wind event

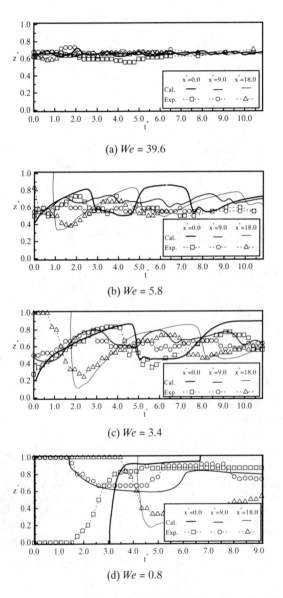

(a) *We* = 39.6

(b) *We* = 5.8

(c) *We* = 3.4

(d) *We* = 0.8

Fig. 5 Time variation of the density interface after the release of wind stress

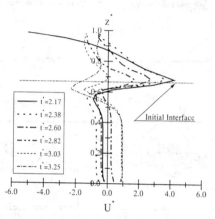

Fig. 6 Computed vertical profiles of horizontal velocity ($We=39.6, x^*=9.0$)

takes place to compensate the leeward flow in $z^*>0.88$. Thus, it is evident that a vertical circulation appears above the initial density interface in the case of strong stratification. This indicates that the influence of the wind stress hardly reach to the lower layer. On the other hand, as time passes after the release of wind forcing ($t^*>2.17$), the vertical distribution of horizontal velocity changes and oscillates. Judging from the above, it is found that vertical mixing between the upper and lower layers hardly takes place even under the influence of the wind stress. Also, the density interface variation after the release of wind forcing is found to behave as an internal seiche.

(b) Case of $We = 5.8$ and 3.4

Figures 4(b) and (c) indicate that the density interfaces begin to change immediately after the wind starts. The density interface at the windward side ($x^*=18.0$) continues to rise and approaches the water surface at $t^*\cong1.1$, while the density interface at the leeward side moves down. As shown in Figs. 5(b) and (c), after the wind stops, the density interface at the windward side, which is reached to the water surface during the wind event, moves down due to the gravitational force, and the density interface at the leeward side turns up. After that, the density interface at each point begins to oscillate. The computed oscillation periods are $T_c/t_0\cong4.0$ and 6.3 for $We=5.8$ and 3.4, while the theoretical oscillation period T_i/t_0 obtained from Eq. (7) are 3.5 and 4.7. The computed values do not coincide with the theoretical ones, differently from the case of $We=39.6$ and 13.1. In other words, the two-layer theory cannot be applied to estimate the oscillation period after the release of the wind stress in these cases because the mixing occurs near the density interface.

Figure 7 shows the time variation of velocity and density difference fields obtained by the numerical experiments in a flume with a 1/7.5 slope at the windward side. The density difference

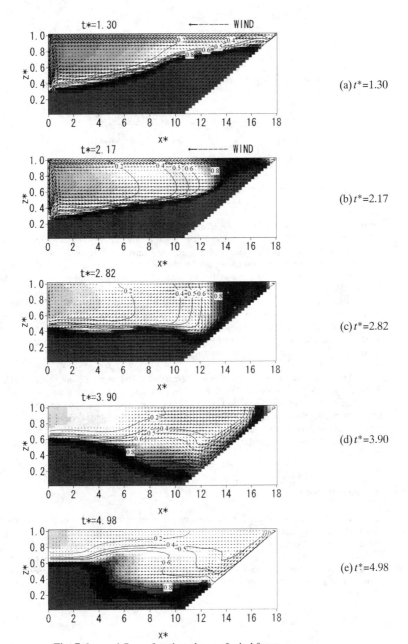

Fig. 7 Internal flow after the release of wind force

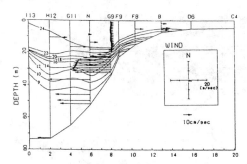

Fig. 8 Observation results of water temperature and flow in Biwa Lake
(cited from Hayami et al.(1996))

between upper and lower layers are plotted every 10 ~ 20% by the contour lines and the shadings. As seen in Figs. 7(a) and (b), the interface of the initial average density between upper and lower layers ($\rho = 0.5$) surfaces at the upwind end of flume, and moves leeward. Therefore, it can be said that the upwelling of bottom water takes place by the wind stress influence. On the other hand, it is found from Figs. 7(c) ~ (e) that the upwelling of bottom water stops after the release of wind force, and the density interface begins to turn down due to the gravitational force. A strong return flow occurs along the slope, as seen in Fig. 7(d). Moreover, a large vortex appears over the slope (11< x^*<13, 0.2< z^*<0.8). This means that the vertical mixing of waters is accelerated. Further, the interval of density difference becomes larger after the wind stops. This indicates that the density interface is unstable and cannot remain sharp. It can be said from this that the vertical mixing causes strongly after the release of the wind stress than during the wind event.

The field observation results of water temperature and internal flow in Biwa Lake by Hayami et al. (1996) are shown in Fig. 8. A northward strong flow (the flow from right to left in this figure) is found to occur near the bottom after the wind stops. It can be expected that return flow appear along the slope although the detail of vertical structure of the flow is not clear because the field data were obtained by means of ADCPs. Comparing Fig. 7(d) with Fig. 8, it may be said that the computed results explain the qualitative characteristics of the field observations of Hayami et al. (1996).

(c) Case of $We = 0.8$

The characteristics of the density interface variation in the case of weak stratification are shown in Figs. 4(d) and 5(d). As seen in Fig. 4(d), the density interface at the windward side rapidly moves up after the start of wind event, and approaches the water surface at t^*=0.76, similar to the case of We=5.8 and 3.4. Also, the density interface at the leeward side reaches to the bottom at t^*=1.1, different from the other cases. In this case, we found from hydraulic experiments that vortex trains are induced by Kelvin-Helmholtz instability, though the numerical model could not reproduce such a phenomenon. On the other hand, after the wind stops, the variation of the density interface is different from other cases and its oscillation is not recognized definitely because the strong vertical mixing of water takes place in case of weak stratification.

As seen in Figs. 4 and 5, the agreement between the numerical and hydraulic results is not bad, and the validity of the present numerical computation model is, to some extent, verified. However, considering the difference of the results between numerical and hydraulic experiments, the k-ε model used in the present study may not represent vertical mixing process precisely because this model assumes turbulences to be isotropic. This discussion remains to be seen.

CONCLUDING REMARKS

In the present study, both numerical and hydraulic experiments were performed to investigate the internal characteristics of a closed two-layer stratified flow system under the influence of the wind stress. According to both numerical and hydraulic results, it was found that the patterns of the two-layer stratified flow systems could be classified mainly into three regimes in terms of the Wedderburn number defined as $We = 2R_i \ h_1 \ /L$. In addition, we also examined the characteristics of the density interface variation and internal flow structure in a two-layer stratified flume before and after the release of wind forcing. Hereafter, we will modify the turbulence model so that the non-isotropic turbulence components can be considered, and the effects of the wind stress on the internal structure of stratified water will be examined in detail.

ACKNOWLEDGEMENTS

Thanks are due to Mr. Tomoki Otani of master student in Osaka University for his assistance in data analysis. This research was partly supported by Grant-in-Aid for Scientific Research of The Ministry of Education, Science, Sports and Culture (Grant No. 11305036).

REFERENCES

Hayami, Y., Fujiwara, T. and Kumagai, M. (1996) : Internal surge in lake Biwa induced by strong winds of a typhoon, Jpn. J. Limnol., 57 4(2), pp. 425-444.

Patankar, S.V. (1980) : Numerical heat transfer and fluid flow, Hemisphere Publishing Corporation Press.

Spigel, R. H. and Imberger, J. (1980) : The classification of mixed-layer dynamics in lakes of small to medium size, J. Physical Oceanography, Vol. 10, pp. 1104-1121.

Thompson, R. O. R. Y. and Imberger J. (1980) : Response of a numerical model of a stratified lake to wind stress, Proc. Int. 2nd Symp. Stratified Flow, pp. 562-570.

Yoon, J. S. (1994) : Hydraulic study on upwelling of anoxic bottom water in a stratified water areas, Dissertation, Osaka Univ., 163p (in Japanese).

The Mary River Estuary -A New Approach to Modeling of Marshes and Over-Bank Areas

Ian P. King[1] and David K. Williams[2]

1. Abstract

The Mary River is a complex estuarial system with multiple channels and a broad flood plain that is subject to seasonal long term flooding during the monsoon season and steady draining during the long dry season. Probably due to geometric changes made at the estuary outlets the tidal prism now progresses more that 20 kilometers upstream and significant salt intrusion now occurs. From a modeling perspective a number of difficulties arise.

The major problem arises from flow transitioning from the main channels to the over-bank areas. These over-bank flood plains allow transfer of water between adjacent sections of the river system. Depending on the season, they are (a) flooded (monsoon), (b) slowly draining (beginning of the dry season) and (c) subjected to tidal influence only (parts of the dry season). In addition to this unusual system there are a several other examples in hydraulic flow modeling where flow transitions of this type occur. For example:

1.   Flow in estuaries with many sandbanks that are exposed at low tide. In this case, at low tide, the flow regime becomes essentially one-dimensional in the channels that are formed. Florida Bay in the United States is an excellent example of this type of situation.

2.   Flow in estuaries surrounded by marshes with numerous channels of various sizes. In this case, in a similar manner to that discussed immediately above, the system becomes a system of one-dimensional channels at low tide and a fully inundated two dimensional system at high tide.

3.   River flows with extensive over-bank areas that are subject to flooding from runoff events. The Mary River system in the Northern Territory of Australia is a typical example of this type of system.

---

[1] Associate, Resource Modelling Associates, c/o University of New South Wales, Manly Vale, Australia, 61-2-9949-4488

[2] Geomorphologist. Department of Lands, Planning and Environment, Natural Resources Division, Palmerston Northern Territory Australia, 61-8-8999-3694

Representation of these system types presents a conceptual problem for model representation. The depth transition from the main channel to the over-bank area can occur over a relatively short distance with relatively steep bed/water depth gradients. Although one-dimensional models can very satisfactorily represent flows in the main channel they cannot adequately represent the two-dimensional over-bank flow regime. Similarly two-dimensional models become burdened by extensive detail required to represent the channel and the transition region to the over-bank area. Geometric representations developed for large regions become impracticably complex.

This paper presents a finite element model formulation for this type of system, where river sections are represented by one-dimensional approximations and the over-bank areas are represented by fully two-dimensional approximations. The model formally integrates these two regimes with a fully mass conserving approach. The methodology, its theoretical basis and finite element implementation will be presented along with simple test cases designed to validate the approach. A more detailed application will show results from the Mary River system described earlier.

2.   Introduction

2.1   The Mary River System

The floodplains of the Lower Mary River lie in close proximity to Darwin and are a valuable cultural and economic resource. The floodplains support a wide variety of plant and animal communities that are dependent on the extensive freshwater system. This area has been extensively degraded by saltwater intrusion and as it continues to expand much more of the freshwater habitat is under threat. The Mary River has a catchment area of 9000 square kilometres and drains northward into Chambers Bay via two tidal tributaries (Sampan and Tommycut Creeks) 80 kilometres to the east of Darwin, see Figure 1. The last 100 kilometres of river form the wetlands and 35 kilometres are under tidal influence. The wetlands cover an area of 3000 square kilometres and comprise a large area of salt marsh and an equally extensive area of fresh water billabongs. The climate is wet dry monsoonal and the average annual rainfall is 1700 millimetres, which falls in the wet season from November to March.

The floodplains are unique amongst Northern Territory coastal floodplains as until recently there had been no significant estuarine development. This lack of estuarine development has not provided a distinct outlet for the annual wet season floodwaters to escape to the sea. Annually this vast expanse of water spreads out over a large area where it slowly drains, evaporates and infiltrates. The floodplains have had several channels formed on them. Evidence suggests that these channels were wider and shallower than those that are forming at present. Flow paths have changed their course on the floodplains over time. It is probable that channel switching occurred concurrently with the formation and breaching of chernier ridges at the previous coastlines. When a chernier ridge forms it can create a barrier that limits the entry of tidal waters onto the floodplain. During this

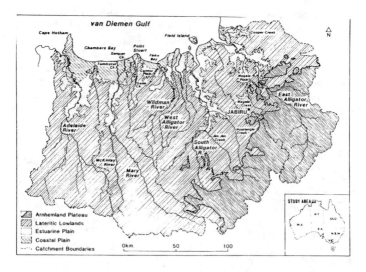

Figure 1 Location Map for Mary River

time the channels can fill with alluvial sediments carried by the wet season floods. If a chernier is breached a new flow pathway is formed and the channels may change their course.

Saltwater intrusion has been increasing in the Lower Mary River wetlands since at least the 1940's. This is evident from an examination of aerial photographs and the water level recording station that later existed at Roonees Lagoon (14 kilometres inland from the coast) from 1958. The main channel of Sampan Creek was at that time narrow and discontinuous. A survey of the channel form of Sampan Creek at Roonees Lagoon in 1963 shows that the width of the stream was 25 metres and the maximum depth was 1.5 metres. The 1994 survey shows that the stream is now 90 metres wide with a maximum depth of 7 metres, see Figure 2.

Associated with the increase in channel dimensions has been a marked increase in tide levels. In 1959 the tidal range at Roonees Lagoon was only small (a maximum of 0.3 metres) and the channel networks were only beginning to branch out onto the floodplains. It is noteworthy that the Roonees Lagoon tide record in 1959 is composed of two superimposed tides, see Figure 3. Current recorded water levels show that the tidal variations at this location are now at a maximum of 4 metres. Figure 4 shows that the top of the main channel's banks are regularly overtopped by spring tides and that tides flow into the tributaries under all conditions.

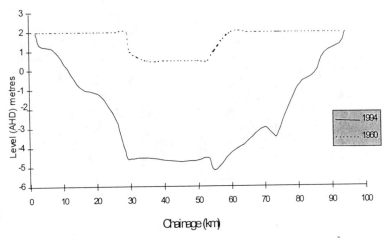

Figure 2 Variation of River Cross Sections at Roonees Lagoon over Time[3].

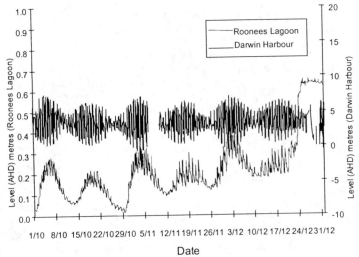

Figure 3 Tidal Stage at Roonees Lagoon and Darwin Harbour in 1959

---

[3] AHD refers to Australian Height Datum

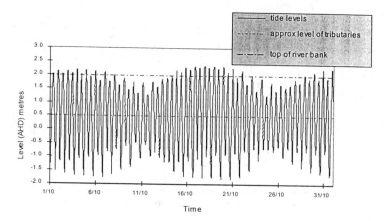

Figure 4. Tidal Stages at Roonees Lagoon in 1993

Tidal variations were not seen at Shady Camp (35 kilometres inland) until 1980 where the maximum variation was 0.1 metres. This variation had increased to 0.5 metres by 1989 and in 1994 had reached a maximum of 2 metres. A barrage was constructed at this location in 1989 to prevent the saltwater from moving further upstream.

Tidal energy is persistent and the existing barriers that were at the coastline have deteriorated with time. Once these barriers are breached their cemented resistant outer layer is severely weakened exposing them to erosional forces. As a result of this erosion more tidal energy is allowed to enter the Sampan and Tommycut Creek systems. This results in an increase in the volume of saltwater that can move into the channel networks. The channels respond by widening and deepening to accommodate the increased volume and this in turn applies more erosional stress on the system. Coupled with the increase in tidal volumes is the annual wet season flooding, the majority of which is borne by Sampan Creek. The additional energy results in the rapid erosion of the channel networks.

At present the mouths of Sampan and Tommycut Creeks experience the same tidal range as the offshore ocean tides and show no sign of reaching equilibrium conditions in the near future as they are still actively eroding. The tributary networks continue to expand and the channels widen and deepen. This allows more saltwater onto the floodplains and the erosion hazard increases. It is not possible to predict what the system may evolve to if left unchecked but the worst case scenario is that Shady Camp will experience much larger tides which will result in the overtopping or sidestepping of the barrage and the invasion of the extensive upstream freshwater habitat by saltwater. Additionally, based on the rate of growth of the tributary network, the area between Sampan and Tommycut Creeks could become a large shallow tidal inlet and the extensive floodplains and freshwater habitats to the west of Tommycut Creek could be invaded.

The growth of the main channel and the tributary network has altered the drainage characteristics of the floodplains after the wet season. When the Lower Mary River was a discontinuous channel with only a small opening to the sea, wet season floods would remain on the plains for long periods of time until the water was eventually removed by a combination of drainage, infiltration and evaporation. Now with an extensive network of tributaries, a continuous larger main channel and a much wider interface with the sea, wet season floods drain more rapidly. Water level records show that in the 1960's the floodplains would be inundated for up to 8 months. Since the mid 1990's the floodplains are only remaining inundated for up to 4 months.

2.2   RMA-2

RMA-2 (King 1978) is part of the RMA suite of finite element models designed for simulation of flow and water quality in rivers, estuaries and coastal systems. It is designed to simulate transient flow and water depth in prototype systems that can be approximated with either the depth-integrated two-dimensional or area-integrated one-dimensional shallow water equations. The model incorporates a fully non-linear solution of the governing equations including terms for convective inertia, pressure gradient, turbulent viscosity, bed and bank friction, wind stresses and Coriolis forces. It has been applied to many estuaries and river systems with results that calibrate well against measured data. An unstructured finite element network of quadrilaterals and triangles permits variable resolution over two-dimensional sections of the system. One-dimensional linear elements may be directly connected to two-dimensional elements to allow a transition in the representation for upstream areas that are part of the tidal prism or flow regime but away from the area of major interest. This methodology has been presented elsewhere (King 1990) and will not be expanded upon here

2.3   Modelling of Flooding and Drying

Many estuarial systems have relatively large bays that incorporate sandbanks or mudflats within them and marshes that form an exterior border. Other systems have relatively narrow channels with extensive adjacent areas that are flooded at high tide. Similarly many river systems have over-bank areas that flood for extreme events. To satisfactorily represent this type of system layout, models must be able to simulate areas of the geometric network that do not remain flooded for an entire tidal or flood cycle. Various methods have been adopted to accommodate this requirement. For example, parts of the system may be dropped out when depth approaches zero or, as adopted in RMA-2, a sub-surface flow analogy may be used where flow across the flood plain is treated as flow through a low porosity region when the water surface elevation drops below the ground level. For further information on this method see Roig and King (1988). These methods work well when the area that remains flooded at all times is extensive and the transition to the drying zones has a gentle gradient. However, in the Mary River system a relatively steep transition occurs between the flowing areas and drying zones. Flow is confined at low water surface elevations and then rapidly changes to over-bank flow as the depth increases. Modeling of this type of system will be the focus of this paper

2.4   Difficulties with Existing Methods

The main difficulty facing the modeler is adequate representation of the two types of flow regime represented by these systems. At low water surface levels flow is one-dimensional, at high water surface levels flow is best represented by a two-dimensional depth integrated approximation. If the modeler chooses to represent the system with a fully two-dimensional approximation, then as flow crosses from the channel to the over-bank region there is a rapid transition from the velocities in the deep channel to different velocity regimes in the shallow over-bank region. For finite element methods, the continuous representation of system variables such as velocity and depth becomes a handicap, as additional elements are required to represent the sloping bank. These elements are then frequently relatively long and narrow and add many dependent variables to the system of equations that must be solved. On the other hand, regular or rectangular finite difference methods cannot adequately represent the small-scale features that these often-meandering channels represent without a very fine grid resolution and curvilinear methods have problems with representation of the numerous channels that are frequently involved.

In summary, what is required is a method that uses a two dimensional approximation for the over-bank areas and a one-dimensional approximation for the channel system. In the sections that follow such a method is described, along with a test case comparing results with those obtained from a standard approach. In later sections of the paper the Mary River System network is presented and results from application of this method are discussed.

3.   *Modeling Method*

3.1   Problem Statement

The objective of this model development is to construct a method that permits a one-dimensional river element to be embedded within a two dimensional network and allow for exchange of flow between these two element types in a manner consistent with the two approximations. Figure 5(a) presents an elevation view of a typical section of river with an over-bank/flood plain area and Figure 5(b) is a conventional finite element representation of this type of system. Note that extra detail is needed to represent the slope of the banks, even for this case of relatively gentle slopes. If the slopes were steeper then the elements would be even less rectangular. Figure 5(c) shows a representation of the system comprised of one-dimensional and two-dimensional elements. The center section is then a series of one-dimensional elements with specified bottom widths and side slopes.

The model representation must be devised so that appropriate boundary conditions are applied at the interface, that is, between the one-dimensional elements and the two-dimensional elements at the top of the bank slope. The conditions that must hold are:

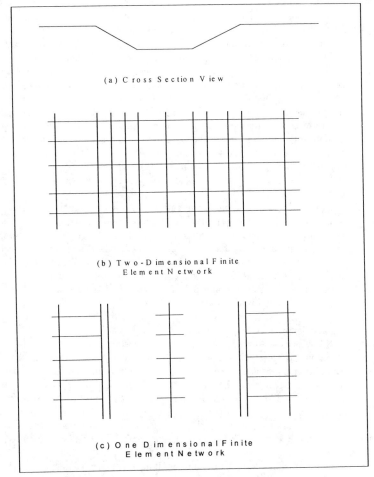

Figure 5   Typical River Section and Model Representations

1   Water surface elevation at centerline of the channel (one-dimensional representation) must equal the water surface elevation at the outer boundary of the two dimensional element.   This is strictly an approximation but given that the rivers are relatively narrow it is considered a reasonable requirement.

2   The specific inflow/outflow rate per unit length from the one-dimensional element must equal the specific outflow/inflow along the boundary of the two-dimensional element.

3.2   Model Equations

Before detailing how these conditions are imposed, it is relevant to explain some of the features of the finite element method as implemented in RMA-2.

1   The model is formulated using the primitive form of the shallow water equations. There are three equations, two equations of momentum and the flow continuity equation. The dependent variables are the velocity components that are approximated with bi-quadratic basis functions and depth that is approximated with bi-linear basis functions. During development of the global set of simultaneous equations, three equations are formed at corner nodes (dependent variables are the two velocity components and depth) and two equations at mid-side nodes (dependent variables are the two velocity components only). The equations and dependent variables thus form a balance. Velocity variables are associated with the momentum equations and depth variables associated with the continuity equation.

2   In order to preserve global conservation of flow, no continuity equation (and thus depth variable) may be dropped from the system.   As a consequence all boundary conditions are applied through the momentum equations. In order to achieve a specified depth on the boundary, external tractions (RHS terms) are applied to the edge of the system until the desired depth is achieved. Note that momentum equation remains active for this case. This condition is thus achieved only in a weak sense.

3   Flow or velocity boundary conditions are applied by eliminating the momentum equation and replacing it with a direct specification for the desired component.

4   RMA-2 uses a simultaneous Newton Raphson process with multiple iterations to solve the global set of non-linear equations derived in the finite element process.   In this form the right-hand side of the global equation set represents the current estimate error and the global matrix is the Jacobian derived from taking derivatives with respect to the dependent variables.

The boundary conditions at the interface between the one-dimensional elements and the two-dimensional elements are applied within the simultaneous process.

This means that a formal structure for all the relationships must be created. For the one-dimensional element, the continuity equation is rewritten, so that the flow crossing the boundary of the two-dimensional region is used as tributary inflow to the one-dimensional region. For each one-dimensional element a linear basis function interpolation of the corner node specific flow is applied along the element boundary.

The one-dimensional continuity equation takes the form:

$$w_s \frac{\partial h}{\partial t} + A\frac{\partial u}{\partial x} + u\frac{\partial A}{\partial x} - (\beta \, h_{n1}u_{n1} + (1-\beta) \, h_{n2}u_{n2}) = 0$$

where: β = a geometric weighting function

h_{n1}, h_{n2} = depth at nodes n1 and n2.

u_{n1}, u_{n2} = velocity normal to the boundary at nodes n1 and n2.

Other variables have the usual definitions.

During solution, the integral process of the finite element method combines the continuity equations for the one-dimensional elements and the two-dimensional elements for the purpose of computing the change of depth.

The momentum equation for the normal component of velocity at the mid-side nodes on the boundary of the two-dimensional region is replaced by a condition that forces the velocity distribution to vary linearly along the element boundary. Thus, with reference to Figure 6;

1.   Flow crossing the boundary line 3-4 is set equal to flow leaving the one dimensional element 1-2.

2.   Elevations at nodes 1 and 3 and 2 and 4 are forced to be equal.

3.   For node 8 the nodal flow normal to the boundary is interpolated from that at nodes 3 and 4.

4.   Flow parallel to the boundary line 4-8-3 is computed from the momentum equation for the two-dimensional element and is uncoupled from the one-dimensional element.

5.   Because of the linear approximation for depth only velocity components are active for node 7, so it is only influenced by the one-dimensional momentum equation.

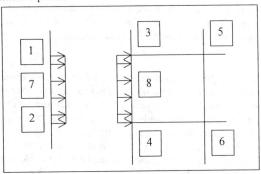

Figure 6   Flows Between Element Types

During implementation Newton Raphson derivatives are applied to reflect sensitivity to changes in tributary flow into the one-dimensional element and to changes in elevation between nodes in the one-dimensional element and the two-dimensional element. This is achieved by creating a pseudo element that connects 1-7-2 and 3-8 4.

3.3   Representational Consequences

When a numerical representation such as this is constructed, it is appropriate to study the consequences of the approximation made.

1.   When the system is such that the flow level is below bank crest. If the over-bank elements are dropped from the system then the solution is as good as the solution of any other one-dimensional system using RMA-2. If the over-bank elements are not dropped and the "marsh porosity" method is used then,

typically, the flow in the over-bank section is very small and the one-dimensional section continues to control the flow.

2. When the system is such that flow is just above bank crest. For this case, flow is still predominantly within the river section, and over-bank flow is driven by the elevation differences upstream and downstream or across the over-bank region. Because the boundary condition at the transition is equality of water surface elevation, momentum from the one-dimensional region to the two-dimensional region is not preserved. In fact, it cannot be, because the one-dimensional approximation inherently only defines momentum along the axis of a river channel. Thus, for the case of the meandering channel in the test case shown below, water flowing into the river from over one bank and out over the other bank only preserves flow continuity and not continuity and momentum. For water surface elevations just above the bank level bed friction will offer significant resistance and it is believed that the error introduced would be small.

3. When the system is such that flow is well above bank crest. For this case it can be expected that velocities in the over-bank region may have significant momentum and the impact of not preserving momentum is more uncertain. It should be pointed out at this time that for high momentum flow normal to a river bank with a relatively steep slope three-dimensional effects could be expected and it is not certain that a two dimensional depth integrated approximation would provide a more refined result.

3.4   Test Case

The test case presented below is designed to compare results from a fully two-dimensional simulation with those obtained from the revised "mixed" method. The system is based on a real stream that meanders across a flood plain that is controlled at the downstream boundary by a bridge crossing. Figures 7 and 8 show the bathymetry and the "mixed" method finite element representation. The fully two-dimensional representation replaces the one-dimensional stream segment with six elements across the trapezoidal section. A number of steady state cases have been investigated.

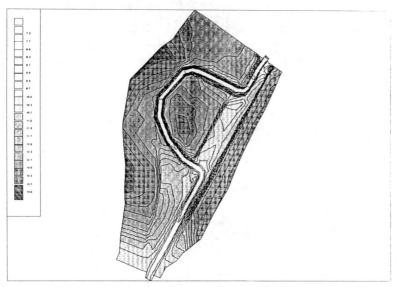

Figure 7   Bathymetry of test case

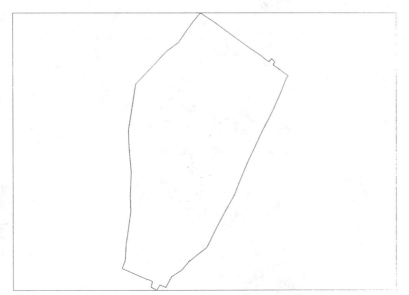

Figure 8 Finite Element Network for Test Case

For low flows (15 m³/s inflow and 8.68 m downstream elevation) flow is confined within the banks. For the high flows (289 m³/s inflow and 12.6 m downstream elevation) a more complex situation arises. Flow is now largely over-bank, short-circuiting the stream section. Water depths in the over bank area are in excess of 3 metres. Figures 9 and 10 illustrate the results from the two methods.

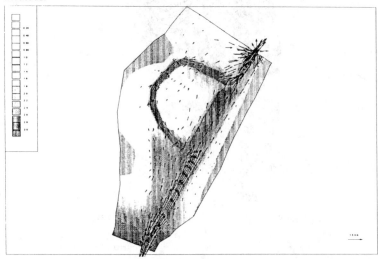

Figure 9 Contours of Water Depth (m) and Velocity Vectors (m/s)
for Fully Two-Dimensional System

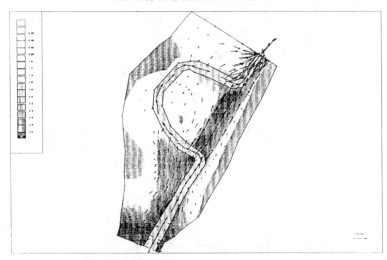

Figure 10 Contours of Water Depth (m) and Velocity Vectors (m/s)
for "Mixed" System

3.5   Discussion

The test case was designed to evaluate the model approximations in a flood flow case where data was available. For this case the riverbed was typically less than 2 metres below bank level and bank slopes were not very steep. Thus it was possible to construct a reasonable approximation with a fully two-dimensional network.

At low flows and low downstream elevations both methods result in flow confined within the stream section and the results are essentially identical. When the over bank area becomes dry the model correctly applies the "marshing method" for the new transition region.

For high flows, the system is now largely over-bank, short-circuiting the stream section. The fraction of flow on the right bank is between 60 and 70% (with the two dimensional results showing the higher value) for both cases. In practice, it is very difficult to design model representations that are identical for this complex situation. For example the different types of approximation in the vicinity of the bridge opening cause variations in head loss in that area and as a result the water depth throughout is a little greater for the fully two-dimensional system. However the overall flow regimes show very similar characteristics. The areas unflooded shown by the darker shading on the edges of the system in Figures 9 and 10 are very similar for each case. It can be expected that results would differ for this relatively extreme case because there is considerable momentum at the bend in the river that is lost in the "mixed" approach.

One significant point to note is that cpu time for the "mixed" method was 1.2 seconds per iterative step and 2.8 seconds for the fully two-dimensional method. This represents a reduction of run time to about 40% thus meeting one of the objectives of this development. Further the "mixed" method model showed better convergence properties as the complex redirection of the velocity vectors in the region where flow goes over-bank was avoided. This reduced the overall run time further.

4.   *Application to the Mary River System*

4.1   System Representation

The Mary River project has been undertaken in a number of phases. First and somewhat optimistically, the downstream section was laid out as a fully two-dimensional network. Experience has shown that large relative differences of depth across an element can cause instability and poor model performance. This model proved very cumbersome due to the large number of elements required to represent the steep river side-slopes. This specific application was the motivation for the "mixed" model development.

The second stage of the project has separated the study into the two main seasonal components.

(a) <u>Dry season in-bank simulation.</u>

As the dry season progresses the flow becomes confined within banks and a one-dimensional approximation is appropriate. Figure 11 shows the finite element representation of the study area. The network consists of 291 elements and 583 nodes. The dashed lines are drawn on 5-km squares to clarify the scale of the network. Studies for this system focused on two aspects:

(1) Distribution of flow between the two main river branches during flood conditions. These simulations serve as precursors to the "mixed" method application and results will be presented in tabular form for comparison purposes.

(2) Evaluation of the impact insertion of submerged weirs on tidal propagation in the estuary. These results will be presented elsewhere and are not the primary interest of this paper.

(b) Wet season bank overtopped simulation.

The newly developed "mixed" model was designed for this situation. During the wet season and at times of extreme tidal range the over-bank sections are extensively flooded and exchange and transport over the flooded sections are of significant interest.

At present a preliminary section of the floodplain has been added to the baseline network presented above. Figure 12 shows a contour view of the two-dimensional network and the bathymetry (refer to the area A in Figure 11 for the location of this detailed section). The dashed lines on this figure are on 1-km squares. The model has been tested for two cases,

(1) Steady state flood flows of 200, 400 and 600 m³/sec applied at the southern boundary and a constant 0.0 metre elevation applied at the northern tidal exits. These levels of inflow were considered characteristic of flows experienced in the area.

(2) Zero net inflow applied at the southern boundary and time varying tidal boundary conditions at the two northern exits of estuary. Specifically the system was simulated for 180 time steps each of 20 minutes for a total period of 60 hours or more than 4 complete tidal cycles. The overall tidal range applied at the boundary for this period was 4.25 metres.

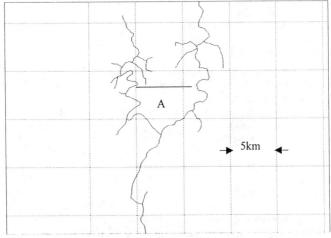

Figure 11 One-Dimensional Finite Element Network for Mary River

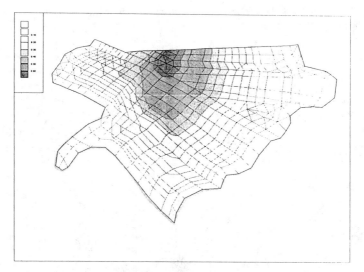

Figure 12 Two-Dimensional Finite Element Network for Mary River
(Elevations in metres)

4.2   Model Results

The steady state case was undertaken to study in a preliminary fashion (no calibration attempted), how the flow splits at times of high flood and how this amount varies as the flow level increases. A comparable simulation was made using the one-dimensional model for the same boundary conditions. Table 1 below shows the flows and proportions for each simulation.

Inflow m³/sec	Percentage outflow eastern exit		Percentage outflow western exit		Outflow eastern exit m³/sec		Outflow western exit m³/sec	
	1-D	2-D	1-D	2-D	1-D	2-D	1-D	2-D
200	68.7	65.7	31.3	34.3	137	131	63	69
400	67.5	60.5	32.5	39.5	270	242	130	158
600	66.6	57.0	33.4	43.0	400	342	200	258

Table 1 Flow Distributions from Steady State Cases

Figure 13 shows the velocity distribution and contours of water depth in the area where flow is crossing the flood plain.

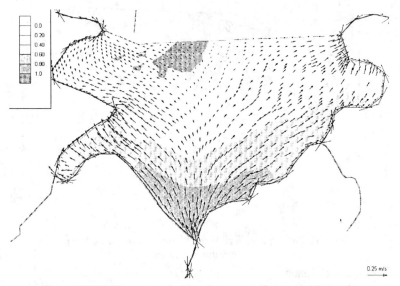

Figure 13 Velocity vectors (m/s) and contours of water depth (m) for
400m^3/sec inflow

Figure 14(a) and (b) show representative velocity vectors for the two-dimensional section of the system during the tidal simulation. Figure 14 (a) shows a time when the tide has just reversed over the flood plain and a section over the centre is dry. To reduce the number of vectors plotted only those at corner nodes are shown in the figure. Figure 14(b) shows a close up view of the velocity vectors (corner and mid-side) focussing on the area overflowing from the river to the flood during the period of maximum flood. From a model development perspective this is the area of most interest. The smooth transition from flow upstream to over-bank flow entering the two-dimensional section is clear.

4.3   Discussion

The results from the steady flow application of the model show how the over-bank area can have an important influence on distribution of flow along the rivers. As the flood flow increases, the flood plain plays an important role in redirecting flow. This must be considered significant when one considers the very limited section of the flood plain included in the present network.

It is important to note that this simulation has successfully modelled a river where the channel bed is 6 metres below bank elevation without instability

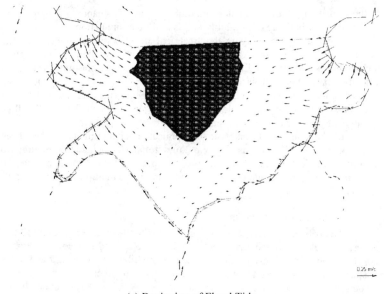

(a) Beginning of Flood Tide

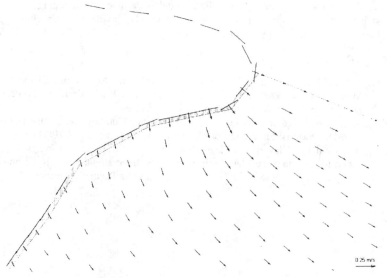

(b) Detail showing velocity vectors at maximum flood

Figure 14 Velocity Vectors (m/s) from Transient Simulation

at the transition. Flow alternates smoothly as the tide changes. There is no evidence of any build up of instability in this multi-day simulation.

The main uncertainty is the treatment of velocity components at the transition. For flow in a normal direction to the one-dimensional section momentum is not preserved, however, the evidence suggests a quite reasonable velocity distribution. This component is driven by continuity of elevation. Similar concerns apply with respect to the component parallel to the direction of one-dimensional flow. It would be quite possible to match the parallel flow components if this was deemed necessary.

In the context of this type of application approximations have been deliberately applied (such as the one-dimensional channel approximation) to simplify a problem that is extremely large. If more detailed information is needed about the flow structure into or from a flood plain then a multi-dimensional model is essential.

5   Concluding Remarks

The two applications, the test case and the limited test section of the Mary River, have demonstrated that the "mixed" model provides a solution that is consistent with overall expectations. The flow distributions (overland and in the channel) agree with results derived from a fully two-dimensional case and the velocity vectors are spatially consistent for the two cases. The model results confirm that overall continuity is maintained. The results also show the expected ebb and flow of the tide with wetting and drying occurring as needed with no instability at the transition. The revised model is capable of simulating river systems with steep banks and is not limited to slowly varying bank transitions characteristic of the fully two-dimensional model.

6.   References

King, I.P. and W.R. Norton, 1978, "Recent Application of RMA's Finite Element Models for Two Dimensional Hydrodynamics and Water Quality," Finite Elements in Water Resources, Pentech Press, London.

King, I.P., 1990, "Modeling of Flow in Estuaries Using Combination of One and Two-Dimensional Finite Elements", Hydrosoft, vol. 3, no. 3, pp. 108-119.

Roig, L.C, and I.P. King, 1988, "Two-Dimensional Finite Element Models for Flood Plains and Tidal Flats," presented at the International Conference on Computational Methods in Flow Analysis, Okayama Japan, September.

Modeling Water Quality for Dredged Material Disposal

Mark Dortch[1], Fellow ASCE, Beth Fleming[1], Member ASCE, and
Barry Bunch[1], Member ASCE

Abstract

Water quality modeling was required to assess impacts on dissolved oxygen (DO) and contaminant exposure resulting from dredged material disposal in contained aquatic disposal (CAD) pits in the Lower Bay of the New York/New Jersey harbors and estuary system. Pollutants could potentially be released into the water column during disposal and/or leached into the water column from bottom sediments placed in the pits before being capped. Biochemical oxygen demand (BOD) and sediment oxygen demand (SOD) associated with freshly dredged and disposed sediments were based on laboratory measurements of dredged material taken from NY-NJ harbor/estuary dredging sites. The CE-QUAL-ICM water quality model was linked to the CH3D-WES hydrodynamic model and applied to an inset grid of the Harbor Apex Model (HAM) for the Lower New York Bay, which encompassed areas under consideration for CAD pits. This paper describes the laboratory measurement methods and results for BOD and SOD, the modeling approach, and example results.

Introduction

Under its Dredged Material Management Program (DMMP), the U.S. Army Engineer District, New York (CENAN), studied disposal options for material dredged to maintain navigation within the NY-NJ harbor/estuary system.

---

[1] U.S. Army Engineer Research and Development Center, Waterways Experiment Station, Vicksburg, MS 39180

The U.S. Army Engineer Research and Development Center, Waterways Experiment Station (WES), was requested by CENAN to assist with various technical analyses of the DMMP. Water quality modeling was required to assess impacts on contaminant exposure concentrations and dissolved oxygen (DO) in Lower NY Bay of the NY-NJ harbor/estuary system resulting from dredged sediment disposal in contained aquatic disposal (CAD) pits and an island confined disposal facility (CDF). To accomplish this requirement, a water quality model was linked to a hydrodynamic model applied to an inset grid for the Lower NY Bay driven by a larger Harbor-Apex Model (HAM). The study area is shown in Figure 1. The inset grid had a smaller domain and finer resolution than the HAM and encompassed areas under consideration for sediment disposal options. Model results were used to evaluate the distribution of potential pollutants and DO as a result of sediment disposal activities. Three sediment disposal operations were modeled: two CAD pit sites during disposal and filling before being capped and an island CDF filled with disposed sediments. Only the CAD pit results are discussed here.

The objective of this effort was to assess spatial concentrations of contaminants and DO resulting from CAD pit operation where pollutants may be released into the water column during disposal and/or leached into the water column from bottom sediments disposed in the pit prior to capping with clean sediments. This study was not intended to determine water quality conditions within the pits since each pit was represented by only a few grid cells, and thus the grid resolution was not fine enough to address such small scale, near-field issues. This paper focuses on the water quality modeling aspects of the study, whereas the hydrodynamic modeling aspects are only briefly described.

Modeling Approach

The three-dimensional water quality model, CE-QUAL-ICM (ICM) (Cerco and Cole 1995), was linked to output from CH3D-WES, a three-dimensional (3D), boundary-fitted, finite difference, hydrodynamic model (Chapman et al. 1996) that was applied to the Lower NY Bay. Hydrodynamics taken from CH3D-WES included circulation as affected by tides, winds, freshwater flows, Coriolis forces, geometric features, and bathymetry. Hydrodynamic output files were provided to the water quality model (WQM) for each disposal option and for each condition of freshwater flows, tides, and winds. Pollutant loadings were introduced in the model at the disposal site location, and the simulations projected water quality concentrations in the Lower NY Bay relative to imposed ambient conditions.

The WQM was used to simulate DO using biochemical oxygen demand (BOD) and sediment oxygen demand (SOD) as sinks. The WQM was not intended to determine occurrence of DO standard violations or what ambient DO will be

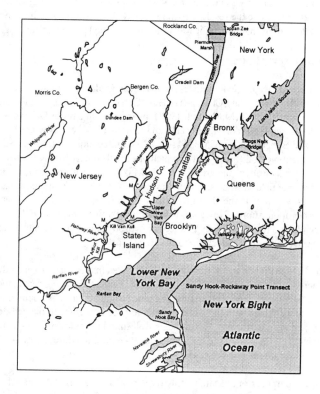

Figure 1. Site map

in the future since ambient DO can vary for reasons independent of disposal operations. Rather, the model provided DO impacts relative to ambient DO conditions. With this approach, it was not necessary to model any sources of DO, nor all the various sinks for DO and sources of BOD, other than the BOD and SOD associated with the disposal pits. This approach greatly simplified the time and cost for conducting the study without compromising the objective.

Contaminants were modeled by introducing a single continuous contaminant source emanating from the pit sites. The contaminant was modeled as a conservative tracer, i.e., it did not react, decay, or settle. Therefore, the model provided an estimate of maximum expected concentrations of contaminants originating from the disposal sites, and the results were used to screen suitability of CAD pit sites. There was no need for WQM parameter estimation or calibration for this analysis since the results depended solely on advection and diffusion driven by the HM output. Of course, it was necessary for the HM to be validated.

Model Description

ICM is a 3D water quality model that has been applied mostly to estuarine and coastal systems, including Chesapeake Bay (Cerco and Cole 1993 and 1994), Lower Green Bay, WI (Mark et al. 1993), inland coastal bays of Delaware (Cerco et al. 1994), the New York Bight (Hall and Dortch 1994), and Newark Bay, NJ (Cerco and Bunch 1997). ICM is based on the finite volume, or integrated compartment method (hence ICM), which has the flexibility to link to structured or unstructured hydrodynamic grids. Since ICM is strictly a water quality model, the hydrodynamic field must be specified through a hydrodynamic model, such as CH3D, as was done in the Chesapeake Bay study (Johnson et al. 1993), as well as most other applications. However, with the finite volume method, ICM can and has been linked to output from a variety of hydrodynamic models, including a finite element hydrodynamic model of Florida Bay.

The ICM model contains multiple state variables that simulate oxygen dynamics as affected by the interaction of phytoplankton, benthic algae, nutrients, and organic matter. The model contains a benthic sediment diagenesis submodel that is dynamically coupled with the water column to produce sediment oxygen demand (SOD) and nutrient fluxes. The model is well suited for study of long-term eutrophication issues, but the model is general enough to address short-term water quality questions, as was the case here.

For the purposes of this study, only three model transport state variables were required: a general, conservative contaminant tracer; DO; and BOD. Model SOD ($g/m^2/day$) was specified as a benthic boundary condition for the computational cells containing the pit, and temperature was specified as a constant throughout the domain. With SOD specified in the model, it was not necessary to use the benthic sediment diagenesis submodel. The diagenesis submodel is required for long-term simulations of eutrophication when SOD varies with time and must be predicted for long-term evolving conditions. For short-term transport issues, it is acceptable to assume that SOD is at steady-state and can be specified rather than simulated. As BOD decayed (i.e., exerted), the amount lost was equivalent to the DO taken up. Ultimate BOD was used to specify sources of BOD originating at the disposal site. Estimates of ultimate BOD were made from laboratory measurements of DO uptake versus time (as explained in the next section) and converted to a loading (mass/day) for model input at the pit as explained in the Model Specifications section.

BOD and SOD Determinations

Freshly dredged sediments are expected to exert oxygen demand when disposed in aquatic environments, such as CAD pits. BOD is the water column DO demand exerted by suspended dredged material dumped at the disposal site, while SOD is the DO demand exerted at the benthic boundary due to deposition of freshly dredged sediments. Both BOD and SOD can have chemical and biological components of oxygen demand due to oxidation of reduced chemical species (e.g., sulfide) and reduction of organic matter, respectively. Freshly dredged sediments obtained from the NY-NJ harbor/estuary waters were tested in laboratories of the Hazardous Waste Research Center (HWRC) at WES to obtain the required estimates of BOD and SOD. Continuously stirred BOD bottle tests were conducted to estimate BOD. Additionally, respirometer studies were conducted which continuously supplied oxygen to the sediment/water system to estimate ultimate BOD. SOD was measured using sealed columns with re-circulated water.

Materials and Methods

Sediment samples were shipped to WES in 55-gallon drums for multiple analyses. The sediments were homogenized and placed in a storage cooler at 4°C until testing. The concentration of sediment was varied by dilution to determine the impact of sediment concentration on oxygen uptake. Concentrations of 1000 mg/L, 2000 mg/L, 3000 mg/L, and 4000 mg/L were evaluated with bottle BOD tests, and 10,000 mg/L and 50,000 mg/L were evaluated with the respirometer tests. All sediment mixtures were prepared using laboratory constructed salt water. The solids concentrations were determined according to methods described in Standard Methods 2540 B (American Public Health Association, American Water Works Association, and Water Pollution Control Federation 1989). The average percent of total solids of the bottom sediment samples was determined to be 44.9%.

Bottle BOD. BOD tests were conducted according to methods described in Standard Methods 5210 B, with modifications (American Public Health Association, American Water Works Association, and Water Pollution Control Federation 1989). Four sets of BOD tests were conducted, one set each for sediment concentrations of 1000, 2000, 3000, and 4000 mg/L. For each set, three replicate controls were run with no sediment and analyzed for DO uptake at hour 120. Dissolved oxygen was measured at 0, 3, 6, 12, 24, 48, 72, 96, 120 hours for the bottles containing sediment. For each time interval that DO was measured, three replicate bottles were analyzed.

The tests involved mixing the sediments with Instant Ocean salt water using an Eberbach Corporation shaker table. The mixed samples were agitated at a rate of

3 shakes per second. The DO was measured using an Orion™ Model 860 dissolved oxygen meter. A BOD bottle was required for each DO measurement at each time interval since the measurement was followed by sacrifice of the sample to avoid the problem of introducing DO when samples are opened and re-capped multiple times.

Ultimate BOD was determined according to the least-squares method presented by Tchobanoglous and Burton (1991). Time series of DO uptake values are used to determine the ultimate BOD and the decay rate, k_1 (day^{-1}). BOD is assumed to follow the first-order decay law

$$\frac{dL}{dt} = -k_1 L \tag{1}$$

where L is BOD remaining (mg/L), and t is time (days). Also $L_n = L_o - y_n$, where L_n is BOD remaining at day n, L_o is the ultimate BOD (mg/L), and y_n is DO uptake after n days (mg/L). Therefore, Equation 1 can be written as

$$\frac{dy}{dt}\Big|_{t=n} = k_1 \left(L_o - y_n\right) \tag{2}$$

Equation 2 is a linear equation of the form $Y = a + b X$, where $Y = dy/dt$, $a = k_1 L_o$, $b = -k_1$, and $X = y_n$. The observed data are DO uptake over time (y_n). The slope dy/dt can be calculated from the observed data, thus providing observations of the dependent variable dy/dt for values of the independent variable, y_n. The sum of the squares of the residuals for the observed and fitted slopes (dy/dt) is minimized. The least squares fit provides estimates of a and b for the general linear equation, from which L_o and k_1 can be calculated.

Respirometer. A BI-1000 Electrolytic Respirometer (Bioscience Inc., Bethlehem, PA) was used to monitor oxygen consumption. Operationally, the respirometer is designed to absorb carbon dioxide as it is produced by microorganisms and introduce oxygen for microbial utilization. The oxygen is supplied at a rate equal to the rate at which oxygen is consumed. Respirometer studies were conducted using sediment concentrations of 0 mg/L (control), 10,000 mg/L, and 50,000 mg/L. The 10,000 mg/L and 50,000 mg/L samples were run in triplicate while the 0 mg/L sample was run in duplicate. Tests were conducted for a period of 10 days.

The results of respirometer studies were used to evaluate ultimate BOD directly through measurement of the accumulated DO uptake (mg/L). The amount of DO uptake by the control samples (which was likely due to a leak in the system) was used to correct the oxygen uptake of the 10,000 mg/l and 50,000 mg/L sediment concentration tests.

SOD Columns. SOD Column studies were conducted in 4 in. (diameter) x 8 in. (height) plexiglass columns in duplicate and included a control. The test columns contained 2 in. of sediment, 5 in. of salt water, and 1 in. of Humco heavy mineral oil to seal off air entry. The column control contained 7 in. of salt water and 1 in. of mineral oil. The DO drop was measured for approximately two days using an Orion model number 084010 DO probe. Water was re-circulated across the DO probe according to manufacturer's instructions at approximately 222 ml/min using a 1:100 Masterflex peristaltic pump. Additionally, this flow helped to maintain a moving boundary layer at the sediment-water interface, thus providing field-like conditions for DO exchange across the interface.

The SOD was determined from

$$SOD = \frac{\Delta DO}{\Delta t} H \tag{3}$$

where,
ΔDO = change in DO, mg/L
Δt = change in time, days
H = water height above the sediments, m

Two SOD column studies were performed utilizing replicate columns (containing sediment and Instant™ ocean water) and a control (containing Instant™ Ocean water only) for each column study.

BOD and SOD Results

Bottle BOD. Replicate DO uptake values for each analysis time interval were averaged and used to calculate ultimate BOD and k_1, which are presented in Table 1. The k_1 and ultimate BOD values were based upon 48-hour analysis of the BOD results. The 48-hour results were used to avoid biasing caused by oxygen depletion at longer elapsed times that was experienced with the relatively high decay rates. In order to provide the BOD loading input required for the model, it was necessary to estimate ultimate BOD normalized to sediment concentrations (per 1000 mg/L sediment). Thus, the normalized ultimate BOD values are also presented in Table 1. For reasons unknown, the normalized ultimate BOD values tended to decrease with increasing sediment concentrations. This indicates that the BOD bottle results did not scale consistently with sediment concentrations as hoped. A representative value of normalized ultimate BOD is required to scale the total ultimate BOD available in a disposal load with a known sediment concentration. Therefore, the 48-hour normalized ultimate BOD values were averaged yielding a value of 2.98 mg/L per 1000 mg/L of sediment. Fortunately, this value agrees closely with the respirometer results. The BOD decay rate of 1.55 day^{-1} is relatively high indicating that the DO demand was mostly chemical in nature.

Table 1. Ultimate BOD and k_1 results			
Suspended Solids Concentration, mg/L	k_1, day^{-1}	Ultimate BOD, mg/L	Normalized Ultimate BOD, mg/L (1000mg/L solids)$^{-1}$
1000	1.1	4.9	4.9
2000	1.7	6.0	3.0
3000	1.9	6.4	2.13
4000	1.5	7.5	1.87
Average	1.55		2.98

Respirometer. The respirometer tests were run for 10 days with higher sediment concentrations than those used in the bottle BOD tests. These tests indicated that 10 days were enough to obtain the approximate ultimate oxygen demand since cumulative oxygen uptake approached a nearly constant value within a few days. The control samples showed an accumulated oxygen uptake of 20 to 30 mg/L which cannot be explained. Although the source of the oxygen demand was not found, there may have been a leak in the system. Through personal correspondence with the respirometer manufacturer, Bioscience, Inc., it was found that oxygen uptake in control samples is not unusual in respirometer systems, due to a number of reasons including temperature changes, leakage around the ground glass joints, and changes in barometric pressure. The manufacturer recommended the general practice of correcting by subtracting the control oxygen uptake from the sample oxygen uptake.

The ten-day respirometer results indicated that sediment mixtures of 10,000 and 50,000 mg/L exerted an average ultimate oxygen demand of approximately 60 mg/L and 170 mg/L, respectively. Subtracting the accumulated oxygen uptake of the control, the accumulated oxygen uptake of the 10,000 mg/L and 50,000 mg/L sediment mixtures are 30 to 40 mg/L and 140 to 150 mg/L, respectively, in 10 days. The accumulated oxygen uptake of the 50,000 mg/l sediment mixture was almost five times greater than that of the 10,000 mg/l sediment mixture, a result that favors the idea of scaling BOD with sediment concentrations. These results indicate a normalized ultimate BOD of approximately 3 mg/L per 1000 mg/L sediment which agrees well with the average normalized ultimate BOD obtained from the bottle BOD tests.

SOD Columns. Sediment concentrations for all SOD tests varied between 174 g/L and 200 g/L. The results of the SOD column studies for run 1 were approximately 0.25 g/m^2/day. For run 2, column 1, estimated SOD was 0.74 g/m^2/day, while column 2 SOD was approximately 0.51 g/m^2/day. The reason that run 1 SOD values were lower is not clear, but problems with the DO sensors were suspected, and temperature changes were greater. To provide conservative estimates of DO uptake, the highest value was used in the model.

Summary of BOD and SOD Results

Based upon the BOD bottle tests, the normalized, ultimate BOD of NY-NJ harbor/estuary sediments is approximately 3 mg/L per 1000 mg/L of sediment. This result was confirmed by the respirometer studies. The SOD of the sediments ranged from 0.25 g/m^2/day to 0.74 g/m^2/day with the higher value used.

Model Specifications

Specifications for the WQM consist of the grid, hydrodynamics, initial conditions and boundary conditions, pollutant loadings, and kinetic parameters. Each of these is discussed below along with methods of model analysis and presentation.

Grid and Model Linkage

WQM grid generation was based on HM grid files that included node coordinates and cell depths. The HM grid file was used to construct a map file relating HM and WQM cells. The WQM grid had a direct, one-to-one cell correspondence with the HM grid (Figure 2), which had a planar density of 95 x 69 (rows and columns) and contained 9 sigma coordinate layers. After discounting land cells, the grid contained 33,039 active cells. A subroutine with the HM computes time-averaged flows across cell faces and vertical diffusivities and writes this information to an output file along with cell volumes at the end of each averaging interval. Following validation of the HM for tides, currents, and salinity, the WQM was linked to the HM. Proper linkage of the HM and WQM was ensured through volume comparisons between the two models and mass conservation tests within the WQM.

Hydrodynamic Input

The WQM was applied to each CAD pit option using four different hydrodynamic input conditions that were expected to exacerbate water quality.

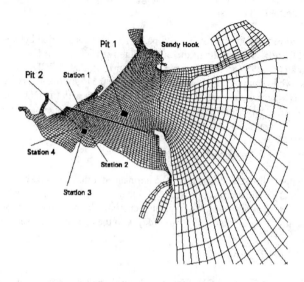

Figure 2. Model domain and grid with pit and plotting station locations

These included: 1) low flow and mean tide; 2) high flow and mean tide; 3) low flow and neap tide; and 4) condition 3 with wind. Conditions 1-3 had no wind. The low and high flows were constant and were the monthly-averaged, freshwater flow that is exceeded 95 % and 5 % of the time, respectively. Repeating mean and neap tides were used. For condition 4, a constant wind of 16 mph from the southeast was applied, which is representative of prevailing summer winds of medium to high magnitude (80 percentile value). This wind direction tends to blow water toward the bay, thus decreasing flushing from the bay. For each of the above four conditions, the HM was run to a dynamic steady-state, and the last full tidal cycle of HM output was saved to disk and used later to drive the WQM. Each WQM run simulated 30 days, at which time the model had reached approximately a dynamic steady-state condition.

Initial and Boundary Conditions

For all runs, the following IC and BC were specified for the three state variables, DO, BOD, and contaminant tracer. A constant DO concentration of 6.0 g/m^3 (or mg/L) was specified for IC and BC throughout the domain and on each flow boundary. A surface boundary condition of no reaeration was specified for DO. A time-invariant SOD was applied along the bottom boundary with a value of zero set throughout the domain except for the cells containing the CAD pits where a value of 0.74 g/m^2/day was specified, which has the same effect as specifying an SOD that is 0.74 g/m^2/day greater than ambient SOD. A constant temperature of 18.0 °C was specified for IC and BC throughout the model domain and boundaries, and surface heat exchange was turned off to keep temperature constant. BOD was specified as zero at all boundaries and initially throughout the domain. Contaminant concentrations were specified as zero initially throughout the domain and on all boundaries.

Loadings

The ultimate BOD loading for each pit site was calculated as follows. The total ultimate BOD load in a hopper dredge is

$$W = L_{os} \, \rho_b \, V \tag{4}$$

where
  W = ultimate BOD loading mass per hopper dump (g)
  L_{os} = ultimate BOD normalized to total solids concentration (g/m^3 (1000 g solids / m^3)$^{-1}$)
  ρ_b = bulk density of sediment in hopper (g/m^3)
  V = volume of sediment and water in hopper (m^3)

Measurements of dredged sediments from the region indicate that on the average $\rho_b = 1.374 \times 10^6$ g/m^3. Typically, hopper dredges used in this region carry about 4,000 cu yds of material, or $V = 3,058$ m^3. Based upon the laboratory studies presented above, $L_{os} = 3$ g/m^3 (1000 g solids / m^3)$^{-1}$. Using these values in Equation 4 results in $W = 12.6 \times 10^6$ g. It is expected that an average of 6 hopper dumps per day will be made at a pit site. Thus, the total BOD mass loading from the hoppers is 75.6×10^6 g/day.

However, about 96% of the dumped material immediately settles to the bottom (Schroeder 1998). The remainder stays suspended for extended periods (about 3 or 4 days) and can exert a BOD demand within the water column. The settling time for the 96% that immediately settles varies between a few minutes for sands to about 25 minutes for fines, but overall, most of the material hits the bottom on the order of a few minutes. It was assumed that 96% of the dumped material is in the water column for 15 minutes. Therefore, with 6 dumps per day, 96% of the dumped material is in the water column for 90 minutes per day, or about 6.25% of the time. The remaining 4% of each hopper load was assumed to remain in the water column indefinitely. With these assumptions, the total BOD loading from the disposal operations is 7.56×10^6 g/day, which was obtained from (0.04 x 75.6

x 10^6) + (0.96 x 0.0625 x 75.6 x 10^6). Thus, a continuous ultimate BOD mass loading rate of 7.56 x 10^6 g/day (7.5 metric tons) was specified in the model for the computational cells corresponding to the CAD pits. This loading rate was divided and distributed evenly throughout each of the nine vertical water column layers residing over the site. For the contaminant loading at the pit, an arbitrary, continuous mass loading rate of 100 kg/day was specified in the bottom layer over the disposal site.

Parameters

Only one water quality kinetic parameter was required for this study which was the BOD decay rate (k_1) equal to 1.55 day^{-1}.

Methods of Analysis and Presentation

Two CAD pit sites were studied, and their locations are shown in Figure 2. Time series of DO, BOD, and contaminant tracer were plotted for the top layer, bottom layer, and mid-depth for each station. Four plotting stations were located in each of four directions around and near the disposal sites, and one station was located off Sandy Hook in the entrance to the Apex. Figure 2 shows the plotting stations for Pit 2. Additionally, a time series plotting station was located immediately above the pit sites. Surface layer contaminant concentrations at day 30 of each simulation were also plotted as contours in the horizontal plane.

Model Results

Results for CAD Pit 2 are discussed since disposal at that site degraded water quality more than Pit 1 site. Of the four hydrodynamic conditions, condition 3 (low flow with neap tide) encountered the lowest DO values and highest contaminant concentrations. Of the six plotting stations, Station 2 had the lowest DO, where DO decreased about 1 mg/L. Station 2 time series of bottom DO and mid-depth contaminant concentration are shown in Figure 3. The contaminant concentration contours (in parts per billion, ppb) for the surface layer after 30 days are shown in Figure 4. The contaminant plume tends to remain close to the NJ shore for this hydrodynamic condition and pit site. Contaminant concentrations were about an order of magnitude higher for Pit 2 than for Pit 1.

Conclusions and Recommendations

The combination of laboratory studies and numerical modeling provided a cost effective means for evaluating water quality impacts of disposal alternatives. Laboratory determinations of normalized, ultimate BOD were consistent using standard BOD bottle analysis and respirometer studies. Ultimate BOD

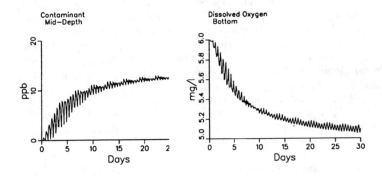

Figure 3. Contaminant and DO concentration versus time at Station 2, Pit 2

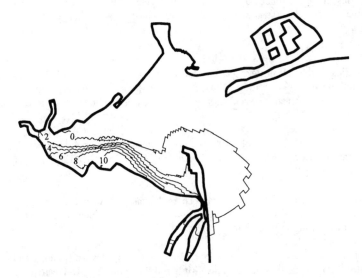

Figure 4. Near-surface contaminant concentration contours (ppb) after
30 days for Pit 2

approximately scaled with sediment concentration, with a value of about 3 mg/L per 1000 mg/L of sediment. A relatively high BOD decay rate was determined indicating the demand was mostly chemical rather than biological, which is expected for bottom sediments containing sulfide and other reduced substances. Using the normalized ultimate BOD value, dredged sediment bulk density, hopper dredge volume, and the number of dredge disposals per day, it was possible to calculate the BOD loading per day at the CAD sites. Pit 2 encountered the lowest DO and highest contaminant concentrations of the two CAD pit sites modeled, where the lowest DO occurred on the bottom and was about 1 mg/L below ambient DO. The location of Pit 2 within a relatively sheltered part of the bay experienced slower flushing compared to the Pit 1 site.

It is recommended that rather than conducting standard 5 day BOD analyses, long-term BOD analyses should be run, where samples are aerated after each DO reading and DO drop between readings are recorded over the long-term (e.g., 10 to 30 days or longer if necessary).

Acknowledgment

This study was funded by the U.S. Army Engineer District, New York. The Chief of Engineers has granted permission to publish these results.

References

American Public Health Association, American Water Works Association, Water Pollution Control Federation. (1989). *Standard Methods for the Examination of Water and Wastewater*, seventeenth edition, American Public Health Association, Washington, DC.

Cerco, C. F., and T. M. Cole (1993). Three-dimensional eutrophication model of Chesapeake Bay, *J. Environmental Engineering*, Am. Soc. Civil Eng., 119(6), 1006-1025.

Cerco, C. F., and T. M. Cole (1994). Three-dimensional eutrophication model of Chesapeake Bay, Technical Report EL-94-4, US Army Engineer Waterways Experiment Station, Vicksburg, MS.

Cerco, C. F., B. Bunch, M. A. Cialone, H. Wang (1994). Hydrodynamics and eutrophication model study of Indian River and Rehoboth Bay, Delaware, Technical Report EL-94-5, US Army Engineer Waterways Experiment Station, Vicksburg, MS.

Cerco, C. F., and T. M. Cole (1995). User's guide to the CE-QUAL-ICM three-dimensional eutrophication model, release version 1.0, Technical Report EL-95-15, US Army Engineer Waterways Experiment Station, Vicksburg, MS.

Cerco, C. F., and B. Bunch (1997). Passaic River tunnel diversion model study, report 5, water quality modeling, Technical Report HL-96-2, US Army Engineer Waterways Experiment Station, Vicksburg, MS.

Chapman, R. S., B. H. Johnson, and S. R. Vemulakonda (1996). User's guide for the sigma stretched version of CH3D-WES," Technical Report HL-96-21, US Army Engineer Waterways Experiment Station, Vicksburg, MS.

Hall, R. W., and M. S. Dortch (1994). New York Bight study, report 2, development and application of a eutrophication/general water quality model, Technical Report CERC-94-4, US Army Engineer Waterways Experiment Station, Vicksburg, MS.

Johnson, B. H., K. W. Kim, R. E. Heath, B. B. Hsieh, H. L. Butler (1993). Validation of three-dimensional hydrodynamic model of Chesapeake Bay, *J. Hydraulic Engineering*, Am. Soc. of Civil Eng., 119, 1, 2-20

Mark, D. J., N. W. Scheffner, H. L. Butler, B. W. Bunch, and M. S. Dortch (1993). Hydrodynamic and water quality model of lower Green Bay, Wisconsin, volume 1, main text and appendices A-E, Technical Report CERC-93-16, US Army Engineer Waterways Experiment Station, Vicksburg, MS.

Schroeder, P. (1998). Personal communication, U.S. Army Engineer Waterways Experiment Station, Vicksburg, MS.

Tchobanoglous, G. and F. L. Burton (1991). *Wastewater Engineering Treatment, Disposal, and Reuse*, Third Edition, Metcalf & Eddy, Inc., McGraw-Hill Publishing Company, New York, NY.

A Framework for Integrated Modeling of Coupled Hydrodynamic-Sedimentary-Ecological Processes

Y. Peter Sheng, Member, ASCE
Civil & Coastal Engineering Dept., University of Florida, Gainesville, FL 32611-6580

ABSTRACT

To solve practical engineering and environmental assessment problems in estuarine and coastal systems, scientists and resource managers have begun to recognize that most problems are interdisciplinary and often involve coupled hydrodynamic-sedimentary-ecological processes. Hence, to conduct integrated study of coastal and estuarine systems, it is essential to develop an integrated modeling system which includes all the important processes. This paper presents the framework of an integrated model for estuarine and coastal systems, with particular emphasis on the coupling methodology between various component models in shallow sub-tropical estuaries.

INTRODUCTION

More than 50% of the U.S. and world population live within 100 miles from the coastline. By 2025, it is expected that more than 75% of the population will live in the coastal zone. It is becoming more and more challenging to balance the various conflicting goals of economic development, maintenance and restoration of ecosystems, sustaining of living resources, and coastal hazard mitigation. Changes to estuarine and coastal ecosystems have been caused by anthropogenic and natural activities. Increased pollutant loadings have led to deteriorated water quality, increased incidences of harmful algal bloom, and loss of fishery habitat. On the other hand, climatic change associated with El Nino-La Nina, severe storms, and long-term global warming have caused noticeable changes in the ecosystems as well. Hence, it is important to develop an integrated modeling and observing system to quantitatively understand and predict the response of ecosystems due to anthropogenic and natural climatic changes.

This paper focuses on the development of an integrated modeling system for a large coastal and estuarine ecosystem - Indian River Lagoon along the east coast of Florida (Figure 1). The Indian River Lagoon extends more than 200 kilometers from Ponce de Leon Inlet at the North to St. Lucie Inlet at the South, with a width varying between 2-10 km and an average depth of 2 m. This lagoon is one of the most biologically diverse ecosystem in the world, with fishery resources in the ecosystem amounting to almost 1 billion dollars per year. In recent decades, however, seagrass and fishery resources have both declined. In order to develop scientifically sound strategy to restore the ecosystem, efforts began at the University of Florida and the St. Johns River Water Management District in 1996 to develop an Indian River Lagoon Pollutant Load Reduction (IRLPLR) model, an integrated modeling system which can be used to predict the response of the ecosystem due to changes in pollutant loading. Field experiments have been planned and conducted to collect synoptic (simultaneous in time and space) data of physical, chemical, and biological parameters. This paper presents an overview of the IRLPLR model. In the paper, coupling methodology in a circulation-wave-sediment model is first discussed, followed by discussion of the coupling methodology in a

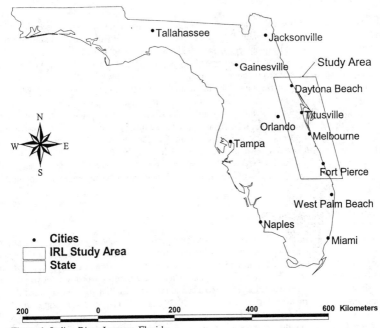

Figure 1. Indian River Lagoon, Florida.

hydrodynamic-sediment-water quality model. Coupling methodology between a water quality model and a light attenuation model is then addressed, followed by a discussion on the coupling between a light attenuation model and a seagrass model. The coupling methodologies are compared in terms of the processes involved, the robustness of results, the type of ecosystems, and computational efficiency.

COUPLING AMONG PROCESSES

The various estuarine processes contain different spatial and temporal scales and there may exist one-way or two-way coupling between any two processes. The coupling among processes depends significantly on the particular estuary. For example, in the relatively deep Chesapeake Bay, wind waves usually do not reach the bottom and hence wave dynamics and sediment transport processes are usually decoupled from nutrient budget considerations. In a shallow sub-tropical estuary such as the Indian River Lagoon, coupling among the various processes is much more pronounced. A few examples of the coupling among processes are:

Coupling Between Atmospheric and Estuarine Processes

Wind fields over the water can generate wind-driven circulation and wind waves which in turn produce wave-induced circulation. During the recent Hurricane Floyd, which went through the Atlantic coastal water in mid September, significant North-South setup in water level was observed (Sheng et al. 1999). Atmospheric heating and cooling can affect the water temperature and circulation. Evaporation and precipitation can significantly affect the salinity distribution in a sub-tropical or tropical estuary. For example, salinity inside Florida Bay can reach 40-50 psu, which is much higher than the ocean salinity of 35 psu, due to a significant net evaporative flux even during the rainy season (Sheng et al. 1995). Even in Indian River Lagoon, during months with high evaporation, the northern part of the lagoon can reach 30 psu in spite of poor tidal flushing (Sheng 1997). In addition, atmospheric deposition can contribute directly to the nutrient budget in an estuary. Hence it is important to incorporate these coupled atmospheric-estuarine processes into an integrated model of an estuarine ecosystem. Evaporation and precipitation must be incorporated into the hydrodynamic model and atmospheric deposition, if determined to be significant, should be included in the nutrient model.

Coupling Between Hydrodynamic and Sedimentary Processes

Bottom sediments in a shallow estuary can be resuspended due to significant bottom shear stress caused by current and wave action during high wind events and/or boat passage. For example, Sheng et al. (1995) observed and simulated the wave-induced resuspension of bottom sediments in Tampa Bay during a strong winter storm. Recent data collected during Hurricane Floyd (Sheng 1999) showed that SSC (suspended sediment concentration) exceeded 250 mg/l in the Indian River Lagoon where the normal SSC is on the order of 20-30 mg/l. Hence for a sediment study in a shallow estuary such as Indian River Lagoon, it is essential to use a wave model in addition to a circulation model and a sediment transport model.

Coupling Among Hydrodynamic, Sedimentary, and Water Quality Processes

In a relatively deeper estuary such as Chesapeake Bay, dissolved nutrients are often released from the bottom sediments during periods when the water column is vertically stratified and the lower layer is anoxic or hypoxic, as shown in Figure 2. In a shallow estuary such as Indian River Lagoon and Tampa Bay, however, bottom nutrients are brought into the water column during high wind events when bottom sediments are resuspended into the water column, as shown in Figure 3. Dissolved nutrients in the interstitial water can be brought into the water column. In addition, before the resuspended sediments are deposited back onto the bottom, nutrients that are absorbed onto the sediment particles can be desorbed into the water, depending on the ambient concentrations of nutrients, pH, DO, and certain ions (Sheng 1999).

Sheng (1994) and Chen and Sheng (1995) presented a coupled hydrodynamics-sediment-water quality model which used a rectangular horizontal grid and a sigma vertical grid. The desorption and absorption/adsorption processes are treated as kinetic processes instead of assuming equilibrium partitioning. A sediment transport model, based on Sheng (1993), which included resuspension, deposition, settling/flocculation, advection, and mixing, was explicitly incorporated into the overall model. Diffusive as well as resuspension fluxes of nutrients were both incorporated into the nutrient model.

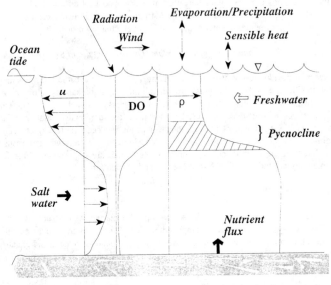

Figure 2. Important parameters and processes related to the development of hypoxia/anoxia in a stratified estuary. (From Sheng 1999).

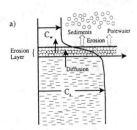

Figure 3(a). Beginning of a resuspension event.(From Sheng 1999)

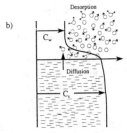

Figure 3(b). Peak of a resuspension event. (From Sheng 1999).

Coupling Among Hydrodynamic, Sedimentary, Water Quality, and Light Processes

Light attenuation processes are much more important in the shallow sub-tropical and tropical Florida estuaries than the deeper temperate estuaries in the northern U.S., e.g., Chesapeake Bay. In shallow sub-tropical estuaries, sediments and nutrients are more readily resuspended which in turn attenuate the light. The increased light attenuation and nutrient concentration will affect the growth of phytoplankton in the water column and the nutrient cycling. Hence there exists nonlinear coupling among the sedimentary, water quality, and light attenuation processes.

Coupling Among Hydrodynamic, Sedimentary, Water Quality, Light, and Seagrass Processes

A seagrass meadow can grow over a much larger portion of the estuary bottom in a Florida estuary than in Chesapeake Bay. Seagrass growth is limited by light, salinity, temperature, water column nutrients, sediment column nutrients, as well as seagrass density (Fong and Harwell 1994). The additional effects of vegetation on hydrodynamics (e.g., the additional vegetation-induced drag on flow in the air and water columns, the additional vegetation-induced turbulent kinetic energy, etc.) and on sediment transport (e.g., increased sediment deposition and reduced sediment resuspension) must be included in the overall model. Other processes that are important in Florida include evaporation and precipitation in the Gulf of Mexico and Florida Strait.

COMPONENT MODELS

An integrated modeling system for a sub-tropical estuarine system such as Indian River Lagoon or Florida Bay should consist of component models of the following processes: circulation, wave, sediment transport, nutrient dynamics, dissolved oxygen, plankton dynamics, light attenuation, and seagrass dynamics. Table 1 shows the component models included in the integrated modeling system being developed for Indian River Lagoon.

TABLE 1. COMPONENT MODELS OF VARIOUS PROCESSES IN IRL.

Component Model	*Model Name*
Circulation Model:	Curvilinear-Grid 3D (CH3D)
Wave Model:	SMB/SWAN/HISWA
Sediment Transport Model:	CH3D-SED3D
Water Quality Model:	CH3D-WQ
Nitrogen Model	
Phosphorus Model	
Dissolved Oxygen Model	
Phytoplankton Model	
Zooplankton Model	
Light Attenuation Model	CH3D-LA
Submerged Aquatic Vegetation(SAV)Model	CH3D-SAV

Circulation Model - CH3D

CH3D is a Curvilinear-grid Hydrodynamic 3D model originally developed by Sheng (1987, 1989, 1997). In the present IRLPLR model, evaporation and precipitation based on measurements are incorporated into the CH3D circulation model. The present model also uses much finer horizontal grid resolution than that used in the Chesapeake Bay model (Sheng et al. 1989, Johnson et al. 1989) and Tampa Bay model (Yassuda and Sheng 1997). The grid for a portion of the IRL is shown in Figure 4. A numerical grid which includes the entire IRL with a total of 20988 (477x44) cells is shown in Figure 5. Another grid extends the IRL grid into the offshore water by 6-10 km. Both simulated and measured water level (Figure 6) show that tidal influence is strong in the southern IRL and diminishes in the northern IRL.

Wave Model

In order to simulate the sediment transport in IRL, it was necessary to include a wave model such that the wave height and period and wave-induced bottom stresses at every grid point can be calculated. A wave model was selected for calculation of wave height and period, based on model tests using the relatively simple SMB model (USAE 1984), the SWAN model (Ris et al. 1996), and the HISWA model (Holthuijsen et al. 1989). Wave-induced bottom stresses are then calculated using a turbulent boundary layer model (Sheng and Villaret 1989) applied to the entire water column. Our model test also indicated that the wave-induced mass transport is negligible compared to the wind-driven or tide-driven residual currents in the IRL. Hence the only effect of waves on sediment

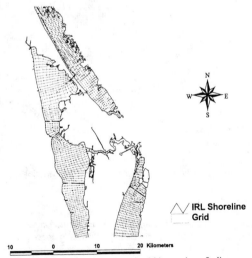

Figure 4. Horizontal curvilinear grid in northern Indian River Lagoon.

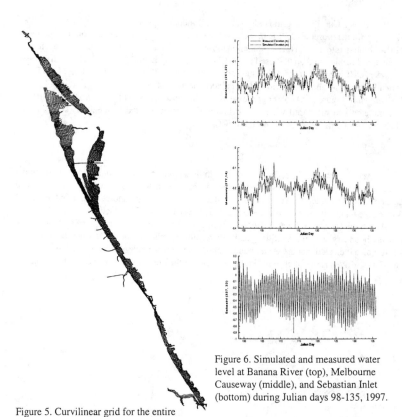

Figure 6. Simulated and measured water level at Banana River (top), Melbourne Causeway (middle), and Sebastian Inlet (bottom) during Julian days 98-135, 1997.

Figure 5. Curvilinear grid for the entire Indian River Lagoon.with 477x44 cells.

transport is through the wave-induced bottom stress and sediment resuspension.

Sediment Transport Model

The sediment transport model is based on the fine sediment transport model developed by Sheng et al. (1991) and Sheng (1993) which includes the following processes: advection, mixing, settling, deposition and resuspension. In addition, flocculation is empirically modeled by the use of a settling velocity which depends on the SSC and turbulence intensity in the water column. Considering the fact that bottom sediments in the IRL are composed of a mixture of fine and coarse sediments, we modeled the sediment transport with two SSC equations, one for the fine sediments and one for the coarse sediments. For the coarse sediments, bottom boundary conditions are similar to those used by van Rijn (1989). Recently the sediment model was coupled to the hydrodynamic model and the wave model for simulation of resuspension of sediments at a site in the northern IRL near Grant, FL which is slightly to the north of Sebastian Inlet. The results indicate that wave induced sediment resuspension was very significant (Figure 7).

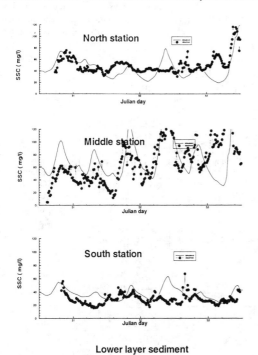

Lower layer sediment

Figure 7. Simulated (solid line) and measured (heavy dots) suspended sediment concentration at a station near Grant in IRL during a storm event in 1999. Resuspension occurred at the beginning and the end of the event of the northern platform.

Water Quality Model

The water quality model developed for Lake Okeechobee (Chen and Sheng 1995) includes a phosphorus cycle and a phytoplankton cycle, while the Tampa Bay model (Yassuda and Sheng 1997) includes a nitrogen cycle, a phytoplankton cycle, and a dissolved oxygen cycle. Both models include the effect of sediment transport on nutrient dynamics through the explicit use of a sediment transport model and the incorporation of resuspension flux of nutrients. In addition, both models include the absorption/adsorption and desorption of nutrients as kinetic processes instead of assuming equilibrium partitioning. In the IRLPLR model, both phosphorus and nitrogen cycles are included since field data indicate that nitrogen concentration is higher in the northern part of the lagoon while phosphorus concentration is higher near the southern part of the lagoon, while nutrient limitation experiments indicate that the limiting nutrients for phytoplankton growth may vary between nitrogen and phosphorus, depending on the location and time. A dissolved oxygen model is included in the model. A phytoplankton model which may include three or one species is being tested with field data to determine how many species will be included in the model. In addition, a zooplankton model is being tested. During the storm event shown in Figure 7, a significant increase in nutrient (N and P) concentrations accompanied the sediment resuspension. The water quality model, therefore, must include the internal loading of nutrients from bottom sediments via resuspension and diffusion. For example, the nitrogen model used in the IRLPLR model as shown in Figure 8 contains both the diffusive flux of dissolved nitrogen species and the resuspension flux of particulate species at the sediment-water interface.

Figure 8. Nitrogen Model included in the IRLPLR model.

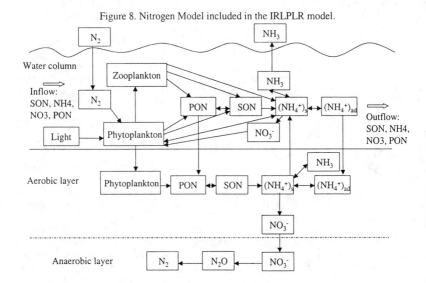

Light Attenuation Model

Light models in use include the simple regression model developed by McPherson and Miller (1993) and the spectral light model developed by Gallegos (1993). Based on the synoptic data collected by us during 1997 and 1998 and other data, we determined the dominant light attenuator for each dataset (Kornick 1998). Presently, we are using a model similar to Gallegos' model as shown in Figure 9.

Seagrass Model

A conceptual seagrass model was developed by Fong and Harwell (1994) with data from Florida Bay. The model includes the growth and turnover of three seagrass species (Thalassia, Halodule, and Seringodium) due to limitation of light, nutrient, temperature, and salinity. The model was modified and applied to Roberts Bay while coupled to a hydrodynamic model, a water quality model, and a regression type light attenuation model (Sheng et al. 1995). In the present IRL study, we consider the growth and turnover of Halodule, as well as epiphytic algae and drifting macroalgae. Extensive experiments are being conducted to determine the fluxes of nutrients between these species and the water column and sediment column.

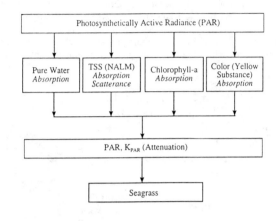

Figure 9. Basic framework of the light attenuation model in the IRLPLR model.

AN INTEGRATED MODELING SYSTEM

The component models are presently being calibrated and verified with extensive data collected in the IRL system during the past few years. The calibrated and verified component models are then combined into an integrated modeling system as shown in Figure 10. The rectangular boxes contain the various component models: circulation model, wave model, sediment transport model, water quality model, light attenuation model, and seagrass model. The elliptical boxes contain the input data required for model initiation/forcing and calibration/verification as well as model output of major model attributes. For example, the water quality model produces output of nutrients, dissolved oxygen, and phytoplankton, while receiving input of flow field (from circulation model) and SSC field (from sediment transport model). The light attenuation model receives input of SSC (from sediment transport model), chlorophyll-a (from water quality model), and producesoutput of photosynthetically active radiance (PAR) at the top of the seagrass meadow. The seagrass model uses the output from the light attenuation model (PAR), water quality model (N & P concentrations), and circulation model (temperature and salinity) and produces the change in seagrass biomass. Single arrows in the chart indicate one-way coupling between two components, while double arrows indicate two-way interaction between two components.

Modeling two-way coupling with the same time and spatial resolutions for both processes may give more robust results, but also requires more computational effort. Hence, appropriate coupling methodology must be developed between the various component models based on both scientific and computational considerations. To achieve that goal, field and/or laboratory data and various statistical methods may be used to examine the potential coupling mechanisms and formulations.

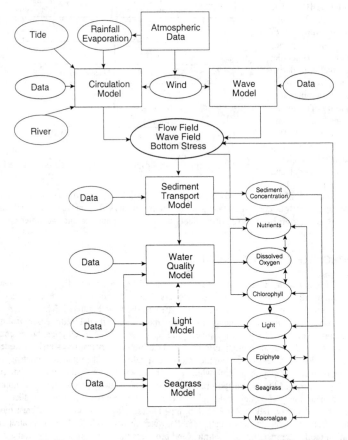

Figure 10. Block diagram of an integrated modeling system - IRLPLR model.

CONCLUSIONS

This paper discussed the rationale and framework for formulating an integrated modeling system for large estuarine and coastal ecosystems with coupled hydrodynamic-sedimentary-ecological processes. Coupling among processes is stronger in shallow sub-tropical estuaries as opposed to deeper temperate estuaries. Using Indian River Lagoon in Florida as an example, we illustrated the requirements for various component models, including a circulation model, wave model, sediment transport model, water quality model, light attenuation model, and seagrass model. Based on our data and model results, it is apparent that an integrated modeling system is necessary to assess the changes to the large aquatic ecosystem due to anthropogenic or natural perturbations. Significant coupling is found to exist among atmospheric, hydrodynamic, sedimentary, water quality, light and seagrass processes. Long-term simulation using the integrated modeling system requires a large amount of data and computer resources. As part of this effort, we have developed a robust GIS model (Sun et al. 1999) of the entire Indian River Lagoon ecosystem to facilitate data management and analysis and comparison with model results. To allow multi-year simulations of large ecosystems with millions of grid cells, it is recommended to develop high-performance parallel computer models (Davis and Sheng, 2000).

ACKNOWLEDGMENT

This work is supported by the St. Johns River Water Management District, Palatka, Florida and the USEPA National Center for Environmental Research and Quality Control.

REFERENCES

Chen, X. and Y.P. Sheng, 1995:"Application of A Coupled 3-D Hydrodynamics-Sediment-Water Quality Model," in *Estuarine and Coastal Modeling, IV*, American Society of Civil Engineers, 325-339

Davis, J.R. and Y. P. Sheng, 2000: "High-Performance Estuarine and Coastal Environmental Modeling," in *Estuarine and Coastal Modeling, VI*, American Society of Civil Engineers.

Fong, P. And M.A. Harwell, 1994: "Modeling Seagrass Communities in Tropical and Sub-Tropical Bays and Estuaries," *Bull. Mar. Sci.* 54(3):757-781.

Galegos, C.L., 1993: "Development of Optical Models for Protection of Seagrass Habitat," in *Proceedings and Conclusions of Workshops on Submerged Vegetation Initiative and Photosynthetically Active Radiation* (L.J. Morris and D. Tomasko, eds.), St. Johns River Water Management District, Palatka, FL, pp.77-90.

Holthuijsen, L.H., N. Booij, and T.H.C. Herbers, 1989: "A Prediction Model for Stationary, Short-Crested Waves in Shallow Water with Ambient Currents," *Coastal Engineering,* 13, 23-54.

Johnson, B.H., K.W. Kim, Y.P. Sheng, and R.E. Heath, 1989: ``Development of a Three-Dimensional Hydrodynamic Model of Chesapeake Bay," *Estuarine and Coastal Modeling*, (M.L. Spaulding, Ed.), ASCE, pp. 162-171.

Ris, R.C., N. Booij, L.H. Holthuijsen, and R. Padilla-Hernandez, 1996: *SWAN Cycle 2 User Manual*, Department of Civil Engineering, Delft University of Technology, Delft, the Netherlands.

Sheng, Y.P., 1987: ``On Modeling Three-Dimensional Estuarine and Marine Hydrodynamics," *Three-Dimensional Models of Marine and Estuarine Dynamics* (J.C.J. Nihoul and B.M. Jamart, Eds.), Elsevier Oceanography Series, Elsevier, pp. 35-54.

Sheng, Y.P., 1989: ``Evolution of a Three-Dimensional Curvilinear-Grid Hydrodynamic Model for Estuaries, Lakes and Coastal Waters: CH3D," *Estuarine and Coastal Modeling* (M.L. Spaulding, Ed.), ASCE, pp. 40-49.

Sheng, Y.P., 1993: ``Hydrodynamics, Sediment Transport and Their Effects on Phosphorus Dynamics in Lake Okeechobee," *Nearshore, Estuarine and Coastal Sediment Transport* (A.J. Mehta, Ed.). In Coastal and Estuarine Studies, **42**, American Geophysical Union, pp. 558-571.

Sheng, Y.P., 1997, "Pollutant Load Reduction Models for Estuaries," in *Estuarine and Coastal Modeling, V,* American Society of Civil Engineers, pp. 1-15.

Sheng, Y.P., H.-K. Lee and K.H. Wang, 1989: ``On Numerical Strategies of Estuarine and Coastal Modeling," *Estuarine and Coastal Modeling* (M.L. Spaulding, Ed.), ASCE, pp. 291-301

Sheng, Y.P., D.E. Eliason, X.-J. Chen, and J.-K. Choi, 1991: ``A Three-Dimensional Numerical Model of Hydrodynamics and Sediment Transport in Lakes and Estuaries: Theory, Model Development, and Documentation," *Final Report for Environmental Research Laboratory*, U.S.E.P.A., Athens, GA.

Sun, W., D. Sun, C. Qiu, J. Davis, Y.P. Sheng, 1999: "Use of GIS in Integrated Study of Coastal & Estuarine Ecosystems," Poster paper presented at the 6th International Conference on Estuarine and Coastal Modeling, New Orleans, LA, Nov. 1999.

U.S. Army Corps of Engineers, Coastal Engineering Research Center, 1984: *Shore Protection Manual*, Volume I and II, 4th Edition, Vicksburg, Mississippi.

Van Rijn, L.C., 1989: *Handbook, Sediment Transport by Currents and Waves.* Delft Hydraulics Report H461, Delft Hydraulics, Delft, the Netherlands.

Yassuda, E.A. and Y.P. Sheng, 1997, "Modeling Dissolved Oxygen Dynamics in Tampa Bay during the Summer of 1991," in *Estuarine and Coastal Modeling, V,* American Society of Civil Engineers, pp. 35-50.

Application of a Barotropic Hydrodynamic Model to
Nearshore Wave-Induced Circulation

Mark Cobb[1] and Cheryl Ann Blain[2]

Abstract

A better understanding of nearshore circulation is crucial to problems concerning coastal zone management such as sediment transport, pollutant transport, water quality, and ship navigation. The two main questions we wish to address in this study are: 1) Is it possible to apply a shelf-scale hydrodynamic model to nearshore environments and 2) How sensitive is the model to changes in its physical parameters (e.g. nonlinear bottom friction coefficient, lateral dispersion coefficient)? ADCIRC-2DDI, a two-dimensional finite element barotropic hydrodynamic model, is used to study the contributions of nearshore nonlinear processes (wave-current interactions, nonlinear bottom stress, advection, lateral diffusion/dispersion) within the context of wave-induced nearshore flow for planar and barred-planar beaches. The mesh resolution necessary for capturing nearshore circulation within the numerical model was determined through a convergence study. The sensitivity of the model to its nonlinear processes is examined to assess their effect on nearshore circulation. Model-data comparisons are made using data from Leadbetter, CA (1980) and the 1990 DELILAH experiment at Duck, NC. These comparisons indicate that alongshore currents forced by wave radiation stress gradients are well simulated by the ADCIRC-2DDI model.

[1] Sverdrup Technology, Inc. (ASGMS), MSAAP Building 9101, Door 136, Stennis Space Center, MS 39529, USA, Tel: 228-689-8541, E-mail: cobb@nrlssc.navy.mil

[2] Oceanography Division (Code 7322), Naval Research Laboratory, Stennis Space Center, MS 39529-5004, Tel: 228-688-5450, E-mail: Cheryl.Ann.Blain@nrlssc.navy.mil

Introduction

In recent years there has been a renewed effort to understand the behavior of wave-driven circulation in nearshore environments, the practical aspects of this research being a better understanding of sediment transport, pollutant transport, and water quality. Advances in computers and modeling techniques make it possible to model coastal areas on a scale of several square kilometers with length scale resolutions of several meters. Important questions concerning the physics of nearshore circulation can now be addressed, in particular regarding the nonlinear processes that affect coastal circulation such as advection, nonlinear bottom friction (with and without wave-current interaction), and lateral mixing (diffusion/dispersion).

In this paper we focus on modeling nearshore environments that are relevant to actual field studies of wave driven nearshore circulation. We apply ADCIRC-2DDI (Luettich et al. 1992; Westerink et al. 1994a), a two-dimensional finite element shelf-scale barotropic hydrodynamic model that solves the equations of continuity and momentum, to the problem of nearshore circulation. Application of the ADCIRC model to such small scales is a first and we find that the model can simulate the elevation and the circulation resulting from small-scale dynamics quite well. ADCIRC is a fully developed model that has the capability of simulating domains with non-uniform grids (Cartesian or spherical coordinates), three-dimensional domains, tides, and baroclinic phenomena. The wave fields that drive the circulation in our simulations were generated using REF/DIF1 (Kirby and Dalrymple 1994) a monochromatic, linear, phase-resolving wave model.

Two types of ideal beaches are considered, planar and barred-planar. In addition, we simulate and compare to alongshore current observations taken at Leadbetter, CA (1980) (a planar beach) and to observations collected Oct 7, 1990 at Duck North Carolina as part of the DELILAH experiment (a barred planar beach). Overall we find that ADCIRC reproduces the respective alongshore current profiles of each experiment quite well. In order to determine the necessary grid resolution for our simulations, we performed a convergence study based on Richardson error estimates (e. g. Blain et al. 1998).

We investigate sensitivity of the alongshore current and surface elevation to the specification of the bottom friction and lateral mixing coefficients. An additional physical mechanism was included in the ADCIRC model in order to capture wave-current interaction; this interaction takes the form of a Grant-Madsen (Grant and Madsen 1979; Signell et al. 1990) approach that enhances the nonlinear bottom friction coefficient. We also investigate the role of advection in the nearshore circulation by utilizing the modular nature of the ADCIRC model, which permits the inclusion of physical processes individually or in combination. Wind and tidal forcing were not considered in this initial study of wave-induced circulation, although such effects will be considered in future studies.

ADCIRC and REF/DIF1 Models

Sea surface elevation and coastal currents are modeled using the fully non-linear, two-dimensional, barotropic hydrodynamic model ADCIRC-2DDI (Luettich et al. 1992). The ADCIRC (ADvanced CIRCulation Model for Shelves, Coasts and Estuaries) model, developed by Luettich et al. (1992) and Westerink et al. (1994a), has a successful history of tidal and storm surge prediction in coastal waters and marginal seas (e.g. Blain et al. 1994; Kolar et al. 1994a; Westerink et al. 1994b; Blain et al. 1998). The depth-integrated shallow water equations, derived through vertical integration of the three-dimensional mass and momentum balance equations subject to the hydrostatic assumption and the Boussinesq approximation, form the basis of the ADCIRC-2DDI model. Following are the set of conservation statements in primitive, non-conservative form expressed in a Cartesian coordinate system:

$$\frac{\partial \zeta}{\partial t} + \nabla_{xy} \cdot (Hv) = 0 \tag{1}$$

$$\frac{\partial v}{\partial t} + v \cdot \nabla_{xy} v + f \times v = -\nabla_{xy} \left[\frac{p_s}{\rho_0} + g(\zeta - \alpha \eta) \right] + \frac{M_{xy}}{H} + \frac{\tau_{sxy}}{\rho_0 H} - \frac{\tau_{bxy}}{\rho_0 H} \tag{2}$$

where t represents time, x,y are the Cartesian coordinate directions, ζ is the free surface elevation relative to the geoid, v is the depth-averaged horizontal velocity vector, $H = \zeta + h$ is the total water column depth, h is the bathymetric depth relative to the geoid, f is the Coriolis parameter, p_s is the atmospheric pressure at the free surface, g is the acceleration due to gravity, α is the effective Earth elasticity factor, η is the Newtonian equilibrium tide potential, ρ_0 is the reference density of water, M_{xy} is the horizontal momentum diffusion/dispersion, τ_{sxy} are the applied horizontal free surface stresses, and τ_{bxy} are the horizontal nonlinear bottom friction terms. This set of equations is considered to be time averaged over a wave period. Bottom stress terms are parameterized using the standard quadratic friction law:

$$\tau_{bxy} = C_f \rho_0 \left(U^2 + V^2 \right)^{1/2} v \tag{3}$$

where C_f is the nonlinear bottom friction coefficient. Lateral mixing due to diffusion/dispersion is represented through a simplified eddy viscosity formulation:

$$M_{xy} = E_h \left(\nabla^2_{xy} (Hv) \right) \tag{4}$$

where E_h is the horizontal eddy viscosity coefficient for momentum diffusion/dispersion. The wetting and drying of computational elements is possible within ADCIRC-2DDI but is not activated. Preliminary simulations invoking the shoreline inundation mechanism demonstrate that for the beach profiles studied in this paper this feature had negligible influence on the nearshore currents of interest. A rigorous derivation of equations (1) and (2) is presented by Kolar et al. (1994b) and will not be repeated here.

Simplified Momentum Equation. In order to isolate wave-driven nearshore flows, tidal and wind forcing are neglected as well as Coriolis effects that are minimal over small beach domains. A spatially variable surface stress forcing is determined from the radiation stress (S_{xx}, S_{yy}, S_{xy}, S_{yx}) gradients of the wave field (Longuet-Higgins and Stewart 1964). The remaining balance is between the local acceleration and advection and forcing through the pressure gradient, a surface wave stress, a bed stress, and lateral mixing. The result is a simplification of the momentum conservation equation (2):

$$\frac{\partial v}{\partial t} + v \cdot \nabla_{xy} v = -\nabla_{xy}(g\zeta) + \frac{M_{xy}}{H} + \frac{\tau_{sxy}}{\rho_0 H} - \frac{\tau_{bxy}}{\rho_0 H} \tag{5}$$

$$\tau_{sx} = -\left(\frac{\partial S_{xx}}{\partial x} + \frac{\partial S_{yx}}{\partial y}\right) \tag{6}$$

$$\tau_{sy} = -\left(\frac{\partial S_{yy}}{\partial y} + \frac{\partial S_{xy}}{\partial x}\right) \tag{7}$$

The wave field that is used to force the ADCIRC-2DDI hydrodynamic model is derived from the linear, monochromatic wave model, REF/DIF1 (Kirby 1986; Kirby and Dalrymple 1994). The monochromatic nature of this model allows propagation of a single wave frequency over irregular bathymetry. The REF/DIF1 model is a phase-resolving, frequency domain model based on the parabolic approximation to the mild-slope equation for water wave propagation.

Numerical solution of the governing equations (1) and (5) is achieved by recasting the continuity equation into a generalized wave continuity equation (GWCE) (Lynch and Gray 1979; Kinnmark 1984) and discretizing using the finite element method. The ADCIRC-2DDI hydrodynamic model solves the GWCE in conjunction with the momentum equations in non-conservative form. The GWCE formulation involves simple rearrangement of the continuity and momentum balances prior to spatial discretization with the result that short wavelengths are successfully suppressed without resorting to non-physical dissipation. The accuracy of this approach is well documented with respect to the solution of various shallow water problems (Lynch and Gray 1979; Luettich et al. 1992; Westerink et al. 1994a; Kolar et al. 1994a,b). A variably weighted three-time-level implicit scheme is applied to linear terms in the GWCE. Advective terms within the GWCE are evaluated at two known time levels. This time discretization results in a system of linear algebraic equations associated with the

GWCE that is solved for the unknown sea surface elevations. In the momentum equations, a Crank-Nicolson two-time-level implicit scheme is applied to all terms except the bottom friction and advective terms, which are treated explicitly. Elevation boundary conditions are enforced within the GWCE equation and normal velocity boundary conditions are enforced in the momentum equations. Westerink et al. (1994c) have shown that solutions to the GWCE equation are insensitive to this standard boundary condition formulation.

Though grids used for this study utilize uniform spacing, the finite element formulation of ADCIRC-2DDI lends itself to tremendous grid flexibility allowing easy incorporation of coastline detail and nodal densities which range over three to four orders of magnitude. A wide gradation of nodal density permits significant resolution of shoreline geometry as well as high levels of refinement near shallow coastal areas and in regions of rapid bathymetric change. Furthermore, the specification of meaningful boundary forcing can necessitate the use of large model domains which cover the continental shelf, coastal regions, and perhaps even include portions of the deep ocean (Blain et al. 1994; 1998). Such domains are readily handled within the context of a variably-graded finite element mesh leaving the discrete problem well within computational limits.

Methodology

The wave heights and directions from the REF/DIF1 wave model are interpolated onto the nodes of the finite element grid using a bicubic spline method. It should be noted that REF/DIF1 applies a criterion of H > 0.78h (wave height H, water depth h) to determine if wave breaking will occur (Kirby and Dalrymple 1994). The radiation stress gradients are then determined for each element; nodal values are assigned the average of gradients at surrounding elements. This averaging is necessary since the first derivative in a finite element scheme is discontinuous at the element boundaries. It should be noted that a finite difference approach to the first derivative produces essentially the same result and that our approach can be applied to an unstructured finite element grid without any additional work. The wave field is held constant throughout the simulation.

The current field is affected by friction due bottom roughness as well as the motion of waves. One well-known approach to account for both frictional effects was developed by Grant and Madsen (1979) and further refined by Signell et al. (1990). This approach involves a positive enhancement of the nonlinear bottom friction coefficient determined from the interaction of the current with the wave orbital motion within a bottom boundary layer. This enhancement of the friction coefficient is what we refer to as wave-current interaction. REF/DIF1 and ADCIRC-2DDI are only one-way coupled; therefore the elevations and currents determined from ADCIRC-2DDI do not affect the REF/DIF1 wave field during the simulation.

An initial grid convergence study is conducted to determine the resolution necessary to capture wave-driven circulation. We apply a Richardson-based error estimation (Blain et al. 1998) to determine if the nodal resolution of our simulations produces converged solutions. All grids in this study are essentially planar and thus the convergence study is performed only for the ideal planar beach. Three uniform grids with respective resolutions of 10 m (7701 nodes), 5 m (30401 nodes), and 2.5 m (120801 nodes) were used ($C_f = 0.004$, $E_h = 0.5$ m^2/s). It was determined that the 5 m grid fell within the asymptotic convergence region and thus suitably converged solutions should result. All uniform grids used in this study therefore have approximately 5 m nodal spacing. For our simulations we employ lateral radiation boundary conditions, a zero offshore elevation boundary condition, and a no-flow boundary condition at the shoreline.

Planar Beaches

As a first step in applying ADCIRC-2DDI to nearshore environments an ideal planar beach is simulated with a slope of 0.009, dimensions of 500 m by 1500 m (x by y) and a nodal spacing of 5 m requiring 60,000 elements and 30401 nodes. These dimensions were chosen to minimize interaction between the nearshore and offshore boundaries as well as influence of the lateral boundaries. The site of a field study in 1980, Leadbetter, CA is also simulated in order to compare modeled alongshore currents to measurements. The Leadbetter beach simulation, also a planar beach, had dimensions of 185 m by 200 m and a nodal spacing of approximately 6.0 m (2048 elements and 1089 nodes). The beach depth profile is essentially flat to within about 85 m of the shore and beyond this point the beach has a slope of approximately 0.05. The wave field used to compute a surface stresses over the planar (Leadbetter) beach had an incident wave height of 1.0 m (0.56 m), incident angle of 30.0° (9.0°), and a wave period of 10.0 sec (14.2 sec). Each simulation is of duration 0.60 days with a time step of 0.25 sec. The simulation length is sufficient to reach a steady state solution[3]. The reader should note that unless it is stated explicitly all simulations were performed *without* wave-current interaction.

In order to understand the effects of nonlinear bottom friction and lateral mixing on nearshore sea surface elevation and alongshore currents we simulate the planar beach with several different values of the nonlinear bottom friction and eddy viscosity coefficients (Figure 1). Uniform bathymetry in the alongshore direction leads to an invariant alongshore current with respect to the width of the beach. We find that the alongshore current is sensitive to changes in the coefficients of friction and eddy viscosity whereas the cross-shore elevation (not shown) is quite insensitive. This is to be expected since friction and lateral mixing are dissipative and diffusive effects that depend strongly on the velocity. The diffusive and dispersive nature of the lateral mixing is illustrated in Figure 1; as E_h increases the sharpness of the alongshore current decreases. Increasing the

---

[3] All simulations in this paper have durations of 0.60 days and time steps of 0.25 seconds.

nonlinear bottom friction coefficient decreases the nearshore alongshore current but leaves the shape of the profile unaffected. The smaller offshore alongshore current increases to a small extent in the region just prior to wave breaking.

Field observations at Leadbetter, CA and REF/DIF1 computed wave heights at the same location (Figure 2(A)) agree reasonably well within the nearshore region (< 84 m). Observed alongshore currents are compared with modeled alongshore currents to determine the effects of advection, nonlinear bottom friction, and lateral mixing. In Figure 2(B) we plot measured alongshore current data from Leadbetter, CA (1980) against two computed profiles of the alongshore current ($C_f = 0.006$, $E_h = 0.5$ m^2/s). The values for C_f and E_h were chosen (based on a parameter study) to minimize the difference between the calculated and observed alongshore current peak. The sensitivity of the alongshore current profile to changes in these parameters can be seen from Figure 1. Material derivative terms, which include advective and local acceleration terms, are not included in the momentum equations for simulation (1) (squares); simulation (2) includes the material derivative terms (solid line only). The exclusion of advection results in better agreement between computed and observed alongshore current peaks. The simulation including advection significantly smoothes and underpredicts the alongshore current peak but is in good agreement with the observed values further offshore. It is clear that the inclusion of advection adversely affects the prediction of nearshore circulation but a definitive explanation of this effect has not been determined. Efforts are being made to analyze the source of this discrepancy including examination of the numerical formulation of the advective terms within the ADCIRC model (Hench and Luettich 1998).

Alongshore profiles of the lateral mixing (M_x/H) and nonlinear bottom friction ($\tau_{by}/H\rho_0$) terms are shown in Figure 3 for simulations 1 (no advection) and 2 (advection). The lateral mixing profiles are quite close offshore but rapidly diverge close to the shore. When advection is included (solid line) the lateral mixing remains negative close to the shore, but without advection (circles) the lateral mixing reaches a well-defined peak and becomes less negative. Including advection significantly increases the magnitude of the nearshore lateral mixing but leaves the other terms (nonlinear bottom friction (Figure 3) and elevation gradients (not shown)) relatively unaffected.

In order to determine the effect of wave-current interaction we applied the Grant-Madsen approach with a constant wave bottom friction coefficient of 0.01 to our planar beach ($C_f = 0.004$, $E_h = 0.5$ m^2/s) (Figure 4). The degree to which the nonlinear bottom friction coefficient increases depends on the magnitude of the computed wave orbital velocity (determined from the REF/DIF1 computed wave field). As seen from Figure 4 wave-current interaction noticeably damps the alongshore current close to the shore and slightly increases the offshore current. This behavior is much like that observed with increased values of the nonlinear friction coefficient (Figure 1). The cross-shore elevation was found to

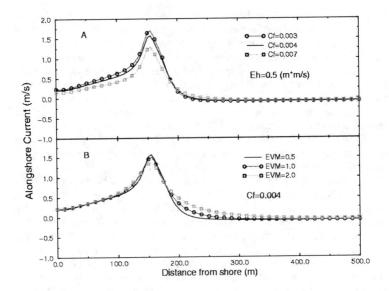

Figure 1. Planar beach (slope = 0.009), transect from x = 0 to 500 m, y = 750 m. (A) Alongshore current (m/s) for E_h = 0.5 m²/s, C_f = 0.003, 0.004, 0.007. (B) Alongshore current (m/s) C_f = 0.004, E_h = 0.5, 1.0, 2.0 m²/s.

be quite insensitive to wave-current interaction (Figure 4). Including wave-current interaction in a simulation of Leadbetter beach damped the alongshore current peak below the observed values. Specification of the wave bottom friction coefficient is still rather subjective and a better understanding of the physical interaction between waves and currents is required in order to select physically meaningful friction coefficients for both waves and mean currents. Two additional considerations pertaining specifically to wave-current interaction are 1) the relevance of a spatially varying wave friction coefficient and 2) actually resolving the bottom boundary layer using three-dimensional simulations. In addition the effect of wave-current interaction on lateral mixing in a region of wave breaking requires some investigation as well. Further study along these lines should lead to more realistic modeling of wave-current interaction.

A very interesting phenomenon observed in the ideal planar beach simulations is a dramatic transition from steady-state to time-dependent circulation patterns. A decrease in diffusion and dissipation prevents the system

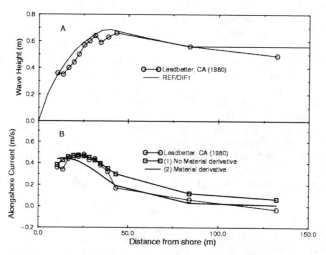

Figure 2. Transect along x data points, y = 100 m (A) Leadbetter, CA (1980) and REF/DIF1 wave heights. (B) Leadbetter, CA (1980) alongshore current, (1) Material derivative terms not included in calculation, (2) Material derivative terms included in calculation.

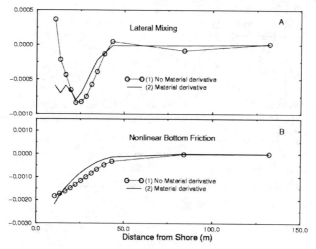

Figure 3. Transect along x data points, y = 100 m (A) Lateral mixing term M_x/H (m/s^2) with and without material derivative terms. (B) Nonlinear bottom friction term $\tau_{by}/H\rho_0$ (m/s^2) with and without material derivative terms.

from relaxing into a steady state and allows complex unsteady circulation patterns to form. As dissipative and diffusive effects (nonlinear bottom friction or lateral mixing) are decreased (values of C_f less then 0.003, and E_h less then 0.3 m^2/s) time-dependent vortices appear in the circulation patterns in contrast to the steady-state circulation patterns previously discussed (Figure 1). The vortices move from one side of the domain to the other in the direction of the alongshore current with a speed of approximately half the alongshore current. In Figures 5 ($C_f = 0.004$, $E_h = 0.5$ m^2/s) and 6 ($C_f = 0.002$, $E_h = 0.5$ m^2/s) the time-independent and time-dependent circulation patterns are plotted in the vicinity of the alongshore current peak (which is in essentially the same location as the alongshore current peak of the time independent case, see Figure 1). Changes in the magnitude and direction of the current occur in the vicinity of and propagate with the vortices. These time-dependent states are completely stable and cycle periodically as a large-scale circulation pattern for the duration of the simulation.

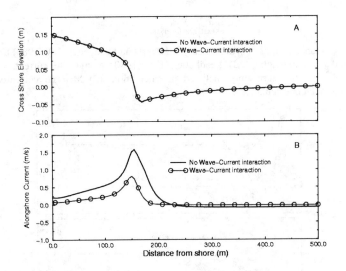

Figure 4. Ideal planar beach ($C_f = 0.004$, $E_h = 0.5$ m^2/s) (A) cross-shore elevation (m). (B) Alongshore current (m/s) profiles with and without wave-current interaction.

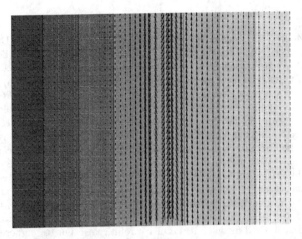

Figure 5. Ideal planar beach <u>time independent</u> velocity field (m/s) with filled bathymetric contours (0.31 meters per contour interval). Alongshore current has a maximum at ~150 m offshore.

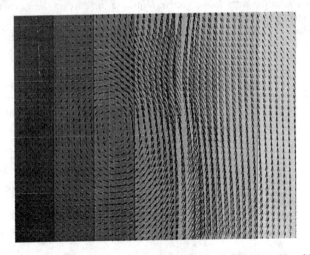

Figure 6. Ideal planar beach <u>time dependent</u> velocity field (m/s) with filled bathymetric contours (0.31 meters per contour interval). Alongshore current has a maximum at ~150 m offshore.

Barred-planar Beaches

Model data comparisons are undertaken using observations at Duck, NC taken October 7, 1990 for the DELILAH experiment (Figure 7). The Duck, NC beach bathymetry presents a more challenging environment for the prediction of nearshore currents. The simulation domain has dimensions of 1695 m by 1587 m, and is discretized to a level of 6.72 m, using 118,944 elements and 59,961 nodes. A cross-shore array with 9 data recording stations provided data on alongshore current and wave heights (Figure 7). The offshore slope of the bathymetry, along the cross-shore array, is approximately 0.01. The REF/DIF1 computed wave field was generated from an incident wave height of 0.53 m, an incident wave angle of -28.0°, and a wave period of 10.72 sec (values determined from the observed wave field). The October 7 data set was chosen for this study of wave-induced circulation because of low wind speeds during that part of the experiment. Figure 8(A) shows a comparison between the October 7, 1990 mean values of the wave height measurements (mean values are determined from data taken between 0400 and 0616 hours) and modeled REF/DIF1 wave heights. For the purposes of this study the observed and modeled wave heights are in reasonably good agreement. The overshoot of wave heights prior to wave breaking is due to the linear nature of the REF/DIF1 model. In Figure 8(B) the ADCIRC alongshore current for $C_f = 0.008$ and $E_h = 0.5$ m^2/s (no advective terms are included in the calculation) is plotted against the maximum, minimum, and mean (averaged over 0400 to 0616 hours) observed alongshore current. For this initial comparison we find that ADCIRC is reproducing the general trends of the data quite well.

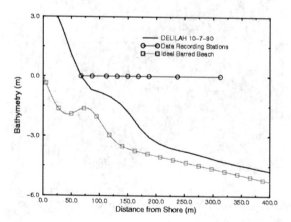

Figure 7. DELILAH 10/7/90 bathymetry, the location of the 9 cross-shore array recording stations, and ideal barred beach bathymetry.

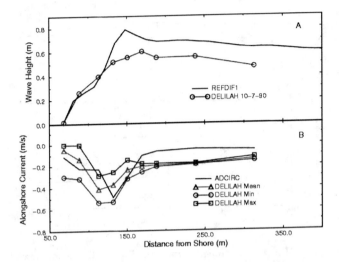

Figure 8. (A) Mean observed wave heights on 10/7/90, 0400-0616 hr plotted against REF/DIF1 computed wave heights. (B) Computed and Max, Min, and Mean Observed 10/7/90, 0400-0616 hr alongshore current.

An additional simulation over an ideal barred beach (Slinn et al. 1998) (Figure 7) is examined in order to better understand how bar and trough bathymetry affect the alongshore current (all material derivative terms are included). Except for differences in bathymetry, the simulation parameters are identical to those of the ideal planar beach ($C_f = 0.004$, $E_h = 0.5$ m^2/s, and an offshore slope of approximately 0.006). The incident wave height was 0.7 m, incident wave angle was 20°, and the wave period was 8.0 sec. As seen in Figures 9 and 10 the alongshore current profile is sharply peaked within the trough close to the shoreline and the wave height has a sharp peak in the trough region. Because the waves are breaking within the trough (sharp peak at 20 m) the alongshore current is confined within the trough. For the case of Duck, NC shown here the wave breaking occurs on the bar structure (Figure 8(A)) but there is no trough to channel the alongshore current. This ideal case provides a baseline for predicting the alongshore currents over more complex bathymetry present during the DELILAH experiment (in particular those with more trough structure).

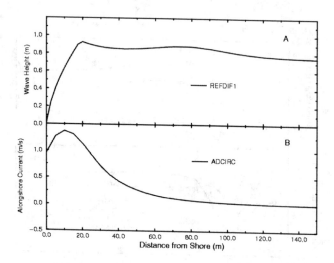

Figure 9. (A) REF/DIF1 wave height (m) and (B) ADCIRC alongshore current (m/s) for the ideal barred beach.

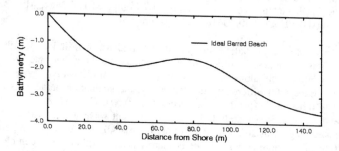

Figure 10. Bathymetry for the ideal barred beach between 0 and 150 m offshore.

Conclusion

The two key issues, 1) whether ADCIRC, a shelf-scale hydrodynamic model, can be applied in nearshore environments and 2) how sensitive is the model to its nonlinear terms, have been addressed in this initial study. The simulations performed demonstrate significant model sensitivity to the nonlinear bottom friction and lateral mixing terms at the small spatial scales required for nearshore dynamics. Alongshore current profiles are damped or smoothed by increasing the nonlinear bottom friction and lateral mixing coefficients, respectively. Secondly a decrease in the nonlinear bottom friction and/or lateral mixing coefficients results in a transition from time-independent to time-dependent circulation. Wave-current interaction is shown to have an effect similar to increasing the nonlinear bottom friction coefficient. We also find that there are significant differences between the alongshore currents predicted by ADCIRC and the observed results for the field cases at Leadbetter, CA and Duck, NC respectively when advection is included; eliminating advection improves comparisons between predicted and observed alongshore current peaks. We are currently investigating the role of advection in nearshore dynamics in addition to implementing improvements to the advection scheme within the ADCIRC model.

In this initial application of ADCIRC-2DDI to nearshore regions we find that the model is quite successful at predicting wave-induced nearshore circulation. Model-data comparisons to observations taken at Leadbetter, CA (1980) and Duck, NC as part of the 1990 DELILAH experiment clearly show that ADCIRC-2DDI is capable of modeling the nearshore circulation of real beaches. Although the advective and wave-current interaction formulations require further refinement we are very encouraged by these initial results. Because ADCIRC is a fully developed three-dimensional finite element model that can incorporate tides, Coriolis forces, and baroclinic effects, the possibility of modeling a wide array of complex nearshore problems is now possible. Future studies will focus on the influence of tide and wind forcing on nearshore wave-induced flows and further investigate wave-current interaction in a range of nearshore environments.

Acknowledgements: The authors gratefully acknowledge the U.S. Army Field Research Facility and Dr. Ed Thornton for use of the DELILAH data set; and Dr. Jane Smith at the Army Corps of Engineers WES for making available the Leadbetter, CA (1980) observational data. Special thanks go to Dr. James Kaihatu (Naval Research Laboratory) and Mr. Erick Rogers (Planning Systems, Inc.) for their wave model results and additionally their involvement in preliminary work and ongoing discussions with regard to nearshore modeling. This work is supported by the Office of Naval Research and the Naval Research Laboratory (BE-35-2-64).

References

Blain, C. A., J. J. Westerink, and R. A. Luettich, "The Influence of Domain Size on the Response Characteristics of a Hurricane Storm Surge Model", *J. Geophys. Res.*, **99**, C9, 18467-18479, 1994.

Blain, C. A., J. J. Westerink, and R. A. Luettich, "Grid Convergence Studies on the Prediction of Hurricane Storm Surge", *Int. J. Num. Methods Fluids*, **26**, 369-401, 1998.

Grant, William D., and Ole Secher Madsen, "Combined Wave and Current Interaction With a Rough Bottom", *J. Geophys. Res,.* **84**, C4, 1797-1808, 1979.

Hench, James L., and R. A. Luettich, "Analysis and Application of Eulerian Finite Element Methods for the Transport Equation", *Proceedings of the 5th International Conference on Estuarine and Coastal Modeling*, edited by Malcolm Spaulding and Allan Blumberg, pp. 138-152, Am. Soc. of Civ. Eng., Reston, VA, 1998.

Kinnmark, I. P. E., *The Shallow Water Wave Equations: Formulation, Analysis and Application*, Ph.D. Dissertation, Department of Civil Engineering, Princeton University, 1984.

Kirby, J. T., "Higher-Order Approximations in the Parabolic Equation Method for Water Waves", *J. Geophys. Res.*, **91**, 933-952, 1986.

Kirby, James T., and Robert A. Dalrymple, "Combined Refraction/ Diffraction Model REF/DIF1, Version 2.5: Documentation and User's Manual", *CACR report No. 94-22*, Center for Applied Coastal Research, University of Delaware, Newark, DE, 171p., 1994.

Kolar, R. L., J. J. Westerink, M.E. Cantekin, and C. A. Blain, "Aspects of Nonlinear Simulations Using Shallow Water Models Based on the Wave Continuity Equation", *Computers and Fluids*, **23**, 523-538, 1994a.

Kolar, R. L., W. G. Gray, J. J. Westerink, and R. A. Luettich, "Shallow Water Modeling in Spherical Coordinates: Equation Formulation, Numerical Implementation, and Application", *J. Hydraul. Res.*, **32**, 3-24, 1994b.

Longuet-Higgins, M. S., and R. W. Stewart, "Radiation Stresses in Water Waves; a Physical Discussion, with Applications", *Deep Sea Research Vol.*, 11, 529-562, 1964.

Luettich, R. A., J. J. Westerink, and N. W. Scheffner, "ADCIRC: An Advanced Three-Dimensional Circulation Model for Shelves, Coasts, and Estuaries, Report 1: Theory and Methodology of ADCIRC-2DDI and ADCIRC-3DL", *Technical Report DRP-92-6*, U.S. Army Corps of Engineers Waterways Experimental Station, Vicksburg, MS 137 pp., 1992.

Lynch, D. R. and W. G. Gray, "A Wave Equation Model for Finite Element Tidal Computations", *Comp. Fluids*, **7**, 207-228, 1979.

Signell, R. P., R. C. Beardsley, H. C. Graber, and A. Capotondi, "Effect of Wave-Current Interaction on Wind-Driven Circulation in Narrow, Shallow Embayments", *J. Geophys. Res.*, **95**, C6, 9671-9678, 1990.

Slinn, Donald N., J. S. Allen, P. A. Newberger, and R. A. Holman, "Nonlinear Shear Instabilities of Alongshore Currents over Barred Beaches", *J. Geophys. Res.*, **103**, C9, 18,357-18,379, 1998.

Westerink, J. J., R. A. Luettich, C. A. Blain, and N. W. Scheffner, "ADCIRC: An Advanced Three-Dimensional Circulation Model for Shelves, Coasts, and Estuaries, Report 2: User's Manual for ADCIRC-2DDI", *Tech. Report*

DRP-92, Department of the Army, 1994a.

Westerink, J. J., R. A. Luettich, and J. C. Muccino, "Modeling Tides in the Western North Atlantic Using Unstructured Graded Grids", *Tellus*, **46A**, 178-199, 1994b.

Westerink, J. J., R. A. Luettich, J. K. Wu, and R. L. Kolar, "The Influence of Normal Flow Boundary Conditions on Spurious Modes in Finite Element Solutions to the Shallow Water Equations", *Int. J. Num. Meth. Fl.*, **18**, 1021-1060, 1994c.

Circulation and sediment transport in the vicinity of the Hudson Shelf Valley

Courtney K. Harris[*] and Richard P. Signell[*]

Abstract

Sediment transport in the Hudson Shelf Valley and on the adjacent Long Island Shelf are evaluated using available data along with a three-dimensional wind-driven circulation model and a one-dimensional sediment transport model. Winds from the northwest drive currents up the Hudson Shelf Valley, while winds from the east produce weaker currents directed down the valley. Consistent with previous studies, sediment transport on the Long Island Shelf is dominated by resuspension during energetic wave events that are correlated with strong winds from the northeast, and net sediment flux is predicted to be towards the southwest along bathymetric contours. Transport of muddy sediments in the Hudson Shelf Valley, however, does not appear to be wave-dominated. These sediments are most likely to be resuspended by energetic currents driven by strong winds from the northwest that are not associated with energetic waves. The strong up-valley flows associated with these winds implies that net sediment flux along the Hudson Shelf Valley is up-valley.

1   Introduction

Study of sediment transport and oceanography of the Hudson Shelf Valley region has been motivated, in part, by the fact that a large amount of sewage sludge has been disposed of in the Christiansen Basin, located at the head of the valley (see, eg. Drake, 1977; Manning et al., 1994; SAIC, 1995). The area is

[*]U.S. Geological Survey; 384 Woods Hole Road; Woods Hole, MA 02543-1598. Email: ckharris@usgs.gov, rsignell@usgs.gov.

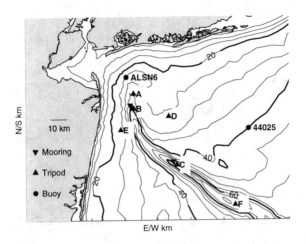

Figure 1: Map of Hudson Shelf valley region, located off-shore of the metropolitan New York area. Bathymetric contours in meters. Locations of NODC buoys, and U.S.G.S. tripods and moorings, deployed in Dec. 1999 also shown.

also interesting because the Hudson Shelf Valley itself is the only topographic depression on the Atlantic Coast that extends across the shelf into water depths as shallow as 25 m (Fig. 1). The shape of the coastline and the bathymetry of the shelf valley must play significant roles in both oceanographic circulation and in sediment transport processes in this region, but the effects of these topographic controls are poorly understood.

The distribution of mud-bound contaminants in the Hudson Shelf Valley suggest movement of fine-grained sediment more than 100 km down-valley from their source on the inner shelf (Buckholtz ten Brink et al., 1996). Sediments in the axis of the Hudson Shelf Valley show decreasing levels of lead and other contaminants with distance from the historic dumpsites, and contain concentrations that are elevated above natural background to water depths of 100 m (Buckholtz ten Brink et al., 1996). The contaminant distribution, however, does not indicate which processes have redistributed the sediments, or indicate the long-term fate of contaminants. Bottom boundary layer measurements of current velocity indicate that mean and energetic currents flow up-valley, and therefore do not provide a direct mechanism for down-valley transport of sediment (Manning et al., 1994).

Previous studies in the Mid-Atlantic Bight indicate that sediment resuspension occurs primarily during energetic wave events (Butman et al., 1979; Lyne

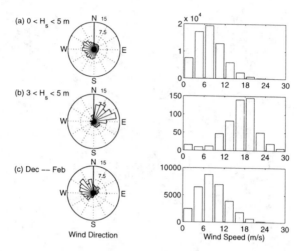

Figure 2: Wind directions and speeds observed by Ambrose Light Buoy. (a) All data, 1989–1999; (b) conditions that occurred when wave height exceeded 3 m; and (c) wind conditions during the winter months.

et al., 1987; SAIC, 1995). Observations from the Ambrose Light Buoy (ALSN6 on Fig. 1; NODC, 1993) indicate that energetic waves coincide with strong winds from the northeast (Fig. 2), and that northeast winds exceed 10 m/s during 5% of the winter (Dec.–Feb.) record. Winds are predominantly from the west, however, and though they do not produce energetic waves, northwest winds that exceed 10 m/s account for 19% of the winter record (Fig. 2). It was thought that noreaster conditions would dominate sediment transport at this site, but the relative contributions of transport of noreaster wave events had not been compared to transport during other types of conditions.

This issue, and the inability for observed up-valley flows to explain the accumulation and down-valley transport of contaminants has motivated a combined observational and modeling study of sediment transport in the vicinity of the Hudson Shelf Valley. Observations of currents and suspended sediment concentrations will be made from a winter 1999/2000 field deployment (see Fig. 1). This data will be used to document suspended sediment transport and to complement conceptual and numerical models that are being developed to account for bathymetric and seasonal influences on circulation and sediment transport in this region.

The present study uses available wind, wave, and sediment data along with circulation and suspended sediment transport models to develop intu-

Wind Speeds (@10m) m/s:	3	6.4	10.6	17.3	
Wind Stress dy/cm^2	0.14	0.62	1.7	6.0	
Wind Directions $°$ T	45	135	225	315	
Orbital Velocities cm/s	4.0	11.5	20.0	32.5	55.0

Table 1: Representative wind speeds, directions, and wave-orbital velocities that span the 1989–1999 observations from the Ambrose Light buoy. Combinations of these waves and winds yield 80 climatic conditions.

ition about transport processes in this region, and to prepare for the field experiment. In particular, the effect of wind forcing and bathymetric steering on oceanographic circulation is predicted, and the ability of these circulation patterns to affect sediment transport is evaluated. Secondly, the types of sediment transport events that are likely to dominate sediment flux are identified, and the magnitude of suspended sediment concentrations and flux are estimated within the shelf valley and on the adjacent Long Island Shelf.

2   Methods

A climatological approach was used to characterize sediment transport and wind-driven circulation in this region. Separate models of circulation and sediment transport were driven using available wind and wave data in the following steps. The details of the individual models are described later.

1. A ten year record of wind and wave data from the Ambrose Light buoy was used to estimate the frequencies of occurrence of 80 combinations of winds and waves. Table 1 provides the values used.

2. Current velocities were calculated for 16 combinations of wind speeds and directions (see Table 1) using the Estuarine and Coastal Ocean Model, ECOM-3D[1], a three-dimensional sigma-coordinate model that calculates oceanographic response to a variety of forces, including winds, tides, and buoyancy (Blumberg and Mellor, 1987).

3. Suspended sediment concentrations and sediment flux were estimated using a one-dimensional model of bottom boundary layer resuspension (Wiberg et al., 1994; Harris and Wiberg, 1997). These calculations were carried out for 80 climatic conditions at sites A–D (Fig. 1).

4. The estimates of suspended sediment flux for each climatic condition were weighted by the observed probability of those conditions to estimate its contribution to sediment flux for sites A–D.

---

[1]Trade names are for identification purposes only, and do not constitute endorsement by the U.S. Geological Survey.

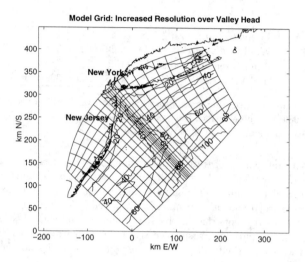

Figure 3: Model grid used for circulation calculations showing every fifth gridline. Bathymetric contours in meters.

3 Circulation Calculations

The circulation model grid includes the Hudson Shelf Valley offshore to the shelf-slope break and extends approximately 200 km to the northeast and southwest of the shelf valley (Fig. 3). The resolution of the model grid was increased in the vicinity of the upper Hudson Shelf Valley, and had a minimum alongshelf grid spacing of 0.5 km over the valley and maximum grid spacing of 6 km at boundaries (Fig. 3). Ten evenly spaced vertical sigma levels were used, and the shelf water was assumed to be of uniform density. Tidal and elevation boundary conditions for the off-shore and northeastern boundaries were obtained by nesting the ECOM-3D model domain within an East Coast ADCIRC model that was driven using identical wind stresses. ADCIRC provides estimates of Kelvin wave effects that would propagate into the model domain due to wind forcing (Luettich et al., 1992).

Steady winds account for approximately 40% of the circulation variance on this shelf (Manning et al., 1994; Mayer et al., 1987). By modeling the wind-driven currents, we determine the degree to which wind forcing and bathymetry influence sediment transport. We expect these results to apply to winter well-mixed conditions of strong wind forcing. The circulation response to each of four wind speeds, and four wind directions (see Table 1) was estimated, assuming the wind forcing to be a steady and uniform wind field.

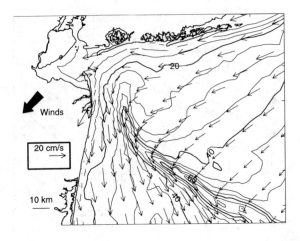

Figure 4: Calculations of current velocity 1 meter above the bed under steady, 17.3 m/s winds from the northeast. (Bathymetry: 5 m contour lines.)

Each wind speed was input from four directions (see Table 1) that were chosen after preliminary calculations were run over a much finer directional mesh, so that a total of 16 circulation runs were done.

Shelf topography exerts a strong control over predicted currents. Under strong winds (17 m/s) from the northeast, for example, shelf bathymetry and the coastline orientation result in near-bed off-shore directed flows of 5–15 cm/s over the shelf valley, while near-bed currents over the Long Island Shelf are predicted to be about 15 cm/s (Fig. 4). Winds from the northeast of this magnitude account for 0.5% of the Ambrose Light winter record. Topographic control is also predicted under 17 m/s winds from the northwest (Fig. 5). Northwest winds of this magnitude are more prevalent than northeast winds, and have been observed in 1.2% of winter observations at Ambrose Light. Under northwest winds, the near bed currents flow up-valley. Predicted near-bed currents reach 30 cm/s in the Hudson Shelf Valley, but are less than 5 cm/s on the Long Island Shelf. Previous observations support this prediction of strong up-valley directed flows during winds from the west (Lavelle et al., 1975; Manning et al., 1994).

Near-bed current velocities were determined from the circulation model for sites A, B, C, and D (Fig. 1) and used as input for the sediment transport model (Table 2). Calculations were done at these sites to evaluate transport

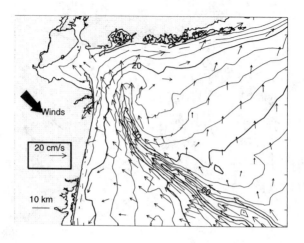

Figure 5: Calculations of current velocity 1 meter above the bed under steady, 17.3 m/s winds from the northwest. (Bathymetry: 5 m contour lines.)

along the Hudson Shelf Valley, and on the adjacent Long Island Shelf. Currents at sites A and B in the Hudson Shelf Valley responded most strongly to the northwest and southeast winds that parallel the valley axis (Table 2). Likewise, winds that paralleled the Long Island Shelf bathymetry (the northeast and southwest winds) created the largest currents at site D.

4   Resuspension calculations

Resuspension calculations were made using a one-dimensional suspended sediment transport model that has compared well against field data from both muddy and sandy shelf sites (Wiberg et al., 1994, 1999). A similar model has been applied to a sandy site on the Mid-Atlantic Bight (Lyne et al., 1987). The calculations include combined wave-current shear stress (Smith, 1977; Grant and Madsen, 1979), density stratification from suspended sediment (McLean, 1992), and bed armoring by winnowing (Harris and Wiberg, 1997). The model requires specification of sediment properties (see Table 3), wave characteristics, and a representative near-bed current velocity. Following Wiberg et al. (1994), the settling velocity and critical shear stress of fine grained sediment (<44 μm) were set equal to 0.1 cm/s and 1.0 dy/cm^2, respectively, to partially

Winds:		Calculated Currents (cm/s):			
Direction	Speed (m/s)	A	B	C	D
(A) NE	3.0	1.3	1.5	1.7	0.6
	6.4	3.2	4.7	3.6	2.7
	10.6	4.7	7.6	6.4	5.9
	17.3	7.4	14.0	13.3	16.2
(B) SE	3.0	1.1	1.2	1.1	0.3
	6.4	4.5	5.5	4.4	1.1
	10.6	7.8	10.2	9.3	2.5
	17.3	18.4	23.5	25.2	6.6
(C) SW	3.0	1.7	1.8	1.7	0.6
	6.4	5.9	7.8	7.7	4.0
	10.6	9.4	12.4	14.2	7.8
	17.3	13.4	18.2	23.5	16.8
(D) NW	3.0	1.1	1.0	0.7	0.3
	6.4	5.8	5.6	3.9	0.6
	10.6	12.0	13.1	11.5	1.1
	17.3	26.1	29.9	26.5	1.8

Table 2: Current velocities calculated 1 meter above the bed at sites A–D for the 16 winds considered.

account for flocculation and bed cohesion.

The calculations estimate velocity and suspended sediment concentration in the bottom boundary layer by balancing downward settling and upward diffusion using an exponential eddy viscosity profile (Wiberg and Smith, 1983) and a reference concentration for the bottom boundary condition (Smith and McLean, 1977). This mass-balance assumes that the suspended sediment field is equilibrated with flow conditions, and neglects flux divergence. These calculations cannot, therefore, be directly used to predict redistribution, erosion, or deposition of sediment. They do, however, provide insight into the processes that suspend sediment at a site, and the magnitudes of suspended sediment concentrations. While results from this model have compared well to measure-

		Site	A	B	C	D
		Water Depth	40 m	55 m	70 m	25m
Size Class:	w_s cm/s	τ_{cr} dy/cm^2	Grain Size Distribution:			
< 44 μm	0.10	1.0	90	60	55	2
44–88 μm	0.23	1.0	5	34	30	30
88–177 μm	0.77	1.6	3	5	9	64
> 177 μm	5.6	2.8	2	1	6	4

Table 3: Water depth of sites for which suspended sediment calculations were completed. Summary of the seven grain size classes that were used for the calculations.

ments made in a variety of field settings (Wiberg et al., 1994, 1999), it can not be tested for the New York Bight until field observations are available.

Resuspension calculations were completed at sites A–D for 80 combinations of wind speed, wind direction, and wave orbital velocity. Table 2 provides the estimated wind-driven current for each combination of wind speed and direction. These currents were used along with the wave orbital velocities from Table 1 to estimate sediment resuspension and flux at the four sites.

Resuspension in the Hudson Shelf Valley: The strength of the driving current velocity affects the vertical distribution of suspended sediment at the Hudson Shelf Valley sites. Along the Hudson Shelf Valley (Sites A–D), currents are predicted to be especially high under conditions of winds from the NW, and lowest under NE winds (Table 2). The importance of the current velocity for driving sediment transport in the Hudson Shelf Valley is illustrated by contrasting two of the 80 conditions; the classic noreaster storm event and a norwester wind event. The noreaster storm event has been shown to dominate shelf sediment transport in many Mid-Atlantic shelf sites, because it combines high winds and energetic waves. Energetic northwest winds, while they occur more frequently, are not associated with energetic waves and therefore it has been assumed that they would not be important for sediment transport.

The calculations, however, indicate that the noreaster is surprisingly less effective at transporting sediment in the Hudson Shelf Valley than energetic winds from the northwest (Fig. 6). While near-bed suspended sediment concentrations are high for the noreaster, the currents predicted under northeast winds do not produce sufficient turbulence to bring sediment up out of the near-bed region (Fig. 6). Suspended sediment concentrations are lower near the bed for the northwesterly event, but the stronger currents that are predicted under northwesterly conditions produce enough shear to mix sediment up into the entire water column. Sediment concentrations outside of the wave boundary layer are therefore orders of magnitude higher for the norwester than they are for the case driven with northeast winds (Fig. 6b). The result is that the depth integrated flux is higher under the case of strong northwesterly winds with moderate waves than it is for the case of a classic noreaster.

Similar results were obtained for the other sites in the Hudson Shelf Valley. Sites A–C all had predictions of high sediment flux driven by high near-bed currents when winds were strong and from the northwest.

Resuspension on the Long Island Shelf: Results from the Long Island Shelf site were somewhat different (Fig. 7). Sediment at this site is predominantly a fine sand, and currents were predicted to be highest under northeasterly winds (Tables 2, 3). Sands are rarely suspended high above the sediment bed when subjected to the shear stresses that normally occur on shelves, even

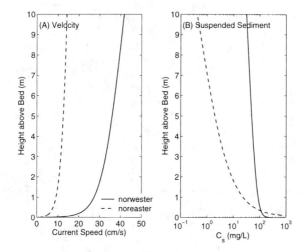

Figure 6: Bottom boundary layer calculations for site A. (A) current velocity and (B) suspended sediment concentration for the classic "noreaster" and "norwester" storms made for Site A. Dashed lines are calculations for energetic waves (u_0=55 cm/s) and strong, north-east winds (17.4 m/s). Solid lines were calculated using weak waves (u_0=11.5 cm/s) and strong northwest winds (17.4 m/s).

under strong current velocities. This implies that on the Long Island Shelf, the amount of resuspension and sediment flux is less sensitive to current speed than was the case in the Hudson Shelf Valley. The NE winds that are associated with the largest waves on this coast are aligned parallel to the Long Island Shelf bathymetry, and these winds therefore drive strong currents. The most significant suspension events are therefore predicted to be energetic waves that are correlated with strong winds from the northeast that also produce large near-bed currents (Fig. 7).

5   Shelf Climatology and Sediment Transport

The significance that each of the climatic states has for transporting sediment was evaluated by estimating the expected flux for each of Sites A–D. The joint frequency for each combination of wind speed, wind direction, and wave orbital velocity was multiplied by the sediment flux calculated for those conditions. Joint probabilities of winds and waves were estimated from the 1989–1999 record of wind speed, wind direction, and wave spectra measured by the Ambrose Light buoy (NODC, 1993). Wave orbital velocities at each of

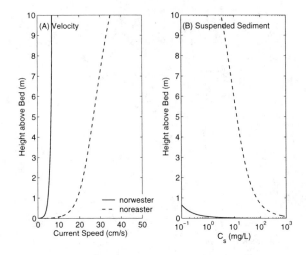

Figure 7: Bottom boundary layer calculations for site D. (A) current velocity and (B) suspended sediment concentration for the classic "noreaster" and "norwester" storms. Dashed lines are calculations for energetic waves (u_0=55 cm/s) and strong, north-east winds (17.4 m/s). Solid lines were calculated using weak waves (u_0=11.5 cm/s) and strong northwest winds (17.4 m/s).

the four sites were estimated from the hourly wave spectra following Madsen (1994). This assumes that the wave spectra measured by the buoy is valid over the region studied. These time series of wave orbital velocities were then combined with the concurrent measurements of wind speed and wind direction to estimate the frequency of occurrence of each of the 80 climatic states summarized in Table 1.

Flux at each site was dominated by a small subset of the 80 climatic conditions with the valley sites being dominated by northwesterly wind events, and the shelf site by noreaster storms. The resultant cumulative flux estimates were therefore up-valley for the three Hudson Shelf Valley sites, and westward on the Long Island Shelf (Table 4). Sediment flux in the Hudson Shelf Valley was dominated by transport during strong winds from the northwest that coincide with low energy waves. For example, at Site A, the climatic condition of low waves (u_o=5 cm/s) and energetic winds (17 m/s) from the northwest account for 60% of the sediment flux predicted for this site, even though these conditions were observed for only 0.44% of the buoy record. Sediment flux predicted during down-valley favorable conditions was 2×10^3, 5×10^3, and 6×10^3 kg/(m year) for Sites A–C, respectively, small fractions of the net flux

Site	A	B	C	D
Cuml. Flux kg/(m yr)	28×10^3	26×10^3	31×10^3	2×10^3
Direction (° T)	4°	353°	310°	248°

Table 4: Cumulative flux magnitude and direction calculated using the estimates of sediment suspension and frequency of occurence for 80 climatic conditions.

predicted for these sites (Table 4). The sandy site on the Long Island Shelf responded most to wave-dominated resuspension during noreaster winds, and the net transport at this site is westward along isobaths. The cumulative flux for this site was estimated to be lower than the other sites, because the sediments here are coarser, and the current velocities were lower.

6   Discussion

The circulation model indicates an interesting response of wind-driven currents to westerly vs. easterly winds so that up-valley flows are favored in the shelf valley. This appears to be a function of topography and coastline orientation, and is consistent both with observations of strong up-valley flows in this region (Lavelle et al., 1975; Manning et al., 1994), and with numerical models that predict enhanced upwelling over submarine canyons (Klinck, 1996). Further study is required to identify the dominant forces that create this pattern.

While it is difficult to infer geologic patterns without a three-dimensional sediment transport model of this region, it is interesting to compare these calculations to present-day sediment texture and contaminant distributions. The fact that muddy sediment has persisted at the head of the valley, but not at equivalent water depths on the adjacent shelves, supports our conclusion that net transport is directed up-valley. The observed migration of contaminants down the valley axis does not contradict these calculations, which predict that down-valley flows would transport 2–6×10^3 kg/(m year) of sediment.

Several potentially important processes were neglected in these calculations, including particle dynamics, fluid mud processes, buoyancy, sediment advection, larger-scale forcing, and time-dependent forcing. These may affect the details of our conclusions, but the sensitivity of the flux of fine-grained material to current speed is a robust conclusion.

The potential error from these assumptions was evaluated in several sensitivity tests. The accepted threshold required to trigger gravity driven sediment flux (10g/L) is exceeded 0.07% of the time. Precluding observations of near-bed suspended concentrations in the Hudson Shelf Valley, it does not appear that fluid mud processes are significant in this region. The sensitivity of our calculations to neglecting tides was tested by adding a liberal estimate of tidal velocity (10 cm/s) to the currents used to drive resuspension calculations. This increased the contribution to net flux experienced during northeast wind conditions from 2% to 12%, but did not qualitatively change the up-valley prediction of net flux. Model runs were done that used a two-layer density structure, with a pycnocline at a depth of 60 m. The density difference between the two layers, 0.4 kg/m^3, was chosen to be similar to observed wintertime buoyancy. Near-bed currents calculated with this density structure were within 4 cm/s of the original values. A final test was run to evaluate how the presence of a mean southwesterly current equal to 10 cm/s on the Long Island Shelf might influence the calculations. While the mean current did impact the circulation calculated for light to moderate winds, the energetic wind cases were unaffected by the presence of a mean current. Because these are the cases that are associated with high suspended sediment transport the suspended sediment calculations were not affected by the presence of a mean current.

7   Conclusions

Climatological analysis of wind and waves coupled with circulation and sediment tranpsort models has yielded insights into the nature of sediment transport in the New York Bight. Sediment flux is predicted to be higher along the Hudson Shelf Valley than on the adjacent Long Island Shelf because sediment is finer in the valley and stronger currents are generated in the valley, particularly under westerly winds. Sediment transport in the Hudson Shelf Valley region appears to be dominated by strong northwesterly wind events. In this area the fine grained sediment is vertically distributed throughout the bottom boundary layer by strong up-valley currents driven by these winds. Transport on the adjacent Long Island Shelf appears to be dominated by noreaster storm events, when strong waves and high current velocities are capable of suspending and transporting the sandy sediments.

Acknowledgements: This paper benefited from discussions with Drs. Brad Butman, Marilyn Buckholtz ten Brink, and Chris Sherwood of the USGS. Sediment size data were provided by E. Mecray (USGS). Hydroqual, Inc. supplied the ECOM-3D model. The authors also appreciate input from two anonymous reviewers of an earlier version of this manuscript.

References

Blumberg, A. F. and Mellor, G. L. (1987). A description of a three-dimensional coastal ocean circulation model. In Heaps, N. S., editor, *Three-Dimensional Coastal Ocean Models, Coastal and Estuarine Sciences*, volume 4, pages 1–16. American Geophysical Union, Washington, D.C.

Buckholtz ten Brink, M., Allison, M., Schlee, J., Casso, M., Lotto, L., and Bopp, R. (1996). Sewage sludge and sedimentary processes in the new york bight: Where does it go from here? In *EOS, Transactions of the American Geophysical Union*, volume 76, page OS14, San Diego, CA. American Geophysical Union, Ocean Sciences Meeting.

Butman, B., Noble, M. A., and Folger, D. W. (1979). Long-term observations of bottom current and bottom sediment movement on the Mid-Atlantic continental shelf. *Journal of Geophysical Research*, 84(C3):1187–1205.

Drake, D. E. (1977). Suspended particulate matter in the New York bight apex, fall, 1973. *Journal of Sedimentary Petrology*, 47(1):209–228.

Grant, W. D. and Madsen, O. S. (1979). Combined wave and current interaction with a rough bottom. *Journal of Geophysical Research*, 84(C4):1797–1808.

Harris, C. K. and Wiberg, P. L. (1997). Approaches to quantifying long-term continental shelf sediment transport with an example from the northern California STRESS mid-shelf site. *Continental Shelf Research*, 17(11):1389–1418.

Klinck, J. M. (1996). Circulation near submarine canyons: a modeling study. *Journal of Geophysical Research*, 101(C1):1211–1223.

Lavelle, J., Keller, G., and Clarke, T. (1975). Possible bottom current response to surface winds in the Hudson shelf channel. *Journal of Geophysical Research*, 80:1953–1956.

Luettich, R. J., Westerink, J., and Scheffner, N. (1992). ADCIRC: an advanced three-dimensional circulation model for shelves, coasts, and estuaries. Report I: Theory and methodology of ADCIRC-2DDI and ADCIRC-3DL. Technical report, U.S. Army Corps of Engineers Waterways Experiment Station, Vicksburg, MS.

Lyne, V., Butman, B., and Grant, W. (1987). Sediment movement along the U.S. east coast continental shelf – II. Modeling suspended sediment concentration and transport rate during storms. *Continental Shelf Research*, 10(5):429–461.

Madsen, O. (1994). Spectral wave-current bottom boundary layer flows. In *Coastal Engineering 1994, Proceedings (24th international conference)*, pages 384–397, Kobe, Japan. Coastal Engineering Research Council/ASCE.

Manning, J., Oey, L., Packer, D., Vitaliano, J., Finneran, T., You, K., and Fromm, S. (1994). Observations of bottom currents and estimates of resuspended sediment transport at the New York Bight 12-mile dumpsite. *Journal of Geophysical Research*, 99(C5):10221–10239.

Mayer, D., Han, G., and Hansen, D. (1987). Circulation in the Hudson shelf valley: MESA physical oceanographic studies in New York Bight, 1. *Journal of Geophysical Research*, 87(C2):9563–9587.

McLean, S. R. (1992). On the calculations of suspended load for non-cohesive sediments. *Journal of Geophysical Research*, 97(C4):5759–5770.

NODC (1993). NOAA Marine Environmental Buoy Database: Atlantic Ocean, disc number 1, ALSN6. Technical report, National Oceanographic Data Center, Silver Spring, MD.

SAIC (1995). Analysis of waves and near-bottom currents during major storms at the New York mud dump site, report #20 of the New York mud dump site studies. Technical report, Science Applications International Corporation, Newport, Rhode Island.

Smith, J. D. (1977). Modeling of sediment transport on continental shelves. In Goldberg, E. D., McCave, I. N., O'Brien, J. J., and Steele, J. H., editors, *The Sea*, volume 6; Marine Modeling, chapter 13, pages 539–577. John Wiley, New York.

Smith, J. D. and McLean, S. R. (1977). Spatially averaged flow over a wavy surface. *Journal of Geophysical Research*, 82(12):1735–1746.

Wiberg, P. L., Drake, D. E., and Cacchione, D. A. (1994). Sediment resuspension and bed armoring during high bottom stress events on the northern California continental shelf: measurements and predictions. *Continental Shelf Research*, 14(10/11):1191–1219.

Wiberg, P. L., Drake, D. E., Harris, C. K., and Noble, M. A. (in review, 1999). The sediment transport environment of the Palos Verdes shelf. *Continental Shelf Research*.

Wiberg, P. L. and Smith, J. D. (1983). A comparison of field data and theoretical models for wave-current interactions at the bed on the continental shelf. *Continental Shelf Research*, 2:147–162.

Inverse Flow Modelling of Waves and Currents
In the Surf Zone

David H. Henderson[1], David Johnson[1] and Graham J. M. Copeland[1]

Abstract

Inverse methods are used to build a 2D horizontal (2DH) numerical model able to assimilate data from a large-scale physical model of the surf zone. The residuals of the governing equations can be used to identify the compatibility between data and equations and to reveal shortcomings in the theory. A particle tracking method is used to simulate sediment transport and morphological change. A one-dimensional vertical model calculates the flow field used for the particle tracking. This 1DV can then be coupled to the flow and wave field from the inverse model to provide a quasi 3d wave resolving description of the time dependent currents for use in the particle tracking module.

Keywords: Direct model, UKCRF data, Inverse model, Particle tracking.

Introduction

Complex flows in the nearshore zone are driven by a number of processes. Coupled wave and current models can describe the behaviour in this zone. These, generally, take account of refraction, diffraction, dissipation and, of course, wave-current interactions. Such effects become important as the water depth decreases and waves break at the shore. As the waves break the excess momentum of the waves (radiation stress) drives the flow. This is turn affects the incoming waves. Two important illustrative and practically important examples, which can be described by such models, are rip currents and wave-flow interactions due to detached breakwaters.

---

[1] Department of Civil Engineering, The University of Strathclyde, 107 Rottenrow, Glasgow G4 ONG, Scotland U.K.

Noda (1974) developed the first two-dimensional finite-difference numerical model that included refraction and wave-flow interaction. Ebersole and Dalrymple (1980) added the effect of the non-linear convective terms. Yoo and O'Connor (1986) considered wave-flow interaction in two dimensions via a phase-averaged model (see below). Three-dimensional models have recently been developed, e.g. Pechon and Teisson (1995). The gaps in our knowledge of the Physics of the system were summarized by Pechon et al. (1997) who identified four deficiencies in recent coastal wave-flow models:

1. The calculated values of wave-driven forcing of the flow are over-estimated.
2. The formulation of the bed shear stress due to waves and flow needs to be improved.
3. Mixing of momentum due to turbulence behind breakwaters is too high
4. Generation of currents by *irregular* waves requires more study.
5. In addition, the parameterisation of mean flow forcing by the radiation stress may need to be altered. This is because the momentum flux from breaking waves (the radiation stress) is not transferred straight to the mean flow, but is advected further shoreward within a region of turbulence on the leading face of the breaking wave (Svendsen, 1984). It is only when this turbulence is dissipated that flow forcing occurs.

In contrast to the solutions provided by direct models, inverse models are able to interpolate objectively between   scattered observations (e.g. of flow) by making use of the system equations (e.g. the momentum or continuity equation) and need no data at boundaries. Therefore, they are useful when only scattered observations are available. Also, the inclusion of data in the solution is an advantage because errors due to the model formulation (discretization, model parameters and boundary conditions) may be as important as the errors in the data.

The development of an inverse model depends on the availability of good data. This data can be gathered in the field or in laboratory models. The United Kingdom Coastal Research Facility (UKCRF) situated at HR Wallingford Ltd[2] allows a wide variety of experiments to be conducted on coastal engineering problems so improving our understanding of coastal processes. Projects carried out there include studies of hydrodynamics: waves and currents, sediment transport, and pollution and mixing processes. In particular, experimental data is now available on rip-current production in the near shore and also on waves and flows around a model of the Elmer breakwater scheme in Sussex, UK. Both cases represent excellent opportunities of testing and demonstrating the use of inverse models in civil engineering.

---

[2] HR Wallingford Ltd, Howbery Park, Wallingford, OX10 8BA, England U.K. http://www.hrwallingford.co.uk/facilities/crf/

A second part of the part of the work has been to develop a way of simulating sediment transport using a particle tracking method within the flow fields generated by the inverse or direct model. A Lagrangian model provides an interesting alternative description of the transport processes to the more conventional Eulerian methods. A primary advantage is that the effect of turbulence on the sediment movement can be included directly with a stochastic velocity component. It is hoped this may provide some interesting insights in a zone of intense turbulence such as the surf zone. A novel approached is used to calculate fluid velocities at the particle location, by using a method of advecting a one-dimensional vertical (1DV) phase resolving intra-wave model with the particle. The 1DV model used is very similar to that of Elfrink *et al.* (1996), solving the momentum equations in the vertical profile and using a k-ε turbulence closure. To resolve intra-wave variables from time mean values, a wave theory is required and here a linear theory is used. The sediment tracking provides a time series of modelled particle trajectories. Morphological change is simulated by tracking a number of particles from points throughout the domain and recording their positions over a period of time. By counting particle occurrence at points throughout the domain at each time step a relative measure of erosion and deposition is made. Areas in which particles occur more frequently at the bed are areas of deposition and conversely areas of lower occurrence are areas of erosion. Concentration profiles may also be calculated by applying the same technique in the vertical.

The 2DH Model

A 2D depth averaged model of wave-current interaction that accounts for current-depth refraction, diffraction, wave-induced currents, set-up and set-down, mixing processes, and bottom friction was formulated by Yoo and O'Connor (1986). Conservation of waves and energy gives the following wave model:

$$\frac{\partial K^i}{\partial t} + \left(U_i + QK^j\right)\frac{\partial K^i}{\partial x_j} + K^j\frac{\partial U_j}{\partial x_i} + R\frac{\partial d}{\partial x_i} - \frac{1}{2kA}\frac{\partial}{\partial x_i}\left(\nabla_j^2 A\right) = 0, \quad (1)$$

$$\frac{\partial A}{\partial t} + \frac{1}{2A}\frac{\partial}{\partial x_i}\left[\left(U_i + QK^i\right)A^2\right] + \frac{S^{ij}}{\rho g A}\frac{\partial U_j}{\partial x_i} + C^A A^2 = 0, \quad (2)$$

and the current is governed by the momentum and mass conservation equations given by:

$$\frac{\partial U_i}{\partial t} + U_j\frac{\partial U_i}{\partial x_j} + \frac{1}{\rho d}\frac{\partial S^{ij}}{\partial x_j} + g\frac{\partial \eta}{\partial x_i} + \frac{C^u}{d}U_i\left|\sqrt{U_i^2 + V_i^2}\right| = 0, \quad (3)$$

$$\frac{\partial \eta}{\partial t} + \frac{\partial}{\partial x_i}\left(dU_i\right) = 0, \quad (4)$$

where K^i is the wave number vector (i represents the index, x or y), U_i is the fluid velocity vector, A is the wave height, η is the set-up or set-down, d is the water depth, ρ is the water density, g is the acceleration due to gravity, and C^U and C^A are the bottom friction coefficients associated with U_i and A respectively. To take account of diffraction the wave number is replaced by a separation factor k, where the relation

$$k^2 = K^2 - \frac{1}{A}\frac{\partial^2 A}{\partial x_i^2} \quad (5)$$

was introduced by Battjes (1968). Also, Q is the group velocity divided by the wave number and R is a depth variation factor for the wave number vector i.e.

$$Q = \frac{\sigma_o}{2k^2}\left(1 + \frac{2kd}{\sinh 2kd}\right) \quad (6)$$

$$R = \frac{\sigma_o k}{\sinh 2kd} \quad (7)$$

$$\sigma_o^2 = gk\tanh kd \quad (8)$$

The radiation stresses, which are responsible for driving currents, are

$$S^{ij} = \frac{1}{2}\left[\left(1 + \frac{1}{Ak^2}\frac{\partial^2 A}{\partial x_i^2}\right)\left(1 + \frac{2kd}{\sinh 2kd}\right)\frac{K_i}{|K|}\frac{K_j}{|K|} + \frac{2kd}{\sinh 2kd}\delta_{ij}\right]\frac{1}{2}\rho g A^2 \quad (9)$$

This system of equations can be used to model wave-current behaviour in the surf zone in conjunction with a wave-breaking criterion. Commonly, this criterion allows the waves to grow to a certain limit before the breaking criteria is imposed. The condition used here is that the maximum wave height is given by $B_h = \gamma d$, where γ is a number (typically 0.78).

Direct Modelling and Physical Data
The wave and current equations (1)-(4) have been solved numerically in a number of ways, e.g. using the alternating derivative explicit scheme for the momentum equations and upstream differencing for the wave equations, see Yoo and O'Connor (1986). In this work, the direct model used a simplified upwinding finite difference scheme for the wave and momentum equations and central differencing was used for the continuity equation.

Techniques for collecting physical data have improved greatly in recent years. The United Kingdom Coastal Research Facility (UKCRF) is a 36m by 23m 3-dimensional experimental wave basin dedicated to the understanding of the near shore environment. Two recent studies are of interest, rip currents and the wave-current interaction around a system of detached breakwaters. Such studies make available large sets of data that direct models make poor use of. This is because only boundary data, which is notoriously difficult to collect, can be used. However, scattered data is easily assimilated into inverse models.

Rip Currents

A study of rip currents on a sinusoidal (cuspate) beach was conducted by
Borthwick et al. (1997). A number of cases were considered involving regular
waves: (a) period 1s, wave height 0.1m, $0°$ incident angle; (b) 1.2s, 0.125m, $0°$;
oblique regular waves (c) 1.2s, 0.125m, $20°$, and random waves (d) 1.2s, 0.125m,
$0°$. During the tests the main characteristics of rip currents were   reproduced.
Nearshore circulation cells and high velocity offshore jets were observed. Some of
this data is shown in Figures 2 and 3 in comparison with direct model solutions.

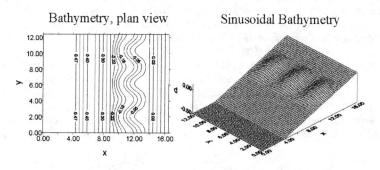

Figure 1

Figure 1 shows the bathymetry used in physical modelling of this system at the UKCRF.
Numerical investigations of this system have shown that direct models can correctly
predict the formation of symmetric counter-rotating circulation cells. However, rip-
current peak velocities are under-estimated as is the offshore extent of the rip currents.
The deficiencies in the direct model solution can be seen by comparing the predicted and
observed flows, see Figure 2.

The peaks observed in the measured waveheights, Figure 3, can be attributed to
the sampling procedure used to obtain the data. This effect is due to the unsteady
motion in the swash zone. There is a clear difference between the distribution of
measured wave heights and the wave heights produced by the direct model, see
Figure 3.   Inverse models should provide a best compromise between data

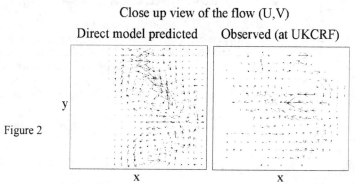

Close up view of the flow (U,V)

Figure 2

(including sampling errors) and calculations (including modelling inadequacies).

Breakwaters

Direct models can also be used to predict wave and current fields around detached breakwaters. However, in this instance it is very difficult to accurately specify boundary conditions at the diffracting structures. Experiments at the UKCRF have also been conducted on the Elmer breakwater scheme by Ilic et al. (1999). The Elmer scheme in Sussex, UK consists of a re-nourished macro tidal beach made up of a shingle beach and a lower sand beach overlying a chalk platform. The beach is protected by a system of eight offshore breakwaters and a terminal groyne, and is two kilometres in length. This system has undergone much research e.g. actual field measurements by Ilic and Chadwick (1995), physical modelling at the UKCRF by Simons et al. (1995), and numerical modelling by Li (1994). Salients were predicted but instead tombolos formed behind the breakwaters. The modelling of the flow and sediment transport is complicated by the fact that the structures themselves are permeable. Wave transmission is thought to play a key roll in the formation of salient/tombolo structures. Studies of this system have concentrated on the behaviour associated with one complete beach cell. That is, the area between two breakwaters with two outlying breakwaters. This has been

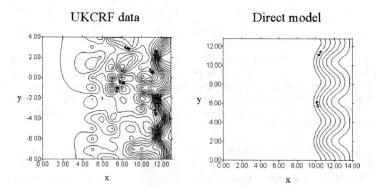

Figure 3: UKCRF waveheight data for a cuspate
beach compared with direct model calculations.

modelled at the UKCRF; see Figure 4 (figures supplied by S Ilic). Wave height and scattered flow measurements are available for fixed bed experiments. It was noticed when studying one breakwater cell; see Figure 4 (bay 3), that a single rotating eddy forms at the entrance to the bay instead of two eddies as expected. This is thought to be due to the partially transmitting property of the breakwaters. Data from mobile bed measurements will soon be available. Again, although good data exists, it is difficult to use it with a direct model. Thus, the idea is to assimilate the data into a solution using an inverse model.

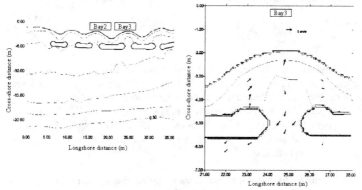

Figure 4: Elmer breakwater scheme

Inverse Modelling

Many inverse methods employ a variational approach in which the object is to find a solution which is close to the observed values whilst also satisfying imposed constraints. In essence, inverse methods seek to create a model in which the discrepancy between available observed data values and their model counterparts is minimized. The classical variational approach involves constructing a system of Euler-Lagrange equations, the solution of which provides the desired mimimization. However, the formulation of the Euler-Lagrange equations is often very difficult and the solution can be computationally intensive. Alternatively, the problem can be addressed by a more accessible method, namely *direct minimization*, Ghil and Mosebach (1978). This method has been used successfully in the modelling of tidal flows in coastal waters, Copeland and Bayne (1998).

Direct minimization methods involve minimizing a cost function by using an iterative numerical algorithm. The model equations, equations (1) – (4), are cast in a finite difference form and then the residuals for each equation are defined. These are then incorporated in a least-squares sense into a cost function. The cost function is a measure of the distance of the calculated solution from the physical solution represented at scattered points by actual data values. The constrained cost function for the 2DH model is given by

$$CC = \sum_{i=1}^{NX}\sum_{j=1}^{NY}\left\{ wmK_{i,j}smK\left[\left(rK_{i,j}^{x}\right)^{2} + \left(rK_{i,j}^{y}\right)^{2}\right] + wmA_{i,j}sA\left(rA_{i,j}\right)^{2} \right.$$
$$\left. + wm\eta_{i,j}sm\eta\left(r\eta_{i,j}\right)^{2} + wmUV_{i,j}smUV\left[\left(rU_{i,j}\right)^{2} + \left(rV_{i,j}\right)^{2}\right]\right]$$
$$+ \sum_{i=1}^{NX}\sum_{j=1}^{NY}\left\{ wdA_{i,j}sA\left(A_{i,j}^{d} - A_{i,j}\right)^{2} + wdUV_{i,j}sdUV\left[\left(U_{i,j}^{d} - U_{i,j}\right)^{2} + \left(V_{i,j}^{d} - V_{i,j}\right)^{2}\right]\right\} \quad (10)$$

where NX and NY are the number of grid points in the x and y directions respectively. The weights on the model and data are signified by wm and wd. Scaling of each variable, where appropriate, is achieved by the factors signified by sm and sd. Residuals of the model equations (1)-(4) are given by rK^{x}, rK^{y}, rA, rU, rV and $r\eta$. Available data values are given by A^{d}, U^{d}, and V^{d}. $smK=T_{off}(gd_0)^{1/2}$ and $smUV=T_{off}$ and $sdA=1/T_{off}$, where the offshore wave period is T_{off} and offshore depth of domain is d_0.

The direct minimization procedure continues by taking the derivatives of the cost function with respect to each variable at grid points (i,j). By illustration, it is instructive to consider a much-reduced system based on the wave model plus set-up (from the U momentum equation). The governing partial differential equations are

$$\frac{\partial\left(QA^{2}\right)}{\partial x} = 0, \quad (11)$$

$$Q\frac{\partial K^{x}}{\partial x} + RK^{x}\frac{\partial d}{\partial x} = 0, \quad (12)$$

$$\rho g d\frac{\partial\eta}{\partial x} + \frac{\partial S^{xx}}{\partial x} = 0, \quad (13)$$

where shallow water assumptions have been made i.e. $Q=(gd)^{1/2}$ and $R=(gd)^{1/2}/2$. In finite difference form we have

$$rA_{i,j} = \frac{Q_{i,j}A_{i,j}^{2} - Q_{i-1,j}A_{i-1,j}^{2}}{dx} = 0, \quad (14)$$

$$rK_{i,j}^{x} = \frac{Q_{i,j}K_{i,j}^{x} - Q_{i-1}K_{i-1,j}^{x}}{dx} + R_{i,j}\frac{\left(K_{i,j}^{x} + K_{i-1,j}^{x}\right)\left(d_{i,j} - d_{i-1,j}\right)}{2dx} = 0, \quad (15)$$

$$r\eta_{i,j} = \rho g d_{i-1,j}\frac{\left(\eta_{i,j} - \eta_{i-1,j}\right)}{dx} + \frac{\left(S_{i,j}^{xx} - S_{i-1,j}^{xx}\right)}{dx} = 0, \quad (16)$$

where

$$S_{i,j}^{xx} = \frac{1}{2}\rho g A_{i,j}^{2}. \quad (17)$$

It is important to properly specify the boundary values of the residuals. In the spatial domain two types of boundary condition are possible:

1. The residuals are zero, Dirichlet boundary conditions. Generally the gradients in this case are non-zero, thus allowing possible material transfer across the boundary. This condition is appropriate for open boundaries and allows variables to be adjusted at the borders towards a solution.
2. The gradients are zero, Neumann boundary condition. This condition is appropriate at closed boundaries since zero gradients imply no adjustment and hence specified boundary values in the variables are maintained e.g. at a no-flow boundary.

The gradients of the constrained cost function (CC) can now be calculated. Note, that account should be taken of occurrences of the variables at locations represented by indices prior and subsequent to (i,j). The gradients are given

$$\frac{\partial CC}{\partial A_{i,j}} = 4Q_{i,j}A_{i,j}\frac{\left(rA_{i,j} - rA_{i+1,j}\right)}{dx}, \quad (18)$$

$$\frac{\partial CC}{\partial K_{i,j}^{x}} = 2rK_{i,j}^{x}\frac{1}{dx}\left[Q_{i,j} + \frac{R_{i,j}}{2}\left(d_{i,j} - d_{i-1,j}\right)\right]$$
$$+ 2rK_{i+1,j}^{x}\frac{1}{dx}\left[Q_{i+1,j} + \frac{R_{i+1,j}}{2}\left(d_{i+1,j} - d_{i,j}\right)\right], (19)$$

$$\frac{\partial CC}{\partial \eta_{i,j}} = 2r\eta_{i,j}\frac{\rho g d_{i-1,j}}{dx} - 2r\eta_{+1i,j}\frac{\rho g d_{i,j}}{dx}. \quad (20)$$

These gradients are used to find the minimised cost function and hence the optimum values for the variables as described below.

Solution Methodology

Minimization of the cost function starts with a first guess of the model variables used to determine the residual values. These residuals are then used to determine the gradients associated with the first guess. The minimization then proceeds iteratively with new estimates for the variables to determine new residual values. The process repeats until the desired convergence criterion is achieved. The solution algorithm comes from a class of gradient descent methods. The simplest method of steepest descent employs the gradients directly to 'nudge' the variables towards the minimum e.g.

$$A_{i,j}^{k+1} = A_{i,j}^{k} - \varepsilon\frac{\partial CC^{k}}{\partial A_{i,j}}, \quad (21)$$

This is, however, not very efficient and frequently results in inappropriate solutions. The model will be developed to use a conjugate gradient algorithm (e.g. CONMIN, Shanno and Phua (1976) or the Brent, Pollack-Ribiere method, Press et al (1992)). The choice of 'first guess' is important. This set of values given to the variables at the out-set of the minimization procedure can determine the effectiveness of the minimization algorithm and the final solution. A good initial guess based on the physical background of the problem is required. Interpolating the available data across the domain can do this.

Particle Tracking
Particle tracking model
The time dependent movement of an individual particle requires the fluid velocities at the particle location. These time dependent velocities are found with a quasi-3d model, by coupling a 1DV model to the horizontal flow field. A computationally efficient alternative to solving a series of 1DV models at every horizontal grid location and interpolating between these for the particle position, is to move the 1DV model with the particle. The assumption is made that the waves have a constant form over the length scale of a particle displacement over one wave period. Formally:

$$\frac{\partial}{\partial t} = -\frac{\sigma}{|K_h|}\frac{\partial}{\partial x_i}, (22)$$

where σ is wave frequency and K_h is the wave number at the particle position, x_i is the horizontal vector in the direction of wave propagation. Within a wave orbit, the movement can therefore be modelled simply in terms of a phase shift. At the end of each orbit the profile is moved and the new wave form centred at this location is used for the subsequent wave orbit. This is illustrated in Figure 5. At each new profile location, the values from the 2DH model are found by linear interpolation between grid points. These new values are used directly in the following iterations of the 1DV profile at this point. This limits the rate at which the profile can move through the changing 2DH hydrodynamic field.

The 1DV model is illustrated in Figure 6. The model domain is between the level of the wave troughs to a distance z_o, where z_o is the roughness length, above the bed. The total depth at the location of the profile is the sum of depth below MWL, and the set-up η_h.

The assumption is made of hydrodynamically rough flow without bedforms. This allows the approximation of roughness length, z_o being set to $z_0=d_{50}/12$, where d_{50} is the median grain diameter, Krecic & Hanes (1996), Soulsby (1997). This can be changed within the model to simulate different bed compositions.

Governing equations
The 1DV model closely follows that of Elfrink et al. (1996). The model equations, neglecting convective terms are:

$$\frac{\partial u}{\partial t} = -\frac{1}{\rho}\frac{\partial p}{\partial x} + \frac{1}{\rho}\left[\frac{\partial \tau_{zx}}{\partial z} - \frac{\partial \overline{\tau}_{zx}}{\partial z}\right], (23)$$

$$\frac{\partial v}{\partial t} = -\frac{1}{\rho}\frac{\partial p}{\partial y} + \frac{1}{\rho}\left[\frac{\partial \tau_{zy}}{\partial z} - \frac{\partial \overline{\tau}_{zy}}{\partial z}\right], (24)$$

where u,v are velocity components in the x and y directions respectively, p is the horizontal pressure, ρ is the density of sea water and τ is the shear stress in the x and y direction as indicated by the subscripts. The overbar on the second shear stress term indicates a phase-averaged value.

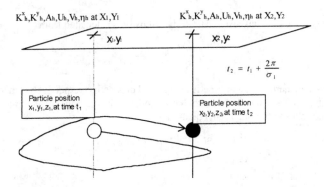

Figure 5. Movement of the profile position. The variables, denoted with a subscript h above from the 2DH model are re-evaluated every wave orbit. U an V are depth averaged velocity components (in x and y direction respectively). K^x and K^y are wave number components, A is wave amplitude and η is water level set-up. σ_1 is the wave frequency at X_1, Y_1. Over each complete cycle the wave form is assumed to be constant.

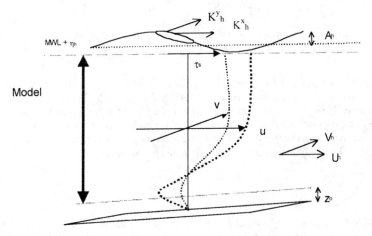

Figure 6. 1DV model domain.(Note: not to scale- height of z_o greatly exagerated)

Using an eddy viscosity approximation the time dependent shear stress term can be expressed as

$$\tau_{zx} = \rho v_t \frac{\partial u}{\partial z},\,(25);\tau_{zy} = \rho v_t \frac{\partial v}{\partial z}\,(26)$$

where v_t is the eddy viscosity. The eddy viscosity is calculated from:

$$v_t = c_1 \frac{k^2}{\varepsilon},\,(27)$$

where k and ε are turbulent kinetic energy and dissipation respectively. k and ε are found with a two equation turbulence model with equations (neglecting convection as with momentum equations):

$$\frac{\partial k}{\partial t} = \frac{\partial}{\partial z}\left(\frac{v_t}{\sigma_k}\frac{\partial k}{\partial z}\right) + v_t\left\{\left(\frac{\partial u}{\partial z}\right)^2 + \left(\frac{\partial v}{dz}\right)^2\right\} - \varepsilon,\,(28)$$

$$\frac{\partial \varepsilon}{\partial t} = \frac{\partial}{\partial z}\left(\frac{v_t}{\sigma_\varepsilon}\frac{\partial \varepsilon}{\partial z}\right) + c_{\varepsilon 1}\frac{\varepsilon}{k}v_t\left\{\left(\frac{\partial u}{\partial z}\right)^2 + \left(\frac{\partial v}{dz}\right)^2\right\} - c_{\varepsilon 2}\frac{\varepsilon^2}{k},\,(29)$$

The pressure terms in equations (23) and (24) are calculated using linear wave theory. The pressure term is therefore:

$$-\frac{1}{\rho}\frac{\partial p_i}{\partial x_i} = A\sigma^2 \frac{\cosh[|K_h|(d+z)]}{\sinh(|K_h|d)}\sin(|K_h|x - \sigma t),\,(30)$$

where A is wave amplitude, σ is frequency, K_h is wave number, d is total depth, and z is the depth of evaluation, measured upwards from the mean water level.

The distribution of the time mean component of the shear stress in the direction of the wave propagation is assumed to be linear, as determined by Deigaard and Fredsøe (1989). The time mean shear stress at the top of the profile due to wave breaking is calculated from the assumption of the surface roller model (*ibidem*). Assuming that the energy dissipation at the bed is negligible compared to that due to wave breaking the surface shear stress, τ_s can be expressed as:

$$\overline{\tau}_s = \frac{\overline{D}}{c} = \frac{1}{c}\frac{dE}{dr},\,(31)$$

where D is the energy dissipation rate per m^2 of sea surface due to the surface roller, and c is the wave celerity. E is the wave energy density, r is distance in direction of propagation. The wave breaking dissipation is calculated from the graidient in total wave energy in the direction of propagation. This assumes that energy losses due to bottom friction are negligible in comparison with the loss due to breaking. Linear wave theory is used for the calculation of the energy gradient. As in the model of Rakha (1998), the bottom shear stress is left as a free parameter. Its value is found by iteration in the model solution with the requirement that the phase-averaged velocities of the 1DV model are equal to U_h and V_h. The vertical component of the free stream velocity is calculated from the continuity equation:

$$\frac{\partial u}{\partial x} + \frac{\partial v}{\partial y} + \frac{\partial w}{\partial z} = 0, \ (32)$$

under the assumption of a constant waveform.

Particle movement

The instantaneous fluid velocity at the particle position is found by adding a turbulent fluctuation to the free stream velocity. The free stream velocity comes directly from the 1DV model. Linear interpolation is used between grid points. The turbulent fluctuations are calculated from the t.k.e. at the particle location, again directly from the 1DV model. Between the bottom grid point and the bed, a logarithmic wall law is assumed so that:

$$|\vec{u}_{h1}| = \frac{u_*}{k} \log\left(\frac{z_1}{z_o}\right), \ (33)$$

where u_* is the shear velocity, u_{b1} is the horizontal components of velocity at the first grid point, k is von Karman's constant (0.4) and z_1 is the height of the first grid point above the bed. Initiation of motion is calculated by evaluating the critical shear stress from an algebraic formulation of the Shield's curve, Soulsby and Whitehouse (1997). The instantaneous bed shear stress, τ_b is:

$$\tau_b = \rho.u_*^2, \ (34)$$

If the instantaneous shear stress exceeds the critical shear stress, particle motion is initiated. Once mobile, the particle moves in one of two ways:
1. In contact with the bed using an approximation based on the work of White and Schulz (1977) and Krecic and Hanes (1996):

$$\frac{d\vec{x}}{dt} = 2\vec{u}_* + \vec{w}_p + \vec{w}_s, \ (35)$$

2. In suspension (once the upward fluid motion exceeds the settling velocity), with assumption that particle inertia and lift forces are neglected:

$$\frac{d\vec{x}}{dt} = \vec{u}_p + \vec{w}_s, \ (36)$$

where u_p is the complete fluid velocity vector, including turbulent component, at the particle position, and w_s is the settling velocity. w_p denotes the vertical component of the local fluid velocity. A more robust treatment of the particle interaction with the flow would be preferable, see for example the work of Krecic and Hane (1996). The approximations above are used at present mainly for computational reasons, as more detailed particle fluid equations must be solved at every time step. Grain interaction is also neglected at present.

Morphological change

The particle position at each time step is saved to give a time series of positions. A particle that is stationary on the bed is counted at that position repeatedly. The particles are assigned to the nearest horizontal grid square for the occurrence counts. A measure of deposition or erosion is made relative to the total number of iterations multiplied by the number of particles started at each grid point. (i.e. a grid area in which the particles spend the whole simulation not moving and no new particles enter). This method does not directly provide a measure of the mass of sediment transferred. A coefficient is required to equate each particle with suitable mass of sediment. This will require some assumptions about the sediment packing at the bed, followed by calibration with field data. Calibration against a Eulerian transport model under well-tested hydrodynamic conditions would also be a suitable method.

Particle tracking results

Figure 7 shows results from a model run with a grain size of 0.5mm. The lower section shows a map of particle occurrence counts. The model was run on tank scale so the results cannot be interpreted as representative of full scale as the sediment grains are not scaled. The upper part of the figure shows an example of a particle track. This expanded view of the domain, seen from above shows the time series of particle x,y positions.

Conclusions

An inverse model of wave-current interaction has been developed. The model is designed to assimilate data from physical modelling experiments carried out at UKCRF. At present a full implementation of the model is yet to be achieved. A more efficient minimization algorithm will improve model performance. Initial results from the particle tracking methodology suggest that it may be a useful alternative to the Eulerian methods in investigating morphological change.

Acknowledgements

This work was supported through funding by the Engineering and Physical Sciences Research Council (EPSRC), grant nr. GR/K48303. The authors would like to thank Alistair Borthwick at Oxford University (rip-currents), Suzanna Ilic at Lancaster University and Andrew Chadwick at Plymouth University (Elmer breakwater scheme) for supplying UKCRF data.

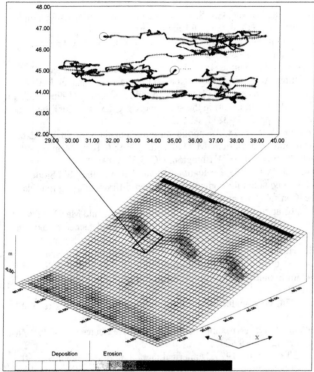

Figure 7. Particle occurrence map showing areas of erosion and deposition. Simulation run with grain size of 0.5mm. Wave and mean flow field calculated with direct model on tank model scale with incident waves directed directly onshore with wave number 0.6. The dark band at the far shoreward edge is due to a numerical boundary effect, and does not represent a zone of high erosion. The top section of the diagram shows particle tracks for an enlarged area indicated of one particle started at grid point 35,45. Crosses shown occur at intervals of approximately 0.04 seconds. Larger crosses indicate greater distance from bed. Model scale: x,1:0.25m ; y,1:0.2m

References

Battjes, J.A., 1968, Refraction of water waves, Journal of waterways, and Harbours Division, ASCE, 94, 437-451.

Borthwick, A.G.L, Y.L.M.Foote, A.Ridehalgh, 1997. Measurements at a cusped beach in the UK-CRF. Coastal Dynamics, 1997.

Copeland G.J.M., 1994. An inverse method of kinematic flow modelling based on measured currents. Proc. Instn. Civ. Eng. Wat., Marit. and Energy; Volume 106, 249-158.

Copeland, G.J.M. and Bayne, G.L.S, 1998, Tidal flow modelling using a direct minimisation method, Coastal Engineering, 34, 129-161.

Deirgaard,R. and Fredsøe,F. (1989) Shear stress distribution in water waves. Coastal Eng. 13(14), 357-378

Ebersole, B.A. R.A. Dalrymple, 1980. Numerical modelling of nearshore circulation, ASCE Proc. Conf. Coastal Engineering, Chapter 163.

Elfrink,B., Brøker,I., Deigaard,R., Hansen,E.A. and Justesen,P. (1996) Modelling of 3D sediment transport in the surf zone. Proc. 25th Int. Conf. on Coastal Eng., vol 4. pp. 3805-3817. ASCE

Ghil, M., Mosebach, R., 1978, Asynoptic variational method for satellite data Assimilation. In: Harlem, M. et al. (Eds.), NASA Tech. Memo. 78063. US Govt. Printing Office, Washington, DC, 3.32-3.49

Ilic, S., Chadwick, A.J., 1995, Evaluation and validation of the Mild Slope evolution equation for combined refraction-diffraction using field data. Proc. Coastal Dynamics 95, ASCE, 149-160.

Ilic S., Pan S., Chapman B., Chadwick A.J., O'Connor B., and MacDonald N.J. 1999 Laboratory measurements of flow around a detached breakwater 2000 scheme. Proc. Coastal Structures '99, Santander, Spain, ASCE.

Krecic, M.R, Hanes,D.M. (1996) An analysis of particle saltation dynamics. Proc. 25th Int. Conf. on Coastal Eng., vol 3. pp. 3847-3859. ASCE.

Li, B., 1994, An Evolution Equation for Water Waves, Coastal Engineering, 23, 227-242.

Noda, E.K., 1974, Wave-induced nearshore circulation, J. Geophys. Res. 79, 4097-4106.

Pechon, P. et. al., 1997, Inter-comparison of wave-driven current models, Coastal Engineering, 31, 199-215.

Pechon, P. and Teisson, C., 1995. Numerical modelling of three-dimensional wave driven currents in the surf zone. In: Proceedings 24th Conference on Coastal Engineering. Am. Soc. Civ. Eng., New York, NY. pp2503-2512.

Press, W.H., Teukolsky, S.A., Vetterling, W.T., Flannery, B.P., 1992, Numerical Recipes in Fortran, 2nd ed., Cambridge University Press.

Rahka, K.A. (1998) A quasi-3d phase-resolving hydrodynamic and sediment transport model. Coastal Eng., 34, 277-311.

Simons, R. R. et al., 1995, Evaluation of the UK Coastal Research Facility, Proc. Coastal Dynamics 95, ASCE, 161-172.

Shanno, D. F., Phua, K. H. 1976, Algorithm 500: minimization of unconstrained multivariate functions. ACM Trans. Math. Software 2 (1), 87-94.

Soulsby, R.L. (1997) Dynamics of Marine Sands, pub. Thomas Telford.

Soulsby,R.L. and Whitehouse,R.J.S.W. (1997) Threshold of sediment motion in coastal environment. Proc. Pacific Coasts and Ports '97 Conf., Christchurch. 1,149-54. University of Canterbury, New Zealand

Svendsen, 1984. http://earth.agu.org/revgeophys/holman01/node2.html

White,B.R. and Shultz,J.C.(1977) Magnus effect in saltation. J.Fluid. Mech. 81, 497-512.

Yoo, D. and B.A. O'Connor, 1986, Mathematical modelling of wave-induced nearshore circulation, Proc. 20th Int Conf. Coastal Eng, 2, 1667-1681.

Three-dimensional Modeling of Temperature Stratification and Density-driven Circulation in Lake Billy Chinook, Oregon

Zhaoqing Yang[1] (Associate Member, ASCE), Tarang Khangaonkar[2] (Member, ASCE), Curtis DeGasperi[3], and Kevin Marshall[4]

Abstract

A three-dimensional (3-D) hydrodynamic model was used in conjunction with a hydrothermal mass balance model to simulate the density driven circulation in Lake Billy Chinook, Oregon. Lake Billy Chinook reservoir, created by the construction of Round Butte Dam, receives flow from three tributaries (Crooked, Deschutes, and Metolius), that have distinct temperature and flow characteristics. The velocity pattern in the lake is dominated by temperature stratification and is also affected by wind stresses. The existing flow conditions are considered ineffective for collection and passage of downstream migrating fish. The immediate goal of the project is to identify the reservoir geometry modification(s) most favorable to the downstream passage of juvenile salmon through the project.

In this paper, we present application of a 3-D hydrodynamic model to evaluate the effectiveness of various proposed flow modification structures on reservoir stratification and circulation. The hydrodynamic model is externally coupled to a long-term seasonal temperature model to provide temperature initial conditions for the hydrodynamic model. This study demonstrated the

---

[1] Senior Engineer. Foster Wheeler Environmental Corporation, 10900 NE 8th Street, Suite 1300, Bellevue, WA 98004.
[2] Program Manager. Foster Wheeler Environmental Corporation, 10900 NE 8th Street, Suite 1300, Bellevue, WA 98004.
[3] Senior Engineer. Foster Wheeler Environmental Corporation, 10900 NE 8th Street, Suite 1300, Bellevue, WA 98004.
[4] Project Manager. Portland General Electric, 121 SW Salmon Street, Portland, OR 97204

effectiveness of these models in identifying the forcing mechanisms for the lake circulation patterns and assisting the design of fish passage facilities.

Introduction

The construction of Round Butte Dam and creation of Lake Billy Chinook on the Deschutes River in Central Oregon affected the downstream migratory fish passage of juvenile salmonids. Fish passage efforts were discontinued in the late 1960s due to ineffective downstream passage after the dam was constructed in 1964. As part of the current re-licensing process, Portland General Electric (PGE) is attempting to re-establish fish passage at Round Butte Dam. The passage problem has been associated with ineffective collection of downstream migrating smolts, whose motion is strongly influenced by the currents. The complex current patterns are attributed to reservoir stratification and mixing of inflows from three tributaries (Crooked, Deschutes, and Metolius Rivers) with distinct temperature characteristics. The Metolius River typically supplies the coldest inflow, which tends to plunges deep into the bottom layer of the reservoir and is withdrawn from the deep powerhouse intake at the forebay. The water from Crooked River is warmer throughout the year due to the influence of thermal springs. The warm Crooked water tends to travel downstream in the surface layer and moves upstream to the Metolius River. These distinct inflow characteristics result in a two-layer circulation system in the reservoir. The temperature stratification in the lake is mainly controlled by solar radiation and inflow temperatures. The reservoir turns over in early winter and similar inflow temperatures result in a fully mixed uniform and cold reservoir during the winter months. The reservoir becomes highly stratified in the summer. The magnitude of the temperature stratification has direct impact on the circulation patterns in the reservoir and therefore affects the downstream fish migration.

A solution to this problem may be to use flow modification structures to reduce the extent of stratification in the lake and modify flow patterns such that flow to the forebay is enhanced. This would be accomplished by designing structures such as curtains or surface intakes. The first step in designing structures to improve the success of downstream salmon passage is the development of a predictive reservoir hydrothermal and hydrodynamic model. Such a model would predict the response of the reservoir to each structural or operational modification. The 3-D numerical model used in this study was the Environmental Fluid Dynamic Code (EFDC) developed at the Virginia Institute of Marine Science (Hamrick 1992). The vertical 2-D Box Exchange Transport Temperature Ecology Reservoir (BETTER) model developed by the Tennessee Valley Authority (Bender et al., 1990) was used to simulate the long-term seasonal temperature distribution and provide temperature initial conditions for the hydrodynamic model. Numerical drifter experiments were also conducted using the Lagrangian particle tracking module in EFDC to investigate the over all water movement in the lake.

This paper presents the application of the hydrodynamic model of Lake Billy Chinook in support of the PGE's studies of the feasibility of fish migration enhancement in Lake Billy Chinook. Results demonstrate the model's ability to reproduce existing flow patterns and temperature. The potential effects of proposed structural modifications such as curtains, reservoir drawdown, and relocation of the intake to the reservoir surface are also presented.

Hydrodynamic Model Setup

Model Description

The numerical model used in this study was the Environmental Fluid Dynamic Code (EFDC) developed at the Virginia Institute of Marine Science (Hamrick 1992). The model is designed for the simulation of flows and transport processes in estuaries and coastal oceans, as well as reservoirs, lakes, and rivers. The model is a time domain, finite difference model. It solves the three-dimensional primitive equations of motion for turbulent flow. Three-dimensional transport equations for the turbulent intensity and length scale as well as temperature, salinity, dye tracer, and suspended sediment can also be solved simultaneously in EFDC. A second-moment turbulence closure model is solved to provide the vertical turbulent eddy viscosity in the model (Mellor and Yamada 1982). Horizontal diffusion is calculated using the Smogarinsky formula (Smogarinsky 1963). The model uses boundary-fitted curvilinear-orthogonal coordinates in the horizontal plane and a sigma-stretched coordinate system in the vertical direction. The computational scheme uses an external-internal mode splitting to solve the horizontal momentum equations and the continuity equation on a staggered grid, using a combination of finite volume and finite difference techniques. For a more detailed description of EFDC, the reader is referred to Hamrick (1992) and Hamrick and Wu (1996).

Model Grid

The Lake Billy Chinook is a 561×10^6 m^3 reservoir consisting of three narrow and deep river tributaries that converge near the forebay. The lake has steep slopes and the maximum water depth is about 110m. The model grid was constructed in a horizontal curvilinear-orthogonal and vertical sigma-stretched coordinate system and contains 493 cells in the horizontal plane (Figure 1). The average grid cell size in the forebay region was about 90m by 90m.

A total of 10 layers were specified in the vertical direction. Because the epiliminion is mainly confined within the upper 30 m, finer vertical layers were specified in the upper 30m of the water column to capture the sharp temperature gradient. The vertical layering scheme had an exponential distribution such that the thickness of the vertical layers increased from the surface to the bottom. The percentage of each layer in the water column is given in Table 1. (Layer 1 corresponds to the bottom layer and the water depth is normalized to 1).

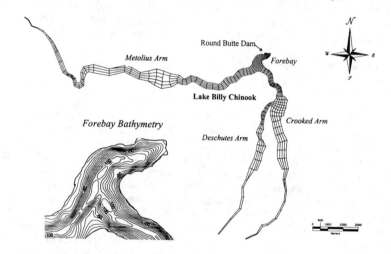

Figure 1: Hydrodynamic Model Grid for Lake Billy Chinook

Table 1: Normalized Vertical Layer Thickness Distribution
for the Lake Billy Chinook Hydrodynamic Model

Layer Number	1	2	3	4	5	6	7	8	9	10
Layer Thickness	0.23	0.18	0.14	0.11	0.09	0.07	0.06	0.05	0.04	0.03

Boundary Conditions

In Lake Billy Chinook, there are no horizontal open boundary conditions. The river inflows and power intake outflow were simulated as sources and sinks in the model system. The river inflow rates and temperatures were uniformly distributed in the water column at each of the three upstream boundaries. Because the power intake is located at the bottom of the reservoir (70m bellow the surface), the power intake outflow was only specified in the bottom layer of the forebay grid cell.

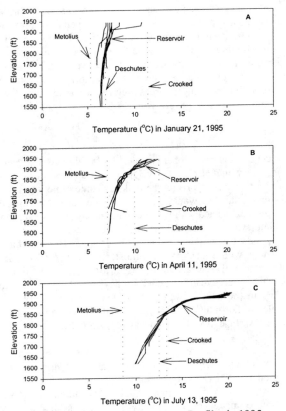

Figure 2: Reservoir Temperature Profiles in 1995

Figure 2 shows the comparisons of the river inflow temperatures and the reservoir temperatures at all observed stations. Below 30m from the water surface, the reservoir temperatures at all stations are very similar in any particular season. In winter, observed temperatures indicate the reservoir is almost fully mixed vertically. Compared to the reservoir water, the Metolius River water is colder and the Crooked River water is much warmer. This pattern has a strong influence on the reservoir temperature near the river inflows to the reservoir. The

Deschutes River temperature is much colder than the Crooked River temperature and is about the same as the mid-layer temperature in the reservoir.

In summer, the temperature stratification in the reservoir is greatly enhanced due to the surface warming. The surface temperature in the reservoir is much warmer. The horizontal temperature distributions at all water levels in the reservoir are almost uniform. The Metolius River temperature is colder than the bottom temperature in the reservoir and the Crooked and Deschutes river temperatures are colder than the surface temperature in the reservoir. All these seasonal differences in the temperature structures result in relatively different seasonal circulation patterns in the reservoir.

Initial Conditions

Any structural and operational changes that would alter the geometry and discharge conditions would affect the development of the stratification patterns in the lake. In order to correctly predict the hydrodynamic response of the lake the thermal behavior of the lake would also need to be correctly simulated. The computational costs associated with long period hydrodynamic model simulations with a small time step (20 seconds) are high. It is more efficient to utilize a thermal balance model to simulate the long-term temperature distribution in the reservoir separately, with a much larger time step (12-hour). The model output from the temperature therefore can be used to provide initial temperature conditions for the hydrodynamic model at any specific time. A review of temperature data in the lake showed that the temperature distribution patterns were uniform across the width of the lake arms and could be approximated with a laterally averaged, vertical, two-dimensional (2-D) representation.

The initial temperature conditions for the hydrodynamic model calculations in this study were obtained from a calibrated seasonal vertical 2-D temperature model (BETTER) that was setup specifically to operate in conjunction with the 3-D hydrodynamic model. The calibration and verification of the temperature model of Lake Billy Chinook is described elsewhere (DeGasperi, *et. al.*, 1999). The temperature model was applied independent of the hydrodynamic model for all simulations for the duration of entire year. The hydrodynamic model was applied for 7-day intervals at specific periods in the year critical to salmon passage using the predicted temperatures as the initial condition.

Model Performance

Because April is an important month for downstream fish passage, April 1995 was selected as the baseline period for model calibration. The bottom roughness, background horizontal diffusion, and vertical eddy viscosity were adjusted so that the general circulation pattern and temperature profiles in the reservoir matched the observations made in the previous studies (Raymond *et al.* 1997; McCollister and Ratliff 1996; Truebe 1996). The model was run for 6 days during each simulation and the model simulation time step was 20 sec. The background horizontal diffusion and background vertical eddy viscosity were adjusted within the typical range reported in the literature such that the model could successfully produce the vertical two-layer circulation in the system. The final adjusted background horizontal diffusion, background vertical eddy viscosity and bottom roughness were 1 m^2/s, 10^{-5} m^2/s and 5 cm, respectively.

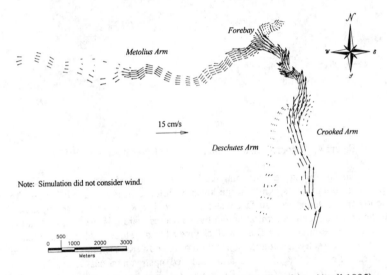

Figure 3: Surface Velocity Distribution for Existing Condition (April 1995)

Figure 3 shows predicted velocity vectors at the surface layer of the reservoir. Figure 4 shows velocity vectors along the vertical longitudinal sections in the Metolius, Deschutes, and Crooked arms of the reservoir. These model results confirm the prevailing understanding of the two-layer behavior of this system. Crooked River water is warmer than the surface water in the reservoir while the Metolius River water is cooler than the bottom water in the reservoir. So the water from the Crooked River enters the reservoir system in the surface layer, and the water from the Metolius River enters the reservoir in the bottom layer. This results in this the observed two-layer circulation pattern. The surface water from the Crooked River moves downstream towards the forebay and continues upstream into the Metolius River.

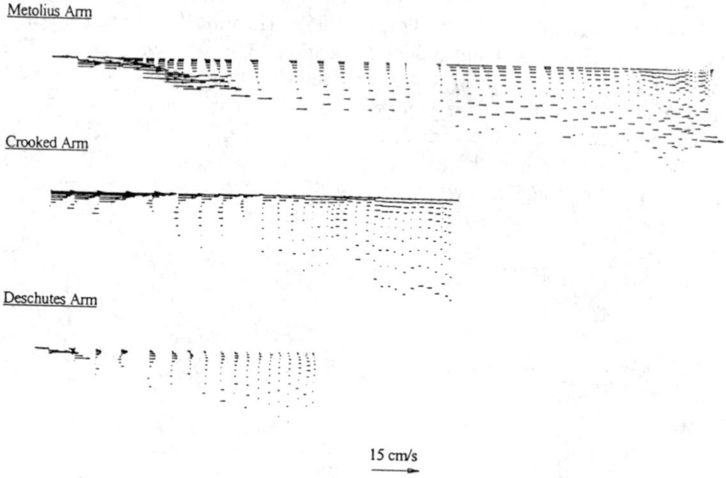

Metolius Arm

Crooked Arm

Deschutes Arm

15 cm/s

Figure 4: Velocity Distribution Along the River Arms (April 1995)

The predicted velocity magnitude in the Crooked and lower Deschutes arms was about 6 cm/s, while in the lower Metolius arm it was about 3 cm/s. In the lower Deschutes arm, surface water from Crooked River also moved upstream to the upper Deschutes arm, although the velocities were relatively small. In the forebay near Round Butte Dam, a clockwise eddy developed, which was mainly due to the geometry effect and the location of the powerhouse intake. The surface upstream flow in the Metolius arm extended upstream about 15km from the Forebay and the cold Metolius River water mainly moved downstream in the bottom layer. Due to the outflow at the powerhouse intake, relatively large bottom velocities at the location of powerhouse intake were observed. In the Crooked River, the two-layer circulation also developed from upstream. The vertical circulation in the upper Deschutes arm was slightly different from that in the Metolius and Crooked arms. There was a reduction in the velocity in the surface layer because the Deschutes River water was slightly cooler than the reservoir surface water but warmer than the reservoir bottom water in April.

Therefore, water from Deschutes River entered the reservoir system in the sub-surface layer with partial Crooked River water overlaying it. It is noted that there is a significant difference between the two-layer circulation in the Metolius and Crooked arms. In the Crooked arm, the surface velocity is much stronger than that in the bottom layer, while in the Metolius arm the bottom velocity is as strong as that in the surface layer. In the Crooked River, the surface water moved downstream towards the forebay, while in the Metolius, an upstream movement of water was predicted.

Lagrangian particle tracking was also conducted to investigate how the water from the Crooked and Deschutes rivers travels through the reservoir. Fifty neutral buoyant water particles were released in the lower Deschutes arm in the entire water column. Two days of particle movement were simulated by running the model for another 2 days using the restart file outputted at the end of the 6-day model run. Figure 5 shows the particle trajectories after 2 days of simulation. As shown in the figure, most of the surface particles moved up the Metolius arm and bottom particles moved in the upstream direction of the Lower Deschutes arm. Only a small fraction of the particles traveled into the forebay.

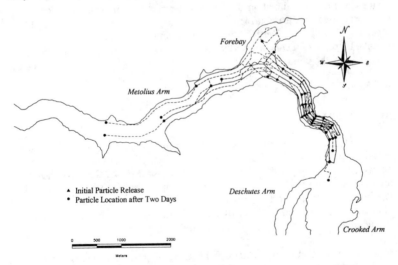

Figure 5: Particle Tracking Trajectories for Existing Conditions (April 1995)

Wind stress is an important factor in the observed circulation patterns in Lake Billy Chinook, especially in its effect on the surface current. Therefore, sensitivity of the model-predicted circulation in the lake to the effects of wind was investigated. Due to the limitation of the wind data in the study area, a designed wind pattern was generated based on statistical analysis of wind data at three wind stations (Forebay, Metolius River, and Deschutes River). At all three stations, the wind directions were mainly aligned in the directions of the reservoir arms. Based on these statistical results, a simple wind pattern given in Table 2 was used in the model to evaluate the wind effect. Wind forcing was assumed constant in the forebay, Metolius arm, and Deschutes/Crooked arms, respectively.

Table 2: Wind Data Used in the Model Sensitivity Study

	Forebay	Metolius River	Deschutes and Crooked Rivers
Wind speed (m/s)	10.0	2.5	2.5
Wind direction (degree)	270.0	247.5	225.0
Current speed without wind (m/s)	0.02	0.03	0.06
Current speed with wind (m/s)	0.08	0.08	0.11

As shown in Table 2, surface velocities in the entire lake increased and aligned in the same direction of the wind stresses. In the Metolius River, the wind-driven downstream currents were superimposed on the top of the upstream reversed flow, thus a vertical three-layer circulation was developed. In the Crooked and Deschutes arms, the surface downstream flows were strongly enhanced by the wind effect. Also, because of the strong wind in the forebay, the clockwise eddy was no longer evident, and the surface water in the forebay moved towards the Round Butte Dam.

Model Applications

As part of the current re-licensing process, PGE is attempting to re-establish fish passage at Round Butte Dam by modification of the reservoir geometry. The hydrodynamic model was applied to the proposed four alternatives to investigate the impact of the proposed reservoir modifications on the circulation patterns and provide data to aid in the selection of the best alternative for the improvement of downstream fish passage. The alternatives considered are shown in Figure 6 and are described below:

- **Alternative 1. Forebay Curtain:** Deep curtain placed in the reservoir forebay. Top of the curtain within 6 to 9 m of the water surface.
- **Alternative 2. Metolius Curtain:** Deep curtain placed in the mouth of the Metolius branch. Top of the curtain within 6 to 9 m of the water surface.
- **Alternative 3. Surface Withdrawal:** Surface withdrawal (1.5m below the water surface) **from the existing intake structure.**

- **Alternative 4. Reservoir Drawdown:** Permanent reservoir drawdown about 24 meters.

All the alternative simulations were conducted for the period of 4/11/95 to 4/17/95 and compared to the base simulation case (existing conditions in April 1995). New temperature initial conditions for each individual alternative run were obtained from independent application of the calibrated Lake Billy Chinook BETTER temperature model .

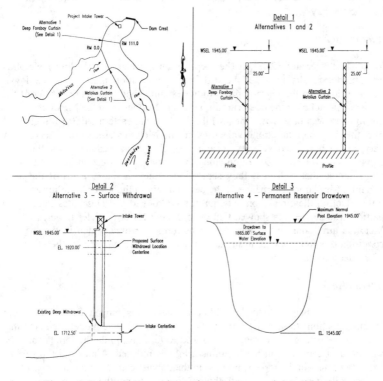

Figure 6: Proposed Four Alternatives for Reservoir Modifications

Alternative 1

To simulate Alternatives 1 and 2 with a curtain, the hydrodynamic model required some coding modifications. A thin barrier in the water column was successfully implemented in the code to simulate a curtain. The surface velocities near and behind the deep curtain in the forebay increased while the surface velocities in the lower arm of Metolius branch slowed down. However, velocity distributions in the Deschutes and Crooked arms remained almost the same compared to the existing condition. Overall two-layer circulation in all three rivers still persisted, except the surface velocities in the Metolius arm decreased. This indicated that the forebay deep curtain effectively reduced surface water movement up the Metolius arm. The particle tracking trajectories also showed that fewer water particles moved up the Metolius arm and more particles moved into the forebay.

Alternative 2

For Alternative 2, where the deep curtain was placed in the lower Metolius arm, model results indicated very different circulation patterns in the lower Metolius arm and forebay compared to the existing condition. At the center of the surface channel over the deep curtain, very strong flow from the Metolius arm to the forebay was observed. Meanwhile, in the southeastern side of the channel over the deep curtain, strong velocities up the Metolius arm were observed. A very strong horizontal velocity shear in the flow was predicted over the deep curtain. The velocity distribution along the vertical cross-section in the Metolius arm indicted that a distinct flow separation occurs at the location of the deep curtain. Similar to Alternative 1, the deep curtain in the lower Metolius arm had little effect on the circulation patterns in the Crooked and Deschutes arms. The particle tracking results showed that even though fewer particles traveled up the Metolius arm, some water particles went further upstream than in the existing condition due to the strong upstream surface velocities over the southeastern end of the deep curtain.

Alternative 3

In Alternative 3, the powerhouse intake was moved from the bottom (522m elevation) to near the surface (585m elevation). The model predicted that much more surface water from the Crooked River moved into the forebay and the surface upstream flow in the Metolius arm was reduced. Almost all the surface velocity vectors in the forebay pointed in the downstream direction because of the surface withdrawal. Compared to the existing condition, the surface downstream flows in Crooked and Deschutes arms increased slightly. Figure 7 shows results of particle tracking simulations for the surface withdrawal option. In comparison with the existing conditions (see Figure 5), particle tracking results for the surface withdrawal alternative showed that most of the surface water particles moved directly to the forebay.

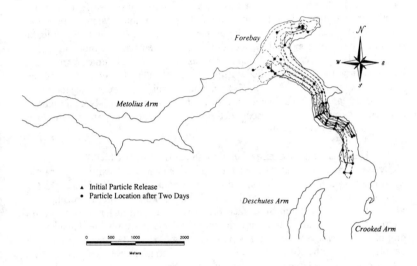

Figure 7: Particle Tracking Trajectories for Alternative 3

Alternative 4

For Alternative 4, many water grid cells for the existing condition become land cells because of the 24 meters permanent drawdown. Lowering of the reservoir also resulted in changes in the river inflow boundary locations. The new locations of the river inflow boundaries were moved farther downstream in all the tributaries. The general horizontal and vertical circulation patterns in the reservoir for alternative 4 were very similar to the existing condition, except the velocity magnitudes were increased because of the decreases of the reservoir cross-sectional area in all three branches resulting from the lower water depth. Also, surface water particles moved further upstream in the Metolius arm due to the increase in upstream surface velocities.

Discussion and Conclusions

A predictive 3-D hydrodynamic model of Lake Billy Chinook was developed to evaluate the feasibility of enhancing downstream fish passage at the Round Butte Dam. The objective of the model application was to develop an understanding of the hydrothermal behavior of the Lake Billy Chinook and to evaluate relative merits and demerits of various alternatives proposed. Each alternative affects the thermal balance of the entire reservoir. Since thermal balance and stratification in a lake are long-term processes, temperature simulations are required for an entire year. To perform these calculations with a 3-D hydrothermal model with a time step of 20 seconds is not effective for preliminary engineering analysis. In this study, the temperature modeling on an annual scale was conducted separately using a 12-hour time step. The hydrodynamic model was applied during specific months of interest over a 6-day period at a time with the initial temperature condition from the long-term temperature model. This de-coupling of temperature and hydrodynamic models allowed the models to be used cost-effectively, supporting engineering and management decision making.

The existing hydrothermal behavior in the Lake Billy Chinook is primarily the result of four different types of forcing: 1) temperature/density stratification, 2) wind stresses, 3) river inflows, and 4) power intake withdrawal. The effect of temperature-related density stratification is the dominant forcing mechanism. Stratification is a function of surface heating/cooling and different river inflow temperatures. Wind stresses are important and can result in enhanced or diminished current velocities, but wind stresses only affect the surface layer. The circulation driven by river inflows is very small due to the relatively small river inflow rates and the large lake cross-sectional areas.

For the existing condition, the hydrodynamic model results show that the two-layer circulation occurs in the Metolius River (upstream surface flow and downstream bottom flow) because the Metolius water is colder than the reservoir water all the time and plunges to the bottom of the reservoir and is withdrawn from the deep powerhouse intake. The warmer inflows from the Crooked and Deschutes Rivers ride on top of the cold Metolius River water and travel up the Metolius arm. In the Deschutes and Crooked arms of the reservoir, the pattern and strength of the vertical circulation depends on the different relationships between the river inflow temperatures and reservoir temperatures.

A screening level evaluation of four alternatives was conducted using neutrally buoyant particles as indicators of potential fish movement. The surface withdrawal, forebay curtain, and Metolius curtain alternatives appear to increase movement of water towards the forebay. Relative to existing condition, considerably higher numbers of drogues end up near the potential collector location in the forebay. Metolius curtain alternative also showed that it would effectively eliminate the undesirable upstream movement of salmonids. The permanent drawdown option resulted in increased velocities in the entire reservoir but did not change the pattern of movement of drogues relative to the existing conditions.

References

Bender, M.D., G.E. Hauser, M.C. Shiao, and W.D. Proctor. 1990. BETTER: A two-dimensional reservoir water quality model. Technical Reference and User's Guide. Tennessee Valley Authority Engineering Laboratory, Norris, Tennessee.

DeGasperi, C., T. Khangaonkar, Z. Yang and K. Marshall. 1999. BETTER Temperature Model of Lake Billy Chinook, Oregon. In: Water Resources into the New Millennium: Past Accomplishments, New Challenges. R. Walton and R.E. Nece (eds). American Society of Civil Engineer's International Water Resource Engineering Conference, August 8-12, 1999, Seattle, WA.

Hamrick, J.M. 1992: A Three-Dimensional Environmental Fluid Dynamics Computer Code: Theoretical and Computational Aspects. Virginia Institute of Marine Science, 317,VA. Special Report 317. 63pp.

Hamrick, J.M. and T.S. Wu. 1996. Computational Design and Optimization of the EFDC/HEM3D Surface Water Hydrodynamic and Eutrophication Models. In: Next Generation Environmental Models Computational Methods. G. Delic and M. F. Wheeler (eds). Proceedings in Applied Mathematics. 87.

McCollister, S.A. and D. Ratliff. 1996. Surface Currents in the Forebay at Round Butte Dam during Spring 1996. Pelton Round Butte Hydroelectric Project. FERC No. 2030. Portland General Electric Company, Portland, OR. 28 pp.

Mellor, G.L. and T. Yamada. 1982. Development of a Turbulence Closure Model for Geophysical Fluid Problems. *Rev. Geophys. Space Phys.* 20: 851-875.

Raymond, R.B., J.M. Eilers, K.B. Vaché, and J.W. Sweet. 1997. Limnology of Lake Billy Chinook and Lake Simustus, Oregon. Final report prepared for PGE, Portland, OR. E&S Environmental Chemistry, Inc., Corvallis, OR.

Smogarinsky, J. 1963. General Circulation Experiments with the Primitive Equations. I. The Basic Experiment. *Mon. Weather Rev.* 91: 99-164.

Truebe, J. 1996. Water velocity measurements in Lake Billy Chinook using acoustic doppler current profiler taken during stratified and unstratified conditions. 1995 Report. Pelton Round Butte Hydroelectric Project. FERC No. 2030. Prepared for Portland General Electric Company, Portland, OR. Lakeside Engineering, Inc.,

Sources of Uncertainty for Oil Spill Simulations

Eric Anderson[1] and Henry Rines[1]

Abstract

The quality of the trajectory predictions for oil spill model systems is subject to the quality of the environmental (wind and current) input data. Wind and current data as used for inputs to these models are typically a combination of direct time series observations and forecasts (especially for winds), and results of numerical simulations (both for wind and water surface currents). Reviews of oil spill modeling systems (most recently Spaulding, 1995) describe the processes modeled by the systems, but typically do not describe in detail how variability in the forcing mechanisms is managed. This paper describes two sources of variability in wind and current forcing data and compares the relative uncertainty associated with these sources of variability. The first source of uncertainty is the variability of the wind field caused by the passage of weather systems. A stochastic trajectory approach samples the wind time history of a wind time series, capturing the spill trajectory forced by the time series, and sums the resulting sea surface prediction fields. A second source of variability is uncertainty of measurement for each wind speed and direction measurement within a particular wind record. Randomization of wind or current vector direction and speed mimics this variation in trajectory simulations. The paper defines the two sources of variability (weather system passage and local measurement error) and attempts to compare the relative contribution of each for example simulation cases. A real spill event is the basis for the discussions, but the focus of the paper is on the variability of the forcing elements (winds and currents), rather than on an exact or detailed analysis of this event.

---

[1] Applied Science Associates, Narragansett, RI 02882

Introduction

Predictions of the movement of oil spills in surface water bodies (e.g. rivers, lakes, estuaries, coastal environments) are dependent upon adequate representations of the environmental forcing. The accuracy of a numerical model's prediction of the surface and/or subsurface movement (advection and dispersion) of pollutants introduced into the environment comes directly from the major forcing mechanisms for the water movement. In a river, the major physical forcing is usually the fresh water flow. For an estuary, the major local forcing may be tidal. In an enclosed lake, wind forced flows often dominate. In coastal environments there is often an interplay of tidal and wind forced movement.

Since many oil spills occur in coastal and estuarine environments, there is a need to adequately describe the tidal and wind-forced movement of spilled oil. From a practical standpoint, the greatest interest in the short term (over the first few days of a spill) is in the prediction of the movement of the oil floating on the surface of the water. The predictions are used for evaluating and selecting response strategies in the management of the effort. It is important for the modeler to describe the most likely event and some measure of the uncertainty of his/her prediction so that the response effort is efficiently allocated. Given exact knowledge, the response effort could be directed to cover exactly those shoreline areas that are to be affected by surface oil. The trajectory modelling effort focuses on a useful description of the modelled future locations of surface oil for the use of the responders in getting deflection and collection equipment ahead of the leading edge of the oil. In practice, the spill trajectory is often computed with an underlying tidal hydrodynamics simulation and a wind-forced oil transport component that is computed directly from a percentage of the wind observations (history) and predictions for the next one to two days. Since the tidal fluctuations are defined for a particular time of study, much of the expected variability in the trajectory estimate comes from the wind forced part of the oil advection scheme. The example used in this paper is based on the North Cape oil spill event on the U.S. east coast for which we have data for a number of coastal wind stations and a reasonable collection of sea surface observations. We explore the uncertainty of model predictions of the observed sea surface movement of oil with various combinations of available coastal wind time series and tidally forced hydrodynamic model simulations. This spill has been modeled in detail as part of the Natural Resource Damage Assessment (NRDA) under contract to the U.S. government. French (1998 a,b,c) describes this modeling effort. The simulations for this paper are simplifications of the release, and so will not exactly replicate the modeling in French (1998 a,b,c). The simulations focus on surface oil trajectory of the main release of oil and are designed to illustrate model response to variation in forcing and resulting uncertainty in the results.

Study Area

Figure 1 shows an overview of the study area. The central estuary is Narragansett Bay, contained within the state of Rhode Island, on the U.S. northeast coast. The spill site is shown by a dark circle with a plus sign above the notation "Block Island Sound". The location of wind record sites used in the analysis are shown by open circles.

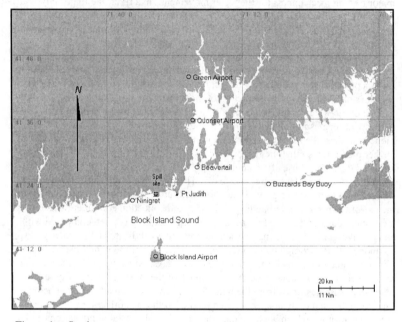

Figure 1:   Study area.

Spill Event

The 5,506 gross ton tank barge *North Cape* ran aground off Moonstone Beach, Narragansett, R.I. during a storm on the evening of January 19, 1996, and began releasing its cargo of No. 2 fuel oil into Block Island Sound (Figure 2) (OPB, 1996). The barge grounded near 41° 21' N, 71° 34' W about 4 hours after the tug *Scandia*, which was towing the barge, suffered an engine room fire about 6 km southwest of Pt. Judith, Rhode Island. The barge was carrying 94,000 barrels or about 4 million gallons of fuel oil. The spill commenced the evening of January 19, 1996, and over the next several days released an estimated 828,000 gallons. The exact release rate and schedule are unknown, but appeared to be related to the tidal stage, with higher release rates at lower tidal levels. The release appeared to cease by late in the day on January 21. The remaining cargo on the barge was successfully lightered over the next week and the barge removed from the beach on January 26, 1996. In the simulations for this paper, the release is assumed to occur at an even rate between 19:00 to 22:00 on 19 January. In reality, the release was uneven in time and over a longer time period. Thus, those simulations are for the initial release of oil the night of 19 January.

Wind Station Data

Figure 1 shows the locations for six wind time series that were collected as part of the study. Block Island Airport, Quonset Airport, and Green Airport are all coastal airports that take hourly observations. The Buzzards Bay Buoy is an offshore meteorological buoy maintained by NOAA/NODC, and also records hourly wind speed and direction. The Ninigret and Beavertail wind observation sites are privately maintained and are part of a network of coastal stations reporting through a wind service dedicated to board sailors: the Wind HotLine (http://www.windhotline.com/). A description of the implementation of these local wind time series for oil spill modeling is contained in Anderson and Atkinson (1996). Figure 2 shows wind stick plots of the reported speed and direction for each of the six stations for the time period January 19-29, 1996. The spill started at approximately 19:30 on January 19, 1996. The individual records show reasonable similarity for the major wind events, with some differences in wind speed and direction attributable to location and perhaps local topography.

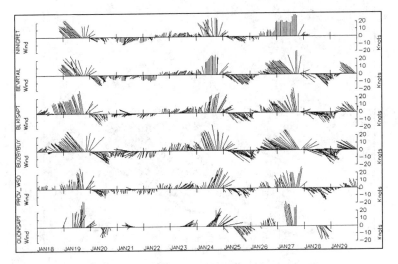

Figure 2: Wind stick plots for the six wind stations.

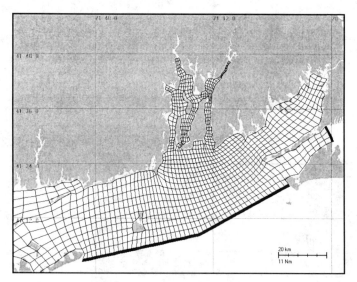

The wind event that brought the vessel ashore shows on the left of the records from mid-afternoon to early evening on 19 January. Quonset Point observations were not available through the night time hours, and this record was not used in the subsequent analysis. The station labeled PROV_WSO in the stick plot figure is located at Green Airport. The Ninigret and Beavertail (top two records) wind records were sampled with eight direction bins and show coarser angular definition than the other records which were recorded with 36 direction bins. The major wind energy, observable as the onshore wind signal from passing weather systems, can be seen in each of the wind records. The passage of three low pressure systems with strong onshore winds (January 19, 24-25, and 27) are shown in all six time series.

Tidal Current Predictions

Figure 3 shows the boundary-conforming grid and simulated maximum. Figure 4 shows the predicted M_2 constituent maximum flood flow vector field for the study area. The M_2 tidal prediction was created by ASA's WQMAP model and used for the prediction of the tidally forced water movement for the subsequent analysis. The semi-diurnal tidal components account for 70 to 85 percent of the tidal energy for the area of interest based on a comparison of predictions and observations for several tidal observation stations in the model domain Petrillo (1981). The flood is from east to west along the shore and bathymetric contours and the ebb west to east. Maximum flood and ebb speeds can reach 0.5 m/s.

Figure 3: Hydrodynamic grid for boundary conformal model.

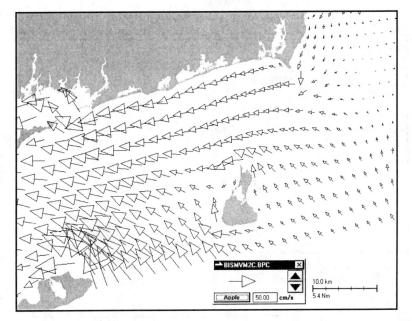

Figure 4: Maximum flood prediction, M_2 tidal constituent.

Simulation Comparisons

Figures 5 and 6 show the predicted 37 hour surface trajectory using the M_2 current field plus 3.5 percent of the wind for wind speed records collected at locations just west of the spill site (Ninigret) and east of the spill site (Beavertail). The black outline is the extent of surface oil observed by a NOAA overflight at the time of the prediction. The initial spill occurred just before high tide and reached the area of Point Judith the next morning. First landfall at Block Island was recorded on 21 January at 0800, about 37 hours after the first release. For all of the following simulations the release schedule is modeled as a uniform release over the first three hours of the spill and the trajectory is followed for 37 hours.

Spaulding et al. (1996) and Anderson and Atkinson (1996) discuss the comparison of model predictions with NOAA overflight observations. The oil spill model used is ASA's OILMAP model, Spaulding et al. (1992a,b). The focus here is on a description of the differences observed by using various wind records in concert with the M_2 tidal prediction.

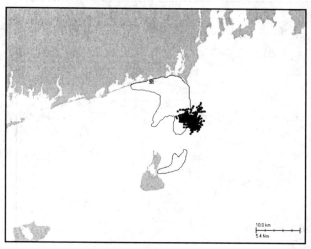

Figure 5:   Trajectory prediction, Ninigret wind, M_2 current forcing,
            37-hour simulation.

Figure 6:   Trajectory prediction, Beavertail wind, M_2 current forcing,
            37-hour simulation.

The combination of an ebb tide and late night winds from west to east forced the oil along the shore from the release point to Point Judith. Using the Ninigret winds (Figure 5), the predictions for the first 12 hours of the spill are very good, adequately describing the shoreline impacts along the coastline toward Pt Judith. The later eighteen hours of prediction lags observations, yielding an underestimate of the time it took for oil to reach Block Island. Using the Beavertail wind (Figure 6), the arrival time at Block Island is improved (still lagging by some seven hours) but the concentration of oil around Pt. Judith and the entrance to the estuary directly to the west of Pt. Judith is not predicted. Figure 7 uses the Ninigret wind for the first 12 hours of the prediction and the Beavertail wind for the later 18 hours for the "base case" simulation. Figure 8 shows the prediction based on the M_2 current and the Block Island Airport wind record. This prediction predictably gives a better estimate of the offshore movement of the surface oil but does not predict the impact on the Pt Judith area. Figure 9 shows the prediction using the M_2 tidal current and the Buzzard's Bay Buoy wind. This simulation again misses the Pt. Judith shoreline impact and overpredicts the offshore movement of the oil.

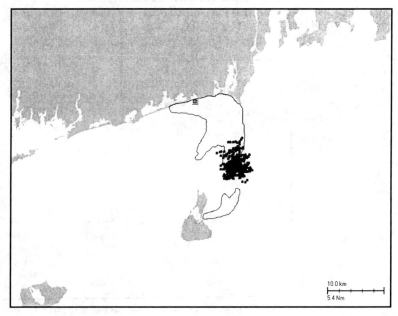

Figure 7: Trajectory prediction, Ninigret-Beavertail wind, M_2 current forcing, 37-hour simulation.

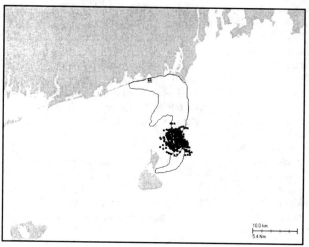

Figure 8:   Trajectory prediction, Block Island Airport wind, M$_2$
current forcing, 37-hour prediction.

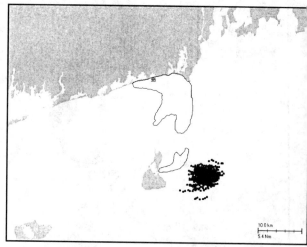

Figure 9:   Trajectory prediction, Buzzards Bay buoy wind, M$_2$
current forcing, 37-hour simulation.

An examination of the local upland topography shows that there is a ridge
extending along the coastline just north of the Ninigret wind station, which
shields both that station and the nearshore waters from northerly winds. It is our
hypothesis that the existence of this ridge and its shadowing effect for offshore

winds explains both the correct prediction of shoreline impacts at Pt. Judith as well as the limited offshore movement and subsequent underestimate of the arrival time of oil at Block Island. These are examples of what Lehr et al. (1999) refer to as a "fourth source of error, spatial variability in the wind." The first source cited was the unpredictability of frontal passages, the second the error in forecasts of weather models, the third that of representing the actual temporal variability of the wind for each element of oil being modelled.

Stochastic Simulations

We have shown a number of hindcast trajectory simulations using the various coastal and offshore buoy wind records available for the time of the incident. For performing these hindcasts, we have time available to collect and compare these records and to make some conjectures about the causes for the differences in the predictions based on them. In fact, the differences in the predictions among the five wind stations shown is not great. With the time constraints in effect during a real spill event, it is necessary to perform a prediction with some measure of the uncertainty in a robust and efficient way.

If we posit a spill from the same location during any January, we can get a representation of the "typical" January spill event from that location. Figure 10 shows probability contours of oil intersecting the water surface generated from 100 separate advection simulations using the Green Airport historical wind speed and direction time series (data from 1988-1995) and the same M_2 tidal hydrodynamics simulation as the simulations above. Each of the 100 simulations begins in January of one of the years for which the wind record has data (1988-1995) and continues for 37 hours. The higher likelihood paths for the oil trajectory show as the darker colors extending to the south and east and south and west of the spill site. The reader can see that the simulations in Figures 5 through 9 all fall within the stochastic contours. In addition, for wind events common to previous January periods, the envelope of possible trajectories is significantly longer. The analysis shows the expected envelope of trajectory simulations taken from the population of wind events sampled at Green Airport over the eight year time span. Turner and Spaulding (1988) reported cn a sensitivity analysis of similar trajectory simulations done off the coast of Alaska using variable wind record length and simulation numbers. They found that the expected "footprint" (outline of the lower probability contour) for analyses stabilized with sampling from a ten year wind time series, and with sufficient temporal samples to include all major weather system passage events.

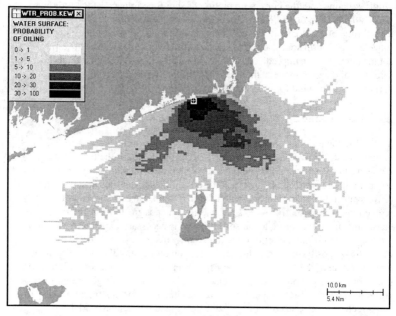

WTR_PROB.KEW ☒
WATER SURFACE:
PROBABILITY
OF OILING

0 -> 1
1 -> 5
5 -> 10
10 -> 20
20 -> 30
30 -> 100

10.0 km
5.4 Nm

Figure 10:   Probability of sea surface contact, Providence Airport January wind
and M_2 current forcing, 100 37-hour simulation.

Figure 11 shows a similar analysis done for 1000 spill scenarios from any
month of the year for the entire eight year wind record. Additional events from
other seasons spread the overall boundary of the "footprint." The 1000
simulations yields about the same sampling interval within the 8x12 month record
as do the 100 simulations for the 8x1 month record of Figure 10. As a general
rule, we should sample the record adequately to have scenarios from each of the
weather system passage events (typically 3-5 days) in every month used in the
wind record. This approach of sampling the historical record of a locally
collected wind time series gives the modeller and the spill response planning
personnel a good picture of the historical variability associated with a modelled
spill event for the particular site. Additionally, the comparison of the predicted
probability contours for the one month forcing (e.g., Figure 10) and the one year
forcing (e.g., Figure 11) shows something of the seasonal characteristics of the
wind time history for that location.

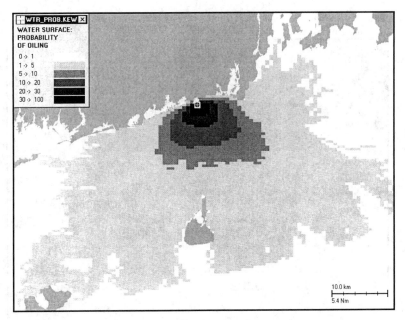

Figure 11: Probability of sea surface contact, Providence Airport year-around wind and M_2 current forcing, 1000 37-hour simulations.

Uncertainty Analysis

Another way to modify the wind forcing for the simulations makes the assumption that there is significant uncertainty about the measured or predicted speed and direction of wind and current forcing. To simulate this uncertainty, another class of particles is generated. For each time step of the simulation we add some uncertainty to each wind and current speed and direction. Figure 12 shows the trajectory outcome of the base case (Figure 7) with the random addition of plus or minus (+/-) 0-30 degrees direction and +/- 0-30% speed for each wind and current vector computation. These values are small as compared with Lehr et al. (1999), but no autocorrelation of these errors is assumed for this simulation. Lehr et al. (1999), indicated that the state of practice for that group, some persistence of speed and direction error is assigned to a particular surface particle for an undefined length of time. The dark gray particles represent these "uncertainty particles" in this presentation, and lead and lag the main body of predicted surface particles. The additional spatial coverage of the uncertainty particles is smaller than that of the weather system passage coverage from the stochastic model simulations. The variability of the local wind and current

measurement uncertainty is smaller than the variability of wind speed and
direction from weather system passage.

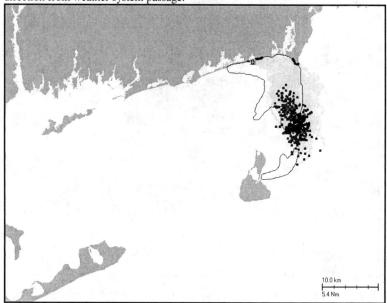

Figure 12:   Trajectory prediction: base case plus uncertainty (wind and current),
             37-hour simulation.

This comparison supports our assumption that in most cases the variability
in the weather system passage as represented by a wind time series record of
adequate length is larger than randomly distributed measurement or prediction
error for wind and current inputs.

Specific Inference

For this particular spill event, the strong onshore winds during the
grounding kept the spilled oil in the highly turbulent mixing regime of nearshore
breaking waves. The high turbulence mixed a large proportion of the oil into the
water column, and there were extraordinarily large mortalities of benthic shellfish
(French, 1998 a,b,c). Figure 13 shows the mass balance for the base case
simulation and the observed high mixing regime with 85% of the oil in subsurface
suspension or solution. If the weather system passage were retarded four hours
the major wind forcing would have been offshore during the large spillage.
Supposing that this offshore wind would keep a good proportion of the spilled oil
out of the wave-induced turbulent mixing nearshore area, the amount of entrained
oil would be markedly reduced. In this particular case, with a large lobster

nursery ground adjacent to the spill location, the ecological impacts resulting from the spill would have been significantly different. Figure 14 shows the standard weathering calculation for the event with the offshore wind pushing oil away from the nearshore turbulent mixing zone. In this simulation the model predicts less than two percent of the oil was mixed into the water column.

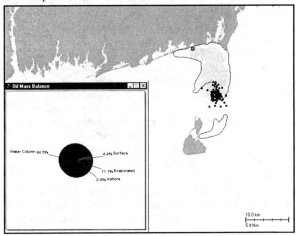

Figure 13: Trajectory and fates prediction: base case (37 hour) with local wave-forced entrainment.

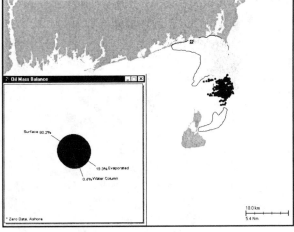

Figure 14: Trajectory and fates prediction: base camp (37 hour) with wind record retarded 4 hours.

Conclusions

The prediction of wind forced movement of surface oil particles in oil spill trajectory modeling has been addressed through the example of a well studied oil spill event in Rhode Island, USA. The predicted 37 hour trajectory using tidal hydrodynamics and wind forcing from five different land and buoy meteorological stations have been presented, as well as a stochastic analysis using an eight year long historical wind time series from a land station, and an uncertainty analysis technique which randomly varies input vector speed and direction. The stochastic analysis is shown both to contain the observed single event predictions and to show additional variability which might occur in another similar event. If trajectory modeling is to be useful for spill response planning input, the variability represented by the passage of weather systems should be included. The uncertainty analysis does not show a similar breadth of predicted trajectory "footprint." The use of local wind time series histories, which include local weather system passages in stochastic analyses appears to be a useful way to depict the possible variation in trajectory forecasts, and thus to aid in the planning for response to spill events.

Additionally, the high local turbulent mixing of breaking waves at the site of the spill, combined with an onshore wind which kept the spilled oil in the wave breaking zone, caused a high oil entrainment rate into the water column for the North Cape spill (French, 1998 a,b,c). Model simulations which change the weather system passage time by 4 hours show a large differential in predicted oil entrainment into the water column. The difference in the modelled oil behavior for this particular study points to the importance of the time of frontal passage to model predictions.

References

Anderson, E.L. and Philip Atkinson, 1996. Implementation of near real time shoreline wind station data for oil spill trajectory modelling: North Cape Spill, Rhode Island. AMOP 1996.

Galt, J.A., 1998 "Uncertainty Analysis Related to Oil Spill Modeling", Spill Science & Technology Bulletin, Vol. 4, No. 4, pp. 231-238, 1998.

French, D. P., 1998a. Estimate of Injuries to Marine Communities Resulting from the North Cape Oil Spill Based on Modeling of Fates and Effects. Report to NOAA Damage Assessment Center, Silver Spring, MD, January 1998.

French, D. P., 1998b. Updated Estimate of Injuries to Marine Communities Resulting from the North Cape Oil Spill Based on Modeling of Fates and Effects. Report to NOAA Damage Assessment Center, Silver Spring, MD, December 1998.

French, Deborah, 1998c. "Modeling the Impacts of the North Cape Oil Spill, Twenty-first Arctic and Marine Oilspill Program (AMOP) Technical Seminar, June 10-12, 1998, West Edmonton Mall Hotel Edmonton, Alberta Canada. p. 387-430

Golob's Oil Pollution Bulletin (OPB), 1996. Grounded barge causes major spill off Rhode Island coast, Vol. VIII, No. 2, January 26, 1996.

Lehr, W, C. Barker and D. Simacek-Beatty, 1999, New developments in the use of uncertainty in oil spill forecasts, Vol. I, Proceedings of the 22nd Arctic and Marine Oilspill Program (AMOP Technical Seminar, 24 June 1999, Calgary, Alberta Canada.Petrillo, A.F., 1981. Southern New England Coastal Sea Bottom Pressure Dynamics. Masters Thesis, Ocean Engineering Dept., University of Rhode Island, 1981.

Spaulding, M. L., E. Howlett, E. Anderson, and K. Jayko, 1992a. OILMAP a global approach to spill modeling. 15th Arctic and Marine Oil Spill Program, Technical Seminar, June 9-11, 1992, Edmonton, Alberta, Canada, p. 15-21.

Spaulding M. L., E. Holwett, E. Anderson, and K. Jayko, 1992b. Oil spill software with a shell approach. Sea Technology, April 1992. P. 33-40.

Spaulding, M.L., 1995. Oil Spill Trajectory and Fate Modeling: State-of-the-Art Review. Presented at the Second International Oil Spill Research and Development Forum, IMO, London, U.K. 23-26, May 1995.

Spaulding, M. L., T Opishinski, E. Anderson, E. Howlett, and D. Mendelsohn, 1996. Application of OILMAP and SIMAP to predict the transport and fate of the North Cape spill, Narragansett, RI. 19th Arctic and Marine Oil Spill Program, Technical Seminar, June 12-14, 1996, Calgary, Alberta, Canada, p. 745-776.

Turner, A. C. and M. Spaulding, 1988 "Influence of Wind Record Length on Oil Spill Trajectory Calculations," 1988. ASA report #1787.

Topographic and Wind Influences on
the Chesapeake Bay Outflow Plume and the Associated Fronts

Patrick C. Gallacher[1], Michael Schaferkotter[2] and Paul Martin[3]

Abstract

Several simulations of the low salinity plume at the mouth of the Chesapeake Bay have been conducted using the Navy Coastal Ocean Model (NCOM). This model is a limited-area model with multiply nested grids and a combination of z and sigma levels in the vertical. The nesting allows us to make high resolution simulations of the head of the plume while simulating the entire Chesapeake and Delaware Bays region at coarser resolution.

The simulations focus in particular on the influences of winds and topography on the plume, especially in the vicinity of Cape Henry and Virginia beach. When the wind is predominately northerly or easterly the plume is compressed along the coast and the southward transport of the plume is enhanced. Southerly winds generate an offshore Ekman transport. This transport pushes the plume offshore and upwelling occurs on the inshore side of the plume causing a wedge of salty continental shelf water to form between the plume and the shore. We also found that the wedge of salty water can form without wind on strong flood tides.

[1] Coupled Dynamics Processes Section, Oceanography Division, Naval Research Lab, Stennis Space Center, MS 70452, gallacher@nrlssc.navy.mil
[2] Sverdrup Technologies, Inc., Stennis Space Center, MS., 70452
[3] Coastal and Semi-enclosed Seas Section, , Oceanography Division, Naval Research Lab, Stennis Space Center, MS 70452

1. Introduction

The low salinity water discharged from the Chesapeake Bay can be identified in remotely sensed images at several frequencies as a plume and a series of fronts that contrast with the generally saltier and clearer continental shelf water. The plume moves outward on the ebb tide and stalls or migrates back into the mouth of the Bay on the flood tide. As the plume moves a series of fronts are generated. The formation and propagation of these fronts strongly affect the transport and distribution of materials on the continental shelf and beyond.

Advances in remotely sensed observations have allowed us to see the surface signatures of these fronts in great detail. Fronts associated with the plume can be seen in remotely sensed observations of sea surface temperature (SST), sea surface salinity (SSS), sea surface color and roughness. We need to understand the dynamics that generate and maintain these fronts in order to interpret the measurements and use them to aid in understanding and forecasting the coastal ocean.

We will report on a series of numerical simulations designed to study the generation and evolution of the fronts associated with the Chesapeake Bay outflow plume. The numerical simulations use a series of nested grids with increasingly high resolution. The large scale coarse grid encompasses the entire Delaware and Chesapeake Bays. This allows us to achieve the realistic large scale circulation in and outside the Bay. The high resolution inner grid allows us to focus on the submesoscale dynamics of the plume and the fronts. In this way a realistic plume develops in response to the riverine, tidal and wind forcing, and we can study the details of the fronts that develop at the head of the plume.

The model is described in section 2. The experiments are detailed in section 3. The results are given in section 4 and discussed in section 5. Conclusions are offered in section 6.

2. The Model

The model used in this study is the Navy Coastal Ocean Model (NCOM) developed by Paul Martin (see Martin, in this volume). NCOM combines a sigma coordinate system on top of a z coordinate system in the vertical. This allows some flexibility in setting up the vertical coordinate system to accommodate the large changes in depth common in continental margins and to have high vertical resolution near the surface to accurately model the vertical mixing. The horizontal grid is orthogonal and curvilinear. The time stepping technique is leapfrog with an Asselin filter to prevent time splitting. The spatial finite difference scheme is centered. The variables are staggered horizontally in a C grid and w is staggered vertically relative to the other variables. In the horizontal several parameterizations are available for the eddy viscosity. The vertical mixing uses a the Mellor-Yamada level 2 or 2.5 scheme with some enhancements (See the discussion by Martin in this volume).

The model allows multiple grids to be nested inside each other, each at higher resolution and smaller time step. At present the coupling between nested grids is one way, from the lower resolution to the higher resolution nest. The prognostic variables are interpolated from the coarse resolution nest to the finer resolution nest at the boundaries of the finer resolution nest. In this set of experiments three grids are used. The locations and resolutions of the grids are discussed in the next section.

The model accommodates realistic bottom topography and irregular coastlines. The bottom topography can be refined in each of the higher resolution grids. The model is forced at the surface by wind stress and heat fluxes. River inflows can be specified and tidal forcing is included.

3. The Experiments

For this set of experiments the model is forced with 8 tidal components, climatological river flows representing the 7 major rivers flowing into the Chesapeake and Delaware Bays and a surface wind stress. The wind stress is composed of a constant southward (toward the south) component of 0.1 dynes/cm^2 and a component of 0.4 dynes/cm^2 that rotates cyclonically with a period of 5 days. The latter is a crude representation of the weather systems that move across the Bays with approximately that frequency. The surface heat flux was set to zero, so the temperature is not a forcing term in these simulations. In these simulations we used a horizontal eddy viscosity composed of a constant minimum value and a variable maximum value computed to keep the grid cell Reynolds' number equal to 10. Using the option defines the maximum velocity that is allowed in the simulation.

Figure 1 Bathymetry of model domain. dx = dy = 2 km

The option of using the Smagorinsky formulation may weaken the horizontal gradients somewhat but it will not qualitatively alter the results.

The model domain extends from north of Delaware Bay almost to Cape Hatteras. It encompasses all of Chesapeake and Delaware Bays and extends eastward to the shelf break (Figure 1). The grid spacing of the coarsest domain is dx = dy = 2 km. The first nested grid includes the portion of the lower Chesapeake Bay south of Cherry Stone Inlet and east of Little Creek Cove. Part of the shelf is also included from north of the Bay mouth to south of French Point. The grid spacing on this first nested grid is dx = dy = 2/3 km (Figure 2). The second nested grid encompasses the mouth of the Chesapeake Bay south of Fisherman's Island, the portion of the lower Chesapeake Bay east of Lynhaven Bay and the part of the shelf at the mouth of the Bay and along Virginia beach to south of Rudee Inlet. The grid spacing on this nest is dx = dy = 2/9 km (Figure 3). The resolution of the bathymetry is approximately 2dx on each grid. Thus the resolution of the bathymetry on the high resolution grid is 400 m.

For the 2 nested grids, 11 sigma levels are used in the vertical with a logarithmic spacing from the surface to the bottom. For the coarsest domain, that included part of the continental slope to depths of greater than 2000 m, 21 levels were used, 11 sigma levels over 10 z levels.

The simulations are started from rest with constant temperature and a salinity field derived from the results of a 2 year spin up on the main grid (See the discussion by Martin, this volume). The simulations are run for 30 days. The results we will discuss are from analyzing the last 15 days of the simulation.

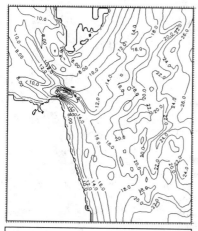

Figure 2. Bathymetry of nest 1. dx = dy = 2/3 km.

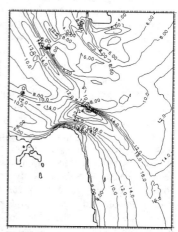

Figure 3. Bathymetry of nest 2. dx = dy = 2/9 km.

4. Results

In the first simulation the winds are set to zero. The plume exits the mouth of the Bay and turns south under the influence of the earth's rotation. The simulated outflow exhibits the classic surface signature of the Chesapeake Bay

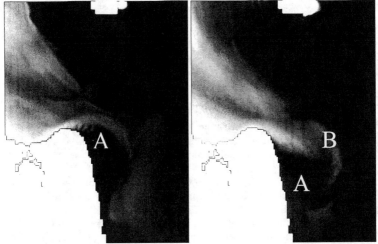

Figure 4. Sea surface salinity from simulation of Chesapeake Bay outflow plume during flood tide with no winds. Darker areas represent high salinity water, lighter areas represent lower

Figure 5. Same as Figure 4 except for the addition of the winds.

outflow plume. The plume forms an inertial bulge just outside the mouth of the Bay and the low salinity water flows south along the coast as a coastally trapped jet (Boicourt, 1973).

During flood tide the plume is pushed back toward the mouth of the Bay and occasionally all the way back into the Bay (Figure 4). When the flood tide is sufficiently strong a wedge of salty water appears east of Cape Henry, inshore of the plume (see point A in Figure 4). This is shelf water that appears between the plume and the shore. Both Figures 4 and 5 are from day 17 of the simulation.

A mass of low salinity water remains on the continental shelf during flood tide. This water propagates down the coast as a coastally trapped internal wave (see point B in Figure 4). In situ and remotely sensed observations have tracked the low salinity water south of the mouth of the Chesapeake Bay (Johnson, 1995) and at Duck North, Carolina (Rennie *et al.,* 1999) and as far south as Cape Hatteras (Berger, *et al.,* 1995).

When the wind forcing is added the plume is pushed offshore by winds southerly or westerly (see point B in Figure 5). Under these conditions the plume appears to dissipate during flood tide and reform on the ebb tide. Southerly winds are favorable for upwelling along the coast. Under these conditions a region of salty shelf water appears east of Cape Henry and inshore of the plume (see point A in Figure 5).

Northerly winds compress the plume against the coast. Under these conditions the plume propagates further south and maintains its integrity. In these wind conditions the plume forms a narrower, more coherent version of the "classic" inertial bulge and coastal current seen in the no wind case.

One purpose of these experiments was to attempt to simulate the formation and behavior of fronts seen in remotely sensed images from airborne real aperture radar (RAR) (Sletten *et al.,* 1999). These frontal signatures in RAR images result from the enhanced steepening and breaking of surface waves when they interact with the converging currents of the fronts (Marmorino *et al.,* 1999). To this end we calculated the horizontal gradient of SSS. We use regions of high SSS gradient as surrogates for convergence.

Fronts form on the ebb tide in the channel, over the steep slope on the north side of the hole in the channel northeast of Cape Henry (see the bathymetry in Figure 3), and east of Cape Henry (Figure 6). These fronts can be clearly seen form on ebb tide and propagate east and south. The fronts bend, kink and break as they propagate.

Adding wind forcing increases the frequency of the frontal formation and the intensity of the fronts (Figure 7). The fronts break apart more frequently with the wind forcing. More fronts are formed east of Cape Henry. During periods of southerly and westerly flow the fronts are advected further to the east in the case with winds

5. Discussion

The inner front of the inshore side of the plume was observed during two Chesapeake Bay Outflow Plume experiments (COPE I in September, 1996 and COPE II in May, 1997) (Marmorino, 1999). An apparently similar plume has been observed near the mouth of the Delaware Bay (Sanders and Garvine, 1996). We had initially hypothesized that the front was a result of upwelling during southerly wind events. However, when all the data were analyzed the wedge was found to exist for several wind directions. The simulations also generated a wedge

Figure 6. Sea surface salinity gradient from simulation of Chesapeake Bay outflow plume during flood tide with no winds. Bright areas represent high salinity gradient, Dull gray areas represent low salinity water. The gray scales have been chosen to enhance the high gradient regions.

Figure 7. Same as Figure 6 except with winds.

for the no wind case.

Marmorino (1999) speculates on several possible explanations for the existence of inshore front. One of these is advection around an anticyclonic eddy centered southeast of Cape Henry that was identified by Harrison *et al.* (1964). An anticyclonic eddy does form in our simulations. It appears to be due to flow separation at Cape Henry. However, Marmorino argues that there is some observational evidence against flow separation. It could also be a residual tidal eddy and the time scale of the eddy in our simulations is tidal. Future simulations and analysis will address these issues. However, in our simulations the most saline surface water inshore of the outflow plume is not associated with the anticyclonic circulation. Rather it appears to form in place which would imply some form of upwelling, although not driven by the local winds.

The SSS fronts that are generated in our simulations look very much like those seen in the RAR images (Sletten *et al.*, 1999). The fronts are generated in approximately the same locations and appear to develop similar structure (kinks, bends, bifurcations etc.) through interactions with topography and other water masses and through nonlinearities in the advection of the fronts themselves. The number of fronts and their intensity increases when wind forcing is included.

6. Conclusions

We conducted several simulations that focused on the Chesapeake Bay outflow plume near the mouth of the Bay. By using a nested model we were able to look at details of the plume while providing the large scale circulation necessary to correctly include the riverine and tidal forcing and the impact of the large scale circulation in the Bay and along the coast.

The simulations developed the classic inertial bulge and coastal jet circulation. We demonstrated the importance of winds in forcing the plume in shore and offshore. The simulations generated fronts that are qualitatively similar to those seen in RAR images that were taken during COPE I and COPE II.

The simulations showed a salt water wedge inshore of the plume east of Virginia Beach and Cape Henry. This front was observed in RAR images and in in-situ observations during COPE I and COPE II. The wedge is better defined during southerly winds when it could be enhanced by Ekman pumping. However, it also appears during strong flood tides with no winds.

Acknowledgements

This work was done at the Naval Research Laboratory at Stennis Space Center, MS. The work was funded by the Office of Naval Research through the Physics of Coastal Remote Sensing Accelerated Research.

References

Berger, T.J., P. Hamilton, R. J. Wayland, J. O. Blanton, W.C. Boicourt, J.H. Churchill, and D. R. Watts, 1995, *A Physical Oceanographic Field Program Offshore North Carolina, Final Synthesis Report*. OCS Study MMS 94-0047, 345pp., U.S. Dep. Of the Inter., Miner. Manage. Serv., Gulf of Mexico OCS Region, New Orleans, La..

Boicourt, W. C., 1973, The circulation on the continental shelf from Chesapeake Bay to Cape Hatteras, Ph.D. thesis, 183 pp., Johns Hopkins Univ., Baltimore, Md.

Harrison, W., M. L. Brehmer and R. B. Stone, 1964, Nearshore tidal and non-tidal currents, Virginia Beach, Virginia, U.S. Army Coastal Engineering Research Center, Tech. Memo., No 5, 20 pp.

Johnson, D. R., 1995, Wind forced surface currents at the entrance to the Chesapeake Bay: their effect on blue crab larval dispersion and post larval recruitment. Bulletin of Marine Science, 57, 726-738.

Marmorino, G.O., T. F. Donato, M. A. Sletten, and C. L. Trump, 1999, Observations of an inshore front associated with the Chesapeake Bay outflow plume, Cont. Shelf Res., submitted.

Rennie, S. E., J. L. Largier and S. J. Lentz, 1999, Observations of a pulsed buoyancy current downstream of the Chesapeake Bay. J. Geophys. Res., 104, 18227-18240.

Sanders, T. M. and R. W. Garvine, 1996, Frontal observations of the Delaware Coastal Current source region, Cont. Shelf Res., 16, 1009-1021.

Sletten, M.A, Marmorino, G.O., Donato, T.F., McLaughlin, D. J., Twarog, E., 1999, An airborne, real aperture radar study of the Chesapeake Bay outflow plume. J. Geophys. Res., 104, 1211-1222.

Modeling Surface Trapped River Plumes: A Sensitivity Study

Jason Hyatt[1] and Richard P. Signell[1]

Abstract

To better understand the requirements for realistic regional simulation of river plumes in the Gulf of Maine, we test the sensitivity of the Blumberg-Mellor hydrodynamic model to choice of advection scheme, grid resolution, and wind, using idealized geometry and forcing. The test case discharges 1500 m^3/s of fresh water into a uniform 32 psu ocean along a straight shelf at 43° north. The water depth is 15 m at the coast and increases linearly to 190 m at a distance 100 km offshore. Constant discharge runs are conducted in the presence of ambient alongshore current and with and without periodic alongshore wind forcing. Advection methods tested are CENTRAL, UPWIND, the standard Smolarkiewicz MPDATA and a recursive MPDATA scheme. For the no-wind runs, the UPWIND advection scheme performs poorly for grid resolutions typically used in regional simulations (grid spacing of 1-2 km, comparable to or slightly less than the internal Rossby radius, and vertical resolution of 10% of the water column), damping out much of the plume structure. The CENTRAL difference scheme also has problems when wind forcing is neglected, and generates too much structure, shedding eddies of numerical origin. When a weak 5 cm/s ambient current is present in the no-wind case, both the CENTRAL and standard MPDATA schemes produce a false fresh- and dense-water source just upstream of the river inflow due to a standing two-grid length oscillation in the salinity field. The recursive MPDATA scheme completely eliminates the false dense water source, and produces results closest to the grid-converged solution. The results are shown to be very sensitive to vertical grid resolution, and the presence of wind forcing dramatically changes the nature of the plume simulations. The implication of these idealized tests for realistic simulations is discussed, as well as ramifications on previous studies of idealized plume models.

---

[1] United States Geological Survey, 384 Woods Hole Road, Woods Hole, MA 02543. Email: jhyatt@usgs.gov, rsignell@usgs.gov. The entire paper with color figures and animations may be viewed at http://smig.usgs.gov/SMIG/features_0300/plumes.html

Introduction

Surface trapped river plumes are important features around the world, carrying freshwater, nutrients, and pollutants into the coastal ocean. In the western Gulf of Maine, river plumes have been linked with the along-coast delivery of red tide organisms (Franks and Anderson, 1992), making prediction of plume behavior a major regional goal. Realistic simulation of surface trapped river plumes presents a particular challenge, because plumes respond rapidly to wind and baroclinic forcing, and often exhibit strong vertical and horizontal density gradients. Numerical calculations of river plume dynamics are therefore particularly sensitive to the advection scheme, grid resolution, and wind forcing used.

Many previous investigations have studied idealized plume responses. Chao and Boicourt (1986) studied the onset of the plume on a flat-bottomed coastal ocean without ambient current, observing a bulge of anticyclonic surface flow at the estuary mouth and a bore intrusion along the shelf. Chao (1988) added a sloping bottom and classified the plumes into four regimes, supercritical, subcritical, diffusive-supercritical, and diffusive sub-critical. The plume in this study is diffusive-subcritical. Oey and Mellor (1992) used a flat bottom domain to examine the unsteady aspects of the plume and front system, noting pulsating detached pools. Kourafalou et al. (1996) also observed the bulge and bore and included wind in the their calculations over a flat bottom domain. Table 1 summarizes the geometry of these studies. Most of these studies resolved the internal Rossby radius with just a few grid cells in the horizontal and the thickness of the river plume with just a few grid cells in the vertical dimension. It therefore seemed a useful contribution would be to examine the role of advection scheme as well as vertical and horizontal resolution on idealized plume cases under both no-wind and wind-forced conditions.

Author, Date	Resolution, Δx, Δy, Δz	Basin
Chao and Boicourt, 1986	3 km, 3 km, 3m	Flat bottom 15 m
Chao, 1988	4 km, 3 km, 2.5 m	Sloping bottom 15 to 25 m
Oey and Mellor, 1992	3 km, 3 km, 2 m	Flat bottom, 20 m depth
Kourafalou et al, 1996	5 km, 5 km, 0.5-2 m	Flat bottom, 5, 10 and 20 m depth

Table 1. Summary of some previous idealized plume studies geometry and resolution

Modeling Framework

We used the Estuary, Coastal, and Ocean Model (ECOM-3D), the HydroQual, Inc.

version of the Blumberg and Mellor sigma-coordinate model (1987). This version allows for easy switching between four advection schemes, including CENTRAL, UPWIND, and the two forms of multi-dimensional positive definite advective transport algorithm (MPDATA) (Smolarkiewicz, 1984, Smolarkiewicz and Clark, 1986). Each of these advection schemes includes advantages and disadvantages, especially in the presence of sharp gradients in scalar variables. The CENTRAL scheme has a long history of use in a variety of domains and is second order accurate with no numerical diffusion (Roach, 1976). However, it introduces numerical dispersion resulting in ripples and subsequent under- and over-shoot of initial values. The UPWIND scheme has no dispersion and is positive definite, but introduces strong numerical diffusion as a result of the first-order truncation. MPDATA is an iterative scheme that attempts to correct the diffusion errors in the UPWIND scheme. After an initial calculation of the scalar field using the UPWIND scheme, an "anti-diffusion" velocity field based on the local first-order truncation error is generated. The UPWIND scheme is then implemented with this velocity field to "bring back" the scalar field that has spread too far during the time step due to the numerical diffusion. This restorative step may be exercised a number of times, which depends upon the users discretion and computational resources. Here SMOLAR_2 refers to the scheme that makes exactly two corrective sweeps for numerical diffusion.

A significant advance was made by Margolin and Smolarkiewicz (1989) who realized that a recursion relation for the antidiffusion velocities could be determined analytically, allowing the benefits of many iterations to be realized in a single correction step (albeit with a more complex determination of the antidiffusion velocities). Here we term this scheme SMOLAR_R, implemented in ECOM-3D by Gomez-Reyes and Blumberg (1995).

The scale of the model domain was chosen to be representative of the Androscoggin-Kennebec river system in the western Gulf of Maine. The base case was run on a grid with a horizontal resolution of 50 x 70 and 13 sigma layers (Figure 1). The grid cells at the coast measured 1.5 x 3 km and linearly increased in size to 3 x 3 km at the eastern open boundary. The bottom sloped linearly seaward from 15 m at the coast out to a depth of 190 m. A freshwater source of 1500 m^3/s flows into a uniform coastal ocean of 32 psu with an ambient southward current of 5 cm/s specified at the northern boundary. The southern outflow boundary had an Orlanski radiation condition on elevation and a no-gradient condition on salinity and temperature.

All cases had a bottom roughness of 0.003 meters. Smagorinsky mixing is employed in the horizontal with a coefficient of 0.05. In the vertical, the Mellor-Yamada level 2.5 scheme was used, with the minimum vertical background mixing held at 5 x 10^{-6} m^2/s. The CENTRAL scheme was always used for advecting momentum.

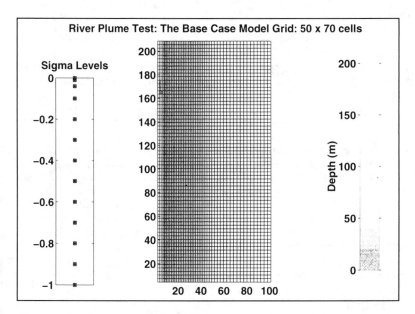

Figure 1. *The Base Case has 13 Sigma Layers, 50 by 70 grid cells, and a sloping bottom of 15 meters at the shore to a depth of 200 meters at the seaward bound. An ambient current of 5 cm/s flows southward.*

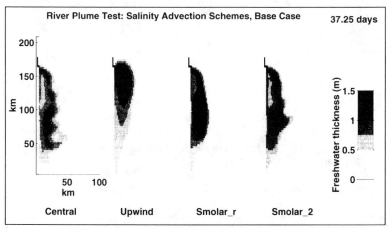

Figure 2. *The original problem of four runs with identical forcing, but different advection schemes yielding very different results.*

Results

Our study was strongly motivated by our initial finding that the base case run with the four different advection schemes led to markedly different results (Figure 2). This led to an analysis of the sensitivities of the run to horizontal and vertical resolution and wind forcing. In each test, the base case was modified and the four advection schemes tested.

Sensitivity to Horizontal Resolution

The initial approach to determining the "correct" solution was to increase the horizontal resolution until all schemes converged on the same answer. The hope was that we could identify which of the original resolution simulations best approximated the convergent solution.

Doubling the horizontal grid resolution, however, did not cause the four schemes to converge (Figure 3). Another doubling of the horizontal resolution also had surprisingly little effect.

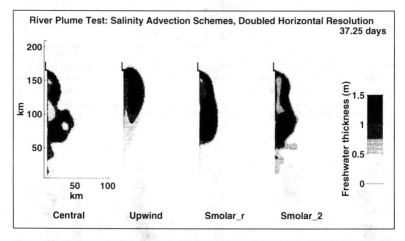

Figure 3. *Same calculations as in Figure 1, but using a grid with doubled horizontal resolution. (Figure 2).*

Sensitivity to Vertical Resolution

Doubling the vertical resolution had a much more dramatic effect on improving convergence of the solutions (Figure 4). It revealed quite clearly that the initial distribution of vertical levels was seriously underresolving the surface trapped plume.

This became especially true as the surface sigma layers expanded in the vertical in the deeper offshore waters.

In order to further explore the effect of vertical resolution, an exponential function for describing the distribution of vertical layers was defined as

$$y = \frac{1 - e^x}{e^\gamma}, x = 0 : \gamma / n : \gamma \tag{1.1}$$

where y are the sigma levels, γ is a shape parameter, and n is the number of vertical layers. Runs of varying vertical resolution were then carried out using the CENTRAL and SMOLAR_R difference schemes. For the runs described herein, γ was held at 3. This shape function is useful in that it provides for a bottom layer of equal thickness for all tests (Figure 7) and allows for a simple direct comparison of runs with any integer number of vertical layers with packing of levels near the surface.

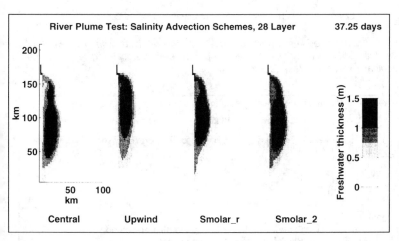

Figure 4. Increased vertical resolution: With 28 layers, the plume structures converge towards the solution most similar to the Base Case SMOLAR_R solution.

The increase in vertical resolution results in a gradual convergence in the model solutions (Figures 5) for the CENTRAL scheme. The base case 13 sigma level spacing was not adequate for the CENTRAL scheme to resolve the plume, whereas the SMOLAR_R at this resolution produced results qualitatively similar to the converged solution. Increasing the vertical resolution results in a convergence of the solution, with the CENTRAL runs showing little change from the 28 layer to the 62 layer run.

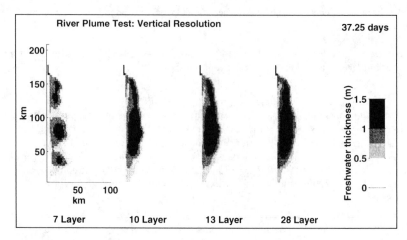

Figure 5. *The eddies where observed to be reduced with increased vertical resolution. Note that this 13 layer run has different vertical segmentation than the original 13 layer run (see Figure 7).*

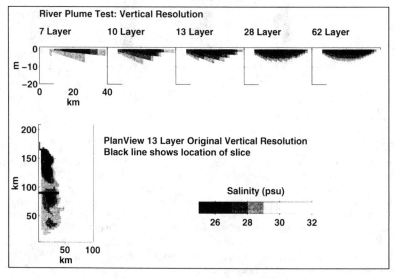

Figure 6. *For the CENTRAL scheme, adequate vertical resolution of the plume is necessary to avoid eddies of numerical origin.*

Figure 6 shows the problem of vertically underresolving a surface trapped river plume. In order to separate sigma-coordinate effects from vertical resolution effects, plume simulations were run at various vertical resolutions over a flat bottom and a similar solution convergence with higher vertical resolution was observed. This was confirmed over a flat bottom of various depths.

Figure 7 compares the vertical resolutions tested here with those used in past studies (Chao and Boicourt, 1986; Chao 1988; Oey and Mellor, 1993). While the different depths, slopes, plume structures and numerical models prohibit direct comparisons, the previous studies appear to be in the range where the choice of advection scheme and vertical resolution negatively affect the solutions.

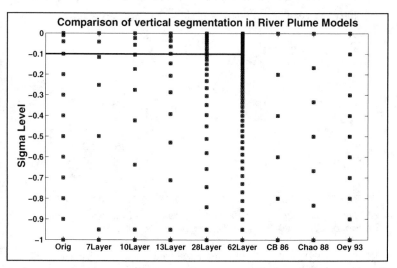

Figure 7. Comparison of vertical segmentation. Orig refers to the base case. CB refers to Chao and Boicourt, 1986. Chao 1988. Oey and Mellor, 1993.

False Dense/Fresh Water Source

Numerical dispersion resulting from advection calculation of scalar variables can result in ripples and over- and under-shoot. In the simulation of a surface trapped river plume, this can represent a significant source of false dense and fresh water where the ambient current encounters the plume and both higher than ambient salinity water and bogus fresh water form in the near-stationary upstream ripples. The dense plume sinks to the bottom and flows along the slope and offshore. This occurs in both the dispersive CENTRAL and SMOLAR_2 schemes, but not in the UPWIND or SMOLAR_R schemes (Figure 8.) Although this dense-water source did not play a dynamically significant role in our study of the surface trapped river plumes, studies interested in bottom boundary layer phenomena should be aware that this seemingly interesting dynamical feature is, in fact, a purely numerical artifact. Also, the additional fresh water can result in artificial freshening of the plume. This effect can easily go unnoticed by total freshwater budgets, a common method of validating a plume model, because the spurious freshwater is exactly offset by the spurious dense water.

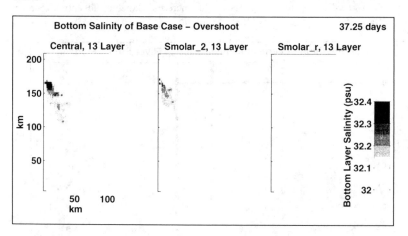

Figure 8. The bottom sigma layer shows the false dense water flowing downstream and downslope. SMOLAR_R does not produce the false dense water.

It is interesting to note that neither increased vertical nor horizontal resolution eliminates this artifact, as dispersive schemes will generate 2 dx ripples upstream of the salinity front regardless of resolution. However, the average bottom layer salinity over the entire domain does come closer to the initial background salinity of 32 with increased vertical resolution, but not horizontal resolution (Figure 9). This comparison is valid because the bottom layer of each of these runs is equally thick (Figure 7).

The observation in Figure 9 also shows an association of the false dense- and freshwater production with the numerical eddy shedding. It is with the 13 Layer CENTRAL run that these eddies disappear (Figure 5) *and* the average bottom layer salinity approaches the background salinity. In addition, the SMOLAR_R and UPWIND schemes produce neither the scalar over- and under-shoots nor the eddies. This association is the subject of future work.

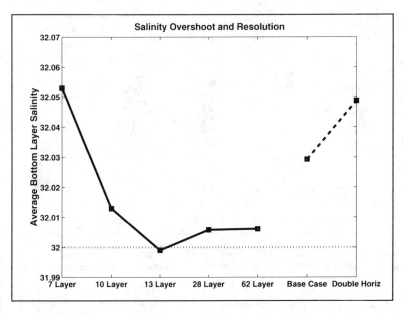

Figure 9. The average bottom layer salinity, a proxy for the amount of false dense water produced, as a function of vertical and horizontal resolution at time = 37.25 days. The solid line shows the dependence on vertical resolution and the dashed line shows the dependence on horizontal resolution.

Sensitivity to wind forcing

While idealized no-wind simulations over month time-scales are interesting, in most regions wind varies significantly in the 2 to 10 day weather band. In order to determine whether the presence of wind forcing affects our conclusions, a sinusoidal north-south wind with a 4 day period and a 5 m/s maximum was imposed. The domain responds to the alongshore wind forcing with periodic upwelling and downwelling, driving the plume on and offshore consistent with Ekman dynamics.

This run demonstrates the dramatic effect of wind on the behavior of surface trapped river plumes (Figure 10). Strong mixing occurs during upwelling wind conditions (Fong, 1997). This mixing can overshadow some of the subtleties caused by changes in advection scheme and resolution.

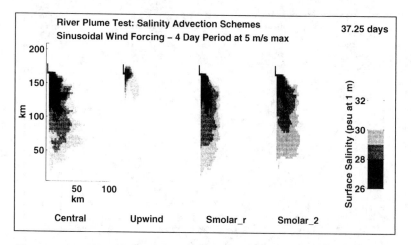

Figure 10. *The effect of wind forcing on plume behavior: wind effects can mask some of the differences caused by advection scheme and resolution (compared to Figure 2).*

Implications to the Realistic Simulation

A model of the Western Gulf of Maine using realistic geometry and forcing was examined in light of the sensitivity analysis. Due to the dominant role of wind mixing on the plume, the differences due to resolution and advection scheme are much more subtle than in the simulations with no wind forcing.

We examined these subtle differences between advection schemes and resolution in the presence of realistic wind forcing. First, we find the elimination of the false dense water source by using the SMOLAR_R advection scheme highly desirable. The modeled biology of red tide includes the process of cell cysts, or seeds, in the sediments germinating and swimming upward in response to environmental factors, including salinity and temperature. The presence of a false dense water source has the definite potential to influence this source of cells.

Next, we wished to test whether or not our optimal advection scheme and resolution combination, SMOLAR_R with normal vertical resolution, provided a significant improvement in the agreement with the data (Figure 11). The plumes appear qualitatively similar, with the SMOLAR_R scheme and CENTRAL scheme with high vertical resolution in good agreement, as expected (two center panels). The CENTRAL scheme with the normal vertical resolution (far left), however, also appears qualitatively similar, again suggesting that advective effects are being overshadowed by wind-driven mixing and other processes. The fourth panel shows dramatically the effect of wind mixing by revealing the simulated salinity field without wind. It is temping to favor the CENTRAL scheme with normal vertical resolution, as this would be the lowest-cost solution. Because our biological growth functions are based on temperature and salinity, however, over- and under-shoots and a dense water source are highly undesirable. We therefore are using SMOLAR_R in our ongoing coupled physical-biological simulations in realistic Gulf of Maine domains.

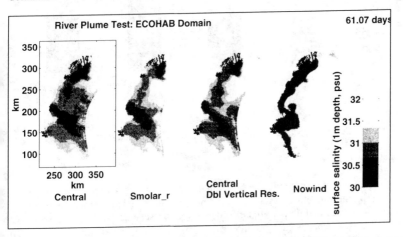

Figure 11. Surface salinity in the Gulf of Maine runs with identical forcing, except for the no-wind case on the right. The no-wind case uses the CENTRAL Scheme.

Computational Expense

Both higher vertical resolution and higher order advection schemes add computational expense to a model simulation. In the case of the idealized domain, using SMOLAR_R at a lower vertical resolution proved much cheaper than using CENTRAL at a higher vertical resolution (Figure 12). The extent of the savings provided by a higher-order scheme depends on the number of vertical levels and the degree of time splitting between internal and external modes.

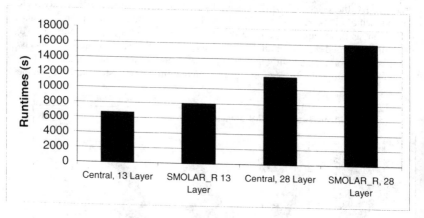

Figure 12. Computational Expense for the idealized river plume domain.

<u>Conclusions</u>

These studies demonstrate that for horizontal and vertical resolutions commonly used in river plume modeling, advection schemes can have a large effect, particularly under low wind conditions. Higher order advection schemes such as the recursive Smolarkiewicz scheme can provide results that can only be obtained with much higher grid resolution and at higher cost. The SMOLAR_R scheme also had the desirable characteristic of eliminating the formation of spurious fresh- and dense-water plume upstream of the river source.

While the advection scheme made a great difference in the no-wind cases, the addition of typical wind forcing greatly modified the plume characteristics, and reduced the sensitivity to the choice of advection scheme. In our realistic simulations with western Gulf of Maine topography and wind forcing, the choice of advection scheme made little qualitative difference in the salinity fields, thus suggesting that the inexpensive CENTRAL scheme might be the optimal choice if false dense water source is not a major concern. In our case, however, consideration of the biological ramifications of over- and under-shoots caused us to choose the more expensive SMOLAR_R scheme. This highlights the fact that numerical models should be designed to allow for a choice of advection schemes with different degrees of accuracy and cost, so that simulations where advection plays a stronger or weaker role can be optimized accordingly.

References

Blumberg, A.F. and Mellor, G.L., 1987, A description of a three-dimensional coastal model, in Three-Dimensional Coastal Ocean Models, N. Heaps [ed], Coastal and Estuarine Sciences, v. 4, p. 1-16.

Fong, D.A., Geyer, W.R., and Signell, R.P., 1997, The wind-forced response on a buoyant coastal current: Observations of the western Gulf of Maine plume. *J. Marine Sys.* 12, 69-81.

Franks, P.J.S. and Anderson, D.M., 1992, Alongshore transport of a toxic phyoplankton bloom in a buoyancy current: Alexandrium tamarense in the Gulf of Maine. *Marine Biology* 112, 153-164.

Gomez-Reyes, E. and Blumberg A.F., Pollutant Transport in Coastal Water Bodies, Computer Modelling of Seas and Coastal Regions II, Computational Mechanics Publications, Southhampton, U.K. 1995, pp. 87-94.

Margolin, L.G. and Smolarkiewicz, P.K, 1989, Antidiffusive velocities for multipass donor cell advection. *Lawrence Livermore National Laboratory*, Report W-7405-Eng-48.

Roach, P.J. Computational Fluid Dynamics. Hermosa Publishers, Albequerque, New Mexico, 1976.

Smolarkiewicz, P.K., 1983, A simple positive definite advection transport scheme with small implicit diffusion. *Monthly Weather Review.* 111, 479-486.

Smolarkiewicz, P.K., 1984, A fully multidimensional positive definite advection transport algorithm with small implicit diffusion. *J. Comptutional Physics* 54, 325-362.

Smolarkiewicz, P.K., and T.L. Clark, 1986, "The Multidimensional Positive Definite Advection Transport Algorithm: Further Development and Applications". *J. Comptutional Physics* 67, 396-438.

Typical and Extreme Responses of Chesapeake Bay and its Coastal Plume to Riverine Forcing

Jerry L. Miller[1], Patrick Gallacher[1], Michael Schaferkotter[2], and Paul Martin[1]

Abstract

The Chesapeake Bay and its coastal low salinity plume exhibit significant spatial and temporal variability even when forced by smoothed (long-term monthly-averaged) river discharge. When forced with realistic daily river discharge, baroclinic structure is markedly increased, currents are stronger and length scales are shorter. Response to a weak hurricane Fran produces a qualitative change in the coastal jet which is driven by the salinity structure.

Introduction

The coastal mesoscale response of a low salinity coastal plume and that of its originating estuary to typical and extreme variations in terrestrial freshwater input have been examined. Such variations have significant implications for transport and fate of dissolved and suspended constituents as well as biota. The Chesapeake system, a mid-sized coastal plain estuary which opens onto a gently sloping wide continental shelf, was the chosen subject of this series of investigations. Contrasts are drawn between response to monthly averaged climatological forcing and forcing associated with rainfall generated by a weak hurricane.

The Model

The Navy Coastal Ocean Model (NCOM), a fully nonlinear three-dimensional primitive equation code, was utilized in its doubly nested mode with intermediate and highest spatial resolution at the mouth of the Bay. NCOM has fully nonlinear primitive equation dynamics and can be forced by river discharge, winds, and/or surface heat flux. It bears resemblence to the Princeton Ocean Model in terms of its set-up and capabilities but uses implicit rather than explicit time stepping. See Martin (1999) for details.

For this series of experiments, NCOM was forced with weak (5 knot) winds which rotated with a 5-day period. No surface heat flux was applied. Radiation conditions were used on the open ocean boundaries. The Chesapeake Bay was forced with fresh water discharge from 10 rivers; the Delaware Bay with two. Surface elevation was adjusted consistent with continuity to account for input of rinerine waters. The model was spun up from rest over a 600 day period during which riverine forcing was held constant at a 50-year mean.

---

[1] Naval Research Laboratory, Oceanography Division, Code 7332, Stennis Space Center, MS 39529.

[2] Sverdrup Technology Inc., Stennis Space Center, MS 39529.

Experiment 1

The first series of runs were conducted with a time-varying annual cycle of "smooth" riverine forcing based on monthly averages calculated from 50 years of discharge data. This monthly forcing was interpolated to each model time step such that transition from month to month was smooth. Runs were initialized on June 9 as that is the date when interpolated monthly means equals the 50-year mean used in the 600 day spin-up. Runs were continued through the end of September.

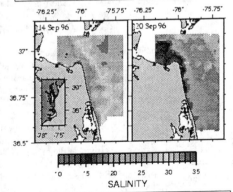

Figure 1a shows simulated sea surface salinity (SSS) from Sept 14. Although we do not expect precise correspondence given the nonlinearity of this system, its structure is comparable in complexity to remotely-sensed SSS (figure 2a) on the same nominal date. (See Miller et al., 1998 for description of the remote sensing technique.) This is noteworthy given the smoothness of the riverine forcing and implies a high level of baroclinic and or barotropic instability. This is likely due to the strong salinity gradients typically observed in such regions. For reference below, we note that the basic structure and salinity contrasts of simulated SSS on Sept 20 (figure 1b) is generally similar to that of Sept 14.

Figure 1a. SSS from 14 Sept 1996 of model run forced by monthly averaged river discharge. Scale on Fig 2.

Figure 1b. SSS from 20 Sept 1996 of model run forced by monthly averaged river discharge. Scale on Fig 2.

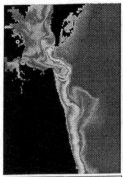

Figure 2. Remotly sensed SSS from 14 (left) and 20 (right) Sept 1996.

Experiment 2

This second series of runs was conducted with much more realistic daily-averaged riverine forcing for June-Sept 1996, the year during which the remotely-sensed images of figure 2 were obtained. As for the monthly forcing of experiment 1, this daily forcing was interpolated to each model time step such that transition from day to day was smooth. Runs were again initialized on June 9 using the same initial conditions described above.

Figure 3a shows simulated sea surface salinity (SSS) from Sept 14. Although a bit more complex, its spatial structure is again comparable to remotely-sensed SSS (figure 2a) on the same nominal date and not markedly different from figure 1a. On Sept 20 (figure 3b), we see a qualitative difference. SSS in the lower Chesapeake Bay is much lower as are salinities in the coastal plume. Southward alongshore velocities (not shown) are also significantly higher corresponding to a strengthening of

Figure 3a. SSS from 14 Sept 1996 of model run forced by daily river discharge. Scale on Fig 2.

Figure 3b. SSS from 20 Sept 1996 of model run forced by daily river discharge. Scale on Fig 2.

the Chesapeake's coastal buoyancy jet.

Discussion

On 8-10 September 1996, Hurricane Fran traversed northward through central Virginia, Maryland, and Pennsylvania producing large amounts of rainfall in the drainage basins of the Chesapeake Bay's major tributaries. While notable in comparison to other summer 1996 events

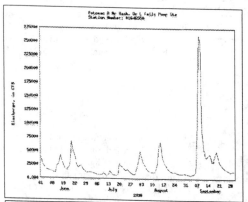

Figure 4. Discharge from the Potomac River, June - Sept 1996.

(figure 4) the pulse associated with this otherwise weak hurricane is rather similar to a typical winter storm in this region. However, this short-lived pulse is approximately 40 times the climatological mean for September. A "back of the envelope" calculation indicates that the baroclinic signal associated which such an event would require about 2 weeks to arrive at the mouth of the Bay if it originated at mid Bay.

The remotely-sensed SSS fields shown in figure 2 reveal that the fresh water pulse arrived at the mouth of the Bay between 14 and 20 Sept. A time series of SSS from experiment 2 reveal that, in fact, there were multiple arrival over a period of a few days - first from the James river which is closest to the Bay mouth and which was first to receive rainfall from the hurricane, followed sequentially by the York, and Potomac. A distinct pulse from the Bay's largest tributary, the Susquehanna, was not identified ; the weakening of the hurricane as it moved northward across the large Susquehanna drainage basin resulted in a broader low-amplitude peak in discharge relative to the rivers (e.g., the Potomac) further south.

Summary

The system was forced with the monthly averaged streamflow to establish nominal seasonal variation in density-driven flow. Day to day variations in forcing can depart substantially from monthly means so short-term simulations forced by observed daily streamflow were conducted. An extreme case corresponding to the rainfall event resulting from Hurricane Fran was examined and is described in the context of typical daily and seasonal variations. Significant response of the density and velocity fields both interior and exterior to the Bay were revealed. "Wet" and "dry" year conditions are to be examined in future experiments by using forcing one standard deviation above and below normal, respectively.

References

Martin, P.J., 2000. A description of the Navy Coastal Ocean Model version 1.0. NRL report in press. Naval Research Laboratory, Stennis Space Center, MS.

Miller, J. L., M. A. Goodberlet, and J. B. Zaitzeff, 1998. Airborne salinity mapper makes debut in coastal zone. EOS, Trans. Amer. Geophys. Union, 79(14): 173-177.

High Performance Estuarine and Coastal Environmental Modeling: the CH3D Example

Justin R. Davis[1] and Y. Peter Sheng[1] [2]

Abstract

To enhance ecosystem management and restoration, managers and scientists are demanding more detailed information on estuarine and coastal ecosystems. To obtain this detailed information, data collection and model simulations must be conducted with finer spatial and temporal resolutions over longer time periods. Presently some 3-D model simulations use horizontal grid spacing on the order of a few meters and more than 10 grid cells in the vertical direction, while the simulation period can be more than one year. To accomplish long-term fine-grid modeling, it is essential to develop high-performance modeling techniques for estuarine and coastal environmental models. This paper presents a preliminary study on the use of parallel techniques to solve governing equations of estuarine dynamics, with the CH3D model as an example. The first technique, shared memory (multi-thread) method, makes use of multiple-CPUs which have access to a single shared memory image, e.g. SGI Origin-2000, to perform computations in various grid zones simultaneously. In the second technique, message passing, CPUs rely on the explicit sending and receiving of messages to communicate with other CPUs. Preliminary results show that using the shared memory technique with a 2-D model, a 950 day simulation of the Indian River Lagoon estuarine system can be performed in one day. The message passing technique, particularly when implemented with a Beowulf Cluster, offers an attractive alternative. To achieve the most user friendly high performance code, macros and scripting languages can be used within the model. Several examples of usage are shown in this paper.

[1]Department of Civil & Coastal Engineering, 345 Weil Hall, P. O. Box 116580, University of Florida, Gainesville, FL 32611-6580.

[2]Member, ASCE.

Introduction

The sophistication of numerical modeling has improved greatly in recent years. Hydrodynamic, sediment, wave, nutrient, light, and seagrass models have all been developed and coupled to further the understanding of estuaries as integrated systems (Sheng, 1998). However, even on current high performance workstations, model execution time increases dramatically when meeting the needs of more complex, integrated models, larger model domains, finer model resolution, and longer simulation periods.

Our focus in high performance estuarine modeling has centered around the use of parallel techniques to solve governing equations of estuarine dynamics. Particularly, several parallel techniques are being developed for CH3D (Sheng, 1989, 1998), a 3-D curvilinear-grid model. The early version of CH3D (Sheng, 1989) included only a hydrodynamics model, but the present model (Sheng, 1998, 2000) includes a sediment transport model and a water quality model, as well. CH3D is written in Fortran and contains more than 30,000 lines of coding. Since CH3D is a well-known and widely used code, we selected CH3D to test the various methods for high performance computing.

Parallel computing can be divided into two predominant communications models, the shared memory model and the message passing model. The shared memory model provides the application programmer with an abstraction of globally accessible data while the message passing model requires explicit message passing constructs. The shared memory model, while easier to implement, does not scale well with the number of processors. The message passing model, on the other hand, scales better because the message passing is explicit and the application programmer can provide for the most efficient communication.

Coding complexity is always an issue with numerical models. Another focus of our modeling effort is to implement the parallel features as generically as possible. Every computer system implements parallel constructs differently, through an extensive use of macros, implementation becomes easier and portability is increased.

In this paper, we discuss the following aspects of high performance estuarine and coastal environmental modeling using the CH3D code as an example:

- Goals of high performance modeling

- Available parallel methodologies

- Preliminary results of the shared memory approach

- Potential use of Beowulf Clusters for environmental modeling

- Use of macros and scripting languages to enhance model functionality

Goals of high performance modeling

During the past two decades, the study of estuaries as integrated systems has advanced from using simple 1-D cross-sectionally averaged models to using fully 3-D models complete with coupled sediment, wave, nutrient, light and seagrass models. Along with the inclusion of more complicated processes into the models, there have been significant increases in simulation domain, grid resolution and simulation period(Table 1).

Table 1: Evolution of grid sizes used with the CH3D model during the past 15 years.

Year	Water Body	Grid Size	Total Grid Cells	Simulation Period (days)
1983	Mississippi Sound (Sheng, 1983)	60x116x4	28,000	5
1988	Lake Okeechobee (Liu, 1988)	31x50x1	2,000	30
1991	Sarasota/Tampa Bay (Peene et al., 1991)	189x50x4	38,000	60
1993	Yellow Sea (Luo, 1993)	50x69x5	17,000	30
1996	Florida Bay (Davis, 1996)	194x148x1	29,000	60
1996	Indian River Lagoon (Sheng et al., 1996)	478x44x4	84,000	60
1997	Tampa Bay (Yassuda and Sheng, 1997)	45x85x8	32,000	120
1999	Charlotte Harbor	63x105x8	53,000	90
1999	Pinellas County	104x185x4	77,000	90
2000	Indian River Lagoon-3x(proposed)	1434x132x4	757,000	365
2002	Indian River Lagoon-5x(proposed)	2390x220x8	4,206,000	365

One such study which exemplifies the trend to larger domains, finer resolution and longer simulations periods is that of the Indian River Lagoon (Sheng et al., 1996), which lies on the Atlantic coast of central Florida (Figure 1) and is one of the most biologically diverse and complex ecosystems in the United States. The lagoon is approximately 200 km long and extends from Ponce de Leon Inlet in the north to Jupiter Inlet in the south. It is 2-4 km wide with an average depth of 2 m outside of the Intercoastal Waterway. The lagoon receives fresh water from numerous natural creeks and rivers and a series of man-made canals.

One horizontal grid system for the Indian River Lagoon is 478x44 with a minimum grid resolution of approximately 100 meters. With such a large grid, 90 day simulations of hydrodynamics, sediments, and nutrients can take several days to execute. Even with such a seemingly fine grid, the model cannot accurately represent the navigation channel, i.e., the Intercoastal Waterway, which runs along the entire lagoon. This channel is on the order of tens of meters wide and approximately 3-4 meters deep. Hence, the interior resolution of the grid needs to be refined to adequately represent the channel (Figure 2). The proposed 5x finer grid system shown in the figure better represents the Intercoastal Waterway. However, because of the grid's immense size (2390x220) and the limitation of presently available, reasonably priced processor technology, it is not feasible to complete a long-term simulation without using parallelism.

Hence, our goal is: Through the uses of parallel algorithms, macros and scripting languages, to obtain a portable, scalable model which can perform simulations in

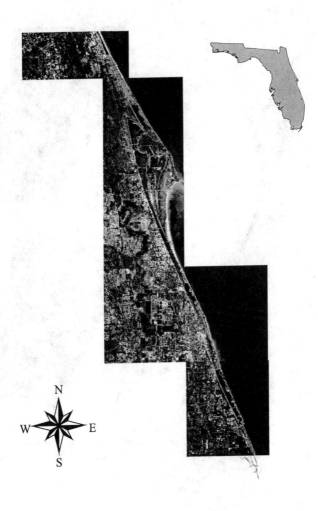

Figure 1: The Indian River Lagoon.

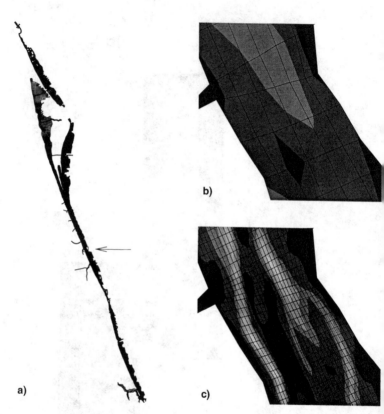

Figure 2: a) The Indian River Lagoon grid system (478x44). b) A closeup of the lagoon showing the crude representation of the Intercoastal Waterway. c) A closeup of the proposed 5x finer grid system (2390x220) showing a better representation of the Intercoastal Waterway.

a reasonable execution time allowing for

- Large simulation areas,

- Finer grid resolutions,

- Longer simulation periods, and

- Inclusion of more complicated processes.

Parallel communication models

There are two predominant parallel communication models, the shared memory model (Figure 3) and the message passing model (Figure 4). The shared memory (also known as multi-thread) model relies on an area of shared memory which can be accessed by an processing element. The message passing model requires that data explicitly be sent from one processing element and explicitly received by another.

Shared communication link

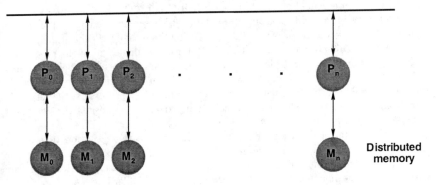

Figure 3: A schematic representation of the shared memory model.

The shared memory model is generally implemented with parallel constructs which are added to the serial source code. These constructs are slightly different between compiler vendors but a standard called OpenMP (http://www.openmp.org) is now being supported by major compiler vendors (SGI, HP, Intel, Sun, Compaq, Absoft, Portland Group).

The message passing model is generally implemented with explicit library calls to send and receive data. Several different libraries exist to facilitate the passing of data, *e.g.* , MPI(Lam, Mpich, ...), PVM, and shmem (SGI proprietary).

CH3D decomposes the governing equations into independent sweeps which allows for easy implementation of the shared memory model (Figure 5). CH3D is composed of approximately 60 procedures(subroutines), the 14 procedures which take

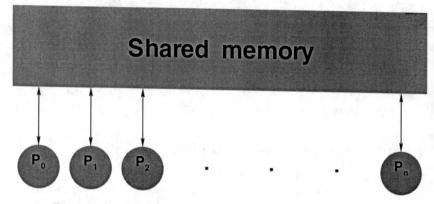

Figure 4: A schematic representation of the message passing model.

the most CPU time have been made parallel using the shared memory model. The model was then executed on our 4 processor SGI Origin 2000. Preliminary execution times of these 14 procedures, attained simulating 3-D circulation and transport in the Indian River Lagoon, are shown in Table 2. Taking the total CPU time of the parallel procedures (0.766 s) and dividing it by the total CPU time of all of the procedures (0.874 s), both for the $n = 1$ case, gives a good approximation % of parallel work being done by the CH3D model, or 87.6%. With significant more effort, the CH3D code could be made nearly 100% parallel. However, with only 4 processors, the relatively little gain making the code 100% parallel is not warranted because of the significant coding effort required.

Computational speedup is defined as the ratio of the execution time of the serial code/procedure to the execution time of the parallel code/procedure. The execution time of the serial CH3D code and the $n = 1$ parallel CH3D code are the same. The maximum speedup is governed by Amdahl's Law (Amdahl, 1967), $S_w = \frac{1}{f_s + \frac{f_p}{n}}$, where S_w is the maximum theoretical wall clock speedup, f_s is the % of serial work, f_p is the % of parallel work $(f_s + f_p = 1)$ and n is the number of processors. For a 100% parallel code, the maximum speedup equals the number of processors. Table 3 illustrates the maximum theoretical speedup for several different cases and shows how little speedup improvement is possible for all but highly parallel codes. For example, the CH3D code executed with 1024 processors can reach a speedup of at most 8.01. Even with a 99% parallel CH3D code, 1024 processors will only yield a speedup of 91.

Based on the percent of parallel work being done by CH3D being 87.6%, the maximum theoretical speedup for the $n = 2$, $n = 3$, and $n = 4$ cases is 1.78, 2.40, and 2.92, respectively, as compared to the measured values of 1.7, 2.14, and 2.43. The actual speedup is somewhat smaller than the theoretically calculated values due to some remaining serial code within the parallel procedures and the inherit serial

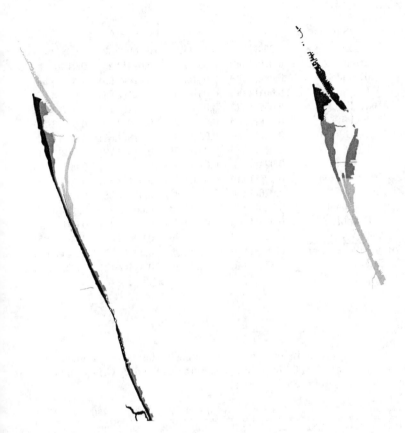

Figure 5: The mapping of the Indian River Lagoon grid system onto 4 processors. The processor distribution of the x-sweep rows and the y-sweep columns are shown on the left and right, respectively.

Table 2: CPU times for the CH3D procedures which have been made parallel using the shared memory model. Times shown are per timestep iteration of the model and are given in seconds, n is the number of processors used and speedup is shown in parenthesis.

Procedure/Subroutine	$n = 1$	$n = 2$	$n = 3$	$n = 4$
Species	0.264(1.00)	0.141(1.87)	0.109(2.42)	0.098(2.70)
Turbulence	0.079(1.00)	0.037(2.13)	0.024(3.26)	0.019(4.18)
Baroclinic(I)	0.064(1.00)	0.034(1.87)	0.024(2.65)	0.019(3.34)
N.L./Diff.(J)	0.061(1.00)	0.031(1.94)	0.021(2.84)	0.017(3.67)
Baroclinic(J)	0.057(1.00)	0.030(1.93)	0.021(2.72)	0.017(3.43)
Salinity	0.050(1.00)	0.026(1.95)	0.019(2.62)	0.016(3.21)
N.L./Diff.(I)	0.045(1.00)	0.024(1.87)	0.018(2.56)	0.014(3.13)
3D Vel.(v)	0.041(1.00)	0.021(1.91)	0.015(2.77)	0.012(3.51)
3D Vel.(u)	0.024(1.00)	0.013(1.93)	0.009(2.76)	0.007(3.67)
Interpolation	0.024(1.00)	0.014(1.75)	0.011(2.30)	0.009(2.74)
Int. Vel.(U)	0.016(1.00)	0.012(1.38)	0.010(1.58)	0.009(1.74)
Dimensionalize	0.015(1.00)	0.009(1.73)	0.006(2.47)	0.005(3.18)
3D Vel.(w)	0.014(1.00)	0.008(1.76)	0.006(2.43)	0.004(3.11)
Int. Vel.(V)	0.013(1.00)	0.008(1.76)	0.006(2.42)	0.004(3.05)
All parallel routines	0.766(1.00)	0.406(1.89)	0.298(2.57)	0.249(3.08)
Total Runtime	0.874(1.00)	0.515(1.70)	0.408(2.14)	0.359(2.43)

Table 3: Theoretical limits of parallel processing performance (Amdahl's Law of maximum speedup). $f_p = 0.876$ corresponds to the preliminary CH3D timing results and n is the number of processors used.

f_p	$n = 2$	$n = 3$	$n = 4$	$n = 6$	$n = 8$	$n = 32$	$n = 256$	$n = 1024$
0.700	1.54	1.87	2.11	2.40	2.58	3.17	3.30	3.36
0.800	1.67	2.14	2.50	3.00	3.33	4.44	4.92	4.99
0.876	1.78	2.40	2.92	3.70	4.28	6.61	7.85	8.01
0.900	1.82	2.50	3.08	4.00	4.71	7.81	9.66	9.91
0.950	1.91	2.73	3.48	4.80	5.93	12.55	18.62	19.63
0.970	1.94	2.83	3.67	5.22	6.61	16.58	29.60	32.31
0.990	1.98	2.94	3.88	5.71	7.48	24.43	72.11	91.18
0.999	2.00	2.99	3.99	5.97	7.94	31.04	203.98	506.18

communication costs necessary to perform parallel processing. As the problem size approaches infinity, f_s approaches 0 and S_w approaches n. Thus, we reduce the serial fraction as the problem size increases.

Another measure of parallel algorithm performance is *efficiency* (the fraction of time that the processors spend performing calculations). It is calculated by dividing the measured speedup by the maximum theoretical speedup(the number of processors). For the current CH3D model using the shared memory approach the efficiencies are 85%, 71%, and 61% for the $n = 2$, $n = 3$, and $n = 4$ cases. However, since not all of the CH3D procedures have been made parallel (only 87.6% of the total runtime is accounted for by parallel procedures), instead of dividing by the number of processors, the speedup can be divided by the maximum theoretical speedup attained by Amdahl's Law to produce a *relative efficiency* of 96%, 89%, and 83%, respectively.

As an example, simulated water elevation during a 950 day simulation of the Indian River Lagoon is shown in Figure 6. This 2-D simulation, consisting of approximately 2.5 million time integration steps using a timestep of 30 seconds, was performed in one day using 3 processors of a SGI Origin 2000. Results of this parallel simulation were identical to those of the corresponding serial simulation.

Potential use of message passing approach with Beowulf Clusters

An attractive alternative to achieve significant speedup of CH3D is the use of the message passing approach with a Beowulf Cluster (http://www.beowulf.org). Beowulf Clusters are composed of Commercial Off The Shelf (COTS) PCs running the Linux operating system connected via a private network. The private network is create by placing the processing nodes behind a head node; thus, only inter-nod processing-related communication consumes network bandwidth. Because many researchers have had success with the Beowulf concept (Sterling et al., 1995; Ridge et al., 1997), a Beowulf Cluster (Figure 7) is in the process of being designed and built to assess the feasibility of the Beowulf concept on estuarine and coastal environmental modeling.

The main advantage of the Beowulf concept is the cost per processing node. One additional processor for the SGI Origin 2000 costs about $10,000 presently, while a COTS 450 MHz Pentium III computer costs about $1,000, a 10x savings. Even though the SGI processor is faster than a Pentium processor, it is not 10x faster. Along with decreased processor cost, memory cost is also much cheaper for a common Pentium based computer.

The main disadvantage of the Beowulf concept, however, is the time needed to modify the code to work with a message passing model. Once modified, the code will need to be analyzed to determine if it is communication limited or processor limited.

Detailed results using the message passing approach and Beowulf Cluster will be reported when available.

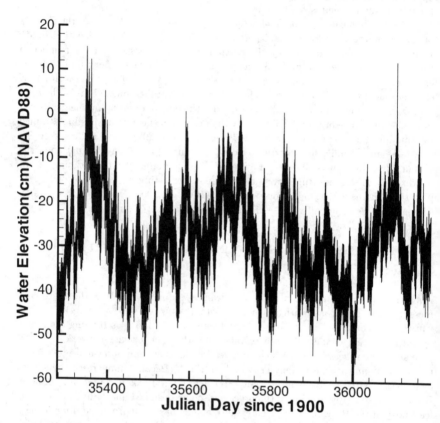

Figure 6: The water surface elevation at Florida Department of Environmental Protection Station #872-16471 (Merritt Causeway East) during a 950 day Indian River Lagoon simulation.

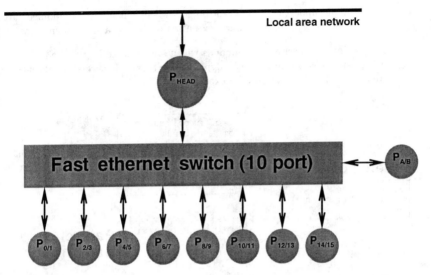

Figure 7: The proposed UF Coastal & Oceanographic Engineering Beowulf Cluster.

Macros and scripting languages

One disadvantage of high performance computing is the complexity which must be added to the code. For example in the shared memory paradigm, parallel constructs must be added to the code to make loops parallel and in the message passing paradigm explicit message passing calls must be inserted into the code.

Our implementation of high performance computing techniques to CH3D, which is written in Fortran, has revolved heavily around the use of macros and shell scripts. These tools allow parallel constructs and message passing interfaces to be hidden within the code so as to make it more more user friendly. The macros are also used to increase code portability. Parallel constructs are implemented in slightly different methods under different compilers. Macros (Figure 8) can be used to hide these differences allowing the programmer to focus on the larger details of the problem. We describe several different macro processors and shell scripts which are used in our model development.

- *cpp* is the standard C preprocessor. It performs text substitutions, manipulations and conditional inclusion as described by the C standard.

- *m4* is a sophisticated macro processor intended as a front end for various programming languages.

- *make* and its associated *makefile* is a utility which controls the compilations and linking of source files into an executable. The *make* utility examines time relationships and updates only relevant objects when source code is modified. We also use the *makefile* to control the execution of the *cpp* and *m4* preprocessors.

- Shell scripts are a series of commands which are executed by the shell. We use shell scripts to generate a *makefile* dynamically; thus, only necessary subroutines are compiled and linked into the executable

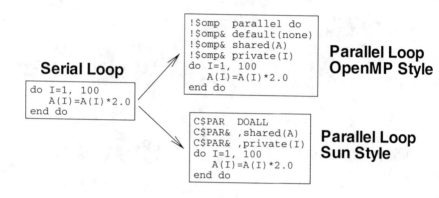

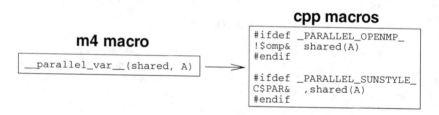

Figure 8: Examples of using macros to implement the shared memory parallel constructs.

Conclusions

This paper discusses several aspects of high performance estuarine and coastal environmental modeling. Goals of high performance modeling as well as a brief discussion of the available parallel models were discussed. As an example of implementation, the CH3D code was made parallel using the shared memory approach. Parallel constructs, which were created using the OpenMP standard implemented via macros, were added to the most time consuming procedures.

Major conclusions are summarized in the following:

- Preliminary tests indicate that the shared memory (multi-thread) method can be used to significantly (3x faster with 4 processors) increase the speed of CH3D, hence allowing more robust processes, finer resolution, larger model domains and longer simulation periods. A 950 day simulation of the Indian River Lagoon estuary system, consisting of 2.5 million time integration steps, was completed in one day using the parallel 2-D model.

- An attractive alternative to high priced shared memory computers is the use of a Beowulf Cluster. Further investigation is underway.

- Macros and scripts, which control both the method and the content of compilation, have been found to provide more readable and portable code.

Acknowledgments

The high performance modeling work is supported by the National Center for Environmental Research and Quality Control, USEPA. The Indian River Lagoon study is supported by the St. Johns River Water Management District. Dr. Y. P. Sheng is the Principal Investigator on both projects.

References

Amdahl, G. (1967). Validity of the single processor approach to achieving large-scale computing capabilities. In *Proceedings of the American Federation of Information Processing Societies*, volume 30, pages 483–485, Washington, DC.

Davis, J. R. (1996). Hydrodynamic modeling in shallow water with wetting and drying. Master's thesis, University of Florida.

Liu, Y. (1988). A two-dimensional finite-difference model for moving boundary hydrodynamic problems. Master's thesis, University of Florida.

Luo, Y. (1993). Three-dimensional moving-boundary hydrodynamic model. Engineer's thesis, University of Florida.

Peene, S., Sheng, Y. P., and Houston, S. (1991). Modeling tidal and wind driven circulation in Sarasota and Tampa Bay. In *Estuarine and Coastal Modeling*, volume 2, pages 357–369. American Society of Civil Engineers.

Ridge, D., Becker, D., Merkey, P., and Sterling, T. (1997). Beowulf: Harnessing the power of parallelism in a Pile-of-PCs. In *Aerospace Applications Conference Proceedings*, volume 2, pages 79–91. IEEE.

Sheng, Y. P. (1983). Mathematical modeling of three-dimensional coastal currents and sediment dispersion: Model development and application. Technical Report CERC-83-2, Aeronautical Research Associates of Princeton, Princeton, New Jersey.

Sheng, Y. P. (1989). Evolution of a three-dimensional curvilinear-grid hydrodynamic model for estuaries, lakes and coastal waters: CH3D. In *Estuarine and Coastal Modeling*, pages 41–49. American Society of Civil Engineers, New York.

Sheng, Y. P. (1998). Pollutant load reduction models for estuaries. In *Estuarine and Coastal Modeling*, volume 5, pages 1–15. American Society of Civil Engineers.

Sheng, Y. P. (2000). Integrated modeling of coupled hydrodynamics-sedimentary-ecological process. In *Estuarine and Coastal Modeling*, volume 6. American Society of Civil Engineers.

Sheng, Y. P., Barry, K., Davis, J. R., Liu, Y.-F., Zhong, J., and Ge, R. (1996). A preliminary hydrodynamics and water quality model of Indian River Lagoon. Final Report of Project #94W241, Coastal and Oceanographic Engineering Department, University of Florida, Gainesville, Florida.

Sterling, T., Savarese, D., Becker, D. J., and Dorband, J. E. (1995). BEOWULF: A parallel workstation for scientific computation. In *Proceedings of the International Conference on Parallel Processing*, volume 1, pages 11–14.

Yassuda, E. and Sheng, Y. P. (1997). Modeling dissolved oxygen dynamics of Tampa Bay during summer of 1991. In *Estuarine and Coastal Modeling*, volume 5, pages 35–58. American Society of Civil Engineers.

Use of Membrane Boundaries to Simulate Fixed and Floating Structures in GLLVHT

Venkat S. Kolluru[1], Edward M. Buchak (Member, ASCE)[1] and Jian Wu[1]

Abstract
The setting up of the grid in any three dimensional hydrodynamic and transport model takes a considerable amount of time even with the use of grid generation software. The effort increases with the complexity of the water body. Refining the grid for proposed design elements such as jetties, breakwaters, and curtain walls becomes especially time consuming. Also, any calibration performed with the original grid needs to be confirmed whenever the original is edited. In order to avoid these problems, an interior membrane boundary condition has been developed and implemented in the three dimensional hydrodynamic and transport model, GLLVHT. The interior boundary is used to restrict the amount of flow through each grid cell. The interior membrane boundary behaves like the permeability factor used in porous media modeling and is applied at the grid cell face. The interior boundary condition has been used to simulate flow around piers, around curtain walls used to bring in bottom cold water to thermal power plant intakes and around break waters. Multiple configurations of such features can be simulated using a single grid.

Introduction
The last several decades have seen the accelerated use of numerical models to predict the circulation and transport and fate of pollutants discharged into river, lake, estuary and coastal water bodies. Many of these models are available in the form of software that includes a built in grid generation tool. The setting up of a grid for a complex water body takes a considerable amount of time even with the use of such grid generation tools. The difficulty increases with the complexity of

---

[1] J. E. Edinger Associates, Inc., 37 West Avenue, Wayne, PA 19087

the water body, especially when such design considerations as jetties, breakwaters and curtain walls are to be considered. Often the modeler needs to evaluate different configurations in a short time to fully participate in the design process. A significant impediment to rapid evaluation is that the model needs to be recalibrated whenever the original grid is modified.

In order to avoid these problems, an interior membrane boundary condition has been developed and implemented in the three dimensional hydrodynamic and transport model called GLLVHT (Generalized Longitudinal, Lateral and Vertical Hydrodynamic and Transport). The membrane boundary resembles permeability property used in ground water modeling. A similar porosity boundary was used in the scheme FAVOUR (Fractional Area-Volume Obstacle Representation) of the 3-D Navier Stokes code called FLOW-3D (Hirt and Sicilian, 1985). In addition to porosity, properties such as turbulence generation, friction etc can also be included in the membrane boundary to represent flow around bridge pilings (Wang and Kim, 1999).

The GLLVHT model is incorporated in the software called GEMSS (Generalized Environmental Modeling System for Surface waters). GEMSS software is developed for the Windows NT/95/98 operating systems by the staff of J. E. Edinger Associates, Inc. The heart of this system is a series of hydrodynamic and transport models. The software is equipped for geographic information system (GIS) operations and includes an automated grid generator, a control file editor, a powerful graphical post processor with quantitative 2-D and 3-D capabilities, and additional support tools such as a meteorological data processor.

The GLLVHT Model
GLLVHT (Generalized, Longitudinal-Lateral-Vertical Hydrodynamic and Transport) is a three dimensional model that computes time-varying velocities, water surface elevations, and water quality constituent concentrations in rivers, lakes, reservoirs, estuaries, and coastal waterbodies. The computations are done on a horizontal and vertical grid that represents the waterbody bounded by its water surface, shoreline, and bottom. The water surface elevations are computed simultaneously with the velocity components. The water quality constituent concentrations are computed from the velocity components and elevations. Included in the computations are formulations for friction, wind shear, turbulence, inflow, outflow, surface heat exchange, sediment transport and water quality kinetics. GLLVHT model properties are shown in Table 1.

The theoretical basis of the model was first presented in Edinger and Buchak (1980) and subsequently in Edinger and Buchak (1985). The GLLVHT model has been peer reviewed and published (Edinger and Buchak, 1995; Edinger, Buchak and McGurk, 1993; Edinger et al., 1998, Kolluru et al., 1998). Version 7.0 of

GLLVHT has been updated with the higher order transport schemes QUICKEST with ULTIMATE and linked with the GEMSS software. The GEMSS modeling kernel is designed in modular fashion for easy coupling of existing as well as user defined models. Currently the GEMSS has five modules: (1) Hydrodynamic Module (**HDM**), (2) Water Quality Module (**WQM**), (3) Sediment Transport Module (**STM**), (4) Particle Transport Module (**PTM**), and (5) Oil and Chemical Spill Module (**OSCM**).

The inputs to these modules are obtained from their respective control files and the outputs are designed for easy uploading into 2-D and 3-D display modules of GEMSS. A detailed description of the model is available on the JEEAI web site www.jeeai.com. The model has had over 40 applications including both peer-reviewed publications and agency-accepted applications.

Interior Membrane Boundary Formulation
The interior membrane boundary condition is specified at the grid cell faces. It is applied in the model as a permeability coefficient that varies between 0 and 1. This coefficient can be applied at any or all of the u, v and w cell interfaces. The membrane coefficients in u, v and w directions are represented in the model as the variables $bmu(n,k)$, $bmv(n,k)$ and $bmw(n,k)$, respectively. These variables are two dimensional arrays and are a function of cell number, n, and vertical layer number, k. The cell number n is related to grid cell location i,j as shown in Figure 1. This figure also illustrates how the membrane boundary concept is applied to a bridge opening. The interior boundary coefficients are shown relative to the velocity vector components at the bridge causeway and through the bridge opening. At the causeways the interior boundary coefficient $bmu(n,k)$ is set to zero.

In the derivation of the finite difference forms of the hydrodynamic equations, each velocity component, $u(n,k)$ and $v(n,k)$ is multiplied by its respective $bmu(n,k)$ and $bmv(n,k)$ in the continuity equations, the momentum equations, and the transport equation. The interior boundary coefficients are carried through the derivation of the surface wave equation used in the solution to the hydrodynamic equations. The $bmu(n,k)$ and $bmv(n,k)$ are set to unity unless there is an interior boundary in which case they are set to zero.

The membrane boundary condition is implemented in the GLLVHT model using the boundary condition control file. The boundary condition control file allows the user to easily change or add floating or fixed structures to the existing model grid.

Applications

The interior boundary has been applied to a wide variety of structures including locks, bridges, jetties, breakwaters and curtain walls. Curtain walls are used to minimize the entrainment of nutrient-rich inflow into the downstream epilimnion in a reservoir (Asaeda et al. 1996). Curtain walls may also be used to reduce the entrainment of heated water from a thermal surface wedge into the intake of a generating station. The curtain wall method of controlling mixing is a technically viable, low cost intake temperature control device. The use of such a curtain wall to reduce the intake temperature at the intake of a power plant in lakes and in a river will be discussed in the succeeding sections.

Flint Creek Lake

Flint Creek Lake (Figure 2) is located at northwest Arkansas near the town of Gentry. The lake has a surface area of 2×10^6 m^2 at a pool elevation of 345.9 m above sea level. The water depths are about 5 to 10 meters in the two arms at the northeast end of the lake, and 10 to 20 meters in the main part of the lake. A thermal generating station uses Flint Creek Lake for open cycle cooling. The station is a single unit plant with a net capacity of 528 MWe and was placed in service in 1978. The Station previously operated with summer and winter generation peaks, but with recently completed engineering improvements, it is anticipated that the full unit loading of 528 MWe net will be maintained 24 hours per day, 7 days per week during summer. This loading represents an overall higher level of continuous output and increased heat rejection to the Lake.

A deep cooling lake like Flint Creek Lake produces a characteristic circulation pattern. Heated water from the discharge flows down-lake on the surface, cooling as it approaches the dam. The cooled water sinks to the deep portion of the lake, and then returns up-lake along the bottom to mix with the heated discharge. The cooled bottom water is available for downstream releases if outlet structures allow access to deep portions of the lake. Larger volumes of cool water are produced during nighttime atmospheric cooling than during the daytime. Buoyancy, near-field intake and outlet dynamics, pumping-induced circulation, and diurnally varying surface heat exchange are dominant processes and need to be included in any 3-D model.

The primary objective of this study was to identify options that minimize intake temperatures so that ancillary station equipment can operate more efficiently. The temperature reduction options need to be assessed for costs, expected heat dissipation improvement, and environmental effects. The curtain wall technique satisfied these requirements and was therefore proposed for further evaluation.

The curtain wall will be located across the mouth of the intake embayment as shown in Figure 2. The wall will extend to a depth of 7 meters below the water surface. Figure 3 shows a cross-section of computed excess temperatures and velocities taken through the intake embayment as indicated in Figure 2. Excess temperature is the increase in temperature above the lake natural water temperature due to the discharge of waste heat from the station. Cross-sections without and with the curtain wall are shown. These cross-sections are from 3-D model simulations for constant meteorological conditions and a steady wind of 3 m s^{-1} from the North-East (NE), a wind direction that nominally brings heat directly from the discharge to the entrance to the intake embayment.

Without the curtain wall, excess temperature in the main portion of the lake are stratified, with values of about 6°C on the surface decreasing to 2°C near the bottom. The flow toward the intake is distributed almost uniformly through the water column. The excess temperatures near the intake approach 5°C resulting in an intake excess temperature of 5.4°C.

With the curtain wall, the excess temperatures stratification in the main portion of the lake is reduced, with values from 6°C to 3°C. The station withdrawal is from the cooler water located near the bottom at the curtain wall opening. Excess temperatures near the intake approach 4°C resulting in an intake excess temperature of 4.2°C. The ideal improvement with the curtain wall under a limiting NE wind for constant meteorological conditions is 1.2°C. Figure 4 shows the excess temperature profiles for these two cases at the profiling point indicated on Figure 2 for two dates, one in early summer and another in late summer. The figure shows that there is less stratification with the curtain wall and bottom temperatures are up to 4°C higher than for the case without the curtain wall. The decrease in stratification is partially due to the downward velocities induced in the main lake by the curtain wall, and partially due to more of the station heat by-passing the intake embayment and moving toward the dam. At the dam, this heat is pulled downward and moves back up the lake near the bottom further decreasing stratification.

The effect of wind speeds on intake temperature predictions and curtain wall performance is illustrated in Figure 5. It shows that intake excess temperatures without and with the curtain wall decrease with increasing wind speed. However, the difference between the two, i.e., the improvement with the curtain wall, stays almost constant up to a wind speed of 3 m s^{-1} and then decreases at higher wind speeds. The average wind speed for the summer months of June 1997 through September 1997 was 1.8 m s^{-1} ± 1.19 m s^{-1}, which indicates that the wind speed was less than 3 m s^{-1} about 68% of the time.

The wind speed effect results from: 1) higher winds leading to greater surface heat exchange; 2) higher winds leading to stronger vertical mixing, and therefore less stratification; and 3) the assumed wind direction being from the northeast, a direction that pushes the discharge water towards the dam thus making better use of the surface area of the lake.

The curtain wall was constructed in the fall of 1999 and continuously recording thermistor data as well as detailed ADCP velocity measurements are currently being evaluated.

Squaw Creek Reservoir

Squaw Creek Reservoir has a surface area of 13.3 million m^2 at a water surface elevation of 236.2 m and is located in central Texas. The Reservoir's volume is 186.8 million m^3 and its mean depth is 14 m. There is little natural inflow to SCR with sufficient additional inflow from Lake Granbury, north of the site to compensate for evaporation from the SCR.

Squaw Creek Reservoir provides once-through cooling water for two nuclear-based, electrical generating units. Each unit generates approximately 1165 MWe of electricity from a reactor thermal output of 3411 MWt. The waste heat load is approximately 2246 MWt for each unit. Cooling water enters the station from the western edge of the SCR about half way up the SCR, picks up waste heat from the condensers and is returned to the Reservoir at its southwestern end.

There is a single deep intake structure for both generating units. Each unit is cooled by 4 circulating water pumps with a capacity of 63.0 m^3 s^{-1}. Normally only 3 pumps per unit are run during the winter (October 1st to June 15th), resulting in a per unit rate of 47.2 m^3 s^{-1}. At full load and summer pumping, Unit 1 and Unit 2 have a condenser temperature rise of 8.5°C; at full load and winter pumping, the condenser temperature rise is 11.4°C. The residence time based on these pumping rates is 17 and 23 days for summer and winter, respectively.

As at Flint Creek Lake, several intake configurations were examined with the goal of reducing intake temperatures. The supplemental cooling options were evaluated in a straight forward manner. First the calibrated 3-D model was used to obtain the annual time-series of full-load temperature rises at the intake for the existing configuration. This temperature rise time series was then added day-by-day to a synthetic record of natural temperatures. The synthetic record was developed from meteorological data and calibrated to reservoir temperatures during periods of reduced station operations. The resulting time series is called the Base Case. Then the intake temperatures were computed with the 3-D model for each option as though the options were in operation in 1998. The reduction for each option was obtained by subtracting the intake temperature from the base case intake temperature. These reductions were subtracted from the long-term time series record of intake temperatures for the Base Case. The resulting time-series of temperature for each option was then summarized in a number of ways, including frequency-duration statistics and the average number of days exceeding 95 F (35 C) at the intake. A detailed description of the cooling water options is described in Buchak et al. (1999).

The first intake temperature reduction option includes a curtain wall to channel flow to relatively unused parts of the SCR (Option 2a, see Figure 6). The total length of curtain wall is about 3000 meters. This option decreases the intake temperature by an average of 0.04°C, an insignificant amount. This result is expected because Option 2a allows access only to shallow surface water at the upper end of the SCR. This water already participates in the surface heat exchange process because heat is carried up-lake through buoyant distribution.

The second option (Option 2b, see Figure 7) is a curtain wall surrounding the intake to allow access to cooler bottom water. The intake curtain wall placed very close to the intake. The two side cells are closed to the bottom, which prevents withdrawal from the surface warm water at these shallow ends. The bottom opening at the central cell is 7 m. Given the full-load circulating flow rate of 138 $m^3 s^{-1}$ the velocity underneath the curtain wall is estimated at 0.2 m s^{-1}. The intake temperature reductions are about 0.3 to 0.6°C. Another simulation shows that a 6-m deep curtain wall reduces the intake temperature by about only 0.3°C.

Since Squaw Creek Reservoir supplies cooling water for a nuclear power plant, utility engineers raised two issues with the curtain wall: (1) the safety-related issue of the curtain wall getting loose from its mooring and clogging the intake and (2) the environmental issue of the NPDES permit being based on the original, full-depth intake design. Modifying this design to draw bottom water would change the entrainment rates for fish eggs and larvae as well as change the impingement characteristics. For these reasons, the curtain wall option was rejected. Separately, however GEMSS was able to simulate use of a series of compressed air bubblers that would bring cool bottom water to the surface and hence to the intake. The use of GEMSS software to simulate air bubblers is presented in a companion paper by Edinger and Kolluru (1999). Continuous time-series temperature data is currently being evaluated to assure the effectiveness of the bubbler system and to confirm the model's predictions.

Charles River
Three dimension thermal transport modeling is being undertaken to find an effective way for discharging heated water from an existing thermal power plant

in the Charles River where salinity gradients exist due to the opening/closing of locks at the new Charlestown Dam (see Figure 8). The Dam is situated downstream of the river where it joins the Boston Bay. The far field model grid for the Charles River and around the power plant intake and discharge is shown in Figure 9 and Figure 10, respectively. A curtain wall is proposed right in front of the power plant intake as shown in Figure 9 using the membrane boundary condition. A sluice exists at the Museum of Science Dam with a solid wall covering the top 1m of the water column. The sluice upper wall is also simulated using the membrane boundary. The plant discharge is kept at the bottom as shown in Figure 9. The blocking of high saline and high temperature water is clearly shown in the lateral-vertical slice taken through the pump intake in Figure 11 and Figure 12, respectively. The results shown here for this application is only for illustrating the use of membrane boundaries in 3-D modeling.

Conclusion

A simple membrane boundary condition is developed for the three dimensional model, GLLVHT. Its application to real engineering flow problems was illustrated by implementing curtain wall to reduce the entrainment of surface heated water into the intake of a thermal power plant. Many different configurations such as breakwater, jetties were simulated using the membrane boundary condition.

Acknowledgement

The authors wish to thank Dr. John E. Edinger who included membrane boundary condition in the GLLVHT model.

References

T. Asaeda, D.G. Nimal Priyantha, T. Saitoh and K. Gotoh, 1996. "A New Technique for Controlling Algal Blooming Using Vertical Curtains in Withdrawal Zone of Reservoirs". Ecological Engineering, 7, pp. 95-104.

Buchak, E. M., V. S. Kolluru, and J. Wu, 1999. Studies in Support of Circulating Water Temperature Reduction Using Three-dimensional Modeling and Frequency-Duration Analysis. Prepared for TU Electric, Glen Rose, Texas. Prepared by J. E. Edinger Associates, Inc., Wayne, Pennsylvania. 18 May.

Edinger, J. E. and E. M. Buchak, 1980. "Numerical Hydrodynamics of Estuaries," Estuarine and Wetland Processes with Emphasis on Modeling, P. Hamilton and K. B. MacDonald eds., Plenum Press, New York, pp. 115-146.

Edinger, J. E. and E. M. Buchak, 1995. "Numerical Intermediate and Far Field Dilution Modelling". Water, Air and Soil Pollution, November, 83, pp. 147-160.

Edinger, J. E., and E. M. Buchak, 1985. "Numerical Waterbody Dynamics and Small Computers". Proceedings of ASCE 1985 Hydraulic Division Specialty Conference on Hydraulics and Hydrology in the Small Computer Age. American Society of Civil Engineers, Lake Buena Vista, FL. Aug. 13-16.

Edinger, J. E., E. M. Buchak and M. D. McGurk, 1993. "Analyzing Larval Distributions Using Hydrodynamic and Transport Modelling". ASCE Third Estuarine and Coastal Water Modelling Symposium. Oak Brook, Illinois, September 8-10.

Edinger, J. E., E. M. Buchak and V. S. Kolluru, 1998."Modeling Flushing and Mixing in a Deep Estuary". Water, Air and Soil Pollution, 102, pp. 345-353.

Hirt, C. W., and J. M Sicilian. 1985. "A Porsosity Technique for the Defnition of Obstacles in Rectangular Cell Meshes". Proceedings of 4th International Conference on Ship Hydrodynamics, National Academy of Sciences, Washington D.C.

Kolluru, V. S., E. M. Buchak and J. E. Edinger, 1998. "Integrated Model to Simulate the Transport and Fate of Mine Tailings in Deep Waters." Proceedings of the Tailings and Mine Waste '98 Conference, Fort Collins, Colorado, USA, January 26-29, pp. 599 – 609.

Edinger, J. E., V. S. Kolluru, 1999. Implementation of Vertical Acceleration and Dispersion Terms in an Otherwise Hydrostatically Approximated Three-Dimensional Model", 6th International Conference on Estuarine and Coastal Modeling", New Orleans, LA, USA, November 3-5.

Table 1   General Properties of GLLVHT Model

Property	Description	Advantage
ΔX, ΔY, ΔZ	Variable from cell to cell. Curvilinear	Fit shorelines precisely, provide more refined grid detail where needed. Each cell has its own orientation for accurate orientation of winds and for mapping. Can map to GIS.
Interior Boundaries	Yes	Representation of interior structures such as breakwaters, marinas, underflow/overflow curtain walls.
Vertical momentum	Included. Relaxes Hydrostatic Approx.	Important for drawdown at outflow structures, mixing devices, and accurate representation of water surfaces in regions of large horizontal velocity changes.
Discharge Momentum	All three directions	Used for proper representation of high velocity discharges.
Time Stepping Solution	Implicit solution over all space on each time step.	Not limited by the Courant wave speed criterion of $\Delta t < \Delta x/(gHmax)^{0.5}$
Coriolis Acceleration	Variable with latitude. Incorporated in implicit part of the time step computations.	Can do large water bodies with large time steps.
Transport Scheme	Quickest, Ultimate	Better prediction of constituent profiles in regions of sharp changes
Turbulence Closure	Higher Order Schemes	Better description of turbulence in regions of rapid changes in bathymetry and around structures. Also at density interfaces.
Wind Speed	Variable through time and across grid	Realistic representation of wind events on a water body.
Surface Heat Exchange	Time varying term by term heat budget	Accurate representation of diurnal variations in heat exchange.
Highest Level Water Quality Model	WQ3DCB coupled with sediment exchange model	More realistic representation of processes taking place. WQ3DCB now includes dinoflagellates with vertical light dependent vertical migration.
Other Supported Routines and Processes	Sediment transport Spill Model Toxics Model Intake Entrainment Model.	Additional routines can be included in a modular fashion and run directly in GLLVHT on a real time basis.

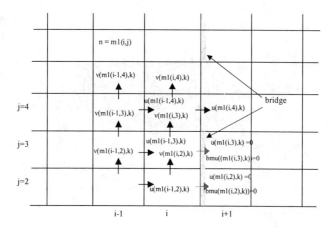

Figure 1   Interior membrane boundary set up for a bridge opening into a canal.

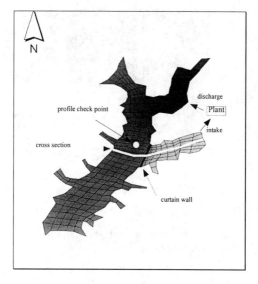

Figure 2   Flint Creek Lake showing the location of the generating station and proposed curtain wall across the mouth of the intake embayment.

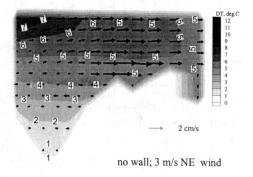

no wall; 3 m/s NE wind

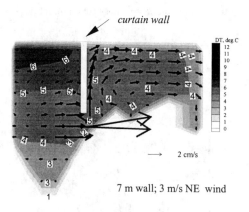

7 m wall; 3 m/s NE wind

Figure 3   Flint Creek Lake cross-section showing excess temperatures and velocities with and without the proposed curtain wall.

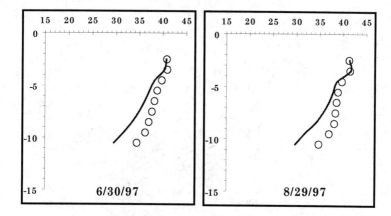

Figure 4   Excess temperature profiles for these two cases at the profiling point. The line represents profile without curtain wall and circle symbols are results with the 7-m curtain wall.

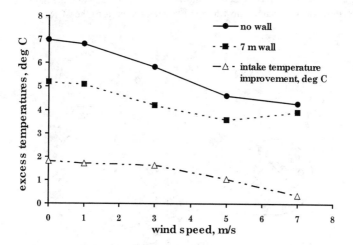

Figure 5   Excess temperature at the intake with no wall, with a 7 meter wall, and the improvement (no wall minus 7 meter wall) as a function of wind speed.

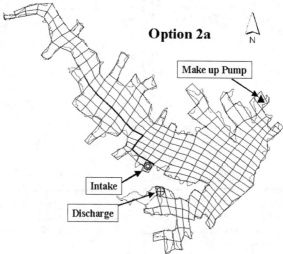

Figure 6   Squaw Creek Reservoir showing general features and the option of diversion walls to channel flow to relatively unused parts of the SCR.

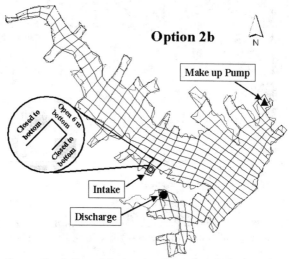

Figure 7   Squaw Creek Reservoir showing the option a curtain wall surrounding the intake and which allows access to cooler bottom water.

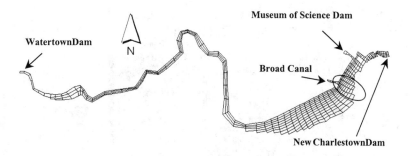

Figure 8   GLLVHT model grid for Charles River.

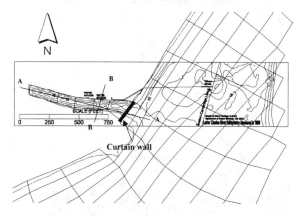

Figure 9   GLLVHT model grid in the vicinity of power plant intake and discharge. Also shown is the location of the curtain wall

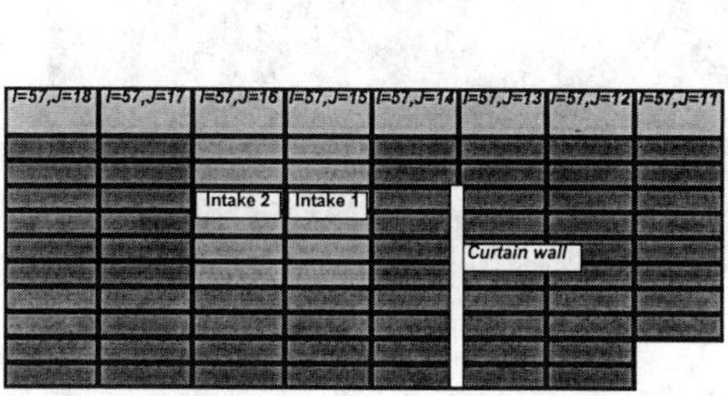

Figure 10   Cross-sectional view along transect AA showing intake location and curtain wall.

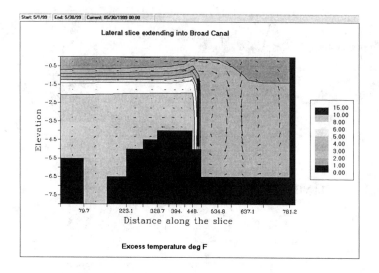

Figure 11 Excess temperature contours along transect AA. The curtain wall inhibits the passage of high excess temperature water into the intake.

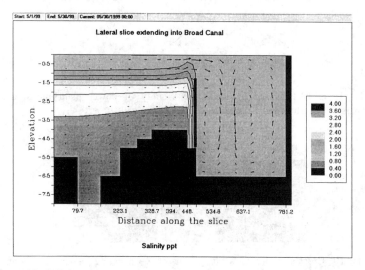

Figure 12 Salinity contours along transect AA. The curtain wall inhibits the passage of high saline water into the intake.

Using GIS as an Interface for 3D Hydrodynamic Modeling

I. Morin[1], K. Hickey[1], and M. Greenblatt[1]

Abstract

A Geographical Information System (GIS) software package was applied as the main interface to the three-dimensional, finite-difference hydrodynamic and transport model HEM-3D. The work was carried out as part of a numerical modeling study in support of an outfall relocation study in an estuary in Nova Scotia, Canada. Circulation and transport of a conservative tracer were modeled to predict effluent mixing and flushing. The GIS application was used throughout the project to provide an integrated environment for processing and viewing the data and model results. Data for the model setup were obtained in the form of digital nautical chart and field measurements at known GPS locations. The model was therefore setup to preserve all geographic references.

The GIS application was also used to view the results of the numerical simulations and to provide spatial analysis capabilities for evaluating the scenarios considered. A number of outputs were generated to illustrate various aspects of the results, such as particle-tracking paths, velocity vector plots, 2-D contour plots of water quality parameters, and 3-D view of the bathymetry. The GIS application provides an integrated environment into which multiple layers of information such as field measurements, numerical modeling results and location of critical habitats can be combined, thus providing a more complete picture of the issues at hand. The visual interface greatly facilitates the interpretation of results and enhance communications with the project's stakeholders.

---

[1] ENSR, 35 Nagog Park, Acton, MA 01720

Introduction

The widespread use and relative low cost of PC-based Geographical Information System (GIS) software, such as ESRI's ArcView, constitute one of its main attractions as an interface to hydrodynamic modeling. While customized modeling applications, such as SMS (2-D Surface Modeling System by Boss International) or MIKE (Danish Hydraulic Institute modeling environment), the GIS environment presents unrivaled functionality for sharing data and combining data from a multitude of sources and disciplines, such as maps of water uses and critical ecological habitats. ESRI ArcView software is widely used and a large amount of information layers are readily available from state and federal agencies as "shapefile" coverages (the ArcView format). The combination of modeling results with other levels of information in an integrated GIS environment offers appealing possibilities for disseminating the results of a study and provides a tool to better understand the issues at hand. Moreover, ArcView like most GIS commercial software enables the user to develop customized routines, or scripts, to analyze and perform operations on geographical information.

Approach

The present work was carried out as part of a hydrodynamic modeling study to relocate an existing industrial effluent discharge in an estuary in Nova Scotia, Canada (Hickey et al., 1999). The study involved the collection of field data, the development and calibration of both a near-field diffuser model, and a far-field three-dimensional hydrodynamic model. The models were used together to identify the optimum location for the proposed effluent discharge. The three-dimensional finite difference hydrodynamic and transport model HEM-3D (Shen et al., 1997) was used to simulate circulation in the estuary and evaluate the transport and flushing of the effluent. The HEM-3D model is composed of a solver and a simple grid generation program, both distributed with their source code. At the present time, HEM-3D does not have a dedicated user interface, therefore all pre- and post-processing tasks must be performed outside of the model using other software.

In this project, ESRI's GIS software ArcView was used to perform pre- and post-processing tasks for HEM-3D. The GIS platform was selected because of its ability to handle spatial data generated during the project and three-dimensional information. A combination of ArcView's standard features and customized programs or "scripts", were used to prepare the input files to the model, particularly the computational grid, and to process and display simulation results. A number of different outputs were produced in ArcView to illustrate the various aspects of the model results: time series, particle-tracking plots, contour plots of water quality components, velocity vectors plots, and three dimensional representations of the bathymetry.

GIS Interface

Preparation of the Computational Grid

Data for the development of the computational grid was obtained in the form of a digital nautical map of the study area. A series of customized ArcView scripts were developed to generate a finite difference grid for the HEM-3D model from the soundings and contours found on the digital nautical chart. The GIS environment greatly accelerated the grid generation process by providing a visual interface to the study area and capabilities to generate a grid file that was compatible with that expected by the hydrodynamic model.

The ArcView Spatial Analyst extension (ESRI) was used to interpolate a regularly spaced grid within specified boundaries (shoreline, upstream and offshore boundaries). The grid file obtained was then modified using a custom script that translated the water depth into the appropriate cell characteristics as required by HEM-3D (land, water, and shoreline identifiers) by considering the depth at any given grid cell with that of its neighbors. The result is a regularly spaced square grid of 1,365 water cells which is exported to ASCII files in a format compatible with that expected by HEM-3D, i.e. a listing of cell indexes with their corresponding depth and UTM coordinates, and a file containing cell type identifiers. Figure 1 shows the computational grid overlaid on the nautical chart of the study area. A number of grids were produced during the course of the project with varying grid resolution and position of the boundaries. These grid modifications, which would otherwise have been very time consuming, were implemented relatively quickly. Moreover, the ability to overly the grid to the nautical chart of the area and interactively modify the grid was a great asset over the "default" HEM-3D grid generator which consists in manually editing the ASCII files.

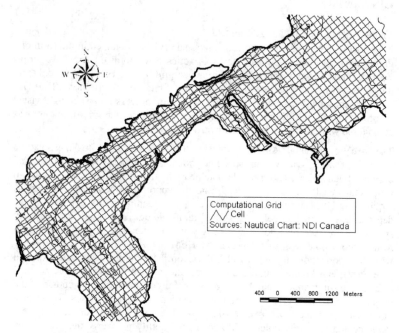

Figure 1:　　　View of the digital nautical chart of the study area showing the computational grid.

Model Calibration

The project included a field study to collect measurements of flow, velocity and water quality parameters at various locations of the study area and is detailed in Hickey *et al.* (1999). Figure 2 shows the location of Acoustic Doppler Current Profiler (ADCP) transects and water quality and tidal height measurements collected during the field effort of December 1998. The position of each measurement location was obtained from a Global Positioning System (GPS) unit installed on the boat during the field survey.

The field measurements were later mapped into the GIS application and used for calibration of the hydrodynamic model by comparing their value with those predicted by the model for the corresponding time during the tide cycle. The GIS environment facilitated the comparison of model results with the field measurements by providing both visual and quantitative basis for the comparison.

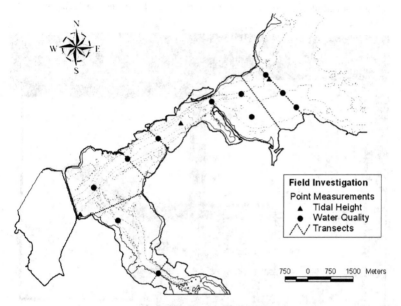

Figure 2:   Field investigation of December 1998. Location of water velocity
              transects, water quality measurements and temporary tidal gauges.

Post-Processing and Presentation of Results

The GIS application was used to view and analyze the results of the
hydrodynamic simulations. Modifications were made to the HEM-3D source code
to output the results in a format that would facilitate their importation into
ArcView. Since each of the output includes the projected UTM coordinates of
each cell of the computational grid, the files can easily be viewed as overlays to
the existing set of data and basemaps. Outputs from the numerical model typically
consisted, for each selected time step, of x, y, and z coordinates for each grid cell,
along with values for predicted variables such as the tracer concentration or the
salinity. Water velocity was given in terms of vector components u and v at each
grid cell and vertical layer of the computational domain.

The GIS environment provided a number of very useful features for analyzing and
displaying the simulation results. Standard tools allow the user to identify values
by clicking at locations on a view of the study area, to query the underlying data
and to produce charts based on the values selected. Figure 3 presents a view of the
ArcView interface where a time series of concentrations in each vertical layer at a

selected location was created by simply clicking at the desired location on the
view of the study area.

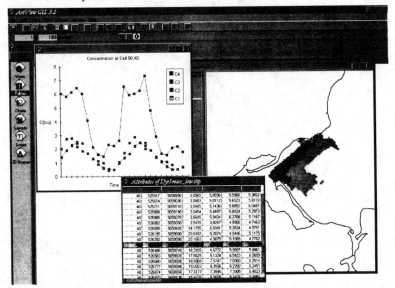

Figure 3:   A view of the GIS environment showing some of the model
              results. The time series shown in the upper left portion of the
              screen was created by querying the results data file (shown at the
              bottom and illustrated to the right) at a given point of the grid.

The tracer concentration and salinity files were imported as point files for each
time period of interest (typically at maximum flood, maximum ebb, low water
slack, and high water slack tides). When displayed in ArcView, the raw output
file from HEM-3D, which contains a sequence of data by time step appears as an
animation. In order to produce contour plots of concentrations, the point files
were interpolated using ArcView Spatial Analyst Extension. The result is a series
of snapshots of the surface or bottom concentration at specific times of the
simulation. An example of a contour plot of tracer surface concentration is shown
in Figure 4. The contour plots are stored as grid files. ArcView has built-in
functions to analyze, compare, and transform grid files. This feature proved
particularly useful in determining the extent of the zone of influence of the
discharge and comparing proposed discharge locations.

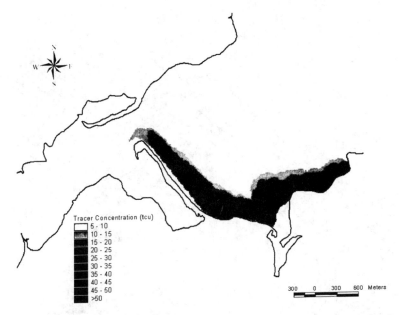

Tracer Concentration (tcu)

- 5 - 10
- 10 - 15
- 15 - 20
- 20 - 25
- 25 - 30
- 30 - 35
- 35 - 40
- 40 - 45
- 45 - 50
- >50

300   0   300   600 Meters

Figure 4:   Example of contour plot. Predicted surface concentration of conservative tracer at maximum flood tide.

Scripts were developed to transform the HEM-3D velocity outputs, given as u and v vector components, into line shapefiles (an ArcView file format) representing the vector magnitude and direction. The result is a series of velocity vector plots for each layer and time period. Velocity vector plots present a visual representation of the circulation in the estuary and, overlaid on the bathymetry or tracer concentration, provide further insight on the transport patterns observed. An example of velocity vector plot is shown in Figure 5, which presents surface water velocity during maximum flood tide.

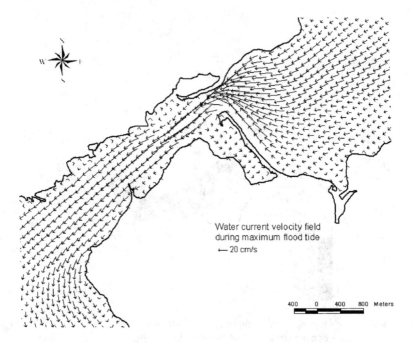

Figure 5:   Example of vector plot. Surface water current velocity at maximum flood tide.

Finally, the 3-D visualization and animation capabilities of the GIS software were also evaluated. The Arcview 3D Analyst extension was used to construct a three-dimensional representation of the study area and bathymetry. The 2D grid file generated during preparation of the computational grid can be easily converted into a TIN surface. The tracer concentration was also displayed on the 3D representation. It is possible to display the tracer concentration at each of the grid cells, using the x, y, and z coordinates, as points in the 3D representation. However, ArcView doesn't yet have the 3D interpolation and solid representation capabilities to display the plume satisfactorily. The ArcView 3D interface was used to provide a quick view of the plume and get an idea of how it evolved in time, but to generate report-quality "mpeg" computer animations of the transport of the effluent plume with the tide, the software EarthVision was used instead since it produces much better representations of three-dimensional objects.

Conclusions

This paper presented some of the possibilities offered by the use of a GIS application as a pre- and post-processing interface to the three-dimensional hydrodynamic model HEM-3D. The use of GIS software, in this case ArcView (ESRI), facilitated the preparation of model inputs, particularly the computational grid. It also enabled the creation of a number of outputs to illustrate the many variables of the system.

The most unique and appreciable gain from the use of a GIS application as the interface to a modeling project is that it provides a single, integrated environment into which can converge the various components of the project: base information, field measurements, numerical modeling, analysis of model results, and communications to stakeholders.

References

Shen, J., Sisson M., Kuo A., Boon J., and Kim S. (1997) Three-dimensional numerical modeling of the tidal York River system, Virginia. Estuarine and Coastal Modeling, Proc. of the Fifth International Conference, Alexandria, Virginia, Oct. 1997.

Hickey, K., Morin, I., Greenblatt, M., and Gong, G. (1999) Three-dimensional hydrodynamic of an estuary in Nova Scotia, Estuarine and Coastal Modeling, Proc. of the Sixth International Conference, New Orleans, Louisiana, Nov. 1999.

Numerical simulation of wave propagation in the entrance of the Tagus estuary.

Filipa S. B. F. Oliveira

Abstract

A two-dimensional simulation of wave propagation is used to investigate wave transformation over the complex bottom geometry of the Tagus estuary entrance. The application of a hyperbolic numerical model based on the mild-slope equation in such large coastal area, where wave transformation is due to the simultaneous occurrence of the processes of refraction, diffraction, reflection and breaking, is done for the first time.

Wave climate analysis is obtained for a vital area, the access to the port of Lisbon, where there is no field data available. Like in other coastal areas worldwide, navigation routes and strong currents are an obstacle to find strategic locations for wave-rider stations.

Model results reveal the sheltering effects of the existent submerged sand bars in the estuary entrance against small frequency wave energy entering through the navigation channel. High frequency incoming waves are less susceptible to dissipate energy through the process of wave breaking over the bars, thus causing higher disturbance into the channel.

1. Introduction

This paper is a contribution to the understanding of the wave propagation phenomenon in a complex coastal area that faces the Atlantic Ocean, the entrance

Laboratório Nacional de Engenharia Civil, Av. do Brasil, 101, 1700-066 Lisboa, Portugal of the Tagus estuary. The Tagus estuary has two very distinct regions: the upper estuary and the lower estuary (Fig. 1). The lower estuary is on itself a natural harbour offering to the port of Lisbon excellent conditions for the implementation of several facilities distributed along the banks of the estuary, mainly in the north bank. The entrance of the estuary is characterised by a navigation channel, about 20 m deep, limited on both sides by submerged sand bars, about 5 m deep: the Bico do Pato to the North, and the Bugio bar, to the South. The sand bars follow the alignment of the navigation channel towards sea side from the cape of S. Julião at the North side, and the Bugio Island at the South side.

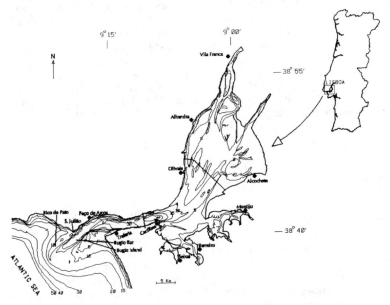

Figure 1 – The Tagus estuary: bathymetry (in metres) and places identification.

The morphology of the entrance has been changing over recent years (Oliveira, 1992): between 1929 and 1985, the Bugio bar has extended 700 m to the North and Bico do Pato has shifted 850 m South–East. An important effort has been done towards the understanding of the physics of the system not only in terms of tidal propagation (e.g. Silva and Oliveira, 1995; Fortunato et al., 1997) and sediment transport (Vicente and Clímaco, 1997), but also in terms of water quality issues, like contamination by heavy metals (e.g. Andreae et al., 1983; Lima et al, 1986) and faecal material (Câmara et al, 1987).

The existence of the navigation route, with a vast amount of traffic (Oliveira, 1992), and strong currents is an obstacle to the installation of wave-rider stations in the entrance of the estuary. Thus, a prognostic wave propagation model, like the one here presented, is crucial to support management decisions with repercussions in the lower estuary.

The numerical model used in this study is based on Copeland's formulation (Copeland, 1985) of the mild-slope equation. The model describes linear and dispersive gravity wave propagation over coastal regions of mild-slope sea bottom, where wave transformation is due to the simultaneous occurrence of refraction, diffraction and reflection. A first analysis of the topography of the estuary entrance, lead to the prediction that breaking would also play a major role in the process of wave transformation over the two existing sand bars. Thus, the process of energy dissipation through wave breaking was included in the model.

The numerical model applied in this study is introduced in section 2, where the legitimacy of its application is also analysed. In section 3, the methodology used for the model application is presented, the tests performed are described and the results obtained are discussed. Finally, section 4 presents the concluding remarks of the present study.

2. Numerical model

2.1 Governing equation

The model is based on a transient or time dependent form of the mild-slope equation. The mathematical formulation is a hyperbolic system of first order equations, the equation of continuity and the equation of motion, and is discretised using a finite differences scheme. This formulation describes linear and dispersive gravity wave transformations from deep to shallow water due to the combined physical processes of refraction, diffraction and reflection. The mild–slope equation derived by Berkhoff (1972, 1976) is based on the principle of conservation of energy under the assumption that the bed is mildly sloping ($\nabla h/kh = O(\varepsilon) \ll 1$, where ε is the wave steepness parameter); that is, the rate of change of water depth is small within a characteristic wave length, and the water motion is irrotational, which allows the introduction of a wave potential function. Energy exchange with the exterior system, like wind action, wave breaking, bottom friction and bed percolation was neglected.

The mild-slope equation for transient conditions can be written as (Booij, 1983)

$$\nabla \left(C\, C_g \, \nabla \eta \right) + (k^2 \, CC_g - \omega^2) \eta - \frac{\partial^2 \eta}{\partial t^2} = 0, \tag{1}$$

where C = wave celerity, $C_g = \dfrac{\partial \omega}{\partial k}$ = group velocity, η = water surface elevation relative to the sea water level (SWL) by passage of a wave, k = local wave number governed by the dispersion relation

$$\omega^2 = gk\tanh(kh),\tag{2}$$

ω = wave angular frequency, h = local water depth and g = gravitational acceleration.

The appropriate harmonic solution for a steady state condition is

$$\eta(x,y,t) = \phi(x,y)\,exp(-i\omega t)\tag{3}$$

where ϕ = velocity potential, so

$$\frac{\partial^2 \eta}{\partial t^2} = -\omega^2 \eta.\tag{4}$$

Substituting (4) in (1) gives

$$\nabla(CC_g \nabla\eta) + \left(k^2 CC_g - \omega^2\right)\left(-\frac{1}{\omega^2}\frac{\partial^2 \eta}{\partial t^2}\right) - \frac{\partial^2 \eta}{\partial t^2} = 0\tag{5}$$

which is the same as

$$\nabla\left(CC_g \nabla\eta\right) - \frac{C_g}{C}\frac{\partial^2 \eta}{\partial t^2} = 0.\tag{6}$$

From this equation, Ito and Tanimoto (1972) derived a hyperbolic system of first order equations:

$$\nabla Q + \frac{C_g}{C}\frac{\partial \eta}{\partial t} = 0\tag{7}$$

and

$$\frac{\partial Q}{\partial t} + CC_g \nabla\eta = 0,\tag{8}$$

where $Q = C_g\,\eta$ = flux (vertically integrated function of particle velocity). The first is the equation of continuity and the second is the equation of motion.

This system contains the governing equations for the hyperbolic model and the first order derivatives are discretised by a finite difference technique:

$$\eta_{ij}^{t+\Delta t/2} = \eta_{ij}^{t-\Delta t/2} - \left(\frac{C}{C_g}\right)_{ij}\left[Qx_{i+1,j}^t - Qx_{ij}^t\right]\Delta t / \Delta x - \left(\frac{C}{C_g}\right)_{ij}\left[Qy_{i,j+1}^t - Qy_{ij}^t\right]\Delta t / \Delta y\tag{9}$$

$$Qx_{ij}^{t+\Delta t} = Qx_{ij}^t - \left(CC_g\right)_{ij}\left[\eta_{ij}^{t+\Delta t/2} - \eta_{i-1,j}^{t+\Delta t/2}\right]\Delta t / \Delta x\tag{10}$$

$$Qy_{ij}^{t+\Delta t} = Qy_{ij}^t - \left(CC_g\right)_{ij}\left[\eta_{ij}^{t+\Delta t/2} - \eta_{i,j-1}^{t+\Delta t/2}\right]\Delta t / \Delta y\tag{11}$$

where t is the instant time, i is the index of the nodes in the x direction and j is the index of the nodes in the y direction for each cell as follows:

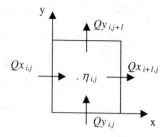

The fluxes in the x and y directions, Qx and Qy, are calculated at a time $\Delta t/2$ ahead of the corresponding values of η and are offset from the locations of η by $\Delta x/2$ and $\Delta y/2$, respectively.

The wave height H, is calculated at each cell from the root-mean-square value of η, $H_{i,j} = 2\left(2\overline{\eta_{i,j}^2}\right)^{\frac{1}{2}}$.

The process of energy dissipation through breaking was introduced in the model through the criterion breaker height to breaker depth ratio. The constant value H/d=0.78 was used to limit wave height growth. Although this criterion is not the most accurate to predict wave breaker height, because important factors like wave steepness are not taken into consideration, it is still very much used in numerical models to predict wave decay in the nearshore region. Despite a reasonable knowledge of the wave breaking phenomenon (e.g. Hamm et al., 1993; Oliveira, 1997), its incorporation in deterministic models of wave propagation remains a task to be largely improved.

2.2 Boundary conditions

To simulate an unbounded domain of coastal zone the model deals with two types of open boundary conditions (BCs): driving boundaries and downwave boundaries. Driving boundaries are considered as the open boundaries through which incoming waves enter the domain of calculation. Downwave boundaries are considered as the open boundaries through which waves leave the domain of calculation.

For the first type, Copeland (1985) introduced what he called an ad-hoc physical approach equivalent to the most simple (first order) radiation BC given by Engquist and Madja (1977), that is

$$\frac{\Delta Qx}{\Delta x} - \left(\frac{\cos\theta}{C}\right)\frac{\Delta Qx}{\Delta t}\bigg|_{x=0} = 0, \tag{12}$$

where θ = angle between wave direction and the direction normal to the boundary.

The reflected or backscattered wave travelling back towards the interior of numerical domain is absorbed at the boundary by calculating its value which is derived by decomposing the flux normal to this boundary, Qx, into the required driving function qx and the reflected wave.

The required driving function at the boundary at a time $t+\Delta t$ is

$$qx_{0,j}^{t+\Delta t} = \eta C_g \cos\theta . \tag{13}$$

The reflected wave at the boundary, defined by QE at a time $t+\Delta t$, can be interpolated from the reflected wave at the previous time step t, calculated at the boundary (location 0,j), $E1$, and at the row adjacent to the boundary (location 1,j), $E2$, as it follows:

$$QE = E1 - (E1 - E2)\, C\, \frac{\Delta t}{\Delta x \cos\theta} \tag{14}$$

where

$$E1 = Qx_{0,j}^t - qx_{0,j}^t \tag{15}$$

and

$$E2 = Qx_{1,j}^t - qx_{1,j}^t . \tag{16}$$

Hence the value of the normal flux to the boundary used in the computation is

$$Qx_{0,j}^{t+\Delta t} = qx_{0,j}^{t+\Delta t} + QE . \tag{17}$$

For the second type, the downwave boundaries, the value of the flux at the boundary is calculated from the flux in the adjacent upwave grid row affected by a time delay τ and a weighting factor AF, both parameters calculated for a certain reflectivity. Copeland (1985) derived these two parameters and as a result the flux at the boundary at an instant t can be written as

$$Qx_{N,j}^t = AF\, Qx_{N-1,j}^{t-\tau} \tag{18}$$

considering the (N,j) location at the boundary and the (N-1,j) location at the adjacent upwave grid row.

2.3 Applicability of the model

The suitability of the numerical model, to simulate wave propagation in the entrance of the Tagus estuary, requires the validation of the assumptions considered during its derivation.

One of the assumptions is the sea bottom mildly sloping, although Booij (1983), comparing results obtained from the mild-slope equation discretised by a finite elementes technique with the full linear equations (3-D model), reported favourably on the use of the mild-slope for quite large slopes, like 1:3 or even the order of unity for certain circumstances. Despite the complexity of the bottom geometry of the entrance of the estuary, the sea bed is generally mildly sloping, with the exception of the areas near the banks, which nevertheless do not violate the validation conditions.

The other assumption regards the legitimacy of the application of the linear wave theory. In a domain where dispersive wave characteristics prevail over nonlinearities, this assumption is satisfied by imposing restrictive values for the wave steepness, H/L, and the Ursell or Stokes parameter defined by Stokes (1847): $H/L \ll 1$ and $Ur = L^2H/d^3 \leq 75$, respectively. For the tests performed in this study, the above conditions are satisfied.

3. Model application

3.1. Numerical domain

Three numerical domains, A, B and C, were established to study wave propagation in the entrance of the estuary. Each numerical domain covers a rectangular area of 12x6 Km^2.

The spatial discretization consists on a regular grid of $\Delta x = \Delta y = 10.0$ m. The grid size was established based on the requirement of a minimum of about 8 nodes per wave length for a correct representation of the physical phenomenon. For each numerical domain the total number of calculation nodes was 720 000.

The three numerical domains were implemented for the direction of the main side of each rectangular area equal to: 210° (S-30°-W), 240° (S-60°-W) and 270° (W), for the numerical domain A, B and C, respectively. These directions were established to obtain offshore driving boundaries normal to the incident waves.

The location of each numerical domain can be done through the identification of a reference point, the lowest left boundary node, of numerical coordinates (x,y) = (0,0). The geographic coordinates of these points are in Table 1.

Domain (numerical coordinates)	Latitude	Longitude
A$_{(0,0)}$	38° 36′ 18″	9° 17′ 24″
B$_{(0,0)}$	38° 36′ 56″	9° 19′ 40″
C$_{(0,0)}$	38° 38′ 54″	9° 23′ 00″

Table 1 – Geographic coordinates of the numerical domains reference point.

The SWL considered was the mean sea level (MSL), 2 metre above the chart datum (CD). For these conditions, the water depth at the entrance of the estuary can be seen in Figs. 2, 3 and 4, for the domain A, B and C, respectively.

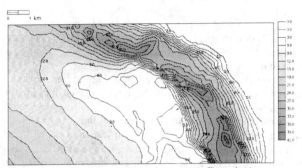

Figure 2 – Bathymetry (isolines in metres). Domain A.

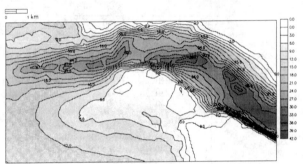

Figure 3 – Bathymetry (isolines in metres). Domain B.

Figure 4 – Bathymetry (isolines in metres). Domain C.

3.2. Numerical tests

Due to the lack of wave data in the offshore side of the Tagus estuary, the wave climate at the entrance of the area of study was estimated based on the following procedure:
1. Transference to offshore of a wave climate measured in the nearshore area of Figueira da Foz, about 170 km north of the Tagus estuary entrance.
2. Transference to the entrance of the area of study, of the offshore wave climate obtained in 1.

Both transferences were made based on the application of a wave ray model, that simulates only the processes of shoaling and refraction, used backwards in step 1 and forwards in step 2.

The wave climate at the offshore side of the entrance of the estuary is presented in Table 2, together with the identification of the tests performed. The wave climate is characterised by the incident wave direction, θ, in degrees, the significant wave height, Hs, in metres, and the zero crossing period, Tz, in seconds.

Case	θ (°)	Hs (m)	Tz (sec)
a1	210	4.50	10.6
a2	210	3.50	9.5
a3	210	2.50	8.5
a4	210	1.50	7.5
a5	210	0.50	6.5
b1	240	8.50	14.7
b2	240	7.50	13.6
b3	240	6.50	12.6
b4	240	5.50	11.6
b5	240	4.50	10.6
b6	240	3.50	9.5
b7	240	2.50	8.5
b8	240	1.50	7.5
b9	240	0.50	6.5
c1	270	7.50	13.6
c2	270	6.50	12.6
c3	270	5.50	11.6
c4	270	4.50	10.6
c5	270	3.50	9.5
c6	270	2.50	8.5
c7	270	1.50	7.5
c8	270	0.50	6.5

Table 2 – Wave climate at the offshore side of the entrance of the estuary and identification of the tests performed.

The numerical tests were performed to simulate the propagation of the incident waves, over the three numerical domains described in 3.1. The stability of the explicit numerical scheme was achieved by imposing a time step, Δt, which guarantees that the domain of dependence of the differential equation is contained in the domain of dependence of the discretised equation, through the Courant–Friedrichs–Lewy (CFL) condition $\left(0 < C\dfrac{\Delta t}{\Delta x} \le 1 \right)$.

3.3. Results and discussion

The numerical results presented in Figs. 5, 6, 7, 8, 9 and 10, were obtained for the cases a1, a4, b1, b8, c1 and c7, respectively, and were chosen for being representative of incident waves with low and high zero crossing period within the range of study, for each of the three incident directions. The results are expressed in terms of the wave amplification factor, which is the ratio H_s/H_{s0}, where H_s is the local significant wave height and H_{s0} is the incident significant wave height.

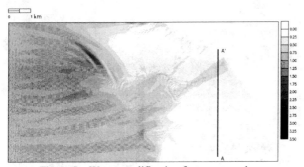

Figure 5 – Wave amplification factor: case a1.

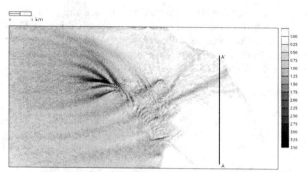

Figure 6 – Wave amplification factor: case a4.

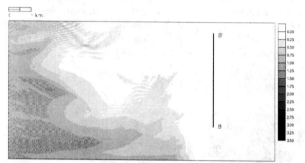

Figure 7 – Wave amplification factor: case b1.

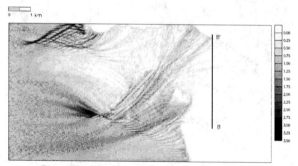

Figure 8 – Wave amplification factor: case b8.

Figure 9 – Wave amplification factor: case c1.

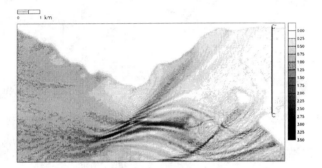

Figure 10 – Wave amplification factor: case c7.

From the 22 tests performed, it was observed that the entrance of the estuary contains 2 zones of energy focus, localised at both sides of the navigation channel, which reveal the influence of the presence of the two submerged sand bars: the Bico do Pato to the North, and the Bugio bar to the South. Such zones are more evident for the high frequency incident wave results, in Figs. 6, 8 and 10, because high frequency waves dissipate less energy through the process of wave breaking than the low frequency waves, in Figs. 5, 7 and 9.

Along the Bugio bar extension (south side of the channel), it can be observed an area of concentration of energy at the western side of the bar, followed by an area of wave decay to the east side: an initial increase of the wave amplification factor due to the process of refraction as waves propagate into shallow water, followed by its decrease due to wave breaking, as waves reach an area of shallow water of reasonable extension.

The north bank of the channel is more exposed to wave disturbance than the south bank and such is due to the sheltering effects of the Bugio bar.

The variation of the wave amplification factor along a section of the navigation channel, 1 km away from the Bugio island into the interior of the estuary, sections A – A', B – B' and C – C', for the domains A, B and C, respectively, is shown in Figs. 11, 12 and 13. The results obtained confirm: the highest exposure of the north bank; the sheltering effects of the Bugio bar in the south bank; and that high frequency waves are responsible for higher disturbance in the navigation channel because they are less susceptible to dissipate energy through the process of breaking over the bars at the entrance of the estuary than small frequency waves.

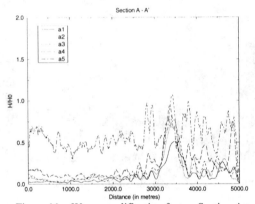

Figure 11 – Wave amplification factor. Section A – A'.

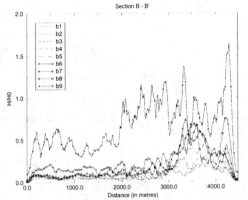

Figure 12 – Wave amplification factor. Section B –B'.

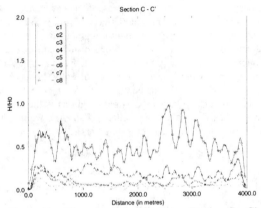

Figure 11 – Wave amplification factor. Section C – C'.

4. Conclusions

A numerical model of wave propagation was used, as a prediction tool, to investigate wave transformation over the complex bottom geometry of the Tagus estuary entrance. The importance of such information, within an area where up to the present date there is no field wave data available due to the difficulty to locate strategically wave-rider stations, is crucial.

The merit of this study is the number of the physical processes simulated numerically and the vast size of the coastal area where the model was applied. In fact, the simulation of simultaneous occurrence of the processes of refraction, diffraction, reflection and breaking over such large area is done for the first time.

Due to the complexity of the bottom geometry a simplistic refraction analysis based on the ray wave theory could never offer a realistic wave climate prediction. Formulations based on the Boussinesq and the mild-slope equations give a more accurate solution to the problem, however, an adequate description of the physics involved is usually compromised by a high computational cost. One of the most applied formulations for large coastal areas is the parabolic approximation of the mild-slope equation, which nevertheless, does not allow an adequate description of the underlying physics for this particular case, as the process of reflection is neglected and the process of diffraction is simulated with limitations. Thus, the advantage of the model applied in this study is that, despite being computationally more expensive than a model based on the parabolic formulation, it offers a more realistic wave climate prediction.

The numerical results reveal two areas of energy concentration in the entrance of the estuary and the sheltering effects of the submerged sand bars against small frequency wave energy entering through the navigation channel.

The major limitation of this study is the no inclusion of current effects. Wave–current interaction can increase or decrease the wave height and period and its implementation in the numerical model is currently being experimented. Another significant limitation is the mathematical modelling of wave decay due to dissipation through breaking. Instead of limiting directly the wave height, a different approach, based on the wave energy equation from where energy dissipation can be estimated, will be implemented in the future.

Acknowledgements
The writer wishes to thank to her colleagues from NPP division for their help in data preparation.

References
Andreae, M.O., Byrd, J.T. and Froelich, P.N. (1983). Arsenic, antimony, germanium and tin in the Tejo estuary, Portugal: modeling a polluted estuary. Environmental Science Technology, vol. 17, pp. 731-737.
Berkhoff, J.C.W. (1972). Computation of combined refraction-diffraction. Proceedings 13th Int. Conf. Coastal Eng., Vancouver, A.S.C.E., pp. 471-490.
Berkhoff, J.C.W. (1976). Mathematical models for simple harmonic linear water waves. Wave refraction and diffraction. Publication no. 163, Delft Hydraulics Laboratoy, The Netherlands.

Booij, N. (1983). A note on the accuracy of the mild-slope equation. Coastal Engineering, vol.7, pp. 191-203.

Câmara, A.S., Silva, M.C., Ferreira, J.G., Rodrigues, A.C., Remédio, J.M., Castro, P.P., Fernandes, T.F. and Machada V. (1987). Sistema de gestão de qualidade de água do estuário do Tejo, Report Projects Tejo 15, Lisbon, Portugal.

Copeland, G.J.M. (1985). A practical alternative to the "mild-slope" wave equation. Coastal Engineering, vol.9, pp 125-149.

Engquist, B. and Majda A. (1977). Absorbing boundary conditions for the numerical simulation of waves. Mathematics Computation, vol.31, no.139, pp. 629-651.

Fortunato, A.B., Baptista, A.M. and Luettich, Jr., R.A. (1997). A three-dimensional model of tidal currents in the mouth of the Tagus estuary. Continental Shelf Research, vol.17/14, pp. 1689-1714.

Hamm, L., Madsen, P.A., Peregrine, D.H. (1993). Wave transformation in the nearshore zone: a review. Coastal Engineering, vol.21, no.1-3, pp. 5-39.

Ito, Y. and Tanimoto, K. (1972). A method of numerical analysis of wave propagation – application to wave diffraction and refraction. Proceedings 13th Int. Conf. Coastal Eng., Vancouver, A.S.C.E., pp. 471-490.

Lima, C., Martin, J.M., Meybeck, M. and Seyler, P. (1986). Mercury in the Tagus estuary: an acute or absolete problem. In Estuarine Processes: an Application to the Tagus Estuary, Secretaria de Estado do Ambiente e Recursos Naturais, Lisbon, Portugal, pp. 413-454.

Oliveira, F.S.B.F (1997). Numerical modelling of irregular wave propagation in the nearshore region. Ph.D. Thesis, Department of Civil Engineering, Imperial College of Science Technology and Medicine, London, U.K.

Oliveira, I.B.M. (1992). Port of Lisbon improvement of the access conditions through the Tagus estuary entrance. Proceedings 23rd Int. Conf. Coastal Eng., New York, A.S.C.E., pp. 2745-2757.

Silva, M.C. and Oliveira, E.M. (1995). EXPO'98: Hydrodynamic modelling of the Tagus estuary. Report 174/95 – NET, Laboratório Nacional de Engenharia Civil, Lisbon, Portugal (in Portuguese).

Stokes, G.G. (1847). On the theory of oscillatory waves. Transactions of the Cambridge Philosophical Society, vol.8, pp. 441-455.

Vicente, C. and Clímaco, M. (1997). Protection of the Bugio fortress. Hydrodynamics and sedimentation studies. Stability of beach protection fill. Report 70/97 – NET, Laboratório Nacional de Engenharia Civil, Lisbon, Portugal (in Portuguese).

INTRUSIONS OF THE LOOP CURRENT OVER THE MS BIGHT

Germana Peggion

ABSTRACT

A numerical model is used to analyze the influence of the DeSoto Canyon in the cross-margin exchanges between the Mississippi Bight (MB) and the Gulf of Mexico (GOM).

It is found that northward excursions of the Loop Current (LC) over the MB are generally compensated by a seaward transport of cooler shelf water, with a significant mass exchange between the low-nutrient, high-salinity Gulf waters and the high-nutrient, low-salinity coastal waters. When warm, salty water from the open Gulf approaches the MB and moves westward along the DeSoto Canyon it tends to become unstable. The simulations clearly indicate the genesis of filaments and mushroom-like features that usually entrain cold and fresh coastal waters. As the deep-water intrusion regresses the small cyclonic eddies are carried away from the coastal region and contribute to the shelf deep ocean exchanges. Genesis and evolution of these features are well documented by satellite imagery.

In agreement with observations and previous studies, the numerical simulations indicate that the DeSoto Canyon is the location at which most of the cross-margin exchanges occur. Northward intrusions at the surface and mid-depth from the open ocean, are usually compensated by an offshore movement of dense shelf water down into the Canyon.

1. INTRODUCTION

The MB is a triangular, broad and flat shelf in the northeastern GOM. It is somewhat isolated by the DeSoto Canyon on the east and the Mississippi Delta

PEGGION
The University of Southern Mississippi
Building 1103, Room 249
Stennis Space Center, MS 39529
Phone (228) 688-2897, Fax (228) 688-7072
germana.peggion@usm.edu

abruptly closes the shelf on the west, and only a narrow, shallow passage is off the river mouth. The southern boundary is open to the Gulf. The shelf break is at approximately 100 m isobath. One unique feature of the MB is that tides are weak, with amplitudes on the order of a few centimeters. Water exchanges across the MB are believed to be driven primarily by wind forcing, buoyancy forces, and interaction with the LC.

The shelf circulation is generally cyclonic and wind-driven over the inner shelf (Dinnel, 1988). Winds have been observed to advect the surface layer, and it is postulated to form up- and downwelling of bottom waters at the shelf break. Monthly wind observations indicate that summer is the only period when upwelling favorable winds occur above 20% of the time (Schroeder and Wiseman, 1985). However, winds experience high variability that may generate a significant residual circulation. Strong events such as storms and hurricanes occur throughout the year. Although major storms are sporadic and quickly pass over a given location, the ocean response can last for many days and the associated energy input may substantially contribute to the cross-margin transfer of mass and momentum.

The Mobile, Pascagoula, and Pearl Rivers are the primary sources of fresh water to the MB. Under favorable wind conditions, water from the Mississippi River may also be advected into the area. Although the percentage of the Mississippi River discharge into the MB is unknown, even a small fraction of the river transport could be a large source of fresh water. River waters are usually confined to thin surface plumes; however, wind and other mixing processes may stir the fresh water down to the lower levels of the water column and create horizontal density gradients that may drive a buoyancy circulation (Brooks and Darnell, 1991).

Intrusion of the LC and associated eddies and filaments over the outer and inner shelf have been documented by satellite images (Schroeder et al., 1985; Vastano et al., 1991). Under such conditions, saline and nutrient-poor waters sweep up the shelf to displace a large portion of the resident water and dominate the shelf circulation for a period of days to a week or more. There are three documented patterns in which the LC interacts with the shelf (Kelly, 1991): 1) LC filaments move along the DeSoto Canyon to a few miles off the coast, 2) outer shelf water entrains into the LC edges, eddies, or filaments, with the LC generally located in deep waters off the continental margin, and 3) diluted open Gulf waters penetrate over the shallow shelf.

The inland movement of Gulf water is generally compensated by a seaward transport of cooler shelf water with a significant mass exchange between the low-nutrient, high-salinity Gulf water and the high-nutrient, low-salinity coastal water. Whether an equivalent flux of momentum occurs during the events is unknown.

The DeSoto Canyon is considered the most likely locus for upwelling onto the shelf. Upwelling may occur either as a result of upwelling favorable winds or interactions with the LC and its extended circulation. As stated previously, upwelling favorable winds are rare. In contrast, the relatively frequent intrusions of the LC may represent an important source of upwelled nutrients (Bogdanov et al., 1968). The objective of this paper is to use a numerical model to understand the effect of the DeSoto Canyon on the cross-margin exchanges between the MB and open Gulf. The next section briefly describes the modeling approach. Section 3 discusses the intrusion of LC over the continental shelf and shelf break of the MB. Section 4 presents the concluding remarks.

2. THE COUPLED MODELING SYSTEM

Two independent models, a coarse-grid resolution for the GOM and a fine-grid resolution for the MB, are executed in parallel and communicate the coupling variables to each other. This approach offers the benefit of modeling coastal environments, taking into account the mutual interactions between the shallow and deep waters, without the computational burden of configuring the basin domain at the high resolution required by the coastal applications.

The Princeton Ocean Model (POM) is the model of choice for the development of the coupling system. POM is a 3-D, free surface, primitive equation model that uses a bottom-following vertical σ-coordinate and a curvilinear orthogonal, horizontal grid and parameterizes vertical mixing with the Mellor and Yamada 2.5 turbulence scheme (Blumberg and Mellor, 1983). The scheme is finite differencing centered in space. The horizontal time differencing is explicit, whereas the vertical mixing is implicit. The model uses a split time step for the external mode computations. Since POM is widely used in the oceanographic community, we refer the reader to Mellor (1996) for the mathematical and numerical formulations.

Although POM can accommodate generalized curvilinear coordinates, a Cartesian coordinate system is used here to simplify the nesting algorithms. We also impose the same number of σ-levels (20) and same stretching factor to both domains. The boundaries of the fine grid coincide with grid lines of the coarse domain, and the ratio of the coarse grid size to the fine grid is an integer.

2.1 The Gulf of Mexico Model Configuration

The GOM grid size is 20 km in both horizontal dimensions. The treatment of the open boundary conditions is based on the Fletcher (1987) method. A constant 30 Sv ($10^6 m^3 s^{-1}$) inflow is prescribed at the southern port. The total outflow at the Florida Strait is nudged to 30 Sv. The experiments described in the next sections are forced by the ECMWF-1993 winds and initialized by the Minerals Management Service (MMS) climatology developed by Dynalysis of Princeton (Herring et al., 1999). Bathymetry and monthly climatological heat fluxes are from Dynalysis of Princeton (Herring et al., 1999). The domain configuration does not include river discharges. The GOM model is spun up from a state of rest.

It is known that the σ-coordinates in the presence of steep topography may lead to errors and model instability. The effects on the pressure force terms are well documented (Haney, 1991). Errors associated with the nonlinear and diffusion terms may also produce spurious heating/cooling in the temperature Eq., and equivalent forcing for the other variables (Martin et al., 1998). To prevent the model solution drift over extensive simulations, the constituent fields are nudged to climatology over a one-year time scale (Welsh and Inoue, 1999). Oey (1994, 1995) discusses the sensitivity of the GOM solutions to POM's physical and numerical parameters.

As expected, winds have little influence in determining the large-scale features (such as ring separation and migration) of the basin model. On the other hand, operational (not-climatological) winds are an important factor for reproducing an appropriate coastal circulation, with more marked currents and jets. It is found that a realistic bathymetry is essential for a correct orientation and more marked northward penetrations of the simulated LC axis. Under these conditions, an almost-permanent cyclonic eddy develops offshore of the Big Bend region. This feature is well documented by TOPEX/ERS-1 satellite images (Biggs et al., 1996).

2.2 The Mississippi Bight Model Configuration

The horizontal grid size of the MB module is 5 km (or nesting ratio =4) (Fig. 1). The model is initialized by the following procedure: 1) the temperature and salinity fields from the GOM one-year solution are interpolated over the MB grid; 2) the MB is run in diagnostic mode for 10 days; and 3) the diagnostic solution is utilized as the initial condition for the MB. Finally, the GOM and MB model are restarted together and run in parallel. This initialization procedure avoids the inconsistencies of interpolating both constituent and dynamical 3-D fields

Table 1: Buoy Stations

St	Longitude	Latitude	w.depth
1	87.95W	29.22N	447 m
2	87.33W	29.75N	101 m
3	87.34W	29.39N	481 m
4	87.34W	29.03N	1299 m
5	86.88W	29.34N	541 m

Fig. 1.: The domain and bathymetry of the high resolution coastal model. (92.1W, 26.5N) and (82W, 30.6N) are the geographical coordinates of the lower-left and upper-righ corner, respectively. Numbers marks the locations of the simulated mooring stations.

The baroclinic mode CFL stability condition of the fine domain restricts the choice of the internal time steps. It is required that the internal time step, Δt_C, of the coarse grid be a multiple of the internal time step, Δt_F, of the fine grid. Surface forcings are interpolated by the GOM model and communicated to the MB model to avoid the memory cost of storing wind and heat fluxes for the fine grid.

2.3. The Nesting Algorithm

The nesting procedure is formulated by assuming that there are two dominant scales: the small, energy-containing features and the large, slowly varying flow. The two-way algorithm develops as follows:

1) From coarse to fine - the basin model determines the location of the inflow and outflow ports at the open boundaries of the coastal model. However, there is a fundamental difference in the numerical treatment of inflow and outflow. While models need to be forced by a priori knowledge of the dynamics at the inflow ports, the outflowing solutions need to leave the domain 'undisturbed'. Therefore, inflow conditions are linearly interpolated and radiation-like schemes are imposed at the outflow areas.

2) From fine to coarse - a feedback procedure from the MB to the GOM model parameterizes and controls the transfer of energy of the sub-scale (relative to the coarse-grid spacing) features. That is, every 96 hours the MB model sends the 3-D data (u, v, T, and S) of the overlapping points. The values are nudged with the GOM solutions. The numerical algorithm is described in Peggion (1999)

3. CIRCULATION PATTERNS FROM THE NUMERICAL SIMULATIONS

The numerical simulations are able to reproduce the interactions between the shallow and deep waters at the shelf break and over the continental slope as observed and described by Kelly (1991). When warm and salty water from the open Gulf approaches the MB and moves along the DeSoto Canyon, it tends to become unstable. The high-resolution nested model clearly indicates the genesis of filaments and mushroom-like features that usually entrain cold and fresh coastal waters (Fig 2a). As the deep-water intrusion regresses, these small cyclonic eddies are carried away from the coastal region and contribute to the shelf-deep ocean exchanges. The genesis and evolution of these features have been observed and documented by satellite imagery (Leben, 1999).

At other times, filaments detach from the LC edge and move northward along the eastern side of the De Soto Canyon, approximately following the 100m isobath (Fig. 2b). At the head of the canyon the front develops meanders in which cold, coastal water is trapped. The cold filaments eventually become cold eddies that intrude into the warm, salty water of the open Gulf. Due to the presence of the cyclonic eddy located off the Florida coast, warm salty water approaches the slope and shelf break primarily from the western side and recirculates backwards to the eastern side of the canyon.

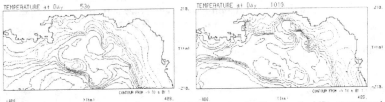

Fig. 2.: Two snapshot images of intrusion of LC filaments and eddies over the MB. a) westward and b) eastward intrusion.

During a one-year simulation, 5 mooring stations have been simulated (Fig. 1 and Table 1). Although the locations are the closest model grid points to the buoys deployed by Science Applications International Incorporated (SAIC) for MMS (Berger, 1997), the model-data comparison is not addressed in the present study. The buoys are configured in two major arrays: stations 1, 3, and 5 (Array A) are approximately along the 500m isobath at the head of the canyon; stations 2, 3, and 4 (Array B) are across the canyon from the shallow to the deep side. One-hour averaged (9 time steps) values of temperature, salinity, and horizontal velocity components are saved at each σ level. The records start Oct. 1.

Table 2: Statistical Analysis

St.	Depth	Temperature		Salinity		U- Velocity		V- Velocity	
1	5	21.42	1.37	36.09	0.08	0.47	0.17	0.10	0.13
	66	18.39	1.05	35.92	0.10	0.33	0.14	6.1 *	0.08
	380	9.84	0.46	35.32	0.04	0.13	0.13	4.3 *	0.04
2	1	16.57	1.34	35.79	0.08	1.1 *	0.16	0.10	0.13
	15	16.17	1.20	35.77	0.07	1.6 *	0.13	8.9 *	0.13
	86	13.08	0.87	35.60	0.08	-1.2 *	0.05	-2.9 *	0.15
3	5	19.58	2.14	35.98	0.12	0.56	0.46	-0.7 *	0.16
	71	15.97	1.52	35.76	0.09	0.35	0.29	-1.2 *	0.19
	410	9.62	0.41	35.31	0.04	5.2 *	0.10	0.1 *	0.04
4	15	22.10	0.65	36.13	0.02	0.29	0.15	3.3 *	0.14
	191	14.00	0.57	35.65	0.03	0.10	0.07	4.9 *	0.05
	1108	5.76	0.10	35.03	0.01	-0.5 *	0.04	-0.0	0.06
5	6	20.26	2.06	36.02	0.11	0.61	0.28	6.1 *	0.29
	80	16.35	1.41	35.78	0.09	0.38	0.19	4.5 *	0.15
	461	9.07	0.41	35.26	0.34	1.6 *	0.06	0.6 *	0.07

Tab. 2.: Mean and standard deviation of temperature (C), salinity (ppt), and horizontal velocity components (m/s) at the model σ-levels 2, 6, and 18. The symbol * indicates values in cm/s. Column 2 reports the water depth values for each station.

Table 2 summarizes the mean and standard deviation at each station for the surface (level 2), middle (level 6) and bottom (level 18) depths. With the exception of station 2 (located in the outer shelf) which is stably stratified, all other stations indicate high variability, especially at the surface and mid-column. Table 3 shows the cross correlation between the levels at each station. Surface and mid-column have high correlation coefficients, while the bottom is usually out of phase with respect to the upper strata. Station 2 presents a homogeneous response throughout the water column.

Table 3: One-year cross correlation between levels

St.	Temperature		Salinity		U- Velocity		V- Velocity	
	L6	L18	L6	L18	L6	L18	L6	L18
1_L2	99.3	-40.0	99.3	-14.0	99.2	-27.9	99.6	-20.8
2_L2	99.6	94.8	99.4	85.5	99.1	80.4	99.6	91.5
3_L2	98.7	-62.2	98.7	-39.0	97.8	-59.5	99.1	-33.8
4_L2	98.4	-99.4	99.4	-98.4	97.9	-59.5	99.1	-33.8
5_L2	98.8	-76.5	99.1	-81.5	99.0	-70.3	98.7	-32.2

Tab. 3.: One-year cross correlation between levels 2, 6, and 18 at each simulated station. Values are scaled by a factor of 100.

We are now analyzing the cross correlation of the two arrays. Table 4 indicates the depth at which the correlation coefficients of Tables 5 and 6 are computed. The depths have been chosen to minimize averaging operations over the σ-levels. Only few time series (marked with an asterisk in Table 4) are the result of averaged values over 2 levels.

Table 4: Depth of Cross Station Correlations

	Array A			Array B		
	1	3	5	2	3	4
D1	5	5	5	15	15 *	14
D2	92	99	111	51	57 *	54
D3	197	212	239	---	240	267

Tab. 4.: Depth of the cross-correlation values of Tables 5 and 6. The symbol * indicates time series averaged over two σ levels.

3.1. Array A: East-Westward propagation

Adjacent moorings have a high cross correlation, which decreases between the edge moorings. Fig. 3a shows the one-year time series at level B (about 95m). Station 3 is generally characterized by a higher variability and more marked oscillations. Although intrusion of shelf waters are found at each station, the coldest and saltiest waters are reported at the middle site, indicating the cross-margin interactions between shelf and deep ocean mainly occur at the head of the canyon. The records clearly indicate that Gulf (warm) water is more likely at the western side of the Canyon, and shelf (cold) water is more likely at the head and eastern side of the topographic feature. The records also show westward preferential pathways, between Stations 3 and 5. Two events, approximately at day 120 and 270, indicate shelf water carried by Gulf water moving westwards over both sites. The excursion of cold and warm mesoscale features is usually recorded at lower depths (Fig. 3b).

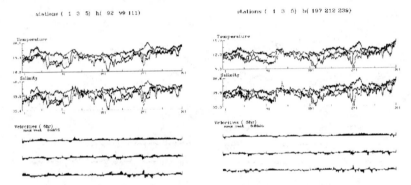

Fig. 3: Time series of the simulated moorings of Array A approximately at a) depth 95m and b) depth 200m. i) temperature, ii) salinity, and iii) velocity vectors.

Table 5: One-year cross correlations at Array A

St.	Temperature		Salinity		U- Velocity		V- Velocity	
	3	5	3	5	3	5	3	5
1_D1	73.3	62.7	40.6	19.8	83.1	83.7	36.2	12.7
3_D1	---	65.7	---	38.8	---	83.9	---	36.2
1_D2	56.5	3.2	41.6	11.9	84.6	85.5	29.8	-39.7
3_D2	---	43.7	---	41.8	---	83.3	---	-48.7
1_D3	57.1	18.5	55.3	33.5	78.0	77.1	44.7	-39.6
3_D3	---	34.0	---	33.4	---	65.4	---	-55.7

Tab. 5.: One year cross correlations between stations 1, 3 and 5 at depth D1, D2 and D3.

3.2. Array B: North-South oscillations

There is a clear separation between the shallow and deep moorings, indicating that station A is on the inner side of the coastal front. The front is generally located along the 200 m isobath, but the axis location experiences north-south oscillations. The time series of the constituent fields at station 3 show the presence of intermediate water, so that mixing is one of the dominate processes over the continental slope. As Fig. 4a illustrates, a few events report interaction of the front with station 2 and 3, while the circulation at station 4 is dominated by the dynamics of the open Gulf. Warm Gulf water has the tendency of intruding over the middle mooring for longer periods. However, at about day 75, cold water moves over the head of the DeSoto canyon and stays there for about a month. The event represents the most southward excursion of the coastal jet in our simulations.

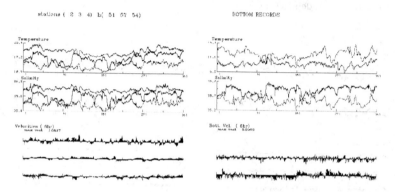

Fig. 4: Time series of the simulated moorings of Array B a) approximately 95m and b) bottom level. Same as Fig. 3.

Table 6: One-year cross correlations at Array B

St.	Temperature		Salinity		U- Velocity		V- Velocity	
	3	4	3	4	3	4	3	4
2_D1	-56.0	-48.8	-35.5	-19.1	13.4	11.4	75.8	27.2
3_D1	---	52.3	---	14.9	---	75.8	---	34.9
2_D2	-28.5	-76.7	-20.1	-58.9	10.8	5.22	57.7	10.6
3_D2	---	37.9	---	4.15	---	79.5	---	39.2
2_D3	---	43.3	---	38.2	---	28.2	---	22.1

Table 6: One-year cross correlations between stations 2, 3 and 4 at depth D1, D2 and D3.

There are few occurrences (such as day 40, 180 and 270 when warm water is moving over the head of the canyon and colder water is recorded at the shelf station. These are upwelling events induced by the interaction of the open Gulf water with the continental slope and shelf break. Bottom colder, but saltier water compensates the warm water at 50 m depth (Fig. 4b).

4. CONCLUSIONS

The MB has large topographic variations. Though wide, it narrows near the DeSoto canyon and is closed by the Mississippi delta to the west leaving only a narrow passage that connects the western and eastern side of the northern shelf. This configuration has a great effect on the along-shore wind driven circulation that dominates the inner shelf (Schroeder et al., 1987). Transport conservation requires high spatial variability of the currents which provides more energy to enhance coastal front instabilities. The instabilities are associated with meander and eddy formations and front breaks, which result in considerable exchange of mass and momentum between shelf and deep ocean.

Due to a general westward tilting of the LC axis, and the presence of a quasi-permanent cyclonic eddy off the Florida Big Bend, intrusions of warm and salty waters are usually from the west. The water flows parallel to, and interacts with, the coastal jet and recirculates backward, carrying cold eddies along the eastern side of the canyon. However, eastern penetration of LC waters have been equally observed and reproduced by our model simulations.

Because of the coarse resolution of the GOM model, we are not able to track the fate of the shelf water entrained in the LC outer edge as it leaves the nested domain. Satellite imagery and oceanographic observations suggest that shelf water does not mix with the surrounding environment; traces of it have been found along the West Florida Coast and in the Florida Strait (Maul, 1997; Ortner et al., 1995).

The DeSoto Canyon only marginally enhances the instability of the coastal jet. The topographic variations speed up the genesis and formation of eddies and meanders and modulates the spatial and temporal scale of the mesoscale variability (Chapman and Gawarkiewicz, 1995). Moreover, topographic features induce an asymmetric bottom stress that may generate substantial time-averaged currents even in the presence of zero-mean fluctuations (Haidvogel and Brink, 1986). This source of energy may also enhance the coastal front instability.

In agreement with observations and previous studies, the numerical simulations indicate that the DeSoto Canyon is the location at which most of the cross-margin exchanges occur. Northward intrusions at the surface and at mid-depth from the open ocean are usually compensated by an offshore movement of dense shelf water down into the Canyon.

Acknowledgements.

This research was supported by the Office of Naval Research under the DoD-EPSCoR program (contract N00014-96-1-0937) through the Ocean Modeling and Prediction Program. Thanks to Drs. A. Lugo-Fernandez and W. Johnson of Minerals Management Service and Drs. J. Herring and R. Patchen of Dynalysis of Princeton for their collaboration and making their data sets available. The numerical experiments were performed at the Naval Oceanographic Office Major Shared Resource Center and at the Mississippi Center for Supercomputing

Research. Thanks to my friend Ms. M. Lyle for her continuing assistance.

5. REFERENCES

Berger, T., 1997: *DeSoto canyon eddy intrusion study. Data Products; March-September 1997.* MMS Contract 1435-01-96-CT-30825. Science Applications International Corporation.

Biggs, D.C., G.S. Fargion, P. Hamilton, and R.R. Leben, 1996: Cleavage of a Gulf of Mexico Loop Current Eddy by a Deep Water Cyclone. *J. Geophys. Res., 101,* 20,629-42.

Blumberg, A.F. and G.L. Mellor; 1983: Diagnostic and prognostic numerical circulation studies of the South Atlantic Bight. *J. Geophys. Res., 88,* 4579-92.

Bogdanov, D.V, V.A Sokolov, and S. Khromov; 1968: Regions of high biological and commercial productivity in the Gulf of Mexico and Carribbean Sea. *Oceanology, 8,* 371-81.

Brooks, J.M. and R.M. Darnell, 1991: Executive Summary. *In Mississippi-Alabama Continental Shelf Ecosystem Study. Data and synthesis.* U.S. Dept. of the Interior. Minerals Managment Service. Gulf of Mexico OCS Region.

Chapman D.C. and G. Gawarkiewicz; 1995: Offshore transport of dense shelf water in the presence of a submarine canyon. *J. Geophys. Res., 100,* 13373-87.

Dinnel, S.P., 1988: Circulation and sediment dispersal on the Louisiana-Mississippi Alabama continental shelf. Ph.D. Dissertation. Louisiana State University. Baton Rouge, LA, 173 pp.

Fletcher, R., 1987: *Practical Methods of Optimization.* Wiley and Sons. 436 pp.

Haidvogel, D.B. and K.H. Brink, 1986: Mean current driven vy topographic drag over the continental shelf and slope. *J. Phys. Oceanogr., 16,* 2159-71.

Haney, R.l., 1991: On the pressure gradient force over steep topography in sigma coordinate ocean models. *J. Phys. Oceanogr., 21,* 610-619.

Herring, H.J., I. Masamichi, G.L. Mellor, C.N.K. Mooers, P.P. Niiler, L.Y. Oey, R.C. Patchen, F.M. Vukovich, and W.J. Wiseman; 1999: *Coastal ocean modeling program for the Gulf of Mexico.* Dynalysis of Princeton. Rep. 115.

Kelly, F.J., 1991: Physical Oceanography and water mass characteristics. *In Mississippi-Alabama Continental Shelf Ecosystem Study. Data and synthesis.* U.S. Dept. of the Interior. Minerals Managment Service. Gulf of Mexico OCS Region.

Leben R.R., 1999: Gulf of Mexico Web Sites. Document available at: http://www.ccar.colorado.edu/leben/gom/websites

Martin P.J, G. Peggion, and K.J. Yip, 1998: *A comparison of several coastal ocean models.* NRL Report 7322-97-9672, 96 pp.

Maul, G.A., 1977: The annual cycle of the Loop Current. Part I: Observations during one year time series. *J. Mar. Sci., 35,* 29-47.

Mellor, G.L., 1996: *POM User's Guide. Document* available at: http://www.aos.princeton.edu/WWWPUBLIC/htdocs.pom/

Oey, L.Y., 1994: Loop Current and eddies: 3-D model experiments and analyses. *Ocean Model. Group Rep.*, **15**, 42 pp. Stevens Inst. of Tech., Hoboken, N.J.

Oey, L.Y.; 1995: Eddy and wind forced shelf circulation. *J. Geophys. Res.*, **100**, 8621-38.

Ortner P.B., T.N. Lee, P.J. Milne, R.G. Zika, M.E. Clarke, G.P. Podesta, P.K. Swart, P.A. Tester, L.P. Atkinson, and W.R. Johnson; 1995: MS River flood waters that reached the Gulf stream. *J. Geophys. Res.*, **100**, 13595-601.

Peggion, G., 1999: A Two-way Nested Model for the Northeastern Gulf of Mexico on a Moderately Parallel Environment. *Procceding of the DoD HPC Users Group Conference*. Monterrey, Ca. June 7-10, 1999. In Press.

Schroeder W.W., S.P. Dinnel, W.J. Wiseman, and W.J. Merrell, 1987: Circulation patterns inferred from the movement of detached buoys in the eastern Gulf of Mexico. *Cont. Shelf Res.*, **7**, 883-94.

Schroeder, W.W. and W.J. Wiseman, 1985: An analysis of the winds (1974-1984) and sea level elevations (1973-1983) in coastal Alabama. *MASGP-84-024*. Mississippi-Alabama Sea Grant Consortium. 102 pp.

Schroeder, W.W., O.K. Huh, L.J. Rouse, and W.J. Wiseman, 1985: Satellite observations of the circulation east of the Mississippi Delta: Cold-air outbreak conditions. *Remote Sens. Environ.*, **18**, 49-58.

Vastano A., C. Barrow, C. Lowe, and E. Wells, 1991: Satellite oceanography. *In Mississippi-Alabama Continental Shelf Ecosystem Study. Data and synthesis*. U.S. Dept. of the Interior. Minerals Managment Service. Gulf of Mexico OCS Region.

Welsh S.E. and M. Inoue, 1999: Loop Current rings and deep circulation in the Gulf of Mexico. Submitted to *J. Geophys. Res.*

The Use of GIS in 3D Hydrodynamic Model Pre- and Post-Processing for Feature-Specific Applications

G. McAllister Sisson[1], John D. Boon[1], and Katherine L. Farnsworth[1]

ABSTRACT

An environmental effects study of a proposed bridge tunnel crossing in the port of Hampton Roads has shown the advantages of combining a GIS (Geographic Information System) with a three-dimensional hydrodynamic-sedimentation model, HEM-3D. The implementation of an interface between the GIS and the spatial framework of the model presented several immediate advantages, most notably the combining of high resolution satellite imagery (10-meter resolution in the horizontal) with client-specified features rendered in a Computer Aided Design (CAD) format compatible with the GIS system. Grids constructed using GIS can be considered "registered" and readily verified visually as to the accuracy of shoreline configuration, bathymetry, and the spatial representation of proposed features in particular. The latter ability is especially important for circulation-altering features such as the bridges and islands designed for possible highway crossings of Hampton Roads in the James River Estuary. The ability to examine and display features at any scale facilitates the assessment of the effects of model input changes in a series of scenario (i.e., hypothetical) test cases. A key advantage of GIS is the ability to overlay spatial data quickly and easily (e.g., high-resolution shorelines and model grid cells). The ability to enhance the commercially-available GIS package has led to the development of an ArcView® GIS extension for a "point-and-click" cell editing capability for rapid Cartesian grid development. Another extension (the commercially-available Spatial Analyst) allows one to generate contour plots depicting bathymetry input as well as to interpolate bi-linearly to assign cell depths from available bathymetric

[1]Virginia Institute of Marine Science, School of Marine Science, The College of William & Mary, Gloucester Point, VA 23062

surveys. The limitations of a Cartesian grid are reduced by 1) the use of triangular half-cells of specified orientation selected to provide a best-fit of the shoreline and 2) a variable cell size allowing increased resolution in areas of keen concern. Whereas most applications thus far has been towards pre-processing of the model, the display of model results for different output state variables, as well as at different layers of the model, is being investigated.

INTRODUCTION

The spatial accuracy in the application of 3D hydrodynamic models has increased dramatically over the past decade due to several trends:

1) additional computer resources, both storage and simulation speed, allow for high resolution computational domains that were once considered impractical
2) with the advent of global positioning systems (GPS) and differential GPS, (DGPS), field observation systems supplying data for model calibration and verification are far superior than former methods
3) due primarily to the internet, accessibility to high resolution shoreline and bathymetry datasets has facilitated use of these sources of model input
4) the widespread usage of CAD/CAM graphic packages allowing model clients to specify feature locations and dimensions with greater precision
5) improved use of datum reference information, both horizontally and vertically, due to sophisticated satellite equipment (as used in NAD-83 horizontal survey) measuring the earth to within 2 meters, and,
6) more suitable usage of map projection coordinates, such as State Plane rather than lat/lon, reducing distortion factors to less than one part in 10,000.

For these reasons, it is extremely advantageous to handle the spatial data associated with hydrodynamic models with an appropriately sophisticated geographic database management tool such as ArcView (ESRI, 1997). However, several other key features of this package exist which can greatly facilitate the modeling effort in the pre- and post-processing stages. The purpose of this paper is to illustrate how the use of ArcView can improve the modeling effort, both in accuracy and in facility of pre- and post-processing tasks. Specific tasks, including Cartesian grid construction, assignment of model cell depths, and displaying results, will be discussed more fully.

PRE-PROCESSING: GRID CELL EDIT USING VIMS ArcView EXTENSION

The integration with ArcView was accomplished with the use of the VIMS ArcView Utility. This custom utility allowed for a user-friendly interface for the manipulation of model files as shown in Figure 1.

The construction of a grid is facilitated by a set of scripts written in AVENUE, ArcView's programming language. These scripts were written to allow a "cell editing" capability whereby the user could specify the addition, removal, or re-classification of a cell or group of cells by highlighting cell positions and designating new cell types. Allowable cell types are shown in the tabs of the pull down cell-editing window shown in Figure 2. Working with a high-resolution shoreline image in the background, the optimal grid alignment is easier to discern. Using the triangular half-cell feature of HEM-3D, the user is able to approximate surface areas more closely by choosing when to use triangles and at which orientation.

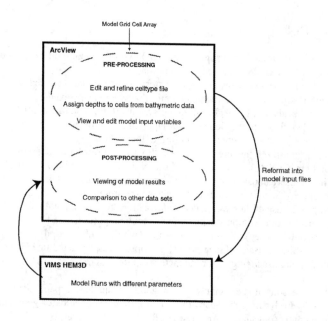

Figure 1. Protocol between hydrodynamic model and ArcView.

INTERNAL OPERATION OF THE CELL EDITING DIALOG

Scripts work with a cell-type theme which is a large array containing grid cell center locations and cell types. The "cell type" refers to the identification of a grid cell as either land, shoreline, or water. This is consistent with the needs for model input. The ArcView project uses both a cell type file and a master file. The master file is a subset of the cell type file, including only those cells which are water cells. The master file contains other attributes of the grid cells, including depth, dimensions, and bottom roughness.

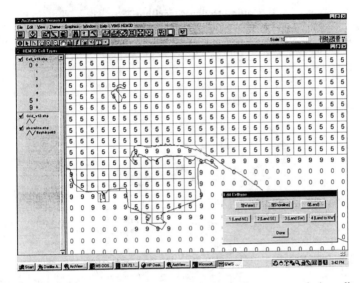

Figure 2. Display of cell type values along a shoreline segment during edit session. Cell types of '1-5' represent water cells whereas '9' represents the shoreline and '0' represents land.

These editing scripts allow for quick changes to the cell type of a grid cell, with a visual check. This leads to a change in the cell type file. However, equally important is the update to the master file, which contains among its records the cell locations and dimensions of each water cell or half-cell. As land cells are edited to become water cells, the master file has records added. As water cells are edited to become land cells, the master file has records deleted. Other modifications (e.g., orienting triangular cells) simply cause the corresponding records in the master file to be updated for cell type and dimension (see Figure 3).

Quadrilateral (full) cells are denoted by a cell type '5', shoreline cells for '9', and land cells by '0'. Cell types '1' through '4' represent the right triangular half-cells whose corners occupy three of the four corners of a quadrilateral cell with dimensions dx and dy. A triangular cell can have only four orientations with its hypotenuse facing land to the northeast (type '1'), southeast (type '2'), southwest (type '3'), or northwest (type '4').

The next step in development of these ArcView scripts is to automate two previously cumbersome protocol requirements for HEM-3D cell input. One of these requirements is that each water cell must have eight neighboring cells, one on each of its four corners and four sides, that consist of either a shoreline cell (type 9) or another water cell – but not a land cell (type 0). Also, a triangular water cell must be bordered by shoreline cells on the two sides away from the triangle and water cells on the two sides along the triangle; the remaining four corners can be bordered by cells of either type (shoreline or water).

Editing Dialog Algorithm

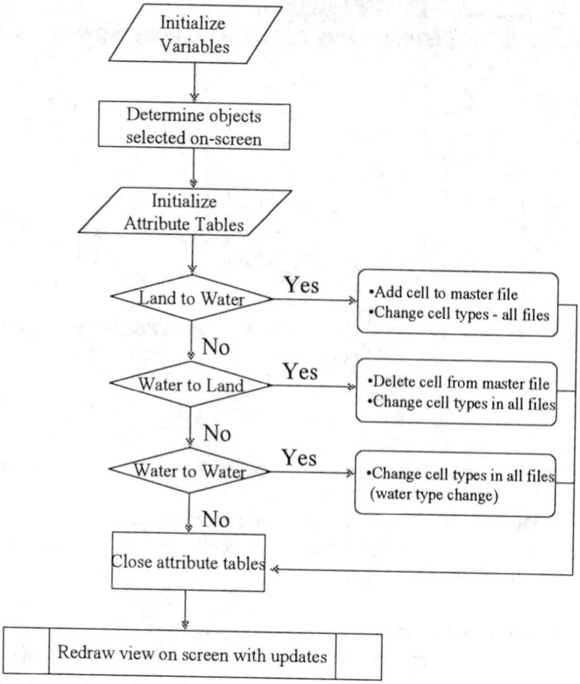

Figure 3. The algorithm employed in the cell edit utility.

ASSIGNING MODEL CELL DEPTHS USING ArcView SPATIAL JOIN

What had previously been a tedious task in the setup of a hydrodynamic model is the assignment of cell depths. This was normally done through a bi-linear interpolation algorithm using either a fixed radial search or a specified number of nearest neighbors.

Use of the ArcView Spatial Analyst module greatly streamlines this process. After the user has specified a shoreline (x,y) and a bathymetry field (x,y,z), he specifies (using the ArcView Spatial Analyst) both the interpolation of the prototype surface and the formation of contours. This is an important intermediate step in that selected contours can be visually compared prior to the final assignment of depths.

The steps involved in assigning model cell depths using ArcView are as follows (see Figure 4):

1) Import as separate themes the bathymetry data (x,y,z), shoreline data (x,y), and cell center locations (x,y). Form a continuous polyline from the shoreline data for use as a shield. Create contours specifying an interval that will result in adjacent contours being everywhere less than the gridlength.

2) Perform the ArcView Spatial Join operation between all cell centers and nearest contours. This creates a merged table including the cell center positions, assigned depths, and distances from these centers to nearest contours ("contour distances").

3) When contour distances have been verified as sufficiently near to cell centers for specified model accuracy, export table in a format readable as model input.

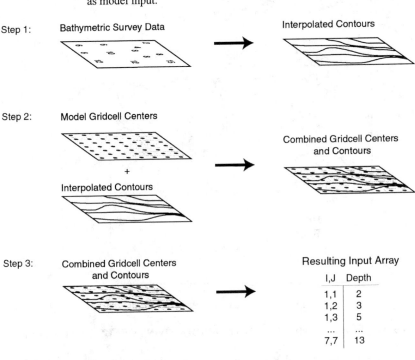

Figure 4. An illustration of the steps in the ArcView Spatial Join of bathymetry contours ad grid cell center locations. The resulting table includes a cell depth for each active horizontal cell in the hydrodynamic model.

EXAMPLE APPLICATIONS

The VIMS hydrodynamic/sediment model HEM-3D was initially applied to the James River in Virginia. Whereas the model domain extended from Norfolk to Richmond, its primary focus was on the Lower James River in the area known as Hampton Roads (see Figure 5). This was the area of concern in the recent evaluation of alternatives for a third highway crossing of Hampton Roads (Boon et al, 1999). It is also the region of the key Newport News frontal system presumed to be vital to the shellfish population in the Lower James (Shen and Kuo, 1998). The original grid of 370 m cell length is shown in Figure 6.

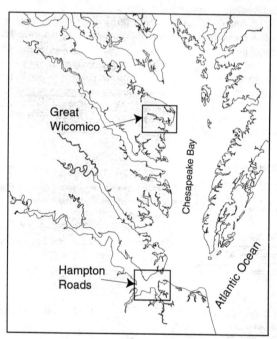

Figure 5. Locations of Hampton Roads (Lower James) and Great Wicomico Rivers.

JAMES RIVER GRID WITH HIGH RESOLUTION IN ELIZABETH RIVER

As interest in the Elizabeth River, a tributary of the lower James River, has intensified, the need for a high-resolution grid of this tributary emerged. A gridlength of 123.3 m (one-third that of the coarse grid) was selected. Figure 7 shows this grid superimposed upon a satellite image with 10-m resolution in the horizontal dimension.

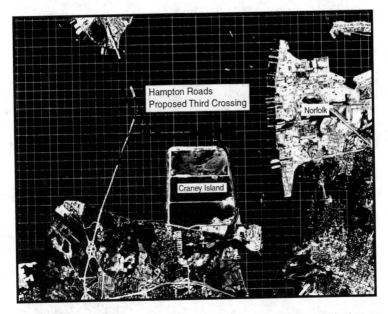

Figure 6. Coarse grid of 370 m cellsize superimposed on satellite image.

Figure 7. Illustration of cell editing session used for high resolution
Elizabeth River portion. A gridlength of 123.3 m is used here.

DISCUSSION

As one investigates the numerous options of the Spatial Analyst capabilty of
ArcView, it is often practical to isolate upon a single criterion. ArcView's ability to
select allows such isolations. Figure 8 displays the 12-foot contour of the Great
Wicomico River bathymetry, which can be quickly compared with a navigation chart.
And because ArcView exports easily to other systems (i.e., grid generator software), it is
easy to position grid cell boundaries at strategic positions, for example, along a channel.

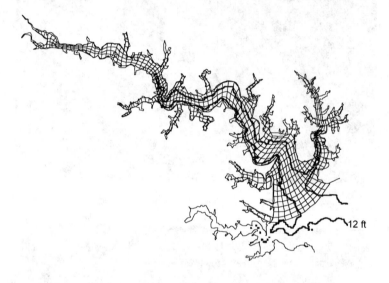

Figure 8. Curvilinear grid for the Great Wicomico River, Virginia with the 3.7 m
(12-foot) contour displayed.

One strength of ArcView is its ability to find unions among different types of
spatial data representations, namely point, line, and areal (i.e., polygonal) files of data
groupings, or themes, as they are known internally. These unions may be either one
dataset within another (i.e., *containment*) or simply the closest that one set approaches
another (i.e., *nearest*). The latter is the criterion for the near intersection of bathymetry
contour lines and the cell center points characterizing a grid. This operation is known as
a *spatial join*, and is incorporated in the ArcView extension known as geo-processing.

The additional role of ArcView as a database management system give it the
deciding edge over the many software systems capable of displaying the spatial
distribution of a variable. Figure 9 shows the distribution of a single variable over the
entire domain, but ArcView is unique in that there is the option in post-run analysis to

zoom into an area of special interest and examine the distribution of values of all the variables of interest.

DISPLAY OF MODEL RESULTS

The ability of ArcView to create a theme relating spatial data to an array of parameter values, such as model output, makes it a viable tool for quick display of this output after a model run. The internal capabilities of ArcView to identify individual cell values as well as to manipulate a view (e.g., zoom, pan, and rotate) make it a very adequate diagnostic tool for post-processing hydrodynamic model results.

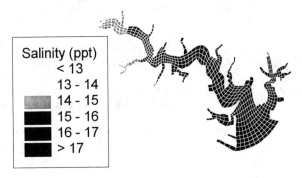

Great Wicomico River Salinity

Salinity (ppt)
< 13
13 - 14
14 - 15
15 - 16
16 - 17
> 17

Figure 9. The Great Wicomico River salinity distribution (surface layer).
The values associated with polygonal cells can be color mapped
and indexed by a legend.

CONCLUSIONS

Although many software packages are available for the pre- and post-processing tasks associated with hydrodynamic modeling, GIS is more likely to become the accepted tool for the management and analysis of spatial data. As computer technology improves and more data becomes available in GIS compatible format, this trend will no doubt continue. Also as related technologies (i.e., surveying and field data collection, visualization and MIS), improve, the need for GIS will expand (Heywood et al, 1998).

ArcView provides a cost-efficient, straight-forward introduction to more elaborate GIS applications which can be considered later. For the present, the learning curve for ArcView is short enough to warrant most application projects to consider this tool for both model analysis and as an interface to other systems. For the hydrodynamic modeler, providing an ArcView interface is an important strategic investment towards future needs to import and export model data to other GIS-based systems.

REFERENCES

Boon, J. D., H. V. Wang, S. C. Kim, A.Y. Kuo, and G. M. Sisson. 1999. Three Dimensional Hydrodynamic Sedimentation Modeling Study: Hampton Roads Crossing, Lower James River, Virginia. A Report to the Virginia Department of Transportation. Special Report No. 354 in Applied Science and Ocean Engineering, VIMS, Gloucester Point, VA 36 p, Appendices.

ESRI (Environmental Systems Research Institute, Inc.). 1997. Getting to Know ArcView GIS: the geographic information system (GIS) for everyone. Redlands, CA. 528 pp.

ESRI (Environmental Systems Research Institute, Inc.). 1996. AVENUE: Customization and Application Development for ArcView. Redlands, CA 260 pp.

Heywood, I., S. Cornelius, and S. Carver. 1998. An Introduction to Geographic Information Systems. Addison Wesley Longman, New York. 279 pp.

Shen, J. and A. Y. Kuo. May/June 1999. Numerical Investigation of an Estuarine Front and Its Associated Eddy. Journal of Waterway, Port, Coastal, and Ocean Engineering. 125(3):127-135.

Three-dimensional curvilinear modelling of wind-induced flows through the North Channel of the Irish Sea

Emma F. Young[1], John N. Aldridge[1] and Juan Brown[1]

Abstract

A three-dimensional non-orthogonal curvilinear model of the Celtic Sea, Irish Sea, North Channel and Malin Shelf has been developed to investigate the wind-induced dynamics of flows through the North Channel of the Irish Sea. This variable resolution approach allows the inclusion of far field effects whilst providing the appropriate fine resolution (of the order of 1 km) to resolve the complex local bathymetry and current structure of the North Channel.

Wind-induced fluxes through the North Channel are presented for a variety of meteorological conditions including a 'big event', and the importance of appropriate representation of far field effects is demonstrated. The structure of the predicted flows both vertically and across sections of the North Channel is discussed and conclusions are drawn on the dominant mechanisms for the transport of material from the Irish Sea. This work has particular relevance to the understanding of factors controlling the flushing of contaminants from the Eastern Irish Sea and fisheries recruitment.

Introduction

Management of the Irish Sea requires that demands imposed by competing activities such as fishing, leisure and industry (e.g. oil and gas extraction and nuclear reprocessing) be balanced. To achieve this, it is necessary to have a sound

---

[1] The Centre for Environment, Fisheries & Aquaculture Science, Pakefield Road, Lowestoft, Suffolk, NR33 0HT, U.K.

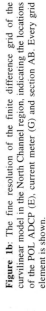

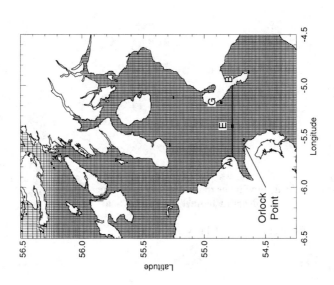

Figure 1b: The fine resolution of the finite difference grid of the curvilinear model in the North Channel region, indicating the locations of the POL ADCP (E), current meter (G) and section AB. Every grid element is shown.

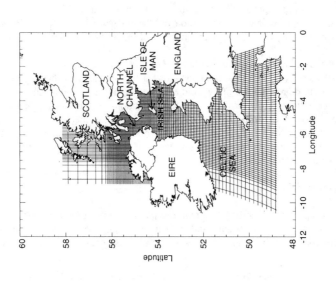

Figure 1a: Finite difference grid of the curvilinear model. Every fifth grid element is shown.

understanding of the circulation and processes controlling fluxes of material leaving or entering the Irish Sea. The North Channel of the Irish Sea provides a convenient constriction across which to measure the exchange of water between the Irish Sea and Malin Shelf. However, the harsh tidal regime in this region, with tidal flows approaching 1.5 m s^{-1}, makes direct observations difficult. Consequently, to understand the region, data have been combined with numerical simulations.

On time scales exceeding several months there is a net northwards flux of Irish Sea water through the North Channel and along the Scottish Shelf (Howarth, 1982; Brown and Gmitrowicz, 1995; Leonard et al., 1997). Modelling studies have suggested that tidal residuals contribute about 0.02 Sv (1 Sv = 1 x 10^6 m^3 s^{-1}) to this northwards flux (Pingree and Griffiths, 1980; Proctor, 1981; Young et al., 1999). However, the low frequency non-tidal flow through the area is highly sensitive to the magnitude and direction of winds over the shelf (Howarth, 1982; Brown and Gmitrowicz, 1995; Knight and Howarth, 1999).

Although the response of the region to wind forcing has been examined in previous modelling studies (Pingree and Griffiths, 1980; Davies and Jones, 1992; Davies et al., 1998), the grids used in these large area models were relatively coarse and insufficient to resolve the complex cross-channel variability in transport revealed by Brown and Gmitrowicz (1995). Unfortunately, the current availability of computer power cannot support a fine grid over a sufficiently large area to adequately represent far field effects. Therefore, a single, variable resolution grid has been developed with high resolution in the North Channel for the adequate representation of the complex topography and current variability, but with a sufficiently large model domain to incorporate the significant influence of far field forcing. The finite difference grid of the curvilinear model is shown in Figure 1a (every fifth grid element) with a more detailed view of the North Channel region shown in Figure 1b (every grid element).

In this paper, the application of a validated large-area, three-dimensional, non-orthogonal curvilinear model to the study of the wind-forced dynamics of flows through the North Channel of the Irish Sea is described. Initially the model was forced with a number of uniform winds of varying orientation with respect to the channel, and the resulting currents and elevation fields will be discussed. In subsequent calculations, the ability of the model to reproduce observed currents for two specific periods in 1994 will be assessed and the importance of the inclusion of far field effects in these simulations will be considered. Finally, the significance of a large storm event on 3-4 February 1994 for the flushing of the Irish Sea will be considered.

Wind-forced dynamics of the North Channel

As tidal currents in the region are strong, it was necessary to include tidal forcing at the open boundary in all calculations in order to obtain the correct background level of turbulence and bed stress. Solutions were generated from initial conditions of zero elevation and motion. For simulations with constant winds, the model was forced until a steady state was reached (4 days). For dynamic winds the wind forcing was ramped over a period of 4 days prior to the start of the required simulation period. This allowed the tidal component of the forcing to reach steady state (after 7 tidal periods; Young et al., 1999) and prevented the generation of persistent inertial oscillations. As the objective of this study was to investigate the meteorological residual, this was obtained by subtracting a purely tidal solution from that with both tidal and wind forcing. For the purposes of this study, fine resolution (approximately 1 km) bathymetry was digitised from Fair Sheets and Admiralty Charts supplied by the U.K. Hydrographic Office (Brown et al., 1999). The numerical method used to solve the three-dimensional hydrodynamic equations was a mixed spectral-finite difference approach, which has been described in detail in the literature (e.g. Davies, 1980; 1987). The development and validation of the model for the two main tidal constituents (M_2 and S_2) is detailed in Young et al., 1999.

Simulations with constant winds

These simulations were intended to provide some insight into the processes by which wind forcing (both local and far field) and the associated elevation gradients influence flows through the North Channel and the Irish Sea. Initial simulations used constant wind forcing of 10 m s^{-1} from four directions; south, west, along channel (from 151°) and cross channel (from 248°). The model showed an essentially linear response to constant wind forcing, thus the response to any wind direction could be constructed from two orthogonal components. Predicted wind-induced fluxes through the North Channel along section AB (Figure 1b) were greatest for an along channel wind (0.26 Sv) and least for a cross channel wind (-0.02 Sv). These results are broadly in agreement with conclusions drawn from observations (Howarth, 1982; Brown & Gmitrowicz, 1995; Knight & Howarth, 1999).

The predicted flows induced by an along channel wind showed little cross channel variability. The predicted surface currents were essentially uniform and in the direction of the wind and the combination of direct wind forcing, topographic steering and pressure gradients associated with the wind-induced surface elevations drove along channel depth-mean and bottom currents. A cross channel wind produced much greater cross channel variability. Although the surface currents were essentially uniform and eastward, the resulting weak pressure

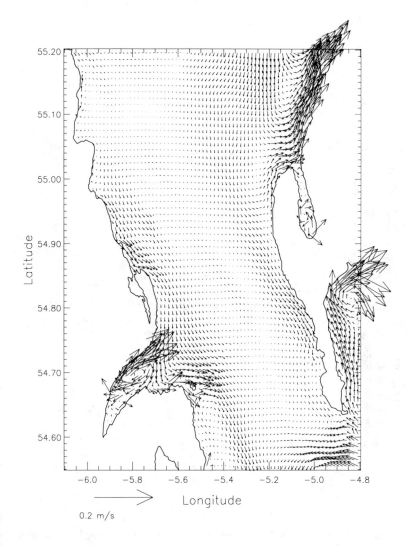

Figure 2: Predicted wind-induced depth-mean currents for a cross channel wind (from 248°) of 10 m s^{-1}.

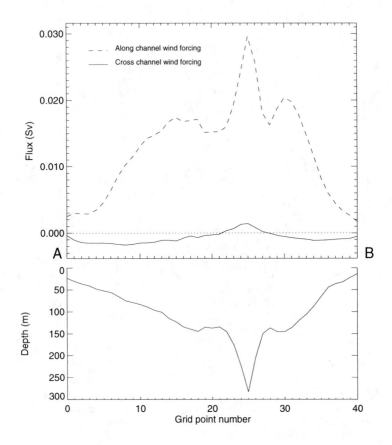

Figure 3: Predicted residual fluxes through section AB (Figure 1b) for along and cross channel winds (from 151° and 248° respectively) of 10 m s^{-1}. The model depths are shown below.

gradients in the North Channel drove a predominantly westward bottom flow with a weak north-westward flow along the centre of the channel. The resultant depth-mean flow (Figure 2) has a complex pattern with southward flow along the margins and northward flow along the centre of the channel. The calculated fluxes through section AB of the North Channel for along and cross channel winds are shown in Figure 3, illustrating the large differences in both the predicted magnitude and the cross channel variability with wind direction.

The importance of adequate representation of far field effects on model predictions of wind-induced flows was investigated by comparing the results of these simulations with those obtained using a smaller model grid. The smaller grid was created by extracting the portion of the original curvilinear grid between the latitudes of 52.13°N and 55.13°N and thus had identical resolution and bathymetry. This ensured that any differences in model predictions were purely due to the limited geographical coverage of the smaller grid. The open boundary of the small grid was forced with tidal information extracted from the original, validated curvilinear model grid and the wind-induced residual flows were extracted as described earlier. The predicted fluxes through the North Channel were markedly different; for example, along and cross channel wind-induced fluxes were 0.07 Sv and 0.02 Sv respectively. This difference is due to a combination of the reduced geographical coverage of the smaller grid and the proximity of the open boundaries to the area of interest (the North Channel). These affect the development and orientation of the predicted pressure gradients, which influences both the magnitude and direction of the predicted fluxes. This is clearly demonstrated by a comparison of the predicted wind-induced surface elevations for an along-channel wind (Figure 4a,b). In addition to the considerable difference in the absolute elevations, the smaller grid predicts north-south elevation gradients in the western Irish Sea unlike the east-west gradients predicted with the large grid.

Simulations with dynamic winds

Whilst uniform winds have generally been used to study the wind-induced dynamics of the Irish Sea, they are unrepresentative of real conditions. Dynamic winds, using three-hourly surface wind fields from the U.K. Met. Office, were applied to two specific periods in 1994. January-February corresponded to a significant outflow event when ADCP (acoustic Doppler current profiler) observations were available from the Proudman Oceanographic Laboratory (Knight, 1995). A second period in September coincided with an observational program undertaken by CEFAS in the North Channel. These point observations are not directly comparable with the model-predicted current average over a 1 km grid square, and also differ with the model predictions in the method of analysis and the inclusion of additional residuals due to density gradients. However, they

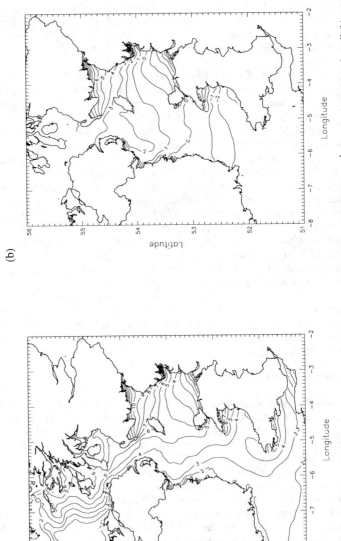

Figure 4: Predicted wind-induced residual elevations for an along channel wind (from 151°) of 10 m s^{-1} using the large (a) and small (b) curvilinear grids.

provide a good test of the model's ability to predict the broad current features and are useful for investigating the importance of appropriate representation of far field effects.

i) January-February 1994; a 'big event'

The northward component of the observed depth-mean flow recorded by the POL ADCP at 54° 46'N, 5° 24'W (site E, Figure 1b) for the period 24 January to 8 February is shown in Figure 5. Observed velocities for the end of January show variable currents, with slightly more southward flow. However, a strong storm event on the 3-4 February generated northward currents of nearly 0.5 m s^{-1}. The strong along channel winds associated with this storm were caused by a low-pressure system whose centre remained over the west coast of Ireland for approximately two days before moving northwards (Figure 6).

The northward depth-mean currents predicted using the large area curvilinear model are generally in good agreement with the observations (Figure 5). The simulation was repeated using a non-spatially varying wind field, specifically the observed winds at Orlock Point (Figure 1b). Although the general trends were predicted, this simulation failed to accurately predict the extremes of the observed currents (Figure 5). A comparison simulation was also run using predicted winds from the Met. Office model interpolated to the location of Orlock Point to obtain a non-spatially varying wind field. This was undertaken to investigate the possibility that inaccuracies in the Met. Office model predictions in the North Channel region, or the effect of using a three-hourly wind field versus hourly observations could be influencing the predictions. However, this showed only minor differences with the results obtained using the observed winds and is not presented. The differences in the model predictions are therefore due to the spatial variability of the wind field. These results show that differences in the far field forcing due to spatial variability in the wind field can have a significant effect.

ii) September 1994

The observed northward residual velocities at current meter G (54° 51.6'N, 5° 13.0'W; Figure 1b) for the period 12-22 September 1994 are shown in Figure 7. This Aanderaa RCM current meter was deployed in 75 m of water at a height of 56 m above the sea bed. The predicted wind-induced residual currents at a depth of 19 m using the spatially varying Met. Office model wind forcing generally agree well with the observed currents. However, the results of the simulation using Orlock Point wind forcing show some significant differences with the observations, in particular the prediction of a southward flow of approximately 0.2 m s^{-1} on the 16-17 September, much stronger than the observed

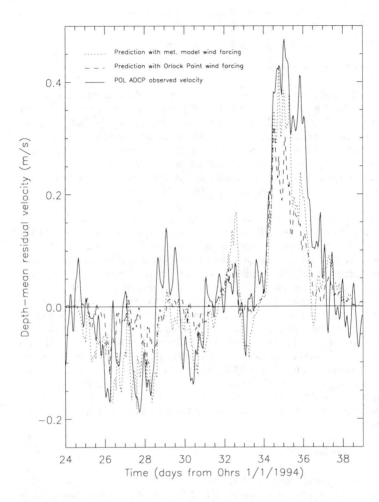

Figure 5: Comparison of model-predicted depth-mean northward residual velocities at site E (Figure 1b) using spatially varying Met. Office model wind forcing and observed winds from Orlock Point, and the observed velocities from a POL ADCP.

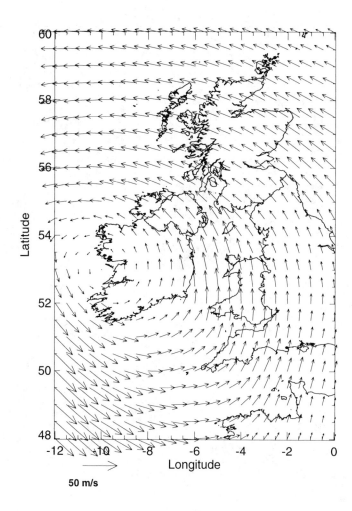

Figure 6: Predicted wind field from the U.K. Meteorological Office model for the 4th February 1994.

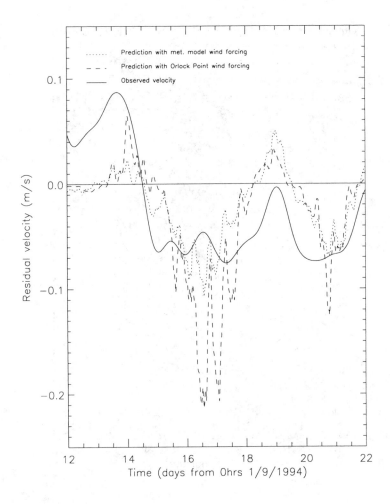

Figure 7: Comparison of model-predicted northward residual velocities at a depth of 19 m at site G (Figure 1b) using spatially varying Met. Office model wind forcing and observed winds from Orlock Point, and the observed velocities from a CEFAS current meter.

flow of about 0.06 m s^{-1}. During this period the observed north-westerly winds at Orlock Point were much stronger than those predicted by the model and also those observed at Ronaldsway on the Isle of Man. This would suggest that the observed winds at Orlock Point were a localised event. Assuming these winds to be applicable to the whole model domain was therefore inappropriate and resulted in the poor predictions shown. This again emphasises the importance of appropriate representation of far field effects in numerical models.

iii)   How significant was the 'big event'?

Whilst the February 1994 event was large, in terms of the flushing of the Irish Sea it is important to place it in the context of average conditions. The model predicted a strong northward mean residual for February of 0.23 Sv. However, this was not purely due to the wind event of 3-4 February; fluxes throughout February had a predominantly northward direction, and if the 3-4 February was neglected the mean wind-induced residual flux was still strongly northward at 0.16 Sv. Using these predictions to estimate a total volume flux for water passing through the North Channel from the Irish Sea in February, and including the tidal residual flux of 0.023 Sv (Young et al., 1999), a value of 6.1x10^{11} m^3 was obtained. This represents the equivalent of 25% of the volume of the Irish Sea, of this 34% passed through on 3-4 February. Thus, while strong wind events generate significant short term fluxes through the North Channel of the Irish Sea, prolonged periods of along channel winds such as observed throughout February 1994 are more significant for the longer term flushing of the Irish Sea.

Summary

Model simulations of wind-induced flows through the North Channel of the Irish Sea have demonstrated the importance of adequate representation of complex bathymetry and current structure whilst maintaining sufficient spatial coverage to incorporate far field forcing. Initial simulations with uniform winds and model grids of differing geographical coverage showed that a limited area grid generates incorrect pressure gradients, particularly in the western Irish Sea. Consequently, both the magnitude and direction of predicted depth-mean currents showed significant differences with those obtained using a large grid. The significance of appropriate representation of far field effects was further demonstrated by comparing observed currents for two periods in 1994 with predictions obtained using spatially and non-spatially varying wind forcing. Whilst the former showed good agreement with observations, the latter did not compare well due to unrepresentative far field forcing. For many oceanographic processes (e.g. larval recruitment, oil spills), the time scales of interest are days to weeks rather than years. For these applications, accurate prediction of the short-term current variability is of more importance than the long term mean transport, which demands appropriate representation of the dynamic wind field.

References

Brown, J., Gmitrowicz, E.M. (1995). Observations of the transverse structure and dynamics of the low frequency flow through the North Channel of the Irish Sea. *Continental Shelf Research*, 15 (9), 1133-1156.

Brown, J., Joyce, A.E., Aldridge, J.N., Young, E.F., Fernand, L., Gurbutt, P.A. (1999). Further identification and acquisition of bathymetric data for Irish Sea modelling. DETR research contract CW0753.

Davies, A.M. (1980). Application of the Galerkin method to the formulation of a three-dimensional non-linear hydrodynamic numerical sea model. *Applied Mathematical Modelling*, 4, 245-256.

Davies, A.M. (1987). Spectral models in continental shelf sea oceanography. In: Heaps, N.S. (Ed.), Three-dimensional coastal ocean models. A.G.U., pp. 71-106.

Davies, A.M., Jones, J.E. (1992). A three-dimensional wind driven circulation model of the Celtic and Irish Seas. *Continental Shelf Research*, 12, 159-188.

Davies, A.M., Kwong, S.C.M., Flather, R.A. (1998). A three-dimensional model of wind-driven circulation on the shelf: application to the storm of January 1993. *Continental Shelf Research*, 18, 289-340.

Howarth, M.J. (1982). Non-tidal flow in the North Channel of the Irish Sea. In: Nihoul, J.C.J. (Ed.), Hydrodynamics of semi-enclosed shelf seas. Elsevier, Amsterdam, pp. 205-241.

Knight, P.J. (1995). Current profile, pressure and temperature records from the North Channel of the Irish Sea from a bottom mounted ADCP and WLR at 54° 46' N 05° 24' W, July 1993 – October 1994. Proudman Oceanographic Laboratory, Report no. 39, 267 pp.

Knight, P.J., Howarth, M.J. (1999). The flow through the North Channel of the Irish Sea. *Continental Shelf Research*, 19, 693-716.

Leonard, K.S., McCubbin, D., Brown, J., Bonfield, R., Brooks, T. (1997). Distribution of Technetium-99 in UK coastal waters. *Marine Pollution Bulletin*, 34, 628-636.

Pingree, R.D., Griffiths, D.K. (1980). Currents driven by a steady uniform wind stress on the shelf seas around the British Isles. *Oceanologica Acta*, 3 (2), 227-236.

Proctor, R. (1981). Tides and residual circulation in the Irish Sea: A numerical modelling approach. Ph.D. thesis, Liverpool University.

Young, E.F., Aldridge, J.N., Brown, J. (1999). Development and validation of a three-dimensional curvilinear model for the study of fluxes through the North Channel of the Irish Sea. *Continental Shelf Research*. In press.

Acknowledgements

The authors greatly appreciate the ADCP data supplied by the Proudman Oceanographic Laboratory and the model wind data provided by the Meteorological Office. This research was supported by the Ministry of Agriculture, Fisheries and Food.

Development of a Long-Term Predictive Model
of Water Quality in Tokyo Bay

Jun Sasaki[1] and Masahiko Isobe[2], M. ASCE

Abstract

A predictive model for hydrodynamic flows and water quality in a semi-enclosed bay is developed. The model consists of two parts, a hydrodynamic model, which is a quasi-three-dimensional baroclinic circulation model including temperature and salinity, and a simple water quality model, which is focusing on reproduction of dissolved oxygen and phytoplankton variation. Considering the capability of long-term computation, an efficient semi-implicit finite difference algorithm is proposed in σ-coordinate system. For the treatment of steepness or abrupt change of bottom slope, such as trenches, a multi-σ-coordinate transformation is introduced. The model is applied to reproduction of flows and water quality in Tokyo Bay during 1994 and 1996, when monitoring data are available, and verified through comparison between computational results and field observation. The model is also applied to the occurrence of *blue tides*, a blued surface phenomenon caused by upwelling of anoxic water, and the subsequent phytoplankton blooming, red tide, initiated by the transport of nutrient-rich bottom waters to the surface.

Keywords: Long-term Simulation. Semi-Implicit, Multi-σ-coordinates, Red Tides, Blue Tides

Introduction

Tokyo Bay is one of the most polluted bays in Japan. It is approximately 50 km in length and 20 km in width, with an average of depth about 15m (Figure 1). The water column of the bay is stratified from late spring to autumn. Phytoplankton proliferates in the upper layer; oxygen depletion occurs in the lower layer. Appearance of anoxic water which causes mortality of benthic animals, such as shellfish and short necked clams, is a very serious problem caused by phytoplankton blooming due to excess loading of nutrients from the rivers contaminated with sewage. In late summer or fall, upwelling of anoxic water sometimes occurs at the head of the bay when the wind direction changes from southerly to northerly. Since the anoxic bottom waters contain hydrogen sulfide, the color of upwelling waters change milky-blue due to oxidation of hydrogen sulfide. This phenomenon is called *blue tide* and sometimes causes mortality of benthic animals in

---

[1]Associate Professor and [2]Professor, Institute of Environmental Studies, The University of Tokyo, 7-3-1 Hongo, Bunkyo-ku, Tokyo 113-0033, Japan.

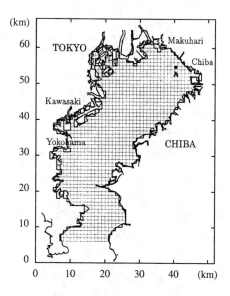

Figure 1: Map of Tokyo Bay with a monitoring site A and numerical model grid.

shallow waters. During blue tides, nutrient-rich bottom waters are also transported to the surface and as a result blooming of phytoplankton often occurs.

Predictive numerical modeling tools to aid in the assessment of various remedies are necessary, and thus, the objective of this study is to develop a long-term predictive model for water quality and to verify the model through comparison between computation and measurements. The model consists of a hydrodynamic model and a water quality model. The hydrodynamic sub-model is a quasi-three-dimensional σ-coordinate baroclinic circulation model which includes temperature and salinity. The water quality sub-model incorporates phytoplankton dynamics, nutrient cycling processes and dissolved oxygen as well as sedimentation processes of particulate organic material.

Hydrodynamic Model

Objective

A three-dimensional baroclinic bay circulation model was developed to study the mechanism of a blue tide phenomenon, upwelling of anoxic water, which sometimes causes mortality of shellfish and damage to its fisheries in late summer in Tokyo Bay (Sasaki *et al.*, 1996). The time scale of blue tides is approximately one week and requires a short time scale to accurately simulate. Spatial and temporal variations of water quality are, however, very complex during summer due to the increase of river discharge and stratification, and it is difficult to give appropriate initial conditions for such a kind of short time scale events. If we start computation from early spring when the water column is well mixed throughout the bay and the water quality is almost uniform, there are few difficulties in giving initial conditions. For this case, the cost of computation increases due to necessity of a long-term simulation, and therefore development of an efficient scheme becomes a main objective of modeling. Our model succeeded in this respect and will be introduced in the following subsections with some modification related to the balance between efficiency and accuracy.

Governing equations in σ-coordinates

The model equations are the primitive equations with the hydrostatic and Boussinesq approximations. The momentum and continuity equations in Cartesian coordinates are given by

$$\frac{\partial \boldsymbol{u}}{\partial t} + \frac{\partial (u\boldsymbol{u})}{\partial x} + \frac{\partial (v\boldsymbol{u})}{\partial y} + \frac{\partial (w\boldsymbol{u})}{\partial z} = f \begin{pmatrix} v \\ -u \end{pmatrix} - g\nabla_h \zeta - \frac{g}{\rho_0} \nabla_h \int_z^\zeta \rho' dz$$
$$+ A_h \left(\frac{\partial^2 \boldsymbol{u}}{\partial x^2} + \frac{\partial^2 \boldsymbol{u}}{\partial y^2} \right) + \frac{\partial}{\partial z} \left(A_v \frac{\partial \boldsymbol{u}}{\partial z} \right) \tag{1}$$

$$\frac{\partial u}{\partial x} + \frac{\partial v}{\partial y} + \frac{\partial w}{\partial z} = 0 \tag{2}$$

where t is time, x and y are the horizontal coordinates, z is the vertical coordinate upward from the still water level, u, v and w are velocities in the x, y and z directions, respectively, $\boldsymbol{u}$ is the horizontal velocity vector ${}^t(u, v)$, ζ is the water surface elevation from the still water level, ρ_0 and ρ' are the constant reference density and deviation from it and the sum of them is the sea water density $\rho(= \rho_0 + \rho')$, A_h and A_v are the horizontal and vertical kinematic eddy viscosities, respectively, f is the Coriolis parameter, and ∇_h is the horizontal differential operator ${}^t(\partial/\partial x, \partial/\partial y)$.

The density of sea water ρ (kg/m^3) can be defined from Eq. (3), the equation of state, as a function of the water salinity S (psu) and temperature T (°C) which will be determined by the transport equations described in the following subsection,

$$\rho = 1000 + 20.99 + \left\{ -4.3 \times 10^{-3}(T - 20) - 0.256 \right\} (T - 20)$$
$$+ \left\{ 2.3 \times 10^{-4}(S - 30) - 1.53 \times 10^{-3}T - 20) + 0.7577 \right\} (S - 30). \tag{3}$$

The resolution of the vertical processes by a constant layer thickness often fails to reproduce the processes in the region of abrupt topography variations or in the surface or bottom boundary layers. To overcome these obstacles σ-coordinate which is originally proposed by Phillips (1957) is often introduced through the transformation

$$\sigma = \frac{z + h}{h + \zeta} = \frac{z + h}{H} \tag{4}$$

where $H = h + \zeta$. In σ-coordinate the water column from the surface ($z = \zeta$) to the bottom ($z = -h$) is transformed into a uniform depth ranging from $\sigma = 1$ to $\sigma = 0$.

The momentum and continuity equations are transformed as follows,

$$\frac{\partial(H\boldsymbol{u})}{\partial t} + \frac{\partial(H\boldsymbol{u}\boldsymbol{u})}{\partial x} + \frac{\partial(H\boldsymbol{v}\boldsymbol{u})}{\partial y} + \frac{\partial(H\dot{\sigma}\boldsymbol{u})}{\partial \sigma} = H\begin{pmatrix} v \\ -u \end{pmatrix} - \frac{gH}{\rho_0}\left\{(\rho_0 + \rho'\sigma)\nabla_h\zeta\right.$$

$$\left. -\rho'(\sigma-1)\nabla_h h - \nabla_h\left(H\int_\sigma^1 \rho'\,d\sigma\right)\right\} + HA_h\left(\frac{\partial^2\boldsymbol{u}}{\partial x^2} + \frac{\partial^2\boldsymbol{u}}{\partial y^2}\right) + \frac{1}{H}\frac{\partial}{\partial \sigma}\left(A_v\frac{\partial\boldsymbol{u}}{\partial \sigma}\right) \quad (5)$$

$$\frac{\partial\zeta}{\partial t} + \frac{\partial(Hu)}{\partial x} + \frac{\partial(Hv)}{\partial y} + \frac{\partial(H\dot{\sigma})}{\partial \sigma} = 0 \quad (6)$$

where $\dot{\sigma}$ is introduced as

$$\dot{\sigma} = \frac{\partial\sigma}{\partial t} + u\frac{\partial\sigma}{\partial x} + v\frac{\partial\sigma}{\partial y} + w\frac{\partial\sigma}{\partial z}$$

$$= \frac{\sigma}{H}\frac{\partial\zeta}{\partial t} + \frac{u}{H}\left(\frac{\partial h}{\partial x} - \sigma\frac{\partial H}{\partial x}\right) + \frac{v}{H}\left(\frac{\partial h}{\partial y} - \sigma\frac{\partial H}{\partial y}\right) + \frac{w}{H} \quad (7)$$

The advection terms, the left-hand side of Eqs. (5), were expressed in the conservation form by using the continuity equation (6).

To solve $\dot{\sigma}$, first the continuity equation (6) is integrated from 0 to 1

$$\frac{\partial\zeta}{\partial t} + \int_0^1 \frac{\partial(Hu)}{\partial x}d\sigma + \int_0^1 \frac{\partial(Hv)}{\partial y}d\sigma = 0 \quad (8)$$

and also integrated form 0 to σ

$$\frac{\partial\zeta}{\partial t}\sigma + \int_0^\sigma \frac{\partial(Hu)}{\partial x}d\sigma + \int_0^\sigma \frac{\partial(Hv)}{\partial y}d\sigma + H\dot{\sigma} = 0 \quad (9)$$

Erasing the term $\partial\zeta/\partial t$ from Eqs. (8) and (9) yields

$$\dot{\sigma} = \frac{1}{H}\left[\sigma\int_0^1 \frac{\partial(Hu)}{\partial x}d\sigma + \sigma\int_0^1 \frac{\partial(Hv)}{\partial y}d\sigma - \int_0^\sigma \frac{\partial(Hu)}{\partial x}d\sigma - \int_0^\sigma \frac{\partial(Hv)}{\partial y}d\sigma\right] \quad (10)$$

which strictly satisfies the kinematic free surface and bottom boundary conditions.

Multi-σ-coordinates transformation

Along the coast of the head of the bay, there are several trenches, with a scale of a few kilometers in the horizontal and 30 meters in the vertical, which were dredged for the reclamation of the foreshore. There is difficulty in implementing σ-coordinate model to the area of this kind of abrupt change of bottom slope. As a remedy for this problem, one of the authors proposed a multi-σ-coordinate model (Sasaki, 1997). In this system, another σ-coordinate system denoted σ_b-coordinate here is attached to the original σ-coordinate system as shown in Fig. 2, through the transformation

$$\sigma_b = \frac{z+h}{h_b - h} = \frac{z+h}{D} \quad (11)$$

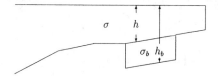

Figure 2: Schematic description of multi-σ-coordinates.

where $D = h_b - h$, in which the water column from the bottom of the trench ($z = -h_b$) to the interface between the trench and the flat bottom ($z = -h$) is transformed into a uniform depth ranging from $\sigma_b = -1$ to $\sigma_b = 0$.

The momentum and continuity equations are transformed as follows;

$$
\begin{aligned}
&\frac{\partial(D\boldsymbol{u})}{\partial t} + \frac{\partial(uD\boldsymbol{u})}{\partial x} + \frac{\partial(vD\boldsymbol{u})}{\partial y} + \frac{\partial(\dot{\sigma}_b D\boldsymbol{u})}{\partial \sigma_b} = fD\begin{pmatrix} v \\ -u \end{pmatrix} \\
&- \frac{gD}{\rho_0}\{-\rho_0 \nabla_{\sigma_b}\zeta - \rho'\sigma_b\nabla_{\sigma_b}h_b + \rho'(\sigma_b - 1)\nabla_{\sigma_b}h\} - \frac{gD}{\rho}\nabla_{\sigma_b}\left(D\int_{\sigma_b}^{0}\rho'\,d\sigma_b + H\int_{0}^{1}\rho'\,d\sigma\right) \\
&+ DA_h\left(\frac{\partial^2 \boldsymbol{u}}{\partial x^2} + \frac{\partial^2 \boldsymbol{u}}{\partial y^2}\right) + \frac{1}{D}\frac{\partial}{\partial \sigma_b}\left(A_v\frac{\partial \boldsymbol{u}}{\partial \sigma_b}\right)
\end{aligned}
\tag{12}
$$

$$
\frac{\partial(Du)}{\partial x} + \frac{\partial(Dv)}{\partial y} + \frac{\partial(D\dot{\sigma}_b)}{\partial \sigma_b} = 0
\tag{13}
$$

Boundary conditions

The tangential stress at the free surface along a horizontal direction is equal to the wind stress and written as follows,

$$
\rho A_v\frac{\partial u}{\partial \sigma} = H\rho_a C_f|\boldsymbol{W}|W_x \qquad \text{and} \qquad \rho A_v\frac{\partial v}{\partial \sigma} = H\rho_a C_f|\boldsymbol{W}|W_y
\tag{14}
$$

where C_f, ρ_a are the wind drag coefficient and the air density, respectively, W_x and W_y are the x and y components of the wind vector $\boldsymbol{W}$ at the altitude of 10m above the sea surface. Kondo (1994) found that C_f under a neutral atmospheric stability condition depends on the wind velocity and is formulated by

$$
\begin{aligned}
10^3 \times C_f &= 1.08W^{-0.15}, & 0.3\text{m/s} \le W &< 2.2\text{m/s} & (15a) \\
&= 0.771 + 0.0858W, & 2.2\text{m/s} \le W &< 5\text{m/s} & (15b) \\
&= 0.867 + 0.0667W, & 5\text{m/s} \le W &< 8\text{m/s} & (15c) \\
&= 1.2 + 0.025W, & 8\text{m/s} \le W &< 25\text{m/s} & (15d) \\
&= 0.073W, & 25\text{m/s} \le W &< 50\text{m/s}. & (15e)
\end{aligned}
$$

At the sea bottom, the quadratic dependence of the bottom stress on the velocity is well recognized:

$$
\rho A_v\frac{\partial u}{\partial \sigma} = H\rho\gamma_b^2 u\sqrt{u^2 + v^2} \qquad \text{and} \qquad \rho A_v\frac{\partial v}{\partial \sigma} = H\rho\gamma_b^2 v\sqrt{u^2 + v^2}
\tag{16}
$$

where γ_b^2 is the bottom friction coefficient and set to be 0.0026 which is commonly accepted in the bay (Ichihara et al., 1980).

At lateral wall boundaries, normal velocities are set to be zero and a free-slip condition is applied for the friction terms.

At the mouth of the bay, the tide level was estimated from the tide table of Meteorological Agency for the entire simulation.

Numerical Scheme

Many three-dimensional circulation models for coastal seas have been proposed since Leendertze et al. (1973) developed a leapfrog explicit model in Cartesian coordinates. Efforts have been made especially for improving efficiency of computation as well as accuracy. The stability of the model is constrained by an external or barotropic mode of motion, including Courant-Friedrichs-Lewy (CFL) condition of the surface gravity waves and the advection terms, and the restriction of the diffusion number. Leendertze and Liu (1977) modified their original model discretizing the vertical diffusion term implicitly to improve the stability of vertical diffusion term. A number of models, such as Princeton Ocean Model (Blumberg and Mellor, 1983) or the model by Lardner and Smoczynski (1990), adopt a mode splitting approach for the fast (external) and slow (internal) modes, in which external or barotropic mode of motion is described by a two-dimensional, vertically-averaged set of equations and solved explicitly in a short time increment and the internal mode allows a much longer time step in explicit. Sato et al. (1993), however, pointed out a disadvantage in accuracy of a mode splitting algorithm. Backhaus (1985) developed an efficient model, which is implicit for the surface displacement and the vertical diffusion term except the bottom friction. Sato et al. (1993) also discretized the surface displacement and the vertical diffusion in fully implicit. While applying to a bay of a steep bottom slope with vertically high resolution, the time increment, however, was found to be severely constrained, maybe by the instability of vertical advection term due to explicit discretization. Therefore, we developed a more robust algorithm which is semi-implicit for the vertical advection in addition to the surface displacement and vertical diffusion terms. This new approach is enabled by using three-point difference scheme for the vertical advection term as shown below.

Here, we introduce the notations (i, j, k) for a spatial grid point of (x, y, z) in difference form. Since we adopt three-point scheme in the vertical, the vertical advection and diffusion terms of Eq. (5) can be written as follows

$$\frac{\partial(\dot{\sigma}\tilde{u})}{\partial\sigma} \Rightarrow \frac{\dot{\sigma}_{i,j,k+\frac{1}{2}}\tilde{u}_{i,j,k+\frac{1}{2}} - \dot{\sigma}_{i,j,k-\frac{1}{2}}\tilde{u}_{i,j,k-\frac{1}{2}}}{\Delta\sigma_k} \tag{17}$$

$$\frac{\partial}{\partial\sigma}\left(A_v\frac{\partial\tilde{u}}{\partial\sigma}\right) \Rightarrow \frac{1}{\Delta\sigma_k}\left(A_{v_{i,j,k+\frac{1}{2}}}\frac{\tilde{u}_{i,j,k+1} - \tilde{u}_{i,j,k}}{(\Delta\sigma_k + \Delta\sigma_{k+1})/2} - A_{v_{i,j,k-\frac{1}{2}}}\frac{\tilde{u}_{i,j,k} - \tilde{u}_{i,j,k-1}}{(\Delta\sigma_{k-1} + \Delta\sigma_k)/2}\right) \tag{18}$$

where $\tilde{u}_{i,j,k-1}$, $\tilde{u}_{i,j,k}$ and $\tilde{u}_{i,j,k+1}$ are the values at the center of the grid whereas $\tilde{u}_{i,j,k-1/2}$ and $\tilde{u}_{i,j,k+1/2}$ are located at the boundary of the grid in a staggered grid arrangement, and thus are evaluated by appropriate interpolation using $\tilde{u}_{i,j,k-1}$, $\tilde{u}_{i,j,k}$, $\tilde{u}_{i,j,k+1}$, which means three-point scheme, such as 1st-order upwind scheme or Lax-Wendroff scheme (Kowalik and Murty, 1993), must be chosen.

Finally, the difference form for x-momentum equation can be expressed by

$$-a_{ijk}\tilde{u}_{ijk-1}^{n+1} + b_{ijk}\tilde{u}_{ijk}^{n+1} - c_{ijk}\tilde{u}_{ijk-1}^{n+1} = d_{ijk}\zeta_{i+1,j}^{n+1} + e_{ijk}\zeta_{i,j}^{n+1} + f_{ijk} \tag{19}$$

where the super-script $^{n+1}$ denotes the value at new time step, and the coefficients a_{ijk} to f_{ijk} are known quantity evaluated at the current time step n. Eq. (19) can be interpreted as a tri-diagonal matrix system of $\tilde{u}$ and thus the Tri-Diagonal Matrix Algorithm (TDMA) is applied to the system and then yields

$$\tilde{u}_{ijk}^{n+1} = p_{uijk}\zeta_{i+1,j}^{n+1} + q_{uijk}\zeta_{i,j}^{n+1} + r_{uijk} \tag{20}$$

where p_u, q_u and r_u are known coefficients evaluated at the current time step n. Similarly $\tilde{v}$ is also formulated as

$$\tilde{v}_{ijk}^{n+1} = p_{vijk}\zeta_{i,j+1}^{n+1} + q_{vijk}\zeta_{i,j}^{n+1} + r_{vijk} \tag{21}$$

On the other hand, finite difference form of Eq. (9) in fully implicit is given by

$$\frac{\zeta_{ij}^{n+1} - \zeta_{ij}^n}{\Delta t} + \sum_k \frac{\tilde{u}_{i+1jk}^{n+1} - \tilde{u}_{ijk}^{n+1}}{\Delta x}\Delta\sigma_k + \sum_k \frac{\tilde{v}_{ij+1k}^{n+1} - \tilde{v}_{ijk}^{n+1}}{\Delta y}\Delta\sigma_k = 0 \tag{22}$$

and recast into a matrix equation of ζ by substituting the expressions of Eqs. (20) and (21) for $\tilde{u}$ and $\tilde{v}$ to give

$$\alpha_{ij}\zeta_{i-1,j}^{n+1} + \beta_{ij}\zeta_{i,j}^{n+1} + \gamma_{ij}\zeta_{i+1,j}^{n+1} + \delta_{ij}\zeta_{i,j-1}^{n+1} + \epsilon_{ij}\zeta_{i,j+1}^{n+1} = \eta_{ij} \tag{23}$$

where α to ϵ and η are known coefficients evaluated at the current time level and thus Eq. (23) is algebraic equations and can be solved by, e.g., successive over relaxation (SOR) method in which a relaxation parameter must be appropriately estimated. The parameter is strongly affected by the horizontal grid size and time increment Δt, and can be determined through the trial and error manner. To overcome this inconvenience, Benqué et al. (1982) or Sato et al. (1993) adopted iterative Alternate Direction Implicit (ADI) algorithm. If the relaxation parameter for SOR method is, however, well estimated, the computation by SOR is found to be much faster than iterative ADI method, and also once the parameter is selected under some fixed horizontal grid spacing and time increment, the same parameter can be applicable to any field if the grid size and time increment are the same.

Once ζ is solved, the horizontal velocities and $\dot{\sigma}$ can be readily calculated by Eqs. (20), (21) and (10). The vertical velocity w in Cartesian coordinates is obtained by Eq. (24) which is equivalent to Eq. (8).

$$w = -\sigma\frac{\partial\zeta}{\partial t} + u\left(\frac{\partial H}{\partial x}\sigma - \frac{\partial h}{\partial x}\right) + v\left(\frac{\partial H}{\partial y}\sigma - \frac{\partial h}{\partial y}\right) + H\dot{\sigma} \tag{24}$$

Transport equations for temperature and salinity

The transport equation of temperature T in σ-coordinate is given by

$$\frac{\partial(HT)}{\partial t} + \frac{\partial(uHT)}{\partial x} + \frac{\partial(vHT)}{\partial y} + \frac{\partial(\dot{\sigma}HT)}{\partial\sigma} = \frac{1}{H^2}\frac{\partial}{\partial\sigma}\left(K_v\frac{\partial(HT)}{\partial\sigma}\right)$$
$$+ HK_h\left(\frac{\partial^2 T}{\partial x^2} + \frac{\partial^2 T}{\partial y^2}\right) + \frac{1}{\rho C_p}\frac{dq(\sigma)}{d\sigma} \tag{25}$$

Table 1: Coefficients for the bulk equations.

s	σ_h $(\mathrm{Jm^{-2}sK^{-4}})$	c_1	c_2	C_E	C_T
0.97	5.8×10^{-8}	1,869	1,233	1.2×10^{-3}	1.5×10^{-3}

$$q(z) = (1 - \beta)\exp(\alpha z) \tag{26}$$

where K_v and K_h are the horizontal and vertical kinematic eddy diffusivities, C_p and q are the specific heat and the short wave radiation, respectively.

At the sea surface, heat balance is expressed by

$$-\frac{1}{H^2}K_v\frac{\partial(HT)}{\partial\sigma} = \frac{Q_s}{\rho C_p} \tag{27}$$

where Q_s is the net surface heat flux penetrating into the water column and given by

$$Q_s = (1 - A)Q_A - Q_B - Q_e - Q_h \tag{28}$$

where A, Q_A, Q_B, Q_e and Q_h are the albedo, the net short wave radiation, the net long-wave radiation, the latent heat and the sensible heat, respectively, and can be estimated by the following standard bulk equations as functions of meteorological variables such as wind speed W, surface water temperature T_w, and air temperature T_a, which are monitored by the Chiba and Tokyo weather stations of Meteorological Agency,

$$Q_B = s\sigma_h\{T_w^4(0.39 - 0.058\sqrt{E_a}) + 4T_a^3(T_w - T_a)\}(1 - 0.83c) \tag{29}$$

$$Q_e = c_1C_E(0.98E_w - E_a)W \tag{30}$$

$$Q_h = c_2C_T(T_w - T_a)W \tag{31}$$

The coefficients for equations (29) to (31) are found in Table 1. Here s is the coefficient of emissivity, relating the properties of radiating surfaces to those of a black body, σ_h the Stefan-Boltzman's constant, E_a vapor pressure (hPa), c cloud amount in fractions of unity, and E_w the saturated vapor pressure at T_w, which is evaluated by

$$E_w = 6.1078 \times 10^{aT/(b+T)} \tag{32}$$

where $a = 7.5$ and $b = 237.3$ according to Kondo (1994).

At the bottom boundary and lateral wall boundary, no flux conditions are specified. Spatial and temporal temperature variation at the mouth of the bay is given as an open boundary condition based on collected field data.

Salinity distribution is determined considering river discharge as well as precipitation and evaporation at the sea surface. The transport equation of salinity S in σ-coordinate is given by

$$\frac{\partial(HS)}{\partial t} + \frac{\partial(uHS)}{\partial x} + \frac{\partial(vHS)}{\partial y} + \frac{\partial(\dot{\sigma}HS)}{\partial\sigma} = \frac{1}{H^2}\frac{\partial}{\partial\sigma}\left(K_v\frac{\partial(HS)}{\partial\sigma}\right)$$

$$+ HK_h\left(\frac{\partial^2 S}{\partial x^2} + \frac{\partial^2 S}{\partial y^2}\right) - RS \tag{33}$$

where R denotes the river discharge introduced at several coastal nodes of the rivers.

Boundary conditions at the sea surface can be described as

$$-K_v \frac{\partial(HS)}{\partial\sigma} = H^2 S(P_{rct} - E_{vap}) \tag{34}$$

$$E_{vap} = C_{ev} W(E_w - E_a) \tag{35}$$

where P_{rct} and E_{vap} are precipitation and evaporation, respectively.

Numerical scheme for scalar quantity

Transport equations for scalar quantity ϕ, such as temperature, salinity and other water quality variables mentioned in the following section, are discretized in a similar manner to momentum equations; the vertical diffusion and vertical advection terms in implicit and the others in explicit. Thus the constraints imposed by the instability of the vertical advection term and the condition of diffusion number disappear. Similar to the case of momentum equations, finite difference form for scalar quantity ϕ can be summarized as

$$-a_{ijk}\phi_{ijk-1}^{n+1} + b_{ijk}\phi_{ijk}^{n+1} - c_{ijk}\phi_{ijk+1}^{n+1} = f_{ijk} \tag{36}$$

where a, b, c and f are known coefficients evaluated at the current time step. Eq. (36) also formulate a tri-diagonal system of equation and can be easily and efficiently solved by the Tri-Diagonal Matrix Algorithm (TDMA).

Estimation for kinematic eddy viscosity and diffusivity

Kinetic turbulent energy transport equation is generally given by

$$\frac{\partial E_t}{\partial t} + u\frac{\partial E_t}{\partial x} + v\frac{\partial E_t}{\partial y} + w\frac{\partial E_t}{\partial v} = A_v\left[\left(\frac{\partial u}{\partial z}\right)^2 + \left(\frac{\partial v}{\partial z}\right)^2\right] + \frac{\partial}{\partial z}\left(A_v\frac{\partial E_t}{\partial z}\right) - \frac{g}{\rho}K_v\frac{\partial\rho}{\partial z} - \epsilon \tag{37}$$

where E_t and ϵ are the kinetic turbulent energy and dissipation, respectively. Consider Eq. (37) for steady state without nonlinear and diffusion terms with the Kolmogorov's hypothesis of similarity

$$A_v\left[\left(\frac{\partial u}{\partial z}\right)^2 + \left(\frac{\partial v}{\partial z}\right)^2\right] - \frac{g}{\rho}K_v\frac{\partial\rho}{\partial z} - A_v^3/l^4 = 0 \tag{38}$$

where l is the mixing length scale. From Eq. (38) the following expressions are obtained (Kowalik and Murty, 1993)

$$A_v = l^2\left[\left(\frac{\partial u}{\partial z}\right)^2 + \left(\frac{\partial v}{\partial z}\right)^2\right]^{1/2}\sqrt{1 + R_i/P_r} \tag{39}$$

$$K_v = A_v/P_r \tag{40}$$

where R_i is the Richardson number

$$R_i = -\frac{g}{\rho}\frac{\partial\rho}{\partial z}\bigg/\left\{\left(\frac{\partial u}{\partial z}\right)^2 + \left(\frac{\partial v}{\partial z}\right)^2\right\} \tag{41}$$

The turbulent Prandtl number P_r is estimated by the method proposed by Kondo et al. (1979) based on the Monin-Obukhov's similarity theory and given by

$$\frac{1}{P_r} = \frac{1}{6.73R_i + 1/(1 + 6.873R_i)} \qquad R_i \leq 1.0 \qquad (42)$$

$$= \frac{1}{7R_i} \qquad R_i > 1.0 \qquad (43)$$

To evaluate the mixing length l, the water column is divided into the surface mixed layer whose thickness L_s is estimated by $L_s = \text{MAX}(\sqrt{\overline{A_v}/f}, h)$ and the bottom boundary layer where $\overline{A_z}$ is the depth averaged vertical eddy viscosity. In the surface mixed layer, l is considered to be proportional to the wind speed W as $l = rW/f$ based on Langmuir circulation hypothesis, which is a very important phenomenon to sustain phytoplankton in the upper layer and lead to blooming, and in the bottom layer, $l = \kappa(z + H)/\{1 + \kappa(z + H)/l_0\}$ following Blackadar (1962).

Water Quality Model

Governing equations

A water quality sub-model was developed to simulate oxygen depletion and phytoplankton dynamics in Tokyo Bay. The model includes the following parameters: phytoplankton (PHY), zooplankton (ZOO), phosphate (PO4), detritus (DET), dissolved oxygen (DO), and irradiance (IRD). In this model, for simplicity, only phosphate is included as a dependent variable among nutrients because the bay has been considered to be phosphate limiting rather than nitrogen limiting except in very shallow waters where denitrification reach high rates. The model is a very standard one and details were described in Sasaki et al., 1998).

The conservation equation for any water quality parameter ϕ is given by a three-dimensional transport equation with a source term S_ϕ in σ-coordinate system as follows

$$\frac{\partial(H\phi)}{\partial t} + \frac{\partial(uH\phi)}{\partial x} + \frac{\partial(vH\phi)}{\partial y} + \frac{\partial(\dot{\sigma}H\phi)}{\partial \sigma} = \frac{1}{H^2}\frac{\partial}{\partial \sigma}\left(K_v\frac{\partial(H\phi)}{\partial \sigma}\right)$$
$$+ HK_h\left(\frac{\partial^2\phi}{\partial x^2} + \frac{\partial^2\phi}{\partial y^2}\right) + HS_\phi \qquad (44)$$

Modeling of phytoplankton processes

The most important dependent variable in the model is phytoplankton and the details of its modeling is outlined here.

According to field data related to phytoplankton species, which were collected by Tokyo Metropolitan Government and Chiba Prefecture, the predominant species is Skeletonema costatum, from spring to early summer and in winter, Heterosigma akashiwo, from late spring to early summer and in autumn, and Thalassiosira sp. in summer. Among them, S. costatum is the most common species in the bay, and consequently S. costatum is chosen as the representative phytoplankton and the property is modeled as a first step.

The source term of the equation for phytoplankton S_{phy} can be summarized as follows

S_{phy} = Growth rate due to photosynthesis

   − Grazing by zooplankton − Mortality − Respiration − Deposition   (45)

The growth rate due to photosynthesis is set to be a function of temperature, salinity and light intensity in a Steele type formula, and phosphate in Monod type formula, in which model parameters were selected based on literature review for $S.\ costatum$ and field data. For the reproduction of phytoplankton blooming, irradiance is a dominant factor, and thus the spatial distribution of irradiance is computed considering the effect of the self shading by phytoplankton on the attenuation coefficient.

Numerical scheme

A split-operator approach was adopted for each equation of water quality. The original equation (44) is, for convenience, decomposed into two parts

$$\frac{\partial(H\phi)}{\partial t} + \frac{\partial(uH\phi)}{\partial x} + \frac{\partial(vH\phi)}{\partial y} + \frac{\partial(\dot{\sigma}H\phi)}{\partial \sigma} = \frac{1}{H^2}\frac{\partial}{\partial \sigma}\left(K_v\frac{\partial(H\phi)}{\partial \sigma}\right)$$

$$+ HK_h\left(\frac{\partial^2\phi}{\partial x^2} + \frac{\partial^2\phi}{\partial y^2}\right) \tag{46}$$

$$\frac{\partial(H\phi)}{\partial t} = HS_\phi \tag{47}$$

Eq. (46) is first solved by semi-implicit approach which corresponds to the method of differencing temperature and salinity equations, and then Eq. (47) is solved successively by the second-order Runge-Kutta method.

Application to Long-term Simulation in Tokyo Bay

The model is applied to long-term reproduction of hydrodynamic processes and water quality in Tokyo Bay. The model domain is divided in a 1km by 1km horizontal grid, with 20 vertical levels. Computation is carried out from Apr. 1 to Sep. 30 in 1994 and 1996 with the time increment of 300s by giving hourly meteorological data provided by Meteorological Agency, which include the surface stresses due to wind. For the initial conditions, the water in the whole domain was at rest, with the specified temperature and salinity of 13°C and 33.5psu, respectively. At the open boundary, predicted tide level obtained from the tide table of Meteorological Agency and the specified temperature and salinity distributions are given at each time step. Comparison of a time variation of temperature and salinity at St. A (see Figure 1 for location) between computation and the field observation of Environment Agency (1995) is performed as shown in Figures 3 and 4. The model reproduces the observation satisfactorily. The abrupt change of the observed surface salinity is due to a problem with the sensors.

Figure 5 shows a time variation of observed DO, computed DO, at St. A. The time variation of dissolved oxygen (DO) concentration reproduces observed data fairly well including DO dynamics at the bottom, which is very important for considering survival of benthic animals and elution of phosphate in the bottom sediments.

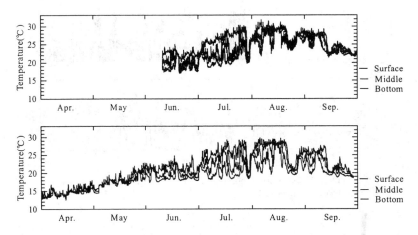

Figure 3: Comparison of measured (upper) and computed (lower) temperature at St. A.

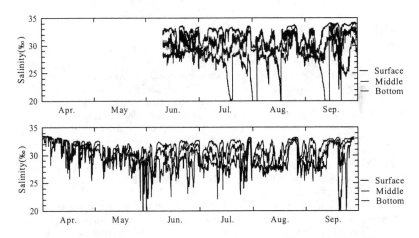

Figure 4: Comparison of measured (upper) and computed (lower) salinity at St. A.

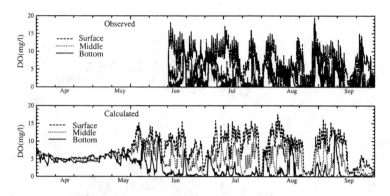

Figure 5: Comparison between the observation of DO and the computational results of DO from April to September in 1994.

The model is also applied to the simulation of water quality in 1996 when the observed data of the time variation of chlorophyll-a are available (Yagi *et al.*, 1997). Figure 6 shows a comparison of computed and observed temperature and chlorophyll-a concentration from April to September in 1996. The temperature distribution shows a good agreement between computation and observation. On the other hand, the reproduction of the variation of chlorophyll-a is not satisfactory and further improvements are recommended.

Simulation of Blue Tides and the Subsequent Red Tides

In late summer or fall, the wind direction sometimes changes from southerly to northerly, and as a result, an upwelling event occurs along the coast at the head of the bay. Since the bottom waters during summer are often anoxic and contain hydrogen sulfide, the color of surface waters sometimes changes milky-blue after upwelling due to oxidation of hydrogen sulfide. This phenomenon is called *blue tide*, *Aoshio* in Japanese, and sometimes causes mortality of benthic animals, especially short necked clams, due to hypoxia.

The bottom anoxic waters were produced not only on the flat bottom, but also in the trenches, and thus two types of simulation were performed; one the upwelling of the waters on the flat bottom and the other upwelling of the waters in the trench shown in Figure 7. The time variation of the surface area of blue tide is shown in Figure 8 for the flat bottom water and in Figure 9 for the trench water. These results were validated through field observation.

After blue tides, due to upwelling of nutrient-rich bottom waters, the concentration of phosphate of surface water increases, and as a result, sudden proliferation of phytoplankton were observed. These events were also reproduced by the numerical model. Figure 10 shows the surface distribution of phosphate during upwelling, and Figure 11 shows surface phytoplankton distribution after the upwelling, which is in good agreement with the observed phytoplankton distribution measured by Chiba Fisheries Experimental Stations.

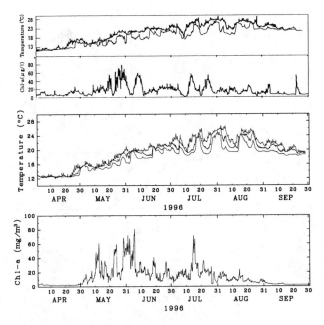

Figure 6: Comparison of the time variation of temperature and chlorophyll-a in 1996 between computation and observation.

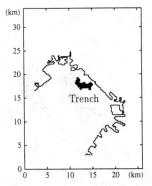

Figure 7: Location of the trench at the head of Tokyo Bay.

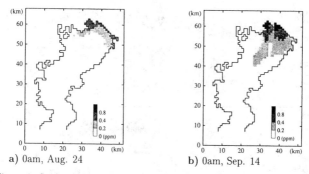

Figure 8: Surface distribution of blue tide due to upwelling from the flat bottom.

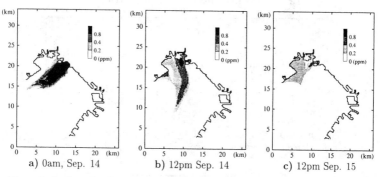

Figure 9: Surface distribution of blue tides caused by upwelling of the trench water.

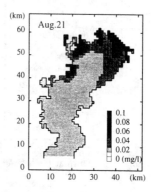

Figure 10: Computed surface distribution of phosphate during the upwelling event.

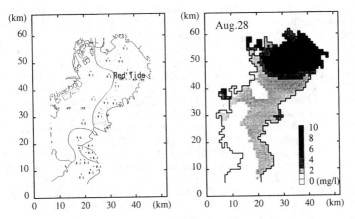

Figure 11: Observed (left) and computed (right) surface phytoplankton distribution after upwelling of nutrient-rich bottom waters.

Conclusion

The comparison between numerical results and the field data of the time variation of temperature and salinity is satisfactory. For spatial distribution, there are not enough data to validate the model.

Computational results of dissolved oxygen dynamics corresponds to the trends in the field. This is because the mechanism of the dynamics of dissolved oxygen is mainly determined by hydrodynamic processes. Phytoplankton dynamics is also rather simple in the bay: the most dominant species of phytoplankton is *S. costatum* and it is adapted to polluted and temperate estuaries and thus its magnitude is mostly limited by the variation of nutrients, in this model phosphate, and as a result, phytoplankton blooming follows after large river discharges with high nutrient concentrations.

For further improvement, we need more field data of water quality, especially chlorophyll-a which can be measured by towing system and nutrients which water sampling is usually required to measure. Motivated by this, starting in April, 1999, the Coastal Engineering Laboratory of The University of Tokyo has been conducting field measurements of various water quality at three stations. The measurements include towing system, velocity field, and weekly nutrient sampling (Koibuchi *et al.*, 1999) The species of phytoplankton in the sampled water is identified. Based on this new data, further improvements will be achieved.

References

Backhaus, J. O. (1985): A three-dimensional model for the simulation of shelf-sea dynamics, *Dtsch. Hydrogr. Z.*, Vol. 38, pp. 165–187.

Benqué, J. P., J. A. Cunge, J. Feuillet, A. Hauguel and F. M. Holly, Jr. (1982): New Method for Tidal Current Computation, *J. Waterway, Port, Coastal and Ocean Division, ASCE*, Vol. 108, No. WW3, pp. 396–417.

Blackadar (1962): The vartical distribution of wind and turbulent exchanges in a neutral atmosphere, *J. Geophys. Res.*, Vol. 67, pp. 3,095–3,120.

Blumberg, F.A. and G.L. Mellor (1983): Diagonstic and prognostic numerical circulation studies of the South Atlantic Bight, *J. Geophys. Res.*, Vol. 88, pp. 4579–4592.

Environment Agency (1995): Report of prediction of blue tide phenomenon in Tokyo Bay, Environment Agency (in Japanese).

Ichihara, M., M. Kyoma and E. Washimi (1980): Estimation of bottom friction coefficient in Tokyo Bay, *Proc. Coastal Eng., JSCE*, Vol. 27, pp. 473–476 (in Japanese).

Kondo, J., Y. Sasano and T. Ishii (1979): Wind-driven current and temperature profiles with diurnal period in the oceanic planetary boundary layer, *J. Phys. Oceanogr.*, Vol. 9, pp. 360–372.

Koibuchi, Y., Y. Ando, H. Ogura, M. Gomyo, J. Sasaki and M. Isobe (1999): Spring observation of water quality by towing system in Tokyo Bay, *Autumn Conference of Oceanogr. Society of Japan* (in Japanese).

Kondo, J. (1994): Meteorology of water environment, *Asakura-shoten* (in Japanese).

Kowalik, Z. and T. S. Murty (1993): Numerical modeling of ocean dynamics, *World Scientific*.

Lardner, R. W. and P. Smoczynski (1990): A vertical/horizontal splitting algorithm for three-dimensional tidal and storm surge computations, *Proc. Royal Soc. London*, Vol. 430 A, pp. 263–283.

Leendertse, J. J., R. C. Alexander and S. K. Liu (1973): A Three-Dimensional Model for Estuaries and Coastal Seas, Vol. 1, Principles of Computation, *The Rand Corporation, R-1417-OWRR*.

Leendertse, J. J. and S. K. Liu (1977): A three-dimensional model for estuaries and coastal seas, Vol. IV, Turbulent Energy Computation, *The Rand Corporation R-2187-OWRT*.

Phillips, N. A. (1957): A coordinate system having some special advantages for numerical forecasting, *Journal of Meteorology*, Vol. 14, pp. 184–185.

Sasaki, J. (1997): On the mechanism of upwelling phenomenon and its effect on blue tides at the head of Tokyo Bay, *Proc. Coastal Eng., JSCE*, Vol. 44, pp. 1111–1115 (in Japanese).

Sasaki, J., M. Isobe, A. Watanabe and M. Gomyo (1996): Discussion on the scale of blue tides in Tokyo Bay, *Proc. Coastal Eng., JSCE*, Vol. 43, pp. 1111–1115 (in Japanese).

Sasaki, J., H. Sanuki and M. Isobe (1998): Numerical simulation of eutrophication in Tokyo Bay, *Proc. Coastal Eng., JSCE*, Vol. 45, pp. 1036–1040 (in Japanese).

Sato, K., M. Matsuoka and K. Kobayashi (1993): Devolpment and application of an efficient method for computation of three-dimensional tidal flows, *Proc. Coastal Eng., JSCE*, Vol. 40, pp. 221–225 (in Japanese).

Yagi, H., Y. Uchiyama, Y. Koibuchi, H. Hinata, S. Miyazaki and K. Nadaoka (1997): Field experiments on properties of water around the head of Tokyo-Bay during the formation of stratification, *Proc. Coastal Eng., JSCE*, Vol. 44, pp. 1076–1080 (in Japanese).

A 3D Shear Dispersion Model Applied to Georges Bank

Zhigang Xu[1], Charles G. Hannah[1] and John W. Loder [1]

Abstract

A three-dimensional model for horizontal dispersion and drift in a spatially- and temporally-varying flow field is described and used to estimate the magnitude and spatial structure of shear dispersion on northeastern Georges Bank. Important dispersion processes associated with vertical and horizontal shears in the seasonal-mean and tidal currents, in combination with random vertical mixing, are represented. Applications to (vertical) line-source releases of suspended sediment and passive tracers are presented, focussed on sensitivity to horizontal and vertical location. On the time scale of the tidal period, vertical shear in the tidal current results in elevated horizontal dispersion in the bottom boundary layer. On longer time and space scales, this dispersion combines with horizontal shears in the mean and tidal flows to enhance the dispersion rates further, particularly near the bank edge.

Introduction

An important factor to horizontal dispersion in the coastal ocean is the occurrence of strong current shears. These include frictionally-induced vertical structure in the bottom boundary layer, horizontal current variations over variable topography and near the coastline, and vertically-varying currents

[1]Fisheries and Oceans Canada, Bedford Institute of Oceanography, P.O. Box 1006, Dartmouth, N.S., Canada B2Y 4A2

associated with horizontal density gradients. In conjunction with smaller-scale current fluctuations that move water and materials across these shears, there can be enhanced levels of dispersion in the direction of the sheared flow, through the mechanism of "shear dispersion" (Taylor 1921; Bowden 1965). For horizontal flows, both vertical and horizontal shears can contribute, the important factors being the shear magnitude and the time scale for transverse mixing across the shear.

In this paper we use a recently developed 3D dispersion/drift model together with realistic seasonal and tidal flow fields from a hydrodynamics model (Naimie 1995; 1996) to explore aspects of shear dispersion on northeastern Georges Bank. The dispersion/drift model, referred to as "spatial bblt", is an extension of a "local bblt" (benthic boundary layer transport) model (Hannah et al. 1995, 1996) to include horizontal variations in the physical environment. Our motivation has been the issue of the fate and impacts of suspended drilling wastes from hydrocarbon activities, but the shear dispersion processes represented in the models are also relevant to other issues such as the cross-bank dispersion rates of nutrients and fish larvae.

In the next section, we provide a brief description of spatial bblt. Then, we apply the model to three sites on Georges Bank to estimate the magnitude and illustrate the spatial structure of shear dispersion. The applications will first be for suspended sediments of relevance to the drilling waste issue, and then for passive tracers restricted to various portions of the water column to explore the vertical structure of the underlying dispersion mechanisms.

Model Description

The key features of the bblt models are the quantitative representations of the combined effects of sheared horizontal flows and mixing in the cross-flow direction, thereby providing quantitative estimates of the rates of shear dispersion (e.g. Taylor 1921) as well as of horizontal drift.

The development and example applications of local bblt, which makes the simplifying assumption of no horizontal variations in the water depth and hydrodynamics, have been described previously (Hannah et al. 1995, 1996, 1998). Other key aspects of the approach are the mathematical partitioning of the sediment load into discrete portions referred to as packets, and the separation of packet movement into horizontal advection by a specified flow field and random vertical mixing with a semi-empirical probability distribution. For the sediment applications described previously, the vertical distribution was based on the Rouse profile (Rouse 1937), providing various degrees of bottom trapping dependent on bottom stress and an effective settling velocity w_s for the overall sediment load. The rate of vertical mixing was specified through a vertical mixing time estimated for the region of interest. The vertically-varying velocity field in local bblt could be taken from current meter data or a 3D numerical hydrodynamics model. With the assumption of no horizontal hydrodynamics variations, the horizontal dispersion in local bblt arises entirely from vertical mixing in combination with vertically-sheared horizontal flow.

On scales of order 10–100 m, the predominant shears in the coastal ocean (away from the immediate vicinity of the coastline) are vertical shears such as those in the bottom boundary layer, so that local bblt should capture the primary shear dispersion processes on small space and time scales. However, the previous bblt applications (e.g. Hannah et al. 1998) have shown that dispersion arising from vertical shear can spread initial (vertical) line-source patches of sediment over a tidal excursion or more within a couple of tidal periods. On Georges Bank, the tidal excursion is order 10 km, and there are substantial horizontal shears on this scale associated with tidal flow over steep topography and a tidal-mixing front (e.g. Loder et al. 1993). This provides the potential for the horizontal dispersion arising from the small-scale vertical shears to combine with larger-scale horizontal shears and provide further enhanced horizontal dispersion, through a combination of shear dispersion processes (e.g. Zimmerman 1976). An additional effect of spatial hydrodynamics variability is that there can be either reduced or enhanced dispersion following a patch of sediment depending on whether the patch moves into a less or more dispersive environment (e.g. Hannah et al. 1998).

The key extension in spatial bblt (beyond local bblt) is the inclusion of horizontal variations in the water depth, currents and bottom stress, so that effects of horizontal shears and horizontal hydrodynamics changes are included. This requires a 4-d flow field which in most cases must come from a 3D numerical model, and involves a large additional computational burden in estimating the local velocity of each packet at each time step. For the applications in this paper, we use the same 3D flow fields as in Hannah et al. (1998); namely, the summer climatological-mean and M_2 and M_4 tidal currents from Naimie's (1995) solution on a triangular finite-element mesh, obtained using a prognostic advanced-turbulence model (Naimie 1996) initialized and forced with observational information on tides, wind and density.

In order to obtain statistically-robust results for the resulting nonlinear problem (the Euler-Lagrange transformation in a spatially- and temporally-structured environment with random vertical mixing), a large number of packets is required, taken as 10^4 in our simulations. To reduce the computational burden, a number of steps are introduced: 1) the packet tracking (or "particle tracking" in numerical modeling terminology), and hence the required mesh and velocity fields, are limited to only a small portion (100-km scale) of the original 3D model mesh (1000-km scale); 2) the simulations are done (essentially in parallel) for subgroups of packets, with different random-number seeds (vertical mixing) for each subgroup; and 3) a sparse-matrix technique (Mathworks 1992) is used to allow for direct computer memory addressing of mesh nodes and elements for the velocity interpolation.

The representation of near-bottom hydrodynamics in spatial bblt closely follows that in local bblt, including a quadratic bottom stress law involving the velocity at the deepest node in the 3D model (approximately 1 m above the actual seafloor), and a logarithmic velocity sublayer below this node (Hannah et al. 1998). For the suspended sediment applications, the Rouse profile is used as in Hannah et al. (1998) with the primary exception that the vertical extent of the profile is restricted to a height approximating the base of the seasonal thermocline, thereby confining the sediment to the bottom boundary layer. For the exploratory passive-tracer applications, the vertical probability distribution is taken to be uniform over the specified interval, with various mixing time scales considered.

The raw output of spatial bblt is the time series of the packets' coordinates, $X(t)$, $Y(t)$ and $Z(t)$, which can be used together with the packet mass to compute various properties of the evolving distribution. These include concentrations in 4-d space and statistics of the distribution. For the presentations in this paper, we use the following statistics:

- $\bar{Z}$, the mean value over the simulation of the height above the seafloor of the vertical center of mass;

- $\sigma_{maj}^2(t)$, $\sigma_{min}^2(t)$, the variances of the horizontal positions of the packets about the horizontal center of mass, in the directions of the major and minor axes of the packet distributions;

- K_{max} and K_{min}, the effective horizontal diffusivities (or dispersion rates) along the major and minor axes, defined as (e.g. Csanady, 1973),

$$(K_{max}, K_{min}) = \frac{1}{2} \left(\frac{\partial \sigma_{max}^2}{\partial t}, \frac{\partial \sigma_{min}^2}{\partial t} \right) ;$$

- D, the overall horizontal diffusivity defined as $D = \sqrt{K_{max}K_{min}}$;

- $\bar{U}$ and $\bar{V}$, the mean (over the duration of the simulation) of the horizontal (east and north) velocity components of the center of mass.

Applications to Suspended Sediment

In this section, we use spatial bblt to examine the drift and dispersion patterns of an initial distribution of suspended sediment at three sites on northeastern Georges Bank (Fig. 1). The focal site is the Edge site investigated previously by Hannah et al. (1998), located near the bank edge and a frontal-zone jet with a water depth of 74 m. The other sites are: a Side site of depth 125 m located 10 km north of the Edge site; and a bank-plateau NEP site of depth 72 m on the Northeast Peak. The Rouse profile is used for the vertical probability distribution, with values of 0.1 and 0.5 cm/s for w_s to explore alternative levels of bottom trapping and approximate flocculated drilling wastes (Muschenheim and Milligan 1996). The maximum height of the profile is set at 34, 17 and 49 m for these respective release sites, based on the observed hydrography and the 3D hydrodynamics model. The duration of each simulation is six M_2 tidal periods. Other specified parameters are 0.005 for the bottom drag coefficient, 0.003 m for the bottom roughness height and the reference height in the Rouse profile, 1 hour for the horizontal advection time step, and 3-4 hour for the vertical mixing time.

Current Structure

The summer currents on northeastern Georges Bank are dominated by the M_2 tidal and seasonal-mean components (Loder et al. 1993) whose spatial structure is illustrated in Fig. 1. The tidal currents have peak amplitudes near 1 m/s and associated excursions near 15 km on the bank plateau, and decrease abruptly (with increasing water depth) across the northern edge of the bank. They also have important vertical structure in the bottom boundary layer, with reduced amplitudes and increased phase leads as the seafloor is approached. The seasonal-mean currents reflect the clockwise gyre driven by tidal rectification and fronts over Georges Bank, with largest values in the northern-edge jet and in the upper water column, and a weak drift across the bank plateau. The vertical shears in the tidal bottom boundary layer and in the frontal-zone jet, and the horizontal shears in the tidal and frontal flows near the bank edge provide the potential for shear dispersion.

Dependences on Release Site and Settling Velocity

Figure 2 shows the drift and dispersion patterns for the two different settling velocities at the three study sites. Further information on the evolving

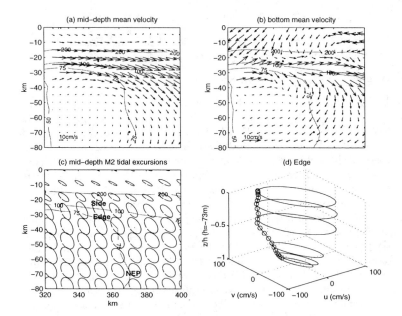

Figure 1: Horizontal distributions of mean velocity at (a) mid-depth and (b) bottom; (c) horizontal distribution of M_2 tidal excursions at mid-depth; and (d) M_2 tidal ellipse profiles at the Edge site.

distributions is provided by the time series plots of selected statistics for the Edge and NEP cases in Fig. 3, and the average values of selected statistics for all cases in Table 1. Substantial differences in both the drift and dispersion rates are apparent, both among sites and with settling velocity. The drift differences are consistent with the spatial current structure in Fig. 1, with highest drift rates at the Edge and Side sites (northern-edge jet) and for the smaller settling velocity (higher center of mass; Fig. 3 and Table 1). The dispersion differences are particularly pronounced among the sites, but there are also substantial differences with settling velocity. The highest dispersion rates occur at the Edge site, while the lowest rates occur at the NEP site. At particular sites, there is generally a higher dispersion rate for the higher settling velocity (reduced sediment height) consistent with the largest vertical shears associated with the tidal currents in the near-bottom region (Fig. 1).

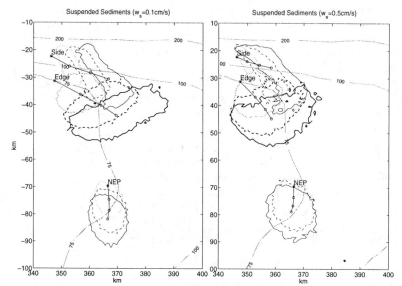

Figure 2: Patch positions and distributions (vertically integrated) at days 1 (dotted), 2 (dashed) and 3 (solid) for the suspended sediments at the Side, Edge (thicker lines) and NEP sites, with $w_s = 0.1$ (left) and 0.5 (right) cm/s. The release sites and the daily positions of the mass centers are marked by circles linked by lines. The contours outline the areas with more than 90% of the packets. The light thin contours indicate bathymetry (m).

Indications of the importance of vertical versus horizontal shears to the dispersion rates can be seen from the statistics for corresponding simulations with local bblt (Table 1), in which the release-site currents and depth are used for all positions. The overall diffusivity D from spatial bblt is close to, or larger than that from local bblt in all cases, pointing to an underlying contribution from vertical shear in all cases, and additional horizontal-shear influences that vary with location. The differences between the spatial and local bblt results are smallest at the NEP site, consistent with the reduced horizontal variations on the bank's plateau than over its sides. (Test cases using the same horizontally-uniform flow field in spatial and local bblt give the same results, confirming that local bblt is a limiting case of spatial bblt). The largest dispersion differences occur at the Edge and Side sites, and involve changes in the overall diffusivity magnitudes and increased anisotropy in the dispersion. Whereas the distribution shapes at the NEP site (Fig. 2) closely resemble the tidal ellipse as expected when vertical shear in the tidal currents is dominant (e.g. Hannah et al. 1996), the shapes are much more irregular at the Edge and Side sites, pointing to the effects of horizontal shear. For example, the major axes at the Edge and Side sites increasingly rotate clockwise and counterclockwise respectively, consistent with the horizontal shear in local mean flow (Fig. 1).

Table 1: Averages (over the 3-day simulations) of statistics for the base suspended-sediment simulations at the three study sites, from the spatial and local bblt models. See text for definitions of the statistics.

Site	w_s (cm/s)	Model	$\bar{Z}$ (m)	K_{max} (m²/s)	K_{min} (m²/s)	D (m²/s)	$\bar{U}$ (cm/s)	$\bar{V}$ (cm/s)
Side	0.1	sbblt	7.5	52	5.1	16	7.5	-1.9
		lbblt	7.2	35	10	19	9.5	-0.6
	0.5	sbblt	2.6	87	9.2	28	4.4	-0.7
		lbblt	1.7	27	4.7	11	2.9	1.2
Edge	0.1	sbblt	15.7	111	19	46	7.6	-4.1
		lbblt	15.6	58	31	34	8.0	-6.8
	0.5	sbblt	9.8	104	34	60	3.5	-0.4
		lbblt	8.5	72	21	34	4.8	-6.9
NEP	0.1	sbblt	23.1	25	13	18	-1.1	-4.3
		lbblt	23.0	22	8.5	14	-1.1	-5.1
	0.5	sbblt	14.9	35	21	27	-1.3	-3.2
		lbblt	15.0	41	20	29	-1.6	-4.3

The time series of the statistics (Fig. 3) further illustrate the nature of the dispersion. The sediment concentration in areas of interest can be estimated from the local packet density, and shows variations over the tidal period due to contraction and expansion of the Rouse profile associated with bottom stress variability, and a longer-term reduction due to the dispersion.

Dependence on Phase of the Tide at Release Time

Besides the effect of horizontal shears on local dispersion, an additional effect of horizontal variability in the hydrodynamics is the potential for drift of the

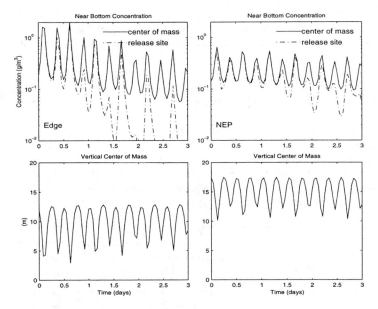

Figure 3: Time series of selected statistics for the $w_s = 0.5$ cm/s cases at the Edge (left) and NEP (right) sites. The panels show near-bottom concentrations (assuming a total mass of 53 tons, Hannah et al. 1998) at the center of mass and release site (upper), and the height of the vertical center of mass (lower).

sediment into a different physical regime. The latter effect can be expected to be particularly important along the northern edge of Georges Bank where topography and current magnitudes change over the scale of the tidal excursion. One approach to examining the importance of this effect is to consider releases at alternative phases of the M_2 tidal flow at a particular site in this area. These different phases can be viewed as representing releases into water columns which undergo tidal excursions in different regions.

Figure 4 shows the drift and dispersion patterns for suspended sediment released at the extreme northern and southern positions of the tidal excursion at the Edge site (at the times when the north-south velocity component changes sign). Statistics for these simulations are presented in Table 2. It can be seen that the different release phases result in large differences in

both drift and dispersion. Release at the extreme northern position (prior
to on-bank tidal flow) results in lower and more isotropic dispersion, and
weaker and more southward drift characteristic of the bank plateau, than at
the extreme southern position (prior to off-bank tidal flow), in which case the
drift is more eastward and the dispersion more anisotropic, characteristic of
the bank edge. Comparison of the statistics (Table 2) with those for the base
(Mid) cases at this site, in which the release phase was intermediate between
the two extremes (Fig. 2), shows higher diffusivities for release between the
northern and southern extremes, particularly for diffusivities along the major
axis. This indicates that there is a cross-bank maximum in the overall shear
dispersion rates in the vicinity of the Edge site, which is consistent with the
pattern among the Side, Edge and NEP sites (Table 1). At the Edge site,
there can also be larger dispersion rates for the lower settling velocity due
to increased vertical and horizontal shears in the frontal-zone jet away from
he seafloor.

Table 2: Averages (over 3-day simulations) of statistics for the spatial bblt
sensitivity cases to tidal phase (excursion position), for suspended sediment
at the Edge site, for different vertical intervals and mixing times (t_m).

w_s (cm/s)	Phase	$\bar{Z}$ (m)	K_{max} (m^2/s)	K_{min} m^2/s	D m^2/s	$\bar{U}$ (cm/s)	$\bar{V}$ (cm/s)
	Northern	15.9	34	15	23	1.6	-3.6
0.1	Mid	15.7	111	19	46	7.6	-4.1
	Southern	15.8	91	21	44	9.6	-3.5
	Northern	10.5	25	9	15	-0.1	-3.6
0.5	Mid	9.8	104	34	60	3.5	-0.4
	Southern	9.3	59	25	38	5.4	- 4.2

Applications to Passive Tracers at Edge Site

In this section we present a set of more idealized simulations for passive tracers, intended to elucidate the vertical structure of the underlying dispersion
processes. The tracers are assumed to be uniformly distributed over (and
limited to) set vertical intervals of the water column following their horizontal drift (i.e. in $\sigma = -z/h$ space, where z is depth below the surface and
h the water depth); specifically over thirds of the column which we desig-
nate as Upper, Middle and Lower. The tracers are released at the Edge site

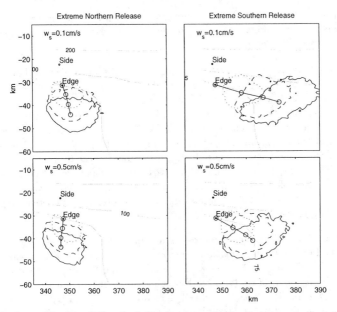

Figure 4: Patch positions and distributions (vertically integrated) at days 1, 2 and 3 for suspended sediment simulations at the Edge site, for releases at two different phases of the tidal current, and $w_s = 0.1$ cm/s (top panels) and 0.5 cm/s (bottom panels). The left panels are for release at the northern extreme of water parcel positions during the tidal excursion, and the right panels for release at the southern extreme.

because of its location in the northern-edge frontal zone. Sensitivity simulations are carried out for vertical mixing times (t_m) of 1-100 hr, with other model parameters as in the suspended sediment cases.

The simulations reveal a dramatic dependence of dispersion on the vertical interval (Fig. 5, Table 3), and a dependence on t_m consistent with theoretical expectations for a combination of oscillatory and mean flow dispersion (Fischer et al. 1979). In the Upper interval, the drift is southeastward across the Northeast Peak consistent with the on-bank spreading of the tidal-mixing front (e.g. Loder et al. 1993), and the dispersion is weak. In the Middle interval, the southward component of drift is reduced consistent with reduced on-bank spreading of the deeper flows but there is greater overall drift associated with the tracer remaining in the tidally-rectified bank-edge flow. The dispersion rates are increased consistent with greater influences of tidal shears in this interval (Fig.1), but remain lower than in the suspended sediment cases (Table 1).

The most remarkable result is the magnitude of dispersion in the Lower interval, with diffusivity magnitudes exceeding those in the sediment cases by a factor of 2–3. It can be seen (Fig. 5) that the dispersion is associated with an overall eastward drift in the Lower interval, and strong effects of the horizontal shear in the mean flow on tracer distribution, including a clockwise rotation of the patch, and spreading both onto the bank plateau and into deeper water along the northern side. It is important to note that the Lower-interval dispersion is highly anisotropic, with large magnitudes (300 m²/s) in the direction of the mean flow and lower (but still substantial) values of 40–70 m²/s in the cross-bank (or cross-frontal) direction for expected mixing times (4-10 hours).

In all three intervals, the diffusivities are maximum for t_m values of the order of the tidal period pointing to the major contributions from tidal shears, but remain substantial for larger values of t_m pointing to important contributions from mean shears (also see Hannah et al. 1996, for a discussion of expectations from theoretical studies). The variation of diffusivity with t_m is most pronounced in the Lower interval, consistent with the vertical tidal shears (and hence oscillatory shear dispersion) being most dominant there.

Summary

The spatial bblt simulations illustrate the importance of spatially-structured shear to horizontal dispersion on Georges Bank, and the need for 4-d representations of the currents and inclusion of cross-flow mixing to obtain quantitative estimates of the dispersion rates. Distinct spatial patterns in the dispersion rates are apparent, which can be related to spatial patterns in the vertical and horizontal shears in the predominant tidal and mean flows. In the bottom boundary layer, vertical shear in the tidal current combines with high levels of small-scale vertical turbulence to provide effective diffusivities of order 10–70 m²/s for even small (horizontally) initial patches. As a result, small patches spread to the scale of the tidal excursion within a tidal period or so, where horizontal shears in the tidal or mean flow can lead to further enhancements of the diffusivities. The latter enhancements have strong spatial variability, resulting in diffusivities of order 100–300 m²/s (on cross-bank scales of 20–40 km) in the direction of the mean flow near the northern edge at the bank, but limited enhancements on the bank plateau.

Table 3: Averages (over 3-day simulations) of statistics from the passive tracer simulations for the Edge site, for different vertical intervals and mixing times.

Vertical Interval	t_m (hours)	$\bar{Z}$ (m)	K_{max} (m²/s)	K_{min} (m²/s)	D (m²/s)	$\bar{U}$ (cm/s)	$\bar{V}$ (cm/s)
Upper	4	68.6	6.1	2.0	3.4	11.6	-10.5
	10	67.9	8.9	3.2	5.3	11.7	-10.6
	100	62.7	5.9	3.4	4.5	12.2	-11.0
Middle	4	44.1	41	9.3	19	16.7	-8.2
	10	44.9	54	8.7	22	16.4	-8.4
	100	46.3	26	3.5	10	16.8	-9.1
Lower	4	16.3	318	41	114	12.4	-2.4
	10	22.8	311	70	148	17.4	-2.9
	100	10.1	101	1.3	11	5.4	-5.3

These diffusivity magnitudes are a factor of a few larger than those reported by Okubo (1974) for comparable patch scales in the upper ocean, pointing to the importance of shear dispersion in the bottom boundary layer. Outside the bottom boundary layer, the passive-tracer simulations indicate much lower diffusivity magnitudes, but significant influences from mean-flow shears at the northern edge.

Observational estimates of dispersion on Georges Bank have been largely limited to the near-surface region (e.g. Flagg et al. 1982; Drinkwater and Loder 2000). These have indicated a strong scale-dependence in the diffusivities as expected, with large diffusivities and pronounced influences of the mean-flow shear on scales of 50–100 km, and influences of near-surface frontal convergence on scales below 20 km. Observational estimates of subsurface dispersion have been limited to the 150–380 m²/s values of Loder et al. (1982) for the central bank, indicating strong dispersion although the precise roles of shear dispersion and chaotic stirring over topography (e.g. Ridderinkhof and Loder 1994) have not been resolved. These observational estimates are not inconsistent with the present estimates, but together they point to the need for subsurface observational studies and complementary modelling and theoretical studies to resolve important spatial structure.

Our observational knowledge of Georges Bank also points to other processes which can modify the estimates provided here based on the summer-mean and tidal flows in combination with vertical mixing. These processes include vertical mixing and vertical shears associated with storm-band forcing of the surface mixed layer; frontal-zone instabilities and convergences providing additional horizontal motions; internal tides and waves affecting vertical processes in the pycnocline; and nonlinear tidal current interactions over irregular topography. Numerical models like the 3D hydrodynamics model and spatial bblt which can resolve and isolate the contributions of various factors are valuable in both elucidating the underlying mechanisms and describing the pronounced spatial structure of shear dispersion in this and other parts of the coastal ocean.

Acknowledgements

We are grateful to Chris Naimie for making available the 3D model field; Elizabeth Gonzalez, Kee Muschenheim and Yingshuo Shen for contributions to bblt model development and applications; Ken Drinkwater and David Greenberg for internal reviews; and the (Canadian) Federal Panel for Energy, Research and Development (PERD) for funding.

References

Bowden, K. F. 1965. Horizontal mixing in the sea due to a shearing current. *J. Fluid Mech.* **21**, 83–95.

Csanady, G.T. 1973. *Turbulent Diffusion in the Environment.* D. Reidel Publishing Company.

Drinkwater, K.F. and J.W. Loder. 2000. Near-surface convergence and dispersion near the tidal-mixing front on northeastern Georges Bank. *Deep Sea Res. II* (in press).

Fischer, H.B., E.J. List, R.C.Y. Koh, J. Imberger and N.H. Brooks. 1979. *Mixing in Inland and Coastal Waters.* Academic Press. 483 pp.

Flagg, C.N., B.A. Magnell, D. Frye, J.J. Cura, S.E. McDowell and R.I Scarlet. 1982. Interpretation of the physical oceanography of Georges Bank. EG&G Environmental Consultants, Waltham, Mass. Final report prepared for U.S. Dept. of Interior, Bureau of Land Management, 901 pp.

Hannah, C.G., Y. Shen, J.W. Loder and D.K. Muschenheim. 1995. bblt: Formulation and exploratory applications of a benthic boundary layer transport model. Can. Tech. Rep. Hydrogr. Ocean Sci.166: vi+52 pp.

Hannah, C.G., J.W. Loder and Y. Shen. 1996. Shear dispersion in the benthic boundary layer. In: Estuarine and Coastal Modeling. Proceedings of the 4th International Conference, edited by M.L. Spaulding and R.T. Cheng, pp. 454-465, ASCE, New York.

Hannah, C.G., Z. Xu, Y. Shen and J.W. Loder. 1998. Models for suspended sediment dispersion and drift. In: Estuarine and Coastal Modeling. Proceedings of the 5th International Conference, edited by M.L. Spaulding and A.F. Blumberg, pp. 708-722, ASCE, New York.

Loder, J.W., K.F. Drinkwater, N.S. Oakey and E.P.W. Horne. 1993. Circulation, hydrographic structure and mixing at tidal fronts: the view from Georges Bank. *Phil. Trans. Roy. Soc. Lond., Ser. A* **343**, 447-460.

Loder, J.W., D.G. Wright, C. Garrett and B.-A. Juszko. 1982. Horizontal exchange on central Georges Bank. *Can. J. Fish. Aquat. Sci.* **39**, 1130-1137.

Mathworks. 1992: Matlab Reference Guide, 548 pp.

Muschenheim, D.K. and T.G. Milligan. 1996. Flocculation and accumulation of fine drilling waste particles on the Scotian Shelf. *Mar. Pollut. Bull.* **32**, 740-745.

Naimie, C.E. 1995. Georges Bank bimonthly residual circulation — prognostic numerical model results. Report No. NML-95-3, Numerical Methods Laboratory, Dartmouth College, Hanover, N.H. 38 pp.

Naimie, C.E. 1996. Georges Bank residual circulation during weak and strong stratification periods — prognostic numerical model results. *J. Geophys. Res.* **101**, 6469-6486.

Okubo, A. 1974. Some speculations on oceanic diffusion diagrams. *Rapp. P.-v. Reun. Cons. int. Explor. Mer* **167**, 77-85.

Ridderinkhof, H. and J.W. Loder. 1994. Lagrangian characterization of circulation over submarine banks with application to the outer Gulf of Maine. *J. Phys. Oceanogr.* **24**, 1184-1200.

Rouse, H. 1937. Modern concepts of the mechanics of turbulence. *Trans. Am. Soc. Civ. Eng.*, **102**, 463-543.

Taylor, G.I. 1921. Diffusion by continuous movements. *Proc. London Math. Soc.* **A20**, 196-211.

Zimmerman, J.T.F. 1976. The tidal whirlpool: a review of horizontal dispersion by tidal and residual currents. *Neth. J. Sea Res.* **20**, 133-154.

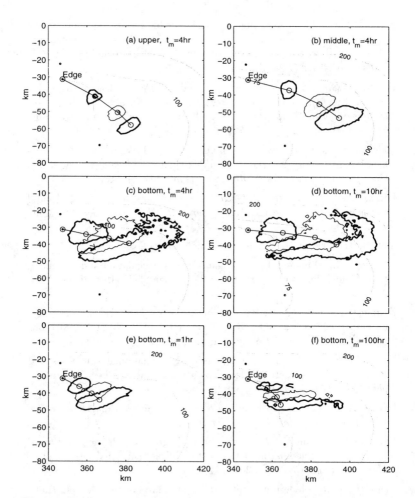

Figure 5: Patch positions and distributions at days 1, 2 and 3 for the release of a passive tracer in different vertical intervals at the Edge site. Cases shown are: (a-c) the Upper, Middle and Lower intervals for $t_m = 4$ hr, and (d-e) the Lower interval for $t_m = 10, 100, 1$ hrs.

Hydrodynamic modeling for the 1998 Lake Michigan coastal turbidity plume event

Dmitry Beletsky[1], David J. Schwab[2], Michael J. McCormick[2],
Gerald S. Miller[2], James H. Saylor[2], and Paul J. Roebber[3]

Abstract

A three-dimensional primitive equation numerical ocean model, the Princeton model of Blumberg and Mellor (1987), was applied to Lake Michigan to simulate hydrodynamic conditions during the 1998 coastal turbidity plume event. A massive turbidity plume in southern Lake Michigan was caused by a strong storm with northerly winds up to 17 m/s during this period. The hydrodynamic model of Lake Michigan has 20 vertical levells, and a uniform horizontal grid size of 2 km. The model is driven with surface momentum flux derived from observed meteorological conditions at 12 land stations in March 1998 and also with surface winds calculated using the mesoscale meteorological model MM5 (Dudhia, 1993) on a 6 km grid. Current observations from 11 subsurface moorings showed that while the model was able to qualitatively simulate wind-driven currents, it underestimated current speeds during strong wind events and in particular an onshore-offshore component of the flow in the area of observations. This may be due at least in part to the significant decrease of modeled current speeds with depth during strong wind events while observed currents showed almost no vertical shear. Hydrodynamic model results using MM5 winds as the forcing function were slightly better than results which were based on objectively analyzed winds.

---

[1]Department of Naval Architecture and Marine Engineering, University of
Michigan; Cooperative Institute for Limnology and Ecosystems
Research/NOAA GLERL and University of Michigan, Ann Arbor, MI, 48105-2845
[2]NOAA Great Lakes Environmental Research Laboratory, Ann Arbor, MI,
48105-2845
[3]Department of Mathematical Sciences, University of Wisconsin-Milwaukee,
Milwaukee, WI, 53201

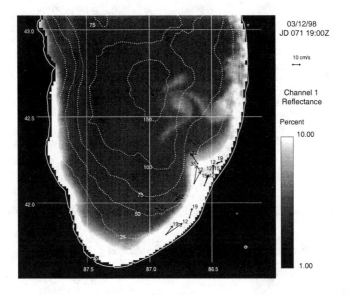

Figure 1. Satellite measurements of surface reflectance in southern Lake
Michigan with currents at various depths observed at 1900 GMT, 3/12/98.

Introduction

Satellite observations of surface reflectivity in Lake Michigan have revealed a
recurrent turbidity plume (Eadie et al., 1996). A 10 km wide plume of
resuspended material extending over 100 km along the southern shore of the lake
was first observed in satellite imagery by Mortimer (1988), and has since been
observed every spring since 1992, when satellite imagery for the Great Lakes
region first became available on a routine basis. The resuspension plume of
March 1998 was one of the largest events of record. Satellite observations (Fig.1)
reveal a well developed plume extending over 300 km of coastline from

Milwaukee, WI to Muskegon, MI with several dominant offshore features originating from the southeastern shoreline. The plume originated around March 10 following several days of intense storms that produced 17 m/s northerly winds and generated waves in the basin over 6 m high (Schwab et al., 2000). Our current understanding is that the initiation of the plume is caused by a major storm with strong northerly winds generating large waves in southern Lake Michigan. The plume appears along the entire southern coastline of the lake. It occasionally veers offshore along the eastern shore of the lake, coincidentally near the areas of highest measured long-term sediment accumulation in the lake.

The recurrent turbidity plume in southern Lake Michigan and associated nearshore-offshore water mass and material exchange is an object of intense study within a multidisciplinary research program called EEGLE (Episodic Events-Great Lakes Experiment) (http://www.glerl.noaa.gov/eegle). The program is jointly sponsored by NOAA (National Oceanic and Atmospheric Administration) and NSF (National Science Foundation). One of the goals of EEGLE is to create a suite of physical and biological models to understand the nearshore-offshore transport of biogeochemically important materials. Currently, a linked system of wave, circulation and sediment transport models is being developed at NOAA Great Lakes Environmental Research Laboratory (GLERL). Preliminary results of application of that system to the March 1998 resuspension event were described in Schwab et al. (2000). In this paper, we will investigate the dynamics of wind-induced circulation and offshore transport in March 1998 in more detail by using current observations and analyzing sensitivity of hydrodynamic model results to meteorological data. We will focus mainly on the large storm of 9-12 March 1998 because the most significant offshore transport occurred during that period.

Hydrodynamic model

A 3-dimensional circulation model for the Great Lakes (Schwab and Beletsky, 1998) is used to calculate lake circulation. The model is based on the Princeton Ocean Model (Blumberg and Mellor, 1987) and is a nonlinear, fully three-dimensional, primitive equation, finite difference model. The model is hydrostatic and Boussinesq so that density variations are neglected except where they are multiplied by gravity in the buoyancy force. The model uses time-dependent wind stress and heat flux forcing at the surface, zero heat flux at the bottom, free-slip lateral boundary conditions, and quadratic bottom friction. The drag coefficient in the bottom friction formulation is spatially variable. It is calculated based on the assumption of logarithmic bottom boundary layer using constant bottom roughness of 1 cm. Horizontal diffusion is calculated with a Smagorinsky eddy parameterization (with a multiplier of 0.1) to give a greater mixing coefficient near strong horizontal gradients.

The Princeton Ocean Model employs a terrain following vertical coordinate system (σ-coordinate). The equations are written in flux form, and the finite differencing is done on an Arakawa-C grid using a control volume formalism. The finite differencing scheme is second order and centered in space and time (leapfrog). The model includes the Mellor and Yamada (1982) level 2.5 turbulence closure parameterization for calculating the vertical mixing coefficients for momentum and heat from the variables describing the flow regime.

The hydrodynamic model of Lake Michigan has 20 vertical levels and a uniform horizontal grid size of 2 km. Model bathymetry is based on the new, high resolution bathymetric data (NGDC, 1996).

Meteorological data 1: objectively analyzed

In order to calculate momentum flux fields over the water surface for the lake circulation model, it is necessary to estimate wind and air temperature fields at model grid points. Meteorological data were obtained from 12 National Weather Service stations around Lake Michigan (Fig. 2). These observations form the basis for generating gridded overwater wind and air temperature fields. Because overland wind speeds generally underestimate overwater values we apply the empirical overland-overlake wind speed adjustment from Resio and Vincent (1977). (See Beletsky and Schwab (1998), and Schwab and Beletsky (1998) for more detail.)

In order to interpolate meteorological data observed at irregular points in time and space to a regular grid so that it can be used for input into numerical circulation model, some type of objective analysis technique must be used. For this study we first used the nearest-neighbor technique (NRST), with the addition of a spatial smoothing step (with a specified smoothing radius). In the NRST technique, we also consider observations from up to three hours before the interpolation time to three hours after the interpolation time. In the nearest-neighbor distance calculations, the distance from a grid point to these observation points is increased by the product of the time difference multiplied by a scaling speed. The interpolation scaling speed is taken as 10 km/hr. Interpolation smoothing distance is 30 km. We found that the NRST technique provided results comparable to results from the inverse power law or negative exponential weighing functions discussed in Schwab (1989).

While nearest neighbor technique was used earlier in the Lake Michigan Mass Balance Study models (Beletsky and Schwab, 1998) and Great Lakes Forecasting System (GLFS) models (http://superior.eng.ohio-state.edu), the GLFS models have subsequently adopted a new geometrically-based technique that appears to provide a more realistic representation of the 2-d wind field than NRST techniques. The approach is called 'Natural Neighbor' interpolation (NTRL) and is based on the Delaunay triangulation of the station observation network (Sibson, 1981, Watson, 1994). According to Sambridge et al. (1995), the method has the following useful properties:

1) the original function values are recovered exactly at the reference points

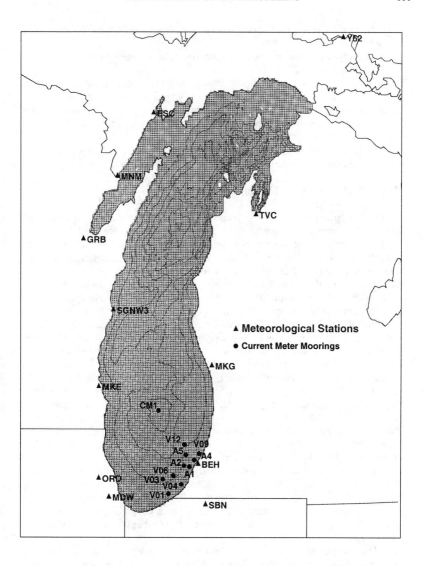

Figure 2. Observations network and 2 km computational grid.

2) the interpolation is entirely local (every point is only influenced by its natural neighbor nodes)

3) the derivatives of the interpolated function are continuous everywhere except at the reference points.

Points 1) and 2) are especially important for the type of network we deal with in the Great Lakes, i.e., not every station is available at every hour and some stations (ships) appear only intermittently. This technique is also advantageous for interpolating data fields for which the spatial autocorrelation function is not well known, such as hourly wind fields.

In this paper we will use both NRST and NTRL interpolation techniques in order to find out which method gives currents that match observations better.

Meteorological data 2: MM5 model based

In addition to objectively analyzed data, we also used meteorological model data as the forcing function in order to compare results obtained by various methods. In order to generate atmospheric forcing fields that take full advantage of the advanced capabilities of modern numerical weather prediction, the Penn State/NCAR 5th generation mesoscale model (MM5) was run for the period 7-10 March 1998. In the hydrodynamic model run, NRST winds during that 4-day period were replaced with MM5 winds. A triply nested domain configuration (54/18/6 km) with two-way interactions (such that exterior domains feel the influence of interior domains and vice versa) was employed, with the innermost nest providing 6 km grid point resolution in an area centered on Lake Michigan. Model initialization and lateral boundary conditions were determined as follows. First guess fields of atmospheric variables (wind, temperature, moisture) were obtained from the National Centers for Environmental Prediction (NCEP) historical archives of global (2.5° latitude by 2.5° longitude) mandatory-level analyses and were then adjusted using a Cressman-type objective analysis of surface and rawinsonde data for all stations within or near the grid domain. These analyses provided boundary conditions on the outermost grid domain throughout the course of the integrations and were used in the four dimensional data assimilation (FDDA) procedure described below.

Vertical sigma levels were arranged such that the model output was available on a total of 23 levels, with a relative concentration at the lowest levels in order to resolve planetary boundary layer structure. The planetary boundary layer was modeled using a high-resolution Blackadar scheme coupled with a 5-layer soil model. Physiographic and land use patterns were back interpolated from a high-resolution data set to the model grids.

The surface and upper-air meteorological analyses described above were incorporated into the simulation using the FDDA technique known as Newtonian Relaxation or nudging. In this technique, an analysis dataset that provides time continuity and dynamic coupling among the various model fields is generated by weakly forcing the model solutions toward three-dimensional gridded analyses of wind, temperature and mixing ratio. In this way, the model solution remains "bounded" by the observations and the horizontal resolution of the observations is effectively enhanced by the added time-dimension.

Current meter data

Current meters were deployed along the east coast of southern Lake Michigan in order to capture nearshore-offshore flow in the vicinity of Benton Harbor, MI (BEH in Fig. 2) during significant northerly wind events. The 1997-98 installation was carried out during a pilot year of the EEGLE program and only 11 moorings were deployed. The 4 central moorings (A1, A2, A4, and A5) were equipped with Acoustic Doppler Current Profilers (ADCP) deployed at 18 (A1 and A4) and 38 m (A2 and A5) depths while the remaining moorings (V01, V03, V04, V06, V09 and V12) deployed at 20 and 60 m depths had 2 Vector Averaging Current Meters (VACM) each at 12m and at 1 m above the bottom (Fig. 1). Observations lasted from October 1997 to June 1998. The mid-lake station (CM1) is a part of an ongoing GLERL monitoring program and had 3 VACM's at 20, 115 and 152 m.

Base model run and comparison with observations

The base model run employs NRST data, and all other runs (NTRL and MM5 data based) will be compared against it. Hourly meteorological data from the 12 stations shown in Fig. 2 were obtained for the period 1-30 March, 1998. Overwater wind and air temperature fields were interpolated to the 2 km grid. Time series of wind speed and direction from a point in the middle of the southern basin (near station CM1 in Fig. 2) are shown in Fig. 3. There are four major wind events in March, two storms with northerly winds (on the 9th and 21st) and two with southerly winds (on the 13th and 27th). In early spring, the lake is thermally homogeneous and density gradients are negligible. Therefore, the circulation model was applied in a barotropic mode with uniform (2˚C) water temperature.

Observation data and model results showed that circulation in Lake Michigan is highly episodic since it is almost entirely wind-driven in early spring. The characteristic wind-driven circulation pattern in a lake consists of two counter-rotating gyres, a counterclockwise-rotating (cyclonic) gyre to the right of the wind and a clockwise-rotating (anticyclonic) gyre to the left (Bennett, 1974). The gyres are separated by a convergence zone along the downwind shore with resulting offshore flow and a divergence zone along the upwind shore with onshore flow. This two-gyre circulation pattern was clearly seen during the two northerly wind events in March in southern Lake Michigan. The computed circulation is illustrated through the use of a snapshot of a computer animation which gives an indication of current magnitude and direction over the previous 48

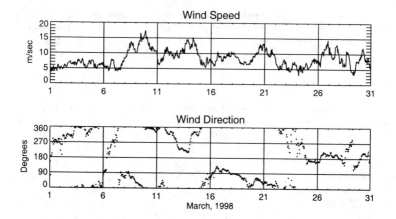

Figure 3. Time series of interpolated wind near station CM1 in southern Lake Michigan for 1-30 March, 1998.

hours (Schwab et al., 2000). The animation depicts the trajectories of passive tracer particles which were introduced into the computed depth-averaged velocity field on 1 March and are traced through the 30 day computational period. The first storm with northerly winds up to 17 m/s on March 9 caused strong along shore southerly currents that converged south of BEH and caused massive offshore flow lasting several days (Fig. 4).

The model qualitatively reproduces the observed large-scale circulation pattern (Fig. 5) although the offshore flow in the model is located somewhat south of the observed convergence zone. This may be explained by the sensitivity of the large-scale lake circulation pattern to the direction and vorticity of the wind field. We were not able to reproduce the spectacular spiral eddy observed in the middle of the lake on March 12 (Fig. 1) which is probably either a result of meandering of the strong offshore jet or direct atmospheric forcing. The last argument seems to be more convincing since there is a strong evidence based on National Weather Service radar data that on March 11 a mesoscale atmospheric vortex was present in southern Lake Michigan. That vortex is almost missing in objectively analyzed winds because of the lack of overlake observations. Monthly mean modeled currents match observed currents very accurately (Fig. 6).

For the purposes of comparison, modeled and observed currents were decomposed into longshore and onshore components. Comparison at nearshore stations at southern (V01) and northern (V09) boundaries of the array of moorings

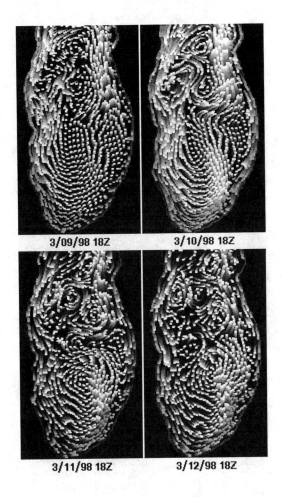

Figure 4. Modeled circulation in southern Lake Michigan, March 9-12 (see text for more explanation).

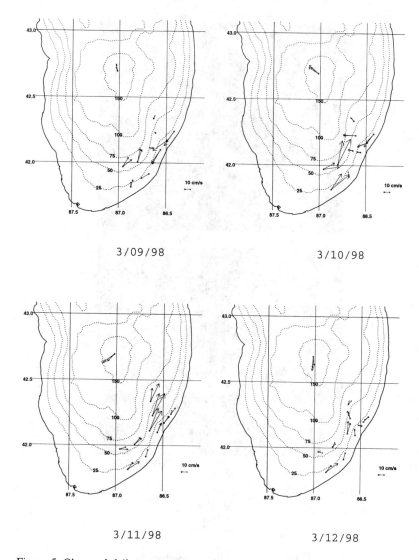

Figure 5. Observed daily averaged currents on March 9-12. Currents experience predominantly counterclockwise rotation toward the bottom.

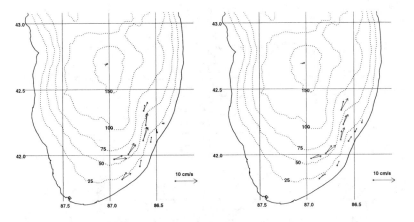

Figure 6. Observed (left) and modeled (right) monthly averaged currents. Currents experience predominantly counterclockwise rotation toward the bottom.

shows good prediction by the model of an offshore flow at 12 m depth but significant underestimation of the longshore flow (Fig. 7). ADCP data (station A1) provided valuable information on vertical current distributions. Observations during March 9-14 (Fig. 8a) showed strong southerly longshore currents (up to 45 cm/s) around March 10 followed by current reversal on March 11 (with northerly currents up to 35 cm/s) and persisting northerly currents for the rest of the period. Model longshore currents also peaked on March 10 at this location although reversed currents were not as strong (up to 10 cm/s). There is also an increase in model current speed around March 14 not seen in observations. The onshore component was also calculated qualitatively correctly (Fig. 8b) but its magnitude was significantly less than in observations. It is interesting to note that while observed currents posess almost no vertical shear, modeled currents showed significant reduction (almost twice) in speed with depth during strong wind events. This demands further investigation of the influence of both bottom friction and vertical turbulent viscosity on model results.

A statistical comparison of modeled and observed currents is presented in the form of the Fourier norms (rms difference). The Fourier norm of time-series of observed current vectors $\mathbf{v_0}$ and computed $\mathbf{v_c}$ is defined as

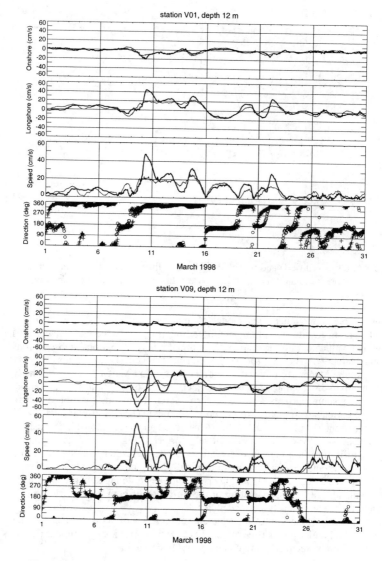

Figure 7. Time-series of modeled (thin line, crosses) versus observed (thick line, open circles) currents at stations V01 and V09.

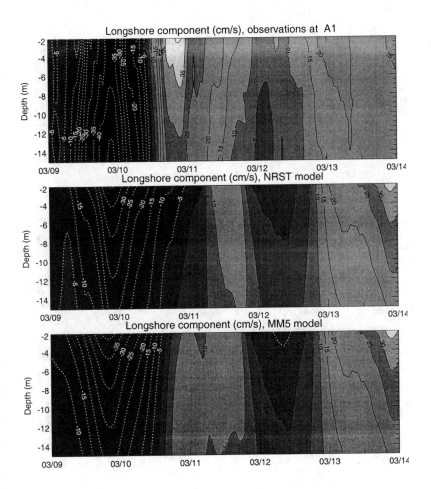

Figure 8a. Time-series of modeled versus observed longshore currents at station A1.

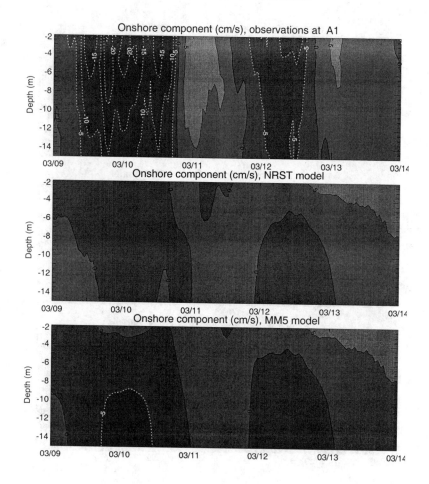

Figure 8b. Time-series of modeled versus observed onshore currents at station A1

$$\|v_o, v_c\| = \left(\frac{1}{M} \sum_{t=\Delta t}^{M\Delta t} |v_o - v_c|^2 \right)^{1/2}$$

We use a normalized Fourier norm:

$$F_n = \|v_c, v_o\| / \|v_o, 0\|$$

In the case of perfect prediction $F_n=0$. In the case $0 < F_n < 1$, model predictions are better than no prediction at all (zero currents). Using F_n not only allows us to use one number for characterization of model skills in predicting vector entities, but also to compare our model results more objectively with previous model results. For example, in one of the earlier modeling exercises, Schwab (1983) calculated $0.79 < F_n < 1.01$ for a comparable barotropic simulation of Lake Michigan circulation on the 5 km grid. Our numbers show significant improvement over this result, from 0.45 to 0.76 (Table 1).

	F_n, range	F_n, mean	CC longshore	CC onshore
NRST	0.45-0.76	0.64	0.77	0.79
NTRL	0.41-0.76	0.63	0.80	0.82
MM5	0.43-0.74	0.58	0.81	0.84

Table 1. Statistical comparison of March 8-10 observed and computed currents (Fourier norm and correlation coefficient).

Sensitivity to meteorological data

Hydrodynamic model runs with NTRL winds yielded currents similar to NRST runs (Table 1 presents only 3-day comparison results but NRST and NTRL numbers were similar for the whole 30-day long comparison). Therefore, NTRL technique can be used now as a reliable alternative to NRST in the EEGLE study (there is evidence that it provides better wind wave predictions with models developed at GLERL).   On the other hand, the MM5 data showed some improvement in model results (Table 1). Figure 8, for example, shows better timing of the nearshore current reversal on March 10-11 and stronger longshore and onshore currents during wind events. The spiral eddy on March 12 is absent in both NTRL and MM5-based results (MM5 runs were recently extended to cover

11-12 March period but showed no indication of the atmospheric vortex).

Although MM5 winds yielded slightly better currents than ones calculated with objectively analyzed winds, there is still room for improvement. Unfortunately, accurate modeling of over-lake atmospheric dynamics can present significant challenges during the early spring period. Analysis shows that there were two important events during the March 9-12 episode: strong winds on March 9-10 that caused the initial sediment resuspension event, and the mesoscale atmospheric vortex that apparently formed on March 11. The first event is a strong cold front with air temperature dropping from 0°C to -10°C. With water temperatures around 2°C this should cause significant instability of atmospheric boundary layer and thus increased wind stress. Unfortunately, as was mentioned earlier, there were no overlake wind observations in southern Lake Michigan during that period. It is possible that the FDDA in the MM5 results is driving the winds towards a low bias in magnitude (since it is based on the available observations, which are practically all land observations) during the storm. Problems with the second event (mesoscale vortex) can be also caused by lack of over lake data. Currently, work is underway to improve MM5 results by experimenting with alternatives to FDDA and also incorporation of radar observations into MM5 runs.

Conclusions

The Princeton ocean model was applied to Lake Michigan to simulate hydrodynamic conditions during the 1998 coastal turbidity plume event. The model is driven with objectively analyzed winds (NRST and NTRL techniques) and also with surface winds from the meteorological model MM5. Comparison with observations showed that the model was able to qualitatively simulate wind-driven currents but underestimated current speeds during strong wind events and in particular the onshore-offshore component of the flow in the area of observations. This may be due at least in part to the significant decrease of modeled current speeds with depth during strong wind events while vertical shear was almost absent in observed currents. Model results with MM5 winds were slightly better than the ones that used objectively analyzed winds (NRST and NTRL). The difference between NRST and NTRL results was minimal. More experiments are underway to study the effects of wind field interpolation, grid resolution, and friction on hydrodynamics in Lake Michigan.

Acknowledgements

This research is supported by the EEGLE (Episodic Events - Great Lakes Experiment) Program sponsored by the National Science Foundation and the National Oceanic and Atmospheric Administration Coastal Ocean Program. This is GLERL Contribution Number 1148.

References

Beletsky, D., and D.J. Schwab. 1998. Modeling thermal structure and circulation in Lake Michigan. Estuarine and Coastal Modeling, Proceedings of the 5th International Conference, October 22-24, 1997, Alexandria,VA, p. 511-522.

Bennett, J.R., 1974. On the dynamics of wind-driven lake currents. J. Phys. Oceanogr. 4(3), 400-414.

Blumberg, A.F. and G.L. Mellor, 1987. A description of a three-dimensional coastal ocean circulation model. Three dimensional Coastal Ocean Models, Coastal and Estuarine Sciences, 5, N.S. Heaps [ed.] Amer. Geophys. Union, Washigton, D.C., pp 1-16.

Dudhia, J., 1993. A nonhydrostatic version of the Penn State-NCAR mesoscale model: validation tests and simulation of an Atlantic cyclone and cold front. J. Atmos. Sci., 46,3077-3107.

Eadie, B.J., D.J. Schwab, G.L. Leshkevich, T.H. Johengen, R.A. Assel, R.E. Holland, N. Hawley, M.B. Lansing, P. Lavrentyev, G.S. Miller, N.R. Morehead, J.A. Robbins, and P.L. Van Hoof, 1996. Anatomy of a recurrent episodic event: a winter-spring plume in southern Lake Michigan. EOS, Trans. Amer. Geophys. Union., 77, 337-338.

Mellor, G.L., and T. Yamada, 1982: Development of a turbulence closure model for geophysical fluid problems. Rev. Geophys. Space Phys., 20(4), 851-875.

Mortimer, C.H., 1988. Discoveries and testable hypotheses arising from Coastal Zone Color Scanner imagery of southern Lake Michigan. Limnol. Oceanogr., 33(2), 203-226.

National Geophysical Data Center, 1996. Bathymetry of Lake Michigan. Report MGG-11.

Resio, D.T. and C.L. Vincent, 1977. Estimation of winds over the Great Lakes. J. Waterway Port Coast. Ocean Div., ASCE, 102, 265-283.

Sambridge,M., J. Braun, and H. McQueen, 1995. Geophysical parameterization and interpolation of irregular data using natural neighbors. Geophys. J. Int., v122, 837-857.

Schwab, D.J., 1983. Numerical simulation of low-frequency current fluctuations in Lake Michigan. J. Phys. Oceanogr. 13(12), 2213-2224.

Schwab, D.J. and D. Beletsky, 1998. Lake Michigan Mass Balance Study: Hydrodynamic modeling project. NOAA Tech. Memo. ERL GLERL-108. NOAA Great Lakes Env. Res. Lab., Ann Arbor, MI, 53pp.

Schwab,D.J., D. Beletsky, and J. Lou, 2000. Modeling a coastal turbidity plume in Lake Michigan. Estuar., Coast. and Shelf Sci. (in press)

Sibson, R., 1981. A Brief Description of Natural Neighbor Interpolation, in Interpreting Multivariate Data, ed. by V. Barnett, John Wiley & Sons, New York, pp. 21-36.

Watson, D. 1994. nngridr - An Implementation of Natural Neighbor Interpolation. Dave Watson Publisher, Claremont, Australia.

Numerical Modeling and Field Experiments of Wind-Induced Currents in Enclosed Bay

Tadashi Fukumoto[1] and Takehiro Nakamura[2]

Abstract

Currents in bays caused by wind, tide and density gradient are fairly complicated. The wind-induced current is normally dominant in an enclosed bay if the mean tidal range and river discharge are very small. Field observations and data analyses have been performed in such an enclosed bay where current and meteorological data have been collected for the last ten years. The current velocities were measured at 5 m below the water surface and at 5 m above the seabed in the center of the enclosed bay. The wind speed and direction were measured near the station of the current measurements. The current velocity variations due to wind contained long period components associated with meteorological variations. There was a time delay about 3 hours between the measured currents and wind. A three-dimensional baroclinic model for shallow water is also applied to reproduce and interpret the wind effects on the vertical variations of currents. The Boussinesq approximation and the hydrostatic pressure distribution in the vertical direction are assumed in the three-dimensional model.

Introduction

Omura Bay, which is located in the western part of Japan, is a complex coastal plain estuary with 320 km² surface area and 16 m average water depth. It is connected with Sasebo Bay by two narrow channels as shown in Figure 1. The inflow of ocean water is small because of these topographic features. The mouth of Omura Bay, which is called Hario-Seto, is very narrow (200 m wide). The currents in the vicinity of the bay mouth of Omura Bay, which is typical of enclosed bays in Japan, are caused mostly by tides. The water near the bay mouth is approximately 30 m deep and vertically well mixed. The inflow of ocean water is relatively small but the peak magnitude of the spring flood currents exceeds 5.0 m/s. On the other hand, in the center and end of Omura Bay, currents are also induced by wind and density gradient. The formation of strong thermal stratification can occur in summer if wind is not strong. This has led to generations of red tide formation. Damages caused by the red tide have often been reported (Iizuka *et al.* 1989). Lately, the developments along the bay, such as

[1] Research Engineer, Nishimatsu Construction Co., Ltd., Technical Research Institute
  2570-4 Shimotsuruma, Yamato, Kanagawa 242-8520, Japan
[2] Professor, Nagasaki University , Faculty of Environmental Studies
  1-14 Bunkyo-machi, Nagasaki, Nagasaki 852-8521, Japan

reclamation for housing, have taken place and the inflow pollutant loads exceed the self-purification rate of Omura Bay. Thus, in order to study the long-term transport and baroclinic flow, field observations are very important and numerical models are needed to simulate not only tidal time-scale variables but also the long-term residual current and transport including wind and stratification effects. The authors have been carrying out both field observations and numerical simulations on the currents, water temperature and salinity in order to comprehend the behaviors of the currents and water quality in Omura Bay (Fukumoto *et al.* 1992, 1995, 1997). Wind-induced currents and wind effects on the thermal stratification are investigated in this paper.

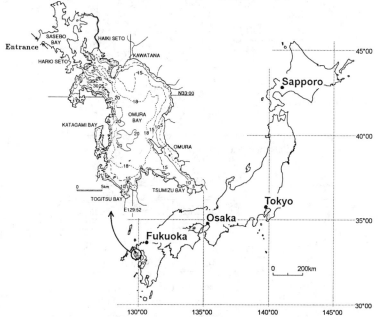

Figure 1 Location and contour map of bottom topography in Omura Bay

Field Observations

Field observations throughout Omura Bay have been carried out over 15 days since 1989. The measurements include water temperature and tidal elevation as well as current velocity and direction using electro-magnetic current meters (ALEC Electronic Co., Ltd. ACM16M and ACM8M). Figure 2 shows the locations of measurement stations. There are 31 time series of current meter data at the surface and bottom layers in ten stations as summarized in Table 1. These instruments were attached at one to three elevations and data were stored for 30 seconds (sampling rate was 1 second) every 10 minutes.

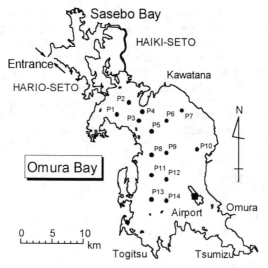

Figure 2 Stations for field observations, showing the locations of the
anemometer gauges (■) and current meters (●)

Table 1 Current measurement from 1989 to 1992

Station	Data block	Duration
P5	P5B1	3/2 - 3/18, '89
P6	P6B1	2/14 - 3/2, '89
	P6S2, P6B2	8/26 - 9/11, '91
	P6S3, P6B3	3/13 - 4/15, '92
	P6S4, P6B4	8/20 - 9/14, '92
P7	P7B1	4/5 - 4/20, '89
P8	P8B1	3/18 - 4/5, '89
P9	P9B1	1/26 - 2/14, '89
	P9S2, P9B2	8/26 - 9/11, '91
P10	P10B1	4/20 - 5/9, '89
P11	P11B1	3/5 - 3/22, '90
	P11B2	3/22 - 4/9, '90
P12	P12B1	1/9 -1/26, '89
	P12B2	1/11 - 1/28, '90
	P12M3	1/31 - 2/14, '90
	P12S4	2/14 - 3/4, '90
	P12S5, P12B5	8/6 - 8/26, '91
	P12S6, P12B6	3/13 - 4/15, '92
	P12S7, P12B7	8/20 - 9/14, '92
P13	P13S1,P13M1,P13B1	7/25 – 8/4, '95
P14	P14S1, P14B1	8/6– 8/26, '91

In the following section, the field observations conducted in winter from 1989 to 1992 are discussed first because stratification effects on currents are small in winter. Then, the field observations in summer in 1991, 1992 and 1995 are described to assess the stratification effects.

Tidal constituents

The time series of the free surface elevation is analyzed to obtain the major tidal constituents (M_2, S_2, K_1 and O_1) using a least-square method. The dotted line in the bottom panel of Figure 3 corresponds to the time series based on the fitted tidal constituents. The free surface oscillations are well described by these four tidal constituents only. Table 2 summarizes the average amplitude and phase of the four tidal constituents for the 18 time series near the bottom layer during 1989 to 1992 listed in Table 1.

The measured currents are separated into the mean and oscillatory constituents. The measured mean current (regarded as M_0) pattern at the surface and bottom layer in the summer and winter were shown in Fukumoto *et al.* (1997). An anti-clockwise horizontal circulation pattern was observed near the bottom layer, which was probably caused by the topography (deeper on the western side) shown in Figure 1.

Table 2 Tidal constituents of free surface oscillations

Tide	Amplitude (cm)	Phase (degrees)
M_2	23.1	326
S_2	10.8	359
K_1	10.3	264
O_1	10.4	262

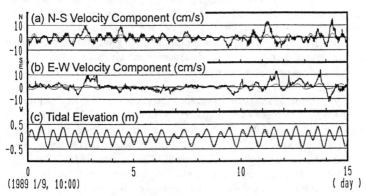

Figure 3 Measured time series of currents and free surface elevation for data P12B1 at station P12

Time series of prototype data on current and wind velocity

The solid lines in Figure 3 shows the measured time series corresponding to the data P12B1 listed in Table 1. The solid lines in the top (a), middle (b) and bottom (c) panels are the measured north-south current, the east-west current and the free surface elevation based on the measured water pressure. The water temperature varied little during these measurements.

The oscillatory constituents of the measured currents are analyzed in the same way as the free surface elevation using the four tidal constituents for each of the north-south and east-west velocities. The sum of the mean (M_0) and tidal oscillatory constituents (M_2, S_2, K_1 and O_1) is simply called the combined tidal velocity in the following. The dotted lines in the top (a) and middle (b) panels of Figure 3 are the combined tidal velocities in comparison with the solid lines representing the measured current velocities. The difference between the measured and tidal velocities is the velocity caused mostly by wind in winter. It is noted that the wind effects on currents are not negligible unlike the wind effects on the free surface oscillations as shown in the bottom (c) panel of Figure 3. This current velocity is correlated with the measured wind speed and direction at the Nagasaki International Airport located on an island in Omura Bay as shown in Figure 2.

The measured and tidal current velocities ($M_0+M_2+S_2+K_1+O_1$) shown in the top (a) and middle (b) panels of Figure 3 are plotted in vector forms in the top (a) and second (b) panels of Figure 4. The remaining current velocity is also plotted in the same vector form in the third (c) panel of Figure 4. The remaining current velocity was given by the difference between the prototype data and tidal current velocity. The measured wind speed and direction are plotted in the bottom (d) panel of Figure 4. The remaining current speed tends to be large at the time of the large wind speed, whereas the current direction tends to be opposite to the wind direction. The current measured at 5m above the bottom may correspond more to return current near the bottom than wind-induced surface current. The time delay is estimated as that, which gives the largest correlation coefficient between the two time series of the current and wind speeds. The average of the time delay between the current and wind speeds was about 3 hours.

Figure 5(a) and (b) show the power spectrum of current speed for data P12B1 at station P12 and wind speed at the airport during the same period. There are peaks at about 5 days, 12 hours, and 80 minutes for the current speed spectrum as shown in Figure 5(a). The peak at about 80 minutes is a natural frequency of seiche in Omura Bay. A semi-diurnal current at the peak of about 12 hours is dominant. The shapes in Figure 5 (a) and (b) for the long-term frequency ranges are very similar and the peaks at about 5 days in the current and wind speed spectra are related. As a result, the current variations of the time scale of days are caused by the corresponding wind variations.

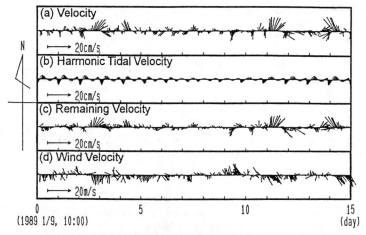

Figure 4 Current and wind speed and directions for data P12B1 at station P12

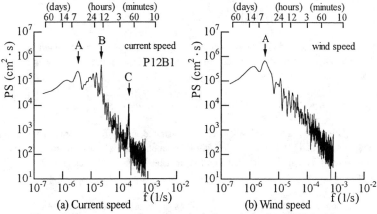

Figure 5 Power spectra of current and wind speeds for data P12B1 with
peaks at about 5 days(A), 12 hours (B) and 80 minutes (C)

Method of analysis on the effect of wind-induced current

The measured current velocity denoted by V_{mes} includes the long-term component V_{long} related to long-term wind effects, the tidal constituent V_{tide} in the time scale of 12-24 hours, and the remaining component denoted by V_{re}. V_{long} is obtained by the analysis of moving average over 25 hours. V_{tide} is obtained by the harmonic analysis using a least-square method.

$$V_{mes} = V_{long} + V_{tide} + V_{re} \tag{1}$$

The remaining component is defined as $V_{re} = V_{mes} - (V_{long} + V_{tide})$. Wind velocity is similarly separated without any tidal constituent.

$$W_{mes} = W_{long} + W_{re} \tag{2}$$

where W_{long} is obtained by the analysis of moving average over 25 hours. W_{re} is the difference between the measured wind speed W_{mes} and the long-term wind speed W_{long}.

The angle shift between the wind and current is estimated as the angle which minimizes the sum of the square of the angle difference between the two time series of the wind and current directions where the current angle is shifted by $\Delta\theta$, where the angle shift $\Delta\theta$ is obtained by minimizing the following quantity

$$VD_{re} = \sum_{i=1}^{n} \left[\theta_{W_i} - \left(\theta_{V_i} + \Delta\theta \right) \right]^2 \qquad (i = 1, 2, \cdots n) \tag{3}$$

where θ_W = wind direction, θ_V = current direction, $\Delta\theta$ = estimated angle shift and n = number of data points. To estimate the degree of the effects of wind on current, the ratio of the magnitude of the current velocity to that of the wind velocity is estimated as

$$WVD = \frac{1}{n} \sum_{i=1}^{n} \left(|V_i| / |W_i| \right) \qquad (i = 1, 2, \cdots n) \tag{4}$$

where $|V|$ = magnitude of the current velocity, $|W|$ = magnitude of the wind velocity, WVD = average ratio between the current and wind speed and n = number of data points.

Long-term and remaining components of the currents

The current and wind velocities are separated into different components using equations (1) and (2). Equations (3) and (4) are applied to the long-term and remaining components. The angle shift of the current is taken to be positive clockwise from the wind direction.

Table 3 summarizes the average values of the angle shift and the ratio between the current and wind speeds. On the average, the difference of the angle shifts for the surface and bottom layers in Table 1 is about 200° for both summer and winter and for the long-term and remaining components. The direction of the bottom current is hence approximately opposite to the surface current direction. Their results are consistent visually with the vectorial time series shown in the third (c) and fourth (d) panels of Figure 4. This tendency is larger in summer than in winter. The angle shift of the long-term component at the surface layer is larger in winter than in summer but the angle shift of the remaining component is larger in summer.

On the other hand, the speed ratio at the surface layer is 1.5-2.0 times larger than the speed ratio at the bottom layer for the long-term component but the ratio for the remaining component is similar at the surface and bottom layers. The speed ratios for the remaining component are larger than those of the long-term components. The effect of wind is strong in the remaining component in winter.

Table 3. Estimated angle shift between current and wind
at the surface and bottom in winter and summer

component	Season	Angle shift in degrees (°)		Speed ratio	
		Surface layer	Bottom layer	Surface layer	Bottom layer
Long-term	winter	89	251	0.0469	0.0211
V_{long}, W_{long}	summer	34	261	0.0305	0.0202
Remaining	winter	54	241	0.0493	0.0317
V_{re}, W_{re}	summer	71	296	0.0399	0.0403

Near-surface and bottom currents during one day in spring and summer

The near-bottom residual currents observed in 1989 and 1990 have been interpreted as return currents caused by wind-induced currents near the surface in the enclosed bay. To examine the validity of this interpretation, currents were measured in March, 1992 at two elevations of 5 m above the seabed and 5m below the water surface at P12 in 20 m water depth which is located in the center of Omura Bay as shown in Figure 2.

Figure 6 shows the measured speeds and directions of the wind velocity W_w, the near-surface current V_{sw} at 5m below the water surface, and the near-bottom current V_{bw} at 5m above the seabed at the interval of 10 minutes on March 18, 1992. The water temperature was approximately 12 ℃ and did not vary vertically. The time delay for both currents were about 3 hours. The angle shift was about 40° for the near-surface current V_{sw} probably due to the Coriolis effect and about 250° for the near-bottom current V_{bw}. The characteristics of the near-bottom current measured in spring 1992 shown in Figure 6 were similar to that measured in winter and spring in 1989 and 1990. Figure 6 clearly shows that the vertical variation of the wind-induced current is important in an enclosed bay even in the absence of stratification because of no flux boundary condition along the bay shoreline.

To examine the effects of stratification on the near-surface and bottom currents the same measurements as those conducted in March 1992 were made in August 1992. Figure 7 shows the measured speeds and directions of the wind W_s, the near-surface current V_{ss} at 5 m below the water surface, and the near-bottom current V_{bs} at 5 m above the seabed at station P12 on August 26, 1992. The CTD measurements on this day indicated the density increment of approximately 2.0 kg/m^3 at 4-6 m below the water surface. The near-surface current responded to the wind in a manner similar to that shown in Figure 6 on the day of no stratification. On the other hand, Figure 7 indicates that the near-bottom current did not always respond to the wind but the tidal

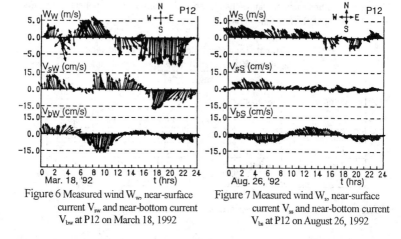

Figure 6 Measured wind W_w, near-surface current V_{sw} and near-bottom current V_{bw} at P12 on March 18, 1992

Figure 7 Measured wind W_s, near-surface current V_{ss} and near-bottom current V_{bs} at P12 on August 26, 1992

effect on the near-bottom current is apparent. The presence of the thermocline appears to reduce the wind effect on the near-bottom current perhaps because a return current may flow above the thermocline.

Vertical variations of currents in summer

The vertical variations of currents and temperatures were measured at station P13 in 20 m water depth for 10 days starting from July 25, 1995. Figure 8 shows the

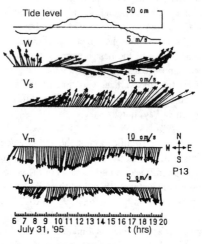

Figure 8 Measured tide level, wind W, near-surface current V_s, and near-bottom currents V_m and V_b at P13 on July 31, 1995

measured time series of the free surface elevation and the speeds and directions of the wind W, the near-surface current V_s at 5 m below the water surface, and the near-bottom currents V_m and V_b at 5 m and 1 m above the seabed, respectively, on July 31, 1995. The measured wind W at the airport was from the west after 9:00 but the wind at station P13 was from the southwest. The high tide occurred at 13:00. The surface current V_s responded to the wind but the near-bottom currents V_m and V_b responded more to the tidal variation.

Figure 9 shows the vertical variations of the water temperature and current speed at P13 measured every two hours using a instrument of water quality attached the current meter (ALEC Electronic Co., Ltd. ACL-1183PDK) on July 31. The current speed was small in the vicinity of the thermocline located at 5-8 m below the water surface, while the current direction was reversed below the thermocline. The thermocline was maintained even though a 5.0 m/s wind continued for approximately 10 hours.

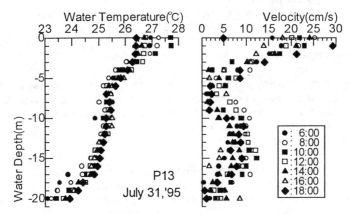

Figure 9 Vertical variations of water temperature (left) and current speed (right) measured every two hours at P13 on July 31, 1995

Figure 10 shows the vertical variation of the current vector measured every one meter vertically at P13 at 8:00 and 18:00 on July 31 as shown in the right side of Figure 9. The number for each arrow indicates the depth below the water surface. The surface currents at 8:00 are effected by wind until 5 m below the water surface (thermocline level). The current at 18:00 within 3 m below the water surface flowed to the northeast in response to the wind but the current in the zone of 4-6 m below the water surface flowed to the north perhaps because of the combined wind and tidal effects. The currents below the thermocline flow to the south mainly because of the tidal effect.

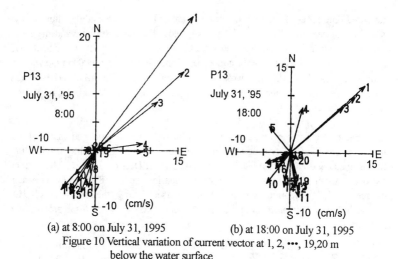

(a) at 8:00 on July 31, 1995 (b) at 18:00 on July 31, 1995

Figure 10 Vertical variation of current vector at 1, 2, •••, 19,20 m below the water surface

Numerical Simulation

The measurements presented above indicate the three-dimensional and unsteady current patterns caused by tide, wind and density gradient in the enclosed bay. A three-dimensional baroclinic model is applied to reproduce and interpret the current data.

Numerical model

The 3-D equations for baroclinic flow in shallow water are solved here by a finite difference scheme. Omura bay is sufficiently large so that Coriolis acceleration effects must be included in the momentum equations. To simplify the governing equations based on incompressible Navier-Stokes equations, the hydrostatic pressure is assumed and the Boussinesq approximation is applied (*e.g.*, Matsuo *et al.* 1993, Fukumoto *et al.* 1997).

The equations integrated over a control volume are discretized by means of a space-staggered grid system (Iwasa *et al.* 1983). A leap-frog and Adams-Bashforth method for the temporal differences and an upwind scheme (Doner-Cell) for the convective terms are also adopted (Shen 1991). For the initial conditions, the water temperature and salinity are assumed to be uniform and the still water state is given. The water temperature and salinity of the ocean water and the measured tide level are given as the boundary conditions at the grid of the bay mouth, Hario-Seto. The discharge and water temperatures of 24 rivers flowing into Omura Bay are also considered. Meteorological heat exchanges are modeled at the water surface. The vertical eddy viscosity and eddy diffusivity are given as a function of Richardson

Number. The Sub-Grid Scale model is used to set the coefficients of eddy viscosity and eddy diffusivity in the horizontal plane (Nakatsuji 1994).

In the follow computations for typical summer conditions, the number of control volumes in a horizontal plane is taken nx x ny = 39 x 26, respectively. The lengths of the horizontal sides are $\Delta x = \Delta y = 1$ km. Where the x axis is taken to be the long (mouth to head) axis of the bay. The number of vertical layers is eleven. The time step limited by the Courant condition is taken as 20 seconds.

Model calibration of simulation and computed current for no wind

First, computation is made for the case of no wind in summer to examine whether the baroclinic model can reproduce the stratified tidal flow in the bay. Figure 11 shows that the computed M_2 tidal ellipse at P6 in 16 m water depth is in fair agreement with the measured ellipse at 5 m above the seabed.

The mean currents tend to flow toward the bay mouth near the surface and toward the bay end near the bottom as expected in the circulation pattern in the stratified bay. An anti-clockwise circulation pattern exists under the middle layer on the northern side of the bay. This mean circulation pattern was also observed in winter and is likely caused by the topography (Fukumoto *et al.* 1997).

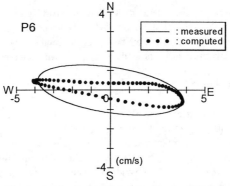

Figure 11 Observed and calculated M_2 tidal ellipses at P6 in summer

Computed wind-induced currents

This numerical model is used to reproduce the field observations made on July 31, 1995 as shown in Figure 8-10. The observed stratified conditions at 6:00 (local time) are used as the initial conditions. The wind is assumed to be 3.0 m/s from the west for 3 hours and then 5.0 m/s from the south for 10 hours.

Figure 12 shows the computed vertical variation of the current vector at P13 at 18:00. The computed current speed decreases downward and becomes small in the vicinity of the thermocline. The current direction was reversed below the thermocline. These computed results are qualitatively similar to the observed current vector shown in the right side of Figure 9.

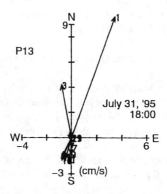

Figure 12 Computed vertical variation of current vector
at 18:00 (local time)

The computed current patterns at 18:00 are shown in Figure 13 for three layers at 1 m (top), 9 m (middle) and 17 m (bottom) below the water surface. The wind-induced current is dominant near the surface but both tidal and wind effects are noticeable in the middle and lower layers in manners similar to those shown in Figures 7 and 8.

Wind effects on stratification

Anoxic water occurred at the bay end below the thermocline in summer (Iizuka and Si 1989, Fukumoto *et al.* 1995). Strong wind accompanying a typhoon was observed to destroy stratification and remove anoxic water. Consequently, additional computations are made for different wind conditions starting from 6:00. Table 4 summarizes the five cases where the wind is assumed to blow from the south, which is the prevailing direction in summer.

Table 4 Wind speed and duration assumed for five cases

case	Wind speed (m/s)	Duration (hr)
1	10.0	6.0
2	15.0	6.0
3	20.0	6.0
4	10.0	12.0
5	10.0	24.0

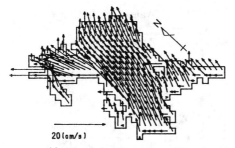

20 (cm/s)

(a) at 1 m below the water surface

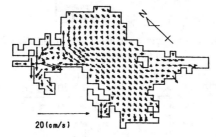

20 (cm/s)

(b) at 7 m below the water surface

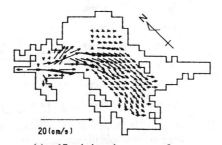

20 (cm/s)

(c) at 17 m below the water surface
Figure 13 Computed wind-induced current patterns

Figures 14 and 15 show the computed and initial vertical variations of the water temperature and horizontal velocities in the x and y directions at station P13 at the end of the assumed wind duration, respectively. The water temperature stratification is reduced with increase in wind speed and duration. The vertical velocity variation increases mostly with increases in wind speed. The water temperature and velocity change little in the vicinity of the thermocline at 5-7 m below the water surface. Their figures suggest that the stratification in Omura Bay may be reduced significantly by wind exceeding 10 m/s provided it blows more than 6 hours. This may be typical in relatively small-enclosed bays in Japan.

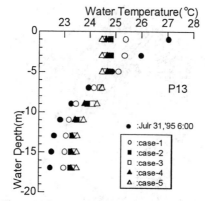

Figure 14 Computed and initial vertical variations of
water temperature for five cases

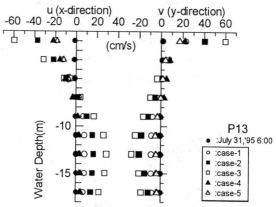

Figure 15 Computed and initial vertical variations of current
velocities in x (left) and y (right) directions

Conclusions

Field observations and a three-dimensional baroclinic model are used to investigate the spatial and temporal variations of currents induced by tides, wind and density gradient in Omura bay. Currents in the absence of stratification are affected by both wind and tides where the wind effect decreases downward from the water surface. Currents in the presence of stratification are more complicated. Wind-induced currents tend to be dominant above the thermocline. Both tidal and wind-induced currents are noticeable below the thermocline. Wind speed of about 5 m/s lasting about 10 hours reduced the stratification in July but did not destroy it completely. Wind exceeding 10 m/s would be required to cause complete vertical mixing.

Acknowledgments

The authors would like to thank the people in the 15 fisheries co-operative associations around Omura Bay. The authors also appreciate the useful advice by Prof. Nobuhisa Kobayashi, Center for Applied Coastal Research, University of Delaware.

References

Fukumoto,T., T.Nakamura, and H.Togashi (1992), " Tidal Residual Current in Omura Bay." *Proc. of 2nd Int. Conf. on Hydr. and Envir. Modeling of Coast. Engrg. and River Waters*, Bradford, 79-87.

Fukumoto,T., A.Tada, T.Nakamura and H.Togashi (1995), " Three-Dimensional Tidal Currents and Water Quality in Stratified Omura Bay." *Proc. of the 4th Int. Conf. on Estuarine and Coast. Modeling*, ASCE, San Diego, 707-721.

Fukumoto,T., A.Tada, T.Nakamura and H.Togashi (1997), " Seasonal Characteristics of the Tidal Current Circulation in Omura Bay." *Proc. of the 5th Int. Conf. on Estuarine and Coast. Modeling*, ASCE, Alexandria, 181-195.

Iizuka,S., H.Sugiyama and K.Hirayama (1989), " Population Growth of Gymnodinium Nagasakiense Red Tide in Omura Bay." in T.Okaichi et al. (eds.): Red Tide: *Biology Environmental Science and Toxicology, Elsevier Science Publishing Co. Inc.*, 269-272.

Iizuka,S. and Si Hong Min (1989), " Formation of Anoxic Bottom Waters in Omura Bay." *Bulletin on Coast. Oceanog., Vol.26, No.2*, 75-86 (in Japanese).

Iwasa,Y., K.Inoue, S.Liu and T.Abe (1983), " Numerical Simulation of Flows in Lake Biwa by Means of a Three-Dimensional Mathematical Model." *Annuals, Disas. Prev. Res. Inst., Kyoto University, No.26 B-2*, 531-542 (in Japanese).

Matsuo,N., K.Inoue and N.Nagata (1993), " Prediction of Algae Blooming and its Control by Bubble Plume in a Eutrophicated Reservoir." *Proc. of 25th IAHR Congr., Vol.5*, Tokyo, 101-108.

Nakatsuji,K. (1994), " Estuarine Circulation and Mass Transport in Osaka Bay." *Lecture Notes of the 30th Summer Seminar on Hydr. Engrg., Committee on Hydr., Course A*, JSCE, 1-28 (in Japanese).

Shen,H. (1991), " Numerical Analysis of Large-Scale Flows and Mass Transports in Lakes." *Doctoral Thesis*, Kyoto University, 1-197.

An Isopycnic Model Study of the Black Sea

Gokay M. Karakas[1], Alec E. James and Alaa M. A. Al-Barakati

Abstract

Miami Isopycnic Coordinate Ocean Model (MICOM) has been used to simulate the seasonal circulation features of the Black Sea. Thermodynamic forcing is interpolated from various climatic data while wind forcing has been described from the Hellermann and Rosenstein (1983) wind stress climatology by using a factor of 2.5 to achieve realistic values. Runoff from 3 major rivers in the northwestern shelf of the basin has been integrated to the model together with the Bosphorus Strait, which is the only source for the saltier Mediterranean waters. Using a horizontal grid resolution of $\sim 0.25^{O}$ on 15 potential density layers in the vertical, the model has been run with monthly forcing data. The results show a well pronounced cyclonic Rim current, consisting of 4 to 6 cyclonic gyres and flowing along the steep shelf slope of the basin, which is the dominant characteristic of the surface circulation throughout the winter months. In summer, on the other hand, the Rim current destabilizes and becomes unrecognizable in the southern and eastern coast of the sea and it is pushed towards the interior of the basin in the northern and western part by the enlarging anti-cyclonic eddies. These anti-cyclonic eddies, most of which are persistent all around the year, are usually confined between the coastline and the Rim current and located near Cape Kaliakra, Cape Baba, and the Bosphorus in the western part, near Batumi and along the Caucasian coast in the eastern part and on both sides of the Crimean Peninsula in the north. While sea surface temperature, roughly increasing towards the south, follows the same pattern as air temperature, surface salinity is easily correlated with the general circulation. A sharp salinity front between low saline

[1] Environmental Technology Centre, Department of Chemical Engineering, UMIST, P.O. Box 88, Manchester M60 1QD, UK

northwestern waters and high saline interior waters is detected particularly in winter months. The Rim current looses its strength in deeper layers of the sea. The model depicted a reversal in the direction of the cyclonic Rim current in the bottom waters of the Black Sea.

1. Introduction

The Black Sea is the largest anoxic basin in the world. Although nearly enclosed, it is connected to the Sea of Azov through the Kerch Strait on the north and to the Sea of Marmara through the narrow Bosphorus on the south where there is an input of saltier Mediterranean water. It is an oval basin with an area of 423 000 km^2 and the average depth of 1240 m (Ross et al., 1974).

Soviet oceanographers were first interested in numerical modeling of the Black Sea in the 1970's (see Marchuk et al. (1975), Dzhioyev and Sarkisyan (1976)). Since then numerous investigations have been made (see Trukhchev and Stanev (1983), Trukhchev et al. (1985), Stanev (1990), Eremeev et al. (1992), Stanev et al. (1995), Oguz et al. (1995), Oguz and Malanotte-Rizzoli (1996), Kiran et al. (1996), Stanev et al. (1997), Rachev and Stanev (1997)). One of the main difficulties encountered in modelling the circulation in the Black Sea is the highly stratified water column caused by an excess of precipitation and river runoff over evaporation. The isopycnic approach in which the vertical co-ordinate system is related to potential density layers seems to be more appropriate for modelling such phenomena rather than the depth or sigma co-ordinate models used previously. To this end the Miami Isopycnic Coordinate Ocean Model (MICOM) is used in the present study where we report the results of model runs forced by seasonal climatological data.

2. Model Description

Ocean general circulation models, until recently, have been developed using cartesian co-ordinates in the vertical. However, when the conservation equations for salt and temperature are written in x,y,z direction, it is very difficult to avoid diffusion of the transport variables in all three directions resulting from mixing between isopynic layers. This is a particular problem in long term simulations. The adiabatic character of fluid motion and eddy diffusion on long time scales has brought about a new type of model that reduces truncation errors introduced when approximating finite-difference equations or numerical errors resulting from physically improper split of transport and diffusion processes into vertical and horizontal components. In contrast to conventional multilevel models for stratified fluids, the newly developed isopycnic models comprise a set of constant potential density layers. Since oceanic mixing processes take place predominantly along constant potential density surfaces, isopycnic models where the dynamic equations are changed from x, y, z to x, y, ρ coordinates, ρ being the potential density, achieve a more accurate representation of the physical system. The reader is referred to Bleck (1978) for the transformation of the equations from the cartesian system. Transport in x, y direction occurs on isopycnic surfaces and transport along the z axis transforms into transport along the ρ axis, which can be totally suppressed so that it would not have any diapycnal component unless required. Accordingly, false heat exchange between warm surface waters and cold bottom layer waters are minimised as well as the horizontal heat exchange along sloping isopycnals (Chassignet *et al.*, 1996; Langlois, 1997).

The model uses 3 prognostic equations. These are a mass continuity equation (1), an equation for the horizontal velocity vector (2), and finally a conservation equation for buoyancy related variables like density, salinity and temperature (3).

$$\frac{\partial}{\partial t_s}\left(\frac{\partial p}{\partial s}\right) + \nabla_s \cdot \left(\vec{v}\frac{\partial p}{\partial s}\right) + \frac{\partial}{\partial s}\left(\dot{s}\frac{\partial p}{\partial s}\right) \ = \ 0 \tag{1}$$

$$\frac{\partial \vec{v}}{\partial t_s} + \nabla_s \frac{\vec{v}^2}{2} + (\zeta + f)k \times \vec{v} + \left(\dot{s}\frac{\partial p}{\partial s}\right)\frac{\partial \vec{v}}{\partial p} + \nabla_a M$$

$$= \ -g\frac{\partial \tau}{\partial p} + \left(\frac{\partial p}{\partial s}\right)^{-1}\nabla_s \cdot \left(\upsilon\frac{\partial p}{\partial s}\nabla_{\bar{s}}\,\vec{v}\right) \tag{2}$$

$$\frac{\partial}{\partial s}\left(\frac{\partial p}{\partial s}T\right) + \nabla_s \cdot \left(\vec{v}\frac{\partial p}{\partial s}T\right) + \frac{\partial}{\partial s}\left(\dot{s}\frac{\partial p}{\partial s}T\right) = \nabla_s \cdot \left(\upsilon\frac{\partial p}{\partial s}\nabla_s T\right) + H_T \tag{3}$$

In these equations $\vec{v} = (u,v)$ is the horizontal velocity vector, p is pressure, $\alpha = \rho^{-1}$ is the specific volume, $\zeta = \partial v/\partial x_s - \partial u/\partial y_s$ is the relative vorticity, $M \equiv gz + p\alpha$ is the Montgomery potential, gz is the geopotential, f is the Coriolis parameter, k is the vertical unit vector, υ is a variable eddy viscosity coefficient, τ is the wind and bottom drag induced shear stress vector, T is any one of the model's buoyancy related variables (temperature, salinity, density) and H_T is the sum of the non-adiabatic sources. Subscripts show the variables that are kept constant during partial differentiation. The model uses the formulation of Gaspar (1988) for the surface mixed layer. Wind-induced stress is assumed to be zero at the bottom of the surface mixed layer, which therefore cannot be shallower than the Ekman layer.

The model used in the present study employs an Arakawa C-grid to distribute the variables such that velocity grid points are located halfway between mass grid points in the x,y direction (Arakawa and Lamb, 1977). Thermodynamic variables and variables of motion are treated as layer variables that are constant within a layer but change across layer interfaces. However, p,z and $\dot{s}\,\partial p\,/\,\partial s$ are characterized as level variables defined on interfaces. A more detailed description of the model, the numerical schemes used and the numerical issues arising from isopycnic modeling can be found elsewhere (see Bleck *et al.* (1989), Bleck and Smith (1990), Smith *et al.* (1990), Bleck *et al.* (1992), Chassignet and Bleck (1993), Chassignet *et al.* (1995), Langlois (1997) and Bleck (1998)).

3. Model Geometry and Density Profile

A model resolution of 0.25^O longitude x 0.25^O *cos* ϕ latitude was chosen for the horizontal plane. Figure 1 shows the bathymetry used in the model which was interpolated from the Scripps topography data set (Pacanowski, 1995) and verified using hydrographic charts (Hydrographic Office, 1996). The model was initialized with transverse-basin meridional density profiles interpolated from the data set of Altman *et al.* (1987).

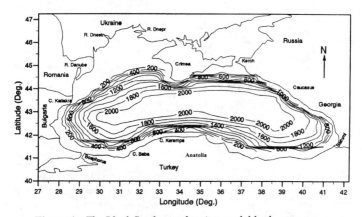

Figure 1. *The Black Sea basin showing model bathymetry.*

The model includes fourteen potential density layers in vertical and a surface mixed layer. In choosing these layers attention was paid to specific vertical features of the basin (e.g. densities corresponding to upper and lower interface of the Cold Intermediate Layer, and upper interfaces of the suboxic and anoxic layers). The base-layer of the model ($\sigma_\theta = 17.23$) is chosen to represent the seabed mixed layer. This is the thickest (~ 400-500 m) bottom layer observed in the world ocean. Ozsoy and Unluata (1997) note that this water is homogenised by convective motions resulting from geothermal heat fluxes. Average salinity values were specified for each layer so that the sharp vertical stratification of the basin was retained. These salinities were interpolated linearly from Altman *et al.* (1987). This data set consists of 70 years of CTD and Nansen bottle measurements. The data comprises monthly averages of temperature and salinity at 22 vertical levels with a resolution of 1° longitude and 0.5° latitude. Table 1 shows the layer densities and salinity fields.

Layer potential density (σ_θ)	Salinity (psu)
13.10	18.02
13.59	18.22
13.85	18.28
14.15	18.41
14.45	18.61
15.05	19.37
15.65	20.19
16.20	20.94
16.45	21.29
16.72	21.65
16.88	21.87
17.05	22.08
17.18	22.25
17.23	22.32

Table 1. *Potential density layers of the model and corresponding salinities.*

4. Model Forcing

The model is forced by mechanical (wind stress), thermodynamic and relaxation functions. Wind stress data is derived from the Hellermann and Rosenstein (1983) (HR) monthly climatologies. It should be noted that these wind stresses are reported to be weaker than expected but are well documented and are thought to reflect the main features of the Black Sea winds in space and in time (Stanev, 1990; Oguz *et al.*, 1995). Oguz and Malanotte-Rizzoli (1996) point out that HR fields are 2-3 times weaker than values reported by meteorological stations. For this reason our HR data values are increased by a factor of 2.5.

Thermodynamic forcing in MICOM is based on the calculation of the thermal balance (B),

$$B = R + H + \varepsilon \tag{4}$$

where, R is the balance of incident solar radiation and radiation emitted by the sea surface, ε is the latent heat transfer due to evaporation and H is the sensible heat transfer, when there is a considerable difference between sea surface temperature and air temperature (Langlois, 1997).

$$H = C_{Pair} E_x (T_s - T_a) \tag{5}$$
$$E_x = \rho_a C_T W \tag{6}$$

where, E_x is an exchange coefficient, ρ_a is mass / volume of the air, C_T is heat transfer coefficient and C_{Pair}, T_s, T_a and W are the specific heat of air, the sea surface temperature, the temperature in the atmospheric boundary layer and the wind velocity respectively.

The latent heat flux is (Langlois 1997):

$$\varepsilon = E_x L (H_u - E_v) \tag{7}$$

where, L, H_u and E_v are the latent heat of vaporization, the specific humidity and the evaporation respectively. The balance of evaporation and precipitation is also taken into account within the model to calculate the salinity of the mixed layer.

The thermodynamic functions are calculated from five sets of forcing data, namely precipitation, wind speed, effective air temperature, net radiation at the sea surface and the atmospheric water vapour mixing ratio. Precipitation and wind speed data were interpolated from Meteorological Office charts (1963). In addition, Oort's (1983) global air temperature statistics were used.

Net surface radiation is calculated from the difference between the incoming short-wave radiation and outgoing long-wave radiation. If t_d is length of day in hours and A_n is noon altitude of the sun in degrees, then the short-wave radiation input in the absence of clouds (Q_{so}) and allowing for the average atmospheric loss is (Bunker $et\ al.$, 1982):

$$Q_{so} = 0.4 A_n t_d \tag{8}$$

The rate at which the radiant energy arrives at the sea surface ($Q_s{}'$) after allowing reduction by cloud cover (C) is:

$$Q_s{}' = Q_{so} (1 - 0.0012 C^3) \tag{9}$$

The amount of short-wave radiation (Q_r) reflected from the sea surface is

$$Q_r = 0.15 Q_s{}' - \left(0.01 Q_s{}'\right)^2 \tag{10}$$

Finally,

$$Q_s(W/m^2) = Q_s{}' - Q_r = 0.85Q_s{}' - 10^{-4}Q_s{}' \qquad (11)$$

where Q_s is the actual short-wave radiation penetrating into the sea (Bunker et al., 1982). The net rate of heat loss by the sea as long-wave radiation to the atmosphere is taken as the world average value of 50 W/m^2 (Bunker, 1976).

The atmospheric water vapour mixing ratio (q; Bunker et al., 1982) is found from

$$q = \frac{0.622e}{p-e} \qquad (12)$$

where e is the pressure of water vapour in the atmosphere, p is the atmospheric pressure. The values for e and p were interpolated from the charts of the Meteorological Office (1963) and Hydrographer of the Navy (1990).

Since river runoff is supposed to have crucial effect on the physical processes of the basin, runoff input from the major rivers on the northwestern shelf (Danube, Dnestr and Dnepr) was added to the grid cells corresponding to their geographical location in the precipitation data. Flow rates are 960, 210 and 210 mm/day respectively and seasonal changes are estimated using rainfall data for the nortwestern region. The Bosphorus Strait, which is the only source for high salinity waters from the Mediterranean was included in the domain as an open boundary where relaxation is applied. Although the relatively coarse resolution of the present model does not give a realistic simulation of the strait, its inclusion helps to reveal the significance of its contribution. In contrast, the Kerch Strait, which connects the Black Sea to the shallow Sea of Azov, is neglected and closed in the domain due to its very low rate of water exchange. The temperature, salinity and pressure data for the Bosphorus were interpolated from surveys conducted during 1986 (Unluata et al., 1990).

5. Results of the Seasonal Model Runs and Discussion

Figure 2 and Figure 3 illustrate the circulation features in January and August which are chosen hereafter to be representative of winter and summer circulation respectively. Note that for clear presentation the maximum current speed in the figures has been restricted to 10 cm/s. The main feature of the circulation is the cyclonic Rim current that flows along the continental slope, and which can be most clearly observed in winter months. This consists 4 to 6 cyclonic gyres extending from west to east. Oguz et al. (1994) report a number of anti-cyclonic eddies, which are confined between the coastline and the Rim current. However, due to the coarse resolution of the present model, these eddies are not observed clearly.

In summer there is a substantial increase in the number of anti-cyclonic eddies found along the Anatolian coast and the Rim current can no longer be distinguished in the southern and eastern Black Sea (see Figure 3). In the north and west of the sea, it is not difficult to track the Rim current throughout the year, although enlarging anti-cyclonic eddies push it towards the interior of the basin. It is worth noting as well that displacement of the Rim current from its route in summer brings about formation of meso to large-scale cyclonic eddies in the eastern part of the basin outside the Rim current. Some opposing vectors are seen in this figure, possibly suggesting the presence of eddies that cannot be resolved.

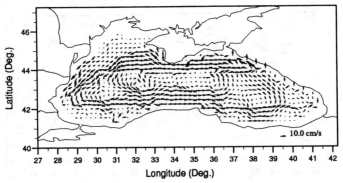

Figure 2. *Surface circulation in January.*

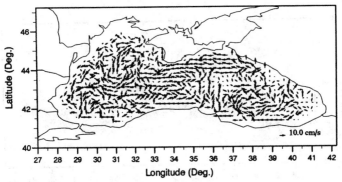

Figure 3. *Surface circulation in August.*

Maximum current velocities in the surface mixed layer usually lie in the range between 20 to 60 cm/s, being in good agreement with hydrographic survey data reported elsewhere (see Oguz et al., 1992). It is also noted that currents are faster in spring and summer months when intensified eddy processes are observed.

The distribution of sea surface temperature in winter follows the pattern of air temperature. As shown in Figure 4, a belt of relatively warmer waters lies along the southern coastline in January. The coldest temperatures are encountered either in the far north of the shallow northwestern shelf or in the western cyclonic gyre of the Rim current. The temperatures, in this month, vary from $6^\circ C$ in the northwestern shelf up to $9.2^\circ C$ in the western Turkish coast.

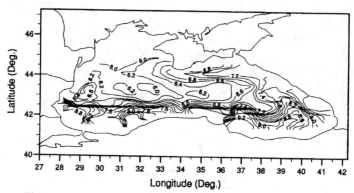

Figure 4. *Surface temperature in January.*

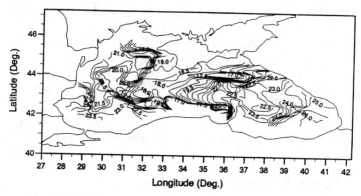

Figure 5. *Surface temperature in August.*

In summer, nevertheless, warmer waters cover a large portion of the eastern and western Black Sea including the northwestern shelf. In August, colder water (typically at about $18^{\circ}C$) covers the central part of the basin and extends south to the Anatolian coast (see Figure 5.). Temperatures as high as $26^{\circ}C$ are encountered in this month in the southeastern part of the basin. Generally sea surface temperatures are in good agreement with observed values (see Oguz *et al.*, 1992 and Oguz *et al.*, 1994)

In winter, the sea surface salinity distribution is easier to correlate with the mixed layer circulation pattern than the temperature. As Figure 6 shows, highly saline waters are trapped in the Rim current and surrounded by fresher waters resulting from river runoff or heavy precipitation. This sharp front is also observed in satellite images along the western shelf/slope boundary when the less saline shelf waters meets with more saline and denser interior waters (Oguz *et al.*, 1992). Furthermore, the salinity contrast between the interior and coastal waters is sharper in the western part of the basin than in the eastern part, due to the input from the Danube, Dnestr and Dnepr in the northwestern shelf. Very low salinity values, less than 17 psu, are found in this shallow region all around the year. In January, low salinity waters are found along the coastline from western side of the Crimean Peninsula to the western part of the Bosphorus. This is supported by salinity observations made along the Anatolian coast and which show that freshwater from the northwestern shelf usually flows cyclonically to the southwest of the basin (Sur *et al.*, 1994). Typical salinity of this belt is characterized as 18 psu in January. In the eastern part of the sea lowest salinity values are encountered around northern Caucasian coast and are as low as 18.1 psu. Salinities usually increase towards the centre of the basin, where the highest values are observed to be around 18.4 psu.

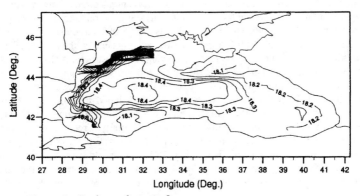

Figure 6. *Surface salinity in January.*

The sharpest surface salinity differences in summer are also found across the front created between the northwestern shelf waters and high saline interior waters (see Figure 7). This contrast is characterised by a difference of around 0.5 psu, which was also observed during HydroBlack'91 (Oguz *et al.*, 1994). Low salinity water masses are also detected along the Caucasian coast and in the easternmost part of the basin, although salinity differences are not as great as those found along the northwestern shelf front. The minimum salinities observed in these regions are around 18.05 psu in August. Maximum values across the sea, however, are recorded around 18.3 psu.

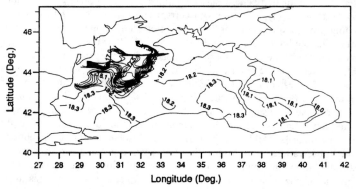

Figure 7. *Surface salinity in August.*

The vertical structure of circulation in the Black Sea is still subject to debate. Oguz *et al.* (1994) note that the results of HydroBlack '91 suggest that even though the large scale components of the general circulation are similar down to 500 m, both large and mesoscale features of the circulation are subject to considerable structural change at deeper levels. The SO90 survey shows not only the disappearance of the Rim Current, the shift of eddy centres, the coalescence of eddies, and the persistence of some components but also structural changes in the deeper layers (Oguz *et al.*, 1993). Bulgakov and Kushnir (1996), for instance, after carrying out some predictions based on theoretical analysis and laboratory tests, conducted special field observations and concluded that an anti-cyclonic counter current lies below the Rim Current at the depth of the main pycnocline in the northern and northwestern parts of the basin. Similarly, Eremeev and Kushnir (1996), based on the measurements conducted during ComsBlack surveys in August 1992 and in October-November 1993, point out that under certain conditions a counter current may be formed under the Rim Current in the same region although even in these recent surveys current velocity profile measurements below 200-250 meters are not claimed to be reliable (Ozsoy and Unluata, 1997).

In agreement with Eremeev and Kushnir (1996) and Bulgakov and Kushnir (1996), results obtained from the present model indicate the existence of a deep rim counter current throughout the year in both intermediate and bottom layers (see Figure 8) and furthermore that the counter current is weak in the intermediate layer but is fully developed in the bottom layer.

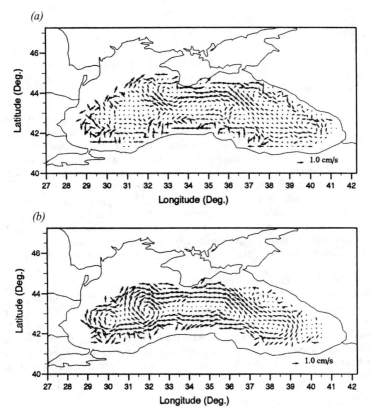

Figure 8. *Circulation in layer (a) $\sigma_\theta = 16.72$ and (b)$\sigma_\theta = 17.23$.*

The seasonal circulation in intermediate and bottom layers are similar, suggesting that the effect of seasonal forcing does not penetrate the strong stratification. Figure 8 presents the circulation in the potantial density layers $\sigma_\theta =$ 16.72 and $\sigma_\theta = 17.23$. Circulation characteristics in the layer $\sigma_\theta = 16.72$ start to deviate from those in the upper layers. Although in the interior of the basin the cyclonic Rim current can still be detected in most months, an opposing current that flows along the coastal slope encircles it. A clearer representation of the anti-cyclonic Rim current found in the bottom layers of the Black Sea is shown in the layer $\sigma_\theta = 17.23$, which characterizes the upper interface of the bottom mixed layer. The present study indicates that the deep Rim counter current consists of four gyres.

6. Conclusions

The seasonal circulation features of the Black Sea have been described. It appears that an isopycnic model, the MICOM, is appropriate and successful in modelling these features.

A well-pronounced cyclonic Rim current is clearly observed flowing along the steep shelf slope of the basin. This is a dominant characteristic of the surface circulation in the winter. In contrast the Rim current destabilizes in summer and becomes unrecognisable in the southern and eastern coast of the sea. It is pushed towards the interior of the basin in the northern and western parts by enlarged anti-cyclonic eddies. These anti-cyclonic eddies, most of which are persistent all around the year, are usually confined between the coastline and the Rim current. While sea surface temperature, roughly increasing towards the south, following the same pattern as air temperature, the surface salinity is easily correlated with the general circulation. A sharp salinity front between low saline northwestern waters and high saline interior waters is seen throughout the year.

The Rim current looses its strength in deeper layers of the sea. The model shows a reversal in the direction of the cyclonic Rim current in the bottom waters of the Black Sea. Although this feature has been suggested by some authors, further modelling studies and surveys are essential to clarify the issue. The contributions of various forcing mechanisms (wind forcing, thermodynamic forcing, river input from the northwestern shelf and seawater exchange through the Bosphorus) to the circulation are examined elsewhere (see Karakas et al. (2000)).

7. References

Altman, E.N., Gertman, I.F., Golubeva, Z.A., 1987. Climatological Fields of Salinity and Temperature in the Black Sea. Sevastopol Branch, State Oceanography Institute, Sevastopol, Ukraine. 109 pp.

Arakawa, A., Lamb, V.R., 1977. Computational design of the basic processes of the UCLA General Circulation Model. Methods in Computational Physics 17, 174-265.

Bleck, R., 1978. Finite Difference Equations in Generalized Vertical Coordinates, Part 1: Total Energy Conservation. Contributions to Atmospheric Physics 51, 360-372.

Bleck, R., Hanson, H. P., Hu, D., Kraus, E. B., 1989. Mixed layer/thermocline interaction in a three-dimensional isopycnal model. Journal of Physical Oceanography 19, 1417-1439.

Bleck, R., Smith, L., 1990. A Wind-Driven Isopycnic Coordinate Model of the North and Equatorial Atlantic Ocean 1: Model Development and Supporting Experiments. Journal of Geopyhsical Research 95, 3273-3285.

Bleck, R., Rooth, C., Hu, D., Smith, L.T., 1992. Salinity-driven Thermocline Transients in a Wind- and Thermohaline-forced Isopycnic Coordinate Model of the North Atlantic. Journal of Physical Oceanography 22, 1486-1505.

Bleck, R., 1998. Ocean modeling in isopycnic coordinates. In: Chassignet. E.P., Verron, J. (Eds.), Ocean Modeling and Parameterization. Kluwer Academic Publishers, pp. 423-448.

Bulgakov, S.N., Kushnir, V.M., 1996. Vertical structure of the current field in the Northern Black Sea. Oceanologica Acta 19(5), 513-522.

Bunker, A.F., 1976. Computation of surface energy flux and annual air-sea interaction cycles of the North Atlantic Ocean. Monthly Weather Review 104, 1122-1140.

Bunker, A.F., Charnok, H., Goldsmith, R.A., 1982. A note on the heat balance of Mediterranean and Red Seas. Journal of Marine Research 40, 73-84.

Chassignet, E. P., Bleck, R., 1993. The influence of layer outcropping on the separation of boundary currents. Part I: The wind-driven experiments. Journal of Physical Oceanography 23, 1485-1507.

Chassignet, E. P., Bleck, R., Rooth, C. G. H., 1995. The influence of layer outcropping on the separation of boundary currents. Part II: The wind- and buoyancy-driven experiments. Journal of Physical Oceanography 25, 2404-2422.

Chassignet, E.P., Smith, L.T., Bleck, R., Bryan, F., 1996. A Model Comparison: Numerical Simulations of the North and Equatorial Atlantic Oceanic Circulation in Depth and Isopycnic Coordinates. Journal of Physical Oceanography 26, 1849-1867.

Dzhioyev, T.Z., Sarkisyan, A.S., 1976. Prognostic Calculations of Black Sea Currents. Atmospheric and Oceanic Physics 12(2), 217-223.

Eremeev, V.N., Ivanov, L.M., Kirwan, A.D., Melnichenko, O.V., Kochergin, S.V., Stanichnaya, R.R., 1992. Reconstruction of Oceanic Flow Characteristics From Quasi-Lagrangian Data 2. Characteristics of the Large-Scale

Circulation in the Black Sea. Journal of Geophysical Research 97(C6), 9743-9753.

Eremeev, V.N., Kushnir, V.M., 1996. The Layered Structure of Currents and Vertical Exchange in the Black Sea. Oceanology 36, 9-15.

Gaspar, P., 1988. Modeling the Seasonal cycle of the Upper Ocean. Journal of Physical Oceanography 2, 514-517.

Hellermann, S., Rosenstein. M., 1983. Normal monthly wind stress over the World Ocean with error estimates. Journal of Physical Oceanography 13, 1093-1104.

Hydrographer of the Navy, 1990. Black Sea Pilot. 12th edition, HMSO, Somerset, 231 pp.

Hydrographic Office, 1996. The Black Sea including the Marmara Denizi and the Sea of Azov. Chart No 2214, Somerset, UK.

Karakas, G.M., James, A.E., Al-Barakati, A.M.A., 2000. Seasonal Circulation in the Black Sea: An Isopycnic Model Experiment. Paper submitted for publication to Deep-Sea Research.

Kiran, N., Yenigun, O., Albek, E., Borekci, O., 1996. Wind-Induced Circulations of the Black Sea. Water Science and Technology 32(7), 87-93.

Langlois, G., 1997. Miami Isopycnic Coordinate Ocean Model: User's Manual, Code Version 2.6. http://www.acl.lanl.gov/CHAMMP/micom.html, 136 pp.

Marchuk, G.I., Kordzadze, A.A., Skiba, Y.N., 1975. Calculation of the Basic Hydrological Fields in the Black Sea. Atmospheric and Oceanic Physics 11(4), 379-393.

Meteorological Office, 1963. Weather in the Black Sea. HMSO, London, 291 pp.

Oguz, T., La Violette, P.E., Unluata, U., 1992. The Upper Layer Circulation of the Black Sea: Its Variability as Inferred from Hydrographic and Satellite Observations. Journal of Geophysical Research 97(C8), 12569-12584.

Oguz, T., Latun, V.S., Latif, M.A., Vladimirov, V.V., Sur, H.I., Markov, A.A., Ozsoy, E., Kotovshchikov, B.B., Eremeev, E.E., Unluata, U., 1993. Circulation in the surface and intermediate layers of the Black Sea. Deep-Sea Research Part I 40(8), 1597-1612.

Oguz, T., Aubrey, D.G., Latun, V.S., Demirov, E., Koveshnikov, L., Sur, H.I., Diaconu, V., Besiktepe, S., Duman, M., Limeburner, R., Eremeev, V., 1994. Mesoscale circulation and thermohaline structure of the Black Sea observed during HydroBlack '91. Deep-Sea Research Part I 44(4), 603-628.

Oguz, T., Malanotte-Rizzoli, P., Aubrey, D., 1995. Wind and thermohaline circulation of the Black Sea driven by yearly mean climatological forcing. Journal of Geophysical Research 100(C4), 6845-6863.

Oguz, T. And Malanotte-Rizzoli, P., 1996. Seasonal variability of wind and thermohaline-driven circulation in the Black Sea: Modeling studies. Journal Geophysical Research 101(C7), 16551-16569.

Oort, A., 1983. Global atmospheric circulation statistics 1958-1973, NOAA, US Government Printing Office, Washington D.C., 180 pp.

Ozsoy, E. and Unluata, U., 1997. Oceanography of the Black Sea: a review of some recent results. Earth-Science Reviews 42, 231-272.

Pacanowski, R., 1995. MOM 2 Documentation User's Guide and Reference. Manual Version 1.0. GFDL Ocean Technical Report No. 3, pp. 175.

Rachev, N.H. and Stanev, E.M., 1997. Eddy Processes in Semienclosed Seas: A Case Study for the Black Sea. Journal of Physical Oceanography 27, 1581-1601.

Ross, D.A., Uchupi, E., Prada, K.E., MacIlvaine, J.C., 1974. Bathymetry and Microtopography of Black Sea. In: Degens, E.T. and Ross, D.A. (Eds.), The Black Sea: Geology, Chemistry and Biology. The American Association of Petroleum Geologists, pp. 1-10.

Smith, L. T., Boudra, D. B., Bleck, R., 1990. A Wind-driven Isopycnic Coordinate Model of the North and Equatorial Atlantic Ocean 2: The Atlantic Basin Experiments. Journal of Geophysical Research 95, 13105-13128.

Stanev, E.V., 1990. On the Mechanisms of the Black Sea Circulation. Earth-Science Reviews 28, 285-319.

Stanev, E.V., Roussenov, V.M., Rachev, N.H., Staneva, J.V., 1995. Sea response to atmospheric variability. Model study for the Black Sea. Journal of Marine Systems 6, 241-267.

Stanev, E.V., Staneva, J.V., Roussenov, V.M., 1997. On the Black Sea water mass formation: Model sensitivity study to atmospheric forcing and parameterizations of physical processes. Journal of Marine Systems 13, 245-272.

Sur, H.I., Ozsoy, E., Unluata, U., 1994. Boundary current instabilities, upwelling, shelf mixing and eutrophication processes in the Black Sea. Progress in Oceanography 33, 249-302.

Trukhchev, D.I., Stanev, E.V., 1983. Numerical Models for Currents in the Western Part of the Black Sea. Oceanology 23(1), 10-13.

Trukhchev, D.I., Stanev, Y.V, Balashov, G.D., Miloshev, G.D., Rusenov, V.M., 1985. Some Unique Features of the Mesoscale Structure of Hydrological Fields in the Western Part of the Black Sea. Oceanology 25(4), 443-446.

Unluata, U., Oguz, T., Latif, M.A., Ozsoy, E., 1990. On the Physical Oceanography of the Turkish Straits. In: Pratt, L.J. (Ed.), The Physical Oceanography of Sea Straits. Kluwer Academic Publishers, Netherlands, pp. 25-60.

The Coastal Marine Demonstration Project

L.J. Walstad[1], G. J. Szilagyi[2], F. Aikman[3], L.C. Breaker[4], F.
Klein[5], J. S. D'Aleo[6] and J. T. McQueen[7]

Abstract

The Coastal Marine Demonstration of Forecast Information to Mariners for
the U.S. East Coast (CMDP) is exploiting existing nowcast and forecast models
for the atmosphere and ocean to provide products in real-time to selected users.
We are also developing new approaches for oceanic data assimilation that will
lead to improved products. This project is a collaboration of university,
government, and industry partners funded by the National Oceanographic
Partnership Program.

This project consists of two overlapping and complementary activities:
operational and research. The operational components have coupled operational
and research models to produce nowcasts and forecasts of relevant marine
quantities in the region west of 70°W and between 32°N and 42°N. In the
atmosphere, these quantities include the surface wind speed and direction,
temperature, precipitation, and visibility. Oceanographic products include sea
surface height, temperature, salinity, currents, and waves. Within the Chesapeake
Bay, water level is forecast. In addition to demonstrating the nowcast/forecast
system, we are incorporating new data sources and improve forcing estimates.

---

[1] University of Maryland Center for Environmental Science Horn Point Laboratory, Cambridge,
MD 21613-0775
[2] Litton-TASC Inc., 4801 Stonecroft Blvd. Chantilly, VA 20151
[3] NOAA/National Ocean Service/Coast Survey Development Laboratory, Silver Spring, MD
20910-3285
[4] NOAA/National Weather Service/National Centers for Environmental Prediction, Washington,
D.C. 20233-9910
[5] Mitretek Systems, 7525 Colshire Drive McLean, VA 22102-7400
[6] Weather Services International, 4 Federal Street Billerica, MA 01821
[7] NOAA/Office of Atmospheric Research/Air Resources Laboratory, Silver Spring, MD 20910-
3285

The research components of the project are primarily directed toward the implementation of new forecast methodology and assessment of our ability to nowcast and forecast. Within Chesapeake Bay, an assimilation of all available tide gauge data is being developed. Future dynamical models for Chesapeake Bay will incorporate temperature and salinity. Our ability to estimate currents in Chesapeake Bay will also be assessed. Wave forecasting for Chesapeake Bay is being examined and will become operational later this year. In the coastal ocean, efforts are underway to assimilate remotely sensed observations of surface elevation and sea surface temperature.

Introduction

The Coastal Marine Demonstration Project (CMDP) is a two-year effort funded by the National Oceanographic Partnership Program and managed by the Office of Naval Research. The partnership is a collaboration of academia, private industry, government, and private citizens. Academic institutions involved are the University of Maryland Center for Environmental Science, Princeton University, and the University of Rhode Island. Private industry partners include Litton-TASC and Litton-Weather Services International (WSI). Government partners include the National Weather Service, the National Ocean Service, the Navy, and the Coast Guard. While products are being made publicly available, a selected group of users are members of this partnership. These users include Maryland and Virginia pilots, the Navy, and the Coast Guard. These partners provide important feedback for the improvement of the products and delivery mechanism.

A major component of the CMDP is the implementation of a state-of-the-art forecast system during two demonstration periods. During demonstrations, model calculation schedules are formalized, more frequent checks are made to ensure that models have run, and model output is provided to forecasters and users in real time who assess the product quality and utility. The first CMDP demonstration began on June 17th and lasted until July 30th 1999. The second demonstration will be held in February and March of 2000. The region of interest is the Chesapeake Bay and surrounding coastal ocean west of 70°W between 32° and 42°N.

Models

Building upon several NCEP operational models, the CMDP devoted significant resources to running and monitoring models that had previously been used experimentally. These models were coupled into a system (Figure 1) and run on a regular production schedule (Figure 2). During the first demonstration, these models were run in accordance with this regular production schedule and output was provided to users in real time. Figure 1 depicts the relationship of the atmospheric and oceanographic models utilized in the demonstration.

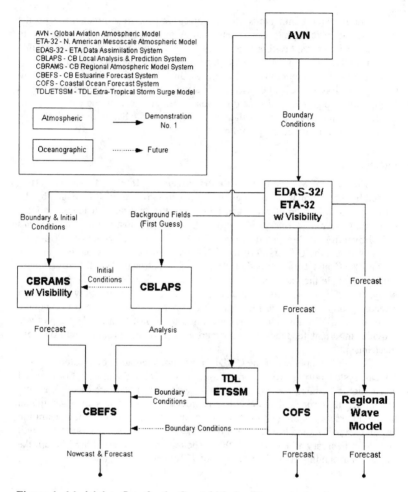

Figure 1. Model data flow for the Coastal Marine Demonstration I.

The operational NCEP mesoscale Eta atmospheric model (Black, 1994) provides input fields for several of the CMDP models. Eta was developed for mesoscale atmospheric phenomena to diagnose these smaller-scale processes that could not be resolved by synoptic scale numerical models such as the Nested Grid Model (NGM) or the Aviation Model (AVN). The "Eta" or "step mountain" coordinate system was implemented in order to remove some of the errors introduced by the sigma coordinate system over steep topographic slopes. Eta presently has a horizontal resolution of 32 km and 45 vertical layers. Eta assimilates data from the Eta Data Assimilation System (EDAS). EDAS generates eight 3-hourly initial states or analyses during each 24-hour period, utilizing a wide range of observed data from satellites, point observation stations, radar and other sources.

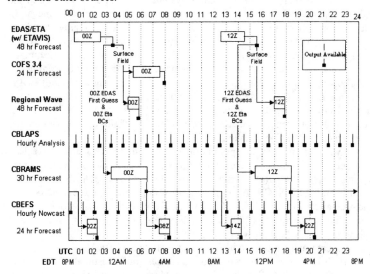

Figure 2. CMDP production schedule for Demonstration I.

The second operational model used during the first demonstration was the NCEP Regional Wave Model. The Regional Wave Model is a 1/4° resolution, third generation wave forecasting system for the U.S. East Coast and Gulf of Mexico that complements predictions of the present operational global wave model that has a grid resolution of 2.5°. The system employs the WAM model Cycle-4 version software package (Gunther et al., 1991). A 1° grid covering the Atlantic Ocean is used to provide boundary conditions for the smaller 1/4° grid for the CMDP. Ocean surface winds used to drive the model are obtained from the mesoscale Eta model (ETA32).

The NWS Extra-Tropical Storm Surge Model (ETSSM) was the third operational model relied upon by the CMDP. The ETSSM provides sub-tidal water level forcing forecasts at the mouth of the Chesapeake Bay to drive the Bay hydrodynamic model discussed below.

The Chesapeake Bay Local Analysis and Prediction System (CBLAPS) (Fuell et al., 1999), the Chesapeake Bay Regional Atmospheric Modeling System (CBRAMS) (McQueen et al., 1999) and the Chesapeake Bay Estuarine Forecast

System (CBEFS) (Gross et al., 1999) comprise the suite of models/analyses covering the Chesapeake Bay region. The Coastal Ocean Forecast System (COFS), a three dimensional ocean circulation model, was implemented for use in the U.S. East Coast waters (Breaker et al., 1999a).

The Local Analysis and Prediction System (LAPS) was developed at the Forecast Systems Laboratory with the objective of producing real-time high-resolution atmospheric forecasts. The LAPS has the ability to integrate data from virtually every meteorological observation system. Some of these data are not utilized in larger scale models (i.e. Eta Data Assimilation System). For example, CBLAPS ingested data from the three Chesapeake Bay Observing System buoys (www.cbos.org). Since the only other over-water wind measurement in the Chesapeake Bay is Thomas Point Lighthouse, these data proved to be very valuable. EDAS fields are used by CBLAPS for the background field estimate. For the first demonstration, CBLAPS produced nowcast fields on a four-kilometer grid centered over the Chesapeake Bay.

RAMS was used as the prediction component of the Chesapeake Bay atmospheric prediction system (CBRAMS). The Regional Atmospheric Modeling System is a non-hydrostatic mesoscale model. CBRAMS utilized a 4 km grid identical to the CBLAPS grid and a 16 km outer grid to extend domain boundaries away from the Bay. EDAS fields are used for initialization and Eta forecast fields provide CBRAMS boundary conditions. In the future, a 3 dimensional CBLAPS field with radar and local mesonet data ingested should provide a more timely and accurate initialization field for CBRAMS.

CBEFS is a water level prediction system designed to estimate the tidal and non-tidal effects. CBEFS is based on the Model for Estuarine and Coastal Circulation Assessment (MECCA). CBRAMS surface wind fields are used to estimate the surface stress needed for CBEFS. The Chesapeake Bay Bridge Tunnel (CBBT) water level gauge provides water level forcing for the nowcast and ETSSM for the forecast. As depicted in Figure 1, we plan to use COFS water level at the bay mouth to force CBEFS, replacing the ETSSM forecast of water level. A future enhancement to CBEFS will be the replacement of the MECCA with a finite element model, allowing forecasts of current, salinity and water temperature. The assimilation of water-level data is being developed and should help to provide more accurate nowcasts and forecasts.

CBEFS builds upon the Chesapeake Area Forecasting Experiment (CAFÉ) (Bosley and Hess, 1997) system. CAFÉ uses Chesapeake Bay Bridge Tunnel and Thomas Point wind measurements for the nowcast and Eta-32 winds for the forecast, whereas CBEFS uses CBLAPS and CBRAMS wind fields respectively. The operational version of CAFÉ is the Chesapeake Bay Operational Forecast System (CBOFS). CBOFS incorporates computer and network changes to make the system reliable (Bosley et al., 1999).

The COFS is based on the Princeton Ocean Model, a hydrodynamic ocean circulation model (Blumberg and Mellor, 1987). COFS is driven at the ocean surface boundary by forecast fluxes of momentum, heat and moisture derived from the Eta model and produces forecasts of currents, temperature and salinity. The spatial resolution of the model varies from 20 km offshore to 10 km nearshore.

A visibility product depicting areas of reduced visibility was also implemented during the first demonstration. Visibility is calculated by applying the Stoelinga and Warner algorithm (Stoelinga and Warner, 1998) to the lowest CBRAMS or Eta model level. The algorithm incorporates cloud liquid water, cloud frozen water, liquid precipitation, and frozen precipitation.

These models provided the basis for the end-to-end CMDP system shown in Figure 3. Model output is sent to the National Oceanographic Data Center (NODC) for archiving and re-distribution, as well as to various groups within NOAA for use and evaluation. WSI pulled model output from NODC and made use of observational data to create valued-added products. Users accessed a web site where products from WSI were displayed and used an HTML web form to send feedback to the developers.

CMDP models were run on three computers at different locations on the production schedule shown in Figure 2. The coastal ocean models were run on the Cray C-90 in Suitland, MD. CBRAMS and CBEFS were run at the Coast Survey Development Lab in Silver Spring, MD and CBLAPS analyses were accomplished at the Sterling Weather Forecast Office (WFO). Eta 48-hour forecasts for periods beginning at 00Z and 12Z daily were available at approximately 03Z and 15Z respectively. Eta output is used to drive 48 hour, 00Z and 12Z Regional Wave forecasts and a 24 hour, 00Z COFS forecast. Eta fields are also used as boundary conditions for 30 hour, 00Z and 12Z CBRAMS forecasts. CBLAPS analyses were run hourly and typically completed by 22 minutes past the hour. CBEFS utilized CBRAMS surface winds to produce 24 hour forecast at 02Z, 08Z, 14Z and 22Z. Hourly CBEFS nowcasts were typically available 7 minutes past the hour.

All model output was stored in GRIdded Binary Format (GRIB) at NODC. The WFOs pulled CBLAPS and CBRAMS GRIB files directly from NODC so they could de-GRIB and display on their own systems using GEneralized Meteorological PAcKage (GEMPAK). This procedure allowed the WFOs maximum flexibility in utilizing CBLAPS and CBRAMS. WSI pulled all GRIB files from NODC except for Eta fields that they receive via the NOAA Family of Services (FOS). WSI received station observations, radar and satellite imagery through FOS and other sources. WSI created products on an SGI O2 workstation using commercially available software.

Products

The primary CMDP products were output fields from the models. WSI was responsible for the graphical display of these fields and also contributed

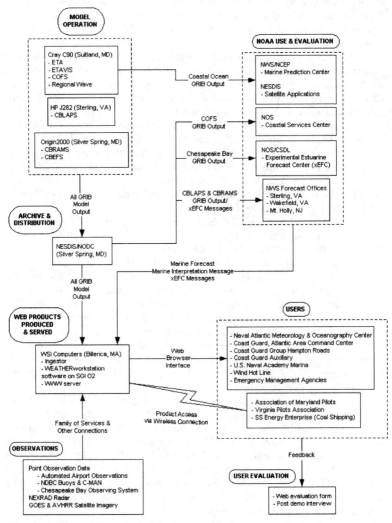

Figure 3. System Overview of the Coastal Marine Demonstration I.

additional products such as radar and satellite imagery, to complete the CMDP suite. These products were made available via the web (http://cmdp.wsicorp.com) as shown in Figure 4. Some users received GRIB files and displayed the model fields on their own systems for added flexibility in analyzing model output.

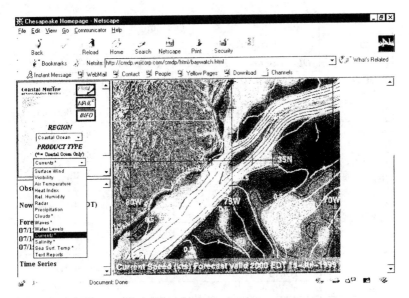

Figure 4. Coastal Marine Demonstration I web based product delivery. Regions and product type are selected by the pull down menu. The displayed field is the surface current in the coastal ocean from the COFS. Streaklines indicate current direction. Contours indicate speed.

An Experimental Estuary Forecast Center (xEFC) was established during the demonstration period to monitor the Chesapeake Bay modeling system and provide limited interpretation of model output. The xEFC issued regular status messages three times a day and special status messages when necessary to keep users informed. During weekdays, xEFC discussions were issued to provide guidance regarding the analyses, nowcasts and forecasts issued for that day.

To organize the information available in the CMDP products, the CMDP region of interest was separated into two areas: 1) the Chesapeake Bay and 2) the surrounding coastal ocean. Mariners operate in a single location and move through the region at a relatively slow rate of speed, thus this division aids them in obtaining the information they need. Table 1 summarizes the products and identifies the sources of these observations or forecasts.

Product	Domain			
	Chesapeake Bay		Coastal Ocean	
	Observations[8]	*Forecast*[9]	*Observations*	*Forecast*
Surface Wind	Land/Water	**CBRAMS**	**Land/Water**	Meso-Eta
Visibility	Land	**CBRAMS**	Land	Meso-Eta
Air Temp	Land/Water	**CBRAMS**	Land/Water	Meso-Eta
Heat Index	Land	**Meso-Eta**	Land	Meso-Eta
Humidity	Land	**Meso-Eta**	Land	Meso-Eta
Radar	NEXRAD	-	NEXRAD	-
Precipitation	Land	Meso-Eta	Land	Meso-Eta
Clouds		-	GOES	Meso-Eta
Waves		-	Water	Regional Wave
Water Levels	NWLON[10] PORTS[11]	CBEFS		
Currents				COFS
Salinity				COFS
Sea Surface Temp	Water		Water, AVHRR	COFS

Table 1. Source of products for Chesapeake Bay and coastal ocean maps.

Samples of the graphics produced are shown in the Figures 4 through 8. Figure 5 shows a 4 km grid CBRAMS surface wind forecast for the lower Chesapeake Bay. Notice the increase in wind speed over water. The 32 km resolution Eta field only has only a few grid points over Chesapeake Bay, making it difficult to reproduce such small-scale patterns. This phenomena and frequent wind reversals over the Bay in the late afternoon were commonly observed during Demonstration I.

The increase in wind speed over water is the result of the differences between the terrestrial and marine boundary layers (McQueen *et al.*, 2000).

[8] Land stations included Automated Surface Observing Systems (Class 1/2 airports) and Automated Weather Observing Systems (Class 3/4 airports). Water stations included National Data Buoy Center's moored buoys, the Coastal Marine Automated Network and the Chesapeake Bay Observing System buoys.

[9] Along with particular Chesapeake Bay forecast fields, hourly analyses or nowcasts were provided as products. This included analyses of surface wind and air temperature from CBLAPS and nowcasts of water levels from CBEFS.

[10] NWLON – National Water Level Observation Network.

[11] PORTS – Physical Oceanographic Real Time System.

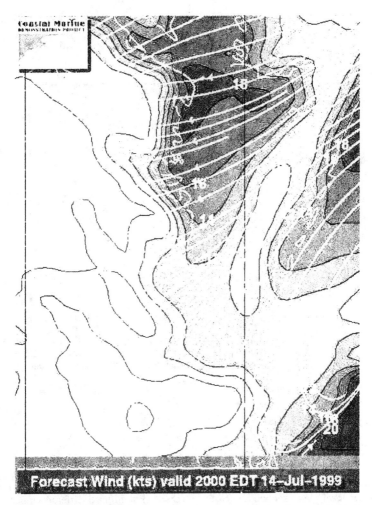

Figure 5. CBRAMS surface winds for July 14, 1999. Streaklines indicate
direction. Contours indicate speed.

These differences include the surface roughness, the heat flux, and the
moisture flux. The boundary layer is not transformed abruptly at the coast.
Rather, there is a gradual transition that is determined by the wind direction and
the particular characteristics of the boundary layers. Wind reversals occur when

northerly winds are present early in the day, and there is differential land-sea heating of the lower atmosphere. This leads to a southerly wind in the late afternoon over the water while over-land winds remain northerly. These differences are significant for mariners in small boats.

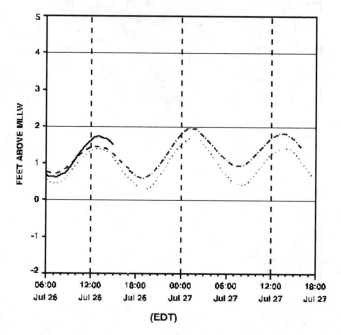

Figure 6. Water level at Lewisetta, MD: CBEFS nowcast (– – –) and forecast (– • – •), astronomical tides (•••), and gauge observations (———).

Figure 6 depicts observed and predicted water levels for 36 hours at Lewisetta in the lower Potomac. None of the gauges within the Bay are used in the assimilation. These observations are only used to assess the accuracy of the water level nowcasts and forecasts. Occasional significant differences occur. These differences are not well understood and are being characterized (Gross *et al.*, 2000). One hypothesis is that because the fetch of the Bay is very short, the surface stress determined through standard relationships may be inaccurate.

Figure 7 is a COFS surface current forecast that indicates the Gulf Stream flow to the northeast. The COFS currents over the shelf of the Mid-Atlantic Bight (MAB) are incorrect. The MAB currents are typically to the southwest except when strongly forced by wind. The COFS currents may be incorrect because of

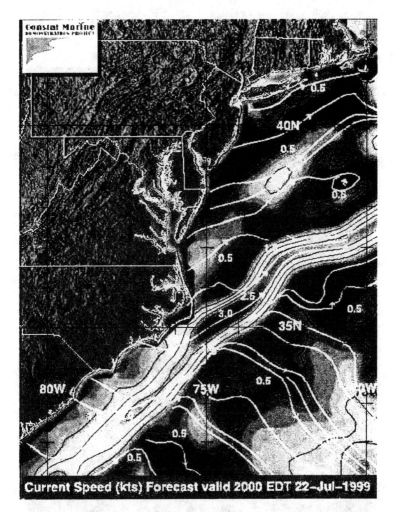

Figure 7. Coastal ocean map of COFS surface current forecast. Streaklines
indicate current direction. Color contours indicate the speed.

insufficient resolution upstream (i.e. north of Cape Cod). This would lead to
incorrect buoyancy input at the northern end of the bight. Off the shelf, the
currents are reasonable. The position of key features of these field estimates are
being assessed (Breaker *et al.*, 1999a). These currents can be important for

anticipating the location of steep and large waves. In Figure 8, the Regional Wave Model forecast is displayed with the direction indicated by quadrant. This product is useful to all mariners and may be combined with the surface wind field to determine areas of large and steep waves.

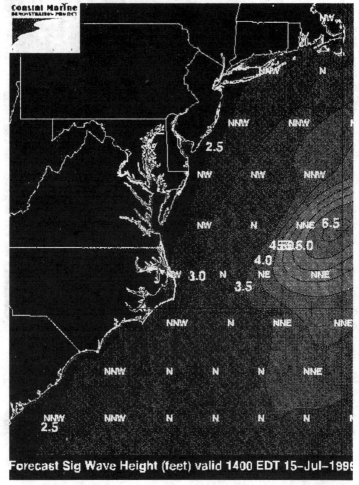

Figure 8. Wave estimates for July 15. Height is in feet as indicated by contours. Direction from which waves are propagating is indicated by letters.

Evaluation

Evaluation of CMDP products was a crucial component of the demonstration and took three forms (Richardson and Szilagyi, 1999; Breaker *et al.*, 1999b). First, model developers assessed model performance and conducted objective reviews of forecasts by comparing predictions with independent measurements. The second form consisted of NOAA organizations, including several National Weather Service Weather Forecast Offices and the Marine Prediction Center, incorporating CMDP products into routine forecasts and messages. The third form involved organizations outside of NOAA, including the Navy and Coast Guard forecasters and marine pilots, who primarily accessed CMDP products through the web site.

NOAA organizations incorporating CMDP products included Weather Forecast Offices (WFOs) from Sterling, VA, Wakefield, VA and Mt. Holly, NJ daily downloaded CBRAMS and CBLAPS 2D surface field GRIB files. WFOs also ingested CBOS buoy data which helped to verify CBLAPS and CBRAMS wind fields. The Coastal Services Center in Charleston, NC compared COFS forecast sea surface temperature (SST) to AVHRR SST imagery on daily basis. The CSC noted the position of features including the Gulf Stream and rings. The NESDIS Satellite Applications Branch assessed the Eta based visibility fields through comparisons with GOES-8 visible satellite imagery. The NCEP Marine Prediction Center evaluated five forecast fields; waves from the Regional Wave Model, visibility, and COFS sea surface temperature, currents and salinities. MPC summarized their observations in daily Marine Interpretation Messages. These analyses are being incorporated into our assessment of the capability of the system for producing useful forecasts.

Our external partners are:

> the Naval Meteorology and Oceanography Center (NLMOC)Norfolk, VA,
> the U.S. Coast Guard Fifth District Headquarters - Portsmouth, VA,
> the U.S. Coast Guard Group Hampton Roads – Portsmouth, VA,
> the U.S. Coast Guard Auxiliary,
> the U.S. Coast Guard Research & Development Center – Groton, CT,
> the Association of Maryland Pilots – Baltimore, MD,
> the Virginia Pilots Association – Virginia Beach, VA,
> the Wind Hot Line – Poquoson, VA,
> the U.S. Naval Academy Sailing Center – Annapolis, MD,
> the Emergency Management Agencies, and
> the SS Energy Enterprise.

These users, with two exceptions, accessed the products through the CMDP web site. The USCG R&D Center pulled COFS GRIB files for input to a drift model. The SS Energy Enterprise did not have internet access on the water, so selected graphical products were faxed directly to the ship. Feedback was gained through an HTML form at the web site along with short user interviews after the demonstration concluded. During the demonstration, 77 user evaluation forms were submitted via the internet. A Maryland pilot and NLMOC submitted 26 and 20 evaluation forms respectively, accounting for over 50% of the total responses. Clearly, the most popular product was surface wind observations and forecasts. Users rated four categories; accessibility, accuracy, display/format and needs met on a scale of poor, fair, good, very good, or excellent. Users rated these categories as good, very good, or excellent for the following percentage of the forms: accessibility – 87%, accuracy – 86%, display/format – 65%, needs met – 62%. The high accessibility number indicated that graphical image size was small enough to be downloaded quickly, and that system reliability was good enough so that when users went to get a product it was there. The perceived accuracy of the products was high. Most criticism of the display/format related to legends, labeling and the use of streamlines rather than wind barbs. The desire for additional products contributed to the needs-met rating. Written comments indicated that the products were desired and that all external partners will actively participate in the second demonstration period in February and March, 2000.

Discussion

In addition to producing real-time products, a real-time system provides a focus upon the comparison of model predictions and observations. This was evident during the CMDP Demonstration I in the coastal atmosphere. Because of the focus upon products, an unusual pattern was noted in the Eta model derived humidity over the southern Bay. Closer inspection revealed that a grid point representing a primarily marine region was functioning as a land point. This was corrected.

We are also gaining a better understanding of coastal phenomena through scrutiny of nowcast and forecast fields with available data. Wind-reversals over the Bay are not observed in over-land winds, nor in the offshore wind observations. These winds are tied to the geometry of the Bay despite the relatively low nearby topography. Only over-water observations or a high resolution numerical model such as the CBRAMS are able to produce estimates of these phenomena. Such circulation features may be an important forcing mechanism for circulation within the Bay. This suggests that simulations of Bay water circulation may require fine-resolution atmospheric simulation models.

The first demonstration was successful. Products were delivered on a regular basis with minor system failures. Extensive feedback was received from both NOAA and external user-partners. While improvements are needed and anticipated, existing products are of sufficient accuracy. There were requests for additional products as well. Feedback from the first demonstration is being incorporated into improvements to the models and delivery system.

We anticipate that systems such as the CMDP system will be developed for the coastal U.S. in the near future. These systems will provide enhanced maritime safety and efficiency as well as improved circulation estimates for coastal atmospheric and oceanographic environments. These improved estimates will clearly be useful for scientific purposes. They may also be useful for improving management strategies where circulation impacts the chemistry, biology, and productivity of the marine environment.

Acknowledgements

This work was funded by a grant from the National Oceanographic Partnership Program to the University of Maryland Center for Environmental Sciences, Litton-TASC Inc., Weather Services International Corp., and NOAA.

References

Aikman, F, J.G.W. Kelley, J. McQueen, T.F. Gross, K. Fuell and G. Szilagyi, 1999: Atmospheric and Oceanographic Analyses and Forecasts for the Chesapeake Bay Region During the Coastal Marine Demonstration. *Third Conference on Coastal Atmospheric and Oceanic Prediction and Processes,* New Orleans, LA, Amer. Meteor. Soc., 180-183.

Black, T., 1994: The New NMC Mesoscale Eta Model: Description and Forecast Examples. Weather and Forecasting, **9,** 265-278.

Blumberg, A.F. and G.L. Mellor, 1987: A description of a three-dimensional coastal ocean circulation model. Three-Dimensional Coastal Ocean Models, 4, ed. By N. Heaps, American Geophysical Union, 208 pp.

Bosley, K.T. and K.W. Hess, Development of an Experimental Nowcast/Forecast System for Chesapeake Bay Water Levels. Proceedings, *Fifth International Conference on Estuarine and Coastal Modeling,* Alexandira, VA, October 22-24, 1997.

Bosley, K.T., T. Bethem and M.J. Evans, January 1999: Migration of the Chesapeake Area Forecasting Experiment (CAFÉ) to an Operational Nowcast/Forecast System: the Chesapeake Bay Operational Forecast System (CBOFS), NOAA/NOS/Oceanographic Products and Services Division Report.

Breaker, L.C., L.D. Burroughs, Y.Y. Chao, D.M. Feit and D.B. Rao, 1999a: NCEP Participation in the Coastal Marine Demonstration Project. Preprints, *Third Conference on Coastal Atmospheric and Oceanic Prediction and Processes,* New Orleans, LA, Amer. Meteor. Soc., 195-200.

Breaker, L.C., B. Balasubramaniyan, A. Brown, L.D. Burroughs, Y.Y. Chao, R. Kelly, H.J. Thiebaux, P. Vukits and K. Waters, 1999b: Results from Phase 1 of the Coastal Marine Demonstration Project: The Coastal Ocean. NOAA Office Note 178, 22 pp.

Fuell, K.K, J.G.W. Kelly and J. McQueen, 1999: The Development of High Resolution Analyzed Wind Fields for the Chesapeake Bay Region. Preprints, *Third Conference on Coastal Atmospheric and Oceanic*

Prediction and Processes, New Orleans, LA, Amer. Meteor. Soc., 184-188.

Gross, T.F., F. Aikman, J. McQueen, K.K. Fuell, K. Hess, J.G.W. Kelley, 2000: Water Level Model Response to Wind Forcing Over the Chesapeake Bay During the Coastal Marine Demonstration Project. *Proceedings, Sixth International Conference on Estuarine and Coastal Modeling,* New Orleans, LA, American Society of Civil Engineers.

Gunther, H., S. Hasselmann and P. Janssen, 1991: The WAM Model Cycle 4. Tech. Rept. 4, Deutsches Klimarechenzentrum, Hamburg, 91 pp.

McQueen, J.T., G.D. Rolph, F. Aikman, J.G.W. Kelley, T. Gross, G. Szilagyi, K.K. Fuell, C. Tremback and J. Titlow, 1999: Development and Evaluation of a Non-hydrostatic Atmospheric Prediction System for the Chesapeake Bay Region. Preprints, *Third Conference on Coastal Atmospheric and Oceanic Prediction and Processes,* New Orleans, LA, Amer. Meteor. Soc., 189-194.

Richardson, S. and Szilagyi, G.J., 1999: User Evaluation of Coastal Marine Demonstration No. 1, Litton-TASC Technical Report, 32 pp.

Stoelinga, M. and T. Warner, 1998: Non-hydrostatic mesoeta-scale model simulations of cloud ceiling and visibility for an east coast winter precipitation event, J. App. Meteor., **38,** 385-404.

A Hydrodynamic Model Calibration Study of the Savannah River Estuary with an Examination of Factors Affecting Salinity Intrusion

Daniel L. Mendelsohn[1], Steven Peene[2], Eduardo Yassuda[3], Steven Davie[2]

ABSTRACT

As part of a channel deepening feasibility study, a comprehensive field program and modeling study of the Lower Savannah River Estuary was conducted. Potential impacts from the deepening were investigated using a 3-D, coupled, prognostic, general curvilinear coordinate, boundary-fitted, hydrodynamic and water quality model system (WQMAP). Salinity increases in the brackish and fresh water, tidal marshes in the upper portions of the estuary due to an upstream shift of the freshwater/saltwater interface is a concern. The present discussion focuses on the factors affecting salinity intrusion in the estuary and model predictions of that intrusion.

A preliminary calibration showed that model performed well in the prediction of tidal amplitude and currents but not as well for salinity and intrusion using a traditional turbulence energy model. The range of predicted diffusivities were not enough to reproduce the development and collapse of stratification observed in the data, nor the large variation in the extent of the salinity intrusion. The relationship observed between mixing and tidal range were used to develop a direct Log Fit relationship between the range and the vertical eddy diffusivity.

Comparisons between model predictions and observations indicate that the model is capturing most of the important physical processes in the estuarine system. It is capable of reproducing complex, transient physical phenomena in this extremely

[1] Applied Science Associates, Inc., 70 Dean Knauss Drive, Narragansett, RI 02882
  Email:mendo@appsci.com
[2] Applied Technology and Management, 350 Cumberland Circle, Suite 2070, Atlanta, GA 30339
  Email:speene@atm-s2li.com
[3] ASATM Brasil, Sao Paulo, Brazil
  Email: eyassuda@asatm.com.br

dynamic system and is in very good agreement with observed values of surface elevation, currents and salinity.

BACKGROUND
The Savannah River, with its head near Hartwell, Georgia at the confluence of the Tugaloo and Seneca Rivers, flows along the South Carolina/Georgia border for approximately 320 km before discharging to the Atlantic Ocean. Depths along the river vary from 16m in the shipping channel to 3-4m in the Back River and upstream of the I-95 Bridge (Figure 1). The river receives freshwater inflow of approximately 400 m^3/s (mean) from its drainage basin, which covers a total area of 27,500 square km within southwestern North Carolina, western South Carolina and Georgia. The river discharges to the Atlantic Ocean at the Georgia/South Carolina border at Savannah.

Savannah Harbor constitutes the lower 34.3 km of the Savannah River from its mouth at Fort Pulaski up to Port Wentworth (Figure 1). This reach of the river, along with approximately 18 km offshore of Fort Pulaski has been dredged in support of navigation since the early 1900s. Since that time the design depth of the channel has increased from 8 m MLW to 12 m MLW, (ATM, 1998a) with varying depths along specific reaches.

Presently the Georgia Ports Authority is examining the feasibility of deepening the harbor to 15 m MLW. Under this proposal the potential impacts of this action upon water quality conditions within the harbor need to be evaluated. The following two specific areas of impact need to be addressed through the application of predictive models:

- Potential salinity increases in the upstream portions of the Savannah River due to a shifting of the freshwater/saltwater interface upstream.
- Potential reduction in the dissolved oxygen levels within the system.

Anthropogenic modifications of the Lower Savannah River Estuary, including the construction of a tide gate and the New Cut channel, between the Middle and Back Rivers, to control sedimentation in the harbor area and channel deepening have altered the temporal and spatial distribution of salinity. Pearlstine (1990) identified this shift in the salinity distribution as one of the more adverse environmental impacts to the various ecosystems within the estuary. This has lead to both the loss of freshwater wetlands within the Savannah National Wildlife Refuge and the reduction of the local striped bass population.

SAVANNAH RIVER DATA ANALYSIS
In 1987, under a *Feasibility Study* (ATM, 1998a), for a previous deepening, the USGS installed continuous monitoring instruments at various locations

throughout the system (Figure 1). These stations monitored specific conductance and water level on 15-minute intervals. The specific conductance, (and therefore the salinity upon conversion) was measured near the bottom for all stations. During this same time period, flow rates were measured at the Clyo gauging station, (not shown) located 100 km upstream. These stations have been maintained from 1987 to the present.

Statistical and multivariate correlation analyses of these historical data (ATM, 1998a) demonstrated that longitudinal intrusion of salinity into the Savannah River is a function of four primary factors:

- Freshwater inflow rate at the headwaters
- Tide range at the mouth
- Offshore mean water level
- Physical geometry of the system (including structures and marsh systems)

The historical data were analyzed in order to understand the behavior and to quantify the relative influence of the various factors affecting salinity intrusion. Figure 2a shows the salinity recorded at Port Wentworth versus the volume flow as recorded at Clyo for 1990, 1992, 1995 and 1996. Figure 2b shows the salinity at Port Wentworth versus the tidal elevation recorded at the NOAA station at Fort Pulaski at the mouth of the River. During the time period covered in the plots, three distinct conditions existed in the estuary:

1) Before March 1991 the tide gate was in operation and New Cut was open.
2) After May 1992, New Cut was closed, the tide gate was out of operation.
3) After March 1994, the 1.2 m deepening had been completed, New Cut was closed, and the tide gate was out of operation.

The influence of the differing conditions can be seen quite clearly in both the salinity versus flow and the salinity versus tidal elevation data plots. In particular, the decommissioning of the tide gate between 1990 and 1992 is clearly visible in both data sets. After the tide gate was taken out of operation, (opened permanently), the mean salinity at Port Wentworth drops considerably.

Referring to Figure 2a it can be seen that through all four years of flow and salinity data there is a clear inverse correlation between the river flow rate and salinity. The difference between the before and after channel deepening data sets (1992 and 1995) is not quite as dramatic but yielded an increase in the mean of about 0.5 ppt.

In the 1990 data set for salinity versus tidal elevation, (Figure 2b), there is a clear positive correlation between the spring/neap range variation and salinity. As the tidal range increases the salinity intrusion into the system also increases. After the decommissioning of the tide gate in 1991, however, the relationship is inverted, so that the maximum observed salinities now occur during the lower neap tide range. This can be seen in the 1992, 1995 and 1996 data sets. Similar analyses performed on data sets from other continuous monitoring stations showed the same results.

The following conclusions were reached from this data analysis:

- A clear impact of the tide gate decommissioning is seen in all of the data sets, with significant alterations in the mean and maximum salinities, along the Little Back River Stations. Alteration in the means and maximums as well as the character of the salinity intrusion on the Front River are also seen.
- Salinity intrusion along the Front River is greater during the neap tide than spring tide range after the tide gate was permanently opened.
- Some impact of the deepening is evident on the Front River with changes in daily mean salinities on the order of 0.5 ppt.

Detailed descriptions of the methodology and the results are presented in the Engineering Appendices of the *Tier I Environmental Impact Statement for the Savannah River Deepening Project,* (ATM, 1998a) in the section entitled *Analysis of the Historic Data for the Lower Savannah River Estuary.*

To better understand the complex dynamics of the system and for the purpose of model calibration, an intensive field-monitoring program to quantify the salinity and water quality conditions within the Lower Savannah River Estuary was conducted. During the summer months of 1997, 14 continuous monitoring stations, with permanently mounted instruments that recorded surface elevation, salinity, and dissolved oxygen (DO) at 15-minute intervals, were deployed. Figure 3 shows the locations of the stations throughout the system. The stations within the navigation channel generally recorded salinity and DO concentrations approximately 1 meter from the surface and 1 meter off the bottom, while those outside of the channel recorded near-bottom concentrations. The instruments were placed near the bottom (1 m) in the reaches outside of the channel, in order to provide the worst-case condition of salinity intrusion and DO concentration.

In order to quantify the vertical structure of the tidal and density-driven flow, two bottom-mounted Acoustic Doppler Current Profilers (ADCP) at stations GPA-04 and GPA-08 and two electromagnetic current meters (S-4) at GPA-10 and GPA-15 were deployed. The ADCPs recorded continuous currents at 1-meter intervals over the water column. The S-4 collected mid-depth current velocities, at 15-minute interval, during at least 30 days. In addition, boat-mounted ADCP transects measured cross-section discharges across the width and depth of the river at critical locations including the confluence of the Front and Back Rivers at Fort Jackson and the upstream split of the Front and Back Rivers south of the I-95 bridge (see Figure 1). These data were used to quantify volume flux at these areas throughout the tidal cycle.

A detailed description of the methodologies utilized, the location of stations, the data collected, and findings from the intensive field monitoring program are presented in the Engineering Appendices of the *Tier I EIS* (ATM, 1998b), *Hydrodynamic and Water Quality Monitoring of the Lower Savannah River Estuary, July-September 1997.*

One of the major findings of the historic data analysis and the 1997 monitoring program was to confirm the importance of the stratification/de-stratification process and the density-driven currents in salinity intrusion. The analysis of the historic data showed that following the decommissioning of the Tide Gate, the maximum salinity intrusion along the Front River, occurred primarily during neap-tide conditions, as opposed to maximums during spring tide when the Tide Gate was in operation. The primary cause of this change appears to be a reduction in the velocities along the Front River above Fort Jackson, during ebb tide following the decommissioning. Since the turbulence energy, which causes mixing, is directly proportional to the tidal velocities, a reduction in the flows causes an associated reduction in turbulent mixing. Therefore, during neap-tide conditions, there is not enough energy to break down the vertical salinity gradient and water masses near the bottom with higher salinity are able to intrude further into the system.

This system of stratification and destratification was again observed in the data from the monitoring program during the summer of 1997. As with the historical data, increased stratification, and salinity intrusion in the Front River, are directly correlated with neap tide conditions. Destruction of stratification and decreased intrusion were related to the spring tide condition. Figure 4 shows the clear development and collapse of stratification at station GPA-04 near Fort Jackson just east of the confluence (or split) between the Front and Back Rivers. This is a critical juncture, and is indicative of, and in conjunction with the river flow rate controls the salinity conditions farther upstream. Also plotted in Figure 4 is the tidal elevation at station GPA-01 just offshore of the Savannah River Entrance. It is clear that the state of stratification in the system is directly related to the tidal range, via the increasing and decreasing current velocity and therefore, turbulent mixing energy.

It is also clear from the observations that the relationship between the tidal range, turbulence and mixing, places the system in a delicate balance between a mixed and a stratified system. An evaluation of the estuary number can be made from the following equation:

$$E_D = \frac{1}{\pi} \frac{u_t^3}{[\Delta\rho/\rho]\, g\, h_0\, u_r} \qquad [1]$$

where

E_D = estuary number
u_t = amplitude of profile-averaged tidal velocity at estuary mouth (m/s)
u_r = river velocity, (flow rate over cross sectional area, m/s)
h_0 = depth at mouth of estuary (m)
$\Delta\rho$ = difference in density between river and sea water (kg/m^3)
ρ = density of fresh water (kg/m^3)
g = gravity (m/s^2)

and from Harleman and Ippen (1967), the ranges of estuary classifications can be determined as:

well mixed conditions, $E_D > 8$,
partly mixed conditions, $8 > E_D > 0.2$,
stratified conditions, $0.2 > E_D$

For the mean Savannah River flow of 328 m^3/s the river velocity is in the range of 0.05 to 0.1 m/s depending on the chosen cross-sectional area of the river. Using a mean, profile-averaged tidal velocity amplitude of 1 m/s derived from the ADCP stationed at GPA-04, fresh water and salt water densities of 996.8 and 1024 (kg/m^3) respectively, and an estuary depth of 14 m the estuary number is in the range of 1.8 to 0.8. In terms of the estuary number, which is the ratio of the energy input by the tidal current to the energy needed for mixing, (Abraham 1988; Thatcher and Harleman 1981; Harleman and Ippen, 1967), the Savannah River estuary can be classified at the low end of partly mixed conditions.

Abraham (1988) suggests that for sufficiently stratified estuaries, turbulence models used in salinity intrusion problems must be capable of reproducing the primarily internal mixing, which is developed from within the region where the mixing occurs, (e.g. the interface), around the slack tides and the primarily external mixing, where energy for mixing is supplied by the solid boundaries, (e.g. the bottom), external to the region where the mixing occurs, when the tidal currents are large. The complexity and importance of this range of mixing regimes is compounded by the large variation, (up to 50%), in the tidal range between the spring and neap tidal cycle.

HYDRODYNAMIC MODEL DESCRIPTION
To model this complex river and estuarine system a 3-dimensional, boundary-fitted coordinate, hydrodynamic, and transport model system was chosen. This system approach uses transformation functions such that all domain boundaries

are coincident with coordinate lines. The transformation equations are applied to a user-defined grid of arbitrarily sized quadrilaterals, mapped to the coastal geometry of the water body in the study area (Figure 3).

The hydrodynamic model (Muin and Spaulding, 1997; Huang and Spaulding, 1995; Swanson et al., 1989) and the mass transport model (Mendelsohn and Swanson, 1991) equations are written and solved on the boundary conforming, transformed grid using the well known finite difference solution technique (Spaulding, 1984; Thompson et al., 1977). An algebraic transformation is used in the vertical to map the free surface and bottom onto coordinate surfaces. The resulting equations are solved using an efficient semi-implicit finite difference algorithm for the exterior mode (2-dimensional vertically averaged) and by an explicit finite difference algorithm for the vertical structure of the interior mode (3-dimensional) (Muin and Spaulding, 1997; Swanson, 1986).

A detailed description of the model with associated test cases can be found in Muin and Spaulding, 1997. The boundary-fitted hydrodynamic and transport models are contained within the Water Quality Mapping and Analysis Program model system, WQMAP, (Mendelsohn et al, 1997).

The boundary conditions are as follows:

- At land, the normal component of velocity is zero
- At open boundaries, the free surface elevation must be specified and salinity specified on inflow. On outflow, salinity is advected out of the model domain
- A bottom stress or a no-slip condition can be applied at the bottom.
- No water or salt is assumed to transfer to or from the bottom
- A wind stress is applied at the surface

There are a number of options for specification of vertical eddy viscosity, A_v, (for momentum) and vertical eddy diffusivity, D_v, (for constituent mass [salinity]). The simplest formulation is that both are constant, A_{vo} and D_{vo}, throughout the water column. More complex formulations include a mixing length model or a full 1 or 2-equation turbulence-closure model, adding the dependence on mixing length and turbulent energy.

APPLICATION TO THE SAVANNAH RIVER

Figure 3 presents the computational grid used in the model simulations. The grid extends from the open boundary offshore of Tybee Island, with a typical longstream/crossstream resolution of 1000x300m, to the USGS gauging station at Clyo approximately 100 km upstream, where the typical resolution is 100x50m. It includes the Front River, South Channel, Back River, Middle River, and the

Little Back River. The extensive marsh areas of the system are represented by storage cells that are attached to secondary tributaries and feeder creeks. The model boundaries were determined from the NOAA digitized shoreline, and incorporated into the WQMAP basemap.

The navigation channel depths utilized for the model calibration were taken from the USACOE 1997 Annual Survey. Depths outside of the maintained channel were taken from an intensive survey conducted by the USACOE in the following reaches: Front River from the Houlihan Bridge to the I-95 Bridge, Middle River, Little Back River and the South Channel.

The NOAA, National Ocean Survey maintains a series of tide gauges along the coast including a station at Fort Pulaski. A time series of surface elevation recorded at this station were used to simulate the offshore water surface elevation variations.

The primary source of freshwater to the Lower Savannah River Estuary is from the Savannah River watershed. The upstream model boundary was placed at Clyo where the USGS maintains a river stage height monitor. Detailed records, both past and present, are available for river flow at this site. Hourly volume flow records were obtained from the USGS for the summer 1997 period and used as input to the model.

CALIBRATION TO THE 1997 FIELD DATA

The primary objective of the model calibration was to be able to capture the dynamics of the system by satisfactorily reproducing the summer of 1997 data set including tidal elevation, currents and salinity. During this process, the grid configuration, time step, and model coefficients (e.g., bottom friction) were adjusted until the computed solution produced the best match to the observed data. The criteria for the desired level of accuracy are variable dependent, including (1) graphical comparison, (2) root-mean-square error (rms), (3) statistical comparison (mean, standard deviation), (4) coefficient of variation, and (5) linear regression.

Field data clearly show that the currents in Lower Savannah River Estuary are dominated by the tides, (barotropic circulation), where the semi-diurnal are the strongest (85 percent of the tidal energy) components. The maximum tidal range is about 3 meters in the entrance and is amplified 25% up to the area near station GPA09, (see Figure 2.2). At GPA10 the tidal wave begins to be damped, most likely due to marsh inundation and super elevation of the mean water level (upstream riverbed elevation increases).

The model was run using 11 layers in the vertical with the 1-equation, turbulence model and mixing length parameterization to obtain the vertical eddy viscosity. The barotropic time step was 10 minutes, although the model was still stable and accurate for 30-minute time step. The advective time step for salinity transport was 0.2 minutes due to the large currents, (up to 2 m/s) in the main channel of the river. For the initial simulations, the only physical parameter adjusted was the bottom friction. The parameter which controls the mixing length has been shown by Muin (1993) to be 0.3. After several simulations it was found that the quadratic law bottom friction coefficient, C_b equal to 0.0015 produced the best fit to current measurements at GPA-04 and GPA08. The surface elevation predictions were relatively insensitive to mixing and bottom friction.

In order to assess the accuracy of the model predicted, surface elevation and currents, the following comparative analysis were performed:

1. Time series comparison between the measured and simulated of water surface elevation and currents.
2. Comparison of the power spectra of the measured and simulated tidal elevation and currents.
3. Comparison of the eight major tidal constituent components in the region (Z0, M2, N2, S2, O1, K1, M4, M6) for the simulated and observed tidal wave and currents.

The use of multiple analysis to quantify the errors between the simulated and measured tides and currents isolates individual weaknesses in the model simulations allowing further tuning of model parameters. The graphical comparisons identify the overall quality of the simulations. The spectral comparisons isolate the ability of the model to simulate the distribution of energy in the various major frequencies; the sub-tidal, and the diurnal, semi-diurnal and higher order harmonics generated within the estuary. The harmonic analysis evaluates the model's ability to replicate the primary astronomical forcing constituents and the amplitude changes and phase shifts in those components due to the passage of the tidal wave through the system. The final calibrated model predictions compared very well with the observed trends and statistics, (surface elevation constituent errors ranged from 2-16% and velocity constituent errors ranged from 2-12% for example). The details of the model calibration to 1997 field data have been covered elsewhere and will not be presented here. The interested reader is referred to ATM, 1998a.

SALINITY CALIBRATION AND VERTICAL MIXING IN THE MODEL
Preliminary simulations, for the calibration, using various constant, vertical mixing coefficients proved incapable of capturing the dynamic range of salinity regimes seen in the data. Although predicted tidal velocity profiles and ranges

were representative of the observations using lower values of the vertical eddy viscosity, (for momentum), constant values for the vertical eddy diffusivity, (for salt) produced either too much stratification for lower coefficients, or a too much mixing for larger values. Too much mixing proved to restrict salinity intrusion, keeping the estuary far fresher than is observed. Test simulations using lower diffusivities predicted salinities in the observed range but without the variability.

Simulations using the turbulence kinetic energy model, (Muin and Spaulding, 1997), to predict the eddy viscosity and diffusivity were also run. As expected, the turbulent energy in the system varied dramatically with both the major semi-diurnal tidal oscillations as well as the spring/neap range variation. An example of the model predicted vertical eddy diffusivity at station GPA-04 over time is plotted in Figure 5. In the figure, the calculated eddy diffusivities have been vertically averaged for clarity. The turbulence model and the predicted vertical eddy viscosities performed extremely well in the prediction of tidal currents. The application and coefficients were calibrated to the ADCP measurements at both stations GPA-04 and GPA-08 where a number of levels were compared, and to the point current meters at stations GPA-10 and GPA-15.

The predicted vertical eddy diffusivities, salinity and salinity intrusion however did not fare so well. The predicted diffusivities were not able to reproduce the development and collapse of stratification observed in the data. Nor was the model capable therefore, of reproducing the large variation in the extent of the salinity intrusion up the Front River. A number of formulations were tested along with a large range of coefficients. The model either produced too much mixing during the critical slack tide periods or not enough mixing during the maximum flood and ebb. This may be due in part to the formulation of the turbulence model and the relative importance of the internally and externally derived mixing during low flow and slack tides. Abraham (1988) suggests that for certain types of stratified flows turbulence models using standard damping functions such as those employed in the present model, may not be capable of reproducing the low water slack conditions. For a detailed analysis of the behavior of turbulence and mixing in stratified tidal flows the interested reader is referred to Abraham (1988).

With the inability of the present formulation of the turbulence energy model to adequately determine the vertical eddy diffusivity and subsequent salinity regimes in the estuary an alternate method was sought. As the primary driving mechanisms for the development and collapse of salinity stratification and intrusion in the estuary and its relationship with the tidal regime is well understood in the present context of the Savannah River it seemed possible that a simple, quantifiable relationship between the two could be developed.

From analysis of the data it is apparent that the behavior of salinity upstream in the estuary is dependent on the conditions observed at GPA-04, (Figure 4). GPA-04 was therefore selected as the indicator of stratification, (or mixing) in the lower estuary. The relationship between the range in turbulence energy, (and subsequent mixing) and the range in current amplitude and therefore the tidal range, were used to develop a direct relationship between the range and the vertical eddy diffusivity. The relationship between the tide range and eddy diffusivity was assumed to be non-linear based on the earlier mixing studies as shown in Figure 5.

NEW VERTICAL MIXING MODEL DEVELOPMENT

A preliminary analysis using the surface and bottom salinities and current profile data at both GPA-04 and GPA-08 was performed to determine the form of the mixing relationship and the range of expected mixing values. Many semi-empirical relationships for the eddy coefficients have been proposed in terms of the gradient Richardson number, (Ri) which relates the local gravity force to the inertial force, where,

$$A_v = A_0 \, f(Ri) \; , \qquad \text{vertical eddy viscosity} \qquad\qquad [2]$$
$$D_v = D_0 \, g(Ri) \; , \qquad \text{vertical eddy diffusivity}$$

and as suggested by Munk and Anderson, (1948) :

$$f(Ri) = (1 + \alpha \, Ri)^{-n} \qquad\qquad [3]$$
$$g(Ri) = (1 + \beta \, Ri)^{-m}$$

and the values of A_0 and D_0 are potentially determined from mixing length theory, (Blumberg, 1986) or arbitrarily calibrated to available data. The gradient Richardson number is given by:

$$Ri = -\frac{2g}{\rho h} \frac{\dfrac{\partial \rho}{\partial \sigma}}{\left(\dfrac{\partial u}{\partial \sigma}\right)^2 + \left(\dfrac{\partial v}{\partial \sigma}\right)^2} \qquad\qquad [4]$$

where
$$\sigma \qquad = \text{vertical coordinate}$$
$$u,v \qquad = \text{velocity vector components}$$

A number of authors have suggested values for the coefficients α, β, n and m, in Equation 3, the most well known being those of the original authors and those of Officer (1976) as given below in Table 1:

Table 1. Values cited for α, β, n and m in the literature, (Blumberg, 1986).

Reference	α	n	β	m
Munk-Anderson (1948)	10	1/2	3.33	3/2
Officer(1976)	1	1	1	2
Bowden-Hamilton (1975)	1	7/4	7	1/4

Figure 6 shows the observed gradient Richardson number, calculated using Equation 4, as a function of time for (a) GPA-04 and (b) GPA08, where ADCP, and near surface and near bottom salinity and temperature data were available. Figure 7 is the filtered observations plotted against the filtered tide range (m). Also shown in Figure 7 are two lines representing linear regressions of the Log (Richardson number) vs. tidal range. The vertical mixing determined from the observed gradient Richardson number and the functional relationships given in Equations 2 and 3 are shown in Figure 8 for GPA-04. Finally, a curve was fit to the D_v vs Tide Range data plotted in Figure 8 using a logarithmic relationship with the form given in Equation 5. This will be called the Log Fit model.

$$Log\ (D_v) = \varphi\ Rng + \delta \qquad\qquad [5]$$

where

Rng　　= running mean tidal range, (m)

φ, δ　　= curve fit coefficients, (1.5 and -6.6 respectively)

Figure 9 shows the data same data and Log Fit curve as plotted in Figure 8 except that the observed mixing values have been time averaged to better highlight the trend in the relationship between the mixing and tide range. It can now be seen that the Log Fit model curve closely follows the observed trend at station GPA-04.

The Log Fit vertical eddy diffusivity calculated using Equation 5, is plotted in Figure 8, labeled 'Curve Fit'. The diffusivity plotted as a function of time during the summer of 1997, is shown in Figure 10. Because the Log Fit is a function of the tidal range and not the elevation, the semi-diurnal response seen in Figure 5 is absent from the curve in Figure 10. This smoother, parameterized version of the vertical mixing has lost semi-diurnal variability but has retained a clear and positive link to energy input to the system via variations of tidal range on the spring-neap cycle.

A comparison of the model predicted salinities and the observed salinities for July - September of 1997, at GPA-04, are shown in Figure 11 and a statistical comparison, for the entire study area, is given in Table 2. In the table, the minimum, maximum, mean and standard deviation (STD) of salinity at stations GPA-04 through GPA-13 are given. Summary statistics of the difference, (model - observations), are also given in terms of the mean absolute error (MAE), the standard deviation of the difference and the root mean square error (RMSE). The average of the MAE over all stations is –0.31 (ppt), with the values indicating that the model predictions are slightly low in the lower part of the estuary and slightly high in the upper portion. The average RMSE of the differences over all the stations is 1.66 (ppt), with a high of 3.26 (ppt) at GPA-04 bottom, decreasing up-estuary to a minimum of 0.21 at GPA-13.

It can be seen from Figure 11 that the model predicted salinities clearly display the both the magnitude and the variation in the range of the values observed. This is true for both the surface and the bottom, which show distinctly different signals based on the mixing regime. The controlling stratification/destratification process is also extremely well reproduced.

DISCUSSION AND CONCLUSIONS
There are several parameters in the foregoing argument that must be calibrated to a particular model application. They are the vertical mixing coefficients in Equations 2, A_0 and D_0, and the constants and exponents used in Equations 3, α, β, n and m. Although some guidance is given in the literature, the values vary substantially, (e.g. Table 1) and the final values must be determined empirically. The pathway to development of the final relationship between D_v and the tide range for the Savannah River application was iterative, where initial values for each of the coefficients were posited, values calculated, the curve fit coefficients φ and δ in Equation 5 determined, and the model run using the new relationship. Predicted values of surface elevation, currents and salinity were compared to observations at each of the 15 GPA stations and statistically evaluated following the process, described in Section 3, for each of the parameter estimations. A minimization of the system wide errors, (differences between the model predicted and observed values), produced the final coefficients for the Log Fit model.

Reviewing the calibration results, both qualitatively (visual inspection of the plotted time series) and quantitatively (statistical evaluation), it is clear that the selected approach performs extremely well. The model is capable of reproducing the trends as well as the magnitudes of salinity and its variation over a long period covering highly variable conditions in terms of both river flow rate and tidal amplitude.

REFERENCES

Abraham, G., 1988. Turbulence and Mixing in Stratified tidal Flows. In: Physical Processes in Estuaries, edited by Dronkers and Leussen. Springer-Verlag.

ATM, 1998a. Tier I Environmental Impact Statement for the Savannah River Deepening Project: Analysis of the Historic Data for the Lower Savannah River Estuary. ATM 98-991.

ATM, 1998b. Tier I Environmental Impact Statement for the Savannah River Deepening Project: Hydrodynamic and Water Quality Modeling of the Lower Savannah River Estuary. ATM 98-991.

Blumberg, A. F. 1986. Turbulent Mixing Processes in Lakes, Reservoirs and Impoundments. In: Physics Based Modeling of Lakes, Reservoirs and Impoundments, ed. William Gray, published by ASCE.

Bowden, K.F. and P. Hamilton, 1975. Some Experiments with a numerical model of Circulation and Mixing in a Tidal Estuary. Estuarine and Coastal Marine Science, 3, pp. 281-301.

Harleman, D.R.F. and A.T. Ippen, 1967. Two Dimensional Aspects of Salinity Intrusion in Estuaries: Analysis of Salinity and Velocity Distributions. Committee on Tidal Hydraulics, Corps of Engineers. U.S. Army, Technical Bulletin no. 13.

Mendelsohn, D.L. Howlett and Swanson, 1996. WQMAP in a Windows Environment, in proceedings of 4th International Conference on Estuarine and Coastal Modeling, ASCE, San Diego, October 26-28, 1995.

Mendelsohn and Swanson, 1992. Application of a Boundary Fitted Coordinate Mass Transport Model. Presented at 2nd International Conference on Estuarine and Coastal Modeling, America Society of Civil Engineers, Tampa, Florida, 13-15 November 1991.

Huang and Spaulding, 1995. 3D Model of Estuarine Circulation and Water Quality Induced by Surface Discharges, published in: Journal of Hydraulic Engineering, Vol. 121, No. 4, pp 300-311, April, 1995.

Muin, M, and M.L. Spaulding, 1997. Application of Three-Dimensional Boundary-Fitted Circulation Model to Providence River, published in: Journal of Hydraulic Engineering, January 1997.

Munk, W.H. and E.R. Anderson, 1948. Notes on a Theory of the Thermocline, J. Marine Research, 1, pp 276-295.

Officer, C.B. 1976. Physical Oceanography of Estuaries. John Wiley & Sons. New York.

Pearlstine, L., P. Latham, W. Kitchens, and R. Bartleson, 1990. Development and application of a habitat succession model for the wetland complex of the Savannah National Wildlife Refuge. Volume II. Final Report to the U.S. Fish and Wildlife Service, Savannah Coastal Refuges. Florida Cooperative Fish and Wildlife Research Unit, Gainesville, Florida. 123 pp.

Spaulding, 1984. A Vertically Averaged Circulation Model Using Boundary-Fitted Coordinates, published in: Journal of Phys. Ocean, May, pp. 973-982.

Swanson et al., 1989. A Three Dimensional boundary-Fitted Coordinate Hydrodynamic model, Part I: Development and Testing, published in: Dt. Hydrog, Z.42, 1989, pp. 188-213.

Swanson, J.C., 1986. A Three Dimensional Numerical Model System of Coastal Circulation and Water Quality, Ph.D. thesis, Ocean Engineering, University of Rhode Island.

Thatcher, M.L. and D.R.F Harleman, 1981. Long Term Salinity Calculation in Delaware Estuary. J. Environ Eng Div ASCE, 107 (EE1): 11-27.

Thompson et al., 1977. Principles of Surface Water Quality Modeling and Control, Harper Collins Publishers, New York.

Table 2   Savannah River Model - Data Comparison for Salinity July - September, 1997

	GPA-04 Bottom	GPA-04 Surface	GPA-05 Bottom	GPA-06 Bottom	GPA-06 Surface	GPA-07 Bottom	GPA-08 Bottom	GPA-08 Surface	GPA-09 Bottom	GPA-10 Bottom	GPA-11 Bottom	GPA-12 Bottom	GPA-13 Bottom
OBSERVATIONS													
Min (ppt)	6.27	2.62	0.10	0.30	0.10	0.00	0.00	0.00	0.00	0.00	0.00	0.00	0.00
Max (ppt)	27.60	21.45	16.40	22.09	14.12	11.00	15.40	13.20	13.00	9.40	5.20	5.50	1.60
Mean (ppt)	18.33	9.64	5.54	10.31	4.10	1.10	3.25	1.67	1.63	1.08	0.13	0.32	0.12
Std (ppt)	3.95	3.78	3.46	5.15	2.20	1.76	3.71	2.37	2.61	1.46	0.44	0.56	0.22
MODEL													
Min (ppt)	2.63	1.02	0.21	0.20	0.04	0.09	0.01	0.01	0.00	0.03	0.00	0.00	0.00
Max (ppt)	28.03	24.57	10.70	19.12	9.33	3.99	12.86	5.67	9.10	5.57	3.25	3.16	1.59
Mean (ppt)	18.20	9.21	3.81	8.87	3.10	1.01	3.86	1.36	2.05	1.42	0.67	0.58	0.23
Std (ppt)	5.79	4.02	2.53	4.95	1.96	0.80	3.01	1.21	2.00	1.18	0.72	0.59	0.28
DIFFERENCE													
MAE (ppt)	-0.13	-0.43	-1.73	-1.44	-1.01	-0.11	0.35	-0.46	0.17	0.27	0.34	0.16	0.03
Std (ppt)	3.40	2.69	2.10	2.44	1.34	1.53	1.76	1.48	1.32	0.82	0.56	0.36	0.21
RMSE (ppt)	3.26	2.62	2.72	2.83	1.69	1.53	1.84	1.57	1.33	0.86	0.65	0.39	0.21

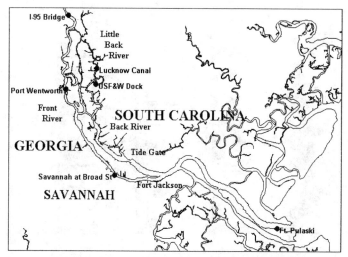

Figure 1. Map of the Savannah River Estuary. Historic water surface elevation and salinity monitoring stations are also shown.

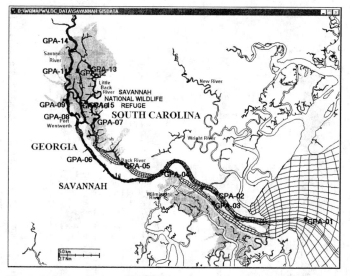

Figure 3. Lower Savannah River Estuary map showing the 1997 long term monitoring stations GPA-01 through GPA-14. The WQMAP boundary-fitted model grid as applied to the estuary is also shown.

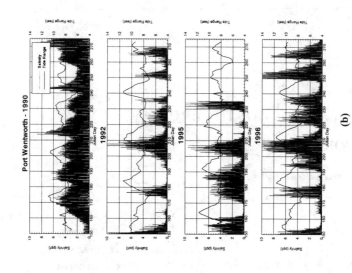

(b)

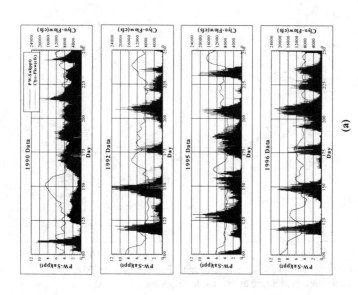

(a)

Figure 2. Historic observations at the Port Wentworth continuous monitoring station for (a) salinity and flow rate, and (b) salinity and tidal range.

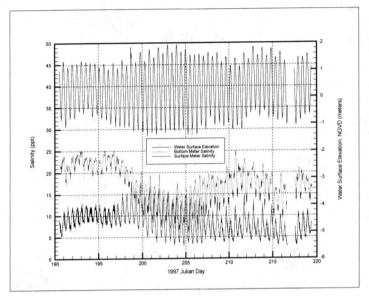

Figure 4. Observed tidal elevation and surface and bottom salinity for station GPA-04 at the Fort Jackson confluence of the Front and Back Rivers.

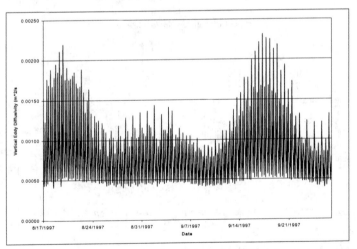

Figure 5. Example model predicted vertical eddy diffusivity at station GPA-04. The eddy diffusivities have been vertically averaged for clarity.

(a)

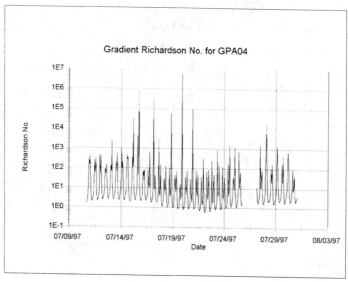

(b)

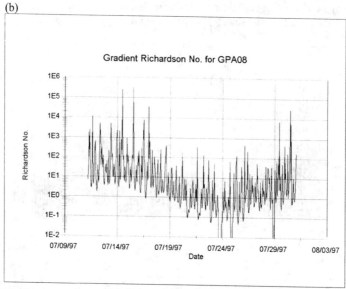

Figure 6. Observed gradient Richardson number calculated from the 1997 summer field data for (a) station GPA-04 and (b) station GPA-08.

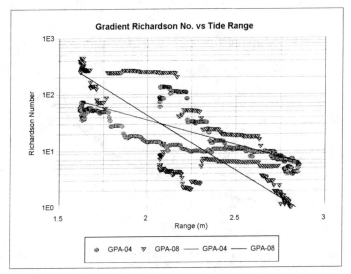

Figure 7. Gradient Richardson number versus tide range for stations GPA-04 and GPA-08.

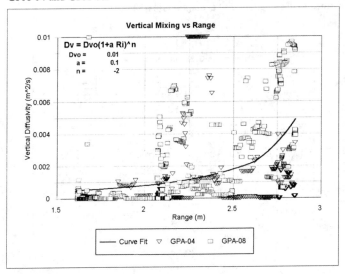

Figure 8. Observer vertical eddy diffusivity plotted as a function of tide range for stations GPA-04 and GPA-08 and compared to "Log Law" curve fit diffusivity.

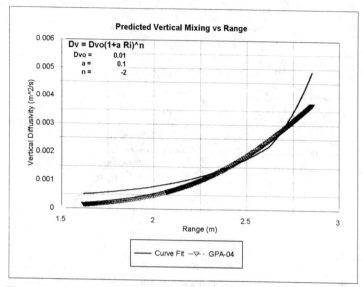

Figure 9. Same as Figure 8 but the observed eddy diffusivities have been smoothed to display the trend at station GPA-04.

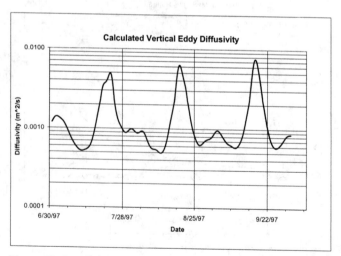

Figure 10. Log Law model predicted vertical eddy diffusivity as a function of time for the 1997 summer simulation period.

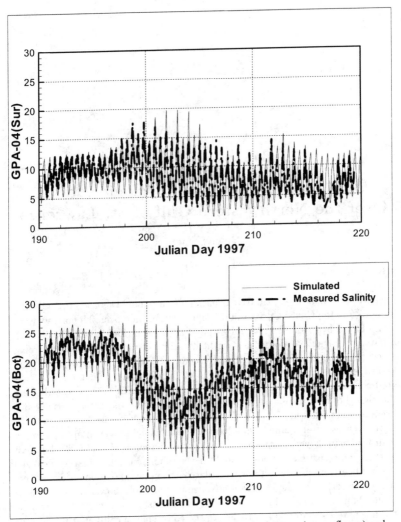

Figure 11. Predicted versus observed salinities for the bottom (upper figure) and surface (lower figure) at station GPA-04.

The Gaspé Current and Cyclonic Motion Over the Northwestern Gulf of St. Lawrence

Jinyu Sheng[1]

Abstract

A three dimensional primitive equation ocean model was used to study the dynamics of the Gaspé Current and cyclonic circulation over the northwestern Gulf of St. Lawrence. The model was first applied to the idealized basin with a flat bottom and a piecewise-straight coastline driven by buoyancy forcing associated with St. Lawrence River discharge. The model reproduces many known features in the early development of the estuary plume and the associated coastal current system. The coastal current, which flows seaward along the right-bounded coast on the northern Hemisphere, is stable initially but becomes unstable quickly with the occurrence of multiple backward-breaking waves as a result of momentum advection. The model was then applied to the realistic basin driven by the same buoyancy forcing. The early development of the plume-current system is qualitatively similar to that in the idealized basin. The combined effects of irregular bathymetry and momentum advection, however, make the intrusion advance considerable slower in the realistic basin than that in the idealized basin. The coastal current in the realistic basin separates from and reattaches to the coast sporadically as it propagates downstream.

Introduction

The Gaspé Current is a buoyancy-driven coastal jet originating in the St. Lawrence Estuary and flowing seaward along the coast of the Gaspé Peninsula

---

[1]Department of Oceanography, Dalhousie University, Halifax, Nova Scotia, Canada, B3H 4J1. email: jinyu.sheng@dal.ca

(Figure 1). Its existence has been confirmed by geostrophic calculation, direct current measurements, and infrared satellite images of sea surface temperature [e.g., El-Sabh, 1976; Tang, 1980; Benoit et al., 1985; Mertz et al., 1988]. Previous studies suggested [Koutitonsky and El-Sabh, 1985] that freshwater runoff from the St. Lawrence River system drives a buoyant estuarine plume in the St. Lawrence Estuary, with two separate coastal currents flowing eastward along the north and south shores, respectively. Combined with the cyclonic circulation of the northwestern Gulf of St. Lawrence, the north shore current in the estuary veers anticyclonically at the estuary mouth and flows southward. It joins the south shore current at the coast, forming the root of the Gaspé Current. The Gaspé Current reaches its highest intensity of about 1 m s^{-1} around the Gaspé Peninsula with a typical width of about 10-20 km in the top 50 m of the water column [Tang and Bennet, 1981]. It apparently leaves the coast at the eastern tip of the Gaspé Peninsula and carries buoyant estuarine waters onto the Magdalen Shallows.

The strength and position of the Gaspé Current varies significantly with

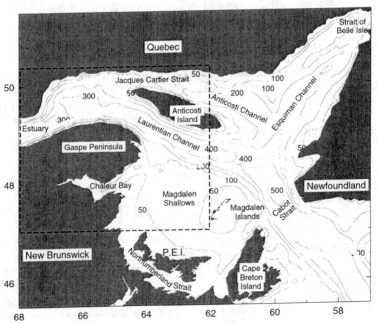

Figure 1. Main bathymetric features of the Gulf of St. Lawrence and a schematic showing the major surface currents. The area marked by the dashed lines is the model domain.

time and space. From satellite infrared images, Tang [1980] was the first to identify the occurrence of instabilities in the Gaspé Current under the summer, high-runoff conditions. The dynamics of the unstable wave development in the Gaspé Current are not yet fully understood. The main objectives of this paper are (1) to study the development of the Gaspé Current and the associated unstable wave features driven by buoyancy forcing associated with the freshwater runoff, and (2) to examine the combined effects of the river discharge and a barotropic westward jet along the Quebec shore on the formation of a large-scale cyclonic motion over the northwest Gulf of St. Lawrence, using a three dimensional eddy-resolving ocean model known as CANDIE [Sheng et al., 1998].

The arrangement of this paper is as follows. The next section summarizes the main features of oceanographic observations over the northwestern Gulf of St. Lawrence in the summer months. This is followed by a description of model configurations used in the numerical experiments. It then presents the model results in an idealized basin that consists of a flat bottom and a piecewise-straight coastline and in a realistic domain of the northwest Gulf of St. Lawrence. The final section is a summary and conclusions.

Review of Oceanographic Observations

Freshwater discharge from the St. Lawrence plays a dominant role in driving circulation over the northwestern Gulf of St. Lawrence (NWGSL). The annual mean discharge of the St. Lawrence is about 0.018 Sv (1 Sv = 10^6 m^3 s^{-1}), with significant seasonal variations. During the period from 1950 to 1984, for example, the monthly mean runoff of the St. Lawrence was about 0.015 Sv from September to March of the next year, but increased to more than 0.03 Sv in May and June [Koutitonsky and Bugden, 1991]. The peak runoff from the St. Lawrence in later spring is most likely to trigger the development of meanders and instabilities in the Gaspé Current during the summer season. Therefore in this paper we only discuss the summer mean climatology of temperature and salinity and use them to initialize the model.

Figure 2 shows the summer surface temperature and salinity over the northwestern Gulf of St. Lawrence (the area marked by the dashed lines in Figure 1). They were gridded using a successive-correction method known as Barnes' Algorithm [Daley, 1991] from the hydrographic observations made at depths of less then 11 m during the summer months (July-September). The results presented in Figure 2 are based on three iterations and an influence radius of about 15 km.

The surface water in summer warms up progressively from the southwestern shallow area to the northern part of the NWGSL, with the coldest water around the estuary mouth and northern part of Jacques Cartier Strait. Along the Gaspé Peninsula, however, the summer surface temperature is relatively uniform. By contrast, the summer surface salinity varies significantly around the estuary mouth and the coastal region of the Gaspé Peninsula, with lower salinity closer to the shore. The combination of strong cross-shore variations in the summer surface salinity and relatively weak variations in the summer surface temperature along the Gaspé Peninsula indicates the predominance of the salinity distribution in driving the buoyancy-driven coastal current, i.e. the Gaspé Current, in the summer season.

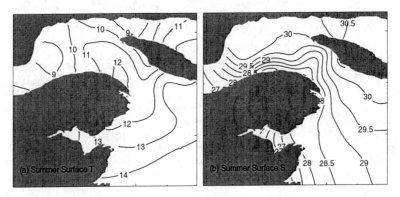

Figure 2. Summer mean surface (0-11m) fields of (a) temperature and (b) salinity gridded from hydrographic data using Barnes' algorithm [Daley, 1991].

Figure 3a presents vertical profiles of summer-mean temperature and salinity. They were obtained by horizontally averaging the three-dimensional gridded summer-mean temperature and salinity over the NWGSL. The water column over this region in summer can be described as a three-layer system: a warm and relatively fresh surface layer, a cold intermediate layer, and a warmer and saltier bottom layer. Figure 3b presents the vertical profile of density calculated from temperature and salinity profiles shown in Figure 3a. The density difference between the surface and at a depth of 250 m is about $5\sigma_t$ (i.e. 5×10^3 kg m^{-3}), which is quite large. By assuming the horizontal scale of motion to be much larger than the vertical scale and ignoring the nonlinear terms, we calculated the baroclinic modes using the method outlined by Kundu [1990]. Figure 3c shows the pressure eigenfunctions of the first three baroclinic modes. The phase speeds c_i of the first three modes are about 0.99, 0.55 and 0.37 m s^{-1}, respectively. The corresponding baroclinic Rossby radii of deformation $r_i = c_i/f$ are about 10, 5.5 and 3.7 km, respectively, where f is the Coriolis parameter.

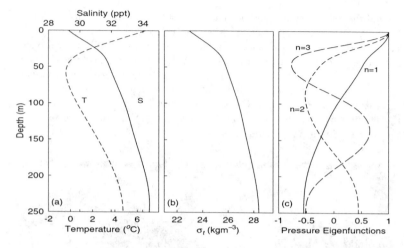

Figure 3. Vertical profiles of (a) temperature and salinity spatially averaged from the gridded hydrographic data over the northwestern Gulf of St. Lawrence, (b) in situ density calculated from the vertical temperature and salinity profiles shown in (a), and (c) pressure eigenfunctions of the first three baroclinic modes calculated based on the vertical density profile shown in (b).

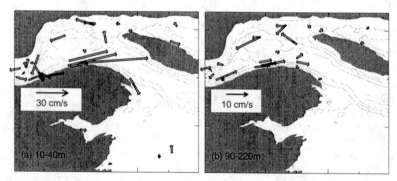

Figure 4. Time mean current-meter observations made during the summer months with record lengths longer than 15 days in the (a) upper-ocean (10-40 m) and (b) lower-ocean (90-220 m) over the northwestern Gulf of St. Lawrence.

Figure 4 shows the time mean current-meter observations made during the summer months with record lengths longer than 15 days in the upper-ocean (10 to 40 m) and in the lower-ocean (90 to 220 m), respectively. The tides were removed before calculating the mean [Gregory et al., 1989]. Horizontal averaging was performed where means were available within 10 km of each other. The summer mean upper-ocean current-meter observations exhibit significant spatial variabilities, particularly near the mouth of the estuary (Figure 4a). Nevertheless, they suggest a transverse flow with typical speed of about 10 cm s^{-1} around the estuary mouth and a strong coastal current (i.e. the Gaspé Current) with the maximum time-mean speed of about 50 cm s^{-1} along the Gaspé Peninsula in the upper-ocean during the summer season. They also suggest a large-scale cyclonic motion in the upper-ocean over the NWGSL region in summer months. The summer mean current-meter observations in the lower-ocean (Figure 4b) are much weaker than those in the upper-ocean, particularly over the estuary mouth and along the Gaspé Peninsula, indicating the surface-intensified characters of the Gaspé Current and the cyclonic NWGSL circulation.

Model Description

The governing equations used in the ocean circulation model are essentially the same as those considered by Sheng et al. [1998], with the exception that Cartesian coordinates are used here.

$$\frac{\partial u}{\partial t} + \mathcal{L}u - fv = -\frac{1}{\rho_o}\frac{\partial p}{\partial x} + \mathcal{D}_m u + \frac{\partial}{\partial z}\left(K_m \frac{\partial u}{\partial z}\right), \tag{1}$$

$$\frac{\partial v}{\partial t} + \mathcal{L}v + fu = -\frac{1}{\rho_o}\frac{\partial p}{\partial y} + \mathcal{D}_m v + \frac{\partial}{\partial z}\left(K_m \frac{\partial v}{\partial z}\right), \tag{2}$$

$$\frac{\partial p}{\partial z} = -\rho g, \tag{3}$$

$$\frac{\partial u}{\partial x} + \frac{\partial v}{\partial y} + \frac{\partial w}{\partial z} = 0, \tag{4}$$

$$\rho = \rho(T, S, p), \tag{5}$$

$$\frac{\partial T}{\partial t} + \mathcal{L}T = \mathcal{D}_h T + \frac{\partial}{\partial z}\left(K_h \frac{\partial T}{\partial z}\right), \tag{6}$$

$$\frac{\partial S}{\partial t} + \mathcal{L}S = \mathcal{D}_h S + \frac{\partial}{\partial z}\left(K_h \frac{\partial S}{\partial z}\right), \tag{7}$$

where u, v, w are east (x), north (y) and vertical (z) components of the velocity, p is pressure, ρ is density, g are the gravitational acceleration, ρ_o is a reference density, and K_m and K_h are the vertical eddy viscosity and diffusivity coefficients. Here $\mathcal{L}$ is an advection operator and $\mathcal{D}_m$ and $\mathcal{D}_h$ are diffusion operators defined as

$$\mathcal{L}q = u\frac{\partial q}{\partial x} + v\frac{\partial q}{\partial y} + w\frac{\partial q}{\partial z}, \qquad \mathcal{D}_{m,h}q = A_{m,h}\left(\frac{\partial^2 q}{\partial x^2} + \frac{\partial^2 q}{\partial y^2}\right) \qquad (8)$$

where A_m and A_h are the horizontal eddy viscosity and diffusivity coefficients, respectively. In this paper, surface wind and surface heat/salinity fluxes were set to zero, the vertical mixing coefficients K_m and K_h were set to 1 cm^2 s^{-1}, and horizontal eddy viscosity and diffusivity coefficients $A_{m,h}$ were determined using the Smagorinsky scheme [Smagorinsky, 1963].

Two types of model domains were used in this study. The first is an idealized basin with a flat bottom and a piecewise-straight coastline (Figure 5a). The dimension of the rectangular estuary is about 40 km wide and 70 km long. The water depth of this idealized basin is 200 m. The second model domain type is a realistic basin with a variable bottom and an irregular coastline, representing the model-resolved bathymetry over the NWGSL region. The water depth at the estuary head in both the idealized and realistic basins was set to 30 m for implementing a source of lower-salinity waters in the model.

The model resolution is about 3.0 km in the east-west direction and 2.3 km in the north-south direction with twenty z-levels in the vertical. The vertical grid spacing in the idealized basin is uniformly 10 m. In the realistic basin the vertical grid spacing is uniformly 10 m from surface to 140 m. Below 140 m the vertical cell boundaries are at depths of 152, 168, 189, 218, 257, and 310 m, respectively.

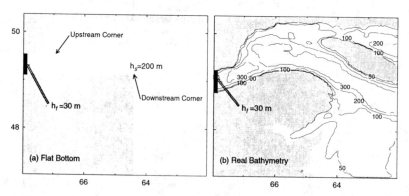

Figure 5. Two types of model domains used in the numerical experiments. (a) The idealized basin of a flat bottom and a piecewise-straight coastline. The water depth of this basin is 200 m except the estuary head region where the water depth is 30 m. (b) The realistic basin with variable bottom and irregular coastline representing the model resolved bathymetry over the area marked by the dashed lines in Figure 1.

At the model lateral closed boundaries, the normal flow, tangential stress of the currents and horizontal fluxes of temperature and salinity were set to zero. The areas with nonzero depths along the eastern and southern boundaries (Figure 5) are the model open boundaries. Along these boundaries the Sommerfeld radiation condition was applied for normal flow, temperature and salinity fields at each model level.

The model was initialized with the summer mean vertical profiles of temperature and salinity shown in Figure 3a with zero initial velocity everywhere except the estuary head. The time step is $\Delta t = 432$ s, or 200 steps per day. To simulate the summer mean St. Lawrence River runoff in the model, we specified the water mass at the estuary head to be 10 ppt fresher than the summer mean salinity at each z-level. We also prescribed a weak barotropic eastward inflow of 2.5 cm s^{-1} at the head to represent the summer mean volume transport of 0.03 Sv from the St. Lawrence River system.

Idealized Basin Experiments

CANDIE was first applied to the idealized basin forced by the St. Lawrence River discharge specified in terms of lower salinity waters and a weak barotropic eastward flow at the estuary head. Shortly after the model initialization, a zonal pressure gradient is created due to the density difference around the estuary head. This zonal pressure gradient drives an eastward surface current that advects the buoyant estuarine water downstream over the whole head region. Due to the Coriolis effect, this surface current veers southward and forms a strong southward jetlike stream along the offshore front of the estuarine plume (Figure 6a). At the north and south shores of the estuary, the Coriolis term in the zonal momentum equation vanishes, leading to a balance between the zonal pressure gradient and the inertial terms in the momentum equation over these two areas (since the dissipative terms are relatively small). Consequently, the buoyant estuarine waters along both shores spread gradually eastward. The southward jetlike stream along the offshore front of the plume turns cyclonically at the south shore, forming a narrow coastal current that flows eastward along the right-bounded coast (Figure 6a). The model-produced onset of the plume-current system in this idealized basin is very similar to that observed in the "dam-break" laboratory experiment [Stern et al., 1982] and produced by numerical models of buoyancy-driven circulation in a channel [Wang, 1985; Chao and Boicourt, 1986].

The coastal current driven by the river discharge is stable initially (Figure 6a). Due to strong horizontal and vertical shears, however, instabilities start to develop around day 10 (Figure 6b). Figure 7 presents the time sequence of surface salinity and flow fields produced by the fully-nonlinear CANDIE model. The plume-current system has become well developed after 20 days of simulation (Figure 7a), with a pool of accumulated estuarine waters and an intense semi-circular anticyclonic eddy at the downstream corner. The surface current inside the anticyclonic eddy abruptly turns cyclonically at the coast to south of the corner, acting as a transition zone connecting the eddy and the coastal current farther south [Chao and Boicourt, 1986].

After 40 days of simulation (Figure 7b), the offshore salinity front of the plume expands further downstream with a small-scale anticyclonic gyre generated inside the plume. The anticyclonic eddy at the downstream corner has grown into a very large size with its center about 50 km northeastward from the corner. The other interesting feature in Figure 7b is the development of several backward-breaking waves along the piecewise-straight coastline, with one well-defined wave about halfway between the estuary mouth and the downstream corner, and two other smaller waves to the south of the corner.

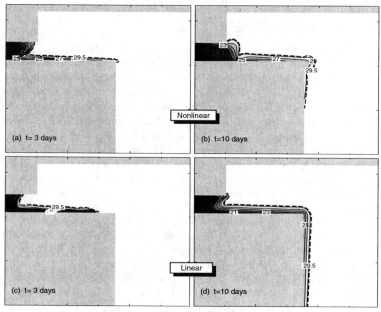

Figure 6. Grayscale images and isopleths of surface salinity fields in the idealized basin produced by the fully-nonlinear model at (a) day 3 and (b) day 10; and those produced by the linearized model (i.e. the momentum advection terms in the model are dropped) at (c) day 3 and (d) day 10. Both numerical experiments were forced by the same river discharge at the estuary head. Darker gray represents lower salinity. The salinity contour interval is 2 ppt. The dashed contour line represents the boundary between the buoyant estuarine and shelf waters.

The well-defined backward-breaking wave has broken upstream and dissipated by day 60, creating a pool of buoyant estuarine waters to the northeast of the upstream corner (Figure 7c). The backward-breaking wave to the south

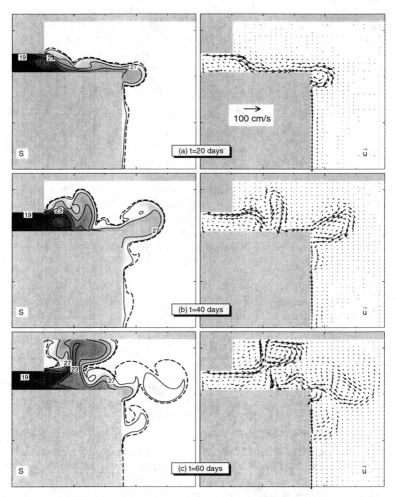

Figure 7. Surface salinity and flow fields in the idealized basin produced by the nonlinear model at (a) day 20, (b) day 40 and (c) day 60. The model was driven by the river discharge at the estuary head. The salinity contour interval is 2 ppt. The dashed salinity contour line represents the boundary between estuarine and shelf waters. Velocity vectors are shown at every fourth grid point.

of the downstream corner enlarges and will break away shortly from the plume-current system (see Figures 7c). Figure 7c also shows that the anticyclonic eddy at the downstream corner has been pinched off from the plume-current system by day 60, with a new anticyclonic eddy created at the downstream corner. This eddy-shedding phenomenon is consistent with the laboratory simulations by Klinger [1994]. He observed that an anticyclonic eddy is always formed when the current, which flows with the coast to its right if one is oriented in the downstream direction on the northern Hemisphere, encounters a sharp convex corner with an angle greater than about 45°. He also found that this eddy grows in size and propagates diagonally away from the corner. As this eddy moves away from the corner, a new anticyclonic eddy is created.

Note that the nonlinear advective terms in the horizontal momentum equations (i.e., $\mathcal{L}u$ and $\mathcal{L}v$ terms) play a predominant role in the development of meanders and instabilities in the plume-current system. Figures 6c and 6d present the surface salinity fields at days 3 and 10 produced by the CANDIE model with the momentum advection terms eliminated. We refer to this type of experiment as the linear case. The overall features of surface salinity and flow in the linear case are similar to those in the nonlinear case (i.e. the results produced by the fully-nonlinear CANDIE model), however, the coastal current in the linear case flows smoothly along the coast without meanders or backward-breaking waves.

Realistic Basin Experiments

The fully nonlinear CANDIE model was applied to the realistic basin forced by the same river discharge as in the idealized basin. All other model parameters are the same as before. Figure 8 shows surface salinity fields at days 3 and 10 in the realistic basin. After 3 days, the buoyant estuarine plume is confined inside the St. Lawrence Estuary with a semicircular salinity front near the mouth. A weak coastal current has just been created over the south shore of the Estuary, advecting the buoyant estuarine waters eastward along the Gaspé Peninsula. A westward intrusion occurs between the salinity front and the south shore near the mouth, very similar to that in the idealized basin.

A comparison of Figure 8a with Figure 6a indicates that the surface salinity and current fields in both idealized and realistic basins have qualitatively similar spatial structures at day 3: a buoyant plume inside the estuary, a semi-cyclonic front around the estuary mouth and a surface-intensified coastal current along the right-bounded coast. Note, however, that the intrusion speed during the first three days of simulation is much slower in the realistic basin (about 0.58 m s^{-1}) than that in the idealized basin (about 1.0 m s^{-1}). The coastal current in the realistic basin is also relatively weaker during this period.

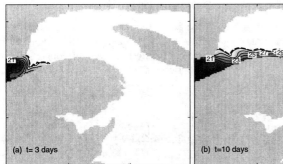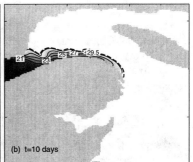

Figure 8. Grayscale images and isopleths of the surface salinity fields in the realistic basin produced by the fully-nonlinear CANDIE model at (a) day 3 and (b) day 10. Otherwise, same as Figure 6.

After 10 days the coastal current in the idealized basin is relatively stable with a small-scale anticyclonic eddy just created at the downstream corner (Figure 6b). By contrast, the coastal current in the realistic basin flows along the irregular coastline without any recognizable separation (Figure 8b). Note that large-scale meanders appear along the outer edge of the coastal current in the realistic basin, indicating that the plume-current system is more unstable in the realistic basin than in the idealistic basin.

Figure 9 shows the evolution of surface salinity and currents in the realistic basin. After 20 days, the meanders along the outer edge of the coastal current grow in size and form several large backward-breaking waves with a typical wavelength of about 50 km. By comparison, the coastal current in the idealized basin is relatively stable at this time with two small-scale backward-breaking waves created along the outer edge of the current (Figures 7a and 9a). The intense anticyclonic eddy appearing at the downstream corner of the idealized basin is not present in the realistic basin. Instead, the coastal current in the realistic basin tends to generate several smaller-scale anticyclonic eddies along the irregular coastline.

The buoyant plume in the realistic basin expands northeastward significantly after 40 days of simulation, with a cyclonic-anticyclonic dipole created to the northeast of the estuary mouth (Figure 9b). The coastal current separates from and reattaches to the irregular coastline sporadically as it propagates downstream. After passing the eastern tip of the Gaspé Peninsula, the current leaves the coast permanently and flows southward onto the Magdalen Shallows. By comparison, the buoyant plume in the idealized basin also expands further eastward but with a small-scale anticyclonic eddy inside the plume and a large-scale backward-breaking wave outside the plume.

At day 60, the buoyant estuarine water in the realistic basin spreads over most of the NWGSL region with an intense anticyclonic eddy to the northeast of the estuary mouth (Figure 9c). A small-scale but well-developed anticyclonic eddy is created over the area between the eastern tip of the Gaspé

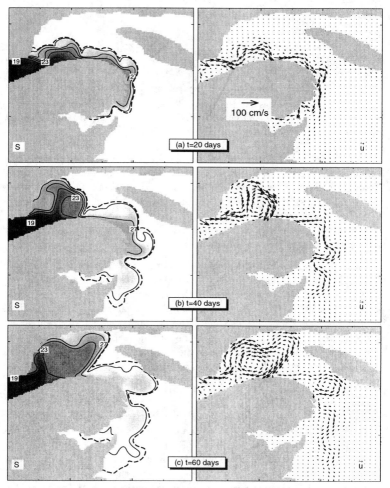

Figure 9. Surface salinity and flow fields in the realistic basin driven by the river discharge at the estuary head at (a) day 20, (b) day 40, and (c) day 60. Otherwise, same as Figure 8.

Peninsula and Anticosti Island. To the south of this eddy several less well developed eddies have been created with a southward drift onto the Magdalen Shallow. By contrast, in the idealized basin a pool of buoyant water tends to separate from the plume-current system.

The model results shown in Figure 9 reproduce many observed features of the Gaspé Current, including the development of backward breaking waves frequently observed in the satellite infrared images [Tang, 1980; Mertz et al., 1988]. As discussed above, however, the buoyant estuarine plume driven by the river discharge expands progressively eastward with time. As a result, a large-scale anticyclonic motion is eventually created over the NWGSL region. This anticyclonic motion produced by the model is, however, contrary to common knowledge of general circulation over this region. Previous hydrographic studies have consistently revealed the presence of a quasi-permanent geostrophic cyclonic motion in this region [Trites, 1972; El-Sabh, 1976; Tang, 1983]. The summer mean current meter observations in the upper ocean (Figure 4) also suggest the existence of this circulation feature. Clearly, other dynamic processes in addition to the St. Lawrence River discharge are involved in the formation of the cyclonic circulation over the NWGSL region. In this section, we consider the role of a barotropic westward jet along the Quebec shore in the formation of the cyclonic NWGSL motion. This is motivated by the findings of Petrie et al. [1988], showing that the Labrador Current enters the Gulf of St. Lawrence through the Strait of Belle Isle with volume transport up to 0.3 Sv. These Labrador Current waters are thought to flow westward along the Quebec shore and are entrained into the cyclonic circulation in the northwest Gulf region.

The fully-nonlinear CANDIE model was then applied to the realistic basin forced by both the river discharge at the estuary head and a barotropic westward jet at the northeast corner of the eastern open boundary. The speed and width of the boundary inflow are about 8 cm s^{-1} and 40 km respectively, corresponding to the volume transport of about 0.3 Sv. The salinity and temperature of this jet are the same as the initial fields. Other model parameters are the same as before. Figure 10 shows the surface salinity and current fields at days 20, 40 and 60.

The addition of the boundary inflow introduces significant variations in the plume-current system and the associated eddy and meander development. Figure 10a shows a weak westward coastal current along the Quebec shore as a direct result of this boundary inflow. The current becomes relatively stronger as it flows through the Jacques Cartier Strait and joins the estuarine outflow at the estuary mouth, forming the root of the surface-intensified Gaspé Current (Figure 10a).

A comparison of the model results in Figures 9b and 10b clearly shows that the westward coastal current along the Quebec shore effectively constrains the eastward expansion of the buoyant estuarine plume, resulting in more estuarine waters advected seaward, and strengthens the Gaspé Current along the Gaspé

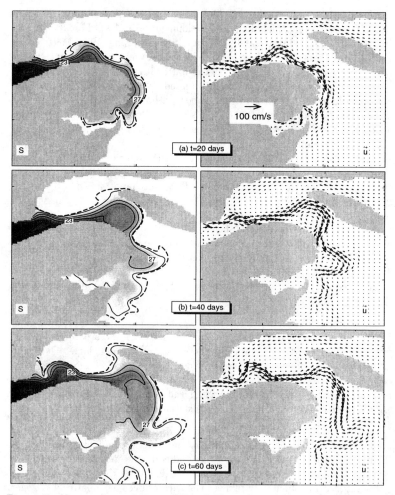

Figure 10. Near surface salinity and flow fields in the realistic basin forced by the river discharge at the estuary head and a barotropic westward inflow at the northeast corner of the eastern open boundary at (a) day 20, (b) day 40 and (c) day 60. Otherwise, same as Figure 9.

Peninsula. Furthermore, the combination of the westward coastal current along the Quebec shore and the eastward Gaspé Current along the Gaspé Peninsula forms a large-scale cyclonic circulation over the NWGSL, agrees qualitatively with the time-mean current observations presented in Figure 4.

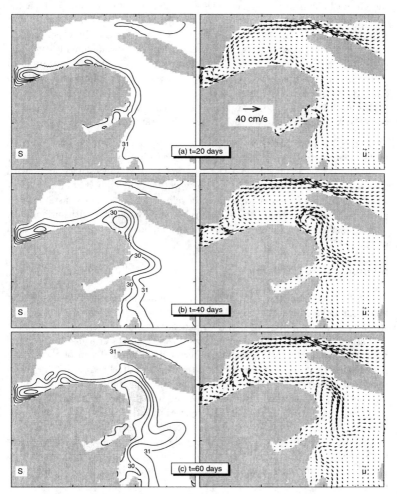

Figure 11. Sub surface salinity and flow fields on the fourth model level centered at 35 m in depth in the realistic basin at (a) day 20, (b) day 40 and (c) day 60. The salinity interval is 0.5 ppt. Otherwise, same as Figure 10.

Figure 10 shows that, initially, the Gaspé Current also flows closely along the irregular coastline. Due to the combining effects of the momentum advection, boundary inflow, and curvature of the coastline, however, a small amount of buoyant estuarine waters are accumulated around the northernmost tip of the Gaspé Peninsula at day 20, with a wavelike feature appearing along the outer edge of the Gaspé Current (Figure 10b). At day 40, a significant amount of estuarine waters are accumulated over the area between the Gaspé Peninsula and Anticosti Island, resulting in an intense anticyclonic eddy over this area (Figure 18b). To the south of this eddy, the Gaspé Current separates from and reattaches to the coast sporadically as it propagates southward. After 60 days, the Gaspé Current separates completely from the coast after passing the northernmost tip of the Gaspé Peninsula, creating an offshore jet that flows onto the Magdalen Shallows with large-scale meanders.

Figure 11 shows the time sequence of salinity and flow fields on the fourth model level centered at 35 m. The sub-surface salinity variations occur mainly along the Gaspé Peninsula and the adjacent coastal and estuary areas with several recognizable pools of lower-salinity waters. These pools of lower salinity waters are highly coherent with the anticyclonic motions. The coastal current along the Quebec shore at this depth is relatively comparable with surface flow. The sub-surface flow along the Gaspé Peninsula, however, is much weaker than at the surface.

Summary and Discussions

This is the first time that a primitive equation model has been used to study the development of the coastal current and the associated eddy and instability generation over the NWGSL region. Two model domains are used in this study. The first is an idealized basin that consists of a flat bottom and piece-wise straight coastlines. The other is the realistic basin of the model-resolved bathymetry over the northwestern Gulf of St. Lawrence. The model results demonstrate that the early development of the plume-current system is qualitatively similar in both the idealized and realistic basins: a buoyant plume is created inside the estuary with a strong eastward surface outflow along the north shore of the estuary. Due to the Coriolis effect, this outflow turns anticyclonically and flows southward along the salinity front of the plume near the estuary mouth. It abruptly turns cyclonically at the south shore to form a surface intensified coastal current that advects the buoyant estuarine water downstream. Due to the combined effects of irregular bathymetry and non-linear momentum advection, however, the coastal current is relatively weaker and the propagation speed of the intrusion nose is much slower in the realistic basin than those in the idealistic basin. Furthermore, an intense anticyclonic eddy is created at the downstream corner in the idealized basin, that grows in size and finally pinches off from the coast with a new eddy generated at the corner. In the realistic basin, however, the coastal current is more unstable with meanders and backward-breaking waves developing along the outer edge of the current.

With the river discharge as the only driving force, however, the buoyant estuarine plume expands successively with time, leading to a large-size anti-cyclonic circulation over the NWGSL region. This is opposite to the quasi-permanent cyclonic gyre observed over this area. The addition of a barotropic westward jet specified at the northeast corner of the east open boundary is efficient to constrain the seaward expansion of the estuarine plume and strengthens the Gaspé Current. The combined effect of the river discharge at the estuary head and the westward coastal current along the Quebec shore results in the formation of a statistically-stable cyclonic circulation over the NWGSL region.

Acknowledgments

The author wishes to thank Michael Dowd, Fraser Davidson, Yuyou Lu, Josko Bobanovic, Richard Greatbatch, Keith Thompson, Dan Wright, and Brad DeYoung for their useful suggestions and comments. I am grateful to Brian Petrie, Ken Drinkwater, and Doug Gregory for making hydrographic data and monthly mean current meter data available to me. Thanks also go to Dan Kelley and Allan Goulding for making the Gri-plotting package work on a DEC Alpha machine. This study was supported by the Natural Sciences and Engineering Research Council of Canada and the start-up funding from Dalhousie University.

References

Benoit, J., M. I. El-Sabh, and C. L. Tang, Structure and seasonal character-istics of the Gaspé Current, *J. Geophys. Res.*, 90, 3225-3236, 1985.

Chao, S. Y., and W. C. Boicourt, Onset of estuarine plumes, *J. Phys. Oceanogr.*, 16, 2137-2149, 1986.

Daley, R., *Atmospheric Data Analysis*, Cambridge University Press, 457 pp., 1991.

El-Sabh, M. I., Surface circulation pattern in the Gulf of St. Lawrence, *J. Fish. Res. Board of Can.*, 33, 124-138, 1976.

Gregory, D. N, O. C., Nadeau, and D. Lefaivre, Current Statistics of the Gulf of St. Lawrence and Estuary, *Can. Tech. Rep. Hydrogr. Ocean Sci.*, 120, 178 pp., 1989.

Klinger, B. A., Baroclinic eddy generation at a sharp corner in a rotating system, *J. Geophys. Res.*, 99, 12515-12531, 1994.

Koutitonsky, V. G., and M. I. El-Sabh, Estuarine mean flow estimation revisited: Application to the St. Lawrence Estuary, *J. Mar. Res.*, 43, 1-12, 1985.

Koutitonsky, V. G., and G. L. Bugden, The physical oceanography of the Gulf of St. Lawrence: A review with emphasis on the synoptic variability of the motion, in *The Gulf of St. Lawrence: Small Ocean or Big Estuary?* J.-C. Therriault, ed., *Can. Spec. Publ. Fish, Aquat. Sci.*, 113, 57-90, 1991.

Kundu, P. K., *Fluid Mechanics*, Academic Press, San Diego, 638 pp., 1990.

Mertz, G., M. I. El-Sabh, and D. Proulx, Instability of a buoyancy-driven coastal jet: the Gaspé Current and its St. Lawrence Precursor, *J. Geophys. Res.*, 93, 6885-6893, 1988.

Petrie, B., B. Toulany, and C. Garrett, The transport of water, heat and salt through the Strait of Belle Isle, *Atmosphere-Ocean*, 26, 234-251, 1988.

Sheng, J., D. G. Wright, R. J. Greatbatch, and D. E. Dietrich, CANDIE: A new version of the DieCAST ocean circulation model, *J. Atm. and Oceanic. Tech.*, 15, 1414-1432, 1998.

Smagorinsky, J., General circulation experiments with the primitive equation. I. The basic experiment, *Mon. Wea. Rev.*, 21, 99-165, 1963.

Stern, M. E., J. A. Whitehead, and B. L. Hua, The intrusion of a density current along the coast of a rotating fluid, blocking waves, *J. Fluid Mech.*, 123, 237-265, 1982.

Tang, C. L., Observation of wavelike motion of the Gaspé Current, *J. Physical Oceanogr.*, 10, 853-860, 1980.

Tang, C. L., and A.S. Bennett, Physical oceanographic observations in the northwestern Gulf of St. Lawrence. Data series BI-D-810–6, Bedford Inst. Oceanogr., Dartmouth, Canada, 127 p., 1981.

Tang, C. L., Cross-front mixing and frontal upwelling in a controlled quasi-permanent density front in the Gulf of St. Lawrence, *J. Physical Oceanogr.*, 13, 1468-1481, 1983.

Trites, R. W., The Gulf of St. Lawrence from a pollution point of view, in *Marine Pollution and Sea Life*, M. Ruivo, ed., FAO, Fishing News Books, London, 59-62, 1972.

Wang, D. P., Numerical study of gravity currents in a channel, *J. Phys. Oceanogr.*, 15, 299-305, 1985.

QUADTREE GRIDS FOR DISPERSION AND INVERSE FLOW MODELS

G. J. M. Copeland[1], R. D. Marchant[1], A. G. L. Borthwick[2]

ABSTRACT
Hierarchical quadtree grids are used in two applications to offer selective grid refinement. Problems of pollutant transport and coastal flow modelling are considered. The first application describes an *adaptive* hierarchical grid-based numerical scheme for predicting localised water-borne pollutant transport in two-dimensions. The prescribed initial pollutant mass is discretised into a set of masslets and the flow domain re-scaled within the unit square. The solution process uses a split operator method in which advection is modelled by discrete masslets and diffusion by finite differences. The quadtree mesh is redefined each step based on the local distribution of masslets. The second application finds a solution to the shallow water equations in a *fixed* quadtree mesh that is refined by the topography. The solution uses an inverse method with direct minimisation of a cost function and allows for data assimilation. When used together, the two models allow simulations to be made of effluent dispersion in coastal waters.

INTRODUCTION

Quadtrees

A quadtree mesh is created by the recursive and selective subdivision of a unit square. Figure 1 shows a coastal topography represented by a quadtree mesh. The criterion for subdivision of a given mesh cell can be defined by the number density of data 'particles' or by the local magnitude of some spatially distributed parameter. In Figure 1 these data 'particles' are a set of location points defining the coastline. Another example is shown in Figure 2. Here a set of 'particles' (or 'masslets') each representing a given mass of material is distributed in space to represent an evolving effluent plume. The resulting quadtree mesh is shown.

---

[1] University of Strathclyde, Department of Civil Engineering, Glasgow G4 0NG, U.K.
   e-mail: g.m.copeland@strath.ac.uk Tel +44 141 548 3252, Fax +44 141 553 2066
[2] University of Oxford, Department of Engineering Science, Oxford OX1 3PJ, U.K.
   e-mail: alistair.borthwick@eng.ox.ac.uk Tel +44 1865273047, Fax +44 1865273010

Pollutant Transport Modelling

Improved accuracy can be achieved in pollutant transport modelling if regions of higher concentration and pollutant fronts are represented by finer grid resolution. Since these regions move, a successful grid refinement scheme must adapt to the changing pattern of pollutant concentration, Figures 2 and 8. An efficient quadtree grid generator has been developed which has been used to produce grid meshes that adapt over time providing selective grid refinement in areas of high pollutant concentration (Borthwick, Marchant & Copeland, 1998, 1999). The calculation of pollutant advection and diffusion uses a split operator method. Diffusion is calculated using an explicit finite difference solution of the diffusion equation in the quadtree mesh. Advection is treated using a Lagrangian particle tracking method. This has the advantages of stability and mass conservation.

Coastal flow modelling

Inverse flow models allow for data assimilation into solutions of the shallow water equations (Copeland 1998; Copeland & Bayne, 1998). This modelling approach has the advantages of providing flow solutions that agree well with field observations and of making best use of available data. Flow models of coastal waters also benefit from selective grid refinement. Regions of complex coastal and bed topography often cause velocity shears that should be resolved. Inverse models may also justify the use of increased resolution in areas of high data density. A grid refinement scheme is required that need not, however, adapt progressively with time. The same quadtree generator has been used to create grids with selectively refined mesh sizes determined by the local density of topographic (coastline) data and moderated by the CFL number. This resulted in fine mesh sizes in shallow water next to the coastline and larger mesh sizes offshore in deeper water.

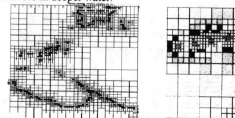

Figure 1. A quadtree mesh representing a coastline in Sepetiba Bay, Brazil.

Figure 2. A quadtree mesh adapted to a moving effluent plume.

The model has been verified with flow around idealised topographic features (a group of islands) and data assimilation tested using the identical twin approach based on results from a direct model solution. Together the two models allow simulations to be made of effluent dispersion in coastal waters with improved local resolution of coastlines and of fronts and plumes.

HIERARCHICAL GRIDS

In order to achieve local grid densities that relate properly to evolving pollutant and flow features, it is necessary to use either unstructured or hierarchical adaptive grids. Hierarchical grids involve recursive spatial decomposition and represent a sensible compromise between control of local mesh refinement, speed of generation, and array storage requirements. Quadtrees have been applied to two-dimensional domains in image processing, e.g. Samet (1982) and increasingly in CFD (e.g. Gáspár *et al*, 1991; Józsa & Gáspár, 1992; Yiu *et al*, 1996). Here, a square quadtree grid generator has been selected because of its robustness and relative ease of generation in comparison with other unstructured grid generation techniques. The quadtree grid generation method is summarised below:

(i)   create a set of seeding points according to the discretised pollutant distribution and/or flow boundaries;

(ii)   rescale the flow domain to fit within the unit square (i.e. grid level 0);

(iii)   input the subdivision level of the finest grid, *m*;

(iii)   divide the square into four panels (i.e. grid refinement level 1);

(iv)   recursively check each panel for seeding points, and, if it contains more than *p* points and is at a level below *m*, then subdivide into four sub-panels;

(v)   carry out grid regularisation (by further subdivision) to ensure adjacent and diagonally touching cells should not be more than one level apart.

Data on each cell are stored in an integer tree array of *n* elements, including pointers to the next cell and information on neighbouring cell levels. Each cell has a unique identification number created by the concatenation of local reference numbers determined from the domain decomposition as it proceeds. Grid regularisation and neighbour finding are undertaken together, starting from the smallest cells (i.e. highest level). Regularisation greatly reduces the possible number of topologies so simplifying the representation of the differenced equations. This advantage outweighs the increase in the total number of cells. Figure 3 depicts the non-regularised level 4 quadtree grid that would be created about a pair of seeding points located at (0.3, 0.3) and (0.35, 0.35) within the unit square. After regularisation the quadtree grid is as indicated in Figure 4.

ADVECTION - DIFFUSION SOLVER

Governing equations

The Fickian species transport equation governing the 2D vertically averaged spatial and temporal distribution of pollutant concentration for isotropic turbulent diffusivity ε (m^{2}s^{-1}) is

$$\frac{\partial C}{\partial t} + U\frac{\partial C}{\partial x} + V\frac{\partial C}{\partial y} - \frac{1}{\rho H}\left[\frac{\partial}{\partial x}\left(\rho H \varepsilon \frac{\partial C}{\partial x}\right) + \frac{\partial}{\partial y}\left(\rho H \varepsilon \frac{\partial C}{\partial y}\right)\right] = S \quad (1)$$

where the depth averaged concentration C units m^{-3} is

$$C = \frac{1}{H}\int_{-d}^{\eta} \overline{c}\,\mathrm{d}z \quad (2)$$

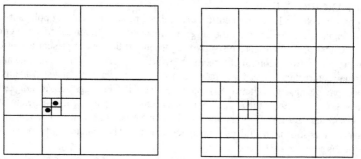

Figure 3. Initial, non-regularised mesh
    • seeding points.

Figure 4. Regularised quadtree mesh

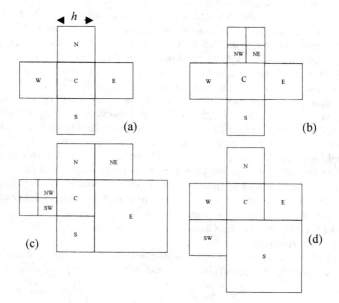

Figure 5. Four examples of the twelve possible cell configurations

in which $\bar{c}$ is the temporal mean concentration averaged over time scale of sub-grid turbulence. Similarly U and V are the vector components of depth-averaged velocities, such that

$$U = \frac{1}{H} \int_{-d}^{\eta} \bar{u}\,dz \quad \text{and} \quad V = \frac{1}{H} \int_{-d}^{\eta} \bar{v}\,dz \tag{3}$$

and the depth-averaged source term is

$$S = \frac{1}{H} \int_{-d}^{\eta} s\,dz \ . \tag{4}$$

The total depth is $H = d + \eta$ where d is the still water depth and η is the free surface elevation. At solid walls located at $x = x_B$ and $y = y_B$, there is no mass flux and so the concentration gradient is zero from Fick's law. Thus,

$$\frac{\partial C}{\partial x} = 0 \quad \text{at} \quad x = x_B \quad \text{and} \quad \frac{\partial C}{\partial y} = 0 \quad \text{at} \quad y = y_B \ . \tag{5}$$

Numerical solver

In a similar manner to Chorin's (1968) operator splitting method, the advection and diffusion processes are treated separately. The advection step is undertaken using Lagrangian particle tracking. The new particle positions are then used as seeding points, about which a regularised quadtree grid is generated. Cell-centre concentrations are determined from the numbers of masslets in each cell multiplied by their individual masses and divided by the mass of water in the cell. The diffusion step is implemented by an explicit forward Euler finite-difference scheme, in which the concentration values are updated, and a new distribution of masslets created according to the nodal concentrations, ready for the next advection step.

Advection

Initially, the mass of pollutant, M, introduced at a source location at each time step is divided equally amongst a prescribed integer number of masslets, N. These are distributed in proportion to a specified initial concentration profile (Gaussian, cone, top hat, etc), and a grid generated with the masslets as seeding points. Cells in regions with non-zero concentrations but which contain no masslets at the end of the preceding time step are each assigned a single masslet. For each cell, the mass of each masslet, m_c, is then revised so that $m_c = M_c / N_c$ where M_c is the cell's pollutant mass and N_c is the number of masslets in the cell. The masslets serve two purposes: one to control mesh density, the other to enable advection to be calculated by Lagrangian particle tracking. The advection step is accomplished by moving masslets with the local flow velocity components using a predictor-corrector step. Cell-centre concentration values are then re-calculated (ready for the diffusion step) by summing the mass contributions from each masslet in the cell.

Diffusion: Finite-difference Scheme

The regularised quadtree grid contains 12 possible cell configurations. For each configuration, the Laplacian (diffusion) operator was discretised using general Taylor expansions. Using the forward Euler scheme to discretise the temporal derivative, an explicit algebraic formula was derived for each cell configuration. The resulting expressions for four example cell layouts depicted in Figures 5(a), 5(b), 5(c) and 5(d) respectively are:

$$\left.\frac{\partial C}{\partial t}\right|_c^n = \frac{\varepsilon}{4h^2}\left(C_w^n + C_n^n + C_e^n + C_s^n - 4C_c^n\right), \tag{6}$$

$$\left.\frac{\partial C}{\partial t}\right|_c^n = \frac{4\varepsilon}{21h^2}\left(5C_w^n + 5C_e^n + 4C_{ne}^n + 4C_{nw}^n + 6C_s^n - 24C_c^n\right), \tag{7}$$

$$\left.\frac{\partial C}{\partial t}\right|_c^n = \frac{4\varepsilon}{75h^2}\left(12C_{nw}^n + 6C_{ne}^n + 8C_e^n + 12C_{sw}^n + 15C_s^n + 13C_n^n - 66C_c^n\right), \tag{8}$$

$$\left.\frac{\partial C}{\partial t}\right|_c^n = \frac{2\varepsilon}{21h^2}\left(9C_n^n + 9C_e^n + 4C_s^n + 3C_{sw}^n + 8C_w^n - 33C_c^n\right), \tag{9}$$

where h is the cell dimension. On updating the concentrations, a new distribution of masslets was calculated. For each cell, the cell centre concentration was assumed to be the mean value over the cell, from which the new mass of pollutant in the cell is computed as M_c. Hence, the number of new masslets in the cell, N_c, is the nearest integer value of $(M_c \, N/M)$. There is an integer round-off error which affects mass conservation. Here, this effect was counteracted by summing the total number of new masslets created per cell, and reassigning each masslet with a new mass $m_c = M_c / N_c$. As in all classical explicit methods, to ensure stability of the scheme a time step criterion must be adhered to, and in this case is determined by the diffusion coefficient and the grid spacing (rather than an advection based criterion, see below). The one employed here is given by Fletcher (1991), and relates to the smallest (i.e. highest level) cells. For the scheme to remain stable across the entire grid, it was necessary to ensure that $\varepsilon \Delta t/h^2 \leq 0.5$. However, in order to reduce numerical diffusion caused by reseeding particles in cells of dimension h at each time step Δt it is required that $h^2/\Delta t \ll \varepsilon$. This is inconsistent with the stability requirement, and so another strategy has been used to reduce numerical diffusion. This strategy ensures that particles are re-seeded as close as possible to their previous locations. It should be noted that random positioning is used if the new number of masslets is greater than the old. It was found that the stability requirement for a small time step could be relaxed somewhat from that of an Euler integration by using the predictor-corrector scheme. This effectively reduced numerical diffusion further by allowing a larger time step to be used. Although there are no stability problems associated with advection of the 'masslets', the accuracy of the advection scheme does depend on the time step. A predictor-corrector method is used to calculate the advection through the varying flow field U(x,y,t), V(x,y,t). The time step determined by the stability conditions discussed above is checked to make sure that the step length does not exceed the local cell dimension. If the step length is too large the time step is subdivided and the advection proceeds as a series of sub-steps.

Initial and Boundary Conditions

One test case and one application are reported here. The test case considers the advection and diffusion of an initial Gaussian concentration profile (see equation 10) in a rotating flow. An initial regular grid of level 5 was employed to position pseudo-masslets according to the specified profile. The pseudo-masslets were then used as seeding points for a revised quadtree grid that replaced the regular grid. The pseudo-masslets were discarded, and a revised initial distribution of masslets created about the quadtree grid. The application considers a continuous discharge into a tidal flow bounded by real topography. In order to simulate a continuous discharge of strength J units s^{-1} (noting that source $S = J/(h^2 d)$ units $m^{-3}s^{-1}$) masslets are injected at a rate of n every time step Δt. The initial mass of each masslet, m_c^i, is therefore $J\Delta t/n$ units. These masslets are released uniformly within an area described by a radius R^i. At solid boundaries, advecting masslets are reflected. For the finite difference diffusion step, the concentration gradient normal to the boundary is set to zero. Advecting masslets are allowed to enter and leave open boundaries depending on the local flow field. Boundary concentration values are determined using linear extrapolation (with a cut-off value of zero concentration to prevent negative concentrations at the boundary).

Results

Advection-diffusion of a Gaussian profile in a rotating flow

An initial Gaussian profile was specified from the analytic solution of the pure diffusion equation as

$$C(x,y,t) = \frac{M_o}{4\pi\varepsilon t}\exp\left[-\frac{(x^2 + y^2)}{4\varepsilon t}\right] \qquad (10)$$

with $M_o = 4\pi$ as initial pollutant mass, $\varepsilon = 0.001$ m²/s and $\Delta t = 0.025$ s. Figure 6 shows concentration plots (as shaded grids and as concentration profiles) at $t = 2.8$ and 3.1 s. Note that $t_o = 2.5$ s so $C(t_o) = 400$ units m⁻². Figure 6 a) shows the quadtree mesh as it adapts over time. The left panel at $t = 2.8$ s and the right panel at $t = 3.1$ s. The method generated 12 intermediate meshes between the two shown, as the Gaussian distribution advected on an anticlockwise circular path. Figure 6 b) shows comparisons between analytical concentrations (full line) and model results (+) at $t = 2.8$ s on the left and at $t = 3.1$ s on the right. Good agreement is achieved.

Application

The method was applied to a tidal embayment (Sepetiba Bay, Brazil; Latitude 23 degrees 0 minutes South, Longitude 44 degrees 0 minutes West) which has a complex coastal topography with several islands, see Figure 7. The overall east to

$t = 2.8$ s $t = 3.1$ s

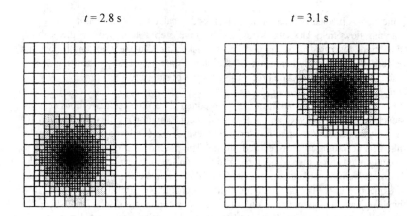

Figure 6 a) Quadtree meshes (at $t = 2.8$ s on the left and at $t = 3.1$ s on the right) adapting to a diffusing Gaussian concentration profile advecting in a rotating flow.

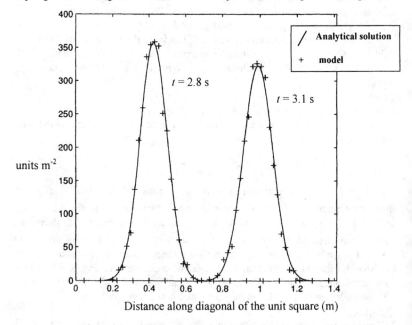

Figure 6 b) Concentration profiles across domain diagonal (at $t = 2.8$ s on the left and at $t = 3.1$ s on the right). Concentration (units m^{-2}) *vs.* distance along the diagonal of the unit square.

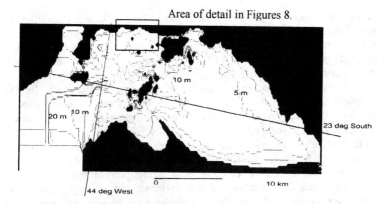

Figure 7. Sepetiba Bay, Rio de Janeiro State. The Bay opens onto the Atlantic Ocean at the south-west corner of the figure.

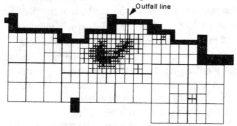

Figure 8 a) plume at 1.5 hrs after high water slack. Typical centreline concentrations are 5000 bacteria per 100ml.

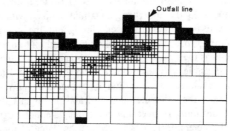

Figure 8 b) plume at 3.5 hrs after high water slack. Typical centreline concentrations are 2500 bacteria per 100ml.

Figure 8 Quadtree meshes generated by the dispersion model representing an evolving effluent plume.

west extent of the computational domain was 50 km and the north to south extent
was 19 km. The coastline was represented by a series of x, y data points which were
used to seed the quadtree representation of the coastal boundaries, see Figure 1, in
which the smallest coastline cell is at level 9 (i.e. 2^{-9} of the maximum domain
dimension which is about 100 m). These coastline seeding points together with the
effluent masslets were used to generate the computational quadtree mesh. Note that
all information was scaled to the unit square and that the part of the square which lay
outside of the rectangular domain shown in these figures was excluded from the
calculations and is not shown in the results. A linear scaling factor σ was defined as
the ratio of the unit square (1 metre) to the longest dimension of the domain, in this
case about 50,000 m. The tidal flow fields were calculated on a regular grid of cell
size 250 m. The current velocities for each masslet position at each time in the
dispersion calculation were delivered from the regular flow grid using linear
interpolation in space and time. The velocities were also scaled by the factor σ, time
remained unscaled. A simulated discharge of sewage effluent was created with a
strength $J = 10^9$ bacteria s^{-1}. Bacterial decay was represented by a $T_{90} = 4$ hrs. Decay
was incorporated into the model calculations by reducing the concentrations in each
cell by a factor of $10^{-\Delta t / T_{90}}$ each time step. Isotropic turbulence was assumed and a
value of $\varepsilon = 0.25$ m^{2}s^{-1} used to represent prototype conditions. This value had to be
scaled by a factor of σ^2 to represent conditions in the model unit square. The adapting
quadtree meshes representing the mid-field plume at 1.5 hrs and 3.5 hrs after high
water slack are shown in Figures 8 a) and 8 b). Typical plume centerline
concentrations are 5000 and 2500 bacteria per 100ml, respectively. Here the smallest
cell is of level 11 which is a cell size of about 25 m.

THE INVERSE QUADTREE SHALLOW WATER SOLVER
Introduction
Unlike direct models, an inverse model of the sort described here does not
integrate forward in time. It calculates a set of flow information throughout an
entire tidal period using the underlying equations coupled with known, sparse field
data. Initially, the field data are used throughout the domain to provide an
approximate starting position for the solver by simple interpolation. This starting
position is known as the first guess. The next stage is to calculate how far the first
guess is from a feasible, physical solution. In the form of a least squares fit, an
error value is calculated; a so called cost function. This depends upon the
differences between the calculated solution and the input data, and also on the
residuals of the discretised governing equations calculated using the first guess.
The object of the inverse solver is to reduce the value of the cost function to a
minimum by iterating progressively towards a better solution.

The Shallow Water Equations

The governing equations

The conserved form of the 2D depth integrated shallow water equations used in the solver are a simultaneous set of three partial differential equations, comprising the continuity equation:

$$\frac{\partial \eta}{\partial t} + \frac{\partial (HU)}{\partial x} + \frac{\partial (HV)}{\partial y} = 0 \tag{11}$$

the x momentum equation,

$$\frac{\partial (HU)}{\partial t} + g(d+\eta)\frac{\partial \eta}{\partial x} + gn^2 \frac{HU\sqrt{(HU)^2 + (HV)^2}}{(d+\eta)^{7/3}}$$
$$- f_c HV + \frac{\partial}{\partial x}\left[\frac{(HU)^2}{(d+\eta)}\right] + \frac{\partial}{\partial y}\left[\frac{H^2 UV}{(d+\eta)}\right] = 0 \tag{12}$$

and the y momentum equation

$$\frac{\partial (HV)}{\partial t} + g(d+\eta)\frac{\partial \eta}{\partial y} + gn^2 \frac{HV\sqrt{(HU)^2 + (HV)^2}}{(d+\eta)^{7/3}}$$
$$+ f_c HU + \frac{\partial}{\partial y}\left[\frac{(HV)^2}{(d+\eta)}\right] + \frac{\partial}{\partial x}\left[\frac{H^2 UV}{(d+\eta)}\right] = 0 \tag{13}$$

where η is elevation, d still water depth, H total water depth, U depth-averaged velocity in x direction, V depth-averaged velocity in y direction, g gravitational constant, n Manning coefficient, and f_c Coriolis frequency. Since the velocity of fluid into or out of a cell is best described as flux through a face edge, and height of fluid in a cell being most easily considered as cell centred, the variables are staggered across the grid, as shown in Figure 9.

Figure 9.   Arrangement of flow variables in a cell.

The discretised forms

For brevity, the discretised forms of these equations are not given here since they are almost identical to those employed by Copeland and Bayne (1998). At present, the model incorporates the continuity equation and simplified x and y momentum equations which omit the convective acceleration terms.

Numerical Methodology

The First Guess

The solution to the shallow water equations found by the inverse method is determined by the data and the first guess. This information along with the data replaces the usual boundary data provided to a direct model.

The Cost Function
This is a series summation of the squared residual values formed by substituting the iterated flow variables into the discretised governing equations. Clearly, if the solution is perfect, the residuals would all be zero. This summation also includes the squared differences between the solution and the available data points, where they exist, to form *The Constrained Cost Function*, (CC). In practice, because the discrete equations are approximate and the field data contain errors, it is not possible to attain a zero value for CC, but a reasonable minimum is sought.

The Gradient Calculations
The cost function comprises functions of the flow variables. Therefore to minimise it through successive changes in the flow variables requires that it is differentiated with respect to every occurrence of each flow variable in the differenced equations. This is not a straightforward task even on a regular Cartesian grid since, for example the flow vector component $U_{i,j}^{k}$ where i, j are x, y coordinates and k is time, will occur in the discretised form of the governing equations several times.

Quadtree Complications
Spatial referencing on a quadtree grid is more complicated than on a regular one, since the straightforward i, j referencing for the x, y coordinates of each cell is no longer possible. Each cell is stored with a list of its surrounding neighbours, possibly of different mesh sizes. The spatial differencing terms become complicated expressions containing interpolations from various surrounding neighbours. In order not to omit any of the terms of the flow variables during differentiation, the cell topology has to be carefully considered. The following three simple topologies illustrate specific difficulties with the quadtree grid.

Velocity interpolations
The topology in Figure 10 shows two smaller cells to the Eastern side of the object cell.

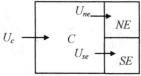

Figure 10. Typical quadtree cell with smaller East cells.

In order that the discretised equations are properly centred across the object cell C, an x-direction velocity must be interpolated on the right hand side of the cell, U_{c}^{East}. In this instance, the velocities of the two Eastern cells are averaged to produce

$$U_{c}^{East} = \frac{U_{ne} + U_{se}}{2} .$$ (14)

As frequently as this topology occurs in a quadtree grid, so too does the case where the right hand cell is larger. This necessitates interpolation involving a

South-Eastern cell to obtain the correct value, as shown in Figure 11. For this particular topology, the value obtained would be calculated as:

$$U_c^{East} = \frac{2U_e + U_{se}}{3} . \qquad (15)$$

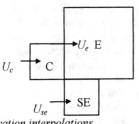

Figure 11. Typical quadtree cell with larger East cell and same size South-East neighbour.

Elevation interpolations

Here, a central value in a non-existent Western cell must be calculated. Figure 12 demonstrates an example of how this is done.

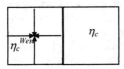

Figure 12. Typical quadtree cell with smaller West neighbours.

The interpolated value here is simply the average of the four smaller cells. If the Western cell is larger, scattered data interpolation is used to determine the value, often requiring as many as four neighbouring cells, and details are omitted here for brevity. In all cases, it must be considered whether any of the neighbouring cells are dry, and if they are then the interpolation formulae must be modified accordingly. Clearly, if the neighbouring cells are of the same dimension as the object cell, no interpolations are necessary. Due to these interpolations, the flow variables of a cell may become involved in the discretised governing equations of several surrounding cells.

Adjusting the Solution

Once the gradients for each flow variable have been calculated, the values are adjusted using a simple nudging (steepest decent) scheme to move them closer towards the final solution. The gradient for each of the flow variables is multiplied by a small value, μ, and then used to give the following simple iterative scheme

$$\eta_{i,new}^k = \eta_{i,old}^k - \mu \frac{\partial CC}{\partial \eta_{i,old}^k} . \qquad (16)$$

The iterative schemes for the U and V components of the current vector are almost identical. The scheme is simple, but convergence is slow, and is not uniform in all three variables. For this reason, further work on the model involves the use of a conjugate gradient algorithm (e.g. CONMIN, Shanno & Phua, 1976).

Results

An idealised domain containing three elliptical islands was chosen to test the model. It was thought that the complex topographies generated by the islands would demonstrate the ability of the quadtree grid to model difficult coastlines, and also that a set of channels would be created where faster flows would occur. The model domain comprised a unit square, subjected to a suitably scaled long wave of amplitude 10^{-3} m. A direct solver of the shallow water equations was used to generate a complete 'tidal' period (6.25 s) of flow vectors around the islands, and then an arbitrary set of 14 grid squares were chosen as data points to run the inverse model, shown in Figure 13. Water level data points were also defined in the same way. The rest of the output from the direct model was then discarded. A first guess flow field was formed by interpolation from these data points, shown in Figure 14. A first guess for the water levels was prepared by the same interpolation method but then reduced by a factor of 0.5 in order to test the model's ability to assimilate water level data. The inverse model was then run for 2000 iterations to develop the complete flow solution, shown in Figure 15. It can be seen that the initial interpolation process results in flow vectors which point directly into land boundaries. After running the inverse model, the flow has correctly taken account of the land boundaries and has developed into a realistic flow scheme. Figure 16 shows a central north-south transect of water elevations across the model domain. The '+' signs indicate cell elevations in the first guess, with the four data points which appear on this transect at clearly much higher levels. The model was run for 2000 and 4000 iterations and the resultant profiles are shown by an 'x' and an 'o', respectively. These results show the convergence of the result towards the water level data.

Future Developments

A conjugate gradient method will be used to improve the speed of convergence. The convective acceleration terms have recently been included into the solver. These terms do not lead to instabilities in an inverse model as they may in a direct model. Even so, it is more difficult to include them than the other terms due to problems of finding good representations of their differenced forms in a quadtree mesh. The inverse flow model and the species transport model will be brought together in the near future to create an integrated package.

CONCLUSIONS

Details have been presented of an adaptive hierarchical grid generator, with local mesh refinement controlled by the level of effluent concentration. The quadtree stores grid information efficiently, while allowing the grid to have fractal-like refinements. In a novel approach, grid regularisation is achieved in a single sweep starting from the finest cells and moving up the quadtree. Good agreement is obtained with analytical solutions to simulations of standard advection-diffusion test cases, one of which is reported here. The adaptive quadtree mesh allows improved resolution of localised features such as sharp pollutant fronts. The application to Sepetiba Bay demonstrates the method is effective at prototype scale and with real topography. The application of a fixed quadtree mesh to the solution of the shallow water equations by a direct minimisation inverse method is also presented. The method provides for data assimilation into an unsteady flow calculation in a mesh with local refinement based on topography. The inverse flow and transport models will be integrated to form a flow and dispersion modelling package. Both models are written in FORTRAN and run on a p.c. (>400 MHz, 128 Mb RAM) in several minutes to several hours depending on the mesh size and number or particles used.

ACKNOWLEDGEMENTS

This work has been supported by the U.K. Engineering and Physical Sciences Research Council through grants GR/L11861 and GR/L11854, and co-investigated by Professor R Eatock Taylor (Oxford University) and Dr A Folkard (Lancaster University). We acknowledge Dr J Józsa of the Technical University of Budapest, Hungary, who substantially contributed to the work with quadtrees. Mr. Scott Couch (Strathclyde University) developed the direct solver of the shallow water equations used in the application to Sepetiba Bay and Prof. Teofilo Monteiro (Fiocruz, Rio de Janeiro) who started the project in this area.

REFERENCES

Borthwick, A.G.L., Marchant, R.D., & Copeland, G.J.M., 1998. *'Adaptive Hierarchical Grid Model of Water-borne Pollutant Dispersion'*, Proc. International Conference on Industry, Technology & Environment, ITE'98, Moscow State University of Technology, Stankin, 15-18 September 1998, Moscow, Russia, pp 11-16.

Borthwick, A.G.L., Marchant, R.D., & Copeland, G.J.M., 1999. *'Adaptive hierarchical grid model of water-borne pollutant'*, submitted to Advances in Water Resources.

Chorin, A.J., 1968. *'Numerical solution of the Navier-Stokes equations'*, Math. Comput., 22, pp 745-762.

Copeland, G.J.M., & Bayne, G.L.S., 1998. *'Tidal flow modelling using a direct minimisation method'*, Coastal Engineering, Elsevier, vol. 34/1-2, pp 129-161.

Copeland, G.J.M., 1998. *'Coastal Flow modelling using an inverse method with direct minimisation'*, Proc. Conf. Estuarine and Coastal Modeling, 1997. Ed. M.L.Spaulding & A.F. Blumberg, Pub. ASCE (ISBN 0-7844-0350-3), pp.279-292.

Fletcher, C.A.J., 1991. *'Computational techniques for fluid mechanics'*, 2nd Edition, Springer-Verlag, New York Berlin Heidelberg.

Gáspár, C., Józsa, J., and Simbierowicz, P., 1991. *'Lagrangian modelling of the convective diffusion problem using unstructured grids and multigrid technique'*, in Proc. 1st Int. Conf. on Water Pollution (Modelling, Measuring and Prediction), 3-5 September, Southampton, U.K.

Józsa, J., and Gáspár, C., 1992. *'Fast, adaptive approximation of wind-induced horizontal flow patterns in shallow lakes using quadtree-based multigrid method'*, in Proc. IX Int. Conf. on Comp. Methods in Water Resources, Denver, USA, 9-11 June.

Samet, H., 1982. *'Neighbor finding techniques for images represented by quadtrees'*, Computer Graphics and Image Processing, 18, pp 37-57.

Shanno, D.F. and Phua, K.H., 1976. *'Algorithm 500 - a variable method for unconstrained nonlinear minimization'*, ACM Trans. Math. Software, 6 (4), pp 618-622.

Yiu, K.F.C., Greaves, D.M., Cruz, S., Saalehi, A., and Borthwick, A.G.L.,1996. *'Quadtree grid generation: information handling, boundary fitting and CFD applications'*, Computers & Fluids, 25(8), pp 759-769.

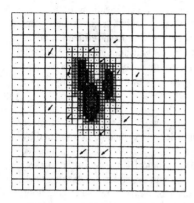

Figure 13 Quadtree grid showing
'data' points.

Figure 14 Interpolated first guess
flow field.

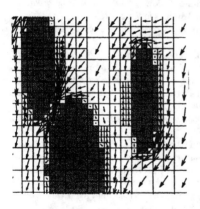

Figure 15 Flow solution.

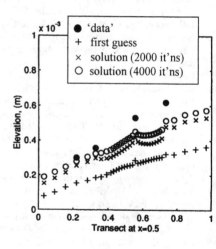

Figure 16 Central north-south
transect showing water levels.

A PARALLEL FINITE VOLUME SCHEME FOR TWO DIMENSIONAL HYDRODYNAMIC FLOWS

Prasada Rao and Scott A. Yost
Department of Civil Engineering
University of Kentucky
Lexington, KY 40506-0281

ABSTRACT

Traditional serial explicit numerical codes, when applied for simulating flows over a large domain for long time periods, require a large amount of computational time. To reduce the simulation time, it is advantageous to make use of a parallel computing technology. A parallel code can speedup an application by working on multiple portions of the computation simultaneously. In this work, which is targeted to the application audience, we discuss the tools that can be used for formulating a parallel code. Then we demonstrate the advantages of using the parallel code by imbedding all the high performance tools into an existing serial two-dimensional finite volume code. The parallel code is based on domain decomposition principles and uses Message Passing Interface (MPI) for all inter-processor communication. Computational results are given to illustrate the efficiency of the parallel code, which simulates flow in transition. On a 16 processor shared memory platform, it has been observed that the parallel code accelerates the transient solution by a factor of 15.5.

INTRODUCTION

Numerical modeling of free surface flows has used a wide variety of explicit schemes both from the family of finite difference and finite volume formulations. For a review of studies done in the field of estuarine modeling using these schemes, we refer readers to the works listed in Spaulding and Blumberg (1997) and Spaulding and Cheng (1995). The popularity that these schemes enjoy among the modeling community can be largely attributed to two factors: (i) simplicity and (ii) robustness for simulating the physics of flow. A limitation in using these explicit schemes is in the choice of time step. From the stability perspective, the magnitude of time step is limited by the CFL stability criteria. For simulations done over a long time period, a small time step requires a huge computational time, taxing the computational resources. A formulation that can accelerate the solution to the desired time period without affecting the accuracy of the solution not only increases the scope in simulating real world problems but also increases the productivity of the researcher. Achieving a higher level of modeling real life flows requires coupling these codes with sophisticated computing environment. This work shows that high performance computing offers a great alternative.

As we discuss in the course of this paper, any existing serial Finite difference (FD) or Finite volume (FV) code can be run in a parallel environment once appropriate high performance computing tools are imbedded. As opposed to sequential codes where the code is run in serial environment (i.e., on one processor/CPU), in the parallel environment the code is run on many processors. Figure 1 depicts the manner in which the input task is divided among 4 processors. By distributing the load equally among all the processors, the time required to complete the input task can theoretically be reduced by a factor of 4. Arriving at an optimal parallel code requires some manual modifications to the sequential code. Depending on the degree of sophistication of the programming tools in the serial code, the parallel programmer needs to rewrite certain portions of the serial code before running it in parallel environment. Relying on the compiler directives, which aids in marginally improving the serial performance, does not automatically generate the efficient parallel version of the serial code. Among the various strategies that can be integrated to the serial code, domain decomposition techniques are unique and are being applied for a wide variety of problems in CFD (http://www.ddm.org). Domain decomposition (DD) algorithms, which are also known as multi-blocking tools, fit naturally into parallel computing and can be run on all high performance architecture machines, including those which have distributed memory. In this approach the entire domain is divided into smaller sub-domains, and a separate processor is assigned to each sub-domain. This paves the way for the computations to be done concurrently. Figure 2 depicts the partitioning of the domain over 2 processors. Various other details associated with a DD formulation are discussed later.

The motivation for performing this work is in showing to our colleagues in the estuarine modeling community the advantages of running an explicit code in parallel environment. This introduction is concluded with a glimpse of the rapid changing technology. Over the past decade computer performance (Table 1) has been improving at a dramatic speed, a trend that is likely to continue in the coming years. It is our belief that future computing in general, and scientific computing in particular, will be performed on these highly parallel machines. Simulating flows in estuaries is one of the challenging problems for application researchers in computational hydraulics, and to this end many serial codes have been formulated. By efficiently porting these codes onto a parallel environment, we know that the capacity to simulate the physics of flow can be greatly enhanced.

In the following sections, we first describe the governing equations and the numerical scheme. We then discuss the domain decomposition methodology, which is the main pillar of our parallel code. The hardware features of the parallel machine on which we test our code is also discussed. The performance parameters to measure the efficiency of the parallel code are then indicated. We finally illustrate these concepts by imbedding them into a two-dimensional serial code that solves the equations for flow in a transition. The focus in this work is equally distributed between the computational savings and the accuracy of the parallel solution.

GOVERNING EQUATIONS

The depth averaged equations that describe the two-dimensional flow in open channels can be written as (Chaudhry 1993)

$$\frac{\partial h}{\partial t} + \frac{\partial}{\partial x}(hu) + \frac{\partial}{\partial y}(hv) = 0 \tag{1}$$

$$\frac{\partial}{\partial t}(hu) + \frac{\partial}{\partial x}\left(hu^2 + \frac{gh^2}{2}\right) + \frac{\partial}{\partial y}(huv) = gh(S_{ox} - S_{fx}) \tag{2}$$

$$\frac{\partial}{\partial t}(hv) + \frac{\partial}{\partial x}(huv) + \frac{\partial}{\partial y}\left(hv^2 + \frac{gh^2}{2}\right) = gh(S_{oy} - S_{fy}) \tag{3}$$

where h is the flow depth, u and v are the depth-averaged velocities in the longitudinal and normal directions, g is the acceleration due to gravity, S_{ox} and S_{oy} are the bed slopes of the channel along the x and y directions, and S_{fx} and S_{fy} are the friction slopes along the x and y axes.

NUMERICAL FORMULATION

The finite volume discretization for the above equations is described in detail in this section. For a review of the finite volume schemes, readers are referred to the work of Hirsch (1996). The scheme we discuss was first proposed by Ni (1982) and is a variant of the more popular Lax-Wendroff formulation. To keep uniformity in the discussion, the above equations are recast in matrix notation as (for convenience, the source terms have been neglected)

$$\frac{\partial U}{\partial t} + \frac{\partial F}{\partial x} + \frac{\partial G}{\partial y} = 0 \tag{4}$$

where the elements of matrices U, F and G are

$$U = \begin{bmatrix} h \\ hu \\ hv \end{bmatrix} \qquad F = \begin{bmatrix} hu \\ hu^2 + \dfrac{gh^2}{2} \\ huv \end{bmatrix} \qquad G = \begin{bmatrix} hv \\ huv \\ hv^2 + \dfrac{gh^2}{2} \end{bmatrix}$$

Starting from the initial time level (n) where all the flow variables are known, the variables at the new time level ($n+1$) for all the interior nodes (see Figure 3) are evaluated as

$$U_I^{n+1} = U_I^n + \delta U_I \tag{5}$$

The corrections at any grid node (δU_I) are related to the flow properties in its surrounding control volumes. For node 1, shown in Figure 3, this can be written as

$$\delta U_I = (\delta U_I)_A + (\delta U_I)_B + (\delta U_I)_C + (\delta U_I)_D \tag{6}$$

where A, B, C, D refer to the volumes surrounding the point 1.

The corrections to a grid node around control volume C, are determined by the two-dimensional distribution formulas,

$$(\delta U_1)_c = 0.25 \left[\Delta U_c - \frac{\Delta t}{\Delta A_c} \Delta F_c - \frac{\Delta t}{\Delta A_c} \Delta G_c \right]$$

$$(\delta U_2)_c = 0.25 \left[\Delta U_c - \frac{\Delta t}{\Delta A_c} \Delta F_c + \frac{\Delta t}{\Delta A_c} \Delta G_c \right]$$

$$(\delta U_3)_c = 0.25 \left[\Delta U_c + \frac{\Delta t}{\Delta A_c} \Delta F_c + \frac{\Delta t}{\Delta A_c} \Delta G_c \right] \tag{7}$$

$$(\delta U_4)_c = 0.25 \left[\Delta U_c + \frac{\Delta t}{\Delta A_c} \Delta F_c - \frac{\Delta t}{\Delta A_c} \Delta G_c \right]$$

where ΔU_c is the finite volume approximation of the equations,

$$\Delta U_c = \frac{\Delta t}{\Delta A_c} \left\{ \begin{array}{l} [\dfrac{F_1 + F_2}{2}(y_2 - y_1) - \dfrac{G_1 + G_2}{2}(x_2 - x_1)] \\[2mm] -[\dfrac{F_3 + F_4}{2}(y_3 - y_4) - \dfrac{G_3 + G_4}{2}(x_3 - x_4)] \\[2mm] +[\dfrac{G_1 + G_4}{2}(x_4 - x_1) - \dfrac{F_1 + F_4}{2}(y_4 - y_1)] \\[2mm] -[\dfrac{G_2 + G_3}{2}(x_3 - x_2) - \dfrac{F_2 + F_3}{2}(y_3 - y_1)] \end{array} \right\} \tag{8}$$

In Eq. 8, ΔA_c is the area of the control volume and Δt is the time step. The unsteady fluxes are computed using

$$\Delta F_c = \Delta f_c \Delta y^l - \Delta g_c \Delta x^l$$
$$\Delta G_c = \Delta g_c \Delta x^m - \Delta f_c \Delta y^m$$

(9)

in which $\Delta f_c = \dfrac{\partial F_c}{\partial U_c}$, $\Delta g_c = \dfrac{\partial G_c}{\partial U_c}$ and

$$\Delta x^l = 0.5(x_2 + x_3 - x_1 - x_4) \qquad \Delta x^m = 0.5(x_3 + x_4 - x_1 - x_2)$$
$$\Delta y^l = 0.5(y_2 + y_3 - y_1 - y_4) \qquad \Delta y^m = 0.5(y_3 + y_4 - y_1 - y_2)$$

(10)

Solutions obtained using Eq. 5 are second-order accurate in space and hence contains numerical oscillations in the vicinity of the shock front. In the present work, the procedure outlined by Jameson et al. (1981) has been used for smoothing the oscillations the instant they are generated.

DOMAIN DECOMPOSITION

As previously mentioned, our parallel code is based on domain decomposition principles. These techniques have been in use for the past 30 years and hence are not new to numerical modelers. In the initial stages these techniques were applied to perform large computations on machines with insufficient memory. To circumvent the memory problem, the domain was divided into sub-domains. Computations were carried out over the smaller domains one at a time, during which time the other sub-domains resided on tape. With the growth in computational hardware, the DD methods were pushed to the background. However, with the advent of parallel computing these methods are again blossoming (Smith et al., 1996). The central theme in these methods is to divide the original mathematical problem to a set of sub problems that can be applied over these smaller regions. A processor is assigned to each sub-domain. To ensure load balancing among the processors, it is important to divide the domain such that each processor performs an equal number of computations.

The domain in explicit FV/FD schemes has an inherent partitioning in it as the grid nodes are separated by a uniform distance. However, before the solution is marched with time it is necessary to ensure that each processor has all the required flow information for solving the equations at all its grid nodes, for the new time level. While this is not a problem for the interior nodes, special treatment is required at the interface boundary nodes of the processors, as some of the variables required for solving the equations at these nodes lie in the domain of other processors. These points are referred to as ghost points. For instance, consider a hypothetical case where the flow domain is divided across two processors, and let the node *(i,j)* in the computational molecule, shown in Figure 4, be located at the boundary between processors 1 and 2. Solving the equations at this node would require the processor 1 to have the flow information at nodes *(i+1,j-1)*, *(i+1,j)* and *(i+1,j+1)*, which are in the domain of processor 2. As evident some inter-processor communication is required at the end of every time step, so as to arrive at the correct solution. This is achieved through the use of Message Passing Interface, MPI (Gropp et al., 1994). MPI is a portable message passing tool. By portability, we mean running the code on different architectures with the same efficiency. The codes written using MPI are portable to all platforms supporting the message-passing model, which includes the distributed-memory parallel supercomputers (MPP's), shared-memory multiprocessors (SMP's) and networks of workstations (NOW's). These codes can be used on virtually all parallel platforms including the IBM SP, Intel Paragon, SGI Origin, DEC Alpha, HP Exemplar, Linux, SUN Solaris and Cluster of workstations running MPI. MPI is fast becoming a standard among the different HPC vendors.

PARALLEL PLATFORM

In the present investigation all the computations have been performed on the HP-Convex Exemplar X-Class Scalable Supercomputer (SPP2200), located at the Center for Computational Sciences in the University of Kentucky campus. While the various details of the machine are listed in Table 2, we would like to briefly discuss its architecture. The Exemplar is a distributed shared memory machine. It has 4 hyper-nodes, with each hyper-node containing 16 processors. While the processors within the hyper-node have a shared memory, the hyper-nodes themselves have access through the network to the distributed memory (Figure 5).

PERFORMANCE PARAMETERS

The two commonly used terms for describing the performance of a parallel code among the engineering community are speedup (*S*) and efficiency (*E*). These parameters give an idea of the performance improvement achieved by parallelizing a code. The speedup is a measure that captures the benefit of solving the problem in parallel and is defined as (Kumar et al. 1994)

$$S = \frac{time\ required\ for\ a\ single\ processor}{time\ required\ for\ p\ processors} \tag{11}$$

Even though ideally one can expect a speedup of *p* on *p* processors, due to communications between processors, the speedup is always less than *p*. The efficiency is a measure of the time the processor spends in the computational phase and is defined as the ratio of speedup to the number of processors (P). It can be expressed as

$$E = \frac{S}{P} \qquad (12)$$

TEST PROBLEM

The definition sketch of the test problem is shown in Figure 6. Of interest, are the flows produced by a gradual transition. As shown in the definition sketch, the initial conditions in the *3.45m* long channel are a flow depth of *0.03m* and a longitudinal discharge of *0.03m³/s*. Starting from $x = 1m$, the cross sectional width of the channel is gradually reduced from *1.2m* to *0.6m* over a distance of *1.45m*. The angle of contraction is *22°*. The change in the cross sectional area causes a flow disturbance in the domain, which starts propagating from the boundaries with time. A Courant number of *0.4* was used in the evaluation of time step (Chaudhry, 1993). The transient simulation was carried out for a total time of *0.5s*.

RESULTS

The number of grid nodes in the x and y directions are *1200* and *64* respectively. The total number of vertices in the flow domain is *76,800,* and the degrees of freedom are *230,400*. The parallel code was run on up to *32* processors on the HP Exemplar. By using *32* processors, we have crossed one hyper-node (i.e., *16* processors), and hence the results also indicate the performance of the code utilizing distributed memory. The performance data of the code is shown in Table 3. By ensuring one-to-one communication among the processors, as the results indicate, the code is well optimized. The number of time steps required for the transient solution to reach a time level of *0.5s* is *277* ($\Delta t = 1.8E\text{-}03$). With an increased number of processors, the efficiency is slightly less than the theoretical values. This can be attributed to the size of the problem. Note that on *32* processors, the number of vertices per processor is only *2400*. This low number offsets the balance between the computation cost and communication cost. Since parallel environments are meant for large problems, the time spent in computation kernels should be large enough. This can be achieved only for large size problems. The computation kernel for explicit schemes, is in general, the routine that computes the flux vector and solves for the unknown variables at the new time level for all the interior grid nodes. In the present formulation, the time spent in this routine is more than 98% of the total time. For finite element formulations, the kernel that requires maximum time is the routine wherein the global matrix is assembled and solved for. Figures 7 and 8 are the plots comparing the speedups and efficiencies with their theoretical values.

To check the reliability of the parallel code in terms of the accuracy of the solution, we have used the solution of the sequential code as the base for comparing the flow profiles. Any discrepancy in the flow profiles of the serial and parallel codes will defeat the whole purpose of running the code in parallel environment.

The transient flow depths at $t=0.5s$ along the boundary wall and centerline of the channel, obtained using both the serial and parallel codes (over 8 and 16 processors), are shown in Figures 9 and 10. As indicated, the flow profiles are in close agreement, which reinforces the reliability of the parallel code in terms of its accuracy. Since the code is explicit, one can theoretically visualize that the solutions should match with one another.

The flow domain was truncated equally among the processors, as indicated by the number of vertices per processor in Table 3. By this we have ensured that the load is fairly well balanced among all the processors. The number of ghost points for the end processors is 64 and for the interior ones is 128. As previously mentioned, the inter processor communication was achieved through the use of MPI. Various tests performed for different flow conditions did not change the over-all trend of the presented results. .

CONCLUSIONS

High sustained performance of a two dimensional parallel finite volume code has been demonstrated in this work. The characteristic features of the parallel code are (i) domain decomposition principles, (ii) MPI for inter-processor communication, and (iii) scalable. On a shared memory platform with 16 processors, the code achieves a speedup of 15.5 and an efficiency of 97%.

REFERENCES

Chaudhry, M.H. 1993. Open Channel Flow, Prentice Hall, NY.

Dongarra, J.J., Duff, I.S., Sorensen, D.C. and van der Vorst, H.A. 1998. Numerical Linear Algebra for High Performance Computers, SIAM.

Gropp, W., Lusk, E. and Skjellum, A. 1994. Using MPI: Portable Parallel Programming with a Message Passing Interface, MIT press.

Hirsch, C. 1996. Numerical Computation of Internal and External Flows, John Wiley & Sons, New York.

Jameson, A., Schmidt, W. and Turkel, E. 1981. "Numerical solutions of the Euler equations by finite volume methods using Runga-Kutta time stepping schemes", *AIAA 14th Fluid and Plasma Dynamics Conference, Palo Alto*, CA, AIAA, pp.81.

Kumar, V., Grama, A., Gupta, A. and Karypis, G., Introduction to Parallel Computing: Design and analysis of algorithms, Benjamin Cummings Inc., 1994.

MPI homepage, http://www.mpi-forum.org/, October 1999.

Smith, B.F., Bjorstad, P. and William Gropp. 1996. Domain Decomposition: Parallel Multilevel Methods for Elliptic Partial Differential Equations, Cambridge University Press.

Spaulding, M.L. and Blumberg, A.F. 1997. Estuarine and Coastal Modeling, Proc. of the fifth International Conference, Alexandria, VA.

Spaulding, M.L. and Cheng, R.T. 1995. Estuarine and Coastal Modeling, Proc. of the fourth International Conference, San Diego, CA.

Table 1. Performance data of supercomputers (Dongarra et al. 1998)

Year	machine	Speed
1979	CRAY –I	160 Mflop/s
1983	CYBER 205	400 Mflop/s
1986	CRAY-2	2 Gflop/s
1990	NEC SX-3	22 Gflop/s
1992	TMC CM-5	130 Gflop/s
1994	VPP-500	205 Gflop/s
1997	Intel ASCI 'Red'	1.8 Tflop/s

Table 2. Hardware characteristics of the SPP Exemplar
Source: http://spp.uky.edu

COMPONENT	DESCRIPTION
Architecture	Cross-bar based symmetry
Number of Processors	64
Number of hypernodes	4
Processors	PA-RISC 8200
Clock Speed/Processor Cycle Time	200 MHz /4.0 ns
Peak performance/processor	800 MFLOPS
Operating System	SPP-UX (similar to Unix)

Table 3. Flow in transitions on Exemplar; fixed size mesh of
76,800 vertices (230,208 unknowns)

Number of Processors	Vertices in the domain	Number of time steps	Time (s)	Speedup	Efficiency (%)
1	76,800	277	3094.2	1.00	100
2	38,400	277	1547	2.00	100
4	19,200	277	772.6	4.00	100
8	9,600	277	387.9	7.98	100
16	4,800	277	199.1	15.5	97
32	2,400	277	117.4	26.4	82.6

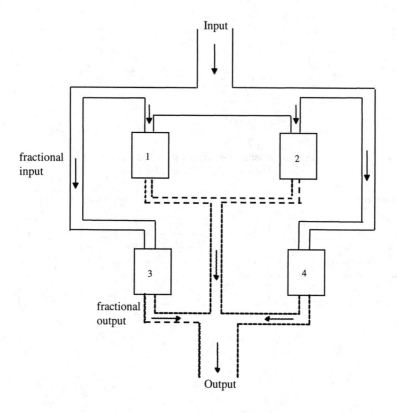

Figure 1. Framework of a parallel formulation (The four processors are
           indicated by the numbered boxes)

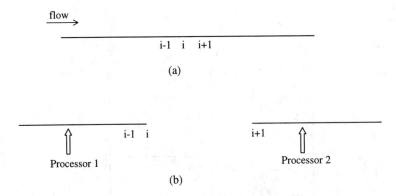

Figure 2. Domain Decomposition over 2 processors for 1d domain
(a) undivided domain (b) domain of individual processors along with the nodes

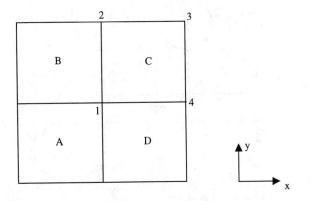

Figure 3. Computational molecule of 2d grid in finite volume discretization

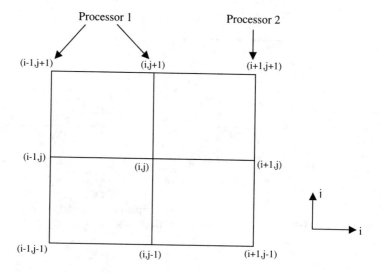

Figure 4. Definition sketch of ghost points
(The ghost nodes for processor 1 are indicated by x)

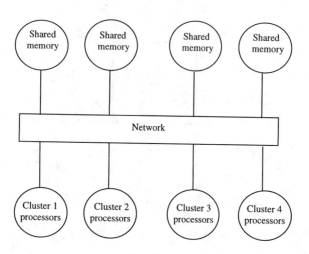

Figure 5. Distributed Shared memory architecture of the Exemplar

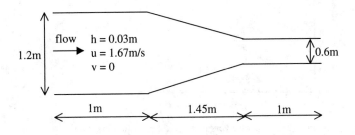

Figure 6. Definition sketch of the test problem

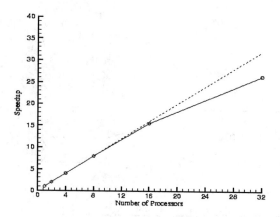

Figure 7. Speedup versus the number of processors
(--- theoretical, — observed)

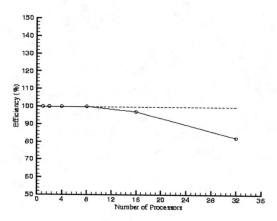

Figure 8. Efficiency versus the number of processors
(--- theoretical, — observed)

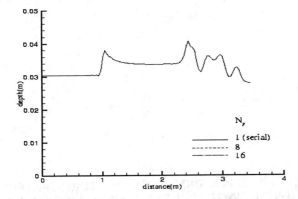

Figure 9. Transient depth profile along the inner boundary wall
(N_P = Number of Processors)

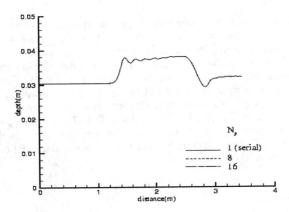

Figure 10. Transient depth profile along the centerline
(N_P = Number of Processors)

INTEGRATING HIGH PERFORMANCE COMPUTING STRATEGIES WITH EXISTING FINITE ELEMENT HYDRODYNAMIC CODES

Scott A. Yost and Prasada Rao
Department of Civil Engineering
University of Kentucky
Lexington, KY 40506-0281

ABSTRACT

Using serial finite-element codes for large scale modeling of estuarine flows is limited by the computational environment. As the modeling domain becomes more complex and larger, it is imperative to move the application platform from serial to parallel architectures. It is our belief that all future scientific computing in general will be on parallel computers. Gaining synergy from the recent advances made in the field of high performance computing, we demonstrate the advantages of running three different serial finite-element codes in a parallel environment. The serial codes, including RMA2, have been ported over to parallel environment using Domain Decomposition principles based on MPI. The tests, up to 32 processors, indicate that these parallelized serial codes are efficient, scalable and portable.

INTRODUCTION

The past 3 decades saw a large-scale application of numerical models for simulating a wide variety of flows. In the field of free surface flows, and estuarine hydraulics in particular, the advantages offered by traditional finite element (FE) formulations are many. For a review of the works done using FE codes, we refer the readers to the papers published in the previous proceedings of the Estuarine and Coastal Modeling conference (Spaulding and Blumberg, 1997 and Spaulding and Cheng, 1995). All these advantages coupled with the rich theory behind FE formulations have prompted many application researchers to use these FE codes for simulating problems of their interest. Over time numerical developers have modified their codes by incorporating the ongoing developments in FE theory. In doing so, they assured that the codes were robust and are capable of simulating the required physics of flow, and hence they have retained their application audience. Code evolution by itself is not new in the field of computational fluid dynamics. Almost every existing code has a long cherishable history. With time the capacity of all these numerical codes to simulate the physics of flow has been enhanced by primarily inserting new subroutines into them. The growth in visualization software also played a significant contribution in enhancing the capabilities of these codes.

Ever since John Backus wrote the first Fortran compiler, the popularity that Fortran enjoys among the engineering community continues to be immense. While the popularity cannot be doubted, it is well known that the code developers in Fortran were not completely satisfied with the programming environment. The alternative of developing a code in object oriented language, although more appealing, always had the risk of not being as popular when compared to its Fortran counterpart among the application audience. So FE codes continued to evolve under the same language banner.

In trying to apply serial FE codes to larger-scale domains, as necessitated by real life applications, a limitation arose in the amount of time spent in the computation kernel. Since all these codes have been written for a serial environment (i.e., over one processor), performing large chunks of computations involving a large amount of data has proven to be tedious job requiring a lot of CPU time. The only way the code can be significantly accelerated is by running it in a parallel environment, over more than one processor. The computational demands of modelers have been the principal driving forces in the development of powerful parallel computers. High processing speed is an essential factor for applications relating to real life simulations.

Relying on the compiler directives does not automatically generate an efficient parallel code, as the compiler technology is not mature enough (Kumar et al., 1994). The parallel programmer needs to rewrite some portions of the serial code and imbed into it the various high performance computing (HPC) tools that are being developed by computer scientists and applied mathematicians. One primary HPC tool that is drawing the attention of many groups is the domain decomposition (DD) technique. According to this, the entire domain is divided into smaller sub-domains, and a separate processor is assigned to each sub-domain. Such approach enables the computations to be done concurrently. Theoretically, the execution time of the serial code can be reduced by a factor equivalent to the number of processors over which the parallel code is run.

Both hardware and software HPC tools have matured considerably over the last decade. The most fruitful years of parallel code development will be in the coming years when HPC tools are used by more application groups. To this end, the objective of this work is two fold:

1. To demonstrate that all the HPC tools can be incorporated into an <u>existing</u> serial code with little effort, eventually arriving at a robust parallel code.
2. To demonstrate the advantages of running a code in parallel environment.

The first objective is a very challenging goal. Rather than developing a parallel code from the beginning (which many times is relatively simpler), we like to show that all existing serial FE codes can be converted into efficient parallel ones. To keep continuity with the discussion, the reader may like to glance through the results of the parallel RMA2 code, presented in the applications section, before continuing with the rest of this discussion. By effectively amalgamating the HPC tools into existing serial codes, we address the major limitation of the serial codes. The programming language of the parallel code continues to be in Fortran (the serial code language), and hence we hope that the parallel code will continue to enjoy the same popularity as the serial one. By this we also want to ensure that the users have a smooth transition while moving from serial to parallel versions of their favorite modeling codes.

In this work, three serial codes are parallelized by imbedding different high performance computing strategies. This aspect alone significantly differs from the previous evolutions of the serial codes by focussing on efficiencies and simplified numerical approximations. This work, while gaining synergy from the recent developments made in the field of high performance computing, targets researchers in the estuarine modeling community. But it also encompasses a general audience, those who are interested in solving any set of partial differential equations.

The outline of the paper is as follows: after presenting the governing equations and the numerical discretization, we discuss in detail the domain decomposition algorithm along with the Message Passing Interface (MPI). Next, we illustrate the features of the parallel machine over which the computations have been carried out. The parallel performance measurements are indicated next. The strategies that we discuss during this document have been implemented in three existing serial FE codes (i) a two dimensional least square finite element code for solving Poisson's equation (ii) a Petrov-Galerkin FE code that solves the three-dimensional turbulent Navier-Stokes equations (Yost, 1995) and (iii) RMA2 code (Donnell and King 1997). In the applications section, we discuss the various performance details of these three parallel codes.

GOVERNING EQUATIONS

Since the first and the last serial FE codes are relatively well known, the reader is referred to the literature for all numerical details; the 2D Least Squares Poisson Equation code is outlined by Jiang et al. (1994), and the RMA2 code is outlined by Donnell and King (1997). The 3D finite element code will be touched on briefly, but more detail can be found elsewhere (Yost, 1995). The well-known Navier Stokes Equations for incompressible flow can be written in tensor notation as follows

$$\frac{\partial u_i}{\partial t} + \frac{\partial u_i u_j}{\partial x_j} = \mathbf{X}_i - \frac{1}{\rho}\frac{\partial p}{\partial x_i} + \frac{\mu}{\rho}\frac{\partial}{\partial x_j}\left(\frac{\partial u_i}{\partial x_j} + \frac{\partial u_j}{\partial x_i}\right) \tag{1}$$

where ρ is the density, μ is the viscosity, p is the pressure, X_i are the body forces and u_i are the velocities. To complete the momentum equations, the continuity equation (incompressibility constraint) is

$$\frac{\partial u_i}{\partial x_i} = 0 \tag{2}$$

These equations are modified to include buoyancy forces (temperature and salinity), a Large-Eddy simulation turbulence model, and a moving free-surface. The details of which are found in the literature.

NUMERICAL FORMULATION

The equations are discretized using 27 node tri-quadratic/tri-linear (Q2/P1) elements, where the velocity is defined at all 27 nodes, and the pressure is defined at the 8 corner nodes. Approximating the dependent variables with the appropriate shape functions,

$$\tilde{u}_i = \sum_{a=1}^{27} N_a u_{ia} \text{ and } \tilde{p} = \sum_{a=1}^{8} M_a p_a \tag{3}$$

multiplying the approximate equations by the associated Petrov-Galerkin weighting functions, and integrating over the entire domain leads to the following residual equations for Eqs. 1 and 2, respectively.

$$R_{u_i} = \int_\Omega (N+P)_a \left[\rho\left(\frac{\partial \tilde{u}_i}{\partial t}\right) - X_i + \frac{\partial \tilde{p}}{\partial x} - \mu\frac{\partial}{\partial x_j}\left(\frac{\partial \tilde{u}_i}{\partial x_j} + \frac{\partial \tilde{u}_j}{\partial x_i}\right)\right]\partial\Omega \tag{4}$$

$$R_c = \int_\Omega M_a\left(\frac{\partial \tilde{u}_i}{\partial x_i}\right)\partial\Omega \tag{5}$$

Time discretization is performed using the Newmark Method (Hughes, 1987). After applying the appropriate initial and boundary conditions, these equations can be solved numerically. Since the system is non-linear, the well known Newton-Raphson routine is used to solve. The general algorithm is

$$\Delta\mathbf{r}_i = -\mathbf{J}^{-1}\mathbf{F}$$
$$\mathbf{r}_{i+1} = \mathbf{r}_i + \Delta\mathbf{r}_i$$

$$(6)$$

where J is the Jacobian matrix (the partial derivative of Eqs. 4 and 5 with respect to the dependent variables), F is the residual vector (Eqs. 4 and 5) and r is the solution vector. There are many nonlinear system solvers available, many of which are more efficient than the one given. The objective of this work precludes the modeler's choice of a system solver; the results will be consistent.

DOMAIN DECOMPOSITION

As previously mentioned, our parallel formulation relies on domain decomposition techniques. With this method, we divide the whole domain into as many parts as the number of processors and run the code simultaneously over all the sub domains. This divide and conquer paradigm is gaining wide acceptability among the different parallel computing groups around the world. The official web site of the domain decomposition school of thought (http://www.ddm.org) contains more details and papers discussing these techniques.

The philosophy underlying a domain decomposition formulation for a finite element code can be written in three steps,

1. Each processor executes the serial code for all the elements lying in its domain.
2. At the end of the execution, each processor makes available the required flow information to its adjacent processors.
3. Each processor writes the required output before the solution is iterated with time.

Figure 1 shows one possible way of dividing the domain across 16 processors. For the solution to be iterated with time, it is a prerequisite that all the processors have all the flow information required for solving the equations. Since some of this information is available on its surrounding processors, some amount of communication is required among the processors. In this work we have utilized MPI, or Message Passing Interface.

Message Passing Interface's goal is to provide a standard for writing message-passing programs. Initiated by the MPI forum in 1992, it is fast becoming a standard among all the vendors who are producing parallel machines. MPI is a portable message-passing standard that facilitates the development and application of parallel applications on a wide range of machines. This standard defines the syntax and semantics of a core of library routines useful to a wide range of users writing parallel programs, both in Fortran 77 or C. Commercial and free public-domain implementations of MPI are widely available. While it is not possible to review the various sites that contain more information about MPI, we would like to refer the readers to the office MPI web site (http://www.erc.MsState.Edu/ labs/ hpcl/ projects/mpi/findex.html), which contains a wealth of information. Codes written using MPI can run on both tightly-coupled massively-parallel machines (MPPs) and on networks of workstations (NOWs).

The MPI standard was developed over a year of intensive meetings and involved over 80 people from approximately 40 organizations. While designing the standards in MPI, the developers incorporated the best features of several systems, rather than relying on one. MPI has roots in PVM, Express, P4, Zipcode and Parmacs (Dongarra et al., 1998).

PARALLEL PLATFORM

In the present investigation all the computations have been performed on the parallel HP-Convex Exemplar, located at the Center for Computational sciences in the University of Kentucky campus. Convex Exemplar, while using scalable parallel processing (SPP) technology, offers a massively parallel processing platform for the application audience. The various hardware details of the machine are listed in Table 1 with more information available at http://spp.uky.edu. The exemplar consists of 4 hyper nodes, each hyper node containing 16 processors. The 16 processors on a hyper node use shared memory while the memory across the hyper nodes is distributed. The shared memory configuration is shown in Figure 2.

In all our runs we have used the modules in Portable Extensible Toolkit for Scientific Computation (PETSc) library (Satish et al, 1999). PETSc library, built at the Argonne National Laboratory, is a powerful set of freely available multi-platform compatible tools for the solution of large-scale problems modeled by partial differential equations. The software uses MPI for all interprocessor communication. An overview of PETSc can be found in Gropp and Smith (1994).

PARALLEL PERFORMANCE MEASUREMENTS

From an engineering point of view, the units by which the performance of a parallel code is measured are:

(i) Speedup: Speedup is the ratio of the time required for the serial to complete the execution to the time required by the parallel code. If t_s indicates the serial time, and t_p denotes the time over p processors, by definition the speedup over p processor machine is given as

$$S_P = \frac{t_s}{t_p}$$

The values of S_p ranges from 0 to P.

(ii)   Efficiency: The efficiency indicates how well the processors are made use
        of in the execution of the parallel code. Indicting efficiency by E, over a p
        processor system, it is given as

$$E = \frac{S_P}{P}$$

The values of E ranges from 0 to 1.

(iii)   Scalability: The scalability of a parallel code is a measure of its capacity in
        yielding increased speedup in proportion to the number of processors
        (Kumar et al. 1994) For a fixed size problem, it has been well documented
        that by increasing the number of processors, the efficiency is reduced. On
        the other hand, one can maintain a constant efficiency by increasing the
        size of the problem with an increased number of processors. A code
        exhibiting this property is known as scalable code. By size of the problem,
        we are addressing the total number of unknowns in the computational
        domain

APPLICATIONS

Problem I
        This problem details the performance of a parallel code that solves the
well known two-dimensional Poisson's equation. The flow equation was
discretized using the Least Square Finite Element (LSFE) approach. While
reviewing the entire LSFE formulation is beyond the scope of this work, those
interested in LSFE literature are referred to the works of Bramble and Schatz
(1970) and Jiang et al. (1994). The reasons that the LSFE approach is enjoying
high success among the heat and mass transfer modelers can be attributed to three
reasons. First, unlike the standard Galerkin and Petrov-Galerkin methods
formulations, LSFE methods lead to symmetric positive definite (SPD) matrix,
even for Navier Stokes equations. A SPD matrix can be solved efficiently using
an iterative solver. As we illustrate next, a typical FE code spends a large chunk
of computational time in the solver kernel. By using an appropriate solver coupled
to a preconditioner, the time spent in this phase can be substantially reduced.
Secondly, a LSFE code is free of any artificial tuning parameters. Thirdly, as
opposed to the Galerkin approach, LSFE method is not subject to Ladyzhenskaya-
Babuska-Brezzi (LBB) condition; thus equal order interpolation functions can be
used for all the variables in the flow equations.

        The flow domain in this test problem is a unit square. The total degrees of
freedom in this domain are 1,484,848. The global matrix was stored in
compressed sparse row format with a block size of 3. The end parallel code was

run using up to 32 processors. Hence, we have crossed one hyper node. In evaluating the efficiency of this code, we have used the conjugate gradient method for solving the equations without any preconditoner. Table 2 documents the observed speedup along with the efficiencies as a function of the number of processors. The number of sub-domains equals the number of processors. This information, along with the theoretical values, is shown in Figures 3 and 4. The residual tolerance of the solution was set to $1x10^{-6}$.

As pointed out earlier, a substantial time is spent in the solver phase. For this data, Figure 5 (over 8 processors), illustrates the time spent in the solver phase for various Krylov subspace solvers. As pointed out by Eriksson et al. (1996), upon selecting an appropriate Krylov solver and preconditioner, substantial savings can be achieved in this phase of the code. The Gustafson style scalability data is shown in Table 3. Note that by maintaining approximately the same number of unknowns per processor, the time spent per iteration is almost linear.

Problem II

Many of the flows in fluid dynamics require the solution of transient Navier-Stokes equations. The original Fortran serial code (Yost, 1995), which is capable of solving the equations presented earlier has been ported over to the HP parallel environment under the PETSc framework. This is a challenging task for a parallel programmer as s/he needs to restructure part of the serial code. One such task is in assigning sub-domains to the different processors along with the associated mapping of flow information from local domain to global domain.

The parallel code was applied for solving the well-known lid-driven cavity flow with a Reynolds number of 100. The number of elements in the domain is 900, number of nodal Points = 8649, and degrees of freedom/node = 4. The global matrix (of size 22305) was stored in compressed sparse row format, of block size 3. The resulting system was solved using the BiCGSTAB (Van der Vorst, 1992) method coupled to a Jacobi preconditioner. The parallel code was run using up to 32 processors, with each processor containing an equal number of elements in its sub-domain. For the time-dependent code our approach has focussed on data decomposition, message passing and on reusing the data structures at each time level. By performing load balancing, inspector/executor optimizations, and data copy elimination, we have achieved high relative performance. The observed speedups are shown in Table 4. By setting a low residual error tolerance on each processor ($1x10^{-8}$), we have ensured that the solutions are similar.

Problem III

RMA2 is a two-dimensional depth averaged serial finite element hydrodynamic model. It computes water surface elevations and horizontal velocity components for 2D free surface flows by solving the Reynolds equations for turbulent flows. The equations are solved by the finite element method using the Galerkin method of weighted residuals. Both steady and unsteady problems can be analyzed using RMA2. The original RMA2 serial code was developed by the Water Resource Engineers in 1973 for the Walla Walla district Corps of Engineers. Subsequent modifications made by the researchers of Resource Management Associates (RMA) and by the Waterways Experimentation Station (WES) resulted in the current version of the code, which is being used by different organizations (Donnell and King 1997). All the evolutions in this code had concentrated on improving its modeling capabilities by imbedding new improved subroutines. The source code is freely available in different public domains.

The source code, which spans over 56 subroutines, has around 15,000 lines in it. We were able to port the code into the PETSc parallel environment, which incorporate various HPC tools. To this end we had to rewrite certain portions in the serial code. The global matrix was stored in compressed sparse row format where only the non-zero values along with their matrix location indicies are stored. This has enabled us to use a variety of Krylov subspace solvers and preconditioners for solving the linearized equations.

The parallel code was run for a sample date file containing 200 elements. This resulted in the number of unknowns to be 1351. Owing to the relatively small size of the problem, we had confined the application to 2 processors. The focus was laid more on the accuracy of the parallel solution, an aspect most crucial to the modeling and design community. Over 2 processors, the domain was divided equally among the processors, as schematically shown in Figure 6. The associated speedups are shown in Table 4. The Kylov solver used is BicGSTAB method coupled to Additive Schwarz preconditioner. The degree of overlap is 2.

To check the accuracy of the parallel solution, we have compared the norm of residual vector ($r = Ax\text{-}b$) as a function of time for the solutions obtained using both 1 and 2 processors. The close agreement of the residual norm profiles, as indicated in Figure 7, indicates the accuracy of the solution obtained using the present formulation.

CONCLUSIONS

In this work, we have demonstrated the advantages of running an existing serial finite element code in parallel environment. Three serial codes have been ported over to parallel environment, by incorporating into them high performance tools, based on Domain Decomposition concepts. All inter processor communication was achieved through Message Passing Interface (MPI). The performance results of all the parallel codes are promising. The observed speedups, up to 16 processors, are almost linear. It is our hope that this work will inspire our colleagues to run their application codes on parallel machines.

REFERENCES

1. Barry F. Smith, Petter Bjorstad and William Gropp. 1996. Domain Decomposition: Parallel Multilevel Methods for Elliptic Partial Differential Equations, Cambridge University Press.
2. Bramble, J.H. and Shatz, A.H. 1970. "On the numerical solution of elliptic boundary value problems by least square approximation of the data", In numerical solution of PDE (edited by Hubbard B.), vol.2, pp.107-133, Academic Press, NY
3. Dongarra, J.J., Duff, I.S., Sorensen, D.C. and van der Vorst, H.A. 1998. Numerical Linear Algebra for High Performance Computers, SIAM.
4. Eriksson, K., D. Estep, P. Hansbo and C. Johnson. 1996. Computational Differential Equations, Cambridge Univ. Press.
5. Gropp, W. and Barry F. Smith. 1994. "Scalable, extensible, and portable numerical libraries", in Proceedings of Scalable Parallel Libraries Conference, 87-93, IEEE, Los Alamitos, CA.
6. Gropp, W., E. Lusk, and A. Skjellum. 1994. Using MPI: Portable Parallel Programming with the message passing Interface, MIT Press.
7. Hughes, T.J. 1987, *The Finite Element Method*, Prentice-Hall Inc., Englewood Cliffs, NJ.
8. Jaing B.N., Lin T.L. and Povinelli, L.A. 1994. "Large scale computation of incompressible viscous flows by least squares finite element method", Comput. Meth. Appl. Mech. Engng 114, 213-231.
9. Kumar, V., Grama, A., Gupta, A. and Karypis, G. 1994. Introduction to Parallel Computing: Design and analysis of algorithms, Benjamin Cummings Inc.
9. Satish Balay, William Gropp, Lois Curfman McInnes, and Barry Smith. 1999. PETSc home page, http://www-fp.mcs.anl.gov/petsc/index.html, September.
10. Spaulding, M.L. and Blumberg, A.F. 1997. Estuarine and Coastal Modeling, Proc. of the fifth International Conference, Alexandria, VA.
11. Spaulding, M.L. and Cheng, R.T. 1995. Estuarine and coastal Modeling, Proc. of the fourth International Conference, San Diego, CA.
12. Donnell, B.P. and Ian King. 1997. Users guide to RMA2 WES Version 4.3.
13. Van der vorst, H.A. 1992. "A fast and smoothly converging variant of BiCG for the solution of non symmetric linear systems", SIAM J. Sci. Stat Comput., 13:631-644.
14. Yost, S.A. 1995. Three-Dimensional Non-Hydrostatic Modeling of Free-Surface Turbulent Flows and Transport of Cohesive Sediment, Doctoral Thesis in partial fulfillment of Doctor of Philosophy Degree, University of Michigan, Ann Arbor, MI.

Table 1. Hardware characteristics of the SPP Exemplar
Source: http://spp.uky.edu

COMPONENT	DESCRIPTION
Architecture	Cross-bar based symmetry
Number of Processors	64
Number of hypernodes	4
Processors	PA-RISC 8200
Clock Speed/Processor Cycle Time	200 MHz /4.0 ns
Peak performance/processor	800 MFLOPS
Operating System	SPP-UX (similar to Unix)

Table 2. Performance data for problem 1

Number of Processors	Time(s)	speedup*	Efficiency(%)
4	2045.6	1.00	1.00
8	1031.0	1.98	0.99
16	535.2	3.82	0.96
32	294.3	6.95	0.86

* speedup is based on 4 processors minimum

Table 3. Gustafson style Scalability for problem 1 (Scaled Speedup)

Unknowns	Processors	Unknowns/ Processor	Time(s)	Time/iteration
184,512	4	46128	86	0.18
371,712	8	46464	120.2	0.18
559,872	12	46656	154.1	0.19
762,048	16	47628	208.1	0.22
1,543,248	32	45414	317.7	0.25

Table 4. Performance data for the test problem 2

Number of processors	Total Time (s)	Speedup	Efficiency (%)
1	3549.5	1.00	100.0
2	1783.1	1.99	99.5
4	905.3	3.92	98.0
8	470.1	7.55	95.3
16	235.0	15.10	94.3
32	118.6	29.92	93.5

Table 5. Performance data for Parallel RMA2
(Number of time steps =26)

Processors	Time(s)	Speedup
1	206.1	1.00
2	103.6	1.99

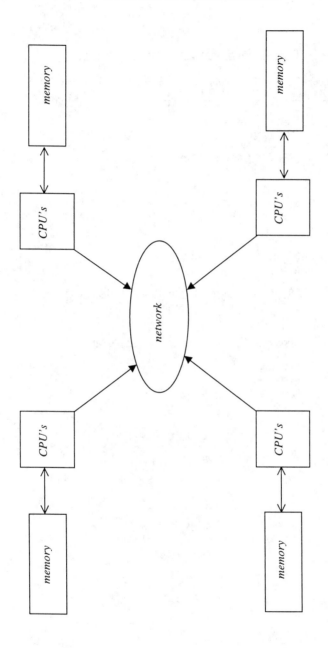

Figure 1. Architecture of the HP Exemplar, located at the University of Kentucky

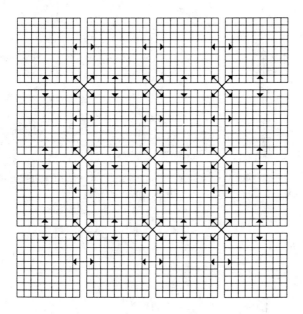

Figure 2. Two dimensional domain decomposition across 16 processors

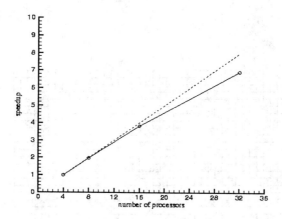

Figure 3. Speedup as a function of number of Processors for Problem 1
(- - - Theoretical, —— Observed)

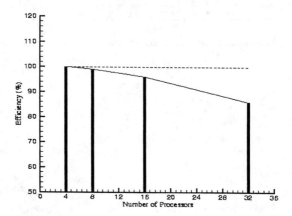

Figure 4.Efficiency as a function of number of Processors for Problem 1
(- - - Theoretical, —— Observed)

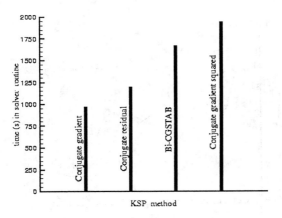

Figure 5. Effect of Krylov solver for problem 1

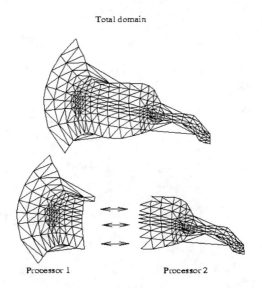

Figure 6: Domain decomposition for RMA2 grid over 2 processors

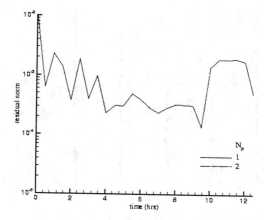

Figure 7. Residual norm as a function of time for the parallel RMA2 (N_p = number of processors)

Lake Michigan Forecast System and the Prospects for a Sediment Transport Model

Yi-Fei Philip Chu[1], Keith W. Bedford[2] and David J. S. Welsh[1]

Abstract

The main focus of the NSF funded EEGLE (Episodic Events –Great Lakes Experiment) project is to study the recurring southern Lake Michigan turbidity plume in cross-margin transport and sediment deposition through intensive field measurements, and modeling programs. The development and implementation of a nowcast/forecast and a sediment transport modeling component for Lake Michigan is critical to the success of this project since the model predictions can be used to assist scientists in cruise planning, shipboard surveys, Lagrangian measurements and data collection programs.

Daily nowcasts of the Lake Michigan Modeling System have been up and running since 1998. Wind fields, water levels, currents and lake surface temperature maps are available on the GLFS web site to assist EEGLE scientists in cruise planning and field sampling. Comparison of predicted water temperatures and buoy observations shows good correlation. However, predicted surface temperatures tend to be half a degree to one and a half degrees lower than the observed data. Short term (6-hour and 24-hour) Lake Michigan forecasts will be achieved early next year by incorporating the forecasted wind fields from a mesoscale meteorological model (MM5) to drive the lake circulation model.

A sediment transport model (CH3D-SED) has been implemented for Lake Michigan. This model is capable of taking multiple sources and sediment size classes as input. Bedload transport can be simulated as well as suspended transport. Preliminary tests show the typical recurring sediment resuspension plume event can be reproduced.

Introduction

The objective of the NSF-funded EEGLE (Episodic Events – Great Lakes Experiment) project is to study impact of the recurring southern Lake Michigan turbidity plumes in cross-margin transport and sediment deposition on the lake's ecosystem through intensive field measurements, data collection and modeling programs. The development and implementation of a

---

[1] Senior Research Associate, Dept. of Civil and Environmental Engineering and Geodetic Science, The Ohio State University, Columbus, OH 43210
[2] Professor and Chair, Dept. of Civil and Environmental Engineering and Geodetic Science, The Ohio State University, Columbus, OH 43210

nowcast/forecast modeling component for Lake Michigan is critical to the success of this project since the model predictions can be used to assist scientists in cruise planning, shipboard surveys, Lagrangian measurements and data collection programs. The information obtained from the nowcast/forecast modeling system is also the required input for sediment transport model.

The Great Lakes Forecasting System (GLFS) is a real-time coastal prediction system that was developed for the daily forecasting of surface water level fluctuations, the horizontal and vertical structure of temperature and currents, and wind waves in the Great Lakes (Schwab and Bedford, 1994). The system uses surface meteorological observations and forecasts from numerical weather prediction models as input. Lake circulation and thermal structure are calculated using a three-dimensional hydrodynamic prediction model (Blumberg and Mellor, 1987). Wind waves are calculated with a parametric wave prediction model (Schwab et al., 1984). Output from the models is used to provide information on the current state of the lake and to predict conditions for the next several days.

The system was originally developed as a demonstration project by Ohio State University (OSU) and NOAA's Great Lakes Environmental Research Laboratory (GLERL). The initial implementation of the system in 1993 and 1994 produced daily nowcasts of system variables for Lake Erie from April to December each year (Schwab and Bedford, 1994). Lake Erie hydrodynamic model calculations with a 2 km horizontal grid run on the CRAY T90 at the Ohio Supercomputer Center (OSC). The products from the system were made publicly available on the World Wide Web from OSU. In 1995, the system began to use mesoscale meteorological forecasts from NOAA's National Meteorological Center (NMC) Eta model to produce 6- and 24-hour forecasts of system variables for Lake Erie. In 1997, nowcasts of lake circulation and thermal structure for Lake Ontario were added to the OSU system and in 1998, Lake Michigan nowcasts were added (Schwab et al. 1999).

Model Description and Data Processing

The hydrodynamic model used for Lake Michigan is the Great Lakes version of the Princeton Ocean Model (POM). The model was configured with 2 km horizontal resolution and 20 vertical layers. The sigma-stretched vertical coordinate layers are more closely grouped near the water surface and lake bed. Internal and external time steps are set at 600 and 12 seconds to satisfy the CFL criteria. A typical 24-hour simulation requires 40 minutes of CPU on a 600MHz DEC ALPHA Linux workstation.

Meteorological data are obtained four times a day from the NOAA Great Lakes Environmental Research Laboratory. Then the data set is checked for missing fields and abnormal values. A nearest-neighbor technique is used to interpolate surface meteorological data to the uniform computational grids. Wind observations are adjusted to account for boundary layer stability and overland versus oversea roughness differences. Water levels at Milwaukee and Muskegon and surface water temperatures at the grid point closest to NOAA buoy 45007 are saved on an hourly basis; all other 2-dimensional water levels and currents and 3-dimensional temperature structures are saved every six hours.

The nowcasts started in mid-March of 1999 and continued through the end of 1999. Examples of interpolated wind fields, computed water levels, currents and surface temperatures are shown below: Julian day 286 (10/13) was chosen because a strong wind event occurred during that particular day. Interpolated wind fields indicate a strong northerly wind at a magnitude of about 25 knots (Figure 1). Water level fluctuates about 20 centimeters between the northern and southern part of the lake (Figure 2). Strong currents also follow the shoreline to the southern part of the basin (Figure 3). All the maps are color-contoured to indicate the magnitude and the arrows in the wind and current plots indicate wind and flow directions. All maps are updated daily and are available on the GLFS website.

Model and Data Comparison

Hourly surface water temperatures at the grid point coinciding with NOAA buoy 45007 were saved for comparison purposes. Hourly temperatures from NOAA buoy 45007 were also extracted and processed. Figure 5 and 6 show both the model results and the buoy observations from Julian days 98 to 266. Buoy observations are dotted lines and solid lines represent model results. Figure 5 shows the model temperature remains unchanged for almost 4 weeks until Julian day 124. Then the model tracks the diurnal cycle and the observation data quite well especially on Julian day 179. Even though the model was not able to hit the peaks with the correct magnitude, the model-predicted temperatures did track the temperature variations relatively well with no lags or phase shift. In addition, time series plots show the model underpredicts the water temperature during the whole simulation period. Table 1 summarizes the correlation coefficient and mean difference on a two-week basis. Correlation coefficient between the model and buoy observation ranges from 0.44 to 0.94 indicating that model predictions track the observation peaks and valleys fairly well except for the first and last two weeks. However, the model-predicted temperatures were consistently lower than the ones observed at buoy 45007 and the underprediction ranges from 0.27 to 1.39 degrees Centigrade.

The most probable cause for the model underprediction is the difference in the depths of the buoy location and the model's sigma coordinate. Buoy 45007 is located in the center of the southern lake Michigan basin, 43 miles southeast of Milwaukee, WI. Even though we increased the model vertical structure to 20 vertical layers, the top layer is still about 6 meters thick due to the depth of the location, while the temperature measurement instrument is only .6 meters below the water surface.

Table 1. Summary of the correlation coefficient and mean difference of surface water temperature

Date	Correlation coefficient	Mean difference (degrees C)
98 - 112	0.444	0.519
112 -126	0.634	1.108
126 - 140	0.849	0.273
140 - 154	0.945	0.843
154 - 168	0.815	0.598
168 - 182	0.921	1.395
182 - 196	0.883	1.040
196 - 210	0.854	1.337
210 - 224	0.867	1.330
224 - 238	0.549	1.154
238 - 252	0.682	0.954
252 - 266	0.498	0.807

Sediment transport model description

The CH3D-SED marine circulation and sediment transport model has been deployed in Lake Michigan. CH3D was initially developed as a stand-alone hydrodynamic code (Chapman et al., 1996). The SED sediment resuspension and transport sub-model was subsequently fully integrated with CH3D, resulting in the CH3D-SED code. CH3D has many features in common with the POM code mentioned above. CH3D predicts 3-dimensional current, temperature, and salinity fields, and 2-dimensional water surface elevation fields. The model solves conservation equations for mass, momentum, and thermal energy using finite difference discretizations on a curvilinear horizontal grid and a sigma layer (terrain following) vertical grid. A mode-splitting solution technique is used in which external (barotropic) and internal (baroclinic) motions are solved separately, allowing an enhanced baroclinic time-step. A level 2 $\frac{1}{2}$ turbulence closure is used. The external mode uses vertically-averaged equations to predict water surface elevations and mean horizontal velocities, then the internal mode re-introduces vertical variations. CH3D requires inputs of time-varying water surface winds and heat fluxes or air temperatures.

The SED sub-model simulates erosion, deposition, bedload transport, and suspended sediment transport for a number of user-specified sediment size classes. The particle size and density are specified for each class, along with initial and boundary conditions for bed and suspended sediment concentrations. Each time-step, CH3D supplies SED with grids of water depth and current. SED returns grids of bottom roughness, modified water depth, and the near-bottom water/sediment mixture density.

The SED model is based on the concept of an active layer at the top of the bed in which erosion, deposition, and bedload transport occur. The conservation of mass for a particular size class in the active layer is given by

$$\rho_s(1-p)\frac{\partial(\beta E_m)}{\partial t} + \tilde{\nabla} \cdot \tilde{q}_b + S_e - S_d - S_F = 0, \tag{1}$$

where ρ_s is the sediment density, p is the bed porosity, β is the size fraction by mass, E_m is the active layer thickness, $\tilde{q}_b$ is the bedload flux vector, S_e is the erosion sink term (movement of active layer particles into suspension), S_d is the deposition source term, and S_F represents the exchange of particles between the active layer and the strata below.

The erosion term is parameterized as

$$S_e = -\beta_2\left(D_V\frac{\partial(\rho_m C)}{\partial z}\right)_{z=a}, \tag{2}$$

where β_2 represents an availability factor for the size class, D_V is the vertical diffusivity for sediment, ρ_m is the total density of the sediment/water mixture, C is the concentration for the size class, and a is a near-bed reference elevation. The deposition term is parameterized as

$$S_d = \left(w_f \rho_m C\right)_{z=a+\Delta a}, \tag{3}$$

where w_f is the settling velocity. The cross-layer exchange term is parameterized as

$$S_F = \rho_s'(1-p)\frac{\partial}{\partial t}\left[\beta'\left(E_m - z_b\right)\right], \tag{4}$$

where the primed parameters are for the same size class below the active layer, and z_b is bed elevation.

Integrating (1) over all n size classes ($\Sigma \beta = \Sigma \beta' = 1$), using (4), and assuming constant ρ_s then gives an equation for the conservation of all active layer sediment:

$$\rho_s(1-p)\frac{\partial z_b}{\partial t} + \Sigma_n\left(\tilde{\nabla} \cdot \tilde{q}_b + S_e - S_d\right) = 0. \tag{5}$$

In the water column, SED conserves suspended sediment mass for each size class at each sigma layer using

$$\frac{\partial(\rho_m C)}{\partial t} + \frac{\partial(\rho_m Cu)}{\partial x} + \frac{\partial(\rho_m Cv)}{\partial y} + \frac{\partial(\rho_m Cw)}{\partial z} - \frac{\partial(\rho_m Cw_f)}{\partial z} = \frac{\partial}{\partial x}\left[D_H \frac{\partial(\rho_m C)}{\partial x}\right] + \frac{\partial}{\partial y}\left[D_H \frac{\partial(\rho_m C)}{\partial y}\right] + \frac{\partial}{\partial z}\left[D_V \frac{\partial(\rho_m C)}{\partial z}\right],$$

(6)

where u, v, and w are the CH3D current components for the sigma layer, D_V is as above, and D_H is the horizontal diffusivity for sediment.

The set of equations (1) –(6) contain several unknown quantities. To permit a solution, the primary unknowns are selected as z_b, the modified β and β' values, and the C values. The remaining unknowns are parameterized using expressions by auxiliary relations as specified in Spasojevic and Holly (1997).

SED uses an implicit numerical scheme that is forward in time and centered in space. The same horizontal grid is used as in CH3D. The solution procedure each time-step is:

1. simultaneously solve (5) and (1) for each sediment class; this gives β, β', and z_b predictions
2. solve (6) for each size class, one at a time, for each sigma layer; this gives $C(z)$ predictions
3. iterate steps 1 and 2 until convergence criteria are met
4. repeat steps 1 to 3 across the horizontal grid

Idealized test case

      The first test case was setup to simulate the annual episode of sediment resuspension along the southern shores in Lake Michigan, a phenomenon which has been well demonstrated by satellite observations. Figure 7 shows a NOAA satellite image of sediment concentration during March and April 1996. The same test case is used here to test the coupled hydraulics and sediment models: the test is described below.
      A 2-km grid with 160 x 250 points is used and the vertical grid has 13 layers. To simplify the test, only uniform wind input and two sediment size classes were used. The circulation and sediment transport are driven by a simplified wind field, which exists in the first day (linearly increases to 15 m/s during the first 18 hours, then remains unchanged until the end of the day) and becomes zero during the rest of the simulation period. Spatially, the wind is from the north and uniformly distributed over the lake. Two typical size classes of sediments (diameter 0.01 mm and 0.1 mm) are used in the simulation. The initial temperature is set to 4 degrees Centigrade throughout the lake. Heat fluxes at bottom and lateral boundaries and the water surface are set to zero; this simplification is reasonable in a short simulation. These initial conditions were setup for the current version of the code and tested on a single processor of the Ohio Supercomputer Center's Origin2000 by running 1008 iterations with a time step of 10 minutes. Figures 8 and 9 show sediment concentration contours in mg/l at a depth of 13 m, plus surface current vectors, for hours 12 and 24 of the simulation. It can be seen that the sediment plume predicted by the idealized CH3D-SED simulation qualitatively reproduces the satellite observed plume.

Conclusions and Future Work

Daily nowcasts of the Lake Michigan Modeling System have been performed since March of 1999. Wind field, water level, current and lake surface temperature maps are available on the GLFS website to assist EEGLE scientists in cruise planning and field sampling. Comparison of predicted water temperatures and buoy observations shows good correlation. However, model-predicted surface temperatures tend to be half a degree to one and a half degrees lower than the observed data. Short term (6-hour and 24-hour) Lake Michigan forecasts will be achieved early next year by incorporating the forecasted wind fields from a mesoscale meteorological model (MM5) into the lake circulation model.

A sediment transport model (CH3D-SED) has been implemented for Lake Michigan. Preliminary tests using uniform wind and two sediment classes show the typical recurring sediment resuspension event can be simulated. Future work will incorporate spatial varying wind forcing and multiple sediment grain sizes to this model. In addition, model results will be verified with sediment concentration measurements from the EEGLE field data.

Acknowledgment

This research is funded by National Science Foundation (NSF) and NOAA Coastal Ocean Process (CoPS). The authors would like to thank Mr. Yong Guo for his assistance with CH3D-SED simulations.

References

Blumberg, A.F., and G.L. Mellor, 1987: A description of a three-dimensional coastal ocean circulation model. Three Dimensional Coastal Ocean Models, Coastal and Estuarine Sciences, 5, N.S. Heaps [ed.] American Geophys. Union, Washington, D.C., 1-16.

Chapman, R.S., Johnson, B.H. and Vemulakonda, S.R., 1996: User's guide for the sigma stretched version of CH3D-WES; a three-dimensional numerical hydrodynamic and temperature model. Technical Report HL-96-21, U.S. Army Engineer Waterways Experiment Station, Vicksburg, MS.

Schwab, D.J., G. Lang, K.W. Bedford and Y.P. Chu, 1999: Recent Development in the Great Lakes Forecasting System (GLFS). Third Conference on Coastal Atmospheric and Oceanic Prediction and Processes, New Orleans, LA, 201-206.

Schwab, D.J., and K.W. Bedford, 1994: Initial implementation of the Great Lakes Forecasting System: a real-time system for predicting lake circulation and thermal structure. Water Poll. Res. J. of Canada, 29(2/3), 203-220.

Schwab, D.J., J.R. Bennett, P.C. Liu and M.A. Donelan, 1984: Application of a simple numerical wave prediction model to Lake Erie. J. Geophys. Res., 89(C3), 3586-3592

Spasojevic, M. and Holly, M. J., 1997: Three-dimensional numerical simulation of mobile-bed hydrodynamics. Technical Report 262, Iowa Institute of Hydraulic Research.

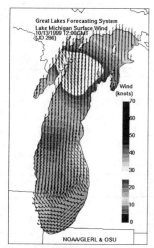

Figure 1. Lake Michigan interpolated Wind fields

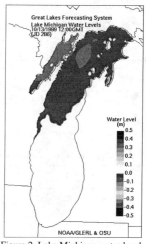

Figure 2. Lake Michigan water levels

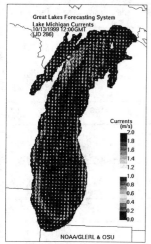

Figure 3. Lake Michigan currents

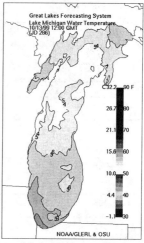

Figure 4. Lake Michigan surface temperatures

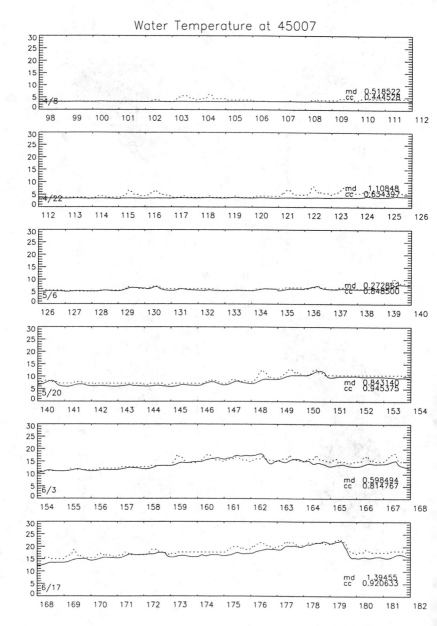

Figure 5. Comparison between measured and computed water temperature at 45007 (Jd98-182)

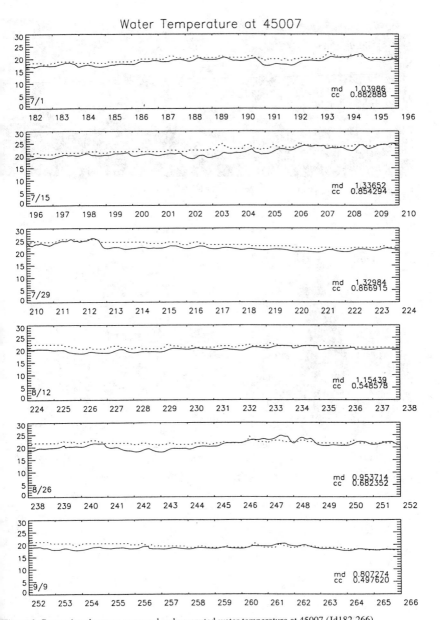

Figure 6. Comparison between measured and computed water temperature at 45007 (Jd182-266)

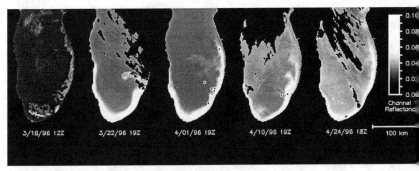

Figure 7. Sediment resuspension event observation by NOAA AVHRR imagery

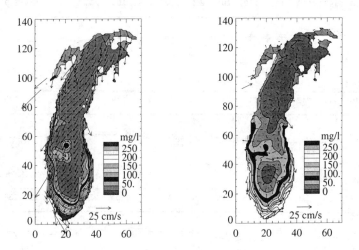

Figure 8.Sediment concentration plot at hour 12 Figure 9.Sediment concentration plot at hour 24

The Tampa Bay Nowcast-Forecast System

Mark Vincent[1], David Burwell,[1] and Mark Luther[1]

Abstract: This paper provides an overview of the recent status, advancements and ongoing research of the Tampa Bay Nowcast-Forecast Modeling System. The system integrates two local real-time oceanographic data acquisition networks (PORTS and COMPS), internet based real-time hydrologic and atmospheric data and forecast products provided by the NWS and USGS. These products are used to drive a three dimensional hydrodynamic circulation model. The ultimate objective is to develop a reliable and accurate tool for providing prompt hindcast, nowcast and forecast circulation model output useful to the maritime, management and coastal communities. Validation of the automated nowcast model indicates good agreement with the PORTS observations, with maximum lags of less than 6 minutes, global rms values of 1.6 centimeters and 100 % of all difference values falling in the range of -6.5 centimeters to 9.5 centimeters. A preliminary audit of the nowcast model during the validation phase indicates that nowcast model products were disseminated to the internet within 15 to 40 minutes of real time.

Introduction

In early 1997 an experimental prototype nowcast-forecast modeling system was developed and tested in Tampa Bay, Florida (Vincent et. al, 1997). The objective of this early phase was to investigate the technical feasibility and merits of producing and disseminating nowcast and forecast model products for use by the maritime and management communities of the bay. During a one year evaluation period, the nowcast forecast system proved reasonably reliable and accurate, however several specific limitations and associated solutions were identified. Implementation of these

[1]Department of Marine Science, University of South Florida, 140 7th Avenue S. St. Petersburg, FL 33701

solutions has lead to substantial revisions of and enhancements to the model system protocols. This paper will provide an overview of the present status of the nowcast and forecast models, as well as results of water level calibration and validation and a performance audit of the automated nowcast model.

Study Area

Tampa Bay is located on the west central coast of Florida, and is the largest port and estuary in the state. It has been classified as a coastal plain estuary with mixed semi-diurnal tides with less than a one meter tide range. The bay covers approximately one thousand square kilometers with a average depth of four meters.

Real-time data acquisition

A total of ten real-time telemetered data acquisition stations are located within the bay. These are comprised of eight Tampa Bay Physical Oceanographic Real-time System (PORTS) stations and two Coastal Oceanographic Monitoring System (COMPS) stations (Vincent et al., 1997). The station locations are provided in Figure 1. Two of the stations, Egmont Key (EGK) and Anna Maria Island (ANM) provide water level, salinity and temperature data for the model open boundary conditions. Wind data from three of the stations, CCUT, VMAN and EGK, are used to provide the surface wind stress forcing.

Real-time or recent river discharge and meteorological data are derived from the US Geological Survey (USGS) and National Weather Service (NWS) ftp internet sites respectively. After proper amplification for ungaged basins and curve number adjustment, the river discharge data are used to provide real-time boundary conditions for a total of 31 river and stream sources within the nowcast model. Regional and local daily meteorological data are used to provide surface mass sink and source fluxes due to precipitation and evaporation.

The Numerical Model

In Tampa Bay, the Blumberg-Mellor ECOM-3D model was initially deployed in work by Galperin et. al, (1992). Salient features of the model include: vertical sigma coordinates, boundary fitted curvilinear grid, a split time step for the solution of the baroclinic 3-D mode and the barotropic 2-D mode, and an embedded second order turbulence closure model to provide vertical mixing coefficients. Details of the model can be found in Blumberg and Mellor (1987) and Blumberg (1990). This new nowcast-forecast phase uses a high resolution grid with 70 by 100 cells in the horizontal and 11 layers in the vertical. The average dimension of the horizontal grid

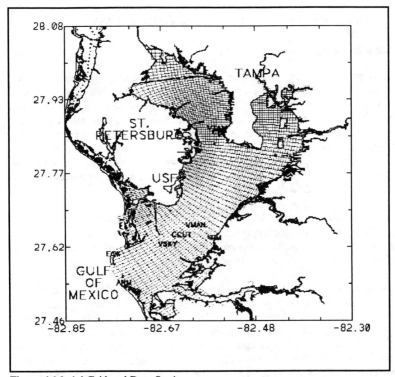

Figure 1 Model Grid and Data Stations

face is on the order of 640 meters. The split time step is 60 seconds for the internal mode and 6 seconds for the external mode. During hundreds of simulations, model input parameters have been tuned to provide good agreement with observations. A noteworthy adjustment of the new higher resolution configuration has been the reduction of the Smagorinsky horizontal diffusivity coefficient (HORCON) to 0.02 from the value of 0.2 used for the old coarser model grid.

Several experiments and investigations have shown that the baroclinic or density driven circulation of Tampa Bay is on the order of ten percent of the total flow (Galperin et al., 1992; Burwell et al., 1999). To account for this important forcing, proper hydrologic source and sink terms must be included. This has been accomplished by the 31 real-time fresh water lateral sources discussed above, as well as the addition of evaporation and precipitation physics to the ECOM-3D model. Recent investigations have shown that these terms do seasonally modulate the bay salinity, thus influencing the baroclinic circulation.

For spill, search and rescue and dispersal studies, passive tracer and an internal Lagrangian particle model have been added (Burwell et al, 1999).

Model Protocols

Three parallel model protocols (hindcast, nowcast and medium range forecast) are currently being tested. The primary engine of the system is the nowcast model protocol. Using real-time data as boundary conditions, this model is updated in time every twelve minutes. Using the output fields from the last nowcast run (restart file), the model is run forward to the most recent boundary conditions received. The current nowcast protocol has been significantly enhanced to provide redundant tidal open boundary conditions (EGK and ANM) as well as a preferred hierarchy of wind stations. This redundancy helps ensure the continued model performance during periods of station failure.

Archived nowcast restart files and the most recent restart files serve as the initialization fields for hindcast and forecast runs respectively. The hindcast protocol has been developed specifically to allow the rapid simulation of hindcast to forecast simulations for purposes of spill response and search and rescue operations.

The forecast model protocols are by far the most difficult, owing to the uncertainty and error of the forecast boundary conditions. The short range forecast protocol uses the National Weather Service (NWS) Nested Grid Model (NGM) Model Output Statistics (MOS) wind fields. The predicted open boundary water level is provided by a signal composed of the tidal harmonics with a super-imposed subtidal component that is a function of the past twenty four hour setup, the setup at the last recorded water level, and the prediction from the NWS Storm Surge Gulf Model (SSGM).

Calibration and Validation of Nowcast Model Water Levels

The nowcast model outputs time series data of water levels, currents and salinities at the grid cells that contain COMPS or PORTS observing stations. This information, which is output every internal time step (1 minute), is used for real-time dissemination as well as archiving for subsequent analysis. Calibration and validation of the nowcast model produced water levels was performed for the periods February 1, 2000 to February 20, 20000, and February 21, 2000 to March 10, 2000 respectively. During both of these phases, the nowcast model protocol was fully automated. Both of these periods were characterized by the passage of strong winter fronts which resulted in large non-tidal water level fluctuations, thereby providing a challenging time frame for skill assessment.

Model output (1 minute frequency) was tested against six minute frequency data obtained from the four PORTS water level stations (i.e. St. Petersburg, Port Manatee,

Old Port Tampa, and Port Tampa). During the calibration phase, Port Tampa experienced two periods of down time, therefore had a reduced data set for comparison.

Various methods were employed to compare the nowcast model and observed PORTS water levels. These included skill assessment tests and as well as cross correlation, cross covariance and difference histograms. Tabulated results of nowcast water level calibration are provided in Table 1. Columns one through four of Table 1 contain the PORTS station abbreviation, the lag of highest correlation, the cross covariance and cross correlation. Columns five through eight contain the mean extrema (R), the rms difference between the model and data (Drms), the rms nondimensionalized by the extrema (Dp=Drms/R), the gain ratio of model extrema to data extrema (Gw), the rms difference of extrema values (Arms), the mean lag of extrema (Lm), and rms extrema lag (Lrms). In these computations, developed by Hess and Bosely (1994), extrema is defined as a parameter maxima or minima that is separated by at least two hours. The summary products of this method are SD (six minute difference skill), SA (extrema amplitude skill) and SL (the extrema lag skill).

The nowcast water levels during the calibration phase were in good agreement with the PORTS data. The cross correlation and cross covariance coefficients exceeded 0.99, with lags of less than 6 minutes. The global gain ratio (Gw) was almost unity (i.e. 0.999)

Time series plots of water level comparisons at each of the four PORTS stations are provide in Figures 2 to 5, which again show the close agreement in phase and amplitude. The phase of model extrema was in excellent agreement with the observations at three of the PORTS stations (i.e. SP, PM, OPT) with less than one minute of lead. Port Tampa (PT), located in a narrow channel at the head of the bay, recorded the worst extrema lag rating (Lm=0.154), which still only equates to a 9.24 minute model lag. All three of the global skill scores (SD, SA, and SL) were very close to the maximum, with scores of 96.2%, 97.2% and 96.3% respectively. The individual station and the global Drms values were all less than 2.3 centimeters, indicating very close agreement between the 6 minute PORTS data and model output.

As required, all tuneable model parameters (friction coefficients etc.) were held constant for the subsequent twenty day nowcast validation period. Similar to the calibration period, the nowcast model was fully automated during this period, requiring no intervention by the model managers.

station	grid cell	num. of Data	lag (hr)	cross covar	cross corr	R	Drms	Dp	num. of extrema	Gw	Arms (m)	Lm (hr)	Lrms (hr)
PM	e45,30	5039	0.000	0.998	0.998	0.473	0.014	0.030	59	1.004	0.011	0.210	0.293
SP	e15,37	5039	0.000	0.998	0.997	0.491	0.015	0.031	60	1.014	0.013	0.093	0.265
OPT	e27,54	5038	0.100	0.998	0.998	0.542	0.013	0.024	61	1.000	0.013	0.048	0.209
PT	e55,66	5038	0.000	0.997	0.995	0.592	0.022	0.038	59	1.012	0.018	0.164	0.255
Global Values							0.016	0.030		1.007	0.013	0.128	0.255
Skill Parameters							SD=97.0%			SA=97.5%		SL=95.9%	

Table 1 Water Level Calibration Statistics

station	grid cell	num. of Data	lag cross (hr)	cross covar	R	corr	Drms	Dp	num. of extrema	Gw (m)	Arms (hr)	Lm (hr)	Lrms
PM	e45,30	4794	0.100	0.996	0.996	0.560	0.021	0.037	54	1.008	0.015	0.098	0.259
SP	e15,37	4764	0.000	0.995	0.995	0.563	0.023	0.041	55	1.009	0.017	0.005	0.245
OPT	e27,54	4792	0.100	0.996	0.996	0.614	0.023	0.037	55	0.975	0.018	-0.009	0.185
PT	e55,66	814	0.100	0.997	0.994	0.645	0.022	0.034	11	1.018	0.015	0.154	0.219
Global Values							0.022	0.038		0.999	0.016	0.039	0.229
Skill Parameters							SD=96.2%			SA=97.2%		SL=96.3%	

Table 2 Water Level Validation Statistics

Output from the model was obtained at one minute intervals and compared to the six minute interval PORTS water level data. Validation statistics are provided in Table 2, which follows the same format described for Table 1.

Cross correlation and cross covariance coefficients exceeded 0.994, with zero lag occurring at three of the stations (SP, PM, PT) and only a six minute lag at station OPT. For the validation time period, the individual station and the global model gain was almost identically unity, indicating there was not any amplification or damping of the model extrema relative to the observation extrema. The six minute PORTS data agreed closely with the model output, with a station averaged Drms value of 1.6 cm. The station averaged mean lag (Lm) indicated a slight model lag of 0.128 hours or 7.68 minutes. The skill scores SD, SA and SL were again very high, with values of 97.0%, 97.5% and 95.9% respectively. Relative to the calibration results, two of the three skill scores were higher (SD and SA), and the average of the three scores increased to 96.80% from 96.56%. All three primary skill scores increased from the validation conducted for the previous model configuration, which had an average skill score of 95.46% (Vincent et al., 1997).

Useful information on the distribution of model errors can be gleaned from histogram plots of model-observation differences. Such information facilitates the identification of extreme failures, if any, which could eliminate certain model configurations. Histograms of the differences between model output and observation data at each of the four PORTS stations are provided in Figure 6 and 7 for the calibration and validation periods respectively. The data for these plots is divided into bins centered from -0.15 meters to 0.15 meters, on 0.01 meter increments. The height of each of the rectangles in the histogram corresponds to the number of values in that interval. A tight unimodal spread centered around the 0.0 meter difference interval indicates that a station has low model - observation differences and is relatively immune from problematic outliers. Such is the case for all four stations in the calibration plot (Figure 6) and the validation plot (Figure 7). For the calibration phase, 94.5% of the differences were within plus or minus 0.045 meters (4.5 cm) and 100% of the differences were within the range of -0.105 meters to 0.145 meters. Agreement was even better for the validation phase, within 97.64% of the differences were within plus or minus 0.045 meters (4.5 cm) and 100% of the differences were within the range of -0.065 meters to 0.095 meters.

Preliminary Audit of Nowcast Model

The previous section detailing calibration and validation indicates that the automated model performs relatively well at reproducing the the water level physics of Tampa Bay. However, this is only part of the challenge in developing an operational system. Specifically, in addition to being accurate, a nowcast (and forecast) system must provide prompt, dependable and useful information. Significant modifications and revisions have been incorporated into the new model protocols in order to immunize them from events of boundary condition station failures, interruption of data transmission, spurious data, power failures and computer shutdown and restarts. The current model protocols have implemented contingencies for these cases.

To test the reliability of the nowcast model protocol, a preliminary performance audit was performed concurrent with the twenty day water level validation phase. During this period, the nowcast model protocol was fully automated. The key measure of the model system reliability was the delay time for the model products to be available via the the project www site (http://ompl.marine.usf.edu/TBmodel). A unix cron job was scheduled to run every six minutes to plot and disseminate the most recent model water levels. This program recorded the computer system date and time, the most recent nowcast model date and time, and the skill assessment of the last twenty four hours of the model forecast. Figures 8 A and B provide a plot of the product delay time versus calender days of the validation and audit period, and a plot of the skill scores (SD,SA, SL) versus days of the validation and audit period. The mean and maximum product delay times during the audit period was 0.22 hours and 0.56 hours. Since this process is called every six minutes, the average and maximum age of the posted internet data is 0.32 hours and 0.66 hours.

As indicated by Figure 8 B, skill scores were very high during this period, with average values for SD, SA and SL of 97.05%, 97.5% and 95.9% respectively.

Summary

A prototype nowcast forecast circulation model was developed and tested for a one year period. Based upon a critical evaluation of the system, the model; model protocols; real-time forcing data and forecast forcing products were revised. The new system incorporates enhanced real-time hydrologic forcing via the inclusion of evaporation and precipitation physics to the model, as well as the addition of 31 real-time freshwater stream and river sources. The nowcast model protocol has also been redesigned to be much more robust, incorporating redundancy where possible. Example nowcast water level comparisons are provided in Figures 9-12. The nowcast model conducts updates every twelve minutes. Calibration and validation of water levels for the automated nowcast model indicated very good agreement with PORTS observation data. For the water level validation, rms values were 1.6 centimeters with lags of less than six minutes. Additional skill assessment of water levels as well as currents is ongoing. The medium range forecast protocol conducts 24 hour forecast simulations every four hours. This protocol is currently automated and is being evaluated for water level and current prediction skill. Model and data products from this system can be reviewed online at ompl.marine.usf.edu

References

Blumberg, A. F. 1990. A primer for ECOM-3D. Prepared for HydroQual, Inc. Mahwah, NJ

Blumberg, A. F. and Mellor, G. L. 1987. A description of a three-dimensional coastal ocean circulation model, in Three-Dimensional Coastal Ocean Models, Heaps, N. S. (Ed.), American Geophysical Union, Washington, DC 1-16

Burwell, D., Vincent, M., Luther, M. and Galperin, B. 1999. Modeling estuarine residence times: Eulerian and Lagrangian. In Proceedings of the 5th International Conference on Estuaries and Coastal Modelling, New Orleans, LA. Nov. 3-5, 1999 In Review

Galperin, B., Blumberg A. F., and Weisberg, R. H., 1992 A time-dependant three-dimensional model of circulation in Tampa Bay, in Proceedings, Tampa Bay Area Scientific Information Symposium

Hess K., and Bosley, K., 1992. Techniques for validation of a model for Tampa Bay. In Proceedings of the 2nd International Conference on Estuaries and Coastal Modeling, Tampa, FL, Nov. 13-15, 1991, 83-94

Hess K., 1994. Tampa Bay Oceanography Project: Development and application of the numerical circulation model, NOAA Technical Report NOS OES 005. NOAA, National Ocean Service, Office of Ocean and Earth Science, Silver Spring, MD. 90 pp.

Vincent, M., Burwell, D, Luther, M., and Galperin, B. 1997.   Real-time data acquisition and modeling in Tampa Bay, in Proceedings of the Fifth International Conference on Estuaries and Coastal Modeling, Alexandria, VA. Oct. 22-24, 1997, 427-440

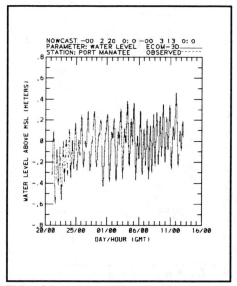

Figure 2 Port Manatee Water Level Validation

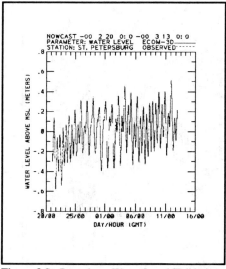

Figure 3 St. Petersburg Water Level Validation

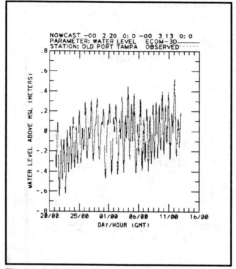

Figure 4 Old Port Tampa Water Level
Validation

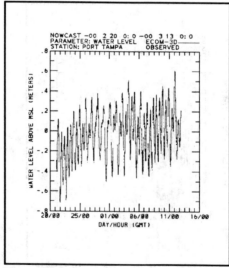

Figure 5 Port Tampa Water Level Validation

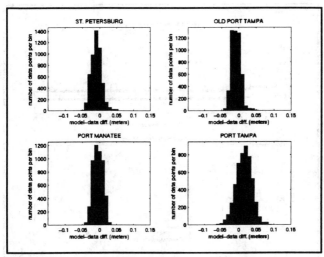

Figure 6 Water Level Validation Difference Histograms

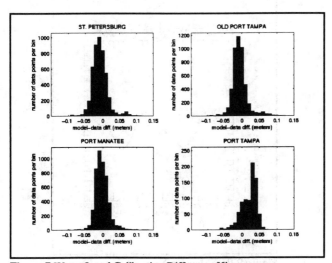

Figure 7 Water Level Calibration Difference Histograms

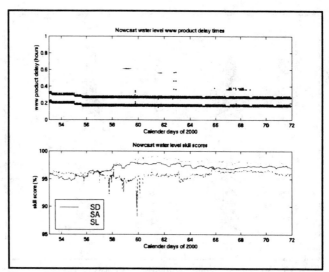

Figure 8 Validation Product Delay and Skill Scores

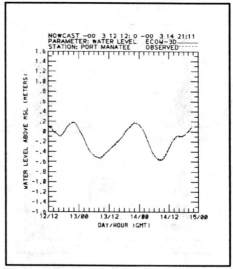

Figure 9 Port Manatee Nowcast Water Levels

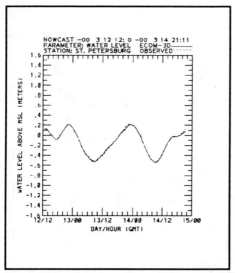

Figure 10 St. Petersburg Nowcast Water Levels

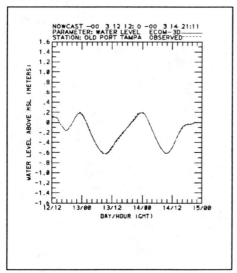

Figure 11 Old Port Tampa Nowcast Water Levels

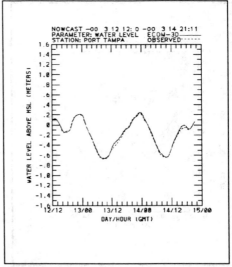

Figure 12 Port Tampa Nowcast Water Levels

NAVIGATION STUDY FOR MATAGORDA BAY, TEXAS

David J. Mark[1], Dennis W. Webb[1], M., and Nicholas C. Kraus[2],Members, ASCE

Abstract

A navigation study was undertaken for Matagorda Bay, Texas for relocating the Gulf Intracoastal Waterway (GIWW). Of the five maintained channels in the bay, the GIWW and the Matagorda Ship Channel have the greatest amount of commercial traffic. In addition to the high traffic volume, pilots must navigate with cross-currents that can exceed 1 m/s at the intersection of the waterway and ship channel. There is a risk of grounding in the shallow waters close to the intersection should a vessel veer from the channel. Consequently, these factors pose potential safety and operational hazards. This paper describes the combined application of a hydrodynamic and vessel-response model to evaluate alternative routings for the GIWW in Matagorda Bay.

Introduction

Situated along the Gulf of Mexico coast, Matagorda Bay is located approximately midway between Corpus Christi and Galveston, Texas. Matagorda Bay is separated from the Gulf by Matagorda Island and Matagorda Peninsula (Figure 1). Until 1963, Pass Cavallo served as the only inlet to the bay. Several attempts were made to stabilize the natural inlet. Because of the great width and shifting channel at Pass Cavallo, this proved infeasible. Through Federal authorization obtained by local interests, in 1963 the U.S. Army Corps of Engineers (USACE) excavated a new inlet through Matagorda Peninsula approximately 4 km north of Pass Cavallo. The Matagorda Ship Channel passes through two jetties and assures a reliable channel for deep-draft access to the bay.

[1]Research Hydraulic Engineer, [2]Research Scientist, U.S. Army Engineer Research and Development Center, Coastal and Hydraulics Laboratory, 3909 Halls Ferry Road, Vicksburg, Mississippi 39180-6199, USA.

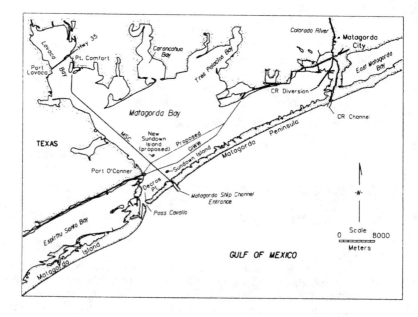

Figure 1. Schematic of study area.

Five maintained channels are presently in active use in the bay, the two most prominent of which are the Gulf Intracoastal Waterway (GIWW) and the Matagorda Ship Channel (MSC). The MSC enters the bay through the artificially cut inlet and runs to Port Lavaca. Sheltered from the Gulf by Matagorda Peninsula, the GIWW extends within the bay from East Matagorda Bay to Espiritu Santo Bay. Presently, the GIWW crosses the MSC about 4 km northward of the bayward side of the MSC entrance, and there is substantial ship and barge traffic at that intersection. Significant shoaling as well as strong currents are also present because of the proximity of the intersection to the MSC entrance.

During peak flood, the current regularly exceeds 3 m/s at the inlet and approaches 1 m/s at the intersection. Strong currents such as these pose a safety and operational hazard for barges and other watercraft. For example, ships entering the MSC during peak flood may have limited rudder control. According to a letter sent to the USACE Galveston District by the U.S. Coast Guard, there were 44 groundings and 382 aids-to-navigation discrepancies near the intersection in 1994. The combination of rapid shoaling, strong wind, and strong tidal current greatly increases the likelihood of grounding by towing vessels. Because of the strong current flowing along it, the MSC entrance is one of the few locations along the coast of Texas where tidal force almost always dominates over wind force. Wind-induced currents, as well as the wind acting on high profiles of ships and empty barges, are also significant factors for navigation in all Texas coastal waters, including at the MSC-GIWW intersection.

Strong currents at the intersection also increase the cost associated with maintaining the GIWW because of frequent dredging, especially in the reach to the northeast of the intersection. Sediment deposited in this reach originates primarily from Sundown Island, a disposal site for material dredged from the GIWW and MSC that now serves as habitat for the endangered brown pelican. Because of the strong currents flowing around the island, the sediment composing this island is being eroded at a relatively fast rate and subsequently being deposited in the channel. As determined by Kraus, Mark, and Sarruff (2000), by moving the GIWW away from the inlet, a new beneficial-uses disposal site can be placed in a less dispersive area, thereby requiring less frequent dredging to maintain the channel.

This study describes the procedures followed in establishing, evaluating, and implementing a numerical long-wave hydrodynamic model for simulating the tide- and wind-induced circulation patterns within Matagorda Bay. After the model was calibrated, three sets of simulations were conducted, for which one set depicted the GIWW with its present location and the other two sets the alternative locations of proposed channels. With the current fields calculated by the calibrated model serving as environmental conditions, navigability was evaluated with a ship/tow simulator. Experienced pilots familiar with the bay operate the simulator's barge console as they would their own vessel. Together with time-varying visual displays of the bay and simulated forcing with the calculated currents, barge response is calibrated to respond similarly as in the bay. After calibration has been achieved, pilots then evaluate the proposed alternatives with the simulator.

Environmental Setting

With a surface area of $1,100 \text{ km}^2$ (Ward 1982), Matagorda Bay is the second largest bay on the Texas coast, after Galveston Bay. Several smaller bays, notably Lavaca Bay, Carancahua Bay, Tres Palacios Bay, and Espiritu Santo Bay directly connect to Matagorda Bay. Depth over much of the natural (non-dredged) bottom of the inner bay and tributary bays runs between 3.0 and 3.7 m mean lower-low water (mllw).

The Colorado and Lavaca Rivers empty into Matagorda Bay. The Colorado River is the larger and discharges into Matagorda Bay through the Colorado River Diversion Channel located at the northeastern end of the bay. From about 1935 to 1991, the Colorado River emptied directly into the Gulf of Mexico. Beginning in 1992, the bulk of the river flow was directed through the diversion channel. A minor amount of water volume can flow through navigation locks on the GIWW and is limited to the amount passing while operating the locks.

Tide along the Texas coast is predominantly diurnal, and mean tide range in the Gulf of Mexico is 0.43 m at the National Ocean Service (NOS) tide gauge located at the Galveston Pleasure Pier on the Gulf of Mexico side of Galveston Island. Secondary NOS gauges at Port O'Conner (located just inside Pass Cavallo) and at Lavaca Bay (located at the north end of the bay) indicate mean and semidiurnal tide ranges are nearly the same, about 0.24 m.

Extreme changes in water level in Matagorda Bay that define navigation-controlling depth are primarily wind-induced and accompany weather fronts that occur from about October to May. The strong sea breeze and fronts experienced along the Texas coast dominate the astronomical tidal forcing and significantly alter the circulation and water level in the shallow bays and estuaries. Harwood (1973) noted that typical strong winds contribute as much as 0.3 m of change in water level, as compared to a typical tidal range of 0.24 m in the bay. Seasonal variability is also substantial as compared to the mean range of tide in Texas bays and estuaries. Further discussion of wind and tides in the area is given by Kraus and Militello (1999).

Pass Cavallo was one of the largest inlets on the Gulf Coast of the United States and was approximately 2.9 km wide when first visited by European explorers. The inlet was navigable to small ships, although the deepest channels shifted frequently and were interrupted by shoals. Prior to cutting of the MSC in 1963, Pass Cavallo was approximately 2.4 km wide. Recent aerial photography and bathymetric surveys show it as approximately 0.6 km wide. Measurements made in the present study show that three-fourths of the discharge now goes through the MSC. If Pass Cavallo were to close, our calculations show that the velocity through the MSC would increase by approximately 10 percent, leading to greater difficulties in navigation.

Sundown Island, which is located near the MSC-GIWW intersection, is a significant natural resource for bird nesting, because its location far from shore in the bay hinders mainland predators from reaching it. Bird species that nest on the island include brown pelicans (endangered species), herons, egrets, ibises, roseate spoonbills, gulls, terns, and black skinners (Texas Department of Transportation 1994). Relocation of the GIWW may require building a new island for nesting.

Hydrodynamic Modeling

The long-wave hydrodynamic processes in the study area were simulated with the ADvanced CIRCulation (ADCIRC) numerical model (Luettich, Westerink and Scheffner 1992). This model is based on a finite-element algorithm for solving the governing equations over complicated bathymetry encompassed by irregular sea and coastal boundaries. The solution algorithm allows for flexible spatial discretization over the computational domain while retaining stability. This flexibility allows larger elements to be specified in open-ocean regions where less resolution is needed, and smaller elements to be specified in the nearshore and estuary areas where hydrodynamic details in channels and around islands must be resolved.

Figure 2 shows the numerical grid with the existing GIWW position as developed for this study. Figure 3 displays the grid in the vicinity of the MSC entrance. The grid encloses Matagorda Bay entirely and incorporates Lavaca, Carancahua, Tres Palacios, and Espiritu Santo Bays, together with several smaller bays. The alongshore open-ocean boundary is situated approximately 65.2 km in the seaward direction from Matagorda Peninsula. Lateral open-ocean boundaries are positioned 91.7 and 69.5 km to the northeast and southwest, respectively, from the MSC entrance. The grid consists of 20,320 elements and 11,575 nodes, and elemental areas ranged from 1.02×10^3 to 1.31×10^8 m^2. The largest element lies on the Gulf boundary and the smallest in the GIWW at its northeast entrance to Matagorda Bay.

Bathymetry specified in the grid was obtained from three sources. The general source was the NOS bathymetric database, which was accessed to define water depths for those nodes residing in the open-ocean portion of the grid. NOS Chart 11319 was the second source and provided the depths for nodes located in Matagorda Bay and in the adjacent bays. Depths measured by the USACE during field surveys conducted in 1997 provided the bathymetry for Pass Cavallo and the MSC entrance.

Calibration was conducted over a 14-day period beginning on YD 243 (31 August 1997) at 0000 Greenwich Mean Time (GMT) and ending on YD 257 (13 September) at 0000 GMT. The model was run with a 4-s time step. Uniform, time-varying wind velocities were applied over the entire model domain, and water-level forcing was specified at the open Gulf boundary. Model accuracy was denoted with measured currents and water surface elevation (Kraus, Mark, and Sarruff 2000). Two current meters were deployed in the vicinity of the MSC-GIWW intersection. One gauge was placed adjacent to the MSC, and the other adjacent to the GIWW.

Calculated water-surface elevation was compared to measurements at the Port Lavaca station, shown in Figure 4. The model generally replicates the measured water surface elevation to within 2 cm. During the strong wind event on YD 265, the model underpredicted the measured water level by 7 cm. Figure 5 compares model-generated current with the measured current at the station in close proximity to the intersection. The calculated current was generally within 5 cm/s of the measurements. Discrepancies occur at about YD 263 where the model underpredicted the current by about 11 cm/s. The phase lag between the computed and measured current time series is attributed to driving the open Gulf boundary with water levels measured approximately 100 km from the study site (from Corpus Christi).

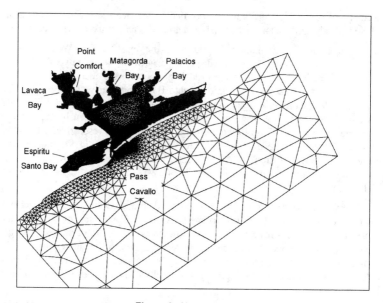

Figure 2. Numerical grid.

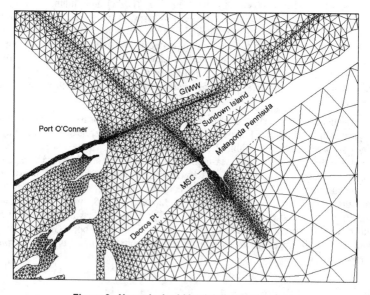

Figure 3. Numerical grid in vicinity of study area.

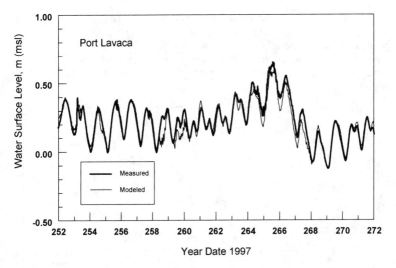

Figure 4. Comparison of model-computed and measured water surface elevation at the Port Lavaca gauge.

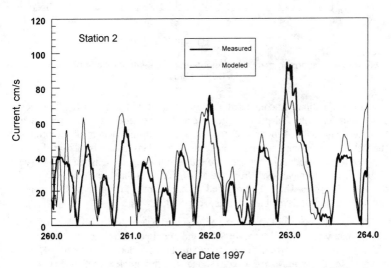

Figure 5. Comparison of model-computed and measured current at Station 2, located near the GIWW-MSC intersection.

To measure the ebb and flood flows exiting and entering the bay, discharge measurements were taken with an Acoustic-Doppler Current Profiler (ADCP) across the inland sides of the MSC entrance and Pass Cavallo. Comparison of model-computed and measured flood flows are presented in Figure 6. The model underpredicted the peak discharge (hour 18) by about 610 m^3/s, or 8 percent of the measured discharge.

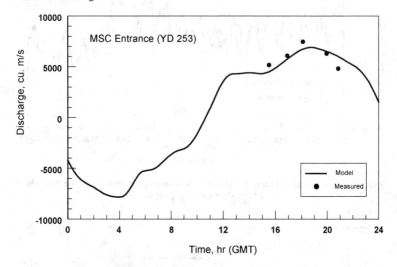

Figure 6. Comparison of model-computed and measured discharge across the MSC entrance.

After calibration was completed, the grid constructed for the existing Matagorda Bay condition was adapted to represent the proposed GIWW alignments. Modifications included increasing nodal resolution in the vicinity of the proposed channels and positioning additional nodes to depict these channels. Channel depths assigned to these nodes are consistent to those composing the existing GIWW (with respect to displacement away from the MSC) for areas that are shallower than the channel. For areas where the proposed channel depth is equal to or less than the ambient conditions, the ambient depths were used.

Because the existing GIWW can be expected to eventually fill with sediment, this channel was removed from the grid, and nodal positions were adjusted in this region such that elemental sizes approximate neighboring elements. Depths assigned to these new nodes correspond to the ambient bathymetry in the surrounding area. For similar reasons, the existing Sundown Island was removed, and a new island was added to the grid to represent a long-term perspective.

With the existing and proposed GIWW-configuration grids, the hydrodynamic model was run to generate current fields for the ship simulator. The model simulated spring-tide conditions with a 12.0 m/s wind blowing in the direction parallel to the MSC; during flood tide, the wind was directed towards the northwest, increasing the tide-induced current entering the bay. During ebb tide, the wind was directed towards the open Gulf. Figure 7 displays the current field for one of the alternative channels under flood tide conditions. Current fields were stored during peak flood and ebb current and subsequently inputted to the ship simulator.

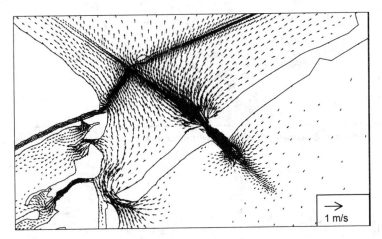

Figure 7. Current vectors for GIWW Alternative 1.

Vessel-Response Model

The vessel-response model is the heart of the simulator used in calculating ship motion as produced by the variety of forces that can be exerted on a vessel. Forces causing ship motion are both environmental and mariner controlled (Figure 8). Environmental forces include current, bank effects, wind, and waves. Mariner-controlled forces include rudder angle, propeller revolution, tugs, and bow and stern thrusters. The six degrees of ship motion are composed of three degrees of motion in both the horizontal and vertical planes (Figure 9). Straight-line motion is calculated by summing forces along a given axis, whereas ship rotation is calculated by summing moments along a given axis.

The three degrees of horizontal motion are surge, sway, and yaw. Surge is either a forward or astern motion and is calculated by summing the forces along the longitudinal axis. Sway is a sideways or crabbing motion calculated by summation

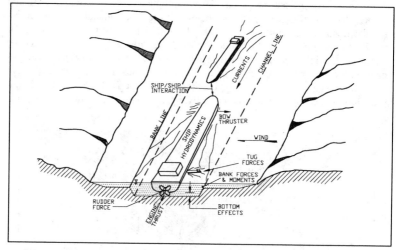

Figure 8. Schematic displaying environmental and mariner-controlled forces.

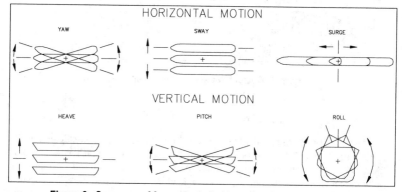

Figure 9. Summary of forces included in vessel-response model.

of forces along the lateral axis. Yaw is the rotation of the vessel calculated by summing the moments about the vertical axis.

Wave action is the primary force causing vertical ship movement. The three degrees of vertical motion are heave, pitch, and roll. Heave is the up and down motion determined by the summation of forces in the vertical direction. Pitch is bow-to-stern rotation computed by summing the moments about the lateral axis.

Roll is the side-to-side ship rotation calculated by summing the moments about the longitudinal axis of the ship. Typically, the design vessel for a navigation project is a large, loaded tanker, bulk carrier, or container ship. Waves, unless fairly high and with a long period, usually have little effect on these ships. However, tows generally do not operate under these conditions because barges tend to separate from them. Therefore, waves were not considered in this study.

Mariners operate the simulator by issuing engine, rudder, and tug-thruster commands. The engine and rudder commands are input to the vessel-response model at the ship's console by either the mariner or a helmsman, whereas tug and thruster commands are input to the model by an operator stationed at the tug console. The model calculates the resultant vessel movement based on these inputs and the environmental conditions. The ship's motion is then shown on the visual and radar displays. If necessary, the mariner responds to this movement by issuing additional commands. This interaction of the mariner and the simulation process is known as "man-in-the-loop." A schematic diagram of the ship/tow simulator is shown in Figure 10.

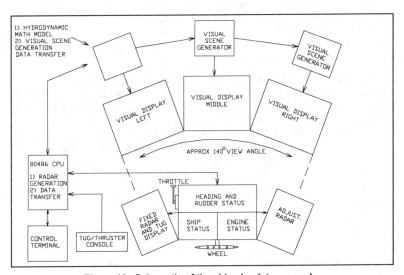

Figure 10. Schematic of the ship simulator console.

Navigation studies are initiated with a reconnaissance trip to the project site. Pilots are interviewed to determine navigation conditions in the project area. Information gathered from the pilots includes: areas where the vessel is subject to severe environmental conditions such as cross-currents, uneven banks, wind, and waves; harbor operating conditions such as vessel transit restrictions, meeting-and-passing restrictions, tug size and availability; and any visual cues or informal navigation aids such as steeples, radio or television towers, and buildings which should appear in the visual scene. Any areas outside of the authorized channel which are as deep as the channel are identified at this time because pilots are not restricted to the authorized channel and sometimes depend on areas outside the authorized channel limits for safe navigation. These areas should be a design consideration for channel deepening projects because some areas may not be deep enough for the vessels transiting the proposed, deepened channels.

During the reconnaissance trip, the study team will ride transits with the pilots. Ideally, the vessels will be close in size to the project design vessel for existing conditions. Pilot commands and ship response to environmental conditions should be observed. This information is used in evaluating the model database. Photographs and video taken during these transits are adapted in developing the visual scene displayed on the simulator monitors.

Databases are generated for the "Maximum Credible Worst Case Scenario," which is defined as the worst condition under which a tow will normally operate. The currents, wind, and waves defining these conditions are evaluated for both the existing and proposed channels. Evaluation is limited to these conditions because of the time and costs associated with real-time simulation testing. The channel design is based on the assumption that if the design is acceptable for extreme condition, it will be acceptable for less severe conditions.

Once the existing-condition databases are prepared, an experienced pilot who is licensed for the project site validates the model. Validation is accomplished by having the mariner evaluate the simulation's visual representation of the project and the vessel's response to the environmental conditions being run. This is an iterative process where the simulator databases are adjusted in response to the mariner's observations during a run. The simulation is then re-run and further modifications are made to the databases, if necessary. The process is continued until the simulation reproduces existing prototype conditions as realistically as possible. If appropriate, those changes made to the existing condition are also represented in the proposed-condition databases.

Simulation evaluation of navigation-channel improvements was accomplished through a rigorous exercise. In order to provide a base with which to compare results, simulations were conducted with the existing channel configuration as well as with the proposed channel. For this study, 12 licensed mariners took part in the evaluation process to minimize personal bias while piloting the tow (Webb 1993a). The existing and proposed channel configurations were simulated under different combinations of transit directions (towards the northeast or the southwest), tide- and wind-current (ebb or flood). Because the mariners' skill at the simulator often increases over the course of a week-long evaluation program, runs are conducted in random order to avoid prejudicing results.

Simulator output includes vessel-track plots and ship-parameter plots. Every five seconds during a simulation, the simulator outputs the control, positioning, and orientation parameters of the ship to a file. These parameters include the horizontal position of the center of the vessel and the vessel's heading, speed, propeller rpm, rudder angle, and rate of turn. The output file is then processed to generate vessel track plots and ship parameter plots. In addition, statistical plots of speed, propeller rpm, rudder angle, drift angle, rate of turn, and port and starboard clearance are also plotted against distance along-track. The distance along-track is calculated by projecting the position of the ship's center of gravity perpendicular to the centerline of the channel and is measured from the beginning of the centerline. For reference, the locations of important landmarks are identified.

Together with a questionnaire completed by the pilots concerning their opinions about the proposed channel improvements, plots tracking the vessel's movement as it transits a channel are the primary means of evaluating a project. Output from the ship simulator showed that the pilots were able to maintain course within the proposed channels, including at the MSC-GIWW intersection. Conversely, their vessels had a tendency to veer from the GIWW channel in close proximity to the intersection. Furthermore, the pilots noted that the simulator responded much like the actual tows while operating under adverse environmental conditions, and that navigation along both proposed alternative channels was improved. It was concluded that either of the proposed channels is acceptable for navigation.

Summary

A navigation study was performed for Matagorda Bay, Texas to investigate alternatives for relocating the GIWW. Presently, pilots must contend with high traffic volume, navigate their craft through strong cross-currents, and risk grounding in the shallow water adjacent to the GIWW if they veer from the channel. These factors pose a potential hazard to vessels and their crew. The alternatives were evaluated through combined modeling of the hydrodynamics and vessel-response model. Output from the ship simulator showed that the pilots were able to maintain course within the two proposed channels, including at the MSC-GIWW intersection. Furthermore, based on a review of questionnaires, all twelve pilots judged that navigation through the proposed channels is superior to navigating the existing GIWW. It was therefore concluded, from the standpoint of navigation, that either of the proposed channels is acceptable.

Acknowledgements

The study described herein and the resulting data presented, unless otherwise noted, were obtained from research sponsored by the US Army Engineer District, Galveston, by the US Army Engineer Research and Development Center (ERDC), Vicksburg, MS. Mr. Fulton Carson of the ERDC assisted in preparing the model grids and figures. We appreciate discussions with Mr. Herbie Maurer of the USACE Galveston District during the course of this study. Permission was granted by the Chief of Engineers to publish this information.

References

Harwood, P. J. W. (1973). "Stability and Geomorphology of Pass Cavallo and its Flood Delta Since 1856, Central Texas Coast." Unpublished M.A. Thesis, The University of Texas at Austin, Austin, TX.

Kraus, N.C., and Militello, A. (1999). "Hydraulic Study of Multiple Inlet System: East Matagorda Bay, Texas." *J. Hydr. Res.*, 125(3): 224-232.

Kraus, N.C., Mark, D.J., and Sarruff, S. (2000). "DMS: Diagnostic Modeling System, Report 1, Reduction of Sediment Shoaling by Relocation of the Gulf Intracoastal Waterway, Matagorda Bay, Texas." Technical Report, U.S. Army Engineer Research and Development Center, Vicksburg, MS.

Luettich, R.A., Jr., Westerink, J.J., and Scheffner, N.W. (1992). "ADCRIC: An Advanced Three-Dimensional Circulation Model for Shelves, Coasts, and Estuaries." Technical Report DRP-92-6, , U.S. Army Engineer Research and Development Center, Vicksburg, MS.

Texas Department of Transportation. (1994). "The Gulf Intracoastal Waterway in Texas." Texas Department of Transportation, Austin, TX.

Ward, G. H. (1982). "Pass Cavallo, Texas: Case Study of Tidal Prism Capture." *J. Waterway, Port, Coastal and Ocean Div.*, 108(WW4): 513-525.

Webb, D. W. 1993a. "Use of Computer-Aided-Drafting (CAD) Software in Ship Simulation." International Conference on Marine Simulation and Ship Maneuverability, MARSIM '93, St. John's, Newfoundland, Canada.

Webb, D. W. 1993b. Ship Navigation Simulation Study, San Juan Harbor, San Juan, Puerto Rico." Technical Report HL-93-17, U.S. Army Engineer Research and Development Center, Vicksburg, MS.

HYDRODYNAMIC MODELING OF A SEA-BREEZE DOMINATED
SHALLOW EMBAYMENT, BAFFIN BAY, TEXAS

Adele Militello[1]

ABSTRACT: Bays and lagoons of the south Texas coast experience strong wind forcing. During the summer, two components of wind are present - a persistent southeast (onshore) wind and the sea breeze. These winds produce a pattern of diurnal strengthening and weakening of the onshore wind. Baffin Bay, Texas, is a shallow, nontidal, wind-dominated embayment with no direct connection to the Gulf of Mexico. Summer onshore wind blows along the bay, inducing significant setup on its western end and setdown at its mouth. Water motion in Baffin Bay under sea breeze forcing was simulated with a two-dimensional finite-difference model. A 1-month-long simulation for July 1995 was conducted during which the wind speed exceeded 10 m/s on 10 days. The diurnal cycle of setup (western end) and setdown (eastern end) along the length of the bay was reproduced by the model. During stronger onshore winds, the water elevation in the western portion of the bay is elevated, and the water level at the bay mouth (eastern end) is lowered. When the onshore wind weakens, water returns from west to east, reducing the surface slope. The bay surface remains relatively flat only during brief intervals when the onshore wind is weak. Peak daily difference in water-surface elevation between the western and eastern ends of the bay, a distance of 30 km, ranged from 0.11 to 0.58 m over the simulation interval. For July 1995, the calculated difference in mean water level was 10 cm higher on the western end of the bay as compared to the eastern end because of the onshore bias in wind direction.

---

1) Research Oceanographer, US Army Engineer Research and Development Center, Coastal and Hydraulics Laboratory, 3909 Halls Ferry Road, Vicksburg, MS 39180-6199. Tel: (601) 634-3099; militea@wes.army.mil.

INTRODUCTION

Wind can be a significant forcing in shallow embayments and dominate over tidal motion at some locations. Where a well-developed and strong sea breeze is present, it produces setup in some regions and setdown in others on a diurnal cycle. Over relatively long time intervals, such as weeks or months, a mean surface tilt can occur in the water body. One implication of this non-horizontal mean surface is potential error in datum elevations. These errors can be passed on to applications such as bottom topography calculations from surveys and referencing water-level gauges to a datum. Baffin Bay, Texas, is one such location where wind forcing dominates the tide. To determine the patterns of water-level variation and circulation in Baffin Bay, a two-dimensional model was applied to simulate a month when the sea breeze was strong.

Baffin Bay is located on the south Texas coast approximately 45 km south of Corpus Christi (Fig. 1). The bay has no direct connection to the Gulf of Mexico, and tidal motion that propagates from the Gulf must traverse a minimum of 75 km of shallow water (typical depth is 1 m in the Laguna Madre) before reaching the bay mouth. Diurnal and semi-diurnal tidal signals are severely attenuated in the Laguna Madre such that the region near the mouth of Baffin Bay was determined to be non-tidal by National Ocean Service (NOS) standards (Gill, et al. 1995). Although Baffin Bay does not experience diurnal or semi-diurnal astronomical tides, the sea breeze forces diurnal water motion. Because the main axis of Baffin Bay is aligned approximately parallel to the cross-shore direction, the sea breeze induces an along-axis water-level tilt with setup on the western end of the bay and setdown on the eastern end, where the bay connects to the Upper Laguna Madre. Response of water level to sea-breeze forcing is particularly notable on the western end of Baffin Bay, at Riviera Beach (Gill, et al. 1995). Setup induced by the diurnal wind varies with wind intensity and ranges from approximately 15 to 30 cm during typical sea breeze wind speeds.

Baffin Bay has three arms on its western end and connects to the Upper Laguna Madre on its eastern end (Fig. 1). Typical depths are 2.3 m in the main body, 1.4 m in the arms, and maximum depth of 3 m. The three arms consist of Alazan Bay, Cayo Del Grullo, and Laguna Salada. Alazan Bay is the largest of the three arms, and its main axis trends along a northeast-southwest axis. Typical depths in Alazan Bay are 1.5 m in the lower reach (closest to the main body of Baffin Bay) and 1 m and shallower in the upper reach. Cayo Del Grullo is a northwest-trending extension of western Baffin Bay, with typical depth of 1.4 m. Laguna Salada, the smallest of the three arms, extends to the southwest from western Baffin Bay. Typical depths in Laguna Salada are 1.5 m in the lower reach and 0.75 m in the upper reach.

Nonliving serpulid worm reefs are present in abundance in Baffin Bay. Reef distribution is most dense in the bay margins and mouths of Baffin Bay and Alazan Bay (Andrews 1964). Reef growth across the mouth of Baffin Bay was so

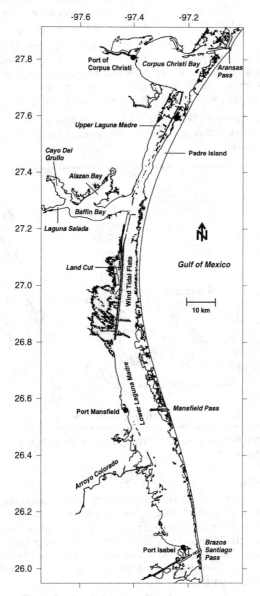

Fig 1. Regional site map of study area.

extensive that the bay mouth is effectively closed (Shepard and Rusnak 1957) (Fig. 2). Tops of reefs at the bay mouth are typically near the water surface or are exposed. Reefs are not found in depths greater than 2.4 m.

Summer winds are composed of two primary components; a persistent southeast wind and the sea breeze. Superposition produces onshore wind that varies in strength over time. Strongest onshore wind occurs when the sea breeze peaks in the afternoon. In the evening, weakening of the sea breeze reduces the strength of the onshore wind, but its direction is generally not reversed.

Wind-forced waves develop as the sea breeze strengthens during the day. At the mouth of the bay, waves with heights approaching 1 m and having periods of 4 or 5 s are common.

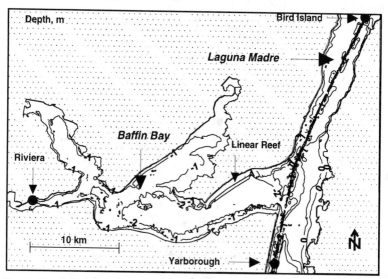

Fig. 2. Monitoring stations and bottom topography for Baffin Bay and southern Upper Laguna Madre, Texas.

MEASUREMENTS

Water level and wind measurements were made at three monitoring stations in the study area; Riviera, Bird Island, and Yarborough (Fig. 2). Water-level gauges were located at all three monitoring stations. Measurements of water level to NOS standards were made every 6 min. The Riviera water-level gauge was located in a marina and sheltered from wind waves. All water-level gauges were set in stilling wells that filtered wind waves. Wind was measured at the Yarborough station every hour.

HYDRODYNAMIC MODEL

Circulation and water level were calculated with a two-dimensional (2D), finite-difference model called M2D (Militello 1998). A 2D modeling approach is appropriate for calculating the hydrodynamics in Baffin Bay because of its irregular shape, very shallow water, and wind-dominated forcing. Although no vertical profiles of the current or salinity were obtained in Baffin Bay for this study, the system can be assumed to be vertically well mixed. Similar shallow embayments on the Texas coast have been documented to be vertically homogenous (Shideler 1984; Ward, et al. 1985; Brown, et al. 1995).

Model Description

M2D is a two-dimensional, finite-difference numerical approximation of the depth-integrated continuity and momentum equations given by

$$\frac{\partial \eta}{\partial t} = -\left(h+\eta\right)\left(\frac{\partial u}{\partial x}+\frac{\partial v}{\partial y}\right) - u\frac{\partial\left(h+\eta\right)}{\partial x} - v\frac{\partial\left(h+\eta\right)}{\partial y} \tag{1}$$

$$\frac{\partial u}{\partial t} = -g\frac{\partial \eta}{\partial x} - u\frac{\partial u}{\partial x} - v\frac{\partial u}{\partial y} + 2a_h\frac{\partial^2 u}{\partial x^2} + fv - C_b\frac{u|U|}{\left(h+\eta\right)} + C_d\frac{\rho_a}{\rho_w}\frac{W^2\cos\left(\theta\right)}{\left(h+\eta\right)} \tag{2}$$

$$\frac{\partial v}{\partial t} = -g\frac{\partial \eta}{\partial y} - u\frac{\partial v}{\partial x} - v\frac{\partial v}{\partial y} + 2a_h\frac{\partial^2 v}{\partial y^2} - fu - C_b\frac{v|U|}{\left(h+\eta\right)} + C_d\frac{\rho_a}{\rho_w}\frac{W^2\sin\left(\theta\right)}{\left(h+\eta\right)} \tag{3}$$

where h = still-water depth, η = deviation of the water-surface elevation from the still-water level, u = depth-averaged current velocity parallel to the x-axis, v = depth-averaged current velocity parallel to the y-axis, g = gravitational acceleration, f = Coriolis parameter, a_h = horizontal coefficient of eddy viscosity, U = total current velocity, C_b = empirical bottom-stress coefficient, C_d = wind drag coefficient, ρ_a = density of air, ρ_w = density of water, W = wind speed, θ = wind direction, and t = time. The convention for wind direction is specified to be 0 deg for wind from the east with angle increasing counterclockwise. In simulations conducted for Baffin Bay, mixing terms were not included in the calculations.

The wind-drag coefficient is expressed in the model as (Hsu 1988)

$$C_{10} = \left(\frac{\kappa}{14.56 - 2\ln W_{10}}\right)^2 \tag{4}$$

where W_{10} = wind speed in m/s at 10 m above the surface. This coefficient is particularly suited for shallow water. Wind velocity is spatially constant over the model domain, but can vary in time.

Cells are defined on a staggered, rectilinear grid and can have variable side length. The momentum equations are solved in a time-stepping manner first, followed by solution of the continuity equation, in which the updated velocities calculated by the momentum equations are applied. The solution scheme is explicit, although calculation of the advective and nonlinear continuity terms specifies water-level and velocity values from two time steps, a necessary condition for stability (Fischer 1959; Kowalick and Murty 1993). Robust mass-conserving flooding and drying algorithms simulated inundation and exposure of wind-tidal flats.

Grid Development

The model domain includes Baffin Bay and a portion of the Upper Laguna Madre extending from approximately the location of the Bird Island monitoring station (northern limit) to the location of the Yarborough station (southern limit), a distance of 32 km. Bottom topography was supplied from available NOS soundings for Baffin Bay and from a 1995 Corps of Engineers survey of the Laguna Madre. Figure 2 shows the bottom topography for Baffin Bay and the southern Upper Laguna Madre. Shallow water at the mouth of Baffin Bay owes to the presence of worm reef deposits.

The numerical grid, Fig. 3, consists of 3,435 cells with side dimensions of 297 and 330 m parallel to the x- and y-axes, respectively. The grid is aligned with its x- and y-axes parallel to East-West and North-South, respectively. Measurements taken at the Bird Island and Yarborough water-level gauges (Fig. 2) were applied

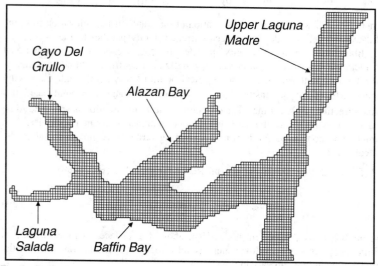

Fig. 3. Baffin Bay computational grid.

as forcing at the northern and southern ends of the grid in the Upper Laguna Madre, respectively. Wind velocity for surface forcing was obtained from the Yarborough monitoring station. Simulations were conducted with a 5-s time step.

Values of Manning's *n* were spatially variable and ranged from 0.03 to 0.08 s/m$^{1/3}$, depending on proximity to reefs. Values of 0.065 and 0.080 s/m$^{1/3}$ were assigned to the reefs because of the increased roughness associated with the deposits and their horizontal and vertical irregularity.

Comparison of Measured and Calculated Water Level

Measured and calculated water levels are compared for the time interval August 3 - 31, 1994 (Julian Days (JD) 215 through 243). Forcing data taken from the Bird Island and Yarborough water-level gauges are shown in Fig. 4 and wind data from the Yarborough anemometer are shown in Fig. 5. The convention for the measured wind direction is 0 deg = North, 90 deg = East. A difference in water-surface elevation of 5 to 10 cm between the Bird Island and Yarborough gauges persisted through the simulation interval. Maximum differences in water level at Bird Island and Yarborough occurred on JDs 232, 233, and 236. On each of these days, the measured peak wind speed exceeded 12 m/s.

Measured and calculated water levels at Riviera in Baffin Bay are compared in Fig. 6 and show general agreement. Daily setup and relaxation of the water surface are well represented, as is the fortnightly tide. The RMS error is 2.5 cm for the 29-day simulation. Between JDs 221 and 241, the best agreement between

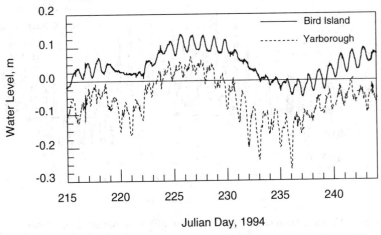

Fig. 4. Measured water level at Bird Island and Yarborough gauges, August 3 - 31, 1994.

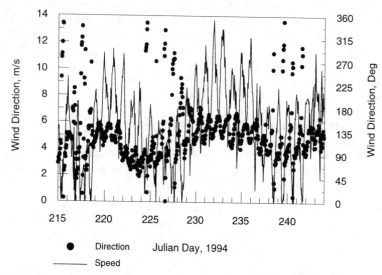

Fig. 5. Measured wind at Yarborough, August 3 -31, 1994.

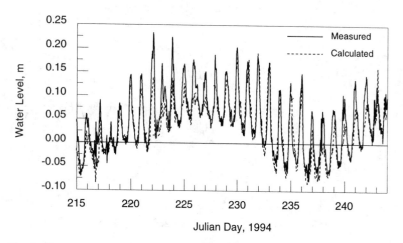

Fig. 6. Measured and calculated water level at Riviera, August 3 - 31, 1994.

the model and measurements occurred from JD 229 through JD 235, and there was less agreement during JD 221 through JD 228 and JD 238 through JD 241. During the time of best agreement, the wind was from the southeast with speed typically peaked at 10 m/s or greater. During the periods of least agreement, the wind speed was generally less than 9 m/s and from the east. Because the model has been demonstrated to calculate wind-induced setup accurately (Militello 1998), the error in water level may owe to representation of the wind field as spatially constant or to the location of the anemometer being approximately 3.5 km west of the barrier island.

WATER LEVEL AND CIRCULATION IN BAFFIN BAY

A 1-month-long simulation of water level and current in Baffin Bay was conducted for July 1 - 31, 1995 (JDs 184 through 213), during which a strong sea breeze was present. Results are applied to investigate the wind-induced setup and setdown, as well as circulation in Baffin Bay, which has not been previously studied or calculated.

Figure 7 plots the wind speed and direction for the simulation interval. Strongest winds, peaking at 15 m/s occurred on JDs 184 through 186 and JDs 203 through 206. During these intervals, the wind was from the southeast. When the wind was weaker, such as on JDs 187 through 199, its direction had a rotary pattern, which is a signature of the sea breeze.

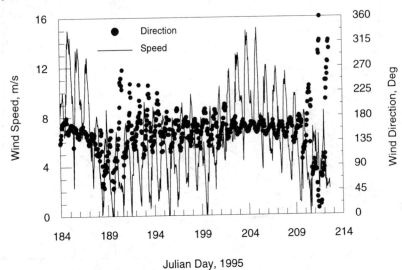

Fig. 7. Measured wind at Yarborough, July 3 - 31, 1995.

Setup and Setdown

The distribution of wind-induced setup and setdown along the main axis of Baffin Bay, including Cayo Del Grullo and Alazan Bay, is shown in Fig. 8 for time JD 184.9 when the wind was strong (15 m/s). Water-surface elevation difference over the length of the bay is 0.5 m, giving an overall surface slope of 1.5×10^{-5}. The surface slope is not spatially constant and changes in slope coincide with changes in depth. For example, the water slopes in Cayo Del Grullo and Alazan Bay are steeper than in the main body of Baffin Bay, where the water is relatively deep.

Time series of water-surface elevation in the Upper Laguna Madre and the terminal ends of Cayo Del Grullo and Alazan Bays are shown in Fig. 9. The diurnal wind forces setup in Cayo Del Grullo and Alazan Bay and setdown in the Upper Laguna Madre. The water-surface elevation difference between Cayo Del Grullo and the Upper Laguna Madre during the peak setup/setdown ranged from 0.11 to 0.58 m over July, 1995, with the difference dependent on the strength of the wind. For the 28-day simulation interval, the mean difference in water elevation was 10 cm over the length of the bay. The extreme reduction in water level seen in Fig. 9 for the Upper Laguna Madre near the end of JD 204 is a drying event.

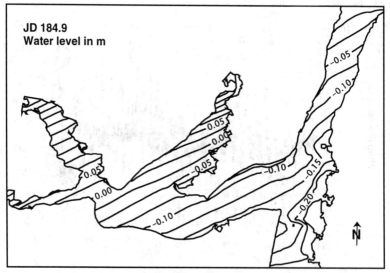

Fig. 8. Calculated setup and setdown in Baffin Bay at time JD 184.9.

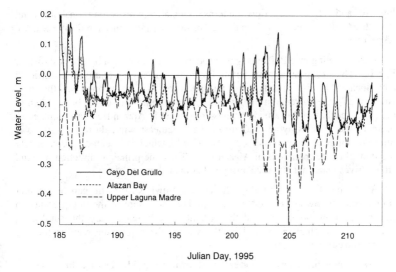

Fig. 9. Calculated water level for Cayo Del Grullo, Alazan Bay, and Upper Laguna Madre.

During times of peak setup, the water-surface elevation is higher in Cayo Del Grullo than Alazan Bay. This difference in elevation owes to the orientations of the bays, with Cayo Del Grullo being forced along its main axis by the sea breeze and Alazan Bay being forced from an angle of approximately 90 deg. When the sea breeze relaxes, the water levels in Cayo Del Grullo and Alazan Bay are approximately equal. Periods of strong sea breeze, such as occurred on JD 185 through JD 186 and JD 201 through JD 206 (Fig. 7), induced sustained differences in water-surface elevation between the arms of Baffin Bay and the Upper Laguna Madre. During these intervals, the weakest wind, i.e., the steady onshore (non-sea breeze) wind, was sufficiently strong to maintain setup across the bay. Militello and Kraus (submitted) have examined the harmonic components of the water level and current produced by sea-breeze forcing in Baffin Bay.

Circulation

To examine the circulation driven by the sea breeze, a period of strong sea breeze was selected so features of the circulation, such as eddies, would be pronounced. The time interval selected ranged from JD 203.45 through JD 204.38, 1995, covering one sea-breeze cycle. During this interval, the sea breeze was strong and peaked at 15 m/s (Fig. 7). Two representative plots,

Figs. 10 and 11, display flow patterns associated with strong onshore wind (measured $W = 12$ m/s) and reduced onshore wind (measured $W = 9$ m/s), respectively.

Water flowing into Baffin Bay in response to the strong sea breeze is shown in Fig. 10. The strongest currents within Baffin Bay occur in a relatively constricted area located at the approximate mid-point of the main body of the bay. Additionally, the currents in the lower Cayo Del Grullo are strong, but weaken in the upper reaches of this arm. Currents are weaker in Alazan Bay, as compared to those in the lower Cayo Del Grullo. The southeast wind during JD 203 was directed along the main axis of Cayo Del Grullo, which is oriented approximately orthogonal to the main axis of Alazan Bay. Thus, the principal direction of wind stress drives water across Alazan Bay rather than along it.

At the mouth of Baffin Bay, the current flowing from the gap in the reef (Fig. 2) moves toward the northwest. The northern shore of the bay presents an impervious flow boundary, and the current is deflected toward the southwest, along the main channel axis. Directly west of the gap in the reef (at the bay mouth) the current is weak.

Water movement toward the east is shown in Fig. 11. During this time, the sea breeze has relaxed, and surface slope of the water creates a gravity-induced flow out of the bay. An east-directed current extends from upper Cayo Del Grullo to the mouth of Baffin Bay. A weaker current originates in upper Alazan Bay and flows into the main current in Baffin Bay.

The eddy northwest of the mouth forms when water moves out of the bay. The northern part (northwest-directed flow) of this eddy persists throughout the sea breeze cycle. The southern half of the eddy disintegrates when water is moved into Baffin Bay by the strong onshore wind. During some portions of the cycle, the size of the eddy spans the width of Baffin Bay. The current in the northern part of the eddy is aligned with the predominant wind direction. Flow entering Baffin Bay through the gap in the reef is pushed toward the northwest by the southeast wind. As stated above, the northern shore forces the flow to be re-directed toward the southwest (during strong onshore wind) and along the main channel of Baffin Bay. The deepest continuous reach of Baffin Bay extends from western Baffin Bay, near the confluence with Alazan Bay, to its mouth. The relatively deep region northwest of the mouth is located in the northern part of the eddy, which is maintained by the southeast wind-induced currents.

A long, linear reef is present near the northern shore of Baffin Bay (Fig. 2). Because this reef is shallow, its presence is expressed through the current patterns. In Fig. 10, the current moves in opposite direction on either side of the reef, with water flowing toward the southwest over the reef top. Figure 11 shows diverging flow over the reef.

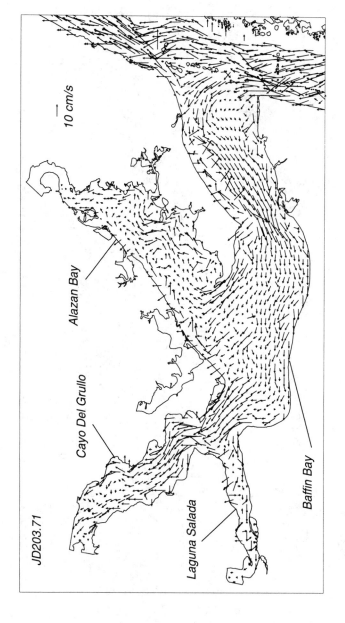

Fig. 10. Calculated current pattern in Baffin Bay at time JD 203.71, measured wind speed equals 12 m/s.

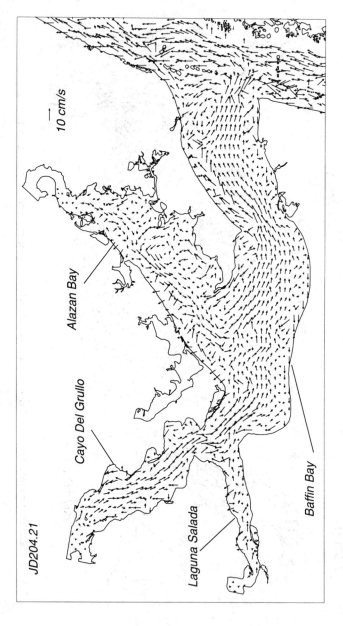

Fig. 11. Calculated current pattern in Baffin Bay at time JD 204.2, measured wind speed equals 9 m/s.

CONCLUSIONS

Two-dimensional circulation modeling was conducted to calculate the water-level distribution over Baffin Bay and to determine circulation patterns under summer wind forcing. Summer winds of the south Texas coast are composed of two primary components, a persistent southeast wind and sea breeze. The intensity of the southeast wind varies with a 24-hr cycle, corresponding to strengthening and weakening of the sea breeze.

Agreement between measured and calculated water level was good, particularly for wind speed exceeding 10 m/s. Thus, the wind-drag coefficient of Hsu (1988) is appropriate for calculating wind-induced motion in shallow (order of 1 m) water.

Water level and circulation in Baffin Bay are wind dominated. Because of the predominant onshore wind direction, the average bay surface is tilted with the western end being at greater elevation than the eastern end. For July 1995, the average difference in water elevation between the west and east ends of the bay, a distance of 30 km, was 10 cm. During strong onshore wind, the water-elevation difference was calculated to be as great as 0.58 m for this particular month. Such strong diurnal forcing by wind could be mistaken for tidal motion. In locations where sea breeze and diurnal tides are present, analysis of water level and current must be carefully conducted to distinguish these forcing mechanisms.

Calculated circulation patterns in Baffin Bay showed currents directed toward the west during strong onshore wind and directed toward the east during weak onshore wind. The eastward-directed flow owes to the sloped bay surface formed during intervals of stronger onshore wind. When the wind relaxes, water piled up on the western end of the bay flows eastward. Thus, circulation in the bay has bi-directional flow, even though the primary forcing (wind) is directed onshore.

ACKNOWLEDGEMENTS

Reviews provided by Dr. Nicholas Kraus, Dr. Norman Scheffner, and two anonymous reviewers improved this paper. Preparation of this paper was supported by the Coastal Inlets Research Program of the U.S. Army Corps of Engineers (USACE). Permission was granted by Headquarters, USACE, to publish this information.

REFERENCES

Andrews, P. B., 1964: Serpulid Reefs, Baffin Bay, Southeast Texas. *Field Trip Guidebook, Depositional Environments, South Central Texas Coast, Gulf Coast*, Association of Geological Societies, 101-120.

Brown, C. A., Kraus, N. C., and Militello, A., 1995: Hydrodynamic Modeling for Assessing Engineering Alternatives for Elevating the Kennedy Causeway, Corpus Christi, Texas. *Proceedings of the 4th International Estuarine and Coastal Modeling Conference,* American Society of Civil Engineers, 681-694.

Fischer, G., 1959: Ein numerisches Verfahren zur Errechnung von Windstau und Gezeiten in Randmeeren. *Tellus,* **2**, 1, 60-76.

Gill, S. K., Hubbard, J. R., and Dingle, G., 1995: Tidal Characteristics and Datums of Laguna Madre, Texas. NOAA Technical Memorandum NOS OES 008, National Oceanic and Atmospheric Administration, Silver Spring, Maryland.

Hsu, S. A., 1988: *Coastal Meteorology.* Academic Press, San Diego, 260 pp.

Kowalik, Z., and Murty, T. S., 1993: *Numerical Modeling of Ocean Dynamics.* World Scientific, New Jersey, 481 pp.

Militello, A., 1998: Hydrodynamics of Wind-Dominated, Shallow Embayments. Doctoral Dissertation, Florida Institute of Technology, Melbourne, Florida.

Militello, A., and Kraus, N. C., Submitted December 1999: Generation of Harmonics by Sea Breeze in Nontidal Water Bodies. *Journal of Physical Oceanography.*

Shepard, F. P., and Rusnak, G. A., 1957: Texas Bay Sediments. *Publications of the Institute of Marine Science, University of Texas,* **IV**, 2, 5-13.

Shideler, G. L., 1984: Suspended Sediment Responses in a Wind-Dominated Estuary of the Texas Gulf Coast. *Journal of Sedimentary Petrology,* **54**, 3, 731-745.

Ward, G. H., Wiersema, J. M., and Armstrong, N. E., 1985: Matagorda Bay: A Management Plan. Report FWS/OBS-82/73, Division of Biological Services, U.S. Fish and Wildlife Service, Washington, D.C.

Tidal inlet circulation: observations, model skill and momentum balances

James L. Hench[1] and Richard A. Luettich, Jr.[1]

Abstract

A combined observational and modeling study of circulation at Beaufort Inlet, North Carolina is underway to elucidate circulation patterns and momentum balances at shallow barrier island inlets. The observational data include fourteen shipboard ADCP cruises, each over a complete semi-diurnal tidal cycle. A high resolution (<50 meter grid spacing) depth-integrated nonlinear barotropic model is used to simulate the circulation. With simple model forcing, preliminary comparisons with semi-diurnal tidal ellipse data are quite favorable.

The contribution of each term in the momentum equations to the overall momentum balance is determined from model elevation and velocity fields. Results show that at maximum ebb the streamwise momentum balance is dominated by advective acceleration, pressure gradient, and bottom friction. In the cross-stream direction the momentum balance is dominated by the centrifugal acceleration, pressure gradient, and (to a lesser degree) Coriolis. Momentum balances were highly localized, often with sub-kilometer length scales.

[1]Institute of Marine Sciences, University of North Carolina at Chapel Hill
3431 Arendell Street, Morehead City, NC 28557 USA

Introduction

Tidal inlets provide an important conduit between estuarine waters and the coastal ocean for the movement of contaminants, nutrients, sediments, and biota. An understanding of these transport processes is based implicitly on knowledge of the circulation. However, a complete understanding of inlet circulation has proved elusive.

Stommel and Farmer (1952) gave the earliest description of tidal inlet kinematics and noted the distinct difference between flood (resembling a potential sink) and ebb (resembling a jet) circulation outside an inlet. The net circulation was described as an offshore directed jet, with inflow along the sides of the inlet. This conceptual picture has been replicated in numerical models of idealized inlets (e.g. Kapolnai, et al. 1996; Wheless and Valle-Levinson, 1996). Progress has also been made in describing the dynamics of inlet flow. Imasato (1983) used an idealized inlet model to make a qualitative assessment of force balances. The most complete analysis to date appears to be Ridderinkof (1988), who computed term-by-term momentum balances at maximum ebb and flood from a model of Wadden Sea inlets.

There is a notable paucity of field data to verify these notions of inlet circulation. Even the classical circulation patterns described by Stommel and Farmer remain essentially unobserved in nature. This is perhaps not surprising given the small spatial scales and difficult field conditions found at tidal inlets. Moored instruments can be buried by rapidly shifting ebb shoals or damaged by trawling activities. Shoaling and breaking waves on the ebb delta often make shipboard measurements unfeasible.

Over the past four years we have been conducting a combined observational and modeling study at Beaufort Inlet, North Carolina to understand the circulation and momentum dynamics at small tidal inlets. In our initial modeling efforts at Beaufort Inlet (see Luettich et al., in press), model results compared well with moored instrument data at 16 elevation and 10 velocity stations distributed within the estuary. Drogue observations during flood tide (Churchill et al., 1999) also provided encouraging comparisons with the model. These studies indicated a high degree of spatial structure in the immediate vicinity of the inlet and have motivated additional observations and an attempt to explain the forcing mechanisms for these flow features. This paper reports on our most recent progress.

Methods

Observations

Beaufort Inlet has a width of about one kilometer at its narrowest point (Figure 1). Depths range from two to ten meters along the ebb delta. The inlet's main channel is dredged for navigation to 15 meters. The flood delta is cut by several connecting sloughs which give way to tidal marsh and an estuarine complex. The astronomical tide at Beaufort Inlet is dominated by the M_2 constituent with maximum flow speeds exceeding 1.25 m/s in the inlet throat. Diurnal tides are relatively weak and are less than 15% of semi-diurnal tides (Klavens, 1983).

Fourteen cruises on the 8 meter *R/V Parker* were conducted in 1998 (Table 1). A boom-mounted 1200 kHz RDI Workhorse broadband ADCP with bottom tracking was used to measure current velocities. Each cruise covered a small region of the inlet and made repeated laps over a complete semi-diurnal tidal cycle (Figure 1). Multiple cruises were made to obtain coverage of the entire ebb delta region. Boat speed was nominally 2.0 m s^{-1} while collecting ADCP data at 0.85 Hz. Ship heading and speed were removed from the ADCP data using the ADCP's internal flux gate compass and bottom-tracking. The instrument's internal pendulum sensors were used to remove ship pitch and roll. The location of each ADCP profile and navigation information were obtained from a NorthStar 951XD DGPS. Vertical bin size was 0.5 meters. ADCP and DGPS data were recorded on a laptop computer with each 13.5 hour survey yielding about 150 megabytes of data.

Cruise No.	Date	Notes
1	Dec. 19, 1997	Incomplete due to poor weather
2	Jan. 11, 1998	
3	Jan. 31, 1998	
4	Mar. 13-14, 1998	
5	Jun. 8, 1998	
6	Jul. 7-8, 1998	
7	Jul. 10, 1998	Incomplete due to poor weather
8	Aug. 5-6, 1998	
9	Aug. 21-22, 1998	
10	Aug. 22-23, 1998	
11	Sep. 16-17, 1998	Incomplete due to poor weather
12	Sep. 20-21, 1998	
13	Oct. 5, 1998	
14	Oct. 11, 1998	
15	Oct. 12-13, 1998	
16	Oct. 16-17, 1998	
17	Oct. 25, 1998	

Table 1. Summary of shipboard ADCP cruises.

The shipboard ADCP data were depth-integrated and ensemble averaged into 100 m horizontal bins. After ensemble averaging, the standard deviation of the ADCP data was about 1 cm s⁻¹ (see Gordon, 1996), and thus sufficiently accurate for studying tidal and tidal residual flows. The resulting short time-series from each cruise were harmonically analyzed for the M_2, M_4, M_6 and Steady tidal constituents. The time series were too short to separate constituents in neighboring frequency bands (e.g. S_2, N_2) or to resolve the diurnal tide (e.g. K_1, O_1). The cruise data were then combined by tidal constituent to form detailed mosaics of the tidal current structure (see Geyer and Signell, 1990).

Modeling

Although the flow at Beaufort Inlet is periodically stratified (primarily due to salinity stratification from estuarine outflow), long-term moored CTD data and ADCP profiles indicate that there are substantial periods when both density variations and vertical shears are small. Considering these observations we have begun modeling the circulation using the barotropic nonlinear two-dimensional version of ADCIRC (Luettich et al., 1992). In the absence of wind and tidal potential forcing, and assuming a constant lateral viscosity, the governing depth-integrated continuity and momentum equations used in the model are:

$$\frac{\partial \eta}{\partial t} + \frac{\partial UH}{\partial x} + \frac{\partial VH}{\partial y} = 0 \tag{1}$$

$$\frac{\partial U}{\partial t} + U\frac{\partial U}{\partial x} + V\frac{\partial U}{\partial y} - fV + g\frac{\partial \eta}{\partial x} - v\nabla^2 U + \left(\frac{C_f \sqrt{U^2 + V^2}}{H} \right) U = 0 \tag{2}$$

$$\frac{\partial V}{\partial t} + U\frac{\partial V}{\partial x} + V\frac{\partial V}{\partial y} + fU + g\frac{\partial \eta}{\partial y} - v\nabla^2 V + \left(\frac{C_f \sqrt{U^2 + V^2}}{H} \right) V = 0 \tag{3}$$

where: $U(x,y,t), V(x,y,t) =$ depth-integrated velocities
$H(x,y,t) = h(x,y) + \eta(x,y,t) =$ total water column
$\eta(x,y,t) =$ water surface vertical displacement from geoid
$f(y) =$ Coriolis parameter
$g =$ gravitational constant
$v =$ lateral eddy viscosity / dispersion coefficient
$C_f =$ quadratic bottom friction coefficient

Equation [1] is transformed into a wave-equation formulation and the resulting coupled system of equations is discretized using a finite element method in space and a finite difference scheme in time (see Luettich et al., 1992).

The Beaufort Inlet model was set up using high resolution bathymetric data from a 1998 NOAA survey (at 85 m track spacing) and supplemented in inshore areas with bathymetry collected with our own research vessel (at 50 m track spacing). We constructed a detailed finite element grid that spanned from the adjacent continental shelf to the estuarine waters (Figure 2). The inlet is adjacent to Cape Lookout and the grid was extended to include the associated shoal region. Nodal spacing ranged from about 2 km offshore to less than 25 m in the estuary; grid resolution in the inlet was 50 m. Prior to numerical solution, the grid was projected into a Cartesian coordinate system (see Kolar et al., 1994). The model was forced at three open boundaries with specified elevations for the M_2, M_4, M_6, and Steady tidal constituents obtained from the larger domain circulation model of Luettich et al. (in press). A constant bottom friction coefficient of 0.0025 was used with a constant lateral eddy viscosity of 7 m^2/s (the minimum for model stability) and the time-step was 2 seconds. The model was spun-up for 6 days to dynamic equilibrium; results from days 6-8 were harmonically analyzed (using M_2, M_4, M_6, and Steady constituents) for comparison with the shipboard ADCP data.

Model / data comparisons

Harmonic constituents from the shipboard ADCP data and the high-resolution circulation model were interpolated to a common uniform grid for comparison. Grid spacing was 200 m (coarser than both the model and observed data) and yielded 405 comparison stations. The spatial density of these observations would be prohibitively expensive to obtain using conventional moored instruments, and provides a unique data set to assess model skill at sub-kilometer scales.

A comparison of the spatial structure of the semi-diurnal tidal ellipses reveals strong similarities between observed and modeled (Figure 3). Velocities are strongest near the inlet throat and highly rectilinear, particularly in the navigation channel. With increasing distance from the inlet, flow decreases and becomes more rotary. Much of the flow is cross-isobath suggesting advection is important in the momentum balance. Tidal ellipse rotation was generally clockwise, except for a nearshore region west of the inlet, as seen in both modeled and (to a lesser degree) observed. Model semi-minor axes were generally slightly smaller than observed, as evident by more rectilinear ellipses. Ellipse orientations (shown as degrees clockwise from North) agreed quite closely and were mostly directed northward.

Momentum flux balances

The circulation field at tidal inlets is sufficiently complex and nonlinear as to elude analytic dynamical description. Here we dissect our model results to assess the contribution of each term in the momentum equations to gain an understanding of the dynamics.

Velocity and elevation fields from the model were used to directly evaluate the momentum terms at each computational node throughout a tidal cycle. The x-y components of each term (e.g. pressure gradient) were then rotated at each node into a local streamwise-normal (s-n) coordinate system aligned with the local velocity vector (Figure 4). The local coordinate system was defined with the s-direction positive in the direction of flow, and the n-direction positive left of the flow. After multiplying by the total water column to obtain momentum fluxes the rotated momentum equations are:

$$H\left[\frac{\partial U_s}{\partial t} + U_s\frac{\partial U_s}{\partial s} + g\frac{\partial \eta}{\partial s} - \nu\nabla^2 U_s + \left(\frac{C_f|U_s|}{H}\right)U_s = 0\right] \qquad [4]$$

$$\quad\;\; \text{(S1)} \quad\;\; \text{(S2)} \quad\; \text{(S3)} \quad\; \text{(S4)} \quad\;\;\; \text{(S5)} \qquad\quad \text{(S6)}$$

$$H\left[\frac{\partial U_n}{\partial t} + U_s\frac{\partial U_n}{\partial s} + g\frac{\partial \eta}{\partial n} - \nu\nabla^2 U_n + fU_s = 0\right] \qquad [5]$$

$$\quad\;\; \text{(N1)} \quad\;\; \text{(N2)} \quad\; \text{(N3)} \quad\; \text{(N4)} \quad \text{(N5) (N6)}$$

where: (S1) streamwise acceleration flux
           (N1) normal (rotary) acceleration flux
           (S2) streamwise (Bernoulli) acceleration flux
           (N2) centrifugal acceleration flux
           (S3, N3) horizontal pressure gradient fluxes
           (S4, N4) horizontal diffusion/dispersion fluxes
           (S5) non-linear bottom friction flux
           (N5) Coriolis acceleration flux
           (S6, N6) errors

With this choice of coordinate system there is, by definition, no normal component to the flow; the Coriolis term is zero in the s-equation, as is the bottom friction term in the n-equation. Moreover, the advective acceleration terms collapse to a single term in each equation; streamwise in the s-equation and centrifugal in the *n*-equation. For our Beaufort Inlet model, momentum flux was conserved to within a few percent after the rotation. The horizontal diffusion/dispersion fluxes were generally an order of magnitude smaller than the other terms, and are omitted from the discussion below.

Streamwise direction

Figure 5 shows contours of streamwise momentum fluxes at Beaufort Inlet near maximum ebb. The flow has reached a point of near steady-state and the transient term (S1) is close to zero in the vicinity of the inlet. The exception is in the navigation channel where the flow is slowing down in the inlet throat, while still slightly increasing in speed in the outer channel.

Near inlet advective accelerations (S2) are quite strong; positive as the flow from the estuary converges toward the inlet throat and negative as the flow decelerates outside the inlet. Localized areas of negative advective acceleration at the tips of headland features are associated with flow separation, and are balanced by local adverse pressure gradients. Along the crest of the eastern ebb delta, a Bernoulli effect in the flow can be seen with acceleration on the crest's west side and deceleration on the east side.

The horizontal pressure gradient (S3) at the inlet is generally negative in the direction of the flow, driving the ebb. Outside the inlet is a region of positive pressure gradient which arises from a Bernoulli effect; the rapid deceleration is balanced by an adverse pressure gradient.

Bottom friction (S5) scales with the square of the velocity and inversely with depth. At Beaufort Inlet, friction effects are greatest in the channels. This term rapidly diminishes in importance away from the inlet, where the dynamics more closely follow linear wave theory.

Normal direction

Figure 6 shows contours of normal direction momentum flux at maximum ebb. The local accelerations in the normal direction (N1) are small, indicating small rates of flow direction change at this point in the tide.

Advective acceleration in the normal direction (N2) can be interpreted as a centrifugal acceleration and this term is largest where the flow from the estuary turns toward the inlet throat. On the east side of the inlet, the centrifugal acceleration is positive since the flow curves in the counter-clockwise direction. Conversely, the west side flow curves clockwise and the centrifugal acceleration is negative. The two streams balance near the inlet center to form a somewhat symmetric pattern. Just outside the inlet the navigation channel turns slightly westward; the centrifugal acceleration gains a clockwise orientation (so the sign changes) and again is balanced by the pressure gradient.

Normal direction pressure gradients (N3) closely balance centrifugal accelerations. At Beaufort Inlet, this balance results in a dome of high water across the inlet throat. Near maximum ebb, model results indicate that the elevation difference from the sides of the inlet to the center is about 3-4 cm.

The Coriolis term (N5) plays a secondary role to the centrifugal acceleration and normal pressure gradients at maximum ebb and is largely confined to the navigation channel. Coriolis acceleration forces the flow against the channel side wall and is balanced by a local pressure gradient.

These results suggest that near maximum ebb, the momentum flux balance can be approximated as:

$$H\left[U_s \frac{\partial U_s}{\partial s} + g \frac{\partial \eta}{\partial s} + \left(\frac{C_f |U_s|}{H} \right) U_s = 0 \right] \tag{6}$$

$$H\left[U_s \frac{\partial U_n}{\partial s} + g \frac{\partial \eta}{\partial n} + f U_s = 0 \right] \tag{7}$$

Discussion

Comparisons between the model and data were most favorable closest to the inlet. Further offshore, the observations are less reliable as the ADCP noise-to-signal ratio increases with weaker circulation. In addition, tidal circulation is weaker offshore and thus more easily contaminated by non-tidal effects (e.g. along-shelf pressure gradients, wind, baroclinicity) not included in the model runs.

Runs with a grid at double this resolution showed only minor differences from the present velocity field. Given this fact and the high resolution bathymetry in the model, we expect many of the discrepancies between the model and observed are due to differences in the forcing. These preliminary model runs were forced with a limited set of tidal constituents (M_2, M_4, M_6, and Steady). In future work we plan to force the model with the entire suite of astronomical constituents and synchronize the harmonic analysis of the model to the times when the shipboard ADCP data were collected. These modifications will permit a more consistent basis from which to compare the model and observations and we expect improvement in these comparisons.

An analysis of the detailed momentum balance spatial structure shows that momentum balances at a tidal inlet can vary dramatically over sub-kilometer scales. The larger scale momentum features in our model are in general agreement with those of Ridderinkhof (1988). However the dynamics of the flow separation regions described here were not present in his model. We attribute this difference to our factor of ten greater grid resolution, and note their similarity to those found around a single headland (Signell and Geyer, 1991). In future work we plan to extend these results to examine time varying momentum balances over a tidal cycle.

Acknowledgments

We thank Brian Blanton and Crystal Fulcher for help setting up and running the model. The crew of the NOAA ship *Whiting* kindly provided the 1998 bathymetric data. The field measurements were aided by: Stephen Bullard, Stacy Davis, Janelle Fleming, Jon Grabowski, Heidi Hook, Niels Lindquist, Jennifer McNinch, Jesse McNinch, Pia Moisander, Joe Purifoy, Glenn Safrit, and Erik Sotka. Funding was provided by NOAA, National Marine Fisheries Service; Office of Naval Research, Naval Research Laboratory; and U.S. Army Corps of Engineers, Coastal Hydraulics Laboratory.

References

Churchill, J. H., J. O. Blanton, J. L. Hench, R. A. Luettich, Jr., and F. E. Werner, 1999. Flood tide circulation near Beaufort Inlet, North Carolina: Implications for larval recruitment. *Estuaries*, 22(4): 1057-1070.

Geyer, W. R., and R. P. Signell, 1990. Measurements of tidal flow around a headland with a shipboard acoustic Doppler current profiler. *J. Geophys. Res.*, 95(C3): 3,189-3,197.

Gordon, R. L., 1996. *Acoustic Doppler Current Profilers Principles of Operation: A Practical Primer, Second Edition for BroadBand ADCPs*. RD Instruments, San Diego, 54 pages.

Imasato, N. 1983. What is tide-induced residual current? *J. Phys. Oceanography*, 13: 1307-1317.

Kapolnai, A., F. E. Werner, and J. O. Blanton 1996. Circulation, mixing, and exchange processes in the vicinity of tidal inlets: a numerical study. *J. Geophys. Res.*, 101(C6): 14,253-14,268.

Klavens, A. S., 1983. *Tidal hydrodynamics and sediment transport in Beaufort Inlet*, North Carolina. NOAA Technical Report NOS 100, Rockville, Md., 119 pages.

Kolar, R. L., W. G. Gray, J. J. Westerink, and R. A. Luettich, 1994. Shallow water modeling in spherical coordinates: equation formulation, numerical implementation, and application. *J. Hydraulic Res.*, 32(1): 3-24.

Luettich, Jr., R. A., J. L. Hench, C. W. Fulcher, F. E. Werner, B. O. Blanton, and J. H. Churchill, Barotropic tidal and wind driven larvae transport in the vicinity of a barrier island inlet. *Fisheries Oceanography*, in press.

Luettich, R. A., J. J. Westerink, N. W. Scheffner, 1992. ADCIRC: An advanced three-dimensional model for shelves, coasts, and estuaries, Report 1: Theory and methodology of ADCIRC 2DDI and ADCIRC 3-DL. *Dredging Research Program Technical Report DRP-92-6*, Coastal Engineering Research Center, U. S. Army Corps of Engineers, Waterways Experiment Station, Vicksburg, MS, 141 pages.

Ridderinkhof, H. 1988. Tidal and residual flows in the Western Dutch Wadden Sea I: numerical model results. *Netherlands Journal of Sea Research*, 22(1): 1-21.

Signell, R. P., and W. R. Geyer, 1991. Transient eddy formation around headlands. *J. Geophys. Res.*, 96(C2): 2,561-2,575.

Stommel, H., and H. G. Farmer, 1952. On the nature of estuarine circulation, Part I (Chapters 3 and 4). *Woods Hole Oceanographic Institution Technical Report Ref. No. 52-88*, 131 pages.

Wheless, G. H., and A. Valle-Levinson, 1996. A modeling study of tidally driven estuarine exchange through a narrow inlet onto a sloping shelf. *J. Geophys. Res.*, 101(C11): 25,675-25,687.

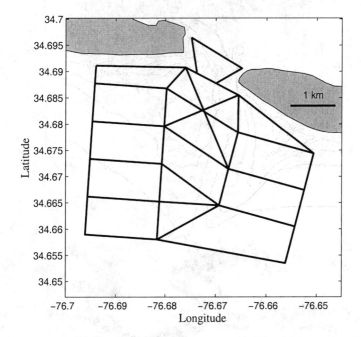

Figure 1. Mosaic of shipboard ADCP survey tracks.

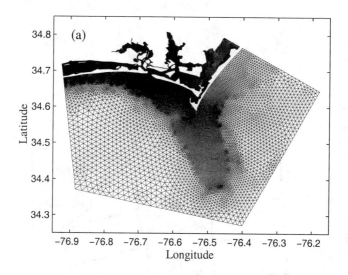

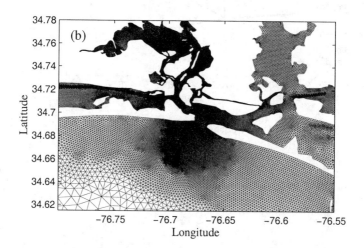

Figure 2. (a) North Carolina coast small domain finite element mesh; (b) Zoom of Beaufort Inlet in finite element mesh. The grid contains 53,505 nodes and 102,228 elements.

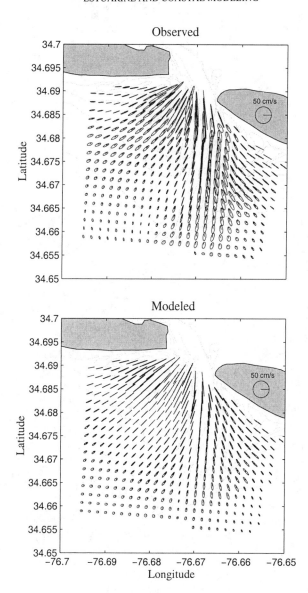

Figure 3. Comparison of depth–averaged semi–diurnal tidal ellipses. Observational data and model results were interpolated to a common uniform grid with 200m spacing.

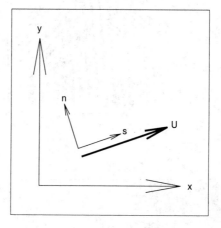

Figure 4. Definition sketch of local rotated coordinate system.

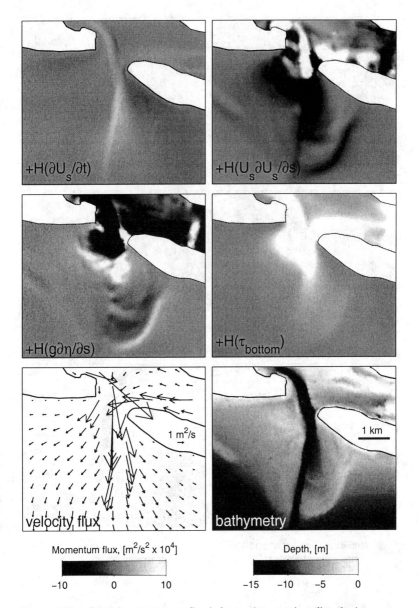

Figure 5. Beaufort Inlet momentum flux balances (streamwise–direction) near maximum ebb. All gradients are relative to the local flow direction.

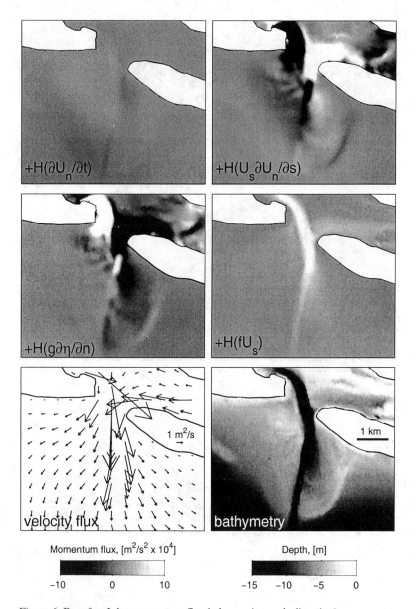

Figure 6. Beaufort Inlet momentum flux balances (normal–direction)
near maximum ebb. All gradients are relative to the local flow direction.

Modeling of Tide-Induced Circulation near a Strait

Masahide Ishizuka[1], Keiji Nakatsuji[2] and Shuzo Nishida[3]

Abstract

    Tide-induced circulation near a strait is clarified by three-dimensional baroclinic flow simulation and field surveys using a VHF oceanic radar, shipboard ADCP and CTD.   According to spatial distributions of surface flow around Kitan Strait from a VHF oceanic radar, the time variations and horizontal structure of a headland eddy can be recognized in the northern part of the Kitan Strait. This eddy is called the Tomogashima Circulation.   Taking the three-dimensional baroclinic flow simulation results into account, the large scale circulation observed by VHF oceanic radar is a barotropic flow induced by tide-induced currents.   The maximum vorticity of the Tomogashima Circulation is -1.0×10^{-3} s^{-1} after maximum northward tidal current.   This transient vortex moves north by strong tidal current and after slack from northward to southward, the scale of the vortex becomes the largest.   The salinity distributions are significantly influenced by tidal stirring in the Kitan Strait.   The residual current is vertically different in the northern area of the Kitan Strait.   In a lower layer, clockwise residual circulation was observed.   On the other hand, in the upper layer, residual circulation does not exist due to southwestward tidal current along the east-coast in Osaka Bay.

[1] Student, Department of Civil Engineering, Osaka University, 2-1 Yamadaoka, Suita City, Osaka 565-0871, Japan. e-mail:ishizuka@civil.eng.osaka-u.ac.jp
[2] Professor, Department of Civil Engineering, Osaka University, Japan. e-mail: nakatsuj @civil. eng.osaka-u.ac.jp
[3] Associate Professor, Department of Civil Engineering, Osaka University, Japan. e-mail: nishida @civil.eng.osaka-u.ac.jp

Introduction

Many of semi-enclosed coastal waters are composed of wide and shallow basins, and narrow and deep straits.   In the former, stratification often develops due to low velocity tidal currents, river water discharge and the heat-balance at the sea surface, while in the latter fluid is vertically well-mixed due to high velocity tidal currents.   We call it a "strait-basin system".   Tidal currents in straits are much stronger than those within the basin.   Therefore, the current in the strait has important roles to not only the three-dimensional flow structure and mass transport processes but also ecosystem in the basin.   Osaka Bay is a typical example of the strait-basin system, which is composed of two narrow straits and one wide basin in the eastern end of the Seto Inland Sea, Japan.   The location of Osaka Bay and its topographical features are shown in Fig. 1 and Fig. 2, respectively.

Previous study (Pingree and Maddock (1977); Tee (1976)) have shown that headland eddies are basically formed by simple separation vortices in the lee of the headland during both flood and ebb tides.   The characteristics of eddies are greatly influenced by complicated topography and the non-linearity of tidal currents near the strait.   Moreover, the spatial distributions of eddies are time-dependent during a tidal cycle.   In order to understand the behavior of headland eddies in detail, high-resolution measurement instruments and numerical simulation are indispensable. Geyer and Signell (1990) measured tidal currents around a headland using a shipboard ADCP in Vineyard Sound, Massachusetts.   Fujiwara et al. (1994) used four ADCPs near the Akashi Strait in Osaka Bay to clarify the mechanism of tidal-jet and vortex-pair driving residual circulation. Fine convoluted jet flow was seen in the north part of the Kitan Strait from a satellite image (Fujiwara et al., 1994).   This structure is about 10 km in diameter.   The existence of large eddy is recognized in the eastern part of the Akashi Strait in the same satellite image, which we call

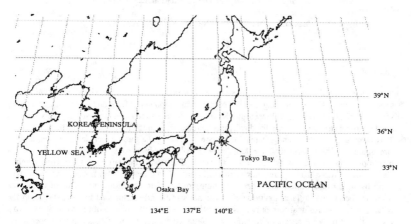

Fig. 1 : Location of Osaka Bay

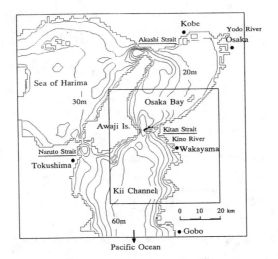

Fig. 2 : Topographical features of Osaka Bay and its surrounding coastal waters.
Two rectangular areas indicate the computer domain for large and small scales

the Okinose Circulation.   Nakatsuji and Fujiwara (1997) clarified the mechanism of the large-scale Okinose Circulation; successively separated vortices from the headland of Awaji Island coalesce into a large-scale eddy.   As long as we observed the satellite image of both circulations, the mechanism of circulation observed in Kitan Strait is different from the Okinose Circulation.

The aim of the present study is to clarify the hydrodynamics peculiar to the coastal waters near Kitan Strait.   Therefore, we carried out field surveys using VHF oceanic radar, shipboard ADCP and CTD in 1998 and 1999.   In addition, three-dimensional baroclinic flow simulation was carried out to complement field results.

Outline of Field Surveys

As shown in Fig. 2, Osaka Bay is composed of a wide basin and two narrow straits.   The basin has a mean depth of 28 m and has an elliptic shape with a long axis of 56 km and a short axis of 28 km.   One of the straits is the Akashi Strait, which connects Osaka Bay to the Sea of Harima, and to the Seto Inland Sea.   Another is Kitan Strait, which connects Osaka Bay to Kii Channel and to the Pacific Ocean.   The width of Kitan Strait is about 11km, and has one main channel Yura Seto, as well as two smaller channels and two islands as shown in Fig. 3.   The width and maximum depth of the three channels are; the Yura Seto: 3.7 km and 150 m, the Naka no Seto: 0.5 km and 40 m and Kada Seto: 0.8 km and 70 m, respectively.   The tidal current through the Kitan Strait reaches about 2 m/s during spring tide.

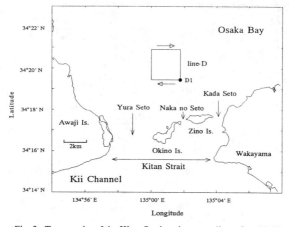

Fig. 3 : Topography of the Kitan Strait and traverse line using ADCP

Field surveys of flow velocities, salinity and temperature were carried out during the period from November 1998 to March 1999 around Kitan Strait. The field survey area and traverse line are shown in Fig. 3.

(1) First survey on 18 December 1998 (spring tide)

Measurements of the current distributions were carried out along the square transect (line-D) by turning around operation with shipboard ADCP (600kHz). Sampling time of data was 15 second and cell thickness was 50 cm. ADCP-equipped vessel whose speed was about 2.7 m/s ran on the line-D and took about 1 hour to come back to the starting point (D1). This measurement was carried out during 8 hours from the maximum northward tidal current (7:00) to the time after the maximum southward current (15:00).

(2) Second survey from 1 February to 4 March 1999

VHF (Very High Frequency) oceanic radar provides long-term variations and horizontal distributions of surface tidal currents. The measurement principal was established by Barrick *et al.* (1977). The radar sites and covering area are shown in Fig. 4. Radio waves transmitted by VHF oceanic radar have a center frequency of 41.900MHz, and the wavelength is 7.16 m. The depth of the measured layer is about 30 cm (Stewart and Joy, 1974). The Yagi type antenna is used and it rotated mechanically. The width of antenna is about 14.4 m long and the height of the radar is about 6 m. The observation distance is about 20 km maximum. The increments and intervals of antenna beam can be easily changed. In the present survey, these conditions were set to a total scanning angle of 100 degree in steps of 10 degree (eleven directions).

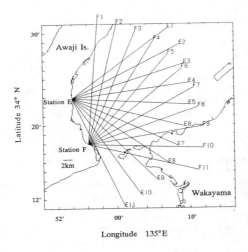

Fig. 4 : Location of VHF oceanic radar stations

Outline of Numerical Simulation

Numerical Model

A three-dimensional baroclinic flow model ODEM (Osaka Daigagu Estuarine Model), is used in the present study, which has been applied to the major Japanese bays, namely, Osaka Bay, Ise Bay and Tokyo Bay (Nakatsuji and Fujiwara (1997), Sugiyama et al. (1997), Choi et al. (1996)). The hydrodynamics of tidal current and density stratification were clarified as compared with field survey results and satellite images. The governing equations used in the model were: the continuity equation, conservation of momentum in three directions, temperature and salinity. Hydrostatic and Boussinesq assumptions were introduced. Strictly speaking, therefore, ODEM is a quasi-3D model. Water density is calculated using Mamaev (1964) formulation as a function of salinity and temperature. The eddy viscosity and eddy diffusivity coefficients in the horizontal direction are determined using the SGS (Sub-Grid Scale) concept (Smagorinsky, 1963) and are functions of space and time. The coefficients of vertical eddy viscosity and eddy diffusivity are calculated every time step as a function of Richardson number. The details of the formulation are shown in the published papers cited above.

The computational domains are shown in Fig. 2. We carried out the 3-D baroclinic flow simulation in two stages. First, hindcasting was performed on a large-scale domain (112 km x 106 km) which covers Osaka Bay and its surroundings coastal waters, the Sea of Harima and Kii Channel. The object of hindcasting simulation is to examine the characteristics in Osaka Bay and to verify the application of ODEM to baroclinic flow, tidal currents and residual currents as well.

Another objective was to obtain the boundary conditions for a small-scale domain simulation. For the second computation, a fine meshed small-scale flow model (57 km x 58 km) was used to study tidal and residual currents in the vicinity of Kitan Strait.

Large-Scale Domain Simulation

Hindcasting was carried out from 5 to 24 August 1997 in the large-scale area shown in Fig. 2. Monthly field surveys of salinity, temperature and water quality were conducted by the Osaka, Hyogo, Wakayama and Tokushima Prefectural Fisheries. For example, the Osaka Prefectural Fisheries have 22 sampling stations in Osaka Bay. The observation points are distributed and the observed data are interpolated onto all numerical grids and used as the initial and boundary conditions in hindcasting. Inflow of salinity and temperature at the open boundaries are also interpolated in space and time using the data observed in August and September 1997. In addition, in order to suppress the artificial rapid inflow of salinity and temperature and large density gradients in the vicinity of open boundaries, the decreasing function derived from Thatcher and Harleman (1972) was adopted. Tidal elevations at the open boundaries were generated from six astronomical constitutions: K_1, O_1, P_1, M_2, S_2 and K_2. Only the river discharge of the Yodo River was used based on time-dependant observation data, and provided by the Yodo River Construction Branch of Ministry of Construction. There are 14 rivers along coast in Osaka Bay. According to analysis of rainfall data in August 1997, discharge volume is very low as compared with that in other year. Therefore we assumed that the total discharge volume could be discharged from Yoko River. Average river discharge in August 1997 was given for other 13 rivers. The meteorological conditions at the sea surface such as winds, rainfall, solar radiation, etc. were also input into hindcasting system as a function of time. The grid spacing was set to 1 km in space with 10 vertical layers of 2, 2, 2, 2, 2, 4, 6, 10, 15 and 15 m in thickness from the surface to the bottom.

Small-Scale Domain Computation

A small-scale computation was performed to examine the generation and development of circulation at the northern area of Kitan Strait and to clarify water exchange through the strait. The horizontal grid spacing of the small-scale model was 250 m and vertical resolutions were the same as the large domain model. There are two open boundaries in the north and in the south of Kitan Strait (see Fig. 2). The semi-diurnal and quarter-diurnal component of tidal elevations and tidal currents at the boundaries were obtained by harmonic analysis of computed results of the large computation in 6:00-18:00 23 August 1997 during a low diurnal inequality. Tidal elevation conditions were given at the boundary in the Kii channel. On the other hand, tidal current conditions were given at the northern boundary to include the large tide-induced residual circulation (the Okinose Circulation).

Boundary Condition in the Kii Channel :

$$\xi = \xi_1 \cos\left(\frac{2\pi}{T_1}t - \theta_1\right) + \xi_2 \cos\left(\frac{2\pi}{T_2}t - \theta_2\right) \tag{1}$$

Boundary Condition in Osaka Bay :

$$U = U_0 + U_1 \cos\left(\frac{2\pi}{T_1}t - \phi_{u1}\right) + U_2 \cos\left(\frac{2\pi}{T_2}t - \phi_{u2}\right) \tag{2}$$

$$V = V_0 + V_1 \cos\left(\frac{2\pi}{T_1} t - \phi_{v1}\right) + V_2 \cos\left(\frac{2\pi}{T_2} t - \phi_{v2}\right) \quad (3)$$

where subscript "1", "2" and "0" means semi-diurnal, quarter-diurnal and residual constituent, respectively. T is a tidal period, ξ is a tidal amplitude, θ is a phase of tide, U is a EW-component velocity and ϕ_u is a phase of EW-component tidal current. The NS-component velocity is defined similarly. Tidal boundary conditions in Kii Channel are shown in Table 1. The boundary conditions were linearly interpolated to the southern boundary using the values of Wakayama and Tokushima sides. As to the boundary conditions of velocity in Osaka Bay, these are calculated at $8 \sim 30$ grid interval in each layer and interpolated. The initial salinity and temperature distributions were interpolated using the results of large-scale domain computation at 6:00 on 23 August 1997. The boundary and initial conditions of the hindcasting system and the small-scale domain computation are summarized in Table 2.

Table 1 : Tidal boundary conditions in the Kii Channel of small-scale domain computation. Phase of tide is Japan Standard Time referred to Akashi Strait.

Tidal current constituents	Wakayama side	Tokushima side
Tidal amplitude (semi-diurnal) : ξ_1	0.531 (m)	0.512 (m)
Tidal amplitude (quarter-diurnal) : ξ_2	0.005 (m)	0.005 (m)
Phase of tide (semi-diurnal) : θ_1	270 (deg.)	268 (deg.)
Phase of tide (quarter-diurnal) : θ_2	332 (deg.)	350 (deg.)

Table 2 : Boundary and initial conditions in hindcasting system for two simulations

Factors	Large-scale computation	Small-scale computation
Computational domain	112 km x 106 km	57 km x 58 km
Horizontal resolution	1 km x 1 km	0.25 km x 0.25 km
Time-step	30 sec	10 sec
Tidal elevation	$K_1, O_1, P_1, M_2, S_2, K_2$	Semi/quarter-diurnal comp.
Tidal velocity	-	Semi/quarter-diurnal comp.
River discharge	Field survey (14 rivers)	10 m^3/s (Kino River)
Density at open boundary	Field survey	Interpolation
Density at initial conditions	Field survey	Interpolation

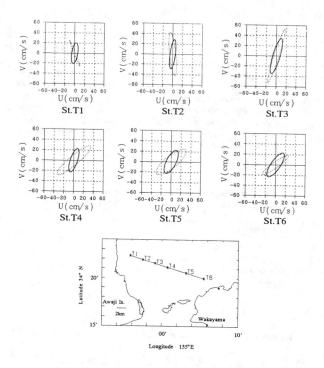

Fig. 5 : Comparison of tidal ellipse in surface layer and observation points:
Solid line represents observation results, while dotted lines are simulation results.

Verification of Simulation Results

For verifying the present simulation, we used harmonic analysis of computed velocity fluctuations in small-scale domain.   Figure 5 shows the comparison of tidal ellipse of the semi-diurnal component (M_2) in a surface layer obtained by small-scale domain simulation and observation.   The observation was conducted by The 3rd Construction Bureaus, Ministry of Transportation from 23 April to 8 May 1978.   The computed and observed tidal ellips are almost same except St. 1 and St. 4.   The difference at St. 1 and St. 4 is supposed to be directly affected by flows due to complicated topography in the Kitan Strait and large-scale vortex generated in the north area of Kitan Strait as well.   One of other reason is that only a semi-diurnal and a quarter component were yield at the boundary condition.   In addition, the observed tidal ellipse calculated

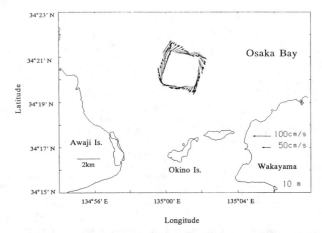

Fig. 6 : Clockwise circulation observed by ADCP in the northern area of the
Kitan Strait during slack time from northward to southward
tidal current at a depth of 10 m on 18 December 1998.

using 15 day's data, while the computed ellipse is based on hindcasting computation of 12 hours.
Synthetically judging, this computation is possible to simulate the observation results.

<u>Results of Simulation</u>

Figure 6 shows the clockwise circulation observed by ADCP in the northern water area of the
Kitan Strait during slack time from northward to southward tidal current.   Since the observation
was performed in a spring tide, the maximum northward current approaches 1.2 m/s.   The
observed circulation may be one of tidal vortex pairs in the vicinity of strait due to the strong tide.
The measured velocities were more than 50 cm/s.

Figure 7 shows the simulated distributions of currents and salinity around the Kitan Strait at
the same time of Fig. 6 at a depth of 1 m.   Here, the tidal current and circulation in the northern
area of the Kitan Strait are mainly discussed.   Therefore, the southern area of computational
domain is not displayed.   Figure 7 shows that vortex pairs appear in the northern area of Kitan
Strait at slack time from northward to southward tidal current, and that the vortex on the right is
larger than the vortex on the left.   The right side circulation is the same as the observed vortex in
Fig. 6.   When the tidal current is slack from northward to southward, the tidal currents start to
flow southward through all cross-sections from the side of Awaji Island to the side of Osaka and
Wakayama.   The sea water around the strait is divided into two water masses by the value of

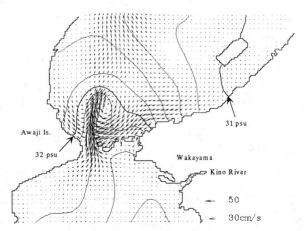

Fig. 7 : Simulated clockwise circulation and salinity distributions around Kitan Strait
        at slack time from northward to southward tidal current at a depth of 1 m on 23
        August 1997.   The contour lines show salinity distributions: the thick lines are
        31 and 32 psu and the contours are drawn every 0.2 psu.

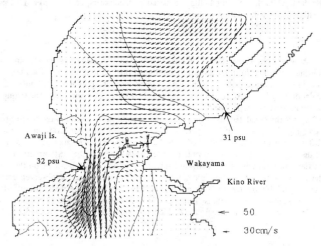

Fig. 8 : Simulated currents and salinity distributions around Kitan Strait
        at slack time from southward to northward tidal current
        at a depth of 1 m on 23 August 1997.

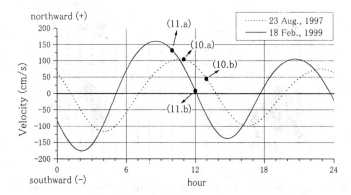

Fig. 9 : Predicted tidal current in Kitan Strait by Maritime Safety Agency, Japan.
: Dotted line is in 23 Aug. 1997; Solid line is in 18 Feb. 1999.

salinity, namely, whether it is less than or more than 32 psu.   The former is water mass remained in Osaka Bay, while the latter is in Kii Channel.   The simulated results show the salinity distributions at sea surface more than 32 psu are widely spread in the inner part of Osaka Bay due to advection of the strong tidal stirring at the end of the northward tidal current.

Tidal current at slack time from southward to northward is shown in Fig. 8.   The salinity contour of 32 psu is distributed in Kii Channel due to the southward tidal current.   It is worth noticing that lower salinity exists only in the southern area of Yura Seto, while the 32 psu line spreads widely in the northern area of the strait.   It is conjectured from the contour line of 32 psu salinity distributions that the northward tidal current flows though the three openings and transports high salinity into Osaka Bay, while the southward tidal current mainly flows though the Yura Seto into the Kii Channel.   These asymmetrical features of the salinity distribution are affected by the difference in the tidal current pattern.

Figure 9 and 10 shows that tidal curve in Kitan Strait and the simulated vorticity distributions at the depth of 1 m on 23 August 1997, when it is (a) at 0.5 hours after maximum northward tidal current and (b) at 0.5 hours before slack time from northward to southward tidal current, respectively.   It is obviously found that three tidal vortex pairs appear in the back of islands. The vorticity becomes higher closer to the extremity of the islands at 0.5 hours after maximum northward tidal current.   The absolute values of the vorticity in the center of circulation are calculated to be 1.0×10^{-3} s^{-1}.   As the northward tidal currents are getting stronger, the center of vorticity moves to the inner area of Osaka Bay and the scale of the vortex is becoming larger. Then, the absolute values of maximum vorticity have been decreased from 1.0×10^{-3} s^{-1} to 5.0×10^{-4} s^{-1}.

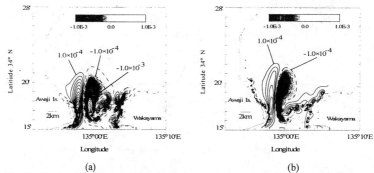

(a)   (b)

Fig. 10 : Simulated vorticity distributions of vortex pairs at the depth of 1m on 23 August 1997;
(a) The left is at 0.5 hours after maximum northward tidal current, (b) the right is at 0.5
hours before slack time from northward to southward tidal current.   The solid lines
represent a counter clockwise (positive) vorticity and the dotted lines do a clockwise
(negative) vorticity, and the contours are drawn every 1.0×10^{-4} s^{-1}.   The light and shaded
colors mean the contours of vorticity and then the darker colors indicate higher vorticity.

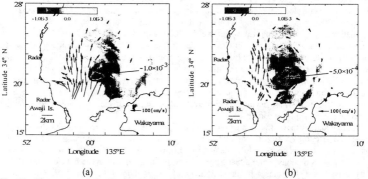

(a)   (b)

Fig. 11 : Observed vorticity distributions and the Tomogashima Circulation using VHF oceanic
radar on 18 February 1999; (a) at 1.0 hours after maximum northward tidal current (left),
(b) at 0.5 hours before slack from northward to southward tidal current (right).
The light and shaded colors mean the contours of vorticity and then the darker colors
indicate higher vorticity.

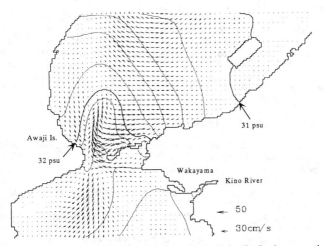

Fig. 12 : Simulated residual circulation and residual salinity distributions around
the Kitan Strait at 1 m depth on 23 August 1997.   The thick solid line
indicates 32 psu of salinity and the contour lines are drawn every 0.2 psu.

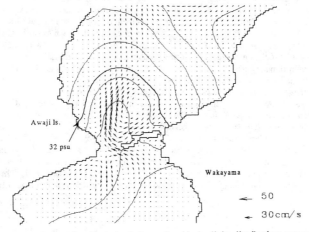

Fig. 13 : Simulated residual circulation and residual salinity distributions around the
Kitan Strait at 25 m depth on 23 August 1997. The thick solid line indicates
32 psu of salinity and the contour lines are drawn every 0.2 psu.

Figure 11 shows the observed vorticity distributions and the Tomogashima Circulation using VHF oceanic radar on 18 February 1999 (spring tide). Tidal curve on 18 February 1999 is shown in Fig. 9. The tidal time is almost the same as simulation results of Fig. 10, although the observation season and tide are different. Since strong stratification does not develop even in the summer, the difference in season may be out of question. The maximum northward tidal current was about 1.6 m/s. When the tidal current in Kitan Strait is maximum northward, the tidal-jet flows into Osaka Bay. At this time, large clockwise and small counterclockwise currents appear in both sides of the northward fast-flowing current. After a quarter tidal period, the tidal current in the strait becomes slack and the clockwise current almost rolled up. From the results of VHF oceanic radar, the headland eddy became largest at 10 km diameter.

Residual Current System and Density Field

Figures 12 and 13 show simulated residual currents and the tide-integrated salinity distribution near the Kitan Strait at a depth of 1 m and 25 m on 23 August 1997, respectively. The residual current is defined as the velocity time average over one tidal cycle; namely from 6:00 to 18:00.

As shown in these figures, the residual current pattern at the upper layer is much different from that at the lower layer. There is no tidal-jet and vortex pair in the northern area of the Kitan Strait in the upper water column. In addition, a distinct vortex pair is observed in the residual current in the southern area of Kitan Strait independent of depth. One of the reasons is the difference in the topography between the upper layer and lower layer. As shown in Fig. 2, the topography at the depth larger than 25 m are simple and symmetric with respect to the cross-section of the Yura Seto and also to the centerline across the Yura Seto. Therefore, four circulations are well developed. Another reason is the effect of southwestward residual current along the east-coast in the Osaka Bay, which is induced by the strong southwest flow on ebb tide. It becomes accelerated when approaching the Kitan Strait. Two small islands located in the east side of the Kitan Strait block the outflow to Kii Channel and makes the direction of the residual current change westward. The velocity reaches to 0.3 m/s. This current changes direction to northward near the centerline of the Yura Seto, and deforms the clockwise circulation. As a result, only one counterclockwise circulation occurs in the northern water area of the Kitan Strait.

It is of worth noting that the westward residual current occurs in both northern and southern water areas in the vicinity of three islands in eastside of the Kitan Strait. It means that most of the water mass transported by southwestward flow along the coastline in the Osaka side moves into Osaka Bay again. According to fishermen, the waste material discharged from the Kino River is sometimes found in coastal waters in Tokushima, which is the opposite shore of Wakayama. This comment is in a good agreement with the residual current pattern near the Kitan Strait. Generally speaking, residual circulation seems to go out of sight in transient flow patterns as in the case of the Okinose Circulation observed in Akashi Strait. In the case of the Kitan Strait, the transient flow pattern forms a circulation by itself. High salinity water flows into Osaka Bay. It is clear that the contour line of 32 psu flows into Osaka Bay, which corresponds to the north end of the residual circulation. The contour line of 32 psu in the lower layers extend larger than that in the upper layer.

Conclusion

The behavior of a headland eddy generated from the Kitan Strait was examined by field surveys and 3-D baroclinic simulation.   The tidal jet flowing into Osaka Bay through Kitan Strait encounters a complex topography and generates large-scale vortices due to shear flow and non-linearity of tidal current.   Such physical phenomena are not unique but rather common in tidal estuaries of a strait-basin system.

After the slack time of the tidal current, the local circulation is originally generated in the extremity of island.   As the tidal currents get stronger, the tidal-jet from the strait generates a moving vortex pair.   In Osaka Bay, the clockwise circulation is larger than the counter-clockwise circulation.   The observed and simulated clockwise circulation is one of the tidal vortex pair and we refer to this circulation as Tomogashima Circulation.   The absolute values of vorticity in the center of Tomogashima Circulation are calculated to be more than $5.0 \times 10^{-4}\,s^{-1}$.   The peak of the vorticity moves to the inner region of the bay and the scale of the vortex becomes larger with divergence.   As a result, the clockwise moving-vortex, Tomogashima Circulation, is generated from Okino Island in the Kitan Strait and can be defined as a headland eddy.

The residual current generate from Kitan Strait into Osaka Bay is vertically different due to the southward tidal current along the east-coast in Osaka Bay.   The residual current appears in both upper and lower layers in the northern area of the Kitan Strait, but they are different. Residual circulation appears only in the lower layer.

The Tomogashima Circulation is a transient flow.   This tidal stirring near the strait plays an important role in water exchange (Imasato, 1983).   The salinity distributions around Kitan Strait are largely influenced by tidal stirring and these transient vortices.

Acknowledgements

This research is partly supported by Grants-Aid for Scientific Research of the Ministry of Education, Science, Culture and Sports (Grants No. A-11305036 and B-09555198) and Kansai International Airport Ltd.

References

[1]   Barrick, D. E., M. W. Evans and B. L. Weber: Ocean surface currents mapped by radar, Science, Vol. 198, No.4313, pp.138-144, 1977.
[2]   Choi, S. Y., K. Nakatsuji, T. Yuasa and J. S. Yoon : Three dimensional modeling of wind-driven residual circulation in Tokyo Bay, Japan, Proceedings of the 2nd International Conference on Hydrodynamics, Chwang, Lee and Leug eds., pp.691-696, 1996.
[3]   Fujiwara, T., H. Nakata and K. Nakatsuji : Tidal-jet and vortex-pair of the residual circulation in a tidal estuary, Continental Shelf Research, Vol.14, No.9, pp.1025-1038, 1994.
[4]   Geyer, W. R. and R. Signell : Measurement of tidal flow around a headland with a shipboard acoustic doppler current profiler, J. Geophysical Reserch, Vol. 95, No. C3, pp.3189-3197, 1990.
[5]   Imasato, N. : What is tide-induced residual current?, J. Physical Oceanography, Vol.13, pp.1307-1317, 1983.

[6]   Mamaev, O. I. : A simplified relationship between density, temperature and salinity of sea water, *Bull. of the Acad. of Sci. USSK Geophys. Ser.*, trans. by I. S. Sweet, No.2, pp.180-181, 1968.

[7]   Nakatsuji, K., and T. Fujiwara : Residual baroclinic circulation in semienclosed coastal seas, *J. Hydraulic Engineering*, ASCE, Vol.123, No.4, pp.362-377, April 1997.

[8]   Pingree, R. D., and L. Maddock : Tidal eddies and coastal discharge, *J. Marine Biological Association. U. K.*, Vol.57, pp.869-875, 1997.

[9]   Smagorinsky, J. : General circulation experiments with primitive equation I. The basic experiment, *Monthly Weather Review*, Vol.91, No.3, pp.99-164, 1963.

[10]   Stewart, R. E. and J. W. Joy : HF radio measurements of surface currents, *Deep-Sea Research*, Vol.21, pp.1039-1049, 1974.

[11]   Sugiyama, Y., F. Takagi, K. Nakatsuji and T. Fujiwara : Modeling the baroclinic Circulation in Coastal Seas under Freshwater Influence, *Proceedings of the 5th International Conference on Estuarine and Coastal Modeling*, ASCE, pp.90-102, 1997.

[12]   Tee, K. T. : Tide-induced residual current, a 2-D nonlinear numerical tidal model, *J. Marine Research*, Vol. 34, pp.603-628, 1976.

[13]   Thatcher, M. L. and D. R. F. Harleman : A mathematical model for the prediction of unsteady salinity intrusion in estuaries, *Massachusetts Institute of Technology*, Report no 144, February 1972.

Application of Quantitative Model - Data Calibration Measures to Assess Model Performance

Malcolm Spaulding[1]
Craig Swanson[2] and Daniel Mendelsohn[2]

Abstract

The success of any model application depends on the credibility of the model predictions. This credibility is gained by careful comparison of model predictions with observations. The model calibration process requires that model parameters be adjusted to best match the available data. The success of the calibration and subsequent validation or confirmation is measured by both qualitative and quantitative means. Qualitative comparisons are usually based on visual comparisons to determine whether the model reproduces observed patterns, in time and in space.

Quantitative calibration measures, on the other hand, are specific statistical analyses that define a measure of goodness of fit between model and data. Such measures include relative error, root mean square error, linear regression analysis, and error coefficient of variation. These measures are usually applied to a series of individual observations or time series of model predicted quantities: tide range, maximum flood and ebb speeds, tidal constituent amplitude and phase, salinity, temperature and any other constituent

Results of the application and verification of ASA's WQMAP three dimensional boundary fitted hydrodynamic model to estimate the circulation and thermal dynamics of Mt.. Hope Bay, RI/MA and salinity intrusion in Savannah River, GA are presented. Observations and discussion of the strengths and weaknesses of using such calibration measures, including suggested guidelines, are presented.

[1] Department of Ocean Engineering, University of Rhode Island, Narragansett Bay Campus, Narragansett, RI 02882 USA, Tel: (401) 874-6666
[2] Applied Science Associates, Inc., 70 Dean Knauss Drive, Narragansett, RI 02882 USA, Tel: (401) 789-6224

843

Introduction

In practical applications of hydrodynamic and water quality models to assist in the design, siting, and operation of engineered systems (e.g. waster water treatment plants, power plants, chemical processing facilities) it is critical to establish the model's ability to accurately predict the system dynamics (Thomann and Mueller, 1987). Once validated the models can be then be used to evaluate alternative developments and to assess their impact on the environment. Through a careful program of model calibration and validation it is possible to develop a sense of the credibility of the model and its ability to make accurate estimates of the impact of the various alternatives under consideration (Dee, 1995).

In assessing model performance or skill a variety of evaluation methods have evolved (Thomann and Mueller, 1987; Lynch and Davis, 1995) to compare model predictions to observations. The evaluation methods used depend on the data collected and its spatial and temporal distribution. The evaluation techniques are normally grouped into qualitative and quantitative methods and can be applied to any one or a number of the observed/model predicted variables.

In the qualitative approaches, model predictions are visually compared to observations. This may take the form of comparing model predicted and observed time series at selected locations. These types of comparisons are typically made for observations from moored instruments (e.g. surface elevation, currents, temperature, and salinity), usually at a very limited number of locations in the study area. These comparisons may be performed with temporally filtered or unfiltered data. Model predicted spatial patterns (2-D plan or section views) can also be compared to observed patterns at selected times. This method is typically used when the data collection method allows large spatial coverage in a short time (e.g. satellite observations or measurements from a rapidly moving vessel). Comparative three dimensional, pattern analyses are rare and rarer still are time dependent, three dimensional spatial comparisons. These types of comparisons are typically limited by the availability of data and effective methods to visualize and compare the results. The qualitative comparative approaches are often quite subjective and rely to a considerable extent on the expertise of the assessor.

Quantitative evaluation methods normally take two forms: time series data analyses and statistical analyses. In the time series analysis approach model predicted time series are compared to observations at selected locations. Comparisons are typically made using power spectra and harmonic analysis methods. Comparison of power spectra allow one to assess the model's ability to correctly represent the distribution of energy in the system, but provide no insight into the phasing relationships. Comparison of tidal harmonic constants derived from observations and predictions provides insight into the model's ability to accurately represent amplitude and phasing of the tides and hence is ideally suited to locations were tidal processes dominate. Temporal filtering methods (e.g. high and low pass) are often used in pre-processing data before employing either qualitative or quantitative methods to focus on discrete frequencies.

Statistical analyses methods are also often used in evaluating model performance given the fact that they are simple to apply and yield well defined quantitative measures of skill. The standard techniques (McCutcheon et al., 1990) applied in coastal modeling include root mean square error(rmse), coefficient of variation (coe)(rmse/mean), relative mean error (rme), and regression analysis (r). These metrics can be evaluated at each observation location. They can also be summed over all observation locations to provide an ensemble or global error estimate for the study area. These evaluation metrics are well documented but care must be taken in their application since their results can be strongly dependent on the number, location, duration, and processing of the observations. McCutcheon et al. (1990) provide guidance on acceptable values of each measure for calibrating and verifying estuarine circulation and water quality models.

In some modeling studies, particularly those that are process focussed, the approach is to study the local momentum, energy, and transport balances on which the governing equations are founded. In these cases the balances (and associated assumptions) inherent in the model are compared to similar balances made using field data. These types of comparison are data intensive and typically limited to one or several critical areas in the study domain.

Model performance or skill assessment is often guided by the proposed application of the model. As an example models that are used in support of navigation in estuaries typically focus on accurate estimation of the sea surface elevations. While the currents are of interest in such applications they are of secondary concern compared to surface elevation prediction. Similarly models used to assess the impact of dredging on circulation often are concerned with the impacts on salinity intrusion. Accurate prediction of the salinity hence is critical.

In the presentation to follow WQMAP, an integrated modeling system for predicting three dimensional circulation and water quality, is first summarized and then applied to predict the salinity intrusion in the Savannah River estuary and the impact of a thermal discharge on temperature structure of Mt. Hope Bay. A variety of qualitative and quantitative evaluation methods, as described above, are applied to assess the model's performance. In the Savannah River case the model is validated using an extensive set of time series observations of sea surface elevation, currents, temperature and salinity collected over the river system. In Mt. Hope Bay the model is compared to a very unique, three dimensional set of temperature time series data. The data set provides a time dependent, three dimensional representation of the thermal field over the simulation period.

WQMAP Model System

WQMAP (Water Quality Mapping and Analysis Program) is an integrated system for modeling the circulation and water quality of estuarine and coastal waters (Mendelsohn et al., 1995; Spaulding et al., 1999). The system allows the user to generate boundary conforming grid systems to represent the study area, to perform simulations with a suite of circulation and water quality models, and to display the model predictions in the form of time series plots, vector and contour plots, and color animations. WQMAP operates on a Pentium personal computer, features a Window's based user interface, and is controlled by pull down menus and point and click operations. The software is designed using a shell-based architecture (Spaulding and Howlett, 1995) making application to any geographic area simple and fast. Selected key features of the software are summarized below.

WQMAP may be set up for any region in the world. At any time, WQMAP operates within a specified geographic location. The user can rapidly switch from one location to another by simply "pointing" the system at the appropriate data set. The number of locations and associated supporting data is limited only by the amount of available disk storage space. This location functionality allows WQMAP to be completely re-locatable and is the core of the shell-based architecture. The geographic location, defined by a base map and a series of GIS layers of vector or raster form, can be at any resolution and at any location in the world and provides the reference for all model applications. GIS maps and layers may be prepared by digitizing from paper charts, by importing from existing GIS databases (e.g. ESRI ArcView, MapInfo, AutoCad) or the user may simply make a map from WQMAP's global database for any region in the world. The simplest implementation of a location is a map that represents land and water features (polygons) that define the water-land interface for the grid generator. More complex coastal features such as marshes and inter-tidal zones may also be introduced.

The embedded GIS allows the user to input, store, manipulate, analyze, and display geographically referenced information. The GIS is fast, user-friendly, interactive, and simple to operate. The GIS is often used to provide input data to the models. It is also helpful in the presentation and interpretation of model predictions. Additional information can be linked to the GIS including charts, graphs, tables, tutorials, bibliographies, photographs, animations, text, or other software (e.g. spreadsheets, word processors).

WQMAP includes three basic components: a boundary fitted coordinate grid generation model, a two/three dimensional hydrodynamics model, and three separate water quality or pollutant transport and fate models. These models are accessed through the system directory. The models are setup through the user interface and supporting data is accessed through the location specific data sets, the GIS, or input through the menu windows. All models are configured for operation on a boundary conforming grid system. They can also be operated on a rectangular grid system, as this is a special case of the boundary conforming grid.

Given the present focus of the paper on skill assessment of the model predicted circulation and transport of salt and heat, an overview of the hydrodynamic model will be presented. Summaries of the boundary fitted coordinate grid generation model and the suite of pollutant transport models are given in Mendelsohn et al. (1995) and Spaulding et al. (1999).

The hydrodynamic model included in the system solves the three dimensional, conservation of water mass, momentum, salt and energy equations on a spherical, non-orthogonal, boundary conforming grid system and is applicable for estuarine and coastal areas (Muin, 1993; Muin and Spaulding, 1996, 1997a,b). The eddy viscosities can be specified by the user or obtained from a one equation turbulent kinetic energy model. The velocities arc represented in their contra-variant form. A sigma stretching system is used to map the free surface and bottom to resolve bathymetric variations. The model employs a split mode solution methodology (Madala and Piaseck, 1977). In the exterior (vertically averaged) mode, the Helmholtz equation, given in terms of the sea surface elevation, is solved by a semi-implicit algorithm to ease the time step restrictions normally imposed by gravity wave propagation. In the interior (vertical structure) mode the flow is predicted by an explicit finite difference method, except that the vertical diffusion term is treated implicitly. The time step generally remains the same for both exterior and interior modes. Computations are performed on a space staggered grid system in the horizontal and a non-staggered system in the vertical. Time is discretized using a three level scheme. Muin and Spaulding (1996, 1997a) provide a detailed description of the governing equations, numerical solution methodology, and in-depth testing against analytic solutions for two and three dimensional flow problems. Additional applications are given in Swanson and Mendelsohn (1993, 1996) and Mendelsohn et al. (1995).

The model output module allows the user to present the results of any simulations performed. The user opens the scenario file of interest and then selects the output desired. As an example for the hydrodynamics model, the user selects the variable of interest (currents, temperature, surface elevation, salinity) and the vertical level (sigma layer) and then animates the results (color contours or vectors) over the model simulation period. The animation control allows forward, step forward, reverse, rewind, and stop options. The animations can also be overlayed on the bathymetric data to allow the user to visualize the relationship between the predicted values (currents, salinity, and temperature) and the depth. The user can also select a section view window (vertical transect along user selected line) that allows the variable of interest to be visualized along the section line at the same time as the plan view is animated. An environmental data window is also available to display time series data for selected model forcing functions with a time marker.

The salt and heat transport are solved using a simple explicit finite difference technique on the boundary conforming grid (ASA, 1997). The vertical diffusion however is represented implicitly to ease the time step restriction caused by the normally small vertical length scale that characterizes many coastal applications. The model includes five options to solve the advective transport: 2 nd order Lax-Wendroff, 1st and 2 nd order upstream differencing, MIE 2 nd order upstream and 2 nd order QUICKEST (Thompson, 1984). These can be selected through the user interface to address the particular problem of interest. The horizontal diffusion term is solved by a centered in space, explicit technique. The solution to the advective diffusion equation has been validated by comparison to one and two dimensional analytic solutions for a constant plane and line source loads in a uniform flow field and for a constant step function at the upstream boundary. The model has also been tested for salinity intrusion in a channel (Muin, 1993).

All model simulations within WQMAP are organized on a scenario basis. The scenario is specified through the user interface (model parameters, terms in governing equations, bathymetric data set, grid system, output file specifications, hydrodynamic data set, contaminant loading, river input, boundary conditions). Once the simulation is complete the results are stored in a scenario output file. Both the input and output files are stored and can be visualized. This approach creates a record of each simulation including both the input parameters and data files employed in the simulation and the associated output. Scenario files can be archived or discarded depending on the need.

Selected applications of WQMAP

To illustrate the use of various model performance or skill assessment methods application of WQMAP to predict the circulation and salinity intrusion in the Savannah River estuary and circulation and thermal dynamics, as impacted by a power plant discharge, during the winter in Mt. Hope Bay are presented below. The Savannah River study was performed as a portion of the Teir I Environmental Impact Statement to determine the potential impacts of a proposed navigation channel deepening project. The two primary objectives were to assess the possible increase in salinity in the upstream, brackish marsh system and the potential for a decrease in bottom water dissolved oxygen concentrations as a result of increased stratification. More detailed information on the Savannah River application can be obtained in ATM (1998). The ultimate goal of the Mt. Hope Bay study was to ascertain whether the power plant effluent is responsible for the observed decline in fisheries resources. The modeling study focused on different times in the year that are critical for various life stages of a variety of fish species. Both maximum temperatures and incremental temperature rises were investigated. The application to winter temperature prediction for Mt. Hope Bay is presented in Swanson et al. (1999).

Savannah River Estuary- Salinity Intrusion

WQMAP's three dimensional hydrodynamic model was applied to predict the circulation and salinity transport in the Savannah River estuary extending from offshore of Tybee Island to the US Geological Survey (USGS) gauging station at Clyo. The study area, including the Front River, South Channel, Back River, Middle River and Little Back River, was represented by the boundary conforming grid system shown in Figure 1. The non-orthogonal, boundary conforming grid allowed concentration of the grids in the central channel of the river and an along to cross channel grid aspect ratio of approximately 8 to 1. The impact of the marsh area on the flows in the upper portion of the river was approximated by the use of water storage reservoirs for selected grids. The reservoir capacity versus sea surface elevation was based on a review of the topographic data for the marsh areas. The bathymetric data for the river and estuarine areas were based on an intensive survey performed by the US Army Corp of Engineers in 1997. The water depth in the central channel is approximately 16 m and extends from the Atlantic Ocean (note closely laterally spaced grid lines in Figure 1) to Onslow Island. Water depths in the coastal ocean offshore of Tybee Island are typically 2 to 4 m and incised by the dredged channel extending into deeper water. The water depths in the south channel and the Back River are typically 3 to 4 m. Depths in the upper portion of the main channel progressively decrease from

8 to 6 m and are below 4 m in Little Back and Middle Rivers and associated inter-connecting channels.

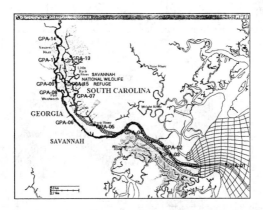

Figure 1.   Savannah River study area showing the location of the
               Georgia Port Authority (GPA) monitoring stations.

Model simulations were performed from July through September 1997 corresponding to the time period of an intensive measurement program. The program included water surface elevation observations collected at 14 stations (GPA-01 to 14), continuous vertical measurements of currents at GPA-04 and 08 using a bottom mounted acoustic doppler current profiler; discrete current velocities measured using an InterOcean S-4 at GPA-10 and 15; surface and bottom salinity time series at stations GPA-04, 06, and 08; and bottom salinity measurements at GPA-03, 05, 07, 09, 10, 11, 12, and 13. The station locations are shown in Figure 1. The model was forced by observed sea surface elevation at the mouth, using phase lagged data from the NOAA/NOS station at Fort Pulaski and USGS observed river discharge (mean = 225 m³/sec over period) for the time period at Clyo. The sea surface elevation was assumed to be uniform across the outer model boundary. The quadratic bottom friction coefficient was set at 0.0015. The model employed 11 layers to represent the vertical structure in the water column. The vertical eddy diffusivities for salt and heat were based on a Richardson number based formulation and calibrated to provide observed stratification and de-stratification of the water column as a function of tidal range. The details are provided in ATM (1998).

Model predictions are compared to observations below. Qualitative comparisons are given first followed by quantitative measures. Each variable is discussed separately. In the interest of space only data from station GPA- 04 is presented since it is in the center of the study area and along the main stem of the river. Comparisons for other locations are summarized.

Figure 2 shows the model predicted and observed sea surface elevation versus time for Julian days 190 to 272. Model predictions are in good agreement with observations at this station correctly representing the amplitude, phasing, diurnal inequality and neap-spring cycle in tidal amplitude. Model performance is equally good at stations in the lower estuary and the main channel but degrades slightly in the upper estuary marsh area due to super elevation of the mean water level and the effects of the marshes acting as storage reservoirs for tidal waters.

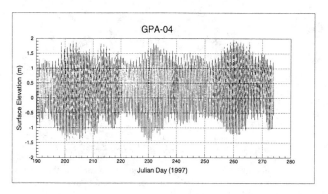

Figure 2.   Model predicted (solid line) and observed (dashed line) water surface
elevation time series at GPA-04.

The sea surface elevation power spectral density for observations and model predictions is shown in Figure 3. Both show that the semi-diurnal components (M_2, N_2, and S_2) dominate the tidal signal and account for over 85% of the total energy. The signature of the M_4 and M_6 and diurnal components are clearly evident in both predictions and observations. The model and data both show little low frequency energy during the application period. A review of the comparisons from other stations shows that the model predictions are as good as at GPA-04 and correctly estimate the amplification of the M_4 and M_6 as one moves from the ocean upstream.

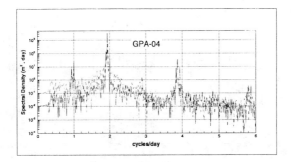

Figure 3. Model predicted (solid line) and observed (dashed line) power spectra of water surface elevation at Station GPA-04.

A harmonic analysis was performed on the model predicted and observed data for all 14 stations. The results for station GPA-0 are shown in Table 1 for the Z_0, O_1, K_1, N_2, M_2, S_2, M_4, and M_6 constituents. These constituents account for the majority of the tidal energy. The agreement between the two is excellent with typical differences in the major constituents of (M_2 and N_2) of 2% in amplitude and 0.48 to 0.74 hr. in phase. For other stations (not shown) the errors are typically the same. Maximum differences occur in the upper reaches of the system or are restricted to small amplitude (<0.1 m) components.

Table 1 Harmonic analysis on water surface elevation of observed and model predicted amplitudes and phases at Station GPA-04.

Tidal Constituents	Amplitude				Phase			
	Data (m)	Model (m)	Diff. (m)	Error (%)	Data (deg)	Model (deg)	Diff. (deg)	Diff. (hr)
Z_0	0.332	0.299	-0.033	10	-	-	-	-
O_1	0.090	0.078	-0.012	13	129	132	3	0.22
K_1	0.176	0.128	-0.048	27	152	139	-13	-0.86
N_2	0.262	0.266	0.004	2	243	222	-21	-0.74
M_2	1.108	1.125	0.017	2	248	234	-14	-0.48
S_2	0.084	0.197	0.113	135	271	261	-10	-0.33
M_4	0.092	0.119	0.027	29	338	326	-12	-0.21
M_6	0.022	0.012	-0.010	45	90	92	2	0.02

Model predicted and observed currents (in the along the channel axis direction) are shown in Figure 4, for the near surface and near bottom stations. The predictions are seen to be in very good agreement with the data. Both are dominated by the tide and show a modest reduction (15%) between the surface and bottom stations. Predictions are equally good at station GPA-08; the only other station where current measurements at the surface and bottom were made.

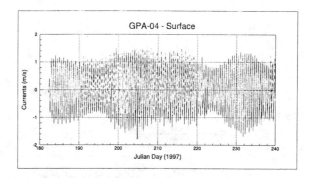

Figure 4. Model predicted (solid line) and observed (dashed line) major axis
current time series for the surface at GPA-04.

The power spectral density for the observed and model predicted values are provided in
Figure 5. The model shows a good ability to predict the distribution of tidal energy correctly
estimating the energy in the M_2, M_6, O_1, and K, constituents but slightly overestimating the

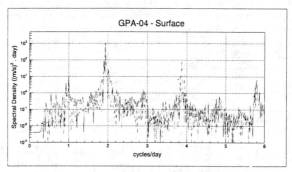

Figure 5. Model predicted (solid line) and observed (dashed line) major axis
current power spectra for the surface at Station GPA-04.

energy for the M_4. Model predicted and observed tidal constituent amplitudes and phases are
provided in Table 2. The differences are 7 to 24% for the major constituent amplitudes and 0.04
to 0.76 hr. for the phase for the surface and bottom stations. The differences increase as the
amplitude of the component decreases. They are also generally higher for the bottom station.
Similar model performance was observed at station GPA-08.

Table 2. Harmonic analysis on major axis currents of observed and model predicted amplitudes and phases for surface and bottom at Station GPA-04.

Tidal Constituent	Amplitude				Phase			
	Data (m/s)	Model (m/s)	Error (%)	Diff. (m/s)	Data (deg)	Model (deg)	Diff. (deg)	Diff. (hr)
(Surface)								
O_1	0.085	0.039	54	-0.046	229	234	5	0.36
K_1	0.107	0.062	42	-0.045	245	243	-2	-0.13
N_2	0.256	0.224	13	-0.032	336	335	-1	-0.04
M_2	0.964	0.895	7	-0.069	327	344	17	0.59
S_2	0.072	0.130	81	-0.058	38	16	-22	-0.73
M_4	0.040	0.189	373	-0.149	51	41	-10	-0.17
M_6	0.069	0.052	25	-0.017	159	211	52	0.60
(Bottom)								
O_1	0.043	0.030	30	-0.013	209	234	25	1.79
K_1	0.061	0.047	23	-0.014	238	242	4	0.27
N_2	0.198	0.169	15	-0.029	332	335	3	0.11
M_2	0.546	0.675	24	-0.129	321	343	22	0.76
S_2	0.067	0.098	46	-0.031	63	16	-47	-1.57
M_4	0.104	0.137	32	-0.033	349	41	52	0.90
M_6	0.035	0.031	11	-0.004	122	200	78	0.90

Model predictions of the discharge (volume flux) were compared to observations for short periods of time during September and October 1997. The observations were based on vessel mounted ADCP data collected during partial tidal cycle cruises at selected river transects. Comparisons of the predictions and observations are shown in Figure 6 for the Front, Back, and Fort Jackson Rivers at

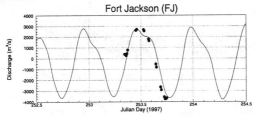

Figure 6. Model predicted (solid line) and observed (closed circles) flux at transects FR3 on the Front River, BR on the Back River, and FJ at Fort Jackson.

the location where Rte 17 crosses the upper estuary. The model is seen to correctly predict the magnitudes and the relative difference in flux for each of the three passages. The model predicts a double peaked flood and single peaked ebb for each of the three transects. It is most pronounced for the Fort Jackson section. This is the result of the interaction between the M_2 and M_4 tidal constituents. This effect is not clearly evidenced in the observations, potentially because of its small amplitude and associated sampling problems.

Model predictions and observations of the salinity at the surface and bottom at GPA-04 are shown in Figure 7 for the study period. The model shows a very good ability to

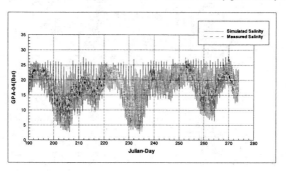

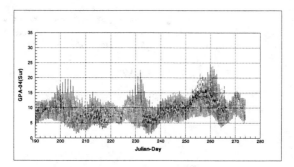

Figure 7.   Model predicted and observed salinity at the surface and
bottom at Station GPA-04.

simultaneously predict both the surface and bottom values. The model has also captured the stratification -destratification cycle over the neap-spring cycle. During neap tidal conditions the tidal currents and vertical mixing are at their minimum and the water column develops a strong (10 to 15 ppt) vertical salinity difference between the surface and bottom. As the currents increase toward their spring state the vertical mixing becomes sufficiently large to overcome the salinity stratification and the river becomes vertically well mixed. The cycle repeats itself as the tidal range decreases and vertical stratification is re-imposed. Observations of salinity data at other stations shows that when the water column becomes stratified penetration of salinity along the bottom of the deeper areas of the river increases substantially. When the system de-stratifies penetration of salinity up the Front River is decreased but increased along the shallower Middle and Back Rivers.

To provide additional insight into the model's ability to predict the salinity structure in the system, detailed statistical comparisons were made (Table 3) for each salinity station. The table presents the minimum, maximum, mean and standard deviation for observations and model predictions. Also shown are the relative mean error, standard deviation, and root mean square error based on comparing model predictions and observations. The relative mean error and coefficient of variation are finally shown at the bottom of the table. A review of the means show that the model does reasonably well in capturing the salinity intrusion (decreasing salinity with distance upstream) at surface and bottom stations. The model also correctly approximates the range of variations in the salinity. The statistical comparisons of the differences also show good model performance as well. For comparison McCutcheon et al. (1990) recommend a relative mean error of 25% and error coefficient of variation of 45% as acceptable levels for model salinity calibration. The mean relative errors are generally less than 1.0 ppt while the standard deviations and root mean square errors are typically less than 2.5 ppt. Both measures decrease as the mean salinity decreases. For stations along the main stem of the river the percent relative mean error and coefficient of variation are 30% or less. These values both increase substantially in other branches of the river and in the upper reaches. Decreases in performance are generally associated with areas where the salinity levels are low (mean less than 2 ppt). As an example the relative mean error for station GPA-12B is 78%. The model predicted (0.32 ppt) and observed (0.58 ppt) mean salinity vary only by 0.26 ppt. In general model performance is best in the main stem of the river and poorer in the smaller less critical Little Back and Middle Back Rivers and the upper reaches of all rivers.

Table 3: Statistical comparisons of model predicted and observed salinity at selected GPA stations along the Savannah River							
	GPA-04B	GPA-04S	GPA-05B	GPA-06B	GPA-06S	GPA-07B	GPA-08B
Observations							
Min (ppt)	6.27	2.62	0.10	0.30	0.10	0.00	0.00
Max (ppt)	27.60	21.45	16.40	22.09	14.12	11.00	15.40
Mean (ppt)	18.33	9.64	5.54	10.31	4.10	1.10	3.25
Std (ppt)	3.95	3.78	3.46	5.15	2.20	1.76	3.71
Model							
Min (ppt)	2.63	1.02	0.21	0.20	0.40	0.09	0.01
Max (ppt)	28.03	24.57	10.70	19.12	9.33	3.99	12.86
Mean (ppt)	18.20	9.21	3.81	8.87	3.10	1.10	3.86
Std (ppt)	5.79	4.02	2.53	4.95	1.96	0.80	3.01
Differences							
MAE (ppt)	-0.13	-0.43	-1.73	-1.44	-1.01	-0.11	0.35
Std (ppt)	3.26	2.62	2.10	2.44	1.34	1.53	1.76
RMSE (ppt)	3.26	2.62	2.72	2.83	1.69	1.53	1.84
RME%	-0.71	-4.48	-31.23	-13.98	-24.52	-8.10	18.92
COV%	18.53	27.93	37.95	23.64	32.67	139.18	54.21

	GPA-08S	GPA-09B	GPA-10B	GPA-11B	GPA-12B	GPA-13B
Observations						
Min (ppt)	0.00	0.00	0.00	0.00	0.00	0.00
Max (ppt)	13.20	13.00	9.40	5.20	5.50	1.60
Mean (ppt)	1.67	1.63	1.08	0.13	0.32	0.12
Std (ppt)	2.37	2.61	1.46	0.44	0.56	0.22
Model						
Min (ppt)	0.01	0.00	0.03	0.00	0.00	0.00
Max (ppt)	5.67	9.10	5.57	3.25	3.16	1.59
Mean (ppt)	1.36	2.05	1.42	0.67	0.58	0.23
Std (ppt)	1.21	2.00	1.18	0.72	0.59	0.28
Differences						
MAE (ppt)	-0.46	0.17	0.27	0.34	0.16	0.03
Std (ppt)	1.40	1.32	0.82	0.56	0.36	0.21
RMSE (ppt)	1.57	1.33	0.86	0.65	0.39	0.21
RME%	-18.96	25.41	30.57	413.14	77.92	94.60
COV%	88.19	80.94	75.99	424.89	111.32	178.01

Mt. Hope Bay – Winter Thermal Dynamics

A summary of the application of WQMAP to predict the winter temperature conditions in Mt. Hope Bay (Swanson et al., 1999) is presented here. Calibration and verification of the model for spring and summer were presented in Spaulding et al. (1999). The present application hence represents an additional validation of the model. Additional information on the application of the model to Mt. Hope Bay can be obtained in Swanson et al. (1998) and supporting documents (Rines and Schuttenberg, 1998). The focus of the presentation below is on validation of the predicted temperature field with particular emphasis on the plume resulting from the power plant discharge.

Mt. Hope Bay is a shallow water (mean depth of 5 m) estuary located in the north-eastern corner of Narragansett Bay, and extends across the state boundary been Rhode Island and Massachusetts (Figure 8). The bathymetry of the area is characterized by a broad shallow area on the north-western side of the bay and a 120 m wide, 10.5 m deep dredged channel progressing from near the mouth of the bay (Mt. Hope Bridge) to the lower reaches of the Taunton River. The major sources of freshwater include the Taunton River (mean discharge -29.7 m3/sec m^3/sec) and the Cole River (0.81 m^3/sec). The Brayton Point power plant (1600 MW) is located on the northern shore of the bay with water intakes on the northeastern and northwestern side of

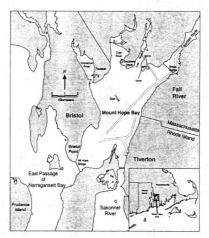

Figure 8.   Mount Hope Bay, showing connecting water bodies and
Brayton Point Station (BPS).

the facility and a discharge, through a venturi structure, on the southern tip of the site. The flow rate and temperature rise vary depending on the number of cooling units in operation but are typically 40 to 50 m^3/sec and 5 to 8° C, respectively. The Lee and Kickamuit Rivers contribute minimal flows to the bay. Circulation in the area is dominated by the semi-diurnal tide with a mean range of about 1.34 m and current speeds in the range of 10-25 cm/sec (Spaulding and White, 1990; Rines and Schuttenberg, 1998). Winds play an important role in mixing in the shallow area on the western side of the bay and density induced circulation is important in the lower Taunton River and the area immediately around its mouth during high flow periods. The water column is generally well mixed although salinity induced stratification is observed near the mouth and in the lower Taunton River when flow rates are large. Stratification in temperature and salinity, particularly for temperature in the northern reaches of the bay, is observed due to the effects of river and cooling water discharges.

WQMAP's three dimensional hydrodynamic model, and embedded thermal model, was applied to the Mt. Hope Bay study area using the boundary conforming grid shown in Figure 9. Environmental heat transfer at the surface included short and long wave solar radiation and

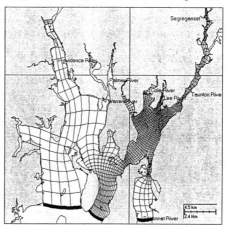

Figure 9.   Model grid used for simulations in Mt.. Hope Bay. The gridded area
extends into substantial portions of Narragansett Bay to more
accurately reflect conditions at the entrance to Mt.. Hope Bay.

sensible and latent heat transfer. The grid provides high resolution in Mt. Hope Bay (200 to 300 m for the bay in general and 50 to 100 m in the vicinity of the power plant intakes and discharge) (insert Figure 9) and coarser resolution for the upper portion of adjacent Narragansett Bay and the upper Sakonnet River. The extension of the grid to upper Narragansett Bay was performed to minimize the impact of the open boundary conditions on model predictions. The model grid also incorporates the Taunton River to the head of the tide. Eleven (11) sigma layers are used to represent the vertical structure. Bathymetric information was obtained from the NOAA/NOS hydrographic survey data for the bay. Editing of this data was done to better represent small-scale features in the vicinity of the power plant intake and discharge systems and the dredged channels in the area.

The bottom friction, eddy viscosities and diffusivities were the same as used in prior calibration and verification studies for the bay (Spaulding et al., 1999; Swanson et al., 1998). Swanson et al. (1998) presents the results of an extensive sensitivity study addressing impact of the specification of these model parameters on predictions.

The winter field measurement program consisted of the deployment of 31 strings (total of 173 thermistors) of Optic Stowaway (Onset Computer Corp., Pocasset MA), fully sealed, internally logging temperature recorders and two moorings where currents, temperature, and salinity were measured (Figure 10). The goal of the thermal measurement program was to accurately characterize the thermal field of the power plant discharge plume and background conditions in the bay. Each string consisted of five thermistors located at about ¼, ½, 1, 2 and 4 meters below the surface. At stations where the water depths were about 4 meters at extreme low tide, the bottom thermistor was placed at 3½ meters. At the deepest stations, in the dredged channel, and in the southern part of the bay, an additional thermistor was added at 6 meters depth. At the 15 stations closest to the BPS discharge, an additional bottom measurement (30 cm from the bottom) was made.

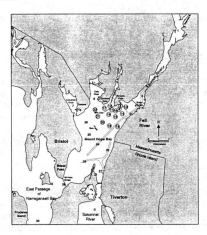

Figure 10. Initial distribution of thermistor strings on February 12 and 16 for the winter 1999 thermal mapping study in Mount Hope Bay. Circles indicate locations of bottom sensors. The Brayton Point current meter station was close to the location of thermistor string 9. The Borden Flats current meter station was on the other side of the channel just to the north of thermistor string 31.

For the winter 1999 deployments, the thermistors recorded temperatures every five minutes. The thermistors have a temperature accuracy and resolution of 0.2 °C and a two-minute response time. The instruments were deployed between 12 and 16 February 1999, and removed after 22 March. Data was first downloaded on 24 February for 26 of the strings at which time one string was missing and replaced. The second download, of 13 strings, took place on 10 March. Retrieval of the strings was delayed by bad weather, and during the period from the last download until retrieval there was a large loss of instruments due both to parting of the wires (suspected to be due to corrosion) and to completely missing buoy/strings. The total data return was 75% for the thermistors deployed.

Two current meter stations were established during this study, one at the Brayton Point station, close to thermistor station 9, and the other at the Borden Flats station at the edge of the channel off Borden Flats, close to thermistor station 31 (Figure 10). Both stations were the same as those used for prior monitoring efforts. Falmouth Scientific 3D-ACM current meters (sampling interval once per hour) with temperature/salinity/depth sensors attached, were used, one near the surface and one near the bottom, at each of the two stations. Water temperature and tidal height data were available from the NOAA/NOS station at Newport, meteorological data from the weather station at Green Airport, river flow rates for the Taunton and Three Mile River from USGS, and power plant data (cooling water flow rate and temperature) for PG&E Generating.

Model simulations were performed from January through March of 1999. This allowed a one month spin up time prior to comparisons with observations. The model open boundaries were forced by tidal elevations based on time lagged data from the Newport NOAA/NOS station and river forcing from USGS data on the Taunton River. Water temperature at the open boundary was based on data from the Newport NOS station and the salinity set constant based on prior observations in the lower Narragansett Bay. Atmospheric forcing was derived from meteorological data (air temperature, wind speed/direction and solar radiation data) from nearby observation stations. The thermal load from the power plant was specified from plant records.

Figure 11 shows a surface plan and vertical section (insert) view plot of model predictions of the temperature field at 1500 on March 18, 1999. The vertical section view is taken along the dashed line in the figure and follows the central axis of the dredged channel. An additional insert is also provided that shows the solar radiation and air temperature versus time. The vertical line on this insert denotes the time of the plan and section views. The plot shows conditions at ebb tide. The impact of the power plant on the temperature is clearly evident from the thermal plume extending from the plant discharge to the southwest. The highest temperatures at the discharge are 10.5 °C. Water temperatures to the east are almost constant (6.5 °C).

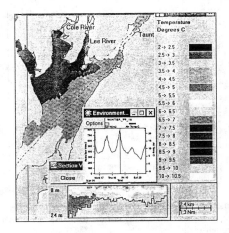

Figure 11. Plan view of model predicted surface temperatures
at 1500 on 18 March 1999. Insert shows vertical
section along dashed line shown in plan view.

The vertical section view shows the results of thermal warming during the day resulting in a
vertical temperature stratification of about 2 °C throughout the bay.

Figure 12 shows a time series comparison between the observed and model predicted
surface temperatures (February 13 to March 22, 1999) at stations 11 through 15. These stations
are located in the vicinity of the power plant discharge and the lower Taunton River (Figure 10).
Similar comparisons were made for all stations and at all depths but are not shown here. In
general the model correctly predicts the atmospherically forced heating and cooling trend over
the period, including the major cooling events in late February and mid March. The model
accurately estimates the mean and higher temperatures but consistently under-predicts
temperatures during the coolest periods. The model, consistent with the data, shows much higher
tidal variability in temperature at Station 11, compared to the other stations, reflecting the impact
of tidal forcing of the thermal plume near the discharge and the associated large thermal
gradients in the plume.

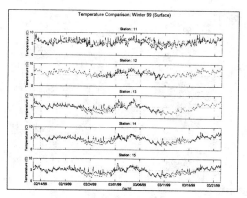

Figure 12. Time series of comparisons between predicted and observed
           temperatures at surface thermistors (0.25 m) for stations 11
           through 15.

A statistical comparison of thermistor observations and model predictions was performed
for all thermistors. The relative mean error (*rme*) varies from 0.00 to 0.22, averaging 0.07, well
below the guidance *rme* of 0.25. The root mean square error (*rmse*) varies from 0.40 to 1.75 °C,
averaging 0.87 °C. The correlation coefficient (*r*) ranges from 0.40 to 0.97, averaging 0.80,
which is close to the guidance level of 0.80. The *ecv* ranges from 0.09 to 0.39, averaging 0.17,
well below the guidance level of 0.45. The model performance for each depth is similar to that
for the means of all observations.

To help understand how these statistical measures vary by location, the relative mean
error and the root mean square error for the surface temperature were contoured at the surface.
The plan view *rme* contours for the surface temperature are shown in Figure 13. The highest
values (approximately 1 °C) are seen in the lower Taunton River, the Lee River, near the
discharge and just south of Mt. Hope. Figure 14 shows a plan view of the contoured *rmse* for the
surface thermistor data. The highest values occur in the Taunton River and near the BPS
discharge with lowest values in the western and southern portions of Mt. Hope Bay. In general
the *rmse* is highest near the discharge where the combination of the tidal currents and the strong
spatial gradient of the plume lead to larger differences between model predictions and
observations than in the remainder of the field.

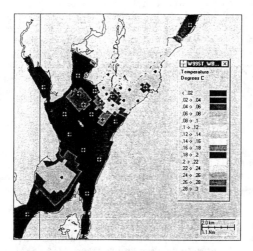

Figure 13. Plan view of the relative mean error (*rme*) between model predictions and observations for the surface (0.25 m) thermistor locations.

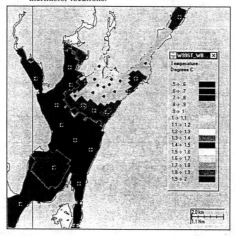

Figure 14. Plan view of root mean square error (*rmse*) between model predictions and observations for the surface (0.25 m) thermistor locations.

Model predictions and animations of the thermistor data both showed that the thermal plume from the power plant occasionally sank below the surface. During these periods the thermal plume, normally confined to the upper several meters of the water column, would sink to mid-depth and on rare occasions impact the bottom. This occurred primarily during times when freshwater flow was sufficient to create a vertical salinity gradient. Both model and data showed that the plume consistently sank in less saline surface waters when the salinity of the cooling water (intake located at the bottom on the eastern side of Brayton Point) increased sufficiently to overcome the buoyant effects of the heat added to the cooling water by the power plant heat exchangers.

Conclusions

WQMAP was applied to predict salinity intrusion in the Savannah River estuary and thermal plume dynamics in Mt. Hope Bay. Model performance in each case was validated by both qualitative and quantitative assessment methods. For the statistically based methods the results have been compared to recommended guidelines proposed by McCutcheon et al. (1990).

In the Savannah River case the hydrodynamic model provided good quality predictions over month long time scales for the sea surface elevations, currents, volume fluxes, and salinity. The predictive performance was best in the central channel of the river and poorest in the smaller branches of the river in the upper part of the estuarine system. Model performance was best for the dominant tidal constituents and poorer for the smaller amplitude constituents. The model accurately captured the stratification and de-stratification of the water column over the neap-spring tidal cycle and reasonably represented the phasing and degree of stratification. The model correctly predicted the mean salinity intrusion and did best in the main stem of the river where the salinity signal was the strongest. Model performance decreased in the upper reaches of the river where the salinity values were quite low and the variability high.

For the Mt. Hope Bay case the model correctly predicted the long term natural heating and cooling of the bay over the simulation period and the increased tidal scale variability in temperature in the vicinity of the plant discharge plume. The mean rmse errors were generally quite low at 0.87° C. An inspection of spatial pattern of the rme and rmse shows that they were largest in areas where the spatial gradients of the temperature were largest and smallest where the gradients were smallest. Model and data showed that the plume sank on occasion. This consistently occurred when the salinity of the cooling water (intake located at the bottom on the eastern and western sides of Brayton Point) increased sufficiently to overcome the buoyant effects of the heat added to the cooling water by the power plant heat exchangers.

In reviewing the two applications presented it is clear that no one skill assessment tool provides a clear, unambiguous evaluation of performance. The qualitative methods are very useful in providing a sense of the patterns of model predictions but are subjective and often open to question. The quantitative based measures give more rigorous and measurable objective criteria but can be significantly impacted by the number, location, duration, and processing of the data. As an example in an area with strong tidal currents and a modest sub-tidal current estimates of the root mean square error can be made using the data (15 minute time increment) and show that model performance is reasonable even though the model may in fact do a very poor job of estimating the sub-tidal currents. It is clear that model performance can vary significantly depending on the location where model and data are compared.   Statistical comparisons in areas of low temporal and spatial variability (open bay) could be excellent while those in areas of high variability (near the thermal discharge) were quite poor. The mix of stations selected for comparison could also significantly impact the global assessment.

It is also possible to have excellent model performance for one variable while the performance for another closely related variable is marginal or even quite poor. As an example model predictions of sea surface elevations can be quite good even when predicted currents vary substantially from the observations. This seeming inconsistency can be explained by the fact that the continuity equation (sea surface elevation) has no directionality while the momentum equations requires accurate prediction of both magnitude and direction.

Acknowledgements:

The application of WQMAP to Savannah River was performed as a team effort with Applied Technology Management and supported by the Georgia Port Authority. The modeling investigation for Mt. Hope Bay was supported by PG & E Generating, Somerset, Massachusetts.

References

Applied Technology Management, Inc. (ATM), 1998. Hydrodynamic and water quality modeling of the lower Savannah River estuary. Prepared by ATM, Gainesville, FL, prepared for the Georgia Ports Authority, GA.

Applied Science Associates, Inc (ASA), 1997. WQMAP user's manual, Applied Science Associates Inc, Narragansett, RI. March 1997, 92 p.

Dee, D., 1995. A pragmatic approach to model validation, in *Quantitative skill assessment for coastal ocean models*, Coastal and Estuarine Studies, American Geophysical Union, Washington, DC, Lynch, D. and A. Davies (Eds.).

Lynch, D. and A. Davies (Eds.), 1995. *Quantitative skill assessment for coastal ocean models*, Coastal and Estuarine Studies, American Geophysical Union, Washington, DC.

Madala, R. V. and S. Piaseck, 1977. A semi-implicit numerical model for the baroclinic ocean, Journal of Computational Physics, Vol. 23, p. 167-178.

McCutcheon, S. C., Z. Dongwei, and S. Bird, 1990. Model calibration, validation, and use, Chapter 5 in Technical Guidance Manual for Performing Waste Load Allocations, Book III:

Estuaries, Part 2: Application of estuarine waste load allocation models, Edited by J. J. Martin, R. B. Ambrose, and S. C. McCutcheon, US Environmental Protection Agency, Office of Water, March 1990.

Mendelsohn, D., E. Howlett, and J.C. Swanson, 1995. WQMAP in a Windows environment. In: Proceedings of the 4th International Conference on Estuarine and Coastal Modeling, October 26-28, 1995, SanDiego, CA., Ed: M.L.Spaulding. Pub. American Society of Civil Engineers. Muin, M., 1993. A three dimensional boundary fitted circulation model in spherical coordinates, Doctoral Dissertation, Ocean Engineering, University of Rhode Island, Narragansett, RI.

Muin, M. and M. L. Spaulding, 1996. Two dimensional boundary fitted circulation model in spherical coordinates, Journal of Hydraulic Engineering, Vol. 122, No. 9, September 1996, p. 512-521.

Muin, M., 1993. A three dimensional boundary fitted circulation model in spherical coordinates, Doctoral Dissertation, Ocean Engineering, University of Rhode Island, Narragansett, R.I.

Muin, M. and M. L. Spaulding, 1997a. Three dimensional boundary fitted circulation model, Journal of Hydraulic Engineering, Vol. 123, No. 1, January, 1997, p. 2-12.

Muin, M. and M. L. Spaulding, 1997b. Application of three dimensional boundary fitted circulation model to the Providence River, Journal of Hydraulic Engineering, Vol. 123, No. 1, January 1997, p. 13-20.

Rines, H. and H. Schuttenberg, 1998. Data report New England Power Company, Brayton Pt Station, Mt. Hope Bay field studies 1997, report to New England Power Company, Applied Science Associates Inc., Narragansett, RI.

Spaulding, M. L. and F. M. White, 1990. Circulation dynamics in Mt. Hope Bay and the lower Taunton River, Coastal and Estuarine Studies, Vol. 38, R. T. Cheng (Ed.), *Residual Currents and Long Term Transport*, Springer-Verlag, New York.

Spaulding, M. L. and E. Howlett, 1995. A shell based approach to marine environmental modeling, Journal of Marine Environmental Engineering, Vol. 1, p. 175-198.

Spaulding, M.L., D. Mendelsohn, and J.C. Swanson, 1999. WQMAP: An integrated, three dimensional hydrodynamic and water quality model system for estuarine and coastal applications, Journal of Marine Technology Society, Special Issue on Advances in Coastal and Ocean Modeling, Fall 1999.

Swanson, J. C. and D. Mendelsohn, 1993. Application of WQMAP to upper Narragansett Bay, RI, Proceedings of the 3rd International Conference on Estuarine and Coastal Modeling, Oak Brook, IL, September 8-10, 1993.

Swanson, J. C., and D. Mendelsohn, 1996. Water quality impacts of dredging and disposal operations in Boston Harbor, ASCE North American Water and Environmental Congress '96 (NAWEC '96), Anaheim, CA, June 22-28,1996.

Swanson, C. J., D. Mendelsohn, H. Rines, and H. Schuttenberg, 1998. Mt. Hope Bay hydrodynamic model calibration and confirmation, prepared for New England Power Company, West Borough, MA , prepared by Applied Science Associates, Inc., Narragansett, RI.

Swanson, J. C., H. Rines, D. Mendelsohn, T. Isaji, and M. Ward, 1999. Mt. Hope Bay winter 1999 field data and model confirmation, Report prepared for PG&E Generating, Brayton Pt, Somerset, MA, prepared by Applied Science Associates, Inc., Narragansett, RI 02882.

Thomann, R. V., and J. H. Mueller, 1987. *Principles of surface water quality modeling and control*, Harper Collins, New York, NY.

Thompson, J. F., 1984. Convection schemes for use with curvilinear coordinate systems- a survey, US Corp of Engineers, Misc. Paper: E-84-4, Vicksburg, MS.

Demonstration of a Nowcast/Forecast System
for Galveston Bay

Richard A. Schmalz, Jr.[1], M. ASCE

Abstract

The National Oceanic and Atmospheric Administration installed a Physical Oceanographic Real Time System (PORTS) in June 1996 in Galveston Bay. Water surface elevation, currents at prediction depth (4.6m below MLLW) as well as near-surface and near-bottom temperature and salinity, and meteorological information are available at six-minute intervals. To complement the PORTS a nowcast/forecast system was initially developed over the Bay using a modified version of the Blumberg-Mellor (1987) three-dimensional hydrodynamic model. The Bay model was used to provide boundary conditions for a one-way coupled, fine resolution Houston Ship Channel model (Schmalz, 1998a). The nowcast component has recently been revised to work directly from the PORTS universal flat file format (PUFFF) files rather than from the PORTS data screen. During the forecast, the National Weather Service's Aviation Atmospheric and Extratropical Storm Surge Forecast Models are used to provide the meteorological and Gulf of Mexico subtidal water level forcings, respectively. The results are presented for a one-month demonstration period, 9 April through 9 May 1999, during which daily 24-hour nowcasts and 36-hour forecasts were performed using both bay and channel models in a pseudo-operational setting. The nowcast and forecast are separately evaluated for water levels, currents, salinity, and temperature. Evaluations of forecast Gulf of Mexico subtidal water level, freshwater inflows, and wind and sea level atmospheric pressure are also considered. Results are summarized and contrasted with a previous one-month evaluation during September 1997 (Schmalz, 1998c). To seek improvement, a pseudo-operational test-bed was developed and several tests performed. An assessment of improvement is made over the first week of June 1999. Finally, plans for additional demonstration periods and system developments are discussed.

1   Oceanographer, National Oceanic and Atmospheric Administration, National Ocean Service, Coast Survey Development Laboratory, Marine Modeling and Analysis Program, 1315 East-West Highway, Rm 7824, Silver Spring, Maryland 20910. (301)-713-2809 x104; e-mail: Richard.Schmalz@noaa.gov.

Introduction

The National Oceanic and Atmospheric Administration installed a Physical Oceanographic Real Time System (PORTS), patterned after Bethem and Frey (1991), in June 1996 to monitor Galveston Bay. Water surface elevation, currents at prediction depth (4.6m below MLLW) as well as near-surface and near-bottom temperature and salinity, and meteorological information are available at six-minute intervals for five, three, and four stations, respectively, as shown in Figure 1 in the format illustrated in Figure 2. To complement the PORTS, a nowcast/forecast system has been designed based on the National Ocean Service (NOS) Galveston Bay three-dimensional hydrodynamic model and the National Weather Service (NWS) Aviation atmospheric model. A three-dimensional sigma coordinate Galveston Bay and near shelf (GBM) model has been developed (Schmalz, 1996) based on a version of the Blumberg and Mellor (1987) model extended to orthogonal curvilinear coordinates. The GBM computational grid in Figure 3 consists of 181x101 horizontal cells (dx = 254-2482m, dy= 580-3502m) with 5 levels in the vertical. The Galveston Bay model (GBM) was initially calibrated to May 1995 astronomical tide (Schmalz, 1998a). To further improve the Bay model, drying/wetting and flux-corrected salinity transport schemes were incorporated. The drying/wetting scheme patterned after Hess (1994) in Tampa Bay is modified for application in Galveston Bay to simulate winter time "Northers", during which northerly winds of up to 40 knots persist over the Bay, associated with cold front passages. A second order van Leer-type upstream-biased transport scheme (Lin et al., 1994) was modified for application to the shallow water to treat the very sharp horizontal salinity gradients in Galveston Bay. Hindcast results for the extended GBM for the October 1994 flood and the January 1995 "Northers" are discussed in Schmalz (1998a). The Galveston Bay model is used to provide Bay wide water level and near entrance current forecasts as well as to directly provide water levels, density, and turbulence quantities to a finer resolution one-way coupled, three-dimensional Houston Ship Channel model (HSCM) to simulate currents within the channel. The refined channel grid was developed in three sections based on the Wilken (1988) elliptic grid generation program patterned after Ives and Zacharais (1987). Each grid section was linked in order to develop the final composite channel grid (see inset Figure 3 and Figure 4) consisting of 71 x 211 horizontal cells (dx=63-1007m, dy=133-1268m) with the same 5 sigma levels as in the GBM. Hindcast results for the extended GBM and coupled HSCM for April 1996 are discussed in Schmalz (1998a).

Nowcasting/Forecasting System Description

The data delivery system consists of an Operational Data Acquisition and Archive System (ODAAS) maintained by the NOS Coast Survey Development Laboratory (CSDL) in which NWS Aviation Model wind/pressure fields are automatically downloaded to CSDL machines. Additional scripts decode NWS Techniques Development Laboratory (TDL) extratropical storm surge water levels at Galveston Pleasure Pier. The NWS Western Gulf River Forecast Center (WGRFC) uploads to CSDL anonymous ftp, three day 6-hour interval forecasted river flow and stage for the Trinity River at Liberty , Texas and Lake Houston Dam near Sheldon, Texas, respectively. In addition, the previous day's hourly discharges at Liberty, Texas on the Trinity River and at Piney Point, Texas on Buffalo Bayou and stage for Lake Houston Dam near Sheldon, Texas are uploaded. A decode script accesses and decodes either the Houston/Galveston PORTS screen (see Figure 2) or PUFFF files every 6 minutes and stores daily station files.

An initial design of a nowcasting/forecasting system compatible with the above ODDAS has been completed. The design concept is modular such that refined hydrodynamic models can be readily substituted for the initial models. To this end, a separate nowcast/forecast program has been developed to establish hydrodynamic model inputs. The program utilizes the following ten step procedure:

1) Setup 24 hour nowcast and 36 hour forecast time periods,
2) Predict astronomical tide,
3) Predict astronomical currents,
4) Read PORTS screen or PUFFF files and develop station time series,
5) Develop GBM subtidal water level signal,
6) Assimilate PORTS salinity and temperature data into GBM and HSCM initial conditions,
7) Establish GBM and HSCM salinity and temperature boundary conditions,
8) Establish GBM and HSCM SST forcings,
9) Establish USGS observed and NWS/WGRFC forecast freshwater inflows, and
10) Establish PORTS based and NWS Aviation Model (AVN) wind and pressure fields.

Details of the above steps may be found in Schmalz (1998b). A one month demonstration using the PORTS Screen as input was performed over September 1997 in hindcast mode (see Schmalz, 1998c). Since the forecasts were performed after the fact, forecast results could be directly assessed against the observations. Due to transmission problems prior to the installation of MCINET, the PORTS screens were not reliably obtained on a six-minute basis, and an hourly sampling interval was used for the nowcasts. Water level comparisons were made by demeaning both model and observations and were not placed on a MLLW datum. Rms water level errors were order 10 cm during the nowcast and order 15 cm over the forecast periods. This initial one month demonstration served as a proof of concept for the above ten step procedure.

In an effort to improve reliability, the PUFFF files were accessed directly along with the instrument control files. Using this approach, the system data interval was decreased from one hour to six minutes. In addition, all water levels are reported on station MLLW datum during the 9 April - 9 May 1999 second one-month demonstration, which was run in a pseudo-operational setting described below.

<u>One-Month Demonstration: 9 April - 9 May 1999</u>

The Galveston Bay Nowcast/Forecast system was established under operational accounts (hgops) and run under the UNIX crontab environment. Each cycle consists of a 24 hour nowcast and 36 hour forecast. Each morning, mail messages were sent to hgops to document the performance of the scripts run under the crontabs. NCAR graphics programs were used to display nowcast/forecast results for both the Bay and Channel hydrodynamic models. Nowcast results were compared with observations while forecast results were contrasted against predictions when available. In addition, separate NCAR graphics programs were run each day to evaluate the previous day's forecasts against the observations. On only one day was the hydrodynamic nowcast/forecast cycle delayed. NWS/AVN and NWS/TDL forecasts were available on 27 and 29 days, respectively, over the 31 day period. WGRFC flow forecasts were not available for only 3 days over the period. Overall system performance was robust and demonstrated that the above ten step procedure could be implemented in a pseudo-operational setting.

Nowcast/Forecast System Evaluation

An IDL graphics program is used to assess the overall system performance separately for the nowcast and forecast component as illustrated in the Figures 5-10. In these figures, the date corresponds to the finish of the nowcast period and the beginning of the 36 hour forecast. Only the first 24 hours are considered in the forecast evaluation. As noted in Figure 5, nowcast water level errors (order 15 cm) are larger than the forecast water level errors (order 10 cm) on the Gulf coast and in lower Galveston Bay. For the mid and upper Bay, forecast water level errors (order 20 cm) are larger than the nowcast water level errors (order 15 cm) as indicated in Figure 6. The magnitude of both the nowcast and forecast water level errors increases from the lower to upper Bay. In Figures 7 and 8 for prediction depth currents, one notes that nowcast and forecast errors are nearly the same in both speed (20 - 30 cm/s) and direction (50 - 70 deg T). Direction errors are large because comparisons are made regardless of current speed. In Figure 9 at Morgans Point, surface temperature and salinity are reasonably represented with errors under 1.5 deg C and 3.5 PSU, respectively. In Figure 10 for wind and sea level pressure, one notes that errors are considerably larger in the NWS/AVN forecasts. Nowcast errors in wind and sea level pressure represent Barnes (1973) interpolation error only. Water level residual forecast errors from the NWS/TDL Extraptropical Storm Surge Model are order 10 cm and range from 4 to 28 cm. NWS/WGRFC riverflow forecast errors are very small (under 1000 cfs) except during the first few days of the evaluation period (9 - 12 April 1999).

Water level rms errors were 5-10 cm larger than the 10 cm target rms error norm. Examination of time series of the observed subtidal water level signal at Galveston Pleasure Pier (used along the entire GBM open boundary) and the simulated total water level signal at this location indicated large oscillations. In addition, forecast time series at Morgans Point exhibited excessive temperature stratification (greater than 5 deg C). Wind magnitude and direction showed dramatic shifts at the nowcast/forecast transition.

Operational Test Bed Results and Improvements

To study the causes of these phenomena an operational test bed was developed by saving selected initial condition and restart files. The initial condition files were used to study astronomical tide dynamics and represent a cold start from climatological density initial conditions adjusted for the given nowcast/forecast cycle thereby resetting the density structure. Wind and subtidal water level forcings are not while river inflows were considered in the tidal dynamics. The restart files are used to resimulate a particularly troubling nowcast/forecast cycle and contain the full range of meteorological forcing (wind, subtidal water level, and river inflow).

To focus on the tidal component of the total water level error, the tidal dynamics were initially tested for the 14 December 1998 nowcast/forecast cycle and then rechecked for the 20 April 1999 nowcast/forecast cycle. Forecast water levels at Galveston Pleasure Pier, while in excellent agreement for 14 December 1998 forecast cycle, were offset by approximately 13 cm from the 20 April 1999 forecast cycle predictions. To seek to reduce this offset, the long period harmonic constituents Sa and Ssa were adjusted along the GBM open boundary to be consistent with Galveston Pleasure Pier values and the order 3 cm open boundary water level offsets were set to zero. Both test nowcast/forecast cycles were repeated and as shown in Figure 11, the forecast cycle results are now in excellent agreement with predictions indicating that the tidal component is now squared away. Note in regions of large Sa and Ssa such as in Galveston Bay, it is necessary to perform seasonal tidal studies to confirm the proper behavior of the these long period tidal constituents.

The restart file was saved for the 26 April 1999 nowcast/forecast cycle during which large oscillations in the water levels primarily during the nowcast cycle occurred. In addition, excessive temperature stratification at Morgans Point was evidenced during the forecast cycle. It was also noted that the wind directions switched dramatically from nowcast to forecast. The following five experiments were conducted to address these issues:

Experiment 0: To demonstrate the ability to recreate the problematic operational nowcast/forecast cycle. The cycle was rerun from the saved restart file, which was the same file used in the operational cycle to start the problematic nowcast/forecast.

Experiment 1: After examining time series files over the entire period, it was evident that at Morgans Point, the temperature stratification gradually increased during every simulation. It appeared that the initial 9 April 1999 nowcast/forecast cycle exhibited the same behavior and that this was due to a problem with the initial conditions at a single grid point. A revised procedure to eliminate excess stratification was used to reinitialize the initial temperature fields, which were then readjusted based on the PORTS measurements. The excessive temperature stratification in the forecast no longer occurs.

Experiment 2: The previous experiment was repeated but with the modifications to the Sa and Ssa long period tidal constituents with zero water level offsets along the open boundary. Nowcast water level results at Galveston Pleasure Pier still exhibit large oscillations. Reasonable temperature stratification is retained at Morgans Point in both nowcast and forecast.

Experiment 3: A three point smoother was used to smooth the nowcast and forecast water level residuals, except at nowcast inception and conclusion. At the initial nowcast point, the water level residuals need to be preserved from nowcast to nowcast and smoothing is not appropriate. At conclusion (the switch from nowcast to forecast) the proper offset to the NWS/TDL Extratropical Storm Surge Model water level is determined and therefore no smoothing should be performed as well. Nowcast water level results are contrasted with those from experiment 0 at Galveston Pleasure Pier in Figure 12. The excessive water level oscillations in the nowcast are greatly reduced.

Experiment 4: Wind directions from the PORTS screen were compared with those contained in the PUFFF files to determine that the wind directions are in meteorological convention. This is not clearly indicated in the PUFFF documentation. Wind directions were corrected and the nowcast water level results show an improvement of 8 cm rms compared with experiment 3 as noted in Figure 13 at Morgans Point in the upper Bay.

In addition to the above improvements, the NWS Aviation Model wind strengths have been reduced by a factor of two. Water level results for the first week in June 1999 are shown in Figures 14 and 15 and exhibit considerable improvement over the 9 April - 9 May 1999 one month demonstration results.

Additional Demonstration Periods and System Developments

The nowcast/forecast system demonstration period has continued since 9 May 1999 to gain further pseudo-operational experience and develop web-site based dissemination. The next focus of the operational test bed experiments is on the improvement of the currents. The recently acquired (obtained jointly by Texas A&M University and NOS through a Sea Grant Project) towed ADCP and CTD measurements over a 4 nautical mile section of the Houston Ship Channel above Redfish Bar will be used to this end. In addition, we will pursue a more formal evaluation of the nowcast/forecast system based on NOS guidelines (1999).

Acknowledgments

Dr. Bruce B. Parker, Chief of the Coast Survey Development Laboratory, conceived of this project and provided leadership and critical resources. Discussions with Dr. Kurt W. Hess on modeling shallow bay systems were of great assistance. Mr. Philip H. Richardson was instrumental in the development of the wind and atmospheric pressure field interpolation procedures and in the forecast system input analysis. Mr. John F. Cassidy enabled extremely efficient performance of the computations via his maintenance and upgrades to the computing systems employed.

References

Barnes, S. L., 1973: Mesoscale objective map analysis using weighted time-series observations, *NOAA Technical Memorandum ERL NSSL-62*, National Severe Storms Laboratory, Norman, OK.

Bethem, T. D., and H. R. Frey, 1991: Operational physical oceanographic real-time data dissemination. *Proceedings, IEEE Oceans 91*, 865 - 867.

Blumberg, A. F., and G. L. Mellor, 1987: A description of a three-dimensional coastal ocean circulation model. *Three-Dimensional Coastal Ocean Models*, (ed. Heaps), American Geophysical Union, Washington, DC., 1 - 16.

Hess, K. W., 1994: Tampa Bay Oceanography Project: Development and application of the numerical circulation model. NOAA, National Ocean Service, Office of Ocean and Earth Sciences, *NOAA Technical Report NOS OES 005*, Silver Spring, MD.

Ives, D. C. and R. M. Zacharias, 1987: Conformal mapping and orthogonal grid generation, Paper No. 87-2057, *AIAA/SAE/ASME/ASEE 23rd Joint Propulsion Conference*, San Diego, CA.

Lin, S-J., W. C. Chao, Y. C. Sud, and G. K. Walker, 1994: A class of the van-Leer transport schemes and its application to the moisture transport in a general circulation model, *Monthly Weather Review*, 122, 1575-1593.

NOS, 1999: NOS procedures for developing and implementing operational nowcast and forecast systems for PORTS, NOAA, National Ocean Service, Center for Operational Oceanographic Products and Services, *NOAA Technical Report NOS CO-OPS 0020*, Silver Spring, MD.

Schmalz, R. A., 1996: National Ocean Service Partnership: DGPS-supported hydrosurvey, water level measurement, and modeling of Galveston Bay: development and application of the numerical circulation model. NOAA, National Ocean Service, Office of Ocean and Earth Sciences, *NOAA Technical Report NOS OES 012*, Silver Spring, MD.

Schmalz, R. A., 1998a: Development of a Nowcast/Forecast System for Galveston Bay, Proceedings of the 5th Estuarine and Coastal Modeling Conference, ASCE, Alexandria, VA, 441-455.

Schmalz, R. A., 1998b: Design of a Nowcast/Forecast System for Galveston Bay, Proceedings of the 2nd Conference on Coastal Atmospheric and Oceanic Prediction and Process, AMS, Phoenix, AZ, 15-22.

Schmalz, R. A., 1998c: Initial Evaluation of a Nowcast/Forecast System for Galveston Bay during September 1997, Proceedings of the Ocean Community Conference, MTS, Baltimore, MD, 426-430.

Wilken, J. L., 1988: A computer program for generating two-dimensional orthogonal curvilinear coordinate grids, *Woods Hole Oceanographic Institution (unpublished manuscript)*, Woods Hole, MA.

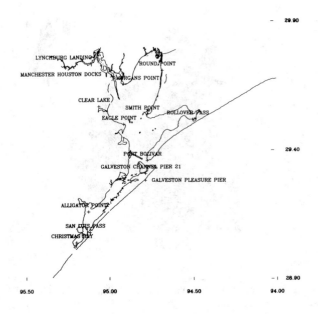

Figure 1. Galveston Bay Base Map. (PORTS station locations are indicated.)

```
            Houston/Galveston PORTS, National Ocean Service/NOAA
                  at 12:48 pm CDT  June 25, 1999
.........................................................................
            TIDES                  :             CURRENTS
Morgans Point        1.7 ft.,Falling: Morgans Point    0.2 kts.(S), 169°T
Eagle Point          1.5 ft.        : Bolivar Roads     0.8 kts.(E), 129°T
Pier 21              1.4 ft.
Bolivar Roads        1.5 ft.,Rising : (F)lood,(S)lack,(E)bb,towards °True
Pleasure Pier        1.6 ft.        :.................................
                                    :                Salinity   S.G.  W.Temp
                                    : Morgans Point    7.2 psu  1.002   84°F
                                    : Eagle Point     14.6 psu  1.008   81°F
                                    : Bolivar Roads   14.7 psu  1.008   82°F
                                    : Pleasure Pier                     83°F
.........................................................................
    METEOROLOGICAL     Wind Speed/Dir              Air Pressure    Air Temp
Morgans Point    11 knots from SSW, gusts to 17   1014 mb,Rising    74°F
Eagle Point      15 knots from S  , gusts to 18   1013 mb,Steady    83°F
Bolivar Roads    16 knots from SSE, gusts to 22   1013 mb,Steady    84°F
Pleasure Pier    22 knots from SSE, gusts to 24   1012 mb,Steady    84°F
.........................................................................
For more information about PORTS,
go to http://www.opsd.nos.noaa.gov/corms_status.html
```

Figure 2. Houston/Galveston PORTS Screen Data Format.

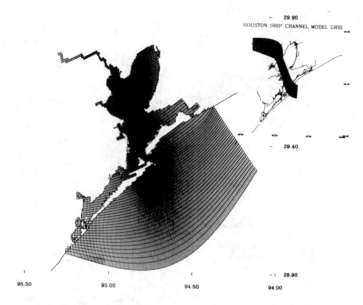

Figure 3. Galveston Bay Model Grid With Fine Resolution Houston Ship Channel Model Grid Inset.

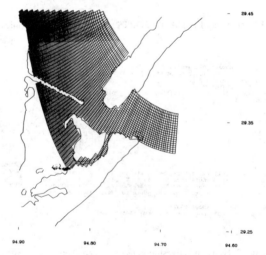

Figure 4. Houston Ship Channel Model Grid Near Galveston Entrance. (Resolution is enhanced by approximately a factor of two over the Galveston Bay Model.)

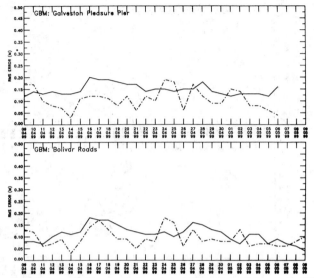

Figure 5. Gulf Coast and Lower Galveston Bay Water Level Nowcast (solid)/Forecast (dashed) RMS Errors.

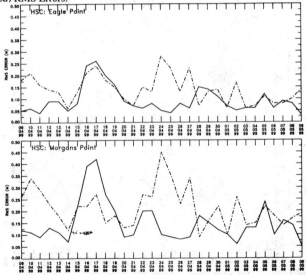

Figure 6. Upper Galveston Bay Water Level Nowcast (solid)/Forecast (dashed) RMS Errors.

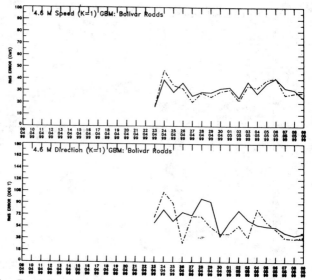

Figure 7. Bolivar Roads Prediction Depth Current Nowcast (solid)/Forecast (dashed) RMS Errors.

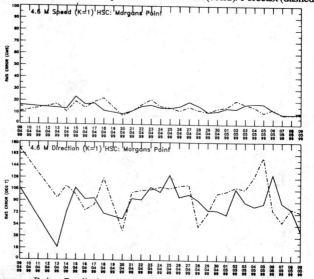

Figure 8. Morgans Point Prediction Depth Current Nowcast (solid)/ Forecast (dashed) RMS Errors.

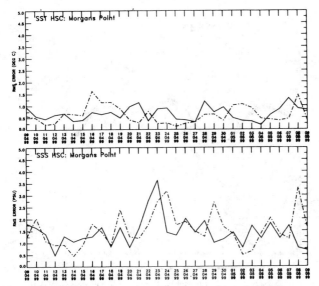

Figure 9. **Morgans Point** Surface Temperature and **Salinity Nowcast** (solid) / Forecast (dashed) **RMS Errors.**

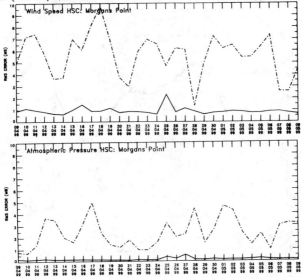

Figure 10. Morgans Point Wind and Sea Level Atmospheric Pressure Nowcast (solid)/ Forecast (dashed) RMS Errors.

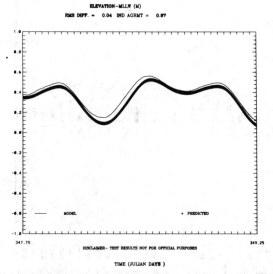

ELEVATION-MLLW (M)

RMS DIFF. = 0.04 IND AGRMT = 0.97

Figure 11a. Tidal Elevations at Galveston Pleasure Pier for 14 December 1998 Forecast Cycle Based on Revised Sa and Ssa and Boundary Water Level Offsets.

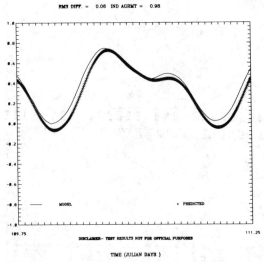

ELEVATION-MLLW (M)

RMS DIFF. = 0.06 IND AGRMT = 0.98

Figure 11b. Tidal Elevation at Galveston Pleasure Pier for 20 April 1999 Forecast Cycle Based on Revised Sa and Ssa and Boundary Water Level Offsets.

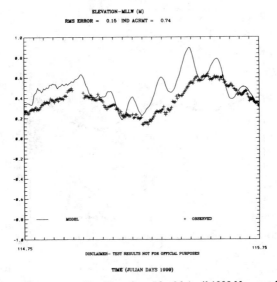

Figure 12a. Galveston Pleasure Pier Water Level for 26 April 1999 Nowcast Cycle: Experiment 0.

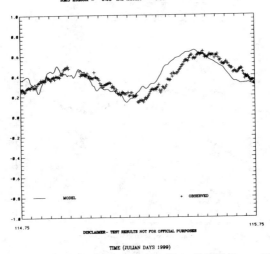

Figure 12b. Galveston Pleasure Pier Water Level for 26 April 1999 Nowcast Cycle: Experiment 3.

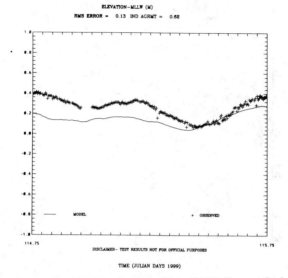

Figure 13a. Morgans Point Water Level for 26 April 1999 Nowcast Cycle: Experiment 3.

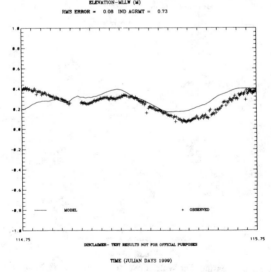

Figure 13b. Morgans Point Water Level for 26 April 1999 Nowcast Cycle: Experiment 4.

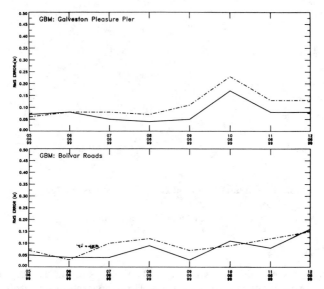

Figure 14. Lower Bay Water Level Nowcast (solid)/Forecast (dashed) RMS Errors in June 1999.

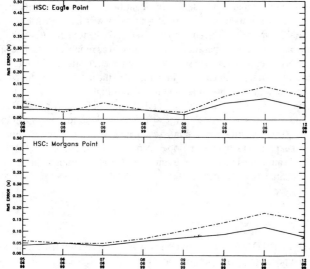

Figure 15. Upper Bay Water Level Nowcast/Forecast RMS Errors in June 1999.

Skill Assessment Methods for NOS Forecast Systems

Kurt W. Hess[1] and Thomas F. Gross[1]

Abstract

NOS has established a set of procedures for assessing the skill of its oceanographic nowcast and forecast systems. The components of skill assessment include the identification of the variables relevant to the intended application, model run scenarios, the comparison statistics or quantities, and the acceptance criteria. User-based skill assessment assumes that a forecast will be most useful when it includes variables specifically needed by the user, and when accuracy is high enough that specific management or operational decisions can be reliably made. The primary user for NOS's nowcasts and forecasts is the maritime navigational community. The primary forecast variables, in terms of importance to the user, are: the magnitude of the water level at all times, the times and heights of high and low water, the speed and direction of the currents at all times, and the times of slack water before flood and ebb. The scenarios describe the type of forcing and the sources of forcing data of forecasts. The statistics are a small set of quantities that are easy to define, compute, and explain and that cover global accuracy and extreme cases. Acceptance criteria include numerical values for the statistics. The application of the method and criteria is illustrated by examples taken from NOS's Chesapeake Bay forecast system.

Introduction

The National Ocean Service (NOS) is developing and implementing nowcast and forecast models for coastal areas as part of its Physical Oceanographic Real-Time

[1]Coast Survey Development Laboratory, Office of Coast Survey, National Ocean Service, 1315 East-West Highway, Silver Spring MD 20910 .

Systems (PORTS). Skill assessment is critical in helping NOS ensure that these models are developed so that they are consistent with user needs, they are implemented in a scientifically sound and operationally robust way, their shortcomings are well understood, and that all products and procedures are authoritative in the face of potential legal challenges. Therefore, NOS has established a set of procedures for developing, evaluating, and implementing nowcast and forecast systems (NOS, 1999). In this context, a *hindcast, nowcast, and forecast* are scientific predictions about the past, present, and future states, respectively, of water levels and/or currents (and possibly other relevant oceanographic variables) in a bay or estuary made by a numerical, statistical, or hybrid model or method.

This paper discusses the approach used to select the skill assessment statistics and criteria, and focuses on the application of these methods to the evaluation of numerical model-based systems. From the wealth of material written on skill assessment (c.f., Lynch and Davies, 1995), NOS, guided by user needs, has selected a small set of quantities that are easy to define, compute, and explain. User-based skill assessment, as this approach is called, assumes that a forecast will be most useful when the forecast includes variables specifically needed by the user, and when accuracy is high enough that specific management or operational decisions can be reliably made. The primary user for NOS's nowcasts and forecasts is the maritime navigational community, including pilots and port operators. The application of the methods and criteria is illustrated by examples taken from NOS's Chesapeake Bay nowcast/forecast system (Bosley and Hess, 1998) and skill assessment (Gross, in preparation). Since the criteria were derived before a full skill assessment could be made, it is expected that some of the tests, criteria, and scenarios will be revised when additional experience is gained.

The Chesapeake Bay nowcast/forecast system consists of a barotropic numerical circulation model driven by winds, coastal water levels, and river flows to produce predictions of water levels of 11 locations around the Bay (Figure 1). For the nowcast, the model uses observed winds at two locations to generate a wind field by spatial interpolation, observed water levels near the Bay's entrance to provide the coastal input, and temporal interpolation of climatological monthly river flows from seven tributaries. For the forecast, the model uses National Weather Service (NWS) forecast winds, a coastal water level that is the sum of an astronomical tide prediction and an NWS prediction of extra-tropical storm surge, and climatological river flows. Twice daily, the system is run in a nowcast cycle followed by a forecast cycle. In the nowcast cycle, the model is initialized with the output of the previous nowcast (approximately 12 hours old) and run for 12 hours to bring the modeled bay to the present. In the forecast cycle, the model is initialized with the output from the just-completed nowcast, and run for 24 hours. A sample of a water level forecast produced by this system is given in Figure 2.

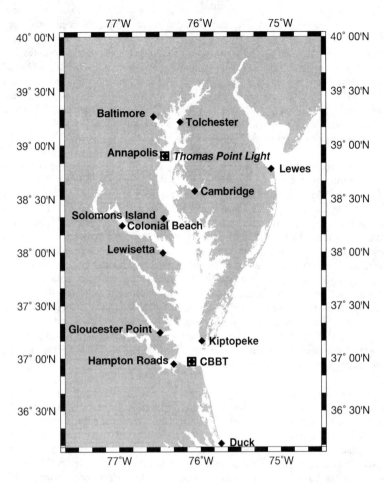

Figure 1. Chesapeake Bay showing the locations of the 11 locations (◆) where nowcasts and forecasts of water levels are produced, and the two wind stations (□) used in the nowcast system.

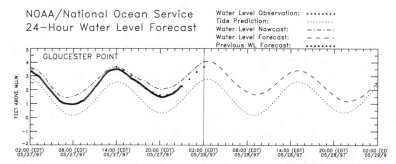

Figure 2. Sample of water level time series output for Gloucester Point, Virginia, in the lower Chesapeake Bay. Water levels are given relative to Mean Lower Low Water (MLLW), which is the NOS chart datum. The vertical line in the center represents the time the forecast was made.

Components of Skill Assessment

Skill assessment can of course be defined in many ways. For our purposes, skill assessment is the (primarily) objective judgement of a model's performance (i.e., its ability to reproduce observed variability and to predict future variability) in relation to (a) a set of standards and (b) other prediction methods. Skill assessment applies both to the model and to the entire nowcast/forecast system, since the availability, quality, and timeliness of input data (from observations and from other models) affects the quality of the nowcast/forecast. The methods discussed are to be applied to a numerical circulation model that has been previously developed; therefore, basic questions about methodology (in the use of numerical models), about mass conservation, etc., will have been settled.

Skill assessment is discussed in terms of four components: (1) the variables relevant to navigation, (2) the model run scenarios, (3) the comparison statistics or quantities, and (4) the acceptance criteria. These are explained in detail below.

Relevant Variables

In terms of importance to navigation, the four primary sets of variables are:

- the magnitude of the water level (relative to MLLW) at all times and locations (for under-keel clearance),

- the times and heights of high and low water at docking/anchorage sites(for planning port arrivals and departures),
- the speed and direction of the currents at all times and locations but especially at channel junctions (for maneuvering), and
- the start and end times of slack water before flood and ebb at all locations but especially at channel junctions (for planning turns in confined areas).

Other potential variables (not addressed here) include water density (for cargo loading capacities) and maximum flood/ebb speed (for maneuvering).

Model Run Scenarios

The scenarios determine the conditions under which the model is run, and are discussed below in the order they would normal occur during model development. The conditions include the type of boundary values used for coastal water level, wind, river flow rate, and water density, and the source of the data. There are five scenarios:

- Astronomical Tide Only
- Test Nowcast
- Test Forecast
- Semi-operational Nowcast
- Semi-operational Forecast

Forecast development usually begins with Astronomical Tide Only simulations because tidal variations are generally dominant, tides may account for a significant part of the error, and because there are extensive data available for validation. Modeled time series can be harmonically analyzed to produce constituent amplitudes and phases for comparison with NOS's accepted values. These values provide information on the model's behavior in frequency space and can also illuminate the role of friction and other non-linear processes. The Test Nowcast is made with observed data with few (or filled in) data gaps. Similarly, the Test Forecast is made with input wind, boundary, and river forecasts that have few (or filled in) gaps. Finally, the Semi-Operational Nowcasts and Forecasts are made in an operational environment (i.e., running one or more times a day) and so will occasionally encounter missing observations and forecasts; the system must be able to handle these conditions without significant loss of accuracy.

Skill Assessment Statistics

Although no single set of statistics can quantify model performance perfectly, we have selected several, easily-calculated quantities that provide relevant information on the important categories of model behavior. Here the error is defined as the predicted value minus the observed value. For a global assessment of errors, we use the series mean (SM), the Standard Deviation (SD), and the Root Mean Square Error (RMSE) of a time series. Calculation of the SM will determine whether any system biases exist. SD and RMSE help when comparing results to those of other model forecast systems. Another statistic is the frequency that errors lie within specified limits. Here the central frequency, or CF(X), is defined as the fraction the errors that lie within the limits of $\pm$X. The concept of the CF, which embodies some of the information captured by the SD and RMSE, is easier to explain to users lacking a background in statistics, as may be the case for pilots and port operators.

The frequency of times of poor model system performance is determined by analyzing the outliers, which are errors which exceed specified limits. The Positive Outlier Frequency, POF(X), is defined as the frequency the nowcast/forecast is higher than the value X. The Negative Outlier Frequency, NOF(X), is defined as the frequency the nowcast/forecast is more negative than -X. A related statistic is the duration of the occurrence of positive or negative outliers. The maximum duration of positive outliers, MDPO(X), is defined as the maximum number of consecutive occurrences when the model's error exceeds X. The maximum duration of negative outliers, MDNO(X), is defined as the maximum number of consecutive occurrences when the model's error is more negative than -X. The MDPO and MDNO must be computed with data without gaps. All the above statistics are termed the Standard Suite and are summarized in Table 1.

Table 1. The Standard Suite of statistics and the Standard Acceptance Criteria that apply to all nowcast and forecast variables. The Criteria are shown in general form; the error value X and duration N are determined by the specific application and variable.

STATISTIC	SYMBOL	CRITERION
Series Mean	SM	none
Root Mean Square Error	RMSE	none
Standard Deviation	SD	none
Central Frequency	CF	$CF(X) \geq 90\%$
Positive Outlier Frequency	POF	$POF(2X) \leq 1\%$
Negative Outlier Frequency	NOF	$NOF(2X) \leq 1\%$
Maximum Duration of Positive Outliers	MDPO	$MDPO(2X) \leq N$
Maximum Duration of Negative Outliers	MDNO	$MDNO(2X) \leq N$

We note that this approach is fundamentally different from that applied to Tampa Bay hindcasts by Hess and Bosley (1992). That approach described skill scores based of the error expressed as a percent of the range of the tide or tidal current. The use of those skill scores remains a valid way of assessing a model's accuracy from a scientific point of view, especially in regions such as Tampa Bay where tides usually dominate and storm surges are relatively rare events. Here we have emphasized the size of the error by itself (i.e., not relative to tide range) because the users are concerned with actual depth of water under the keel.

The SD and the RMSE have limited use in our approach because model errors were not found to be normally distributed. For example, for nowcast water levels at Baltimore, Maryland (Figure 3), a higher number of errors cluster around zero and a higher number of large positive errors are seen to occur than in a Gaussian distribution. Calculation of the Chi-square value confirm a significant departure from normality. This departure is especially important in the description of outliers, which would be underestimated by a normal distribution, but are critical in navigation.

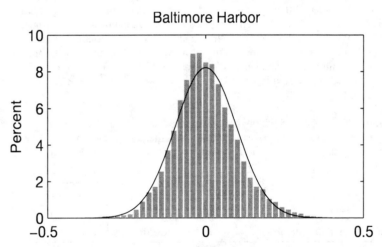

Figure 3. Histogram of percentage of water level nowcast errors within 0.02-m intervals for 365 days of hourly water levels at Baltimore. Error is defined as the predicted minus the observed. The solid line is the Gaussian curve with the same standard deviation as the errors. The Chi-square test indicates that the errors depart significantly from the Gaussian distribution.

Another statistic determines the frequency of the 'worst case' forecast situation, which (from a model-based forecaster's viewpoint) is when the user would have been better off using the astronomical tide water level prediction rather than the model-based forecast. The Worst Case Outlier Frequency, WOF(X), is defined here as the frequency of occurrence of the situation when the model-based forecast error magnitude was greater than X and either (1) the simulated value of water level is greater than the astronomical tide and the observed value is less, or (2) the simulated value of water level is less than the astronomical tide and the observed value is greater.

Standard Acceptance Criteria

Once the statistic has been calculated, its value can be compared to some pre-established standard. For a nowcast or forecast at a particular station to be approved for release to the public, the statistics related to model performance at that station must (a) meet or exceed all standard acceptance criteria, or (b) meet or exceed most of the criteria and be granted a waiver. A waiver may be needed to allow for the wide variety of coastal areas and their dynamics, for changes in the priorities of users, and because a forecast is not likely to be as accurate as a nowcast.

The standard acceptance criteria have a generalized form and are expressed in terms of a representative error (which determined by the specific variable and application) and a threshold frequency or duration value. For example, the standard criteria for the distribution of errors, given a representative error X, are $CF(X) \geq 90\%$, $POF(2X) \leq 1\%$, and $NOF(2X) \leq 1\%$. This means that 90% of all errors must fall within $\pm X$. The specific threshold frequency values (90%, 1%, and 1%) are somewhat arbitrary, but they approximate the bell-shaped distribution of errors and are simple enough to be readily understood by most navigation community users. The limit of 1% of the time is equivalent to about 88 hours per year (about $3\frac{1}{2}$ days). While the system may be inaccurate during those few extreme events during a typical year, mariners will recognize that most weather-related forecasts are likely to be less dependable and will plan accordingly. Note that for a Gaussian (normal) distribution, the condition that $CF(X) = 90\%$ implies that $SD = 0.606X$ and $POF(2X) = 0.048\%$. However, as explained above, the errors are not perfectly Gaussian.

The standard acceptance criteria for the error duration (defined as the number of consecutive occurrences of an outlier), given a representative error X, are $MDPO(2X) \leq N$ and $MDNO(2X) \leq N$, where N is the limit on the duration (N may or may not have time units). Figure 4 shows a typical histogram of water level outlier events and their durations. The general criteria on error distribution are required for the assessment of nearly all variables, and are collectively called the Standard Acceptance Criteria (Table 1).

Water Level Time Series

Before the statistics can be computed, several series of observed and modeled

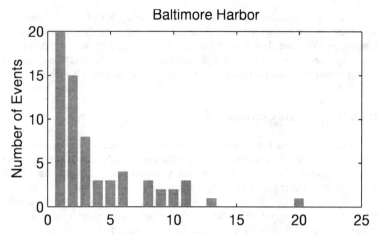

Figure 4. The number of occurrences of errors greater that 0.2 m for values of duration (hr) for outliers in the year-long water level nowcast for Baltimore.

variables must be generated. Actual values of the selected variables are obtained by *in situ* observation and by running the forecast model. All time series of water level data are to consist of at least 365 days of hourly values; a full year of values is required so that the impact of annual changes in astronomic tides and seasonal weather conditions will be included in the evaluation. For hourly water levels, the time series is referred to as WL_1. For the Test Forecast and the Semi-operational Forecast, WL_2 is a series which is created by collecting all forecasts valid at a specific projection time (i.e., +00 hr, +06 hr, +12 hr, +18 hr, and +24 hr) from all 24-hr forecasts made each day. For example, consider the WL_2 series consisting of 6-hour forecasts at a specific location. In the first forecast run of the first day, the predicted water level at 6 hours from the start of the forecast will become the first value in the series. The second value in the series will be the 6-hour forecast water level predicted in the second forecast run of the day, and so on for each forecast in the first day. The number of forecasts per day is M. The next value in the series will be the predicted 6-hour water level in the first forecast of the second day, and so on.

Another series is required for the high and low waters, and consists of consecutive values of either the height or time of either high or low water. Precise times and heights may be found by polynomial interpolation of the hourly water levels.

Error (X) and Duration (N) Values for Water Levels

The Standard Suite of statistics and the Standard Acceptance Criteria (Table 1) are common to all scenarios for water levels. The criteria are related specifically to the marine navigational community by the selection of the representative error X and the duration N. For the time series of water levels (WL_1), the representative error X is 15 cm and maximum outlier duration N is 24 hours. This means that 90% of the errors will be within ±15 cm, 1% of the errors will exceed 30 cm, and 1% of the errors will be more negative than -30 cm. It also means that the maximum number of consecutive hours that a nowcast or forecast is in error by more than 30 cm is 24. These values for X and N were chosen because of pilots' expressed need for that level of accuracy in water depths to avoid groundings and for planning approaches to U.S. harbors with sufficient time. The requirements are summarized in Table 2.

Table 2. The representative error (X) and duration (N) values (see Table 1) for acceptance of hourly water levels (WL_1), constant projection time forecasts of water levels made M times per day (WL_2), heights of high and low water (HHW, HLW), and times of high and low water (THW, TLW).

VARIABLE	X	N
WL_1	15 cm	24
WL_2	15 cm	1+M
HHW, HLW	15 cm	3
THW, TLW	30 min	3

For forecast water levels (WL_2), the representative error is 15 cm, again in keeping with the pilot's expressed requirements. The maximum duration of outliers, N, was chosen as follows to be analogous to the 24 hours for WL_1. Suppose the number of forecasts per day, M, is 2 and consider the forecast for +06 hr. The first +06 hr forecast of the day is valid at 0600 GMT, the second is valid at 1800 GMT, and the first forecast of the next day is valid at 0600 GMT of the second day. Therefore, it takes three forecasts to cover 24 hours, so N = M+1.

In addition, there are requirements for the height of high water (HHW), the time of high water (THW), the height of low water (HLW), and the time of low water (TLW). For the heights and times of high (low) water, the maximum error is 15 cm for amplitude and 30 minutes for time. For duration of maximum outliers, the duration, N, is the number of consecutive high (low) waters, and is limited to 3. This means that high and low water values can exceed limits at most on 3 consecutive high or low tides. Requirements on these are also given in Table 2.

Statistics which have no specific acceptance requirements (SM, SD, and RMSE) still are necessary for scientific model evaluation. Also, it is expected that as nowcasts/forecasts become widely disseminated that some of the X and N values selected will need to be revised. Accuracy requirements may be altered after further experience with present-day modeling capabilities. For example, although the nowcast is required to meet $CF(X) \leq 90\%$, the requirement for forecast accuracy may be reduced. Thus the 6-, 12-, 18-, and 24-hr forecasts may be required to meet, say, $CF(X) \leq 85\%$, 80%, 75%, and 70%, respectively. The outlier frequencies could be treated in a similar way.

Additional Criteria for Water Levels

In addition to the Standard Acceptance Criteria, there are three additional criteria for evaluating water levels. The first is a requirement for the Test Nowcast and the Semi-operational Nowcast on the worst case outlier frequency, WOF(30cm):

$$WOF(30 \text{ cm}) \leq 0.5\%. \tag{1}$$

This requirement is relatively severe because it is important to develop a water level forecast system that is nearly always better than the astronomical tidal prediction during each extreme event.

Another requirement is for the comparison of forecast skill using three methods: the model-based forecast, the astronomical tidal prediction added to a persisted non-tidal component (i.e., based on an observed offset during some time interval before the forecast is made), and the astronomical tidal prediction alone. If the model-based forecast is not an improvement over the other two forecasts, the model should not be implemented. The requirements for water level error and the height and time of high/low water error, are

$$
\begin{aligned}
CF_{astronomical} &\leq CF_{persistence} \leq CF_{model} \\
POF_{astronomical} &\geq POF_{persistence} \geq POF_{model} \\
NOF_{astronomical} &\geq NOF_{persistence} \geq NOF_{model} \\
MDNO_{astronomical} &\geq MDNO_{persistence} \geq MDNO_{model} \\
MDPO_{astronomical} &\geq MDPO_{persistence} \geq MDPO_{model,} \\
WOF_{astronomical} &\geq WOF_{persistence} \geq WOF_{model.}
\end{aligned}
\tag{2}
$$

Another requirement applies specifically to the Astronomical Tide Only simulations, and requires a comparison of the tidal constituents (37 amplitudes and phases) obtained by harmonically analyzing the modeled water levels with the

NOS accepted values; there is no specific criterion to be met, but the results should give additional insight to how the model performs.

Results for Chesapeake Bay Water Levels

The above statistics and criteria for water levels were applied to the output of the Chesapeake Bay nowcast/forecast system presently under development (Bosley and Hess, 1998). The statistical results for Solomons Island, Maryland, are shown in Table 3 (see Gross [in preparation] for details). At this location, the model passes the acceptance criteria for the WL_1, HHW, and HLW in the Astronomical Tide Only and Test Nowcast scenarios, and passes the acceptance criteria for the WL_2 in the Test Forecast scenario for the +00, +06, and +12 hr projections. The model fails many of the criteria for THW and TLW; the large negative value of SM indicates that the phase of the predicted astronomical tide is occurring too early in the model. In the comparison tests, the Semi-operational Forecast is better than Astronomical Tide Only at all projection times, and is better than persistence (Astronomical Tide plus Persisted Observation) beyond 6 hr. The persistence forecast up to 6 hr may be better than the model-based forecast because the model, which at present does not assimilate water levels, does not usually have a zero error at +00 hr. Another point is that the Semi-operational Nowcast, since it may have missing data, was expected to be less accurate than the Test Nowcast. However, the data show that it may be more accurate: a possible explanation is that data is often missing during extreme events, which is the time the model's performance may be the worst.

A sample of the results for CF at all stations is shown in Table 4. The model system does best in the Astronomical Tide Only scenario, where it is quite accurate throughout except in the extreme upper Bay (Tolchester) and at the two stations located up rivers (Lewisetta and Cambridge). In the Test Nowcast, the model is better at locations closer to the entrance, where the ocean water level boundary forcing is applied, and poorer in the upper Bay (Baltimore and Tolchester), where winds are at least as important as the coastal non-tidal forcing. For the Test Forecast, many of the lower Bay stations do not meet the acceptance criteria; this indicates that the forecast of the coastal non-tidal water level may be the source of the inaccuracy.

Summary

Skill assessment procedures are necessary to determine the accuracy of nowcast and forecast variables. The approach developed here is called 'user-based' because it was specifically tailored to users in the marine navigation community (primarily pilots and port operators). The method consists of a small set of easily-computable statistics and their corresponding acceptance criteria. The statistics describe the frequencies of occurrence of a representative error and of the outliers, and the duration of the outliers. The approach was applied to Chesapeake Bay water

Table 3. Preliminary skill assessment quantities for Solomons Island, Maryland, using hourly values from 1998. Values of SM, RMSE, and SD for water levels and high/low water heights are in meters and for times of high/low water in minutes. NOF, CF, POF, WOF, MDPO, and MDNO are percentages. MDPO and MDNO percentage is duration divided by N from Table 2. Na means not applicable.

Variable		SM	RMSE	SD	NOF	CF	POF	MDPO	MDNO	WOF
Criterion		none	none	none	≤1.0	≥90.0	≤1.0	≤100.	≤100.	≤1.0
Scenario: Astronomical Tide Only										
WL1		0.028	0.058	0.051	0.0	99.8	0.0	0.0	0.0	0.000
HHW		0.018	0.046	0.043	0.0	100.0	0.0	0.0	0.0	Na
HLW		0.052	0.064	0.037	0.0	100.0	0.0	0.0	0.0	Na
THW		-33.866	35.069	9.108	1.2	39.0	0.0	0.0	33.3	Na
TLW		-25.947	27.793	9.960	0.0	68.6	0.0	0.0	0.0	Na
Scenario: Test Nowcast										
WL1		0.000	0.085	0.085	0.0	92.3	0.1	16.7	0.0	0.047
HHW		-0.004	0.087	0.087	0.0	92.0	0.1	33.3	0.0	Na
HLW		0.020	0.073	0.071	0.0	95.0	0.0	0.0	0.0	Na
THW		-34.283	40.503	21.568	7.2	34.6	0.7	33.3	66.6	Na
TLW		-30.340	37.653	22.299	7.7	46.9	0.1	33.3	100.0	Na
Scenario: Semi-Operational Nowcast										
WL1		0.005	0.083	0.082	0.2	93.6	0.2	16.7	29.2	0.000
HHW		0.002	0.083	0.083	0.0	92.6	0.0	0.0	0.0	Na
HLW		0.024	0.072	0.068	0.0	94.9	0.0	0.0	0.0	Na
THW		-34.804	41.642	22.863	7.4	36.6	0.8	33.3	100.0	Na
TLW		-29.143	38.140	24.604	8.2	53.9	0.4	33.3	66.7	Na
Scenario: Test Forecast										
WL2	+00hr	0.008	0.089	0.089	0.0	90.8	0.0	0.0	0.0	0.000
WL2	+06hr	-0.012	0.074	0.073	0.0	95.8	0.0	0.0	0.0	0.000
WL2	+12hr	-0.003	0.087	0.087	0.0	91.7	0.1	33.3	0.0	0.000
WL2	+18hr	-0.031	0.101	0.096	1.1	88.4	0.0	0.0	66.7	0.000
WL2	+24hr	-0.016	0.118	0.117	1.6	81.6	0.3	33.3	100.0	0.000
HHW		-0.013	0.090	0.089	0.1	91.1	0.1	33.3	33.3	Na
HLW		-0.001	0.076	0.076	0.6	95.5	0.0	0.0	33.3	Na
THW		-34.159	39.028	18.877	6.4	38.5	0.1	33.3	66.7	Na
TLW		-26.722	34.191	21.330	6.4	57.2	0.1	33.3	100.0	Na
Scenario: Semi-Operational Forecast										
WL2	+00hr	0.008	0.081	0.081	0.0	93.2	0.2	33.3	0.0	0.000
WL2	+06hr	0.004	0.092	0.092	0.0	90.1	0.4	33.3	0.0	0.000
WL2	+12hr	0.010	0.108	0.108	0.6	84.0	0.6	33.3	66.7	0.195
WL2	+18hr	0.012	0.120	0.120	0.8	80.2	0.6	33.3	66.7	0.196
WL2	+24hr	0.012	0.135	0.134	1.8	75.1	1.8	100.0	66.7	0.588
HHW		0.008	0.106	0.106	0.4	87.5	0.4	33.3	33.3	Na
HLW		0.019	0.086	0.084	0.0	90.7	0.0	0.0	0.0	Na
THW		-28.306	34.494	19.713	4.7	56.0	0.0	0.0	66.7	Na
TLW		-23.483	31.006	20.247	2.7	64.9	0.8	66.7	33.3	Na
Forecast Method: Astronomical Tidal Predication										
WL1		-0.003	0.157	0.157	3.4	68.5	2.4	145.8	133.3	0.000
HHW		-0.188	0.247	0.160	23.4	37.4	0.4	100.0	433.3	Na
HLW		-0.194	0.246	0.151	23.4	40.6	0.3	66.7	433.3	Na
THW		-0.060	22.142	22.142	0.6	87.6	1.3	66.7	33.3	Na
TLW		-3.163	26.418	26.228	2.8	81.9	1.3	33.3	66.7	Na
Forecast Method: Persisted Observation Plus Astronomical Prediction										
WL2	+00hr	0.000	0.000	0.000	0.0	100.0	0.0	0.0	0.0	0.000
WL2	+06hr	-0.007	0.071	0.071	0.1	95.5	0.3	66.7	33.3	0.141
WL2	+12hr	-0.001	0.117	0.117	0.8	83.1	1.7	33.3	66.7	0.847
WL2	+18hr	-0.008	0.155	0.155	2.8	72.0	3.4	66.7	66.7	1.836
WL2	+24hr	0.000	0.176	0.176	4.1	68.1	4.8	100.0	100.0	2.975
HHW		-0.003	0.122	0.122	0.7	83.3	2.5	66.7	33.3	Na
HLW		-0.006	0.119	0.119	1.5	83.2	1.2	33.3	33.3	Na
THW		0.002	22.103	22.103	0.6	87.3	1.5	66.7	33.3	Na
TLW		-3.171	26.426	26.235	2.8	81.9	1.3	33.3	66.7	Na

Table 4. Central Frequency (CF) percentage for 365 days of hourly water levels at 11 Chesapeake Bay stations. The water levels were generated for the Astronomical Tide Only, Test Nowcast, and Test Forecast (at +6 hr and +12 hr) scenarios. Present acceptance criteria is CF ≥ 90%; locations not passing are shaded. Locations, in order of increasing distance from the entrance, are Chesapeake Bay Bridge Tunnel (CB), Hampton Roads (HR), Kiptopeake Beach (Ki), Gloucester (Gl), Lewisetta (Le), Colonial Beach (Co), Cambridge (Ca), Annapolis (An), Baltimore (Ba), and Tolchester (To). The system does well in Astronomical Tide Only everywhere, and in Test Nowcast in most locations except the extreme upper Bay (Ba, To) and up the rivers (Co, Ca). The accuracy in the Test Forecast is reduced in the lower Bay (CB, HR, Ki, and Gl) due to errors in the forecast of the non-tidal water level.

WL Series	CB	HR	Ki	Gl	Le	Co	So	Ca	An	Ba	To
Tide Only	100	99	100	99	100	96	99	91	99	98	90
Nowcast	100	98	99	98	96	88	92	82	90	87	87
Fcst +06h	86	86	88	88	98	92	95	86	90	83	85
Fcst +12h	83	84	84	83	93	93	91	79	92	87	88

levels when the model-based nowcast/forecast system was run under five scenarios. It is anticipated that further experience with the application of the methods will lead to refinements in the statistics and the numerical criteria.

Conclusions

- Standard statistics that describe Gaussian distributions of errors were not useful for the acceptance criteria because errors in water levels were found to be significantly non-Gaussian.
- The astronomical tide must be simulated accurately because, in the absence of significant meteorological events, the tides are the dominant forcing and may have a large contribution to the error.
- Model nowcasts could be improved with higher spatial resolution that would better simulate circulation in the tributaries, and by data assimilation.
- At +00 hr and +06 hr, the persistence forecast (astronomical tide plus an observed offset) may be better than the model-based forecast because the model, which at present does not assimilate water levels, does not usually have a zero error at +00 hr.

- The Semi-operational Nowcast, although it may have missing data, was not always less accurate than the Test Nowcast. The most likely explanation is that data is often missing during extreme events, which is precisely the time the model's performance may be the worst.
- Ultimately, system forecast accuracy is limited by the accuracy of the forecasts for winds and coastal non-tidal water levels, which drive the system.

Acknowledgments

The skill assessment methodology was developed by a joint NOS team consisting of Drs. Kurt Hess, Frank Aikman, and John Kelley from the Coast Survey Development Laboratory (CSDL); Dr. Kathryn Bosley and Thomas Bethem from the Center for Operational Oceanographic Products and Services (CO-OPS); and Sonny Richardson, a contractor for CO-OPS. CSDL's Dr. Bruce Parker and Lloyd Huff provided significant components of the methodology.

References

Bosley, K. T., and K. W. Hess, 1998: Development of a Experimental Nowcast/forecast System for Chesapeake Bay Water Levels. **Proceedings, 4th International Conference on Estuarine and Coastal Modeling**, Alexandria, VA, October 20-22, 1997. 413 - 426.

Gross, T. F., (in preparation): Skill assessment of the CAFE Model. pp 43.

Hess, K. W., and K. T. Bosley, 1992: Techniques for validation of a model for Tampa Bay. **Proceedings, 2nd International Conference on Estuarine and Coastal Modeling**, Tampa FL, November 11-13, 1991. 83 - 94.

Lynch, D. R., and A. M. Davies (eds), 1995: **Quantitative Skill Assessment for Coastal Ocean Models**. American Geophysical Union, Washington DC. pp 510.

National Ocean Service, 1999: NOS Procedures for Developing and Implementing Operational Nowcast and Forecast Systems for PORTS. **NOAA Technical Report** NOS CO-OPS 20. 33 pp.

Coastal Wave Measurement and Forecast System : Preliminary Results and Model Selection

Michael S. Bruno[1], Member, ASCE, Kelly L. Rankin[1], and Thomas O. Herrington[1]

Stevens Institute of Technology recently established the New Jersey Coastal Monitoring Network (CMN). This system provides real-time observations and archived records of shallow water (5m) wave, water level and meteorological conditions (wind speed and direction, temperature, atmospheric pressure), as well as digital images of the beach and nearshore, at three locations that span the State's ocean shoreline. This information is disseminated via the Internet through the Davidson Laboratory's web site: http://www.dl.stevens-tech.edu. The system is designed to provide real-time information to local, State, and Federal emergency management personnel, and long-term records of wave, weather conditions and shoreline response for use by the coastal scientific community. The observational network is currently being combined with an existing wave transformation model and deepwater wave forecasts in order to create a wave measurement and forecasting system covering the entire New Jersey and Long Island coastal ocean region. As part of this effort, we first examined the wave transformation patterns throughout the region. Synoptic measurements of incident deepwater wave conditions at NOAA buoys 44025 and 44009 and nearshore conditions at three shallow water wave gauges were employed in the analysis. These measurements, combined with the local wind observations at each measurement location, were employed to examine several factors of concern to the eventual forecasting scheme. Factors that were identified as being critical to the success of the modeling effort include the spatial variability in the wave climate associated with the sheltering of incident deepwater waves from the north, the highly variable contribution of fetch-limited windsea, and wave dissipation through white capping, bottom friction and breaking.

---

[1] Stevens Institute of Technology, Hoboken, New Jersey 07030

Introduction

Understanding and predicting the intensity of coastal storms and the resulting nearshore wave and surge climate are necessary to mitigate the potential loss of life and property damage often associated with these storms. The fact that over 50% of the US population lives within 50 miles of the coast lends urgency to the need for real-time coastal information and short-term forecasts for emergency management operations, pre- and post-storm hazard assessment, and mitigation. Except for a few short-term, high-resolution coastal experiments (e.g., DELILAH, DUCK94, SandyDuck97, SWADE, and presently, SHOWEX), the current observation system lacks the necessary spatial and temporal resolution to capture localized weather phenomenon such as sea breezes, severe weather outbreaks, intensifying storm systems and strong wind-wave interactions. In particular, the impact of local wind fields in the coastal zone on strongly refracted low frequency swell and the interaction of local wind sea on waves is poorly understood (ONR, 1995). Masselink and Pattiaratchi (1998) found that moderate sea breezes (average wind speed of 5.7 m/s) occurring in western Australia can significantly modify the incident offshore wave climate as it propagates toward the coast. The sea breeze system was found to directly force the nearshore wave field and induce a diurnal cycle of nearshore change by causing an increase in the wave height, a decrease in the wave period and an increase in suspended sediment transport. Measurements obtained by Herrington et. al., 1998 adjacent to the downdrift shoreline of a tidal inlet in New Jersey revealed that an ambient inlet-directed alongshore current was strongly enhanced during colinear wind, wave and current events and strongly suppressed during periods when the wind and wave fields were nearly orthogonal. All of the observations indicate the importance of the local wind fields in modifying the nearshore wave and current regime.

In recognition of the scientific community's need for high-resolution, long-term meteorological and coastal dynamics measurements, and the emergency management community's need for real-time information regarding the condition of the coast during storm events, the State of New Jersey initiated in 1998 the development of the Coastal Monitoring Network (CMN). In its final form, the network will provide real-time observations at three shallow water locations along the New Jersey shoreline, and 12-to-72 hour forecasts of wave conditions throughout the region. The information will be freely accessible to users via the Internet. In the following, we describe the observational network and the results of an initial analysis of the data, undertaken with the aim of identifying the optimal configuration of the wave forecasting model.

Description of Project Region

The Atlantic Ocean shoreline of New Jersey is bordered by Long Island, NY, to the north and the Delaware Bay to the south. The coastline is generally oriented north to south for 80 km from Sandy Hook to Barnegat Inlet and northeast to southwest for 120 km from Barnegat Inlet to Cape May (Figure 1). The continental shelf gradually slopes eastward from the coastline reaching a depth of 100 m below sea level approximately 150 km offshore. Within 5 km of the coast the bottom is covered with numerous shoals and sand waves. The shelf contains several submarine canyons, the most significant of which is the Hudson Canyon, which extends southeastward from the entrance of New York Harbor to the edge of the continental shelf. It is as much as 30 m deeper than the surrounding ocean bottom at the western end, and reaches depths of nearly 1000 m near the eastern edge.

The southern and northern ends of the state can experience dramatically different weather conditions. The southern extent of the state is surrounded by water and is removed from the influence of frequent storms tracking east-northeast across the Great Lakes into eastern Canada. The northern portion of the state is well within the influence of the northern storm track. This difference in climate leads to significantly different wind fields along the northern and southern portion of the coast. Extratropical storms moving along the coast from the late fall through early spring have the most significant impact on the coast. These storms, known as northeasters, are characterized by sustained winds from the northeast for durations typically greater than 24 hours. The coastal geometry and offshore bathymetry play a significant role in the impact of northeasters. Southwesterly propagating waves directly impact the southern coastline between Barnegat Inlet, approximately in the center of the coast, and Cape May, located at the southern tip of New Jersey. Depending on the location of the storm, the northern portion of the coastline can experience a less energetic north and northeast wave climate due to the shadowing effect of Long Island to the north. During the summer and fall, the entire coast is typically subject to less energetic windsea from the southeast, and occasional south and southeasterly swell generated by tropical cyclones. The incident wave climate interacts with the offshore bathymetry, generating complex wave refraction and diffraction patterns, especially near the Hudson Canyon, resulting in a significant variation in the wave height distribution along the coast. Our experience indicates that a 2 m variation in wave height along 25 km of coast is not uncommon.

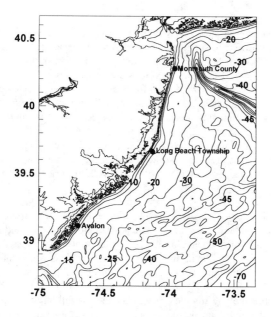

Figure 1: Bathymetry of project area; depth contours in meters

Existing Near Real-Time Coastal Data Acquisition Systems

The National Data Buoy Center (NDBC) of the National Weather Service (NWS) currently maintains two moored buoys and one Coastal-Marine Automated Network (C-MAN) station offshore of New Jersey (Figure 2). All three stations obtain measurements of wind speed, wind direction, wind gusts, barometric pressure, air temperature, water temperature, wave height, and wave period on an hourly basis. The two moored buoys additionally measure wave direction. Buoy 44025 is located approximately 50 km east of Long Branch, NJ, in 40 m water depth, and buoy 44009 is located 40 km southeast of Cape May, NJ, in 30 m water depth. The Ambrose Light C-MAN station is located at the entrance of New York Harbor, 9 km east of Sandy Hook, NJ. In addition to the wave measurements obtained by the NDBC, the Field Wave Gauging Program of the US Army Corps of Engineers currently supports shallow water directional wave measurements offshore of Long Branch, NJ and Westhampton, NY. The Long Branch wave gauge is located in a water depth of 8 m, approximately 11 km south of Sandy Hook and the Westhampton gauge is located in 10 m water depth, approximately 130 km east northeast of Sandy Hook (Figure 2). Each gauge records wave height, peak wave period, peak wave direction and water elevation on an hourly basis.

The NWS maintains two weather observation stations along the coast of New Jersey; one in Atlantic City and one in Belmar. The Atlantic City station is located 70 km north of Cape May and the Belmar station is located 25 km south of Sandy Hook (Figure 2). Each station measures wind speed, wind direction, wind gust, barometric pressure, and air temperature on an hourly basis. Nearshore tide data is recorded hourly at Cape May, Atlantic City and Sandy Hook by the National Ocean Service (NOS) of NOAA. Of the three tide stations, the Atlantic City gauge is the only one located on the open ocean.

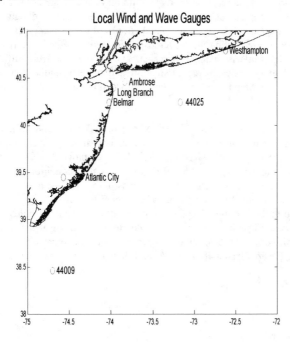

Figure 2. Location of existing wind and wave gauges in the region

Existing Wave Forecasting Systems

The global WAve Model (WAM) was developed through an international collaborative effort and was subsequently freely distributed to the scientific community. The model is currently employed throughout the world in the forecast of sea state conditions. The third-generation WAM model has been operational since 1994 at the U.S. Navy Fleet Numerical Meteorology and Oceanography Center. This model is run on a 1-degree spherical (latitude/longitude) grid with 25 frequency components and 24 directional bands at a 20-minute time step. Surface wind fields obtained from the Naval Operational Global Atmospheric Prediction System (NOGAPS) are used to drive the model. The model provides forecasts up to six days in advance twice daily (0000 GMT, 1200 GMT). NOAA also has an operational global WAM model and a high-resolution model for the East Coast and Gulf of Mexico. The B-Grid for the high-resolution model extends from 65 degrees west to 98 degrees west and from 15 degrees north to 45 degrees north at a resolution of 0.25 degrees. The directional spectrum is discretized using 25 frequency components and 12 directional bands. The surface wind fields are obtained from NOAA's National Center for Environmental Prediction (NCEP) atmospheric models. Significantly, the global WAM model neglects shallow water effects such as refraction-diffraction by underwater topography, energy dissipation due to depth-induced breaking and bottom friction, and energy redistribution due to nonlinear three-wave interactions.

The NWS Marine Prediction Center (MPC) provides forecast information and warnings concerning marine meteorological and oceanographic conditions for the north Atlantic and Pacific Oceans. The MPC employs the Navy and NOAA wave models to provide a 48-hour forecast of sea state conditions four times per day for offshore areas along both U.S. coastlines. In the north Atlantic, the forecasts are provided in the offshore area extending from 31 degrees north to 67 degrees north, and from 65 degrees west to within 25 nautical miles (43 km) of the U.S. coast. This information is provided to National Weather Service coastal forecast offices, for use as guidance material in the preparation of narratives describing the anticipated 48-hour sea conditions in the affected areas.

Coastal Monitoring Network Station Locations

In order to monitor the spatial variation in the coastal wind and wave climate, it was determined that three CMN stations would be utilized to provide coverage of the entire shoreline region. To maximize the usefulness of the measured data, the stations were sited distant from existing nearshore wave and tide gauges. Additional considerations included infrastructure needs, such as a reliable power supply and shelter for the data acquisition computers. Based on all of the factors identified, the following sites were selected (Figure 3): southern Monmouth County, located 35 km south of Sandy Hook; Long Beach Township, located 16 km south of Barnegat Inlet; and Avalon, located 26 km north of Cape May.

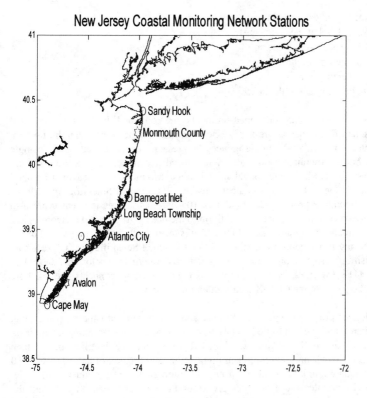

Figure 3. Location map showing positions of CMN nodes

Coastal Monitoring Network Instrumentation

The required   measurements and frequency of sampling were identified by focusing on the end uses of the data. The system requirements were defined by the desire to create a system that would generate a long-term database for the analysis of regional coastal processes, as well as provide near-real time data for emergency management operations and the rapid assessment of the impact of episodic storms on the coast. In order to provide real-time data, the system was required to measure the parameters of interest at high frequency and provide the data to the end users in a rapid and easily accessible way. In addition, the system was required to allow for remote access of the data acquisition program so that sampling intervals and data transfer can be changed if needed. Based on a benefit/cost analysis, the following types of measurements were selected :

- Nearshore wave height and wave period
- Water level at the open coast
- Wind speed and direction
- Barometric pressure
- Air Temperature
- Digital Imagery of the beach

Nearshore wave and tide measurements are obtained by bottom mounted Paroscientific digiquartz pressure sensors located in a mean water depth of 4.6 m. In order to minimize flow interference, the pressure sensors are attached to single point bottom mounted moorings. Wind speed and direction measurements are obtained by R.M. Young model 05103 Wind Monitors mounted on Campbell Scientific, Inc. UT3 Meteorological Towers. A Campbell Scientific, Inc. Model 107 temperature probe, and a model CS105 barometric pressure sensor are also mounted on each UT3 tower. In addition to the oceanographic and meteorological measurements, a digital image of the beach is taken by a digital camera. All of the monitoring instruments are hard wired to a Campbell Scientific, Inc. CR10X data acquisition system mounted in a UV-stabilized fiberglass enclosure attached to each UT3 tower. A double armored data/power cable connects the offshore pressure gauge to the CR10X data acquisition system.

Datalogger programming, data retrieval and storage, image capture, processing, and data transfer is controlled locally by a Pentium computer running a windows-based program (PC208W) developed by Campbell Scientific, Inc. which facilitates logger programming, communication, and transfer of data. The CRX10 continuously monitors wind speed and direction signals at a frequency of 1 Hz. Every 10 minutes the datalogger processes the measured data to calculate the average wind speed and direction as well as the highest wind gust over the 10 minute sampling interval. Air temperature and barometric pressure are recorded as point measurements every 10 minutes. Ten minutes before the top of the hour, the CRX10 powers up the Paroscientific pressure gauge and records data over a 10 minute sampling interval. This relatively short sampling interval was chosen because of concerns regarding the sea state variability over larger time intervals. The 10-minute samples are adequate to describe the frequency spectrum at the site, and offer the advantage of providing wave records at a rate consistent with the meteorological measurements.

At the top of every hour the PC208W software accesses the datalogger and transfers the recorded data first into local memory and then - via a phone modem connected to a local internet provider - to a web-based server located at the Davidson Laboratory in Hoboken NJ. Every five minutes, the computer accesses the digital camera, which automatically snaps and transfers a digital image to the server. The base station computer is equipped with a backup power supply and is configured to store the previous 100 hours of recorded data. Data and programs resident on the CR10X are stored either in non-volatile flash memory or battery-backed RAM. The memory can store 62,000 data points in two Final Storage areas. The CR10X is powered by a rechargeable 9.6 VDC battery connected to an AC powered 12 volt charger/regulator. During power outages, the battery can supply 12 amp hours of power, enough to operate the system for at least 3 months. The data transferred from the base station computer is received by a Sun Ultra Workstation located at the Davidson Laboratory. The pressure time series is then processed through standard spectral analysis procedures and the significant wave height, peak wave period, and average water elevation are determined. The full data set of wind speed, wind direction, air temperature, and barometric pressure is transferred and archived on the Sun Workstation; however, only the most recent 10 minute averages are posted on the web page. The local computer at each CMN site can be accessed from the Davidson Laboratory workstation so that changes in instrument sampling rates, data acquisition, and data archiving can be programmed remotely. At the workstation level, the data is analyzed for outliers, archived and posted in a tabular format on the World Wide Web at www.dl.stevens-tech.edu, providing anyone with internet access real-time images and information regarding the conditions along the coast.

Preliminary Data Analysis

In an effort to ascertain the dominant factors influencing wave generation and transformation in the region (e.g., bathymetry, coastal geometry, local winds), an analysis of the meteorological and wave/water level data obtained to-date was conducted. As we anticipate utilizing predictions of deepwater wave characteristics as boundary conditions for the eventual wave forecasting system, we began our analysis with an examination of the wave transformation patterns indicated by the wave records at NOAA buoy 44009 and the shallow water gauge

at Avalon. Figure 4 illustrates the significant wave height at each gauge and the
ratio of the two wave heights, for the period covering Julian Day 220 through 245
(August 8 – September 2, 1999). Note the dramatic variability in wave height
associated with the passage of frontal systems - with a period of approximately 2
to 4 days. The large wave heights experienced toward the end of the time period
are associated with Hurricane Dennis, which stalled to the south of the region
during this time. Note the semi-diurnal variability in wave height at the shallow
water gauge at Avalon at this time – a clear indication of the depth-limited nature
of these large (greater than 2 m) waves at the Avalon site. The transformation
coefficient (the ratio of the two wave heights) exhibits a high degree of variability.
This variability can be attributed at least in part to the time lag in wave
growth/decay at the two locations due to the passage of weather systems across
the region, as well as to the travel time of swell between the two gauges. Local
winds also play a role, as we shall see later. It is interesting to note that the
transformation coefficient is influenced to a large extent by the magnitude of the
wave height, with minimum values recorded during periods of large wave heights.
This is likely a manifestation of the increased wave dissipation (via white capping,
breaking, and bottom friction) experienced in the shallow water regions between
the two gauges during these high wave events.

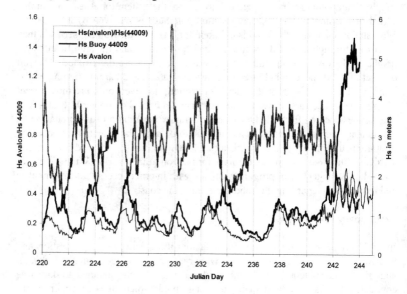

Figure 4 : Significant wave heights and transformation coefficient

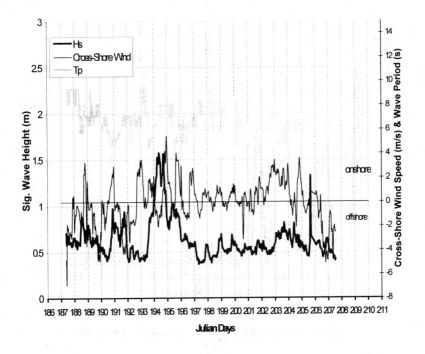

Figure 5: Time history of significant wave height, cross-shore wind, and peak wave period

The role of local wind wave generation can be assessed by a detailed examination of the wind and wave variability at the Avalon station. Figure 5 illustrates the time history of significant wave height and cross-shore wind speed for the period from Julian Day 187 to 208 (July 6 – 27, 1999). The peak wave period is also shown. Note that although the figure demonstrates the presence of the expected correlation between high onshore winds and increased wave heights, there are several instances where the two appear to be uncorrelated, an indication of the influence of remotely-generated swell. This is further illustrated in the variations in peak wave period – low wave periods (indicative of local wind sea) are associated with high onshore winds whereas high wave periods (swell waves) tend to dominate during time periods of low onshore, or offshore winds. Interestingly, there appears to be a rather strong diurnal component to the wind and wave variability, particularly during the period from Julian Day 197 to 204. In order to further examine this variability, we computed the frequency spectra for the

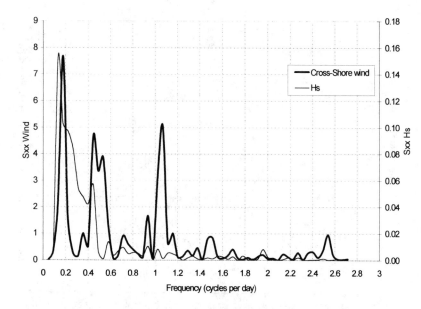

Figure 6: Spectra of cross-shore wind and sig. wave height

cross-shore wind speed and the significant wave height time series for the full 22-day time period. The spectra, illustrated in Figure 6, indicate that as expected a significant portion of the variability in both the cross-shore wind speed and the significant wave height lies within the frequency band associated with the passage of frontal systems – 0.5 to 0.2 cycles per day (2 to 5 days). However, there are also distinct peaks in both spectra at frequencies around the diurnal frequency (period of one day), a clear indication that the sea breeze can be an important factor in the nearshore wave characteristics at the site, as was found by Masselink and Pattiaratchi (1998) at a location on the western coastline of Australia.

In order to illustrate the potential influence of sheltering by Long Island, we present in Figure 7 a time history of significant wave height at the three gauges in the northern area of the model domain – buoy 44025 and the Westhampton and Long Branch gauges - over the period from Julian Day 160 to 165 (June 9 – 14, 1999). Note the dramatic change in relative wave heights at the three gauges late in the day on Julian Day 160. As shown in Figure 8, this change was associated with an equally dramatic change in the wind speed and direction – to a northeast wind unfavorable for wind wave growth at Westhampton because of sheltering.

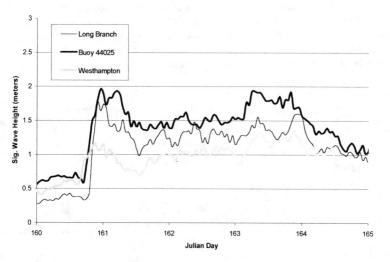

Figure 7 : Sig. wave height at Buoy 44025, Westhampton and Long Branch

Figure 8 : Wind speed and direction at Buoy 44025

As an indication of the need for high spatial and temporal resolution in the eventual wave forecasting system, and as evidence of the robust nature of the CMN instrumentation and data transmission, we present in Figure 9 the time histories of atmospheric pressure, wind speed, wind direction, water depth and significant wave height over a 26-day period that included the influence of Hurricane

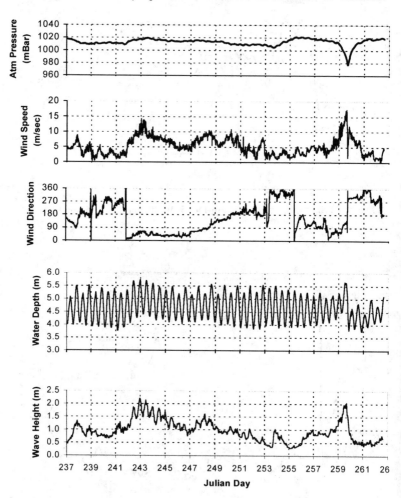

Figure 9 : Measurements at Avalon during Hurricanes Dennis and Floyd

Dennis (peak on Julian Day 243 - August 31, 1999) and the passage of Hurricane Floyd on Julian Day 259 (September 16, 1999). Of particular note is the dramatic pressure drop and associated 180-degree wind shift as the eye of Hurricane Floyd passed over the Avalon station. Clearly, the timing of the hurricane's passage (at near-neap tide and during a falling stage of the tide) greatly limited the magnitude of the storm surge in this area.

Wave Forecasting Model

The analysis described here has demonstrated the need for a forecasting algorithm capable of including local wind wave generation, white capping, shallow water dissipation and breaking. The model must include the large-scale topographic features (e.g., Long Island) that play a major role in the wave generation processes in the region, and must possess the high spatial and temporal resolution necessary to accurately resolve the wind fields and wave growth associated with fast-moving frontal systems, extratropical storms and tropical cyclones.

A number of shallow-water wave propagation models have been developed that can simulate with varying degrees of success the various nearshore wave transformation processes. For wave propagation in open water where the processes of wind-wave growth, shoaling and refraction are dominant, spectral models based on the wave energy density conservation equation are typically used. The global WAM model used to forecast waves in deep water could, in principle, be applied to nearshore regions. The computational time would however be prohibitive because of the high-grid resolution and fine time step required to resolve details of wave transformation in shallow water. We plan to employ the SWAN model (Booij et al., 1999) as our wave transformation algorithm in combination with the WAM deepwater wave forecasts. The SWAN model takes into account the effects of local wave generation due to wind; shoaling and refraction due to varying water depth; energy dissipation due to wave breaking and bottom friction; cross-spectral energy transfer due to nonlinear wave-wave interactions; and wave-current interaction.

In order to minimize the computational effort, the model assumes stationary conditions over the computational area. This assumption is acceptable for most coastal applications since the propagation time from the seaward boundary of the model to the coastline is typically much smaller than the time scale of the variations in the offshore wave boundary conditions (which, in our case will be provided by WAM deepwater forecasts).

The computational domain extends from Long Island to Delaware Bay. The seaward model boundary will be determined from the position of the WAM forecasts. The bathymetry will be generated using data obtained from the

GEOphysical DAta System (GEODAS) from the NOAA National Geophysical Data Center.

Directional wave spectra obtained from WAM forecasts will be used to initialize the model. As discussed earlier, the WAM model is currently run at the U.S. Navy Fleet Numerical Meteorology and Oceanography Center and at NOAA, using surface wind fields obtained from two different atmospheric models. We intend to initially compare predictions from both WAM models with measurements obtained at NOAA buoys 44025 and 44009. Based on these comparisons, we will select the WAM model to be employed in the operational use of the SWAN model.

Extensive calibration/validation of the complete wave forecast model will be performed using measured data at the three CMN stations. The model will initially be run using both measured buoy data and WAM forecasts as boundary conditions. This will help to isolate uncertainties associated with possible errors in the offshore boundary conditions and the wave propagation model.

References

Booij, N., R. C. Ris, and L. H. Holthuijsen, A third-generation wave model for coastal regions. 1. Model description and validation, *J. Geophys. Res., 104*(C4), 7649-7666, 1999.

Herrington, T. O., M. S. Bruno, M. Yavary, and K. L. Rankin, Nearshore processes adjacent to a tidal inlet, *Proceedings of 28th International Conference on Coastal Engineering*, pp. 3331-3344, Am. Soc. of Civ. Eng., Reston, Va., 1998.

Masselink, G. and C. B. Pattiaratchi, Sea breeze climatology and nearshore processes along the Perth metropolitan coastline, Western Australia, *Proceedings of 28th International Conference on Coastal Engineering*, pp. 3165-3177, Am. Soc. of Civ. Eng., Reston, Va., 1998.

ONR, Summary of the Workshop on Coastal Waves, cheyenne.rsmas.miami.edu/showex, , Arlington, Va., Nov.17-18, 1995.

EFFECTS OF BED COARSENING ON SEDIMENT TRANSPORT

Craig Jones[1] and Wilbert Lick[1] (member)

Abstract

As sediments of different sizes are eroded and deposited, the erosion rates of the surficial sediments can change significantly with time due to either coarsening or fining of this layer. The effects of this coarsening/fining on sediment transport are here examined by means of numerical calculations. Examples are flow and transport in a straight channel, in a channel with a rapid expansion, and in the Lower Fox River.

Introduction

Erosion rates of bottom sediments are generally highly variable throughout a sediment bed and can change significantly not only in the horizontal direction (e.g., from soft, cohesive, easily erodible sediments in shallow, near-shore, slow-flowing areas of a river to coarse, non-cohesive, difficult-to-erode sediments in deeper, rapidly flowing parts of a river) but also in the vertical direction (from unconsolidated, easily erodible sediments at the sediment-water interface to very compact, difficult-to-erode sediments at depth). Due to combined erosion and deposition, the properties and especially erosion rates of the surficial layer of a sediment bed can also change significantly with time. For example, if coarse, non-cohesive sediments are deposited on a bed of finer sediments or if the fine fraction of a sediment bed is washed away leaving coarse sediments behind, the erosion rate of the sediment bed can be greatly reduced. This is called bed coarsening or armoring. Conversely, if fine non-cohesive sediments are deposited on a bed of coarser sediments, the erosion rate of the sediment bed can be increased (fining).

In order to accurately determine erosion rates of relatively undisturbed sediments from field and laboratory reconstructed sediments, a unique flume, called Sedflume (McNeil et al, 1996), has been developed and tested which can measure sediment erosion rates (for both cohesive and non-cohesive sediments) as

---

[1]Department of Mechanical and Environmental Engineering, University of California, Santa Barbara, CA 93106

a function of depth in the sediments and at high shear stresses. The models described here are extensions of previous models and have been modified so as to include this information and data from Sedflume.

In the present paper, these models are used to illustrate the process of bed coarsening and its effects on sediment transport. Illustrative examples are those for sediment transport in a straight channel, transport in a channel with an expansion region, and transport in the Lower Fox River.

Bed Coarsening in a Straight Channel

As a simple example to illustrate bed coarsening and its effects, a calculation was made of sediment transport in a steady-state flow in a straight channel including coarsening of the sediment bed. Erosion rates were obtained from Sedflume data corresponding to an average particle size at the sediment surface. Both two-dimensional and three-dimensional models were used in the study in order to illustrate the similarities and differences between the two types of models.

Hydrodynamic and Sediment Transport Models

In the two-dimensional hydrodynamic model (Ziegler and Lick, 1986; Gailani et al., 1991; Lick et al., 1998), the conservation of mass and momentum equations are vertically integrated with the assumption that velocity differences in the vertical are small. This is valid for well-mixed systems. A hydrostatic pressure assumption is used. The conservation equations were solved by an explicit, finite-difference procedure.

Although the water column is assumed to be well-mixed, it is still necessary to solve for the vertical sediment profile since the larger particles may not be evenly dispersed. By assuming a quasi-steady distribution of sediments in the vertical and setting the downward flux due to sediments settling through the water column equal to the upward flux due to turbulent diffusion, the concentration as a function of distance from the bed, C(z), can be obtained and is

$$C(z) = C_B e^{-\frac{w_s z}{\varepsilon_v}} \qquad (1)$$

where C_B is the concentration at the sediment bed (mg/L), w_s is the settling speed (cm/s), and ε_v is the vertical diffusivity (cm^2/s). The diffusivity is calculated from

$$\varepsilon_v = 0.067 h u^* \qquad (2)$$

which is an empirical relationship based on river observations (Fischer et al., 1979), where h is the local water depth in cm and u^* is the shear velocity in cm/s.

The three-dimensional hydrodynamic model used for this study is the time-dependent, estuarine and coastal circulation model of Blumberg and Mellor (1987). A hydrostatic pressure assumption is used in the momentum equation. The conservation of mass and momentum equations were then solved using a finite difference procedure. The vertical diffusivity is assumed constant in the vertical for this case and is calculated from Eq. (2).

Erosion and Deposition Rates

Erosion rates, E, and critical shear stresses for erosion, τ_{ce}, were determined from laboratory measurements with Sedflume (Roberts et al., 1998). In that investigation, sediment cores with particle sizes ranging from 5 to 1350 μm of pure quartz were prepared; E and τ_{ce} were then determined for each particle size for shear stresses from 0.2 to 3.2 N/m^2. In the present calculation, only data for sediments larger than 200 μm were used. These erosion rates are shown as a function of particle size in Fig. 1. In the calculations, it was assumed that E and τ_{ce} were determined by the D_{50} of the top layer of the sediment bed.

The top layer is considered to be a layer of mixing between the depositing and eroding sediments and will be referred to as the active layer. Borah et al. (1982) have defined the maximum thickness of the active layer as twice the largest particle size present. Annular flume and Sedflume observations have shown that this is a reasonable thickness. This active layer is only formed when the original underlying sediments are eroding whether the eroding sediments are cohesive or non-cohesive. When the underlying sediments are not eroding, the deposited sediments simply form a new discrete layer.

Deposition begins below the critical shear stress for suspension, τ_{cs}. The values for τ_{cs} for this case, sediments larger than 200 μm, were determined from

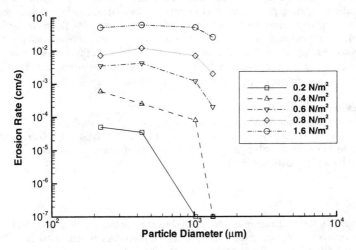

Figure 1. Erosion rates vs. particle size from laboratory studies on quartz particles (Roberts et al, 1998). Results for shear stresses of 0.2, 0.4, 0.6, 0.8, and 1.6 Nm^2 are shown.

Van Rijn's (1984) method for non-cohesive sediments. For each size class of sediments, the deposition rate is calculated from

$$D = p \, w_s \, C \qquad (3)$$

where p is the probability of deposition and w_s is the settling speed of each size class. The probability is 1 when $\tau < \tau_{cs}$, and is 0 for $\tau > \tau_{cs}$. More complex methods have been presented for the calculation of the probability function, but none have conclusively been demonstrated to accurately model deposition (Krone, 1962; Gessler, 1967). C_B is used for the concentration C in the two-dimensional model, while the concentration in the bottom numerical cell is used in the three-dimensional calculation. The settling speed for each sediment size class is determined from Cheng's (1997) formula:

$$w_s = \frac{v}{d}\left(\sqrt{25 + 1.2d_*^2} - 5\right)^{1.5} \qquad (4)$$

where v is the kinematic fluid viscosity (cm^2/s), d is the particle diameter (cm), d_* is the non-dimensional particle diameter calculated from $d_* = d[(s-1)g/v^2]^{1/3}$, s is the specific density (g/cm^3), and g is the acceleration of gravity (cm/s^2).

The final sediment bed flux is the erosion rate minus the deposition rate with a positive flux being from the bed. Although erosion and deposition are calculated separately, it is important to note that in a situation with simultaneous erosion and deposition the deposition of sediments will vary the D_{50} of the active layer. This variation will cause the erosion rate to vary thereby coupling the erosion and deposition. This situation takes place when the shear stress is greater than the τ_{ce} of the bed, as determined by Sedflume measurements, and less than the τ_{cs} of one or more of the suspended sediment size classes.

Model Parameters

In the present calculation, it is assumed that the channel is 1 m wide, 1 m deep, and 10 m long. The grid spacing is 0.5 m and the time step is 0.01 seconds. The flow rate is 0.52 m^3/s; this produces a velocity of 52 cm/s. The shear stress is constant and is calculated from

$$\tau = C_f u^2 \qquad (5)$$

where $C_f = 0.003$; τ is then 0.8 N/m^2. A free slip condition is imposed along the sidewalls of both the three-dimensional and two-dimensional models. The vertical diffusivity is 20 cm^2/s throughout the channel.

The model uses three size classes of sediment with properties as shown in Table 1. The bed has an initial D_{50} of 556.7 μm, an initial erosion rate of 0.01 cm/s, and a τ_{ce} of 0.32 N/m^2. The bulk density of the sediment bed is 1.8 g/cm^3.

Particle Size (μm)	Initial Bed Percentage	w_s (cm/s)	τ_{ce} (N/m^2)	τ_{cs} (N/m^2)
222	33	1.96	0.25	0.28
432	34	4.64	0.3	0.4
1020	33	10.25	0.425	1.7

Table 1. Sediment parameters.

Results and Discussion

Examination of Table 1 shows that, at 0.8 N/m^2, all sediments will erode but only the 1020 μm sediment will deposit. Therefore, sediments will initially erode everywhere while the largest size class will begin to settle out downstream. This downstream deposition causes an increase in the percentage of 1020 μm particles in the active layer and therefore an increase of the D_{50} of the active layer. This causes a decrease in erosion rate as the bed begins to coarsen.

Fig. 2 shows a time-dependent plot of the concentration and particle size of the active layer at the end of the channel. The particle size increases over time until it reaches 1020 μm. At small time, the concentration increases and then decreases rapidly. This is due to the fact that erosion rates are relatively high at small time. As the bed coarsens, the concentration levels off to its steady state value. Fig. 3 shows the erosion and deposition rates at the end of the channel as a function of time. As the D_{50} increases, the erosion rate decreases. Also, as the concentration increases and then levels off, the deposition rate also increases until it is equal to the erosion rate. At steady state, the erosion rate equals the deposition rate.

In the steady state, only sediments in the first 2 m of the channel show continued net erosion. Deposition of coarse sediments is insufficient to coarsen the bed sufficiently so that net erosion ceases. After 2 m, the active layer in the bed has almost fully coarsened to an average particle size of 1020 μm, and the erosion rate then equals the deposition rate.

The downstream steady state concentration at the bed can be simply determined by equating the deposition and the erosion rates for the largest particles ($E=w_sC$). Solving for the concentration of the 1020 μm sediment with an erosion rate of 0.007 cm/s and a settling speed of 10.25 cm/s gives 1,229 mg/L for the concentration at the sediment bed. By integrating the concentration profile in the vertical with 1,229 mg/L as the bottom concentration, an average concentration of 22.7 mg/L is calculated for the 1020 μm sediment. The two-dimensional model also calculates a 22.7 mg/L average concentration for the 1020 μm sediment. This is approximately 1/3 of the total sediment concentration with the 222 μm and 432 μm sediments making up the other 2/3.

A characteristic time may be analytically determined by assuming that the bed fully coarsens when 0.2 cm (active layer thickness) of 1020 μm sediment deposits onto the surface of the bed. Dividing 0.2 cm by the maximum deposition rate

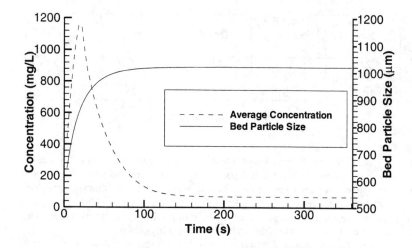

Figure 2. Average sediment concentration and bed particle size as a function of time at a point 10 m downstream for the straight channel bed coarsening example.

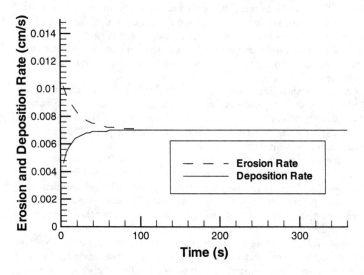

Figure 3. Erosion and deposition rates as a function of time at a point 10 m downstream for the straight channel bed coarsening example.

of 0.007 cm/s gives a characteristic time of 29 s. Fig. 2 shows that 2/3 of the coarsening takes place in about 30 s.

A three-dimensional model was also used with the same parameters to further examine the coarsening phenomena. In this calculation, the coarse sediment size fraction is concentrated in the bottom portion of the water column due to its high settling speed. A critical length for the vertical distribution of the coarse sediment can be analytically determined so that the vertical grid can be constructed with sufficient detail. The exponential describing vertical sediment distribution, Eq. (1), yields a critical length of ε_v/w_s, so that the critical length for this case is 2 cm. To capture the vertical profile, a grid size of 1 cm was used for the first 5 cm from the bed, then a grid size of 5 cm was used for the next 95 cm.

The results for the steady state were similar to those for the two-dimensional case except that an average concentration of 95 mg/L was calculated (37% higher than the two-dimensional case). This is due to the still somewhat coarse 1 cm grid used near the bottom.

For the three-dimensional calculation, the time to steady state is also longer by an order of magnitude than that for the two-dimensional calculation. The three-dimensional model calculates time-dependent diffusion in the vertical instead of assuming an instantaneous steady state profile as in the two-dimensional model. For a depth of 100 cm and a diffusivity of 20 cm²/s, a diffusion time of 250 s is calculated in the vertical ($t_d = h^2/2\varepsilon_v$). This is the same order of magnitude as the coarsening time (250 s), and the vertical diffusion time is therefore primarily responsible for the transient in this case.

Straight Channel with an Expansion Region

To further investigate the effects of coarsening and differential settling of sediments, the flow and transport through a straight channel with an expansion into a larger straight channel was modeled. In this calculation, the fluid shear is first constant and then drops off rapidly in the larger channel; in this expansion, the sediments settle out differentially and are graded according to size. A two-dimensional hydrodynamic model has been used to illustrate this process.

Model Parameters

The narrow part of the channel has a width of 2.5 m and a length of 10 m; this expands into a channel with a width of 8.25 m which extends 30 m downstream. The depth of the entire channel is 1 m. The grid spacing is 0.5 m in the downstream and 0.25 m in the cross-stream and the time step is 0.01 seconds. A flowrate of 2 m³/s was assumed. Shear stresses were calculated from Eq. (5).

The same sediment size classes were used with the same initial percentages making up the sediment bed as in the previous case. The bed has an initial D_{50} of 556.7 μm, an initial erosion rate of 0.01 cm/s, and a τ_{ce} of 0.32 N/m². The bulk density of the sediment bed is 1.8 g/cm³.

Results and Discussion

A flow rate of 2 m^3/s produces a constant velocity of 74 cm/s in the upstream channel; this rapidly falls off to about 23 cm/s in the downstream part of the channel. After the flow exits the upstream channel, it is non-uniform with the higher velocities in the center of the channel and low-speed eddies just to the sides of the exit flow from the upstream channel. Fig. 4 shows a pronounced decline in the shear stress from a value of 1.6 N/m^2 upstream to 0.16 N/m^2 in the downstream region. The shear stress in the upstream region is large enough to erode sediments there, but is also small enough to allow the coarse particles to settle out ($\tau < \tau_{cs}$). The upstream region reaches a steady state condition similar to that in the previous straight channel example. The concentration comes to a constant value of 220 mg/L in the upstream region once the active layer has stabilized to a D_{50} of 1020 μm (Fig. 5).

After the flow moves through the expansion, the velocity and shear stress drop rapidly, and the smaller particle sizes begin to settle out as the shear stress falls below the τ_{cs} for each size class. Fig. 5 shows the D_{50} of the bed decreasing rapidly as the finer size classes deposit just past the expansion. The D_{50} decreases to a minimum value of 255 μm and then begins to increase again to the initial value of the bed (556.7 μm) as all of the suspended material has deposited out. The shear stresses in Fig. 4 are seen to fall well below τ_{ce} about 5 m outside the expansion; there is therefore no further erosion downstream of this point.

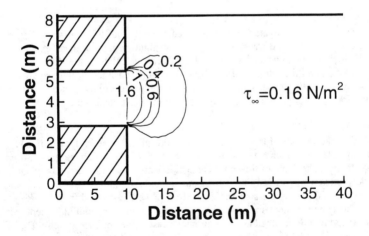

Figure 4. Shear stresses in N/m^2 resulting from a 2 m^3/s flow rate. Downstream shear stress is 0.16 N/m^2.

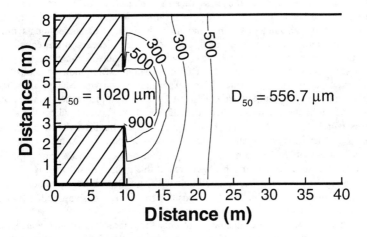

Figure 5. Steady state bed particle sizes in μm. Upstream bed particle size is 1020 μm and downstream particle size is 556.7 μm.

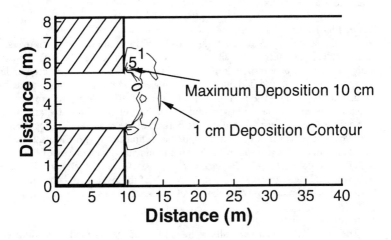

Figure 6. Net change in the sediment bed thickness in cm after 15 minutes of 2 m³/s flow. Maximum deposition is 10 cm and maximum erosion upstream (not shown) is 10 cm.

The net change in bed thickness after 15 minutes (Fig. 6) further reveals the nature of the sediment bed. In the upstream region, there is erosion at the entrance to the channel that is not shown in the plot. Coarse sediments from upstream armor the entrance channel, and erosion and deposition are equalized. As stated earlier, when the shear stress begins to drop outside the expansion, the finer sediment size classes begin to settle out. The most pronounced deposition (10 cm) takes place to the sides of the expansion region where there are small eddies and the shear stresses are lowest. A band of 1 cm deposition can be seen downstream of the expansion. This coincides with the location where the shear stress falls below 0.4 N/m^2 and the 432 μm sediment begins to settle out. The location where the shear stresses fall below 0.28 N/m^2 does not produce a pronounced band of 222 μm particle deposition because the 222 μm particles have been diffused over a much larger area at this point. However, Fig. 5 does show a region where the D_{50} does fall below 300 μm due to the deposition of the 222 μm size class. The thickness of deposition in this region is approximately 0.2 cm.

The present example illustrates not only the importance of bed coarsening but also the importance of differential settling to the formation of a graded sediment bed. Bed coarsening actively decreases erosion rates so that the higher flow regions show only the coarse size class of sediments. The drop in shear stress outside the expansion allows the smaller size classes to deposit out. Without multiple size classes of different transport characteristics, this behavior would not be reproduced.

Lower Fox River

To illustrate an application of the two-dimensional model to a more realistic case, a calculation has been made for the Fox River in Green Bay, Wisconsin. Extensive studies of transport in this river (with erosion rates based on annular flume measurements and without bed coarsening) have been previously made by Gailani et al. (1991). The portion of the Lower Fox River from DePere Dam to Green Bay is approximately 11 km long with a maximum width of 0.8 km. Over 20 Sedflume cores were taken over this portion of the river in 1994 (McNeil et al.,1996). These cores have been compiled in conjunction with EPA and USGS studies to provide an accurate picture of the sediment characteristics of the river. A 280 m^3/s flow rate (a 99.7 percentile flow rate) over the DePere dam was assumed; sediment concentrations at the mouth of the river were compared with USGS data for similar flow rates for model validation.

Sediment Transport

The two-dimensional vertically averaged hydrodynamic model was utilized. The Fox River is a shallow wide river that generally can be considered well mixed in the vertical. The vertical sediment profile was assumed to always be at steady state, and was calculated from Eq. (1).

The initial erosion rates were determined from the Sedflume study. For this test case, the erosion rates were averaged over the depth of the core. These values for erosion rates are utilized until deposition occurs on the top layer of the sediment. If the original sediment is eroding ($\tau > \tau_{ce}$), then an active layer is formed; otherwise the newly deposited material forms a discrete layer. Erosion rates and τ_{ce} for newly deposited sediments are determined from available particle sizes from Sedflume measurements on reconstructed Fox River sediments (Jepsen et al., 1996) and average values from the field study. The erosion rate and τ_{ce} for any newly deposited sediment from 10 to 300 µm are interpolated from these values.

Deposition rates were determined from Eq. (3) and settling speeds from Eq. (4). The τ_{cs} used in the deposition calculation was determined from Van Rijn's (1984) method for sediments larger than 200 µm. For smaller sediments τ_{cs} is assumed to be equal to the τ_{ce} for newly deposited sediments (less than 1 day old). This assumption is supported by observations made by Krone (1962), Mehta and Partheniades (1975), and Van Rijn (1984). All sediment size classes are assumed to be disaggregated particle sizes with no flocculation occurring.

Model Parameters
Grid sizes are 90 m by 30 m. A constant flow rate over the DePere Dam of 280 m^3/s was used. For this flow, the East River, as seen in the downstream portion of Fig. 7, also has an inflow of approximately 28 m^3/s. The shear stress is calculated from Eq. (5); a no-slip boundary condition is imposed along the shoreline and a time step of 2 seconds was used.

The sediment size classes consist of the three sediments shown in Table 2.

Particle Size (µm)	Initial Percentage	w_s (cm/s)	τ_{ce} (N/m^2)	τ_{cs} (N/m^2)
10	By Region	0.057	0.10	0.10
30	By Region	0.508	0.18	0.18
300	By Region	1.680	0.28	0.32

Table 2. Sediment parameters for the Fox River.

A flow with a sediment concentration of 75 mg/L enters the river at both the DePere dam and the East River. These concentrations are typical for this flow rate according to USGS measurements. Since the actual particle size distribution is unknown, it is assumed that this distribution is 45% 10 µm, 45% 30 µm, and 10% 300 µm particles. This distribution is chosen due to the difficulty in keeping significant concentrations of 300 µm sediments suspended over the dam and through the slower flowing East River.

The sediment bed for this case is constructed from two cores obtained in the 1994 survey. The first core was obtained from a fine-grained, generally low flow, depositional region of the river. It has an average disaggregated particle size of 20 µm and was used in the low shear stress regions of the river. The second core was obtained from a high velocity, mid-channel region of the river. It has an average disaggregated particle size of 273 μm and was used to describe the sediments in the high shear stress regions of the river. Both cores had an in-situ τ_{ce} of 0.25 N/m². This is due to the large particle size of the 273 μm core and the cohesive nature of the 20 μm core. The newly deposited sediments, because of their low density, are assumed to have a lower τ_{ce} than the original consolidated Sedflume core shown in Table 2. This is consistent with the Roberts et al. (1998) data for low density, newly deposited sediments.

Results and Discussion

A 280 m³/s flow rate at the upstream DePere Dam produces maximum velocities of 59 cm/s in the narrow portion downstream (Fig. 7). Fig. 8 shows the corresponding shear stresses for this flow rate. In the upstream portion of the river, the maximum shear stress is 0.4 N/m², with much of the mid-channel showing shear stresses between 0.2 and 0.4 N/m². As the river narrows, the flow accelerates producing shear stresses up to 1.2 N/m².

This calculation was run until the sediment concentrations throughout reached a steady state. The time to steady state for this case was 36 hours. Fig. 9 shows the suspended sediment concentrations at steady state throughout the river. In the upstream portion of the river, most of the suspended sediments present are from the flow over the dam. The suspended sediments from the dam have mostly settled out by about 5 km down the river. The concentration begins to rise dramatically in the downstream portion of the river. This is due to the high shear stresses producing large erosion rates in this part of the river. In this region, a maximum concentration of 440 mg/L was calculated. The USGS sampling station is near the center of the downstream end of the river where it flows into Green Bay. At this point and for this flow rate, concentrations from 50 to 100 mg/L were recorded. The present calculation produced a 53 mg/L concentration at the USGS sampling station.

Amounts of erosion and deposition are shown in Fig. 10. In the upstream region, the only erosion was just below the dam and in the first narrow portion of the river. Both of these areas had shear stresses over 0.25 N/m² and a maximum erosion of about 1 cm. The near-shore regions of the wider sections of the river show net deposition and D_{50} up to 30 μm. These areas had shear stresses below 0.1 N/m² and a maximum deposition of 0.6 cm. The high shear stress regions at the end of the river have a maximum erosion of 30 cm. As this section of the river widens towards the mouth at Green Bay, the shear stress drops rapidly to about 0.2 N/m², and the 300 μm particles previously eroded now settle out. This section has a maximum deposition of 25 cm and a D_{50} of 300 μm.

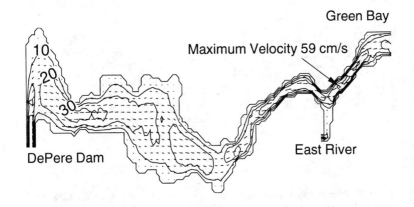

Figure 7. Velocities in cm/s and velocity vectors in the Lower Fox River. Flow rates are 280 m³/s at the DePere Dam and 28 m³/s from the East River.

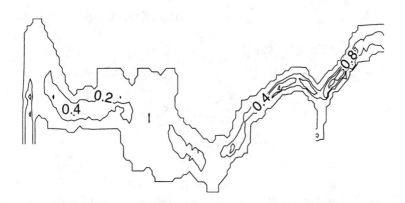

Figure 8. Shear stresses in N/m² on the Lower Fox River. Flow rates are 280 m³/s at the DePere Dam and 28 m³/s from the East River.

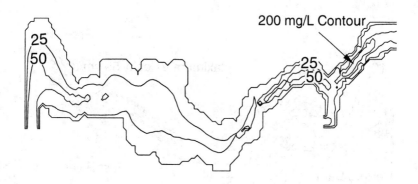

Figure 9. Average suspended sediment concentration in mg/L on the Lower Fox River. Flow rates are 280 m^3/s at the DePere Dam and 28 m^3/s from the East River.

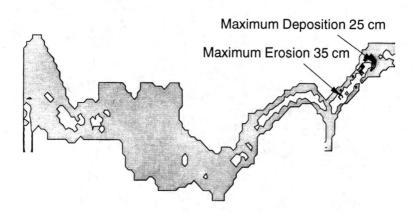

Figure 10. Net change in the sediment bed thickness in cm after 36 hours. Flow rates are 280 m^3/s at the DePere Dam and 28 m^3/s from the East River.

Summary and Concluding Remarks

In natural sediment systems, areas of consistently high shear stress generally show larger average particle sizes at the sediment surface. This is due to the high flow eroding away the smaller particles that can be suspended, and the larger particles, which are unable to be suspended, settling out of the flow. The first two cases presented here serve to illustrate the modeling of this phenomenon. In the straight channel example, a moderate flow rate turns a sediment bed with a 556.7 μm particle size into a bed armored by a layer of 1020 μm particles. This process is illustrated using measured erosion rates for quartz sediments. The time scales for the process can be verified by comparison with physically significant characteristic times. These times are the time to deposit an active layer of 1020 μm sediment when the time for vertical diffusion is relatively short, and the time for vertical diffusion when it is relatively long. The expansion example shows that the model behaves in a realistic manner with more complex geometries. The bed coarsens in the higher shear stress region upstream, and then becomes graded as smaller particles settle out in the expansion region. This grading is what can be observed in realistic cases.

The model was then applied to the Lower Fox River to illustrate a more realistic application. In regions where the flow velocities and shear stresses are high, the model predicts an increase in the average particle size at the sediment surface and therefore decreased erosion rates. In regions where the flow velocities and shear stresses are low, the model predicts finer deposited sediments. These predictions are consistent with observations.

The present models are extensions of previous hydrodynamic and sediment transport models. In general, the modifications include variation of erosion rates horizontally and with depth, different size classes of sediments, bed coarsening, consolidation of newly deposited sediment, and bed load. The present calculations illustrate the effects of bed coarsening on sediment transport. Additional calculations are being made to illustrate the effects of the other processes.

Acknowledgement

This research was funded by the U.S. Environmental Protection Agency.

References

Blumberg, A.F. and G.L. Mellor, 1987, A description of a three-dimensional coastal ocean circulation model, Three-Dimensional Coastal Ocean Models, N. Heaps, Ed., 1-16, American Geophysical Union.

Borah, D.K., C.V. Alonso, and S.N. Prasad, 1982, Routing graded sediments in streams: formulations, *J. Hydr. Engrg., ASCE*, 108(12), pp.1486-1503.

Cheng, N.S., 1997, Simplified settling velocity formula for sediment particle, *J. Hydr. Engrg., ASCE*, 123(2), pp. 149-152.

Gailani, J.Z., C.K. Ziegler, and W. Lick, 1991, The transport of suspended solids in the Fox River, *J. Great Lakes Res.* 17, pp. 479-494.

Gessler, J., 1967, The Beginning of Bedload Movement of Mixtures Investigated as Natural Armoring in Channels, W.M. Keck Laboratory of Hydraulics and Water Resources, California Institute of Technology, Translation T-5.

Jepsen, R., J. Roberts, and W. Lick, 1997, Effects of bulk density on sediment erosion rates, *Water, Air, and Soil Pollution,* Vol. 99, pp. 21-31.

Krone, R. B., 1962, Flume Studies of the Transport of Sediment in Estuarial Shoaling Processes," Final Report to San Francisco District, U.S. Army Corps of Engineers, Washington, D.C.

Lick, W., Z. Chroneer, C. Jones, and R. Jepsen, 1998, A predictive model of sediment transport, *Estuarine and Coastal Modeling,* pp. 389-399.

McNeil, J., C. Taylor, and W. Lick, 1996, Measurements of erosion of undisturbed bottom sediments with depth, *J. Hydr. Engrg., ASCE,* 122(6), pp. 316-324.

Mehta, A.J. and E. Partheniades, 1975, An investigation of the depositional properties of flocculated fine sediments, *J. Hydr. Res.* 12(4), pp. 361-609.

Roberts, J., R. Jepsen, D. Gotthard, and W. Lick, 1998, Effects of particle size and bulk density on erosion of quartz particles, *J. Hydr. Engrg., ASCE,* 124(12), pp. 1261-1267.

van Rijn, L.C., 1984, Sediment transport, Part II: Suspended load transport, *J. Hydr. Engrg., ASCE,* 110(11), pp. 1612-1638.

Ziegler, C.K., and W. Lick, 1986, A Numerical Model of the Resuspension, Deposition, and Transport of Fine-Grained Sediments in Shallow Water, Univ. of Ca., Santa Barbara Report ME-86-3.

Mixing in a Small Tidal Estuarine Plume

M. Pritchard [1], D.A. Huntley [1] & T.J. O'Hare[1]

Abstract

Features inside a radially spreading river plume discharge in the English Channel are compared to those predicted by a simple generic model that simulates radial spreading from a constant source. X-band radar imagery used to scale model output predicted the presence of an internal bore lagging the surface front of the plume by approximately 200m. High-resolution measurements of temperature, salinity and velocity recorded inside the plume outflow found no evidence of the internal bore in its predicted position. However, temperature contours showed an abrupt shallowing of the interfacial region some 40m behind the surface front. Super-critical Froude numbers present in this region of the plume suggest the bore may form closer to the leading front than predicted by the model. Field measurements show the initial value chosen for the mixing parameter (β) included in the frontal boundary condition is underestimated by about a factor of 2. The required increase in β reduced the lag length scale of bore formation to one where critical Froude numbers were detected inside the plume. Further estimates of entrainment velocity in the frontal region showed convergence and downwelling at the surface front that is followed by a region of upwelling. A 'caterpillar track' type circulation is thought to be responsible for mixing and entrainment in the head region of the plume. This head remains a reasonably constant size due to any increase in across frontal velocity over the ebb tidal cycle being matched by increased entrainment and mixing.

Introduction

River plume dynamics have been studied in the past through extensive modelling exercises (see O'Donnell, 1993). However, as yet, few observational campaigns have provided the high-resolution data necessary to resolve the internal features and mixing hypothesised by their output. Recent field experiments made by O'Donnell *et.al.* (1998) observed convergence and downwelling in the frontal zone of a plume outflow. Field results are compared with a model proposed by Garvine (1974) which examined small-scale frontal dynamics. The results presented by O'Donnell *et.al.* (1998), although interesting, failed to quantify mixing parameters in the frontal region. However, mixing is thought to be intense and site specific to the plume front. Lukitina & Imberger (1989) using a high-

---

[1] Institute of Marine Studies, University of Plymouth, Drake Circus, Plymouth, Devon, PL4 8AA, UK.

resolution data set identified a highly turbulent regime within the frontal region, but their instrumentation was static and could not follow frontal development and mixing over a tidal cycle.

To fully understand mixing and dispersion of freshwater plumes and thereby aid model development, detailed spatial and temporal field measurements are required to quantify boundary conditions and parameterise mixing. This will hopefully encourage improvements in plume model boundary conditions used to forecast the dispersion of conservative pollutants and sediment in the coastal zone. The aim of this paper is to present some results from recent field measurements and a modelling exercise that examined the sensitivity of a radial spreading river plume model to *in situ* derived mixing parameters.

Theoretical Background

The model chosen for this exercise was one developed by Garvine (1984). This effectively simulated a radial spreading river plume unbound by coastal features. After an initial start up, the source flow was assumed constant at all times, Coriolis force and wind-stress were ignored and the receiving coastal water was assumed stationary. The method of characteristics was employed to solve the non-linear wave dynamics of the unsteady flow. Full model derivation and methods of finding solutions may be found in Garvine (1981, 1984).

The model basically sets two regimes, one of steady non-critical flow, where the Froude number, *Fr* is <1 and the second as a front boundary where *Fr* >1 as shown in classical hydraulic theory on hydraulic jumps and bores (Stoker, 1957). The governing equations of the model are:

mass:
$$\frac{\partial c^2}{\partial t} + \frac{1}{r}\frac{\partial}{\partial r}\left(rc^2 q\right) = 0 \qquad (1)$$

momentum:
$$\frac{\partial q}{\partial t} + \frac{\partial}{\partial r}\left(\frac{q^2}{2} + c^2\right) = 0 \qquad (2)$$

r = Radial distance from spreading source (m)
t = Time From release (s)
c = Phase speed of internal wave on the interface = $\sqrt{g'd}$ (ms⁻¹)
g' = Reduced gravity $(\Delta\rho/\rho)g$ (ms⁻²)
d = Buoyant layer depth (m)
q = Fluid velocity (ms⁻¹)

The mass continuity and momentum equations (1&2) are hyperbolic with a pair of characteristic lines given locally by:

$$\frac{dr}{dt} = q \pm c \qquad (3)$$

where a change in radius with change in time is equal to radial water velocity $\pm$ long internal wave speed. The corresponding characteristic equations along the lines given by (3) are:

$$dq \pm 2dc = \mp \frac{cq}{r} dt \qquad (4)$$

The $\pm$ families represent non-linear internal gravity waves that propagate upstream and downstream respectively, at phase speed c relative to the buoyant layer and at an absolute wave speed $q \pm c$ relative to fixed co-ordinates. The model is non-dimensionalised by introducing scaled variables using q^*, d^* and r_0 as a reference velocity, depth and length scale (schematic diagram is shown in Fig 1):

$$R \equiv r/r_0 \qquad\qquad T \equiv q^* t / r_0$$
$$Q \equiv q / q_0 \qquad\qquad C \equiv c / q^*$$
$$D \equiv d / d^* = C^2 \qquad\qquad F = Q / C$$

The frontal boundary conditions are derived in Garvine (1981). He assumes that the magnitude of the vertical entrainment velocity, q_e (see equation 8) is linearly proportional to the relative velocity of the lower layer, Q_a and decays exponentially away from the front towards the source. The vertical direction of entrainment may be selected as positive (upward), zero or negative (downward). For the limit of a surface front where the buoyant layer depth tends to zero, the frontal 'jump' boundary conditions are:

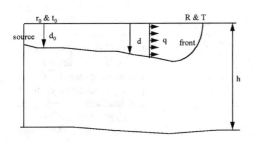

Fig 1: A schematic diagram of the radial spreading plume model (adapted From Garvine. 1984).

$$Q_a = F_{al} \, C_f \tag{5}$$

$$Q_f = F_{al} \, (1 - S_e \beta) C_f \tag{6}$$

Where:

Q_f = Fluid speed behind the front
C_f = Phase speed behind the front
Q_a = The speed of the front relative to the ambient water.

$F_{al} = [2\beta(\tilde{d} + S_e^2 \beta)]^{-1/2}$ = The Froude number based on Q_a and C_f
β = Positive dimensionless constant that indicates the decay scale or fraction of the front where significant turbulent mass exchange (entrainment) and momentum exchange (interfacial friction and entrainment of momentum) occurs.

$\tilde{d}$ = Positive constant that specifies the ratio of interfacial friction and entrainment coefficients.
S_e = Entrainment factor i.e. 0 = No entrainment; 1 = upward entrainment: -1 = downward entrainment.

The Froude number that describes the state of flow behind the front, $F_f = Q_f / C_f = F_{al} (1 - S_e \beta)$ is held constant. Equations (5) and (6) are the two jump conditions of Garvine (1981) adapted to describe the mass continuity and horizontal momentum balance in the frontal region of a radially spreading plume. These are combined with the plus family branch of (4). This relates, Q_f and C_f to the interior flow state Q and C, at the point where the plus or downstream propagating wave of (3) passes through the front and intersects the previous R step. Stoker (1957), Garvine (1981, 1984) and O'Donnell (1993) provide full discussions of the derivation of boundary conditions.

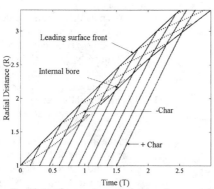

Fig 2: Plot of sample calculation from model in dimensionless form. Leading surface front and internal bore are shown with respect to R-T space and time. $\beta = 0.15$; $F_{al} = 2^{1/2}$ $S_e = -1$.

Results from model runs showed the formation of an internal bore behind the leading front. As the method of characteristics cannot solve discontinuities in a variable properly, Garvine (1984) fitted a second jump condition. Garvine (1984) suggests the internal bore forms as a result of internal waves that propagate downstream from the source coalescing with upstream propagating waves that are reflected from the leading front. A plot of the characteristics plane as shown in Fig 2, describes a steady-state radial spreading regime that is preceded by a radial ring where the leading edge forms the front and the trailing edge an internal hydraulic jump or bore.

Our field experiments discussed in the following section were designed to test the hypothesis of a trailing internal bore that forms behind a leading surface front in a small radially spreading river plume. In addition, the mixing in the plume is quantified, as previous model sensitivity tests showed the magnitude of the mixing parameter is an important factor in determining the existence and / or position of the internal bore.

Field Site

Fig 3 shows the site chosen to compare observational data to the radial spreading model was the Teign Estuary, Teignmouth, Devon, U.K. The most physically pronounced plumes are discharged from the estuary into Lyme Bay between October and March in response to high precipitation. A minimum threshold runoff rate of approximately 12 m^{3}s^{-1} seems to be the lower limit for a definable plume to form. When runoff is sufficient, a radial shaped brackish water plume, roughly symmetrical about the mouth develops on the ebb tide. Plume formation and spreading was captured in a series of X-band radar images recorded in 1995. A radar station located on the

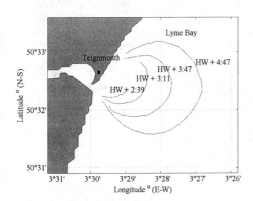

Fig 3: Map of study area including projections of plume development as estimated from X-band radar imagery. Imagery was made available courtesy of M. Tenorio and N. Ward, SOC.

shoreline effectively mapped the developing plumes frontal boundaries and produced a time series of the developing plume on 2 ebb tidal cycles. Thus, by comparing time stack images of plume area development through an ebb tidal cycle it was possible to estimate the rate of radial spreading. This speed was then used to 'seed' the radial spreading model through the scaling arguments. Spatial maps of the plume development through time are shown in Fig 3. Colour contrast due to sediment load and damped sea surface roughness were present inside the confines of the plume front.

The tidal regime at Teignmouth is semidiurnal with a mean range at the mouth of the estuary of 3.5 m. The local coastal waters receiving the discharge have low amplitude recti-linear (NE-SW) tidal currents rarely exceeding 0.25 ms^{-1} and 0.12 ms^{-1} on Spring and Neap tides respectively. Initial formation of the front appears due to a converging coastal tidal current and outflow from the estuary. Radial spreading between the 2nd and 4th quarter of the estuary's ebb cycle is relatively unaffected by the coastal tidal currents. Small residual currents have little impact on plume development and spreading dynamics. Estimates of the Kelvin number as defined by O'Donnell (1993) showed that Earth's rotation is unimportant near to the estuary mouth.

Initial model conditions were scaled to radial spreading rates calculated from the X-band radar imagery, CTD casts and velocity measurements recorded in 1995. Regression analysis of incremental increases in area calculations made from X-band radar imagery gave a steady state critical radial flow velocity of $q* = 0.13$ms^{-1} and we set $Q_0 = 1.2$ and $F_{al} = 2^{1/2}$ as in Garvine (1984). $S_e = -1$ was chosen from field observations made by O'Donnell $et.al.$ (1998). Simpson's (1982) estimate of

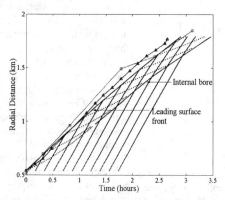

$\beta = 0.15$ the dimensionless mixing coefficient, is derived from laboratory experiments. β describes the ratio of the mean speed of gravity current propagation to the overtaking speed of fluid into the head. In context of the model any increase in β intensifies and spatially lengthens the mixing scale in the frontal zone.

Fig 4: A comparison between modelled characteristics plane and observed radial spreading rates from the 3rd and 7th March 95. □ = Spreading rate 3rd March 95; Δ= Spreading rate 7thMarch. β = 0.15; F$_{al}$= 2$^{1/2}$ S$_e$ = -1.

Fig 4 illustrates the numerical output as a characteristics plane plot scaled to plume spreading rates calculated from X-band radar imagery. For the two-recorded spreading events on the 3rd March 1995 and the 7th March 1995, good agreement is shown between the observed and modelled rate of spreading. The scaled model output predicts the formation of an internal jump or bore lagging the leading front by a distance of some 200 m.

Fig 5 illustrates the predicted interfacial depth scaled by CTD profiles recorded in the plume near the source   (d^*=2.5 m). The abrupt shallowing of the interface from the deeper head region to the sub-critical plume layer behind the internal jump has a predicted length scale of approximately 30 m.

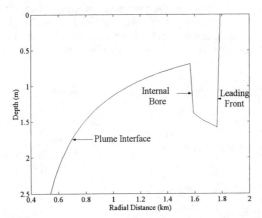

Fig 5: Modelled output of the internal interfacial features that show the formation of an internal hydraulic jump 200m behind the leading surface front. $\beta = 0.15$; $F_{al} = 2^{1/2}$ $S_e = -1$.

Field Study & Results

To test the model hypothesis of internal bore formation and the magnitude of mixing in the frontal zone we carried out multiple survey transects on several tidal ebb spreading cycles of the Teign plume. We used a high frequency (8Hz) logging instrument package consisting of the Estuarine Thermistor Spar (ETS), a vertical array of 15 Fastip thermoprobe thermistors fitted to a 4m aluminium tube (Sturley & Dyer, 1990). On the ETS we mounted 2 Valeport CT probes, 3 Valeport electro-magnetic current meters and Druck pressure transducers. The ETS was deployed from a small boat and used to survey the plume in the autumn of 1997 and 1998 during which a Trimble DGPS with local correction was used to calculate position. All instrument records were subject to 1 Hz low-pass filtering. Spectral coherence filtering between pressure tranducers and all the deployed instruments was used to remove the signals due to surface waves and boat movement. Seasonal temperature contrast between coastal water and the cooler river outflow water was used as the predominant tracer of water type (temperature/salinity correlation: R^2= 0.95). Contoured water column temperature from several across frontal transects recorded through temporal and spatial development of the plume is presented in Fig 6a to 6c. The temperature profiles through the plume outflow reveal the abrupt contrast in the temperature signal of the plume and ambient seawater. Spatially, temperature and salinity changes were detected within a few metres when entering the plume. This abrupt frontal convergence was followed by a steeply descending isotherm layer that extended to a depth of 2 –3m. Fig 6b and 6c show no evidence of an internal bore 200 metres behind the leading front, as was the case in all the longer frontal transects. However, the contoured isotherms did appear to ascend to a shallower depth approximately 40m behind the leading front. This appeared to conform to the form of gravity head feature as observed in the experiments of Britter & Simpson (1978). This head feature was evident in nearly all our observations.

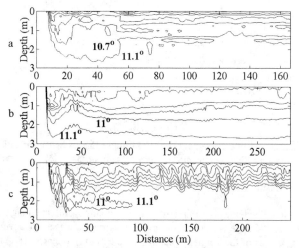

Fig 6a-Fig 6c: Temperature (°C) versus spatial distance (m) contour plots from thermistor outputs of 3 plume front transects. Plots show the deepening of the interface behind the leading front and formation of the gravity head.

Development of flow dynamics inside the plume is illustrated through four sets of across frontal velocity profiles in Fig 7a to Fig 7d. Velocity measurements recorded along each transect at 3 levels (0.95m, 1.85m, 2.9m) in the water column were grouped specific to a time and spatial locality after local high water. The individual transects in a group were then spatially aligned with respect to each other and averaged together. This produced a mean flow for each level in the water column at a progressive distance from the source and time after local high water.

Plate 1. Photograph of spreading plume and convergence line of the Teign plume outflow on the 25th November 1998.

Negative flow (-ve) represents flow towards the front with respect to ambient seawater (non-plume water). The velocity data shows significant flows ranging from 0.1ms^{-1} to 0.25 ms^{-1} towards the front at the 1m level. Below this level flow is less intense but still forced towards the front. In the instance where flow is towards the front at all depths, the observations suggest a deeper surface layer extending beyond the depth of the lower current meter.

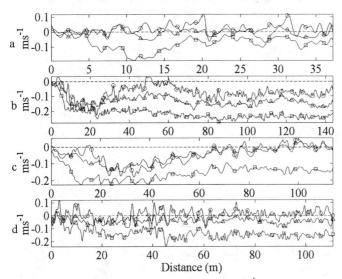

Fig 7a-Fig 7d: Group averaged frontal velocity profiles (ms^{-1}) versus distance (m) of time developing flow at □ = 0.95m; Δ = 1.85m O = 2.9m entering the plume in the first 5m on the left-hand side of the diagram. Fig 7a –HW + 2:42; Fig 7b-HW + 3:36; Fig 7c-HW + 5:12; Fig 7d-HW + 5:42.

The most intensive flow recorded in these observations is mid ebb corresponding to maximum outflow from the estuary and then slightly subsides towards the end of the plume spreading cycle. The across frontal flow was convergent with a distinct foam line present at the boundary as shown in Plate 1. Steeply sloping isotherms indicated the abruptness of the convergence with frontal slopes ranging from 0.13 - 0.47.

To quantify hydraulic jump conditions for model comparisons we calculated the densimetric Froude number, Fr:

$$Fr = \frac{u}{\sqrt{g'h}} \tag{7}$$

u = Surface layer velocity relative to ambient water (ms^{-1})
g' = Reduced gravity ($\Delta\rho/\rho$)g (ms^{-2})
h = Depth of surface layer (m)

Critical flow ($Fr \geq 1$) was evident on immediate encounter with the frontal region, symbolic of a typical hydraulic jump. Interestingly, the sites of critical flow were very localised as shown in Fig 8a – Fig 8c.

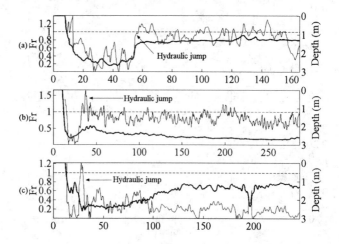

Fig 8a–Fig 8c: Froude number versus spatial distance for the 3 temperature contour plots shown in Fig 6a to Fig 6c. Interface depth (thick line) is plotted to indicate the position of the gravity head.

A running average smoothing of the data gives a spatial resolution of the critical flow sites as approximately 2-3m wide at the leading front. The quiescent sub-critical flow region often present behind the leading front was considered a result of the deepening of the pycnocline that subsequently increased the interfacial wave speed thus reducing the Fr number in the head.

Sharp peaks appeared in the Fr spatial series inside the plume. The position of the peaks and suggestion of jump conditions corresponded to the region behind the apparent gravity head illustrated in Fig 6. Also, the fluctuating critical / sub-critical Fr numbers seen inside the plume in Fig 8a and Fig 8b suggest the presence of multiple hydraulic jumps inside the plume. The main internal jump appeared inside the plume much closer to the surface front than predicted by the model.

To quantify the downwelled flux of brackish water at the front we calculated the vertical entrainment velocity, q_e, for each respective frontal transect. Fig 9 illustrates an example of the observed vertical entrainment velocity, q_e, calculated in a moving reference frame (overbar) from continuity.

$$\frac{\partial}{\partial \overline{x}}\left(\overline{u}D\right) = q_e\left(\overline{x}\right) \tag{8}$$

$\overline{x}$ = Distance from the front (m) $\overline{u}$ = Velocity relative to the front (ms^{-1})
D = Depth of interface (m) q_e = Entrainment velocity (ms^{-1})

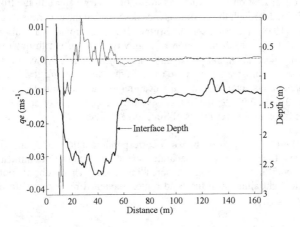

Fig 9: Entrainment velocity (ms^{-1}) versus distance (m) for the temperature transect shown in Fig 6a. Interfacial depth is plotted to indicate regions of down and upwelling.

When encountering the front, brackish surface water is initially forced down at a rate of up to 0.05 ms^{-1} emphasising the dynamic nature of the foremost convergent zone. Immediately behind the convergence observations indicated a smaller upwelled flux of water inside the gravity head. This suggested the previously downwelled water at the front was being entrained back into the plume in and behind the gravity head.

Localised entrainment coefficients, E and mixing coefficient, β were estimated within the realm of the leading front. The results presented are mean values calculated in the region of the vertically sloping interface i.e. from the surface to maximum horizontal depth of the head. E, as defined by Ellison and Turner (1959), Garvine (1981) and Huzzey (1982) describes the ratio of entrainment velocity to interfacial shear velocity.

$$\frac{q_e}{\overline{u}_a} = E \qquad (9)$$

$\overline{u}_a$ = Velocity of front relative to ambient water (ms^{-1})
E = Entrainment coefficient

Further quantification of mixing and influx of fresher water into the frontal region was estimated through the β parameter as described in the experiments of Britter & Simpson (1978) and Simpson & Britter (1979). β provides a bulk estimate of mixing in the gravity head region of a gravity flow current through the ratio of overtaking speed to frontal propagation speed. Estimates of β were calculated for each cross frontal transect from the cross frontal velocity profiles. Values derived from calculations centred around the head region ranged from 0.08 to 0.83 with a mean of 0.37 ± 0.07 (95% C.I.) This value is approximately double that quoted from the laboratory experiments of Simpson & Britter (1979).

Fig 10 shows a comparison between corresponding values of E and β calculated for survey transects from 2 separate spreading cycles. Entrainment increased linearly with mixing through a spreading event. These observations suggest mixing and entrainment in the head region remained proportional to the outflow velocity.

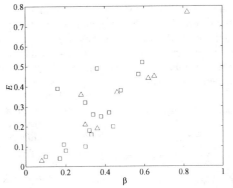

Fig 10: Entrainment coefficient (E) versus mixing coefficient (β). $\square = 25^{th}$ Nov 98; $\Delta = 26^{th}$ Nov 98.

Fig 10 shows a comparison between corresponding values of E and β calculated for survey transects from 2 separate spreading cycles. Entrainment increased linearly with mixing through a spreading event. These observations suggest mixing and entrainment in the head region remained proportional to the outflow velocity.

Thus in context of a constant Fr, if g' decreases and $h \approx$ constant, then u the surface layer velocity must increase. So to maintain volume continuity in the head of the plume outflow velocity must increase, as our observations suggest.

Discussion

The results from the observational data presented in this paper show discrepancies between the theoretical and observed internal features of radially spreading river plumes. Temperature and velocity records failed to identify an internal hydraulic jump in or around the predicted site. However, consecutive temperature profiles made inside the plume outflows did suggest the presence of the gravity head as observed in laboratory simulations (Britter & Simpson, 1978).

Group averaged velocity profiles near to the front illustrate the development of flow inside the plume thorough a spreading cycle. Flow towards the front increased as the ebb tide from the estuary intensified. This eventually subsided towards local low water. The convergence zone at the front was the site of persistent downwelling where surface water was subducted down in to the water column. The mean estimated entrainment coefficient in the frontal zone for the 2 surveys discussed was 0.28, slightly greater than those found by Huzzey (1982). This region was marked by a narrow vertical band of super-critical flow followed by a region of deeper sub-critical flow. Repeated observations showed super-critical Fr numbers behind the head region as well as an upward component of velocity. Super-critical flow and upwelling possibly correspond to a jump as predicted in the Garvine (1984) model. Putting results into context of the modelled flow, we found that the leading front advanced with a Fr number $(O)1.3$, the same order of magnitude as selected for the modelled runs. However, the initial mixing coefficient $\beta = 0.15$, or the extent of turbulent exchange used in the frontal region of the model, was under estimated. Our field observations showed β to be of order 0.37, which is slightly higher than the maximum value recently calculated by O'Donnell (1997) in the Connecticut plume. A positive increase in β basically intensifies and lengthens the region of turbulent exchange following the leading front. Thus as β increases F_f increases. If Q_f remains constant then through Equation 6, C_f decreases. Physically, the increased turbidity and mixing decreases g' and interfacial wave speed 'c'. Hence, Q_a the resultant flow velocity at the front relative to ambient water is reduced through the relationship in Equation 5. The reduction in propagation speed reduces the spatial scale where downstream and upstream interfacial waves coalesce and the internal jump thus forms nearer to the leading front as energy is dispersed through turbulent kinetic energy which fuels local mixing rather than interfacial wave generation. To maintain the same spreading rates as observed in the X-Band radar imagery requires an increase in flow speed behind the front to main continuity. This would compensate for the extra loss in mass due to the elevated value of β. Fig 11 shows how a sequential increase in β and Q retains the speed but reduces the lag distance of the internal jump to some 50 m behind the leading front. This distance is the same order to that of the rear of the head region in the Teign plume where we observed super-critical Fr values.

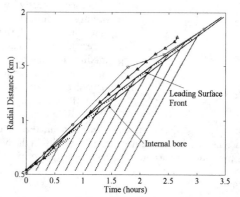

Fig 11: Characteristics plot of plume spreading with modified mixing coefficient (β) and an increase in the initial release speed Q_0. O = Spreading rate 3rd March 95; Δ= Spreading rate 7th March 95.

These results may indicate the presence of rotational circulation in the front /head region as observed by Lukitina & Imberger (1987). As the moving head of the gravity current advanced, brackish water from the surface layer was entrained down at the relatively steep sloping interface of the convergence zone. Downwelled water is then swept back under the plume but is not fully mixed thus retaining some of its buoyancy and thus begins to rise. This upward flux combined with the mean forward motion of the plume creates shear, producing instabilities possibly akin to the Kelvin-Helmoltz billows observed in the laboratory by Britter and Simpson (1978). Brackish water that is not fully mixed then begins to rise finding stability at some equilibrium level. This component or parcel of water would eventually be convected towards the front by the surface outflow current. Thus mixing in the frontal region may be conceptualised as a 'caterpillar track' type process where the mixing zone is conveyed outwards from the source leaving a diluted oscillating wake in the general body of the plume. At the end of the ebb cycle the source flow is terminated and effectively cuts off the mixing mechanism.

To conclude, the current study has presented a comparison between theoretical concepts and observational data of river plume dynamics. The numerical methods employed to explore internal mixing take a very simplistic approach to a complex system that develops in time and space. Entrainment and mixing in the front appears more intense than anticipated from laboratory experiments and remains proportional to outflow velocity. The distance that the predicted internal jump forms behind the leading front is considerably less when mixing (β) is increased in the leading frontal boundary. An increase in initial outflow velocity is thus required to maintain a similar spreading rate to that predicted from the X-band radar.

The physical description of mixing in the front proposed from our results requires further verification from field studies and the possibility of multiple jumps created by this mixing mechanism is not modelled by the method of characteristics and would require a different modelling approach.

Acknowledgements

The presented work was supported by a Plymouth University and PERC studentship. The authors would like to thank Miguel Tenorio and Nick Ward of Southampton Oceanographic Centre for the use of their X-band radar imagery.

References

Britter R.E. & Simpson J.E., 1978. Experiments on the dynamics of a gravity current head. *J. Fluid Mech.* 88, 223-240.

Ellison T.H. & Turner J.S., 1959. Turbulent Entrainment in Stratified Flows. *J. Fluid Mech.* 6, 423-448.

Garvine R.W. 1974. Dynamics of Small Scale Oceanic Fronts. *J. Phys. Oceanogr.* 4, 557-569.

References continued

Garvine R.W. 1981. Frontal jump conditions for models of shallow, buoyant surface layer dynamics. *Tellus*. 33, 301-312.

Garvine R.W. 1984. Radial Spreading of Buoyant, Surface Plumes in Coastal Waters. *J. Geophys. Res.* 89, 2, 1989-1996.

Huzzey L.M. 1982. The Dynamics of a Bathymetrically Arrested Estuarine Front. *Estuarine, Coastal and Shelf Science*. 15, 537-552.

Luketina D.A. & Imberger J., 1987. Characteristics of a Surface Buoyant Jet. *J. Geophys. Res.* 92:5, 5435-5447.

Luketina D.A. & Imberger J., 1989. Turbulence and Entrainment in a buoyant surface plume. *J. Geophys. Res.* 95, 12,619-12636.

O'Donnell J., 1993. Surface Fronts in Estuaries: A Review
Estuaries, 16:1, 12-39.
References Continued

O'Donnell J., 1997. Observations of near-surface currents and hydrography in the Connecticut River plume with the surface current and density array. *J. Geophys. Res.* 102:11, 25,021-25,033.

O'Donnell J., Marmorino G.O. & Trump C.L., 1998. Convergence and Downwelling at a River Plume Front. *J. Phys. Oceanogr.* 28, 1481-1495.

Simpson J.E & Britter R.E., 1979. The dynamics of the head of a gravity current advancing over a horizontal surface. *J. Fluid Mech.* 94, 477-495.

Simpson J.E, 1982. Gravity currents in the laboratory, atmosphere and ocean. *Ann. Rev. Fluid Mech.* 14, 213-234.

Stoker J.J. 1957. Water Waves. Interscience, New York, USA.

Sturley D.R.M., Dyer K.R., 1990. The Estuarine Thermistor Spar: an instrument for making thermal profiles in shallow water. *The Hydrographic Journal*, 55:13-21.

Improving Coastal Model Predictions through Data Assimilation

J. Craig Swanson[1] and Matthew C. Ward[1]

Abstract

The application of numerical models and the collection of remotely sensed and in-situ data in the coastal environment has dramatically increased over the past few decades. The current state of data-model coupling is to calibrate the model to local conditions through both qualitative and quantitative comparisons between data observations and model predictions. The application of data assimilation techniques has the potential to improve model accuracy and provide guidance on the efficiency of the monitoring effort in terms of instrument placement and sampling frequency.

The current application focuses on the temporal and spatial assimilation of temperature data, collected in Mt. Hope Bay, MA, from a series of 30 thermistor chains, for the period of May-June 1997 into the WQMAP hydrodynamic and water quality modeling system. Assimilating the complete in-situ temperature record resulted in a reduction of the average root mean square error by 80%, over the case of no data assimilation. However, further tests show that by strategically choosing five spatially separate thermistor records for assimilation the average RMS error is reduced by as much as 50%.

The predictive capability of the data assimilation scheme was also tested where the thermistor data was continuously assimilated for one day only. This resulted in a two-stage response. The first was an initial rapid increase of the RMS error on the order of 12 hours. The second was a longer term increase of the RMS error towards that of the case of no data assimilation on the order of days.

[1]Applied Science Associates, Narragansett RI, 02882

Introduction

The increase of long-term coastal monitoring programs provide unique opportunities for the development of techniques to assimilate these data into coastal modeling systems. Both remotely sensed (from aircraft and satellite) and in-situ (from moored or moving platforms) data sets offer opportunities. The combined model-data assimilation platform has the potential not only to increase the model performance but also to provide insight and guidance for the monitoring program itself. We have addressed these issues through the assimilation of long term in-situ temperature measurements from Mount Hope Bay, MA, into the WQMAP hydrodynamic and water quality modeling system (Mendelsohn, et. al, 1995). The in-situ temperature data was collected to characterize the thermal discharge from the Brayton Point Station (BPS) powerplant. The overall project goal was to determine the plant's effects upon the bay through a long-term thermal mapping and modeling program.

In this paper, we first present a description of Mt. Hope Bay, the thermal mapping and modeling program, and the baseline numerical simulation where no data assimilation was performed. This is followed by a presentation of the data assimilation methodology. The technique is then applied to several spatial and temporal data assimilation variations. The results of these variations are compared with the in-situ thermistor data and the baseline numerical simulation.

Mount Hope Bay

Mt. Hope Bay is a shallow water estuary with a mean depth of 5 m located on north-eastern Narragansett Bay and extends across the state boundary between Rhode Island and Massachusetts (Figure 1). The bathymetry is characterized by a broad shallow area on the north-western side of the bay and a 120 m wide, 10.5 m deep dredged channel progressing from near the mouth of the bay (at the Mt. Hope Bridge) to the lower reaches of the Taunton River. The Taunton River is also the major source of fresh water having a mean discharge of 29.7 m^3/s.

Circulation in the area is dominated by the semi-diurnal tide with a mean range of 1.34 m and current speeds in the range of 10-25 cm/s (Spaulding and White, 1990; Rines and Schuttenberg, 1998). Winds play an important role in mixing in the shallow area on the western side of the bay and density induced circulation is important in the lower Taunton River and the area immediately around its mouth during high flow periods. Stratification in temperature, particularly in the northern reaches of the bay, is observed due to the effects of the cooling water discharges from the BPS, and enhanced solar heating in the shallow northern reaches. The BPS powerplant (1600 MW) is located on the northern shore of the bay with water intakes on the northeastern and northwestern side of the facility and a discharge, through a venturi structure, on the southern tip of the site. The flow rate and temperature rise vary depending on the number of cooling units in operation but are typically 40 to 50 m^3/sec and 5 to 8° C.

Mt. Hope Bay Thermal Mapping and Modeling Program

PG&E Generating contracted Applied Science Associates (ASA) to characterize the thermal discharge from the BPS powerplant and to determine its effect upon the bay. This required seasonal thermal monitoring to map the plume in three dimensions and the development of a calibrated seasonal hydrothermal model. At this time, four thermal mapping studies have been completed for the periods of May-June 1997, August-September 1997, September 1998 and February-March 1999. Calibrated hydrothermal models have also been developed for the latter three periods (Rines and Schuttenberg, 1998; Rines, 1999; Swanson, et. al., 1998). The model used in this study was that calibrated to the August-September 1997 data set. Tables 1 and 2 presents a summary of temperature calibration statistics computed at three locations within Mt. Hope Bay, Borden Flats, Brayton Point and Gardners Neck (Figure 1).

Table 1. Statistics of temperature observations and model predictions for model calibrated to August-September 1997 data set.

Location	Observations			Model Predictions		
	Mean (°C)	Standard Deviation (°C)	Coefficient of Variation (%)	Mean (°C)	Standard Deviation (°C)	Coefficient of Variation (%)
Borden Flats - Surface	23.96	0.86	4	23.25	0.82	4
Borden Flats - Bottom	23.05	0.75	3	23.27	0.60	3
Brayton Point - Surface	24.54	0.89	4	25.87	1.95	8
Brayton Point - Bottom	23.56	0.79	3	23.99	0.90	4
Gardners Neck - Surface	23.97	0.74	3	25.50	1.10	4
Gardners Neck - Bottom	23.32	0.72	3	25.50	1.10	4

Table 2. Statistical comparison of temperature observations and model predictions for model calibrated to August-September 1997 data set.

Location	Relative Mean Error (%)	Root Mean Square Error (°C)	Correlation Coefficient	Error Coefficient of Variation (%)
Borden Flats - Surface	3	0.90	0.79	4
Borden Flats - Bottom	1	0.56	0.73	2
Brayton Point - Surface	5	2.08	0.58	8
Brayton Point - Bottom	2	0.75	0.74	3
Gardners Neck - Surface	6	1.73	0.66	7
Gardners Neck - Bottom	9	2.42	0.39	10

The thermistor-based thermal mapping studies consisted of thermistor strings distributed radially around the powerplant's outfall (Figure 2) and selected sites elsewhere in the bay. This distribution provides high resolution in the area surrounding the outfall and decreases in resolution toward the mouth of the bay. Each study used approximately 30 buoy-mounted thermistor strings with four to six thermistors per string, located at depths of 0.25, 0.5, 1, 2, 4 and 6 m, recording temperature at five minute intervals.

The modeling platform used for these studies was WQMAP, an integrated system for modeling the circulation and water quality of estuarine and coastal waters (Mendelsohn, et. al., 1995). The system allows the user to generate boundary conforming grids to represent the study area, to perform simulations with a suite of circulation and water quality models, and to display the model predictions in the form of time series plots, vector and contour plots and color animations. The embedded hydrothermal model solves the unsteady three-dimensional conservation of mass, momentum, salt and energy equations on a spherical coordinate system using a split mode methodology (Madala and Piasek, 1977). Muin and Spaulding (1996, 1997) provide a detailed description of the governing equations, numerical solution methodology and testing against analytic solutions for two and three-dimensional flow problems. Mendelsohn (1998) provides a similar description of the thermal model.

Baseline Numerical Simulation

The data assimilation study was conducted for the May-June 1997 field survey. A baseline simulation in which no data was assimilated was run in order

Figure 2. Thermistor string locations within Mt. Hope Bay.

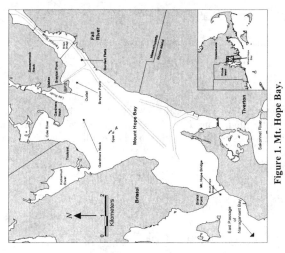

Figure 1. Mt. Hope Bay.

to obtain baseline statistical performance parameters. The computational grid (Figure 3) consisted of 11 sigma layers in the vertical and had a horizontal resolution of 200-300m in most of the bay and high resolution 50-100m in the vicinity of the powerplant. A statistical comparison between the model predicted temperature and thermistor measurements resulted in a maximum relative mean error of 9%, a maximum error coefficient of variation of 5%, minimum and maximum correlation coefficients of 0.39 and 0.79, respectively, and an average root mean square error (RMS) of 1.4°C. Figure 4 shows the RMS error between the baseline simulation and thermistor data at the surface, areas where no bars are present indicate that the thermistor data was incomplete relative to the length of the simulation. The RMS error is greatest in the region directly surrounding the outfall (thermistor strings 8, 9, 10, 11 and 12) and decreases as approach the mouth of the bay. Figure 5a-b presents a plan view of contoured thermistor data and model predicted temperature at the surface. The reasons for this error distribution becomes clear by examining the time series of the model predicted temperature and thermistor data at the surface near the outfall (Figure 6) and near the mouth of the bay (Figure 7). The motion of the thermal plume is governed by the semi-diurnal tide and capturing the proper phasing and extent of the plume in the near field surrounding the powerplant significantly impacts the results. However, the temperature data collected from the thermal mapping program should prove useful in adjusting the model to capture the near-field dynamics and improving the predictions in the far-field.

Data Assimilation Methodology

In order to distribute the discrete temperature observations over the computational domain a spatial interpolation scheme was required. The technique chosen for this study is based upon the method of successive correction as presented by Moore, et. al. (1987). Using this scheme, the model forecast is combined with the temperature observations within a region of influence determined by a correlation length scale. The new temperature at any grid cell is that predicted by the model plus a weighted mean of observational errors within the correlation region. The data assimilation scheme is then expressed as

$$T_i^{AN} = T_i^f + \frac{\sum_{k=1}^{N} \alpha_k \left(T_k^0 - T_k^f\right)}{\alpha_p + \sum_{k=1}^{N} \alpha_k}$$

where the weighting coefficient α_k is defined as

$$\alpha_k = \exp - \left\{ \left(\frac{x_i - x_k}{a_x}\right)^2 + \left(\frac{y_i - y_k}{a_y}\right)^2 \right\}$$

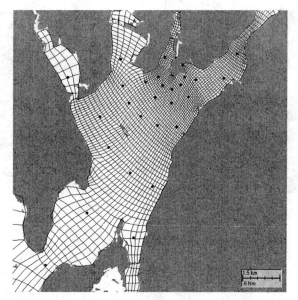

Figure 3. Computational grid for Mt. Hope Bay application.

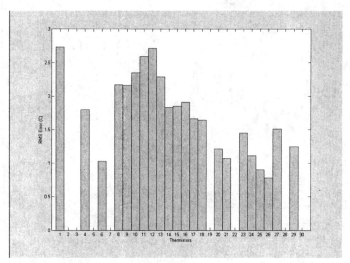

Figure 4. RMS error between baseline simulation and thermistor data at surface.

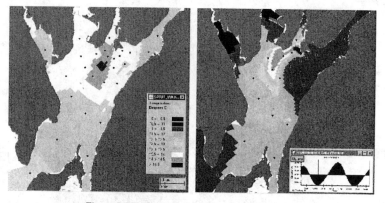

Figure 5. Plan view of surface temperature.
A) contoured thermistor data
B) model predicted temperature

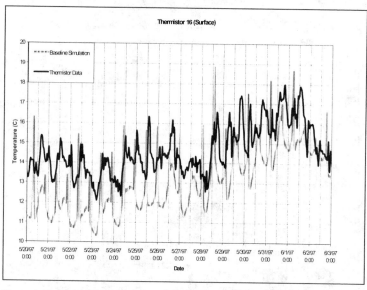

Figure 6. Model predicted temperature and thermistor data at the surface near the out-fall.

if $|x_i-x_k| < a_x$ and $|y_i-y_k| < a_y$ otherwise the value is zero, where a_x and a_y are the correlation length scales in the longitudinal and latitudinal directions, respectively, T_i^{AN} is the analyzed temperature within a grid cell, T_i^f is the model predicted temperature within a grid cell, T_k^0 is the observed temperature at the thermistor locations, and T_k^f is the model predicted temperature at a grid cell coincident with the thermistor observations. The term α_p is the weight assigned to the model predictions and was assigned the value of one. This corresponds to the new temperature, when the grid cell corresponds to the thermistor location and no other thermistors are within the correlation radius, being the average of the observation and model prediction.

The correlation length scale was kept constant for this study versus applying the above equation a number of times with successively smaller correlation length scales as in previous applications of this technique in the analysis of meteorological data for numerical weather forecasting (Cressman, 1959; Barnes, 1964; Lorenc, 1986). The correlation length scale was determined by developing a zero phase correlation matrix for all of the thermistor data at each depth and plotting it versus the corresponding distance matrix. The majority of the results with a correlation coefficient above 0.8 were clustered within a distance of 0.5 NM (0.93 km) (Figure 8). This distance was chosen as the correlation length scale in both the longitudinal and latitudinal direction.

In order to determine the effectiveness of the scheme, two types of tests were developed. The first directly addresses the question of improving the model predictions, in a hindcasting and/or nowcasting sense, and studying the efficiency of the monitoring program. This was accomplished by varying the spatial distribution of thermistor strings while continuously assimilating the data and included cases where all the thermistor strings were used, only the even numbered strings, only the odd numbered strings, only strings at the open boundary and outfall (11 and 29), and five strategically chosen strings (6,11, 13, 23, and 29). Refer to Figure 2 for thermistor string locations. The second tested the scheme's predictive capability by assimilating the data from all the thermistor strings and from only the outfall and open boundary for one day then ceasing the assimilation.

Spatial Test Results

Figure 9 shows the RMS error between the model predictions and thermistor data at the surface, for the baseline simulation and the case where all 30 thermistor strings were continuously assimilated. As in the baseline simulation, the error is greatest in the area surrounding the outfall. However, the maximum error is now 0.68°C relative to 2.73°C for the baseline simulation. The effect of assimilating all the thermistor data reduced the mean RMS error by 81% from

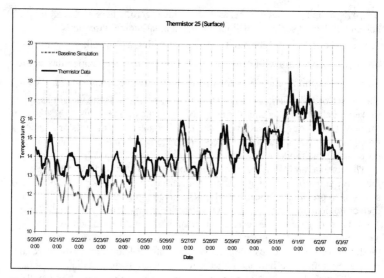

Figure 7. Model predicted temperature and thermistor data at the surface near the mouth of the bay.

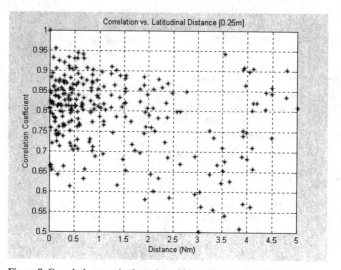

Figure 8. Correlation matrix clustering within a distance of 0.5NM (0.93 km).

1.71°C to 0.32°C. A time series plot of the temperature in the region of the outfall at the surface (Figure 10) demonstrates the data assimilation scheme's ability to improve model predictions in capturing the appropriate plume dynamics.

The results for the cases where only data from the even and odd numbered thermistor strings were assimilated was similar to those for the case where all of the thermistor data was used (Figure 11). The reduction in RMS error relative to the baseline simulation was 69% and 71% for the even and odd numbered thermistor string cases, respectively. This is an important step in evaluating the efficiency of the monitoring program, since the result of using 15 thermistors is almost equivalent to using all 30, the long-term cost of such a program could be dramatically reduced.

The next step was to test model predictions by assimilating data from only representative thermistor strings located near the outfall and open boundary. The improvement was primarily seen in the region directly in front of the outfall, while those on the periphery saw little improvement (Figure 12). The resulting reduction in the mean RMS error at the surface for this case was 35% relative to the baseline assimilation.

With the results of the other four cases we decided to investigate the impact of assimilating the data from five strategically located thermistor strings (6, 11, 13, and 29). The result was a fairly uniform RMS error distribution throughout the bay (Figure 13), a desirable result, with a mean RMS error of 0.84°C at the surface. So, by reducing the number of thermistor strings by a factor of six, a 51% reduction in the mean RMS error relative to the baseline simulation was achieved. Table 3 summarizes the RMS error at the surface for all six cases.

Table 3. RMS error at the surface for all spatial test cases.

Case	Thermistor Strings	Maximum RMS Error (°C)	Minimum RMS Error (°C)	Mean RMS Error (°C)	Percent Mean RMS Reduction (relative to baseline simulation)
Baseline Simulation	0	2.73	0.78	1.71	----
All Thermistors	30	0.68	0.14	0.32	81
Even Thermistors	15	2.48	0.16	0.52	69
Odd Thermistors	15	1.27	0.15	0.48	72
Strategically Located Thermistors	5	2.66	0.16	0.84	51

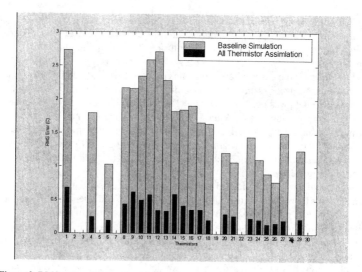

Figure 9. RMS error at the surface for continuous assimilation of all 30 thermistor strings.

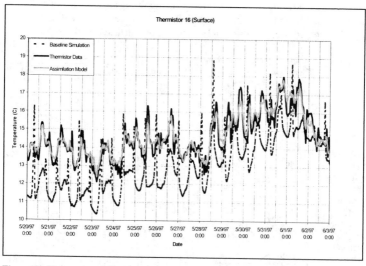

Figure 10. Time series of the temperature in the region of the outfall at the surface for continuous assimilation of all 30 thermistor strings.

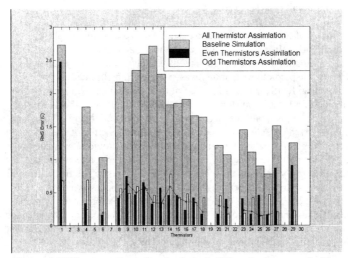

Figure 11. RMS error at the surface for the assimilation of only even and odd numbered thermistor strings, respectively.

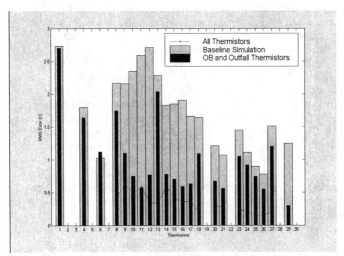

Figure 12. RMS error at the surface for continuous assimilation of thermistor strings in the vicinity of the open boundary and outfall.

Predictive Capability Test Results

This series of tests was designed to examine the forecasting capability of the data assimilation scheme. As neither, any time dependent term was included in the data assimilation scheme nor any feedback included to modify the model coefficients, the predictive capability was expected to be limited. Figure 14 shows the mean RMS error as a function of time for four cases: 1) the continuous assimilation of all the thermistor data; 2) the assimilation of all the thermistor data; 3) only thermistor strings located near the outfall and the open boundary for one day; and 4) the baseline simulation. The mean RMS error of the continuous assimilation of all the data and the baseline simulation provide lower and upper bounds of the predictive capability tests, respectively. For cases where data was assimilated for only one day, there is a two-stage response. The first is an initial rapid increase of the error occurring on the order of 12 hours, which is also the dominant tidal period for Mt. Hope Bay. The second is a longer-term decay towards the baseline simulation on the order of days. This can be attributed to the data assimilation scheme effectively raising the bulk temperature of the bay and is clearly a function of the number of thermistor data sets assimilated.

Summary

The application presented in this paper focused on the temporal and spatial assimilation of temperature data, collected in Mt. Hope Bay, MA, from a series of 30 thermistor strings, for the period of May-June 1997. The data assimilation scheme used proved successful in improving the model predictions by reducing the RMS error by as much as 80% in a hindcast/nowcast sense. This result is useful in situations when an environmental monitoring-modeling program is required for regulatory concerns or a better understanding of the environment. The data assimilation scheme also proved useful in studying the efficiency of the monitoring program. By reducing the number of thermistor strings from 30 to 5, a 51% reduction in the mean RMS error relative to the baseline simulation was achieved. Using this technique has the potential to dramatically reduce the long-term cost of future programs of this type and still provide the error reduction advantages of data assimilation.

The predictive or forecasting capability of the data assimilation scheme was also tested where the thermistor data was continuously assimilated for one day only. This resulted in a two-stage response. The first is an initial rapid increase of the RMS error on the order of 12 hours. The second is a longer-term increase of the RMS error towards that of the case of no data assimilation on the order of days.

Further testing of the data assimilation scheme used is warranted. One test would be to recalibrate the model to the period of the thermistor deployment first

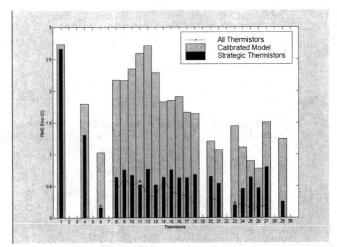

Figure 13. RMS error at the surface for the assimilation of 5 strategically located thermistor strings.

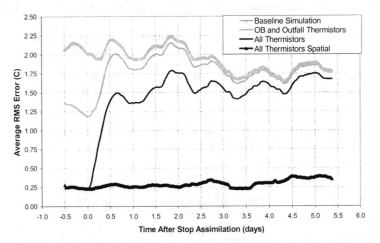

Figure 14. Mean RMS error at the surface as a function of time for predictive capability tests, baseline simulation and the continuous assimilation of all thermistor data.

and then perform the data assimilation. The use of overflight data offers a different type of assimilation problem, but offers more complete spatial coverage. A combination of remote and in-situ data offers an additional line of testing to determine an optimal balance.

Acknowledgements

This work was partially funded by the National Aeronautic and Space Administration through a grant to Brown University. Discussions with Professor John Mustard of Brown have been invaluable in the development of the work reported here.

References

Barnes, S.L., 1964. A technique for maximizing details in numerical weather map analysis, Journal of Applied Meteorology, Vol. 3, pp. 369-409.

Cressman, G.P., 1959. An operational objective analysis system, Monthly Weather Review, Vol. 87, pp. 367-374.

Lorenc, A.C., 1986. Analysis methods for numerical weather prediction, Quarterly Journal of the Royal Meteorological Society, Vol. 112, pp. 1177-1194.

Madala, R.V., and S.A. Piacsek, 1977. A semi-implicit numerical model for baroclinic oceans, Journal of Computational Physics, 23, pp. 167-178.

Mendelsohn, D.E., E. Howlett, J.C. Swanson, 1995. WQMAP in a Windows Environment, Fourth International Conference on Estuarine and Coastal Modeling, ASCE, San Diego, October 26-28, 1995.

Mendelsohn, D.L., 1998. Development of an estuarine thermal environmental model in a boundary-fitted, curvilinear coordinate system, Applied Science Associates Internal Document.

Moore, A.M., N.S. Cooper, and D.L.T. Anderson, 1987. Initialization and data assimilation in models of the Indian Ocean, Journal of Physical Oceanography, Vol. 17, pp. 1965-1977.

Muin, M., and M. Spaulding, 1996. Two-dimensional boundary fitted circulation model in spherical coordinates, Journal of Hydraulic Engineering, Vol. 122, No. 9, pp. 512-521.

Muin, M., and Spaulding, 1997. Three-dimensional boundary-fitted circulation model, Journal of Hydraulic Engineering, Vol. 123, No. 1, pp. 2-12.

Rines, H. and H. Schuttenberg, 1998. Data report New England Power Company, Brayton Point Station, Mount Hope Bay field studies for 1997. Report to New England Power company. Applied Science Associates, Inc., Narragansett, RI.

Rines, H., 1999. Mapping temperature distributions in Mount Hope Bay, Fall 1998 deployment. Report submitted to PG&E Generating Company, ASA project 96-076. 44pp.

Spaulding, M.L. and F.M. White, 1990. Circulation dynamics in Mt. Hope Bay and the lower Taunton River. Coastal and Estuarine Studies, Vol. 38, R.T. Cheng (Ed.), Residual Currents and Long-Term Transport, Springer-Verlag, New York, Inc., 1990.

Swanson, J.C., D. Mendelsohn, H. Rines, H. Schuttenberg, 1998. Mt. Hope Bay hydrodynamic model calibration and confirmation. Applied Science Associates report to New England Power Company.

Parameter Estimation for Subtidal Water Levels Using An Adjoint Variational Optimal Method

Aijun Zhang[1], Eugene Wei[1], and Bruce Parker[1]

Abstract

In this paper, the two-dimensional Princeton Ocean Model (POM) (Blumberg and Mellor, 1987; Mellor, 1998) is implemented to simulate wind-driven subtidal water levels along the U.S. East Coast. The model is forced by a 48km ETA Data Assimilation System (EDAS) analyzed wind field. An optimal adjoint variational data assimilation technique is presented to assimilate the subtidal water levels sampled along the entire Atlantic Coast of the U.S. into the numerical model. In the optimal data assimilation procedure, the subtidal water level misfit is defined as the cost function. The gradient of cost function is determined by the adjoint model. Limited memory Broyden-Fletcher-Goldfarb-Shanno (BFGS) quasi-Newton (Liu and Nocedal, 1989) method for large scale optimization is implemented to minimize the cost function. The data assimilation system was tested in ideal twin identical experimental cases in which the pseudo-observations are generated by numerical model with predefined wind drag coefficients (C_d). The results show that the wind drag coefficients can be recovered from subtidal water levels very accurately by using this adjoint optimal data assimilation system.

1. Introduction

In recent years, data assimilation techniques based on optimal control methods have been developed and widely applied in meteorologic and oceanographic fields. As early as 1970's, the adjoint approach with the governing equations as strong constraints was described by Sasaki (1970a,b) which gave a framework that is readily

---

[1]Coast Survey Development Laboratory, NOS/NOAA, 1315 East-West Hwy., Silver Spring, MD 20910; E-mail: aijun.zhang@noaa.gov

applicable to a set of steady or unsteady state equations. Yu and O'Brien (1991) used a variational adjoint method in a one-dimensional vertical model to estimate the wind stress drag coefficient and the oceanic eddy viscosity profile from observed velocity data. Schwab (1982) used the inverse method to estimate wind stress from water level fluctuations. Das and Lardner (1991) implemented the adjoint method to

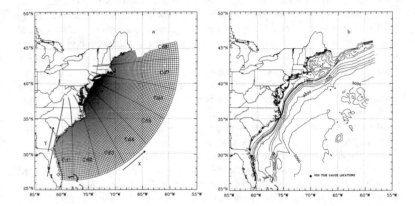

Fig. 1.   U.S. East Coast subtidal water level forecast system, (a) computational grid; (b) tide gauge locations and bathymetry.

determine the bottom friction coefficient and water depth which are position dependent parameters from periodic tidal data. Panchang and O'Brien (1989) applied adjoint state formulation to a one-dimensional hydraulic model to determine the bottom friction factor for a tidal river. Chu et al. (1997) determined open boundary conditions with an optimization method, in which a multiperturbation method is proposed to determine the gradient of cost function. Thus, his method is independent of the ocean model and can be easily applied to linear and nonlinear ocean models. But for the case of a large number of control variables, the multiperturbation method for determining the gradient of the cost function is time consuming.

With POM being widely used by more and more oceanographers (Aikman et al 1996; Mellor and Ezer 1991; Ezer and Mellor 1997) and variational adjoint techniques being successfully applied in many fields, it is possible and necessary to develop an adjoint model of POM in order to efficiently perform data assimilation.

For the purpose of water level forecasting, a subtidal water level forecast system for the U.S. East Coast is under development which uses the two dimensional barotropic Princeton Ocean Model with orthogonal curvilinear model grid (Fig.1).

Grid resolution ranges from 5 to 32 km. The model is forced by a 48km EDAS analyzed wind field. An objective of this system is to forecast subtidal water levels along the East Coast and to provide water level forecasts as the open boundary condition for other regional or bay forecasting systems. With the ability to acquire real-time water gauge observations along the coast, it is now feasible to assimilate real time observed water levels into a numerical model, using the optimal control data assimilation technique in order to improve water level nowcasts and subsequent forecasts. From the sensitivity experiments of subtidal water level simulation, we found that the surface wind plays a predominant role in causing the subtidal water level variation in this area. Thus wind drag coefficients are chosen as the control variables in the data assimilation. The idea is based on the assumption that the forecast wind field is not accurate enough (and its resolution is also too coarse compared with the ocean model grid resolution), hence causing the water level misfits between observed and simulated water levels. For convenience we use a change in wind drag coefficients to represent (and correct for) systematic "errors" in the wind field. However, it is possible that there really could be errors in the wind drag coefficient, since its behavior as wind speed increases is still not well understood, especially at high wind speeds. A problem with the wind drag coefficient can also occur if the effect of atmospheric stability is not included in its formulation, in which case, changing air and water temperature could affect wind stress in an unaccounted for way. Thus, by assimilating the observed subtidal water level into a numerical model, it is possible to improve the wind field by adjusting the wind drag coefficients. In reality, surface wind drag coefficients are position-dependent parameters. In this paper, for the sake of simplification, the forward model is linearized, and the wind drag coefficients are assumed as constant or piecewise constants. Pseudo-observations generated by the numerical model with predefined wind drag coefficients are used in a twin identical experiment. The real observed subtidal water levels are obtained from total observed water level by a 30-hour low-pass Fourier filter.

2. Forward Numerical Model

2.1 Fully Two-dimensional POM Governing Equations

The governing equations of two dimensional POM are given as follows:

$$\frac{\partial \eta}{\partial t} + \frac{\partial UD}{\partial x} + \frac{\partial VD}{\partial y} = 0 \tag{1}$$

$$\frac{\partial UD}{\partial t} + \frac{\partial U^2 D}{\partial x} + \frac{\partial UVD}{\partial y} - F_x - fVD + gD\frac{\partial \eta}{\partial x} = \frac{1}{\rho}\left(\tau_{sx} - \tau_{bx}\right) \tag{2}$$

$$\frac{\partial VD}{\partial t} + \frac{\partial UVD}{\partial x} + \frac{\partial V^2 D}{\partial y} - F_y + fUD + gD\frac{\partial \eta}{\partial y} = \frac{1}{\rho}\left(\tau_{sy} - \tau_{by}\right) \tag{3}$$

where H, τ_s and τ_b are water depth at rest, wind stress and bottom friction, and D=H+η total water depth, g the acceleration due to gravity, f the Coriolis parameter, ρ the water density. And the horizontal viscosity and diffusion terms F_x and F_y are defined as

$$F_x = \frac{\partial}{\partial x}\left[H2A_M \frac{\partial U}{\partial x} \right] + \frac{\partial}{\partial y}\left[HA_M \left(\frac{\partial U}{\partial y} + \frac{\partial V}{\partial x} \right) \right] \tag{4}$$

$$F_y = \frac{\partial}{\partial y}\left[H2A_M \frac{\partial V}{\partial y} \right] + \frac{\partial}{\partial x}\left[HA_M \left(\frac{\partial U}{\partial y} + \frac{\partial V}{\partial x} \right) \right] \tag{5}$$

where A_M, the vertically integrated horizontal eddy viscosity, is defined by the Smagorinsky formula

$$A_M = C\Delta x \Delta y \frac{1}{2} |\nabla V + \nabla V^T| \tag{6}$$

C, a non-dimensional parameter, is set to be 0.2 in this study; Δx and Δy are the grid spacings in the x and y directions for each grid cell.

Table 1. RMS Errors and Correlation Coefficients at 15 East Coast Locations

Station	Mean_ob (cm)	Mean-M (cm)	RMS (cm)	RMS (demean) (Cm)	Correla	Correla (Demean)
Newl	5.13	-1.08	10	8	0.80	0.91
Brid	6.41	2.98	12	11	0.92	0.91
Montk	5.54	4.86	8	8	0.90	0.90
Willets	7.53	4.65	15	15	0.91	0.91
S H.	6.76	-2.51	11	9	0.79	0.87
Atlantic	8.03	-0.45	13	9	0.77	0.89
Lewes	7.54	-1.89	13	10	0.62	0.84
Kipt.	7.83	-1.48	13	8	0.53	0.80
Glou.	7.93	1.36	12	9	0.67	0.77
CBBT	8.36	-1.37	14	10	0.55	0.79
Duck	9.38	-0.59	14	10	0.52	0.68
Cape	7.79	-1.01	12	8	0.28	0.54
Spring	10.6	0.47	13	8	0.60	0.75
Char.	11.5	0.24	14	8	0.55	0.73
Mayport	13.2	2.27	16	11	0.71	0.75

First, the fully nonlinear two-dimensional POM is used to evaluate numerical model subtidal water level simulation at 15 selected locations. The model simulates a period from Sep. 1 to Oct. 31, 1996. During this period, there were three hurricanes that occurred in Sep. 16-18, Oct. 7-9, and Oct. 16-20, respectively. The EDAS analyzed wind fields were used as the surface forcing. Wind stress is calculated using a formula developed by Large and Pond (1981). The results (Fig. 2) show that the variation trends of simulated water levels with time match that of observations fairly well at most locations most of the time. Mean values, correlation coefficient and Root Mean Square (RMS) of the differences between the observed and the simulated subtidal water levels are presented in Table 1. Note that there are mean value

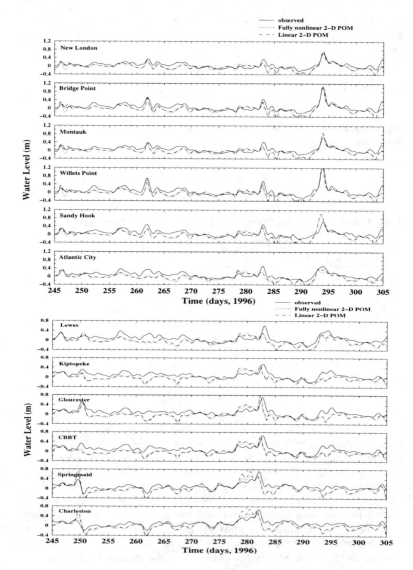

Fig.2.   Comparison of observed and simulated subtidal water levels without data assimilation at NOS tide gauge locations.

differences between the observed and the simulated subtidal water levels. The simulation mean value is lower than that of observation at most locations. The demeaned RMS errors range from 8 cm to 15 cm, the correlation coefficients between observed and simulated subtidal water levels vary from maximum 0.91 at Willets Point and New London to minimum 0.54 at Cape Hatteras. There are differences between the observed and simulated subtidal water levels even during the weak wind period. These differences are because the 2-D model forced by surface wind was run for only two months, so some low-frequency signals in the observed subtidal water levels which are caused by other factors were not represented in the simulated subtidal water levels. This includes steric effects on water level due to water temperature change, which are not included in the model, as well as Gulf Stream effects on the southern stations.

2.2 Linearized Governing Equations

To simplify the adjoint model development, the fully nonlinear two-dimensional POM is linearized by: (1) neglecting the variation of water level surface elevation η relative to water depth (D=H), which is reasonable for the depths over most of model regime; (2) neglecting the horizontal advection and diffusion terms (F_x and F_y); (3) linearizing the bottom friction terms (which is also reasonable for the depths over most of regime) with a constant bottom friction coefficient, $C_b = 1.0 \times 10^{-3}$ Thus, the linearized 2-D POM governing equations are as follows,

$$\frac{\partial h}{\partial t} + \frac{\partial HU}{\partial x} + \frac{\partial HV}{\partial y} = 0 \tag{7}$$

$$\frac{\partial HU}{\partial t} - fHV + gH\frac{\partial h}{\partial x} + C_d|U_.|\cdot U_. - C_t U = 0 \tag{8}$$

$$\frac{\partial HV}{\partial t} + fHU + gH\frac{\partial h}{\partial y} + C_d|U_.|\cdot V_. - C_b V = 0 \tag{9}$$

The model is integrated from rest, and radiational open boundary conditions are used on the southern boundary. The results of the linearized POM (Fig. 2) demonstrate that, for subtidal water level simulation, the linearized 2-D POM gives results similar to those of the fully nonlinear 2-D POM. Therefore, it is reasonable to implement the linearized 2-D POM to simulate subtidal water levels for this coastal ocean area.

3. Adjoint Equations of Linearized 2-D POM

The adjoint method is increasingly being implemented in data assimilation, model tuning, and model sensitivity analysis. It provides an efficient method for calculating the gradients of cost function with respect to control variables (especially in the case of large number of control variables). However, the adjoint equations depend on the forward equations, and derivation of adjoint equations is complicated. Zhu et al. (1997a) simply described the construction of an adjoint model for the 2-D barotropic model of Institute of Atmospheric Physics of China (IAP) and applied it to estimate the open boundary conditions of a tidal model (Zhu et al., 1997b). Ralf and Kaminski (1996) and Ralf (1997) proposed a helpful and basic textbook on how to construct adjoint code from forward model code using a tangent linear compiler (TLC). However, in this particular application, it is difficult to determine what the formula is to calculate the gradient from the adjoint variables. In some occasions, TLC may even generate redundant statements in the adjoint code. Here, as an example for the linearized 2-D POM with well-posed initial and open boundary conditions , the procedure for deriving adjoint equations is presented in continuous notation.

In oceanic numerical modeling we are often confronted with some unknown or undetermined parameters, e.g., initial conditions, open boundary conditions, friction coefficients, wind drag coefficients, etc. In this study, we are concerned with nontidal elevation data assimilation. Since wind stresses play the most important role in nontidal elevation simulation, the surface wind drag coefficient C_d is chosen as the control variable.

The basic procedure in the variational adjoint method consists of minimizing a cost function that represents the misfits between observed data and model output. This minimization is performed subject to the strong constraints of satisfying the governing equations. The constraint minimization involves Lagrange multipliers and leads to additional equations (known as adjoint equations) from which Lagrange multipliers are determined. The model state variables and Lagrange multipliers are used to compute the cost function and its gradient from which the cost function is minimized to obtain the optimal control variables.

In this variational problem the cost function is defined as,

$$J = \frac{1}{2} \iiint_{xyt} W(h - h_o)^2 \, dx \, dy \, dt \tag{10}$$

where h_o and h are observed and simulated elevations and W is the weighting factor. The variational problem is to minimize cost function J subject to equations (7)-(9). Introducing Lagrange multipliers λ_h, λ_u, λ_v for the constraint governing equations (7), (8), (9) (Lawson et al, 1995), the first variation of the cost function J can be written as

$$\delta J = \iiint_{xyt} [\ (h - h_o) \quad + \lambda_h \delta(\frac{\partial h}{\partial t} + \frac{\partial HU}{\partial x} + \frac{\partial HV}{\partial y})$$
$$+ \lambda_u \delta(\frac{\partial HU}{\partial t} - fHV + gH\frac{\partial h}{\partial x} + C_d|U_w| \cdot U_w - C_b U) \tag{11}$$
$$+ \lambda_v \delta(\frac{\partial HV}{\partial t} + fHU + gH\frac{\partial h}{\partial y} + C_d|U_w| \cdot V_w - C_b V) \ ] dx \, dy \, dt$$

The corresponding adjoint variables are defined as follows: λ_h is the adjoint variable of h, λ_u is the adjoint variable of U, and λ_v is the adjoint variable of V, respectively.

Considering the specific case defined above, using wind drag coefficients as the only control variables, Eq. (11) can be rewritten as, after applying the chain rule and integrating by parts,

$$
\begin{aligned}
\delta J = &\iiint_{xyt}\left[(h-h_o)-\frac{\partial \lambda_h}{\partial t}-g\frac{\partial H\lambda_u}{\partial x}-g\frac{\partial H\lambda_v}{\partial y}\right]\delta h\,dxdydt \\
&+\iiint_{xyt}\left(-H\frac{\partial \lambda_h}{\partial x}-H\frac{\partial \lambda_u}{\partial t}-C_b\lambda_u+fH\lambda_v\right)\delta U\,dxdydt \\
&+\iiint_{xyt}\left(-H\frac{\partial \lambda_h}{\partial y}-H\frac{\partial \lambda_v}{\partial t}-C_b\lambda_v-fH\lambda_u\right)\delta V\,dxdydt \\
&+\iiint_{xyt}\left(\lambda_u|U_w|U_w+\lambda_v|U_w|V_w\right)\delta C_d\,dxdydt \\
&+\iint_{xy}\left(\lambda_h\delta h+H\lambda_u\delta U+H\lambda_v\delta V\right)dxdy\,\Big|_0^T \\
&+\iint_{xt}\left(gH\lambda_v\delta h+H\lambda_h\delta V\right)dxdt\,\Big|_0^Y \\
&+\iint_{yt}\left(gH\lambda_u\delta h+H\lambda_h\delta U\right)dydt\,\Big|_0^X
\end{aligned}
\tag{12}
$$

Since the initial conditions and boundary conditions are well-posed for this specific case described earlier, we have

$$
\delta h\,\big|_{t=0}=\delta U\,\big|_{t=0}=\delta V\,\big|_{t=0}=0
$$
$$
\delta h\,\big|_{x=0}^{x=X}=\delta U\,\big|_{x=0}^{x=X}=\delta V\,\big|_{x=0}^{x=X}=0
$$
$$
\delta h\,\big|_{y=0}^{y=Y}=\delta U\,\big|_{y=0}^{y=Y}=\delta V\,\big|_{y=0}^{y=Y}=0
$$

Therefore, the last three terms of (12) are zero if the initial values (t=T) of λ_u, λ_v, and λ_h are forced to be zero. By forcing the coefficients of the non-control variables (δh, δU, δV) to zero, we obtain adjoint equations as follows,

$$
H\frac{\partial \lambda_u}{\partial t}-fH\lambda_v+H\frac{\partial \lambda_h}{\partial x}+C_b\lambda_u=0
\tag{13}
$$

$$
H\frac{\partial \lambda_v}{\partial t}+fH\lambda_u+H\frac{\partial \lambda_h}{\partial y}+C_b\lambda_v=0
\tag{14}
$$

$$
\frac{\partial \lambda_h}{\partial t}+g\left(\frac{\partial H\lambda_u}{\partial x}+\frac{\partial H\lambda_v}{\partial y}\right)=h-h_o
\tag{15}
$$

and the function increment becomes

$$
\delta J = \iiint\left(\lambda_u|U_w|U_w+\lambda_v|U_w|V_w\right)\delta C_d\,dxdydt
\tag{16}
$$

The above process showed that cost function minimization has resulted in new equations (13)-(15) which are called adjoint equations. And the adjoint variables are calculated by integrating the adjoint equations. The gradient of the cost function with respect to the control variable, wind drag coefficient C_d, can be computed by

$$G = \iiint_{xyt} (\lambda_u |U_w| U_w + \lambda_v |U_w| V_w) \, dxdydt \qquad (17)$$

It is shown from (17) that, if C_d varies spatially and temporally, the gradients of cost function with respect to C_d can be computed by integrating the adjoint model once regardless of the number of control variables.

By comparison with 2-D forward POM equations, it is seen that the adjoint equations (13)- (15) are similar to their original model equations (7)-(9). The differences between forward and adjoint equations are: no wind stress forcing terms are included in adjoint equations because the wind drag coefficient is defined as control variable; the data misfit $(h-h_o)$ is added to adjoint equation as a forcing term; the friction terms in the adjoint equations and the corresponding governing equations have opposite signs. This implies that the adjoint equations have to be integrated backwards in time starting from zero values at the final time step.

An iterative scheme for a 24-hour data assimilation is given in Fig.3.

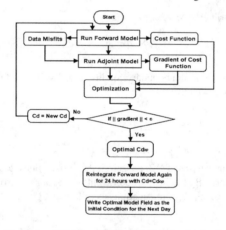

Fig. 3. Flow Chart for a 24-Hour Water Level Data Assimilation Run

4. Identical Twin Experiments

In order to check the performance of adjoint optimal data assimilation technique, identical twin experiments are usually considered. In twin experiments, the pseudo-observed water level data are generated by the numerical model. This is the best situation for data assimilation since the pseudo-observational data contain the same dynamics as the numerical model and are not contaminated by any error. In this study, the pseudo-subtidal water levels are generated by integrating the model with a constant or piecewise constant predefined wind drag coefficient over the entire domain and hourly subsampled at 15 selected locations. With a known initial condition from the model spin-up and the radiation formula implemented on southern open boundary, three cases are devised in the one- and 30- day data assimilation simulations. These are:

Case 1: pseudo-observations contain no errors,

Case 2: pseudo-observations at all locations are contaminated with a random white noise which has zero mean value and a standard deviation 0.029 m,

Case 3: same as Case 2 except the white noise standard deviation is doubled to 0.058 m.

4.1 One-day Data Assimilation

One-day pseudo-observations at 15 selected locations are used to recover the wind drag coefficient in this one-day data assimilation experiment (for convenience, the value of C_d appeared in this paper, including figures and tables, is multiplied by 10^3). The forward model is integrated from rest with a true C_d as 10.0. In the data assimilation procedure, the first guess C_d is 20.0. The variation of the optimal C_d, the gradient of the cost function, and the cost function are plotted as a function of the iteration number in Fig. 4 and the final values are listed in Table 2. For Case 1, the optimal C_d converges to the true value after three iterations and the corresponding gradient and cost function are less than 10^{-8}. For Cases 2 and 3, the optimal C_ds tend close to (but not exactly converge to) their true solutions after four iterations and the cost functions are much greater than that of Case 1. By comparison of the data misfits (the difference between the model results and the observations) and the noise added the observations, it is found that the curves of the misfits and noise for each station are very similar. This indicates that the value of the cost function is primarily contributed by the noise. The noise results in the optimization line search routine failing to find a better converging direction for optimizing procedure. Thus the optimization procedure could not recover the true C_d. This experiment indicates that observation errors can affect the optimal results.

4.2 30-day Continuous Data Assimilation

Table 2. Final optimal Cd, gradient and cost function

	Optimal Cd	Gradient	Cost function
Case 1	9.99998	-3.197×10^{-8}	7.0×10^{-10}
Case 2	10.085	2.07×10^{-6}	0.138
Case 3	10.170	-4.88×10^{-6}	0.553

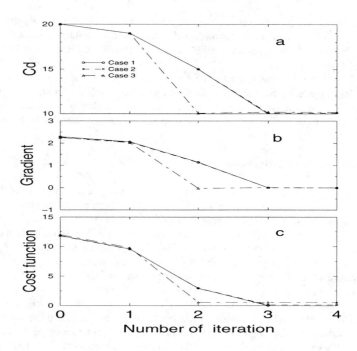

Fig.4. The variation of parameters with number of iteration in the one-day identical
       twin data assimilation: (a) the optimal C_d; (b) the gradient of the cost function;
       and (c) the cost function.

4.2 30-day Continuous Data Assimilation

The three cases described above are repeated in a 30-day continuous data
assimilation. The true wind drag coefficient C_d is set as a function of time, but
constant within each day. The forward model starts from rest and creates a restart file
at the end of each day as the initial condition for next day's data assimilation. The
first guess C_d of the first day is 20.0 and the optimal C_d of the previous day is used as
the first guess C_d for the next day's data assimilation. The time series of the final
optimal C_d, the gradient of the cost function, and the cost function are plotted in Fig.
5. The maximum of difference between true and final optimal C_d, the maximum of

gradients and the maximum of the cost functions from each case are presented in Table 3. For Case 1 with no observational error, the data assimilation procedure recovers the optimal C_d to its true value. The cost function and its gradient are less than 10^{-9} and 10^{-7}, respectively.

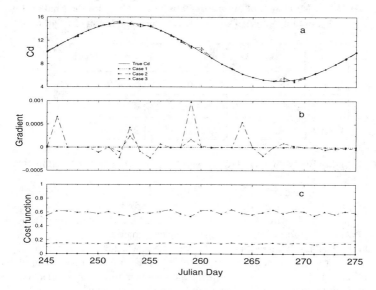

Fig.5 Time series of parameters in the 30-day identical twin data assimilation: (a) the optimal C_d; (b) the final gradient of the cost function; and (c) the final cost function.

For Case 2, the recovered optimal C_d is also close to its true value. However, the cost function is much greater than that of Case 1. As described in the one-day simulation, the value of the cost function is primarily contributed by the random noise added to the pseudo-observations. By adding the white noise to the pseudo-observations, the forward model can not dynamically simulate the random noise so that the final optimal C_d can not coverage to the true C_d.

For Case 3, by doubling the standard deviation of random noise, most of the optimal C_ds are also close to the true values. However, the differences between optimal C_ds and true C_ds are much greater than that of Cases 1 and 2. The cost function and its gradient are the greatest among three cases.

For the continuous data assimilation cases, the restart field is generated by the previous day forward model integration. If there is a difference between the optimal

C_d and its true C_d in the previous day, the restart field with the optimal C_d will be different from true initial field of this day. Then the optimal C_d may not converge to its true value in this day data assimilation.

Table 3. The Maximum of difference between the true and the final optimal C_d, the maximum of cost Functions and maximum of Gradients

	$Max\|Cd_{optimal} - Cd_{true}\|$	$Max\|Gradient\|$	$Max\|Cost\ Function\|$
Case 1	2.0×10^{-5}	3.16×10^{-8}	8.5×10^{-10}
Case 2	0.37	3.4×10^{-5}	0.16
Case 3	0.74	9.77×10^{-4}	0.64

4.3 One-day Identical Twin Experiment with Eight Control Variables

The wind drag coefficient should be time and position dependent. This is not simply because we use the Large-Pond formulation for C_d, in which C_d depends on wind speed, which itself is time and position dependent. In finding optimal C_ds as part of a data assimilation scheme, we are allowing the "improvement" in C_d to represent (and correct for) a systematic error in the wind field, which varies in space and time.

In order to test the performance of the adjoint data assimilation system for the multi-control variables case, the experiment with eight piecewise constant control variables is devised and performed. In this experiment, the entire domain is divided into eight pieces, and the wind drag coefficient in each piece is constant and used as one control variable (see Fig.1). The true solution of C_d is expressed as

$$C_{d\,m} = 10.0 + 5.0 \sin\left[\frac{2\pi}{30.0}(m-1)\right] \qquad m = 1, 2, \cdots 8 \qquad (18)$$

here m denotes the index of control variable in each subregion. The initial guesses for all of C_d are set to be 20.0. The result of this experiment shows that, in the first several iterations, most of the optimal C_d components (Fig.6) converge towards their true solutions very fast. The optimal C_d values for all of the eight components are very close to their true values after about 10 iterations. The maximum difference between the true solution and the final optimal C_d for all eight control variables is 1.5×10^{-4}, which appeared at the eighth control variable component. The cost function and the norm of the gradient (Fig.7) drop rapidly in the first three iterations. The convergence criterion $\|G\| < 10^{-7}$ is satisfied after 30 iterations, and the corresponding values of the cost function and the norm of the gradient are 1.4×10^{-11} and 3.2×10^{-8} respectively.

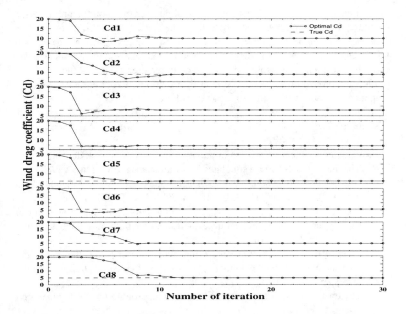

Fig.6. Variation of the optimal C_d components with the number of iterations for one day identical experiment with eight control variables.

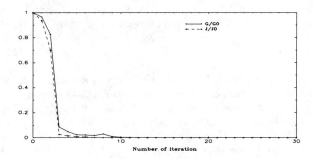

Fig.7. Variation of the norm of the gradient $\|G\|/\|G_0\|$ and the cost function J/J_0 with the number of iterations one day identical experiment with eight control variables. Where J_0 and $\|G_0\|$ are initial values of the cost function and its gradient.

In order to examine the effects of the initial guesses of C_d on the optimization procedure, different initial guesses are tested. The results show that no matter how far the first guesses are away from their true solutions, the cost function and the norm of the gradient drop rapidly in the initial several iterations, and the optimal C_ds are very close to their true solutions after about ten iterations. After that, the optimization procedure adjusts the optimal C_d slowly, and all of the eight optimal C_d components gradually converge to their true solutions.

5. Summary

From the experiments described in previous sections, we conclude that:

a. The linearized 2-D POM simulates the subtidal water level fairly well for this coastal ocean. The demeaned RMS errors and correlation coefficients are 9 cm and 0.87 at Sandy Hook and 10 cm and 0.79 at CBBT, respectively.

b. For perfect pseudo-observations, the adjoint data assimilation system can exactly recover the true wind drag coefficient for both of the one and eight control variable experiments.

c. Since water level observational errors have an impact on the wind drag coefficient recovery, the optimal C_d may not converge to its true value if the errors are significant.

d. The errors in initial field play a very important role in recovering the wind drag coefficient.

Acknowledgments

We would like to thank Dr. Yuanfu Xie of NOAA's Forecast System Laboratory for his assistance and great discussion in the construction of the adjoint model.

References

Aikman, F A., III, G. L. Mellor, D. Sheinin, P. Chen, L. Breaker, K. Bosley and D. B. Rao, 1996. "Towards an operational nowcast/forecast system for the U.S. East Coast". Modern Approaches to Data Assimilation in Ocean Modeling, 61, P. Malanotte-Rizzoli, Ed., Elservier Publishers, 347-376.

Blumberg, A. F. and G. L. Mellor, 1987. "A description of a three-dimensional coastal ocean circulation model." Three Dimensional Coastal Ocean Models. N. S. Heaper, Amer. Geophys. Union, 1-16.

Chu, P., C. Fan, and L. L. Ehret, 1997. "Determination of open boundary conditions with optimization method. J. Atmos. And Oceanic. Tech., Vol. 14, 723-734.

Das, S. K. and R. W. Lardner, 1991. "On the estimation of parameters of hydraulic models by assimilation of periodic tidal data." J. Geophys. Res., Vol.96, C8, 15187-15196.

Ezer, T. and G. L. Mellor, 1997. "Data assimilation experiments in the Gulf Stream Region: how useful are satellite-derived surface data for nowcasting the subsurface fields". J. Atmos. Oceanic Tech., 14, 1379-1391.

Large, W. G., and S. Pond, 1981. "Open ocean momentum flux measurements in moderate to strong winds." J. Phys. Oceanogr., Vol.11, 324-336.

Lawson, L. M, Y. h. Spitz, E. E. Hofmann, and R. B. Long. (1995). "A data assimilation technique applied to a predator-prey model. Bulletin of Mathematical Biology, 1995, 57, 593-617.

Liu, D. and J. Nocedal, 1989. "On the limited memory BFGS method for large scale optimization." Mathematical Programming, B 45, 503-528.

Mellor, G. L., 1991. "User's guide for a three dimensional, primitive equation, numerical ocean model." Princeton University.

Mellor, G. L. and T. Ezer, 1991. "A Gulf Stream model and an altimetry assimilation scheme". J. Geophys. Res., 96, 8779-8795.

Panchang, V. G. and J. J. O'Brien, 1989. "On the determination of hydraulic model parameters using the adjoint state formulation." Modeling Marine Systems. Edited by A. M. Davies, CRC Press, Boca Racon, 6-18.

Ralf, G. and T. Kaminski, 1996. "Recipes for adjoint code construction." Technical Report 212, 1996, p35. Max-Planck-Institut fur Meteorologie.

Ralf, G., 1997. "Tangent Linear and Adjoint Model Compiler. Users Manual, MIT, 53p.

Sasaki, Y., 1970a. "Some basic formalisms in numerical variational analysis." Monthly Weather Review, Vol.98, No.12, 875-883.

Sasaki, Y., 1970b. "Numerical variational analysis formulated under the constraints as determined by longwave equations and a low-pass filter." Vol.98, No.12, 884-899.

Schwab, D. J., 1982. "An inverse method for determining wind stress from water level fluctuations." Dyn. of Atmos. And Oceans., 6, 251-278.

Yu, L. and J. J. O'Brien, 1991. "Variational estimation of the wind stress drag coefficient and the oceanic eddy viscosity profile." J. Phys. Oceanogr., Vol. 21, 709-719.

Zhu, J., Q. C. Zeng, D. J. Guo and C. Liu, 1997a. "The Construction of adjoint model of IAP baratropic model and its second-order adjoint model." Science in China. 27(3), 277-283. (In Chinese)

Zhu, J., Q. C. Zeng, D. J. Guo and C. Liu, 1997b. "Estimation of the open boundary conditions of tidal model from coastal tidal observations by the adjoint method." Science in China, 1997, 27(5), 462-468. (In Chinese)

J1.19 DEVELOPMENT OF THE HIGH RESOLUTION, DATA ASSIMILATING NUMERICAL MODEL OF THE MONTEREY BAY.

I. Shulman [1], C. -R. Wu [1], J. K. Lewis [2], J. D. Paduan [3] L. K. Rosenfeld [3] S. R. Ramp [3], M. S. Cook [3], J. C. Kindle [4], D. -S. Ko [4]

Abstract

The development of a high-resolution, data assimilating model of Monterey Bay area (MBa model) is one of the tasks of the project "An Innovative Coastal-Ocean Observing Network" (ICON) sponsored by the National Oceanographic Partnership Program (NOPP). The objectives of this task are: 1) development of a fine-resolution, hydrodynamic model of the Monterey Bay area capable of resolving the temporal and spatial scales of corresponding oceanic processes and bringing together the unique oceanographic data sources available in this area; 2) development of technology for coupling the Monterey Bay area model with the Pacific West Coast model; and 3) development and implementation of data-assimilating capability for HF radar-derived surface velocity fields into the hydrodynamic model of the Monterey Bay area. Different scientific issues related to model initialization, forcing, open boundary conditions, and model spin-up are discussed. The results of model development, validation and verification, as well as HF radar data assimilation issues are presented.

2. Introduction

Our previous data assimilation experiments with the numerical model and HF radar (CODAR) data in the Monterey Bay area are presented in Lewis et al. (1998). The results of simulations with the Monterey Bay regional model show that the proposed data assimilation approach (Lewis et al., 1998) results in significant increases in the linear and nonparametric rank correlations between the model predicted and CODAR data when compared to the case when no assimilation of CODAR data is performed. The data assimilation technique allows the model to reproduce the trends and variations of the Doppler current data.

[1]COAM, Univ. of Southern Mississippi, Stennis Space Center, MS 39529.
[2]Scientific Solution, Inc., Kalaheo, HI
[3]Naval Postgraduate School, Monterey, CA
[4]Naval Research Laboratory, Stennis Space Center, MS

At the same time, the numerical experiments revealed some problems. Some of the problems are associated with the HF CODAR data, and others with the model used:

a. The amplitudes of CODAR-predicted currents showed significant differences compared with acoustic Doppler current profiler (ADCP) data (Paduan and Rosenfeld, 1996). Moreover, differences between the currents sensed in the same grid at the same time by two different HF systems can be fairly large with respect to the magnitude of the current, on the order of 10-15 cm/s. This suggested that assimilation of Doppler current data should be performed in the sense of the qualitative characteristics of the current direction, trends, and circulation features.

b. It was shown that the radar currents have significant divergences and unrealistic spatial variations of divergence. This suggested the need for additional preprocessing (before use in data assimilation) of CODAR data (Lipphardt et al., 2000) or special treatment of CODAR data during the numerical assimilation process (as, for example, introducing a constraint on the magnitude of the divergence of the current field).

c. In the previous study, the results from the global tidal model of Schwiderski were used as reference values to specify the model open boundary values of sea surface height. The model simulations showed unrealistic currents close to the boundary. This suggested that additional sources of data are needed on the open boundaries as, for example, available observations blended together with the output from a larger-domain, extended area model. These types of data sources would include other physical processes dominant in this area (e.g., the California Current). CODAR measurements are influenced by these processes; in this case, the assimilation of radar currents into the model resulted in the generation of transient waves, which could not propagate out of the model domain.

d. The data assimilation technique employed in the previous study was based on the application of a pseudo-shearing stress over the surface layer of the model. This pseudo-shearing stress depended on the differences between the model-predicted velocities and the velocities observed by the HF Doppler system. The approach is a nudging technique, with the nudging coefficient chosen as a solution of a special optimization problem. The observational error covariance matrix as well as the model error prediction covariance matrix were not modeled and taken into account in this previous data assimilation scheme.

In our present study, the model domain (Fig.1) was extended to the north and south to take into account the research interests and needs of the partners of the ICON/NOPP project. We use an orthogonal, curvilinear grid. In this case, we set up the cross-shelf open boundaries to be orthogonal to the isobaths of bathymetry in order for the flow to be almost perpendicular to the open boundaries. The Pacific West Coast Model (Clancy, et al., 1996) is used for estimation of open boundary conditions for our coastal model. Finally, the error coariance matrixes are taken into account in the CODAR surface currents data assimilation schemes.

In section 3 we outline the model along with the forcing and formulations. In section 4 we present the results of model runs over 1995 and compare these with observations. Section 5 describes the data assimilation approach, and section 6 presents a discussion of and future plans for model development.

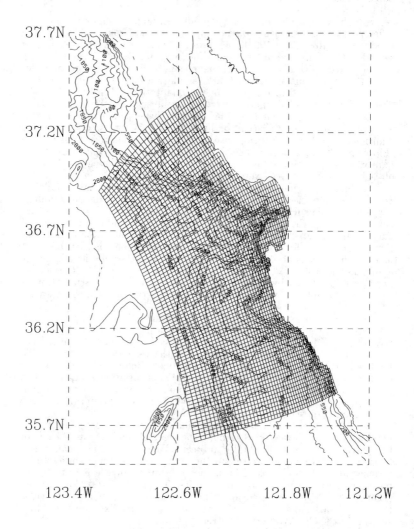

Fig. 1. Model grid and bathymetry

3. Implementation of the Monterey Bay Area Model.

A fine-resolution, sigma coordinate version of the Blumberg and Mellor hydrodynamic model (1987) has been implemented in the Monterey Bay area. This three-dimensional, free surface model is based on the primitive equations for momentum, salt, and heat. It uses the turbulence closure sub-model developed by Mellor and Yamada, and the Smagorinsky formulation is used for horizontal mixing. Additional information on the model can be found in Blumberg and Mellor (1987). The bathymetry and model grid are presented on Fig.1. An orthogonal, curvilinear grid is used for this study. Cross-shelf open boundaries of the model (northern and southern) are approximately orthogonal to the isobaths of bathymetry in order for the flow to be almost perpendicular to the cross-shelf open boundaries. The model has variable resolution in the horizontal, ranging from 1-4 km, with finer resolution around the Bay. Thirty vertical sigma levels were chosen. The Pacific West Coast Model (PWC) is playing a significant role in the initialization and specification of the open boundary conditions for the Monterey Bay model. The PWC (Clancy, et al., 1996) is based on the Blumberg and Mellor model (explicit, sigma coordinate version). The domain extends seaward to 135°W longitude, and from 30°N to 49°N in latitude. The PWC model has a horizontal resolution of 1/12° (around 10 km) and 30 vertical sigma levels. The model includes seven major rivers, and is forced with 12-hourly FNMOC NOGAPS/HR hybrid wind. There is a relaxation of the model sea surface temperature (SST) to the composite observed SST data. An important feature of the PWC model is a coupling to a 1/4°, global Navy Layered Ocean Model (NLOM) which has an assimilating capability for altimeter sea surface height observations. At the initial stage of the development of the Monterey Bay area model, the 12-hourly FNMOC NOGAPS winds are being used. Eventually, the wind forcing provided by NRL's Coupled Ocean/Atmosphere Mesoscale Prediction System (COAMPS) will be used (www.nrlmry.navy.mil/~coamps/). Also, heat fluxes derived from the COAMPS runs are planned for surface forcing of the MBa model.

4. Model Simulations Results.

The model was initialized with a horizontally-constant vertical profile of temperature and salinity based on summer conditions in the Bay. The model was forced with FNMOC NOGAPS 12-hourly surface wind stresses and coupled at the open boundaries to the Pacific West Coast model. The Blumberg and Mellor model uses the so-called splitting technique, where the separation of the vertically-integrated governing equations (barotropic, external mode) and the equations governing vertical structure (baroclinic, internal mode) is introduced. Boundary conditions are formulated for the barotropic and baroclinic modes separately and then adjusted to take into account the different truncation errors for those modes (Blumberg and Mellor [1987]).

Test Case 1. The Monterey Bay area model has been run for one year, from August of 1994 to the end of August 1995. Only the barotropic information from the PWC (sea surface height and transports) was used to specify open boundary conditions for the Monterey Bay area model. The barotropic vertically-averaged velocities on the open boundaries of the Monterey Bay area model were estimated by using the Flather formulation (1976):

$$\bar{u}_n = \overline{u_n^o} + (g/H)^{\frac{1}{2}}(\eta - \eta^o) \tag{1}$$

where u_n is the outward normal component of the velocity on the open boundary at time t (vertically averaged values will be denoted by overbars); $\overline{u_n^o}$ is the outward normal component of the velocity on the open boundary at time t estimated from the PWC model results; η is the model sea surface elevation calculated from the continuity equation and located half of a grid inside of the open boundary; η^o is the PWC model sea surface elevation on the open boundary of the Monterey Bay area model; H is the water depth on the open boundary, and g is a gravitational constant. For the baroclinic mode, information from the PWC model was not used. Alternatively, radiation conditions for layer velocities and advectional boundary conditions for temperature and salinity were used for the baroclinic mode (advected values were from the same single profile used for model initialization). Results of the simulations are presented on the ICON modeling web pages (*www.oc.nps.navy.mil/~icon*). The model reproduced a strong upwelling and the correct position for cold water formation near Pt. Sur, CA. Also, similar to the observations (Rosenfeld et al., 1994), an offshore spreading of a cold water tongue can be seen in the southern portion of the Monterey Bay area. In August, the model produces a strong upwelling and the expected direction of currents on approaching the Bay from deep water. Absent, however, is the background cyclonic circulation within Monterey Bay as seen in 1994 and described by Paduan and Rosenfeld (1996). Also, there is a lack of warm water in the northern part of the Bay as well as an extended cold water formation area in the northern part of the domain. At the southwest corner of the open boundary of the model, a cyclonic circulation was generated and persisted during most of the model run. This flow field seems to follow a bathymetric feature on the open boundary (see Fig.1). However, we tend to attribute this circulation to some artificial instabilities caused by the coupling scheme to the PWC model as well as the use of a single summer T-S profile for estimating the baroclinic open boundary conditions during the entire 12 months of the simulation.

Test Case 2

In this test case, the model was run from June 1994 to the end of 1995. The barotropic as well as baroclinic information from the PWC model results were used to specify open boundary conditions for the MBa model. For the barotropic mode, the open boundary conditions in formulation (1) were used. At the same time, an adjustment procedure was used to balance the records of transport from the PWC model with the records of sea surface elevation. The available outputs from the PWC model have daily records of sea surface elevation and transports; they were spatially interpolated to the MBa grid by using bavariate interpolation, and were linearly interpolated to the MBa model time step in order to form η^o and $\overline{u_n^o}$ in the formulation (1).

As a result, there is a lack of continuity between the total transport through the open boundaries estimated from $\overline{u_n^o}$ and the total change in the sea surface elevation of the modeling area estimated from η^o. To balance these records, any difference was added to the values of $\overline{u_n^o}$ along the open boundaries, those additions were inversely proportional to the water depth at any particular location on the open boundaries. For temperature and salinity on the open boundaries, the advectional boundary conditions were used, but advected values were calculated from the PWC profiles of temperature and salinity, interpolated to the MBa model grid.

The observed 1995 hydrographic conditions (K.M. Sakuma et al, 1996) are reproduced significantly better in the results of this second test case simulation. This is obviously due to the use of baroclinic signal from the records of the PWC model results. For example, in Test Case 2, cool plumes of upwelled water in May -June 1995 extended far north, up to the northern open boundary of the Monterey Bay model, which was not seen in the results of Test Case 1. Also, the ranges of the Test Case 2 August model temperatures are much closer to the observed ranges than in Test Case 1. Moreover, the cyclonic circulation seen in the first test case at the southwest corner of the open boundary of the model is almost absent in the results of Test Case 2. The results of Test Case 2 were compared to the observed data from the MBARI 1 (M1) station, located at $(36^o44' N, 122^o02' W)$. On Fig.2, the time series for 1995 of observed (40 hours low pass filtered data) and model temperatures for M1 location are presented. The model results reproduced the observed trends in temperature for different depths. The computed correlations between observed and model temperature time series for different levels (see Fig.2) are larger than 0.8. At the same time, the model variability is lower then the observed one. We note two situations of high frequency variations in the observations around Julian days 250 and 330. These oscillations (periods of around 2-3 days) are attributed to an oceanic front moving back and forth past the sensor array.

On Fig. 3, the observed and model power density spectra (PDS) for the M1 station location are presented. The solid line is the model PDS. The dased line represents the observed PDS when all observed data presented on Fig.2 were used. The dash-dotted shows the observed PDS when data at two time periods (around 250 and 330 days) were excluded from calculations. The model shows the same frequency for the maximum in the spectra (around 0.11 cycles per day) as the observations. At the same time, the energies at that frequency are lower compared to the observations.

With regard to currents, as noted in Ramp et al., 1997 and K.M. Sakuma et al., 1996, the region featured a northward flow over the continental slope off the Monterey Bay ($36.75^o N$). This northward flow is associated with the California Undercurrent predominated at 200 m. On Fig.4 we present a plot of the model alongshore velocity component to 500 m depth at the cross-shelf section situated around $36.75^o N$. The northward flow over the continental shelf (dashed contours represent negative values) is evident in the model results.

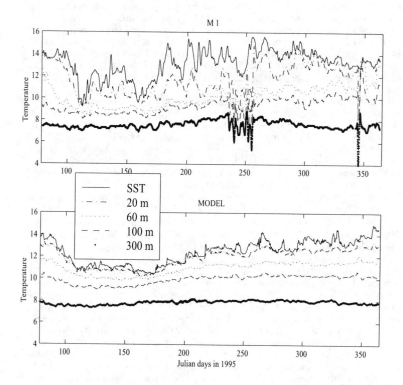

Fig. 2. Time series of temperature at M 1 station (top panel - observations, bottom panel - model)

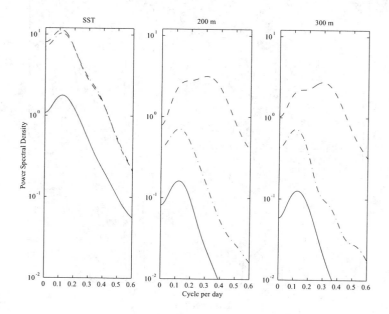

Fig. 3. Observed and model predicted power density spectra

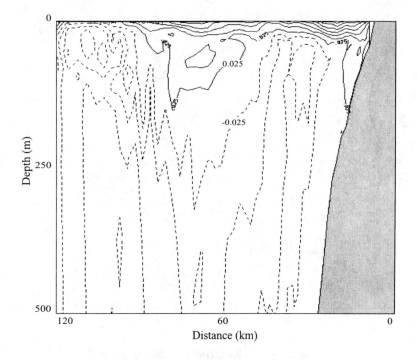

Fig. 4. Plot of alongshore velocity component (m/s, solid contours represent southward flow) up to 500 m depth at the cross-shelf section situated around 36.75 $^{\circ}$ N

5. Implementation of a HF Current Data Assimilating Capability.

5.1 Data assimilation schemes

Below, we will follow the notations and terminology used in (Cohn et al., 1998). Let w^t represents the true state of the oceanic variable w at the model grid point, and let w^f represents the forecast first guess,

$$w^f = w^t + \epsilon^f \tag{2}$$

where ϵ^f represents the forecast error. Let w^o represent the observations available at the analysis time, assuming it is related linearly to the state variables,

$$w^o = Hw^t + \epsilon^o \tag{3}$$

where H is the interpolation operator, and ϵ^o is the observation error. Under the assumption that there is no correlation between forecast error and observation error, the optimal values of analyzed field w^a can be derived from the minimum variance estimation:

$$w^a = w^f + K(w^o - Hw^f) \tag{4}$$

$$K = P^f H^T (H P^f H^T + R)^{-1} \tag{5}$$

Here the matrix K is the Kalman gain, which depends on the forecast error covariance matrix, P^f, and the observation error covariance matrix, R.

$$P^f = \langle (\epsilon^f - \langle \epsilon^f \rangle)(\epsilon^f - \langle \epsilon^f \rangle)^T \rangle \tag{6}$$

$$R = \langle (\epsilon^o - \langle \epsilon^o \rangle)(\epsilon^o - \langle \epsilon^o \rangle)^T \rangle \tag{7}$$

There are two schemes employed in the present research, the OI (Optimal Interpolation) and PSAS (Physical-space Statistical Analysis System) schemes (Cohn et al., 1998). The OI scheme solves equations (4) and (5) directly. On the other hand, the PSAS algorithm first solves the quantity q through a linear system

$$(H P^f H^T + R)q = w^o - Hw^f \tag{8}$$

and then the analyzed state w^a is obtained from the equation

$$w^a = w^f + P^f H^T q \tag{9}$$

The error covariance matrixes P^f and R play a major role in the data assimilation schemes described above. They determine the "blending" of model and observational data. Despite the tremendous amount of research done on estimation of P^f and R during recent years, this subject remains open and we are planning to investigate different approaches for covariance matrixes estimation. In this paper, in our preliminary surface velocity data assimilation experiments covariance matrixes P^f and R were derived from the estimates of horizontal covariances of the observed HF surface current data.

5.2 Covariance matrixes

Estimates of the horizontal covariance for CODAR surface velocities are presented on the web site of the ICON project ($www.oc.nps.navy.mil/\sim icon$). These derived covariances of CODAR surface velocities were approximated by using following assumptions:

a). The covariance matrix for the streamfunction $C_{\varphi\varphi}$ depends only on the distances between points in x and y directions. This covariance is a rotated Gaussian function with different length scales in the x and y directions.

b). The corresponding covariance matrix for the velocity components (C_{uu} and C_{vv}) can be derived from $C_{\varphi\varphi}$ with the assumption that the flow is completely nondivergent (see Daley, 1991).

Let us consider a Gauss function:

$$cov_{i,j} = a \exp\{-b[((\xi_i - \xi_j)/R_1)^2 + ((\zeta_i - \zeta_j)/R_2)^2]\} (10)$$

where (i,j) are two locations with coordinates (ξ_i, ζ_i) and (ξ_j, ζ_j) (ξ, ζ is an orthogonal coordinate system); a and b are unknown parameters; R_1 and R_2 are length scales. Suppose that $C_{\varphi\varphi}$ can be represented by the Gaussian function (10), when axis ξ is the physical axis x rotated counterclockwise with the angle θ. Using Helmholtz's theorem and assuming that the flow is completely nondivergent (geostrophic), the covariance functions for velocity components (C_{uu} and C_{vv}) can be easily derived (see Daley (1991)). These covariance functions depend on five unknown parameters (a, b, R_1, R_2 and θ), which can be determined from the best fit to the observed covariances. The preliminary results of the best fit show that the surface velocity covariances have a seasonal cycle, with the orientation of the major axis (with a larger length scale) changing from season to season throughout the year (Fig. 5). In the preliminary data assimilation experiments described below, these derived covariance functions are used for the forecast error covariance matrix, P^f. The observation error covariance matrix, R, is given by a diagonal matrix with normalized values.

5.3 Surface velocity data assimilation experiment.

Rosenfeld et al. (1994) observed predominantly alongshore southward flow in the upper 70 meters around the Monterey Bay area during the upwelling season. However, the flow is reversed occasionally in accordance with variability in the wind fields. For example, on August 15, 1995, a northward flow appears inside the Monterey Bay, turns westward at 36.85° N, and forms a cyclonic eddy centered at 36.75° N, 122.0° W (Fig. 6, top panel). This cyclonic eddy has not been reproduced by the model (Fig. 6, bottom left panel). In the data asimilation experiment, the surface velocity from from CODAR data on August 15, 1995 was assimilated into the model over the entire month in accordance with the PSAS scheme described in section 5.1, with the covariance matrixes chosen according to section 5.2. The purpose of this experiment is to assess how well the surface velocity assimilation can reproduce eddy-type features. The results after 22 days of

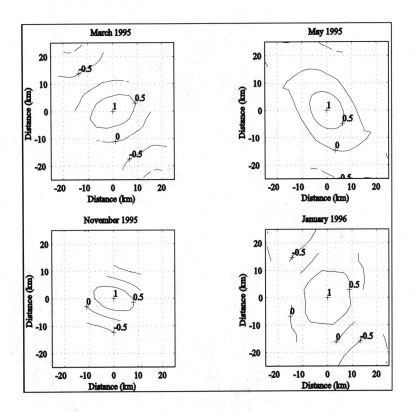

Fig. 5. CODAR surface velocity covariances

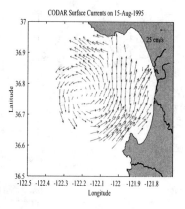

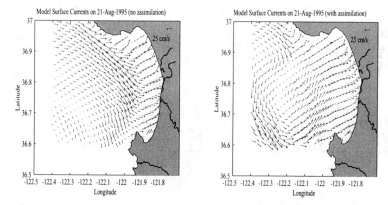

Fig. 6. Observed and model predicted surface velocities in the twin data assimilation experiment

assimilation are shown on Fig. 6, bottom right panel. The cyclonic eddy centered at 36.75° N, 122.25° W which has similar position and trajectory as in observed data is evident in the model results with data assimilation. Due to the fact that in the runs without assimilation and with assimilation (correspondingly, bottom left and right panels of Fig.6) all model forcings were the same, we can conclude that the evident cyclonic eddy is the result of the data assimilation.

6. Discussion and Future Model Development Plan.

The results of model simulations during 1995 reproduced many of the hydrographic conditions observed in the Monterey Bay area. These include cool plumes of upwelled water extending from north to south and seaward of Monterey Bay during May - June 1995, meandering, alongshore ocean front between the upwelled water and the warmer water of the California current, and the northward flow over the continental slope off the Monterey Bay associated with the California Undercurrent predominated at 200 m. Also, the model does well in reproducing the mean water temperatures at a given depth, and it reproduces a peak in energies at 0.11 cpd. The computed correlations between observed and model temperature time series for different levels (see Fig.2) are larger than 0.8. At the same time, the model variability and model energies are lower than the ones observed. This deficiency may be a result of a lack of surface heat fluxes at the air-sea interface of the MBa model. As a result, the model will not include water temperature oscillations associated with sensible, evaporative, and radiative heat fluxes resulting from the passages of atmospheric fronts. As mentioned before, we plan to employ a high-resolution atmospheric forcing from the COAMPS predictions for 1999. In this case, the atmospheric parameters from the COAMPS system will be used to incorporate the various heat fluxes into the MBa model. However, we may find that, even with the inclusion of air-sea heat fluxes, the model produces smaller-than-observed energies in temperature fluctuations. This could require the implementation of a data assimilation technique for satellite sea surface temperatures (SST's) into the model.

In section 5, the assimilation of CODAR surface currents includes only assimilation into the surface layer of the model. The dynamics of the MBa model is used to transfer the surface currents corrections to the subsurface layers. Our future plans include investigation of alternative approaches for projecting surface information into subsurface layers (see, for example, Shulman and Smedstad, 1998).

7. Acknowledgments

This work was supported by the National Oceanographic Partnership Program. We would like to thank M. Carnes of the Naval Oceanographic Office for the help with the optimum interpolation program code. Computer time was provided in part by grants of HPC time from the DoD HPC Centers at: U. S. Naval Oceanographic Office, Stennis Space Center, Mississippi; US Army Corps. of Engineers Waterways Experiment Station (CEWES).

8. References.

Blumberg, A., and G. L. Mellor, 1987: A description of a three-dimensional coastal ocean circulation model. In *Three Dimensional Coastal Models* , N.S. Heaps, Ed., Coastal and Estuarine Sciences, 4, Am. Geophys. Un., 1-16, 1987.

Clancy R.M.,P.W. deWitt, P.May and D.-S. Ko, 1996 "Implementation of a coastal ocean circulation model for the west coast of the United States", in *Proceedings of the American Meteorological Society Conference on Coastal Oceanic and Atmospheric Prediction*, Atlanta, G.A. January 28 - February 2, 72-75.

Cohn S.E., A. da Silva, Jing Guo, M. Sienkiewicz, D. Lamich, 1998: Assessing the effects of data selection with the DAO Physical-space Statistical Analysis System. *Monthly Weather Review*, 126, 11, 2913-2926.

Daley R., 1991: *Atmospheric Data Analysis*, 457 pp., Cambridge Univ. Press.

Lewis J. K., I. Shulman, and A. F. Blumberg, 1998: Assimilation of CO-DAR observations into ocean models. *Continental Shelf Research*, 18, 541-559.

Lipphardt B. L., A. D. Kirwan, C. E. Grosch, J. K. Lewis and J. D. Paduan, 2000: Blending HF radar and model velocities in Monterey Bay through normal mode anlysis, *J. Geophys. Res.*, 105, 3425-3450.

Paduan J.D. and L.K. Rosenfeld, 1996: Remotely sensed surface currents in Monterey Bay from shore-based HF radar (Coastal ocean Dynamics Application Radar). *J. Geophys. Res.*, 101, 20,669-20,686

Ramp S. R., L. K. Rosenfeld, T. D. Tisch and M. R. Hicks, 1997: Moored oobservations of the current and temperature structure over the continental slope off central California. 1. A basic description of the variability, J. Geophys. Res.,102, 22877-22902.

Rosenfeld, L.K., F.B. Schwing, N. Garfield, and D.E. Tracy, 1994: Bifurcated flow from an upwelling center: A cold water source for Monterey Bay, *Continental Shelf Research*, 14, 931-964.

Sakuma K.M., F.B. Schwing, K. Baltz, D. Roberts, H.A Parker, S. Ralston, 1996: The physical oceanography off the Central California Coast during May - June 1995. U. S. Dep. Commer., NOAA Tech. Memo., NOAA-TM-NMFS-SWFSC-232, 146 pp.

Shulman I., O. M. Smedstad, 1998: Data assimilation experiments with the Navy Layered Ocean Model. In *Estuarine and Coastal Modeling*, Ed. Malcolm L. Spaulding, 752 - 765.

Modeling Residence Times: Eulerian vs Lagrangian

David Burwell[1], Mark Vincent[1], Mark Luther[1], Boris Galperin[1]

ABSTRACT

Tampa Bay Florida is the setting for this numerical modeling study of residence times in a coastal plain estuary. The study compares the residence times by two different methods, Eulerian concentration based, and Lagrangian particle tracking using neutrally buoyant particles advected by the Eulerian flow field with an added Markovian dispersion. The results show a doubling of overall residence time for the Lagrangian approach when compared to the Eulerian. The Lagrangian spatial maps of residence time show strong gradients in residence time throughout the bay and likely areas of poor bay flushing.

Introduction

The issue of residence times arises in many studies of Tampa Bay, and other estuaries. Specific applications include larval dispersion, chemical species fate, hazardous material handling and cleanup, and changes in residence time due to fresh water withdrawal from the rivers feeding the bay. Historically, the important concept of estuarine residence time has been variously defined, and usually relied on bulk approximations that were generally spatiotemporally invariant. This research suggests and compares two distinct methodologies for determining estuarine residence times, and is a first step in examining the residence time in Tampa Bay. Of the two methods compared, the first (Eulerian) is a concentration based modeling effort, where the bay is initialized at 100 percent concentration

---

[1]Department of Marine Science, University of South Florida, 140 7th Avenue South, St. Petersburg, FL 33701

in a numerical model and residence time is defined to be the time it takes for each cell's vertically integrated normalized concentration to fall below $1/e$ of the original vertically integrated concentration. The second method is a particle tracking method (Lagrangian) where each model grid cell is initialized with a particle density of 10 particles per grid cell that is uniform in sigma and centered on the cell center. The model outputs the total number of particles in every grid cell each day, by summing the particles in a given grid cell at every internal time step. This total is then normalized by the 14400 particles per grid cell per day that represents no-flow conditions. Residence time is defined to be the time (in days) it takes for each grid cell to fall below $1/e$ of the original normalized daily value without regard to which grid cell a particle started in or the position of the particles in a given grid cell. This paper will describe the data systems used, the numerical circulation model, the particle tracking subroutine, the passive tracer advection scheme, and several comparisons of the two methods. Although the research here is confined to Tampa Bay, the method should be robust enough to be applied to estuarine studies in general.

Data Systems

The modeling effort in this study attempted to capture the dynamics of the bay in as realistic a manner as possible, by using very high resolution environmental data inputs to the model whenever possible. The two systems in the bay for gathering data are the Tampa Bay Physical Oceanographic Real Time System (PORTS) and the Coastal Ocean Monitoring and Prediction System (COMPS).

PORTS

Tampa Bay is the home of the original PORTS installed in 1992 by the National Ocean Service (NOS). It has been housed at and maintained by USF since shortly after its inception. Tampa Bay PORTS is a real-time data acquisition and dissemination system that consists of four tide stations with anemometers scattered around the bay and three current meter stations at various points along the main ship channel. The data are archived both by USF and NOS and were used in this study for validation of model residual fields, as well as, for some of the meteorological inputs to the model.

COMPS

USF also houses and maintains COMPS which consists of several near real time tide stations with various meteorological sensors along the

West Florida coast as well as several offshore buoys with current meters and meteorological packages.

Two of the COMPS tide stations are located near the mouth of Tampa Bay, these stations serve as real-time input to the Now-cast Forecast model that runs at USF, and models the bay in near real time. These systems of monitoring stations allow for integration of model and data sets for many types of benchmark studies, including long time scale historical runs with real data sets. The availability of these long time series at several locations also allow for validation of modeled parameters over long averaging times. (See Vincent et al., 1997 for details on the PORTS/COMPS locations and measured parameters.)

Circulation Modeling System

The Estuarine and Coastal Ocean Model (ECOM) (Blumberg-Mellor, 1987; Galperin et al., 1992a, 1992b) is a version of the Princeton Ocean Model (POM) as modified and adapted for Tampa Bay applications by Galperin, Vincent, and Burwell (Vincent et al., 1997). ECOM uses an orthogonal curvilinear grid in the horizontal and a surface and bottom following sigma coordinate system in the vertical. Vertical turbulent mixing is parameterized by an embedded second moment Mellor-Yamada turbulence closure model and horizontal diffusion is provided by a Smagorinsky formulation. For this study of Tampa Bay, a 70 x 100 grid with 11 sigma layers is used. The internal three-dimensional mode and external two-dimensional mode use respectively a 60 second and 6 second split time step. The advection scheme is second-order and central difference and is non-diffusive while the coefficient for the Smagorinsky diffusion is set near the lower limit for numerical stability.

A passive tracer subroutine was added to the model, as well as a particle advection and tracking subroutine. The passive tracer subroutine mimics the routine used for the advection and diffusion of salt and heat, but is uncoupled from density calculations and has no feedback mechanism to the flow dynamics. The particle tracking subroutine uses a tri-linear interpolation calculation in model x, y, and sigma (Zhang, 1995) to determine the velocity at each particle position at each internal time step (60 seconds). This short time step allows for a simple Euler forward difference particle advection scheme. The dispersion in the particle tracking subroutine is a Markovian random walk, which is in essence Rodean's (Rodean, 1996) random displacement equation (no. 8.43) with eddy diffusivity constant, and with the magnitude of this constant eddy diffusion based on recent measurements. This modification of the Fokker-

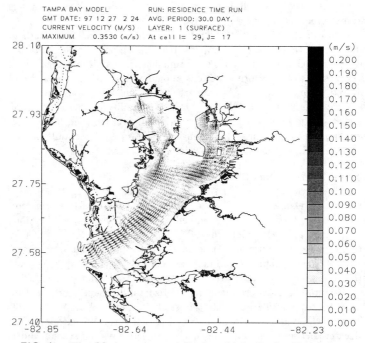

FIG. 1.   The 30 day averaged Surface Velocity for December.

Plank dispersion models the horizontal sub-grid scale dispersion effectively, as recent data shows (Vincent et al., 1997). It can however lead to net mean drift. The sigma component of the dispersion was set to zero for this study, as only vertically integrated particle densities were considered.

The model has been calibrated to the bay geometry and dynamics and does a reliable job of now-casting water levels and currents, in near real time in the bay (Vincent et al., 1997). The residual (30 day averaged) circulation has a classic two layer flow structure (Pritchard, 1956) in the deeper parts of the bay (Figure 1) tapering toward a transverse structure (Wong, 1994) near the edges (Figure 2).

Model Performance

Validation of the model's ability to capture residual (monthly averaged) velocity fields is shown in Figures 3 and 4. The Tampa Bay

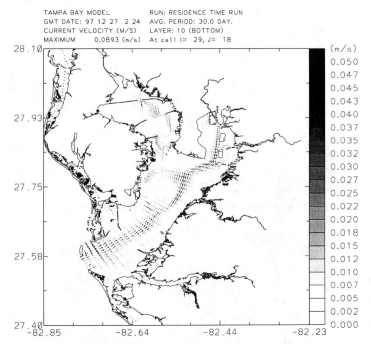

FIG. 2.   The 30 day averaged Bottom Velocity for December.

PORTS Acoustic Doppler Current Profiler (ADCP) data archive for the last three months of 1997 were compared with the monthly averaged velocity profile from the model grid cell containing that particular PORTS current meter. The data (and model) show that, at this location (near the geographical center of the bay), the mean flow is out of the bay near the surface and into the bay at depth. Note the close agreement in absolute residual velocity and the persistent nature of the vertical flow structure from month to month.

Eulerian Monthly Averages

A time series of plots were derived from the model output for both the normalized monthly averaged concentration and the normalized monthly averaged particle density. Figures 5 and 6 indicate the concentration contours for the passive tracer added to the model. These figures

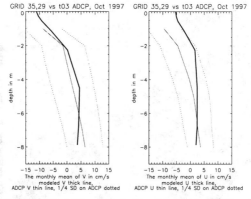

FIG. 3.   The Vertical Structure of horizontal velocities for Oct.

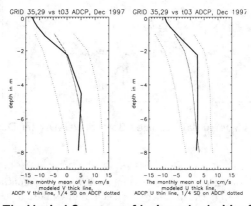

FIG. 4.   The Vertical Structure of horizontal velocities for Dec.

show a progression of lower concentration starting near the mouth of the
bay (Figure 5 left panel), and moving toward the head of the bay (Figure
5 right panel and Figure 6 left panel). Dark regions in each plot show
un-flushed areas as a monthly average from one month following the tracer
release (Figure 5 right panel), until all regions of the bay fall below an
e-folding concentration level (Figure 6 left panel). The bay was initialized
on 10/1/97 with a uniform concentration of a passive tracer. After that
time the tracer input to the rivers and open boundary was maintained at

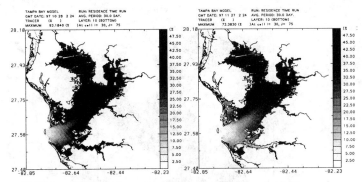

FIG. 5.   Averaged concentration for October and November.

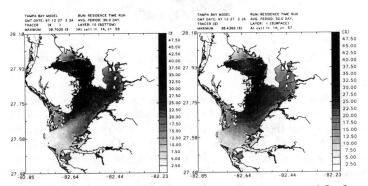

FIG. 6.   Averaged concentration for December Bottom and Surface.

zero concentration. It is clear here that the main pathway for flushing is via the open boundary along the navigational channel, and that the lower bay southwest of the Sunshine Skyway bridge causeway, is quickly flushed, while regions near the northern head of the bay take of the order of 60 days longer to reach 1/e of their original concentration. The right panel in Figure 6 is included to show that the surface signature of this monthly average is similar to that of the bottom, due to the well mixed nature of this estuary.

Lagrangian Monthly averages

The Lagrangian series of monthly plots (Figures 7 - 11) tell a

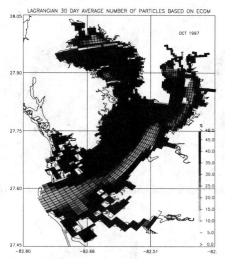

FIG. 7.   The 30 day average particle density for Oct.

similar but more detailed story of the flushing pathways in the bay. The particles are carried by the flow field into convergence zones and persistent eddies, so that the bottleneck at the Sunshine Skyway bridge causeway turns into a particle trap. The residual velocity fields suggest that there should be regions of very long flushing times, and the monthly averaged concentrations implied that the upper bay should be slower to flush. Due to the numerical diffusion inherent in the Eulerian method, however, the details were smoothed and underestimated in time. The Lagrangian fields give a more detailed (and realistic) picture of the flushing. It should be noted here that the Lagrangian plots have been spatially smoothed with a 5X5 boxcar for graphical purposes. The October 1997 through February 1998 monthly averaged plots are included for completeness.

Bay Wide Residence Times

The difference in the two methods on a bay wide average is shown in Figures 12 and 13. The Eulerian plot (Figure 12) shows the expected exponential decay of the normalized concentration in the bay. The steep drop to below $1/e$ of the original concentration can be in part explained via the numerical diffusion inherent in the Eulerian approach, which is a well known problem for grid-based models (Zhang, 1995).

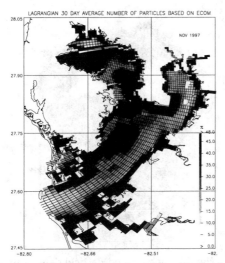

FIG. 8. The 30 day averaged particle density for Nov.

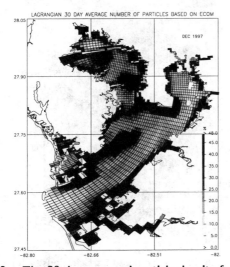

FIG. 9. The 30 day averaged particle density for Dec.

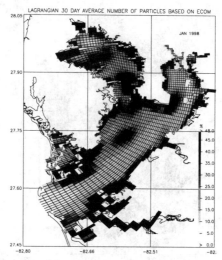

FIG. 10.   The 30 day averaged particle density for Jan.

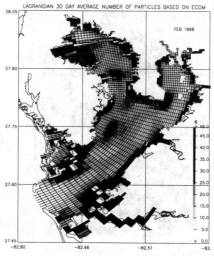

FIG. 11.   The 30 day averaged particle density for Feb.

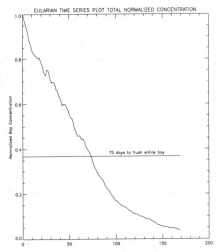

FIG. 12.   The Bay Wide Normalized Eulerian Residence Time.

The Lagrangian plot (Figure 13) shows a more gradual decline toward zero due to the non-dispersive nature of the particle tracking method that relies on the grid structure only for interpolated velocity, and utilizes a random walk to simulate dispersion in the near field. The slow flushing could also be attributed somewhat to the clumping of particles around flow field convergence zones, away from divergent areas which could leave the divergent areas underrepresented from a particle concentration point of view.

Eulerian Maps

The Eulerian spatial map of grid cell by grid cell residence times (Figure 14) shows the time in days until the last $1/e$ crossing for each grid cell in the bay. Recall that the modeled bay had an initial uniform concentration (100 percent) on 10/1/97. This plot appears smoothed in space, due to the numerical diffusion inherent in the method. The low residence times occur near the mouth (order 20 days), and high residence times in the mid to upper bay on the north side (order 80 days). This plot hints at the structure due to the deeper channel running approximately NE/SW from the mouth of the bay toward the eastern lobe of the head of the bay. It also shows an accumulation behind the bridge causeways near the mouth of the bay but is too diffuse to see details.

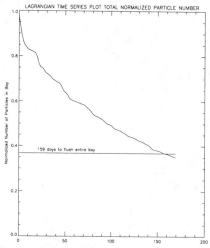

FIG. 13.   The Bay Wide Normalized Lagrangian Residence Time.

Lagrangian Maps

The Lagrangian plot (Figure 15), shows the e-folding time for each grid cell from an initial normalized particle density of 100 percent, also initialized on 10/1/97. Due to the highly variable nature of the particle density, these results are derived from the last crossing of the 7 day smoothed time series for each grid cell. The plot has been further smoothed with a 5X5 boxcar in space for graphical purposes. The major difference in this plot is the strong gradients in flushing time. This structure is possible due entirely to the non-diffusive nature of the particle tracking method. The mouth of the bay region flushes very quickly in agreement with the concentration method, however, there is a distinct region of low residence times along the deep navigational channel from the mouth up to the eastern head of the bay. This result is in agreement with previous observations by Galperin et al., (1992b) that the residual advection of higher salinity waters from the shelf along the deep channels is the dominant long term transport mechanism that maintains the salinity distribution in Tampa Bay. The sides of the bay show much longer (order 100 days) residence times implying that particles swept toward the bay edges experience low flow and are piled up behind the restriction of the Sunshine Skyway bridge causeway. This acts as an impediment to

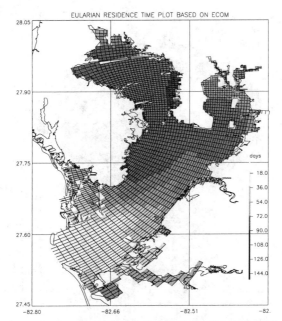

FIG. 14.   The Bay Wide Eulerian Residence Time Map.

outgoing particles, as well as providing an eddy structure causing particle deposition.

Conclusion

Tampa Bay, Florida has been an ideal site for this study. With an existing model, ECOM, two different real time data systems, COMPS and PORTS, the data for boundary conditions and for model validation were readily available, and the residual flow fields accurately captured.

The residual circulation in the bay appears to be a mix of classical two layer flow over the deeper regions, tapering toward a transverse structure near to the edges. The major avenues for flushing are the deep navigation channels running approximately NE/SW through the bay from the mouth. Residence times in these areas are short, on the order of 15 days to one month, while outside this region, particularly near the bay edges and near persistent eddies, the residence time increase is dramatic, up to over three months. The Eulerian fields seem to capture the overall

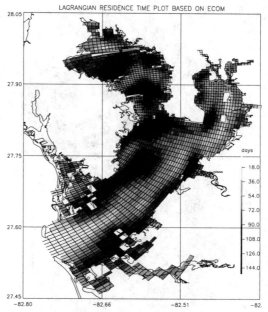

FIG. 15.   The Bay Wide Lagrangian Residence Time Map.

structure of the variation in residence time over the bay during the period studied. The map, however, shows considerable diffusion compared to the Lagrangian map and underestimates residence times in the bay by up to a factor of two, in some areas. The goal of quantifying residence times in a spatially varying map of the estuary is achieved in this study by two methods, and the Lagrangian method captures the structure in more detail than the Eulerian.

It should be noted that, the diffusion in the Eulerian scheme could be reduced via higher order diffusion scheme, and tuning the diffusion coefficient. However some numerical diffusion is inherent in grid based models, and the diffusion coefficient used is near the lower limit for numerical stability. Future research with different schemes could quantify the inherent numerical diffusion. The current scheme however approximates a lower bound on the residence time for the entire bay (Figure 12). The Lagrangian bay wide residence time (Figure 13) could be viewed as an upper bound for the residence time in the bay during this period. However, in the Lagrangian scheme, the effect of possible longterm drift due to the choice of constant eddy diffusivity in the random displacement equation, and the sensitivity to the number of particles used, have not been calibrated to long term data. Future studies should include a passive tracer released in the bay for calibration of these parameters.

Acknowledgments

The authors would like to thank the reviewers for their comments, which helped to refine this work.

References

Blumberg, A. F. and Mellor, G. L. 1987. A description of a three-dimensional coastal ocean circulation model, in Three-Dimensional Coastal Ocean Models. Heaps, N. S. (Ed.), American Geophysical Union, Washington, DC 1- 16.

Galperin, B., Blumberg A. F., and Weisberg, R. H., 1992a. A time-dependent three-dimensional model of circulation in Tampa Bay, in Proceedings, Tampa Bay Area Scientific Information Symposium 2, Tampa, FL, February 27 - March 1, 1991, 77 -98.

Galperin, B., Blumberg A. F., and Weisberg, R. H., 1992b. The importance of density driven circulation in well mixed estuaries: the Tampa experience. In Proceedings of the 2nd International Conference on Estuarine and Coastal Modeling, Tampa, FL, November 13-15, 1991, 332 -343.

Pritchard, D. W., 1956. The Dynamic Structure of a Coastal Plain Estuary. J. Marine Res., 33-42, 1956

Rodean, H. C., 1996. Stochastic Lagrangian Models of Turbulent Diffusion. Meteorological Monographs, V.26, No. 48, American Meteorological Society, 84pp, 1996.

Vincent,M., Burwell, D., Luther, M., Galperin, B., 1997. Real-Time Data Acquisition and Modeling in Tampa Bay. In Proceedings of the 5th International Conference on Estuarine and Coastal Modeling, Alexandria, VA, October 22-24, 1997, 427 - 440.

Wong, K.-C., 1994. On the Nature of Transverse Variability in a Coastal Plain Estuary. J. Geophys. Res. 99(C7), 14209-14222, 1994.

Zhang X., 1995. Ocean Outfall Modeling - Interfaced Near and Far Field Models with Particle Tracking Method Doctoral Dissertation, Massachusetts Institute of Technology, Cambridge, MA. 159, 1995.

PEARL RIVER ESTUARY POLLUTION PROJECT (PREPP) –
An INTEGRATED Approach

Jay-Chung Chen, Ophelia Lee, Chan Wai Man
Center for Coastal and Atmospheric Research[1]
Hong Kong University of Science and Technology
Clear Water Bay, Hong Kong

Introduction

The Pearl River Delta region, being part of the South China Sea, has experienced tremendous growth in the past couple of decades. The increased social and economic activities brought in serious deterioration of the water quality due to pollutants discharged into the Pearl River estuary system. The objectives of the Pearl River Estuary Pollution Project (PREPP) are to study the relative flux of toxic pollutants, sediment and nutrients entering Hong Kong waters from the Pearl River. The outcome of the study is to provide information for international scientific community, Hong Kong and Chinese environmental policy makers and the Hong Kong public community so that a concerted effort will be made to reduce the impact of the Pearl River pollution on Hong Kong water quality and ecological components of the ecosystem.

Background

The Hong Kong waters are seriously disturbed by industrial, agricultural and domestic pollution. From time to time, the coastal environment of Hong Kong has been the site of red-tide outbreaks and other harmful algal blooms; there have also been instances of fish kills and disruption of fish farming activities. These significant and unpredictable events can be regarded as marine environment crises. They indicated that the balance of the natural marine ecosystem is being perturbed. In view of this, the Hong Kong has laid down policy and legislation to address the quality of the environment with regard to air quality, water quality, noise, solid waste disposal.

Pearl River, being one of the largest rivers in the world, contributes to the marine pollution in Hong Kong. The Pearl River catchment covers 453,690 km^2 and annual nutrient loading in 1989 was estimated at COD (12.05 billion tonnes) and total ammonia, nitrate and nitrate of 45.3 tonnes per year. The Chinese governmental agencies at the several levels have monitored basic water quality parameters in the Pearl River and coastal area since early 1980 as part of the National Monitoring Network of Marine Environmental Pollution (Dong and Zen, 1993). In Hong Kong, the Environmental Protection Department has an extensive water quality monitoring network throughout about changes in parameters including e.g. nutrients in the water column and heavy metals in the seabed. Broom and Ng (1995) used the WAHMO model, a hydrodynamic model to investigate the influence of the Pearl River on Hong Kong waters. The concluded that "water quality in Hong Kong is heavily influenced by polluting loads, discharge and hydrodynamic characteristics of the Pearl River".

---

[1] Home page of Center for Coastal and Atmospheric Research: http://ccar.ust.hk

An integrated approach

In 1998, the Hong Kong University of Science & Technology proposed a 3-year integrated study of the water quality of the Pearl River Estuary. This interdisciplinary project would integrate marine biology, chemistry, hydrology, environmental sciences together by deploying various cutting edge technologies for <u>measurements</u>, <u>processing</u> and <u>information dissemination</u>. A schematic of the system is shown in Figure 1

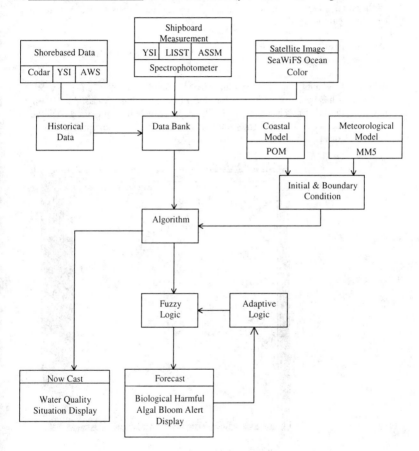

Figure 1 Schematic of the PREPP

The schematic illustrates the relationships of the various components such as the data sources, the mathematical models, the algorithm and the output (the now cast display and the forecast display). In order to give a now cast display, the system must have the capability and the infrastructure to ingest real time data from the various sources. Can one imagine that the now cast will give information that is of yesterday ? So the real time capability becomes a critical requirement of the infrastructure. With the availability of the real time data as the boundary condition of the mathematical model, the 'predictive logic' of the system can then forecast hazardous events such as the harmful algal bloom alerts.

Measurement
- ### *Shorebased data*

In order to understand the coastal situation of the South China Sea, a network of instruments has been installed in various locations. Shorebased SeaSonde HF (high frequency) radars for ocean surface current mapping have been procured from the CODAR Ocean Sensors Ltd. The Codar radar measures the ocean current on the horizontal surface area and output current display as shown in Figure 2

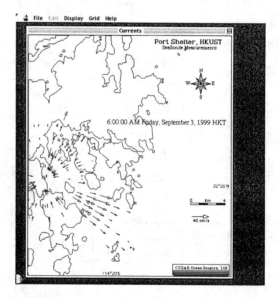

Figure 2.   Codar output indicating the ocean current

Our Codar radars were configured to a preset resolution of 3 kilmeteres and 32 grid range cells giving us a coverage of 96 Km. We have three sites (1) Castle Peak, Tai O, Macau for the Codar equipment. The Macau and Castle Peak sites were chosen to have an overlap coverage so that the radial current from the two sites will form a total vector current map at real time on an hourly basis. The Castle Peak site also serves as the central site which provides communication link with the two remote sites.

While the Codar radars allow us to understand the current of a surface area, we also deploy other equipments for the vertical distributions and other parameters. We have acquired Multi-parameter water quality monitor from the YSI Massachusetts and the configuration of our YSI monitoring system allow us to install various sensors for measuring different parameters. These sensors include (1) dissolved oxygen probe, (2) pH probe, (3) turbidity probe, (4) chlorophyll probe to measure these parameters at three different locations (1) Hu Men (2) Heng Men (3) MoDao Men.

Apart from the above, there are Automatic Weather Stations (AWS) installed in five strategic locations extending to the boundary of China, namely at (1) the campus of HK University of Science & Technology (2) Castle Peak (3) Hu Men (4) Heng Men (5) MoDao Men. Among the five sites, site (1) and (2) are located in Hong Kong while the other three are located in Guangdong province of China. The sensors mainly ingest wind speed, wind direction, humidity and temperature which are factors relating to the biological and chemical conditions of the coastal environment. The locations of these shorebased sensors are shown in Figure 3.

- *Shipboard measurement*

In order to validate and complement the shorebased data, shipboard measurements were conducted for the wet and dry seasons during the project period. We have chartered a ship with a displacement of 996 tons with a cruise team of 40 people. The ship has various types of onboard measurements. The cruise last from 10-15 days taking samples from over one hundred locations along the Pearl River.

The multi-parameter water quality monitors from the YSI Massachusetts were deployed in the cruise for collecting parameters such as water temperature, salinity, dissolved oxygen, pH value, turbidity, chlorophyll. While the shorebased equipment can only take stationary measurement, the shipboard equipment provides measurement from various depths and various locations of the coastal area. There were other equipments such as the Submersible laser-scattering instrument (LISST-ST) that were used to measure the particle size distribution, settling rate and pressure. The acoustic suspended sediment monitor (ASSM-2) was used to measure the vertical suspension sediment concentration profile in the water column. The spectrophotometer in the water column also measured nutrients (NO_3^--N, NO_2^--N, NH_3-N, DTN, SiO_3^{3-}-Si, PO_4^{3-}-P, DTP) in the water

- *Satellite images*

Satellite data acquisition system was installed to collect real-time satellite image of the ocean color and sea-surface temperature. Our High Resolution Picture Transmission (HRPT) station was installed and became operational on November 21, 1994 for the reception of the SEA-viewing WIde Field-of-view Sensor (SeaWiFS) ocean color data and Advanced Very High Resolution Radiometer (AVHRR) transmissions from the NOAA polar orbiting satellites. There are also plans to acquire satellite images from the Chinese HY-1 satellite that will be launched sometimes in the year 2000.

Processing

Data from both the shorebased measurement and the shipboard measurement were integrated together with historical data that we have obtained from other government environmental agency and exported to a database, which is a *real-time* GIS (Geographical Information System) based system.

Most of the current GIS applications are static or associated attributes are to be remained constant over time. In the applications, the need to make a timely and appropriate decision corresponding to the data that are collected and updated responsively is undoubtedly important. Thus, the GIS system was designed with a web-based architecture to fetch real-time data for a dynamic decision support system and fuzzy algorithm. Conceptually, our system can be seems as a real-time GIS or an application of integrating GIS in a real-time setting. The Active Server Page, Perl script and C/C++ are selected for implementing the real-time GIS which was dynamically linked to an external database and web server. Scientists or operators

can use our GIS system for data ingest monitoring through internet browser. Figure 4
illustrates the real-time data ingest monitoring screens.

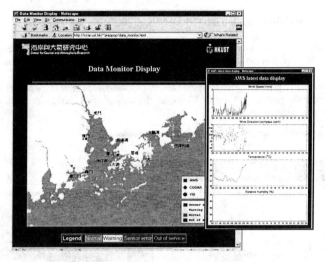

Figure 4.   GIS data ingest monitoring display

As for the real-time data management of the in-situ measurement data ingest process,
the GIS has to match several system requirements. The system configuration should
be joined with real-time data acquisition interface and a real-time kernel. The
framework of the integrated real time dispatching/fetching information system is
exhibited in the following figure.

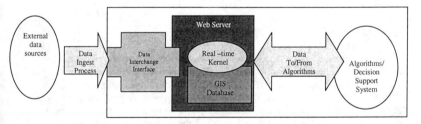

Figure 5.   System architecture

In Figure 5, the external data source, data interchange interface, real-time interface
(for monitoring and reporting), GIS software, web server, data repository are the
principal components. External data source includes historical data, in-situ
measurements data ingest and simulated data from the POM models. On site data are
multiplexed with data-logger equipment and transmitted to our web server via dial-up
or internet line and incorporated into the data repository. The data interchange

interface is to convert the ingest data into a pre-defined format and to ensure the data quality.

Information dissemination

Once the GIS is invoked, it updates and fetches data to our fuzzy algorithm and decision support system to classify the water quality. The results are displayed on our web interface for monitoring operation. In the end, the GIS database is then automatically invoked again to update the results and store the data for adaptive logic system at the back end. Figure 6 illustrates the GIS interface design for water quality display in different regions.

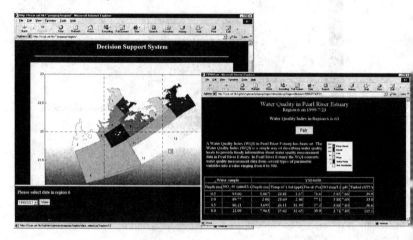

Figure 6. GIS interface design for water quality display in different regions.

Algorithm

The core of the processing is to define a set of algorithm and display water quality situation; this is the now cast output which is to tell the scientists what is currently happening to the water quality. The software will compare the various parameters that affect the quality water quality of the same region with the previous years. The algorithm is a rule-based and case-based reasoning. It uses a set of rules to obtain a preliminary answer for the given phenomena, it then draws analogies from cases to handle exceptions to the rules. Having rules together with case as not only increases the domain coverage, it also allow innovative ways of doing case-based reasoning. The same rules that are sued for rule-based reasoning are also used by the case-based component to do case indexing and case adaptation.

The sophisticated component is the forecasting of water quality which is basically a composite index of the various parameters. The ultimate goal of this index is to let the users of the system know whether the water quality is very good, average or poor. Since the user of this index will comprise of a big population including non-scientists such as fishermen, businessmen, policy makers, this index should reflect the science and yet comprehensible since it may affect the day to day decision of the users.

There are a number of indices for different purposes, as fishermen's concern and swimmers' concern are quite different. Therefore we need to have different indices for various user groups. Each index has a pre-defined set of scientific parameters, which are classified into different fuzzy sets. There are two types of fuzzy sets : (1) one is the scientific parameters and (2) the event related parameters such as seasonal and location. These sets are characterized by a membership function. The fuzzy logic is a set of algorithm (or rules) to define the relationship of these fuzzy sets with respect to the index. In other words, the parameters will contribute to a discrete value of the membership function which may be insignificant for any alert or warning, however, a composite of these discrete values will create an index which carries a meaningful message.

The adaptive logic is the fine tuning of the system. The initial set of algorithm is a good baseline. As time goes, the system has the "self-learning" capability to build new rules and relationship of parameters.

Summary

This paper has introduced an integrated approach as a solution to a pollution monitoring in the Pearl River Estuary. In this project, a sophisticated system has been developed to integrate various formats of data from different devices and scientific models. Without this infrastructure, the data will remain as 'islands' of data and fail to provide the real time capability. The objective of this project is to establish a common platform for water quality data ingest, a standard architecture to process the data and also a internet environment to disseminate the information which is useful to the industry, the scientific community, the policy makers and even to the general public.

Acknowledgement

The present study was supported by the Pearl River Estuary Pollution Project funded by the Hong Kong Government/Hong Kong Jockey Club, and by the 863/818-09-01 granst from the Ministry of Science and Technology of China.

References

Bennet, S 1998, *Real-time Computer Control: An Introduction*, Prentice-Hall, UK.

Broom, M.J. and Ng. A.K.M. 1995. *Water quality in Hong Kong and the influence of the Pearl River. pp. 193-213.* Proceedings of the Symposium on hydraulics of Hong Kong Water 28-29 November 1995.

Dong, D. and Zen, X. 1993 *Sutdy on regularity of sea-entering pollutants flux from Pearl River. pp131-138.* International Workshop on The Development of Strategies for Pollutino Control in the Pearl River Delta 15-19 February 1993.

Zeng, T and Cowell, P.J. 1998, *A GIS-based modelling of Coastal Hazard due to Climate Change*, Proceedings of International Conference on Modeling Geographical and Environmental Systems with Geographical Information System

Implementation of Vertical Acceleration and Dispersion Terms In An Otherwise Hydrostatically Approximated Three-Dimensional Model

John Eric Edinger (Member, ASCE)[1] and Venkat S. Kolluru[1]

Abstract

The acceleration and dispersion terms in the vertical direction are developed for the 3-D hydrodynamic and transport model GLLVHT. With these additional terms, the model was extended to simulate the effects of a compressed air bubbler on the vertical temperature structure of a reservoir used as a source of cooling water for a thermal power plant. The bubbler was proposed as a method to reduce power plant intake temperatures. Two bubbler applications were simulated: a "bubbler curtain" used to exclude the thermal plume from the intake and a deep bubbler to mix cooler bottom water. The results show that temperature reduction can be obtained effectively if the bubbler is operated on an intermittent basis, which allows for the buildup of cooler hypolimnetic water.

Introduction

Most current 3-D hydrodynamic models simplify the Navier-Stokes equation in the vertical direction by use of the Boussinesq approximation, i.e., the hydrostatic relationship. However, the vertical acceleration terms become important in the near field of an issuing momentum jet. Typical examples where the vertical terms become important include thermal discharges in shallow depths, bubblers and intakes (Schladow, 1993).

GLLVHT Model Description

The theoretical basis of the three-dimensional, Generalized, Longitudinal-Lateral-Vertical Hydrodynamic and Transport (GLLVHT) model was presented in Edinger

---

[1] J. E. Edinger Associates, 37 West Avenue, Wayne PA 19087

and Buchak (1980). GLLVHT provides three-dimensional, time-varying simulations of rivers, lakes, impoundments, estuaries and coastal water bodies. The GLLVHT model has been peer reviewed and published (Edinger and Buchak, 1995). The GLLVHT model has over 40 applications including many water quality modeling and cooling water discharge cases. A previous assumption in this model, as in almost all other free surface hydrodynamic models is that the hydrostatic approximation is valid over the space and time scales used to model the prototype.

Fundamentals of the Momentum Balances

There has been much interest in including vertical momentum in the hydrodynamic relationships to eliminate the hydrostatic approximation. Inclusion allows handling problems of downward withdrawal at outlet structures, hypolimnetic pumping devices in stratified reservoirs, bubbler mixing systems, and the vertical momentum of high velocity discharges. The complete momentum formulation is as follows:

The horizontal x-momentum balance is taken as:

$$DU/Dt = -1/\rho_o \ \partial P/\partial x + Smx \quad\quad\quad (1)$$

The horizontal y-momentum balance is taken as:

$$DV/Dt = - 1/\rho_o \ \partial P/\partial y + Smy \quad\quad\quad (2)$$

and the vertical z-momentum balance is taken, with z positive downward as:

$$DW/Dt = g - 1/\rho_o \ \partial P/\partial z + Smz \quad\quad\quad (3)$$

where:

$D(\ )/Dt$	=	sum of the local acceleration of momentum, and advection and dispersion of momentum
U,V,W	=	velocity components in each of the x, y and z directions.
g	=	gravitational acceleration
ρ_o	=	density
P	=	pressure.

Smx, Smy, Smz are the specific momentum in each direction due to stuctures, bubblers, diffusers, etc. The vertical specific momentum, Smz, is incorporated in the vertical momentum relationship similar to the method used to incorporate the horizontal momentum in GLLVHT as shown in Edinger and Buchak (1995).

The numerical solution to the momentum balances depends on evaluating the horizontal pressure gradients from the vertical pressure distribution as it depends on density, water surface elevation, and the vertical component of momentum Smz. The vertical momentum balance can be written as:

$$SM_z = g - 1/\rho\ \partial P/\partial z \tag{4}$$

where:

$SM_z = DW/Dt - Smz$, and is the specific momentum in the vertical direction including the vertical acceleration terms (DW/Dt). The numerical forms of the horizontal momentum relationships require the horizontal pressure gradient as developed from the pressure. The pressure is:

$$P = g \int_{z'}^{z} \rho\ dz - \rho \int_{z'}^{z} SM_z\,dz \tag{5}$$

Differentiating Equation (5) and using Leibnitz's rule gives the horizontal pressure gradient in the x-direction as:

$$1/\rho_o\partial P/\partial x = g\partial z'\ /\partial x - g/\rho_o \int_{z'}^{z} (\partial\rho/\partial x)\ dz - \int_{z'}^{z} (\partial SM_z/\partial x)\ dz \tag{6}$$

where z' is the water surface elevation. The first term on the RHS is the barotropic surface slope, the second term is the baroclinic slope, and the third term is the addition to the horizontal pressure gradient introduced by the vertical specific momentum. Inclusion of the acceleration of vertical velocity in the evaluation of SM_z eliminates the hydrostatic approximation from the numerical relationships.

A relationship similar to (6) can be written in the y-direction. The vertical momentum of a discharge can be included in the evaluation of specific momentum SM_z as $QVz/(\Delta x\Delta y\Delta z)$ where Q is the discharge flow rate, Vz is the vertical velocity component of the discharge, and $\Delta x\Delta y\Delta z$ is the volume of the finite difference cell into which the discharge is placed. The vertical momentum of a discharge thus changes the horizontal pressure gradient and causes horizontal flows to be accelerated inward toward the momentum source. The vertical momentum terms are included in the forcing functions, Fx and Fy, used for the

GLLVHT solution as given in Edinger and Buchak (1995) and the numerical solution proceeds as given there.

Implementation of the Bubbler Algorithm

The vertical momentum computations described in the previous section were extended to simulate an air bubbler in a waterbody. The bubbler is introduced in the GLLVHT model as a continuous source discharging air in the vertical direction towards the water surface. The vertical velocity of air that enters into the specified computational cell is computed using analytical solution given by Cederwall and Ditmars (1970) for a single bubbler. It is given by

$$V_z = \left[\frac{25 g Q_0 H_0 \left(1 + \lambda^2\right)}{24 a^2 \pi \left(H_0 + H\right)} \right]^{1/3} x^{-1/3} \tag{7}$$

where

Q_0 = volume of air discharged at atmospheric pressure in, $m^3 s^{-1}$

H_0 = piezometric head equivalent of atmospheric pressure, m

H = depth above the air source, m

α = coefficient of entrainment

λ = ratio of the lateral spread of density deficiency to momentum

x = distance from the virtual source.

The virtual source as in single-phase plume analysis is usually located below the real source of finite dimension.

To test the algorithm, the bubbler was placed in the center of a numerical wave tank grid. The wave tank 3-D model grid consists of 30 cells in x and y directions of size 50 m and 20 layers in the vertical direction of thickness 0.5 m. The value of α, λ, H_0 was set at 0.2, 0.065 and 10.32 m, respectively. The value of Q_0 was set at 0.6 $m^3 s^{-1}$. A vertical slice through the bubbler location is shown in Figure 1. This figure shows the horizontal entrainment of ambient water due to the vertical motion of air introduced into the wave tank. The velocity pattern close to the bubbler location and near the surface is shown in Figures 2 and 3, respectively. These figures clearly show the bubbler flow characteristics as predicted by (Fanneløp and Sjøen, 1980).

Application of the Bubbler Algorithm

Squaw Creek Reservoir (SCR), shown in Figure 4, provides once-through cooling water for two electrical generating units. The SCR is located in central Texas and has a surface area of 13.3 million m^2, a volume of 186.8 million m^3 and mean depth of 14 m. Each unit generates approximately 1165 MWe of electricity from a reactor thermal output of 3411 MWt. The waste heat load is approximately 2246 MWt for each unit. Cooling water enters the station from the western edge of the SCR about half way up the SCR, picks up waste heat from the condensers and is returned to the SCR at its southwestern end.

There is a single deep intake structure for both generating units. Each unit is cooled by 4 circulating water pumps with a capacity of 63 m^3 s^{-1}. Normally only 3 pumps per unit are run during the winter (October 1st to June 15th), resulting in a per unit rate of 47 m^3 s^{-1}. At full load and summer pumping, Unit 1 and Unit 2 have a condenser temperature rise of 8.5°C; at full load and winter pumping, the condenser temperature rise is 11.4°C. The residence time based on these pumping rates is 17 and 23 days for summer and winter, respectively.

Several intake configurations were examined with the goal of reducing intake temperatures. Bubblers were evaluated as an option using the GLLVHT model; other options included a fabric curtain wall (see the companion paper by Kolluru et al., 1999), cooling towers, discharge structure modifications and relocations, and plant efficiency improvements. A detailed description of the application of GLLVHT 3-D model to the SCR can be found in Buchak et al., 1999.

The bubblers simulated by the model included two types: a "bubbler curtain" used to exclude the thermal plume from the intake and a deep bubbler to mix cooler bottom water. For the former case, the bubbler was more like those used as surface debris separators used at the entrance to harbors. The vertical rise and subsequent bi-directional flow away from the plume establishes a current that effectively traps debris in the harbor.

However, air bubblers are used more frequently in lakes to mix low DO hypolimnetic water with high DO surface water ("bubble plume destratification"). In the application described here, the destratification process was used to bring cool hypolimnetic water to the surface where it could be drawn into the intake. Both the "bubbler curtain" and the "bubble plume destratification" approaches were proposed and subsequently evaluated in the 3-D simulations described here.

In the first series of simulations, a set of six bubblers was located around the intake at the bottom with a very large total airflow rate of 1.7 m^3 s^{-1} as shown in Figure 4. Figure 5 (option 3h) shows the longitudinal slice through the bubbler curtain and the associated velocity pattern. Especially interesting is the retardation of the thermal wedge, which was the primary objective. The average reduction in excess temperature during the simulation time period was 0.3°C as shown in Figure 6. Because the bubblers eventually mix the Squaw Creek lake, the intake temperature reduction is not a steady value, but begins with a large improvement that eventually causes the lake to completely or partially destratify. Simulations showed that the bubblers could be operated intermittently to reduce long term mixing. Results are shown in Figure 7 (option 3i). The main feature of the intermittent scheme is an immediate reduction in intake temperature of 1.4°C, followed by a slight increase as the Squaw Creek lake restratifies while the bubbler is off. There follows a secondary improvement of 0.7°C when the bubbler is again turned on.

A second series of simulations was developed based on anecdotal information that an existing bubbler located mid-depth halfway between the discharge and intake (Figure 8) reduced intake temperatures. The bubbler in this case was thought to deliberately mix cooler bottom water with the warmer surface layer. Destratification of the SCR would presumably depress intake temperatures until the cool bottom water was depleted.

A detailed comparison of the intake temperature record against meteorologically-generated surrogate water temperatures showed insignificant reductions. However, the perception that the bubbler destratified the SCR was reinforced by thermistor data (see Figure 8 for locations of the thermistor chains). Fundamentally, there was no direct way of comparing the prototype behavior for the "with" and "without" cases. The 3-D model provided such a method and was used to help sort these conflicting data sets.

Figure 9, Figure 10 and Figure 11 show computed profiles for a sequence of days at three stations prior to, immediately after and a about week after activation of the bubbler. The Base Case profiles shown on the figures are the simulation without the bubblers; two airflow rates were also simulated. The lower airflow is the rate of the existing bubbler. Note that is it partially effective and in fact mixes the water column from its placement depth to the surface. The proposed higher airflow rate is more effective. Figure 13 shows that both the anecdotal information as well as the thermistor data are both correct: the existing bubbler is capable of partially destratifying the SCR, but intake temperature effects are small. This bubbler will be moved to deeper water next summer in a field test of its effectiveness as well as of the model's predictive ability.

Conclusion

A new approach is developed to include vertical acceleration terms in the 3-D hydrodynamic flow equations. It is currently implemented in the GLLVHT 3-D model and applied to simulate bubblers in a lake which has been used as a way of reducing power plant intake recirculation temperature during summer months.

Acknowledgement

The authors wish to thank Edward M. Buchak, who managed the Squaw Creek Reservoir project, developed the frequency-duration analysis, and Dr. Jian Wu, who ably performed the complex application simulations described here.

References

Buchak, E. M., V. S. Kolluru, and J. Wu. 1999. Studies in Support of Circulating Water Temperature Reduction Using Three-dimensional Modeling and Frequency-Duration Analysis. Prepared for TU Electric, Glen Rose, Texas. Prepared by J. E. Edinger Associates, Inc., Wayne, Pennsylvania. 18 May.

Cederwall, K. and J. D. Ditmars, 1970. Analysis of Air-Bubble Plumes, W. M. Keck Lab. of Hydraulics and Water Resources, California Institute of Technology, KH-R-24.

Edinger, J. E. and E. M. Buchak. 1980. "Numerical Hydrodynamics of Estuaries", Estuarine and Wetland Processes with Emphasis on Modeling, P. Hamilton and K. B. MacDonald eds., Plenum Press, New York, pp. 115-146.

Edinger, J. E. and E. M. Buchak. 1995. "Numerical Intermediate and Far Field Dilution Modelling". Water, Air and Soil Pollution 83:147-160. Kluwer Academic Publishers, The Netherlands. November.

Fanneløp, T. K. and K. Sjøen, 1980. "Hydrodynamics of underwater blowouts", Norwegian Maritime Research, No.4, pp. 17-33.

Kolluru, V.S., E. M. Buchak and J. Wu, 1999. "Use of Membrane Boundaries to Simulate Fixed and Floating Structures in GLLVHT", presented at the 6th International Conference on Estuarine and Coastal Modeling, New Orleans, November 3 – 5.

Schladow, G. S., 1993. "Lake destratification by bubble-plume systems: design methodology", Journal of Hydraulic Engineering, Vol. 119, No. 3, March.

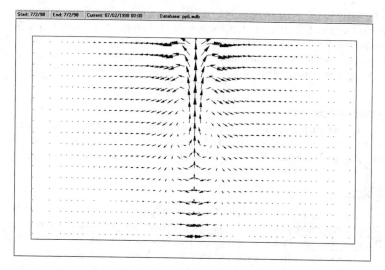

Figure 1   Cross sectional velocity vector plot for a bubbler placed at the bottom.

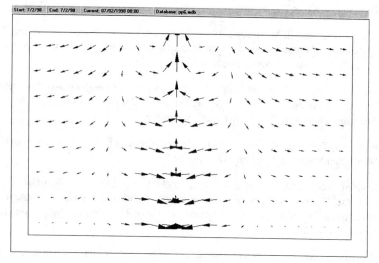

Figure 2   Velocity pattern produced by the bubbler close to its location.

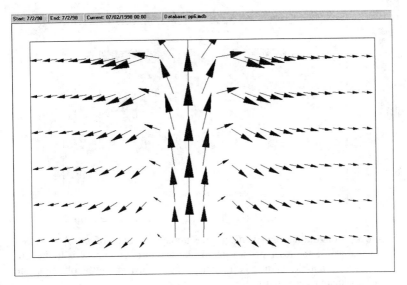

Figure 3   Velocity pattern produced by the bubbler close to the water surface.

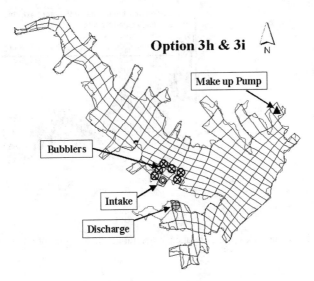

Figure 4   Bubblers arrangement around the intake structure.

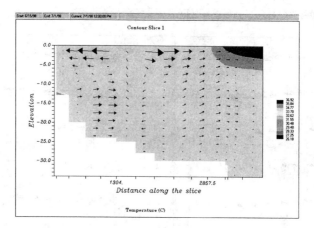

Figure 5   Longitudinal and vertical slice through a series of bubblers placed
          across the reservoir near the intake structure. The intake is to the left,
          the bubbler curtain is just left of center in this view looking up the
          main reservoir arm.

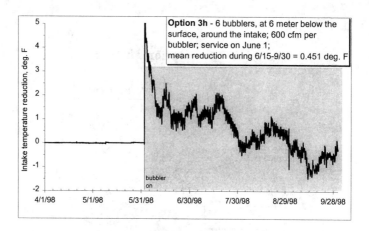

Figure 6   Time series plot of intake temperature reduction for option 3h.

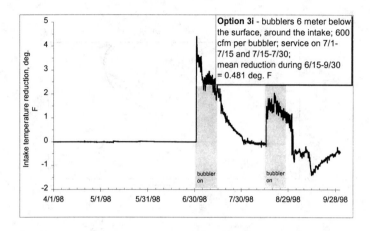

Figure 7   Time series plot of intake temperature reduction for option 3i.

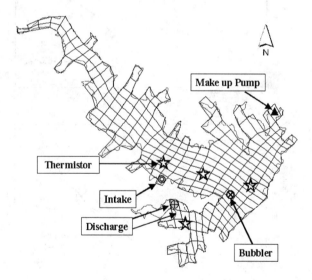

Figure 8   Location of thermistors and Bubbler #1.

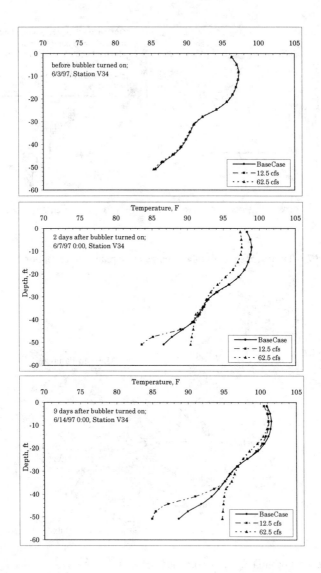

Figure 9   Temperature profiles at V34 during the 1997 simulation.

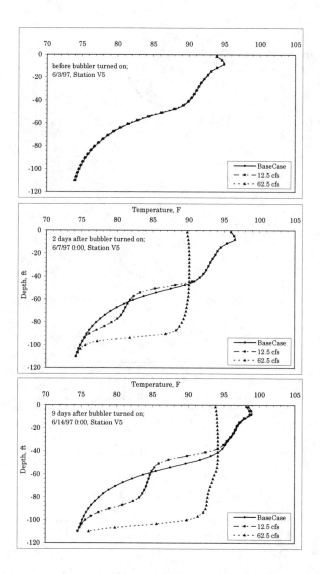

Figure 10   Temperature profiles at V5 during the 1997 simulation.

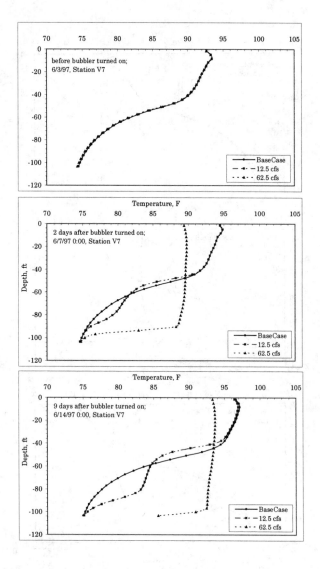

Figure 11   Temperature profiles at V7 during the 1997 simulation.

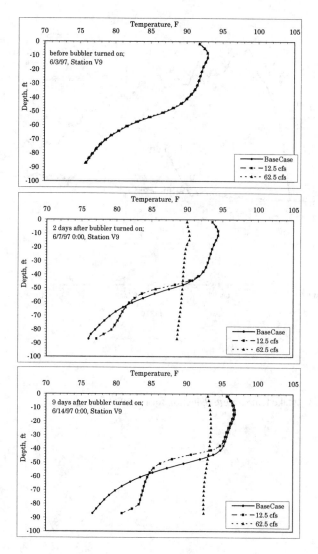

Figure 12   Temperature profiles at V9 during the 1997 simulation.

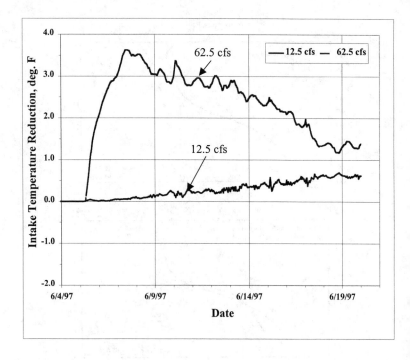

Figure 13   Intake temperature reduction for the 1997 simulation.

MODELING HYDRODYNAMIC AND SEDIMENT PROCESSES IN MORRO BAY

Zhen-Gang Ji[1], M. ASCE, Michael R. Morton[1], M. ASCE, and
John M. Hamrick[1], M. ASCE

ABSTRACT

Morro Bay is located on the central coast of California. Because of sedimentation, the bay has lost more than a quarter of its volume over the last 100 years. At low tides, more than 60% of the bay area emerges and becomes dry. The purpose of this study was to develop a coupled hydrodynamic and sediment model for the Morro Bay National Estuary Program. The Morro Bay model, developed based on the Environmental Fluid Dynamics Code (EFDC), contains 1609 curvilinear grid cells with resolution varying from 50 to 110 meters. The hydrodynamic calibrations included a predictive, dynamically coupled simulation of salinity, and temperature. The primary model hydrodynamic forcings included open boundary tidal elevations, freshwater inflows from two creeks, and surface wind stresses. The measured data for model calibration include tidal elevation, current velocity, water temperature, and salinity at 6 locations for 31 days, from March 9 to April 10, 1998. Comparison between model results and data indicates that the model results match the data reasonably well at the 6 data stations for tidal elevation, current velocity, salinity and temperature. The model also simulates the wet and dry processes in Morro Bay realistically. Using the calibrated model, a 160-day simulation was conducted to determine whether locations of sediment deposition were qualitatively in agreement with historical surveys. The model results indicate that sediment deposition occurred in the same locations as observed in the bay.

---

[1] Tetra Tech, Inc., 10306 Eaton Place, Suite 340, Fairfax, VA 22030

INTRODUCTION

Morro Bay is a natural embayment located on the central coast of California, about 96 km north of Point Conception and about 160 km south of Monterey Bay (Figure 1). It is a shallow lagoon, approximately 6.5 km long in the north-south direction and about 2.8 km wide in the east-west direction at its maximum width. Because of sedimentation, the bay has lost more than a quarter of its volume over the last 100 years. At low tide, more than 60 percent of the bay area emerges and becomes dry. Water quality concerns include excessive levels of bacteria, nutrients, and heavy metals. The purpose of this study was to develop a coupled hydrodynamic and sediment model for sedimentation control and water quality management in Morro Bay.

Historical bathymetric surveys have indicated that Morro Bay is becoming shallower as the result of sedimentation. The three sources of sedimentation in the bay are sediment from upland watershed areas entering the bay via Chorro Creek and Los Osos Creek, littoral transport from Estero Bay, and aeolian transport of sand from the barrier sand spit. The sediments from these sources are not distributed uniformly throughout the bay. Deposition from littoral sources is primarily confined to the entrance channel. About 90 percent of all sedimentation occurring in the entrance channel is the result of littoral transport being intercepted by the harbor entrance; the remaining portion is attributed to riverine and aeolian sources. The U.S. Army Corps of Engineers has maintained dredging records for Morro Bay since 1949. The average sedimentation rate in the navigation channels over the time period 1944-1987 was 87,900 m^3/yr (USACE 1991).

Deposition from Los Osos Creek and Chorro Creek is primarily confined to the central and south bay areas, as well as the delta and lower creek channels. The riverine deposition consists of silt, clay, and fine sands. A study by Philip Williams & Associates (1988) concludes that Morro Bay is gradually changing from a shallow estuary to an intertidal mudflat with subsequent transition to a salt marsh and then a meadow. The overall accretion of the central and back-bay areas over the years has led to a decrease in tidal prism and a corresponding reduction in the ebb tide flow. This in turn has reduced the carrying capacity on ebb tide, resulting in increased net deposition in central Morro Bay. Most of the sediment loading from the creeks is episodic and occurs during major flood events. Since the sediment from the two creeks is the primary source of sedimentation, this study will focus on the transport and deposition of sediment discharged from Chorro Creek and Los Osos Creek. The resuspension and redistribution of sediment deposited on the floor of the bay over the past years will not be included in the modeling.

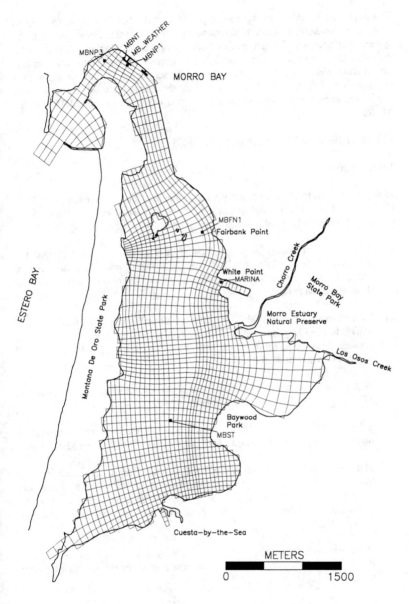

Figure 1. Morro Bay model grid and monitoring stations.

Thorough understanding of the hydrodynamics of the bay is vital to the Morro Bay study. Without a detailed understanding of how water moves through the system, any analysis of water quality issues would be incomplete. In this study, data on tides, stream flow, temperature, salinity, currents, winds, and atmospheric information were compiled to provide a data set for calibration or input to a hydrodynamic model of Morro Bay. The sediment transport and sediment deposition processes in the bay were simulated using the calibrated Morro Bay Model.

FIELD OBSERVATIONS AND ANALYSES

Data Sources

In this study, the model calibration period spanned 31 days, from March 9, 1998 to April 9, 1998. During this period, the following field-measured data were available for model external forcings or for model-data comparison:

1) Tidal gauge recorders were installed at two locations - MBNT near the entrance to Morro Bay and MBST near the southern end of the bay (Figure 1). The tidal elevation data were averaged over 10-minute intervals throughout the period.

2) Current data were available at three locations - MBNP1, MBNP3, and MBFN1. The sampling depths of the current meters were near water surface, varying from 6.0 to 7.4 meters at MBNP1, 5.5 to 7.3 meters at MBNP3, and 3.0 to 4.0 meters at MBFN1.

3) Water temperature data were available at five different locations - MBNT, MBNP1, MBNP3, MBST, and Marina.

4) Salinity values were determined from the conductivity and temperature measurements at four locations - MBNP1, MBNP3, MBST, and Marina.

5) Meteorological data at MB_WEATHER included air temperature, relative humidity, solar radiation, wind speed and direction, and rainfall at one-hour intervals. As shown in Figure 1, MB_WEATHER was located right on the water surface of the bay, and was representing the meteorological conditions realistically.

6) Hourly stream flow measurements were available from a gauge at Canet Road, which is located about 5.5 km upstream from the mouth of Chorro Creek.

7) The shoreline of Morro Bay was determined from several sources, including 1:100,000 scale U.S. Geological Survey Digital Line Graph Data, and an AutoCAD drawing of Morro Bay prepared by Philip Williams & Associates (1988).

The data were processed to provide continuous and simultaneous boundary conditions for the hydrodynamic model during the calibration period. The open ocean boundary was defined at the mouth of Morro Bay and extended a short distance into Estero Bay. Two sources of freshwater entered the bay representing Chorro Creek and Los Osos Creek. More detailed descriptions of the data sources are provided by Tetra Tech (1999a).

Freshwater Discharges

Freshwater inputs to the modeled system consist of discharges from two streams, Chorro Creek and Los Osos Creek, as well as rainfall. The watershed area contributing to Chorro Creek at its mouth is about 111 km^2, and the watershed area at the mouth of Los Osos Creek is about 60 km^2. Tetra Tech (1999a) estimated the historical flow rates at the mouths of Chorro Creek and Los Osos Creek from 18 years of flow records at the Canet Road, and developed regression equations with the help of the HEC-1 modeling results:

$$Q_{ChorroMouth} = 0.8845 \, Q_{Canet}^{1.094}$$

$$Q_{LosOsosMouth} = 0.007427 \, Q_{Canet}^{1.559}$$

where flow rates are in cubic feet per second. The historical record at the Canet Road gauge indicates that Chorro Creek is an extremely "flashy" stream. It is common for flow rates to change dramatically over a period of a few hours. Therefore, hourly inflow time series were developed for the model in order to preserve the high-frequency resolution during storm events. No estimates of groundwater underflow or septic system flows were available for incorporation into the hydrodynamic model.

Sediment Yield

Sediment yield and transport analyses on the major tributaries to Morro Bay was conducted to evaluate the source, sizes, and quality of sediment loading to the bay (Tetra Tech 1999a). This study indicated that the majority of sediment being transported to Morro Bay from watershed and riverine processes were from Los Osos Creek and Chorro Creek. Los Osos Creek and Chorro Creek supplied 14 % and over 80 % of the sediment load respectively. Data used in this assessment were 18 years of flow records at the Chrro Creek. The HEC-1 model was used to estimate the relationship between flows at gaged and ungaged locations, and the sediment yield. The following regression equations of sediment loading versus discharge were derived:

$$SedLoad_{ChorroMouth} = 0.005256 \, Q^{2.212}_{ChorroMouth}$$

$$SedLoad_{LosOsosMouth} = 0.032981 \, Q^{2.118}_{LosOsosMouth}$$

where sediment loads SedLoad are in tons per day.

EFDC MODEL AND MODEL SETUP

Environmental Fluid Dynamics Code (EFDC)

The Environmental Fluid Dynamics Code (Hamrick 1992) is a general-purpose modeling package for simulating three-dimensional flow, transport, and biogeochemical processes in surface water systems, including rivers, lakes, estuaries, reservoirs, wetlands, and coastal regions. The EFDC model was originally developed at the Virginia Institute of Marine Science for estuarine and coastal applications and is considered public domain software. In addition to hydrodynamic and salinity and temperature transport simulation capabilities, EFDC is capable of simulating cohesive and noncohesive sediment transport, near field and far field discharge dilution from multiple sources, eutrophication processes, and the transport and fate of toxic contaminants in the water and sediment phases. Special enhancements to the hydrodynamic portion of the code include vegetation resistance, drying and wetting, hydraulic structure representation, wave-current boundary layer interaction and wave-induced currents, refined modeling of wetland marsh systems, controlled flow systems, and nearshore wave-induced currents and sediment transport. The EFDC model has been extensively tested and documented for more than 50 modeling studies, including study of tidal intrusion and its impact on larval dispersion in James River estuary (Shen et al 1999), simulation of estuarine front and its associated eddy (Shen and Kuo, 1999b), investigation of bottom shear stress in estuaries (Kuo el at 1996), hydrodynamic and sediment transport modeling in the middle Atlantic Bight (Kim el at 1997), and hydrodynamic and water quality modeling in Peconic Bay (Tetra Tech 1999b). The model is presently being used by a number of organizations, including universities, governmental agencies, and environmental consulting firms.

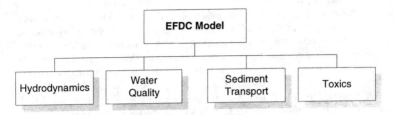

Figure 2. Primary modules of the EFDC model.

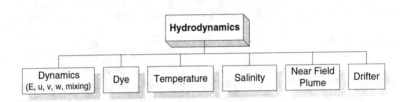

Figure 3. Structure of the EFDC hydrodynamic model.

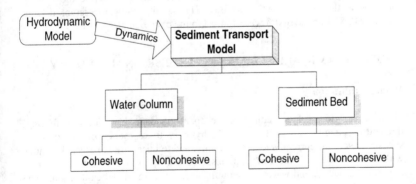

Figure 4. Structure of the EFDC sediment model.

The structure of the EFDC model includes four major modules: (1) a hydrodynamic model, (2) a water quality model, (3) a sediment transport model, and (4) a toxics model (Figure 2). As shown in Figure 3, the EFDC hydrodynamic model itself is composed of six transport modules, including dynamics, dye, temperature, salinity, near field plume, and drifter. Various products of the dynamics module (i.e., water depth, velocity, and mixing) are directly coupled to the water quality, sediment transport, and toxics models. The schematic diagram for the sediment transport model is shown in Figure 4. Details on the EFDC model are documented by Hamrick (1992) and Hamrick and Wu (1997).

Model Grid

The model grid (Figure 1) contains 1609 horizontal curvilinear grid cells and a single vertical layer. Also shown in Figure 1 are the six stations at which tides, velocity, temperature, and/or salinity were measured. The typical cell size varies from 50 to 80 meters in the X-direction and 50 to 110 meters in the Y-direction.

External Forcings

The Morro Bay hydrodynamic circulation model is driven by atmospheric forcings, tributary inflows, and open boundary conditions. Atmospheric forcings are composed of the meteorological data mentioned previously and include atmospheric pressure, air temperature, precipitation, relative humidity, solar radiation, cloud cover, wind speed, and wind direction. As discussed previously, the tributary inflows include the two major freshwater sources to the bay, Chorro Creek and Los Osos Creek, which discharge both freshwater and sediment into the bay. Detailed information on the flow rate and sediment loadings is presented in a sediment loading study by Tetra Tech (1998). Open boundary conditions include the tidal elevation at MBNT, which was used to prescribe the forcing condition at the entrance to Morro Bay. The salinity and temperature data at stations MBNP1 and MBNP3 were applied to the open boundary.

HYDRODYNAMIC AND SEDIMENT PROCESSES IN MORRO BAY

Model Calibration

Model calibration included adjusting the open boundary tidal elevation forcing, the bottom roughness, and the local bathymetry to reproduce observed water surface elevations, current velocities, salinity, and temperature at several locations within Morro Bay. Time series of observed water surface elevation, salinity, horizontal velocity, and water temperature were used for model calibration. The consistency between model predictions and observations was determined qualitatively by comparison plots for the various parameters.

Based on the availability of the measured data, the period of model-data comparison was 31 days, between March 9 and April 9, 1998. During this period, the atmospheric wind is relatively small and uniform, typically varying from 2 to 4 m/s. The model results were compared with measured data in time series and in scatter plots. The model-data comparison from March 9, 1998 to April 9, 1998 at Station MBNT is shown in Figure 5. The corresponding (I, J) location is (38, 75) on the computational domain. In Figure 5, the first panel is the tidal elevation, the second panel is velocity in the major tidal axis direction, the third panel is temperature, and the fourth panel is salinity. As expected, the modeled tidal elevation (solid line) matches the data (dotted line) almost perfectly, since the measured tidal elevation at MBNT is used as the open boundary condition. The modeled velocity has an amplitude of 40 cm/s, and there is no measured velocity at MBNT. The modeled temperature is also consistent with the data, and the model realistically simulates the semi-diurnal temperature variations caused by the tides. There is no measured salinity data at MBNT. The model indicates strong semidiurnal variation from 24 to 32 ppt. Figures 6 and 7, which are similar to Figure 5, show model-data comparisons for Stations MBNP3 and MBST, respectively. These figures indicate that the modeled tidal elevation, velocity, temperature, and salinity match the measured data satisfactorily. Model results at the other three stations, MBNP1, MBFN1 and Marina, are consistent with the data as well. Station MBST is in the south of the bay and is far way from the open boundary, therefore the salinity at MBST is lower and more sensitive to the inflows from the two creeks. The largest discrepancy is at station Marina, where the observed salinity is consistently lower than the model result. Model sensitivity tests reveal that when a small amount of fresh water inflow is added at the end of the little channel, salinity at Marina changes significantly. This is an indication that the salinity discrepancy at Marina could be due to a small source of freshwater entering the marina that is not accounted for in the model, which should have minimum impact on the overall salinity distribution in the bay.

Since tidal velocity is one of the major factors that determine sediment transport and deposition processes, it is necessary to examine and compare the modeled and measured velocity carefully. Figure 8 is a scatter plot of the measured velocity at MBNF1. The velocity is rotated into the major tidal axis direction so that the maximum U-component and the minimum V-component are shown. In Figure 8, the velocity toward the open boundary can be more than 90 cm/s, and the velocity toward the bay can be more than 80 cm/s. The corresponding velocity from the Morro Bay Model is shown in Figure 9. Comparing Figure 8 with Figure 9, it is clear that generally the model simulated amplitudes of the outward velocity and the inward velocity well, except for a few samplings. Since the wind speed was small and there was no storm events during the modeling period, the exceptions are probably the result of the fact that the model has a single vertical layer and does not represent the vertical circulation patterns of the bay. Adding additional model layers might allow better representation of tidal velocities in the harbor area of Morro Bay. The discrepancies in tidal currents between the model and the data might be also caused by uncertainty in the very localized bathymetry, which cannot be represented in the model without resorting to extremely fine-scale grid cell sizes. In general, reproducing tidal currents is a more difficult task because, compared with tide heights, tidal currents are more sensitive to bathymetry, giving rise to a higher degree of uncertainty in field measurements. Scatter plots of velocities at MBNP1 and MBNP3 also show good consistency between the model velocities and the measured velocities. In summary, the model simulated the tidal velocity reasonably well.

Hydrodynamic and Sediment Processes in the Morro Bay

Successful calibration of the Morro Bay hydrodynamic circulation model indicates that the model can be applied to investigate hydrodynamic processes, and sediment transport and deposition processes in Morro Bay.

Figure 10 shows the 30-minute averaged water depth at low tide. The small plot in the upper right corner indicates the tidal elevation on the west open boundary. The water depth is indicated by shading. The grayscale bar at the bottom of the figure shows the water depth, ranging from less than 21 cm to more than 700 cm. In the Morro Bay model, grid cells are switched to "dry" when the water depth is less than 15 cm and to "wet" when the water depth is more than 20 cm. As the result of this wetting-and-drying scheme, the areas shown in Figure 10 are considered dry when the water depth is less than 21 cm. Figure 10 reveals clearly that the majority of the bay becomes dry at low tides and only the main channel area remains wet, with water depth ranging from less than 1 meter in the south to more than 7 meters at the entrance of the bay. This model result is also consistent with the measured low-water tide line.

A 160-day simulation of sediment transport and deposition in Morro Bay was conducted, from November 1, 1997 to April 9, 1998. As would be expected, model results indicate a strong correlation between sediment deposition and storm events. In the first 33 days of the simulation, there was little sediment deposition on the floor of the bay due to the lack of inflows from the two creeks. Figure 11 shows the 1-day averaged bed sediment thickness (mm) on day 128 (in reference to November 1, 1997). The large storm event around day 100, as indicated in the small plot in the upper right corner, brought a large volume of sediment into the bay from Chorro Creek and Los Osos Creek, and sediment started to deposit in the main channel. On day 128, a significant portion of the channel had sediment

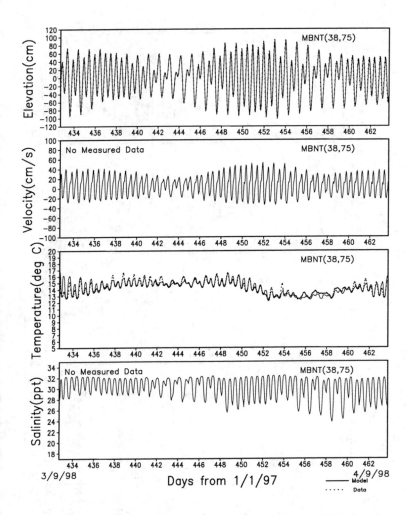

Figure 5. Modeled (solid line) and measured (dashed line) tidal elevation, velocity, temperature, and salinity at MBNT.

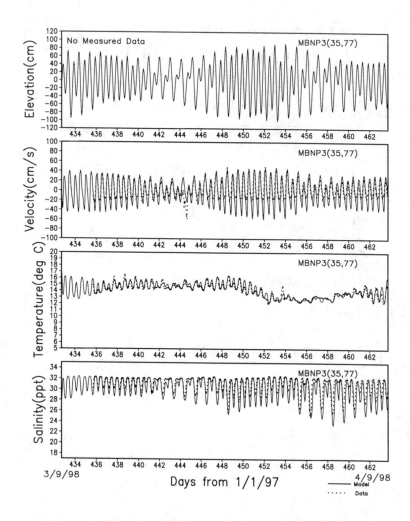

Figure 6. Modeled (solid line) and measured (dashed line) tidal elevation, velocity, temperature, and salinity at MBNP3.

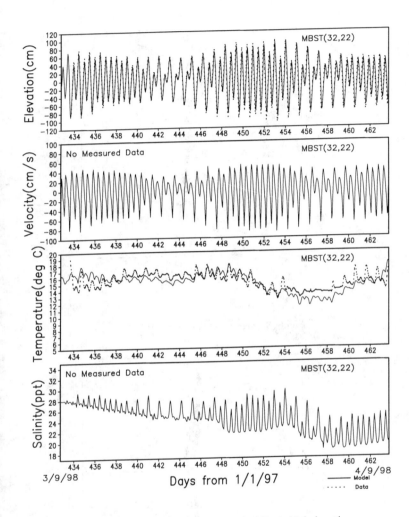

Figure 7. Modeled (solid line) and measured (dashed line) tidal elevation, velocity, temperature, and salinity at MBST.

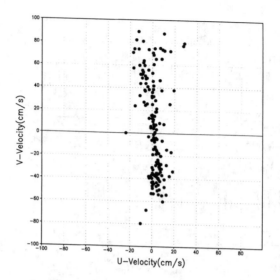

Figure 8. Measured velocity at MBFN1.

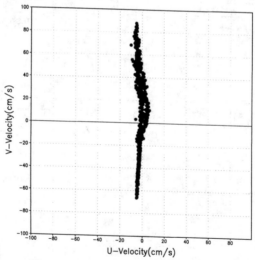

Figure 9. Modeled velocity at MBFN1.

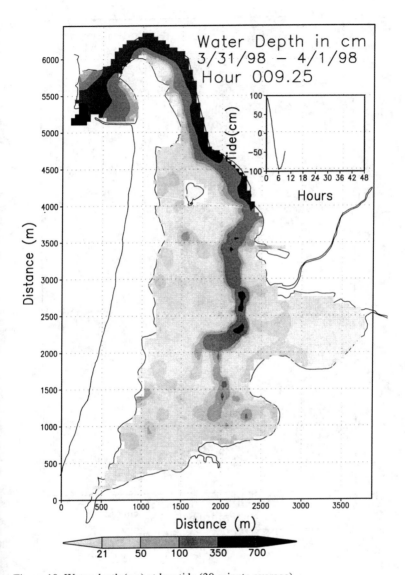

Figure 10. Water depth (cm) at low tide (30-minute average).

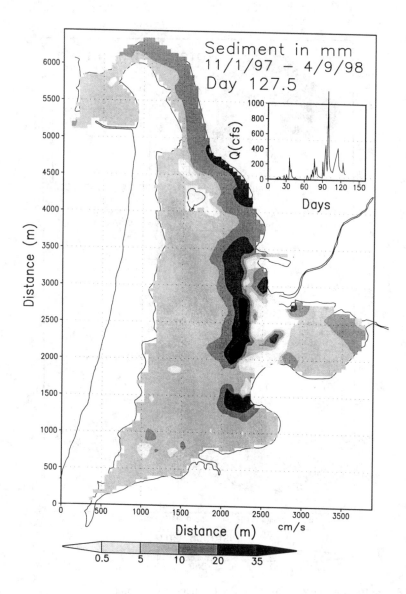

Figure 11. Bed sediment thickness (mm) at day 128.

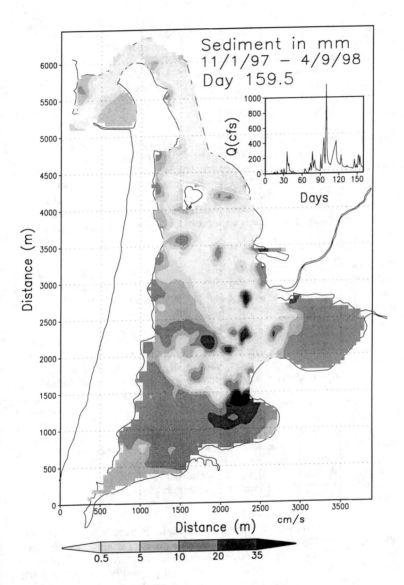

Figure 12. Bed sediment thickness (mm) at day 160.

deposition of more than 35 mm. By the end of the simulation on day 160 (Figure 12), much of the deposited sediment in the channel had been resuspended and transported out of the bay or onto the adjacent mudflat areas, as the result of tidal current forcing. Figure 12 also indicates that sediment from the two creeks was generally deposited in the southern part of the bay on the tidal mudflat areas, the same as the historic records reported by Tetra Tech (1998)

SUMMARY AND CONCLUSIONS

The Environmental Fluid Dynamics Code (EFDC) model was used in this study. The EFDC model is a three-dimensional and time-dependent model with the capability of simulating large wet/dry variability of tidal areas. The measured data for model calibration included tidal elevation, current velocity, water temperature, and salinity at six locations for 31 days, from March 9 to April 9, 1998. The primary model hydrodynamic forcings included open boundary tidal elevations, freshwater inflows from Chorro Creek and Los Osos Creek, and surface wind stresses.

Comparison between the model results and measured data indicates that the model results match the data reasonably well at the six data stations for tidal elevation, current velocity, salinity, and temperature. The model also simulated the wet/dry variability of Morro Bay realistically.

Using the calibrated model, a 160-day simulation was conducted to determine whether locations of sediment deposition are qualitatively in agreement with historical surveys. The modeled sediment deposition occurred in the same locations as observed in the bay. The modeling results indicate that during storm events, large amount of sediment from Chorro Creek and Los Osos Creek are discharged into the Morro Bay, and are immediatly deposited near the mouths of the creeks and on the bottom of the channel nearby. During high tide, especially during spring tide, the seposited sediment are resuspened and transported either out of the bat or onto the adjacent muflat. However, it should be noted that since there are no sediment data available for model calibration, the sediment processes discussed in this paper should only be considered preliminary, and further studies are needed to verify the sediment resuspension and transport processes presented here.

The calibrated EFDC hydrodynamic model of Morro Bay can be used as a tool for quantifying the hydrodynamic characteristics and examining the transport processes in the bay. The hydrodynamic processes in Morro Bay are relatively difficult to model since large areas become "dry" mudflats during low tide. However, the EFDC model has been designed to simulate this wetting-and-drying process. The numerical model can be used as a tool to aid further hydrodynamic and water quality studies and to guide field data collection programs. Time scales associated with traditional water quality and risk assessment issues are on the order of seasons and years, which are much larger than those associated with hydrodynamic circulation. Spills of toxic contaminants often pose an immediate environmental threat and require responses and remediation actions to be taken within a few tidal cycles (one-half to two days). Sedimentation issues require even longer time scales on the order of decades. These important environmental studies usually require information on flows at different time scales, and a calibrated hydrodynamic model can provide circulation and transport information for the entire range of time scales.

ACKNOWLEDGEMENTS

The development of the hydrodynamic model of Morro Bay was one part of a more comprehensive study conducted by Tetra Tech, Inc. for the Morro Bay National Estuary Program (MBNEP). The authors would like to acknowledge the following for their roles in the overall program: Ms. Melissa Moonie, MBNEP Program Director; Ms. Katie Kropp, MBNEP Scientific Coordinator; Mr. Thomas Grieb, Tetra Tech program manager; Mr. Kim Brown, Tetra Tech project manager; Mr. Hank Fehlman, Tetra Tech project engineer responsible for watershed runoff and sediment loading studies; and Mr. Ron Munson, Tetra Tech project scientist responsible for nutrient and bacteria studies. The authors would also like to thank the anonymous reviewer for some valuable comments.

REFERENCES

Hamrick, J. M. (1992). *A three-dimensional Environmental Fluid Dynamics Computer Code: theoretical and computational aspects. Special Report 317.* The College of William and Mary, Virginia Institute of Marine Science, Williamsburg, Virginia. 63 pp.

Hamrick, J. M., and Wu, T. S. (1997). "Computational design and optimization of the EFDC/HEM3D surface water hydrodynamic and eutrophication models." *Computational Methods for Next Generation Environmental Models*, G. Delic, and M. F. Wheeler, Ed., Society for Industrial and Applied Mathematics, Philadelphia. pp 143 – 156.

Kim, S. Wright, D. L., Maa, J. and Shen, J. (1997). "Morphodynamic responses to extratropical meteorological forcing on the inner shelf of the middle Atlantic Bight: wind wave, currents and suspended sediment transport." *Estuarine and Coastal Modeling*, Proceedings of the 5th International Conference, ASCE, 456-466.

Kuo, A.Y., Shen, J. and Hamrick, J. M. (1996). "The effect of acceleration on bottom shear stress in tidal estuaries." *Journal of Waterway, Port, Coastal, and Ocean Engineering*, ASCE, 122 (2), 75-83

Philip Williams & Associates. (1988). *Sedimentation processes in Morro Bay, California.* Prepared for Coastal San Luis Resource Conservation District. Philip Williams and Associates. June.

Shen, J. Boon, J., and Kuo, A. Y. (1999). "A numerical study of a tidal intrusion front and its impact on larval dispersion in the James River estuary, Virginia." *Estuary*, 22(3A), 681-692.

Shen, J. and Kuo, A.Y. (1999). "Numerical investigation of an estuarine front and its associated eddy." *Journal of Waterways, Ports, Coastal and Ocean Engineering*, ASCE, 125 (3), 127-135.

Tetra Tech. (1998). Sediment Loading Study. Prepared for Morro Bay National Estuary Program, Los Osos, California. Tetra Tech, Inc., Lafayette, California.

Tetra Tech. (1999a). Morro Bay National Estuary Program: Hydrodynamic Circulation Model. Tetra Tech, Inc., Lafayette, California.

Tetra Tech. (1999b). *Three-Dimensional Hydrodynamic and Water Quality Model of Peconic Estuary.* For Peconic Estuary Program, Suffolk County, NY. Tetra Tech, Inc., Fairfax, Virginia.

USACE. (1991). *Morro Bay Harbor, California, Feasibility Report.* Volume II. U.S. Army Corps of Engineers, Los Angeles District.

AN OCEAN MODEL APPLIED TO THE CHESAPEAKE BAY PLUME

Paul J. Martin[1]

Abstract

A three-dimensional, baroclinic, coastal ocean model is described and applied to a simulation of the Chesapeake Bay outflow plume. The model has a free surface, uses an Arakawa C grid, is leapfrog in time with an Asselin filter to suppress time splitting, and uses second-order, centered spatial finite differences. The propagation of surface waves and vertical mixing are treated implicitly. A choice of the Mellor-Yamada Level 2 or Level 2.5 turbulence models is provided for the parameterization of vertical mixing. The model's horizontal grid is orthogonal curvilinear. The vertical grid uses sigma coordinates for the upper layers and z-level (constant-depth) coordinates for the lower layers. The transition from the sigma to the z-level grid can occur at any depth, although at least one sigma layer is required at the surface to allow for changes in surface elevation. The inclusion of a source term in the model equations can simplify input of river and runoff inflows. The model has the capability for including an arbitrary number of nests.

The main model domain for the Chesapeake Bay plume simulations consists of a region of 200 by 400 km extent encompassing both the Chesapeake and Delaware Bays, with a horizontal grid resolution of 2 km. Some simulations were conducted with nests in the area of the mouth of the bay to provide increased resolution in this area.

Idealized experiments were performed to look at the response of the Chesapeake Bay outflow plume to tide and wind forcing. As noted by other investigators, the tidal forcing has a significant effect on the salinity structure in the mouth of the

---

[1] Naval Research Laboratory, Stennis Space Center, MS 39529. Email: martin@nrlssc.navy.mil

Bay. An applied wind stress that rotates clockwise with a period of several days illustrates the response of the outflow plume to changing wind direction.

Introduction

This paper discusses the formulation of a coastal ocean model and its application to idealized simulations of the Chesapeake Bay (ChB) outflow plume.

The ocean model has been structured for use in the Coupled Ocean-Atmosphere Mesoscale Prediction System (COAMPS), which was developed by the Naval Research Laboratory (NRL) in Monterey, CA (Hodur, 1997). The model uses dynamic memory allocation so that it doesn't need to be recompiled for different simulations and all variables are passed through subroutine argument lists to facilitate nesting, i.e., the same subroutines are used to calculate different grids. This structure allows for an arbitrary number of nested grids (nesting in the ocean model is currently one-way only). Within COAMPS, all the ocean and atmospheric grids are run within the same Fortran program.

In its physics and numerics, the ocean model is very similar to the Princeton Ocean Model (POM) of Blumberg and Mellor (1987), but with a few differences. The vertical grid uses a combination of sigma and z-level (fixed-depth) coordinates, a source term is provided in the numerical equations to simplify input of river and runoff inflows, and the free surface is implicit.

The model is applied to the simulation of the ChB outflow plume. The behavior of the ChB outflow plume is being investigated in the Coastal Remote Sensing (CoRS) Project at NRL, which has conducted several observational experiments near the mouth of ChB.

There has been a significant amount of research done to investigate the circulation of ChB. A sampling of the references available includes observations (Goodrich and Blumberg, 1991; Valle-Levinson and Lwiza, 1995, 1997; Valle-Levinson 1995; Valle-Levinson et al., 1998), simulations with idealized domains and forcing (Chao and Boicourt, 1986; Chao, 1988a, 1988b, 1990; Valle-Levinson et al., 1996), and modeling with realistic domains with realtime forcing (Blumberg and Goodrich 1990; Johnson et al. 1993).

The purpose of the ChB simulations conducted in this study was to see how well the model could simulate the salinity structure near the mouth of ChB and the general behavior of the outflow plume. Mean river outflows representative of those in ChB and realistic tidal forcing are used. An idealized, rotating wind is used to look qualitatively at the behavior of the outflow plume in response to changing wind direction.

Description of Ocean Model

Model Physics

The ocean model is free surface and employs the hydrostatic, Boussinesq, and incompressible flow approximations. The model employs the standard equations for a baroclinic, free-surface model (Blumberg and Mellor, 1987) with the addition of a source term in the equations to allow specification of local sources and sinks (Martin, 2000). Boundary conditions at the surface are the wind stress and heat and (effective) salt flux. At the bottom, a quadratic drag law is used for the momentum equations and the heat and salinity fluxes are set to zero.

Two equations of state are provided, the polynomial equation of state of Friedrich and Levitus (1972) and the formula of Mellor (1991), which is used in POM. For horizontal mixing, the model offers a choice of the Smagorinsky (1963) or grid-cell Reynolds number (Re) parameterizations. For the grid-cell Reynolds number scheme, the eddy coefficients are computed as the maximum of a specified minimum value and a value computed based on a maximum grid-cell Re.

For vertical mixing, the model offers a choice of the Mellor-Yamada Level 2 (MYL2) or Level 2.5 (MYL2.5) schemes (Mellor and Yamada, 1974; Mellor and Yamada, 1982; Mellor, 1996). The surface roughness and, for the MYL2.5 scheme, the surface turbulent kinetic energy flux can be specified as discussed by Craig and Banner (1994). The model also provides an option for the use of the Richardson-number-based mixing enhancement scheme of Large et al. (1994), which provides for weak mixing at the edge of a turbulent boundary layer for Richardson numbers above the normal critical value up to a value of 0.7.

Model Numerics

The model uses second-order, centered, spacial finite differences on a C grid. Advective and diffusive transports are done in a conservative flux form. The temporal differencing is leapfrog with an Asselin (1972) filter to suppress time splitting. The horizontal grid is orthogonal curvilinear as implemented in POM and discussed in Blumberg and Herring (1987). A grid-curvature correction for horizontal momentum transport is provided for advection but not for mixing.

The model uses a combined sigma/z-level vertical grid. Figure 1 illustrates the different ways this grid can be set up: (a) a mainly z-level grid with a single sigma layer at the surface to take up surface elevation changes, (b) and (c) a mixed (hybrid) grid with sigma layers down to a specified depth and z-levels below, and (d) with sigma layers used all the way to the bottom everywhere. Note that in water depths less than the depth of the transition from the sigma grid to the z-level grid, the grid is the same as a standard sigma-coordinate grid.

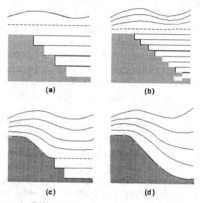

Figure 1. Different ways the vertical grid can be set up

The combined sigma/z-level grid offers some flexibility. Sigma coordinates can be used in the shallow water, e.g., on the shelf, and z-levels, which are more robust to steep bathymetry, can be used in the deeper water (Fig. 1c). For a mainly z-level grid, the option of being able to use several sigma layers to take up surface elevation changes (Fig. 1b) allows high vertical resolution to be used near the surface in conjunction with relatively large surface elevation changes. With only a single sigma layer to take up surface elevation changes, the thickness of the upper layer would be limited by the maximum depression of the free surface. The combined grid also allows comparisons to be made between sigma and z-level coordinates to assess the extent of differences resulting from the use of the different vertical coordinates.

The free surface is treated using an implicit scheme. The time-level weightings for the implicitly treated terms, i.e., the surface elevation gradients in the momentum equations and the transport gradients in the depth-averaged continuity equation, can be specified by the user. An even weighting between the old and new time levels, which minimizes damping, is used for the results presented here.

Shrinkwrapping is used to eliminate calculations over land on the left and right sides of the domain, and the calculations proceed through the model domain in x-z slices to try to optimize the use of high-speed cache memory.

Simulation of Chesapeake Bay Outflow Plume

Idealized model simulations were conducted to look at the response of the ChB outflow plume to tide and wind forcing. The main focus of NRL's CoRS Project is the behavior of the plume and salinity fronts near the mouth of the bay. However, some spin up experiments were conducted to see how well the model could simulate the salinity distribution within the bay itself, starting from a condition of a uniform salinity of 35 psu.

Figure 2. Main model domain for ChB simulations

Figure 2 shows the main model domain used for the simulations and the bathymetry used for the 2-km horizontal grid. The domain consists of both the Chesapeake and Delaware Bays and the overall dimensions of the domain are 200 by 400 km. The bathymetry is based on data from the National Ocean Service (NOS) (Jerry Miller, personal communication). The depths on the model grid were smoothed with two passes of a nine-point box filter. Some cross sections are identified in Fig. 2 that correspond to sections observed over a number of years and reported in Whaley and Hopkins (1952) and Stroup and Lynn (1963).

River Forcing Only

The initial experiment used constant river inflows of 1100, 100, 100, 400, 100, and 200 m³/s for the Susquehanna, Patapsco, Patuxent, Potomac, York, and James Rivers in ChB, respectively (2000 m³/s total), and 600 m³/s from the Delaware River in Delaware Bay. A very weak, southward-directed wind stress of 0.02 dynes/cm² was applied to provide a weak alongshore current at the coast. Simulations conducted without this weak wind stress indicated that it had little effect on the salinity distribution within ChB. There was no tidal forcing for this case. The initial condition was a uniform salinity of 35 psu. Temperature variability was not considered (the temperature was set to a constant 20°C). Radiation conditions were used at the open boundaries, with a relaxation of the elevation and the normal, depth-averaged velocity at the open boundary near the coast to the interior values with a timescale of 5 d to allow the development of an along-shore flow.

The horizontal grid resolution was 2 km. A uniformly-spaced sigma coordinate grid of 10 layers was used down to 50 m depth, and 10 uniformly-stretched z-levels were used from 50 m to the bottom (2000 m). The timestep was 360 s. Horizontal mixing was by the grid-cell Re scheme with a minimum value of 20 m²/s and a grid-cell Re of 50. The MYL2 scheme was used for vertical mixing. The simulations are quite sensitive to the background vertical mixing values used outside of the turbulent boundary layers. Large values cause too much vertical mixing and too weak a salinity stratification within ChB as noted by Johnson et al. (1991). A value of 0.1 cm²/s was used for both momentum and salinity.

It takes about a year for the salinity distribution within the bay to stabilize. Figures 3a and 3b show cross sections of the 90-d averaged salinity and normal velocity from the model simulation at the mouth of ChB looking into the bay after 2 y. The range of salinity at the bay mouth from the simulation (19-32 psu) is larger than that of the climatology of Stroup and Lynn (1963) (annual mean range about 23-30 psu) and the model's vertical stratification is a bit higher.

The model's inflow occurs over the main channel on the south side of the bay mouth. The strongest outflow occurs near the surface on both the north and south sides of the bay mouth, but there is outflow near the surface across the entire section and outflow from surface to bottom on the north side.

Table 1 lists inflow (upbay), outflow (downbay), and net transports at the sections indicated in Fig. 2. These transports are averaged over 90 d, but without tides or other time-varying forcing, the temporal fluctuations are small. Note that the net transports are equal to the river inflows between the particular section and the head of the bay. The model's inflow transport at the mouth of ChB (at Section II), which reflects the gravity-driven estuarine circulation, is 7000 m³/s. The upbay transports decrease towards the head of ChB, though the ratio of the inflow transport to the net transport remains fairly constant up through Section VI. The transports for a simulation with zero wind stress were similar.

Table 1.
Model-predicted inflow (upbay), outflow (downbay), and net transports at several ChB cross sections (see Fig. 2) after 2 y for a simulation with only river forcing. Inflows are defined to be positive and outflows are defined to be negative. Transports are in m³/s.

section	inflow	Outflow	net inflow	inflow/net
II	7000	-9000	-2000	3.5
IV	5820	-7620	-1800	3.2
VI	4580	-5880	-1300	3.5
X	1750	-2950	-1300	1.5

The model-predicted transports at the mouth of ChB in Table 1 can be compared with transports from Valle-Levinson et al. (1998) in Table 2, which were derived from data taken at the mouth of ChB. Valle-Levinson et al. measured currents across the mouth of ChB repeatedly over two tidal cycles on four separate occasions by cruising back and forth across the bay mouth with an acoustic Doppler current profiler (ADCP). Table 2 lists their measurements of the inflow and outflow sub-tidal currents and their estimates of the magnitude of the river and wind setup/setdown outflows at the time of the measurements.

Table 2
Measured sub-tidal inflows and outflows and estimated flows due to rivers and wind setup/setdown at the mouth of ChB from Valle-Levinson et al. (1998). Inflows are positive and outflows are negative. Transports are in m³/s.

	Sept 96	Nov 96	Feb 97	May 97
Measured inflow	3000	8000	14000	4000
Measured outflow	-15000	-7000	-3000	-12000
Net flow	-12000	1000	11000	-8000
Estimated river flow	-4000	-3000	-2000	-2000
Estimated flow from wind setup/setdown	-7000	0	11000	-6000

If the sub-tidal inflows and outflows in Table 2 are barotropically adjusted to give as a net flow the estimated river flow at the time (i.e., to remove the transport due to wind setup or setdown in ChB, which is assumed to be primarily barotropic

(Valle-Levinson et al., 1998)), the inflows are quite consistent (Table 3), with a range of 6000 to 7500 m³/s. The inflow from the model simulation (Table 1) is within this range. The barotropic correction used in Table 3 to adjust the inflow and outflow transports can be compared with the estimates of the flow due to wind setup/setdown by Valle-Levinson et al. in Table 2. They are generally similar.

Table 3
Adjustment of sub-tidal inflows and outflows from Valle-Levinson et al. (1998) to provide a net outflow equal to the estimated river outflow. Transports are in m³/s.

	Sept 96	Nov 96	Feb 97	May 97
Adjusted inflow	7000	6000	7500	7000
Adjusted outflow	-11000	-9000	-9500	-9000
Net flow	-4000	-3000	-2000	-2000
Required correction	-8000	4000	13000	-6000

The salinity range in the model simulation at Section IV (Fig. 2) is 18-26 psu, which is similar to the annual mean range of 19-25 psu in Stroup and Lynn (1963). At Section VI, the model's salinity range of 16-22 psu, is slightly higher than the annual mean range of 14-20 psu in Stroup and Lynn (1963).

River and Tidal Forcing

Another spin-up experiment was conducted in which tidal forcing was used. River inflows were as before. The tidal elevation and depth-averaged currents for 8 tidal constituents (K_1, O_1, P_1, Q_1, N_2, M_2, S_2, K_2) were applied at the open boundaries. The tidal data were obtained from a calculation of the east-coast tides conducted by the Coastal Engineering Research Center, Vicksburg, MS (Norm Scheffner, personal communication), using the ADCIRC model (Westerink et al. 1994). The time required for the salinity distribution in the bay to stabilize was about 1.5 y, i.e., a bit longer than for the case without tides.

Figures 3c and 3d show cross sections of the time-averaged (90-d) salinity and normal velocity at the mouth of ChB for the tidal forcing case. Compared with the case with no tidal forcing (Fig. 3a and 3b), the salinity for the tidally forced case shows significantly larger horizontal gradients. The horizontal salinity gradients are also larger than in the Stroup and Lynn (1963) climatology. This is reflected by the maximum surface salinity on the north side of the bay mouth, which is 30 psu, being higher than in the climatology, which shows 24, 25, 27, and 29 psu for the four seasons. There are often times when the observed surface salinity on the north side of the bay mouth is 30 psu or more with large gradients across the bay mouth (Whaley and Hopkins, 1952; Stroup and Lynn, 1963; Miller et al., 1999). However, there are also times when it is observed to be in the mid 20's and occasionally less, which accounts for the lower value in the climatology.

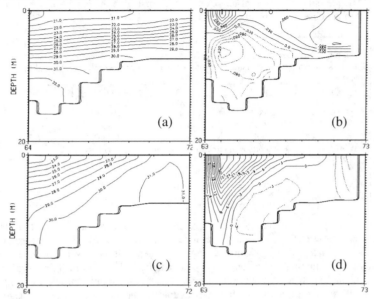

Figure 3. Ninety-day average of salinity (a) and normal velocity (b) at the mouth of ChB after 2 y for river forcing case. (c) and (d) are similar averages for the case with tidal forcing.

The mean normal velocity for the tidally forced case (Fig. 3d) is also quite different from the non-tidal case (Fig. 3b). The outflow is now confined to the south side of the bay mouth, rather than extending across the entire bay mouth, and extends down to about 15 m depth. The inflow occurs on the north side of the main channel below the outflow and extends to the surface on the north side of the bay mouth, which is consistent with the presence of high salinity water on the north side.

The time-averaged currents in Fig. 3d show some agreement with the structure of the sub-tidal currents measured by Valle-Levinson et al. (1998), i.e., maximum outflow near or at the surface on the south side and inflow on the north side of the main channel. Some differences are that the model's inflow in the main channel is much weaker and the Valle-Levinson et al. observations show outflow near the surface over a small channel on the north side of the bay mouth, which would affect the salinities in that area. A factor in the disagreement may be the limitation of the grid resolution used. Neither of the channels in the bay mouth is properly resolved, i.e., the deep channel on the south side is much too shallow (15 m versus a true depth of more than 25 m) and the smaller channel that exits on the north side is not resolved at all. Another factor is that the conditions at the bay mouth are observed to be highly variable.

Table 4
Model-predicted inflow (upbay), outflow (downbay), and net time-averaged transports at the ChB cross sections (see Fig. 2) after 2 y for a simulation with river inflows and tidal forcing. Transports are in m^3/s.

section	inflow	Outflow	net inflow	inflow/net
II	5980	-7980	-2000	3.0
IV	3805	-5605	-1800	2.1
VI	3095	-4395	-1300	2.4
X	750	-1950	-1200	0.5
Transports with reduced horizontal mixing of salinity.				
II	6005	-8205	-2000	3.0
IV	3755	-5555	-1800	2.1
VI	3670	-4970	-1300	2.8
X	2135	-3335	-1200	1.8

Table 4 shows time-averaged transports at the sections indicated in Fig. 2. The mean inflow at the bay mouth was reduced from about 7000 m^3/s without the tides to about 6000 m^3/s with tides. A consequence of using velocity-dependent horizontal mixing coefficients is that adding tides significantly increases horizontal mixing. With the tidal forcing there was a significant decrease in the salinities and the gravity circulation in the mid and upper bay (the salinity range at Section VI decreased to 10-14 psu). It was considered that the strong tidal currents were causing too much mixing due to larger explicit mixing coefficients and to increased numerical mixing effects from the advection scheme.

Another simulation was conducted using minimum horizontal mixing coefficients for salinity reduced by a factor of four. This didn't change the flow at the bay mouth much, but did increase the gravitational circulation in the upper bay (Table 4) and the salinity range at Section VI increased to 11-17 psu. Since, with the reduced mixing, the upper-bay salinities are still low, relative to the climatology, it is likely that the mixing is still too strong.

Idealized wind forcing

A simulation was conducted with a time-varying, idealized surface wind stress to look at the response of the low-salinity plume flowing out of ChB to changes in wind direction. This simulation included the river outflows and tidal forcing of the simulation discussed in the previous section. The wind stress was parameterized as a constant, southward-directed stress of 0.1 dyne/cm^2 plus a clockwise, rotating component with a magnitude of 0.4 dynes/cm^2 and a period of

10 d. This simulation was initialized with the salinity distribution obtained at the end of the tidal forcing simulation. Note that wind directions referred to here refer to the direction in which the surface wind stress is acting, i.e., the direction in which the wind is blowing.

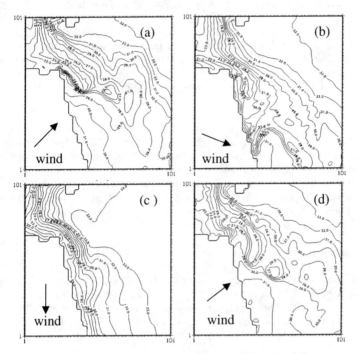

Figure 4. Surface salinity on the nested grid for the clockwise rotating wind case. The current wind direction is indicated by the arrow.

A nested grid (with one-way nesting) was used in the area of the mouth of ChB to provide increased resolution of the outflow plume. A 3:1 grid-nesting ratio was used; hence, the horizontal resolution on the nested grid is 2/3 km. The vertical grid used on the nest is the same as that used on the main grid.

The simulation was run for about 100 days to allow the model to develop a fairly regular cycle in response to the rotating wind. The response of the plume to the changes in wind direction is similar to that observed and discussed by Chao (1988b) in his numerical simulations with idealized estuary geometries.

Figure 4a shows the surface salinity on the nested grid as the direction of the surface wind stress has rotated around to the northeast after a period of westward and northward directed winds. The near-surface, Ekman flow driven by these winds has advected the plume away from the coast and has caused the upwelling of relatively high salinity water (31 psu) to the surface near the coast.

Figure 4b shows the surface salinity on the nested grid about 2 days later as the direction of the surface wind stress has rotated around to the east-south-east. The surface, wind-driven flow is now advecting the plume back towards the coast. The nose of a relatively sharp plume front can be seen against the coast near the southern boundary of Figure 4b. This plume front propagates southward along the coast.

Figure 4c shows the surface salinity on the nested grid as the direction of the surface wind stress has rotated around to the south. The onshore, wind-driven flow has advected the plume against the coast. The width of the plume at this time is about 6-7 km, which is consistent with observations (Boicourt, unpublished observations). Figure 4d shows the surface salinity when the surface wind stress has rotated back around to the northeast. A crescent-shaped plume of relatively low salinity water has been extruded on the ebb and flood of the tide.

Acknowledgments

This work was sponsored by the Office of Naval Research through the Coastal Ocean Model Development for Coupled Air-Sea Modeling Task of the Naval Ocean Modeling and Prediction Program (Program Element 602435N) and through the Naval Research Laboratory Coastal Remote Sensing Program.

References

Asselin, R., 1972: Frequency filter for time integrations. *Mon. Wea. Rev., 100,* 487-490.

Blumberg, A. F., and H. J. Herring, 1987: Circulation modeling using orthogonal curvilinear coordinates. *In Three-Dimensional Models of Marine and Estuarine Dynamics,* J. C. Nihoul and B. M. Jamart, Ed., Elsevier Oceanography Series, Number *45.*

Blumberg, A. F., and G. L. Mellor, 1987: A description of a three-dimensional coastal ocean circulation model. In *Three-Dimensional Coastal Ocean Models,* edited by N. Heaps, American Union, New York, 208 pp.

Blumberg, A. F. and D. M. Goodrich, 1990: Modeling of wind-induced destratification in Chesapeake Bay. *Estuaries, 13,* 236-249.

Chao, S.-Y., and W. C. Boicourt, 1986: Onset of estuarine plumes. *J. Phys. Oceanogr., 16,* 2137-2149.

Chao, S.-Y., 1988a: River-forced estuarine Plumes. *J. Phys. Oceanogr., 18,* 72-88.

Chao, S.-Y., 1988b: Wind-driven motion of estuarine Plumes. *J. Phys. Oceanogr., 18,* 1114-1166.

Chao, S.-Y., 1990: Tidal modulation of estuarine plumes. *J. Phys. Oceanogr., 20,* 1115-1123.

Craig, P. D., and M. L. Banner, 1994: Modeling wave-enhanced turbulence in the ocean surface layer. *J. Phys. Oceanogr., 24,* 2546-2559.

Friedrich, H., and S. Levitus, 1972: An approximation to the equation of state of seawater suitable for numerical ocean models. *J. Phys. Oceanogr., 2,* 514-517.

Goodrich, D. M. and A. F. Blumberg, 1991: The fortnightly mean circulation of Chesapeake Bay. *Estuarine, Coastal and Shelf Science, 32,* 451-462.

Hodur, R. M., 1997: The Naval Research Laboratory's Coupled Ocean/Atmosphere Mesoscale Prediction System (COAMPS). *Mon. Wea. Rev., 125,* 1414-1430.

Johnson, B. H., K. W. Kim, R. E. Heath, B. B. Hsieh, and H. L. Butler, 1993: Validation of Three-Dimensional Hydrodynamic Model of Chesapeake Bay. *J. Hydraulic Eng.*, *119*, 2-20.

Large, W. G., J. C. McWilliams, and S. Doney, 1994: Oceanic vertical mixing: a review and a model with a nonlocal boundary layer parameterization. *Rev. Geophys.*, *32*, 363-403.

Martin, P. J., 2000: *A description of the Navy Coastal Ocean Model Version 1.0.* NRL Report in press. Naval Research Laboratory, Stennis Space Center, MS 39529.

Mellor, G. L., and T. Yamada, 1974: A hierarchy of turbulence closure models for planetary boundary layers. *J. Atmos. Sci.*, *31*, 1791-1806.

Mellor, G. L., and T. Yamada, 1982: Development of a turbulence closure model for geophysical fluid problems. *Geophys. and Space Phys.*, *20*, 851-875.

Mellor, G. L., 1991: An equation of state for numerical models of oceans and estuaries. *J. Atmos. and Ocean Tech.*, *8*, 609-611.

Mellor, G. L., 1996: *User's Guide for a Three-Dimensional, Primitive-Equation, Numerical Ocean Model.* Princeton University, 39 pp.

Miller, J. L., M. A. Goodberlet, and J. B. Zaitzeff, 1998: Remote sensing of salinity in the littoral zone. *EOS, 79*, 173 & 176-177.

Smagorinsky, J., 1963: General circulation experiments with the primitive equations. I: The basic experiment. *Mon. Wea. Rev.*, *91*, 99-164.

Stroup, E. D., and R. J. Lynn, 1963: *Atlas of Salinity and Temperature Distributions in Chesapeake Bay 1952-1961 and Seasonal Averages 1949-1961.* Report 63-1, Chesapeake Bay Institute, John Hopkins University, 410 pp.

Valle-Levinson, A., 1995: Observations of barotropic and baroclinic exchanges in the lower Chesapeake Bay. *Cont. Shelf Res.*, *15*, 1631-1647.

Valle-Levinson, A., and K. M. M. Lwiza, 1995: The effects of channels and shoals on exchange between the Chesapeake Bay and the adjacent ocean. *J. Geophys. Res.*, *100*, 18551-18563.

Valle-Levinson, A., J. M. Klinck, and G. H. Wheless, 1996: Inflows/outflows at the transition between an estuary and the coastal ocean. *Cont. Shelf Res.*, *16*, 1819-1847.

Valle-Levinson, A., and K. M. M. Lwiza, 1997: Bathymetric influences on the lower Chesapeake Bay hydrography. *Journal of Marine Systems*, *12*, 221-236.

Valle-Levinson, A., C. Li, T. C. Royer, and L. P. Atkinson, 1998: Flow patterns at the Chesapeake Bay entrance. 1994. *Cont. Shelf Res.*, *18*, 1157-1177.

Westerink, J.J., R.A. Luettich, C.A. Blain, and N.W. Scheffner, 1994: *ADCIRC: An Advanced Three-Dimensional Circulation Model for Shelves, Coasts and Estuaries; Report 2: Users Manual for ADCIRC-2DDI.* Dredging Research Program Technical Report, U.S. Army Engineers Waterways Experiment Station, Vicksburg, MS., 156 pp.

Whaley, H. H., and T. C. Hopkins, 1952: *Atlas of the Salinity and Temperature Distribution of Chesapeake Bay 1949-1951.* Report 52-4, Chesapeake Bay Institute, John Hopkins University.

Simulation of the Oil Trajectory and Fate on the Diamond Grace Spill in Tokyo Bay

K. Okamoto [1]

Abstract

It is important to predict the process of oil dispersion on the sea surface flowing out from a tanker when an accident and spill occurs.

As oil spill fate model, "OILMAP" model software by Spaulding et al. was developed as a shell based approach allowing separation between the model application and the data and information, so that this system is extremely powerful and flexible. Another major point in this model has been the integration of interactive, limited function geographic information system (GIS) within the models.

Here, the real oil-spill accident had been happened on July 2nd, 1997 in Tokyo Bay, which is semi-enclosed bay. The oil tanker Diamond Grace had run aground at central in the bay and the spill amount was 1500K litters.

In this paper, the object is to apply the real oil-spill accident. The major point in this paper is to use the current data by 3-dimensional multi-level model that is formulated by Finite Differential Method.

The model results are compared against the Diamond Grace spill in the Tokyo Bay. We discuss the accuracy and the limitation for this applying.

1. Oceanic Architecture and Engineering, Nihon University
   Narashinodai, Chiba 2748501, Japan

1. Introduction

From the viewpoint of prevention of the sea disaster, it is important to predict the process of oil dispersion on the sea surface flowing out from the tanker attacked some accidents.

The oil spill fate model can mainly be divided into two types. One is for the environmental assessment before constructing the facilities for oil industry. It is necessary to combine with all process related to the oil spill fate into model. Another type of oil spill fate model is for real time prediction during an oil spill accident. Quickly response is needed to oil spill accident, usually by using PC level computer. It is not necessary to include the minor processes.

As oil spill fate model, "OILMAP" model software by Spaulding et al. (1993) was developed as a shell based approach allowing clear separation between the model application and the data and information, so that this system is extremely powerful and flexible. Another major point in this model has been the integration of interactive, limited function geographic information system (GIS) within the models.

Here, the real oil spill had been happened on July 2nd, 1997 in the Tokyo Bay, which is semi enclosed bay. The oil tanker Diamond Grace had run a ground at central in the bay and the spill amount was estimated 1500K litters.

In this paper, the object is to apply this system to real oil-spill accident. The major point in this paper is to use the current data by using 3-dimensional multi-level model that is formulated by Finite Difference Method.

2. Modification of "OILMAP" system

OILMAP system predicts the trajectory and fate of oil including drift, spreading, evaporation, emulsification, entrainment, and shoreline stranding. The concentration distribution of dispersed oil in the water column is also predicted in space and time. The simulation allows for instantaneous or continuous spills with a constant rate. Each spillet is transported and weathered independently.

OILMAP also provides a current analysis tool, which allows the user to sketch surface current field representations from some 2-dimensional vector plots of currents. This tool is actually used when observed currents data are obtained. The currents are only needed at high tide.

Here, this tool in OILAMP could not be used to reproduce the tidal currents in Tokyo Bay. This is the reason why this tool is not used to calculate a current field, but to interpolate only high tide and to make low tide as symmetry, such as Figure 1. Therefore, the tidal currents must be derived from multi-level model simulation performed by Okamoto et al. (2000).

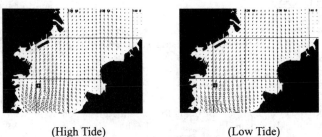

(High Tide)   (Low Tide)
Figure 1   Tidal Currents by Current Tool in OILMAP

3. Multi Level Model for Tidal Currents

The coordinate system for multi level model is shown in Fig.1. The surface level is expressed by h_1 and the bottom level is h_b.

Basic equations are expressed the following equation of motion and continuity equation, and the following boundary conditions are needed.

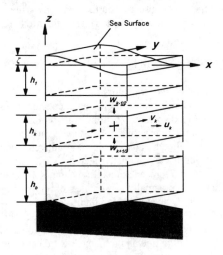

Figure 2   Coordinate System

(1) Momentum Equation

$$\frac{\partial u_k}{\partial t} + \frac{\partial (u_k)^2}{\partial x} + \frac{\partial (u_k v_k)}{\partial y} + (uw)_{k-1/2} - (uw)_{k+1/2}$$

$$- fv_k + \frac{1}{\rho_k}\frac{\partial p_k}{\partial x} - A_x\left(\frac{\partial^2 u_k}{\partial x^2}\right) - A_y\left(\frac{\partial^2 u_k}{\partial y^2}\right) - A_z\left(\frac{\partial u_k}{\partial z}\right)_{k-1/2} + A_z\left(\frac{\partial u_k}{\partial z}\right)_{k+1/2} = 0 \qquad (1)$$

$$\frac{\partial v_k}{\partial t} + \frac{\partial (u_k v_k)}{\partial x} + \frac{\partial (v_k)^2}{\partial y} + (vw)_{k-1/2} - (vw)_{k+1/2}$$

$$+ fu_k + \frac{1}{\rho_k}\frac{\partial p_k}{\partial y} - A_x\left(\frac{\partial^2 v_k}{\partial x^2}\right) - A_y\left(\frac{\partial^2 v_k}{\partial y^2}\right) - A_z\left(\frac{\partial v_k}{\partial z}\right)_{k-1/2} + A_z\left(\frac{\partial v_k}{\partial z}\right)_{k+1/2} = 0 \qquad (2)$$

(2) Continuity Equation

$$\frac{\partial \zeta}{\partial t} + \frac{\partial}{\partial x}\sum_{k=1}^{b} u_k h_k + \frac{\partial}{\partial y}\sum_{k=1}^{b} v_k h_k = 0 \qquad (3)$$

$$w_{k-1/2} = w_{k+1/2} - \frac{\partial u_k h_k}{\partial x} - \frac{\partial v_k h_k}{\partial y} \qquad (k \geq 2) \qquad (4)$$

where h = height of each level, u,v,w = x,y,z velocity components, t = time, f = Coriolis parameter, ρ = density of seawater, p = pressure, A_x, A_y, A_z = eddy viscosity in x, y and z direction, ζ = tide level.

(3) Boundary Condition

Land boundary condition must be used no-slip condition.
At sea surface, below wind stress are given.

$$A_z\left(\frac{\partial u_k}{\partial z}\right)_{1/2} = \frac{\rho_a}{\rho}\gamma_a^2 W_x \sqrt{W_x^2 + W_y^2} \qquad (5)$$

$$A_z\left(\frac{\partial v_k}{\partial z}\right)_{1/2} = \frac{\rho_a}{\rho}\gamma_a^2 W_y \sqrt{W_x^2 + W_y^2} \qquad (6)$$

At sea bottom, below friction stress are given.

$$A_z\left(\frac{\partial u_k}{\partial z}\right)_{n+1/2} = \gamma_b^2 u_n \sqrt{u_n^2 + v_n^2} \qquad (7)$$

$$A_z\left(\frac{\partial v_k}{\partial z}\right)_{n+1/2} = \gamma_b^2 v_n \sqrt{u_n^2 + v_n^2} \qquad (8)$$

where W_x, W_y = x and y components of wind vector, ρ_a = density of air, γ_a^2 = friction coefficient at sea surface, γ_b^2 = friction coefficient at sea bottom and the suffix n express the lowest level of multilevel model.

Here, these equations are discretized by Alternating Direction Implicit (ADI) method under appropriate boundary conditions.

4. Application of the Model to Tokyo Bay Spill

At first, tidal currents simulation must be performed before oil spill simulation.

4.1 Tidal Current Simulation

Here, the parameters for tidal current simulation are shown in Table 1. Map of Tokyo Bay is shown in Figure 3. The calculation domain and observation points (1 & 2) are denoted in Figure 4. From comparison with computed results on tidal ellipse and observed ones, shown in Figure 5, good agreements are obtained. Results of the tidal currents of maximum ebb tide at 1st level are presented in Figure 6, and also the observed ones of tidal currents are shown in Figure 7. These results show good agreement. Obtaining the good results on tidal ellipses and tidal movements, this simulation by multi-level model has sufficiently reproduction of the current of Tokyo Bay.

Table 1   Parameters for Current Simulation

Horizontal Mesh Size		1000[m]	
Vertical Mesh Division	10 Level Division		
	Depth of 1st Level	1[m]	
Friction Coefficient at Bottom		0.0026	
Drag Coefficient		0.0016	
Horizontal Eddy Viscosity		500[m²/sec]	
Vertical Eddy Viscosity		0.0001[m²/sec]	
Time Interval		180[sec]	
Coriolis Parameter		8.42×10^{-5} [sec^{-1}]	
River Discharge	① Edogawa	20[m³/sec]	
	② Arakawa	30[m³/sec]	
	③ Sumida	30[m³/sec]	
	④ Tama	50[m³/sec]	
	⑤ Tsurumi	50[m³/sec]	
Open Boundary		Amplitude	Phase Delay
	Jogashima	38[cm]	146[deg]
	Iwaifukuro	35[cm]	152[deg]

Figure 3　Tokyo Bay
(x : spilled point, ①−⑤:rivers)

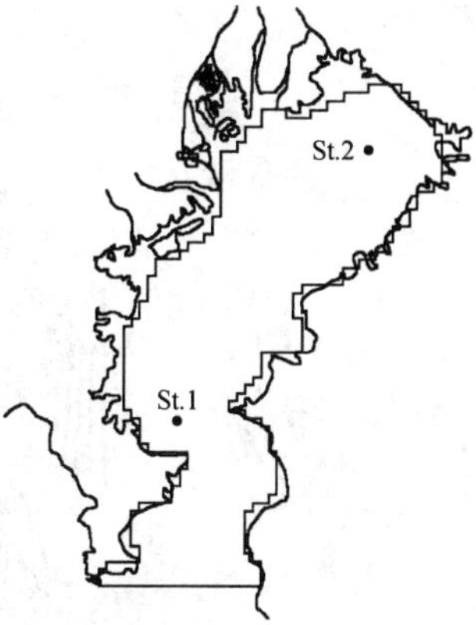

Figure 4　Calculation Domain and
Observation Points

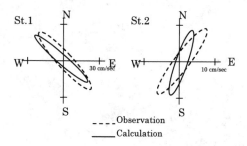

Figure 5　Current Ellipse

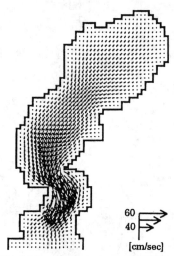

Figure 6   Computed Tidal Current of Maximum Ebb Tide at 1st Level
by Multi-Level Model

Figure 7   Observed Tidal Current of Maximum Ebb Tide by Maritime Safety
Agency, Japan (Unit: [knot]=1852/3600[m/s])

4.2 Oil Spill Simulation

To compare the performance of this modified OILMAP system for spill event; the model was applied to hindcast the Tokyo Bay spill. Outline of this spill is presented in Table 2.

Table 2   Outline of the Tokyo Bay Spill

Spill Time	10:05, July 2nd, 1997
Spill Site	35° 22.99' N 139° 42.44' E (estimated, see Figure 2)
Ship Name	Diamond Grace (256999 [DWT])
Loading Oil	25700 [ton]
Spill Amount	1500 [Klitter] (estimated)
Spill Oil	Light Crude Oil (Umm Shaif)

Simulation Parameters

The parameters for spill simulation were presented in Table 3. The spill model was run with 15 minutes time step and 1000 spillets were used to represent the spill. Here, the currents obtained from a multilevel model are used as an input to OILMAP system whit no interpolating, and the selected output files of current results were used at the corresponding time according time step. The horizontal turbulent diffusion coefficient, used in the oil transport calculation, was $500[m^2/sec]$. This value was decided by horizontal mesh size of the current analysis that was 1000[m] square.

Table 3   Parameters for Spill Simulation

Simulation Duration	24 [hour]
Spill Duration	2 [hour] (estimated)
Horizontal Diffusion Coefficient	$500 [m^2/sec]$
Wind Factor	3.0 [%]
Wind Angle	20 [deg.]
Spreading Coefficient	1.5
Mouse Coefficient	0.000002
Time Interval	15 [min.]
Air Temperature	25 [°C]
Water Temperature	23 [°C]
Observed Wind Station	Yokohama

Oil transport is usually calculated by vector addition of the current plus 3-4[%] of the instantaneous wind speed with angle 3[%] to the right of the wind. The drift factor and angle are based on Reed et al.'s (1990). Here, the wind data was observed at Yokohama station, existing near spilled point. This wind data was recorded by 1[hour] time step, shown in Figure 8.

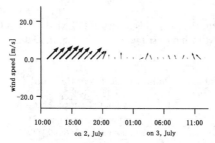

Figure 8   Observed Wind Data at Yokohama Station

Spreading of oil was performed using MacKay et al. (1980) modified formulation, which was based on Fay's (1971) three-regime formulation. Evaporative exposure method was using MacKay and Matsugu's (1973) algorithm. A mouse formulation and emulsification algorithm is based on MacKay et al. (1980,1982).

Observed Spilled Area

In order to compare with simulation results and observed ones in spill simulation, observed spill area is shown in Figure 5, by Maritime Safety Agency, Japan. Now there are three of spilled areas on 19:00, 23:00, 2nd and 09:00, 3rd, July.

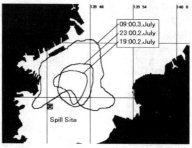

Figure 9   Observed Spilled Area

Simulation Results

Simulation results are presented as 3 cases; (1) using calculated tidal currents by multi-level model shown in this paper, (2) using observed tidal currents by tidal current tool in OILMAP and (3) under no tidal currents. Here, in the following figures on simulation results, black circle ● shows spillet and shade area surrounding spillet denotes oil mass including 0.1[kg/m^2].

(1) Simulation Result using Calculated Currents by Multi-Level Model

The comparisons with simulation results and observed ones were presented in Figure 10 to 13, in case of using calculated tidal currents by multi-level model.

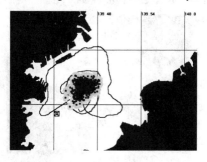

Figure 10   Simulation result at 19:00 on 2 July, using by Multi-Level Model

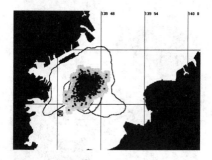

Figure 11   Simulation result at 23:00 on 2 July, using by Multi-Level Model

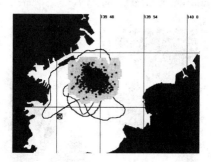

Figure 12   Simulation Result at 09:00 on 3 July, using by Multi-Level Model

(2) Simulation Results in case of using Tidal Current Tool in OILMAP

The comparisons with simulation results and observed ones were presented in Figure 13 to 15, in case of observed currents by tidal current tool in OILMAP

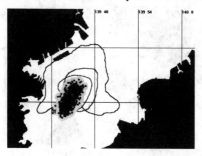

Figure 13　Simulation result at 19:00 on 2 July, by current tool in OILMAP

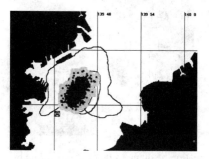

Figure 14　Simulation result at 23:00 on 2 July, by current tool in OILMAP

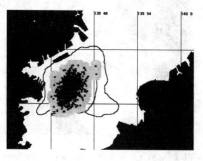

Figure 15　Simulation Result at 09:00 on 3 July, by current tool in OILMAP

(3) Simulation Result under No Tidal Currents

The comparisons with simulation results and observed ones were presented in Figure 16 to 18, under no tidal currents.

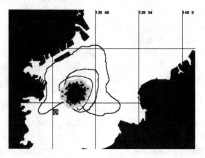

Figure 16   Simulation Result at 19:00 on 2 July, under No Tidal Currents

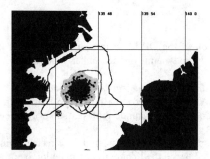

Figure 17   Simulation Result at 23:00 on 2 July, under No Tidal Currents

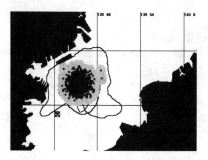

Figure 18   Simulation Result at 09:00 on 3 July, under No Tidal Currents

Discussion

From the comparison with simulation results and observed area in case of (1) as present method, it seems simulation results are over estimated until 23:00 on 2nd July, but good agreement is obtained at 09:00 on 3rd. As fencing operation against spilled oil had been actually done, it is considered spilled area was reduced, so that good agreements are obtained for these simulation results.

From the comparison with calculated area in case of (2) by tidal current tool and one in case of (1), movement of spilled area looks like straight line and southern part of the area moves slowly. This is the reason why tidal current by tidal current tool moves along straight line. Moreover, front part of calculated area does not reach the observed one.

From the comparison with calculated area in case of (3) under no currents and one in case of (1), calculated area cannot be represented at all the time and the results are underpredicted.

Therefore, the modified OILMAP method expressed in this paper was pointed out more sufficient.

5. Conclusions

A modified version of the OILMAP trajectory model was used with current from a multilevel hydrodynamic model rather than from OILMAP's current analysis tool.

Tidal current simulation obtained sufficient reproduction.

Modified OILMAP was applied to the Tokyo Bay spill. This simulation showed the importance of the accurate representation of the current field in the spill.

In the future, oil spill simulation considered the effects of wind driven currents should be recalculated, since the tidal current simulation by multi-level model in this paper did not consider the effects.

Acknowledgement

I owe special thanks to Prof. Malcolm L. Spaulding, developer of the OILMAP system. I would also like to thank to Mr. T. Sato, in Maritime Safety Agency, Japan.

References

Chart of Tidal Streams in Tokyo Wan (No.6216) (1989). Published by Maritime Safety Agency, Japan.

Fay, J. A. (1971). Physical Processes in Spread of Oil on a Water Surface. In *Proc. Joint Conference on Prevention and Control of Oil Spills*, American Petroleum Institute, pp.463-7

Mackay, D. & Matsugu, R.S. (1973). Evaporation Rates of Liguid Hydrocarbon Spills on Land and Water, *Can. J. Chem. Engng.* 51, 434-9

Mackay, D., Buist, I., Mascarenhas, R. & Paterson, S. (1980). Oil Spill Processes and Models. Environmental Protection Service, Canada, Report EE-8

Mackay, D., Shui, W., Hossain, K., Stiver, W., Mcgurdy, D. & Paterson, S. (1982). Development and Calibration of an Oil Spill Behavior Model, U.S. Coast Guard, CG-D-27-83

Okamoto, K., Ono, T. & Saijo, O. (2000). Simulation of Tidal Currents and Diffusion taking into Density Stratification around Very Large Floating Structure, *J. Structural and Construction Engineering*, Transaction of AIJ, No.528, pp.189-195

Reed, M., Turner, C. Odulo,A., Isaji, T., Sorstrom, S. & Mathisen, J.P. (1990). Field Evaporation of Satellite-Tracked Surface Drifting Bouys in Simulating the Movement of Spilled Oil in the Marine Environment. Report to US Department of Interior, Minerals Management Service, MMS 90-0050

Spaulding, M.L., Anderson, E.L., Isaji, T. & Howlett, E. (1993). Simulation of the oil Trajectory and Fate in the Arabian Gulf from the Mina Al Ahmadi Spill, Marine Environmental Research, 36, pp79-115

Spaulding, M.L. & Chen, A. (1994). A Shell Based Approach to Worldwide Oil Spill Modeling, J. Advanced Marine Technology Conference, 11, pp127-141

A Multi-Estuarine Model in Long Island Sound

Frederick E. Schuepfer, M., ASCE[1]
Guy A. Apicella[2]
Robert O'Neill[1]

Abstract

The three-dimensional finite-element mathematical model, RMA-10, was applied to Bridgeport and Black Rock Harbors, two Connecticut estuaries along Long Island Sound. This multi-estuarine model for Bridgeport and Black Rock Harbors and the adjoining section of Long Island Sound has multilayered capability in the vicinity of the freshwater discharges, i.e., wastewater treatment plants (WTP) and river inflows. One WTP outfall is presently positioned in each harbor. The model includes Ash Creek, a third estuary, simulated for purposes of overall tidal prism accuracy.

To provide data sets for calibrating the model, dye, salinity, temperature, tidal stage, and velocity were measured in the harbors. During two neap tide dye tracer surveys, dye was released continuously for five days and was measured, along with salinity, horizontally and vertically in the receiving water bodies. Two water level recorders and two velocity current meters were deployed and bathymetric data were collected to be used in the model calibration process and in the development of model geometry. Using these measurements, the model was then calibrated for stage, velocity, flow, and conservative substances such as salinity and dye. Model calibration parameters were varied over justifiable ranges until the best fit between observed and computed results was obtained. The primary objective of the study was to develop a model that can be used as a planning tool for assessing potential WTP outfall modifications and other water quality management strategies.

[1]Mathematical Modeler, Lawler, Matusky & Skelly Engineers LLP (LMS), One Blue Hill Plaza, Pearl River, NY 10965
[2]Director of Modeling, Lawler, Matusky & Skelly Engineers LLP (LMS), One Blue Hill Plaza, Pearl River, NY 10965

Introduction

The development of a hydrodynamic model can be viewed as a continuous decision-making process, with several universal, key decision points present in all model applications. Choice of numerical method, degree of spatial and temporal refinement, translation of prototype characteristics vis a vis observed data, and model extent and assumptions are all separate but related individual decisions. The choices made determine the ultimate degree of success of the modeling effort. In this paper, along with the description of a modeling study of Bridgeport and Black Rock Harbors, the decision-making process is described.

The three-dimensional finite-element mathematical model, RMA-10, was applied to Bridgeport and Black Rock Harbors (Lawler, Matusky & Skelly Engineers LLP [LMS] 1999a), two estuaries along Long Island Sound shown in Figure 1 . The harbors lie generally to the south of the City of Bridgeport, Fairfield County, in southwestern Connecticut. As shown in Figure 1, the Bridgeport East Side WTP is located near a small basin to the east of Bridgeport Reach. The Bridgeport West Side WTP is situated to the northwest of Cedar Creek. The study area also includes these tributaries to the harbors: a portion of the Pequonnock River and the entire Cedar Creek. Significant branches of the Pequonnock - Yellow Mill Channel, Lewis Gut, and Johnsons Creek - are also included. Ash Creek and its tributary, Rooster River, are modeled due to the proximity to Black Rock Harbor. To properly treat boundary effects, a portion of Long Island Sound out to approximately 4.0 km (2.5 mi) from the shoreline is also within the model bounds.

The RMA-10 model has been used extensively in various studies (Resource Management Associates [RMA] 1996). The hydrodynamic model was applied to the Sacramento River to examine velocity fields around bends (King 1985). The effect of navigation channel widening on saltwater intrusion was investigated in San Francisco Bay (King and Rachiele 1989). To address reservoir water quality issues, RMA-10 and its water quality counterpart were applied by LMS to the Kensico Reservoir in New York (LMS 1995a). The model was used to analyze power plant thermal effects in Mount Hope Bay, Rhode Island (LMS 1995b). Thermal power plant modeling with RMA-10 was also performed for the Delaware Estuary (PSE&G 1999). In studying combined sewer overflows into a number of water bodies in New York City, LMS also applied the two-dimensional predecessors to RMA-10 (LMS 1992, 1994a, 1994b).

For the purposes of this study, the RMA-10 source code was enhanced to include the capability of simulating a temporally varying tidal elevation gradient across the boundary.

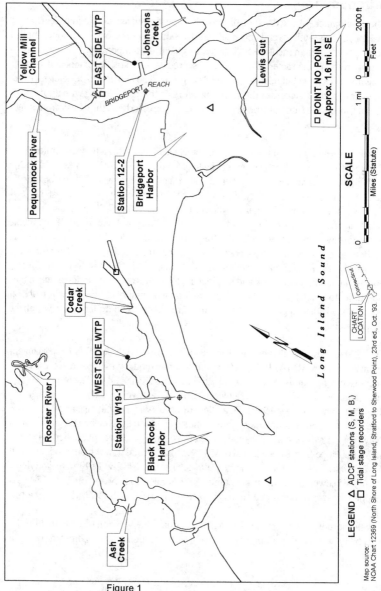

LEGEND △ ADCP stations (S, M, B.)
         □ Tidal stage recorders

Map source:
NOAA Chart 12369 (North Shore of Long Island, Stratford to Sherwood Point), 23rd ed., Oct. '93.

Figure 1
Bridgeport WTP Facility Plan Study Area

Motivation for Developing a Model

The primary objective of the model development is to produce a planning tool that can be used for assessing potential wastewater treatment plant (WTP) outfall modifications and other water quality management strategies. When properly applied, the model can project the effects of hypothetical changes to WTP discharge locations on transport and dilution.

The many uses of the waterways lend added importance to this goal. Nearby in Ash Creek, shellfish beds are cultivated. A section of the Stewart B. McKinney National Wildlife Refuge, containing the largest unditched tidal high marsh in Connecticut (USF&WS, no date), is located along Lewis Gut, a tributary to Bridgeport Harbor, shown in Figure 1. Pleasure, Long, and Seaside Park Beaches, public recreational facilities along Long Island Sound, are located near Bridgeport and Black Rock Harbors.

The complex hydrography and topography of this area would sharply limit the usefulness of applying simpler, analytical solutions. Furthermore, based on a literature search going back to 1980 and earlier no suitable three-dimensional hydrodynamic model had yet been developed for these water bodies. Consequently, on behalf of the City of Bridgeport Water Pollution Control Authority the model described in this paper was developed.

Modeling Decisions

Three of the crucial decisions in the shaping of this model were:

- multilayered versus vertically mixed
- a single multi-estuarine model versus two, single-estuarine models
- finite element as opposed to finite difference method

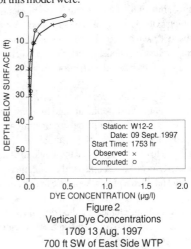

Figure 2
Vertical Dye Concentrations
1709 13 Aug. 1997
700 ft SW of East Side WTP

Each element in the model has a designated number of layers, varying in this model between one and five. Based on the results of the dye surveys showing substantial vertical stratification (as shown, for example, in Figure 2 for a location approximately 700 ft southwest of the Bridgeport East WTP discharge), three-dimensional elements were used in the harbors, the regions of greatest interest in this study. Two-dimensional,

depth-averaged elements were used in
Johnson Creek and Lewis Gut, where little
dye was observed and flow is believed to
be low relative to Bridgeport Harbor.
Two-dimensional elements were also used
to model Long Island Sound. The assumption of complete mixing and vertical
uniformity in this portion of the model is
supported by velocity magnitude, salinity,
and dye profiles in nearby Black Rock
Harbor which are fairly constant vertically.
A graph of dye showing this vertical
characteristic is shown in Figure 3 for a
location 2,200 ft southwest of the West
Side WTP discharge.

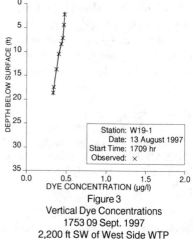

Station: W19-1
Date: 13 August 1997
Start Time: 1709 hr
Observed: ×

Figure 3
Vertical Dye Concentrations
1753 09 Sept. 1997
2,200 ft SW of West Side WTP

In modeling the study area, two smaller
models, one for each harbor, could have been developed instead of the one comprehensive model that was constructed. Smaller models would have had faster run times.
However, the study area was simulated using one comprehensive model rather than two
smaller models for three reasons. First, the potential transfer of wastewater between
the East Side and West Side WTPs and associated change in discharges from the plants
would be analyzed comprehensively using a single model. Second, one comprehensive
model allows for the potential simulation of inter-harbor transport, an important capability for modeling such constituents as nutrients. Third, having a single, larger model
which extends into Long Island Sound for a significant distance provides a more rigorous treatment of boundary conditions than would two smaller models. The single
comprehensive model is capable of predictively evaluating extensions of the treatment
plant outfalls beyond the harbor entrances.

The study area includes a variety of water bodies, including rivers, embayments, and
open water, with a peninsula and two breakwaters. In plan view, elements can be
three- or four-sided, with curved or straight sides. Thus, model elements were developed to take advantage of the flexibility allowed by the finite element method and were
tailored to closely represent these widely varying shapes. Given the presence of such a
wide variety of water body and complementary landform shapes, use of a finite element
model enabled a precise, realistic treatment of the land-water model boundary.

While the finite element method does not intrinsically guarantee local mass conservation,
this property can be checked, and conservation of mass was verified for this model
during steady state conditions.

Qualitative Review of Data

Two velocity surveys conducted by Ocean Services, Inc. (OSI), were conducted throughout the lunar period, including a spring tidal phase (LMS 1999b). The maximum observed velocity at the Black Rock Harbor Acoustic Doppler Current Profiler (ADCP) station was 0.19 m/s (0.61 fps), while the maximum at the Bridgeport Harbor ADCP was 0.30 m/s (1.00 fps) (OSI 1997). Predicted maximum velocities for Bridgeport Harbor are 0.3 to 0.4 m/s (1.1 to 1.2 fps) (Ellison 1997); no predicted maximum is available for Black Rock Harbor.

Collected salinity and dye data show that a significant degree of stratification exists near the WTP discharges, where dye was released, and gradually diminishes towards Long Island Sound. The gauged flow for the Rooster River was used as a basis for modeled flows for the Pequonnock River and the Ash Creek/Rooster River waterway. The maximum modeled flow for the Pequonnock River was 1.33 m^3/s (46.9 cfs), while that of Ash Creek, 1.22 m^3/s (43.0 cfs). There are no significant freshwater inflows recorded for Cedar Creek. In the majority of vertical profiles, the greatest concentrations of dye are at or near the surface and diminish to varying degrees with increasing depth. Minimum salinities are generally at or near the surface and increase with increasing depth. Occasional exceptions occur near the discharge boil, where transient, atypical concentration patterns may occur, i.e., subsurface dye concentration exceeds that at the surface. These exceptions may also be attributed to wind- or turbulence-induced boat movement out of the plume during a profiling.

Discharge Characteristics

Continuously measured WTP effluent flows and daily temperatures generally measured once a day on weekdays were obtained. Temperatures are spot measurements taken each weekday between 1030 and 1100. Since no information was available regarding salinity intrusion into the effluent discharge outfall, the salinity of the WTP discharge is assumed to be zero.

Both East Side and West Side WTPs utilize a surface discharge. Locations of the discharges are shown in Figure 1. The West Side WTP discharges in close proximity to a small cove, while the East Side WTP discharges into a small basin on the eastern shore of Bridgeport Harbor. A small structure deflects the effluent flow towards the head end of the basin, assuring a more complete mixing in the basin before the effluent reaches the harbor.

Model Application

There are three principal steps involved in developing the RMA-10 model geometry: element layout, vertical layering, and model node elevations. The first step consists of dividing the study area into elements, as shown in Figure 4. A sufficient number and

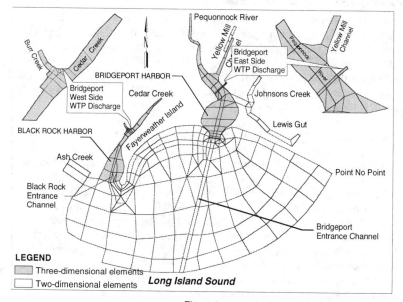

Figure 4
Finite Element Model Network

distribution of elements was included to yield the degree of refinement needed for the
model application of WTP effluent dilution analysis. Rows of elements were tailored to
conform to the most significant bathymetric features such as navigational channels and
bathymetric contour patterns. Fayerweather Island and the Bridgeport Harbor breakwater are modeled using curved-sided elements. Relatively small elements were used to
provide enough detail to attain model solution convergence at these locations. Since
observed dye concentrations were below background concentration in Lewis Gut and
Johnsons Creek, it is assumed that limited tidal exchange occurs, and few elements
were used to represent these channels. Ash Creek was modeled as two simple rectangular elements so as to account for tidal prism effects. Therefore, available topographic
and hydrographic information was obtained so that depth and surface area for Ash
Creek could be depicted as accurately as possible (U.S. Geological Survey [USGS]
1970; R. Bimble, Town of Fairfield Parks Department, pers. commun., 1997; K.
Mulvey, Town of Fairfield Conservation Department, pers. commun., 1997). Ash
Creek was modeled as far upstream as Black Rock Turnpike. As shown in Figure 4,
the study area is represented by a model network of 849 surface nodes and 250
surface elements.

Definition of model node elevations involved first combining bathymetric data from three sources: a bathymetric survey performed by OSI, United States Army Corps of Engineers (USACE) ship channel and turning basin condition surveys (July, August and September 1997), and data collected by the National Oceanic and Atmospheric Administration (NOAA) used to develop navigation charts (1932 - 1989). Each of these data sets have unique advantages, and the judicious use of all three yielded accurate model geometry. The OSI data was best for some of the small branches as well as main channels, and the measurements are relatively recent. The USACE data consists of large data sets along the main navigation channels as well as the Bridgeport Harbor turning basin, and that data set is also relatively recent. The NOAA data has the most comprehensive coverage of the three data sets; these data span both harbors and the relevant portion of Long Island Sound. However, the NOAA source contains some older data which for some areas, such as the navigational channels, are no longer valid.

To provide the necessary tidal elevations for the model boundary, measurements collected at Point No Point, the location shown in Figure 4, were used. Results from a model of Long Island Sound developed by NOAA (1994b) were used to estimate spatial variations in water surface elevation along the boundary. Tidal elevation is specified at the eastern end of the boundary and is computed across the remainder of the boundary using the spatial gradient.

Water quality data collected by USGS were obtained for the Rooster River at Fairfield, Connecticut. Four values each of temperature and specific conductivity were available for July through October 1997 (USGS 1997), and boundary condition values were interpolated from this data set for each model timestep. Salinity was calculated from specific conductivity using the following formula (Texas Instruments 1979):

$$Salinity \ (ppt) = -100 \ln \left(1 - \frac{Specific \ Conductivity \ (\mu mhos)}{178,500}\right)$$

Boundary salinities were first estimated to be 28 ppt from the Long Island Sound Study data (NOAA 1994b). The values were then adjusted to 25 ppt to obtain a good fit with LMS survey data near the boundary. The model will enforce boundary conditions during flood tide (i.e., flow entering model elements). During ebb, the model computes concentrations at the boundary nodes using the advection diffusion equation.

As stated previously, WTP salinities were assumed to be essentially zero for modeling purposes, and USGS data for Rooster River were used to interpolate salinities for Ash Creek and Pequonnock River over the duration of the model simulations.

Both dye surveys were conducted during neap tide conditions and involved a continuous release of Rhodamine WT, a fluorescent dye, for 5 days. The target effluent dye concentration at the West Side WTP outfall was 16 µg/l, for the East Side 30 µg/l. The objective of the dye release operation was to maintain a constant effluent concentration so that each measured dye concentration in the waterways is a direct measurement of effluent dilution. Temporal variations in released dye mass were measured and simulated in the RMA-10 calibration runs.

In formulating the model boundary condition for dye it was assumed that dye which leaves the study area across the Long Island Sound boundary does not return. The value of two times the standard deviation (approximately 95% confidence) of observed background (i.e., pre-release) fluorescence measurements for the Black Rock Harbor survey is 0.09 µg/l, and that for Bridgeport Harbor 0.19 µg/l. Since model computed concentrations near the boundary are substantially lower than two times the standard deviation of background fluorescence/dye, it appears that dye would not have been detected at the model boundary. Also, dye measured at the harbor/Long Island Sound interfaces was approximately less than or equal to two times the standard deviation. Therefore, assuming a boundary dye concentration of zero on flood tide is reasonable.

To properly simulate initial conditions, a test run of a repeated tidal cycle was executed and yielded virtually identical hydrodynamic results at the end of each cycle. This result is indicative of rapid spinup from flat water surface conditions.

Model Calibration and Verification

To develop a reliable model, ranges for model parameters were determined from relevant data and literature. The model was then calibrated and verified by adjusting model parameters within these ranges to attain agreement with measured hydrodynamic data.

To provide data sets for calibrating the model, two tide gauges were deployed from 7 August 1997 through 17 September 1997 by OSI at the locations shown in Figure 1. Two acoustic Doppler current profilers (ADCPs) were operational from 7 or 8 August 1997 through 17 September 1997. Two dye dilution surveys were performed during neap tide conditions. The first survey was executed in Black Rock Harbor on 11-15 August 1997; dye was released through the Bridgeport West WTP discharge outfall. The second survey was performed in Bridgeport Harbor on 8-12 September 1997; dye was released through the Bridgeport East WTP discharge outfall. During each survey, Rhodamine WT dye was released continuously for five days. Horizontal transects were performed in which dye was measured at a depth of 1.5 ft. Vertical profiling of dye, conductivity, and temperature was also done at prescribed sampling stations.

The model was calibrated for velocity magnitude and direction, flow, and stage using the 11-15 August 1997 survey and was verified for these variables using the 8-12 September 1997 survey. The model was calibrated for salinity and dye for the West Side WTP (Black Rock Harbor) using the 11-15 August 1997 survey and was calibrated for the same constituents for the East Side WTP (Bridgeport Harbor) using the 8-12 September 1997 survey.

Parameters which were adjusted to calibrate the model are:

- Manning roughness coefficient
- turbulent eddy coefficient
- eddy diffusion coefficient
- number of layers
- timestep
- wind effects

During calibration, the model timestep was reduced in successive runs until no further improvement in observed-computed velocity and dye comparisons were attained. The effect of further element size reduction near the Bridgeport East WTP will be assessed in future work.

In the course of the velocity and stage calibration, the spatial gradient of water surface along the open boundary was varied, and the final value is 0.187 ft/mile. This value is comparable to a gradient computed in the nearby East River in New York City (LMS 1994a), which had a simulated maximum neap gradient of 0.180 ft/mile.

Friction is mathematically described in RMA-10 using either the Manning roughness coefficient, used in this study, or the Chezy coefficient. Values used in the model are comparable to those found in the literature (Chow 1959, Weisman et al. 1989). Turbulent exchange coefficients were varied during the calibration. Values were selected that would produce the most accurate solution and also would achieve a stable, convergent solution. Values used for the Manning roughness coefficient were 0.025 to 0.030. Vertical turbulent exchange coefficients were set to 8.5 x 10-5 m^3/sec (0.003 ft^3/sec). The vertical eddy coefficients for diffusion were set to 9.29 x 10-4 m^2/sec (0.01 ft^2/sec). Horizontal eddy coefficients are calculated internally by the model as a function of depth and velocity and were adjusted by the modelers by means of scaling factors varied during calibration.

Observed versus computed tidal stage results are seen to agree closely for both Black Rock and Bridgeport Harbors. The calibration results are accurate for both tidal range and timing of high and low waters. Computed water surface elevations at maxima and minima are frequently less than observed values at those times by 0.03 to 0.06 m (0.1-0.2 ft). As NOAA predicted stage values support model results, this difference is suggestive of a possible, slight inconsistency in vertical benchmark elevations.

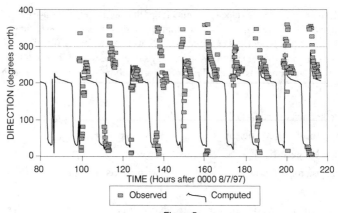

Figure 5
Velocity Direction 11-14 Aug. 1997, Bridgeport Harbor at Bin 9.5 m From Bottom

For the Bridgeport Harbor ADCP data, velocities at the lowest measured height (referred to as a "bin") of 11.5 ft (3.5 m) above the bottom (where total water depth was 39 ft (11.9m) at water surface elevation of 0 ft NGVD) frequently were directed in a predominantly western direction. Approximately 3,000 of the 5,700 velocity measurements (52.6%) taken at this depth have directions varying between 250° and 300° clockwise from true North. Approximately 28% of the measurements at this depth fall between 150° and 250° (southwestern direction) or 20° to 70° (northeastern direction), the expected velocity orientation parallel to the navigational channel. In contrast, velocity directions measured at 18.0 ft (5.5 m), 24.6 ft (7.5 m), and 31.2 ft (9.5 m) above the bottom showed a much more bidirectional, northeast-southwest primary axis which decreases with increasing depth. For example, observed and computed directions for the 31.2 ft (9.5m) bin are shown in Figure 5. It appears that this velocity pattern is a characteristic feature of Bridgeport Harbor hydrodynamics. As shown in Figure 1, the ADCP was deployed in a relatively deep region adjacent to a shoal and breakwater. The abrupt increase in depth near the ADCP causes flow near the bottom to be deflected in a predominantly western direction, instead of moving in the expected tidal orientation of southwest-northeast. This situation could engender the westerly flows seen in the observed data, especially during ebb tide. Because the velocity magnitude decreases with increasing depth, the flow through the lowest bin constitutes a small portion of the total flow. In any event, to calibrate the model for overall transport through the harbor, it was decided to focus on the total cross-sectional flow rather than velocity at the ADCP location. While it would have been preferable to quantify the flow at such a section using more than one ADCP location, the ADCP position, which is outside of the main navigational channel, is believed to be representative of the average cross-sectional velocity.

Comparisons were made for total flow, shown in Figure 6, as well as flow for each of the four measurement depths. The sign convention is that positive flow represents flood flow. Comparisons for specific depths show significant differences at some times. However, as close agreement is generally attained at the bin nearest the surface, and as most of the flow is transmitted at this depth, the total flow summing all bin flows shows good agreement between observed and computed.

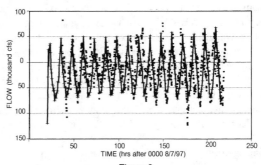

Figure 6
Bridgeport Harbor Observed vs. Computed Flow
(Positive flow represents flood tide)

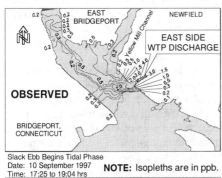

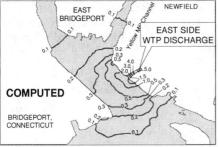

Figure 7
Horizontal Dye Comparison

For Bridgeport Harbor, horizontal dye contour plots show smoother computed dye contours relative to observed results for both nearfield and farfield. One of the 16 horizontal survey comparisons done for this harbor is shown in Figure 7. This difference is attributed to small-scale flow characteristics, e.g., piers, moorings and eddy currents, that are not duplicated by the model. A second reason for the smoother model result is that the horizontal resolution of the model results, i.e., distance between nodes, is typically 350 ft or more, whereas observed data has a resolution on the order of 10 to 20 ft. Nevertheless, the model accurately simulates the effluent discharge plume in key aspects. The surface areas of many of the isopleths are quite comparable for computed versus observed. Given a target Bridgeport East WTP discharge concentration of 30 ppb, the range of dilution factors within the observed and computed plumes are similar. Most importantly, the observed-computed agreement is sufficient for the needs of this broad-scale analysis. From the background survey for Bridgeport Harbor, the value of two times the standard deviation for dye is 0.19 µg/l. When this value is used to define a confidence interval around (plus or minus) observed data, computed results compare well to corresponding ranges of dye concentration. The average background dye concentration for the Bridgeport Harbor survey is 0.223 µg/l, suggestive of the variability of observed data.

One significant characteristic of the observed dye distribution is that in many distributions dye is observed in higher concentrations along the western shore, opposite the discharge location. This effect is due to the presence of piers and watercraft across the harbor from the East Side WTP effluent discharge point that impede flushing along the west shore. To simulate this effect, shoreline friction was added in the Bridgeport Harbor anchorage area; changes in this parameter resulted in impeded flushing. Model sensitivity to wind forcing was also tested; it was found that adding wind to the model based on Igor Sikorsky Memorial Airport data did not improve the model results. Due to the presence of breakwaters at Bridgeport Harbor and the sinuosity of Black Rock Harbor that limits fetch, it is inferred that wind is not a significant factor in harbor hydrodynamics.

Seven of the 118 vertical dye profiles for Bridgeport Harbor are shown in Figure 8. Although variations in dye concentration over depth are generally well depicted, model results are at times somewhat less stratified compared to observed concentrations. Horizontal and vertical turbulent exchange and diffusion coefficients were adjusted to obtain realistic distributions of dye. These coefficients were lowered as much as possible to attain realistic spatial variation without compromising model solution convergence. However, this result may be partially due to the single layer model element in the outlying sound region of the model. A few negative computed dye concentrations were noted near the WTP discharges due to imperfect model convergence. However, the occurrence of these negative values were kept to a very low level (0.5% of computed vertical dye concentrations) by parameter adjustment during the model calibration phase.

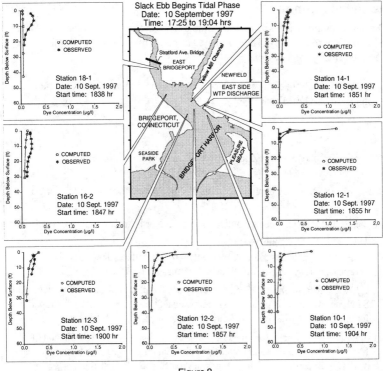

Figure 8
Vertical Dye Comparison

Future Uses of the Model

Application of the model is planned to evaluate dilution characteristics of the harbors with respect to the WTP effluent discharges. Possible effluent discharge modifications, designed to increase dilution, will be tested. In particular, lengthening of the outfall to discharge at greater depths both in and out of the harbors will be simulated. Since vertical momentum is neglected in RMA-10 (hydrostatic assumption), if a higher degree of detail is desired the plume model CORMIX will be used as part of a second stage of model projections. A further, possible use of the RMA-10 model could be to estimate potential for inter-harbor transport of pollutants and nutrients.

Acknowledgment

This study described in this paper was commissioned by the City of Bridgeport Water Pollution Control Authority. The model used in this study was originally developed by Resource Management Associates.

The authors wish to gratefully acknowledge the assistance of Mr. James Guderian, graphic artist; and Dr. Michael Skelly, partner.

REFERENCES CITED

Bimble, R. Telephone conversation between R. Bimble and F. Schuepfer, 18 November 1997.

Chow, V.T. 1959. Open Channel Hydraulics. New York: McGraw-Hill Book Co. 680 p.

Ellison, B. (ed.) 1997. Reed's Nautical Almanac: North American East Coast 1997. Thomas Reed Publications, Inc.

King, I.P. 1985. Strategies for finite element modeling of three-dimensional hydrodynamic systems. Adv. Water Resources (8):69-76.

King, I.P., and R.R. Rachiele. 1989. Simulation of the Impact of Channel Widening in San Francisco Bay Using a Three Dimensional Stratified Flow Model. Prepared for the Waterways Experiment Station, U.S. Army Corps of Engineers.

Lawler, Matusky & Skelly Engineers (LMS). 1992. Task 4.0 Receiving Water Quality Modeling, East River Water Quality Facility Planning Project. Performed for New York City Department of Environmental Protection (NYCDEP)

Lawler, Matusky & Skelly Engineers (LMS). 1994a. Flushing Bay Water Quality Facility Planning Project Receiving Water Modeling. Performed for NYCDEP.

Lawler, Matusky & Skelly Engineers (LMS). 1994b. Subtask 5.3 Receiving Water Modeling, Newtown Creek Water Quality Facility Planning Project. Performed for NYCDEP.

Lawler, Matusky & Skelly Engineers LLP (LMS). 1995a. Task 5.5 Summary Report: Reservoir Water-Quality Modeling, Kensico Water Pollution Control Study. Contract CRO-223. Performed for NYCDEP.

Lawler, Matusky & Skelly Engineers LLP (LMS). 1995b. Mount Hope Bay Modeling Study. Prepared for New England Power Company.

Lawler, Matusky & Skelly Engineers LLP (LMS). 1999a. Effluent Dilution Study of Bridgeport's East Side and West Side Wastewater Treatment Plant Discharges: Subtask 5.3 Model Effluent Dilution of Existing Outfalls and Alternative Outfalls/Diffusers. Prepared for the Bridgeport Water Pollution Control Authority under subcontract to Kasper Group, Inc.

Lawler, Matusky & Skelly Engineers LLP (LMS). 1999b. Effluent Dilution Study of Bridgeport's East Side and West Side Wastewater Treatment Plant Discharges: Subtask 5.2 Dye Dilution and Hydrographic Surveys. Prepared for the Bridgeport Water Pollution Control Authority under subcontract to Kasper Group, Inc.

Mulvey, K. Telephone conversation between K. Mulvey and F. Schuepfer, 18 November 1997.

Ocean Surveys, Inc. (OSI). 1997. Final Report: Bridgeport Facilities Plan Outfall Study. Prepared for Diversified Technologies Corporation.

Public Service Electric & Gas (PSE&G). 1999. Salem Generating Station 316(a) Demonstration.

Resource Management Associates (RMA). 1996. RMA-10V Reference Guide: A Finite Element Model for three Dimensional, Density Stratified Flow.

Texas Instruments, Inc. 1979. 1977 Year-Class Report for the Multiplant Impact Study of the Hudson River Estuary. Prepared for Consolidated Edison Company of New York, Inc.

United States Fish & Wildlife Service (USFWS). No date. Stewart B. McKinney National Wildlife Refuge (pamphlet).

United States Geological Survey (USGS). 1970. Bridgeport Quadrangle (Connecticut- Fairfield County), 7.5 Minutes Series (Topographic).

United States Geological Survey (USGS). 1997. Temperature and Specific Conductance Measurements at Rooster River, Fairfield CT. Facsimile transmission from Bruce Davies, USGS, sent 4 December 1997.

Weisman, R.N., G.P. Lennon, and F. Schuepfer. 1989. Resistance coefficient in a tidal channel. Presented at ASCE Estuarine and Coastal Modeling Conference, Newport, RI, 15 November 1989.

3D Hydrodynamic Model of an Estuary in Nova Scotia

Ken Hickey, Isabelle Morin, Marcia Greenblatt, and Gavin Gong[1]

Abstract

A study of hydrodynamics in an estuary in Nova Scotia was performed through a field investigation and a three-dimensional modeling evaluation. Complex hydrodynamic flow patterns observed in the estuary, such as seasonally dependent stratification, regions with tidally reversing flows, and regions with lateral circulation patterns, necessitated the use of a three-dimensional numerical model. The study was motivated by a requirement that a major discharger identify an acceptable submerged diffuser location somewhere in the study area. The study focused on identifying an optimal outfall location from an effluent mixing and flushing perspective. A variety of locations including an inner harbor, a harbor narrows, and a coastal embayment were evaluated. The hydrodynamic modeling application resulted in predictions of effluent mixing and flushing from a hypothetical submerged diffuser at several alternative locations. Results of the hydrodynamic study provided a basis for assessment of the waterbody's assimilative capacity.

The three-dimensional, finite-difference model HEM-3D (Shen *et al.*, 1997) was selected and applied to perform hydrodynamic simulations. The modeling was supported by a field program designed and implemented to quantify hydrodynamics through measurement of water levels, tidally reversing flows, and water quality profiles throughout an eight mile reach including an estuarine river, a harbor, and a coastal embayment. Calibration of the hydrodynamic model included comparison of model predictions to synoptic tidal flowrate measurements collected at flood and ebb tide. A GIS package, ArcView (ESRI), was used to generate model bathymetry, to generate a model grid, and to illustrate model predictions (Morin *et al.*, 1999). Model predictions were expressed primarily as effluent dilution contours in the near-surface layer.

---

[1] ENSR, 35 Nagog Park, Acton, MA 01720

The model was successful in simulating hydrodynamic conditions in the system and in quantifying differences between alternative outfall locations in terms of predicted effluent dilution and flushing. Model results also identified the utility of hold-and-release discharge scenarios to improve effluent flushing. The hydrodynamic model will serve as a foundation for assessment of potential water quality and biological impacts related to the proposed outfall. Eventually, an acceptable outfall location will be selected based on hydrodynamic factors and other factors including construction costs, waterbody uses, and potential for ice damage.

Introduction

A study of potential locations for a submerged outfall within an estuarine system was necessitated by a legal settlement stipulating that a major discharger terminate use of its present surface outfall location. Selection of an alternative outfall location requires evaluation of a set of complex factors including assimilative capacity of the water body, public opinion, construction costs, and potential for ice damage. As one component of the outfall location project, a preliminary hydrodynamic study was performed to evaluate the potential for effluent mixing and flushing at alternative locations throughout the estuary. This paper describes the preliminary hydrodynamic study.

Figure 1 shows a map of the estuarine study area located on the north coast of Nova Scotia. The estuary receives freshwater inflow from three rivers, known as the East, Middle, and West Rivers and has a watershed area of approximately 1020 km^2. The East River has unrestricted, tidally-reversing flows and receives approximately one-half of the watershed's freshwater inflow. The Middle and West Rivers are separated from the harbor by a reservoir causeway and flow is restricted to a narrow sluice near Abercrombie Point. As a result, strong tidally reversing flows are observed at the sluice location.

The estuary may be described as consisting of an Inner Harbor, a Narrows, and an Embayment, as labeled in Figure 1. The study area extends from the East River to a point in the Embayment along a line between Cole Point and MacKenzie Head (see Figure 1). The Embayment is approximately 2 km long and 2 km wide and is generally 1 to 6 meters deep except in a deeper narrow navigation channel extending from the mouth of the harbor. A navigation channel runs southwesterly from the harbor mouth to the reservoir causeway, a distance of 5.4 km. The channel is 400 to 600 meters wide and between 8 and 14 meters deep. The Narrows begin at the harbor mouth and extend inward for 3 km. At that point, the estuary broadens into the Inner Harbor until it reaches the reservoir causeway and is approximately 2 km long and 2 km wide. The Inner Harbor is characterized by extensive shallows, with depths of approximately 3 meters, and with a navigation channel forged through the center.

Approach

A preliminary evaluation of hydrodynamics in the study area was performed to assess mixing and flushing characterizations throughout the system and to support future, more robust evaluations. Two project constraints, a timeline of four months and a scope of work including only one field survey, were influential in designing the program. The approach selected for this application featured a synoptic field survey and numerical modeling evaluation to support a preliminary evaluation of the feasibility of alternative outfall locations.

The field program was designed to collect a synopsis of hydrodynamics in the system through an intensive one-week survey. The field survey included measurement of water velocities and tidal flowrates, continuous measurement of water elevations, and water quality profile measurements were collected throughout the system.

The need for a three-dimensional numerical model was identified based on the complexity of the estuarine system and level of precision required to achieve project goals. Hydrodynamics in the estuary feature lateral circulation patterns and seasonally dependent vertical stratification. Project goals require simulation of the fate and transport of a buoyant plume with sufficient precision to support outfall location decision-making and, ultimately, to satisfy the concerns of local and regulatory communities. A survey of available three-dimension models was performed and the Virginia Institute of Marine Science (VIMS) Hydrodynamic-Eutrophication Model (HEM-3D) was selected. The HEM-3D model was selected based on several factors including strength of computational approach, cost-effectiveness (i.e., it is a public domain model), and availability of a water quality component to support future applications.

Field Measurements

Field measurements were successfully collected throughout the study area during the week of December 9-15, 1998. The goal of the field survey was to collect a synopsis of hydrodynamic condition in the estuary sufficient to support preliminary hydrodynamic modeling. The field survey featured boat-based Acoustic Doppler Current Profiler (ADCP) measurements throughout the study area, continuous water level measurements at three locations, and water quality profile measurements throughout the study area. Figure 2 shows the locations of all field data collection activities. The field survey focused on measurement of total tidal flowrates and water velocities across transects throughout the study area (shown as black lines in Figure 2) using a boat-based ADCP. Boat-based ADCP measurements were collected over a 5 day period at various tidal stages and provided sufficient data to support a strong hydrodynamic model calibration.

Model Set-up

The model grid was developed from an digital version of a nautical chart and consisted of a Cartesian grid of equal spacing in the x- and y-directions of 40 meters, for a total of 1,365 cells in the horizontal plane. The vertical dimension is modeled in a sigma-coordinate system and is divided into 4 layers, each representing a quarter of the water column.

Figure 3 presents the model domain, including the model grid and model boundaries. Open water boundary conditions exist to the West, South and East. A time series of water surface elevations was specified along the East open boundary to the seaward boundary beyond the Embayment. The tidal-driver was applied as a continuously repeating "average tide". The average tide consisted of a single M2 tidal component with a 12.4 hours period and an average tidal amplitue of 0.4 meters, based on measurements collected during the field survey.

Tidally-reversing flows were experienced throughout the study domain. As a result, flow boundaries at the causeway to the West and in the East River to the South were tidally-reversing. These flows were represented in the model as a sinusoidal function with period and amplitude selected based on field survey measurements.

A screening level near-field diffuser model, UDKHDEN (Muellenhoff et al., 1985), was applied and embedded in the HEM-3D model framework. UDKHDEN is a 1985 refinement of the model originally developed in 1979 by Teeter and Baumgartner under support of the U.S. Environmental Protection Agency. Results of near-field modeling indicated a buoyant effluent plume reaching the surface within 20 meters of the discharge point. The effluent discharge was therefore specified in HEM-3D as a source in a single surface cell (dimensions: 40 meters by 40 meters) above the proposed diffuser location. The effluent was modeled as a conservative tracer with specified discharge concentrations.

Model Calibration

The model was calibrated to water velocity, tidal flowrate, and tidal water level measurements collected during the field survey. Model predictions and field measurements of water surface elevations and tidal flowrates across seven transects in the study area (see Figure 2) were compared during the calibration process. Figure 4 presents a comparison of model predictions vs. field measurements at the mouth of the harbor. Model predictions of both surface water levels and tidal flowrates are well-matched to field measurements.

The model calibration process was straightforward because initial model simulations resulted in model predictions relatively close to field measurements. Thus, major adjustments of modeling parameters was not necessary. Key model parameter settings are presented in Table 1. Experimentation was performed to identify the duration required for initial model set-up. Specifically, the period of time required for the effluent plume to establish a "steady-state" pattern within the system was determined to be 10 days. After 10 days of simulation, effluent mass in the system was neither increasing nor decreasing and the pattern of effluent transport was the same from one simulated day to the next.

Figure 5 presents model predictions of water velocities throughout the system during peak flood tide. The predicted normal water velocities at seven transects of the study area were compared to velocities measured using the boat-based ADCP as part of the calibration process. Both the magnitude and the cross-sectional distribution of the water velocities predicted by the model were well-matched to field measurements. Good agreement was also observed between predicted salinities throughout the water column and field measurements of salinity profiles.

The calibrated model was applied to predict the transport of the existing effluent plume. The existing effluent plume is released on the water surface from a point in the southwest corner of the Embayment (see Figure 1). The present plume is visible and aerial photographs have documented its location at various tide stages. Model predictions of effluent transport from the present discharge location featured a complete reversal of effluent flow direction with each tide cycle and were well-matched to observed effluent transport. Clearly, comparison of model predictions to aerial photographs provide at best a semi-quantitative validation. The strong correlation observed between model predictions and observed effluent transport was useful, however, in that it provided an additional confirmation that the model was performing properly.

Application

The modeling application was designed to obtain a preliminary determination of optimal outfall locations in terms of mixing and flushing objectives. All locations within the model domain were considered as potential outfall locations. Four hypothetical outfall locations were selected for evaluation and are identified in Figure 6 as locations A, B, C, and D. At each alternative outfall location, effluent flow and constituent concentrations were simulated.

Figure 7 presents predicted water surface effluent dilutions contours at Location A during peak flood tide. Peak flood tide provides the worse-case tidal condition because the plume is least well-flushed and tends to be most extensive. Figures 8, 9, and 10 present predicted water surface effluent dilution contours at Locations B, C, and D, respectively, during peak flood tide. The effluent plume is predicted to be most extensive at Location A and least extensive at Location D. Predicted effluent plumes are predicted to be less extensive and are predicted to provide greater effluent dilution with distance seaward (i.e., moving from Location A to D). Thus, the hydrodynamic modeling evaluation identified a location in the Embayment as most favorable for locating a submerged outfall.

Conclusion

The three-dimensional hydrodynamic modeling study was successful in supporting a preliminary evaluation of alternative outfall locations in the study area. The study concluded that the Embayment is relatively well-flushed, the Narrows are moderately well-flushed, and the Inner Harbor is relatively poorly flushed. The hydrodynamic modeling evaluation identified a location in the Embayment as most favorable for locating a submerged outfall. Conversely, two other evaluations performed in support of the outfall location process have recently reached opposite conclusions. Specifically, construction costs and the potential for ice damage are decreased (i.e., minimized) with distance into the harbor, while effluent mixing and flushing is worsened. The outfall location process is ongoing and a compromise between these conflicting factors is being sought.

The HEM-3D model performed well and was a good choice for this application. Additional tasks are necessary to bring the outfall re-location process to a successful conclusion. Additional tasks should include collection of a validation data set, preferably during summertime stratified conditions. Also, improvement of model simulation of vertical gradients and near-shoreline areas through increasing the number of model cells and other enhancements. This will result in better simulation of critical factors such as near-surface effluent dilution and extent of effluent shoreline attachment.

Tables and Figures

Table 1: Selected model parameters.

Parameter	Value	Source
Tidal amplitude (East)	0.40 m	Average for December 1998
Tidal period (East)	12.4 hrs	Average for December 1998
Flow at causeway (West)	293 m³/s	Estimated from reservoir area.
Flow East River (South)	260 m³/s	December 1998 measurements.
Discharge Location	Present, A, B, C, D	
Effluent discharge	Avg: 66,000 m³/day	Typical plant operating
	Max: 110,000 m³/day	conditions (1998)
Wind magnitude	3.5 m/s	Atlantic Climate Centre Data, NS
Wind direction	SW (235°)	Atlantic Climate Centre Data, NS

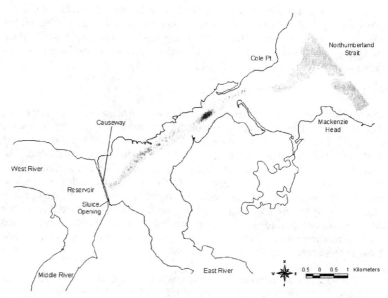

Figure 1: Map of estuary study area.

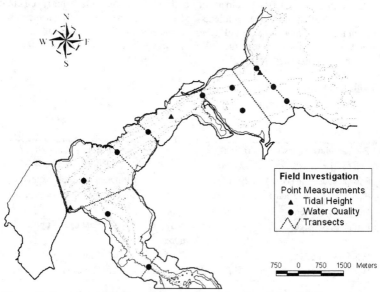

Figure 2: Locations of field measurement activities (December 1998)

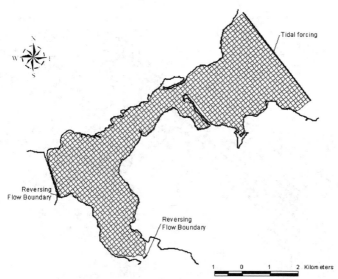

Figure 3: View of the study area showing the computational grid and open water boundaries.

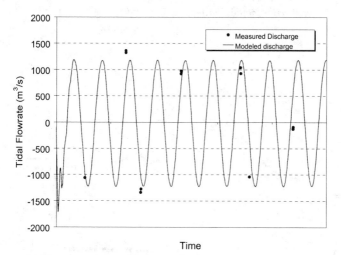

Figure 4: Model calibration. Comparison of model predictions and field measurements of tidal flowrate at the Habor Mouth.

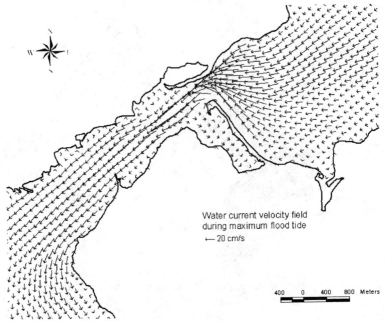

Water current velocity field
during maximum flood tide
← 20 cm/s

400   0   400   800 Meters

Figure 5: Predicted velocities during peak flood tide.

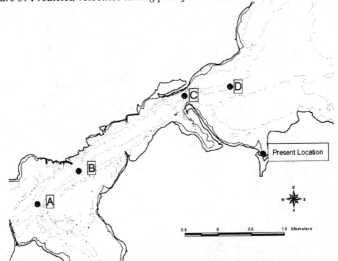

Figure 6: Present and alternative effluent discharge locations.

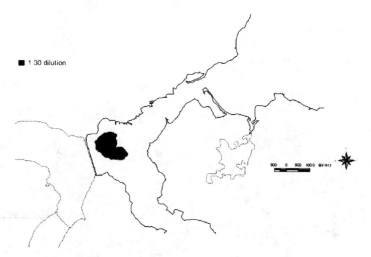

Figure 7: Predicted dilution for location A during peak flood tide under maximum plant operating conditions.

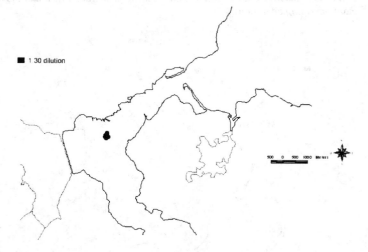

Figure 8: Predicted dilution for location B during peak flood tide under maximum plant operating conditions.

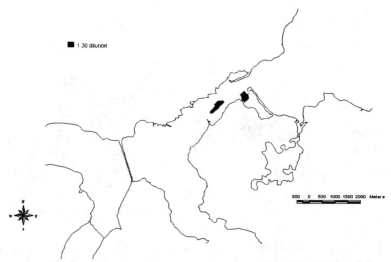

Figure 9: Predicted dilution for location C during peak flood tide under maximum plant operating conditions.

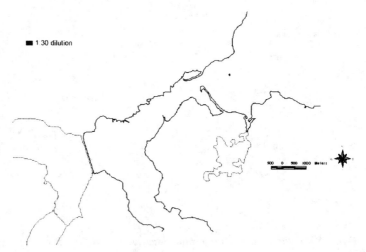

Figure 10: Predicted dilution for location D during peak flood tide under maximum plant operating conditions.

References

Shen, J., Sisson M., Kuo A., Boon J., and Kim S. (1997) Three-dimensional numerical modeling of the tidal York River system, Virginia. Estuarine and Coastal Modeling, Proc. of the Fifth International Conference, Alenxandria, Virginia, Oct. 1997.

Morin, I., Hickey, K., and Greenblatt, M. (1999) Using GIS as an Interface for 3D Hydrodynamic Modeling. Estuarine and Coastal Modeling, Proc. of the Sixth International Conference, New Orleans, Louisiana, Nov. 1999.

Muellenhoff, W.P., A..M. Soldate, Jr., D.J. Baumgartner, M.D. Schuldt, L.R. Davis, and W.E. Frink (1985) Initial Mixing Characteristics of Municipal Ocean Discharges. Volume I – Procedures and Applications. Report No. EPA/600/3-95/073a. Environmental Research Laboratory, Office of Research and Development, U.S. Enviromental Protection Agency. Narragansett, RI.

Modeling the Circulation in Penobscot Bay, Maine

Huijie Xue[1], Yu Xu[1], David Brooks[2], Neal Pettigrew[1], John Wallinga[1]
1. School of Marine Sciences, University of Maine, Orono, ME 04469-5741.
2. Department of Oceanography, Texas A& M University, College Station, TX 77843-3148.

Abstract

Penobscot Bay, with approximate dimensions 50 x 100 km, is the largest estuarine embayment along the Maine coast. It can be characterized by two deep channels on its eastern and western sides, which are separated by several islands and a shoal region in the middle of the Bay. Subtidal circulation in Penobscot Bay is influenced by winds, fresh water discharge from the Penobscot River, and the southwestward Maine Coastal Current flowing pass the mouth of the Bay. The Princeton Ocean Model was adapted to Penobscot Bay to simulate the circulation for the spring and summer of 1998. Observed winds at nearby Matinicus Rock and realistic river discharge rates were used to force the model. Open boundary conditions were specified using the results from a Gulf of Maine climatological model. Simulations were somewhat sensitive to the mixing coefficient in the model. When a background viscosity of 5×10^{-6} m^2/s was used, the model reproduced the observed three-layer structure in the outer western bay with outflows near the surface and the bottom and inflows in the middle of the water column. In contrast, a two-layer estuarine like circulation was found in the outer eastern bay with outflows in the upper water column and inflows in the lower water column.

1. Introduction

Penobscot Bay (Figure 1), where the Penobscot River watershed drains into the sea, is the largest estuarine embayment in the Gulf of Maine, the second largest on the U.S. east coast. There are two deep channels on the eastern and western side of the bay, separated by several islands and a shoal region in the middle. Penobscot Bay has been historically and remains a very important fishery ground. Approximately 70 species of fin and shellfish are harvested in the bay and on the surrounding shelf, and the bay itself accounts for roughly 50% of the lobster landings for the entire state of Maine (Conkling, 1996).

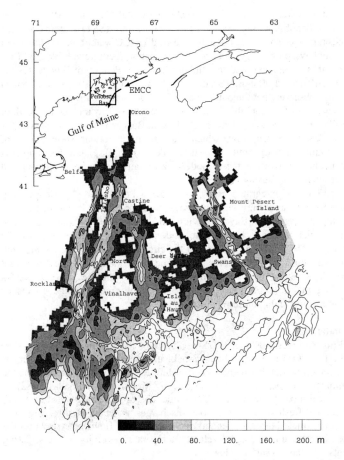

Figure 1. A location map of Penobscot Bay and the bathymetry of the region.

Knowledge of the circulation is fundamental to the understanding of Penobscot Bay as an ecological system. Based on the data of short-term moorings deployed during the summer of 1969, 1970, and 1974 (Normandeau Associates, 1975), Humphreys and Pearce (1981) found a counterclockwise circulation around Islesboro in the case of southerly winds. On the contrary, under similar southerly wind conditions Burgund (1995) found a clockwise circulation around Islesboro in the summer of 1994 using visually tracked surface drifters.

The southwestward flowing Eastern Maine Coastal Current (EMCC) passes the mouth of Penobscot Bay. Sea surface temperature (SST) patterns as seen from Landsat images sometimes indicate the cold EMCC water entering the lower bay that may have profound effects on the circulation (pettigrew, 1998). On the other hand, it is often suggested that the outflow from Penobscot Bay causes the EMCC to be deflected offshore (Brooks, 1994; Bisagni et al, 1996). Pettigrew et al (1998) also observed that the outflow overrides the coastal current.

Previous efforts to model the circulation in Penobscot Bay include the work of Fidler (1978), Humpherys and Pearce (1981), and Burgund (1995). These models lacked the density driven flow, which was the most serious shortcoming of applying simple hydrodynamic models to Penobscot Bay that has apparent estuarine nature. In addition these models had relatively coarse resolution and poor representation of the bay geometry,

As a part of an integrated effort to create a Penobscot Bay Geophysical Infomation System describing the ecological characteristics of Penoscot Bay, we have developed a Penobscot Bay circulation model that simulates time-dependent, three-dimensional velocity, temperature and salinity in the bay and on the adjacent shelf. The principal objective of the circulation model is to identify and characterize features such as fronts and eddies and their associated temporal variability. It is hoped that the model will provide valuable information for understanding marine ecological conditions, and for predicting the effects of marine resource use and management activities.

2. The Model

The Princeton Ocean Model (POM) (Blumberg and Mellor, 1987) is used in this study. The POM is a three-dimensional, fully nonlinear, primitive equation, ocean circulation model that includes complete thermal dynamics. It contains the second moment turbulence closure scheme of Mellor and Yamada (1982) to provide vertical mixing coefficients. The model can be forced with surface wind, surface heat and fresh water fluxes, river discharges, and boundary forcing from the open ocean (including sea level and tidal and subtidal flows) and solve for the time-dependent, three-dimensional velocity, temperature, salinity, vertical mixing coefficients, and the sea surface elevation.

An orthogonal curvilinear grid with 151x121 discrete points covers Penobscot Bay and the adjoining Blue Hill Bay, a domain size roughly 100 km by 80 km (Figure 2). The grid size varies from less than 400 m inside the Bay to about one kilometer offshore. There are 15 sigma levels in the vertical.   The model's offshore boundary is placed on the shelf, about 30km seaward from the mouth of Penobscot Bay so that the coastal current may enter the study area and flow along the shelf portion of the model domain.

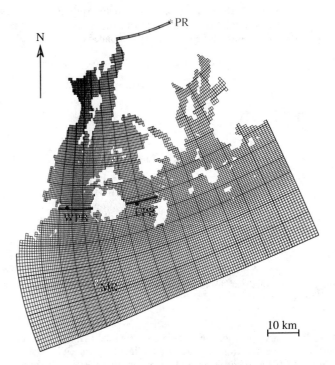

Figure 2. Orthogonal curvilinear grid of the Penobscot Bay circulation model. Asterisks mark the C-MAN station at Matinicus Rock (MR) and the Penobscot River inflow (PR). Dots show mooring locations in the western Penobscot Bay (WPB) and the eastern Penobscot Bay (EPB), respectively. In addition, observations were made along the sections across the outer western and eastern bay (heavy curves) on 29 April 1998.

This paper presents model simulations of the period between April and September 1998. This period was chosen since lobster larvae usually settle in Penobscot Bay during summer. Hourly winds from the C-MAN meteorological station at Matinicus Rock and the daily Penobscot River discharge rates were used to force the model (Figure 3). The spring freshet of 1998 was somewhat unusual with two peaks in April separated by a period of about two weeks. The river discharge rose again in June and July reflecting a relatively wet summer. Winds were strong but variable in the month of April. Southerly components became dominant starting in mid-May and continued throughout the summer except for two northeasters in the end of June and August. Two iepisodes are highlighted in

Figure 3 (vertical lines). The first corresponds with the April hydrographic survey during which winds were from southwest, and the second is a rather strong southeasterly wind event when Penobscot Bay responded quite differently.

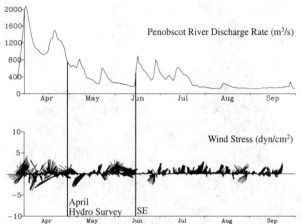

Figure 3. Penobscot River discharge rate and wind stress at Matinicus Rock from April to September 1998.

Open boundary conditions, including both the tidal (M_2 only) and the subtidal forcing, were obtained from a larger scale climatological Gulf of Maine model (Xue et al, 2000). A gravity wave radiation condition was applied to the velocity component perpendicular to open boundaries. An upwind-advection scheme was applied to temperature, salinity, and the velocity component parallel to the boundary so that, in case of inflow, the boundary conditions derived from the climatological Gulf of Maine model were imported by inward velocities.

3. Comparisons of the Model Results with Observations
3.1. Time series

Model simulations were performed for the period from April to September 1998. Results were averaged over the M_2 tidal cycle and saved to illustrate the subtidal circulation. During the same time period, two moorings with downward looking Acoustic Doppler Current Profilers (ADCPs) were deployed in outer Penobscot Bay (see Figure 2). The ADCP in the western bay (WPB) was configured for 8m depth bins, giving a total of 11 bins between 10 and 90m. Four out of the eleven bins are shown in Figure 4a. The ADCP in the eastern bay (EPB) was configured for 2m depth bins. Due to instrument malfunction, only the top three bins of data centered at 8, 10, and 12 m were determined useful, and two of which are shown in Figure 4b.

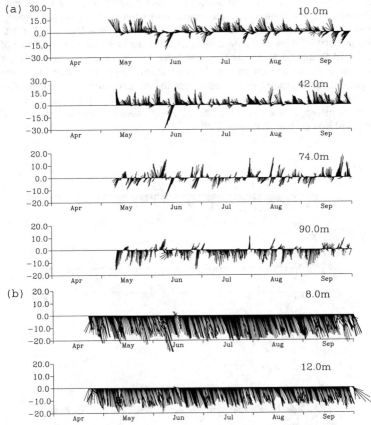

Figure 4. The west (a) and the east (b) Penobscot Bay ADCP tidal residual velocity time series. The northward current component is plotted along the ordinate as a positive, and the eastward component is plotted from its time origin in the positive direction along the abscissa. The depths of the bins are indicated.

In the western bay, velocities were consistently northward (indicating inflows) between 10 and 70 meters except around 14 June when the strong southeasterly wind event caused southward flows throughout the water column. Velocities near the bottom were predominantly southward (indicating outflows). Although velocities were not measured in the top 10 m, from hydrographic data Pettigrew (1998) showed a surface layer (from the surface down to about 5m) of southward Penobscot River outflow, which led to his postulate of a three-layer structure in the outer western bay. In contrast, the upper layer currents in the eastern bay were steadily southward except for the mid-June event (Figure 4b). The 1997 ADCP record showed that the outflows extended from the surface to about 30 m depth, whereas below 30 meters the currents reversed and flowed northward into the bay (Pettigrew, 1998).

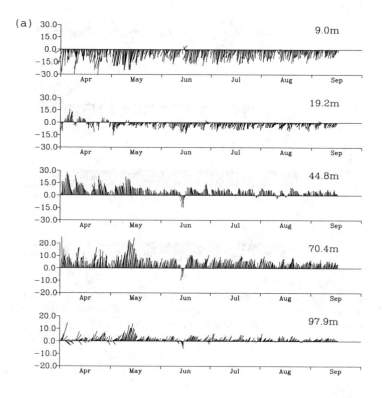

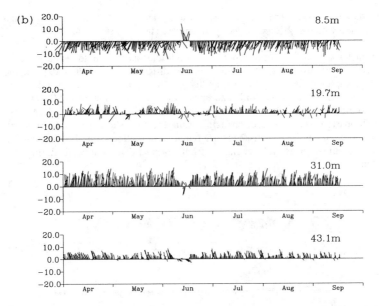

Figure 5. (a) and (b) are model predicted velocity time series at the WPB and the EPB mooring site, respectively, from the experiment with v equal to 5×10^{-5} m^2/s.

Two model experiments were carried out with different values of background viscosity, v. In the POM, the background viscosity is added to the viscosity determined by the Mellor-Yamada closure scheme (Mellor and Yamada, 1982), and it becomes predominant when the local Richardson number is large. Figures 5 and 6 show model predicted velocity time series at the mooring sites in the experiment with v equal to 5×10^{-5} m^2/s and 5×10^{-6} m^2/s, respectively. Differences between the two experiments are larger on the western side (Figures 5a and 6a). Southward flows in the upper water column tend to be weaker. On the other hand, flows near the bottom change from predominantly northward to predominantly southward by reducing the background viscosity from 5×10^{-5} m^2/s to 5×10^{-6} m^2/s. The eastern bay (Figures 5b and 6b) is less sensitive to the change of v except near the bottom where the northward flows become more intermittent.

When the smaller background viscosity was used, the model reproduced the three-layer structure in the western bay and the two-layer structure in the eastern bay postulated by Pettigrew (1998) (Figure 6). However, the surface southward flows reached more than 20 m on the western side but less than 20 m on the eastern side. The former was too deep, and the latter too shallow when compared with the observed velocity time series shown in Figure 4.

We are continuing the sensitivity study with different model parameterizations and open boundary conditions and hope to achieve better agreement of layer thickness. Despite the differences in layer thickness, the model captured the temporal variability on the synoptic scale, especially the abnormal response to the mid-June southeasterly event. Southward flows were found throughout the water column on the western side, whereas the surface flows on the eastern side reversed, changing from southward to northward. Although not observed, model results suggested that bottom flows on the eastern side reversed direction as well, changing from northward to southward during this event.

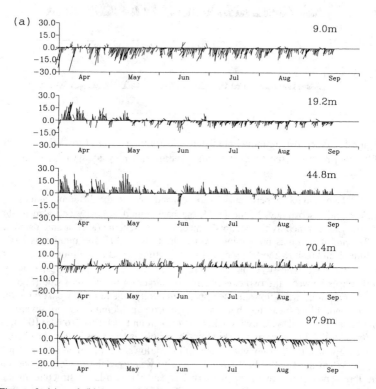

Figure 6. (a) and (b) are modeled velocity time series at the WPB and the EPB mooring site, respectively, from the experiment with ν equal to 5×10^{-6} m^2/s.

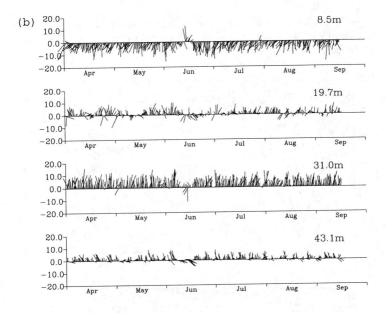

3.2. Spatial variability

It is clear from both the observed and the modeled velocity time series that the subtidal circulation in Penobscot Bay can be highly variable, especially in the western bay. The circulation also exhibits rich spatial variability. Two instances were selected to demonstrate the strong contrasts between the circulation patterns. The first corresponded to the hydrographic survey on 29 April. Model results were compared with the tidally-averaged, ADCP and CTD cross-sections made in the outer western and eastern bay (see Figure 2 for locations of the sections). The second was sampled during the strong southeasterly wind event from 13 to15 June.

During the hydrographic survey, wind was from the southwest and light (Figure 3). Cross-sectional distributions of velocity and density in the outer bay are shown in Figure 7. In the western bay, isopycnals were almost flat near the surface and the surface density was rather low. During this time of the year, temperature doesn't vary much and density gradients arise from salinity effects associated with the river outflow. Isopycnals in the mid-water column tilted downward toward the east. The corresponding along-channel velocity (Figure 7a) shows that outflows occurred on the eastern side. On the western side of the channel, flows were northward on the top and southward near the bottom. The model reproduced these patterns in the density and velocity (Figure 8a and 8b)

although the outflows on the eastern side of the channel were much too strong.
Surface density in the model was slightly lower than observed and the bottom
density in the model was higher.

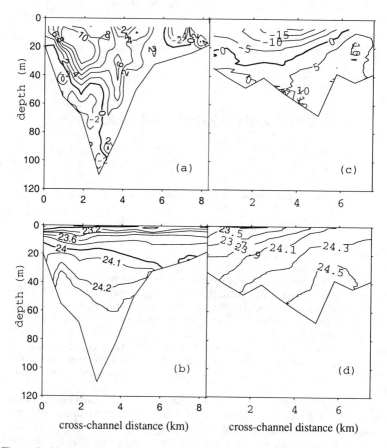

Figure 7. Observed along-channel velocity in cm/s (upper panels) and density in
kg/m³ (lower panels) in the outer western bay (a and b) and in the outer eastern
bay (c and d) on 29 April 1998. Positive (negative) velocities indicate northward
(southward) flows, i.e., inflows (outflows).

In the eastern bay, the observed along-channel velocity (Figure 7c)
suggests that outflows occurred above 30m depth and inflows near the bottom and

to the eastern side of the channel. Isopycnals tilted upward toward the east with the lowest density in the upper western side of the channel (Figure 7d). Again the model appeared to reproduce the basic features of the velocity and density (Figure 8c and 8d). However, the tilt of the isopycnals was not as steep and outflows didn't reach sufficient depth. These findings are consistent with the time series results discussed in the previous section.

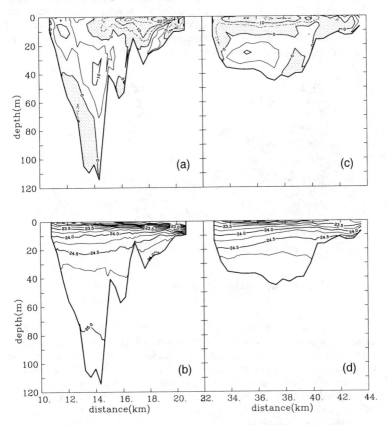

Figure 8. Model simulated along-channel velocity in cm/s (upper panels) and density in kg/m³ (lower panels) in the outer western bay (left panels) and in the outer eastern bay (right panels) on 29 April 1998.

Figure 9 shows the modeled surface velocity and salinity on 29 April 1998. Due to the westerly component of the wind, surface flows on the shelf were eastward. Below the surface (*e.g.*, at 10m) the EMCC still moved southwestward on the shelf at about 15 cm/s (not shown). Inside Penobscot Bay, southwesterly winds drove the northward flows west of Islesboro. The Penobscot River outflow was mostly confined to the east of Islesboro (as seen from the surface salinity) which appeared to form a semi-enclosed clockwise circulation around Islesboro, except for the area between Islesboro and North Haven where the surface currents were northeastward, i.e., downwind. In the outer western bay, the surface currents moved northeastward on the western side and southeastward on the eastern side. The surface inflows on the western side came from the coastal water west of Penobscot Bay. Driven by the wind and the river outflow, surface currents in the outer eastern bay moved southeastward.

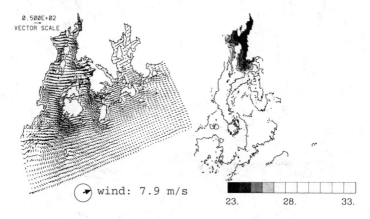

Figure 9. Modeled surface velocity and salinity on 29 April 1998.

During the strong southeasterly wind event between 13 and 15 June, circulation patterns in the bay were quite different. Figure 10 shows the along-channel velocity and density distributions at the same western and eastern outer bay cross-sections. In the outer western bay, southward flows, although weak at the surface, extended throughout the water column on the western side of the channel. Northward flows, driven by the strong wind, filled the eastern and shallower part of the channel and were quite strong. Similarly, in the outer eastern bay, outflows were on the western side of the channel whereas inflows on the eastern side. Overall, the water column was less stratified during this time of the year compared with that at the end of April during the spring freshet. Although not shown, the surface had been warmed by several degrees, but the surface salinity was higher.

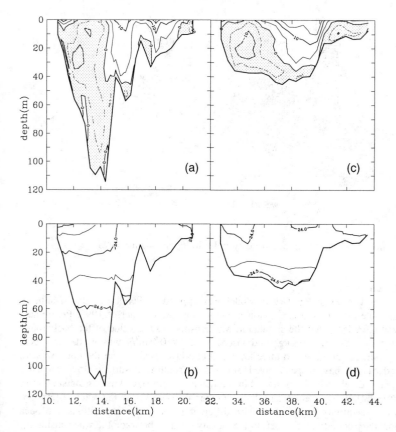

Figure 10. As in figure 8 except for 14 June 1998.

Figure 11 shows the corresponding surface circulation and salinity during the 14 June event. Driven by the strong southeasterly wind, the surface water moved northward into the bay. However, the northward flows were confined mostly to the eastern side of the channels. The circulation pattern around Islesboro was rather complicated. Most of the Penobscot River outflow moved along the channel west of Islesboro. The discharge rate on 14 June was not much lower than that on 29 April, but the surface salinity in the bay was much higher because 29 April was at the end of the spring freshet (see Figure 3).

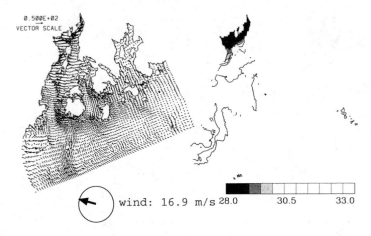

Figure 11. Modeled surface velocity and salinity on 14 June 1998

4. Summary

The Princeton Ocean Model was applied to Penobscot Bay, Maine to simulate the subtidal circulation in the spring and summer of 1998. Preliminary results indicate that the simulation was sensitive to the value of the background viscosity. When a background viscosity of 5×10^{-6} m²/s was used, the model reproduced the observed three-layer and two-layer structure in the outer western and eastern bay, respectively. However, the southward surface flow reached a greater depth than the observed in the outer western bay, whereas the southward surface flow was weaker and shallower than the observed in the outer eastern bay. More sensitivity experiments with different model parameterizations and open boundary conditions that set up the density contrast between the water inside the bay and the coastal water are needed to obtain a better agreement between the observed and the model predicted layer thickness.

The model responded to the synoptic scale forcing and produced temporal and spatial variability resembling the observations, in particular for the 29 April and the 14 June instances. In addition, the model also showed a semi-enclosed clockwise circulation around Islesboro in the case of southwesterly wind, consistent with the Burgund's (1995) finding. Additional observations are needed to verify the circulation pattern and the associated variability in the inner bay. In the future, a multi-year model simulation and a Lagrangian trajectory study will be carried out, with the goal of providing insights into the interannual variability of lobster settlement in the Penobscot Bay region.

5. Acknowledgements

This study was supported by the NOAA Sea Grant project R-CE233 and NOAA NESDIS Penobscot Bay Marine Resource Collaborative to the University of Maine and Texas A & M University.

6. References

Bisagni, J. J., D. J. Gifford and C. M. Ruhsam, 1996. The spatial and temporal distribution of the Maine Coastal Current during 1982. *Cont. Shelf Res.*, 16, 1-24.

Blumberg, A. F., and G. L. Mellor, 1987. A description of a three dimensional coastal ocean circulation model. In *Three-dimensional Coastal Ocean Models*, Vol. 4, N. Heaps ed., Amer. Geophys. Union, 1-16.

Brooks, D. A., 1994. A model study of the buoyancy-driven circulation in the Gulf of Maine. *J. Phys. Oceanogr.*, 24, 2387-2412.

Burgund, H. R., 1995. The currents of Penobscot Bay, Maine: Observations and a numerical model. Department of Geology and Geophysics, Yale University. 71 pp.

Conkling, P. W., 1996. The marine ecology of Penobscot Bay. In *Penobscot the Forest, River and Bay.* D. D. Platt edit, Island Institute, Rockland, ME, 32-44.

Fidler, R. B., 1979. An approach for hydrodynamic modelling of Maine's estuaries. MS thesis, University of Maine.

Humphreys, A. C. and B. R. Pearce, 1981. A hydrodynamic investigation of the Penobscot estuary. Technical Report, Dept. Civil Engineering, University of Maine, 81 pp.

Mellor, G. L., and T. Yamada, 1982. Development of a turbulence closure model for geophysical fluid problems. *Rev. Geophys. Space Phys.*, 20, 851-875.

Normandeau Associates, Inc., 1975. Environmental survey of upper Penobscot Bay, Maine. Unpublished Report.

Pettigrew, N. R., D. W. Townsend, H. Xue, J. P. Wallinga, P. J. Brickley, and R. D. Hetland, 1997. Observation of the Eastern Maine Coastal Current and its offshore extensions. *J. Geophys. Res.*, 103, 30,623-30,640.

Pettigrew, N. R., 1998. Vernal circulation patterns and processes in Penobscot Bay: Preliminary Interpretation of Data. A final report for year 1 of the Penobscot Bay Experiment. University of Maine, 23 pp.

Xue, H., F. Chai, and N. R. Pettigrew, 2000. A model study of the seasonal circulation in the Gulf of Maine. *J. Phys. Oceanogr*, 30, 1111-1135.

A 3D Modeling Study of the Estuarine System: An Application to the Satilla River

Lianyuan Zheng[1] and Changsheng Chen[2]

Abstract

The estuarine system of the Satilla River, Georgia was studied using the Blumberg and Mellor coastal ocean circulation model (so-called Ecom-si). A three-dimensional (3D) technique of the wet/dry point treatment was incorporated into the model for the simulation of flooding and drying processes over inter-tidal salt marshes. Driven by the M_2 tidal forcing at the open boundary in the inner Georgia shelf and real-time freshwater discharge at the upstream end of the river, the model successfully simulated the tidal current and salinity in the river as well as water transport flushing onto and draining from salt marshes. A good agreement was found in tidal ellipse and salinity between the model and observations at available mooring stations and on along-river CTD transects. The numerical experiments suggested that the spatial distribution of tidal-averaged currents in the river is driven by tidal rectification over a variable bottom topography and the baroclinic pressure gradient associated with the along-river distribution of salinity. The fact that the model-predicted tidal current in the case with a fixed river boundary was 50% smaller than the observed tidal current implied that the inter-tidal zone must be taken into account in order to simulate accurately the water movement in Georgia estuaries.

[1] Ph D Student, [2] Professor, Department of Marine Sciences, The University of Georgia, Athens, GA 30602.

Introduction

Understanding Georgia's complex estuarine ecosystem processes requires that scientists have a quantitative estimation of the water transport associated with intertidal salt marshes. Previous field measurements have provided a long-term interdisciplinary data set in major Georgia Rivers. These data have shown that the maintenance of marine resources in Georgia estuaries was manifested through the nonlinear interaction of physical and biological processes. Because of the lack of knowledge about the spatial and temporal distributions of water currents, however, our understanding of the impact of these processes on the marine environment in Georgia estuaries was still limited in a qualitative aspect. It is imperative that we develop a full three-dimensional, prognostic numerical model if we are to quantify the complex physical, biological and chemical interaction processes in Georgia estuaries and if we are to provide managers with a scientific and visual tool with which to make strategic decisions.

The Georgia estuary system is characterized by a series of barrier island complexes and extensive salt marshes (Figure 1). It is more practical and realistic to start a modeling experiment at one of the rivers and then expand it to cover all of Georgia's estuaries. This strategy will help us test and calibrate the model. For this reason, the Satilla River was chosen as an initial site for our model experiments.

The primary physical processes in the Satilla estuarine system include freshwater discharges, tidal mixing, and meteorological forcings (wind and rainfall). This river originates in the coastal plain and is characterized by high amount of dissolved organic C content (Beck *et al.* 1974; and Windom *et al.* 1975). The salinity level in rivers, which is determined by the mixing of fresh water and oceanic-intruded salt water, varies semi-diurnally, seasonally and annually in tidal-induced currents and mixing, amount of freshwater discharges, precipitation against evaporation, and wind intensity and directions (Blanton 1981, 1991; Blanton *et al.*1994). Previous field measurements and laboratory experiments implied that the temporal variation of primary production in the Satilla River was influenced significantly by salinity levels of river water and by marsh soil (Morris and Haskin 1990). The dramatic decline in commercial fishery stocks in Georgia's inner shelf and estuaries is related to increases in local salinity levels.

The M_2 tidal current accounts for about 90% of the along-river current variation in the Satilla River (Blanton 1996). To a certain extent, it controls the spatial and temporal variation of biological and chemical materials through river-ocean or intertidal water exchanges and tidal mixing (Bigham 1973; Dunstan and Atkinson 1976; Manheim *et al.* 1970; Pomeroy *et al.* 1993; Verity *et al.* 1993; and

Windom *et al.* 1975). Since tidal currents and mixing are usually asymmetric over a tidal cycle, a quantitative estimation of the tidal-cycle averaged net flux of water, nutrients, and inorganic and organic matter from salt marshes to rivers is critically important in order to understand the driving mechanism for the maintenance and variability of this highly productive estuarine ecosystem.

The Georgia coast is dominated by northwesterly, northerly or northeasterly winds in winter and by southeasterly or southwesterly winds in late spring and summer (Chen *et al.* 1999). Seasonal wind variations have a direct influence on the recruitment of blue crab and white shrimp, two of the most important commercial species in the inner shelf and estuaries of Georgia (Blanton *et al.* 1994). In general, mature female blue crabs move downstream out of the river in spring and spawn in the inner continental shelf inside the 30-m isobath from April through August (Archambault *et al.* 1990). The zoeae stay in the nursery ground on the shelf until they grow to the postlarval stage, or megalope (McConaugha 1988; Sulkin *et al.* 1980). The postlarval crabs are advected into rivers in late summer and early fall and grow there until the following spring. Since the survival rate of the zoeae and larvae relies critically on the short-term and seasonal variations in direction and intensity of the wind in spring through summer and because the growth stage of these two species takes place in both inner shelf and rivers, it is impossible to understand the recruitment mechanism for blue crab and white shrimp in the Satilla estuary without knowledge of the physical processes controlling the water exchange between estuaries and shelf.

Precipitation in the Satilla River area is generally low in autumn and winter and high in spring through summer. The rainfall-induced freshwater input tends to enhance vertical stratification in the river and hence reduce tidal mixing.

Recently, we developed a flooding/drying treatment technique and incorporated it into the Ecom-si. This modified model was used to simulate the 3D tidal current in the Satilla River with the inclusion of the intertidal zone. The spatial and temporal distributions of salinity also were simulated by adding the real-time freshwater discharge at the upstream end of the river. The model results were compared with observations taken along river transects and at a fixed anchoring station. Good agreement was found in tidal ellipse and salinity, which suggests that our model results are robust.

The remaining sections of this paper are organized as follows. The numerical model and flooding/drying treatment methods are described in the next section, followed by tidal and salinity simulations, and the concluding summary.

The Numerical Model

The numerical model used in this study was a so-called Ecom-si model developed originally by Blumberg and Mellor (1987). This is a three-dimensional (3D), primitive equations, coastal ocean circulation model incorporating with the Mellor and Yamada's (1982) level 2.5 turbulent closure and a free surface for the simulation of vertical mixing and surface gravity waves. The numerical grids were generated using an orthogonal curvilinear coordinate transformation in the horizontal and a σ–coordinate transformation in the vertical. The primitive equations were solved using a semi-implicit finite-difference method with a second-order mass conservation (Casulli, 1990). A detailed description of the Ecom-si is given by Blumberg (1994). This model has been widely used in the coastal ocean (Chen and Beardsley (1995), Chen et al. (1995), Chen et al. (1997), Chen and Xie (1997), Chen et al. (1999), etc.).

Applying the Ecom-si to simulate the water transport over the intertidal zone requires a proper treatment of flooding and drying processes over tidal cycles. To achieve this, two traditional approaches were applied: (1) the moving boundary method and (2) the wet/dry point treatment methods, respectively. In the first method, the model's lateral boundary is defined as an interface position between land and water where the water depth and flux vanish. Since this boundary varies with every time step over a tidal cycle, the numerical computational domain (or model grid) must be re-adjusted (or re-generated) at each time step (Johns, 1982, Lynch and Gray, 1980, Shi, 1995). This method, however, would prove cumbersome for generating model grids, particularly in an estuarine area where complex coastal lines and numerous island barriers exist.

In the second method, the model grids cover the entire intertidal zone and the moving boundary with no water flux is determined by the so-called wet/dry point judgement method. In the z-coordinate model, the equations at dry points were treated simply by assuming the vertical eddy viscosity to be an infinite value so that velocities and sea surface automatically vanish (Leendertse, 1970, 1987; Flather and Heaps, 1975; Casulli and Cheng, 1991, 1992; Casulli, and Gettani, 1994; and Cheng et al, 1993). This approach, however, requires a significant modification of the original code when a σ–coordinate transformation is used. Alternatively, we added a small critical layer h_c to total depth in the code and use it to determine the wet or dry points in the σ–coordinate transformation system. As the total depth at a grid point is larger than h_c, this point is treated as a wet point. Otherwise it remains as a dry point. This approach is like adding a porous medium with a thickness of h_c to the model (Ip. Et al., 1998), which numerically prevents the occurrence of a zero total depth in the σ-coordinate system from computation. The advantages of this method are: (1) simplicity and (2) mass conservation with no sacrifice of the σ–coordinate in making a perfect match with the irregular bottom topography in the vertical. The h_c should be chosen based on sensitivity studies of the velocity and sea surface. In our model experiment, h_c was chosen as 5 cm. Our numerical experiments also show that h_c can be as low as to 1 cm with no occurrence of numerical instabilities.

The numerical computational domain covers the entire Satilla River including of the maximum flooding area associated with intertidal processes (Fig. 2). The horizontal grid points are 148 (along-river)×141 (cross-river) with a resolution of 100 m in the main stream of the river to 2500 m near the open boundary in the inner shelf. Eleven σ-levels were used in the vertical, which corresponded a vertical resolution of less than 0.5 m in most areas inside the river. The model was driven by a semi-diurnal M_2 tidal forcing (sea surface elevation plus phase) at the open boundary. The harmonic constants of the M_2 tidal forcing were specified using the tidal elevations and phases predicted by the inner shelf South Atlantic Bight (SAB) tidal model (developed and calibrated by Chen et al., 1999). The initial distribution of salinity was specified to be 35 PSU, a typical salinity value of oceanic water on the SAB shelf. This is consistent with our assumption that the salinity anomaly in the Satilla River is caused by freshwater discharge and precipitation. When the tidal current and surface elevation reached a quasi- equilibrium state, the real-time freshwater discharge was injected into the model at the upstream end of the river. The bottom topographic data used in this study were digitized directly from the NOAA chart and depth in the inner shelf came from inner shelf model of Chen et al. (1999). The time step used in the numerical computation was 41.4 seconds (1080 for a M_2 tidal cycle).

Tidal Simulation

Tidal current and surface elevation reached a quasi-equilibrium state after 6 model tidal cycles. The model successfully simulated the flooding and drying processes over the intertidal zone in the Satilla estuary. Water flushed onto the saltmarsh area during the flood tide period at a maximum speed of 70 cm/s at the edge of the main channel of the river and reached a 2-m elevation line at slack high water (Fig. 3a). Water drained back into the main channel of the river when the tidal current reversed during the ebb tide (Fig. 3b). The model predicted a strong advection process over a tidal cycle with conservative mass during computation.

The tidal current ellipses were calculated at each grid point using the model-predicted, time series current data after 8 model tidal cycles. A comparison was made between model-predicted and observed tidal ellipses at two selected mooring sites (Fig. 4 and Table 1). At site A, the differences in major axis, minor axis, orientation and phase between model-predicted and observed tidal currents were 0.5 cm/s, 1.3 cm/s, 1.1° and 0.4°, respectively. Similar differences also were shown at site E except in orientation. The relatively large difference of 8.9° found in orientation at this site was probably due to interpolation of the bottom topography based on the old NOAA geometric map, where the depth at site E was

10.3 m (whereas, in our model, it was 9.2 m). The geometric data in the Satilla River must be updated if an accurate simulation is to be achieved.

The synoptic distribution of near-surface residual current vectors is shown in Fig. 5a. The maximum velocity, 38 cm/s, was found at 5 km upstream and reached 25 cm/s at the transient location between the river channel and salt marsh. Two residual gyres controlled by bottom topography were found near the river's mouth. Several convergent and divergent zones were found in the river. It is likely that these contributed directly to the spatial distribution of suspended sediment there. The residual flow in the Satilla River was generated mainly by tidal rectification over varying bottom topography (Chen *et al.*, 1995; and Loder, 1980). Inertial effects around the bent coastline of the main river channel and the asymmetric features of tidal currents, which are due to the flooding/drying processes over the intertidal zone, also contributed to the formation of the complex residual current pattern inside the river (Fischer *et al.*, 1979; and Geyer, 1993).

The importance of the flooding/drying processes in simulating tidal currents in the Satilla River was examined by conducting a model experiment with a fixed lateral river boundary without inclusion of the intertidal zone. In this case, the model-predicted amplitudes of tidal current were about 43.2% and 57.5% smaller than observed values at sites A and E (Fig. 6 and Table 1). In the model, the maximum tidal current occurred 72 minutes earlier than it did both observation stations. Because no flux was allowed at the fixed lateral boundary, the distribution of the residual flow (Fig. 5b) was significantly different from that shown in Fig. 5a for the case, which included the intertidal zone. Also, its magnitude was much smaller. The fact that the model with a fixed lateral river boundary significantly underestimated the tidal current in the Satilla River suggests that the flooding/drying processes over the intertidal zone in estuaries must be taken into account when estimating water transport in the river.

Salinity Simulation

The salinity simulation shown here was conducted using the 1995 Land-Margin Ecosystem Research (LMER) hydrographic survey along the Satilla River and an anchoring station. Two surveys were made along the river on April 7 and April 15, 1995. The observed salinity taken on April 7 was first interpolated into the model grids as the initial condition of salinity for our restart run after tidal currents and elevation reached a quasi-equilibrium state. The model was then restarted and run by adding the real-time freshwater discharge at the upstream end of the river. After running 9 additional days, the model-predicted salinity was output for direct comparison with salinities observed along the river on April 15.

Fig. 7 shows a model-data salinity comparison along the river at different measurement times. The model-predicted salinity was vertically well mixed along the river, which was in good agreement with observations. The distribution of salinity contours also coincided well between model and observations except for a 1-km shift in the location. For a given location, the model-predicted salinity was slightly higher than the observation. This difference was not surprising since the hydrographic survey on April 15 was made during a spring tidal period, and in the model, only the M_2 tidal constituent was included. The previous observation and modeling experiments showed that the maximum tidal velocity was about 10 to 15% greater in the spring tide than in the M_2 tide. If a speed of 10 cm/s was added to our model-predicted tidal current, it could advect the salinity contour about 1 to 2 km upstream during the flood tidal period. This suggests that the location shift in salinity contours found between model and observation is caused by tidal advection during a spring tide.

We also compared the model-predicted salinity with observations taken at an anchoring site shown in Fig. 1 (Fig. 8). The salinity observed at that site exhibited a significant variation over a M_2 tidal cycle: 15 PSU at slack high water (SHW) and under 4 PSU at slack low water (SLW) on April 15 and a varying range of 20 PSU on April 16. The vertical gradient of salinity was largest at SHW and smallest at SLW. The model-predicted salinity at that site varied considerably over a tidal cycle, with 17 PSU at SHW and 3 PSU at SLW. The model showed a vertically well-mixed structure of salinity during flood tide and weak stratification during ebb tide. On April 16, the wind blew northward at a speed of 0.5 to 1 m/s in the morning and then rotated clockwise in the afternoon. Although this wind was too weak to significantly alter the shelf circulation, it could directly affect the water transport in the Satilla River since it is very shallow. Taking both wind and spring-tide effects into account, we conclude that our model is capable of simulating the salinity in the Satilla River.

The significant salinity variation over tidal cycles was clearly evident in the synoptic salinity distribution shown in Fig. 9. The amplitude of the near-surface current was larger in the ebb tide than in the flood tide due to the compensation from the outward freshwater discharge-induced buoyancy flow. As a result, the near-surface salinity was lower at SLW than at SHW. This salinity anomaly increased as the distance to the freshwater source decreased. It was about 9 PSU near the river mouth but increased to 12 PSU at a point of 16 km upstream.

The residual circulation in the river was modified significantly after the freshwater discharge was added (Fig. 5c). The current through the water column was almost uniform in the flood tide but exhibited a strong vertical shear in the ebb tide. The residual gyres found near the river mouth, which were predicted in the case with tides only, were replaced by a strong outward buoyancy flow when the freshwater discharge was added.

Conclusion

The tidal current and salinity in the Satilla River were simulated using the Blumberg and Mellor coastal ocean circulation model (so-called Ecom-si). A three-dimensional (3D) technique of the wet/dry point treatment was incorporated into the model to estimate the water transport over intertidal salt marshes due to tidal flooding and draining. Driven by the M_2 tidal forcing at the open boundary in the inner Georgia shelf and real-time freshwater discharge at the upstream end of the river, the model successfully simulated the tidal current and salinity in the river as well as water transport flushing onto and draining from salt marshes. A good agreement was found in tidal ellipse and salinity between model and observation at available mooring stations and on an along-river CTD transect.

The numerical experiments suggested that the spatial distribution of the mean current in the river be driven by complex physical processes associated with tidal rectification over variable bottom topography, asymmetric tidal current distribution over flooding and draining, inertial effects around the bent coastal line, and a freshwater discharge-induced baroclinic pressure gradient.

The intertidal zone must be taken into account in simulating the water transport in Georgia estuaries. The model with a fixed lateral river boundary not only caused a significant underestimation of the tidal current but also failed to provide a realistic residual circulation in the Satilla River.

It should be pointed out here that our current model did not include spring tides. The effects of spring tides on salinity variations were found to be less significant over the inner shelf of the SAB (Chen et al., 1999), but it must have a remarkable influence on the spatial distribution of salinity and other biological variables inside the river. Two tidal constituents of S_2 and N_2 are being added to our estuarine model for the spring tide simulation.

Acknowledgments

This research was supported by the Georgia Sea Grant College Program under grant numbers NA26RG0373 and NA66RG0282. We thank Drs. Jack Blanton and Merryl Alber for providing the current meter data at two selected mooring sites and for allowing us to use the CTD data taken during the 1995 hydrographic survey along the Satilla River. Without their kind help, the model-data comparison would have been impossible. Encouragement from Dr. Mac Rawson was greatly appreciated. Finally, we want to thank George Davidson for his editorial help on this manuscript.

References

Archambault, J. A., E. L. Wenner, and J. D. Whitaker, 1990: Life history and abundance of blue crab, Callinects sapidus Rathbun, at Charleston Harbor, South Carolina, *Bulletin of Marine Science*, **46**, 145-158.

Beck, K. C., J. H. Reuter, and E. M. Perdue, 1974: Organic and inorganic geochemistry of some coastal plain rivers of the southeastern United States. *Geochim, Cosmochim. Acta*, **38**, 341-364.

Bigham, G. N., 1973: Zone of influence-inner continental shelf of Georgia. *Journal of Sedimentary Petrology*, **43**, 207-214.

Blanton, J. O., 1981: Ocean currents along a nearshore frontal zone on the continental shelf of the southeastern U.S. Journal of Physical Oceanography, **11**,1627-1637.

Blanton, J. O., 1991: Circulation processes along oceanic margins in relation to material fluxes. *Ocean Margin Processes in Global Changes*, edited by R. F. C. Mantoura, J. M. Martin, and R. Wollast, John Wiley & Sons Ltd., 145-163.

Blanton, J. O., F. Werner, C. Kim, L. P. Atkinson, T. Lee, and D. Savidge, 1994: Transport and fate of low-density water in a coastal frontal zone. *Continental Shelf Research*, **14**, 401-427.

Blumberg, A. F. and G. L. Mellor, 1987: A description of a three-dimensional coastal ocean circulation model. *Three-Dimensional Coastal Models*, Amer. Georphys. Union, 1-16.

Blumberg, A. F., 1994: A primer for Ecom-si. *Technical Report of HydroQual, Inc.*, 66 pp.

Casulli V., and E. Cattani, 1994: Stability accuracy and efficiency of a semi-implicit method for three-dimensional shallow water flow. Computers & Mathematics With Application, **27**, 99-112

Casulli, V., 1990: Semi-implicit finit-difference methods for the two-dimensional shallow-water equations. *Journal of Computational Physics*, **86**, 56-74.

Casulli, V., and R. T. Cheng, 1991: A semi-implicit finite-difference model for three-dimensional tidal circulation, *Proceeding of 2nd International Conference on Estuarine and Coastal Modeling*, edited by Spaulding et. al., ASCE, Tampa Florida, 620-631.

Casulli, V., and R. T. Cheng, 1992: Semi-implicit finite-difference methods for three-dimensional shallow-water flow. *International Journal for Numerical Methods in Fluids*, **15**, 629-648.

Chen, C, R. C. Beardsley, and R. Limeburner, 1995: A numerical study of stratified tidal rectification over finite-amplitude banks, Part II: Georges Banks. *Journal of Physical Oceanography*, **25**, 2111-2128.

Chen, C. and L. Xie, 1997: A numerical study of wind-induced near-inertial oscillations over the Texas-Louisiana shelf. *Journal of Geophysical Research*, **102**, 15583-15593.

Chen, C. and R. C. Beardsley, 1995: A numerical study of stratified tidal rectification over finite-amplitude banks, Part I: symmetric banks. *Journal of Physical Oceanography*, **25**, 2090-2110.

Chen, C., D. A. Wiesenburg, and L. Xie, 1997: Influences of river discharge on biological production in the inner shelf: A coupled biological and physical model of the Louisiana-Texas shelf. *Journal of Marine Research*, **55**, 293-320.

Chen, C., L. Zheng, and J. O. Blanton, 1999: Physical processes controlling the formation, evolution, and perturbation of the low-salinity front in the inner shelf off the southeastern United States: A modeling study. *Journal of Geophysical Research*, **104**, 1259-1288.

Cheng, R. T, V. Casulli, and J. W. Gartner, 1993: Tidal, residual, intertidal mudflat (TRIM) model and its applications to San Francisco Bay, California. *Estuarine, Coastal and Shelf Science*, **36**, 235-280.

Dunstan, W. M. and L. P. Atkinson, 1976: Sources of new nitrogen for the South Atlantic Bight. In *Estuarine Processes*, edited by M. Wiley, **1**, 69-78, Academic Press, New York.

Fischer, H. B., E. J. List, R. C. Y. Imberger, and N. H. Brooks, 1979: Mixing in Inland and Coastal Waters. Academic Press, New York, 483 pp.

Flather, R. A., and N. S. Heaps, 1975: Tidal computations for Morecambe Bay. *Geophysical Journal of the Royal Astronomical Society*, **42**, 489-517.

Geyer, W. R., 1993: Three-dimensional tidal flow around headlands. *Journal of Geophysical Research*, **98**, 955-966.

Ip, J. T. C., D. R. Lynch, and C. T. Friedrichs, 1998: Simulation of estuarine flooding and dewatering with application to Great Bay, New Hampshire. *Estuarine, Coastal and Shelf Science*, **47**, 119-141.

Johns, B., 1982: The simulation of a continuously deforming lateral boundary in problems involving the shallow water equations. *Computers and Fluids*, **10**, 105-116.

Leendertse, J. J., 1970: A water-quality simulation model for well-mixed estuaries and coastal seas: principles of computation, *Rand Corporation Report RM-6230-RC*, Vol. I.

Leendertse, J. J., 1987: Aspects of SIMSYS2D, a system for two-dimensional flow computation, *Rand Corporation Report R-3572-USGS*.

Loder, J. W., 1980: Topographic rectification of tidal currents on the sides of Georges Bank. *Journal of Physical Oceanography*, **10**, 1399-1416.

Lynch, D. R., and W. G. Gray, 1980: Finite element simulation of flow in deforming regions. *Journal of Computational Physics*, **36**, 135-153.

Manheim F. T., R. H. Meade, and G. C. Bond, 1970: Suspended matter in surface waters of the Atlantic continental margin from Cape Cod to the Florida Keys. *Science*, **167**, 371-376.

McConaugha, J. R., 1988: Export and reinvasion of larvae as regulators of estuarine decapod populations. *American Fish. Soc. Symposium*, **3**, 90-103.

Mellor, G. L. and T. Yamada, 1982: Development of a turbulence closure model for geophysical fluid problem. *Review of Geophyics*, **20**, 851-875.

Morris, J. T. and B. Haskin, 1990: A 5-yr record of aerial primary production and stand characteristics of spartina-alterniflora. *Ecology*, **71**, 2209-2217.

Pomeroy, L. R., J. O. Blanton, G. A. Poffenhofer, K. L. V. Damm, P. G. Verity, H.L. Windom, and T. N. Lee, 1993: Inner shelf processes. In *Ocean Processes: U.S. Southeast Continental Shelf*, edited by D. W. Menzel, 9-43.

Shi, F., 1995: On moving boundary numerical models of coastal sea dynamics. Ph D thesis, Ocean University of Qingdao, China.

Sulkin, S. D., W. V. Heukelem, P. Kelly, and L. V. Heukelem, 1980: The behavioral basis of larval recruitment in blue crab Callinects-Sapidus Rathbun-A laboratory investigation of ontogenetic changes in geotaxis and barokinesis. *Biological Bulletin,* **159**, 402-417.

Verity, P. G., J. A. Yoder, S. S. Bishop, J. R. Nelson, D. B. Craven, J. O. Blanton, C. Y. Robertson, and C. R. Tronzo, 1993: Composition, productivity and nutrient chemistry of a coastal ocean planktonic food web. *Continental Shelf Research*, **13**, 741-776.

Windom, H. L., W. M. Dunstan, and W. S. Gardner, 1975: River input of inorganic phosphorous and nitrogen to the southeastern salt marsh estuarine environment. *Mineral Cycling in Southeastern Ecosystems*, edited by F. G. Howell, J. B. Gentry, and M. H. Simth, ERDA Symposium Series, 309-313.

Table 1: Comparison of model-predicted and observed tidal ellipse parameters

Site	Cases	U_{major} (cm/s)	U_{minor} (cm/s)	θ_{orien} (deg)	Phase (deg)
	Observed	53.4	-1.6	4.2	156.7
A	FB model	23.1	-0.2	7.7	118.8
	MB model	53.9	-0.3	5.3	156.3
	Observed	47.8	0.8	121.3	150.0
E	FB model	27.5	0.2	128.7	118.3
	MB model	46.1	1.2	130.2	153.3

Note: Station A is located at 81° 32.28′ W, 30° 59.58′ N; Station E is at 81° 30.84′ W, 30° 59.16′ N; FB model — Fixed boundary model; MB model — Moving boundary model; U_{major} — amplitude of the major axis; U_{minor} — amplitude of the minor axis; θ_{orien} — orientation of the major axis; Phase — time of maximum current (Greenwich phase in degree, 1 degree = 2.07 minutes). The orientation is measured counterclockwise from the east. The positive sign in U_{minor} indicates a counterclockwise rotation of the current vector.

Captions

Fig. 1: Georgia coastline (left) and location of Satilla River (right). Solid squares are the locations of two current meter stations. Solid circles are the longitudinal transects of CTD observed stations. Solid triangle is the anchor station.

Fig. 2: The geometry of the Satilla River and the numerical model domain (enclosed by thick solid line). Dashed line is the boundary of the intertidal zone (defined by the 2-meter elevation line).

Fig. 3: Synoptic distributions of the near-surface tidal current vectors of the M_2 tide at slack high water (a) and slack low water (b) with inclusion of salt marshes.

Fig. 4: Comparison between observed (top) and computed (low) current ellipses at two current meter stations (marked at Fig. 1) in the case with inclusion of salt marshes.

Fig. 5: Synoptic distributions of the near surface residual currents averaged over a M_2 tidal cycle for the case with (a) moving lateral river boundary; (b) fixed lateral river boundary; and (c) moving lateral river boundary plus freshwater discharge.

Fig. 6: Comparison between observed (top) and computed (low) current ellipses at two current meter stations (marked at Fig. 1) in the case with fixed lateral river boundary (without inclusion of the intertidal zone).

Fig. 7: Comparison between observed (top) and computed (low) longitudinal distributions of salinity. The measurement was conducted on April 15, 1995 during the spring tidal period. The model was initialized using the CTD data taken on April 7, 1995, and run for 9 days until April 15, 1995. Dots in the upper panel represent the points of measurement. Contour interval was 1 PSU.

Fig. 8: Comparison of observed (top) and computed (low) temporal changes of salinity versus depth over a tidal cycle during spring tide on April 16,1995. Dots in the upper panel represent the points of measurement. During the measurement, the surface was selected as the origin of the coordinator ($z = 0$). Shadow area included the variation of the surface elevation, rather than the real bottom topography. Contour interval is 4 PSU.

Fig. 9: Synoptic distributions of the model-predicted near surface salinity during slack high water (top) and slack low water (low) on April 15, 1995.

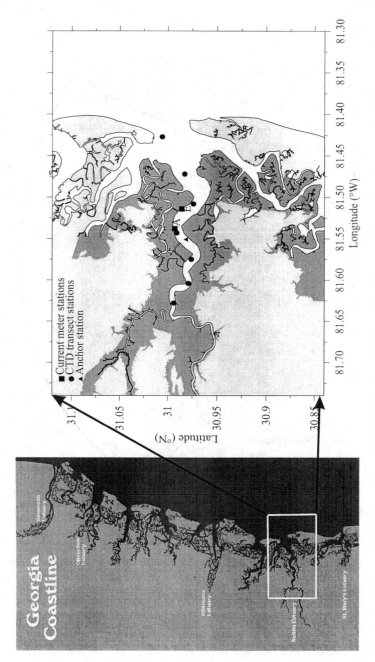

Fig. 1: Georgia coastline (left) and location of Satilla River (right). Solid squares are the locations of two current stations. Solid circles are the longitudinal transects of CTD observed stations. Solid triangles is the anchor station.

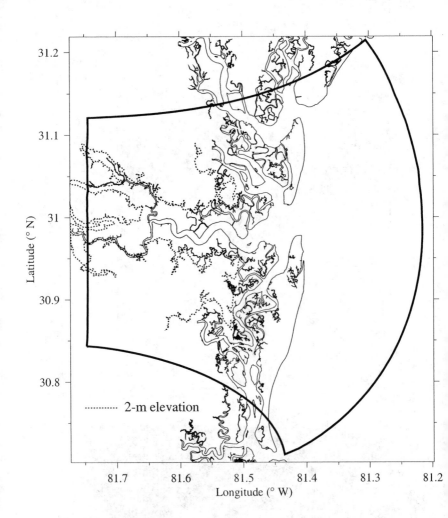

Fig. 2: The geometry of the Satilla River and the numerical model domain (enclosed by thick solid line). Dashed line is the boundary of the intertidal zone (defined by the 2-meter elevation line).

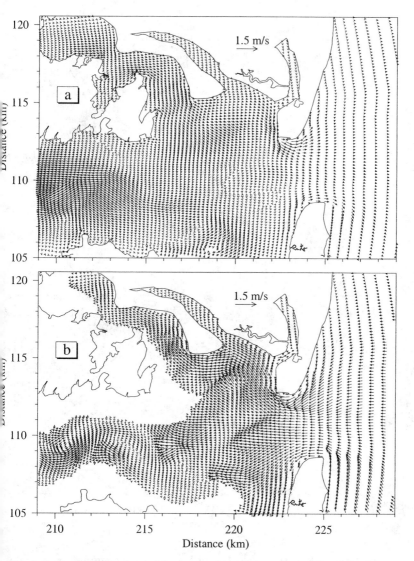

Fig. 3: Synoptic distributions of the near-surface tidal current vectors of the M_2 tide at slack high water (a) and slack low water (b) with inclusion of salt marshes

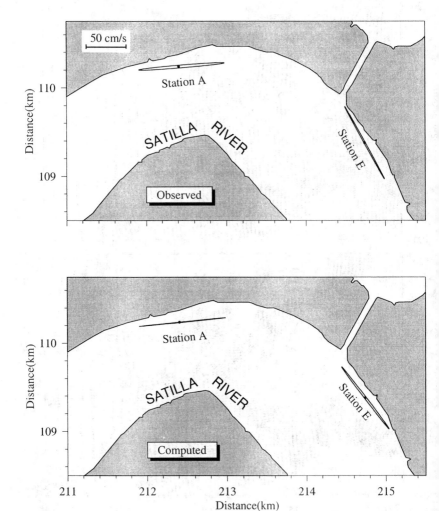

Fig. 4: Comparison between observed (top) and computed (low) current ellipses at two current meter stations (marked at Fig. 1) in the case with inclusion of salt marshes

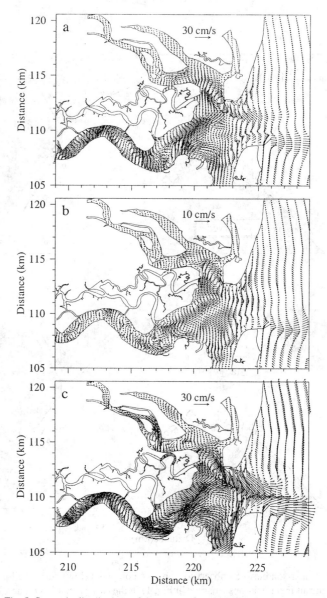

Fig. 5: Synoptic distributions of the near surface residual currents averaged over a M_2 tidal cycle for the cases with (a) moving lateral river boundary; (b) fixed lateral river boundary; and (c) moving lateral river boundary plus freshwater discharge.

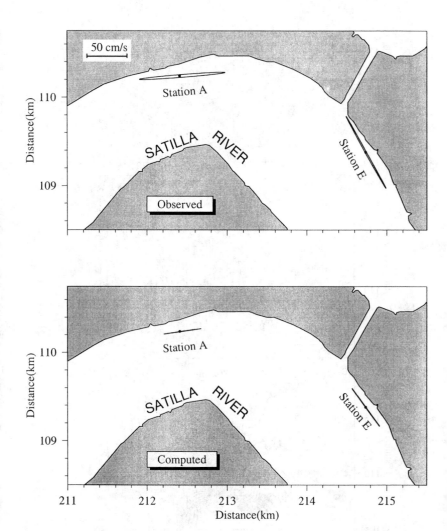

Fig. 6: Comparison between observed (top) and computed (low) current ellipses at two current meter stations (marked at Fig. 1) in the case with a fixed lateral river boundary (without inclusion of the intertidal zone).

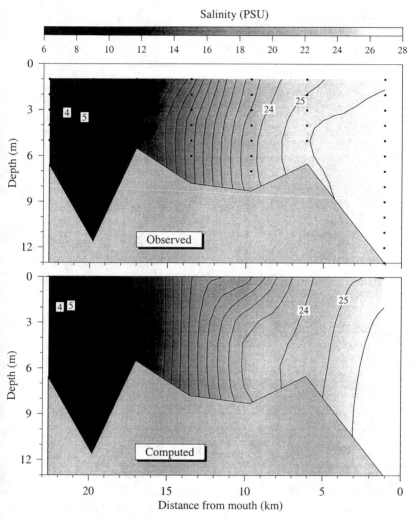

Fig. 7: Comparison between observed (top) and computed (low) longitudnial distributions of salinity. The measurement was conducted on April 15, 1995 during the
spring tidal period. The model was initialized using the CTD data taken on April 7,
1995 and run for 9 days until April 15, 1995. Dots in the upper panel represent the
points of measurment.Contour interval was 1 PSU.

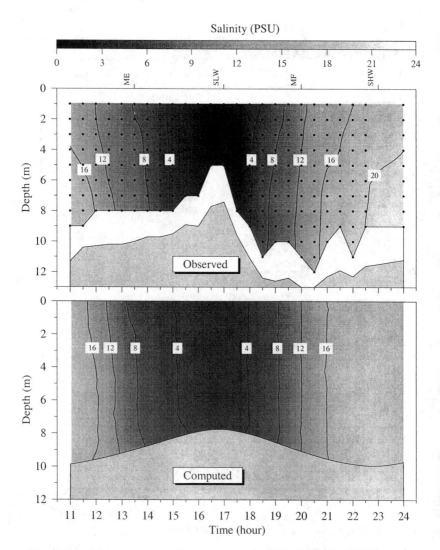

Fig. 8: Comparsion of observed (top) and computed (low) temporal changes of salinity versus depth over a tidal cycle during spring tide on April 16, 1995. Dots in the upper panel represent the points of measurment. During the measurement, the surface was selected as the origin of the coordinator ($z = 0$). Shadow area in the upper panel included the variation of the surface elevation, rather than the real bottom topography. Contour interval is 4 PSU.

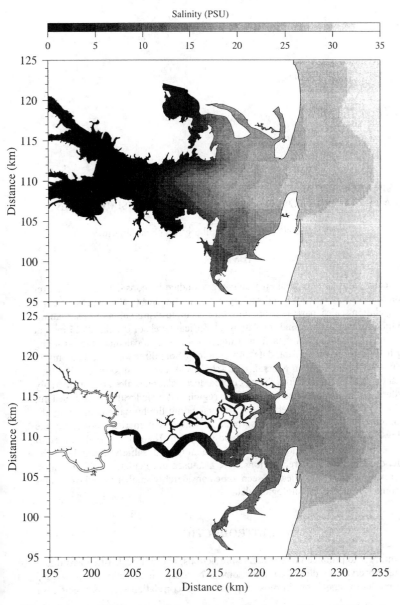

Fig. 9: Synoptic distributions of the model-predicted near surface salinity during slack high water (top) and slack low water (low) on April 15, 1995

SPATIAL DECORRELATION SCALES IN COASTAL REGIONS: A PRECURSOR TO MORE ACCURATE DATA ASSIMILATION

James K. Lewis[1], Alan F. Blumberg[2], B. Nicholas Kim[2]

ABSTRACT

Scales of variability in shelf waters were studied by considering the decorrelation scales for temperature and salinity in the Yellow Sea (water depths < 95 m). Decorrelation scales, parameters required for accurate assimilation of observations into models, were considered for August at vertical levels of 0-1 m, 20-25 m, and 40-50 m. Climatological data from August were used to calculate the e-folding length scale based on all data, data only in a longshore direction, and data only in a cross-shelf direction. In general, cross-shelf scales are seen to be smaller than longshore scales. In addition, the calculated decorrelation scales are considerably smaller than those calculated for a broader region that includes the Yellow Sea but also covers areas with deeper water. Understanding the limitations of using climatological data to determine decorrelation scales, a numerical model for the Yellow Sea was developed, calibrated, and validated. The model was then used to provide temporal "snapshots" of the temperature and salinity throughout the model domain, and these realizations were also used to calculate e-folding scales. The model-determined scales are seen to be considerably smaller than those scales determined from the climatological data.

1. INTRODUCTION

In order to produce more accurate predictions, ocean models often assimilate various observations during computations. The assimilation technique may vary from model to model, but in most cases a spatial correlation scale is specified *a*

[1] Scientific Solutions, Inc., P.O.Box 1029, Kalaheo, HI 96741
[2] HydroQual, Inc., 1 Lethbridge Plaza, Mahwah, NJ 07430

priori in the assimilation scheme. The spatial correlation scale (or decorrelation scale as it is often called) is used in the assimilation methodology to limit the spatial extent of the influence of a given observation within the model domain. In general, the spatial correlation scale in deep waters is of the order of the baroclinic Rossby radius, and this can vary up to hundreds of kilometers. But we expect the correlation scale to be considerably smaller in coastal waters where salinities, temperatures, and currents can have significant spatial (and temporal) variations. As we improve our ability to model coastal regions, we must adapt our assimilation schemes to handle the dynamic environments over continental shelves. This will necessarily include determining appropriate decorrelation scales for coastal regions. Moreover, there can be distinct differences in correlation scales in littoral areas in a longshore direction as opposed to a cross-shelf direction. Future assimilation schemes may be able to take advantage of such differences while ingesting observations into a model, but we must first specify longshore and cross-shelf correlation scales.

To consider correlation scales in a coastal environment, we will use the regions of the Yellow Sea and the Bohai as shown in Fig. 1. The area is relatively shallow, with depths no greater than 93 m. There can be considerable temperature gradients across the region, and the various river inflows can induce large salinity gradients. Moreover, the Yellow Sea is a site of relatively large internal tides. The lateral structure of such internal waves can result in its own associated spatial decorrelation scale.

In this study, we will consider only spatial decorrelation scales, and we will confine our work to the August time period. August is the period of greatest freshwater outflow in the region of interest (Chen, 1993), and others have previously considered summer-time spatial decorrelation scales in the general region. Chu et al. (1997) determined decorrelation scales for temperature (T) using July, August, and September data from the northern East China Sea and the Yellow Sea but excluding the Bohai. Chu et al. used all climatological data but employed a space-time sorting method so that the data better represented temporal snapshots of the temperature fields. They considered data out to the 200 m isobath, so their scales differ from those in this work in that the deeper water domain of the northern East China Sea was considered. Chu et al. (1997) found spatial decorrelation scales for summer of 251 km for the surface temperatures, 169 km for T at 50% of the water column depth, and 157 km for T at 80% of the water column depth.

As pointed out by Lie (1999), the results of the work by Chu et al. (1997) may represent basic dimensions of patterns on a relatively large scale, but it does not satisfactorily describe coastal structures of smaller scales. Lie (1999) listed some of the major limitations of the data set used by Chu et al. (1997) for determining correlation scales. These include the lack of data from the Bohai as well as problems with the spatial and temporal distributions of the data. Moreover, the data does not have the capability to resolve structure in coastal regions with scales of less than 100 km.

In this work, we consider another source of T and salinities (S) in the Yellow

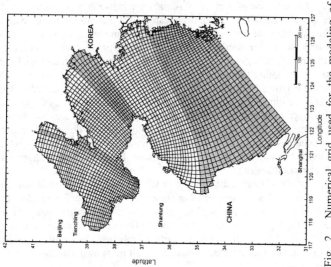

Fig. 2. Numerical grid used for the modeling of the circulation of the region. Grid dimensions range from about 7 to 21 km. The shaded transects are those used to calculate cross-shelf e-folding scales.

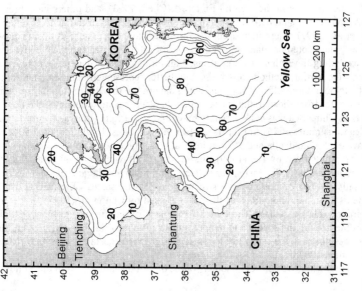

Fig. 1. The Yellow Sea and associated depth contours (m).

Sea, those found in the *Marine Atlas of the Bohai Sea, Yellow Sea, and East China Sea* (Chen, 1993). The data from this atlas are given at depths of 0, 10, 20, 30, 50, and 100 m. These data for August were used to calculate the e-folding spatial linear correlation for temperature and salinity representative of the two mixed layers of the water column: at the surface and at 50 m. The data were also used to calculate the e-folding spatial correlation scales at 20 m, the depth at which the pycnocline in the region tends to be the greatest and we can expect the strongest internal tides.

Obviously, correlation space scales determined from climatological data for a given month will tend to misrepresent the actual lengths of the scales. Too few data points in space and time can lead to a climatology that is highly variable (with too small of a decorrelation scale) while a good sampling of data from a highly variable region can lead to a climatology that is not variable enough (with too large of a decorrelation scale). The over-estimated (under-estimated) scale results in an observation impacting an area larger (smaller) than it should during the data assimilation process in a model. Chu et al.'s (1997) space-time sorting process attempts to correct for this problem, but even their technique does not provide a true snapshot of spatial variability of a given parameter. And since observations are assimilated into ocean models which represent sequential snapshots of the ocean environment, spatial decorrelation scales that are too large (small) will tend to damp (enhance) flow patterns.

In an attempt to address this problem, we produced an ocean circulation model of the area shown in Fig. 1, calibrated the model, and used snapshots in time of the model temperature and salinity to calculate spatial decorrelation scales. The model grid is shown in Fig. 2, and simulations provide estimates of the spatial structure of temperature and salinity associated with tidal forcing and internal tides at any given time. Scales determined in this manner may provide our best means of estimating decorrelation parameters for use in assimilation schemes for numerical ocean models. At the very least, scales determined from observations should be compared to scales calculated from calibrated hydrodynamic models.

2. E-FOLDING SCALES, AUGUST CLIMATOLOGY

The surface temperature data for August as given by Chen (1997) are shown in Fig. 3. Maximum temperatures range from ~27°C near the mouth of the Yellow Sea to about 21°C off the central Korean peninsula. Temperatures tend to increase from the southwest toward the northeast, with the exception being in the Bohai Sea. Temperatures in the deeper waters of the region are relative cold, ranging from ~8°C in the central part of the Yellow Sea to ~13°C near the mouth of the Yellow Sea. The surface salinity structure for August is also shown in Fig. 3. Minimum surface salinities occur near the mouths of the major rivers of the region, with a low of ~19°/oo near the southwestern edge of the region. The

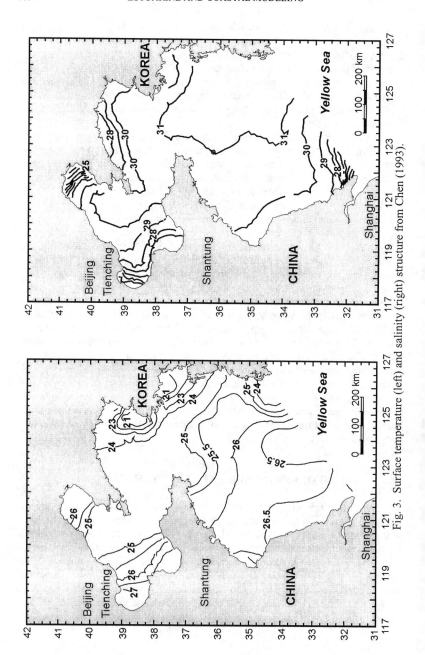

Fig. 3. Surface temperature (left) and salinity (right) structure from Chen (1993).

maximum surface salinity is ~32°/oo in the southeastern part of the area. Salinities in the lower parts of the water column are less variable, ranging from about 32-33.5°/oo.

The data from Chen (1997) were used to calculate e-folding scales (the distance at which the linear correlation falls to e^{-1}) for temperature and salinity at 0, 20, and 50 m. This calculation was first performed using all data at a given vertical level. The calculation was then repeated using data at a given vertical level but for a set water depth ±5 m. This second calculation is used to provide an estimate of the decorrelation scale in a longshore direction. A third e-folding scale was calculated using data from transects crossing the region from the coastline to deeper water (i.e., in a cross-shelf manner). These transects are shown in Fig. 2 and tend to run from the northeast toward southwest.

The various scales resulting from the correlation functions are summarized in Table 1. For T, the e-folding scales tend to be larger at the surface of the water column than at 20 and 50 m. At the surface, the T scale considering the entire domain is 140 km. But the scale for T in the longshore direction is larger, 190 km. The T scale in the cross-shelf direction is even larger, ~230 km.

At 20 m, the temperature scales are 100-105 km for all data and data in the longshore direction. The T scale in the cross-shelf direction is again larger at 130 km. At 50 m, the T scales are 90 km for all data and data in the longshore direction. Once again, the T scale in the cross-shelf direction is larger, this time being

Table 1. Spatial e-folding correlation scales (km) for salinity and temperature based on (top) the data from Chen (1993), (middle) a snapshot in time of T and S from the circulation model, and (bottom) a snapshot in time of T and S from the circulation model in which there was a nudge of T and S to their initial values.

	Temperature			Salinity		
Depth	All Domain	Longshore	Cross-Shelf	All Domain	Longshore	Cross-Shelf
0 m	140	190	230	210	>320	165
20 m	105	100	130	270	>320	195
50 m	90	90	110	>320	>320	170

	Temperature			Salinity		
Depth	All Domain	Longshore	Cross-Shelf	All Domain	Longshore	Cross-Shelf
0-1 m	140	75	120	130	70	115
20-25 m	85	230	160	200	220	210
45-50 m	80	175	150	160	170	200

	Temperature			Salinity		
Depth	All Domain	Longshore	Cross-Shelf	All Domain	Longshore	Cross-Shelf
0-1 m	130	70	120	110	70	115
20-25 m	100	275	215	215	220	185
45-50 m	110	240	170	160	170	180

110 km.

For salinity, the e-folding scales are greater than those of temperature, and they tend to be longer in the longshore direction and shorter in the cross-shelf direction. At the surface, the S scale considering all the domain is ~210 km. The scale for S in the longshore direction is large, never falling to e^{-1} in 320 km. The T scale in the cross-shelf direction is ~165 km.

At 20 m, the salinity scale is 270 km for all data and never falls below e^{-1} in the longshore direction. The S scale in the cross-shelf direction is again smaller at 195 km. At 50 m, the T scales for all data and data in the longshore direction do not fall below e^{-1} in 320 km. Once again, the T scale in the cross-shelf direction is smaller, this time being ~170 km.

3. THE HYDRODYNAMIC MODEL

The hydrodynamic model used in this work is a semi-implicit, z-level adaptation of the model of Blumberg and Mellor (1987). The model is based on the hydrostatic, shallow water, primitive equations for momentum, salt, and heat. Sub-grid scale turbulence is specified using the schemes of Mellor and Yamada (1982) in the vertical and Smagorinsky (1963) in the horizontal. For additional information on the model, the reader is referred to Blumberg and Mellor (1987).

This particular version of the Blumberg-Mellor model uses a semi-implicit solution scheme for solving for the surface height field (Casulli and Cheng, 1992). The original Casulli and Cheng implicit solver uses the water velocities at a previous time level for the Coriolis term. Such a formulation requires a very small time step to overcome numerical stability considerations. The implicit solution scheme used in this study uses both the past and future velocities to determine the Coriolis effect, iterating to obtain a solution at each time step.

Another modification of the original Blumberg-Mellor model is the use of the hybrid z-level coordinate system discussed by Lewis et al. (1994). In this scheme, water velocities are calculated using the hydrostatic momentum equations in a vertically averaged mode using actual bathymetry. These velocities are saved, and the model is then executed for the same time step in a z-level mode to calculate the three-dimensional field of currents using a step-wise approximation to the bathymetry. The transports calculated using the step-wise approximation to the bathymetry are then adjusted (barotropically) to match the transports calculated in the vertically average mode. Since the latter mode uses actual bathymetry, this adjustment allows longwaves to propagate at appropriate speeds across the domain of the model.

The horizontal grid used by the model is shown in Fig. 2, and the distribution of vertical levels in presented in Table 2. The initial temperature and salinity structure used in the model was developed from the data for August at 0, 10, 20, 30, 50, and 100 m as given in Chen (1993). Fig. 3 provides examples of the data used for the initial conditions. Fig. 5 gives an indication of vertical structure.

Table 2. Layer characteristics (m) for the model of the Yellow and Bohai Seas domain.

Layer	Top of Layer	Bottom of Layer	Layer Thickness
1	0	1	1
2	1	2	1
3	2	4	2
4	4	6	2
5	6	8	2
6	8	10	2
7	10	12	2
8	12	16	4
9	16	20	4
10	20	25	5
11	25	30	5
12	30	35	5
13	35	40	5
14	40	45	5
15	45	50	5
16	50	55	5
17	55	60	5
18	60	70	10
19	70	80	10
20	80	93	13

The model uses results from the tidal modeling of Woo et al. (1994) to specify the tidal surface height amplitudes and phases along the open boundaries of the model for the M_2, S_2, K_1, and O_1 tidal constituents. The model uses this sea level height information and the linearized equations of motion to specify the velocities along the open boundaries. Open boundary conditions of this type have at times been formulated without regard to the fact that longwave energy can be generated within the model domain and should be radiated out through the open boundaries. Such "clamped" boundary conditions can often lead to erroneous solutions (Lewis et al., 1994), and we wish to apply a condition that is better adapted for allowing longwave radiation to leave the model domain.

For this study, we use an optimized approach for specifying the open boundary condition. Available information on the open boundary (the sea level heights from the tidal model of Woo et al., 1994) is combined with the physics of the model represented by the energy flux on the open boundary. The optimized open boundary condition is derived from the optimization problem

$$\min_{\eta} (J = 0.5 \, g \int_s (g H)^{1/2} (\eta - \eta^o)^2 \, ds) \quad \text{and} \quad -g \int_s H \, \eta \, u_n \, ds = P_t$$

where S is the open boundary, P_t is the energy flux on the open boundary, u_n is the vertically-averaged outward normal velocity, η is the model sea surface elevation on the open boundary, H is depth, g is the acceleration due to gravity, and η° are the reference values of the sea level elevation on the open boundary (i.e., the information from the Woo et al. tide model). To solve the above, we use the regularization approach to obtain the following minimization problem:

$$\min_{\eta} [\ 0.5\ (P_t + g \int_S H\ \eta\ u_n\ ds)^2 + \gamma\ 0.5\ g \int_S (g\ H)^{1/2}\ (\eta - \eta^\circ)^2\ ds],$$

where γ is a parameter of regularization, chosen according to Shulman (1997) to provide the maximum of the entropy integral. The solution has the form

$$\eta = \eta^\circ + \lambda_t\ u_n\ (g/H)^{1/2}$$

where

$$\lambda_t = -\ (P_t + g \int_S H\ \eta^\circ\ u_n\ ds)\ /\ (g^{1/2} \int_S H^{3/2}\ u_n^2\ ds + \gamma).$$

Thus, in the linearized momentum equations, the pressure gradient term due to horizontal sea level height variations uses $\eta^\circ + \lambda_t\ u_n\ (g/H)^{1/2}$ and the next model-predicted η in the interior of the model domain. In this way, any longwave radiation other than that generated directly by η° can be radiated out through the open boundary according to the $\lambda_t\ u_n\ (g/H)^{1/2}$ term.

The vertical structure of currents at the open boundaries are specified using a Sommerfeld (1949) radiation condition:

$$\partial U/\partial t \pm C\ \partial U/\partial n = 0$$

where U is the velocity at a given vertical level, n is the coordinate perpendicular to the open boundary, and C is the propagation speed at the given depth as determined from vertical density variations and normal mode theory (Veronis and Stommel, 1956).

The boundary conditions used for salinity and temperature are the horizontal and vertical structure of temperature and salinity at the open boundary from the data in Chen (1993). An advective flux condition is used at the open boundary at a given grid cell, with advection of the next interior grid cell's temperature and salinity if the flow is out of the model domain. When the current switches to flowing into the model domain, the T or S that is advected into the model is slowly modified over a period of ~3 hrs from the last T or S that was advected out of the model at a given grid cell to the boundary condition T and S at that grid cell.

4. MODEL SIMULATIONS

Model-predicted sea level variations due to tides were compared to the six tidal stations listed in Table 3. The predictions at the tidal stations were generated using the known M_2, S_2, K_1, and O_1 tidal constituent phases and amplitudes at the six stations. The comparisons of the predicted and observed sea level variations are shown in Fig. 4. The model predicts the phases well, but in some cases the model-predicted tides are somewhat too large in magnitude and in some cases they are somewhat too small. However, overall the model does a good job in predicting the general magnitude, phase, and propagation of the tides in the region.

Table 3. Tidal stations at which model-predicted sea level variations are compared in Fig. 4.

Station Number	Station Name	Latitude (°N)	Longitude (°E)
1	Futau Wan	36.10	120.53
2	Soya To	37.23	126.17
3	Kaiyo To	39.07	123.15
4	Yotoa	38.78	121.13
5	Offing in Pwok Hai	38.75	118.13
6	Offing in the Liautung Gulf	38.97	119.43

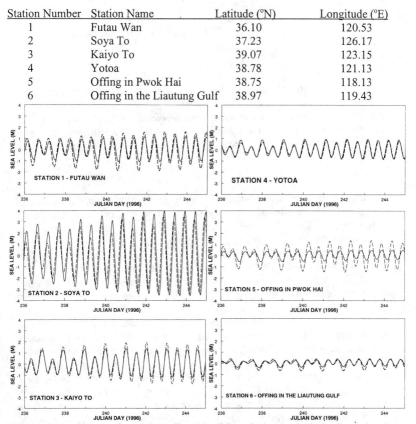

Fig. 4. Comparisons of model-predicted (solid) and observed (dashed) sea level variations due to tides at the six stations listed in Table 3.

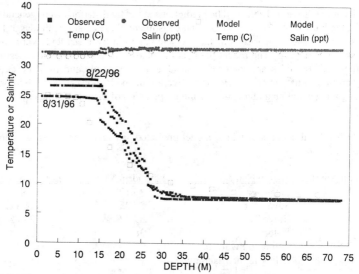

Fig. 5. Observed temperature and salinity profiles during August 1996. The open circles and squares represent the T and S values based on climatological data from Chen (1993) at the model grid location where the August 1996 data were collected.

Current measurements and some salinity and temperature data were collected within the study area during August 1996. The data were collected at approximately 37°N, 124°E in ~74 m of water. Some of the T-S profiles are shown in Fig. 5 along with the initial values of temperature and salinity for each level for the model grid cell at which the observations were collected. The climatological salinity data from Chen (1993) matches the observed salinity almost exactly. The climatological temperature profiles also match the observed temperatures, especially during the latter part of the observational program at the end of August 1996.

Observed currents at 10, 18, 38, and 54 m are shown in Fig. 6. These currents have been 3-hour low-passed filtered. Currents in the north-south direction tend to dominate the flow, with decreasing magnitude with depth. The flow has a distinct tidal variability. The model was used to simulate the August 1996 time period, and the currents from the model grid cell corresponding to where the 1996 observations were made are shown in Fig. 7. The forcing data and the model physics are able to reproduce much of the characteristics of the observed flow. There are some phase differences between the observed and predicted currents,

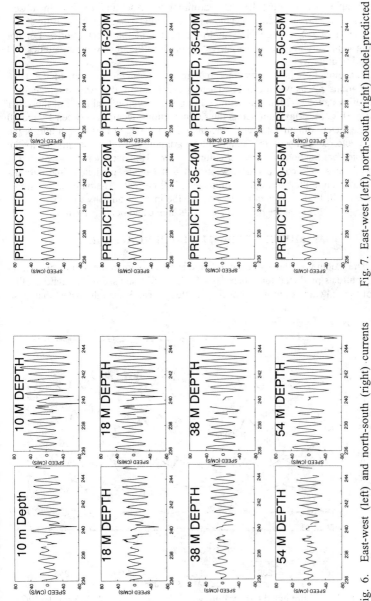

Fig. 7. East-west (left), north-south (right) model-predicted currents at 37°N, 124°E, August 1996 (Julian days).

Fig. 6. East-west (left) and north-south (right) currents observed at 37°N, 124°E, August 1996 (Julian days).

but this appears to be a problem in determining the exact time frame of the observations. The currents were collected only as ancillary data for an acoustic study, and there is a 1-2 hr uncertainty as to the timing of the observations. For not having wind forcing, the model currents reproduce the observed magnitudes and variability quite well.

The model-predicted variations of temperature at the location of the 1996 observations are shown in Fig. 8. There were little in the way of temperature fluctuations above ~10 m and below ~30 m, in the well-mixed parts of the water column (Fig. 5). Maximum temperature fluctuations occurred between 12 and 20 m. The water column depth at this location has a range of up to 4+ m during the tidal cycle, so some of the temperature variations may be due strictly to the expansion/contraction of the water column depth. However, some may be due to internal tides. To consider this, we repotted the model-predicted tidal currents after subtracting the barotropic component of the current. The baroclinic currents are shown in Fig. 9. It is quite easy to see the internal tides by the distinct 180° shift in phase between the baroclinic currents above and below ~14 m.

5. MODEL-PREDICTED E-FOLDING SCALES

The model was allowed to run for an eleven day simulation period, and the T-S structure as predicted for Julian day 240.0, 1996, was used to calculate another set of spatial correlation functions for the study area. We might expect that these

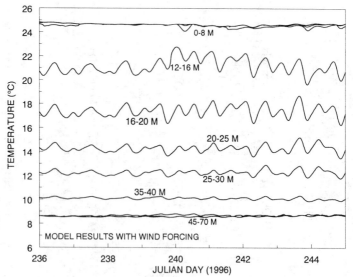

Fig. 8. Model-predicted temperature variations at 37°N, 124°E, August 1996.

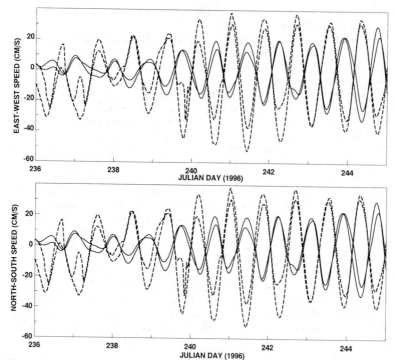

Fig. 9. Baroclinic component of model-predicted east-west and north-south currents at 37°N, 124°E during August 1996. Dashed lines - currents for levels of 4-6, and 10-12 m. Solid lines - currents for levels at 16-20, 25-30 m.

correlation functions could be different from those calculated using the climatological data from Chen (1993) since the temperature and salinity fields from the model simulation will reflect the dynamics of the region in terms of tidal forcing, associated mixing, and the spatial variability related to any internal tides. The resulting e-folding scales are summarized in Table 1. The model-determined correlations are seen to be quite different from those based on climatology.

At the surface, all the model-derived scales for temperature and salinity are considerably smaller than those previously calculated, with the maximum model-derived decorrelation scale being ~140 km. For temperatures at the pycnocline and in the bottom mixed layer, model-derived scales tend to be larger than those at the surface and those calculated using the data from Chen (1993). Moreover, model-derived longshore temperature scales tend to be larger than the cross-shelf temperature scales. The longest decorrelation scales for temperature occur in the region of the pycnocline.

By contrast, the model-derived spatial scales for salinity at the pycnocline and in the bottom mixed layer tend to be smaller than those calculated using the climatological data. Model-derived decorrelation scales at the depth of the pycnocline tend to be longer than those at the surface and lower mixed layers, ~200-220 km. The longshore e-folding scale at the pycnocline is slightly greater than the cross-shelf e-folding scale. The opposite is true for the model-derived longshore and cross-shelf scales in the lower mixed layer. The scales in the lower mixed layers are somewhat smaller than those in the pycnocline.

In the real world, fluxes of freshwater, heat, and salt along all the boundaries of the domain would act to maintain the general character of the T-S fields. But these fluxes are not included in the model simulation due to a lack of knowledge of the exact balance of such fluxes for the model domain. As a result, the spatial variability in the fields of temperature and salinity may be reduced due to mixing over time. This could lead to overestimates of the lengths of the spatial decorrelation scales. To address this possibility, a second simulation was made which included a slight nudge of the temperatures and salinities to their initial values. The nudge acted over a time scale of 3 days. It was found that this allowed for significant variations of temperature and salinity due to internal tides while eliminating any long-term trends of the average temperature and salinity at a given grid cell. Again, the model was allowed to run for an eleven day simulation period, and the T-S structure as predicted for Julian day 240.0, 1996, was used to calculate a third set of spatial correlation functions for the study area. The spatial e-folding scales from these correlations are summarized in Table 1.

6. CONCLUSIONS

By considering the three decorrelation scales at the surface, pycnocline, and lower mixed layer, we can chose length scales that would be appropriate for data assimilation schemes for temperature and salinity. A conservative choice is the minimum of the length scales, as this would keep observations from unduly influencing regions during the assimilation process that are too far away to be represented by the observation. Based on the climatological data from Chen (1993), the minimum spatial decorrelation scales for temperature for August would be 140 km at the surface, 100 km at 20 m, and 90 km at 50 m. The corresponding minimum spatial decorrelation scales for salinity for August are larger, being 165 km at the surface, 195 km at 20 m, and 170 km at 50 m. Our scales for temperature are quite a bit smaller than those determined for summer by Chu et al. (1997), which were 251 km for the surface, 169 km at 50% of the water column depth, and 157 km at 80% of the water column depth. Chu et al. (1997) considered data from July, August, and September, and this wider window of time apparently tends to increase spatial scales of variability. For example, Chu et al. show the average mean temperature field for their summer period, and these data only range from about 24-28°C. On the other hand, the spatial variability for surface temperature indicated in Fig. 3 is considerably greater, ranging from approximately 21-27°C.

The minimum model-predicted scales for temperature are 75 km for 0-1 m, 85 km for 20-25 m, and 80 km for 45-50 m. For salinity, the minimum model-predicted scales are 70 km for 0-1 m, 200 km for 20-25 m, and 160 km for 45-50 m. For the simulation that included a nudge of the temperature and salinity to initial conditions, the minimum model-predicted scales for temperature are 70 km for 0-1 m, 100 km for 20-25 m, and 110 km for 45-50 m. For salinity, the minimum model-predicted scales are 70 km for 0-1 m, 185 km for 20-25 m, and 160 km for 45-50 m. These spatial scales are very similar to those from the simulation without nudging. Overall, these dynamically-determined spatial scales for temperature and salinity during August are considerably smaller than those calculated using climatological data either from Chen (1993) or from Chu et al. (1997). The use of the smaller decorrelation scales would be the more appropriate choice in circulation models since the larger scales will tend to decrease spatial variability and, thus, damp out circulation features such as internal tides.

ACKNOWLEDGMENTS

This work was supported through a contract between HydroQual, Inc., and the Office of Naval Research, Naval Ocean Modeling and Prediction program. Thanks are forwarded to Dr. G. Caille who provided the Yellow Sea current data.

REFERENCES

Blumberg, A. F., and G. L. Mellor, 1987: A description of a three-dimensional coastal ocean circulation model. *Three Dimensional Coastal Models*, Coastal and Estuarine Sciences, 4, N. S. Heaps (ed.), Amer. Geophys. Union Geophysical Monograph Board, 1-16.

Cassulli, V., and R. T. Cheng, 1992: Semi-implicit finite difference methods for three-dimensional shallow water flow. *Int. J. Numer. Methods in Fluids*, Vol. 15, 629-648.

Chen, D. (ed.), 1993: *Marine Atlas of the Bohai Sea, Yellow Sea, and East China Sea*. China Ocean Press, Bejing, China.

Chu, P. C., S. K. Wells, S. D. Haeger, C. Szczechowski, and M. Carron, 1997: Temporal and spatial scales of the Yellow Sea thermal variability. *J. Geophys. Res.*, 102 (C3), 5655-5667.

Lie, H.-J., 1999: Comment on "A parametric model for the Yellow Sea thermal variability" by P. C. Chu et al. *J. Geophys. Res.*, 104 (C3), 5459-5461.

Lewis, J. K., Y. L. Hsu, and A. F. Blumberg, 1994: Boundary forcing and a dual-mode calculation scheme for coastal tidal models using step-wise bathymetry. In *Estuarine and Coastal Modeling III: Proceedings of the 3rd International Conference*. (M. Spaulding et al., eds.) Am. Soc. Civil Eng., 422-431.

Mellor, G. L., and T. Yamada, 1982: A hierarchy of turbulence closure models for planetary boundary layers. *J. Atmos. Sci.*, 31, 1791-1896.

Shulman, I., 1997: Local data assimilation in specification of open boundary conditions. *J. Atmos. Oceanic Techol.*, 14, 1409-1419.

Smagorinsky, J., 1963: General circulation experiments with the primitive equations, I. The basic experiment. *Mon. Wea. Rev.*, 91, 99-164.

Sommerfeld, A., 1949: *Partial Differential Equations: Lectures in Theoretical Physics, Vol. 6.* Academic Press, New York.

Veronis, G., and H. Stommel, 1956: The action of variable wind stresses on a stratified ocean. *J. Mar. Res.*, 15, 43-75.

Woo, S. B., Y. B. Kim, B. H. Choi, and K. T. Chung, 1994: Improved tidal charts of the Yellow Sea and the East China Sea. *Korea Society of Coastal Ocean Engineering Abstracts*, 96-101.

Numerical Modeling Experiments of the East (Japan) Sea Circulation and Mesoscale Eddy Generation based on POM-ES

Young Jae Ro

Abstract

The numerical model, POM-ES is developed to understand the physics of the circulation of the East (Japan) Sea. Numerical modeling experiments are designed to answer major dynamical questions including basin-wide circulation system, topographic control of major currents, meso-scale eddy activities, formation of the polar front, response to in-outflow conditions.

Important conclusions could be drawn that 1) topographic control is a primary factor in defining the present current systems, 2) the influence of the monthly wind stress is prominent along the coastlines of Russia, Korea and Japan, 3) major current systems (TWC, EKWC, NKCC, Liman and Primoriye) are well reproduced, 4) the baroclinic conditions are of crucial importance in the process of the instability of the Tsushima Warm Current along the Japan Island and are leading to subsequent meandering of the off-shore axis of the TWC and generation of the meso-scale warm eddy into the Ulleung Basin.

Introduction

The East (Japan) Sea is drawing keen international attentions from broad spectrum of scientists, diplomats, and defense officers for its geopolitical s¹ituation, peculiar scientific assets recognized as miniature ocean for its basin-wide circulation pattern, boundary currents, polar front, meso-scale eddy activities and deep water formation. In order to understand oceanographic phenomena occurring in the East Sea, it is necessary and inevitable to understand the physical conditions such as currents and circulation of sea water and their variability. Yet, it is extremely difficult to monitor that condition by field observations due to *e.g.*

Professor of the Dept. of Oceanography, Chungnam National University, Taejon, Korea
royoungj@hanbat.cnu.ac.kr

technical difficulties, ship-time charges, and man-hour budgets. The numerical modeling capability has been greatly improved with reliable model code such as POM, skill assessment techniques, more field observational datasets for accurate forcing, graphic visualization of model outputs in their pre- and post-processings

The circulation pattern in the East (Japan) Sea has been of major interests for its peculiar gyre, a western boundary current and its separation (Uda, 1934; Moriyasu, 1972; Yurasov and Yarichin , 1991). This feature is a well- recognized phenomena in the Kuroshio and Gulf Stream in Pacific and Atlantic Oceans respectively. In relation to the gyre system in the East Sea, the formation of the East Korea Warm Current (EKWC) has brought up with many numerical experiments (Kawabe, 1982a.b; Yoon. 1982a.b; Seung, 1992; Seung and Nam, 1992; Wang et al.. 1999). The branching mechanism of the Tsushima Warm Current (TWC) has also been of interest in association with the formation of the EKWC. Recent field observations provide new evidence of the two components of the TWC, i.e., in and offshore ones (Hase and Yoon, 1999). The offshore component of the TWC is very vulnerable to the baroclinic instability and the meandering of this component would quickly generate a meso-scale eddy into the Ulleung basin (Ro, 1999a). Understanding the role of these sequential processes provided with a new idea to explain the formation of the EKWC, whereas most of previous investigators failed to understand.

The Ekman transport due to wind stress over the sea surface plays an important role in the circulation of the upper waters in the East (Japan) Sea. Many investigators (Sekine, 1991; Seung, 1992; Kim and Yoon, 1996) have carried out numerical experiments to see the response of different datasets of wind stress. Among those, Hogan and Hurlburt (1999) did the most extensive analyses using 7 different datasets. They demonstrated that the sensitivity of the model responses to various wind forcing exist with varying degree, yet hard to quantify due to the lack of precise knowledge of the circulation in the East Sea itself.

The most prominent feature seen on the AVHRR imagery is the polar front existing between warm and cold water masses in the south and north of approximate 39 deg. N. The formation of the polar front is a key to understanding of the circulation gyre in the entire basin, since it is closely related to current system involving the TWC, EKWC, NKCC, Liman and Primoriye Currents.

The POM-ES with eddy resolving grid system and realistic bottom topography is used to understanding major questions addressed above. The modeling experiments are designed to reproduce basic circulation features such as dominant current system of the TWC along the Japan Island, the East Korea Warm Current (EKWC), the Liman, Primoriye and North Korea Warm Current (NKCC), separation of the EKWC, the formation of polar front, meso-scale eddy activities in recognized regions such as Ulleung Basin, Sockcho and Donghan Bay, Yamato Rise, and off the Vladivostok area. This study pays special attentions to focus on the generation of the meso-scale eddies and their responses to various forcing functions and initial conditions of hydrography by the model POM-ES.

Model Configuration

The Princeton Ocean Model (POM, code 97') was used in this study. Detailed descriptions of the POM code can be referred to POM97 user manual (Mellor, 1998) and Blumberg and Mellor (1987). The POM is a time dependent hydrodynamic using sigma coordinate with Smagorinsky type lateral friction (Smagorinsky, 1963), complete thermodynamics and second order turbulence closure submodel (Mellor and Yamada, 1982). It solves 3-D components of velocity, T, S and density, elevation, and turbulent kinetic energy. Model can be started at cold and/or hot condition.

The POM-ES was setup for the domain of 34° to 46° N and 128° to 140° E in Fig. 1 with the realistic bottom topography with 1/10 degree horizontal resolutions (dx = 8.34 to 9.57, dy= 11.2 km) in Fig. 2 and 10 vertical sigma levels are placed at 0.0025, 0.00875, 0.01875, 0.0375, 0.0625, 0.1, 0.1875, 0.375, 0.625, 0.875, respectively. The POM-ES has one southern open boundary through the Korea and Tsushima Straits and two eastern open boundaries through the Tsugaru and Soya Straits. The seasonally varying transport of water volume through the southern boundary is specified with 4 Sv in August and 2 Sv in Feburary which is balanced by the outflows through the Tsugaru (75 %) and the Soya Straits (25 %), respectively.

Modeling Schemes

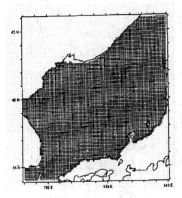

Fig.1 Model Domain and the Grid system for the POM-ES.

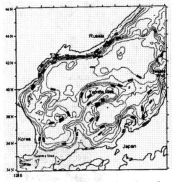

Fig.2 Bottom topography of the East (Japan) Sea.

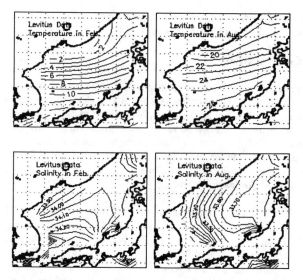

Fig. 3a Horizontal distribution of surface temperature and salinity for winter and summer in the East (Japan) Sea from

Fig. 3b Horizontal distribution of surface temperature and salinity for winter and summer in the East (Japan) Sea GDEM datasets,

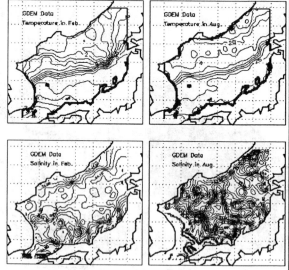

Total of 18 different sets of experiments (Table 1) were setup with various conditions of bottom topography, T-S dataset of Levitus(1994) and GDEM(1998) seen in Fig. 3, Wind dataset (Fig. 4), inflow and outflow magnitudes, and seasonally varying inflow condition for open boundary. Fig 3b show the GDEM T-S distributions in the surface with complicated eddy structures with 1/6 deg. resolution in comparison with Fig 3a of Levitus datasets.

The purpose of the modelling experiments for this study is to understand the response of the model to various forcing conditions and barotropic/baroclinic processes responsible for the meso-scale eddy generations. Thus, the experiments were designed under 3 categories W, T, E which represent experiments for wind-driven, topography induced, and eddy generating circulation in the East (Japan) Sea. The POM-ES was initially spinned up for three years with climatological data of annual mean T-S and wind stress. When the model reached pseudo-steady state (total kinetic energy of the model domain converges within ± 0.1 %), the POM-ES is forced by monthly mean of wind stress and seasonally varying transport of water volume. Surface temperature and salinity values are relaxed to monthly mean T-S with time scale of 30 days.

Table 1 Experimental schemes with various conditions.

Exp. Scheme	No	inflow	Wind	TS	Mode	mode	Major Result
W	01	0	1	f	4	B	Anti-clockwise gyre
	02	2	1	f	4	B	Japan Coast Warm Current
	03	3	8	f	4	B	stronger JCWC
T	04	2	N	f	4	B	same as 03
	05	3	N	f	4	B	stronger JCWC
	06	3	n	f	4	B	no JCWC, weak EKWC
	07	3	n	f	4	B	Much stronger JCWC
	08	3	n	f	4	B	No JCWC, yes EKWC
	09	3	n	8	4	B	Real topography, no EKWC
E	10	3	n	8	4	C	No NKCC
	11	1	n	y	3	C	Long run 100 days
	12	1	n	y	3	C	no EKWC
	13	1	n	y	3	C	stronger EKWC
	14	1	n	y	3	C	stronger EKWC
	15	2	n	y	3	C	very long run 720 days
	16	4	n	y	3	C	very long run 1000 days
	17	2	n	8	3	C	numerous eddies generated
	18	4	n	8	3	C	numerous eddies generated

N : none, f: fix T=18, S=34 psu, y : annual mean

W : wind-driven, T : Topographically controlled, E : eddy generating

Results

Wind-driven Circulation

Response of surface currents to two sets of wind stress data (Hellerman and Rosenstein (HR), 1983; Na and Seo (NS), 1998) were examined. Wind-driven circulation is well developed along the Korean and Japanese coastline. Their magnitude of currents in general is below 20 cm/sec. The responses of the model to two sets of HR and NS are different, since the horizontal resolution of HR and NS are 2 and 0.5 deg. respectively, and NS has higher windstress curl field. To see the differences between two datasets with other conditions of T-S and inflow of 2 Sv fixed, the increased southward transport along the Korean coast by about 20 % were

Fig. 4. Horizontal distribution of wind stress of HR and NS in February over the East Sea,

obtained in Februrary cases seen in Fig. 5.

Topographic Control of the circulation and current system

In the East (Japan) Sea, topography plays an important role in defining basin-scale circulation as well as local currents system. The East (Japan) Sea is divided in three major basins such as Japan Basin in the northern part of the Sea with fairly deeper and flat bottom, Ulleung Basin in the southwestern part and Yamato Basin in the southeastern part. It is well known that in each basin, there exist basin-scale

Fig. 5 Example of circulation experiments with NS wind stress in February.

circulation gyres and recognized locations where meso-scale eddies are most frequently observed. In contrast to basins, there also exist sea mountains in the area off the Vladivostok. Around Yamato Rise and Dok-Island, very steep bottom slope play significant roles in producing meso-scale eddies. These are the areas well recognized for the meso-scale eddy generations. Different results are presented in Figs. 6 where the strength of major currents and size and locations of meso-scale eddies vary.

Circulation patterns controlled by the topography in the East (Japan) Sea are evidently seen in Fig. 7 where two hypothetical cases of flat bottom (A and C) and a realistic bottom (B) cases show strikingly different patterns. EKWC are well developed in A and C, while only TWC is developed in a realistic bottom with barotropic case (B). This result imply the process of EKWC development in association with baroclinic forcing. This is closely related with the onset of the baroclinic instability to be discussed in following section.

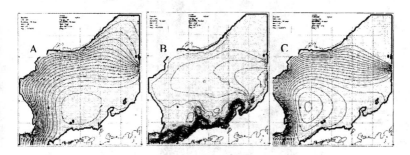

Fig. 7 Example of circulation experiments with hypothetical bottom topography, A: flat (200m) B: realistic, C: flat (1000m)

Circulation Gyre System

s There exist various schemes of the circulation gyre by many investigators such as Uda(1934), Moriyasu(1972), Yurasov and Yarichin(1991). These schemes reflect contemporary knowledge of the current system in the East Sea based on observations and speculations. Yet, they are evolving with time of new findings such as separation points of the EKWC, locations of meso-scale eddies, formation of polar front, magnitudes and variabilities of currents such as the TWC, EKWC, Primoriye Currents, etc. The basic feature is a cyclonic gyre in the northern part of the East Sea which is in contrast to an anticyclonic gyre existing in the southern part with separation of the EKWC around 38 deg. N and flows out of the Tsugaru

Strait. Theses features are well reproduced in Fig. 6 in which dominant current system such as TWC along the Japan Island, the EKWC along the Korean Peninsula, the Liman/Primoriye Current and North Korea Warm Current (NKCC) are well reproduced. In response to various forcing conditions such as in-outflow magnitudes (varying from 2 to 4 Sv. seasonally), two datasets of T-S such as Levitus and GDEM, two datasets of wind stress such as HR and NS, the results are all varying to some degrees. Fig. 8 shows a very extreme case where the very strong NKCC flows down to 35 deg. N which prohibit the development of the EKWC (Kim and Legekis. 1987). This case was obtained with abnormally colder conditions along the Primoriye region which led to the development of stronger Liman/Primoriye Current which flows southward along the Korean coast and joins the Tsushima Warm Current.

Fig.8 An example of circulation
experiments showing an
extreme case where the NKCC
dominate the EKWC.

Meso-scale Eddy Activities

Meso-scale eddy activities in the East (Japan) Sea are extremely pronounced features seen on satellite IR Imageries. It seems that they can be formed any where

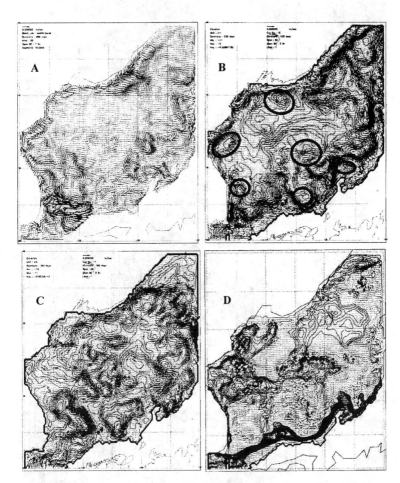

Fig. 6 Examples of circulation experiments with different forcing conditions. A: 13, B:15, C:17, D:18 where numbers indicate the experiments in Table 1.

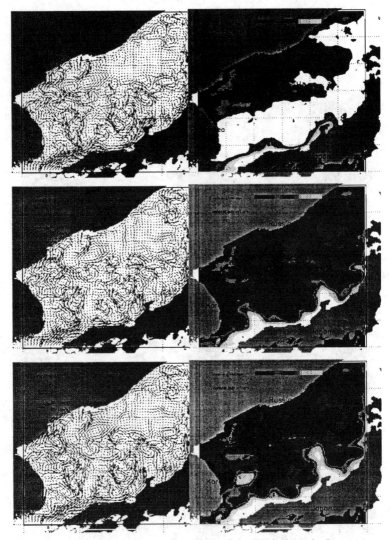

Fig. 9 Onset of the baroclinic instability of the TWC along the Japanese Coast. Right panel show the distributions of sea surface heights for cases before, onset and after the baroclinic instability. Upper, middle and lower panels show the cases with 20-day intervals. Left panel show the corresponding distributions of current vectors for the same cases.

in the Sea. Yet, there are several recognizable locations such as Ulleung Basin, Sockcho and Donghan Bay, Yamato Rise, off the Vladivostok area, etc where the meso-scale eddies are most frequently observed. In Fig. 6B, these eddies are well reproduced and are marked in ellipses. The characteristics and formation mechanism vary for each eddies in that baroclinic origins are associated with Ulleung and Yamato Basins, while eddies in the region off the Vladivostok are in many times barotropic. Eddies over Yamato rise are believed to be formed with baroclinic and bottom topographic interactions where the steepest bottom exist. The detailed descriptions of the eddy formations in the East (Japan) Sea can be referred to Ro(1999b).

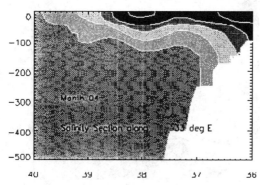

Fig. 10 Cross-sectional distribution of salinity along 133 deg. E where it shows the two cores of maxima corresponding to two components of TWC.

Onset of the baroclinic Instability of the TWC

Of the various numerical experiments, the most striking and significant results are seen in Fig. 9 where the onset of the baroclinic instability of TWC is demonstrated. The offshore component of TWC becomes unstable so that the meandering of its axis take place and it further develops into isolated meso-scale eddy. In model experiments, two regions off Oki Island and Noto Peninsula are clearly recognized for the locations of the onset of instability. The baroclinic instability process can be understood in taking account of vorticity balance. The planetary vorticity gets bigger with northward flow to which the relative vorticity should be balanced. This would be further strengthened by topographic beta effect where bottom depths get deeper offshore. So in principle, the TWC along the Japan

Island is considered to be vulnerable to the baroclinic instability, when it is considered with geometry and bottom topography. Yet, the baroclinic characteristics are strengthened in summer season with more buoyancy in the upper layer so that the instability seems more frequently observed. These phenomena are observed in intensive field observations where the two components of the TWC were revealed recently (Hase et al, 1999). The model results are in a very good agreement with this observation in Fig. 11. Fig. 12 shows the time evolution of characteristics of the flow into the Ulleung Basin from the Oki Island area in terms of velocity, vorticity and temperature. During the onset of the instability period, it is obvious that the negative vorticity is increased by as much as 100 %.

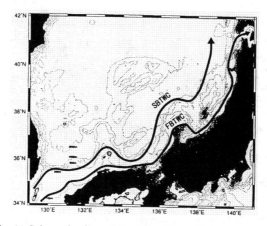

Fig. 11 Schematic view of Tsushima Warm Current system with two components, Figure is reproduced from Hase, Yoon and

The onset of the instability of TWC off the Oki Island plays an important role, since it provides more buoyancy forces into Ulleung Basin so that the development of EKWC take advantage of this energy supply. In the model, such a process is clearly demonstrated. The hypothesis of the three branch system in TWC has been of arguing issue. Based on the modelling experiments in this study, I suggest that the middle branch is a transient phenomenon and the hypothesis of three branch should be reconsidered. Based on these numerical experiments and field observations, a new hypothesis is proposed for the generation of the East Korea Warm Current such that the onset of the instability of the Tsushima Warm Current along the Japan Island is a necessary condition.

Summary and Conclusion

This study carried out diverse numerical modelling experiments to understand the general circulation characteristics in the East (Japan) Sea by using POM-ES. The experiments are based on three schemes of wind-driven (W), topography controlled (T), and eddy generation (E). Following conclusions could be drawn :

1) bottom topography is one of the most important factors for the general circulation in the East (Japan) Sea and eddy generation as well,

2) numerous meso-scale eddies across entire basin were reproduced with different dynamical characteristics (cyclonic or anti-cyclonic, shallow or deep, coastal or open water, short or long life) in various locations such as Ulleung basin, Donghan Bay, Oki Island, Yamato Rise, off Vladivostok area, etc.

3) Liman and Primorye cold current was well developed under some extreme conditions to reach down to 35°N along the Korean coastline,

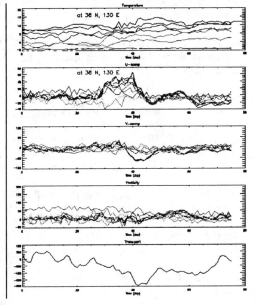

Fig. 12 Time evolution of flow characteristics into the Ulleung Basin from the Oki Island area in terms of velocity, vorticity and temperature. Eight curves in the figures represent points at 36 to 40 deg N with 0.5 deg interval along 130 deg. E.

4) The onset of the instability of the branch of the Tsushima Warm Current along the Japanese coast plays a crucial (necessary) role in the generation of the EKWC by providing sufficient heat flux into the Ulleung Basin. Thus a new hypothesis is proposed for the generation of the EKWC.

Acknowledgement

This study is supported by the academic research fund of Ministry of Education through KIOS (KIOS-97-M-10).

References

Blumberg A.F. and G.L. Mellor, 1987. A description of a three-dimensional coastal ocean circulation model., in Three-Dimensional Coastal Ocean Models, Coastal Estuarine Sci., edited by N. Heaps., AGU, Washington, D.C., Vol. 4, 1-16.

Hase, H., J-H Yoon and W, Koterayama, 1999. The current structure of the Tsushima Warm Current along the Japanese coast, J. Oceanogr. Soc. Japan, 55:217-235.

Hellerman, S. and M. Rosenstein, 1983. Normal monthly windstress over the world ocean with error estimates. J. Phys. Oceanogr., 13, 1093-1104.

Hogan, P. and H,E, Hurlburt, 1999. Impact of different wind forcing on Circulation in the Japan/East Sea, pp 124-127, Proceeding of CREAMS'99, International. Symposium held in Fukuoka, Japan, Jan. 26-28, 1999

Kawabe, M. (1982a). Branching of the Tsushima Current in the Japan Sea, Part I: Data analysis. J. Oceanogr. Soc., 38: 95 - 107.

Kawabe, M. (1982b). Branching of the Tsushima Current in the Japan Sea, Part II: Numerical experiment. J. Oceanogr. Soc., 38: 183 - 192.

Kim, K. and R. Legekis, 1987. Branching of the Tsushima Current in 1981-83. Progress Oceanogr., 17:265-276

Kim. C.-H. and J.-H. Yoon (1996). Modeling of the wind-driven circulation in the Japan Sea using a reduced gravity model. J. of Oceanogr., 52: 359-373.

Levitus, S. and T. Boyer, 1994. World Ocean Atlas Vol. 4: Temperature, NESDIS, NOAA, U.S. Dept. of Commerce, Washington DC.

Mellor, G. L. and T. Yamada, 1982. Development of a turbulence closure model for geophysical fluid problems. Rev. Geophys. Res., 20:851-875.

Mellor, G. L. , 1998. Users guide for a three-dimensional, primitive equation, numerical ocean model, Princeton University, May, 1998.

Moriyasu, S. 1972. The Tsushima Current. In Kuroshio - Its physical Aspects, ed. by H. Stommel and K. Yoshida, Univ. Tokyo, Japan, 353-369.

Na, J. Y. and J. W. Seo, 1998. The sea surface winds and heat flux in the East Asian Marginal Seas, 55 pp, Dept, of Earth and Marine Sciences, Hanyang University, Korea,

Ro. Y.J.. 1999a. Numerical Experiments of the Meso scale Eddy Activities in the East (Japan) Sea. pp 116-120, Proceeding of CREAMS'99, International. Symposium held in Fukuoka, Japan, Jan. 26-28, 1999

Ro. Y.J.. 1999b. Numerical modeling of the Meso-scale Eddies in the East (Japan) Sea, submitted to J. Korean Soc. of Oceanogr.

Sekine, Y. (1991). A numerical experiment on the seasonal variation of the oceanic circulation in the Japan Sea. In: Oceanography of Asian Marginal Seas (Ed., K. Takano) Elsevier Oceanography Series, 54:113-128.

Seung, Y.H. (1992). A simple model for separation of East Korean Warm Current and formation of the North Korean Cold Current. J. Oceanol. Soc. Korea, 27: 189-196.

Seung, Y.H. and S.Y. Nam (1992). A numerical study on the barotropic transport of the Tsushima Warm Current. La mer, 30: 139-147.

Smagorinsky, J., 1963. General circulation experiments with the primitive equations, I. The basic experiment, Mon. Weather Rev., 91:16053-16064.

Uda, M., 1934. The result of simultaneous oceanographical investigation in the Japan Sea and its adjacent waters in May and June, 1932. J. Imp. Fish. Exp. Sta., 5, 57-190(in Japanese with English abstract).

Wang, K, G. Fang and Y.J. Ro, 1999. Barotropic Numerical Studies on the driving mechanism of the East Korea Warm Current, Studia Marina Sinica vol 41.

Yurasov, G. and B. G. Yarichin, 1991, Currents of the Japan Sea, Vladivostok, FEB of Academy of Science, USSR 176 pp.

Yoon, J. H., 1982a, Numerical experiment on the circulation in the Japan Sea, Part I : Formation of the East Korean Warm Current. J. Oceanogr. Soc. Japan, 38: 43-51.

Yoon, J. H., 1982b, Numerical experiment on the circulation in the Japan Sea, Part III: Formation of the nearshore branch of the Tsushima Current. J. Oceanogr. Soc. Japan, 38, 119-124.

Simulating Morphodynamical Processes on a Parallel System

Hartmut Kapitza and Dieter P. Eppel[1]

Abstract

The simulation of morphodynamical processes with a 3–dimensional numerical model requires high horizontal and vertical resolution in the area of interest. Since running the whole model domain with high resolution is not feasible a nested approach is suggested. An estimate on the required amount of computer resources calls for applying parallel systems with distributed memory. This paper explains the philosophy of transfering a 3–dimensional hydro–numerical model aiming at applications for morphodynamics onto a parallel machine.

Introduction

Morphodynamics describes the change of bathymetry under the joint action of wind, waves, and currents, and the subsequent reaction of the flow to this change. Figure 1 shows a schematic of the processes involved.

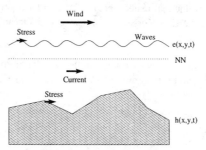

Figure 1: Processes participating in an integrated model approach for morphodynamical changes.

---

1   GKSS Research Center, Max–Planck–Str. 1, D–21465 Geesthacht, Germany

Wind stress and tide drive the current. The wind–generated waves influence the atmospheric flow by changing the surface roughness. Both, waves and current generate a stress on the bathymetry. If this stress is larger than a certain value (critical shear stress) which depends on the characteristics of the seabed material the sediment can be mobilized and erosion sets in. Below the critical shear stress, only deposition of suspended material is active. Since several of the key parameters influencing the bottom shear stress are of stochastic nature (waves, material) the critical shear stress parameterization should be used in a statistical sense. For more details see Soulsby (1997). A comprehensive simulation system for morphodynamics is composed of submodels for wind, waves, currents, and sediment transport.

There are many areas in the world where such changes are of environmental and economical relevance. Being able to predict morphodynamical changes could result in considerable savings of tax money by assessing the impact of seadefence activities, e. g. dredging to keep navigation channels open or estimating in advance the influences of construction works like training walls or wave breakers.

Nesting Approach

On one hand, many of these applications need a horizontal resolution in the meter range and a vertical resolution of a few centimeters. A typical example is a navigation channel, which is often only 50 m wide or even less. In order to resolve such a feature one needs at least 2 grid points, i.e. grids coarser than this value don't even see the channel. In order to gain results which are physically reasonable one needs grid point numbers in the order of 10 to resolve a channel like that. Another, even more demanding example for an application is the scouring effect of bridge pillars (typical scale of 1 m or less). On the other hand, the model domain has to have a reasonable size in order to avoid unphysical boundary effects influencing the interior results. Therefore, running the whole domain with the same high resolution is not feasible. This calls for using a nested model approach, where a hierarchy of models is stacked onto each other. The coarsest model is driven by boundary values from either measurements or model results of larger scales, or a combination of both. All subsequent, finer grids are driven on the boundaries by the results of the coarser grid using established methods like for instance the nudging technique (Davies, 1973).

For estimating the amount of computer resources we restrict ourselves to the current model. The results are directly transferable to the other modules of the integrated model. Let n_k be the number of grid points in the k–th level of refinement, and A a fraction of cells in k to be refined for level $k+1$. A refinement is defined as a splitting of a cell in 4 sub–cells (horizontal or 2D refinement) or 8 sub–cells (3D refinement). A 2.5D refinement is also thinkable where in the vertical only the cells in the lower half of the water column are split into two cells. A=0.5 means that half of the cells of each grid are to be refined in order to reach the next grid level. Let further C be the increase of grid points for level k to k+1. C is given by

C = 4A : 2–dimensional refinement
C = 6A : 2.5–dimensional
C = 8A : 3–dimensional

The total number of grid points for k levels can be expressed as a geometrical series and is a function of A:

$$N_k(A) = \frac{C^{k+1} - 1}{C - 1}$$

As an example consider a domain size of 80 x 160 km, a depth of 20 m, using a basic grid of horizontal and vertical grid sizes of 800 m and 2 m, respectively. The resulting number of cells in the coarsest grid is 200000. With 6 levels of refinement the resulting resolution is 12.5 m in the horizontal and 3 cm in the vertical. The results for different refinement strategies using A=0.5 is summarized in Table 1:

Dimension	N/n0	M pts	M pts/proc
2 D	127	25.4	0.05
2.5 D	1093	218.6	0.43
3 D	5461	1092.2	2.13

Table 1: Grid point estimate for different refinement strategies. For an explanation see text.

Table 1 shows in the first column the refinement strategies (2–, 2.5–, 3–dimensional), in the second column the total number of grid points as a multiple of the coarsest grid, in the third column the total number, and in the last column the number of cells per processor based on a 512–processor machine like for example a Cray–T3E. The results show that a purely 2D refinement with 25 million points might be just manageable on a powerful machine with large memory. All other refinement strategies are way beyond reach of today's single processor machines. But one can also see that even the full 3D refinement is feasible on a machine of the T3E–class with just above 2 million points per processor. Table 1 shows impressively the need for parallel computing.

Parallelization Strategy

When a problem is to be solved on a parallel computer there are usually two kinds of parallelization strategies: One distributes the different parts of the model to different processors, for example let one processor do advection, another is busy with diffusion, and so on. This strategy has the disadvantage of a difficult load balancing among the individual processes. The other approach is to split the data into small chunks and let each processor have one of them at a time. This strategy is often referred to as domain– or data–decomposition (Chan, 1994). We are following the latter strategy since it is much easier to distribute the load evenly among processors.

One aim of this project is to end up with a program which is able to run on any kind of computer architecture, that means there is no explicit assumption on the number of processors available (it could be as well just one). Therefore, each individual grid is split into *sub–domains*, which are defined as a rectangular arrangement of grid cells. An important property of a sub–domain is its global definition of the cell indices within the grid, which are used for navigating. One sub–domain is the smallest portion of a grid which can be run on a single processor, but one processor can accommodate more than one sub–domain if necessary. The model at hand (TRIM3D) is a finite difference code developed by V. Casulli and co–workers at Trento University, Italy. The serial code runs very efficiently on a single workstation. For details on the numerics of the model see Casulli & Stelling, 1998. Since TRIM3D computes some of the terms implicitly in the vertical direction the splitting of the domain into sub–domains is restricted to the horizontal, i. e. all of them contain the full water column. Each sub–domain is surrounded by a layer of *ghost* cells in order to make the communication with its neighbouring sub–domain possible. The ghost cells are filled with the content of a neighbouring sub–domain. The width of the ghost cell layer depends on the size of the discretization stencil used in the numerical implementation of the partial differential equations describing the flow. In our case a width of one grid cell is enough.

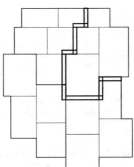

Figure 2: Distributing 17 sub–domains (thin lines) onto 4 processors (medium lines). The external ghost cells of all sub–domains of 1 processor are displayed with thick lines.

Figure 2 shows a sample splitting of a single grid into 17 sub–domains distributed on 4 processors. For one processor the external ghost cell layer is shown, the filling of which means communication between processors. On a distributed memory system some kind of message passing library has to be used for this process. For convenience, the MPI–library (message–passing interface) is used as a standard interface to implement all inter–processor communication (for details on MPI see Snir et al., 1997). All internal ghost cells (not shown) are filled by simple memory to memory copies. Note that sub–domains do not have to have matching boundaries. One sub–domain can have more than one neighbour on a particular face. The neighbouring sub–domains are found automatically by checking four conditions for validity as displayed in Figure 3. The numbered double–arrows point to the cells whose indices are to be compared. The numbers refer to the relation shown in the figure. Note that in order for this algorithm to work the indexing of all sub–domains of a particular grid has to be *global* as noted above.

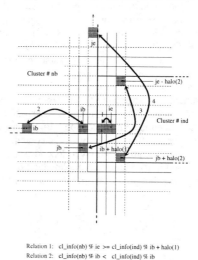

Relation 1: cl_info(nb) % ie >= cl_info(ind) % ib + halo(1)
Relation 2: cl_info(nb) % ib < cl_info(ind) % ib
Relation 3: cl_info(nb) % jb <= cl_info(ind) % je - halo(2)
Relation 4: cl_info(nb) % je >= cl_info(ind) % jb + halo(2)

Figure 3: Schematic description of the algorithm for auto–matically finding the neighbouring sub–domains of *ind*. See text for details.

All the information on the neighbouring sub–domains needed for communication is stored in an entry of a globally available information table.

Testing the Parallelization

A simple test problem was created for a basic test of the communication among the sub–domains on a machine with an arbitrary number of processors. For each inner grid point a simple iteration formula was applied which computes the new value of the grid point by taking the average of all 26 neighbours. Ten iterations were performed. Table 2 shows the results of some test runs using 8 sub–domains of different size (columns) ranging from 5 x 5 x 5 to 200 x 100 x 100. All runs on the T3D were done with 8 processors, i.e. each sub–domain is located on a private processor. Using smaller number of processors was not feasible for this test since the largest sub–domain size just fits into a single processor. Therefore, traditional concepts like speedup and efficiency do not make sense for this particular test. All performance times (initialization phase, communication, computation) are given in seconds per processor on a Cray–T3D system. Additionally, the Mflops rate is given as well as for a particular sub–domain size (50 x 50 x 50) some comparisons with other machines.

Cl.-Size:	5 5 5	10 10 10	20 20 20	30 30 30	40 40 40	50 50 50	100 100 100	200 100 100
Initial.	0.32	0.41	0.34	0.31	0.33	0.37	0.6	0.5
Commun.	0.03	0.03	0.04	0.08	0.10	0.11	0.5	0.6
CPU	0.003	0.015	0.10	0.96	0.88	1.66	15.5	33.3
Mflops	11.5	18.3	21.5	7.9	20.3	21.4	18.1	16.8
Sun-Mflops						7.2		
J90-Mflops						33.6		
C90-Mflops						230.7		

Table 2: CPU times per processor (seconds) of a simple iteration test on 8 processors of a Cray–T3D and some other single–processor platforms for different sub–domain sizes. The results are split into initialization, communication and computation parts of the program.

From this table several interesting facts are recognizable:

· The initialization time is almost independent of the sub–domain size. Since this time value mostly contains the setup and information generation it is proportional to the number of sub–domains which was constant.

· The communication time increases somewhat with the sub–domain size but only by about one order of magnitude, while the sub–domain size increases by a factor of 16000. This indicates that a large amount of the communication time is dictated by the startup of communication calls, which is also a function of the sub–domain number. Only a relatively small increase is due to the increasing message length.

· The time for computation matches fairly well the increase in sub–domain size. Therefore, the ratio of computation to communication becomes larger with larger sub–domain size.

· The Mflop rate sort of saturates around a value of approximately 20.

However, there is a very pronounced performance loss for a sub–domain size of 30 x 30 x 30. This is attributed to cache effects and therefore depends very much on the platform being used. One needs to be aware of this effect which makes performance tests with different sub–domain sizes necessary.

The next test is done with a model bathymetry similar to a step pyramide (see Figure 4). Four grids are stacked onto each other with horizontal resolutions of 800, 400, 200, and 100 m, respectively. All grids have 16 x 16 x 10 grid points. The whole stack is split into 4 sub–domains for each grid resulting in a total of 16 sub–domains all of the same size (so load balancing is just a matter of distributing the sub–domains evenly). In the outermost area a water depth of 10 m is assumed. At each grid boundary the depth decreases by 1 m.

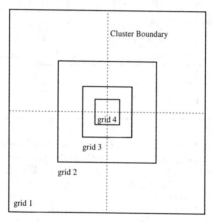

Figure 4: Bathymetry and sub–domain boundaries (dashed) of the grid hierarchy used for the implementation test.

The run is started at rest with a constant water level. The boundary value for water level all around drops by 1.35 m within 1 time step of the coarsest grid (120 seconds) and is kept constant thereafter. This simulates the sudden breakdown of a wall around a rectangular basin. This test mainly served the purpose of making sure that all the data exchanges between sub–domains work right and that the results of the parallel implementation correspond to the ones of the original serial code. Therefore the signal in the model was set to an unrealistic high value.

The results of the parallel code running on a Cray–T3D were compared to the results of the original serial code on a Sun–Ultra 1 workstation and were found to be identical up to 10 significant digits (running both with 64 bit precision). The differences beyond that level can be attributed to round–off resulting from the different order of adding the sub–domain contributions to global dot products used for the conjugate gradient method to solve for the water level. When this part of the computation was turned off the results were identical up to the full range of 13 digits.

Table 3 shows the timing results for 1 time step of the coarsest grid (which is actually a total of 15 time steps since the time steps of the grids are decreased in the same way as the spatial resolution) again split into initialization, computation and communication, and run on the T3D with 1, 2 and 4 processors, respectively. The results show an initalization time almost independent on the number of processors. Also the communication time stays in the same order of magnitude. A first rough inspection indicates that the computation part scales nicely with the number of processors. This is shown more precisely in Figure 5 presenting the CPU times as a function of processors. The ideal speedup is indicated by the diagonal line. Figure 5 shows a more than ideal speedup for 2 processors. The reason for this behaviour is not quite clear.

Nprocs:	1	2	4
Initial.	1.4	1.2	1.3
Comput.	8.9	4.2	2.3
Communic.	0.6	1.0	0.8

Table 3: CPU times in seconds per processor of the step pyramide test with 4 grids (16 sub–domains) using 1, 2 and 4 processors.

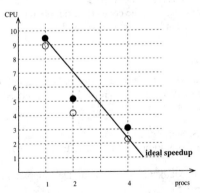

Figure 5: CPU times as function of processor number. The diagonal line indicates the ideal speedup. Circles are just computation times, bullets are computation plus communication.

Conclusions and Outlook

It has been shown that for the small scale simulation of morphodynamical processes the use of the nested model domain approach demands the implementation on a massively parallel computer. The task was split into sub–tasks using the data–decomposition technique. In order to increase the flexibility of the model code the domain was decomposed into rectangular sub–domains which are subsequently distributed among the available processors. Application to simple test problems make sure that the parallel code gives the same results as the serial code and that the nesting of several grid levels works properly. The following points need to be considered in future work:

- An automatic load balancing algorithm is necessary. So far, the sub–domains are distributed manually among the processors. A more advanced technique of load balancing is presented by Chien et al. (1995).

- The parallelization has to be done for all parts of the integrated model (wave model, atmospheric model). The serial codes of those parts are available. For the atmospheric model there is already a parallel version available (Ashworth et al., 1997). Up to now only the flow model is parallelized in the sub–domain philosophy.

- The coupling among the grid levels is one–way (top–down). A two–way nesting will allow some part of the sub–grid scale processes of a coarse mesh (which need to be parameterized) to be expressed explicitly by values of the finer grid.

- The communication has to be optimized, which could mean using different algorithms for some parts of the code (for example the conjugate gradient algorithm is a good candidate for a closer look).

References

Ashworth, M., F. Fölkel, V. Gülzow, K. Kleese, D. P. Eppel, H. Kapitza and S. Unger, 1997: Parallelization of the GESIMA Mesoscale Atmospheric Model. Parallel Computing 23, 2201–2213

Casulli, V., and G. S. Stelling, 1998: Numerical Simulation of 3D Quasi–Hydrostatic, Free–Surface Flows. J. Hydraulic Eng. 124, 678–686

Chan, T. F., and T. P. Mathew, 1994: Domain Decomposition Algorithms., Acta Numerica, 61–143

Chien, Y.P., F. Carpenter, A. Ecer, and H. U. Akay, 1995: Load–Balancing for Parallel Computation of Fluid Dynamics Problems. Comput. Meth. Appl. Mech. Engrg. 120, 119–130

Davies, H.C., 1973: On the Lateral Boundary Conditions for the Primitive Equations. J. Atmos. Sci. 30, 147–150

Snir, M., S. W. Otto, S. Huss–Lederman, D. W. Walker and J. Dongarra, 1997: MPI: The Complete Reference. MIT Press, Cambridge, MA, 336 p.

Soulsby, R., 1997: Dynamics of Marine Sands. Thomas Telford Publications, London, 249 p.

MORWIN – Collaborative Modeling of Coastal Morphodynamics

Rainer Lehfeldt[1], Volker Barthel[2]

Abstract

MorWin is an ongoing research project within the framework of coastal zone management activities. It is concerned with the marine processes that shape the Baltic Sea coast of Germany and the setup of an infrastructure for collaborative modeling using internet technology. The interdisciplinary research team has established a web-based working environment for data management, numerical modeling and documentation.

Numerical studies regarding cause and effect of wide area coastal evolution and sea bed changes in the vicinity of barrier islands or within shipping channels are carried out. The essential data base for longterm modeling of coastal morphodynamics is compiled from field and hindcast data in cooperation with coastal engineers, environmental data providers and local authorities.

Morphodynamic models which continuously update the model bathymetry due to sediment accretion or erosion are instantiated. Results from a 1D-longshore sediment transport study near Rügen island for a 5 year period from 1993 to 1997 are presented which show the sensitivity of morphodynamic models to wind and wave boundary conditions. These results will be used to drive the computationally expensive morphodynamic 2D-area models in small domains around coastal inlets and sandbanks with intricate topographic features.

Introduction

The ongoing research project **MorWin** (**Mor**phodynamic Modeling of **Win**d-Influenced Flats - Internet Based Collaborative Project Handling in Coastal Engineering) was initiated by the German Coastal Engineering Research Council (KFKI). Funding is provided by the German Federal Ministry of Education, Science, Research and Technology in an attempt to establish numerical models as multi-user prognostic tools within the frame-work of coastal zone management activities. It is a joint effort of universities, federal and provincial coastal

[1] Research Scientist, Federal Administration of Waterways and Navigation, Hindenburg Ufer 247, 24106 Kiel, Germany, tel: ++49.431.3394.765, email: lehfeldt@wsd-nord.ki.shuttle.de
[2] Manager, German Coastal Engineering Research Council (KFKI), Am Alten Hafen 2, 27472 Cuxhaven, Germany, tel: ++49.4721.567.385, email: vbarthel@cux.wsd-nord.de

authorities to study the sediment transport processes along the Baltic Sea coast of Germany with focus on the evolution of coastal inlets and wind influenced sandbanks.

There are two distinct work packages concerning a) the extension of CFD methods to reflect actual changes in bathymetry during simulation runs and b) the utilization of state-of-the-art information and communication technology methods to facilitate and optimize the application of such morphodynamic models. The main objectives of the **MorWin** project are

- Gaining insight in sediment transport processes near Rügen island
- Providing the data base for prognostic modeling
- Instantiating hydro- and morpho-dynamic numerical models
- Creating a prototype framework for collaborative project handling in coastal engineering

Because of the complexity of this task, an interdisciplinary research team was established. The collaboration partners are numerical modelers, sediment transport specialists, information and communication technology experts, coastal engineers and representatives from several administrations concerned with the coastal zone. New forms of sharing and organizing work packages are necessary for members of a project-team who remain in different institutions [12] across the nation. Distributed project handling relies on the information and communication technology of the internet. All documents and data are stored on web-servers and accessed via common internet browsers, thus bridging the gap between different hard- and software platforms.

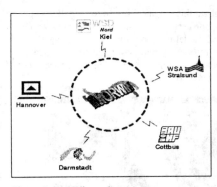

Figure 1: MorWin project partners

The project partners (Figure 1) are the Institute for Computer Science in Civil Engineering at Hannover University, the Institute for Computer Science in Civil Engineering at Brandenburg Technical University in Cottbus, Institute for Water Resource Research at Darmstadt University and the Federal Administration of Waterways and Navigation, Directorate North in Kiel (WSD-Nord) with local offices in Stralsund. The provincial authority for Nature and Environment (StAUN) in Rostock participates in a consulting function.

CFD-activities concentrate on morpho-dynamic models [2] which continuously update the model bathymetry due to sediment accretion or erosion. These models are very sensitive to wind and wave boundary conditions and require comprehensive data acquisition, statistical analysis and preprocessing in order to provide consistent input data. Different concepts of modeling individual hydrodynamic processes due to wind forcing, waves and wave-current interaction have recently been proposed [4] in the literature. Implementations in commercial software packages of Hydraulic Research Wallingford and Danish Hydraulic Institute are being used in the MorWin project for model inter-comparison. Some emphasis is put on improvements of the models developed by the project partners.

However, a greater challenge is seen in using modern information and communication technology to establish a collaborative modeling environment for CFD [14]. The redesign of proven software for internet client/server applications and the design of an information model used for handling bulk input and output data of hydro- and morphodynamic models fall into this category.

MorWin modeling environment

Java technology is used for all components of the cross-platform graphical user interface developed in the project. Navigation and searching in the project's information base is supported by point-and-click selection methods on interactive maps. According to categories (Figure 2) such as bathymetry, beach profiles, water levels, wind, currents, waves or sediment the available data are displayed with geographical reference. Each data type is equipped with specific visualization and analysis methods. After loading the individual data sets on the desktop, all relevant views available can be activated. Bathymetry data, e.g., can be shown as contour plot or wire mesh. Depth profiles along interactively selected cross sections can also be displayed. Unions of data sets can be formed in preparation of input data for automatic grid generation which is also implemented in this user interface. The handling of time series of scalar or vector quantities is realized in the same way with methods of statistics and special plots such as stick diagrams available for these data types.

The concept of meta data is applied throughout this graphical user interface. It is explicitly used when specific queries are generated by the user to find data sets complying with given specifications. Due to the underlying logical model of a point-based data storage [13], the integration of time and space information into one system is straight forward. An object oriented data base system POET [15] stores the data proper on decentralized web servers. This feature eliminates the need for central data pools.

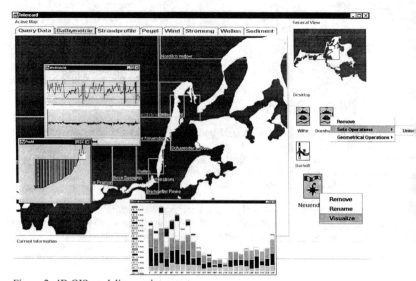

Figure 2: 4D GIS modeling environment

A common user interface for combined visualization of **geo-spatial** data (3D GIS) and **time-dependant** data of both time series at singular locations and 2D-area modeling results has been developed. This greatly supports model validation [10], data valuation and especially illustrating cause and effects of hydro- and morphodynamic phenomena. Many routine tasks relevant in coastal engineering are automatically performed within the 4D GIS modeling environment presented. Similar integrated systems such as COASTMAP [16] have recently been published. However, the key idea to provide an identical suite of **online services** has not been addressed which provides utilities on the desktop of every team member for data visualization, analysis and processing.

Java script methods and hypertext pages are used in a complementary way for **online information** on progress within the collaborative modeling project. By adding to the categorization of physical phenomena (bathymetry, water level, currents, waves, wind, sediment) the related principal areas of activities (data, modeling, assessment) the matrix in Figure 3 is created. Each icon is linked to a complete and up-to-date record of information concerning the selected "activity per phenomenon". These web pages can be maintained on different web servers and typically represent available meta information for a chosen topic. They also give access to all data archives, an internal bulletin board for team members, and propose contact persons to external visitors.

The scalable icon matrix may hold many more categories from biology, chemistry, etc. when applied to environmental or coastal zone management projects. Integrating additional phenomena into the collaborative modeling activity helps to get a better notion of inter-relationships. Figure 3 indicates the input data necessary for morphodynamic modeling, summarizes the modeling efforts for each data type and collects assessment papers of each activity. It documents and presents the work done within the 4D GIS modeling environment and can also be viewed as a work flow instrument for the project management. Each topic opens up with a summary page which can be regarded as online "executive abstracts".

Apart from formalized meta data according to Dublin core standard [3] there are charts, photographs, animated simulation results, scanned documents, etc. provided on the web-server which hold additional information concerning the engineering task. Since the number of objects maintained on the project servers grows continuously, the evolving information system offers search mechanisms in addition to the standard navigation provided by web browsers. An automatic search engine [5] has been implemented which supports keyword search.

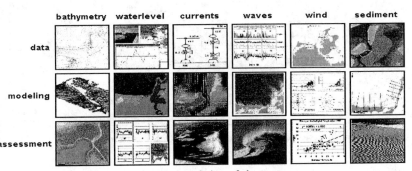

Figure 3: Collaborative modeling; Integrated view of phenomena

The different areas of activities are related to project team members representing the communities of a) data provider, b) CFD modelers and c) the "public" which is generally most interested in the assessment of engineering studies. Accessing this integrated view of phenomena on the desktop facilitates feedback and communication about the collaborative task. At any time, the complete status of the project with all its different work packages can be reviewed. It is helpful to make the working process transparent, it guarantees continuous quality control and adds to further acceptance in the public.

Application to German coastal region at the Baltic Sea

The MorWin study domain (Figure 4) is a micro-tidal coastal region in the southern Baltic Sea where shorelines are shaped by wind-generated waves and currents. Littoral drift accumulates material into large sand flats which are above sea level and dry most of the time but which can be flooded up to 1.5m during extreme events. Sand transported southward along the island of Hiddensee and eastward along the Zingst coast forms the sand flat "Bock" located in the inner part of the Gellen bay. It extends over approximately 10x3km^2 in the vicinity of the northern access channel to the port of Stralsund. This waterway is maintained at a depth of 4.5 meters and requires continuous dredging.

The Darss-Zingst peninsula and the island of Hiddensee belong to barrier islands shielding a shallow lagoon system called "Boddengewässer" from the Baltic Sea. Rather straight coastlines extend in West-East direction for 30km and in North-South direction for 20km along the Gellen bay with water depths <10m in most of the area. Significant water level changes in the Baltic Sea caused by seiching and local wind set-up enter through several coastal inlets with different amplitudes. The resulting pressure gradients can lead to local flow reversals within the lagoon system and to a backward facing delta [12] at Gellen inlet. Sediment entering from the outer bay then nourishes the spit at the southern tip of Hiddensee island and plugs the navigation channels.

The natural channel system is morphodynamically very active. In 1879, shipping routes of the northern access to the port of Stralsund still followed given passages (Figure 5). In order to facilitate navigation for larger vessels, local authorities decided to dredge straight

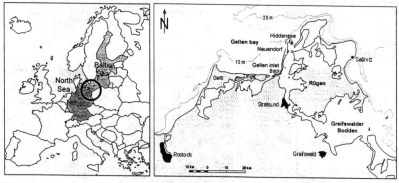

Figure 4: MorWin project domain. Coastal area at Baltic Sea in Germany

waterways at the turn of the century. These are shown on the map of 1986. The dredged material was mainly deposited on the sand bank Bock forming a dam permanently 1.5m above mean sea level which has grown into another barrier island. Continuous dredging of mostly coarse material has been carried out from 1927 to 1944 with an average of $250000m^3$/yr. Until 1970, about $150000m^3$ of fine sand have annually been removed from the waterways. Maintenance dredging in the last 10 years amounts to volumes in the range of 40000 to 130000 m^3/yr.

The diurnal tidal signal of the southern Baltic Sea is in the micro-tidal range of only 20cm. Morphodynamically significant events in the coastal region of Darss-Zingst peninsula and Hiddensee island depend both on the filling state and the local wave and flow conditions. **Water level records** at the gauge of Neuendorf (Figure 6) show a low frequency variation with amplitudes of about 40cm during the 5 year period from 1993 through 1997. These gradual changes result from large scale meteorological conditions and affect the filling state of the Baltic Sea altogether by additional water masses exchanged with the North Sea. Superposition of local wind set-up often causes extreme singular water levels [1] with high water of >100cm above mean sea level which usually have a history of about 2 weeks'

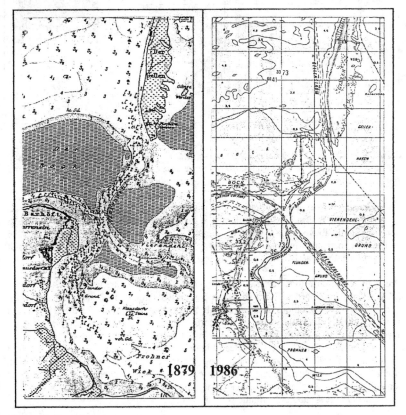

Figure 5: Dredging in channel system

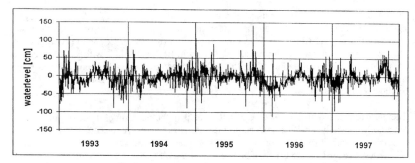

Figure 6: Water level time series at gauge Neuendorf, 1993 - 1997

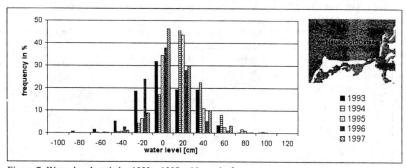

Figure 7: Water level statistics 1993 – 1997 at Neuendorf

duration. Together with an increased wave action they can adversely affect coastal protection works and harbor installations. In those cases sediment is moved across the sand flats and within the channel system depending on the alternating current direction.

The frequency of occurrence (Figure 7) of water levels taken at time intervals of $\Delta t = 6$ hours shows the majority of events during the analysis period as expected near mean sea level or slightly above in the range of 0 - 20cm. There are two exceptions in 1993 and 1996 shifted systematically towards lower water levels in the order of 20 - 40cm which is a result of predominantly easterly winds over long time intervals.

Prototype **wave data** in Gellen bay are available from a wave rider buoy which is located 28 km north of Zingst at Darss sill. Data from 1993 through 1997 have been used for analysis of wave conditions and preparation of boundary conditions for littoral transport modeling. The wave statistics for 1997 (Figure 8) show events of significant wave heights reaching maximum values of 3m. The majority of events lie in the range of 0.4 – 0.8 m. The distribution of wave directions clearly identifies the sectors ENE-E and WSW-WNW as principal directions of wave attack. These are the directions of the longest fetch with according large wave heights. Few events occur from the southern sectors because of shielding effects by the barrier islands and small fetch lengths. An analysis for 1993 through 1996 gives similar distributions for all years.

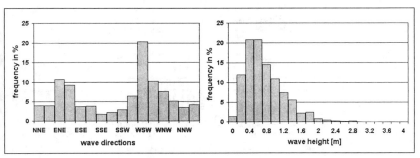

Figure 8: Wave data at Darss sill buoy, 1997

In order to get a global view of the hydrodynamics in the Gellen bight, the repository of topography and hydrographic and meteorological records on a MorWin project server [13] was used to instantiate regional 2D coastal models. The **flow models** operated by the MorWin partners are validated with water level time series of 10 gauges along the coast and within the lagoon system for the entire simulation period. Water levels and current parameters at the open boundaries with the open sea and the wind field over the entire model domain are taken from the operational Baltic Sea circulation model [6] of the Federal Maritime Agency (BSH) in Hamburg.

Morphodynamic models account for the effect of waves being the main driving force of currents and sediment transport. Therefore, different **wave models** [11] [17] which are being used by the individual MorWin partners have been instantiated for the study domain. A wave atlas modeling exercise has been carried out in order to get an impression of the wave fields in the entire Gellen bay corresponding to the wave statistics at the particular location cited in Figure 10. Validation was initially based on near shore wave records off Zingst and is continued at the time of writing with new data from additional locations off

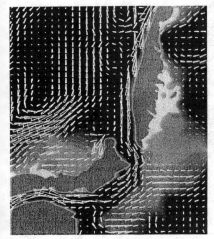
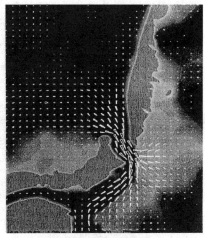

Outflow situation 12.5.1997 10:00   Inflow situation 12.6.1997 03.00
Fig. 9: Flow simulation at Gellen inlet [8]

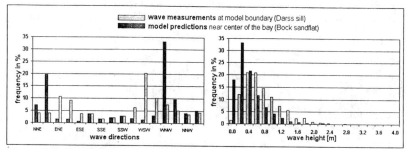

Figure 10: Wave propagation into Gellen bay, 1997
Hiddensee and Bock sandbank.

Two examples from the regional 2D coastal model illustrate typical flow fields in the vicinity of Gellen inlet (Figure 9). The outflow situation on Dec. 5 1997 is characterized by currents which clear the channel of the inlet and continue northward along the coastline of Hiddensee island. Topographically influenced circulation patterns occur in the Gellen bay and in the lagoon system east of Hiddensee where extended areas have fallen dry. Except for the inlet, still water is observed in most of the area 15 hours later. Due to a water level difference at the inlet, there are strong in-going currents with considerable near bottom velocities entering the lagoon system. This flow expands immediately behind the narrow gateway onto the spit. Given a sediment laden water body, this plot would indicate the formation of a sandbank.

According to results from the wave modeling, wave directions measured at the buoy location on Darss sill are transformed due to diffraction and refraction effects. The wave statistics are exemplarily shown in Figure 10 for a reference location near the coastal inlet indicated in Figure 11. In general, measured WSW wave directions transform into WNW and NW directions in the inner bay. Waves from ENE change direction to NE and NNE. The statistical view of the transformed wave heights shows a distinct wave attenuation as they propagate into the bay.

This completes the hydrodynamic systems analysis of Gellen bight which was carried out in close cooperation of numerical modelers at the participating universities and coastal engineers from regional administrative offices in charge of the study domain. The continuous assessment of data used for numerical modeling and instantaneous feedback on simulation results within the interdisciplinary research team has led to a data base of high quality for morphodynamic modeling. The available data are ready for model input on the web site and cover a time span of 5 years which contains a large number of events that appear to be significant for coastal processes.

Case study longshore sediment transport modeling in Gellen bight

Longshore sediment transport rates have been computed with the COSMOS-2D longshore model of HR Wallingford [14] [15] for 10 profiles along the coastline of Darss-Zingst and Hiddensee as indicated in Figure 10. Longshore models can only be applied for straight coastlines. Therefore, the immediate vicinity of the Gellen inlet cannot be studied by such

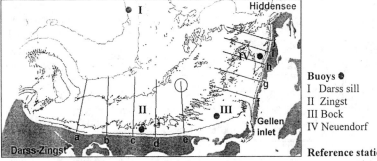

Buoys ●
I   Darss sill
II Zingst
III Bock
IV Neuendorf

Reference station ○

Figure 11: Profiles for long shore modeling at Darss-Zingst peninsula and Hiddensee island

an approach. However, any 2D area morphodynamic model for such regions of interest must be operated with reliable sediment transport boundary conditions. The results from 1D longshore modeling will be used as input data for small scale 2D area modeling and for comparison purposes concerning larger scale morphodynamic models [7].

The profiles extend about 7 km into the bay. The required input parameters are time series of wave height, wave direction, wave period, water level, wind speed and direction to be specified at the seaward end of the profiles. Initial attempts to use archived wave data from the circulation model of the Baltic Sea were abandoned because the grid resolution of the wave model used by the data provider is too coarse in the coastal region of interest. Therefore, boundary data were generated for each profile at 6 hours intervals for the intended 5 years study period based on the wave atlas exercise and the deep water buoy at Darss sill.

The simulation period 1993 –1997 covers several typical situations for the Gellen bay region. First hints of significant sediment transport scenarios are the water level recordings in Figure 6. These indicate different filling states of the Baltic Sea caused by the preference directions of wind forcing together with a correlated prevalent wave climate. The cross shore distributions of longshore sediment transport rates of each simulation year in Figure 12 are distinctively different at the shown locations. Panel b and e correspond to the west-east coast of Darss-Zingst (Figure 11) where the flat shallow beach profiles are dissipative [18] in contrast to the steep profiles located along Hiddensee in north-south orientation shown in the panels f and i which belong to the reflective type.

Darss-Zingst peninsula:
According to Figure 12, westerly transport is significant at profile b and almost negligible at profile e in the years 1993 and 1996. Westward littoral transport depends on NE waves along Darss-Zingst which arrive with smaller wave heights as NNE waves in the vicinity of Gellen inlet. This explains the smaller amplitudes in panel e. Predominant NE waves imply predominantly easterly winds which are in keeping with the water level statistics (Figure 7) for these years. Significant eastward transport is indicated for 1994 and 1997 which can be attributed to westerly waves resulting from westerly winds associated with an enhanced filling state of the Baltic. During 1995, there are several extreme events in spring and fall which have an impact on the sediment transport rates of this year.

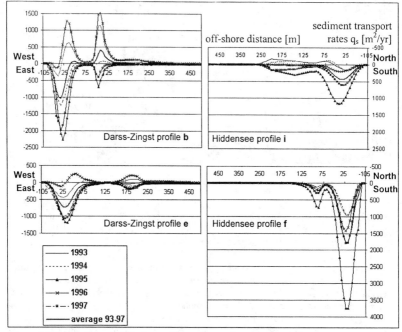

Figure 12: Cross shore distributions of longshore sediment transport rates

Hiddensee island:

There is a small tendency of northward transport at locations close to the northern profile i. Transport near the inlet appears to be directed southward under all conditions. The impact of storm high waters in 1995 exceeds all other results by a factor of 3. According to the 5-year average distributions also shown for all locations, these extreme events considerably influence the local mean values. Longterm studies based on such mean values may tend to overestimate the computed transport rates.

Summing up the total transport across a profile for each time step and adding to it the previous sum gives the time series of cumulative transport (Figure 13) for the entire simulation period. Along Darss-Zingst peninsula, the sediment balance is near equilibrium at profile b. Near the Gellen inlet at profile e, there is continuous eastward transport amounting to $200 \times 10^3 m^3$ for 1993 to 1997. Along Hiddensee island the total southward transport amounts to $400 \times 10^3 m^3$ for this period. Extreme storm high water events can be detected from the cumulative plots as large changes within few time steps. These results will be further used as basis for long term prediction of the evolution of the coastline for the sake of coastal protection planning and dredging strategies for the navigation channels.

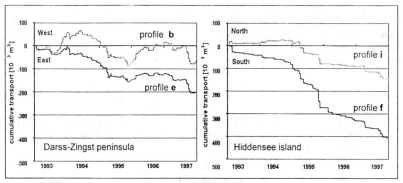

Figure 13: Cumulative sediment transport 1993 - 1997

Conclusions

In this pilot project an attempt is made to install tools and perform modeling in an internet based working environment for data management, modeling systems and information exchange. The area of interest is a coastal region in the Baltic Sea with a negligible tidal motion and the focus is on wind-generated flats nourished by wave-generated littoral drift with an influence on wide-area coastal protection and navigability of harbor access channels.

Project results from 3 years indicate that

- the concept of a distributed working environment based on Java technology works well if everybody involved obeys the rules of making all results available for discussion firstly within the project and later on outside,
- data mining, data assessment by modelers as well as by local coastal engineers and data distribution play a major role in a successful modeling exercise,
- the selective utilization of a suite of 1D- and 2D-models for various aspects of the study and the respective boundary value exchange is an important feature for this project, and
- the definition of high sediment transport scenarios in a non-tidal environment for predictions of long-term evolution of the coastline will have to take into account the results of several model studies as shown above.

Coastal zone management uses the knowledge of the coastal processes as dealt with in this study only as a basis. In addition, many more disciplines such as geology, ecology, biology and socio-economic sciences need to be taken into account. It is evident that a working environment as presented here is the ideal basis for a functioning integrated coastal zone management.

Acknowledgements

The funding for this ongoing research is provided by the German Federal Ministry of Education, Science, Research and Technology under grant BEO52 03 KIS 3120. In addition to the quoted sources data have been acquired from the Federal Maritime Agency (BSH) in

Rostock and the German Meteorological Service in Offenbach (DWD). The authors thank Frank Sellerhoff, Axel Schwöppe and Dr. Peter Milbradt at Hannover University for contributions concerning the collaborative working environment and 2D area modeling of currents and waves. Long shore modeling has been carried out in cooperation with Dr. Howard N. Southgate and Dr. Andy H. Peet at Hydraulic Research Wallingford.

References

[1] Baerens Ch et al.: Zur Häufigkeit von Extremwasserständen an der deutschen Ostsee-küste, Teil I: Sturmhochwasser. Spezialarbeiten aus der AG Klimaforschung des Meteoro-logischen Instituts der Humboldt-Universität zu Berlin, No. 8, 1994

[2] DeVriend H et al., 1993. Approaches to long-term modelling of coastal morphology: a review. Coastal Engrg, 21, pp 225-269

[3] Dublin Core Metadata Element Set: http://purl.oclc.org/dc

[4] European Commission (ed), 1995. MAST G8M Coastal Morphodynamics. Final Overall Meeting, Gdansk/Poland.

[5] ht://Dig -- Internet search engine software: http://www.htdig.org

[6] Kleine E, 1994. Das operationelle Modell des BSH für die Nordsee und Ostsee. Bundesamt für Seeschiffahrt und Hydrographie, Hamburg.

[7] Lehfeldt R, Milbradt P, 2000. Longshore Sediment Transport Modeling in 1 and 2 Dimensions. Accepted at International Conference on Hydroscience and - Engineering, ICHE4, Seoul.

[8] Lehfeldt R, Barthel V, 1999. Numerische Simulation der Morphogenese von Wind-watten. Die Küste, 60, pp 257- 276

[9] Lehfeldt R, Schleider O, 1998. Morphodynamic Modeling of Wind Influenced Flats - Internet Based Collaborative Project Handling in Coastal Engineering. Milestone 1997. http://morwin.wsd-nord.de

[10] Lehfeldt R, Holz P, 1992. Model Validation in X-Environment. Proc. 9th Int. Conf. Computational Methods in Water Resources, Denver, pp 733-740

[11] Milbradt P, 1995. Zur Mathematischen Modellierung großräumiger Wellen- und Strömungsvorgänge (Dissertation) Institutsreihe des Inst. f. Bauinformatik, Universität Hannover.

[12] Molkenthin F and Holz K-P, 1998. Working Process in a Virtual Institute. Hydro-informatics '98. Proceedings of the 3rd International Conference on Hydroinfor-matics, Copenhagen: Balkema. pp 941-948

[13] MorWin web site http://morwin.wsd-nord.de

[14] Nairn RB, Southgate HN, 1993, Deterministic profile modelling of nearshore processes. Part 2. Sediment transport and beach profile development. Coastal Engrg 19, pp 57-96

[15] POET, 1998. http://www.poet.com/products/oss/oss.html

[12] Reinhard H, 1953. Der Bock, VEB Geographisch.Kartographische Anstalt Gotha.

[13] Sellerhoff F, to appear. Ein punktbasiertes Informationsmodell in Anwendung auf das Küsteningenieurwesen (Dissertation) Institutsreihe des Inst. f. Bauinformatik, Universität Hannover.

[14] Sellerhoff F, Lehfeldt R, Milbradt P, 1998. Model Validation in Web Environment - Progress in Distributed Modeling, in K.-P. Holz et al. (eds): Advances in Hydro-Science and–Engineering, Vol III, ICHE98, Cottbus/Berlin. Center for Computational Hydroscience and Engineering, The University of Mississippi.

[15] Southgate HN, Nairn RB, 1993. Deterministic profile modelling of nearshore processes. Part 1. Waves and currents. Coastal Engrg 19, pp 27-56

[16] Spaulding ML, Sankaranarayanan S, Erikson L, Fake T and Opishinski T, 1997. COASTMAP, An integrated system for monitoring and modeling of coastal waters: Application to Greenwich Bay. Proc. 5th International Conference on Estuarine and Coastal Modeling, Alexandria, pp 231-251

[17] SWAN Simulating WAves Nearschore, TU Delft. http://swan.ct.tudelft.nl

[18] Van Rijn LC, 1998. Principles of Coastal Morphology. Aqua Publications, Amsterdam, The Netherlands.

An Improved Formulation for the Bed Shear Stress in Morphodynamic Simulations

Andreas Malcherek [1]

Abstract

Morphodynamic models are usually consisting of a depth integrated hydrodynamic module which is coupled with Exners bed evolution equation. The sediment transport rates are calculated using classical sediment transport formulas which need the bed shear stress from the hydrodynamics as an input. Therefore a good representation of the bed shear stress is important for long term simulations of morphodynamics.

In this paper a formulation for the bed shear stress is presented which takes into account the effect of the depth averaged velocity as well as its horizontal gradients. The influence of the latter is very important when steep bottom gradients are present.

This approach is applied in a long term simulation of the Elbe estuary. At its mouth changes of the wadden sea bed topography in the order of one meter per year are observed which are occurring at a steep side wall of a secondary channel.

Introduction

The simulation of morphodynamics in estuaries and coastal areas is because of its economical merit on one hand and of its scientific complexity on the other an outstanding field of research and development. The Federal Waterways Engineering and Research Institute (BAW) is the central advisory institute for the German waterway authority. It is therefore responsible to provide the scientific support for optimal sediment management procedures for waterway maintenance.

The hydrodynamic analysis of estuarine systems using numerical models is the main approach to support engineering decisions on a scientific basis. Although the application of three dimensional models to engineering problems is increasing and well validated cases are documented, it is restricted due to large computational costs. Therefore in long term simulations of morphological changes principally two dimensional depth averaged models are applied [3].

Here a verification of the simulated free surface elevation can be regarded as the technical standard. Hitherto this validation standard is not achieved in morphodynamic modeling because a lot of general as well as specific problems have not been solved yet.

[1]Research Engineer, Federal Waterways Engineering and Research Institute (BAW), Coastal Department, Wedeler Landstr. 157, 22559 Hamburg, Germany

General problems are on the one hand mainly connected with the heterogeneity of the material itself resulting in a broad variability of processes and phenomena.

From a hydrodynamical point of view an insufficient representation of the flow field can spoil a morphodynamic simulation. One reason for this lack of quality is the fact that the calibration of the model is often done only with respect to the water levels. In order to achieve good results for the free surface elevation, the bottom roughness as well as the turbulent dispersion coefficient are used for the calibration, because both physical processes are connected with energy dissipation. This ambiguity may result in an over- or underestimation of the bed shear stress which is crucial for the morphodynamics.

From a theoretical point of view it is possible to determine the bed shear stress exactly when the three dimensional velocity field in the bottom boundary layer is known. But this is only the case using the direct simulation technique which up to now is not applied to such complex situations as they occur in sediment transport problems [7].

Because of this and historical reasons empirical bed shear stress formulations are applied in three dimensional as well as depth averaged modeling. Standard formulations used for engineering purposes are Chezy's, Mannings, Stricklers, and Nikuradse's law, the three latter can be derived assuming that the vertical velocity distribution is obeying a logarithmic law.

None of these formulations process information about horizontal velocity gradients as they are important at steep bottom slopes. As a matter of fact there are bed load formulas which take into account the effect of gravity on bottom slopes [5]. But the effect on the bed shear stress is usually neglected in morphological models. This papers attention is therefore dedicated to an improved formulation for the bed shear stress which is derived from the viscous stress tensor.

Morphology at the Mouth of the Elbe Estuary

The Elbe estuary is located at the southern coast of the North Sea and connects the harbor of Hamburg with the remaining world. Its tidal range is generally between 2.1 and 3 m. The estuary reach of approximately 140 km is bounded at the upstream end by the Weir of Geesthacht. The estuary is 15 km wide at the mouth and 300 m upstream at the weir. The minimum low water depth in the navigation channel is approximately 13.6 m (1992) up to the harbor of Hamburg. The mean annual river flow rate at the weir is 700 m^3/s.

The morphological situation at the mouth of the estuary can be characterized as follows (figure 1). The navigation channel is touching the harbor of Cuxhaven in a southbound curve. The tidal flats which are part of the German wadden sea National park are supplied with tidal water volume by a large tidal channel called the Klotzenloch. A secondary channel the so-called Medem Rinne has twisted in the seventies from the main channel. A strong erosion can be found at the north side walls of the channel with a counterpart deposition at its southern side. This northbound movement of the channel is related to a certain risk, that the Medem Rinne will connect with the Klotzenloch forming a stable bypass to the Elbe waterway. An upper limit for the deflection of the sediment transport by secondary currents can be estimated using Engelunds formula [2]

$$\tan \delta = 7 \frac{h}{R}$$

as 1.5 degrees. Therefore secondary currents can not be the reason for the shift of the channel.

The sediments in this area are non-cohesive fine sands. Only at some places a high silt content can be found. For the following simulations a uniform grain size distribution with a median value of $d_{50} = 0.15\ mm$ is taken. The slopes at the northern side walls of the Medem Rinne are very steep. The curvature of the Medem Rinne is low, the corresponding curve radius is everywhere larger than 2 km, therefore secondary currents are not expected to play any role in the morphology of the system.

The Numerical Model TRIM-2D

The hydrodynamical model TRIM-2D [1] is used in this project. This code solves the two-dimensional Saint-Venant equations using a finite difference scheme on a staggered Arakawa-C-grid. Because the time stepping for the free surface is semi-implicit and the advection terms are solved applying the Lagrange technique the code is very stable. For morphodynamic simulations Exners bed evolution equation

$$(1 - n)\frac{\partial z_B}{\partial t} + \operatorname{div} \vec{q_s} = 0 \tag{1}$$

is modeled whereby $z_B(x, y, t)$ is the vertical coordinate of the bottom, n is the porosity and $\vec{q_s}$ (in $[m^2/s]$) is the sediment transport rate. The Saint-Venant and Exners equation are solved every time step.

Hydrodynamics and morphodynamics are coupled by the sediment transport rate q_s which is under saturated transport conditions the sediment transport capacity. The following studies are performed using the van Rijn formula [8]

$$q_S = 0.053\sqrt{\frac{\varrho_S - \varrho}{\varrho}gd_{50}^3}\,D_*^{-0.3}T^{2.1}$$

T is the dimensionless bed-shear parameter defined as

$$T = \frac{\tau_B - \tau_c}{\tau_c}$$

and the critical bed shear stress τ_c is calculated according to Shields. D_* is the dimensionless grain size parameter

$$D_* = \left(\frac{\varrho_s - \varrho}{\varrho}\frac{g}{\nu^2}\right)^{1/3}d_{50}$$

The scalar transport rates are transformed to two-dimensional fluxes assuming that the transport direction is parallel to the depth averaged flow direction. Afterwards the gravitational effect of slopes is considered by changing the vectorial bed load transport rate to:

$$\vec{q_S} = \vec{q_S} - \beta\operatorname{grad} z_B\|q_S\| \tag{2}$$

According to *Koch* and *Flokstra* [4], [9] the parameter β lies between 0.6 and 1.3.

An Improved Formulation for the Bed Shear Stress

As many other bed load transport formulas the one of van Rijn needs the bed shear stress as an input calculated from the hydrodynamical part of the code. For a Newtonian Fluid the shear stress acting in a certain sectional area can be determined as the projection of the viscous stress tensor

$$P = \varrho\nu \left(\frac{\partial u_i}{\partial x_j} + \frac{\partial u_j}{\partial x_i} \right)_{(i,j)}$$

onto its normal. For the bed this can be done by multiplying the stress tensor with the normal

$$\vec{n_B} = \frac{1}{\sqrt{1 + \left(\dfrac{\partial z_B(x,y)}{\partial x} \right)^2 + \left(\dfrac{\partial z_B(x,y)}{\partial y} \right)^2}} \begin{pmatrix} \dfrac{\partial z_B}{\partial x} \\ \dfrac{\partial z_B}{\partial y} \\ -1 \end{pmatrix}$$

of the bottom: $\vec{\tau_B} = -P\vec{n_B}$. The result can easily be calculated as

$$\vec{\tau_B} = \frac{\varrho\nu}{\sqrt{1 + \left(\dfrac{\partial z_B(x,y)}{\partial x} \right)^2 + \left(\dfrac{\partial z_B(x,y)}{\partial y} \right)^2}}$$

$$\begin{pmatrix} \dfrac{\partial u}{\partial z} + \dfrac{\partial w}{\partial x} - 2\dfrac{\partial u}{\partial x}\dfrac{\partial z_B}{\partial x} - \left(\dfrac{\partial u}{\partial y} + \dfrac{\partial v}{\partial x} \right) \dfrac{\partial z_B}{\partial y} \\ \dfrac{\partial v}{\partial z} + \dfrac{\partial w}{\partial y} - \left(\dfrac{\partial v}{\partial x} + \dfrac{\partial u}{\partial y} \right) \dfrac{\partial z_B}{\partial x} - 2\dfrac{\partial v}{\partial y}\dfrac{\partial z_B}{\partial y} \\ 2\dfrac{\partial w}{\partial z} - \left(\dfrac{\partial w}{\partial x} + \dfrac{\partial u}{\partial z} \right) \dfrac{\partial z_B}{\partial x} - \left(\dfrac{\partial w}{\partial y} + \dfrac{\partial v}{\partial z} \right) \dfrac{\partial z_B}{\partial y} \end{pmatrix}$$

The components of the shear stress in x- and y-direction can now be determined through the multiplication with the respective unit tangent vectors. Neglecting all the products of derivatives and the vertical velocity w this equation collapses to the well known form

$$\vec{\tau_B} = \varrho\nu \begin{pmatrix} \dfrac{\partial u}{\partial z} \\ \dfrac{\partial v}{\partial z} \\ 0 \end{pmatrix}$$

which is true for a plane bed. In depth averaged models the vertical velocity profile is finally approximated using the assumption of a logarithmic velocity profile.

In our case the horizontal velocity profiles at steep bottom gradients seem to have an enormous effect on the morphodynamics of the system. Therefore we cannot neglect the terms including products of flow velocity and bottom gradients.

The vertical velocity gradients are estimated using Stricklers law. Further it is assumed that the horizontal bottom velocity gradients are proportional to the gradients of the vertically averaged velocities, i.e. the horizontal components of the shear stress can be calculated as

$$\vec{\tau_B} = \begin{pmatrix} \dfrac{\varrho g}{k_{Str}^2 h^{1/3}} \overline{u}\sqrt{\overline{u}^2 + \overline{v}^2} - \varrho\nu_M \left(2\dfrac{\partial \overline{u}}{\partial x}\dfrac{\partial z_B}{\partial x} + \left(\dfrac{\partial \overline{u}}{\partial y} + \dfrac{\partial \overline{v}}{\partial x} \right) \dfrac{\partial z_B}{\partial y} \right) \\ \dfrac{\varrho g}{k_{Str}^2 h^{1/3}} \overline{v}\sqrt{\overline{u}^2 + \overline{v}^2} - \varrho\nu_M \left(\left(\dfrac{\partial \overline{v}}{\partial x} + \dfrac{\partial \overline{u}}{\partial y} \right) \dfrac{\partial z_B}{\partial x} - 2\dfrac{\partial \overline{v}}{\partial y}\dfrac{\partial z_B}{\partial y} \right) \end{pmatrix} \tag{3}$$

Two effects are determining the value of the morphological viscosity parameter ν_M. First it compensates the replacement of the exact velocity gradients at the bottom by the depth averaged velocities. Second the derivatives are replaced in a numerical model by finite differences. When compared to their exact values the gradients of the bottom and the velocity are much smaller when calculated with finite differences. This must be compensated by the morphological viscosity parameter ν_M. Unfortunately ν_M can not be a constant because the gradients are more or less well represented at different positions by finite differences.

Simulation Results for the Elbe Estuary

The Elbe estuary is discretized using quadrilateral elements with a grid size of 50 m. This results in a total number of over 320 000 elements for the whole estuary. The simulation is carried out over five months choosing a time step of two minutes. The model is validated comparing the simulation results for the free surface elevation with 23 tidal gauges located over the whole reach of the estuary.

In order to validate the morphodynamics, the coefficient ν_M and and the coefficient β for the gravitational transport are calibrated. The following results are obtained using the values $\beta = 15$ and $\nu_M = 7.5 \ m^2/s$. Therefore the parameter for the gravitational transport is much higher than the values reported in literature. This discrepancy can be explained by the fact that the bottom slope is calculated with linear finite differences. On a finite discretization this leads to an underestimation of the bottom slope in equation (2) which has to be compensated by a higher parameter for the gravitational transport.

Fig. 3 compares the results for the bed shear stress in the Medem Rinne using the Strickler law and the extended formulation (3). It can be seen that the shear stress calculated with Stricklers law is zero at the slopes because of vanishing velocity. This behavior can also be observed all the time of the tidal cycle. As a result the sediment transport rates are also zero because the actual shear stress does not exceed the critical shear stress. Therefore no change of the bathymetry can be found at the northern edges of the Medem Rinne.

The results for the bed shear stress using the extended formulation show in the deeper parts of the estuary the same behavior as Stricklers law. But higher bed shear stresses can be seen at the northern edge of the Medem Rinne and at the southern bank of the Elbe waterway. Therefore it can be expected that the sediment transport rates are now higher at steep bottom gradients.

The simulated bottom evolution after five months is shown in figure 4. A high erosion takes place at the northern edge of the Medem Rinne. This material is deposited in the channel. Here the simulation agrees well with the natural morphodynamics. A large amount of erosion can also be seen at the southern bank of the Elbe estuary which cannot be found in the measurements. In reality this bank is protected by groines which are as a matter of fact unerodable. Unfortunately the model does not contain an algorithm for the bed load transport over rigid beds although stable ones can be found in literature [6].

In the shipping channel a high accumulation of sediment is simulated which stem from the side walls. This discrepancy can further be explained by the fact that dredging is done in this area.

Conclusions

A new formulation for the bed shear stress is derived which takes into account the influence of the vertical as well as the horizontal velocity profile. The vertical velocity gradients are modeled using the assumption that the velocity profile is logarithmic.

The horizontal gradients are replaced by the gradients of the depth averaged velocity whereby the difference between both is compensated by a new parameter. It is dependent on the grid resolution because of the underestimation of the bottom as well as the velocity gradients using finite approximations. This parameter can be used to calibrate the results for the morphological development at steep bottom slopes.

The application of the new shear stress formulation in a model of the Elbe estuary improved the results for the morphological developments significantly.

References

[1] Casulli, V. Semi-implicit finite difference methods for the two-dimensional shallow-water equations. *J. Comp. Phys.*, 86:56–74, 1990.

[2] Engelund, F. Flow and bed topography in channel bends. *J. Hydr. Div., ASCE*, 100(11):1631–1648, 1974.

[3] Jia, Y. and Wang, S.S.Y. Numerical Model for Channel Flow and Morphological Change Studies. *J. Hydr. Eng.*, 125 (9):924–933, 1999.

[4] Koch, F.G. and Flokstra, C. Bed Level Computations for Curved Alluvial Channels. Reprint from the XIXth Congress of the IAHR, New Dehli, India HE45/92.01, Électricité de France, Direction des Études et Recherches, Laboratoire National d'Hydraulique, Chatou, 1981.

[5] Kovacs, A. and Parker, G. A new vectorial bedload formulation and its application to the time evolution of straight river channels. *J. Fluid Mech.*, 267:153–183, 1994.

[6] Malcherek, A. Application of Bed Evolution Models Over Loose and Rigid Areas. *International Journal of Sediment Research*, 12(2):291–299, December 1997.

[7] Moin, P. and Mahesh, K. Direct Numerical Simulation: A Tool in Turbulence Research. *Annu. Rev. Fluid Mech.*, 30:539–578, 1998.

[8] Rijn, L.C. van. *Principles of Sediment Transport in Rivers, Estuaries and Coastal Seas*. Aqua Publications, Amsterdam, 1993.

[9] Tanguy, J.-M., Zhang, B., and Hamm, L. A New Lax-Wendroff Algorithm to Solve the Bed Continuity Equation With Slope Effects. In Spaulding, M.L., Cheng, R.T., Bedford, K, Blumberg, A., and Swanson, C., editors, *Estuarine and Coastal Modeling, Proceedings of the 3th International Conference in Chicago 1993*. ASCE, 1993.

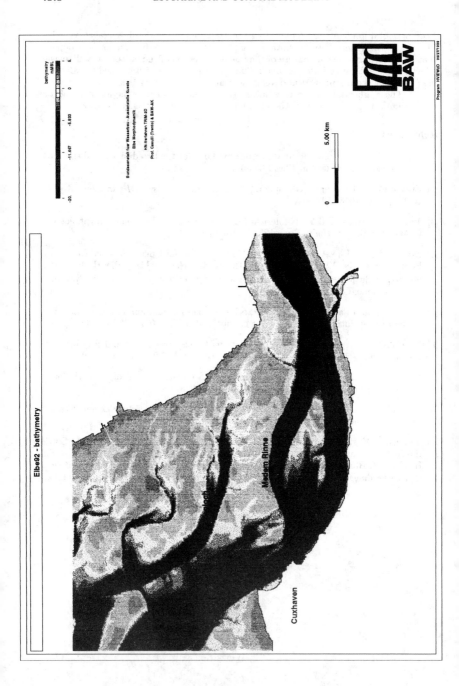

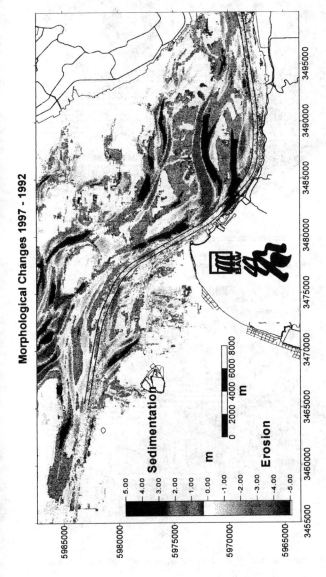

Figure 2: Morpological changes during 5 years

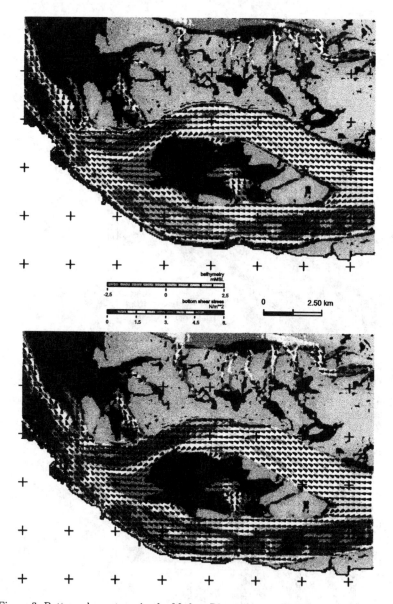

Figure 3: Bottom shear stress in the Medem Rinne. Above: Strickler law, below: improved formulation

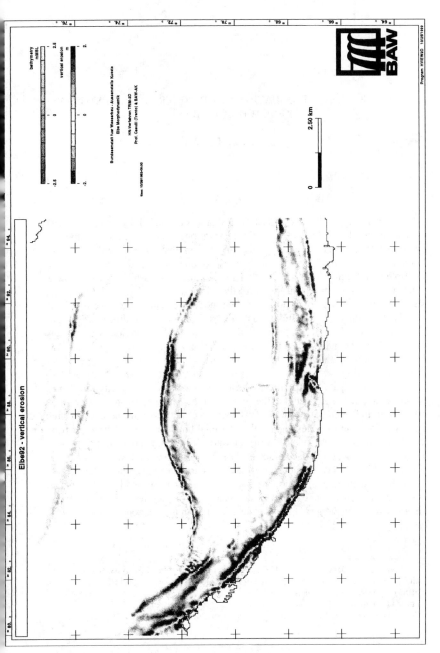

Figure 4: Morphological changes after five months simulation time

MODELING TIDAL CIRCULATION IN A BARRIER-ISLAND ESTUARY: APALACHICOLA BAY

W. Huang[1], K. Jones[2], T.S. Wu[3], H. Sun[4]

ABSTRACT

Apalachicola Bay is a shallow estuary surrounded by several barrier islands in the Florida Panhandle. Field observations indicate that tidal amplitude in the east boundary is considerably larger than that in the south and west boundaries. The circulation in the bay is complex due to the interactions of multiple tidal forcing from east, south, and west directions, while the river flow is discharged into the bay from the north direction. In this study, a calibrated and validated, three-dimensional hydrodynamic model was applied to investigate the circulation induced by multiple tidal forces from different directions in this shallow estuary. In order to focus on the tidal-induced circulation, we used the observed tidal boundary conditions for the period of May-July of 1993 and the historic minimum daily river inflow, but removed wind forcing. Snapshots of model simulations at high, ebb, low, and flood tidal conditions were used to characterize the spatial distributions of surface elevations and velocity in the bay. Model simulations indicated that the currents in the bay were strongly affected by the spatial and temporal variations of the surface elevations in the bay. At high tide, the higher surface elevation in the east tidal boundary drove bay water from east to the west. At ebb tide, majority of water at mid bay moved out of the bay following the steepest surface gradient through West Pass. At low tide, the lower water level in the East Pass caused strong eastward currents in the eastern region of the bay. At flood tide, the lowest water level occurred at mid bay. As a result, flood tidal currents from east, south, and west tidal boundaries and the southward river discharge converged at mid bay, and resulted in mixing zones in the estuary.

[1] Civil Engineering Department, FAMU-FSU College of Engineering, 2525 Pottsdamer St., Tallahassee, FL 32310-6046. Phone (850)-410-6199;E-mail: whuang@eng.fsu.edu
[2] PBSJ Engineering Inc., Tallahassee, FL
[3] Florida Department of Environmental Protection Agency, Tallahassee, FL.
[4] Department of Geological and Marine Sciences, Rider University, Lawrenceville, NJ.

INTRODUCTION

Apalachicola Bay is a barrier island estuarine system located in the Florida panhandle. The bay was formed by deltaic processes of the Apalachicola River and is situated on a prominent point of land that extends onto the Gulf of Mexico continental shelf (Figure 1). It is a highly productive estuarine system, which supports a diverse and abundant commercial and recreational fishery. It provides important nursery and feeding grounds for large varieties of commercial and non-commercial fish and shellfish. (Livingston 1984). The bay has the third largest catch of shrimp statewide and accounts for 90 percent of Florida's oyster harvest and 10 percent of the nation's. The total aquatic area of Apalachicola Bay, including East Bay, St. Vincent Sound, and St. George Sound covers over 450 km^2 (112,000 acres). The study area extends from the western tip of St. Vincent Sound called Indian Pass east to the eastern tip of Dog Island. It is bounded to the south by St. Vincent Island, Little St. George and St. George Island and Dog Island. The study area is approximately 63 km long and 12 km wide at the widest point. The bay is a shallow estuarine system with an average 2 – 3 m depth. It is connected to the Gulf through four natural openings (Indian Pass, West Pass, East Pass, and Lanard Reef), and one man-made pass (Sikes Cut). Sikes Cut was dredged by the United States Corps of Engineers in 1954, and continues to be maintained on a regular basis. West Pass and Sikes Cut connect Apalachicola Bay with the Gulf of Mexico, Indian Pass is the opening into St. Vincent Sound, and East Pass is the large pass from the gulf into St. George Sound. The southward freshwater discharge from Apalachicola River is perpendicular to the long axis the estuary, acting like a buoyant jet discharged into cross flows affected by tidal forcing. Understanding the characteristics of tidal circulation is essential for studying the dynamics of freshwater discharge and the transport processes in the bay.

The circulation dynamics in Apalachicola Bay is complex due to the effects of multiple tidal forces from different directions, with East Pass in the east, West Pass and Sides Cut in the south, and Indian Pass in the west. It is the goal of this study to characterize the tidal circulation in this shallow barrier island estuary. Results from this study would be helpful for the environmental study and water resource management in this important estuary.

TIME SERIES OBSERVATIONS OF SURFACE ELEVATIONS

Apalachicola bay is in the region of the Gulf of Mexico where the tide signal changes from one that is mixed semi-diurnal/diurnal in the east to the one where the semi-diurnal signal is less strong, ultimately changing to a predominately diurnal tide further west. Observations of surface elevations at East

Pass, West Pass, Indian Pass, and S397 gage are given in Figure 2 for the period of 7/2/1993-7/5/1993. Because of tidal variations at the five tidal openings east, south, and west directions, there is a complicated interaction of the water levels in the bay. Tidal amplitudes is strongest in the east with the energy diminishing as the wave moves west due to the friction in the shallow bay. Observations of surface elevation (Figure 2) show that the tidal amplitude decreases from the east opening at East Pass to the west opening at Indian Pass. At the maximum high tide (at 12 hr), the surface elevation at East Pass is about 12 cm higher than that of West Pass, and 21 cm higher than that of Indian Pass. At the minimum low tide (at 20 hr), surface elevations at both Indian Pass and West Pass are 18 cm higher than that of East Pass. The maximum high and low tides are approximately in phase at the openings of East Pass, West Pass, and Indian Pass. However, the secondary high tide in East Pass (at 26 hr) has about a 3-hr lead compared to the tides in the West Pass and Indian Pass. The tidal signal at tidal gage S397 in the bay shows a pattern similar to that in the East Pass, but is lagged by about 2 hrs. Because there are considerably differences in the maximum daily tidal amplitude and the phases of the semi-diurnal tide between the tidal signals in the east and the west openings, an east-west horizontal gravity gradient is formed and varies following the changes of tidal forcing in the ocean openings. The maximum horizontal gravity gradient occurs approximately in the diurnal maximum high and minimum low tide.

HYDRODYNAMIC MODEL DESCRIPTION

In order to investigate the circulation in the Apalachicola Bay, the Princeton Ocean Model (POM) by Blumberg and Mellor (1987) was applied to Apalachicola Bay. It is a semi-implicit, finite-difference model that can be used to determine the temporal and spatial changes of surface elevation, salinity, temperature, and velocity in response to wind, tide, buoyancy, and Coriolis forces. The model solves a coupled system of differential, prognostic equations describing conservation of mass, momentum, heat and salinity at each horizontal and vertical location determined by the computational grid. This model incorporates a second-order turbulence closure sub-model that provides eddy viscosity and diffusivity for the vertical mixing (Mellor and Yamada, 1982). This model has a history of successful applications in other estuaries; for example, Oey et al., (1985a,b,c) for the Hudson-Raritan estuary, Blumberg, and Goodrich, (1990) for Chesapeake Bay, Galperin and Mellor, (1990a,b) for Delaware Bay, River and adjacent continental shelf, and Blumberg and Galperin, (1990) for New York Bight. In all of these studies, the model performance was assessed via comparisons with data and a confidence has been established that the model realistically reproduces the predominant physics. The model is capable of simulating time-dependent wind and multiple river inputs, and a variety of other forcing conditions. An important feature of the present model is the use of a horizontal orthogonal, curvilinear coordinate system that allows one to better represent coastline irregularities in Apalachicola Bay system. The computational model grid for Apalachicola Bay is given Figure 2. Details of model descriptions were discussed by Blumberg and Mellor (1987), and the enhanced version of the curvilinear coordinate formulation is given by Blumberg and Galperin (1990). Major governing equations in the model are given below.

Continuity equation

$$\frac{\partial}{\partial \zeta_1}(h_2 U_1) + \frac{\partial}{\partial \zeta_2}(h_1 U_2) + h_1 h_2 \frac{\partial W}{\partial z} = 0, \tag{1}$$

Momentum equation

$$\frac{\partial}{\partial t}(h_2 U_1) + \frac{1}{h_1}\frac{\partial}{\partial \zeta_1}[(h_2 U_1)U_1] + \frac{1}{h_2}\frac{\partial}{\partial \zeta_2}[(h_2 U_1)U_2] + \frac{\partial}{\partial z}[(h_2 U_1)W]$$

$$+ U_2 \left(\frac{2U_1}{h_1}\frac{\partial h_1}{\partial \zeta_2} - \frac{U_1}{h_2}\frac{\partial h_2}{\partial \zeta_2} - \frac{U_2}{h_1}\frac{\partial h_2}{\partial \zeta_1} \right) - U_2 f h_2$$

$$= -\frac{1}{\rho_0}\frac{h_2}{h_1}\frac{\partial P}{\partial \zeta_1} + \frac{\partial}{\partial z}\left[\overline{-(h_2 u_1)w} \right] + F_1 h_2 \tag{2}$$

$$\frac{\partial}{\partial t}(h_1 U_2) + \frac{1}{h_1}\frac{\partial}{\partial \zeta_1}[(h_1 U_2)U_1] + \frac{1}{h_2}\frac{\partial}{\partial \zeta_2}[(h_1 U_2)U_2] + \frac{\partial}{\partial z}[(h_1 U_2) W]$$

$$+ U_1 \left(\frac{2U_2}{h_2}\frac{\partial h_2}{\partial \zeta_1} - \frac{U_2}{h_1}\frac{\partial h_1}{\partial \zeta_1} - \frac{U_1}{h_2}\frac{\partial h_1}{\partial \zeta_2} \right) + U_1 f h_1$$

$$= -\frac{1}{\rho_0}\frac{h_1}{h_2}\frac{\partial P}{\partial \zeta_2} + \frac{\partial}{\partial z}\left[\overline{-(h_1 u_2')w'} \right] + F_2 h_1 \tag{3}$$

U_1 and U_2 are the horizontal velocities and W is the vertical velocity calculated from continuity. ζ_1 and ζ_2 are horizontal curvilinear orthogonal coordinates, z is the vertical coordinate, h_1 and h_2 are metric coefficients, P_{atm} is the atmospheric pressure, and f is the Coriolis parameter. The terms F_1 is related to the horizontal mixing processes and is parameterized as horizontal diffusion terms. The Reynolds stresses $\overline{u_1'w'}$ and $\overline{u_2'w'}$ are evaluated using the level 2½ turbulence closure model of Mellor and Yamada (1982) modified by Galperin et al. (1988)

The salinity and temperature equations:

$$\frac{\partial(S,T)}{\partial t} + \frac{\partial U_1(S,T)}{\partial \zeta_1} + \frac{\partial U_2(S,T)}{\partial \zeta_2} + \frac{\partial W(S,T)}{\partial z}$$

$$= A_H\left[\frac{\partial^2(S,T)}{\partial \zeta_1^2} + \frac{\partial^2(S,T)}{\partial \zeta_2^2}\right] + \frac{\partial}{\partial z}\left[K_v\frac{\partial(S,T)}{\partial z}\right] \tag{4}$$

where S is the salinity and T is the temperature. K_v is the eddy diffusivity for salt and temperature, which is calculated from a second order turbulent model (Mellor and Yamada, 1982). Density is a function of temperature and salinity calculated from the equation of state. The horizontal viscosity and diffusivity coefficients A_H are calculated according to the Smagorinsky (1963) formulation where the coefficient c is set to 0.05 for both parameters (Equation 3).

$$A_m, A_H = c\Delta x\Delta y\left[\left(\frac{\partial U_1}{\partial \zeta_1}\right)^2 + \left(\frac{\partial U_2}{\partial \zeta_2}\right)^2 + \frac{1}{2}\left(\frac{\partial U_1}{\partial \zeta_1} + \frac{\partial U_2}{\partial \zeta_2}\right)^2\right]^{\frac{1}{2}} \tag{5}$$

MODEL CALIBRATION AND VERIFICATION

The three-dimensional hydrodynamic model of Apalachicola Bay was previously calibrated and validated by Huang and Jones (1998). The model was calibrated for the period of June, and validated for the period of July –November of 1993 by using field observations of hourly surface elevation, salinity, and velocity at several stations in the bay. Model coefficients (bottom drag coefficient, bottom roughness, horizontal diffusion and viscosity, surface wind drag coefficient, time step, vertical and horizontal grid) were selected to minimize the difference between model predictions and observations. The model employs the horizontal grids as given in Figure 2 and five vertical layers. The time step used in model simulation is 2 minutes for internal mode. The model predictions of hourly water level and salinity compare well with observations in the model calibration. Model predictions of surface elevation at station S397 of mid bay match well with field observations with a correlation value of 0.99.

TIDAL CIRCULATION MODELING

The validated hydrodynamic model was used to study the tidal circulation in the bay during a tidal cycle on July 2 of 1993. The residual non-tidal water levels were negligible since no strong wind was observed during the study period. Since the tidal-induced circulation can be better observed in the minimum wind and river forcing conditions, we used zero wind condition and minimum historic river inflow (1939-1993) of 157 m^3/s in this study. Observed water levels were specified at the ocean boundaries. A 60-day spin-up simulation was done using observed tidal boundary conditions to setup appropriate initial conditions. Model simulations given below provide snapshots of velocity and surface elevation field throughout the bay at high tide (12:00, EST), ebb tide (16:00), low tide (20:00), and flood tide (24:00) (Figure 3).

High tide (12:00)

High tide (Figure 4) is a short period of slack water as the water level changes from rising to falling. The tide was in phase, and the velocities were generally low at all bay openings. However, due to the difference of tidal amplitude in the east and west boundaries and the shallow water friction, the surface elevation gradually decreased from East Pass, through mid bay, to the west tidal boundaries at West Pass and Indian Pass. The difference of surface elevation was about 12 cm between East Pass and West Pass, and about 20 cm between East Pass and Indian Pass. The strongest gradient of surface elevation was located near tidal gage S397 where the width is reduced by one-third due to the existence of an island. The horizontal gravity gradient results in westward currents in the bay. Surface velocity was about 40 cm/s in the St. George Sound, and about 20-30 cm/s in the St. Vincent Sound. Strongest currents were near the tidal gage S397. Weaker currents were in the western mid bay where the westward tidal currents from the St. George Sound met the southward currents discharged from that Apalachicola River. Due to the westward current effects, the southward river discharge deviated to the west and southwest directions. The bottom currents generally moved in the same directions as the surface currents with smaller speed. Due to the high water level in the bay and the river mouth, currents from river discharge were relatively weak.

Ebb tide (16:00)

At ebb tide (Figure 5), water levels everywhere in the estuary were falling at the fastest rate of the tidal cycle with the eastern side of the bay falling faster than the western side. Water was leaving the bay through all the tidal openings. At ebb tide, the water level at all tidal openings in the bay were approximately equal, and the maximum water level was located in mid bay. This resulted in the stronger gravity gradient and currents from mid bay to the West Pass because West Pass opening is closer to mid bay than Indian Pass and East Pass. Currents in the St. G. Sound were in eastward direction towards the East Pass, while currents in the St. V. Sound moved to the Indian Pass in the west direction. The divergence of southeastward currents to the West Pass in mid bay and the eastward currents to the East Pass in the St. G. Sounds caused a slack zone between mid bay and St. G. Sound where tidal currents were almost equal to zero. Surface velocity was about 1 m/s in the East Pass, 20-30 cm/s in St. G. Sound, 30-40 cm in mid Bay, 1 m/s in West Pass, and 50 cm/s in Indian Pass. Bottom currents were less than 10 cm/s in most of the areas of the bay except the tidal openings.

Low tide (20:00)

At low tide (Figure 6), the water level at all the boundaries reached the minimum level in the same time. The maximum water level in the bay was -0.1 m in mid bay, with its contour line approximately following the strongest currents from the Apalachicola River discharge. Water levels at ocean openings were lower than that at mid bay, which was about -0.35 m near East Pass and -0.2 m near West Pass and Indian Pass. In St. V. Sound water moved to the Gulf through the Indian Pass, while currents in the vincinity of West Pass and the western mid bay exited from the bay through the West Pass. The strong gravity gradient between mid bay and the East Pass caused the strong eastward currents in St. G. Sounds and eastern mid bay, which was about $20 - 60$ cm/s dependent on the locations. Due to the low water level in the bay, currents from river discharge became stronger. The surface currents resulting from the river discharge near the river mouth were about 0.8-1.0 m/s, which were much stronger than those in the high and ebb tides. The southward currents from the river stretched south almost 8 km away from the river mouth, and then split into the eastward and westward currents due to the block by the barrier island. Bottom currents were weak in mid bay and St. V. Sound, but became stronger in the St G. Sound with speeds of 20-30 cm/s.

Flood Tide (24:00)

At flood tide (Figure 7), water level roses faster in the East Pass than west openings. The surface elevations in all the tidal boundaries were higher than that in the bay. The minimum water level was about -0.2 m to -0.25 m in the mid bay, while the surface elevation was about 0.0 m at the East Pass, -0.1 m at West Pass and Indian Pass. Due to the lower water level in mid bay, the surface currents from Apalachicola River into the bay were strong. The southward surface currents from the river discharge extended south about 12 km away from the river mouth near the south barrier island. Currents from all the tidal boundaries moved towards the estuary. The flood tidal currents from East Pass and the river discharge formed a convergence zone in the area between mid bay and St. G. Sound. Surface currents near East Pass were about 50 cm/s, and gradually decreased as they approached to the west of the St. G. Sound. Part of currents moved from the St. G. Sound to the northern part of the bay. The flood currents from the west tidal boundaries (Indian Pass, West Pass, and Sikes Cut) caused another mixing area in the western mid bay. Surface velocities near the south barrier island were weak as a result of the mixing between the southwestward river discharge and the northeastward tidal currents from the West Pass. Bottom currents were generally weak except the area near East Pass, where the bottom currents were about 20-30 cm/s. The tidal currents from the man-made opening at Sides Cut were strong at high tide, which were about 70 cm/s at the surface and 30 cm/s at the bottom. This shows that ocean salt water could enter the bay during the flood tide, mix with bay water, and transport to other areas of the bay from the changes of tidal circulation during a tidal cycle.

CONCLUSION

In this study, a three-dimensional hydrodynamic model was applied to investigate tide-induced circulation in the barrier-island estuary of Apalachicola Bay. The model was previously calibrated and validated with field observations. Model simulations were performed by using observations at tidal boundaries and minimum historic river flow. In order to focus on tidal circulation, wind forcing was not included in this study. Model simulations were used to characterize tidal circulation in this shallow estuary with multiple openings. Field observations of surface elevations indicated that there were considerable differences between the surface elevations in the east and west tidal forcing boundaries. This resulted in an east-west direction horizontal gravity gradients when tidal forcing varied in the boundaries. Snapshots of model simulations at high, ebb, low, and flood tidal conditions were used to characterize the spatial distributions of surface elevations and velocity in the bay.

Results from this study indicate that tidal circulation in the bay is strongly affected by the spatial and temporal variations of surface elevation in the bay resulting from the multiple tidal forces from different directions. East Pass and West Pass are the strongest tidal boundaries. A small man-made opening also causes interactions of bay and ocean waters. At high tide, the higher surface elevation in the east tidal boundary drives bay water from east to the west. At ebb tide, majority of water at mid bay moves out of the bay through the West Pass. At low tide, the lower water level in the East Pass causes strong eastward currents in the eastern region of the bay. At flood tide, the flood tidal currents from all boundaries and the river discharge converged and caused mixing zones in the mid bay where the lowest water level occurs.

ACKNOWLEDGEMENT

The authors are grateful to the anonymous reviewers' valuable comments and suggestions. Data used in this study were provided by Northwest Florida Water Management District. Model simulations were performed in the SGI 2000 computer at the Super Computer Center of the Florida State University.

REFERENCES

Blumberg, A. F. and B. Galperin, 1990. On the Summer Circulation in New York Bight and Contiguous Estuarine Waters, In *Coastal and Estuarine Studies*, **Vol. 38**, R. T. Cheng (Ed.), Residual Currents and Long-term Transport, Springer - Verlag New York Inc.

Blumberg, A. F., and D. M. Goodrich, 1990: Modeling of Wind-induced Destratification in Chesapeake Bay. In: *Estuaries*, **13**, 3.

Blumberg, A. F., and G. L. Mellor, 1987: A description of a Three-dimensional Coastal Ocean Circulation Model. In: Three-dimensional Coastal Ocean Models, Coastal and Estuarine Sciences, Vol 4. (Heaps, N.S., Ed). Washington, E.D., AGU, 1-16.

Conner, C., A. Conway, B. A. Benedict, and B. A. Christiensen. 1982. Modeling the Apalachicola System: A Hydrodynamic and Water Quality Model with a Hydrodynamic and Water Quality Atlas of Apalachicola Bay. Technical Paper No. 23, for the Florida Sea Grant College. 87pp.

Huang, W. and W.K. Jones, 1998. Hydrodynamic modeling of circulation and salinity of Apalalchicola Bay for low flow seasons. Final technical report of Civil Engineering Department, Florida State University, Submitted to Northwest Florida Water Management District.

Jones, W.K. and M.R. Mozo, 1993. Apalachicola Bay freshwater needs assessment program, geophysical data collection program. Volume 1, Northwest Florida Water Management District, Water Resources Special Reports, 93-5,

Jones, W.K. and M.R. Mozo, 1994. Apalachicola Bay freshwater needs assessment program, geophysical data collection program. Volumes 2-4,

Northwest Florida Water Management District, Water Resources Special Reports, 94-1,94-6,94-7.

Jones, W.K. and G. Rodriguez, 1995. Apalachicola Bay freshwater needs assessment program, geophysical data collection program. Volumes 5-6, Northwest Florida Water Management District, Water Resources Special Reports, 95-1,95-2.

Mellor, G. and T. Yamada, 1982. Development of a turbulent closure model for geophysical fluid problems. Review of Geophysics and Space Physics. Vol. 20. No.4. Pages 851-875, November.

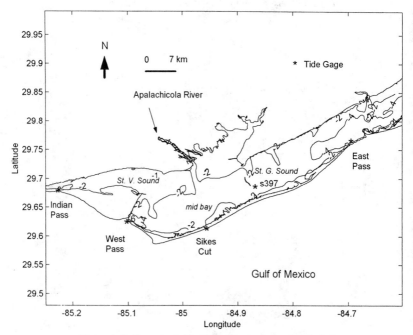

Figure 1, Bathymetry (m) and tidal gages in Apalachicola Bay

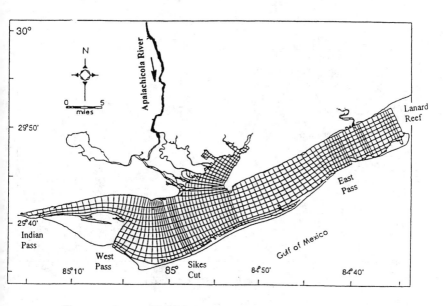

Figure 2. Hydrodynamic Model Grid System.

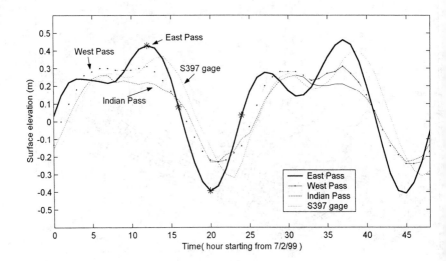

Figure 3.   Time series observations of surface elevations

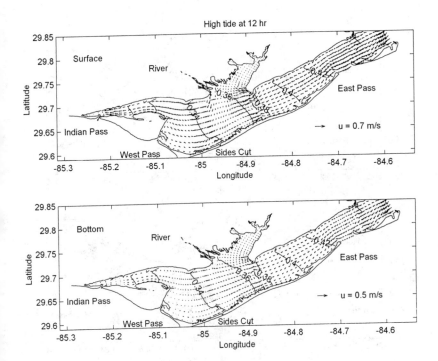

Figure 4.   Water level and currents at high tide: a) surface, b) bottom.

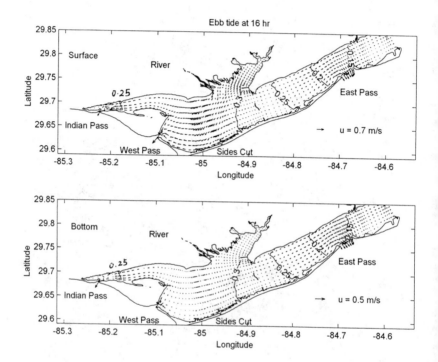

Figure 5.   Water level and currents at ebb tide: a) surface, b) bottom.

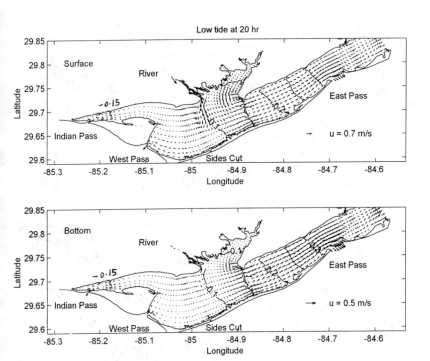

Figure 6.   Water level and currents at low tide: a) surface, b) bottom

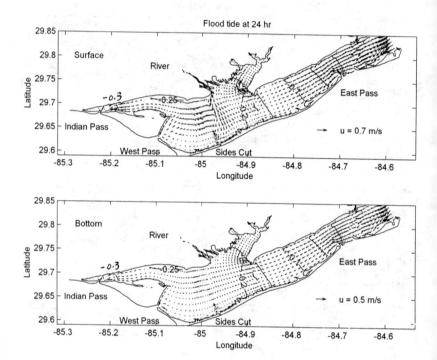

Figure 7.   Water level and currents at flood tide: a) surface, b) bottom

A PC-based Visualization System
for Coastal Ocean and Atmospheric Modeling

Hongqing Wang[1,2], Kai-Hon Lau, Wai-Man Chan and Lai-Ah Wong
Hong Kong University of Science and Technology

Abstract
This paper provides a summary of graphical representations of multi-dimensional data and introduces a powerful visualization system running on *PC-Windows NT operating system.* In addition, we shall also present examples of using the system to visualize multi-dimensional data generated by two well-known numerical models: one for coastal ocean modeling using the Princeton Ocean Model (POM) and the other for atmospheric modeling using the Fifth generation PSU/NCAR Mesoscale Model (MM5).

Key words
Scientific visualization, numerical simulation, multi-dimensional data, computer graphics, personal computer

Introduction
Geophysical research is often based on analyses of large sets of physical parameters generated by integration of numerical models or through observations / analyses over long periods of time. These time-evolving, multi-parameter data sets are huge, and can be conceptualized as a single five-dimensional data set as shown in Figure 1. (For simplicity, we shall henceforth refer to such data sets as *model* data, but the discussion applies equally well for data sets collected through observations or analyses.)

With the amount of data increasing exponentially, a fundamental question for modelers and data analysts is *"What is the most efficient and most easily understandable way to examine these data sets?"* It is difficult to use traditional methods designed for two-dimensional data sets to extract scientific information from these multi-dimensional data sets. It is a substantial challenge for the scientific visualization community to develop more sophisticated analysis tools and display techniques for these data sets.

---

[1] Corresponding author address: Dr. Hongqing Wang, Center for Coastal and Atmospheric Research, Hong Kong University of Science and Technology, Clear Water Bay, Hong Kong, China. (Email: hqwang@ust.hk).
[2] Permanent affiliation: Laboratory for Severe Storm Research, Peking university, Beijing 100871, China.

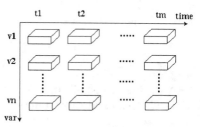

Figure 1: Conceptual structure of a five-dimensional data set
(Every block can represent an one, two or three-dimensional grid)

Computer techniques and hardware have been advancing very rapidly. In recent years, several powerful scientific visualization systems have been developed. For example, VIS5D (Hibbard 1989) is a powerful and popular visualization system for the geophysical community now. However, most of these sophisticated visualization systems (including VIS5D) were developed for *UNIX workstations* and very few for the *Personal Computer* (PC). This is mainly because of the higher computational power and friendlier software development environment on the workstations a few years back. The situation has changed very much since then. Now, the PC's *operating systems* (OS)-especially *Windows NT* - and CPU speed can rival their workstation counterparts. On the OS side, the Windows NT environment has support for multi-processors, powerful development and networking tools. *Graphics accelerators* are also available on the market to allow for a complete set of fast graphics tools like pipelined rendering and full texture mapping. The performance of a properly equipped PC compares very well with mid-range graphics workstation but at a significantly lower price. In addition, software packages that were only available on workstations are now available on PCs as well. A typical example is *OpenGL*. It was the premier environment for developing interactive two- and three-dimensional graphic applications and was the basis of most rendering algorithms on UNIX workstations. Since 1993, OpenGL has been the most widely visualization tool in the industry, and it is now available on Windows NT. On the CPU side, the speed of the Intel clips has also increased very significantly so that compute-intensive algorithms like volume rendering can easily be performed on a PC. Putting all these advances together, a PC-based solution has certainly become more and more attractive when one wants to consider a powerful state-of-the-art visualization system.

In this paper, we will first provide a summary of graphical representations of multi-dimensional data and then introduces a new PC-based visualization system running on the Windows-NT environment. To demonstrate some of the capability of this system, we shall also present examples of two applications to examine simulated data sets from two well-known geophysical models: POM (for coastal ocean modeling) and MM5 (for atmospheric modeling).

Graphical representations of multi-dimensional data

Geophysical variables can be divided into *scalars* and *vectors*, and they can be two-, three-, four-, or five-dimensional in depth. The graphical representations can be different for data with different depths.

1) **Two-dimensional data**: *Contour plots* are commonly used to depict variations in well-behaved two-dimensional scalar variables (e.g., surface temperature at one time). For more complex data sets, *grey-scale* or *color* plots can be used. *Arrows, barbs* and *streamlines* are used to show vector variables. These traditional representations are still the predominant ways for presenting two-dimensional information (including two-dimensional data subsets extracted or reduced from data sets with more dimensions).

2) **Three-dimensional data**: For three-dimensional scalar variables (e.g., temperature within the entire model domain at one time), contour or color plots for selected horizontal and vertical *cross-sections* are often used to help understand the information structure. *Arrows, barbs* and *streamlines* are similarly used to depict the variation of three-dimensional vector variables after reducing their dimension through cross-sections or averaging. The cross-sections have to be compared off-line and many sections are needed to provide the proper understanding of the data. In contrast, *iso-surfaces* of interpolated values and *volume renderings* are better ways to depict the spatial distributions of these three-dimensional variables. *Vertical profiles* are also useful. In vertical profiles, values associated with a fixed vertical data column in the domain are graphed, with the vertical axis showing the variation at different vertical levels and the horizontal axis showing the variation of the parameter.

3) **Four -dimensional data**: There are many three-dimensional data grids in each four-dimensional data set (e.g., temperature within the entire domain at all model times). Typically, they represent the spatial and temporal evolution of one physical variable. To examine the spatial distributions at a certain time, one can use the *three-dimensional data visualization* presentations described above. Since the multi-dimensional data are time-sequences of three-dimensional data sets, *animations* of the three-dimensional representations are particularly useful for depicting the evolution of the parameter. They are also for highlighting and comparing differences in the data set between adjoining time-steps. The idea behind *vertical profiles* can be expanded to vertical profile *evolution plots*, with contours or color plots showing the variation of the parameter at different vertical levels (vertical axis) and time (horizontal axis).

4) *Five-dimensional data*: Five-dimensional data (e.g., temperature and velocity within the domain at all times) are typically used to describe the spatial and temporal evolution of many variables during some physical processes. There are several ways to present the relationship between different variables. For example, an iso-surface is usually drawn in one color. It can also be colored according to the value of another variable at the same space-time location. Such an iso-surface has two information attributes: one is the spatial distribution of the original variable and the other is its relation to the coloring variable.

In addition, there are two special ways to represent four- or five-dimensional data sets: *meteogram* and *trajectory*. Meteogram shows the time evolution of a parameter at one single location. In this case, the horizontal axis is associated with time and the vertical axis indicates the variation of the variable. Trajectory is a sophisticated way to represent vector variables; it depicts the time evolution of one parcel as it travels in the three-dimensional flow field. It is also possible to plot the meteogram of a parcel as it travels along its simulated trajectory. Complex algorithms for tracing particles in forward and backward directions are needed.

Vertical coordinate systems, Map projections, and Geographical information: In earth science, data sets come in different vertical coordinate systems (e.g., equally or unequally spaced *height / pressure / potential temperature levels*) and in many different map projections (e.g., *Mercator, Lambert conformal* or *azimuthal stereographic* projections). A good visualization system should be able to (i) assimilate data from all these different coordinate systems and map projections, (ii) convert data accurately between these representations, and (iii) present data in any of these representations. In addition, the system should be able to display geographical information that are helpful for understanding common geophysical problems, but may or may not be available in the model data set (e.g., coastlines and the earth topography).

A visualization system based on PC-Windows NT Operating System

Features of our new PC-Windows NT based visualization system will be outlined here. This system is designed to manage and analyze very large multi-dimensional data sets. It can produce complex, three-dimensional multivariate images from these data sets, and also has powerful time animation tools for these images. The system allows users to change viewpoints easily so that researchers can "fly into the data" and "observe" the processes in motion from any perspective they choose. Many of its functions are similar to the popular *Vis5D* (Hibbard 1989), but it also has a number of new functions. Since it works on PCs under Windows 98/NT/2000 operating system, we named it as PC-Vis5D.

Similar to Vis5D, PC-Vis5D also has a suite of software tools built into the systems. These include tools to manage data as two- and three-dimensional grids, as trajectories, as images, and as collections of data without any spatial order. Many of these graphics tools were developed gradually as plug-ins as the system was used to analyze / visualize new aspects of some model data set. The data structure is optimized for management of very large data sets spanning many physical variables and multiple time intervals. Data-management tools include handles for reading in(or taking in) and archiving data, libraries of routines for accessing the data archives, conversion tools between different data formats, etc. We also have a convenient on-line HTML help system, and a powerful movie player that can play monocular and stereo movies made during any PC-Vis5D session.

Moreover, PC-Vis5D includes a variety of *commands* for analyzing data in their file structures. These include:

1) re-sampling data to a different spatial or temporal resolution;
2) re-sampling data to a different map projection or different vertical coordinate;
3) transforming data to a moving frame of reference;
4) combining or manipulating data using common arithmetic operators;
5) interpolating non-uniform data (e.g., irregular sounding observations) onto a uniform grid;
6) deriving trajectories from grids of vector components (new algorithms for tracing particles are added: e.g., tracking many particles at the same time);
7) creating images from grids;
8) creating new variables by entering in a formula of known variables and operators.

These *commands* can be used to produce animated sequences of three-dimensional images from the data. The *visual elements* of these images may include:

1) shaded relief topographical maps with physical and political boundaries;
2) time-sequence of satellite or radar images can be mapped onto flat or topographical map near the bottom of three-dimensional box;
3) trajectories drawn as shaded ribbon, which may be opaque or semi-transparent, and may be long, tapered or with length proportional to tracer speed;
4) iso-value contour surface of three-dimensional scalar variables, which may appear smooth with natural shading or semi-transparent;
5) iso-value contour lines, drawn either on topographical surface or on any specified surface in the ocean or atmosphere (e.g., constant pressure or constant density surface);
6) colored slices, may be opaque or semi-transparent;
7) three-dimensional translucent volumes with opacity proportional to selected scalar physical variables;
8) vector fields can be shown in arrowheads, wind-barbs or streamlines;
9) individual data values can be inspected at specific probe time and locations;

10) vertical profile graph and evolution contour for any specified data column in data set ;

11) meteogram for showing time evolution of selected parameters at any specified location.

These images can be assembled as animated sequences to show the *evolution* or a *rotating view at any single time* of selected parameters within a data set. The animated sequences can be loaded into memory so that they can be viewed smoothly on-screen or be recorded directly onto videotape through the video output port of the computer or the graphics card.

All visual elements can be shown in *monocular* or *stereoscopic* viewing mode. Stereoscopic viewing enhances the visual experience by displaying images with true depth perception. It is done by displaying simultaneously the *left- and right-eye views* of a "would-be human user". The user can "fly-in" and examine the data by controlling the viewpoint and the scale of the depicted spatial region in real-time. The three-dimensional images can also be saved into animated stereoscopic sequences and played by our *stereo-movie* player. PC-Vis5D also supports the *virtual reality environment* for real-time stereoscopic display.

In addition to saving images in a number of standard image file formats (e.g., SGI RGB, Windows bitmap, Targa TGA etc), PC-Vis5D can also save three-dimensional images in *VRML (Virtual Reality Modeling Language)* format. The *VRML* format allows for interactive visualization through the internet using standard browsers like the Netscape Navigator.

Our PC-Vis5D system also provides a variety of ways to enhance viewers comprehension of the images. The images are generated in color, and the color of each physical variable is controllable by the user. The user can also control the scale and viewpoint, or have the user viewpoint rotate through in any user-desired rotation angles. In volume renderings, the user can also clip out selected portions of the display to allow for a clearer, dissected image. Depth information is somewhat ambiguous for three-dimensional images in the monocular viewing mode; moving the viewpoint or using stereoscopic viewing can help reduce this ambiguity.

PC-Vis5D also supports a scripting facility. Users can control the entire PC-Vis5D session or part of a session with a *text* file of *Tcl (Tool command language)* commands. This allows an identical set of operations to be performed for different data sets, which makes it easier to compare simulations from different experiments, or for generating movies with preset timings.

PC-Vis5D currently works with following systems:

1) Personal Computers (PCs) with Intel processors: operating systems are Windows 98 and Windows NT4.0/2000.
2) Digital Alpha workstation: operating system is Windows NT4.0/2000.
3) Silicon Graphics visual workstations: operating system is Windows NT4.0/2000.

In all cases, at least 64MB of memory is recommended and at least 16-bit color display is required. For stereoscopic viewing mode, one also needs high-performance three-dimensional graphics accelerator and view glasses with infrared emitter (e.g., Crystal-eyes of StereoGraphics Corporation).

Applications

In the Hong Kong University of Science and Technology (HKUST), the PC-Vis5D system is used to study outputs from our coastal ocean and atmospheric prediction system. We are designing a coupled coastal ocean and atmospheric prediction system for the South China region based on POM and MM5. Here, we shall use outputs from two case studies from our coastal and atmospheric system to demonstrate some of the functions of the PC-Vis5D system.

1) Visualization of MM5 output

Typhoon LEO formed in the South China Sea at 10 UTC on April 26 1999. During the formative stage, it was difficult to pinpoint the position of the cyclone and the center of LEO was relocated a number of times during the next 2-3 days. After that, the storm strengthened very rapidly. The maximum wind estimated for LEO at 06 UTC on April 30 was about 100 knots, up from a maximum of 45 knots 24 hours before. As LEO moved closer to the coast of Guangdong, strong upper level westerly winds acted to shear off the organized convection, resulted in an exposed eye and the cyclone dissipated rapidly before it make landfall about 50 km east of Hong Kong. Most of the aforementioned features were predicted successfully by our MM5 system. The output for each model run contains 11 three-dimensional variables (temperature, cloud, ice, rain, three wind components, etc.) and 26 two-dimensional variables (terrain topography, surface pressure and temperature, rainfall, etc.). The horizontal grids have 63x81 points in Mercator Projection, and there are 27 vertical levels in the three-dimensional fields. The MM5 model outputs at 1-hour interval and the PC-Vis5D data is of size about 128 MB.

Figure 2a depicts the 36-hour forecast of Typhoon LEO valid at 00 UTC May 1 1999. The wind fields at the surface and 200 hPa level are shown by arrows(bottom) and streamlines(top), respectively. The horizontal slice shows the surface wind speed, and the iso-surface shows the cloud water mixing-ratio at 0.2 kg/kg. The figure clearly shows the location, organization and intensity of the storm. It also shows the shearing-off of the convection towards the northeast, which led to the subsequent rapid dissipation of the system over water. Figure 2b shows particle trajectories (as lines and small balls) at the same time, with coloring proportional to the vertical motion field: upward going and downward going particles. The topography is also shown. The spiral motion around the eye is clearly shown by the trajectories. The intense updraft in the rain band and the outflow towards the northeast are also depicted very clearly. It would be difficult to visualize these features without this new visualization tool.

2) *Visualization of Coastal Model*

Pearl River is the 16th (3rd) largest river in the world (China) with an annual averaged discharge rate of about 10,000 m^3 per second. Its drainage area covers some of the most densely populated regions in the world. Environmental pollution is an important problem for this river system, and we are building an extensive monitoring system in the Pearl River Estuary. It is hoped that the system will eventually be able to assimilate multi-disciplinary observations and numerical experiments, and be used as an early warning system for storm surge and red tide. To have a better understanding of the circulation patterns in the Pearl River Estuary, we have developed our own Pearl River Estuary model based on POM. The model data consists of 5 three-dimensional variables (salinity, temperature and three flow components) at 16 vertical levels, with horizontal grid size of 222x242.

In Figure 3, PC-Vis5D was used to visualize (a) the summer and (b) the winter circulation pattern at the 5m level and the corresponding temperature and salinity distributions. A vertical section is also shown in Figure 3b across Ling-Ding-Yang. It can be seen that the area seaward from the Dan-Gang Islands is strongly influenced by the Guangdong coastal current that flows from northeast towards southwest and passes the mouth of Ling-Ding-Yang in the winter circulation pattern. In additional, the river plume hugs the western shore of the Pearl River estuary, generally moves in southwest direction, and eventually meets the coastal current. We also found that there was a rather different flow regime in summer period, possibly due to much higher discharge rates and excessive river outflow of the Pearl River.

Summary

The rapid increase in computational power has enabled researchers to perform intensive scientific computations on workstations and PCs. However, the large amount of analyzed or modeled data makes it increasingly more difficult to extract and present the relevant scientific information. Sophisticated visualization tools are needed to help depict the important scientific contents. In this paper, we introduced a new powerful visualization tool, PC-Vis5D that is based on the PC-Windows NT environment. Examples of using the system to visualize model data generated by two well-known numerical models, POM and MM5 were presented to demonstrate some of the capability of our PC-Vis5D system.

Acknowledgment: This research was supported by the Pearl River Estuary Pollution Project funded by the Hong Kong Government/Hong Kong Jockey Club, by the 863/818-09-01 grant and the 95-s3-3 grant from the Ministry of Science and Technology of China. The authors wish to express their appreciation to Dr. Bill Hibbard of university of Wisconsin's Space Science and Engineering Center, Prof. Jay-Chung Chen of CCAR/HKUST, Mr. Don Middleton and Dr. Bill Kuo of NCAR for their encouragement and support in this effort.

References

1. Gilbert Camara, Lubia Vinhas, et al., Multidimensional Data visualization in Meteorology, *Sixth workshop proceedings on Meteorological Operational system ECMWF*, November 1997, 4:115-118.

2. Hibbard W., Computer generated imagery for 4-D meteorological data. *Bull. Amer. Meteor. Soc.* 1986, 67:1362-1369

3. James, A. S., and Thomas V. P., Visualizing Meteorological data. *Bull. Amer. Meteor Soc.* 1990, 71:1012-1020

4. A. F. Hasler, H. Pierce, and K. R. Morris. 1985. Meteorological Data Field "in Perspective". *Bull. Amer. Meteor. Soc.* 1985, 66:795-801

5. Hibbard W., and D. Santaek., Visualizing large data set in the earth science. *IEEE computer.* 1989, 22 (8):53-57

6. William Ribarsky, The times they are a-changing: PC graphics moves in. *IEEE-Computer Graphics and Application*, 1998 May/June, 6:20-25

7. Philip C. Chen, Portable and Desktop 3D Weather data Visualization systems in UNIX and non-UNIX Environments, *proceedings on Meteorological Operational system of ECMWF*, November 1997, 7:195-201.

8. Wang Hongqing, Zhang Yan, Tao Zuyu, Chen Shoujun, multi-dimensional data sets Visualization in meteorology, *Process in Natural Science*, Vol.8, No.6, 1998, 6:742-747.

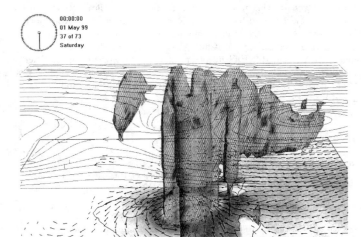

00:00:00
01 May 99
37 of 73
Saturday

(a) Cloud Water, ground and 200mb wind field, ground pressure

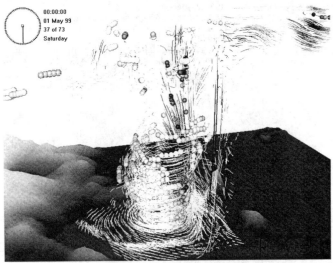

00:00:00
01 May 99
37 of 73
Saturday

(b) Typhoon LEO air particle trajectories

Figure 2. Typhoon LEO visualization

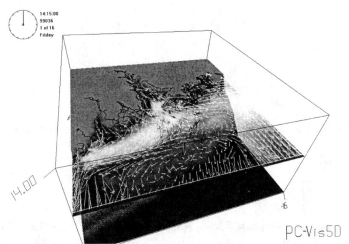

(a) the tidal circulation at 5m below sea surface and distribution of salinity at 8m below sea surface in the summer period.

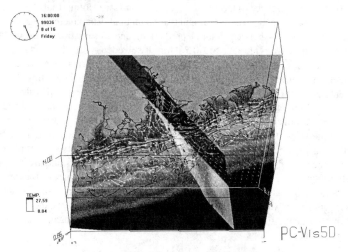

(b) the tidal circulation at 5m below sea surface, salinity contour at 8m below sea surface and distribution of temperature surface across the Ling-Ding-Yang vertically in the winter period.

Figure 3. Ocean Model Visualization

Calibration Performance of a Two-Dimensional, Laterally-Averaged Eutrophication Model of a Partially Mixed Estuary

James D. Bowen[1], A.M. ASCE

Abstract

Eutrophication modeling of the estuarine portion of the Neuse River, North Carolina was conducted to predict the water quality improvement associated with a 30 percent nutrient loading reduction to the estuary. An existing two-dimensional, laterally-averaged model (CE-QUAL-W2) was applied to predict water quality conditions in the lower 80-km of the estuary. During model calibration it was found that the extent to which the model explained observed variability in water quality parameters varied widely. Correlation coefficients between model predictions and observations were 0.94 for salinity, 0.87 for nitrite+nitrate, and 0.76 for dissolved oxygen, but only 0.39 for chlorophyll-a. The relatively poor chlorophyll-a calibration performance was ascribed to the model's inaccuracy in predicting the timing and location of algal blooms. Good agreement was observed, however, between cumulative frequency distributions of chlorophyll-a observations and predictions. Predicted cumulative frequency distributions of chlorophyll-a concentrations matched observations to within a factor of two for frequencies between 0.01 and 0.99, although the maximum observed values were much higher than model predicted maximums. These peak chlorophyll values are of particular concern to the regulatory community as they result in water quality standard violations. A sensitivity analysis, performed to explore the model's capability to predict these highest concentrations, indicated that changes in algal growth parameters alone would not increase predicted maximum concentrations to observed values. It appears from model results that residence time is an important factor in limiting the maximum predicted chlorophyll-a concentrations.

Introduction

North Carolina's Neuse River Estuary has experienced the adverse effects of nutrient enrichment for decades. It is widely believed, however, that these problems

---

[1]Assistant Professor, Department of Civil Engineering, University of North Carolina at Charlotte, Charlotte NC 28223, (704) 547-3130

have worsened lately as a result of increases in human and livestock population and agricultural activity in the watershed. In an attempt to reverse the ecological and water quality trends, the State of North Carolina has drafted regulations aimed at reducing nitrogen loading to the estuary by 30 percent. The State has also funded water quality modeling and monitoring research to investigate the relationship between nutrient loading and estuarine water quality. As part of this effort, eutrophication modeling of the estuary is being performed. In this article we report on some aspects of the models's ability to simulate observed spatial and temporal distributions of key water quality parameters.

Previous studies of water quality in the Neuse River estuary have largely been observational. Long-term monitoring studies have shown that nutrient concentrations generally decrease downstream, although at different rates, such that the time-averaged N:P ratio decreases downstream (Christian et al., 1991; Paerl et al., 1995). Nitrate concentrations in the upper estuary are generally more than an order of magnitude higher than the lower estuary. Temporal variability in both nutrient concentrations, chlorophyll-a concentrations, and primary productivity are significant, with variations occurring both seasonally and associated with loading events from high river flows (Rudek et al., 1991; Boyer et al., 1993; Mallin et al. 1993; Paerl et al., 1995). The monitoring data suggests that limitation of phytoplankton productivity may result from low temperature, light, or nutrients at certain times and places within the estuary, and that accumulation of chlorophyll biomass is also affected greatly by changing water residence times.

The Neuse Estuary is considered to be a hyper-eutrophic estuary, and has experienced serious blue-green algae blooms in the oligohaline portion of the estuary since at least the 1970's (Hobbie and Smith, 1975; Paerl, 1987). Unfortunately, these problems seem to have spread and intensified in the last few years. In 1995, heavy summer rains followed by a prolonged period of density stratification lead to widespread algal blooms and anoxia (defined here as DO concentration < 2 mg/l), and massive fish kills (Paerl and Pinckney, 1996). An outbreak of the toxic algae *Pfiesteria piscicida* (Burkholder et al., 1995) also occurred, and was considered to be the cause of many of the fish deaths. A similar scenario occurred in 1996 after the passage of Hurricane Fran. Anoxia and fish kills occurred along the entire estuary. A smaller, although significant fish kill occurred in the middle estuary in late July, 1998 (Leuttich et al., in press), again following a period of anoxia.

Nutrient bioassays have suggested that a 30% reduction in nitrogen loading might be sufficient to produce a noticeable decrease in algal productivity (Paerl, 1987). Based upon this work and additional expert assessment, the state adopted the 30% nutrient reduction level as the initial goal of its management strategy for the Neuse River. The state's management strategy also includes, however, an effort to develop and calibrate a eutrophication model of the Neuse Estuary, and revise the 30% reduction target, if warranted. The ultimate objective of this modeling study is to determine the water quality consequences of nutrient loading reduction. Of particular interest is the estuary's predicted response to a load reduction of 30%. Two previous transport model studies have been completed for the Neuse River Estuary. Lung (1988) used WASP to investigate the factors related to blue-green algae blooms. Robbins and Bales (1995) performed a circulation and transport study of the Neuse Estuary using a two-dimensional, vertically-averaged model. This model predicted that although the general circulation pattern followed that expected for a partially mixed estuary, lateral variations in water velocities could at times be significant in specific areas of the middle estuary.

In this article we report on the calibration performance of a two-dimensional laterally-averaged model of the Neuse River Estuary. This model has been previously calibrated to data from 1991 that had extensive hydrodynamic information, but very limited water quality information (Bowen and Hieronymus, in press). Here we make use of the extensive hydrodynamic and water quality monitoring data on the Neuse Estuary now being collected (Luettich et al., in press). These data allow us to look carefully at the degree to which the model simulates observed spatial and temporal distributions of nutrients, dissolved oxygen, and chlorophyll-a. Through this analysis we are able to assess, to some extent, the utility of the model with regard to environmental management of the Neuse Estuary.

Description of the Study Site

North Carolina's Neuse River drains approximately 16,000 km^2, and empties into the southwestern corner of the Pamlico Sound. In the river's headwaters is the Raleigh-Durham metropolitan area, with nearly a million inhabitants. Approximately 1.5 million people live in the entire river basin. The lower basin is used intensively for agricultural and livestock production. The estuary occupies the lower 80 km of the river (Figure 1). The estuary gradually deepens downstream, with cross-sectional average depths ranging from approximately 4 m near New Bern to 6 m at the mouth.

Surface water loadings to the estuary come from two rivers and nine creeks, as well as from direct runoff and precipitation. Three of these water bodies (Neuse River, Trent River, Swift Creek) account for more than 90% of the total watershed area (Bales and Robbins, 1999). None of the remaining creeks have a watershed area more than 1.1% of the total. Direct runoff to the estuary accounts for 3.9% of the watershed area.

For the purposes of this study, the Neuse River Estuary between Streets Ferry and Oriental can be divided into three sections, as was done in studies of benthic nutrient and dissolved oxygen fluxes (NC DWQ, 1998). The upper estuary, from Streets Ferry to just upstream of Upper Broad Creek, is oligohaline. Velocities and salinities in this section are strongly affected by riverine flushing (Giese et al., 1985). The upper estuary is considered eutrophic and has experienced numerous blooms of blue-green algae (Hobbie and Smith, 1975; Paerl, 1987).

The middle estuary is much wider and generally shallower than the upper estuary. It runs southwesterly to Cherry Point, where the estuary makes a perpendicular bend to the northeast. Cherry Point is considered the downstream extent of the middle estuary.

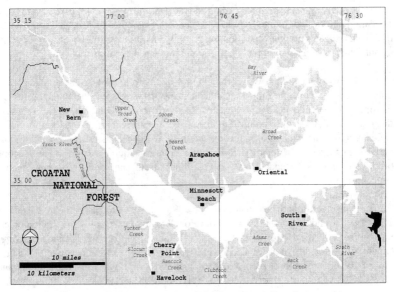

Figure 1. Neuse River Estuary, North Carolina

The lower estuary is generally wider and deeper than the other sections. The middle and lower estuary reaches are considered to be mesohaline and mesotrophic. Recently dinoflagellate blooms have plagued these sections of the estuary (Rudek et al., 1991; Mallin and Paerl, 1994).

Methods and Procedures

The Neuse River Estuary is a relatively long, narrow, partially-mixed estuary. Significant longitudinal and vertical variations in water quality conditions have been observed (e.g. Christian et al. 1991). These variations are expected to be more significant than lateral variations. While lateral variations in circulation patterns have been observed (e.g. Robbins and Bales, 1995), they act primarily as a source of lateral mixing rather than as an important factor in determining estuarine flushing. In addition, stakeholder concerns regarding model uncertainty and data needs suggested that the model should be able to simulate observed spatial patterns, but should also be as simple as possible. For these reasons, a two-dimensional, laterally-averaged approach was chosen, using the water quality model CE-QUAL-W2 (Cole and Buchak 1995).

CE-QUAL-W2 (W2) is a coupled hydrodynamic/water quality model. Its water quality algorithms incorporate 23 constituents (W2 Version 3), 18 of which are used in this application. These 18 constituents are linked to one another to simulate: 1) phytoplankton uptake and release of nutrients and CO_2 through photosynthesis and respiration; 2) remineralization of carbon and nutrients through phytoplankton mortality, exudation, and water column respiration; 3) consumption, production, and transport of dissolved oxygen through respiration, reaeration, and photosynthesis; and 4) recycling of nutrients and consumption of DO through sediment diagenesis.

In this application of W2, certain modifications were made to the standard water quality routines. First of all, three separate algal groups were used: 1) dinoflagellates and diatoms, 2) chlorophytes and cryptophytes, and 3) blue-green algae. The sediment diagenesis model was also modified to include both labile and refractory organic matter state variables, as well as an aqueous sediment oxygen demand (i.e., sulfide) state variable, and a sediment denitrification feature. Denitrification was modeled as a second order rate process dependent on water column nitrate and sediment organic matter concentrations.

A nineteen-month period in 1997 and 1998 was simulated using the model (Table 1). Model boundary conditions were developed using data from the U.S. Geological Survey (river flows, Neuse River bathymetry), the North Carolina Division of Water Quality (water quality of Neuse and Trent Rivers, wastewater treatment plant loadings, the U.S. Weather Service (meteorologic forcings at Cherry Point Naval Air Station), and the UNC Institute of Marine Sciences (elevation, water quality). Water quality measurements were collected weekly from surface and bottom waters at 16 mid-river stations divided nearly uniformly between upper, middle, and lower regions of the estuary (Luettich et al., in press). This monitoring also included measurements at fewer times and places of primary productivity, algal accessory pigments, and benthic nutrient and dissolved oxygen fluxes. Model spatial resolution (Table 1) was established by comparing results of salinity simulations for various numbers of Neuse River model segments ranging from 35 to 140 (Bowen and Hieronymus, in press). Temporal resolution was set according the model's auto time-stepping routines.

The two years simulated, 1997 and 1998, were very different hydrologically. All seven months in 1997 (May - December) had Neuse River flows below the monthly average. Two relatively high-flow events, which occurred in early June and early August, brought the flow only up to the average monthly values. In contrast, 1998 was an exceedingly wet year. Springtime flows were significantly above average. Runoff decreased markedly after April, and the remainder of 1998 was characterized by streamflows below monthly averages.

Calibration of the model was performed first on the hydrodynamic aspects of the model and then on water quality. Parameters adjusted during hydrodynamic calibration included the maximum and minimum eddy viscosities and diffusivities, and horizontal momentum and mass dispersion coefficients. Water quality calibration focused on specification of the algal growth parameters that quantify nutrient, temperature, and light limited growth rates.

Table 1. Model Application Summary

Characteristic	Value
Modeled Region	Streets Ferry - Oriental, North Carolina
Model Time Period	June 1, 1997 - December 31, 1998
Horizontal Grid Resolution	62 segments, length approx. 1 km.
Vertical Grid Resolution	18 layers, 0.5 m thick
Temporal Resolution	approx. 6 min.
Temporal Resolution - upstream flow boundary conditions	daily flows and nutrient concentrations, hourly temperatures, weekly data for other water quality constituents
Temporal Resolution - downstream elevation boundary conditions	water levels every 15 min., water quality profiles bi-weekly
Meteorologic Forcing Data	hourly air and dewpoint temperatures, hourly cloud cover, and wind speed and direction, daily precipitation

Results

The relatively dry conditions in the Summer of 1997 and the very wet conditions in early 1998 produced significant salinity variations in the normally mesohaline middle portion of the Neuse Estuary. By November 1997, salinities in the bottom waters near Cherry Point reached as high as 17 PSU (Figure 2). By March of 1998 salinities dropped to less than 4 PSU, and did not recover to the values of the previous summer until very late in the year. Top to bottom differences in salinity were quite variable. These patterns were also seen in the model predictions (Figure 2), although the model generally seems less dynamic with regard to stratification as compared with the observed data. Considering the estuary as a whole, however, the model's predictions agree well with observations (Bowen and Hieronymus, in press). The hydrodynamic model predictions explained 88.5% of the observed variation in salinity, while normalized mean (predicted - observed) and mean absolute errors were -7.9% and 20.8% respectively (Table 2).

Observed chlorophyll-a concentrations in the surface waters of the Neuse Estuary were also quite variable spatially and temporally. Over the nineteen-month time period simulated (June '97 - December '98), concentrations above the water quality standard value of 40 µg/l were observed in April, June, and August of 1998 in the middle and lower estuary (Figure 3). In 1997, the summertime peak chlorophyll concentrations were generally lower and were located farther upstream. Late in 1997, chlorophyll-a concentrations increase in the lower estuary until the high flows of early 1998 lower residence time sufficiently to limit phytoplankton accumulation (Figure 3), even though nitrate concentrations at this time are very high throughout the estuary (Leuttich et al., in press).

Predicted chlorophyll-a concentrations show this same seasonal pattern (Figure 3). Comparisons of chlorophyll-a model predictions against observations at a single location indicate that although the model predicts the overall seasonality of the chlorophyll-a distributions reasonably well, it generally misses the timing of the peak by at least a month (Figure 3, Figure 4). In addition, the magnitude of the peak chlorophyll-a concentration is not always well predicted. For instance, the observed data show peaks of chlorophyll-a exceeding 50 ug/l at two times during 1998, but the predictions show peaks of significantly lesser magnitude (Figure 3).

Quantitative comparisons between observed and predicted concentrations indicate that the degree of model fit varies widely between key water quality parameters. Correlation r^2 values ranged from a high of 88.5% for salinity to a low of 15.1% for chlorophyll-a (Table 2). These values correspond to correlation coefficients of 0.94 for salinity, 0.87 for nitrite-nitrate, 0.76 for dissolved oxygen, and 0.39 for chlorophyll-a.

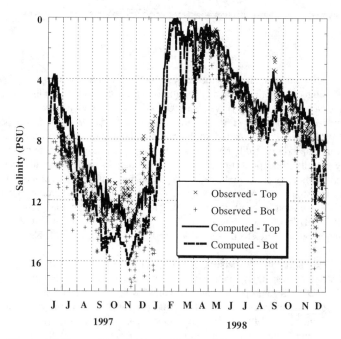

Figure 2. Observed (symbols) and Predicted (lines) Salinities in the Neuse River Estuary at Cherry Point, North Carolina.

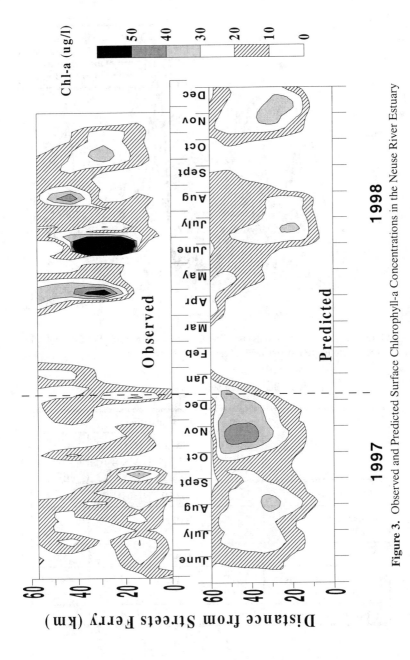

Figure 3. Observed and Predicted Surface Chlorophyll-a Concentrations in the Neuse River Estuary

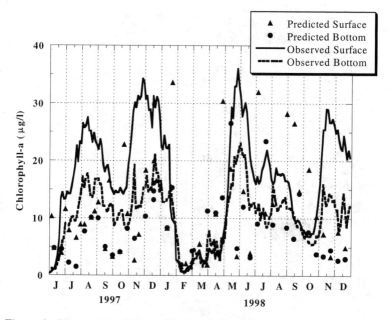

Figure 4. Observed (symbols) and Predicted (lines) Chorophyll-a Concentrations in the Neuse River Estuary near Cherry Point, North Carolina

Table 2. Summary of Model Calibration Performance

Calibration Statistic	Constituent			
	Salinity (PSU)	Nitrite-Nitrate (mg/l)	Dissolved Oxygen (mg/l)	Chl-a (µg/l)
Mean Error (P - O)	-0.48	-0.029	0.047	0.027
Normalized Mean Error (%)	-7.9	-10.5	0.7	0.3
Mean Absolute Error	1.26	0.087	1.25	5.0
Normalized Mean Absolute Error (%)	20.8	31.2	17.7	64.4
Correlation r^2 (%)	88.5	75.7	57.2	15.1

Other quantitative calibration measures followed a similar pattern, with salinity predictions showing the smallest errors, chlorophyll-a the highest, and dissolved oxygen and nitrate-nitrite falling somewhere in between (Table 2). Normalized mean errors were significantly lower than mean absolute errors, with the mean errors being no greater than 10.5% while the absolute errors were as high as 66% (Table 2).

From a regulatory perspective, the ability of the model to predict the timing and location of extreme values is less important than the ability to properly capture the statistical distribution of observed values. Chlorophyll-a cumulative frequency distributions for model predictions and observations were compared to examine the model's performance in this regard. Over nearly the entire range of cumulative frequencies, the model predictions agreed with the observations to within a factor of two (Figure 5). The model generally under predicted concentrations in the lower cumulative frequency range and over predicted concentrations in the upper frequency range, but it was only the most infrequent, bloom events that were badly under predicted by the model. For these cases, having probabilities of less than 1% (cumulative frequencies greater than 99%) the model predictions leveled off at approximately 45 µg/l while the observations continued increasing with decreasing probability to well over 100 µg/l (Figure 5).

Discussion

While the water quality model's predicted chlorophyll-a frequency distribution matched the observations over much of the frequency range, it badly under predicted the peak chlorophyll-a concentrations. Furthermore, the shape of two distributions was distinctly different. The observed frequency distribution looked approximately log-normal (linear distribution when plotted log-c vs. normal probability), while the model predictions had a distinct upper bound (Figure 5). Clearly some of this difference could be attributed to differences in the intrinsic spatial averaging of model predictions versus the observed data. The observations are grab samples of approximately one liter, while the model's predictions can be considered averages over the much larger segment volumes. Nonetheless, the comparison raises the following question: Could the model's peak chlorophyll-a predictions be increased by varying the existing algal growth rate parameters in the model? A numerical experiment was conducted to investigate this question.

A sensitivity analysis was performed to see how model predictions of peak chlorophyll-a concentrations would be increased by changes in algal growth rate parameters. A related question of interest was how the changes in growth parameters might change the shape of the cumulative frequency distributions. Four additional model runs were conducted, each using an alternate set of algal growth parameters designed to increase algal growth rate or decrease algal losses from the water column. In each case, parameters were varied by a similar amount (50%), and all three algal groups (diatoms and dinoflagellates, chlorophytes and cryptophytes, blue-green algae) were changed in

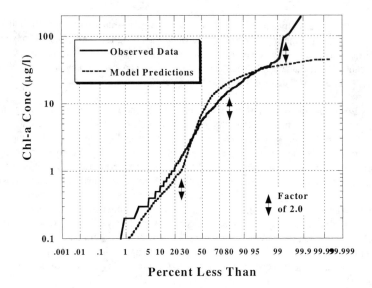

Figure 5. Cumulative Frequency Distributions of Predicted and Observed Chlorophyll-a Concentrations in the Neuse River Estuary from June 1, 1997 through December 31, 1998

an identical manner. Algal growth was adjusted by: 1) increasing the maximum, unlimited growth rate, 2) decreasing the nitrogen half-saturation constant for growth, and 3) decreasing the light requirement for growth. In a fourth case, algal water column losses were decreased by lowering the algal settling velocity. In this initial investigation of the model's ability to simulate algal blooms, no attempt was made to consider possible interactions between the various parameters describing algal growth.

Only one of the four cases, when tested against the earlier base case model prediction and the observed data, had a significantly higher predicted peak chlorophyll-a concentration. Increasing the maximum, unlimited growth rate increased the peak value from about 45 µg/l to just over 50 µg/l (Figure 6). The other three test cases did not increase the peak predicted chlorophyll concentration at all, although there were some minor increases in concentrations at lower cumulative frequencies. Of these test cases, the lower light requirement case seemed to have the largest impact (Figure 6). It is unclear at this time if combinations of parameter changes might have produced a larger effect. The shape of the predicted frequency distribution was identical in every model test case. As before, each model run was characterized by a concave downward frequency distribution, whereas the data appeared to be log-normally distributed (Figure 6).

The fact that the only way to increase predicted chlorophyll-a maximum was to increase maximum growth rate suggests that limitation of biomass accumulation is controlled by factors other than light, nitrogen availability, or settling. This is reasonable in instances where blooms occur during periods of high nutrients, temperature, and light, such as in the upper estuary in Spring and Summer. In this case, peak concentrations are probably mostly dependent on water residence time, rather than growth limitation. Thus the model's inability to simulate the chlorophyll peaks may be related to limitations of the transport model. Although the model well represented the overall mixing and flushing within the system, as evidenced by the relatively low error in mean salinity (Table 2), it may be less accurately predicting circulation in the upper estuary, where the largest blooms occur. Blooms that occur in Fall and Winter in the lower estuary are likely to be

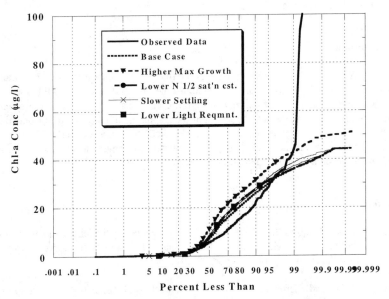

Figure 6. Cumulative Frequency Distributions of Observed Chlorophyll-a Concentrations and Predicted Concentrations for Five Model Runs with Various Algal Growth Parameterizations.

more dependent on light, nutrients, and temperature limitation, and thus better represented by this model, but these blooms generally don't reach concentrations as high as those in Spring and Summer up-estuary. A second possible explanation for the lack of sensitivity is that algal growth in the upper estuary is phosphorus limited rather than nitrogen limited, thus changes in nitrogen half-saturation constants would not be expected to change algal biomass predictions. In fact, there is some evidence of phosphorus limitation in the upper estuary, warranting further investigation of the model's sensitivity to changes in the phosphorus half-saturation constants for algal growth.

The concave downward cumulative frequency distribution is a persistent feature of these and other eutrophication model predictions of chlorophyll-a concentrations (e.g. Hydroqual and Normandeau 1995). Whether this is due to the inherent spatial averaging of these models is unclear at this time. If so, then spatial averaging of the observed data set might be warranted. On the other hand, since eutrophication models are often used in a regulatory context, it may be inadvisable to spatially average monitoring data if this averaging would not be done when determining compliance to water quality standards. In either case, the discrepancy between the observed and predicted distributions is an important issue with regard to the calibration and use of eutrophication models, and justifies further research.

Conclusions

We conclude from this investigation that eutrophication model calibration performance can be a mixed bag. Prediction of the time and place of algal blooms is problematic. Predictions of typical concentrations and seasonality of response match observed patterns quite well. Prediction of the magnitude of blooms fails for only the rarest of events, but for these cases, model under predictions can be severe. Because of the wide variations in model skill, use of these models for environmental management should be accomplished in a thoughtful and selective fashion.

References

Bales, J.D. and Robbins, J.C. 1999. A dynamic water-quality modeling framework for the Neuse River Estuary, North Carolina. U.S. Geological Survey, Water-Resources Investigations Report 99-4017.

Bowen, J. D., and Hieronymus, J. In press. The Neuse River Estuary's response to reduced nutrient loading: Predictions and an uncertainty analysis using a mechanistic eutrophication model. UNC Water Resources Research Institute Technical Report Series, N. C. State Univ., Raleigh, NC.

Boyer, J.N., Christian, R.R., and Stanley, D.W. 1993. Patterns of phytoplankton primary productivity in the Neuse River estuary, North Carolina, USA. *Mar. Ecol. Prog. Ser.*, 97, pp. 287-297.

Burkholder, J.M., Hobbs, C.W., and Glasgow, H.B. Jr. 1995. Distribution and environmental conditions for fish kills linked to a toxic ambush-predator dinoflagellate. *Mar. Ecol. Prog. Ser.* 124, pp. 43 -61.

Christian, R.R., Boyer, J.N., and Stanley, D.W. 1991. Multi-year distribution patterns of nutrients within the Neuse River Estuary, North Carolina. *Mar. Ecol. Prog. Ser.*, 71, pp. 259-274.

Cole, T.M., and Buchak, E.M. 1995. *CE-QUAL-W2: A two-dimensional, laterally averaged hydrodynamic and water quality model, version 2.0, Draft User Manual.* Instruction Report EL-95-1, Waterways Experiment Station, Vicksburg, MS.

Giese, G.L, Wilder, H.B., and Parker, G.G. Jr. 1985. Hydrology of major estuaries and sounds of North Carolina. U.S.G.S. water-supply paper 2221, 105 pp.

Hobbie, J.E., and Smith, N.W. 1975. Nutrients in the Neuse River Estuary. University of North Carolina Sea Grant Program. Report UNC-SG-75-21. Raleigh, NC. 183 pp.

Hydroqual and Normandeau. 1995. *A water quality model for Massachusetts and Cape Cod Bays: Calibration of the Bays Eutrophication Model (BEM.* MWRA Enviro. Quality Dept. Tech. Rpt. Series No. 95-8. Massachusetts Water Resources Authority, Boston, MA., 402 pp.

Luettich, R.A. Jr., McNinch, J.E., Pinckney, J.L. Alperin, M., Martens, C.S., Paerl, H.W., Peterson, C.H., and Wells, J.T. In press. Neuse River Estuary MODeling and MONitoring Project: Monitoring Phase - 12/31/98. UNC Water Resources Research Institute Technical Report Series, N. C. State Univ., Raleigh, NC.

Lung, W.S. 1988. The role of estuarine modeling in nutrient control. *Water Sci. and Technol.*, 20(6/7), pp. 243-252.

Mallin, M.A., and Paerl, H.W. 1994. Planktonic trophic transfer in an estuary: Seasonal, diel, and community structure effects. *Ecology*, 75(8), pp. 2168-2184.

North Carolina Division of Water Quality (NC DWQ), Water Quality Section, Environmental Sciences Branch. 1998. Neuse River Internal Nutrient Loading Studies, 1996-1997. Internal Report, North Carolina Division of Water Quality, Raleigh, NC, 50 pp.

Paerl, H.W. 1987. Dynamics of blue-green algal blooms in the lower Neuse river, NC: Causative factors and potential controls. Report No. 229, Water Resources Research Institute, UNC, Raleigh, NC.

Paerl, H.W., Mallin, M.A., Donahue, C.A., Go, M., and Pieirls, B.L. 1995. Nitrogen loading sources and eutrophication of the Neuse River estuary, North Carolina: direct and indirect roles of atmospheric deposition. Report No. 291, UNC Water Resources Research Institute, N. C. State Univ., Raleigh, NC.

Paerl, H.W., and Pinckney, J.W. 1996. Hypoxia, anoxia and fish kills in relation to nutrient loading in the Neuse River estuary: Why was 1995 a "bad" year? *Water Wise*, 4(2), May 1996. UNC Sea Grant Program, Raleigh, NC.

Robbins, J.C., and Bales, J.D. 1995. Simulation of hydrodynamics and solute transport in the Neuse River estuary, North Carolina. U.S. Geological Survey Open-File Report 94-511, Raleigh, NC. 85 pp.

Rudek, J., Paerl, H.W., Mallin, M.A., and Bates, P.W. 1991. Seasonal and hydrological control of phytoplankton nutrient limitation in the lower Neuse River Estuary. *Mar. Ecol. Prog. Ser.*, 75, pp. 133-142.

Modelling the Humber Estuary catchment and coastal zone

Roger Proctor, Jason Holt (CCMS-POL[1]),
John Harris, Alan Tappin (CCMS-PML[2]) and
David Boorman (CEH-IH[3])

Abstract

The Humber Estuary is one of the largest UK estuaries and the major UK freshwater input to the North Sea (250 m^3s^{-1} annual average). It has a catchment area of 24,000 km^2 drained by two major river systems, the Trent and Ouse, and several minor rivers. These river systems and the estuary itself are subject to extensive industrial and sewage inputs, and agricultural runoff. The steep chemical gradients and complex dynamics within the estuary itself can result in major transformations in the amount, chemical nature and timing of the flux of these and other materials to the adjacent North Sea coastal waters. The UK Land Ocean Interaction Study (LOIS) (1993-2000), combining extensive physical and biogeochemical measurements with an integrated modelling programme, was established to examine the transport and fate of contaminants (including nutrients, metals and micro-organic species) through the land - sea boundary. Three regimes are modelled: the catchment system (comprising land surface and in-stream components), the estuary, and the coastal zone, with interaction between them through a one-way coupling from catchment to coastal ocean. The inputs and the fate of suspended matter

[1] Centre for Coastal and Marine Sciences, Proudman Oceanographic Laboratory, Bidston Observatory, Wirral CH43 7RA, UK. Tel: +44 151 653 8633. Email: rp@ccms.ac.uk, jholt@ccms.ac.uk

[2] Centre for Coastal and Marine Sciences, Plymouth Marine Laboratory, Prospect Place, Plymouth PL1 3DH, UK. Tel: +44 1752 633100. Email: adt@ccms.ac.uk, jrwh@ccms.ac.uk

[3] Centre for Ecology and Hydrology, Institute of Hydrology, Wallingford, Oxon OX10 8BB. Tel: +44 1491 838800. Email: bdd@institute-of-hydrology.ac.uk

and nitrate are examined through the modelled system with emphasis on the component entering the coastal zone.

Introduction

The United Kingdom lies on a wide continental shelf bordering the north Atlantic Ocean and possesses a coastal zone of great physical and ecological diversity, on which lie many large centres of population, commerce and industry (Environment Agency, 1999). Natural factors have caused the physical environment of the coastal zone to change markedly with time (Pye and French, 1993). These changes include, for example, cliff erosion ,coastal reclamation and estuarineinfilling by sediment. These changes are caused by factors including fluctuations in sea level, wind-wave climate and tidal regime. The physical changes are accompanied by parallel changes in the biological character of the coastal zone, and in both cases the extent of the changes can be modified by anthropogenic perturbation. Human impacts have been significant over the last 200 years due to agricultural, urban and industrial development (Davidson *et al.*, 1995; Environment Agency, 1999). A fragmentory approach to the understanding of coastal zones is no longer sufficient if environmental problems are to be satisfactorily addressed. In other words, systematic investigations of the functioning and interactions of both coastal zone physics and ecology is required. The UK developed the Land-Ocean Interaction Study (LOIS) (NERC, 1994) in response to this requirement.

LOIS is a collaborative, multi-disciplinary study undertaken by scientists from the Natural Environmental Research Council (NERC) and Universities, employing novel instrumentation and modelling techniques to integrateterrestrial, marine and atmospheric sciences in studies of the dynamics of coastal environments. The broad aim of LOIS is to gain an understanding of, and an ability to predict, the nature of environmental change in the coastal zone of the UK. Here, the coastal zone is defined as the region extending from coastal plains draining into the sea and from the lower reaches of rivers affected by tidal motion, to the continental slope at the margin of

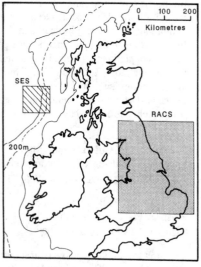

Fig. 1 – LOIS Study areas. SES = Shelf Edge Study, RACS = River, Atmosphere, Coast Study

the Atlantic Ocean. The programme has focussed on understanding the processes active in two areas, one centred on the east coast of Britain - the River, Atmosphere and Coast Study (RACS), and the other on the shelf edge west of Scotland - the Shelf Edge Study (SES), Fig. 1. The RACS covers an area from Berwick-on-Tweed in Scotland to Great Yarmouth in England, approximately 500 km of coastline, taking the discharge of several rivers, the largest of which is the Humber Estuary (Fig. 2).

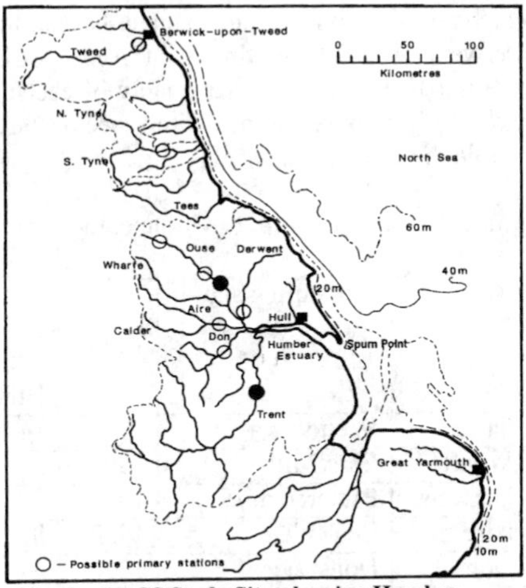

The objectives of RACS focus on three main areas

• to estimate contemporary fluxes of materials, including water, fine particles, biogeochemically important elements and representative contaminants, into, through and out of the coastal zone

• to identify and quantify the physical, chemical and biological transformation processes within river basins, the coastal zone and the atmosphere that govern such fluxes

• to develop models of these fluxes and transformation processes which can

Fig. 2 – RACS Study Site, showing Humber Estuary and catchment . Solid circles represent the tidal limits – Naburn Weir on the Ouse, Cromwell on the Trent.

provide the basis for integrated models of the land-atmosphere-coastal system capable of predicting the effects of future environmental change.

The scientific priorities of the three components (Rivers, Atmosphere and Coasts) have been developed to ensure their integration with programmes focussed on the Humber and Tweed river / estuary / plume systems.

This paper reports on modelling in the RACS associated with the Humber catchment, and is organised into three sections, dealing with each of the main modelling and measurement programmes, namely, the catchment and rivers, the estuary, and the coastal zone. The object of the paper is to demonstrate the integration of these three models in tracing the fate of components from catchment to the open sea.

The models

1) Catchment and rivers model (CR)

i) Model overview

The catchment area draining to the Humber Estuary is of approximately 24 000 km^2, and generates a mean freshwater flow to the North Sea of ca. 250 m^{3}s^{-1} (Jarvie et al., 1997). The region represents approximately one fifth of the area of England and it contains several large, industrially important, metropolitan areas . The Humber Estuary is formed by the confluence of two rivers, the Trent and Ouse, which drain the southern and northern parts of the Humber basin, respectively. The River Ouse catchment is divided into five sub-basins, the Don, Aire, Wharfe, Ouse and Derwent. The UK Environment Agency of England and Wales (UK-EA) monitor the flow and water quality of each of these rivers at a suitable site close to but above the tidal limit (see Table A), and at many upstream sites within the catchments.

Table A (source Natural Environment Research Council, 1998)

River	Monitoring site	Area km^2	% of area draining into Humber	Flow m^3 s^{-1}	% of flow entering Humber
Ouse	Skelton	3315	14	49	20
Wharfe	Flint Mill	759	3	18	7
Derwent	Buttercrambe	1586	7	17	7
Aire	Beal Weir	1932	8	36	14
Don	Doncaster	1256	5	16	6
Trent	North Muskham	8231	34	90	36
Total of six rivers above		17079	71	225	90

Although the six catchments represent only 71% of the drainage area of the Humber Estuary, they represent some 90% of the freshwater flow to the estuary . An overview of the water quality and loads of these rivers is given in Robson and Neal (1997).

The linked catchment – river models are dynamic, and incorporate process-based descriptions of the transport and transformations of simulated constituents. The catchment model had two linked components. The first component, termed the catchment delivery model, represents the land surface. Data required by the model include climatic inputs, such as rainfall and potential evapotranspiration, spatial data describing elevation, soils and land-use, and information describing agricultural management, such as the use of fertilisers. This model employs a spatially distributed representation of processes within the catchment. The second component, termed the in-stream water quality model, is a 1-D model that represents the major channel network of the catchment. Its inputs are both the output from the catchment delivery model and discharges from point sources of pollution, e.g. from sewage treatment works and industrial discharges. The model represents transport through the channel network, and in-stream physical, chemical and biological processes. Data from the UK-EA monitoring network has enabled the models to be calibrated and validated, both at a site near to the tidal limit, and for smaller upstream catchments. The tidal limit is well defined on these rivers, usually by a weir, and these provide convenient boundaries between the river and estuary models.

In addition to modelling the flow of water , the models also simulate the transport and transformations of suspended particles, , nitrate, nitrite, ammonia/ammonium, particulate nitrogen, and two micro-organics (lindane and atrazine). The in-stream component also models dissolved oxygen (DO), biochemical oxygen demand (BOD) and dissolved and particulate zinc. Both models use a daily time step which is adequate for representing the dynamics of the systems and the variability in fluxes.

ii) The catchment delivery model

The catchment delivery model (Cooper and Naden, 1998) represents each catchment as a set of hydrological response units (HRU) which are determined automatically from topography, as defined by a 50 m digital elevation model (Morris and Flavin, 1990), after specifying a maximum HRU area. Within the HRUs, the soil is given a three component vertical structure (upper and lower soil zones and a "groundwater" zone). Precipitation arrives at the soil surface and can either infiltrate into the upper soil zone or become surface runoff. Water in the upper zone can evaporate, percolate, contribute to evapotranspiration, or move laterally; similar transfers occur in the other zones. Physical, chemical and biological characteristics of the zones are derived from spatially distributed data sets of soils and land use. In both cases two data sets have been combined to provide the required data. For soils these are HOST classification (Boorman et al., 1995) which gives a spatial description of 29 soil types based on soil properties and substrate geology and the SEISMIC (Hallett et al., 1993) soil property data set. Land use information came from combining the NERC Institute of Terrestrial Ecology land cover data set (Fuller, 1993) with data describing crop types in parishes provided by the UK Ministry of Agriculture, Fisheries and Food. Other information, for example describing agricultural management were derived from best practice guides. Within the model there is no lateral coupling between cells and therefore it is only required to represent the various combinations of soil and land use, and not their spatial distribution. All lateral flows are transferred directly to the HRU outlet.

iii) The in-stream water quality model

The in-stream water quality model QUESTOR (QUality Evaluation and Simulation TOol for River-systems, Eatherall et al., 1998) was developed partly in response to the requirements of the LOIS programme, and based on earlier river modelling studies (e.g. Whitehead et al., 1997). It requires the river network to be represented as a series of reaches which are delimited mainly by inputs from tributaries, point sources and abstractions. In practice typical reach lengths were approximately 500 m. Within each reach there is perfect mixing, and physical, biological and chemical processes are described by first order rate equations.

The input to the channel network from tributaries, diffuse sources, was provided by output from the catchment delivery model. Although the location of effluent discharges was known, observed data describing flow rates and effluent quality were rarely available. In most cases default values had to be used, based on a classification of discharge type. As the default values were derived from observed data within the study area, errors associated with the use of such values were small at the large catchment scale, although caution would be required in interpreting results at a smaller scale.

Figure 3 shows an example simulation of flow, dissolved oxygen, biochemical oxygen demand, and nitrate and for the River Ouse catchment (see Fig. 2). It will be noted that this is a relatively clean river with high DO, at or close to saturation for most of the time, and low nitrate.

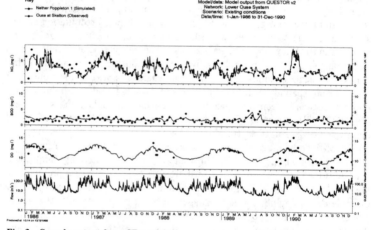

Fig. 3 – Sample output from CR model

2) The Humber Estuary model (HE)

Material that is collected from the catchment and transported seaward by a river system generally enters the sea through an estuary. Estuaries do not, however, act simply as conduits, and are recognised as major modulators of the flux of material from land to sea. This is because there are large gradients in master variables , which within the macrotidal systems of north-western Europe, are also subject to complex and energetic dynamics generated by the combination of extensive tidal oscillations and unidirectional river flows. The result is that the multivariate biogeochemical signal that enters the sea may differ considerably from that which entered from the river (Millward and Turner, 1995).

The physico-chemical complexity of estuaries such as the Humber is obvious, but an understanding dependent on weakly defined detail would be illusory. Taking this view, the model used to simulate fluxes through the system is deliberately simple, without compromising its treatment of those aspects thought to be directly pertinent to net axial transport. The model is sectionally averaged, that is, the only spatial dimension treated explicitly is along the axis of the estuary, but includes branching where the estuary itself branches. To allow robust treatment of seasonal and annual variations, it is also tidally averaged. This means that transport generated by sectional heterogeneity and tidal flow variations must be explicitly represented in the model. In general terms each branch of the estuary is represented by a set of advection-dispersion equations of the form:

$$\frac{\partial c_n}{\partial t} = -\frac{\partial (u_n c_n)}{\partial x} + \partial (k_n \, \partial c_n / \partial x) / \partial x + f_n (c_1, c_2, \dots c_s)$$

Where c_n is the concentration, u_n the axial velocity and k_n the axial dispersion of the nth of s (chemical, biological etc.) species, and

$$f_n (c_1, c_2, \dots, c_s) = \sum_i c_i g_{i,n} (c_1, c_2, \dots, c_s) - c_n \sum_i g_{n,i} (c_1, c_2, \dots, c_s)$$

in which the $g_{i,j}$ represent transfers from species i to species j. In principle, the velocities, dispersions and transfers are all functions of time, space and the local species concentrations. The model has been set up, and these equations solved semi-implicitly, in the ECoS3 modelling shell, based on generic templates developed for use with the software (Harris and Gorley 1998; for other examples of ECoS based models of estuaries see Pham *et al.* 1997, Liu *et al.* 1998).

Three branches of the overall tidal system are represented, the Humber Estuary itself, from the confluence of the Ouse and the Trent to Spurn Point (Fig. 2), the Ouse, to the tidal limit at Naburn Weir, and the Trent, as far as its tidal limit at Cromwell Lock.. Each is represented by a one-dimensional space, and, in each case, closed boundaries are used at the landward limit and open boundaries at the seaward limit. Fluxes determined at the open boundaries of the Ouse and Trent define inputs across the closed boundary of the Humber, while concentrations at this closed boundary define those at the open boundaries of the two rivers. Inputs from the rivers, from either the CR model or observations, are defined as fluxes across the closed boundaries at the tidal limits. Concentrations at the open boundary at the mouth of the Humber are estimated from observations.

The model is being verified and calibrated against results from 27 axial surveys of the Humber and tidal Ouse, undertaken ars from 1994 to 1996 as part of the LOIS programme (Uncles *et al.* 1998a,b,c). This calibration is ongoing, the interim results presented here are for demonstration purposes only.

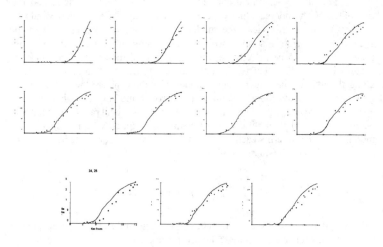

Fig. 4 – Measured surface salinity (dots) during axial surveys of the Humber and Ouse during 1995 (Uncles et al., 1998). EcoS model results (line) with constant axial dispersion of 250m²s⁻¹.

Three basic patterns of flow and dispersion are used to define the transport of material through the estuary. That associated with solutes, that associated with suspended particles and that associated with mobile bed sediment. The passage of materials through the estuary is then defined by the way in which they partition between these three. Neglecting variation in volume (since the model is averaged over tides), the net flow of water, and solutes through the estuary is determined by cubature (see, e.g. Harris *et al.* 1984). Axial dispersion is estimated from observed axial profiles of salinity. An underlying simplicity is suggested by the good representation obtained by using one, constant, axial dispersion (Fig. 4). The advection of suspended particulate matter (SPM) is represented by two additive terms; a seaward component reflecting the effect of riverine flow, and an landward component reflecting the tendency of processes including tidal asymmetry and salinity circulation to transport particles up-estuary when associated with the tendency of particles to sink. The former is represented as a constant proportion of the net water velocity, and declines seawards with reducing river velocity. The latter is depicted by a function of, amongst other things, tidal range and salinity that declines landwards with reducing tidal velocity. Where these two balance (i.e. the net axial velocity of the particles is zero) defines the position of the region of maximum turbidity, which occurs in this estuary, as in other similar systems, towards the limit of saline intrusion (Uncles *et al.* 1998a).

The axial dispersion of particles is described by a similar function in which both terms are positive (the exact form of the functions used will be discussed and justified elsewhere). The result is a model requiring the estimation of four parameters and two boundary conditions. Again, even a preliminary fit appears to give an adequate description of the available data (Fig. 5).

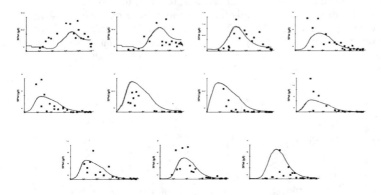

Fig. 5 – Measured surface particle concentrations (dots) during axial surveys of the Humber and tidal Ouse in 1995 (Uncles et al, 1998). EcoS model results (line).

The cycling of nitrogen within the estuary has been modelled as an integral part of the overall cycling of carbon. The nutrient constituents and the transfers between them within the estuarine model are illustrated in Fig. 6.

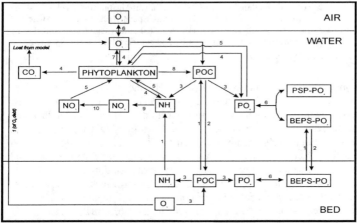

Fig. 6 – Nutrient concentrations, and their transfers, within the EcoS-based estuarine model. POC = particulate organic carbon; PSP = permanently suspended particles; BEPS = bed exchangeable particles. The numbered transfers are: 1, resuspension; 2, sinking; 3, remineralisation; 4, respiration; 5, uptake; 6, exchange; 7, photosynthetic production; 8, mortality; 9, nitrification; 10, oxidation.

The same state variables and exchanges between them are included in the CR model (QUESTOR), except here there is no explicit modelling of the bed and no carbon cycling. Whilst it is acknowledged that processes such as photosynthesis and the oxidation of ammonium and nitrite (Owens 1986), as well as the distribution of particulate organic carbon, are markedly affected by the distribution of suspended particles, resulting in axial distributions of constituents that differ markedly from those to be expected for inert solutes, it would appear that for nitrate in the Humber Estuary, its concentrations within river inputs are sufficiently large that signals from internal cycling are swamped, and the passage of nitrate through the estuary is essentially conservative, as shown in Fig. 7.

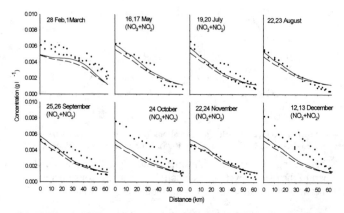

Fig. 7 – **Measured and simulated concentrations of nitrate in the water column of the Humber Estuary during 1995. Trent Falls is at 0 km, Spurn Head at 63 km. Measured concentrations (ca. 1m depth) (filled circles). Modelled values (solid line and dashed line) when treating nitrate as a non-conservative and conservative tracer respectively.**

3) Coastal zone model (CZ)

The outputs of nitrate and SPM from the HE model are applied as input forcing functions to the CZ model. These inputs are then advected and dispersed by the circulation derived from local tidal and wind-driven flow and far field effects extending over the whole of the North Sea. The model used in these simulations is essentially that used by Holt and James (1999a,b) to successfully model the temporal and spatial structure of temperature and suspended particulate matter in the southern North Sea during 1989. Briefly, the model solves the non-linear baroclinic 3-dimensional shallow water equations formulated in geographical coordinates. The equations are written in finite differences on an Arakawa 'B' grid in the horizontal with a depth-following sigma coordinate in the vertical. Vertical mixing is parameterised through the Mellor-Yamada (1974) level 2.5 turbulence closure scheme as modified by Galperin et. al. (1988). Particular attention is paid to the advection of momentum and scalars and the Piecewise Parabolic Method (PPM: James, 1996) is used to maintain horizontal gradients with minimal diffusion. The model is applied to the southern North Sea on the same coarse grid (approx. 20km, see Fig. 8) used by Holt and James (1999a,b) as the emphasis here is on the dispersive fate of the Humber discharge, not the dynamics of the plume itself. Open boundaries across the North Sea and the English Channel are handled through a radiation condition for elevations and currents incorporating a 10-component tidal signal and the residual elevation and currents at hourly intervals from the UK operational storm surge forecasting model archive (Smith, 1994). Meteorological forcing (surface pressure, 10m winds, air temperature, relative humidity, cloud cover) was obtained from European Centre for Medium-range Weather Forecasting (ECMWF) and processed at six-hourly intervals on a 1 degree grid (T. Smyth, Plymouth Marine Laboratory, personal communication).

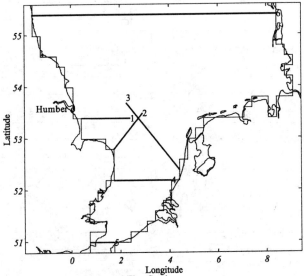

Fig. 8 - Model domain and Flux Sections

Both nitrate and SPM are treated as conservative tracers in the simulation. SPM includes erosion / deposition potential with SPM (C) concentration transferred between the lowest model level and the bed mass (B) by resuspension and deposition following Puls and Sundermann (1990):

$$\frac{\partial C}{\partial t} = \frac{\varepsilon}{\Delta z}(\tau / \tau_{ero} - 1), \qquad \tau > \tau_{ero}, B > 0$$

$$\frac{\partial C}{\partial t} = -w_s \frac{\partial C}{\partial z}$$

$$\frac{\partial C}{\partial t} = -\frac{wC}{\Delta z}(1 - \tau / \tau_{dep}), \qquad \tau < \tau_{dep}$$

$$\frac{\partial B}{\partial t} = \frac{\partial C}{\partial t} \Delta z \qquad B > 0$$

where τ (= $\rho c_D |u_b|^2$) is the bed stress, and $\Delta z = H \Delta \sigma$, H= total depth, ρ=density of water, c_D is the bottom drag coefficient, u_b is the near bed velocity, τ_{dep}=0.1 Nm^{-2}, τ_{ero}=0.41 Nm^{-2} and w_s=0.05 mms^{-1}.

The model is run for 1995 with daily loadings of nitrate and SPM computed by the HE model input as point source loadings (Fig. 8). These are shown as the top series in Fig. 9 (left and right) respectively. Both series show a similar signal, with maximum load to the sea (0.5 tonnes / day for nitrate, 11 tonnes / day for SPM) occurring late January / early February 1995.

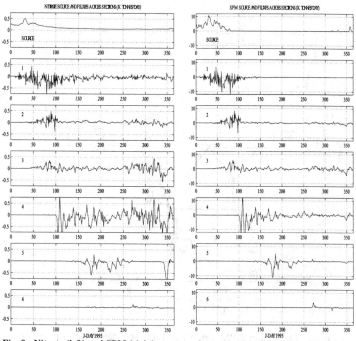

Fig. 9 - Nitrate (left) and SPM (right) source and cross-sectional flux (K tonnes / day)

After this winter pulse, nitrate concentrations reduce to a near constant value of 0.1 tonnes / day and SPM to near zero. Depth integrated horizontal distributions of nitrate and SPM at Julian Day 21, 91, 154 and 315 are shown in Figs. 10 and 11. The nitrate shows that the circulation during 1995 has a predominantly southerly component, pushing both nitrate and SPM southwards. The developing plume crossing eastwards across the southern North Sea (Julian Day 91) is a persistent feature due to the predominantly west/northwest winds which drive an anticlockwise circulation around the southern North Sea during the winter / spring months. The plumes of SPM and nitrate persist during the summer, with decreasing concentrations, especially in SPM which settles out in deeper water. However the dominant motion in this period is southward, through the Dover Straits and into the English Channel. This is driven by an atypical seasonal pattern - a persistent southward flow through the Dover Straits (mean southward flow ~ 0.04ms^{-1}). SPM passing through this open boundary is considered lost to the system.

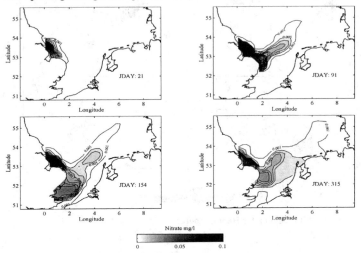

Nitrate mg/l

Fig. 10 – Nitrate concentration for Humber source

To ascertain for how long the input signal remains in the system, fluxes across 6 sections (marked in Fig. 8) are shown in Fig. 9 for nitrate (left) and SPM (right) respectively. Sections 1-3 show similar behaviour for nitrate and SPM, especially up to Julian Day 150 because of strong mixing from tidal and wind-driven currents. During this period SPM acts as a passive tracer and patterns are dictated by the variability in the meteorological forcing and not the source function. However, the advective progression of the source through the sections is clear, with the source term taking 50 days to cross the southern North Sea (section 3) and 100 days to reach section 4 and 150 days to reach the southern boundary of the English Channel. Trace amounts of nitrate are evident at the northern boundary after Julian Day 100. Evidence of SPM reaching the northern boundary in 275 days correlates with a similar timescale reported by Holt and James (1999b). This date is later than for nitrate because SPM settles to the seabed during summer, and is only resuspended with the increasing storminess of autumn. Large fluctuations seen at section 4 correspond to the north-south oscillation of the high concentration patch in the Thames estuary. Oscillations are larger in nitrate because of the continuous low level discharge from the Humber throughout the year. Flux across section 5 indicates that most SPM passed into the English Channel between Julian days 160-250. It is unfortunate that no observations were made other than outside of the nearshore zone of the Humber with which to validate the coastal ocean model but the description of SPM movement in 1995 is similar to the validated simulation of Holt and James (1999b).

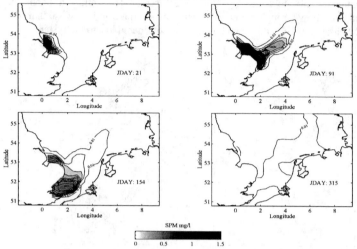

Fig. 11 – SPM distribution for Humber source

Summary

The methodology for linking catchment, river, estuary and the coastal zone, as developed during the LOIS programme, have been demonstrated. Three different kinds of model can be combined to chart the transport of nitrate and SPM from catchment to the coastal ocean, where it is diluted and advected over most of the southern North Sea. The year 1995 is anomalous in that persistent southerly winds during the summer months resulted in a significant proportion of SPM passing out of the North Sea into the English Channel, normally the Channel will act as a source for SPM (Eisma and Kalf, 1987).

References

Boorman, D.B., Hollis, J.M. and Lilly, A. 1995. Hydrology of soil types: a hydrologically-based classification of the soils of the United Kingdom. Wallingford, Oxfordshire. *Institute of Hydrology Report No. 126.* 134 pp.

Cooper, D.M. and Naden, P.S., 1998: Approaches to delivery modelling in LOIS, *Science of the Total Environment*, 210/211, 1-6, 483-498.

Davidson, N.C., Laffoley, D.A. and Doody, J.P. 1995: Land-claim on British Estuaries: Changing Patterns and Conservation Implications. In: *Coastal Zone Topics: Process, Ecology & Management.* 1. The Changing Coastline. N.V. Jones (ed). Joint Nature Conservancy Council, Peterborough. pp. 68-80.

Eatherall, A., Boorman, D.B., Williams, R.J. and Kowe. R., 1998: Modelling in-stream water quality in LOIS, *Science of the Total Environment,*.

Eisma, D. and Kalf, J. 1987. Distribution, organic content and particle size of suspended matter in the North Sea. *Netherlands Journal of Sea Research*, 21: 265-285.

Environment Agency. 1999: *The State of the Environment of England and Wales: Coasts*. The Stationary Office, London. 201 pp.

Fuller, R.M., 1993: The land cover map of Great Britain, *Earth Space Reviews,*2, 13-18.

Galperin, B., Kantha, L.H., Hassid, S. and Rosatti, A. 1988: A quasi-equilibrium turbulent energy model for geophysical flows. *Journal of Atmospheric Sciences*, 45: 55-62.

Hallett, S.H., Jones, R.J.A. and Keay, C.A, 1993: SEISMIC: a spatial environmental information system for modelling the impact of chemicals, *"Environmental modelling: the next ten years"* Eds. Stebbing, A.D.R., Travis, K., and Matthiesson, P. pp.40-49.

Harris, J.R.W., Bale, A.J., Bayne, B.L., Mantoura, R.F.C., Morris, A.W., Nelson, L.A., Radford, P.J., Uncles, R.J., Weston, S.A. & Widdows, J. 1984: A preliminary model of the dispersal and biological effect of toxins in the Tamar estuary, England. *Ecological Modelling*, 22: 253-284.

Harris, J.R.W. & Gorley, R.N. 1998: *Estuary quality templates for ECoS3*. Report available from CCMS Plymouth Marine Laboratory. 47pp.

Holt, J.T. and James, I.D. 1999a: A simulation of the southern North Sea in comparison with measurements from the North Sea Project. Part 1: Temperature. *Continental Shelf Research*, 19: 1087-1112.

Holt, J.T. and James, I.D. 1999b: A simulation of the southern North Sea in comparison with measurements from the North Sea Project. Part 2: Suspended Particulate Matter. *Continental Shelf Research*, 19: 1617-1642.

James, I.D. 1996: Advection schemes for shelf sea models. *Journal of Marine Systems,* 8: 237-254.

Jarvie, H.P., Neal, C. and Robson, A.J., 1997: The geography of the Humber catchment, *Science of the Total Environment*, 194/195, 1-6, 87-99.

Liu, Y.P., Millward, G.E. & Harris, J.R.W. 1998: Modelling the distribution of dissolved Zn and Ni in the Tamar Estuary using hydrodynamics coupled with chemical kinetics. *Estuarine, Coastal and Shelf Science*, 47: 535-546.

Mellor, G.L. and Yamada, T. 1974: A hierachy of turbulence closure models for planetary boundary layers. *Journal of Atmospheric Sciences*, 31: 1791-1806.

Morris, D.G. and Flavin, R.W., 1990: A digital terrain model for hydrology, *"4th International Symposium on Spatial Data Handling (Zurich)"* Anonymous pp.250-262.

Millward, G.E. and Turner, A. 1995: Trace metals in estuaries. In: *Trace Elements in Natural Waters*. ed. B Salbu, E Steinnes. CRC Press, Boca Raton, pp. 223-245.

Natural Environment Research Council. 1998: *Hydrological Data United Kingdom: Hydrometric register and statistics 1991-1995*, Wallingford, Oxfordshire. Institute of Hydrology.Natural Environment Research Council, 1994. Land-Ocean Interaction Study (LOIS). Implementation Plan. Natural Environment Research Council. 61pp.

Owens, N.J.P. 1986: Estuarine nitrification: a naturally occurring fluidized bed reaction? *Estuarine, Coastal and Shelf Science*, 22: 31-44.

Pham, M.K., Martin, J.M., Garnier, J.M., Li, Z. & Boutier, B. 1997: On the possibility of using the commercially available ECoS model to simulate Cd distributions in the Gironde estuary (France). *Marine Chemistry*, 58: 163-172.

Puls, W. and Sundermann, J. 1990: Simulation of suspended sediment dispersion in the North Sea. pp 356-372 in *Residual currents and long term transport*, ed. R.T. Cheng. New York, Springer verlag, 544pp.

Pye, K. and French, P.W.1993: Targets for Habitat Re-creation. *English Nature Science Series*. English Nature, Peterborough. 85 pp.

Robson, A.J. and Neal, C. 1997: A summary of regional water quality for Eastern UK rivers, *Science of the Total Environment,*194/195, 1-6, 15-37.

Smith, J.A. 1994: Operational storm surge model data archive. *Proudman Oceanographic Laboratory Report 32*, 34pp.

Uncles, R.J., Easton, A.E., Griffiths, M.L., Harris, C., Howland, R.J.M., Joint, I., King, R.S., Morris, A.W. & Plummer, D.H. 1998a: Concentrations of suspended chlorophyll in the tidal Yorkshire Ouse and Humber Estuary. *Science of the Total Environment,* 210-211: 367-375.

Uncles, R.J., Howland, R.J.M., Easton, A.E., Griffiths, M.L., Harris, C., King, R.S., Morris, A.W., Plummer, D.H. & Woodward, E.M.S. 1998b: Concentrations of dissolved nutrients in the tidal Yorkshire Ouse River and Humber Estuary. *Science of the Total Environment,* 210-211: 377-388.

Uncles, R.J., Easton A.E., Griffiths, M.L., Harris, C., Howland, R.J.M., King, R.S., Morris, A.W., Plummer, D.H. 1998c: Seasonality of the turbidity maximum in the Humber-Ouse Estuary, UK. *Marine Pollution Bulletin* 37:206-215.

Whitehead, P.G., Williams, R.J. and Lewis, D.R. 1997: Quality simulation along river sytems (QUASAR): model theory and development. *Science of the Total Environment*, 194/195, 447-456.

Young, R.A., Onstad, C.V., Bosch, D.D. and Anderson, W.P. 1989: AGNPS: a non-point source pollution model for evalluating agricultural watersheds, *Journal of Water Resources, Planning and Management*, 44, 168-173.

DEVELOPMENT OF A WASTE LOAD ALLOCATION MODEL FOR THE CHARLESTON HARBOR ESTUARY - PART III: PROJECT APPLICATION

Eduardo A. Yassuda[1], Steven J. Peene[2], Steven R. Davie[2],
Daniel L. Mendelsohn[3], Tatsu Isaji[3]

ABSTRACT

The development of a waste load allocation model within the Charleston Harbor Estuary was conducted to further the understanding of the estuary as an integrated system, and to provide a tool for a quantitative assessment of various management practices. This paper presents the development of a management tool for waste load allocation within the Charleston Harbor Estuary. The working tool is capable of accurately isolating the cumulative impacts of the various point-source discharges on the dissolved oxygen levels during critical conditions of water temperature, river flow, and tidal regime.

Using this integrated tool, the net dissolved oxygen deficit due to the cumulative impacts of all point source discharges was determined longitudinally along the primary tributary in the system, the Cooper River. In addition, the model was utilized to isolate the individual contributions of dissolved oxygen deficit from the primary point source discharges.

1   ASATM Brasil
    Al.Piunas, 140
    Sao Paulo, SP 06430-170
    Brazil
    eyassuda@asatm.com.br

2   Applied Technology & Management (ATM)
    3350 Riverwood Parkway, Suite 2070
    Atlanta, GA 30339
    speene@atm-s2li.com

3   Applied Science Associates (ASA)
    70 Dean Knauss Drive
    Narragansett, RI 02882
    mendo@appsci.com

INTRODUCTION

According to the USEPA guidance for water quality-based decisions, one of the tools for implementing State water quality standards is the TMDL (Total Maximum Daily Load) process. A technically based TMDL constitutes the mechanism for integrating the management of both point and nonpoint pollution sources, and it provides the legal support to decisions for attaining and maintaining water quality standards. The basis for the development of a TMDL is a thorough understanding of the site-specific relationships between pollution sources and instream water quality conditions.

The development of a waste load allocation model within the Charleston Harbor Estuary was conducted to further the understanding of the estuary as an integrated system, and to provide a tool for a quantitative assessment of various management practices. This working tool is capable of accurately isolating the cumulative impacts of the various point-source discharges on the dissolved oxygen levels during critical conditions of water temperature, river flow, and tidal regime.

Study Area

The Charleston Harbor Estuary is located along the southern coast of South Carolina. The estuarine system extends over approximately 3,300 km^2 and consists of 3 primary tributaries: the Cooper, Wando, and Ashley Rivers (Figure 1).

Historically, the Ashley, Wando, and Cooper Rivers were all tidal sloughs with limited freshwater inflow and extensive tidal marshes. Presently, the Ashley and Wando Rivers remain tidal sloughs with varying levels of urban development along their reaches. Over its history, the Cooper River has undergone extensive anthropogenic modifications. In the 1930 , the Santee-Cooper Project created 2 freshwater lakes by diverting the flow from the Santee River and using the naturally high topographic relief at the upper end of the Cooper River for the construction of the Pinopolis Dam. The flow diverted to the west branch of the Cooper River, through the Pinopolis Dam, altered its characteristics from a tidal slough to a riverine system with a significant average freshwater discharge (440 m^3/s). The increased freshwater discharge with its associated sediment load created problems in the lower harbor (Kjerve, 1976). In 1985, a portion of the freshwater inflow to the Cooper River was diverted back to the Santee River and the freshwater flows were reduced to an average of 140 to 170 m^3/s.

The primary mechanism providing hydrodynamic forcing to the system is the tides propagating into the Harbor through the Charleston Harbor Entrance. The tidal excursions range from 1.5 m on average up to 1.8 m during spring tide conditions. These tides extend and amplify up the Wando and Ashley Rivers. On the Cooper River, the tides are felt as high as at the Pinopolis Dam, but there is